# CHAOS II

## Hao Bai-Lin
### Institute of Theoretical Physics, Beijing

**World Scientific**
*Singapore • New Jersey • London • Hong Kong*

*Published by*

**World Scientific Publishing Co. Pte. Ltd.**
P O Box 128, Farrer Road, Singapore 9128

*USA office:* World Scientific Publishing Co., Inc.
687 Hartwell Street, Teaneck, NJ 07666, USA

*UK office:* World Scientific Publishing Co. Pte. Ltd
73 Lynton Mead, Totteridge, London N20 8DH, England

The editor and publisher are grateful to the authors and the following publishers of the various journals and books for their assistance and permission to reproduce the selected articles found in this volume:

Academic Press (*J. Combinatorial Theory*); American Association for the Advancement of Science (*Science*); American Chemical Society (*Accounts of Chem. Res.*); American Meteorological Society (*J. Atmos. Sci.*); American Physical Society (*Phys. Rev., Phys. Rev. Lett., and Rev. Mod. Phys.*); Cambridge University Press (*Quart. Rev. Biophysics*); Ecological Society of America (*Ecology*); Elsevier Science Publishers (*Physica D and Phys. Lett.*); IEEE (*IEEE Trans. Circuits and Systems*); Macmillan Magazines Ltd (*Nature*); Oxford University Press (*IMA J. Math. Applied in Medicine and Biology*); Pergamon Press (*Collected Papers of L. D. Landau*); Plenum Publishing Corp. (*J. Stat. Phys.*); Scientific American (*Sci. American*); Sigma Xi Scientific Research Society (*Am. Scientist*); Springer-Verlag (*Commun. Math. Phys. and Lecture Notes in Phys.*); Systems Dynamics Society (*Systems Dynamics Review*).

**CHAOS II**

Copyright © 1990 by World Scientific Publishing Co. Pte. Ltd.

ISBN 981-02-0095-1
ISBN 981-02-0096-X pbk

Printed in Singapore by Singapore National Printers Ltd.

# Foreword

The First Edition of this introduction and reprint volume was compiled by the end of 1983 and appeared in early 1984. Since then great progress has been made in the observation and understanding of chaotic phenomena. The interest on chaos has been spreading into many fields of science and engineering. Indeed, now it is no longer necessary to convince people that chaos does exist in mathematical models, in laboratory experiments, and in Nature. Instead, people try to answer much deeper questions such as how to compare two strange attractors, how to compare theory with experiments, using various "invariant" characteristics, or how to reconstruct the dynamics from experimental data.

The need for an update of this volume has been felt for some time. However, it has become a much more difficult job as compared to that in 1983. There are so many new concepts and techniques that should be explained within more or less the same length of the Introduction. In rewriting the Introduction, I tried to perserve the comprehensive yet elementary style, and to pay more attention to some recent development, e.g., some aspects of applied symbolic dynamics, characterization of chaotic attractors by using experimental time series and thermodynamic formalism of describing multifractals. The selection of reprinted papers has not been easy either. Among the mostly positive book reviews on the First Edition there was a criticism that some original papers had not been selected. I must admit that it is even truer in this new edition. This volume is not thought to be an archive of milestones in the vast Empire of Chaos, but rather to be a practical introduction for newcomers. While the general layout remains the same, only less than one-third of the selected papers have been kept, the majority being replaced by more recent research articles or reviews. I must apologize to all those whose papers have not been included merely due to limitations in space. The Bibliography has been extended to more than 2200 titles and limited basically to the first half of 1989.

The updating has been accomplished during my visit to the University of Texas at Austin. The hospitality of the Center for Nonlinear Dynamics, the Center for Statistical Mechanics and Complex Systems, and the Institute for Fusion Studies of the University is gratefully acknowledged. I would like also to thank the help of Ms. Zhang Shu-yu, who, in particular, has maintained a computerized bibliography. The collaboration with Dr. K. K. Phua and Ms. Kim Tan, Editor-in-Chief and Editor of World Scientific, respectively, has been a very pleasant experience.

# Foreword to the 1984 Edition

Chaos is a rapidly expanding field of research to which mathematicians, physicists, hydrodynamisits, ecologists and many others have all made important contributions, but in the end it is a newly-recognized and ubiquitous class of natural phenomena and thus belongs to the realm of physics. This volume is designed mainly for physicists with a standard mathematics education, and does not pretend to mathematical rigour in the Introduction but instead tries to rely more or less on physical intuition. However, a demanding reader can easily find precise formulations in the reprinted papers or trace them through the Bibliography.

The better understood part of chaos is essentially classic. Quantum mechanics, in spite of its probabilistic interpretation, happens to be more deterministic, i.e., less chaotic, compared to its classical counterpart. In this volume we shall put aside the problem of quantum chaos except for citing a number of references in the Bibliography.

Stochastic behaviour in classical Hamiltonian, in particular, conservative systems, has become a well-shaped chapter of mathematics and there have been several excellent reviews and books. Chaotic phenomena in dissipative systems are closer to the heart of the physicist, but we need some notions formulated in the study of Hamiltonian systems. That is why a short chapter on the KAM theorem and stochasticity in classical dyanamical systems is included. But in the main, this volume deals with dissipative systems.

The idea to compile an Introduction and Reprints volume on chaos was suggested by Drs. K. K. Phua and K. Young. Dr. K. Young gave valuable advice on the organization of this book. However, the author alone takes responsibility for any possible mistakes in the Introduction and any bias in the choice of reprinted papers. It is absolutely impossible to include all important publications on chaos in a single volume. The author apologizes to all those whose papers did not find a place in this book.

The Bibliography in this volume is based on a computerized bibliography list maintained by Ms. Zhang Shu-yu of the Institute of Physics, Academia Sinica, from which preprints and papers published after 1983 were deleted. If any papers have been overlooked, we hope the authors would send us their papers for inclusion in future editions of the bibliography.

The author expresses his gratitude to all persons mentioned above and to many more colleagues not mentioned who have helped and taught him so much on life in the vast Empire of Chaos.

# To the Reader

This volume consists of three parts. The Introduction and the reprinted papers are divided into chapters under the same headings, indicating their rough correspondence. The Bibliography is subdivided into "Books and Conference Proceedings, and Collections of Papers" and "Papers including Reviews". A reprinted paper is referred to in the Introduction as, e.g., "Hénon, 1976, Paper 11 in this volume", while a title in the second section of the Bibliography is referred to as, e.g., "Feigenbaum (1980a)". Entries in the "Books" section are referred to as, e.g., "Moser, B1973". Since it is impossible to cite all papers on a particular topic, in most cases only some of the earliest and latest references are indicated in the Introduction.

This volume is not supposed to be read from the beginning to the end. Those coming across chaos for the first time are recommended to skip papers of historical significance and to start with the two popular articles by Crutchfield, Farmer, Parkard and Shaw (Paper 1 in this volume), and by Jensen (1987, Paper 2 in this volume), as well as May (1976, Paper 6 in this volume). Several review papers in Chapter 10 contain an introductory part which may well serve the same purpose.

# CONTENTS

## PART ONE: INTRODUCTION

## PART TWO: REPRINTED PAPERS

# PART THREE: BIBLIOGRAPHY ON CHAOS

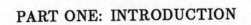

PART ONE: INTRODUCTION

# Chapter 1
# WHAT IS CHAOS?

In 1831, Faraday observed shallow water waves in a container, vibrating vertically with a frequency $\omega$, and discovered subharmonic motion with frequency $\omega/2$. Lord Rayleigh later repeated this experiment and devoted a paragraph to its explanation in the classic treatise *Theory of Sound* (first edition, 1877). 150 years after Faraday, this experiment was redone with modern data acquisition and analysing technique (Keolian *et al.*, 1981; it contains some historical references). What was unusual in it?

Many experimental apparatus may be viewed as frequency transformers. One feeds in some signal with frequencies, say, $\omega_1$ and $\omega_2$, and analyzes the output. If the output contains the same frequency components, the system is essentially linear. If the system is nonlinear, then one may find frequency combinations such as $n\omega_1 \pm m\omega_2$, where $n$ and $m$ are integers. This is, in fact, a simple consequence of trigonometry. If we have a linear Ohm's law:

$$V = RI,$$

where $V$ is voltage, $R$ is resistance, and $I$ is an alternating current, e.g.,

$$I = A\cos\omega_1 + B\cos\omega_2,$$

no other frequency can appear in $V$. However, if the law contains a nonlinear term, say,

$$V = R_0 I + R_1 I^2,$$

then, in addition to $\omega_1$ and $\omega_2$, $V$ must contain $2\omega_1$, $2\omega_2$, and $\omega_1 \pm \omega_2$, owing to the trigonometric relations

$$\cos\omega_1 \cos\omega_2 = \frac{1}{2}[\cos(\omega_1 + \omega_2) + \cos(\omega_1 - \omega_2)],$$

etc. One cannot get subharmonics $\omega_i/2$ in the spirit of linear mode analysis, since everything is expressed through simple sines and cosines. Moreover, these new frequency components are always present, no matter how small is the nonlinearity, i.e., the coefficient $R_1$. In other words, the appearance of harmonics is not a threshold phenomenon.

Contrary to what has been said, the subharmonic observed by Faraday and Rayleigh was a kind of threshold phenomena: they appeared all of a sudden when the nonlinearity reached a certain level. Therefore, the existence of both the threshold and the subharmonic calls for highly nontrivial explanation. People returned to this old experiment, because it had become clear then that subharmonics usually appear as the first tone in the overture to chaos. This understanding was the result of a sequence of initially unrelated events.

In 1963 meteorologist E. N. Lorenz published his numerical observation on a simplified model of thermal convection (see Paper 12 in this volume), a model which now carries his name. He discovered that in this completely deterministic system of three ordinary differential equations, for some parameter range all nonperiodic solutions were bounded but unstable, i.e. they underwent irregular fluctuations without any element of randomness introduced from the outside. In 1971 D. Ruelle and F. Takens (Paper 5 in this volume) coined the term "strange attractor" for dissipative dynamical systems, though they were unaware of the Lorenz model as an example. They further suggested a new mechanism for the onset of turbulence. Li and Yorke (1975) seemed to be the first to introduce the term "chaos" into the literature to denote the apparently random output of some mappings, although the use of the word "chaos" in physics dates back to L. Boltzmann in another context, not unrelated to its present usage. In an excellent review published in 1976 (Paper 6 in this volume), R. May called attention to the very complicated dynamics including period-doubling and chaos in some very simple population models. Then came the discovery by Feigenbaum (Paper 7 in this volume) of the scaling properties and universal constants in one-dimensional mappings, as well as the introduction of renormalization group idea, triggering an upsurge of research interest among scientists.

There has been another course of events leading to the domain of chaos, namely, the study of nonintegrable Hamiltonian systems in classical mechanics. By the end of the last century the development of celestial mechanics and the foundation of statistical mechanics had posed a number of deep problems in classical dynamics. However, the success of relativity and quantum theory as well as the rapid progress of modern technology had absorbed the attention of almost all physicists, these difficult problems were left to mathematicians for calm study for more than half a century. Their efforts crystallized in the formulation of the so-called KAM theorem on nearly integrable Hamiltonian systems in the early 1960's (see Chapter 2). Numerical studies on what happens when the conditions of the KAM theorem fail have revealed an abundance of random motion in nonintegrable systems (see MacKay and Meiss, B1987, for reprinted papers).

The phenomena related to the occurrence of randomness and unpredictability in completely deterministic systems have been called "dynamical stochasticity", "deterministic chaos", "self-generated noise", "intrinsic stochasticity", "Hamiltonian stochasticity" and so on by various authors. We prefer the word "chaos" as a laconic expression and use it mainly in the context of dissipative systems. However, there is as yet no generally accepted definition for chaos. In fact, there have been some rigorous definitions for various manifestation of stochasticity in dynamical systems. For example, the existence of positive topological entropy defines "topological chaos", something definitely chaotic but not necessarily observable. A positive Lyapunov exponent or positive metric entropy certainly means more, however, it is not always easy to fit a model into the mathematical framework or to extract the characteristics from experiments (see Chapter 8 for more).

For practitioners in physical sciences and engineering it is appropriate to use a *working definition of chaos*. If the following ingredients are present:

1. The underlying dynamics is deterministic;

2. No external noise has been introduced;

3. The seemingly erratic behaviour of individual trajectories depends sensitively on small changes of initial conditions;

4. In contrast to a single trajectory, some global characteristics, obtained by averaging over many trajectories or over long time, e.g., a positive Lyapunov exponent, does not depend on initial conditions;

5. And, when a parameter is tuned, the erratic state is reached via a sequence of events, usually including the appearance of one or more subharmonics;

then one may well be dealing with chaos. The first two points are easy to check for theoretical models, but not so apparent in experimental situations. In the latter case one has to invoke techniques to distinguish chaos from random noise (see Chapter 8).

Actually there are at least two different levels of stochasticity, related to two distinct classes of deterministic equations in physics.

On the fundamental "microscopic" level we have the law of dynamics, represented by Newton's equations in classical mechanics. In the absence of external time-dependent forces, they describe reversible movement of conservative systems. Does stochasticity and the necessity of statistical description occur spontaneously, when the system gets complicated enough? This is the old problem of the relation between mechanics and statistical mechanics. In the hands of mathematicians, as part of ergodic theory, dynamical system theory and qualitative theory of differential equations, significant progress has been reached in understanding this problem. We shall touch briefly on the essentials in Chapter 2, because in studying dissipative systems many concepts have been borrowed from the well-shaped mathematics of stochasticity in Hamiltonian systems.

On the other hand, threre are many macroscopic equations in physics, the first example being the Navier-Stokes equtions describing the velocity field **v** of a fluid:

$$\frac{\partial \mathbf{v}}{\partial t} + (\mathbf{v}\nabla)\mathbf{v} = -\frac{1}{\rho}\nabla p + \nu\nabla^2\mathbf{v}, \tag{1}$$

where $p$, $\nu$ and $\rho$ denote pressure, kinematic viscosity and density, respectively. For another example one many take the reaction-diffusion equations describing chemical reactions in an inhomogeneous medium:

$$\frac{\partial x_i}{\partial t} = f_i(x_1, \cdots, x_n) + D_i\nabla^2 x_i, \quad i = 1, \cdots, n, \tag{2}$$

where the reaction kinetics is represented by the usually nonlinear functions $f_i$ and the diffusion caused by spatial inhomogeneity is described by the second term, $D_i$ being the diffusion constant of the $i$-th component $x_i$.

These nonlinear (due to the $(\nabla v)v$ term or the functions $f_i$) and dissipative (due to the $\nu$ or $D_i$ terms) evolution equations are irreversible in nature. Although they can be derived, in principle, from microscopic dynamics by making statistical assumptions or coarse-graining at certain steps of the derivation, the equations themselves, supplemented with appropriate boundary and initial conditions, are manifestatively deterministic. It is a basic experimental fact that under certain conditions a fluid or a reacting system many undergo transitions into states of more and more erratic motion, which in turn necessitates a statistical description. This leads to another long-standing problem in physics, namely, the problem of turbulence. Does the onset of turbulence follow from such macroscopic deterministic equations as (1) or (2)? And, furthermore, can one characterize the state of developed turbulence starting from these same equations? To a certain extent the recent upsurge of interest in chaos was roused by the hope to understand these questions. We shall devote Chapter 3 to the problem of turbulence.

While equations (1) and (2) are partial differential equations, i.e., systems with infinite degrees of freedom, many models studied so far are low-dimensional, e.g., one- or two-dimensional mappings, ordinary differential equations with three or more variables, etc. A moral drawn from recent studies on chaos in low-dimensional systems consists of the belief that at least the onset of turbulent behaviour in dissipative systems, no matter how large is the original phase space, may be described by motions on attractors of much lower dimension. In fact, it is dissipation that realizes the contraction of description in a natural way: a vast number of modes die out due to dissipation, only those spanning the attractors need be taken into account in modelling the system. This kind of simplification cannot take place in Hamiltonian systems owing to the preservation of phase volume (Liouville's theorem). Moreover, dissipation is unavoidable in most experiments. This explains why chaos in dissipative systems has attracted more and more attention.

To conclude this introductory chapter we emphasize that chaos is not to be equated simply with disorder. It is more appropriate to consider chaos as a kind of order without periodicity. Within generally chaotic regimes one can discover patterns of ordered motion interspersed with chaos at smaller and smaller scales, provided sufficiently high resolving power is reached in numerical or laboratory experiments. Instead of the usual spatial or temporal periodicity, there appears some kind of scale invariance which opens up the possibility for renormalization group considerations in studying chaotic transitions (see Chapter 4).

# Chapter 2
# KAM THEOREM AND STOCHASTICITY IN CLASSICAL HAMILTONIAN SYSTEMS

As mentioned before, stochasticity in Hamiltonian systems has become a well-shaped chapter of mathematics and there are many excellent reviews and books (see, e.g., Moser, B1973; Casati and Ford, B1977; Lichtenberg and Liberman, B1983) and a recent introductory and reprints volume (MacKay and Meiss, B1987). This chapter is included since some notions formulated for Hamiltonian systems remain useful for understanding chaotic behaviour in dissipative systems.

To begin with we recall a few concepts from analytic mechanics. The motion of a classical conservative system of $N$ degrees of freedom with a Hamiltonian function

$$H = H(p_1, \cdots, p_N; q_1, \cdots, q_N) \tag{3}$$

is described by the Hamilton's canonical equations

$$\dot{q}_i = \frac{\partial H}{\partial p_i}, \quad \dot{p}_i = -\frac{\partial H}{\partial q_i}, \quad i = 1, \cdots, N. \tag{4}$$

If there exist successive canonical transformations changing $p_i$, $q_i$ into a new set of canonical variables $J_i$, $Q_i$ such that in terms of these new variables the Hamiltonian function depends only on the $J_i$, but not on the $Q_i$, i.e., all the $Q_i$ become so-called cyclic variables:

$$H = H(J_1, \cdots, J_N), \tag{5}$$

then the corresponding canonical equations

$$\begin{aligned}
\dot{Q}_j &= \frac{\partial H}{\partial J_i} = \Omega_i(J_1, \cdots, J_N), \\
\dot{J}_i &= -\frac{\partial H}{\partial Q_i} = 0
\end{aligned} \tag{6}$$

can readily be integrated to give

$$\begin{aligned}
Q_i(t) &= \Omega_i t + Q_i(0), \\
J_i(t) &= J_i(0).
\end{aligned} \tag{7}$$

Now going back to the old variables, one gets $2N$ combinations of $\{p_i, q_i\}$ and $t$:

$$\begin{aligned}
Q_i(0) &= Q_i(p_1(t), \cdots, p_N(t); q_1(t), \cdots, q_N(t)); t), \\
J_i(0) &= J_i(p_1(t), \cdots, p_N(t); q_1(t), \cdots, q_N(t)),
\end{aligned} \tag{8}$$

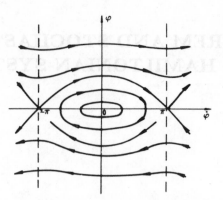

Figure 1: Phase plane of the mathematical pendulum.

which do not change with time. In other words, we have solved the equations of motion completely and obtained $2N$ constants (or "integrals", or "invariants") of the motion. Such Hamiltonian systems are said to be *integrable*. In fact, the existence of $N$ independent integrals of motion suffices to make the system integrable.

A qualitative picture of the motion of an integrable system looks much like that of a system of coupled oscillators, as can be seen from (7). To keep the motion in a finite region of the phase space, the linearly growing $Q_i t$ terms must appear as arguments of periodic functions. When $N = 1$ the motion can be visualized as rotation around a circle of radius $\sqrt{2J(0)}$ with constant angular velocity $\Omega$; therefore the bounded motion of a system with one degree of freedom is always periodic. When $N = 2$, there are two radii determined by $J_1(0)$ and $J_2(0)$ with angular velocities $\Omega_1$ and $\Omega_2$, so the motion is confined to a two-dimensional toroidal surface or a 2-torus. Besides periodic motion a new possibility appears. When the ratio $\Omega_1/\Omega_2$ happens to be an irrational number, the motion can no longer be periodic: the trajectory winds up the 2-torus densely and endlessly. This kind of motion is called *quasiperiodic* (or conditionally periodic in the Russian literature). In general, the motion of an integrable Hamiltonian system with $N$ degrees of freedom is quasiperiodic and is confined to an $N$-torus. Therefore, integrable systems have nothing to do with the requirements of statistical mechanics, since the dimension of the constant energy surface $(2N - 1)$ is larger than that of the torus whenever $N > 1$, and the trajectory can in no way fill up the energy surface, not to mention the equal probability assumption of microcanonical ensemble.

Two questions arise immediately:

1. Are there many integrable systems among all Hamiltonian systems?

2. What happens with the qualitative picture of motion when the system is made slightly nonintegrable, i.e., when the Hamiltonian becomes

$$H = H_0 + V,\qquad(9)$$

where $H_0$ is integrable and $V$ contains a small parameter?

The answer to the first question is definitely negative (Siegel, 1941, 1954). It was found that integrability is an exceptional property for Hamiltonian systems whenever the number of degrees of freedom gets larger than two. Integrable systems are so rare that in general it is impossible to approximate a nonintegrable Hamiltonian system by a series of integrable ones. This statement is to be compared with irrational numbers which can always be approached from both sides by sequences of rationals, because rational numbers are dense on the number axis, though having zero measure.

The answer to the second question is provided by the KAM theorem, first enunciated by Kolmogorov in 1954 and completely proved by Arnold and Moser in the early 60's. We have included the paper of Kolmogorov in this volume (Paper 3) for its historical significance. The proof of KAM required a successful treatment of the small divisor problem in perturbative solutions to the classical many-body problem. The mathematical prerequisite goes beyond that of an average physicist, so we confine ourselves to a loose formulation of the theorem and then turn to its physical implications.

KAM proved that provided the following conditions hold: (a) The perturbation $V$ causing nonintegrability in (9) is small (we ignore the precise formulation of smallness); (b) The Hamiltonian function $H$ is smooth enough; (c) The frequencies $\Omega_i$ of the unperturbed *nonlinear* Hamiltonian $H_0$ satisfy the noncorrelated or nonresonance condition

$$\frac{\partial(\Omega_1, \cdots, \Omega_N)}{\partial(J_1, \cdots, J_N)} \neq 0, \tag{10}$$

then the motion is still confined to an $N$-torus except for a small set (proportional to the smallness of $V$) of initial conditions which may lead to wandering motion on the energy surface. These $N$-tori, now called KAM-tori, KAM surfaces, or KAM curves, if seen in plane sections, may be slightly distorted compared to that of the $V = 0$ case; nevertheless the qualitive picture of the motion remains much the same as in the unperturbed integrable system.

If we follow the exceptional trajectories mentioned in the KAM theorem, then a qualitatively new phenomenon appears for systems with $N > 2$ degrees of freedom. This follows from the fact that the boundary of the constant energy surface must be of dimension $2N - 2$ and when $N < 2N - 2$, the $N$-dimensional KAM torus cannot serve as boundaries dividing the energy surface into regions impenetrable for the wandering trajectories. Therefore, when $N > 2$ these trajectories may wander along the whole energy surface and give rise to a new mechanism of randomness called *Arnold diffusion* (Arnold, 1964), a phenomenon that must be delt with in constructing particle accelerators and magnetically confined nuclear fusion devices. Physically this new possibility remains small as far as the KAM conditions hold.

It is appropriate to summarize briefly at this point: Hamiltonian systems with $N = 1$ are all integrable; the overwhelming majority of systems with $N \geq 2$ becomes nonintegrable; for $N > 2$ Arnold diffusion may show up.

What happens when one violates the condition of KAM theorem? This appears to be a very difficult problem, just to cite such a competent mathematician as Arnold: "Nonintegrable problems of dynamics appeared inaccessible to tools of modern mathematics" (Arnold, 1963a). Still mathematicians were able to tell the qualitative picture of how the KAM-tori are destroyed and modern computers are of much help to visualize this process. The first computer study was reported

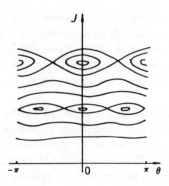

Figure 2: Resonance zones of coupled oscillators.

by Hénon and Heiles (1964) and then followed by intensive study of J. Ford and coworkers and many others.

Let us try to grasp the essentials without involving mathematics. We start with the mathematical pendulum described by the simple nonlinear differential equation

$$\ddot{\phi} + \omega^2 \sin \phi = 0. \tag{11}$$

The angle $\phi$ and the angular velocity $\dot{\phi}$ span a two-dimensional phase plane. Due to periodicity of the motion we can consider only an infinite strip $-\pi \le \phi < \pi$ in the phase plane (Fig. 1). Among the equi-energy curves shown in Fig. 1, there is one connecting the points $(-\pi, 0)$ and $(\pi, 0)$. This *separatrix* divides the strip into three regions with different types of motion: oscillation in the central region and rotation in opposite sense in the upper and lower regions.

In Fig. 1 the stable equilibrium point $(0,0)$, surrounded by ellipses, is an elliptic point or a center, while the unstable equilibrium point $(\pi, 0)$ is a hyperbolic or *saddle point*. We see that the separatrix leaves one saddle point along the "unstable" direction and enters the other saddle point along the "stable" direction and *vice versa*. Actually, $(\pi, 0)$ and $(-\pi, 0)$ correspond to the same point and the stable and unstable directions of the separatrix intersect at the saddle point.

Now take a system of two coupled oscillators in its nearly integrable regime. A section of the phase space in terms of $J_i$, $Q_i$ would appear as if assembled from Fig. 1, i.e., there are regions corresponding to motion of different frequencies called *resonance zones*, see Fig. 2. When the coupling gets stronger and the conditions of KAM theorem begin to break down, these resonance zones tend to overlap and the original separatrices become *stochastic layers* of finite width (see Fig. 3). At the same time, some of the originally simple closed curves split into successions of elliptic and hyperbolic points at smaller scales. The overlap of resonance zones can be cast into a quantitive criterion for the appearance of stochasticity (the *Chirikov criterion*, see, e.g., the review by Chirikov, 1979).

In general, the KAM torus looks much like these figures if intersected by planes in the phase space. The KAM curves can be classified according to the "distance" of their underlying frequency ratio from rationals. The harder an irrational ratio is

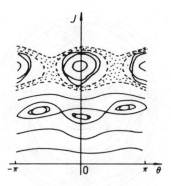

Figure 3: Stochastic layers.

approximated by rationals, the longer it persists during the violation of the KAM conditions. The most persistent KAM curves are those described by the golden mean and other "noble" numbers. The destruction of KAM curves shares the general feature of going from Fig. 2 to Fig. 3. Even after destruction the remains of KAM-tori may still act as obstacles for diffusion of wandering trajectories across the phase space. These obstacles have Cantor-set like structure: KAM-tori now become *Cantori* (for Cantor set see Chapter 8).

In higher dimensions, stable and unstable directions seen in Fig. 1 generalize to "stable manifolds" and "unstable manifolds". A transversal intersection of stable and unstable manifolds is called a *homoclinic point*, if they result from one and the same saddle point, and a *heteroclinic point* if they come from different saddles. The existence of a single homoclinic point implies the presence of an infinite number of homoclinic points, because both stable and unstable manifolds are invariant under the motion, their intersection will be repeated by the dynamics infinitely many times what may lead to very intricate picture as shown schematically in Fig. 4. Homoclinic and heteroclinic points play the role of organizing centers for chaotic motion and are important notions applicable to dissipative systems as well. In dissipative systems, the unstable manifolds of the unstable periodic points outline the attractor (or put pedantically, the attractor is the "closure" of unstable manifolds), while the stable manifolds, traced back in time, outline the boundary of the basin of attraction of the corresponding attractor. In Hamiltonian systems, due to the conservation of phase volume, there are no attractors.

The degree of randomness of classical motion is closely related to its ergodic property, which belongs to a domain of significant progress in the last few decades[1]. Since one encounters the concepts of ergodicity when characterizing the attractors in dissipative systems, a few words may be in order.

A system is said to be ergodic on its energy surface if time averages along a trajectory are equal to ensemble averages over the whole energy surface. Ergod-

---

[1] An elementary introduction to modern ergodic theory can be found in J. L. Lebowitz and O. Penrose, *Physics Today*, 1973, February, but nothing was said there about K-systems. We also recommend another *Physics Today* paper by J. Ford (April, 1983).

Figure 4: Intersections of stable and unstable manifolds (schematic).

icity alone means very little in randomness: two neighbouring points may remain correlated all the time. The next step on the ladder of randomness is called mixing: any initial region on the energy surface evolves into filaments which cover the whole surface when time goes on. Correlation between initial neighbours must decay with time, but nothing is required as regards the decay rate. If the motion suffers local orbital instability, i.e. any nearby trajectories go apart exponentially, then the correlation also decays exponentially. Such system is said to be a *K-flow* (after Kolmogorov). The separation rate of neighbouring points averaged along the trajectory determines the *K-entropy* of the system. A system is a K-flow if it possesses positive K-entropy. This serves as another criterion for stochasticity along with the Chirikov resonance overlap criterion, which often give identical results in simple cases. We shall return to the notion of entropy in Chapter 8. Skipping the highest level of randomness, namely, the Bernoulli flow, we only recall that every higher step on the ladder of ergodicity implies the lower ones, but not *vice versa*. Usually, K-flow is the common case one encounters in "Hamiltonian chaos".

Since quantum chaos is closely related to non-integrable Hamiltonian systems and it will not be touched in this volume[2], a few remarks may be in order. Chaos is essentially a classical phenomenon. It is not typical for quantum mechanics. For time-independent systems the problem reduces to what has been known in energy level distributions of compound nuclei and eigenvalue distributions of random matrices of certain types, the essential new understanding is only a small number of degrees of freedom may lead to non-Poisson distributions of energy levels in non-integrable systems. In time-dependent systems, quantum chaos is mostly the reminiscence of classical chaos. In scattering problems there may exist chaotic behaviour without classical analog. For example, the phase shift may be as "random" as the distribution of the imaginary parts of the roots of the Riemann $\varsigma$-function, but there is no sensitive dependence on initial conditions.

To conclude this chapter let us emphasize that instability does not mean collapse of the system, but opens the way to intrinsic stochasticity. A deeper thought on

---

[2]Quantum non-integrability and quantum chaos will be the subject of the forthcoming volume 4 of *Directions in Chaos*, edited by D-H Feng and J-M Yuan and to be published by World Scientific.

the schematic parallel

$$stability \longleftrightarrow determinism$$
$$instability \longleftrightarrow randomness$$

would help the reader to get rid of the traditional prejudice that classical mechanics is fully deterministic and to recognize stochasticity as an ubiquitous and intrinsic property of nonintegrable Hamiltonian systems.

# Chapter 3
# THE PROBLEM OF ONSET OF TURBULENCE

As we mentioned in Chapter 1, turbulence has been a long-standing problem in physics. The difficulty is rooted in the simultaneous presence of many, many length scales, or in other words, in the lack of a single characteristic length. This can be seen from the intuitive picture of a turbulent fluid: nested and interpenetrated eddies of all scales, from macroscopic down to "molecular". In this respect the problem of turbulence bears similarity to the problem of continuous phase transitions, where length scales, ranging from the correlation length, which approaches infinity at the transition temperature, down to the atomic scale, all play an essential role. Perhaps this explains why people like L. D. Landau and K. G. Wilson who contributed so much to the understanding of phase transitions also thought about turbulence. However, there is an important difference that makes the physics of turbulence even more difficult than phase transitions, namely, the coexistence of large scale "coherent" structures with erratic motion at much smaller scales.

Just as in the case of phase transitions, the key to understanding turbulence may be hidden in the onset mechanism, as pointed out by L. D. Landau almost fifty years ago: "...the problem may be in a new light if the process of initiation of turbulence is examined thoroughly" (Paper 4 in this volume). We would like to make it clear from the very beginning that chaotic phenomena in dissipative systems, at least for the time being, are relevant only to the onset mechanism of turbulence, i.e. to the stage of weak turbulence. It says very little about fully developed turbulence which is of primary importance in engineering. In addition, chaos as it is treated in this volume concerns mainly erratic behaviour in time evolution, whereas turbulence necessarily involves stochasticity in the spatial distribution as well.

Figure 5: A stable fixed point.

Figure 6: An unstable fixed point.

Real turbulence occurs in three-dimensional space. However, most mathematical models and experimental situations studied so far are confined to finite or low-dimensional geometry, e.g., fluid instability between rotating cylinders (the Couette-Taylor instability) or thermoconvective instability in small boxes (the Rayleigh-Bénard instability). Developed turbulence must involve a great number of fluid motion modes, but the onset of turbulence may stem from the loss of stability of only a few modes. Finite geometry together with dissipation just provides the mechanism to suppress many irrelevant modes what makes the experimental results closer to predictions based on simple theoretical models.

It has been realized for a long time that turbulence might be a sophisticated regime of nonlinear oscillation in continuous media. L. D. Landau and E. Hopf attributed the onset of turbulence to the appearance of an increasing number of quasiperiodic motions resulting from successive bifurcations in the system. Being an extension of the Hopf bifurcation idea, the first steps of this process can easily be understood geometrically.

When the Reynolds' number $R$, which represents the relative importance of nonlinear to dissipative terms in the Navier-Stokes equations (1), remains small enough, the fluid motion is laminar and stationary, corresponding to a stable *fixed point* in its phase space. A stable fixed point acts as an *attractor*, i.e., it attracts all nearby initial points towards itself (see Fig. 5). Now, let the Reynolds' number be increased infinitesimally larger than the first critical value $R_{c1}$ where the fixed point loses stability and begins to repel all nearby trajectories (Fig. 6). Since a small change in $R$ cannot cause such drastic consequence as inverting the direction of all flows in the whole phase space, the neighbourhood of the fixed point may become repelling, but it must remain attracting with respect to regions located far enough. Local repulsion and global attraction of the flow implies the formation of a closed curve around the now unstable fixed point and the curve attracts all nearby flows (see Fig. 7). This closed curve is called a *limit cycle* and corresponds to periodic motion of the system.

The process of generating a limit cycle from a fixed point is called a *Hopf bifurcation*. Repeated use of the above arguments would reveal the nature of the next bifurcation when the limit cycle loses stability and becomes repelling at another critical value $R_{c2}$. There would appear an attracting closed tube, i.e. a 2-torus, around the unstable limit cycle. The motion becomes quasiperiodic if the two fre-

Figure 7: A compromise appearance of a limit cycle.

quencies on the torus are incommensurable. Landau and Hopf allowed this process
to continue infinitely and identified the final state with an infinite number of incommensurable frequencies as fully turbulent. This was called the *Landau-Hopf route*
(or scenario) to turbulence.

At present, we do not know any reasonable mathematical model which follows
the Landau-Hopf route to turbulence. The Landau-Hopf route requires successive
appearance of new incommensurable frequencies in the power spectrum, which remains discrete all the time as Reynolds' number increases. Turbulent spectra in
laboratory experiments do develop a few independent frequencies, then turn into
broad noisy bands (Gollub and Swinney, 1975). There is no mechanism for sensitive
dependence on initial conditions in Landau-Hopf scheme, but the details of turbulent states do depend on initial conditions sensitively. In addition, the Landau-Hopf
picture ignores an important physical phenomenon – frequency locking. In fact, in
nonlinear systems new incommensurable frequencies cannot appear without interacting with each other. Nearby frequencies tend to get locked, which will diminish
the number of independent frequencies. The above remarks exclude the Landau-
Hopf route as an onset mechanism for turbulence from both the theoretical and the
experimental points of view.

In 1971 Ruelle and Takens (Paper 5 in this volume) showed that the Landau-
Hopf route is unlikely to occur in nature. It is enough to have four consecutive
bifurcations to get into a state of erratic motion described by interweaving trajectories attracted to a low-dimensional manifold in the phase space called a *strange
attractor*. They identified the motion on strange attractors with turbulence. Their
scheme may be summarized as: fixed point → limit cycle → 2-torus → 3-torus →
strange attractor (turbulence). A few years later, in collaboration with Newhouse,
these authors succeeded in reducing the scheme to: fixed point → limit cycle → 2-
torus → strange attractor, i.e., quasiperiodic motion on a 2-torus may lose stability
and give birth to turbulence directly. This so-called *Ruelle-Takens route* to turbulence seems to be more consistent with recent hydrodynamical experiments, but it
is less well understood in theoretical models except for some nice results on circle
mappings. We shall devote Chapter 7 to transitions from quasiperiodic motion to
chaos.

Perhaps this is the right place to insert a loose discussion on a few frequently
used terms such as attractor, attracting set, strange attractor, and chaotic attrac-

tor, etc. First of all, these notions are applicable to dissipative systems only. Any of these is a collection of points in the phase space; all of them are invariant under the dynamics. An attracting set has a fundamental neighbourhood and a basin of attraction, any point in the basin will eventually move (be attracted) into the fundamental neighbourhood and will never get out. An attracting set may consists of several pieces, not all of them are attracting. An attractor may not have an open basin of attraction, but it is made of one irreducible attracting piece. An attractor may be thought as an experimental object where all observed points accumulate as time goes on. A strange attractor, as defined by Ruelle and Takens, must exhibit sensitive dependence on initial conditions, a property related to the dynamics, not only the geometry. For more rigorous definitions we recommend the nice booklet by Ruelle (B1989). However, not all authors use exactly the same definition. For example, Grebogi and coworkers attribute geometric aspects (Cantor-set like structure, nowhere differentiable, etc.) to the word "strange", while characterize dynamic property of sensitive dependence on initial conditions as being "chaotic". In this sense they introduced the notion of strange, nonchaotic attractors, obtained mostly by quasiperiodic forcing (see, e.g., Grebogi et al., 1984; Ding et al., 1989 a and b).

The notion of stable and unstable manifolds, which have been mentioned in Chapter 2, applies also to dissipative systems. At any point of the phase space the linear space tangent to a trajectory may be decomposed into an expanding and a contracting subspace, with, probably, a neutral subspace. If this decomposition carries over uniformly to the whole attractor and there is no neutral directions, the attractor is said to be *hyperbolic*. If not, it is a nonhyperbolic attractor. In particular, tangent contacts of stable and unstable manifolds lead to nonhyperbolic attractor.

The work of Ruelle and Takens played an eye-opening role. More and more people now believe that at least the problem of turbulence onset can be settled within the framework of Navier-Stokes equations and many new routes to turbulence have been suggested, leading to a situation of "all routes lead to turbulence", among which the most thoroughly studied are the *period-doubling route* of Feigenbaum and the *intermittent route* of Pomeau-Manneville. These two routes are actually twin phenomena. The next Chapter 4 on one-dimensional mappings is at the same time an introduction to the period-doubling route and Chapter 6 will deal with the intermittent route, so we shall not go into details here. We note also that there has been suspicion as regards the relevance of attractors to turbulence (Crutchfield and Kaneko, 1988).

In conclusion we would like to point out that turbulence has become a general concept, related to many branches of natural sciences and not less important than the concept of order. New terms such as solid state turbulence, chemical turbulence, acoustic turbulence or optical turbulence have been emerging into the literature. Subtle measurements at liquid helium temperatures, laser Doppler velocimetry and modern data acquisition technique have brought the study of turbulence back into physics laboratories. The idea of scaling and universality as well as renormalization group arguments which have proved so successful in understanding phase transitions are now being adapted to chaos and turbulence. In one word, turbulence should not be considered as a specific problem in hydrodynamics. It should attract the attention of physicists, because "That is the central problem which we ought to solve some day, and we have not"[3].

_____

[3]R. P. Feynman, R. B. Leighton, and M. Sands, *The Feynman Lectures on Physics*, Vol. I, pp.3-10, Addison-Wesley, 1963.

# Chapter 4
# UNIVERSALITY AND SCALING PROPERTY
# OF ONE-DIMENSIONAL MAPPINGS

It often happens in physics that one-dimensional models are either too trivial or too specific to be extended to higher dimensions, but chaos in dissipative systems offers a lucky and instructive exception. The reason is very simple: dissipation plays a global stabilizing role against local orbital instability and causes the volume representing the initial states in phase space to contract in the process of evolution. This contraction makes the phase volume approach one-dimensional objects in some of its sections and thus enables higher dimensional systems to enjoy some of the universal properties of one-dimensional mappings.

One-dimensional mappings of an interval into itself are simple enough to be accessible to certain analytical tools and are not very time-consuming in numerical study. At the same time they are rich enough to show many of the scaling and universal properties of chaotic transitions observed in higher dimensional systems. Therefore, we shall treat them in more detail than we did in previous chapters.

### Unimodal Maps

Consider a real interval $I$ and a nonlinear function $f$ which transforms a point $x$ of $I$ into some point $x'$ in the same interval $I$. This is called a *map of the interval* the interval $I$

$$f: I \to I. \tag{12}$$

In general, the function $f$ may depend on a parameter $\mu$. We can choose an arbitrary initial point $x_0 \in I$ and iterate it using $f$:

$$x_{n+1} = f(\mu, x_n), \quad n = 0, 1, 2, \cdots. \tag{13}$$

Formula (13) can be rewritten as

$$x_n = f^n(\mu, x_0), \quad n = 1, 2, 3, \cdots, \tag{14}$$

where $f^n$ denotes the $n$-th iterate (not derivative!) of $f$:

$$f^n(\mu, x) \equiv \underbrace{f(\mu, f(\mu, \cdots f(\mu, x) \cdots))}_{n \text{ times}}. \tag{15}$$

A more convenient notation for the kind of nested functions (or functional composition) in (15) is to thread the functions by using small circles ∘. For example, the composition of four functions $f$, $g$, $h$, and $k$ can be written as

$$f(g(h(k(x)))) \equiv f \circ g \circ h \circ k(x). \tag{16}$$

It will be quite rewarding to get accustomed to using ∘ instead of writing many nested parentheses. In general, the property of the sequence $\{x_i, i = 0, 1, 2, \cdots\}$ depends on the function $f$ and on the choice of $x_0$ and $\mu$. We confine ourselves to those functions $f$ which have only one maximum on the interval $I$. Without loss of generality one can rescale $f$ and $x$ in such a way that

1. The maximum is located at $x = 0$, $f'(\mu, 0) = 0$, and $f(\mu, 0) = 1$;

2. $f(\mu, x)$ is monotonically increasing when $x < 0$, and monotonically decreasing when $x > 0$. This kind of mappings has been called *unimodal*. We further require that

3. in the neighbourhood of $x = 0$, $f$ can be expanded as

$$f(\mu, x) = 1 - ax^z + \cdots, \tag{17}$$

   where $z = 2, 4, 6, \cdots$, etc.

For concreteness we shall refer to the *logistic map*

$$x_{n+1} = 1 - \mu x_n^2 \tag{18}$$

as the representative of unimodal mappings. In this case $I = [-1, +1]$, $\mu \in (0, 2)$, and $z = 2$. Sometimes it is more convenient to rescale $x_n$ by letting $\mu x_n$ be the new $x_n$. The logistic map then reads

$$x_{n+1} = \mu - x_n^2. \tag{19}$$

The interval becomes $I = [-\mu, \mu]$, but nothing changes with the parameter $\mu$. We note that, in general, when $x$ runs over the whole interval $I$, $f(\mu, x)$ does not fill up the interval $I$ (as long as $\mu < 2$), hence the name *endomorphism* (endo = internal) of the interval in some mathematical literature.

## Stability of Fixed Points

New let us return to the sequence $\{x_i, i = 0, 1, 2, \cdots\}$. Usually after a few hundred transient points, it settles into one of two kinds of stationary patterns: periodic or aperiodic. In what follows we shall assume that the transients have died away. A periodic pattern of period $p$: $x_{i+p} = x_i$, $x_{i+k} \neq x_i$ for all $k < p$ and all $i$, is also called an orbit of period $p$ or a $p$-cycle for the mapping $f$. The particular case $p = 1$ corresponds to a *fixed point* for $f$:

$$x^* = f(\mu, x^*). \tag{20}$$

Naturally, each point from a $p$-cycle of $f$ must be a fixed point of the $p$-th iterate of $f$, i.e.

$$x_i = f^p(\mu, x_i), \quad i = 1, 2, \cdots, p. \tag{21}$$

A standard question to be asked about a fixed point is its stability, i.e., if $x_n$ is chosen very close to the fixed point $x^*$:

$$x_n = x^* + \epsilon_n,$$

what happens with the next iterate $x_{n+1} = x^* + \epsilon_{n+1}$ ? If

$$|\epsilon_{n+1}/\epsilon_n| < 1, \tag{22}$$

the fixed point $x^*$ is stable. The stability condition (22) is equivalent to the requirement

$$|f'(\mu, x^*)| < 1. \tag{23}$$

Similarly, a $p$-cycle is stable if

$$|(f^p)'| \equiv \left| \prod_{i=1}^{p} f'(\mu, x_i) \right| < 1, \quad i = 1, 2, \cdots, p \tag{24}$$

where the chain rule of differentiation has been used.

The most favourable case for stability appears when

$$\left| \prod_{i=1}^{p} f'(\tilde{\mu}, x_i) \right| = 0, \quad i = 1, 2, \cdots, p \tag{25}$$

then one has quadratic convergence towards the fixed point. The stability condition (24) holds for a finite interval of $\mu$, called a *periodic window*, whereas (25) takes place only at one particular value $\tilde{\mu}$ somewhere in the middle of the periodic window. This $\tilde{\mu}$ value corresponds to a *superstable period*, which serves as the representative of all $p$-cycles from the same periodic window. With our conventions on $f$ we can say that any cycle containing the point $x = 0$ must be superstable, since $f'(\mu, 0) = 0$ leads to (25) for all $p$. This suggests an idea to determine the superstable value $\tilde{\mu}$ by solving the fixed point equation (21) together with (25), but we will mention a better method soon.

## Period-Doubling Cascade

Equipped with a desk calculator, one can easily find the first periodic windows for the logistic map (18):

$$
\begin{aligned}
p &= 1 & 0 &< \mu < \mu_1 = 0.75 \\
p &= 2 & \mu_1 &< \mu < \mu_2 = 1.25 \\
p &= 4 = 2^2 & \mu_2 &< \mu < \mu_3 = 1.3680989 \cdots \\
p &= 8 = 2^3 & \mu_3 &< \mu < \mu_4 = 1.3940461 \cdots \\
& \cdots & & \cdots .
\end{aligned} \tag{26}
$$

This is a *period-doubling bifurcation cascade* with period $p = 2^n$ which quickly converges to an aperiodic orbit at $n = \infty$, the value $\mu_\infty = 1.401155 \cdots$ being approached as a geometric progression, namely,

$$\mu_n \approx \mu_\infty - \frac{A}{\delta^n}, \quad \text{as } n \to \infty, \tag{27}$$

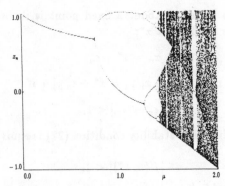

Figure 8: Bifurcation diagram of the logistic map.

where $\delta = 4.66920\cdots$ is a universal (for unimodal mappings with $= 2$) constant, first discovered by Feigenbaum (Paper 7 in this volume).

What happens when $\mu$ gets larger than $\mu_\infty$ is more interesting from the viewpoint of chaos. In the parameter range $(\mu_\infty, 2)$ there exists an infinite number of periodic windows immersed in the background of aperiodic regime. If one examines more carefully the distribution of points in the aperiodic regime, then one sees the iterates jumping between $2^n$ subintervals (or islands) of the interval $I$ with $n$ decreasing from $\infty$ to 0 when $\mu$ goes from $\mu_\infty$ to 2. This is the so-called *reversed*, or *period-halving bifurcation sequence* of chaotic bands, which looks much like a washed-out mirror image of the direct period-doubling bifurcation sequence with respect to $\mu_\infty$. This overall structure can be seen clearly from a *bifurcation diagram*, obtained by plotting a few hundred stationary outputs of the iteration (18) versus the parameter $\mu$, see Figure 8. We call the attention of the reader to a characteristic feature of this kind of bifurcation diagrams: there are many dark lines going through the chaotic zones and they may become sharp boundaries of the latter. The equations for all these lines can be written down explicitly by using the same mapping (13) to define a family of functions $\{P_n(\mu)\}_0^\infty$ of the parameter:

$$P_{n+1}(\mu) = f(\mu, P_n(\mu)), \quad n = 0, 1, \cdots, \tag{28}$$

with the initial function $P_0(\mu) = C$ (see Fig. 1 of Paper 9 in this volume).

## Universal and Scaling Properties

If what has been described above is restricted only to a specific mapping, e.g., (18), it would be worth no more than a rare bird in the mathematical zoo. However, this kind of bifurcation structure with its universal numerical characteristics (such as the Feigenbaum constant $\delta$) appears very frequently in nonlinear mathematical models and real experiments. In what follows we list some more useful notions and results related to the universal and scaling property of one-dimensional mappings.

1. A unimodal map can have at most one stable period for each parameter value $\mu$. It may have no stable period at all for many $\mu$ values. The necessary

condition for $f$ to have at most one stable period was found by Singer (1978) and consists in the *Schwarzian derivative* [4] $Sf$ of $f$ being negative on the interval $I$:

$$SF(x) \equiv \frac{f'''(x)}{f'(x)} - \frac{3}{2} \left( \frac{f''(x)}{f'(x)} \right)^2 < 0. \tag{29}$$

The condition (29) is not sufficient for the stable period to exist. It is just this insufficiency that opens the possibility for chaotic orbit to appear: even when $Sf < 0$ one can get different aperiodic sequences starting from different $x_0$ and never reach a stable period.

In fact, Singer's theorem says a one-dimensional map with $n$ critical points can have at most $n + 2$ stable periodic orbits, the number 2 comes from possible contribution of the end points. If the derivative $f'$ is not too close to zero at the end points, it is often safe to say that a map with $n$ extrema may possess at most $n$ stable periodic orbits.

2. The set of all parameter values which give birth to chaotic orbits possesses a positive measure on the $\mu$-axis, although no such $\mu$ values form an interval. Moreover, to classify an orbit as chaotic one must show that all consecutive points approach a continuous distribution with respect to $dx$. Computers are of no use in proving this kind of statements, and one must appeal to rigorous mathematics. Proofs for some particular parameter values have been known since long, but a general proof for certain classes of mappings appeared only in 1981 (Jakobson, 1981). Since then some stronger results have been proved, e.g., now we know that maps with positive Lyapunov exponents do have positive measure on the parameter axis[5]. These mathematical relsults make us feel safer, although the proofs are beyond the reach of most non-mathematicians.

3. There was a theorem proved in 1964, which became known to physicists much later (Stefan, 1977) and which has been frequently mentioned in the literature:

*Sarkovskii theorem* (Sarkovskii, 1964): consider the following ordering of integers

$$3 \prec 5 \prec 7 \prec 9 \prec \cdots \prec 3 * 2 \prec 5 * 2 \prec 7 * 2 \prec 9 * 2 \prec \cdots$$
$$\prec 3 * 2^2 \prec 5 * 2^2 \prec 7 * 2^2 \prec 9 * 2^2 \prec \cdots 3 * 2^n \prec 5 * 2^n \prec 7 * 2^n$$
$$\prec 9 * 2^n \prec \cdots \prec 2^n \prec \cdots \prec 32 \prec 16 \prec 8 \prec 4 \prec 2 \prec 1.$$

(the symbol $\prec$ means "precede"). If $f$ is an unimodal mapping and has a point $x$ leading to a $p$-cycle, then it must have a point leading to a $q$-cycle for every $q$ that follows $p$ in the sense of the above ordering.

It must be emphasized that the Sarkovskii theorem is only a statement concerning periodic orbit at a fixed parameter value $\mu$. It says nothing about the stability of the orbits, nor about the measure, i.e., the observability of these periods. Unaware of the work of Sarkovskii, many equivalent formulations or corollaries of this theorem were suggested in the intervening years, among which the Li-Yorke theorem "Period 3 implies chaos" (Li and Yorke, 1975), etc.

The classification and ordering of periodic orbits has been a well-studied problem with an extensive literature. In particular, the classification and enumeration of

---

[4]For more on the Schwarzian derivative, see, e.g., E. Hille, *Ordinary Differential Equations in the Complex Domain*, Chapter 10, Wiley, 1976.

[5]M. Benedicks, and G. Carleson, preprint.

stable periodic orbits for unimodal maps has been completely solved. We will touch the problem when describing symbolic dynamics in the second half of this chapter.

4. The *renormalization group equation* and the *universal scaling factor* $\alpha$. Generally speaking, behind any renormalization group arguments there always figures some geometry with infinitely-nested self-similar internal structure (a kind of fractal geometry, see Mandelbrot, B1982). One-dimensional mappings provide us with a simple example.

The very instructive discussion on scaling propert of one-dimensional mapping in Feigenbaum's first paper (Paper 7 in this volume) was illustrated by several not well-proportioned figures, so we redraw them in Fig. 9. Only the vicinity of a certain fixed point is shown in these figures. A part of $f^{(2^n)}$ at the exact superstable parameter $\tilde{\mu}_n$ is given in Fig. 9(b). If we draw $f^{(2^{(n-1)})}$ at the same parameter value $\tilde{\mu}_n$, it represents the period-doubled regime shown in Fig. 9(a), where locally one has a few superstable 2-cycles as outlined by the square boxes. Now increase $\mu_n$ from the situation of Fig. 9(b), until a new period-doubled superstable regime appears at the next parameter value $\tilde{\mu}_{n+1}$, as shown by the smaller boxes in Fig. 9(c). A comparison of the two hatched boxes in Figs. 9(c) and 9(a) shows what happened was a rescaling and change of signs in both the $x$ and $y$ directions. This process repeats itself with $\mu$ increasing.

To put the above geometrical observation into a mathematical frame, let us introduce an operator $T$ to represent this period-doubling, $\mu$-shifting and rescaling proceduce starting from some fixed $\mu_n$, i.e.,

$$T f(\mu_n, x) = \alpha f(\mu_{n+1}, f(\mu_{n+1}, x/\alpha)),$$
$$T^2 f(\mu_n, x) = \alpha^2 f^2(\mu_{n+2}, f^2(\mu_{n+2}, x/\alpha^2)), \tag{30}$$
$$\cdots$$

Feigenbaum gave some plausible arguments to conjecture the existence of a universal, i.e., independent on the starting function $f$, limit

$$\lim_{k \to \infty} \alpha^k f^{2^k}(\mu_{n+k}, x/\alpha^k) = g(x). \tag{31}$$

This conjecture was proved later (Lanford, 1982b; Eckmann and Wittwer, 1987). It follows from the definition (15) of $f^n$ that

$$f^{2^k}(\mu_{n+k}, x/\alpha^k) = f^{2^{k-1}}(\mu_{n+k}, f^{2^{k-1}}(\mu_{n+k}, x/\alpha^k)). \tag{32}$$

Taking the $k \to \infty$ limit on both sides of (32) results in the renormalization group equation of Feigenbaum:

$$g(x) = \alpha g(g(x/\alpha)). \tag{33}$$

The normalization of $f(\mu, 0) = 1$ and the superstable condition (25) imply the following boundary conditions for (33):

$$g(0) = 1, \quad g'(0) = 0. \tag{34}$$

It is worth mentioning that conditions (34) alone do not determine the solution of (33) uniquely. At least the behaviour of $g(x)$ near $x = 0$ should be given, e.g., the series expansion

$$g(x) = 1 + Ax^z + Bx^{2z} + Cx^{3z} + \cdots \tag{35}$$

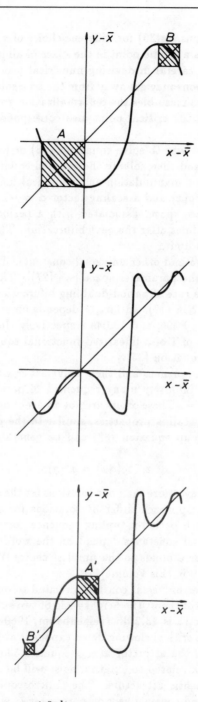

Figure 9: (a). A segment of $f^{(2^{n-1})}(\tilde{\mu}_n, x)$, $\tilde{\mu}_n$ is the superstable point for $f^{(2^n)}$. (b). A segment of $f^{(2^n)}(\tilde{\mu}_n, x)$. (c). A segment of $f^{(2^n)}(\tilde{\mu}_{n+1}, x)$ to be rescaled and inverted to get (a) again.

will give different solutions to (33) for different choice of $z$.

Equation (33) defines a saddle point in the space of all functions leading to $g(x)$. Therefore, one must be careful in devising numerical procedure to calculate $g(x)$, because the approximations may get away from the true solution after converging to it initially. This situation resembles the renormalization group theory of continuous phase transitions where the critical point also corresponds to an unstable saddle point.

Actually the simplest way to solve equations (33) or (36) consists in substituting the expansion (35) and then solving the resulted equation for $A, B, \cdots$ etc., by combined use of algebraic manipulation and numerical languages. The results will be a universal function $g(x)$ and a scaling factor $\alpha = -2.5029 \cdots$. All characteristic lengths in the phase space, associated with a periodic orbit, will be scaled down approximately $\alpha$ times after the next bifurcation. This ratio becomes exactly $2.5029 \cdots$ in the $n \to \infty$ limit.

5. Convergence rate $\delta$ and other universal constants. The Feigenbaum constant $\delta$ is a scaling factor in the parameter space [see (27)]. The cited value $4.66920 \cdots$ refers to the convergence rate of period-doubling bifurcation sequences in quadratic mappings, i.e., for $z = 2$ in (17). In fact, $\delta$ depends on $z$ smoothly and its values for $z = 4, 6, 8$ are 7.248, 9.296, and 10.048 respectively. In general, $\delta$ is defined as the eigenvalue in excess of 1 of a linearized functional equation obtained from the renormalization group equation (33).

6. Period-$n$-tupling sequences and their universal constants. In the bifurcation diagram one can select infinitely many sequences of periodic orbits with periods growing as $n^k$, $n = 3, 4, \cdots$. These orbits are not adjacent on the parameter axis, but they do enjoy universal scaling properties, similar to the period-doubling cascade. The renormalization group equation (33) can be generalized to period-$n$-tupling sequences:

$$\alpha^{-1} g(\alpha x) = g^n(x). \tag{36}$$

When $n \geq 4$, (36) may have more than one solutions for the same $n$ and $z$. One needs better initial values to approach different solutions for various period-$n$-tupling sequences. Actually, each period-$n$-tupling sequence correponds a basic symbolic word $W$, all the universal constants depend on the word: $\delta_W$, $\alpha_W$, and $\kappa_W$, etc. The Feigenbaum case corresponds to the simplest choice $W = R$ (Zeng, Hao, Wang and Chen, 1984, Paper 9 in this volume).

There are many other universal constants related to one-dimensional unimodal mappings. We list some of them: the height ratio of two subsequent period-doubled peaks in the power spectra is 13.2 db (Feigenbaum, 1980b, c, d, 1981; Nauenberg and Rudnick, 1981), the critical slowing down exponent at $\mu_n$ is $\Delta = 1$ (Hao, 1981), the fractal dimension of the attractor at $\mu_\infty$ equals $0.538 \cdots$ (Grassberger, 1981), etc. Another exponent $\kappa$ related to external noise will be mentioned in Chapter 9.

7. The *crises* of chaotic attractors. The "microscopic" structure of achaotic attractor does not depend continuously on parameter $\mu$, whereas the general shape of the attractor changes smoothly with $\mu$ except for certain values where abrupt changes take place. This phenomenon was called crisis of the attractor and explained by Grebogi, Ott and Yorke (1982, see also Paper 13 in this volume), as the result of collisions of the attractor with an unstable orbit. Actually all the band-merging

points in the bifurcation diagram occur at collisions with unstable orbits left by the main period-doubling bifurcations from the other side of $\mu_\infty$. As we shall see below, maps with crises are described by symbolic sequences of $\rho\lambda^\infty$ type. The sudden change or disappearance of chatic attractors at "collisions" with other unstable objects certainly plays a more important role in higher dimensional systems.

## Aspects of Symbolic Dynamics

There exists a very powerful method to describe periodic and chaotic orbits in one-dimensional mappings systematically, namely, the method of *applied symbolic dynamics*. In a sense, everyone who enters the field of chaotic dynamics should start with symbolic dynamics. We give the essence of this beautiful theory in a few pages. We will use the logistic map as an example, although most of the results are extendable to maps with multiple humps and valleys.

We divide the phase space of a one-dimensional map, i.e., the interval $I$, into several subintervals on which the mapping function $f$ is monotone, either increasing or decreasing, and label each subinterval by a letter. For the logistic map (18) or (19) we have a *L*eft branch $L$ and a *R*ight branch $R$, divided by the *C*entral or *C*ritical point $C$.

*The number, symbol, and inverse function correspondence.* Taking an initial point $x_0$, we get a numerical orbit by iterating the function $f$ (we omit the parameter $\mu$ for simplicity). Replacing each number in the orbit by a letter $L$ or $R$ according to which side of $C$ it falls, we juxtapose the numerical orbit by a symbolic orbit or symbolic sequence:

$$x_0, \quad x_1 = f(x_0), \quad x_2 = f(x_1), \quad \cdots, \quad x_{n-1} = f(x_{n-2}), \quad x_n = f(x_{n-1}), \quad \cdots,$$
$$\sigma_0 \quad \sigma_1 \qquad \sigma_2 \qquad \cdots \quad \sigma_{n-1} \qquad \sigma_n \qquad \cdots,$$
$$(37)$$

where $\sigma_i$ may be either $R$ or $L$ (or $C$ if the number happens to be exactly at the critical point). It is useful to name a symbolic sequence by the initial point $x_0$, i.e., to introduce a convention

$$x_0 = \sigma_0\sigma_1\sigma_2\cdots\sigma_{n-1}\sigma_n\cdots. \tag{38}$$

Can we reverse the iteration process and write, e.g.,

$$x_0 = f^{-n}(x_n)?$$

No and yes. No, because the inverse function $f^{-1}$ is multivalued. Yes, if we look at the iterates more carefully. In fact, no problem arises with the forward iterations: each $x_i$ in the first line of (37) chooses its own monotonic piece of $f$ according to the symbol $\sigma_i$ in the second line. Therefore, we should have written

$$x_0, \quad x_1 = f_{\sigma_0}(x_0), \quad x_2 = f_{\sigma_1}(x_1), \quad x_n = f_{\sigma_{n-1}}(x_{n-1}), \quad \cdots,$$

attaching a subscript to each monotonic branch of the function $f$. With this done, we can safely write the inverses as

$$
\begin{aligned}
x_0 &= f_{\sigma_0}^{-1}(x_1) \\
&= f_{\sigma_0}^{-1} \circ f_{\sigma_1}^{-1}(x_2) \\
&= \cdots \\
&= f_{\sigma_0}^{-1} \circ f_{\sigma_1}^{-1} \circ f_{\sigma_2}^{-1} \circ \cdots f_{\sigma_{n-1}}^{-1}(x_n).
\end{aligned} \tag{39}
$$

Now the notations have become too cumbersome. To make one's life easier, let us name a monotone branch simply by its subscript, i.e., to use the symbols $\sigma_i$ also as function names:

$$
\sigma_i(y) \equiv f_{\sigma_i}^{-1}(y). \tag{40}
$$

For the logistic map (19) this means solving the quadratic equation $y = \mu - x^2$ and writing explicitly

$$
\begin{aligned}
R(y) &= \sqrt{\mu - y}, \\
L(y) &= -\sqrt{\mu - y}.
\end{aligned} \tag{41}
$$

Now the last line in (39) may be written as

$$
x_0 = \sigma_0 \circ \sigma_1 \circ \sigma_2 \cdots \sigma_{n-1}(x_n). \tag{42}
$$

A comparison of (37), (38) and (42) shows the "number – symbol – inverse function" correspondence. One-dimensional maps of the interval are sometimes called non-invertible. We see this is not very precise, since they can be inverted once symbolic dynamics removes the multi-valueness.

The *ordering of symbolic sequence* is based on the property of monotone functions: a monotonically increasing function $f$ preserves the order of its arguments, i.e., $x > y$ implies $f(x) > f(y)$, while a monotonically decreasing function $f$ reverses the order, i.e., from $x > y$ it follows that $f(x) < f(y)$. A functional composition of many monotone functions will act as an increasing function if it contains an even number or none of decreasing functions; it will be a decreasing function, if there is an odd number of decreasing components. Therefore, one may assign an even (or $+1$) parity to each increasing piece such as $L$, and an odd (or $-1$) parity to each decreasing piece such as $R$. The parity of a composite function is the product of parities of its components.

The convention (38) allows us to assign the order of the initial points to the corresponding symbolic sequences. This ordering, in fact, does not depend on the numbers. When comparing the order of two symbolic sequences

$$
\Sigma_1 = \Sigma^* \sigma \cdots
$$

and

$$
\Sigma_2 = \Sigma^* \tau \cdots,
$$

where $\Sigma^*$ denotes their common leading string and $\sigma \neq \tau$, we start from the natural order

$$
L < C < R. \tag{43}
$$

If $\Sigma^*$ is even, then the order of $\sigma$ and $\tau$ in the sense of (43) determines the order of $\Sigma_1$ and $\Sigma_2$; if $\Sigma^*$ is odd, then $\sigma > \tau$ implies $\Sigma_1 < \Sigma_2$.

The symbolic sequence starting from the iterate of $C$ plays a key role in symbolic dynamics and has been given a special name *kneading sequence*[6]. Clearly, not any arbitrary symbolic sequence may become a kneading sequence at a certain parameter value. A sequence must be *shift-maximal* in order to be a candidate for kneading sequence, i.e., it must be greater or equal to all its shifts. For unimodal maps all kneading sequences except for $L^\infty$ and $R^\infty$ start from $RL$, i.e.,

$$K = f(C) = RL \cdots. \tag{44}$$

*Admissibility condition* for symbolic sequences. A symbolic sequence is said to be admissible, if it can be reproduced by the map. With our convention (38) and (44) we see that kneading sequences may be taken as the most natural parameter for a unimodal map. Since $K$ is the iterate of $C$ and equals to the maximum of the map, any iterate of a point other than $C$ cannot be greater than $K$. Given an admissible sequence $\Sigma$ one can throw away any number of leading symbols, the remaining part must also be admissible. Therefore, the admissibility condition for $\Sigma$ consists in that all its shifts must not be greater than the kneading sequence $K$.

*Periodic window theorem.* A symbolic sequence that corresponds to a super-stable periodic orbit, must contain the letter $C$ [see what has been said after (25)]. A superstable kneading sequence $\Sigma C$ occurs only at a single parameter value $\tilde{\mu}$, but one may perturb $\tilde{\mu}$ slightly to change the letter $C$ into either $R$ or $L$, thus extending $\Sigma C$ to a window

$$[(\Sigma C)_-, \Sigma C, (\Sigma C)_+], \tag{45}$$

where

$$(\Sigma C)_+ = \text{the larger one of } \{\Sigma R \text{ and } \Sigma L\},$$
$$(\Sigma C)_- = \text{the smaller one of } \{\Sigma R \text{ and } \Sigma L\}.$$

For unimodal maps the larger one is odd, and the smaller one even. The periodic window theorem says: if $\Sigma C$ is admissible, so do $\Sigma R$ and $\Sigma L$. This is a consequence of continuity consideration and the natural order (43). The simplest example of this theorem is the period-doubling cascade (26). In fact, it can be described by the following sequence of triples:

$$(L, C, R) \; (RR, RC, RL) \; (RLRL, RLRC, RLRR) \; \cdots. \tag{46}$$

It is easy to see that all these triples have a signature $(+, 0, -)$, if one assigns a parity 0 to the letter $C$. Only the $-$ parity end of a triple may undergo period-doubling.

*Construction of median words.* Given two admissible superstable sequences $\Sigma_1 < \Sigma_2$, it is easy to construct the shortest superstable period $\Sigma^*$, included in between. One extends both sequences into windows according to the periodic window theorem, then compares the upper sequence $(\Sigma_1 C)_+$ of the smaller sequence with the lower sequence $(\Sigma_2 C)_-$ of the larger one. If these coincide, then $\Sigma_1 C$ and $\Sigma_2 C$ are close neighbours on the parameter axis. In fact, the latter is just the period-doubled regime of the former. If not, then their common leading part $\Sigma^*$ gives the answer[7]. Starting from two words, say, $C$ and $RL^{98}C$, one can construct

---

[6]Milnor and Thurston, 1977 preprint; see Milnor and Thurston, 1988.

[7]The method of MSS(1973) to construct median words via so-called harmonics and antiharmonics of words has become obsolete.

a table of all the admissible words of lengths not exceeding 100. In the Appendix of Metropolis, Stein and Stein (1973, Paper 8 in this volume) a table was given for $n \leq 11$. Symbolic sequences, as ordered in these tables, were called U-sequences by MSS, U standing for universal.

The *number* $N^*(n)$ *of different superstable periodic sequences* of given length $n$ is associated with many interesting enumeration problems. In particular, it may be calculated from the following recursion formula[8]:

$$2^n = 2 \sum_{\{d:d|n'\}}^{1 \leq d \leq n'} 2^k d N^*(2^k d),$$ (47)

where $k$ and $n'$ are determined from the decomposition $n = 2^k n'$ with $k \geq 0$, $n'$ odd, and $d|n'$ denotes a factor $d$ that divides $n'$. For more aspects of this problem and tables see Section 3.9 of Hao (B1989).

The *generalized composition rule*. The whole period-doubling cascade (46) may be obtained by applying a pair of substitutions

$$\begin{aligned} R &\to RL, \\ L &\to RR, \end{aligned}$$ (48)

to the first triple $(R, C, L)$. This hints on the possibility to consider more general substitutions of the following form

$$\begin{aligned} R &\to \rho, \\ L &\to \lambda, \end{aligned}$$ (49)

where $\rho$ and $\lambda$ are symbolic strings, made of $R$ and $L$. Indeed, rigorous conditions may be formulated, that allow one to get many longer admissible sequences from a shorter one by making substitutions like (49); for these conditions see Paper 10 in this volume.

The generalized composition rule includes the *∗-composition* of Derrida, Gervois and Pomeau (1978) as a particular case. If one chooses $\rho$ and $\lambda$ for the substitutions (49) from one and the same periodic window, i.e., let

$$\begin{aligned} R &\to \rho = (\Sigma C)_+, \\ L &\to \lambda = (\Sigma C)_-, \end{aligned}$$ (50)

and applies them to an admissible sequence $\Pi$, then the result is just what was denoted as $\Pi * \Sigma$ by Derrida *et al.* For convenience one could add to (50) a substitution rule for $C$, namely,

$$C \to \Sigma C.$$

We note that a symbolic sequence, decomposable into ∗-composition of shorter primitive words, has a characteristic fine structure in the power spectrum. A word $\Sigma$, representing a periodic orbit in the one-band chaotic zone, "∗-multiplied" by $R$

---

[8]Zheng Wei-mou, unpublished note.

from the right, i.e., $\Sigma * RC$, becomes its period-doubled regime. The period-doubling cascade (26) may be expressed as

$$R^{*n}C, \quad n = 0, 1, \cdots.$$

The same word, *-multiplied by $R$ from left, represents a similar orbit in the 2-band chaotic zone with twice the period. The hierarchy structure of the entire bifurcation diagram, schematically denoted as $[R, RL^\infty]$, may be described as $R^{*n} * [R, RL^\infty]$, for $n = 1, 2, \cdots$. Fine structures conforming to the *-composition and U-sequences have been observed in real and computer experiments.

A *standard chaotic map.* The logistic map (18) or (19) at parameter $\mu = 2$ has been understood more or less thoroughly. We list some of its properties.

1. It is a *surjective* map or a complete map, i.e., the interval $I$ now maps onto the whole interval.

2. The kneading sequence is $RL^\infty$, the largest among all possible sequences, made of $R$ and $L$. Its lower left corner is an unstable fixed point, described by $L^\infty$, the smallest among all possible sequences. All possible symbolic sequences are present between these two extremes. They are as many as real numbers in the interval $[0, 1]$. Any change of initial value will lead to a new type of symbolic sequence. This is just another way of saying that the dynamics has sensitive dependence on initial conditions.

3. It corresponds to a band-ending point in the bifurcation diagram.

4. It possesses *homoclinic points* and infinitely many *homoclinic orbits.*

5. It is a *crisis point* (Grebogi, Ott and Yorke, 1982), since the chaotic attractor collides with an unstable period 1 orbit and changes abruptly.

6. If one keeps iterating, then the points $\{x_i\}$ obey a continous distribution $\rho(x)$. For the logistic map this distribution may be written down explicitly (Ulam and von Neumann, 1947): $\rho(x) = 1/\pi\sqrt{1 - x^2}$.

7. It is an intersection point of all $P_n(\mu)$, $n \geq 2$ [see (28)].

Referring to the standard chaotic logistic map, we can characterize a *class of chaotic maps.* Namely, any periodic window $(\lambda, \rho|_C, \rho)$, where $\rho$ and $\lambda$ satisfy the conditions of the generalized composition rule (49), has a chaotic counterpart with kneading sequence $K = \rho\lambda^\infty$. It is as chaotic as the $\mu = 2$ logistic map in the following sense:

1. Although it cannot be a surjective map, but the $|\lambda|$-th iterate $f^{|\lambda|}$, where $|\lambda|$ denotes the number of letters in the string $\lambda$, is locally surjective: the part restricted to the subinterval

$$(f^{|\lambda|}(C) = \lambda^\infty, f(C) = K = \rho\lambda^\infty), \tag{51}$$

does map onto itself. In fact, there are $|\lambda|$ such subintervals, obtainable by applying the map to the above subinterval.

2. Any symbolic sequence in between $L^\infty$ and $RL^\infty$ in the surjective logistic map, may be put in correspondence with a sequence in between $\lambda^\infty$ and $\rho\lambda^\infty$ by substitutions $R \to \rho$ and $L \to \lambda$. They are as many as real numbers.

3. It corresponds to a band-merging point in the bifurcation diagram.

4. It possesses homoclinic points and homoclinic orbits.

5. It is a crisis point where the unstable orbit is described just by $\lambda^\infty$.

6. Although the chaotic attractor splits into several small islands, points within an island do satisfy a continuous distribution.

7. It is an intersection point of a subset of $\{P_n(\mu)\}$.

*Word-lifting technique.* Now we have seen the importance of superstable periodic kneading sequences $\Sigma C$ and "eventually periodic" sequences $\rho\lambda^\infty$ which represent a class of chaotic maps. Suppose we are given a concrete map, say, the logistic map (19), how to find the parameter value where one can observe the map with a given kneading sequence? The answer comes from a little trick, namely, the word-lifting technique.

We first consider the case of a periodic sequence $\Sigma C$.

With our convention (38) we can break the infinite sequence

$$C|\Sigma C \cdots$$

just after the first letter $C$ and write in accordance with (42)

$$f(C) = \Sigma(C), \tag{52}$$

where $C$ is the number corresponding to the critical point and $\Sigma(y)$ must be understood as a composite function in the sense of (42). Eq. (52) is an equation for the unknown parameter. Take, for example, the three different period 5 orbits $RLRRC$, $RLLRC$ and $RLLLC$ from the MSS table. We have to solve the following three equations

$$\begin{aligned}
f(C) &= R \circ L \circ R \circ R(C), \\
f(C) &= R \circ L \circ L \circ R(C), \\
f(C) &= R \circ L \circ L \circ L(C).
\end{aligned} \tag{53}$$

For the map (19) we have $C = 0$ and $f(0) = \mu$. Using the explicit expressions (41) for the inverse functions, we have

$$\begin{aligned}
\mu &= \sqrt{\mu + \sqrt{\mu - \sqrt{\mu - \sqrt{\mu}}}}, \\
\mu &= \sqrt{\mu + \sqrt{\mu + \sqrt{\mu - \sqrt{\mu}}}}, \\
\mu &= \sqrt{\mu + \sqrt{\mu + \sqrt{\mu + \sqrt{\mu}}}}.
\end{aligned} \tag{54}$$

This kind of equations can be solved by transforming into iteration schemes (Kaplan, 1983). For instance, the first one in (54) becomes

$$\mu_{n+1} = \sqrt{\mu_n + \sqrt{\mu_n - \sqrt{\mu_n - \sqrt{\mu_n}}}}.$$

With a suitable choice for $\mu_0$, say, $\mu_0 = 2$, the three equations in (54) lead to $1.62541\cdots$, $1.86078\cdots$, and $1.98542\cdots$, respectively. One should not try to get rid of the square roots in (54) by squaring them; all the three equations will become one and the same $P_5(\mu) = 0$ [see (28)]. When one deals with long periods, it will cause serious numerical problem: $P_{26}(\mu) = 0$ has more than one million real roots, all populated in the interval $(1.401\cdots, 2)$!

For an infinite kneading sequence $K = \rho\lambda^\infty$ one proceeds similarly and writes

$$f(C) = \rho \circ \lambda \circ \lambda \circ \lambda \circ \cdots. \tag{55}$$

We remind the reader that $\rho$ and $\lambda$ themselves must be understood as composite functions of their constituent letters — inverse functions. Denoting the result of the infinite composition of $\lambda$ by $\nu$ and defining $\nu$ recursively, we get a pair of equations

$$\begin{aligned} f(C) &= \rho(\nu), \\ \nu &= \lambda(\nu). \end{aligned} \tag{56}$$

As an example, we take the period 3 window $(RLR, RLC, RLL)$. Its chaotic zone ends at $K = RLL(RLR)^\infty$. For the logistic map (19) the parameter is determined from

$$\begin{aligned} \mu &= R \circ L \circ L(\nu), \\ \nu &= R \circ L \circ R(\nu). \end{aligned}$$

After transforming into a pair of iterations, we get $\mu = 1.79032749\cdots$ and $\nu = 1.74549283\cdots$. This is the exact location of crisis first studied by Grebogi, Ott and Yorke (1982). The numerical subinterval $(\nu, \mu) \in I$ corresponds to the symbolic range $(\lambda^\infty, \rho\lambda^\infty)$. It is precisely the subinterval (51) where "local surjectivity" takes place.

*Symbolic dynamics analysis of symmetry breaking.* As an example to keep symmetry of the dynamics we take the antisymmetric cubic map

$$x_{n+1} = Ax_n^3 + (1 - A)x_n, \tag{57}$$

which has three monotone branches, to be denoted by $R$, $M$, and $L$, respectively, and two critical points $C$ and $\overline{C}$. We have the natural order:

$$L < \overline{C} < M < C < R.$$

Map (57) does not change under the transformation $x \to -x$, i.e., under symbolic transformation

$$\begin{aligned} R &\longleftrightarrow L, \\ C &\longleftrightarrow \overline{C}, \\ M &\longleftrightarrow M. \end{aligned} \tag{58}$$

The last line means keeping $M$ unchanged. A superstable orbit must contain either $C$ or $\overline{C}$ or both. Suppose it contains only $C$, i.e., it looks like $C\Sigma C\cdots$, where $\Sigma$ is a string made of $R$, $M$ and $L$. This orbit must be asymmetric due to the fact that it does not contain $\overline{C}$. By applying to it the symmetry transformation (58) we get another asymmetric orbit $\overline{C}\Sigma\overline{C}\cdots$. They are the two symmetrically located asymmetric orbits.

Now it is clear that a symmetric orbit must be of the form $C\Sigma\overline{C}\Sigma C\cdots$, the simplest case is $\Sigma$ being blank: $C\overline{C}\cdots$. This is the symmetric period 2 one sees in the bifurcation diagram of map (57). We can disturb $C$ and $\overline{C}$ slightly to bring them into $R$ and $L$, or into $M$ and $M$, respectively. The triple $(MM, C\overline{C}, RL)$ describes the whole window of symmetric period 2. One can apply the symmetry transformation (58) to it to get another triple $(MM, \overline{C}C, LR)$. However, these two triples are the same, because they can be brought one into another by cyclic permutation. If we take the last term $RL$ or $LR$ of these two triples, we can extend them into another two triples $(RL, R\overline{C}, RM)$ and $(LR, LC, LM)$, which cannot be brought one into another by cyclic permutation. These are the two asymmetric period 2; only one of them may be realized at one time. If we check the parity of these triples, then the first two have signature $(+, 0, +)$, thus cannot undergo period-doubling, while the last two have signature $(+, 0, -)$, capable of period-doubling. The corresponding band-merging point is $RL(RM)^\infty$ or $LR(LM)^\infty$, where the full symmetry restores. This explains a phenomenon known in the literature as "suppression of period-doubling by symmetry breaking" (Swift and Wiesenfeld, 1984; Bryant and Wiesenfeld, 1986; Wiesenfeld and Tuffilaro, 1987; Pieranski, 1988). Moreover, we see that only even periods are capable of undergoing symmetry breaking, but not all even periods can do so. The selection rule follows from symbolic dynamics. For more details we refer to Zheng and Hao (1989).

Symbolic dynamics originated in topological theory of dynamical systems in the 1930's and was later used as a tool in ergodic theory by Bowen, Sinai, Ruelle and others. For physicists, symbolic dynamics is nothing but a mathematical realization of coarse-graining or reduction of description. The method of symbolic dynamics is expected to play more role in physics (see, e.g., the Preface of J. Ford to Alekseev and Yakobson, 1981). For a brief summary of applied symbolic dynamics see Paper 10 in this volume. More detailed exposition may be found in the monograph (Hao, B1989) and in a recent review[9].

---

[9]Zheng Wei-mou, and Hao Bai-lin, "Applied symbolic dynamics", in *Directions in Chaos*, vol. 3, World Scientific, 1990.

# Chapter 5
# BIFURCATION AND CHAOS IN HIGHER DIMENSIONAL SYSTEMS

Most physical processes are described by ordinary or partial differential equations, so we are more concerned with chaotic behaviour in such higher dimensional systems. Discrete mappings of a plane or an annulus occupy an intermediate place and often provide the clues for understanding chaotic transitions in higher dimensional systems. In what follows we first list some mathematical models encountered frequently in the current literature. Then we shall say a few words about analytical and numerical methods to study these systems, with emphasis on the latter, since one has to rely heavily on numerical experiments in studying higher dimensional systems.

Higher dimensional systems may be divided into three categories: discrete mappings of two or more dimensions (difference equations of second order or higher); ordinary differential equations (autonomous, nonautonomous and time-delayed); and partial differential equations. We cite examples from each category. The literature on these systems has become so numerous that we are only able to cite a few early ones and indicate some of the latest ones.

## Two- and Higher-Dimensional Mappings

In contradistinction to one-dimensional mappings, higher-dimensional mappings may be conservative (volume-preserving) as well as dissipative (volume-contracting), invertible as well as noninvertible, depending on the parameters in the model. As the first example one should mention the well-studied *Hénon map* (Paper 11 in this volume):

$$x_{n+1} = 1 - \mu x_n^2 + y_n$$
$$y_{n+1} = bx_n$$

(59)

with the Jacobian

$$J = \frac{\partial(x_{n+1}, y_{n+1})}{\partial(x_n, y_n)} = -b.$$

It is an invertible transformation when $b \neq 0$ and goes back to the logistic map (18) when $b = 0$. It preserves area for $b = 1$ and corresponds to a dissipative system for $b < 1$. The case $a = 1.4$, $b = 0.3$ has been studied in great detail by Hénon and many others, (see, e.g., Simo, 1979; Marotto, 1979). The location of periodic orbits up to period 6 in the parameter plane can be determined analytically (Huang, 1985, 1986).

The essential difference from one-dimensional map may be summarized as follows. First, even for one and the same parameter value, the character of stationary output of (59) depends on the initial point. In other words, the $(x, y)$ plane divides into basins, and starting from different basins one may be led to different periodic or aperiodic orbits. Second, for certain parameters and initial points, the iteractions of (59) converge to an attractor with self-similar internal structure (see Figs. 3-6 in Paper 11). This was the first strange attractor known to have fractal structure. We shall return to this point in Chapter 8. Third, it was proved that there exist intersections of stable and unstable manifolds in Hénon's model, i.e. homoclinic and heteroclinic points give rise to chaos (see, e.g., Misiurewicz and Szawc, 1980; Qian and Yan, 1986). From recent literature on the Hénon map (59) we indicate Alligood and Sauer (1988), Liu (1988), Paramio (1988), and Cvitanovic, Gunaratne and Procaccia (1988).

It is interesting to note that the piecewise linear counterpart of the Hénon map (Lozi, 1978)

$$\begin{aligned} x_{n+1} &= 1 - \mu|x_n| + y_n, \\ y_{n+1} &= bx_n, \end{aligned} \tag{60}$$

allows analytical treatment in great detail. In particular, the existence of a strange attractor has been proved for some parameter value (Misiurewicz, 1980), its stable and unstable manifolds have been constructed (Tél, 1982a and b, 1983b).

Another well-known two-dimensional map is the so-called *standard mapping* (Zaslavsky and Chirikov, 1972; Chirikov, 1979):

$$\begin{aligned} x_{n+1} &= x_n + y_{n+1} \\ y_{n+1} &= y_n - \frac{B}{2\pi} \sin 2\pi x_n. \end{aligned} \tag{61}$$

Being an example of two-dimensional Hamiltonian system, this conservative mapping occurs in many physical applications such as the motion of a charged particle in toroidal magnetic field, or the Frenckel-Kontorova model of modulated structures in solid state physics (see Chapter 10). In analogue to the Hénon map, one can introduce a coefficient $0 \leq b \leq 1$ and add a constant $A$ to get from (61) the *dissipative standard map* (Feigenbaum, Kadanoff and Shenker, 1982; Schmidt and Wang, 1985)

$$\begin{aligned} \theta_{n+1} &= \theta_n + \rho_{n+1}, \\ \rho_{n+1} &= b\rho_n + A - \frac{B}{2\pi} \sin 2\pi\theta_n. \end{aligned}$$

In the extremely dissipative limit $b = 0$, it reduces to the circle map (79), to be discussed in Chapter 7.

Instead of listing other two- and higher-dimensional mappings we confine ourselves to a few remarks. First, some mathematical results on one-dimensional maps can be "lifted" to mappings on $R^n$ as outlined in Collet, Eckmann and Koch (1981a). Higher dimensional mappings are richer in their behaviour, e.g., basin dependece and transitions from quasiperiodic motion to chaos may take place. Second, period-doubling bifurcations do occur in area-preserving mappings of the plane, but the universal convergence rate happens to be $\delta = 8.7210 \cdots$. Third, complex mappings present a rich class of two-dimensional mappings. Their graphic representations, e.g., the Julia and Mandelbrot sets, are beautiful creations of modern computers (see, e.g., Peitgen and Richter, B1986).

## Ordinary Differential Equations

Ordinary differential equations displaying chaotic behaviour may be subdivided into three groups: autonomous systems with three or more variables, non-autonomous systems with two or more variables, and time-delayed systems of at least one variable.

A system of ordinary differential equations is called *autonomous*, when there is no explicit time dependence on the right-hand side of the standard form:

$$\frac{dx_i}{dt} = f_i(x_1, \cdots, x_n), \quad i = 1, \cdots, n. \tag{62}$$

A classical example of autonomous system exhibiting chaos is the Lorenz model (Lorenz, 1963, Paper 12 in this volume):

$$\begin{aligned} \dot{x} &= -\sigma(x - y), \\ \dot{y} &= -xz + rx - y, \\ \dot{z} &= xy - bz. \end{aligned} \tag{63}$$

It was obtained by truncating the partial differential equations describing thermal convection between two infinite planes. There are three parameters in this model: $r$ is the ratio of the Rayleigh number to its first critical value, $a$ is the Prandtl number, and $b$.

The Lorenz model is one of the simplest truncated hydrodynamical systems. If one considers planar flow of a fluid with periodic boundary conditions imposed in both directions, then the original partial differential equations (the Navier-Stokes equations) can be transformed into an infinite system of ordinary differential equations for the Fourier coefficients of various hydrodynamical quantities. By truncating this latter system one gets various finite systems of ordinary differential equations. Truncation of partial differential equations has become a minor industry. Truncated equations with 4, 5, 6, 7, 14, 33 or even more modes have been suggested (see, e.g., Curry, 1978; Franceschini and Tebaldi, 1981; Franceschini *et al.*, 1988). We call attention to a newly proposed systematic way of truncating infinite-dimensional systems using the notion of *inertial manifold*, which, in principle, contain all the interesting attractors (Foias *et al.*, 1988).

The behaviour of all the abovementioned models is very complicated, so Rössler tried to construct models as simple as possible, but still exhibiting chaotic behaviour. The simplest of his models contains only one quadratic nonlinear term:

$$\begin{aligned} \dot{x} &= -(y + z), \\ \dot{y} &= x + ay, \\ \dot{z} &= b + xz - cz. \end{aligned} \tag{64}$$

Among other autonomous systems displaying chaotic behaviour we mention a 7-mode model for the Belousov – Zhabotinskii reaction (see Argoul *et al.*, Paper 33 in this volume), the double-diffusive or two-component Lorenz model (see, e.g., Knobloch and Weiss, 1981; Velarde and Antoranz, 1981), a 40-mode system for Gunn instability in semiconductors (Nakamura, 1977, 1979), and the coupled Brusselators (Schreiber and Marek, 1982; Sano and Sawada, 1983; Lahari, 1988). (Brusselator is the name for a hypothetical model of trimolecular reaction with an autocatalytic step, which shows rich temporal and spatial structures when a diffusion term is added.)

We digress to explain why at least three variables are required to allow chaotic behaviour. This can best be seen from the example of period-doubling as the first step leading to chaos. Since the trajectory has no reason to change drastically at an infinitesimal increase of the parameter value, and yet the period $T$ doubles to $2T$, the only imaginable picture is an almost imperceptible splitting of the original orbit. If this splitting takes place in a plane, then there must be at least one point where the trajectory intersects itself and thus violates the uniqueness of the solution. However, usually the systems of ordinary differential equations under consideration are good enough to ensure the validity of the uniqueness theorem. Therefore, splitting of the orbit without self-intersection can take place only in three and higher dimensional space. The subsequence bifurcations and the development of chaotic trajectories follow the same principle.

Now let us return to *nonautonomous* ordinary differential equations. It is well-known that a nonautonomous system can be made autonomous by adding one or more variable, so at least two variables are required for a nonautonomous system to show chaotic behaviour. In fact, the only kind of nonautonomous systems studied in detail up to now consists of nonlinear oscillators driven by an external periodic force.

The forced van der Pol equation

$$\ddot{x} - k(1 - x^2)\dot{x} + x = b\lambda k \cos(\lambda t + \varphi) \tag{65}$$

seemed to be the first example exhibiting stochastic behaviour, reported long before the modern jargon of chaos and strange attractor has emerged (Cartwright and Littlewood, 1945). From recent papers on (65) we indicate Ding (1988a), Rajasekar and Lakshmanan (1988).

The forced anharmonic oscillator (Huberman and Crutchfield, 1979; Herring and Huberman, 1980)

$$\ddot{x} + k\dot{x} - \beta x + \alpha x^3 = b\cos\omega t, \tag{66}$$

being a particular case of the Duffing's equation

$$\ddot{x} + k\dot{x} + f(x) = g(t), \tag{67}$$

where $g(t)$ is a periodic function of $t$, and $f(x)$ a nonlinear function of $x$, appears in many physical applications, e.g., motion of a dislocation line in a supersonic field.

The forced mathematical pendulum

$$\ddot{x} + k\dot{x} + \sin x = \alpha \cos \omega t \tag{68}$$

and the parametrically excited pendulum

$$\ddot{x} + k\dot{x} + (A + \alpha \cos \omega t)\sin x = 0 \tag{69}$$

represent another class of physical models, encountered, say, in describing Josephson junctions in microwave cavity (see Chapter 10).

The forced Brusselator

$$\dot{x} = A - (B+1)x + x^2 y + \alpha \cos \omega t,$$
$$\dot{y} = Bx - x^2 y, \tag{70}$$

provides at the present time one of the most thoroughly studied models by combined use of various numerical methods. All well-known routes to chaos and the U-sequence of Metropolis-Stein-Stein (see Chapter 4) as well as Farey sequences of mode-locked regimes (see Chapter 7) have been observed in this model. For a recent review, see Section 5.7 in Hao (B1989).

In principle, every autonomous system can be extended to a nonautonomous one by adding a periodic force. From such models we mention only the forced Lorenz model (Aizawa and Uezu, 1982b; Saravanan *et al.*, 1985).

Most periodically forced systems can be viewed as coupled systems of one non-linear and one linear oscillators. Usually, in most part of the parameter space, one of the two oscillators dominates. Complicated regimes of oscillations may occur only as compromise between the two. This point of view allows for a more intuitive interpretation of various regimes of oscillation, including period-doubling and chaos. Having the driving frequency as a control parameter at hand opens the possibility to reach very high frequency resolution by using various stroboscopic sampling techniques (see below). These merits distinguish the periodically driven equations from purely autonomous systems in numerical studies.

Whenever analytical solutions are known for a nonlinear oscillator, one may add a special kind of periodic kicks in the form of a train of zero width pulses (the "Dirac comb")

$$\sum_{n=0}^{\infty} \delta(t - nT),$$

where $\delta(x)$ is the Dirac delta-function, to get a two-dimensional map exactly. Then it may be studied in great detail without using too much computer time. A good example was given by Gonzalez and Piro (1983). Recently, a prototype limit cycle oscillator

$$\dot{r} = sr(1 - r^2),$$
$$\dot{\theta} = 1, \tag{71}$$

has been studied in this way (Ding, 1988a). The bifurcation structure in its parameter space resembles the van der Pol oscillator (65).

Now comes the last group of ordinary differential equations — the *time-delayed equations*. One of the first such systems studied from the viewpoint of chaotic dynamics was a physiological model due to Mackey and Glass (1977):

$$\dot{x}(t) = \frac{ax(t-T)}{1 + x^n(t-T)} - bx(t),$$

which may have a strange attractor of rather high dimension (Farmer, 1982a). A seemingly simple equation like

$$\tau \dot{x}(t) + x(t) = 1 - \mu x(t-1) * x(t-1) \tag{72}$$

may possess very complicated bifurcation and chaos structure (Li and Hao, 1989), except for the limiting case $\tau = 0$ when it degenerates into the logistic map (18). Formally the time delay may be written as an infinite order differential operator acting on the function $x(t)$:

$$x(t-1) = exp\left(-\frac{d}{dt}\right)x(t) = \sum_{n=0}^{\infty}\frac{(-1)^n}{n!}\frac{d^n}{dt^n}x(t).$$

Now it becomes clear that a time-delayed equation in one variable actually corresponds to an infinite system of ordinary differential equations. One encounters this kind of equations in ecological models and in models describing optical bistable devices (Ikeda, 1979; Ikeda and Matsumoto, 1987; Le Berre *et al.*, 1986, 1987).

### Partial Differential Equations

Many ordinary differential equations listed above are truncations of partial differential equations. Broadly speaking, any study of turbulence in hydrodynamic and reaction-diffusion equations deals with chaos in partial differential equations. However, direct study of chaos in partial differential equations is still in its beginning stage, because it inevitably involves the interplay of temporal and spatial behaviour and must be very time-consuming in numerical study. We put aside the Navier-Stokes and reaction-diffusion equations and only mention a few other lines of research.

A first example of period-doubling and chaos in a system of partial differential equations was reported by Moore *et al.* (1983). Actually the equations in *(t,x,z)* were replaced by a finite difference scheme on a mesh and the essential nature of the solution was shown not to depend upon altering the mesh interval. Oscillations, period-doublings and chaos were observed in the time evolution of certain global characteristics of the solution. Futher studies on chaos in partial differential equations may be found in Edelen (1985), Wolfe and Morris, (1987), Yorke *et al.* (1987).

Another interesting development connects the two extremes of nonlinear equations: those exhibiting chaos and those having solitons. It is well-known that partial differential equations solvable by the inverse scattering technique are integrable systems. The existence of an infinite number of integrals of motion guarantees the creation of such "stable" objects as solitons and precludes the possibility for chaotic behaviour. However, integrability is a very subtle property which can easily be broken by periodic perturbations. Chaotic behaviour has been observed in the perturbed sine-Gordon equation (Bishop *et al.*, 1983; Overman II *et al.*, 1986), the perturbed nonlinear Schrodinger equation (Nozaki and Bekki, 1983a, 1986) and the closely related Ginsburg-Landau equation (Moon *et al.*, 1983; Doering *et al.*, 1987), see also Bishop *et al.*, 1986, 1988. There is a recent review on such systems (Abudulaev, 1989).

Chaotic behaviour in the Kuramoto-Shivashinsky equation, describing wave phenomena in excitable media, has roused some interest, see, e.g., Nicolaenko *et al.* (1985), Toh (1987), Konno and Soneda (1988).

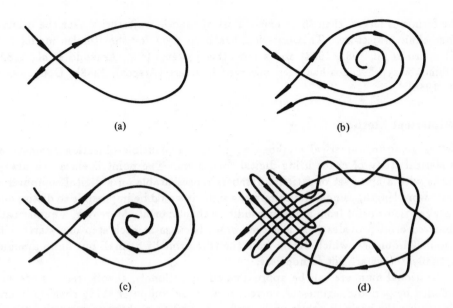

Figure 10: (a). Unperturbed separatrix. (b), (c) and (d). Perturbed separatrices in dissipative systems.

## Analytical Methods

Now we turn to the methodological aspects in studying higher dimensional systems. As regards analytical tools, there are very few mathematical results ready for practitioners to use: no *a priori* criterion for chaos, no classification and enumeration theorem for attractors, etc. Still we would like to call attention to two methods due to Melnikov and Shilnikov respectively, both centered on the existence of homoclinic points or homoclinic orbits.

The *Melnikov's method* applies to near-integrable systems subjected to dissipative time-dependent (usually periodic) perturbations. We mentioned in Chapter 2 that a separatrix becomes a stochastic layer under perturbation which makes the system nonintegrable, but still Hamiltonian. This is not the general rule for dissipative perturbations, as it can be seen from Fig. 10. An unperturbed separatrix shown in (a) may be perturbed into one of other three cases, among which only case (d) leads to an infinite number of homoclinic intersections. The criterion for case (d) consists in the alternating sign of the Melnikov function — the distance between stable and unstable directions of the trajectory. Detailed description of Melnikov's method may be found in Guckenheimer and Holmes (B1983); among recent papers we refer to Grunedler (1985), Salam (1987), and Schector (1987). Ling and Bao (1987) suggested a numerical implementation of the method.

The *Shilnikov's method* applies to stationary points of saddle-focus type, when there is a homoclinic trajectory passing through the saddle-focus. The theorem says roughly that if the only real eigenvalue and the real part of a pair of complex eigenvalues of the linearized system are of opposite sign and the absolute value of the former is larger, then there exists a set of chaotic trajectories near the original homoclinic trajectory. This criterion has been used for the Rössler model (64) (Gaspard *et al.*, 1983, 1984) and a few other systems (e.g., Arneodo *et al.*, 1982). "Shilnikov-type" chaos has been observed in lasers (Arecchi, 1988b; Dongoisse *et al.*, 1988).

## Numerical Methods

Before going to numerical methods we discuss a plausible objection against any numerical study of chaos using digital computers. The point is since one always works with a finite set of rational numbers representable on a digital computer of finite word length, using finite computing time, it would be impossible to distinguish a long periodic orbit from a quasiperiodic or chaotic orbit. Moreover, a numerically observed orbit provides only a *pseudo-orbit*, because at each step one starts with a rounded number, which differs from the true orbit by a small but finite amount. Actually, things are not so hopeless.

Irrational numbers can be approximated by rationals, chaotic regions are surrounded by periodic regimes. A correct strategy in computer study consists in first identifying the periodic orbits with confidence and then characterizing the "aperiodic' motion. The systematics of periodic orbits alone tells much about the nature of the nearby (in parameter space) aperiodic motion. In recent years, many methods of characterizing chaotic motion have been worked out and most of hese methods invoke the tangent space at each point of the trajectory. In addition, it has been proved that each chaotic pseudo-orbit is "shadowed" by a true chaotic orbit (see, e.g., Hammel, Yorke and Grebogi, 1987; Palmore and McCauley, 1987).

We postpone the problem of characterizing the attractors to Chapter 8 and concentrate on how to reach high frequency resolution in numerical experiments. We shall compare the resolving power in terms of the recognizable order $p$ of subharmonics with respect to the natural fundamental frequency of the system under study: $p = 2^n$ for period-doubling sequences, etc.

Direct observation of the trajectories is a method with the least resolution, but it enjoys the merit of having more physical intuition. In this way it is difficult to resolve subharmonics higher than $p = 32$. Sometimes it is useful to draw stereoprojections of a trajectory in some three-dimensional subspaces of the phase space.

The plotting of Poincaré maps, i.e. the intersection points of a given trajectory with a fixed surface in the phase space, offers an effective means to explore the nature of the motion. A practical problem consists in matching the precision of the interpolation scheme used to locate the intersection point with that of the integration algorithm. In this respect a clever suggestion by Hénon (1982) deserves to be mentioned. In practice, high precision calculationΩof Poincaré maps may be quite time-consuming when one is close to a bifurcation point due to critical slowing down. It is better to switch to some periodic orbit following algorithm, based on Newton-Raphson iterations, to achieve quicker convergence.

For both autonomous and non-autonomous systems, one should extract as much as possible information from the Poincaré maps, e.g., rotation numbers and symbolic sequences for periodic orbits. For instance, it has been shown that the systematics of periodic orbits in the Lorenz system (63) is given basically by symbolic dynamics of three letters (Ding and Hao, 1988).

For periodically driven systems stroboscopic sampling at fundamental and subharmonic frequences provides the means to reach very high frequency resolution at the expense of longer computing time. Period-doubling sequences up to $p = 8192$ has been identified in systems of ordinary differential equations (Hao and Zhang, 1982). Some precautions must be taken in using the subharmonic stroboscopic sampling method, e.g., to cope with the accumulation of round-off errors or to distinguish transient or intermittent behaviour from chaotic behaviour.

Power spectrum analysis in the frequency domain deserves a few more comments. Let the sampling interval be $\tau$ and the total sampling time for a single spectrum be $L = N\tau$, $N$ being the number of sampled points. The parameters $\tau$ and $L$ determine two frequencies: $f_{max} = 0.5/\tau$ is the maximal frequency one can measure using the given sampling interval; $\Delta f = 1/L$ is the frequency difference between two adjacent Fourier coefficients. In order to eliminate effectively the aliasing phenomena[10] one has to take $f_{max} = kf_0$, where $f_0$ is the fundamental frequency of the physical system and $k$ is a multiplier of the order of 4 to 8. We aim at resolving the $p$-th subharmonic of $f_0$ and wish the subharmonic peak to be formed by $s$ points in the spectrum, i.e., $f_0/p = s\Delta f$. Putting together all the abovementioned relations, we get

$$p = \frac{N}{2ks}. \tag{73}$$

Note that this is a relation independent of $\tau$ and $f_0$. Taking $N = 8192$, $k = 4$, $s = 8$, we have $p = 128$. This is the limit of resolution in power spectrum analysis on a medium-size computer. In principle, power spectrum is not invariant under coordinate transformation: measurement of voltage or current may lead to different peaks in the spectrum (see the beginning of Chapter 1). Nevertheless, power spectrum analysis is quite useful in telling different periods embedded in the chaotic bands. The presence of broad-band noise in the spectra is still the first symptom for chaos in computer and laboratory experiments.

---

[10]See, e.g., J. N. Rayner, *An Introduction to Spectral Analysis*, Pion Ltd., 1971.

# Chapter 6
# INTERMITTENT TRANSITIONS

The term intermittency has been used in the hydrodynamical theory of turbulence to denote random burst of turbulent motion on the background of laminar flow. However, its machanism slightly differs from what is called intermittency in the context of temporal chaos. In an open fluid the laminar and turbulent regions are divided by irregular and moving boundaries. The flow velocity sampled at a given point near such boundary will show alternating laminar or turbulent behaviour when the point enters or leaves the laminar region as the boundary moves. In one word, temporal intermittency is caused by spatially changing boundaries separating different flow regimes. However, in this volume we use the term intermittency exclusively to refer to random alternations of chaotic and regular behaviour in time evolution without involving any spatial degrees of freedom.

Among the various routes to chaos mentioned in Chapter 2, period-doubling and intermittency are in fact twin phenomena. Inspecting the bifurcation diagram of a one-dimensional mapping or the "phase diagram" showing the bifurcation and chaos structure of a system of ordinary differential equations, one sees many chaotic regions contained in between pairs of type $RL^{n-1}$ and $RL^n$ periods. (For the meaning of these letters see Chapter 4.) If one enters a given chaotic region from the $RL^{n-1}$ side, then the transition takes place via period-doubling bifurcations. If one approaches the same chaotic region from the $RL^n$ boundary, the intermittent transition shows itself. The same may be said towards any period-doubling cascade, described by a nonprimitive word $W$ and born from a tangent bifurcation. Both period-doubling and intermittency are described by the same renormalization group equation (33) (see, however, remark 3 below) with different boundary conditions and are well understood by now. Recommending the reader to consult the reprinted papers and reviews for details, we only touch briefly a few points related to the intermittent transition.

1. The mechanism of intermittency in one-dimensional mappings has been fully understood. It occurs in the neighbourhood of a tangent bifurcation, whereas period-doubling is related to pitchfork bifurcations. For figures and explanation of these bifurcations see May (1976, Paper 6 in this volume). Take, for example the logistic map (18). When the parameter $\mu$ is kept very close to, but still less than $\mu_c = 1.75$, its third iterate $f^3(\mu, x)$ has three humps or valleys that come very close to the bisector in the $f^3$ versus $x$ diagram, but does not touch it, leaving thereby very narrow corridors between the curve $f^3(\mu, x)$ and the bisector (see Fig. 4 in Paper 15). Every time when one iterate point falls near the entrance to one of these

corridors, it will take many iterations to pass through the corridor. When $\mu$ is very close to $\mu_c$, this looks much like a segment of convergent iteration to the would-be fixed point at $\mu_c$ and leads to the "laminar phase" of iteration. Having passed one corridor, the point makes a number of large-step jumps before falling again to the entrance of one or another corridor. These jumps constitute the random bursts of the "turbulent phase". The nearer $\mu$ is to $\mu_c$, the longer the averaged laminar time. Simple mean-field arguments give the passage time $t$

$$t \propto (\mu_c - \mu)^{-(1-1/z)}, \tag{74}$$

where $z$ denotes the order of maximum in the mapping (17). For the usual case $z = 2$, we have $t \propto (\mu_c - \mu)^{-1/2}$ which is to be compared with the divergence rate of correlation length $\xi$ near a continuous phase transition point $T_c$

$$\xi \propto |T_c - T|^{-\nu} \tag{75}$$

with its mean-field value $\nu = 1/2$.

2. Intermittent transitions had been observed in computer experiments on ordinary differential equations before its mechanism was explained by Pomeau and Manneville. In periodically forced systems, intermittency associated with tangent bifurcations of not very long period can be easily distinguished from chaotic or transient behaviour by their specific subharmonic stroboscopic sampling diagrams. Another way to explore the intermittent transitions consists in constructing the $(P_i, P_{i+n})$ map near a tangent bifurcation of period $n$, where $P_i$ may be any of the variables under study. If there were an exact period $n$, one would get $n$ points on the bisector of the $(P_i, P_{i+n})$ diagram, but at the intermittent transition we can only see the "spirit" of the would-be fixed points, i.e., $n$ clusters along the bisector together with a few points scattered away from it. This is a very effective method to recognize intermittency in practice.

Intermittent transitions have been reported in many laboratory experiments, e.g., in chemical turbulence, thermoconvective instability and nonlinear electronic circuits. We would like to point out that since there are infinitely many periodic windows born from tangent bifurcations embedded in the chaotic regions of the parameter space, the intermittent transitions are typical phenomena to be observed in numerical and laboratory experiments.

3. A few words about the renormalization group equation for intermittent transitions. We know from Chapter 4 that a periodic window in one-dimensional mapping remains stable whenever the absolute value of its dervative does not exceed one [see (24)]. In fact, the limit $(f^n)' = -1$ corresponds to period-doubling bifurcations, whereas $(f^n)F' = +1$ indicates an intermittent transition. It is has been known that intermittent transitions can be described by the same renormalization group equation of Feigenbaum

$$g(x) = \alpha g(g(x/\alpha)), \tag{76}$$

with a natural change of boundary conditions

$$g(0) = 0, \quad g'(0) = 1. \tag{77}$$

[cf. (34)]. However, equation (76) is not as "deep" as its counterpart for period-doubling. It was based on simple scaling invariance of counting the number of steps required to pass a corridor when two steps were combined into one and the width of the corridor was readjusted at the same time. In fact, one may combine $n$ steps into one step and get

$$\alpha^{-1}g(\alpha x) = g^n(x) \tag{78}$$

instead of (76). The number $n$ does not enter the universal constant $\alpha$ at all, while in the Feigenbaum case (78) describes period-$n$-tupling and all the exponents depend on $n$. Nevertheless, Hirsch, Nauenberg and Scalapino (1982, Paper 16 in this volume) were able to find an exact solution to (76) and (77). Their solution has been extended to give all the intermittent exponents and universal functions $g(x)$ for general $z$ (Hu and Rudnick, 1982, Paper 17 in this volume, and 1986). Their method can be applied to (78) as well.

4. Our last remark concerns the effect of external noise on intermittency: a small amount of noise increases the averaged passage time (Hirsch *et al.*, 1982 a and b; Eckmann *et al.*, 1981).

# Chapter 7
# TRANSITION FROM QUASIPERIODICITY
# TO CHAOS

The transition from quasiperiodicity to chaos has become a hot subject since 1982. This is provoked by the interest to elucidate whether there are universal aspects in the Ruelle-Takens route to chaos, which was less understood theoretically in comparison with the period-doubling and intermittent routes.

As mentioned in Chapter 3, Ruelle and Takens (1971, Paper 5 in this volume) proposed an alternative to the Landau-Hopf picture of infinitely increasing number of incommensurable frequencies for the onset mechanism of turbulence. They showed that quasiperiodic motion on a 4-torus, i.e., with 4 incommensurable frequencies, is in general unstable and can be perturbed into a strange attractor corresponding to turbulent motion. A few years later, in collaboration with Newhouse they sharpened the above result to that of a 3-torus, i.e., quasiperiodic motion with three incommensurable frequencies is in general unstable and can be perturbed into chaotic motion. Let us denote quasiperiodic motion with $n$ incommensurable frequencies by $Q_n$. The Landau-Hopf route to chaos can be expressed as $Q_n \to$ chaos when $n \to \infty$. Ruelle and Takens first suggested the possibility of $Q_3 \to$ chaos, and then sharpened it to $Q_2 \to$ chaos.

In the original paper of Ruelle and Takens, and in Newhouse, Ruelle and Takens (1978), the perturbation required to destroy $Q_3$ was rather specific, so in subsequent years the question has been raised as whether stable $Q_3$ exists in dynamical systems as a generic, i.e. typical, possibility. Indeed, stable $Q_3$ has been shown to exist in several mathematical models (Grebogi et al., 1983a, 1985b; Takavol and Tworkowski, 1984) and in hydrodynamical experiments (Gollub and Benson, 1980; Fauve and Libchaber, 1981). Even $Q_4$ and $Q_5$ has been observed (Walden et al., 1984). A recent high-resolution experiment may shed more light on this problem (Cumming and Linsay, 1988).

Quasiperiodic orbits do exist in plenty in maps of the interval, but thetransition from quasiperiodicity to chaos becomes fully fledged in maps of a circle onto itself (Arnold, 1965; Zeng and Glass, 1989):

$$theta_{n+1} = f(\theta_n) = \theta_n + A - \frac{B}{2\pi} \sin 2\pi\theta_n \quad \delta(\text{mod } 1), \tag{79}$$

which is the limiting case of a two-dimensional annular map. Shenker (1982), Feigenbaum et al. (1982), Rand et al. (1982), and Ostlund et al. (1983) studied the $Q_2 \to$ chaos transition by using renormalization group idea and discovered univer-

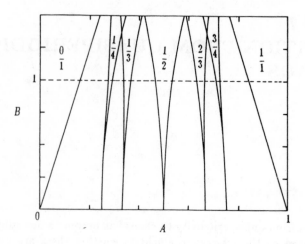

Figure 11: Frequency-locking zones in the parameter plane of the circle map (schematic).

sal characteristics for this transition, when the irrational rotation number $(\sqrt{5}-1)/2$ (the golden mean) is approached through an infinite sequence of rational fractions $F_n/F_{n+1}$, where $F_n$ are the Fibonacci numbers.

In order to explain the notion of *rotation number*, we first look at the $B = 0$ case of (79). Now it is a linear map and describes a rigid rotation by an angle $A$ at each step:

$$\theta_{n+1} = \theta_n + A \quad (\text{mod } 1). \tag{80}$$

It has only two regimes: periodic for $A$ rational, i.e, $A$ is the ratio of two integers $p$ and $q$, and quasiperiodic for $A$ irrational. $A$ is the "bare" rotation number of the map (80). It is sufficient to restrict $A$ to the interval $(0,1)$ due to taking (mod 1). Since rational numbers are isolated and have zero measure on the interval, quasiperiodic motion overwhelms the interval. Nonzero $B$ introduces nonlinearity into the system and causes each periodic regime to persist for a wider range of $B$. This widening of periodic regime leads to tongue-shaped zones of *frequency-locking* or *mode-locking* regimes, called also *Arnold tongues* (see Fig. 11). In fact, frequency-locking is a generalization of resonance phenomenon to nonlinear oscillator systems. For a review on quasiperiodic transition to chaos from an experimentalist's point of view we recommend Glazier and Libchaber (1988, Paper 18 in this volume). When $B \neq 0$, $A$ is no longer the rotation number. One has to calculate the rotation number from the averaged rotation rate per step. If we drop the (mod 1) in (79) and write

$$\theta_{n+1} = F(\theta_n),$$

$F(\theta)$ is called the *lift* of $f(\theta)$. The rotation number (or winding number) $W$ is defined as

$$W = \lim_{n \to \infty} \frac{F^n(\theta_0) - \theta_0}{n}.$$

In the $B - A$ plane there exists a *critical line* $B = B_c$ (in general case it may be a curve), below which the order of occurrence of the frequency-locking regions is given by the increasing succession of rational numbers $p/q$ whose denominators are less than or equal to $n$, e.g., for $n = 5$ we have the 5-th order *Farey sequence*:

$$0/1, \; 1/5, \; 1/4, \; 1/3, \; 2/5, \; 1/2, \; 3/5, \; 2/3, \; 3/4, \; 4/5, \; 1/1.$$

(The case $n = 4$ is shown schematically in Fig. 11.) A rotation number $p/q$ means it is a period $q$ orbit and the $q$ points are being visited in $p$ turns. If in an experiment one has seen two frequency-locked regimes described by $p/q$ and $p'/q'$, then most probably one can find another frequency-locked tongue in between, whose rotation number is the *Farey sum* of the two, namely,

$$\frac{p}{q} \oplus \frac{p'}{q'} = \frac{p + p'}{q + q'}.$$

It is easy to check that if

$$\frac{p}{q} < \frac{p'}{q'},$$

then

$$\frac{p}{q} < \frac{p + p'}{q + q'} < \frac{p'}{q'}.$$

All rational numbers between 0 and 1 may be arranged on a *Farey tree* in the following way. One starts with $0/1$ and $1/1$ and then form the Farey sum $1/2$. Put these numbers in their natural order, then take the Farey sum of every neighbouring pair, one gets a binary tree. Every rational number will have a unique site on this tree. Each Farey member has a continuous fraction representation, a symbolic representation, and an "address" representation [see, e.g., Section 4.4 in Hao (B1989)]. Many scaling properties of circle maps are associated with regularities of the Farey tree. For more on Farey tree, see, e.g., Cvitanovic *et al.* (1985). To deal with three and more competing frequencies, one needs generalizations (the Farey triangle) of the Farey construction (Hu and Mao in Hao, B1987; Kim and Ostlund, 1986; Maselko and Swinney, 1987).

If one varies the parameters along a continuous line in the $A - B$ plane, one would encounter many segments where the rotation number remains constant, while in between these frequency-locked regions rotation number may change. This staircase-like curve is called a *devils's staircase* (Aubry, 1983b; Bak, 1982; 1986).

Below the critical line the rotation number is an invariant property of the map, which does not depend on the choice of the initial point $\theta_0$, i.e., for a fixed set of parameters all possible initial values lead to one and the same number $W$.

Above the critical line, various tongues may overlap and give rise to complicated periodic, quasiperiodic, and chaotic motions. These different types of motion may coexist in a certain parameter range. In particular, at a given point in the parameter plane there may exist a *rotation interval* or rotation set (Newhouse, Palis and

Takens, 1983) $[\rho_-, \rho_+]$. Rotation set is a closed interval (Ito, 1981). For any number in between $\rho_-$ and $\rho_+$ there exists an orbit, i.e., an initial value, which has that number as its rotation number. In addition, there may exist orbits without a rotation number. The existence of a nontrivial rotation interval, i.e., $\rho_- \neq \rho_+$, implies chaos (Boyland, 1986; Casdagli, 1988) and positive topological entropy. It is an empirical fact that the two end points of a rotation interval are always mode-locked, i.e., associated with some tongues (Casdagli, 1988). In principle, rotation interval may be calculated from experimental data (Gambaudo *et al.*, 1984; MacKay, 1987).

Along the critical line, frequency-locked regime overwhelms quasiperiodic one. The latter has a zero measure, but a finite fractal dimension $D_0 = 0.87\cdots$, which is believed to be universal for many smooth circle maps and differential equations with two competing frequencies (Jensen, Bak and Bohr, 1984, Papers 19 and 20 in this volume).

# Chapter 8
# CHARACTERIZATION OF CHAOTIC ATTRACTORS

We mentioned in Chapter 5 that the strategy in studying the bifurcation and chaos "spectrum" of a physical system consists in first identifying various periodic regimes and then trying to characterize the chaotic motion on the attractors. Both in the laboratory and on computers it is very difficult to distinguish a purely chaotic motion from a quasiperiodic one or from a motion perturbed by external noise by simply looking at a finite sequence of data. Moreover, although mathematicians have prepared such nice notions as hyperbolic attractor, Axiom A system or Smale's horseshoe, they are not directly applicable to most systems of physical interest. Consequently, the term "strange attractor" is now being used in the literature very liberally. Anyway, we need some quantitative means to recognize, characterize and classify attractors. In particular, it is desirable to extract invariant characteristics from a single time series of sampled points.

Power spectrum analysis is useful in telling quasiperiodic motion from periodic, and in identifying high order periodicities embedded in chaotic bands by the fine structures in spectra. Power spectra of chaotic attractors associated with period-doubling distinguish themselves by sharp peaks on the background of broad-band noise, but it makes little difference between chaotic attractors of other types. We indicate a few papers on scaling property of power spectra (Feigenbaum, 1980; Collet *et al.*, 1981; Nauenberg and Rudnick, 1981; Wolf and Swift, 1981; Yoshida and Tomita, 1987; Dumont and Brumer, 1988) and turn to more sophisticated ways of characterizing attractors, among which Lyapunov exponents and various definitions for dimension are widely used.

Lyapunov exponents, dimension, and entropy (in a mathematical sense) are closely related notions. For hyperbolic attractors (see Chapter 3) they are backed up by a few mathematical theorems (see the review by Eckmann and Ruelle, 1985, Paper 23 in this volume), but for nonhyperbolic attractors, which are more common in practice, one has to rely on plausible conjectures and numerical experience (see, e.g., Grassberger *et al.*, 1988; Politi *et al.*, 1988). We devote this chapter to a brief summary of the most important concepts and techniques.

## Fractal and Information Dimensions

The phase space volume of a dissipative system contracts in the process of evolution and the motion is confined to attractors in the long time $t \to \infty$ limit. For dissipative systems the dimension $D$ of the attractor is usually lower than the dimension of the original phase space. Using any reasonable definition of dimension, the dimension of trivial attractors can easily be calculated to be integers: $D = 0$ for fixed points, $D = 1$ for limit cycles, $D = 2$ for 2-torus, etc. However, the dimension of chaotic attractors often turns out to have noninteger value, yet what is an object with noninteger dimension?

Hausdorff introduced a generalized definition of dimension as early as in 1919. It can be explained very simply in the following way. Take a usual, regular geometrical object, say, a cube, and double its linear size in each spatial direction. We get a cube whose volume is eight times larger than the original one, because $2^3 = 8$. In general, taking an object of dimension $D$ and increasing its linear size in each spatial direction $l$ times, one would have its volume increased to $k = l^D$ times of the original. Inverting this simple relation, we get a new definition for dimension:

$$D = \frac{\log k}{\log l}. \tag{81}$$

Now, we have freed ourselves from the restriction of $D$ being an integer.

A precise definition of Hausdorff dimension requires the notion of Hausdorff measure (see, e.g. Mandelbrot, B1982), but formula (81) is enough for our practical needs. We begin with the simplest example of a geometrical object which has noninteger dimension – the *Cantor set*.

Take the interval $(0, 1)$ and delete the central one-third $(1/3, 2/3)$, then repeat this operation with respect to the remaining segments again and again, *ad infinitum*. What is the dimension of the limiting set of points thus obtained? Let points left in the interval $(0, 1/3)$ constitute our geometrical object under consideration. Increasing its linear size by a factor $l = 3$ would yield two copies $(k = 2)$ of the same object, therefore

$$D = \frac{\log 2}{\log 3} = 0.6309 \cdots. \tag{82}$$

Actually, many strange attractors have a Cantor set like structure (see, e.g., the frequently cited Figs. 3-6 in Hénon (1976, Paper 11 in this volume)). To calculate the dimension numerically, a box-counting algorithm may be used. One divides the phase space into small boxes of linear size $\epsilon$, and counts the number $N(\epsilon)$ of such boxes that contain at least one point of the orbit. Then, the dimension is calculated by taking the limit

$$D_0 = \lim_{\epsilon \to 0} \frac{\log N(\epsilon)}{\log(1/\epsilon)}. \tag{83}$$

To be precise, (81) or (83) are the *Kolmogorov capacity*. The meaning of the subscript 0 will become clear later. In practice, one counts $N(\epsilon)$ for a sequence of finite $\epsilon$'s and obtains $D_0$ from the slope of the $\log N(\epsilon)$ versus $\log(1/\epsilon)$ plot.

Thus far in counting $N(\epsilon)$ no attention has been paid to the possible inhomogeneity of the attractor, i.e., no matter how many times the orbit travels through a given box, the box is counted only once. To correct this inexactitude one introduces a weight according to how frequently a box is visited. If the $i$-th box is visited with a probability $p_i$, then the definition (83) is replaced by

$$D_1 = \lim_{\epsilon \to 0} \frac{I(\epsilon)}{\log(1/\epsilon)}, \tag{84}$$

where

$$I(\epsilon) = -\sum_{i=1}^{N(\epsilon)} p_i \log p_i, \tag{85}$$

$N(\epsilon)$ being the total number of boxes visited. If every box is visited with equal probability, i.e., $p_i = 1/N(\epsilon)$, then (85) reduces to $I(\epsilon) = \log N(\epsilon)$ and $D_1 = D_0$. In general $D_1 \leq D_0$. $D_1$ is called the *information dimension*.

## Correlation Dimension. Time-delay Method of Reconstructing the Phase Space

A more convenient notion of *correlation dimension* was introduced by Grassberger and Procaccia (1983, Paper 22 in this volume). It has the merit to be readily calculable from experimental data, using the *time-delay method* of reconstructing the phase space. Since this method has become the basis for processing experimental time series, we describe it in some detail.

Suppose we use a digitized instrument and measure the time variation of a physical variable (velocity, temperature, or whatever quantity) for sufficiently long time. The data are sampled at equally spaced time instants $t_i = i\tau$ ($\tau$ is the sampling interval) and recorded as a *time series*

$$x_1, x_2, x_3, \cdots, x_N \tag{86}$$

where the total number $N$ is large enough, say, $N = 10^8$. For simplicity we consider the case of scalar $x_i$ (like temperature). The basic idea is: due to nonlinear inter-actions in the system these $x_i$ contain information on other variables as well and one should be able to extract this information from a single time series like (86). A simple example is a phase portrait of the attractor. If one has measured separately the time series of $x$ and its derivative $\dot{x}$, then the $x - \dot{x}$ curve is just what is required. In the absence of $\dot{x}$ one may try to approximate the derivative by taking differences of $x_i$ and $x_{i+p}$, where $p$ is an integer that gives the delay in units of the sampling interval $\tau$. This idea of using time-delay was suggested by D. Ruelle, first realized by Packard *et al.* (1980), and justified mathematically by Takens (1981), using the embedding theorem of Whitney.

In most experimental situations one does not know the dimension of the under-lying phase space (it may be infinite) and one believes that the motion is confined to an attractor of much lower dimensionality. Therefore, one first tries to construct a vector space of finite but high enough dimension $m$ (the *embedding dimension*) from the scalar series (86). A simple way to construct these $m$-dimensional vectors is to pick up consecutive points from (86) and regroup them as follows:

$$
\begin{aligned}
\mathbf{y}_1 &= \left(x_1, x_{1+p}, x_{1+2p}, \cdots, x_{1+(m-1)p}\right), \\
\mathbf{y}_2 &= \left(x_2, x_{2+p}, x_{2+2p}, \cdots, x_{2+(m-1)p}\right), \\
&\cdots \quad \cdots \\
\mathbf{y}_i &= \left(x_i, x_{i+p}, x_{i+2p}, \cdots, x_{i+(m-1)p}\right), \\
&\cdots \quad \cdots
\end{aligned}
\tag{87}
$$

One gets $M$ vectors, $M$ is of the same order as $N$ when $N$ is large enough.

Next, a distance $|\mathbf{y}_i - \mathbf{y}_j|$ between two vectors is introduced. One may use the Euclidean distance, i.e., the square root of the sum of squares of the component differences. However, in practice it is more convenient to define the distance between two vectors by the largest difference of their corresponding components. If the distance of two vectors is smaller than a predefined constant $\epsilon$, we say that these two vectors are correlated. Count the number of all correlated pairs among the $M$ vectors and normalise its maximal value to one, we get so-called *correlation sum* (or *correlation integral*):

$$
c(\epsilon) = \frac{1}{M^2} \sum_{i,j}^{M} \theta\left(\epsilon - |\mathbf{y}_i - \mathbf{y}_j|\right),
\tag{88}
$$

where the step function $\theta(x)$

$$
\theta(x) = \begin{cases} 1 & \text{if } x > 0, \\ 0 & \text{if } x \leq 0, \end{cases}
$$

does the job of counting. The correlation dimension $D_2$ is defined as

$$
D_2 = \lim_{\epsilon \to 0} \frac{\log c(\epsilon)}{\log \epsilon}.
\tag{89}
$$

If $\epsilon$ is chosen too large, then all vectors would be correlated and one would get $c(\epsilon) = 1$ and $D_2 = 0$. If $c(\epsilon)$ scales as

$$
c(\epsilon) \propto \epsilon^\nu,
\tag{90}
$$

when varying $\epsilon$, then $D_2 = \nu$. If $\epsilon$ is too small, then external and instrumental noise would come into play. Since random noise acts in every dimension, $\nu$ would approach the embedding dimension $m$. Only when the scaling relation (90) holds for a clear range of $\epsilon$ (see Fig. 12), it makes sense to calculate $D_2$.

## Fundamental Limitations of Dimension Calculation

Computer programs implementing the time-delay method to calculate correlation dimension and Lyapunov exponents are now available from many sources (see, e.g., Paper 26; Parker and Chua, B1989) and people in all walks of life are measuring dimensions. There has appeared a real danger of overdoing the job if one is not aware of the limitations of the method. Recently, Eckmann and Ruelle discussed

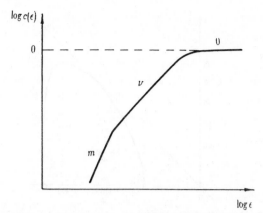

Figure 12: The scaling range in determining $D_2$ (schematic).

this problem[11]. Their arguments are simple and convincing, so we reproduce the essentials here.

Assuming that one has reconstructed the vectors $\mathbf{y}_i$ in the embedding space and one counts the number $N(\epsilon)$ of correlated pairs in a small ball of radius $\epsilon$, the existence of a dimension $D$ means that $N(\epsilon)$ scales with $\epsilon$ as

$$N(\epsilon) = const \times \epsilon^D, \tag{91}$$

(one should have used $D_2$, but the difference is not essential for a qualitative discussion). One may even determine the *const* in (91) from the following consideration. We enlarge $\epsilon$ to the size $L$ of the attractor, then all possible pairs of vectors are "correlated" and we know the total number of pairs is $M(M-1)/2 \approx M^2/2$, where $M$ is of the order $N$, the length of the time series, i.e.,

$$N(L) = const \times L^D \approx M^2/2. \tag{92}$$

Solve (92) for *const*, we get

$$N(\epsilon) \approx \frac{M^2}{2} \left(\frac{\epsilon}{L}\right)^D. \tag{93}$$

Eq. (93) remains valid as long as the inequalities

$$N(\epsilon) \gg 1 \quad \text{and} \quad \epsilon \ll L \tag{94}$$

hold. Taking logarithm of the first inequality and using (93), we get

$$2\log_{10} M - \log_{10} 2 + D \log_{10}\left(\frac{\epsilon}{L}\right) > 0.$$

(It is a little more convenient to use common logarithm.) Neglecting $\log_{10} 2$ for $M$ large enough and assuming $L/\epsilon$ is of the order 10, we are led to

$$D < 2\log_{10} N. \tag{95}$$

---

[11]J-P Eckmann, and D Ruelle, "Fundamental limitations for estimating dimensions and Liapunov exponents in dynamical systems", IHES preprint 1989.

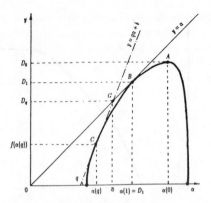

Figure 13: A typical $f(\alpha)$ curve and graphical calculation of $D_q$.

If the calculated dimension is well below $2 \log_{10} N$, where $N$ is the total number of points in the original time series, and, in addition, there is a clear scaling range in the $\log c(\epsilon)$ versus $\log \epsilon$ plot and the slope is smaller than the embedding dimension $m$ and does not change with the latter, the result may be reliable. To have a practical test one may scramble the data points and pass them through the same routine. If the "scrambled dimension" is well above the value calculated before, one may be safe.

One may put forward a even simpler argument against too high values of calculated dimension. Suppose the "attractor" has dimension $D = 30$ and we want to probe it using Monte Carlo idea. The crudest probe requires sampling two points in each direction and that makes $2^{30} \approx 10^9$ points, a value not always reached in the published calculations.

### Generalized Information Dimensions and
### Thermodynamic Formalism

We have attached numerical subscripts to the fractal, information and correlation dimensions, because they happen to be particular cases of the *generalized information dimension* (Hentschel and Procaccia, 1983, Paper 21 in this volume):

$$D_q = \frac{1}{q-1} \lim_{\epsilon \to 0} \frac{\log \sum_{j=1}^{N(\epsilon)} p_j^q}{\log \epsilon}, \qquad (96)$$

for $q = 0$, 1, and 2, respectively. Although the box-counting algorithm to compute the dimension is very time-consuming and becomes impractical when the dimension of the underlying phase space is larger than two or three (Greenside *et al.*, 1982), the definition (96) is given in the spirit of box-counting. However, there are other ways to define $D_q$.

At a certain level of resolution, a *strange set* may seem to be made of $K$ pieces, each having a length scale $l_i$ and a probability $p_i$ for some event to take place on it, e.g., to be visited by a trajectory. At the next, finer level of resolution, one sees that each piece splits into $K$ smaller pieces with length and probability scaled down in the same way. With knowledge of $(K, l_i, p_i)$, $D_q$ may be calculated from the following sum rule or partition function (Halsey *et al.*, 1986, Paper 24 in this volume):

$$\sum_{i=1}^{K} \frac{p_i^q}{l_i^{(q-1)D_q}} = 1. \tag{97}$$

Another definition of $D_q$ introduces two more exponents $\alpha$ and $f(\alpha)$:

$$D_q = \frac{1}{q-1} \left[ \alpha q - f(\alpha) \right]. \tag{98}$$

First of all, let us try to explain the meaning of $\alpha$ and $f(\alpha)$ in an elementary way. Suppose we have a lattice in a $d$-dimensional space (think about a 3-dimensional crystal) and we have measured the value $p_i$ of some quantity at each lattice site. Now we want to transform a lattice sum $\sum_i \cdots$ into an integral over the continuum. We can make use of the well-known relation in solid state physics

$$\frac{1}{(2\pi)^d} \sum_i \cdots = \frac{1}{V} \int \cdots \rho(\mathbf{r}) d^d \mathbf{r}, \tag{99}$$

where $V$ is the volume. Expressed in terms of the linear size $l$, one has $V = l^d$, where $d$ is the dimension of the space. If the density does not contain any singularity, this is all what one needs. However, if $\rho$ has singularities, concentrated in lower dimensional regions, say, at some points or along some lines, then one must be more careful. Although the region have zero volume compared to $V$, their contribution to the integral may be finite due to the presence of singularity. For example, for singularities along a line, one should add a one-dimensional integral with $d = 1$ to the right-hand side of (99). What is said refers to singularities concentrated on regular geometric regions, i.e., regions with integer dimensions. If a singularity of "strength" $\alpha$, i.e., it diverges as $\epsilon^{-\alpha}$ when $\epsilon$ goes to zero, is populated on a fractal object, then one has to introduce a dimension $f$ instead of $d$ to count for its density. Surely, $f$ depends on $\alpha$.

The meaning of $\alpha$ and $f(\alpha)$ becomes more apparent if we sketch the derivation of (98) from (96) as follows. A strange set may be thought as composed of points of different colours. We label these colours by an index $\alpha$. Eq. (96) may be viewed as a time average along the trajectory (the sum over $j$). Now we transform it into an ensemble average over colours. Divide the volume into cells of linear size $\epsilon$. Suppose that when the size goes to zero points of colour $\alpha$ contribute to the probability $p$ a singularity

$$p \propto \epsilon^\alpha,$$

(we use the same $\alpha$ to denote the strength of the singularity). In order to average over the volume, we imagine points of different colours to be packed into subsets of different dimensions, i.e., instead of a single $d$, we use a $f(\alpha)$ for each colour $\alpha$ and write the corresponding volume element as $\epsilon^{f(\alpha)}$. Now we can replace the sum over

$j$ by an integral over $\alpha$:

$$\sum_{j} \Longrightarrow \int d\alpha' \rho(\alpha') \frac{\epsilon^{q\alpha'}}{\epsilon^{f(\alpha')}},$$

where we have placed primes on the integration variable and introduced a density for colours $\rho(\alpha)$ which normalizes to one and drops out from the final result. The integral must be estimated in the limit $\epsilon \to 0$. The most significant contribution comes from that $\alpha$ which makes the exponent of

$$\epsilon^{q\alpha - f(\alpha)}$$

a minimum, i.e., this $\alpha$ is determined from the solution of

$$\begin{aligned}
\frac{d}{d\alpha'}[q\alpha' - f(\alpha')] &= 0, \\
\frac{d^2}{d\alpha'^2}[q\alpha' - f(\alpha')] &> 0,
\end{aligned} \tag{100}$$

that is,

$$\begin{aligned}
\frac{d}{d\alpha} f(\alpha(q)) &= q, \\
\frac{d^2}{d\alpha^2} f(\alpha(q)) &< 0.
\end{aligned}$$

Taking the integral only at the maximum of the integrand and substituting the result back into the box-counting formula (96), we get (98). Sometimes, one introduces a new notation $\tau_q$ and write (98) as

$$\tau_q \equiv (q-1)D_q = \alpha q - f(\alpha), \tag{101}$$

which may be taken as a Legendre transformation between the two sets of thermodynamic quantities $D_q$ versus $q$ and $f(\alpha)$ versus $\alpha$. One can go back and forth between these two curves, using the relations, given above. We indicate a simple graphic construction to get $D_q$ from $f(\alpha)$ (Fig. 13): draw a tangent to $f(\alpha)$ in Fig. 13, its slope is $q$ while its intersection with the bisector gives $D_q$. This analogue with thermodynamics may be developed further. In particular, there may be phase transitions in this formalism (see Fig. 14). For recent reviews on the thermodynamic formalism, see, e.g., Badii (1989) and Amritkar and Gupte[12].

## Lyapunov Exponents from Governing Equations

The fractal dimension reflects only the overall effect of stretching and contraction of phase volume in reducing the latter to an object of lower dimensionality. In order to describe stretching and contraction in more detail, we need the notion of Lyapunov exponents and we shall see that the dimensions are in general connected to these exponents. In addition, Lyapunov exponents furnish us with a classification scheme of attractors.

[12]R. E. Amritkar, and N. Gupte, "Multifractals", in *Directions in Chaos*, vol.3, ed. by Hao Bai-lin, World Scientific, 1990.

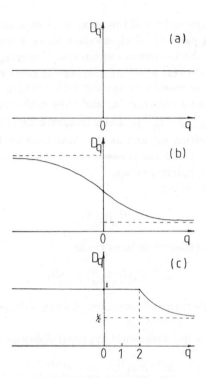

Figure 14: Different possibilities for $D_q$ (schematic). (a). A single-scale homogeneous Cantor set. (b). A multi-scale Cantor set. (c). A "phase transition".

We first consider the case when the governing equation is known, and turn to the calculation of Lyapunov exponents from experimental time series later. Suppose we have an autonomous systems of ordinary differential equations

$$\dot{x}_i = f_i(x_1, x_2, \cdots, x_n), \quad i = 1, 2, \cdots, n. \tag{102}$$

Given any point $x_0$ in the phase space, Eq. (102) may be linearized in the vicinity of this point:

$$x_i = x_{0i} + \delta x_i \quad i = 1, 2, \cdots, n$$

to yield a linear, but nonautonomous (depending on $x_0$) system of equations

$$\frac{d}{dt}\delta x_i = \sum_{j=1}^{n} U_{ij}(x_0)\delta x_j, \quad 1 = 1, 2, \cdots, n, \tag{103}$$

where

$$U_{ij}(x_0) = \frac{\partial f_i}{\partial x_j}\Big|_{x=x_0} . \tag{104}$$

The eigenvalues of matrix (104) determine the local stability of Eq. (102) in the neighbourhood of $x_0$: if the real parts of some eigenvalues happen to be positive numbers, then nearby trajectories will run away at an exponential rate. In principle, one can average the real parts of all eigenvalues along a trajectory to get a set of global characteristics — the Lyapunov exponents. However, there is another more intuitive way to calculate $n$ real numbers playing the same role.

Let us introduce a local coordinate system with origin at $x_0$ by choosing at random $n$ small vectors coming out from $x_0$ and then orthogonalizing them. Denote these vectors by $\{e_0^{(1)}, e_0^{(2)}, \cdots, e_0^{(n)}\}$. Then integrate Eq. (102) one time step forwards to reach a new position $x_1$, and at the same time let $\{e_0^{(i)}\}$ evolve according to the linearized Eq. (103) to form $n$ new vectors with origin at $x_1$. Now we can calculate a number of goemetrical ratios:

1. There are $n$ length ratios:

$$|e_1^{(i)}|/|e_0^{(0)}|;$$

2. And $n(n-1)/2$ parallelogram area ratios:

$$|e_1^{(i)} \wedge e_1^{(j)}|/|e_0^{(i)} \wedge e_0^{(j)}|;$$

3. Several intermediate ratios of geometric objects of dimensions 3 through $n-1$, if $n > 3$;

4. At last, there is one $n$-dimensional hyperparallelepiped volume ratio:

$$\frac{|e_1^{(1)} \wedge e_1^{(2)} \wedge \cdots \wedge e_1^{(n)}|}{|e_0^{(1)} \wedge e_0^{(2)} \wedge \cdots \wedge e_0^{(n)}|}.$$

The exterior product $\wedge$ generalizes the vector product $\times$ to higher dimensional spaces.

Being geometrical quantities, all these ratios are real numbers. Repeating this process step by step along the trajectory and taking the long time average of these numbers we get a set of global characteristics of the attractor. Based on the multiplicative ergodic theorem of Oseledec (1968), it was proved by Benettin *et al.* (1976) that provided the vectors $\{e_0^{(i)}\}$ are taken at random, the length ratios converge to the maximal Lyapunov exponent $\lambda_1$, The parallelogram area ratios converge to the sum of the two largest Lyapunov exponents $\lambda_1 + \lambda_2$, ..., the $n$-dimensional hyperparallelepiped volume ratio converges to the sum of all Lyapunov exponents $\lambda_1 + \lambda_2 + \cdots + \lambda_n$. The last quantity must be equal to the contraction rate of a volume element along the trajectory, i.e.,

$$\lambda_1 + \lambda_2 + \cdots + \lambda_n = -\mathrm{div}\mathbf{f}, \tag{105}$$

where $\mathbf{f} = (f_1, f_2, \cdots, f_n)$ is the vector field on the right-hand side of (102). In practice, Eq. (105) serves as a numerical check for the correctness of the computation. In stating the results above, we have introduced an ordering for all Lyapunov exponents:

$$\lambda_1 \geq \lambda_2 \geq \cdots \geq \lambda_n. \tag{106}$$

We mention in passing that the step-by-step orthogonalization appears to be crucial in numerical calculation, since the initially independent vectors tend to evolve into a subspace of lower dimensionality due to phase volume contraction and they become parallel to each other (see, e.g., Shimada and Nagashima, 1979).

Kaplan and Yorke (1979) conjectured that the Kolmogorov capacity $D_0$ [see Eq. (83)] is related to the Lyapunov exponents by

$$D_0 = j + \frac{1}{|\lambda_{j+1}|} \sum_{i=1}^{j} \lambda_i, \tag{107}$$

where $j$ is the number of Lyapunov exponents which assures the non-negativeness of the sum in (107), i.e.,

$$\sum_{i=0}^{j} \lambda_i \geq 0.$$

In the extreme case $j = 0$ or $n$, define $D_c = 0$ or $n$ respectively. Numerical experiments seem to be in favour of (107). In general, the Lyapunov exponents converge fairly rapidly and provide an effective way to calculate the dimension $D_c$ in comparison to the box-counting algorithm.

The largest Lyapunov exponent alone is of much use in numerical work, since the appearance of a positive exponent signals the onset of chaos. It was suggested by many authors to put forward a qualitative classification of attractors according to the signature of the Lyapunov exponents. For example, $(-, -, -)$ corresponds to fixed point with $D = 0$; $(0, -, -)$ to limit cycle with $D = 1$; $(0, 0, -)$ to 2-torus with $D = 2$; and $(+, 0, -)$ to strange attractor with $D = D_c$. In this scheme a signature $(+, +, 0, -)$ would correspond to strange attractor displaying *hyperchaos* (Rössler, 1979). Since the signature changes with control parameters, they provide also a means to classify transitions between various regimes. Transition from $(+, 0, -)$ to $(+, +, 0, -)$ has been observed in Rayleigh-Bénard convection (Sano and Sawada, 1985) and in semiconductors (Stoop *et al.*, 1989).

## Peculiarities of One-Dimensional Maps

Dimension and Lyapunov exponent for one-dimensional maps have some peculiarities worth mentioning. First, there is only one Lyapunov exponent, sometimes called also Lyapunov number,

$$\lambda(\mu) = \lim_{k \to \infty} \frac{1}{k} \sum_{i=0}^{k-1} \log |f'(\mu, x_i)|. \tag{108}$$

Stable periodic orbits correspond to $\lambda < 0$, chaotic orbits to $\lambda > 0$.

Second, $\lambda = 0$ at all bifurcation points, at the accumulation point $\mu_\infty$ of period-doubling and period-$n$-tupling sequences, as well as at tangent bifurcations. Grassberger (1981) showed that at $\mu_\infty$, where the attractor has a Cantor set structure, the Hausdorff dimension is universal, with value $D_c = 0.538 \cdots$. In fact, the whole $D_q$ versus $q$ and $f(\alpha)$ versus $\alpha$ spectra may be calculated for period-doubling and period-$n$-tupling sequences (Zeng and Hao, 1986). On the other hand, due to critical slowing down, one always gets a cluster of points near every finite bifurcation point $\mu_k$, no matter how long the iteration is being carried on. This cluster acquires an "operational" dimension $D = 2/3$, if measured by using a box-counting algorithm (Wang and Chen, 1984), although the true dimension of a periodic orbit must be zero. In fact, there exists a scaling function $D(k, \epsilon)$, where $k$ numbers the bifurcation and $\epsilon$ is the box size, and the two values $2/3$ and $0.538 \cdots$ appear at two different limits (Hu and Hao, 1983):

$$
\begin{aligned}
\lim_{k \to \infty} \lim_{\epsilon \to 0} D(k, \epsilon) &= 0.666 \cdots, \\
\lim_{\epsilon \to 0} \lim_{k \to \infty} D(k, \epsilon) &= 0.538 \cdots.
\end{aligned}
\tag{109}
$$

## Lyapunov Exponents from Time Series

When the governing equations on attractors of low dimensionality are not known, one has to extract Lyapunov exponents from experimental time series. The calculation is based on *recovering the linearized dynamics* in the reconstructed phase space of embedding dimension $m$. One starts from the list (87) of reconstructed vectors $\{y_i\}$. After $k$ time steps, a given vector $y_j$ becomes vector $y_{j+k}$. If the time interval $k\tau$ ($\tau$ is the sampling interval of the time series) is small enough, the time evolution should be described by the linearized dynamics, i.e., by a linear relation

$$
y_{j+k} = A y_j,
\tag{110}
$$

where $A$ is a $m \times m$ matrix. While components of $y_j$ and $y_{j+k}$ are known, the matrix $A$ may be parametrized by $m \times m$ unknowns (by preliminary sorting of the original time series and suitable choice of $k$, $m$ unknowns are sufficient to parametrize $A$). In order to determine these unknowns, one collects all neighbours of $y_j$ in a small ball of radius $\epsilon$ and describe their time evolution by the same matrix $A$. In this way, one gets more equations than there are unknowns in $A$. This overdetermined system may be solved by standard least-square procedure. Then the Lyapunov exponents are calculated by averaging the eigenvalues of $A$ over many time steps. In principle, one can get not only the largest, but also a few next exponents. This method was suggested by Sano and Sawada (1985), Eckmann and Ruelle (1985, Paper 23 in this volume), and realized, e.g., by Eckmann *et al.* (1986, Paper 26 in this volume). To make the method more efficient, a few subtle points must be taken into account; one may consult, e.g., the work of Conte and Dubois (1988, Paper 28 in this volume).

A problem closely related to the recovering of dynamics is *noise reduction*. The knowledge of the linearized dynamics allows one to go beyond the traditional noise reduction technique, e.g., by filtering the data. We indicate the papers by Farmer and Sidorowich (1987), Kostelich and Yorke (1988, Paper 27 i n this volume). The notion of shadowing orbit may provide another way of noise reduction (Hammel, Yorke and Grebogi, 1987).

## Characterization of Chaotic Attractors using Unstable Periodic Orbits

Dimensions and entropies are global invariant of the motion on chaotic attractors. Two attractors, having the same dimension or entropy, may still be different in their dynamics. The spectra of Lyapunov exponents do provide more detailed description of stretching and contraction in the phase space, but they are also global characteristics, obtained by averaging over the whole trajectory, and hard to calculate (we have in mind the calculation of many exponents, not just the first ones). However, these global invariants should be connected to local invariant properties of the dynamics, the latter may provide a more refined characterization of the motion.

Since strange attractors are the closures of the unstable manifolds of unstable periodic points and periodic orbits of short periods may be extracted from a chaotic orbit, it was natural to characterize chaotic motion by unstable periodic orbits (Auerbach *et al.*, 1987; Guranatne and Procaccia, 1987). Periodic orbits are dense on a strange attractor, but they are necessarily unstable. They are topological invariants, their eigenvalues are metric invariants (Cvitanovic, 1988). Global characteristics like dimension and the $f(\alpha)$ spectrum may be calculated from the eigenvalues (Grebogi, Ott and Yorke, 1987c, 1988; Auerbach *et al.*, 1988), topological and metric entropy may be estimated from the number and frequency of appearance of periodic cycles. Unstable periodic orbits have been extracted from experimental time series to get an estimate for the topological entropy (Lathrop and Kostelich, 1989; see also an earlier attempt by Kahlert and Rössler, 1984). Cvitanovic (1988) has suggested that further comparison of experiment with theory should be carried out by listing unstable periodic cycles and their eigenvalues. In fact, the enumeration of all possible periodic orbits is closely related to symbolic dynamics.

# Chapter 9
# SCALING FOR EXTERNAL NOISE

We have emphasized the intrinsic nature of chaos in Chapter 1 and throughout the chapters thus far. However, external noise is an unavoidable factor in computer experiments (round-off errors) and laboratory experiments (thermal fluctuations of the ambience). A fuller understanding of chaos requires the inclusion of external noise into the theoretical framework.

Actually, in the theory of chaotic phenomena external noise plays a more constructive role than just to be an undesirable but inescapable disturbance from the outside. Only with the external noise taken into account shall we have a full parallel with the theory of continuous phase transitions which has enriched physics with so many new concepts since the mid 1960's.

Phase transition occurs at certain critical value of a control parameter such as the temperature $T$ or the pressure $p$ when the latter is subjected to a slow and progressive variation. Many thermodynamical quantities show singular behavior when $T - T_c \to 0$ where $T_c$ denotes the critical point. Chaotic transition takes place at critical value $\mu_c$ of the control parameter such as the Reynold's number and shows up as singularity in $|\mu - \mu_c|$. The emergence of a new phase may be described by the appearance of nonzero "order parameter", say, macroscopic magnetization in ferromagnetic phase transitions. Chaotic state can be characterized by certain "disorder parameter" as well. It was thought by some authors that the Lyapunov exponent may be taken as the disorder parameter. In fact, the Lyapunov exponent corresponds to the reciprocal of correlation length in phase transitions as we shall see soon. One must take some other statistical characteristic of the chaotic motion on the strange attractor (entropy or the like) to be the disorder parameter.

Moreover, a new ordered phase may appear either spontaneously at the critical point $T_c$, or may be induced by suitable external fields. For example, the external magnetic field plays the role of an "ordering field" in magnetic phase transitions. In phase transition theory the ordering field, coupled to the order parameter, is sometimes called a dual field. Does there exist a "disordering field" for chaotic transition? Chaotic states are characterized by spontaneous appearance of seemingly random motion showing noisy broad bands in the power spectra. Therefore, external noise may well play the role of the disordering field.

To include the external noise, one adds a noisy source term to the nonlinear map (13) and thereby changes it into a discrete Langevin equation:

$$x_{n+1} = f(\mu, x_n) + \sigma \xi_n, \tag{111}$$

where $\xi_n$ are random numbers obeying certain statistical distribution with, e.g., the constraints

$$\overline{\xi_n} = 0, \quad \overline{\xi_n \xi_m} = \delta_{nm}.$$

The coefficient $\sigma$, being a measure of the noise strength, can be compared with the external magnetic field in magnetic phase transitions.

An important notion in phase transition theory is the correlation length $\xi$ which diverges at the critical points as

$$\xi \propto |T - T_c|^{-\nu}, \tag{112}$$

where $\nu$ is a positive universal exponent. So far in this volume, chaotic behaviour has been considered only in connection with time evolution of nonlinear systems (except for a few words in Chapter 4 on partial differential equations). From the discussion in the last chapter we know that when the Lyapunov exponents $\lambda$ are negative, the influence of initial conditions decays quickly and the system approaches an asymptotic state independent of the initial conditions, whereas when at least one $\lambda$ is positive, nearby orbits will run away exponentially and any minor change in initial conditions will be amplified with time. Therefore, $\lambda$ can be taken as a measure of time correlation. Since $\lambda$ passes through zero at bifurcation points, in particular, at the limiting point $\mu_\infty$ of a period-doubling sequence, one expects a relation similar to (112) to hold:

$$\lambda_{-1} \propto |\mu - \mu_c|^{-t}. \tag{113}$$

Starting from the discrete Langevin equation (111) the whole arsenal of critical dynamics in phase transition theory[13] can be brought to bear on chaotic transitions. We have included two pioneering papers along this line into this volume (Crutchfield and Huberman, 1980, Paper 29; Crutchfield, Nauenberg and Rudnick, 1981, Paper 30; see also Haken and Mayer-Kress, 1981 a and b; Haken and Wunderlin, 1982, on Champman-Kolmogorov equation related to the discrete Langevin equation). Shraiman, Wayne and Martin (1981) derived a scaling relation for $\lambda$ in the presence of external noise:

$$\lambda(\mu_\infty - \mu, \sigma) = (\mu_\infty - \mu)^t \Phi\left(\frac{\sigma^\theta}{(\mu_\infty - \mu)^t}\right), \tag{114}$$

where $\Phi$ is a universal function and

$$
\begin{aligned}
t &= \frac{\log 2}{\log \sigma} = 0.4498069 \cdots, \\
\theta &= \frac{\log 2}{\log \kappa} = 0.366739 \cdots,
\end{aligned}
\tag{115}
$$

$\kappa = 6.6190 \cdots$ being a new exponent related to the scaling of external noise. Putting $\sigma = 0$ in (114), we are led to

$$\lambda(\mu_\infty - \mu, 0) \propto (\mu_\infty - \mu)^t \tag{116}$$

---

[13]For a review of critical dynamics, see, e.g., P. C. Hohenberg, and B. I. Halperin, Rev. Mod. Phys. **49**(1977), 435.

which has the desired form of Eq. (113) and was obtained without involving external noise (Huberman and Rudnick, 1980).

The scaling relation derived in Paper 30 looks slightly different:

$$\lambda(\mu_\infty - \mu, \sigma) = \sigma^\theta L \left( \frac{\mu_\infty - \mu}{\sigma^{\theta/t}} \right). \tag{117}$$

It suffices to redefine the scaling function by letting

$$xL(1/x) \equiv \Phi(x), \tag{118}$$

with $x = \sigma^\theta/(\mu_\infty - \mu)^t$, to see that Eq. (117) is identical to (114).

Now we are prepared to summarize the role of external noise in chaotic dynamics as follows. First, external noise will wash out the details in high order bifurcation sequences including both the period-doubling cascades and the inverse sequences of chaotic bands. In order to see one more bifurcation one must decrease the noise level by a factor of $\kappa$. This explains the meaning of the noise exponent $\kappa$. Second, the presence of external noise makes the bifurcation diagram fuzzy and lowers the accumulation point $\mu_\infty$ slightly as it can be expected from the role of noise as a disordering field. Third, the Lyapunov exponent passes through zero at various bifurcation points, but this is no longer true when external noise is added. For more details see review papers, e.g., Crutchfield et al., (1982). The level of external noise, hidden in an experimental time series, may be estimated, in principle, in the time-delay method of reconstructing phase space by looking at the limit of the scaling range (see Fig. 12 in Chapter 8).

The role of external noise has been studied in detail on the example of the logical map (for a recent paper, see, Bene and Szépfalusy, 1988) and the Lorenz model (see, e.g., Zippelius and Lucke, 1981; Nicolis and Nicolis, 1986). There have been reports on the effect of noise on chaotic attractors, strange and non-strange (see, e.g., Ott and Hanson, 1981; Rechester and White, 1983, Franaszek, 1987). The influence of noise on the $Q_2 \to$ chaos transition (see, e.g., Feigenbaum and Hasslacher, 1981; Kajanto and Salomma,1985) and the intermittent transition (se, e.g., Eckmann et al., 1981; Hirsch et al., 1982, Paper 16 in this volume; Landa, 1987) has also been studied. In addition, there has appeared a monograph on chaos in systems with noise (Kapitaniak, B1988).

# Chapter 10
# OBSERVATION AND EXPERIMENTAL STUDY OF CHAOS

Chaotic phenomena are not just a motley collection from the mathematical labyrinth, but a wide class of natural events found in the physical world. Generally speaking, chaos happens more frequently than order, just as there are many more irrational numbers than the rationals. Chaos can be thought of as a new regime of nonlinear oscillations, as a compromise between competing periodicities, as overlap of resonances, as accumulation of many instabilities, as the prelude to turbulence. Therefore, it is not difficult to imagine various experimental situations generating chaotic behaviour.

Moreover, Nature is always richer than mathematical models, Experimental study of chaos has at least a twofold mission: to verify the theoretical understanding gained from model studies and to bring about new physics by challenging the existing theory with unexpected findings, not to mention the prospect for technological development at our present level of knowledge. From this view-point laboratory observations can be roughly divided into three groups: those almost coinciding with theoretical expectations, e.g., experiments using nonlinear circuits; those showing much more variations and complications, only some facets of which resemble one or another aspect of theoretical models, e.g., experiments on hydrodynamical instabilities; and those occupying an intermediate state, e.g., most of the experiments on acoustic, optical, chemical, and solid-state turbulence.

In what follows various kinds of experiments will be listed briefly along with a few references to the bibliography, including some theoretical analysis and proposals which come under the same heading.

## Forced Vibration of Shallow Water in a Finite Container

Historically this was the first experiment where transition to chaos would have been discovered, had Faraday had the idea of period-doubling route to chaos. In fact, Faraday observed subharmonic component $f_0/2$ in the shallow water wave in a container forced to vibrate vertically at frequency $f_0$. Lord Rayleigh noticed, repeated and wrote about this experiment in his *Theory of Sound*, §69b. Modern data accquisition techniques helped to reveal much longer bifurcation sequences and the onset of chaos (Keolian *et al.*, 1981). The experimental findings looked like departure from period-doubling. For example, taking the driving period to be 1, a sequence

$$1, 2, 4, 12, 14, 16, 18, 20, 22, 24, 28, 35$$

has been observed. However, the power spectrum given for period 14 showed clearly a $2 \times 7$ fine structure, testifying to it being a secondary period 7 embedded in a chaotic band of period 2.

## Bifurcation and Chaos in Nonlinear Circuits

The most complete and beautiful results on chaos have been obtained on nonlinear circuits, because the experimental conditions can be precisely controlled and the circuits can be well represented by ordinary differential equations with only a few variables.

Earlier work on nonlinear circuits (Gollub *et al.*, 1978, 1980) were carried out on systems of many coupled nonlinear oscillators. Complicated combinations of periods and chaos were indeed observed, but could not be fitted into simple systematics as has now been done.

Most of later experiments deal with a single nonlinear oscillator driven by a periodic signal (Linsay, 1981; Buskirk and Jeffries, 1985, Paper 31 in this volume). Period-doubling bifurcations, tangent bifurcations, intermittency, crisis of the attractors, and the effect of external noise have been studied. In particular, the Feigenbaum constants $\delta$ and $\alpha$, the noise scaling exponent $\kappa$, and the ratio of successive period-doubling peaks in the power spectra were measured to be in good accord with that of the one-dimensional unimodal maps.

It is interesting to note that a simple autonomous nonlinear circuit with a piecewise linear resistor (Chua's circuit, Matsumoto, 1984) develops a Shilnikov-type (see Chapter 5) chaotic attractor ("the double scroll"). This model has been studied in detail both experimentally and theoretically (see, e.g., Chua *et al.*, 1986).

It seems desirable to carry out a detailed study on transitions from quasiperiodicity to chaos using nonlinear circuits, but probably more than one nonlinear oscillators should be used for this purpose.

## Bifurcation and Chaos in Mechanical or Electromechanical Systems

This is another group of experiments where results in keeping with theory can be expected. Some of these systems may serve as classroom demonstration of chaotic phenomena (see, e.g., Bremer, 1987). We add also that both the Lorenz and Duffing equations can be modelled by simple mechanical systems (see Sparrow, B1982, and Guckenheimer and Holmes, B1983, respectively). Chaotic vibration of mooring offshore structures is a problem of technical importance (Thompson, 1983; Thompson and Stewart, B1986; Liaw, 1988). The deformation of a long elastic string (the Euler elastica) is described by the same equation of the mathematical pendulum [Eq. (68) with friction $k = 0$], with time $t$ being replaced by the arc length $s$. This correspondence provides interesting analogue of temporal chaos in space, in particular, homoclinic intersections may be easily visualized (El Naschie and Al Athel, 1989).

## Onset of Hydrodynamical Instabilities

Low temperature technique, laser Doppler velocimetry, and modern data acquisition systems have brought the study of turbulence back to physical laboratories. However, at the present time most of these studies still concentrate on instabilities in finite containers where the decay of many irrelevant modes of motion makes the results closer to theoretical predictions. As regards the more important shear flow instabilities, little progress has really been achieved with the help of the philosophy of chaos, because infinite degrees of freedom must be involved to interpret the experiments.

Thermoconvective instability of fluid in a container heated from below with free surface (the Bénard instability) or without space in between the fluid and the upper plate of the container (the Rayleigh-Bénard instability) has now become a classical set-up to study the onset mechanism of turbulence. Ahlers first conducted the experiment at low temperatures on liquid helium in 1974. Libchaber and Maurer (1980) observed the successive appearance of quasiperiodic regime with two incommensurable frequencies, frequency locking, and period-doubling. For a recent review, see Libchaber (1987).

Instabilities of Couette flow between rotating cylinders show much more variations, some but not all of which may be put into the framework of the Ruelle-Takens route. Especially when the outer cylinder also rotates or stops suddenly, many more very complicated patterns have been recorded. Most of these experimental findings still await a better theoretical interpretation. We refer to Swinney (1988) for a recent summary.

## Acoustic Turbulence

Subharmonic components have been known for a long time in the spectra of cavitation, i.e., bubble formation in liquid subjected to local pressure decrease due to, say, high power supersonic radiation. It was natural to look for transitions to chaos in these systems (Lauterborn and Cramer, 1981; Cramer and Lauternborn, 1982). Subharmonic generation has been also observed in supersonic absorption in liquid helium (Smith *et al.*, 1982, 1983). Bifurcations and chaos can be observed in simple air acoustic systems and loudspeakers (Kitano *et al.*, 1983), but usually it is difficult to follow the whole cascade in detail due to high level of external noise (see, e.g., Zong and Wei, 1984; Wei *et al.*, 1986; Miao *et al.*, 1987). For recent development in acoustic chaos, see Lauterborn and Parlitz (1988), Holzfuss and Lauterborn (1989).

## Optical Turbulence

Lasers provide a typical class of nonlinear oscillator systems. The Maxwell-Bloch equations, describing a unimodal laser in the semiclassical approximation, reduce to a system of three ordinary differential equations. Haken (1975) showed that this system can be transformed into the Lorenz model (63) under certain conditions, not realistic for a laser but relevant to random spiking in the superradiance regime. However, it has been realized later that Lorenz type of chaos may exist in lasers as well (Weiss *et al.*, 1984, 1986).

Ikeda (1979) predicted theoretically that chaos may appear in an optical bistability device using a ring cavity. This effect has been observed first in so-called hybrid bistable devices (Gibbs *et al.*, 1981; Hopf *et al.*, 1982, 1986). Hybrid means that part of the optical output is delayed by an electronic circuit, or opto-acoustic delay line, or by keeping for a while in a microprocessor, and then fed back to the input signal. The characteristic periods in these systems, being determined by the delay time, are quite low as compared to optical frequencies. Therefore, some people are reluctant to count the effect as optical turbulence. All optical bistability devices have also been designed to show optical chaos (Nakatsuka *et al.*, 1983; Harrison *et al.*, 1983, 1984).

The study of bifurcations and chaos in optical bistable devices has led to detailed exploration of delayed differential equations, which we have mentioned in Chapter 5 (see, e.g., Le Berre *et al.*, 1986, 1987, 1989; Zhang *et al.*, 1985, 1987; Li and Hao, 1989).

Optical chaos has been observed also in other experimental settings. We have included a recent review by Harrison and Biswas (1986, Paper 34 in this volume; see also, Harrison and Uppal, 1988).

## Chemical Chaos

The concentrations of intermediate products in some specially designed chemical reactions may oscillate with time. The most throughly studied reaction of this type, the Belousov-Zhabotinskii reaction, consists of some 20 elementary reactions. During the last two decades these systems have served as prototype for self-organization phenomena, i.e., spontaneous occurrence of temporal or spatial structures or both, when energy and matter flows are supported from the outside. In so-called continuously-flow stirred tank reactor (CSTR), the reacting system may be considered as spatially homogeneous and thus described by systems of ordinary nonlinear differential equations. Reports on experimental observations of chemical chaos began to emerge since 1977 (Schmitz *et al.*, 1977; Yamazaki *et al.*, 1978; Hudson *et al.*, 1979, 1981). Many theoretical models have been suggested to describe chemical turbulence, but most of them failed to capture all the experimental findings. Only quite recently, a system of 7 ordinary differential equations have been shown to exhibit all the experiemntally observed behaviour (see the review by Argoul *et al.*, 1987, Paper 33 in this volume).

In order to keep the reaction in steady state, some reactant and products must be supplied or removed from the system constantly. The flow velocity of these chemicals can be taken as a control parameter and the concentration of certain intermediate products, say, the concentration of $Br^-$ ions in the Belousov-Zhabotinskii reaction, is to be monitored. At very small flow velocity the system approaches thermal equilibrium without showing complicated time behaviour. At very large flow velocity the chemicals rush through the container, having no time to get into reaction. Therefore, only at intermediate flow velocities one can expect complicated dynamics, including multisteady states, simple or relaxation oscillations, intermittency, period-doubling and chaos, hysteresis, etc. It has been proved experimentally that such dynamical behaviour was entirely caused by chemical reactions, not by the hydrodynamics of the flow.

Strange attractors, reconstructed from experimental data of the Belousov-Zhabotinskii reaction, happens to have sufficient low dimensionality (Roux *et al.*, 1980, 1981) and the ordering of the observed periods fits into that of the $U$-sequence (Simoyi *et al.*, 1982). Therefore, we see that in chemical turbulence there are many features in common with other nonlinear systems. Moreover, besides nonlinear circuits, chemical systems are easy to control and the dynamics involves a quite different frequency range as compared to that of nonlinear circuits. All these have made chemical turbulence an interesting subject, drawing more attention. Indeed, there exists a wide literature on chemical chaos. We mention only some new experiments in electrochemical reaction (Albahadily *et al.*, 1989; Basset and Hudson, 1989; Schell and Albahadily, 1989), and experiments using continuously fed unstirred reactors (CFUR) which are capable to show chaotic spatio-temporal patterns (Tam *et al.*, 1988).

## Chaos in Solid State Physics

Coupled or driven oscillators, capable to show chaotic behaviour in suitable parameter range, provide frequently used models in solid state physics. Competition of various commensurable and incommensurable periods in solids makes chaos an intrinsic source of spatial disorder (or to be more precise, order without periodicity). From numerous theoretical models and experimental set-ups we mention only a few.

It has been known since 1977 that in radio-frequency driven Josephson junctions used as parametric amplifiers noise grows anomalously with the gain level. The equivalent noise temperature reaches 50000K whereas the device is kept at 4K. Since such a high noise level cannot be explained by any known noise source and its amplification, Huberman *et al.* (1980) attributed it to the intrinsic dynamics of the junction. In the resistively shunted junction model the current equation reads

$$C\dot{V} + V/R + I_c \sin\varphi = I_r \cos\omega t + I_0, \tag{119}$$

where $R$, $C$, $I_c$, $I_r$, and $I_0$ denote the resistance, capacity, critical current, radio-frequency current, and dc bias, respectively. The phase difference across the junction is given by the Josephson equation

$$\dot{\varphi} = 4\pi eV/\hbar.$$

In terms of dimensionless quantities these two equations lead to the following equation of a forced damped pendulum

$$\ddot{\varphi} + a\dot{\varphi} + \sin\varphi = i_0 + i_r \cos\Omega t. \tag{120}$$

Early study on chaos in Josephson junctions were conducted on simulators of Eq. (120) (Yeh and Kao, 1982, 1983; Yeh *et al.*, 1984; He *et al.*, 1984, 1985; Kao *et al.*, 1986). Reports on experimental study began to appear since 1984 (Gubakov *et al.*, 1984; Octavio and Nasser, 1984; Cronemeyer *et al.*, 1985). Among reviews on chaos in Josephson junctions we mention an early paper by Beasley and Huberman (1982) and a recent one by Pedersen (1988).

Now let us turn to spatial chaos as a source of disorder. In solid state physics there are many cases of competing periods leading to the formation of new periodic or disordered structures. For example, a monolayer of inert gas, adsorbed to a graphite substrate, may form a two-dimensional lattice, which does not necessarily match the lattice of the substrate. If the interaction between the monolayer and the substrate is weak enough, the adsorbed layer keeps its own periodicity. Conversely, when they interact strongly, the monolayer lattice is forced to match that of the substrate. In between these two extremes phase-locking and spatial chaos may occur. This process resembles the behaviour of forced nonlinear oscillators with time evolution replaced by spatial ordering.

The simplest model for modulated structure was proposed by Frenckel and Kontorova in the 1930's. Suppose a one-dimensional chain of atoms is put in a cosine potential of the substrate, the $n$-th atom being at $x_n$. The energy of the system may be written as

$$F = \sum_n \frac{1}{2}(x_{n+1} - x_n - a)^2 + \frac{\mu}{(2\pi)^2} \cos 2\pi x_n, \tag{121}$$

where the period of the substrate field is taken to be 1, and $\mu$ measures the coupling. The equilibrium positions of atoms are determined by solving

$$\frac{\partial F}{\partial x_n} = 0,$$

that is

$$x_n - x_{n-1} - (x_{n+1} - x_n) - \frac{\mu}{2\pi} \sin 2\pi x_n = 0. \tag{122}$$

Let $y_n = x_n - x_{n-1}$, we have

$$\begin{aligned} y_{n+1} &= y_n - \frac{\mu}{2\pi} \sin 2\pi x_n, \\ x_{n+1} &= x_n + y_{n+1}. \end{aligned} \tag{123}$$

This is precisely the standard mapping (61), given in Chapter 5. Now it is natural to expect periodic, quasiperiodic and chaotic solutions depending on the parameter $\mu$. When the characteristic wave number of the modulated structure varies continuously, commensurable phases may occur repeatedly at rational values, leading to the devil's staircase, similar to that in circle maps. Similar situation appears in equilibrium configuration of spin systems (see, e.g., Ananthakrishna et al., 1987). Ruelle (1982) went even further to pose the question whether there exist turbulent crystals. We recommend the reader to consult the reviews by Aubry (1983a), Fisher and Huse (1982), and Bak (1982, 1986).

Chaos due to instabilities in semiconductors has also attracted more attention. We refer to a few recent papers (Landsberg et al., 1988; Stoop et al., 1989).

## Microwave Ionization of Hydrogen Atom

Ionization of hydrogen atom in high Rydberg state in a microwave cavity has been considered as experiment showing "quantum" chaos (Bardsley and Sundaram, 1985; Bayfield and Pinnaduwage, 1985; Van Leeuwen et al., 1985). However, the phenonmenon is essentially a manifestation of classical chaos with, perhaps, some "quantum signature", as a few cautious authors put it. For theoretical explanation, see, e.g., Casati et al. (1986), Zheng and Reichl (1987), Gontis and Kaulakys (1987), Blumel et al. (1988), Jensen et al., (1989).

## Chaos in Geophysics

Besides the long-standing problem of weather forecasting, atmosphere and ocean dynamics provides a wide arena for chaotic dynamics. There have been discussions on the Southern Oscillation (the El Nino phenomenon, see, e.g., Hense, 1986; Vallos, 1986), on large and mesoscale motions in the Pacific Ocean (Osborne *et al.*, 1986). The debate on the existence of climatic attractors was essentially concentrated on the reliability of dimension calculation (Nicolis and Nicolis, 1984; Grassberger, 1986b; Essex *et al.*, 1987). Geomagnetic dynamo models furnish many examples of ordinary differntial equations, exhibiting chaotic behaviour (see, e.g., Miura and Kai, 1986). For more on geophysical and climate dynamics we refer to Ghil, Benzi and Parisi (B1985).

## Chaos in biology and ecology

We have included two reviews on choas in biology (Olsen and Degn, 1985, Paper 35 in this volume) and in ecology (Schaffer, 1985, Paper 36 in this volume). Therefore, we add only two remarks.

The impressive experiment with periodically stimulated beating of cultured chicken cardiac cells (Guevara *et al.*, 1981, 1982; Glass *et al.*, 1983), displaying phase-locking, period-doubling and chaos, has been followed by many interesting studies on the dynamics of the heart (e.g., Goldberger *et al.*, 1985, 1986). It was modelled by a periodically forced oscillator and attempts have been made to connect the chaotic dynamics with cardiac arrhytmias and clinic practice (Goldberger and West, 1987; Courtemanche *et al.*, 1989.)

Chaotic behaviour in neural networks and in electroencephalographic (EEG) signals have drawn much attention in recent years (see, e.g., Rapp *et al.*, 1985; Babloyantz *et al.*, 1985, 1986; Xu and Li, 1986). In particular, measurements of human EEG signal dimensions are subject to the limitations that we have discussed in Chapter 8. There have been some speculations on the possible role of chaos in brain activity (Harth, 1983; Skarda and Freeman, 1987).

## Chaos in Economy

Any attempt to apply chaotic dynamics to social science must be quite speculative at present. We have included a review by Chen (1988, Paper 37 in this volume), where one may find more references.

PART TWO: REPRINTED PAPERS

**1. What is Chaos?**

# Chaos

*There is order in chaos: randomness has an underlying geometric form. Chaos imposes fundamental limits on prediction, but it also suggests causal relationships where none were previously suspected*

by James P. Crutchfield, J. Doyne Farmer, Norman H. Packard and Robert S. Shaw

The great power of science lies in the ability to relate cause and effect. On the basis of the laws of gravitation, for example, eclipses can be predicted thousands of years in advance. There are other natural phenomena that are not as predictable. Although the movements of the atmosphere obey the laws of physics just as much as the movements of the planets do, weather forecasts are still stated in terms of probabilities. The weather, the flow of a mountain stream, the roll of the dice all have unpredictable aspects. Since there is no clear relation between cause and effect, such phenomena are said to have random elements. Yet until recently there was little reason to doubt that precise predictability could in principle be achieved. It was assumed that it was only necessary to gather and process a sufficient amount of information.

Such a viewpoint has been altered by a striking discovery: simple deterministic systems with only a few elements can generate random behavior. The randomness is fundamental; gathering more information does not make it go away. Randomness generated in this way has come to be called chaos.

A seeming paradox is that chaos is deterministic, generated by fixed rules that do not themselves involve any elements of chance. In principle the future is completely determined by the past, but in practice small uncertainties are amplified, so that even though the behavior is predictable in the short term, it is unpredictable in the long term. There is order in chaos: underlying chaotic behavior there are elegant geometric forms that create randomness in the same way as a card dealer shuffles a deck of cards or a blender mixes cake batter.

The discovery of chaos has created a new paradigm in scientific modeling. On one hand, it implies new fundamental limits on the ability to make predictions. On the other hand, the determinism inherent in chaos implies that many random phenomena are more predictable than had been thought. Random-looking information gathered in the past—and shelved because it was assumed to be too complicated—can now be explained in terms of simple laws. Chaos allows order to be found in such diverse systems as the atmosphere, dripping faucets and the heart. The result is a revolution that is affecting many different branches of science.

What are the origins of random behavior? Brownian motion provides a classic example of randomness. A speck of dust observed through a microscope is seen to move in a continuous and erratic jiggle. This is owing to the bombardment of the dust particle by the surrounding water molecules in thermal motion. Because the water molecules are unseen and exist in great number, the detailed motion of the dust particle is thoroughly unpredictable. Here the web of causal influences among the subunits can become so tangled that the resulting pattern of behavior becomes quite random.

The chaos to be discussed here requires no large number of subunits or unseen influences. The existence of random behavior in very simple systems motivates a reexamination of the sources of randomness even in large systems such as weather.

What makes the motion of the atmosphere so much harder to anticipate than the motion of the solar system? Both are made up of many parts, and both are governed by Newton's second law, $F = ma$, which can be viewed as a simple prescription for predicting the future. If the forces $F$ acting on a given mass $m$ are known, then so is the acceleration $a$. It then follows from the rules of calculus that if the position and velocity of an object can be measured at a given instant, they are determined forever. This is such a powerful idea that the 18th-century French mathematician Pierre Simon de Laplace once boasted that given the position and velocity of every particle in the universe, he could predict the future for the rest of time. Although there are several obvious practical difficulties to achieving Laplace's goal, for more than 100 years there seemed to be no reason for his not being right, at least in principle. The literal application of Laplace's dictum to human behavior led to the philosophical conclusion that human behavior

CHAOS results from the geometric operation of stretching. The effect is illustrated for a painting of the French mathematician Henri Poincaré, the originator of dynamical systems theory. The initial image (*top left*) was digitized so that a computer could perform the stretching operation. A simple mathematical transformation stretches the image diagonally as though it were painted on a sheet of rubber. Where the sheet leaves the box it is cut and reinserted on the other side, as is shown in panel *1*. (The number above each panel indicates how many times the transformation has been made.) Applying the transformation repeatedly has the effect of scrambling the face (*panels 2–4*). The net effect is a random combination of colors, producing a homogeneous field of green (*panels 10 and 18*). Sometimes it happens that some of the points come back near their initial locations, causing a brief appearance of the original image (*panels 47–48, 239–241*). The transformation shown here is special in that the phenomenon of "Poincaré recurrence" (as it is called in statistical mechanics) happens much more often than usual; in a typical chaotic transformation recurrence is exceedingly rare, occurring perhaps only once in the lifetime of the universe. In the presence of any amount of background fluctuations the time between recurrences is usually so long that all information about the original image is lost.

46

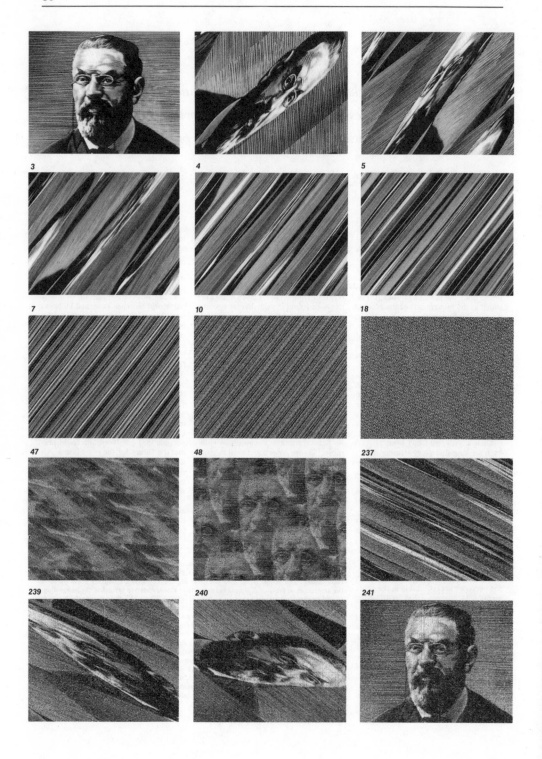

3

4

5

7

10

18

47

48

237

239

240

241

was completely predetermined: free will did not exist.

Twentieth-century science has seen the downfall of Laplacian determinism, for two very different reasons. The first reason is quantum mechanics. A central dogma of that theory is the Heisenberg uncertainty principle, which states that there is a fundamental limitation to the accuracy with which the position and velocity of a particle can be measured. Such uncertainty gives a good explanation for some random phenomena, such as radioactive decay. A nucleus is so small that the uncertainty principle puts a fundamental limit on the knowledge of its motion, and so it is impossible to gather enough information to predict when it will disintegrate.

The source of unpredictability on a large scale must be sought elsewhere, however. Some large-scale phenomena are predictable and others are not. The distinction has nothing to do with quantum mechanics. The trajectory of a baseball, for example, is inherently predictable; a fielder intuitively makes use of the fact every time he or she catches the ball. The trajectory of a flying balloon with the air rushing out of it, in contrast, is not predictable; the balloon lurches and turns erratically at times and places that are impossible to predict. The balloon obeys Newton's laws just as much as the baseball does; then why is its behavior so much harder to predict than that of the ball?

The classic example of such a dichotomy is fluid motion. Under some circumstances the motion of a fluid is laminar—even, steady and regular—and easily predicted from equations. Under other circumstances fluid motion is turbulent—uneven, unsteady and irregular—and difficult to predict. The transition from laminar to turbulent behavior is familiar to anyone who has been in an airplane in calm weather and then suddenly encountered a thunderstorm. What causes the essential difference between laminar and turbulent motion?

To understand fully why that is such a riddle, imagine sitting by a mountain stream. The water swirls and splashes as though it had a mind of its own, moving first one way and then another. Nevertheless, the rocks in the stream bed are firmly fixed in place, and the tributaries enter at a nearly constant rate of flow. Where, then, does the random motion of the water come from?

The late Soviet physicist Lev D. Landau is credited with an explanation of random fluid motion that held sway for many years, namely that the motion of a turbulent fluid contains many different, independent oscillations. As the fluid is made to move faster, causing it to become more turbulent, the oscillations enter the motion one at a time. Although each separate oscillation may be simple, the complicated combined motion renders the flow impossible to predict.

Landau's theory has been disproved, however. Random behavior occurs even in very simple systems, without any need for complication or indeterminacy. The French mathematician Henri Poincaré realized this at the turn of the century when he noted that unpredictable, "fortuitous" phenomena may occur in systems where a small change in the present causes a much larger change in the future. The notion is clear if one thinks of a rock poised at the top of a hill. A tiny push one way or another is enough to send it tumbling down widely differing paths. Although the rock is sensitive to small influences only at the top of the hill, chaotic systems are sensitive at every point in their motion.

A simple example serves to illustrate just how sensitive some physical

---

**Laplace, 1776**

"The present state of the system of nature is evidently a consequence of what it was in the preceding moment, and if we conceive of an intelligence which at a given instant comprehends all the relations of the entities of this universe, it could state the respective positions, motions, and general affects of all these entities at any time in the past or future.

"Physical astronomy, the branch of knowledge which does the greatest honor to the human mind, gives us an idea, albeit imperfect, of what such an intelligence would be. The simplicity of the law by which the celestial bodies move, and the relations of their masses and distances, permit analysis to follow their motions up to a certain point; and in order to determine the state of the system of these great bodies in past or future centuries, it suffices for the mathematician that their position and their velocity be given by observation for any moment in time. Man owes that advantage to the power of the instrument he employs, and to the small number of relations that it embraces in its calculations. But ignorance of the different causes involved in the production of events, as well as their complexity, taken together with the imperfection of analysis, prevents our reaching the same certainty about the vast majority of phenomena. Thus there are things that are uncertain for us, things more or less probable, and we seek to compensate for the impossibility of knowing them by determining their different degrees of likelihood. So it is that we owe to the weakness of the human mind one of the most delicate and ingenious of mathematical theories, the science of chance or probability."

---

**Poincaré, 1903**

"A very small cause which escapes our notice determines a considerable effect that we cannot fail to see, and then we say that the effect is due to chance. If we knew exactly the laws of nature and the situation of the universe at the initial moment, we could predict exactly the situation of that same universe at a succeeding moment. But even if it were the case that the natural laws had no longer any secret for us, we could still only know the initial situation *approximately*. If that enabled us to predict the succeeding situation with *the same approximation*, that is all we require, and we should say that the phenomenon had been predicted, that it is governed by laws. But it is not always so; it may happen that small differences in the initial conditions produce very great ones in the final phenomena. A small error in the former will produce an enormous error in the latter. Prediction becomes impossible, and we have the fortuitous phenomenon."

---

OUTLOOKS OF TWO LUMINARIES on chance and probability are contrasted. The French mathematician Pierre Simon de Laplace proposed that the laws of nature imply strict determinism and complete predictability, although imperfections in observations make the introduction of probabilistic theory necessary. The quotation from Poincaré foreshadows the contemporary view that arbitrarily small uncertainties in the state of a system may be amplified in time and so predictions of the distant future cannot be made.

systems can be to external influences. Imagine a game of billiards, somewhat idealized so that the balls move across the table and collide with a negligible loss of energy. With a single shot the billiard player sends the collection of balls into a protracted sequence of collisions. The player naturally wants to know the effects of the shot. For how long could a player with perfect control over his or her stroke predict the cue ball's trajectory? If the player ignored an effect even as minuscule as the gravitational attraction of an electron at the edge of the galaxy, the prediction would become wrong after one minute!

The large growth in uncertainty comes about because the balls are curved, and small differences at the point of impact are amplified with each collision. The amplification is exponential: it is compounded at every collision, like the successive reproduction of bacteria with unlimited space and food. Any effect, no matter how small, quickly reaches macroscopic proportions. That is one of the basic properties of chaos.

It is the exponential amplification of errors due to chaotic dynamics that provides the second reason for Laplace's undoing. Quantum mechanics implies that initial measurements are always uncertain, and chaos ensures that the uncertainties will quickly overwhelm the ability to make predictions. Without chaos Laplace might have hoped that errors would remain bounded, or at least grow slowly enough to allow him to make predictions over a long period. With chaos, predictions are rapidly doomed to gross inaccuracy.

The larger framework that chaos emerges from is the so-called theory of dynamical systems. A dynamical system consists of two parts: the notions of a state (the essential information about a system) and a dynamic (a rule that describes how the state evolves with time). The evolution can be visualized in a state space, an abstract construct whose coordinates are the components of the state. In general the coordinates of the state space vary with the context; for a mechanical system they might be position and velocity, but for an ecological model they might be the populations of different species.

A good example of a dynamical system is found in the simple pendulum. All that is needed to determine its motion are two variables: position and velocity. The state is thus a point in a plane, whose coordinates are position and velocity. Newton's laws provide

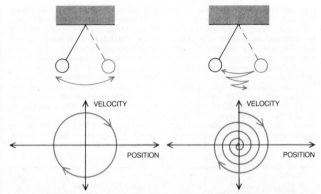

STATE SPACE is a useful concept for visualizing the behavior of a dynamical system. It is an abstract space whose coordinates are the degrees of freedom of the system's motion. The motion of a pendulum (*top*), for example, is completely determined by its initial position and velocity. Its state is thus a point in a plane whose coordinates are position and velocity (*bottom*). As the pendulum swings back and forth it follows an "orbit," or path, through the state space. For an ideal, frictionless pendulum the orbit is a closed curve (*bottom left*); otherwise, with friction, the orbit spirals to a point (*bottom right*).

a rule, expressed mathematically as a differential equation, that describes how the state evolves. As the pendulum swings back and forth the state moves along an "orbit," or path, in the plane. In the ideal case of a frictionless pendulum the orbit is a loop; failing that, the orbit spirals to a point as the pendulum comes to rest.

A dynamical system's temporal evolution may happen in either continuous time or in discrete time. The former is called a flow, the latter a mapping. A pendulum moves continuously from one state to another, and so it is described by a continuous-time flow. The number of insects born each year in a specific area and the time interval between drops from a dripping faucet are more naturally described by a discrete-time mapping.

To find how a system evolves from a given initial state one can employ the dynamic (equations of motion) to move incrementally along an orbit. This method of deducing the system's behavior requires computational effort proportional to the desired length of time to follow the orbit. For simple systems such as a frictionless pendulum the equations of motion may occasionally have a closed-form solution, which is a formula that expresses any future state in terms of the initial state. A closed-form solution provides a short cut, a simpler algorithm that needs only the initial state and the final time to predict the future without stepping through intermediate states. With such a solution the algorithmic effort

required to follow the motion of the system is roughly independent of the time desired. Given the equations of planetary and lunar motion and the earth's and moon's positions and velocities, for instance, eclipses may be predicted years in advance.

Success in finding closed-form solutions for a variety of simple systems during the early development of physics led to the hope that such solutions exist for any mechanical system. Unfortunately, it is now known that this is not true in general. The unpredictable behavior of chaotic dynamical systems cannot be expressed in a closed-form solution. Consequently there are no possible short cuts to predicting their behavior.

The state space nonetheless provides a powerful tool for describing the behavior of chaotic systems. The usefulness of the state-space picture lies in the ability to represent behavior in geometric form. For example, a pendulum that moves with friction eventually comes to a halt, which in the state space means the orbit approaches a point. The point does not move—it is a fixed point—and since it attracts nearby orbits, it is known as an attractor. If the pendulum is given a small push, it returns to the same fixed-point attractor. Any system that comes to rest with the passage of time can be characterized by a fixed point in state space. This is an example of a very general phenomenon, where losses due to friction or viscosity, for example, cause orbits to be attracted to a smaller region of the state space with lower dimension. Any such region is called an attractor. Roughly speaking, an attractor is what the behavior of a system settles down to, or is attracted to.

Some systems do not come to rest in the long term but instead cycle periodically through a sequence of states. An example is the pendulum clock, in which energy lost to friction is replaced by a mainspring or weights. The pendulum repeats the same motion over and over again. In the state space such a motion corresponds to a cycle, or periodic orbit. No matter how the pendulum is set swinging, the cycle approached in the long-term limit it is the same. Such attractors are therefore called limit cycles. Another familiar system with a limit-cycle attractor is the heart.

A system may have several attractors. If that is the case, different initial conditions may evolve to different attractors. The set of points that evolve to an attractor is called its basin of attraction. The pendulum clock has two such basins: small displacements of the pendulum from its rest position result in a return to rest; with large displacements, however, the clock begins to tick as the pendulum executes a stable oscillation.

The next most complicated form of attractor is a torus, which resembles the surface of a doughnut. This shape describes motion made up of two independent oscillations, sometimes called quasi-periodic motion. (Physical examples can be constructed from driven electrical oscillators.) The orbit winds around the torus in state space, one frequency determined by how fast

the orbit circles the doughnut in the short direction, the other regulated by how fast the orbit circles the long way around. Attractors may also be higher-dimensional, since they represent the combination of more than two oscillations.

The important feature of quasi-periodic motion is that in spite of its complexity it is predictable. Even though the orbit may never exactly repeat itself, if the frequencies that make up the motion have no common divisor, the motion remains regular. Orbits that start on the torus near one another remain near one another, and long-term predictability is guaranteed.

Until fairly recently, fixed points, limit cycles and tori were the only known attractors. In 1963 Edward N. Lorenz of the Massachusetts Institute

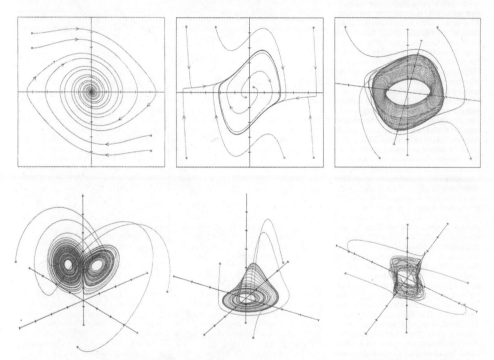

ATTRACTORS are geometric forms that characterize long-term behavior in the state space. Roughly speaking, an attractor is what the behavior of a system settles down to, or is attracted to. Here attractors are shown in blue and initial states in red. Trajectories (*green*) from the initial states eventually approach the attractors. The simplest kind of attractor is a fixed point (*top left*). Such an attractor corresponds to a pendulum subject to friction; the pendulum always comes to the same rest position, regardless of how it is started swinging (*see right half of illustration on preceding page*). The next most complicated attractor is a limit cycle (*top middle*), which forms a closed loop in the state space. A limit cycle describes stable oscillations, such as the motion of a pendulum clock and the beating of a heart. Compound oscillations, or quasi-periodic behavior, correspond to a torus attractor (*top right*). All three attractors are predictable: their behavior can be forecast as accurately as desired. Chaotic attractors, on the other hand, correspond to unpredictable motions and have a more complicated geometric form. Three examples of chaotic attractors are shown in the bottom row; from left to right they are the work of Edward N. Lorenz, Otto E. Rössler and one of the authors (Shaw) respectively. The images were prepared by using simple systems of differential equations having a three-dimensional state space.

84

of Technology discovered a concrete example of a low-dimensional system that displayed complex behavior. Motivated by the desire to understand the unpredictability of the weather, he began with the equations of motion for fluid flow (the atmosphere can be considered a fluid), and by simplifying them he obtained a system that had just three degrees of freedom. Nevertheless, the system behaved in an apparently random fashion that could not be adequately characterized by any of the three attractors then known. The attractor he observed, which is now known as the Lorenz attractor, was the first example of a chaotic, or strange, attractor.

Employing a digital computer to simulate his simple model, Lorenz elucidated the basic mechanism responsible for the randomness he observed: microscopic perturbations are amplified to affect macroscopic behavior. Two orbits with nearby initial conditions diverge exponentially fast and so stay close together for only a short time. The situation is qualitatively different for nonchaotic attractors. For these, nearby orbits stay close to one another, small errors remain bounded and the behavior is predictable.

The key to understanding chaotic behavior lies in understanding a simple stretching and folding operation, which takes place in the state space. Exponential divergence is a local feature: because attractors have finite size, two orbits on a chaotic attractor cannot diverge exponentially forever. Consequently the attractor must fold over onto itself. Although orbits diverge and follow increasingly different paths, they eventually must pass close to one another again. The orbits on a chaotic attractor are shuffled by this process, much as a deck of cards is shuffled by a dealer. The randomness of the chaotic orbits is the result of the shuffling process. The process of stretching and folding happens repeatedly, creating folds within folds ad infinitum. A chaotic attractor is, in other words, a fractal: an object that reveals more detail as it is increasingly magnified [see illustration on page 53].

Chaos mixes the orbits in state space in precisely the same way as a baker mixes bread dough by kneading it. One can imagine what happens to nearby trajectories on a chaotic attractor by placing a drop of blue food coloring in the dough. The kneading is a combination of two actions: rolling out the dough, in which the food coloring is spread out, and folding the dough over. At first the blob of food coloring simply gets longer, but eventually it is folded, and after considerable time the blob is stretched and refolded many

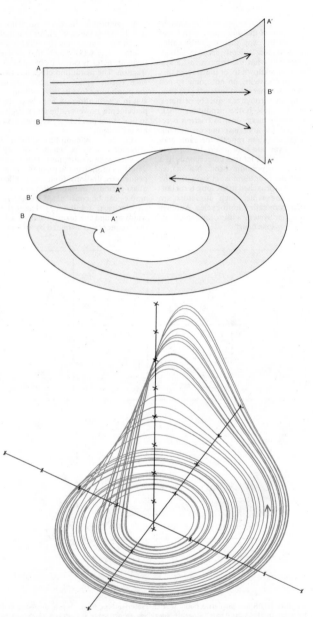

CHAOTIC ATTRACTOR has a much more complicated structure than a predictable attractor such as a point, a limit cycle or a torus. Observed at large scales, a chaotic attractor is not a smooth surface but one with folds in it. The illustration shows the steps in making a chaotic attractor for the simplest case: the Rössler attractor (bottom). First, nearby trajectories on the object must "stretch," or diverge, exponentially (top); here the distance between neighboring trajectories roughly doubles. Second, to keep the object compact, it must "fold" back onto itself (middle): the surface bends onto itself so that the two ends meet. The Rössler attractor has been observed in many systems, from fluid flows to chemical reactions, illustrating Einstein's maxim that nature prefers simple forms.

51

times. On close inspection the dough consists of many layers of alternating blue and white. After only 20 steps the initial blob has been stretched to more than a million times its original length, and its thickness has shrunk to the molecular level. The blue dye is thoroughly mixed with the dough. Chaos works the same way, except that instead of mixing dough it mixes the state space. Inspired by this picture of mixing, Otto E. Rössler of the University of Tübingen created the simplest example of a chaotic attractor in a flow [*see illustration on preceding page*].

When observations are made on a physical system, it is impossible to specify the state of the system exactly owing to the inevitable errors in measurement. Instead the state of the system is located not at a single point but rather within a small region of state space. Although quantum uncertainty sets the ultimate size of the region, in

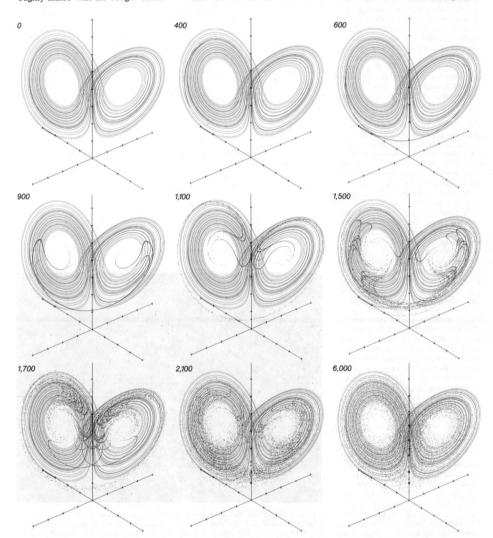

**DIVERGENCE** of nearby trajectories is the underlying reason chaos leads to unpredictability. A perfect measurement would correspond to a point in the state space, but any real measurement is inaccurate, generating a cloud of uncertainty. The true state might be anywhere inside the cloud. As shown here for the Lorenz attractor, the uncertainty of the initial measurement is represented by 10,000 red dots, initially so close together that they are indis-

tinguishable. As each point moves under the action of the equations, the cloud is stretched into a long, thin thread, which then folds over onto itself many times, until the points are spread over the entire attractor. Prediction has now become impossible: the final state can be anywhere on the attractor. For a predictable attractor, in contrast, all the final states remain close together. The numbers above the illustrations are in units of 1/200 second.

practice different kinds of noise limit measurement precision by introducing substantially larger errors. The small region specified by a measurement is analogous to the blob of blue dye in the dough.

Locating the system in a small region of state space by carrying out a measurement yields a certain amount of information about the system. The more accurate the measurement is, the more knowledge an observer gains about the system's state. Conversely, the larger the region, the more uncertain the observer. Since nearby points in nonchaotic systems stay close as they evolve in time, a measurement provides a certain amount of information that is preserved with time. This is exactly the sense in which such systems are predictable: initial measurements contain information that can be used to predict future behavior. In other words, predictable dynamical systems are not particularly sensitive to measurement errors.

The stretching and folding operation of a chaotic attractor systematically removes the initial information and replaces it with new information: the stretch makes small-scale uncertainties larger, the fold brings widely separated trajectories together and erases large-scale information. Thus chaotic attractors act as a kind of pump bringing microscopic fluctuations up to a macroscopic expression. In this light it is clear that no exact solution, no short cut to tell the future, can exist. After a brief time interval the uncertainty specified by the initial measurement covers the entire attractor and all predictive power is lost: there is simply no causal connection between past and future.

Chaotic attractors function locally as noise amplifiers. A small fluctuation due perhaps to thermal noise will cause a large deflection in the orbit position soon afterward. But there is an important sense in which chaotic attractors differ from simple noise amplifiers. Because the stretching and folding operation is assumed to be repetitive and continuous, any tiny fluctuation will eventually dominate the motion, and the qualitative behavior is independent of noise level. Hence chaotic systems cannot directly be "quieted," by lowering the temperature, for example. Chaotic systems generate randomness on their own without the need for any external random inputs. Random behavior comes from more than just the amplification of errors and the loss of the ability to predict; it is due to the complex orbits generated by stretching and folding.

It should be noted that chaotic as

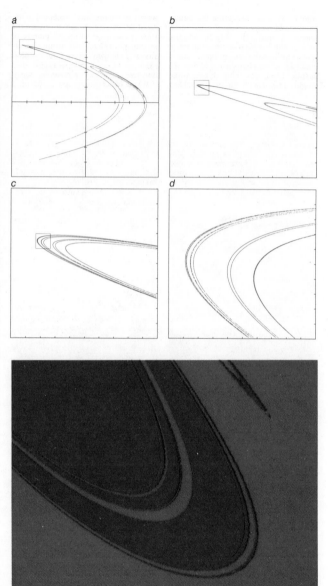

CHAOTIC ATTRACTORS are fractals: objects that reveal more detail as they are increasingly magnified. Chaos naturally produces fractals. As nearby trajectories expand they must eventually fold over close to one another for the motion to remain finite. This is repeated again and again, generating folds within folds, ad infinitum. As a result chaotic attractors have a beautiful microscopic structure. Michel Hénon of the Nice Observatory in France discovered a simple rule that stretches and folds the plane, moving each point to a new location. Starting from a single initial point, each successive point obtained by repeatedly applying Hénon's rule is plotted. The resulting geometric form (a) provides a simple example of a chaotic attractor. The small box is magnified by a factor of 10 in b. By repeating the process (c, d) the microscopic structure of the attractor is revealed in detail. The bottom illustration depicts another part of the Hénon attractor.

well as nonchaotic behavior can occur in dissipationless, energy-conserving systems. Here orbits do not relax onto an attractor but instead are confined to an energy surface. Dissipation is, however, important in many if not most real-world systems, and one can expect the concept of attractor to be generally useful.

Low-dimensional chaotic attractors open a new realm of dynamical systems theory, but the question remains of whether they are relevant to randomness observed in physical systems. The first experimental evidence supporting the hypothesis that chaotic attractors underlie random motion in fluid flow was rather indirect. The ex-

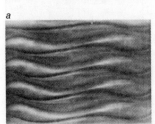

a

b

c

d

EXPERIMENTAL EVIDENCE supports the hypothesis that chaotic attractors underlie some kinds of random motion in fluid flow. Shown here are successive pictures of water in a Couette cell, which consists of two nested cylinders. The space between the cylinders is filled with water and the inner cylinder is rotated with a certain angular velocity (a). As the angular velocity is increased, the fluid shows a progressively more complex flow pattern (b), which becomes irregular (c) and then chaotic (d).

periment was done in 1974 by Jerry P. Gollub of Haverford College and Harry L. Swinney of the University of Texas at Austin. The evidence was indirect because the investigators focused not on the attractor itself but rather on statistical properties characterizing the attractor.

The system they examined was a Couette cell, which consists of two concentric cylinders. The space between the cylinders is filled with a fluid, and one or both cylinders is rotated with a fixed angular velocity. As the angular velocity increases, the fluid shows progressively more complex flow patterns, with a complicated time dependence [see illustration on this page]. Gollub and Swinney essentially measured the velocity of the fluid at a given spot. As they increased the rotation rate, they observed transitions from a velocity that is constant in time to a periodically varying velocity and finally to an aperiodically varying velocity. The transition to aperiodic motion was the focus of the experiment.

The experiment was designed to distinguish between two theoretical pictures that predicted different scenarios for the behavior of the fluid as the rotation rate of the fluid was varied. The Landau picture of random fluid motion predicted that an ever higher number of independent fluid oscillations should be excited as the rotation rate is increased. The associated attractor would be a high-dimensional torus. The Landau picture had been challenged by David Ruelle of the Institut des Hautes Études Scientifiques near Paris and Floris Takens of the University of Groningen in the Netherlands. They gave mathematical arguments suggesting that the attractor associated with the Landau picture would not be likely to occur in fluid motion. Instead their results suggested that any possible high-dimensional tori might give way to a chaotic attractor, as originally postulated by Lorenz.

Gollub and Swinney found that for low rates of rotation the flow of the fluid did not change in time: the under-

lying attractor was a fixed point. As the rotation was increased the water began to oscillate with one independent frequency, corresponding to a limit-cycle attractor (a periodic orbit), and as the rotation was increased still further the oscillation took on two independent frequencies, corresponding to a two-dimensional torus attractor. Landau's theory predicted that as the rotation rate was further increased the pattern would continue: more distinct frequencies would gradually appear. Instead, at a critical rotation rate a continuous range of frequencies suddenly appeared. Such an observation was consistent with Lorenz' "deterministic nonperiodic flow," lending credence to his idea that chaotic attractors underlie fluid turbulence.

Although the analysis of Gollub and Swinney bolstered the notion that chaotic attractors might underlie some random motion in fluid flow, their work was by no means conclusive. One would like to explicitly demonstrate the existence in experimental data of a simple chaotic attractor. Typically, however, an experiment does not record all facets of a system but only a few. Gollub and Swinney could not record, for example, the entire Couette flow but only the fluid velocity at a single point. The task of the investigator is to "reconstruct" the attractor from the limited data. Clearly that cannot always be done; if the attractor is too complicated, something will be lost. In some cases, however, it is possible to reconstruct the dynamics on the basis of limited data.

A technique introduced by us and put on a firm mathematical foundation by Takens made it possible to reconstruct a state space and look for chaotic attractors. The basic idea is that the evolution of any single component of a system is determined by the other components with which it interacts. Information about the relevant components is thus implicitly contained in the history of any single component. To reconstruct an "equivalent" state space, one simply looks at a single component and treats the measured values at fixed time delays (one second ago, two seconds ago and so on, for example) as though they were new dimensions.

The delayed values can be viewed as new coordinates, defining a single point in a multidimensional state space. Repeating the procedure and taking delays relative to different times generates many such points. One can then use other techniques to test whether or not these points lie on a chaotic attractor. Although this representation is in many respects arbitrary, it turns out that the important properties of an attractor are preserved by it and do not depend on the details of how the reconstruction is done.

The example we shall use to illustrate the technique has the advantage of being familiar and accessible to nearly everyone. Most people are aware of the periodic pattern of drops emerging from a dripping faucet. The time between successive drops can be quite regular, and more than one insomniac has been kept awake waiting for the next drop to fall. Less familiar is the behavior of a faucet at a somewhat higher flow rate. One can often find a regime where the drops, while still falling separately, fall in a never repeating patter, like an infinitely inventive drummer. (This is an experi-ment easily carried out personally; the faucets without the little screens work best.) The changes between periodic and random-seeming patterns are reminiscent of the transition between laminar and turbulent fluid flow. Could a simple chaotic attractor underlie this randomness?

The experimental study of a dripping faucet was done at the University of California at Santa Cruz by one of

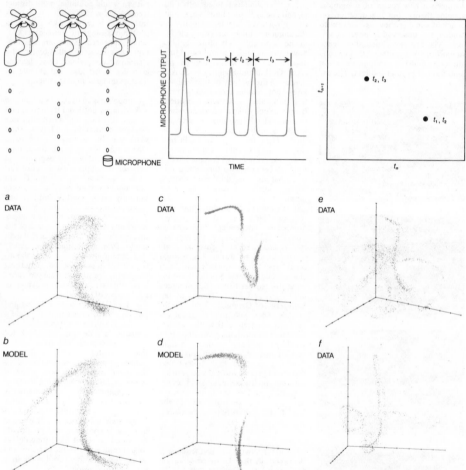

**DRIPPING FAUCET is an example of a common system that can undergo a chaotic transition. The underlying attractor is reconstructed by plotting the time intervals between successive drops in pairs, as is shown at the top of the illustration. Attractors reconstructed from an actual dripping faucet (*a, c*) compare favorably with attractors generated by following variants of Hénon's rule (*b, d*). (The entire Hénon attractor is shown on page 53.) Illustrations *e* and *f* were reconstructed from high rates of water flow and** **presumably represent the cross sections of hitherto unseen chaotic attractors. Time-delay coordinates were employed in each of the plots. The horizontal coordinate is $t_n$, the time interval between drop $n$ and drop $n-1$. The vertical coordinate is the next time interval, $t_{n+1}$, and the third coordinate, visualized as coming out of the page, is $t_{n+2}$. Each point is thus determined by a triplex of numbers ($t_n, t_{n+1}, t_{n+2}$) that have been plotted for a set of 4,094 data samples. Simulated noise was added to illustrations *b* and *d*.**

us (Shaw) in collaboration with Peter L. Scott, Stephen C. Pope and Philip J. Martein. The first form of the experiment consisted in allowing the drops from an ordinary faucet to fall on a microphone and measuring the time intervals between the resulting sound pulses. Typical results from a somewhat more refined experiment are shown on the preceding page. By plotting the time intervals between drops in pairs, one effectively takes a cross section of the underlying attractor. In the periodic regime, for example, the meniscus where the drops are detaching is moving in a smooth, repetitive manner, which could be represented by a limit cycle in the state space. But this smooth motion is inaccessible in the actual experiment; all that is recorded is the time intervals between the breaking off of the individual drops. This is like applying a stroboscopic light to regular motion around a loop. If the timing is right, one sees only a fixed point.

The exciting result of the experiment was that chaotic attractors were indeed found in the nonperiodic regime of the dripping faucet. It could have been the case that the randomness of the drops was due to unseen influences, such as small vibrations or air currents. If that was so, there would be no particular relation between one interval and the next, and the plot of the data taken in pairs would have shown only a featureless blob. The fact that any structure at all appears in the plots shows the randomness has a deterministic underpinning. In particular, many data sets show the horseshoelike shape that is the signature of the simple stretching and folding process discussed above. The characteristic shape can be thought of as a "snapshot" of a fold in progress, for example, a cross section partway around the Rössler attractor shown on page 51. Other data sets seem more complicated; these may be cross sections of higher-dimensional attractors. The geometry of attractors above three dimensions is almost completely unknown at this time.

If a system is chaotic, how chaotic is it? A measure of chaos is the "entropy" of the motion, which roughly speaking is the average rate of stretching and folding, or the average rate at which information is produced. Another statistic is the "dimension" of the attractor. If a system is simple, its behavior should be described by a low-dimensional attractor in the state space, such as the examples given in this article. Several numbers may be required to specify the state of a more complicated system, and its corresponding attractor would therefore be higher-dimensional.

The technique of reconstruction, combined with measurements of entropy and dimension, makes it possible to reexamine the fluid flow originally studied by Gollub and Swinney. This was done by members of Swinney's group in collaboration with two of us (Crutchfield and Farmer). The reconstruction technique enabled us to make images of the underlying attractor. The images do not give the striking demonstration of a low-dimensional attractor that studies of other systems, such as the dripping faucet, do. Measurements of the entropy and dimension reveal, however, that irregular fluid motion near the transition in Couette flow can be described by chaotic attractors. As the rotation rate of the Couette cell increases so do the entropy and dimension of the underlying attractors.

In the past few years a growing number of systems have been shown to exhibit randomness due to a simple chaotic attractor. Among them are the convection pattern of fluid heated in a small box, oscillating concentration levels in a stirred-chemical reaction, the beating of chicken-heart cells and a large number of electrical and mechanical oscillators. In addition computer models of phenomena ranging from epidemics to the electrical activity of a nerve cell to stellar oscillations have been shown to possess this simple type of randomness. There are even experiments now under way that are searching for chaos in areas as disparate as brain waves and economics.

It should be emphasized, however, that chaos theory is far from a panacea. Many degrees of freedom can also make for complicated motions that are effectively random. Even though a given system may be known to be chaotic, the fact alone does not reveal very much. A good example is molecules bouncing off one another in a gas. Although such a system is known to be chaotic, that in itself does not make prediction of its behavior easier. So many particles are involved that all that can be hoped for is a statistical description, and the essential statistical properties can be derived without taking chaos into account.

There are other uncharted questions for which the role of chaos is unknown. What of constantly changing patterns that are spatially extended, such as the dunes of the Sahara and fully developed turbulence? It is not clear whether complex spatial patterns can be usefully described by a single attractor in a single state space. Perhaps, though, experience with the simplest attractors can serve as a guide to a more advanced picture, which may involve entire assemblages of spatially mobile deterministic forms akin to chaotic attractors.

The existence of chaos affects the scientific method itself. The classic approach to verifying a theory is to make predictions and test them against experimental data. If the phenomena are chaotic, however, long-term predictions are intrinsically impossible. This has to be taken into account in judging the merits of the theory. The process of verifying a theory thus becomes a much more delicate operation, relying on statistical and geometric properties rather than on detailed prediction.

Chaos brings a new challenge to the reductionist view that a system can be understood by breaking it down and studying each piece. This view has been prevalent in science in part because there are so many systems for which the behavior of the whole is indeed the sum of its parts. Chaos demonstrates, however, that a system can have complicated behavior that emerges as a consequence of simple, nonlinear interaction of only a few components.

The problem is becoming acute in a wide range of scientific disciplines, from describing microscopic physics to modeling macroscopic behavior of biological organisms. The ability to obtain detailed knowledge of a system's structure has undergone a tremendous advance in recent years, but the ability to integrate this knowledge has been stymied by the lack of a proper conceptual framework within which to describe qualitative behavior. For example, even with a complete map of the nervous system of a simple organism, such as the nematode studied by Sidney Brenner of the University of Cambridge, the organism's behavior cannot be deduced. Similarly, the hope that physics could be complete with an increasingly detailed understanding of fundamental physical forces and constituents is unfounded. The interaction of components on one scale can lead to complex global behavior on a larger scale that in general cannot be deduced from knowledge of the individual components.

Chaos is often seen in terms of the limitations it implies, such as lack of predictability. Nature may, however, employ chaos constructively. Through amplification of small fluctuations it can provide natural systems with access to novelty. A prey escaping a predator's attack could use chaotic flight control as an element of surprise to evade capture. Biological evolution demands genetic variability; chaos provides a means of structuring random changes, thereby providing the possibility of putting variability under evolutionary control.

Even the process of intellectual progress relies on the injection of new ideas and on new ways of connecting old ideas. Innate creativity may have an underlying chaotic process that selectively amplifies small fluctuations and molds them into macroscopic coherent mental states that are experienced as thoughts. In some cases the thoughts may be decisions, or what are perceived to be the exercise of will. In this light, chaos provides a mechanism that allows for free will within a world governed by deterministic laws.

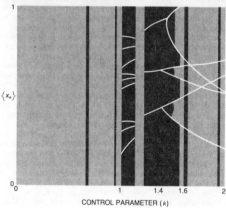

CONTROL PARAMETER ($k$)

TRANSITION TO CHAOS is depicted schematically by means of a bifurcation diagram: a plot of a family of attractors (*vertical axis*) versus a control parameter (*horizontal axis*). The diagram was generated by a simple dynamical system that maps one number to another. The dynamical system used here is called a circle map, which is specified by the iterative equation $x_{n+1} = \omega + x_n + k/2\pi \cdot \sin(2\pi x_n)$. For each chosen value of the control parameter $k$ a computer plotted the corresponding attractor. The colors encode the probability of finding points on the attractors: red corresponds to regions that are visited frequently, green to regions that are visited less frequently and blue to regions that are rarely visited. As $k$ is increased from 0 to 2 (*see drawing at left*), the diagram shows two paths to chaos: a quasi-periodic route (from $k = 0$ to $k = 1$, which corresponds to the green region above) and a "period doubling" route (from $k = 1.4$ to $k = 2$). The quasi-periodic route is mathematically equivalent to a path that passes through a torus attractor. In the period-doubling route, which is based on the limit-cycle attractor, branches appear in pairs, following the geometric series 2, 4, 8, 16, 32 and so on. The iterates oscillate among the pairs of branches. (At a particular value of $k$—1.6, for instance—the iterates visit only two values.) Ultimately the branch structure becomes so fine that a continuous band structure emerges: a threshold is reached beyond which chaos appears.

# Classical Chaos

## Roderick V. Jensen

A wide variety of natural phenomena exhibit complicated, unpredictable, and seemingly random behavior. Common examples include the turbulent flow of a mountain stream, the changing weather, and the swirling patterns of cream, slowly stirred, in a cup of coffee. The paradigm for this class of macroscopic phenomena is the problem of turbulent flow in fluids (Fig. 1). Additional examples of complex, irregular behavior occur in the dynamics of molecules and atoms in a gas or charged particles in a plasma. These microscopic systems define another class of important physical problems which raise a disturbing question: How can the deterministic and reversible motions of individual particles give rise to the irreversible behavior of the system, as described by statistical mechanics and thermodynamics?

Although physics has made monumental strides in the last hundred years, theoretical descriptions of these complex phenomena have remained outstanding unsolved problems. The difficulty lies in the nonlinear character of the mathematical equations which model the physical systems: the Navier-Stokes equations for fluid flows and Newton's equations for three or more interacting particles. Since these equations do not generally admit closed-form analytical solutions, it has proved extremely difficult to construct useful theories that would predict, for example, the drag on the wing of an airplane or the range of validity of statistical mechanics. However, in the last ten years considerable progress has been made, using a unique synthesis of numerical simulation and analytical approximation.

The key to the recent progress has been the use of high-speed digital computers. In particular, high-resolu-

*New methods for studying chaotic behavior make the unpredictable more understandable but also raise disturbing fundamental questions*

tion computer graphics have enabled the "experimental" mathematician to identify and explore ordered patterns which would otherwise be buried in reams of computer output. In many cases the persistence of order in irregular behavior was totally unexpected; the discovery of these regularities has led to the development of new analytical methods and approximations which have improved our understanding of complex nonlinear phenomena.

This novel approach, which combines numerical "experiments" with mathematical analysis, has given rise to a new interdisciplinary field called nonlinear dynamics. The work done in this field has been applied not only to problems in physics but also to a wide variety of nonlinear problems in other scientific fields, such as the evolution of chemical reactions (1), the feedback control of electrical circuits (1), the interaction of biological populations (2), the response of cardiac cells to electrical impulses (3), the rise and fall of economic prices (4), and the buildup of armaments in competing nations (5). In this article I will limit myself primarily to physical problems. However, I hope that readers will recognize the applicability of these methods to their varied fields, since the difficulties in solving nonlinear equations are common to every branch of science.

Nonlinear dynamicists use the word "chaos" as a technical term with a precise mathematical meaning to refer to the irregular, unpredictable behavior of deterministic, nonlinear systems (6). Contrary to what Isaac Newton may have believed, the deterministic equations of classical mechanics do not imply a regular, ordered universe. Although most modern physicists and gamblers would concede that dynamical systems with large numbers of degrees of freedom, such as the atmosphere or a roulette wheel, can exhibit random behavior for all practical purposes, the real surprise is that deterministic systems with only one or two degrees of freedom can be just as chaotic.

Traditionally, the fundamental problems associated with the origins of chaos in turbulent flows, the microscopic foundations of statistical mechanics, and the appearance of random behavior in a variety of other fields have been avoided by using the argument that so many particles and degrees of freedom are involved that it would not be humanly possible to describe these

*Roderick V. Jensen is an associate professor of applied physics at Yale University. He is a graduate of Princeton University (A.B. in physics 1976, Ph.D. in astrophysical sciences 1981), where his dissertation research was devoted to the statistical description of chaotic dynamical systems with applications to plasma physics. His current research is concerned with the role of chaos in the foundations of statistical mechanics and the investigation of chaotic behavior in quantum systems. This work is supported by an Alfred P. Sloan Fellowship and a Presidential Young Investigator Award from the National Science Foundation. Address: Mason Laboratory, Department of Applied Physics, Yale University, Yale Station, New Haven, CT 06520.*

92

Figure 1. When motion becomes chaotic, the results are unpredictable and sometimes disastrous. In classical dynamics, the behavior of turbulent fluids has proved extremely difficult to predict—as we know, for example, from weather forecasting. But new insights about the nature of chaos have revealed an underlying structure that is common in many natural systems and even in human social behavior. These insights have been applied to such problems as the evolution of chemical reactions, the control of electrical circuits, the growth of biological populations, the response of cardiac cells to electrical impulses, the rise and fall of economic prices, and the buildup of armaments. (Photograph © Four By Five.)

phenomena from first principles. However, the discovery of much simpler systems which can nevertheless exhibit behavior as complicated as these standard examples means that we no longer have to throw up our hands in despair. Using the computer as a laboratory apparatus to study these simple systems, we can begin to explore and understand chaotic, irregular, and unpredictable phenomena in nature.

In this review I will concentrate on phenomena which are well described by classical physics and, consequently, on problems of "classical chaos." Unfortunately, the question of chaos in quantum physics remains controversial. At present, "quantum chaos" is a poorly characterized disease for which we have only identified some of the possible symptoms. Both an unambiguous definition as well as the very existence of quantum chaos remain open problems. In contrast, we have a clear understanding of the symptoms and causes of classical chaos, if only a partial understanding of the cure.

I will start by examining in detail two deceptively simple nonlinear dynamical systems which exhibit a transition from regular, ordered behavior to chaos.

These examples will graphically illustrate the irregular, unpredictable, but nevertheless deterministic behavior we call chaos. Then, after formulating a precise definition of classical chaos, I will attempt to dispel the longstanding psychological prejudice which insists on a distinction between deterministic and random behavior by showing that the chaotic behavior of deterministic dynamical systems can be indistinguishable from a random process.

This deeper understanding of chaos will lead, finally, to a slightly more philosophical discussion of where classical chaos really comes from and what it is good for. We will see that investigations of nonlinear dynamical systems have suggested partial answers to some of the fundamental problems of turbulence and statistical mechanics which were first formulated in the nineteenth century. However, this research has also raised new questions, more profound than those they have answered, relating to twentieth-century problems arising from Gödel's incompleteness theorem, the theories of algorithmic and computational complexity of modern computer science, and the principles of quantum mechanics (7).

Figure 2. Many dynamical systems can be approximated by the logistic map (equation 1), which then predicts, for example, the size of a changing biological population, the fluctuation of economic prices, or the dynamics of a periodically kicked and damped nonlinear oscillator. Equation 1 defines an inverted parabola, plotted here for four different values of $a$. Once an initial value, $x_0$, is specified, the evolution of the system is fully determined. One can find the values of $x$ at succeeding time-steps by tracing the colored lines on the appropriate graph: from $x_0$ vertically to the parabola for $x_1$, then horizontally to the 45° line and vertically back to the parabola for $x_2$, and so on for succeeding values of $x_{n+1}$. When $a$ is less than 1, as in the graph at the upper left, all initial conditions converge to 0 (the population becomes extinct, the price falls to zero, etc.). When $a$ is increased to a value between 1 and 3, as in the upper right, almost all initial conditions are attracted to a fixed point. When $a$ is larger than 3, however, the fixed point becomes unstable; at $a = 3.2$ there are two fixed points between which the value for $x$ eventually oscillates. As $a$ continues to increase, there can be more and more fixed points, and for many values of $a$, as when $a = 4$, the values for $x$ wander over entire intervals in an apparently random fashion.

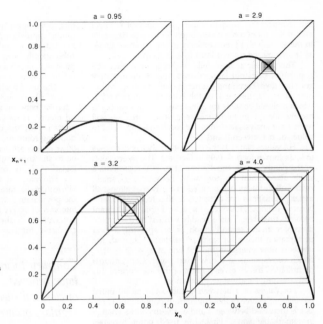

# Examples of chaotic dynamical systems: The logistic map

Perhaps the simplest example of a nonlinear dynamical system is the celebrated logistic map. This system is described by a single difference equation

$$x_{n+1} = ax_n(1 - x_n) \tag{1}$$

which determines the future value of the variable $x_{n+1}$ at time-step $n + 1$ from the past value at time-step $n$. The time-evolution of $x_n$ generated by this single algebraic equation exhibits an extraordinary transformation from order to chaos as the parameter $a$, which measures the strength of the nonlinearity, is increased.

Although nonlinear difference equations of this type have been studied extensively as simple models for turbulence in fluids, they also arise naturally in the study of the evolution of biological populations. In fact, the review article on the logistic map by the biologist Robert May (2) is a historical milestone in the modern development of nonlinear dynamics. Therefore, for illustrative purposes we will examine the use of the logistic map as a crude model for the annual evolution of a single biological population, $x_n$, say that of gypsy moths in the northeastern United States, which exhibits wild and unpredictable fluctuations from year to year. However, we could equally well consider the evolution of economic prices determined by a nonlinear "cobweb" model with nonmonotonic, backward-bending supply and demand curves (4) or the dynamics of a periodically kicked and damped nonlinear oscillator.

Writing equation 1 in a slightly different form, $x_{n+1} = ax_n - ax_n^2$, we see that it is a simple quadratic

equation, with the first term linear and the second term nonlinear. When the original population $x_0$ is small (much less than 1 on a normalized scale, where 1 might stand for any number, such as 1 million individuals), the nonlinear term can initially be neglected. Then the population at time-step (year) $n = 1$ will be approximately equal to $ax_0$. If $a > 1$, the population increases. If $a < 1$, the population decreases. Therefore, the linear term in equation 1 can be interpreted as a linear growth or death rate which by itself would lead to exponential population growth or decay. If $a > 1$, the population will eventually grow to a value large enough for the nonlinear term, $-ax_n^2$, to become important. Since this term is negative, it represents a nonlinear death rate which dominates when the population becomes too large. Biologically, this nonlinear death rate could be due to the depletion of food supplies or the outbreak of diseases in an overcrowded environment.

As emphasized in May's review article, the dynamics of this map and the dependence on the parameter $a$, which measures the rate of linear growth and the size of the nonlinear term, are best understood using a graphic analysis. Consider the graphs of $x_{n+1}$ versus $x_n$ (called "return maps") displayed in Figure 2 for four different values of $a$. Equation 1 defines an inverted parabola with intercepts at $x_n = 0$ and 1 and a maximum value of $x_{n+1} = a/4$ at $x_n = 0.5$. Using these return maps, we can get a qualitative understanding of the dynamics of the logistic map without performing any calculations. The successive values of the populations can be determined simply by tracing lines on these graphs. Just start your pencil at an initial $x_0$ and move vertically to the parabola to get $x_1$. At this point you could return to the horizontal axis to repeat this procedure using the value of $x_1$ to get $x_2$, but it is more convenient simply to trace horizontally to the 45° line and then vertically to the parabola again, as shown by the colored lines in each graph.

This graphic analysis tells us that if the normalized population starts out larger than 1, then it immediately goes negative, becoming extinct in one time-step. Moreover, if $a > 4$, the peak of the parabola will exceed 1, which makes it possible for initial populations near 0.5 to become extinct in two time-steps. Therefore, we will restrict our analysis to values of $a$ between 0 and 4 and to values of $x_0$ between 0 and 1.

For values of $a < 1$, the population always decreases to 0, as shown for $a = 0.95$ in Figure 2. The intersection of the parabola with the 45° line at $x_n = 0$ represents a stable fixed point on the map. Because $a$ is small, perturbation theory can be used to verify that almost all initial populations are attracted to this fixed point and become extinct. However, for $a > 1$ this fixed point becomes unstable. (This is readily verified by tracing the dynamics in the second graph or by applying a local perturbation theory for small populations.) Instead, the parabola now intersects the 45° line at $x = (a - 1)/a$, which corresponds to a new fixed point. Conventional perturbation theory gives no hint of the existence of this nonvanishing steady state population.

For values of $a$ between 1 and 3 almost all initial populations evolve to this equilibrium population. Then, as $a$ is increased between 3 and 4, the dynamics change in remarkable ways. First, the fixed point becomes unstable and the population evolves to a dynamic steady state in which it alternates between a large and a small population. A time-sequence converging to such a period-2 cycle is displayed in Figure 2 for $a = 3.2$: the population eventually cycles between two points on the parabola, $x_n \sim 0.5$ and $x_n \sim 0.8$, in alternate years. For somewhat larger values of $a$ this period-2 cycle becomes unstable and is replaced by a period-4 cycle in which the population alternates high-low, returning to its original value every four time-steps. As $a$ is increased, the long-time motion converges to period-8, -16, -32, -64,... cycles, finally accumulating to a cycle of infinite period for $a = a_{inf} \sim 3.57$.

This sequence of "period-doubling bifurcations" in the long-time, steady state behavior of the logistic map is clearly displayed in Figure 3. The graph shows the steady state values of the population as a function of $a$ between 3.5 and 4. For $a \leq 3$ only a single steady state value of $x = (a - 1)/a$ would be displayed. For $a > 3$, we get two steady state values, then four, then eight, and so on. Each bifurcation in Figure 3 thus represents a doubling of the number of steady state values and a doubling of the time-steps in a period.

The range of $a$ over which a single cycle is stable decreases rapidly as the period of the cycle increases, which accounts for the rapid accumulation of cycles with larger and larger periods. In fact, having observed this period-doubling sequence in numerical experiments, Feigenbaum was able to prove, using a remarkable application of the renormalization group, that the intervals over which a cycle is stable decrease at a geometric rate of $\sim 4.6692016$. The tremendous significance of this work is that this rate and other properties of the period-doubling bifurcation sequence are universal in the sense

that they appear in the dynamics of any system which can be approximately modeled by a nonlinear map with a quadratic extremum (8). Feigenbaum's theory has subsequently been confirmed in a wide variety of physical systems such as turbulent fluids, oscillating chemical reactions, nonlinear electrical circuits, and ring lasers (1).

The investigation of period doubling in nonlinear dynamical systems provides a superb example of the interplay between numerical "experiments" and analytical theory. However, this sequence of regular periodic orbits is only the precursor to chaos. Since the period-doubling route to chaos has been the subject of several other review articles and texts (2, 8–10), I will now move on to still larger values of $a$, where the dynamics of the logistic map are truly chaotic.

For many, if not most, values of $a > 3.57...$ the bifurcation diagram shows that the long-time behavior of the population is aperiodic and ranges over continuous intervals of $x$. As I will demonstrate, the evolution of populations in these continuous intervals is indistinguishable from a random process, even though the

---

*Contrary to what Isaac Newton may have believed, the deterministic equations of classical mechanics do not imply a regular, ordered universe*

---

logistic map is fully deterministic in the sense that there are no "random" forces and the future is completely determined by the initial condition, $x_0$.

However, we also find windows of periodic behavior embedded in this chaotic regime. The most prominent window corresponds to a period-3 cycle for $a \sim 3.83$, in which the population increases in two successive years and decreases in the third. Moreover, as $a$ is increased within this window of stability, the period-3 cycle can also be seen to exhibit period-doubling bifurcations to period-6, -12, -24, ... cycles. In fact, between $a_{inf}$ and $a \sim 3.83$ there are windows of stability for every integer period, which terminate in a period-doubling cascade back to chaos. Although the windows of stability for most of the higher-order cycles are too narrow to be seen in Figure 3, a period-5 and a period-6 cycle can be readily discerned.

It is a remarkable mathematical fact that, although these intervals of stability are dense throughout the range of $a$, it is not correct to conclude that the set of values of $a$ for which the motion is truly chaotic is negligibly small. On the contrary, this set has been proved to have a nonvanishing measure (11). In other words, if the exact evolution of $x_n$ looks chaotic, then it probably is; we are not necessarily being deceived by a very long, but periodic, cycle. In particular, the irregular dynamics for $a = 4$, which deterministically spans the entire unit interval, is easily shown to meet the definitions of both a chaotic and a random process formulated later in this article.

Another striking feature in the bifurcation diagram is the dark streaks which mark the upper and lower boundaries and crisscross the chaotic domain. The dark streaks represent values of $x$ which are more probable and visited more often during the chaotic evolution. These ordered structures were discovered "experimentally" in high-resolution graphs, like Figure 3, displaying hundreds of thousands if not millions of iterations of the logistic map. Once discovered, their explanation was found to be simple (12). The streaks are located at the future values of the "critical" population, $x_0 = 0.5$. The upper bound of values for $x$ is determined by the heights of the inverted parabolas, $x_1 = a/4$, as diagrammed in Figure 2, and the lower bound and all the interior streaks in Figure 3 by the subsequent iterates. The reason that populations have a higher probability of passing through

---

*High-resolution computer graphics have enabled mathematicians to identify ordered patterns which would otherwise be buried in reams of computer output*

---

values near the trajectory of $x_0 = 0.5$ is that the slopes of the parabolas on the return maps (Fig. 2) vanish there, which tends to compress nearby trajectories. Moreover, the intersections of these dark streaks in Figure 3 correspond to "crises" in the chaotic dynamics, where disjoint intervals of chaotic orbits collide to form larger regions, and they have been a topic of recent research (13). The most spectacular crisis is readily visible at $a \sim 3.68$.

The discovery and explanation of such regular structures in the chaotic domain is not just an amusing exercise for experimental mathematicians; rather, an understanding of these probability distributions has important practical applications. Since an analytical description of the chaotic evolution of individual initial conditions is impossible, the best we can hope for is a statistical theory which predicts the likelihood of the variable $x_n$ taking on any particular value. In this case the "order in chaos" which is apparent in Figure 3 plays an important role in delineating the range of validity and the structure of statistical descriptions. For example, in applying this analysis to the evolution of biological populations, we see that for conditions corresponding to $a \sim 4$ the population will fluctuate in an apparently random fashion, over the entire range, but is most likely to lie at either the maximum or minimum values.

### And the standard map

Our second example of a nonlinear dynamical system which exhibits a transition from regular to chaotic behavior is the standard map (14), described by a pair of nonlinear difference equations

$$x_{n+1} = x_n + y_{n+1} \tag{2}$$
$$y_{n+1} = y_n + k \sin x_n \tag{3}$$

which map the values of the two variables $x_n$ and $y_n$ at time-step $n$ into $x_{n+1}$ and $y_{n+1}$ at time-step $n + 1$. In this case the parameter $k$ in equation 3 controls the magnitude of the nonlinearity.

This map can be used to describe a large number of physical systems. It provides, for example, an approxi-

mate description of the one-dimensional motion of a charged particle perturbed by a broad spectrum of oscillating fields, where $x_n$ and $y_n$ denote the position and velocity of the particle at a discrete time $t = n$ and $k$ is a measure of the electric field amplitude. It also arises naturally as an approximate description of general one-dimensional, nonlinear oscillators subject to periodic perturbations (hence the name "standard map").

As the nonlinear parameter, $k$, is increased, the evolution of this map exhibits, like the logistic map, a dramatic transformation from regular, predictable motion to chaotic, statistical behavior. As a consequence, detailed numerical and analytical investigations of this classical mechanical system have played, and continue to play, an important role in studies of the microscopic foundations of classical statistical mechanics.

The simplest physical system described by this pair of coupled, nonlinear difference equations is a rigid rotor, such as the one depicted in Figure 4, which is subject to sudden kicks at regular time intervals. In this case the variable $x_n$ corresponds to the angle of the rotor and $y_n$ to the angular velocity immediately after the $n$th kick, and equations 2 and 3 are just Newton's equations for this classical mechanical system. The kick can be either forward or backward, depending on the sign of $\sin x_n$, and the maximum strength of the kick is determined by the size of the nonlinear parameter, $k$.

Equations 2 and 3 provide an exact, deterministic description of the evolution of the "phase-point" $(x_n, y_n)$ in the two-dimensional $x$–$y$ "phase-space" which is uniquely determined by the initial condition $(x_0, y_0)$. For example, if we set $k = 0$ and look at the motion of the rotor at stroboscopic intervals of time, then the angular velocity would remain constant at $y_n = y_0$ and the angle $x_n$ would increase by $y_0$ each unit of time. A graph of the point $(x_n, y_n)$ in the $x$–$y$ phase-space of this dynamical system would show a sequence of dots lying in a straight, horizontal line of constant $y_n$. The first graph in Figure 5 shows a computer-generated "phase-space portrait" (also known as a Poincaré section) for several values of $y_0$ with $k = 0$. (For convenience we have taken advantage of the natural periodicity of the angle $x$ to restrict the range of $x$ to the interval $[0, 2\pi]$ by evaluating equation 2 modulo $2\pi$.) In fact, an analytical solution which describes this regular behavior for the linear difference equations (linear when $k = 0$) can easily be determined. However, for nonzero $k$ the standard map is no longer integrable and does not admit closed-form analytical solutions for $x_n$ and $y_n$ at an arbitrary time $t = n$. In these cases we must rely heavily on intuition derived from numerical "experiments" to develop new methods of analysis.

We can exploit several symmetries which significantly reduce the complexity of the analysis. The first symmetry is the fact that the map is naturally periodic in $y$ with period $2\pi$. (If we increment $y$ by $2\pi$ on both sides of equation 3, its value remains unchanged.) We have already noted that $x$ is an angle variable which is only defined modulo $2\pi$. Therefore, for the purposes of graphic analysis it is convenient to evaluate both equations 2 and 3 modulo $2\pi$ so that the evolution of $x_n$ and $y_n$ is restricted to a square in the $x$–$y$ phase-space with sides of length $2\pi$.

The graphs in Figure 5 show phase-space portraits

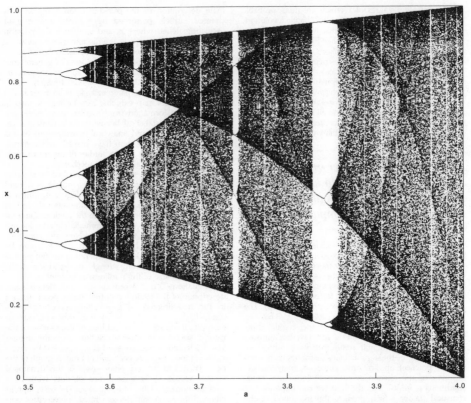

Figure 3. A paradigm in the field of nonlinear dynamics, this bifurcation diagram shows the long-time behavior of the logistic map for values of $a$ between 3.5 and 4. The graph is generated by numerically iterating the map for different values of $a$ and plotting several hundred successive values of $x_n$ after initial transients have died out. The result is a graphic display of the underlying structure of chaos—patterns that show an orderly progression from regular to chaotic behavior in any system that can be modeled by equation 1.

Where the long-time evolution of the map converges to a periodic cycle of period $N$, the diagram shows $N$ discrete values for $x$ (at $a = 3.5$, for example, there are four discrete values for $x$; the system eventually settles down to a periodic oscillation among those four values). Where the evolution is chaotic, the values of $x$ cover continuous intervals, and the darkness of the shading represents the relative probability that $x$ will visit a particular region.

of the restricted dynamics for increasing values of the nonlinear parameter, $k$. This series of graphs clearly shows the transition from regular to irregular, chaotic behavior as the strength of the kicks is increased. (The speckled region in the graph for $k = 2$ is covered by a single orbit.) These graphs of the restricted motion are extremely useful, since initial conditions that show regular or chaotic behavior in the restricted phase-space will also exhibit regular and chaotic behavior, respectively, in the unrestricted dynamics.

A second important property which simplifies the mathematical analysis is that the standard map (like most equations describing nondissipative, classical mechanical systems) is Hamiltonian, and the evolution of $x_n$ and $y_n$ preserves areas in the $x$–$y$ phase-space. This is

easy to see if we think of the map as a coordinate transformation from variables $(x_n, y_n)$ to variables $(x_{n+1}, y_{n+1})$. From elementary calculus we know that under a change of variables infinitesimal areas and volumes stretch or contract by the magnitude of the Jacobian determinant, $J$, of the coordinate transformation. A careful evaluation of $J$ for the standard map shows that it is equal to 1. This property permits the application of a wide variety of mathematical methods and theorems for Hamiltonian systems which have been developed over the last hundred years in the study of celestial mechanics (15).

If $J$ were less than 1, the long-term dynamics would converge to an attracting set in phase-space with zero area. This could be a point or a line or a more complex manifold in two or more dimensions called a "strange attractor" (6, 16). In our simple example, if we added dissipation (friction) to the rotor by replacing equation 3 by

$$y_{n+1} = \lambda y_n + k \sin x_n \qquad (4)$$

with $\lambda < 1$ decreasing the velocity each time-step due to friction, then in the absence of any kicks, $k = 0$, every initial condition would evolve to the attracting set defined by the line $y = 0$. However, for large enough values of $k$ the kicks can overcome the friction, and the attracting set can be much more complicated. For example, the upper diagram in Figure 6 shows the outline of the "strange attractor" for $\lambda = 0.1$ and $k = 8.8$ (17). The

*We could always imagine that in principle, by stirring very carefully, we can separate the cream from the coffee after it has been thoroughly mixed*

reason this attracting set is considered to be strange is that if we magnify a section of any strand of the attractor, we find that it is composed of many strands, which in turn are composed of many strands, ad infinitum. The second diagram in Figure 6 is a magnification of a section of this attractor showing this "self-similar" structure on a smaller scale.

The structure and dimension (which is not necessarily an integer) of these fascinating "fractal" attractors are the subjects of much current research in nonlinear dynamics. This research has many possible applications, including the description of chaotic behavior in dissipative systems such as turbulent flows, chemical oscillators, or neural networks (1); the interested reader should refer to the excellent review article by Ed Ott (16) and the beautiful book by Benoit Mandelbrot (18).

Returning to the nondissipative standard map, we note that in the absence of an attractor, the phase-point $(x_n, y_n)$ can in principle wander anywhere in the available phase-space. However, we have seen that when $k = 0$, the evolution of the phase-point is confined to a horizontal line. For nonzero $k$ the angular velocity is perturbed by kicks and ceases to be a constant of motion. We might then expect the phase-point to explore all of phase-space. Nevertheless, the phase-space portraits in Figure 5 clearly show that this is not necessarily the case.

This "experimental" result is further substantiated by a remarkable theorem for Hamiltonian systems known as the Kolmogorov, Arnold, Moser (or KAM) theorem (15, 19). This theorem states that if you take an integrable Hamiltonian system (such as the standard map with $k = 0$) and add a nonintegrable perturbation ($k \neq 0$), then for sufficiently small perturbations approximate constants of motion will survive and the evolution of the dynamical system will remain regular (if somewhat distorted) for most initial conditions. Although the rigorous mathematical proof of this theorem requires that the perturbation be extremely small (Ian Percival has described its magnitude as comparable to the "gravitational pull of a micro-organism in Australia"), we can see

that in Figure 5 the phase-space orbits remain quite regular for fairly large values of $k$ and that the perturbation must become rather large before the evolution of a single phase-trajectory begins to fill large regions of phase-space, as it does in the graph for $k = 2$.

In practice this transition from mostly regular behavior to global chaos as $k$ is increased has tremendous physical significance. For example, numerical experiments show that for small values of $k$ the angular velocity and kinetic energy of the kicked rotor may increase and decrease but remain confined to a restricted range of values for all time. However, for large values of $k$ the velocity and energy can wander over all of phase-space. If in this case we remove the restriction to velocities on the interval $[0, 2\pi]$, we find that the rotor's velocity and energy can wander to arbitrarily large values. Despite the fact that there are no "random" forces at play, this diffusion in energy appears for all intents and purposes to be a random walk. Since the standard map also provides a model for the interaction of charged particles with a broad spectrum of oscillating electrical fields, this deterministic diffusion in energy provides an important means of heating high-temperature, low-density fusion plasmas where "random" particle collisions are too rare to mediate in the irreversible transfer of energy from the fields to the particles (19).

The numerical experiments indicate that this transition from confined to diffusive motion occurs for $k_c \sim 1$. This observation has led to the development of a series of approximate theories of ever increasing sophistication and accuracy for predicting the critical perturbation strength for the onset of global chaos in general nonlinear systems. At present the best theoretical prediction (20) for the standard map is $k_c \leq 63/64 = 0.984375$, which is very close to the best numerical estimate (21) of $k_c \sim 0.971635406$.

The chains of elliptical "island" structures which proliferate at $k \sim 1$ (Fig. 5) play a very important role in determining this transition to global stochasticity. These regular structures in the nonlinear dynamics result from resonances between the motion of the nonlinear oscilla-

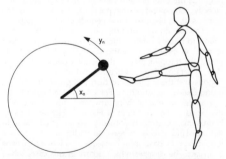

Figure 4. The simplest physical system described by the standard map (equations 2 and 3) is a rigid rotor subject to periodic kicks. The term $x_n$ represents the angular position of the rotor at the time of the $n$th kick; $y_n$ represents the angular velocity of the rotor just before the $n$th kick. The strength and direction of the kicks are determined by the nonlinear term, $k \sin x_n$.

Figure 5. Phase-space portraits (Poincaré sections) for the standard map, shown here for four different values of the nonlinear parameter $k$, are analogous to the bifurcation diagram for the logistic map (Fig. 3), in that they make it easier to see transitions from regular to chaotic evolution. These figures are generated by numerically iterating the standard map for several different initial conditions $(x_0, y_0)$ and plotting several hundred of the succeeding points $(x_n, y_n)$ in the $x$–$y$ phase-space. The graph at the upper left shows the regular, integrable dynamics for $k = 0$ (which corresponds, in Fig. 4, to a kick of zero strength, so that the angular velocity of the rotor is constant). When $k$ is increased to 0.5, the trajectories for various initial conditions are still regular and nearly integrable, as guaranteed by the Kolmogoroff-Arnold-Moser theorem for small values of $k$. A mixture of chaotic and regular trajectories appears when $k = 1$. The graph for $k = 2$ is dominated by chaotic evolution: a single trajectory can wander over large regions of phase-space, although some islands of stability persist.

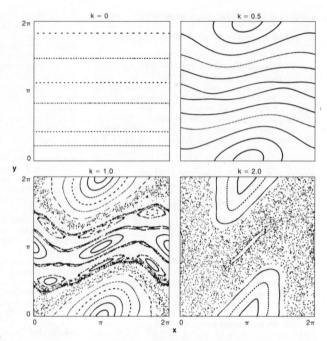

tor (the rotor) and the periodic perturbation. At the center of each "island" is a phase-point $(x, y)$ which recurs after $q$ iterations of the map, where $q$ also happens to be the number of islands in a chain spanning the distance from $x = 0$ to $2\pi$. The reader can readily verify that the point $(\pi, 0)$, which lies at the center of the large island when $k = 0.5$ and $k = 1$, is a fixed point with period 1 of equations 2 and 3. (Remember that the map is periodic, so the bottom half of this large elliptical island can be found at the top of the graph.) It is also easy to check that the phase-point $(\pi, \pi)$, which lies at the center of the smaller two-island chain across the center of the graph for $k = 1$, is a periodic orbit of period 2; the intermediate values of the rotor angle and velocity correspond to the point $(2\pi, \pi)$, which is the same as $(0, \pi)$ because of the periodicity of the map.

The islands surrounding these periodic orbits correspond to nearby orbits which are trapped in nonlinear resonances. Since these trapped orbits will also oscillate within the trapping region, the periodic perturbation will also generate island structures within these regular regions and these islands in turn will grow still smaller island chains, ad infinitum. For $k \lesssim 1$ these higher-order resonances are extremely narrow, and only a few can be discerned at the resolution of Figure 5. However, using Hamiltonian perturbation theory we find that the individual island chains increase in width as $k^{q/2}$, so we would expect catastrophic consequences when $k$ exceeds 1. In fact, the disaster which occurs in this case, when large numbers of resonances interact, is the onset of global chaos (22).

For $k > 1$ the approximate constants of motion are destroyed for most initial conditions, and the corresponding phase-space trajectories are no longer confined to smooth curves but can wander throughout large

regions of phase-space (like the orbit for $k = 2$ in Fig. 5). In the next section I will show that these orbits exhibit the same local instability and extreme sensitivity to initial conditions as the irregular orbits of the logistic map and that the irregular dynamics meet the conditions required by the definition of chaos. Unfortunately, few rigorous mathematical results are available at present for moderately realistic physical models like the standard map; however, since the map can be easily iterated for many millions of time-steps, the numerical evidence can be very convincing. In fact, one numerical study reported the results of a calculation with as many as $10^{12}$ iterations of the standard map (23).

One of the difficulties faced by a rigorous mathematical analysis is that the phase-space is often divided into both regular regions (inside resonant island structures) as well as chaotic regions for most realistic systems. In particular, the standard map already exhibits bands of chaotic orbits for very small values of $k$, although the KAM theory guarantees that they are very narrow. These increase in size as $k$ increases until $k$ exceeds $k_c$, after which the chaotic regions expand until they consume most of phase-space. For example, the bands of chaos are too narrow to be seen in Figure 5 when $k = 0.5$ but begin to appear at $k = 1$ and dominate the phase-space at $k = 2$. Moreover, periodic orbits with stable island structures may persist in the chaotic regime. For example, Figure 5 shows that an island of stability persists around the fixed point at $(\pi, 0)$ for $k = 2$; however, it is eventually washed away by the chaotic sea when $k$ exceeds 4 (19).

## Chaos, determinism, and chance

The graphs of the irregular dynamics generated by the logistic and standard maps provide a picture book of chaos. Like many nonlinear systems in nature, these mathematical models exhibit behavior which appears to be random despite the fact that the equations of motion are fully deterministic. But if the motions are fully determined and the systems are relatively simple, where does this complex behavior come from? What are the symptoms that allow us to identify chaos when we see it? And what are the real differences, if any, between such deterministic chaotic behavior and random processes? To describe more clearly this disease called classical chaos, we must delve a little deeper into the mathematical theory of dynamical systems.

We have already defined what we mean by deterministic behavior in dynamical systems; namely, their evolution is completely determined by the initial conditions and the equations of motion prescribed by the laws of physics. But what do we mean by random behavior? Our intuitive notion of a random or chance process, such as the roll of a die, the flip of a coin, or the spin of a roulette wheel, is a process which exhibits irregular behavior that is not determined by any laws and defies prediction (24). However, this concept would not be very useful if it were not for the fact that statistical properties of these systems, such as the average behavior over time or after many repetitions, are very well described by the calculus of probabilities and the so-called laws of chance (24). Therefore, the traditional definition of an idealized random or, more precisely, stochastic process is a dynamical system which can be described only in terms of average properties determined by an appropriate probability distribution.

Statistical methods based on the calculus of probabilities and the mathematical theory of stochastic processes have been successfully applied to a wide variety of physical problems. The most spectacular example is the theory of classical statistical mechanics developed by Maxwell, Boltzmann, Gibbs, and Einstein; it provides the physical foundation for the theory of thermodynamics, which accurately describes much of the macroscopic world. However, we now have two antithetical descriptions of the evolution of molecules in a gas or atoms in a solid, one deterministic and the other probabilistic; and one must wonder, as did the founders of statistical mechanics, what is the connection between them.

Attempts to reconcile the probabilistic laws of statistical mechanics with the deterministic laws of classical mechanics gave birth to a new branch of mathematics, called ergodic theory, which provides a means of classifying different deterministic dynamical systems with irregular behavior (19, 25, 26). In particular, this classification scheme defines symptoms for a hierarchy of different classes of random-like behavior, "statistical diseases," of increasing severity.

Dynamical systems with the mildest disease are called ergodic (25, 26). These are systems that come near almost every possible state over time but do so in a regular manner. For example, the evolution of the standard map for $k = 0$ is completely described by equation 2, since the angular velocity, $y_n$, is a constant of motion. If the initial angular velocity, $y_0$, is an irrational multiple of $2\pi$, then the angle variable, $x_n$, will eventually cover the entire interval $[0, 2\pi]$ in an ordered and predictable way. This system is merely ergodic.

Although there has been considerable confusion in the physical literature, ergodicity is not sufficient to justify the application of the probabilistic methods of statistical mechanics, since ergodicity alone does not assure that nonequilibrium distributions evolve toward

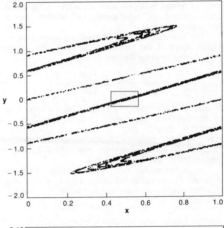

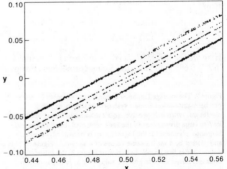

Figure 6. Under certain conditions, the long-term evolution of a chaotic dynamical system can converge to an attracting set in phase-space, which is called, in two or more dimensions, a "strange attractor." For example, if we add dissipation (friction) to the rotor depicted in Figure 4, so that equation 4 replaces equation 3, the diagram at the top shows the strange attractor for the damped, kicked rotor when $\lambda = 0.1$ and $k = 8.8$. The long-term evolution of this dissipative system has been traced by numerically iterating the map for $10^4$ time-steps. The attractor is considered "strange" because of the self-similar structure, which is maintained on all scales and which gives this object a noninteger, "fractal," dimension. The diagram at the bottom magnifies the region enclosed by the colored rectangle in the diagram above, showing some of the self-similar fine structure in the central strand of this attractor.

100

equilibrium (25, 26). However, dynamical systems with a more severe disease, the so-called Kolmogorov systems or K-systems, are irregular enough to rigorously justify a statistical description (26, 27).

K-systems exhibit the mathematical property known as "mixing" with "positive Kolmogorov-Sinai entropy." The "mixing" behavior is a precise character-ization of what you observe when you stir cream in your coffee, although many nonlinear dynamicists prefer the example of rum and Coke (28). "Positive Kolmogorov-Sinai entropy" is an essential technical condition which is difficult to verify directly for a given dynamical system. However, in practice this means that the dynamical system exhibits extreme sensitivity to initial conditions, so that two trajectories started at nearby initial conditions diverge at an exponential rate. This rate is measured by the "average Liapunov exponent," which is equivalent to the Kolmogorov-Sinai entropy and can be easily computed (29, 30). Because of this extreme sensitivity to initial conditions, the evolution of deterministic K-sys-tems defies long-time prediction (like the weather), since small errors or uncertainties in the initial conditions give rise to time-evolutions which are completely different.

We can now define chaos as the behavior of deter-ministic dynamical systems which exhibit these symp-toms of mixing behavior with a positive Kolmogorov-Sinai entropy or, equivalently, a positive average Liapunov exponent.

For example, for one-dimensional maps (like the logistic map) of the form $x_{n+1} = F(x_n)$, the average Liapunov exponent is defined to be

$$\lambda = \lim_{N \to \infty} \frac{1}{N} \sum_{n=1}^{N} \ln \left( \left| \frac{dF}{dx}(x_n) \right| \right) \qquad (5)$$

Figure 7 shows a graph of $\lambda$ versus $a$ for the same range of the nonlinear parameter, $a$, as displayed in the bifurcation diagram (Fig. 3). Here we clearly see a correspondence between chaotic motion and positive values of the average Liapunov exponent and between periodic orbits and sharp dips in $\lambda$.

In particular for $a = 4$, the average Liapunov exponent can be calculated exactly by taking advantage of a remarkable coordinate transformation. If we define a new variable

$$y_n = (2/\pi) \operatorname{Sin}^{-1}(\sqrt{x_n}) \qquad (6)$$

then the logistic map, equation 1, transforms to the "tent map"

$$y_{n+1} = \begin{cases} 2y_n & 0 \le x_n \le 0.5 \\ 2(1 - y_n) & 0.5 \le x_n \le 1 \end{cases} \qquad (7)$$

Here we see that $\ln |dF/dx(x)| = \ln 2$ for all $x$, so that $\lambda = \ln 2 \sim 0.693 > 0$. Since the Kolmogorov-Sinai entropy is

invariant under coordinate changes (26), this proves that the logistic map with $a = 4$ is a K-system and thereby meets our definition of a chaotic dynamical system. It can also be rigorously shown that the logistic map is a K-system and therefore chaotic for many values of $a > a_{inf} = 3.57...$, which is consistent with the numerical evi-dence displayed in Figure 7.

The average Liapunov exponent can also be calcu-lated for dynamical systems in higher dimensions, like the standard map, although the algorithm is more complicated than that for one-dimensional maps (30).

## The root of the disease lies in the mathematical pathologies of the real numbers

For example, a computation of the average Liapunov exponent for the standard map shows that for orbits in the regular regions of the phase-space of Figure 5, $\lambda \sim 0$; in the irregular regions, $\lambda > 0$. Unfortunately, very few realistic systems have been rigorously proved to be K-systems. Consequently, the justification for classifying much irregular behavior as chaos depends on the accu-mulation of numerical evidence and on experience with a few idealized mathematical models which are known to be K-systems.

Using this technical definition of chaos, we now see that chaotic dynamical systems can exhibit many of the attributes of idealized random systems; namely, their evolution is unpredictable because of their extreme sen-sitivity to initial conditions, and their average properties can be described using statistical methods. However, when we observe irregular phenomena in nature, such as turbulent flow in fluids, we don't always perform averages over time or over an ensemble of initial condi-tions. Rather, we often observe a single realization of the dynamical process evolving from a specific (though imprecisely known) initial condition which nevertheless

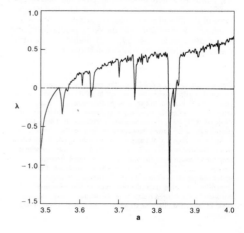

Figure 7. The average Liapunov exponent ($\lambda$, equation 5) defines in precise mathematical terms a system's sensitivity to initial conditions: when $\lambda$ is positive, small changes in initial conditions lead to large divergences in the long-term evolution. The average Liapunov exponent for the logistic map is numerically computed and plotted here for the same values of $a$ shown in Figure 3. This graph verifies that chaotic behavior in Figure 3, the bifurcation diagram, corresponds to positive values for $\lambda$, whereas regular (periodic) behavior corresponds to negative dips on this graph.

appears as random as a sequence of coin flips. Is it possible that deterministic but chaotic dynamical systems can also account for the random appearance of individual realizations of these physical systems? The answer is yes. Using the definition of a random sequence provided by algorithmic complexity theory (7), we will see that the evolution of a chaotic dynamical system can be indistinguishable from a sequence of coin flips and that these completely determined systems can be as irregular as any idealized random system. (This latter conclusion begs the question of whether any idealized random systems exist in the world of classical physics and whether the apparent randomness we observe and exploit in statistical theories is just the chaotic behavior of some underlying deterministic dynamical system.)

Algorithmic complexity theory defines the complexity, $K_N$, of a sequence of $N$ numbers as the length of the shortest computer program that can generate the sequence $(7, 31, 32)$. This length is conveniently measured

---

*Under chaotic conditions the use of pesticides, price controls, or arms control agreements will not necessarily yield the desired outcomes*

---

in terms of the number of bits of information required to input the program, which is proportional to the number of lines of FORTRAN (or any other programming language) plus the number of bits required to specify any numerical inputs or parameters in the program, such as the number of elements in the sequence, $N$. In particular, the minimum program size required to generate a sequence of numbers of length $N$ is at least $\log_2 N$, since this is the number of bits required to specify the length of the sequence in binary notation. Moreover, if we consider binary sequences of 0s and 1s so that an output sequence with $N$ elements corresponds to $N$ bits of information, then the maximum value for $K_N$ is of the order of $N$, since the computer program can simply read the $N$-bit sequence as input and then output the same sequence. (The programming commands to copy the input add only a constant contribution to $K_N$, which varies for different computers but is negligibly small in the limit of large $N$.)

A random sequence is defined to be a sequence with maximal complexity, $K_N \sim N$. A nonrandom sequence can be generated by a shorter program which takes advantage of any order or regularity in the sequence. For example, a sequence consisting of all 1s, corresponding to a sequence of coin flips where heads appears every time, can be generated by the computer program "Print 1, $N$ times," which can be programmed with $\sim \log_2 N$ bits of information. However, a sequence of 1s and 0s with no apparent order, which is most efficiently generated by simply making a copy of it, has maximal complexity, $K_N \sim N$.

This definition of a random sequence, which arose in the work of Kolmogorov, Chaitin, and Solmonov in information theory $(7, 31, 32)$, is in complete agreement with our intuitive concept of a random sequence. Certainly a sequence with any apparent order, such as consecutive 1s, would not be considered to be very random, whereas a sequence that has no regular patterns and can be specified only by a program of length $\sim N$ is likely to meet our intuitive criteria for randomness. In fact, for infinite sequences Martin-Löf has proved that these random sequences will satisfy every conceivable statistical test for randomness (32).

What, then, is the complexity of the time-sequences generated by chaotic dynamical systems? Consider for convenience the one-dimensional map on the unit interval

$$x_{n+1} = 2x_n \quad \text{Mod } 1 \qquad (8)$$

which is closely related to the tent map and consequently to the logistic map via the coordinate-transformation equation 6. Using equation 5, the average Liapunov exponent for equation 8 is easily determined to be $\ln 2 > 0$; so this map is a K-system and therefore chaotic. Now, if we examine the action of this chaotic dynamical system on initial conditions represented in binary, $x_0 = 0.101001110100111...$, then the multiplication of $x_0$ by 2 just shifts the "binary" point to the right and the Mod 1 throws away any integer part of $x$ to the left of the "binary" point. Therefore, successive iterations of this "register shift" simply read off successive binary digits in the initial condition. In particular, if we call "heads" when the value of $x_n > \frac{1}{2}$ (i.e., the leading digit is 1) and "tails" when $x_n < \frac{1}{2}$ (in which case the leading digit is 0), then the evolution of the map will generate, from every initial condition, a sequence of heads and tails which resembles the tossing of a coin. But when will these sequences appear random? The answer is again provided by Martin-Löf, who also proved that almost all initial conditions on the unit interval have a random binary-digit sequence (32). Therefore, the deterministic shift map will almost always generate a random sequence which is indistinguishable from the outcome of an idealized coin toss. Moreover, the same conclusions can be generalized to the tent map (and equivalently the logistic map) and all other chaotic dynamical systems. (Of course if you try to implement equation 8 on a digital computer, only short sequences can be studied, because the shift map quickly runs into the precision limits of the computer, which represents initial conditions with only $\sim 30$ or 60 binary digits in single and double precision, respectively.)

## Is physics conquering chaos, or chaos undermining physics?

The definitions and examples in the previous sections show that nonlinear dynamical systems can exhibit all the attributes of an idealized random process. Moreover, the theory of algorithmic complexity reveals that the origins of chaotic behavior in nonlinear dynamical systems and perhaps in nature itself lie in the randomness of almost all real numbers.

In other words, chaotic dynamical systems are mathematical models which "read" initial conditions. They are like the compulsive librarians in Borges's Library of Babel (where books containing every possible combination of letters are shelved), who read every word and character in the books under their care, whereas regular or nonchaotic systems are like the

Figure 8. On the New York Stock Exchange, the use of computers has decreased the frenzied shouting on the floor while increasing the volume of trading—but prices seem more volatile than ever. Even if the prices of stocks are completely determined by initial conditions—that is, if the system is mechanistic—the behavior of the market on a given day might still satisfy the mathematical definition of chaos. There would be no faster way to compute the outcome than to watch the market itself perform on that day. (Photograph © Four By Five.)

casual readers, who just read the titles and skim the text (33). The unpredictability of chaotic dynamical systems arises from the fact that slight errors or changes in the initial conditions correspond to different books in the library which tell different stories.

More generally, if nonlinear models describing the evolution of biological populations, economic prices, armament stockpiles, or turbulent flows in fluids can exhibit chaotic behavior, then we may be incapable, in practice, of predicting the behavior of these systems or their response to external forces, since any errors or perturbations will grow exponentially. For example, under chaotic conditions the use of pesticides, price controls, or arms control agreements will not necessarily yield the desired outcomes (Fig. 8).

Another manifestation of the unpredictability of chaotic dynamical systems is that the time-evolution is computationally irreducible (7, 34). There is no faster way of finding out how a chaotic system will evolve than to watch its evolution. The dynamical system itself is its own fastest computer. For most books in the Library of Babel the only way you can appreciate the contents is to read the entire book to the end. (Unfortunately, most, in fact almost all, of the books appear to be gibberish and make very uninteresting reading. However, somewhere in the library is a collection of books which contains the complete past and future history of the universe.)

Chaotic dynamical systems are also like football games. Even with the largest imaginable digital computer you could not predict the outcome with certainty. The players themselves provide the fastest analog computation of the evolution of this dynamical system. Because of the complexity and unpredictability of chaos, direct numerical simulations of football games and turbulent flows are likely to remain impractical with even the largest supercomputers. However, we can nevertheless compute reliable odds or probabilities for the outcomes of these processes. As a consequence, probabilistic and statistical theories provide a natural description of average properties of chaotic dynamical systems. (An entertaining account of how several well-known chaos theorists used their knowledge of nonlinear dynamics to improve their odds at roulette is provided in ref. 35.)

One of the most surprising properties of chaotic dynamical systems is that these deterministic models are often very simple. The realization that complex behavior does not require complex mathematical models is one of the most significant contributions of nonlinear dynamics. Since simple models can yield complex, irregular behavior, we can actually hope to develop theoretical descriptions of a wide variety of apparently random, unpredictable natural phenomena using mathematical models which exhibit deterministic chaos. However,

although recent progress has suggested partial solutions to the nineteenth-century problems of the origin of turbulence in fluids and the microscopic foundations of statistical mechanics, many old problems remain, and some new and very profound questions have been raised.

For example, among the old problems, the discovery of chaos has not miraculously solved the problem of turbulence in fluids. But we now have new methods of

---

*There is no faster way of finding out how a chaotic system will evolve than to watch its evolution. The system itself is its own fastest computer*

---

characterizing turbulent behavior, such as the measurement of the average Liapunov exponent or the fractal dimension of the strange attractor associated with turbulence; and we have a much better understanding of why the theoretical and numerical description of the evolution of turbulent flows is so difficult (36).

Moreover, although chaos explains how average properties of nonlinear dynamical systems can exhibit an irreversible approach to thermodynamic equilibrium, it does not account for why individual systems in nature also appear to exhibit the irreversible evolution mandated by the second law of thermodynamics. Since the equations of motion of classical mechanics are deterministic and invariant under time-reversal, we could always imagine that in principle, by stirring very carefully, we can separate the cream from the coffee after it has been thoroughly mixed. All that nonlinear dynamics tells us is that this reversal of time-evolution will be extremely difficult, since any errors or uncertainties will guarantee failure. The missing ingredient required for a complete justification of the foundations of classical statistical mechanics is an argument for why such errors are inevitable. (Certainly, any numerical simulation of the evolution and reversal of a chaotic dynamical system will fail to recover the initial state, because whenever the machine rounds off a number it automatically introduces a slight change in the system which gives a completely

## What are the real differences, if any, between deterministic chaotic behavior and random processes?

different and unpredictable result.) A provocative discussion of the relationship of chaos to the second law of thermodynamics can be found in Prigogine and Stengers's *Order out of Chaos* (37).

While confronting these remaining problems, nonlinear dynamics has also identified some new, uniquely twentieth-century problems which may account for some of these failures. The first problem is that, since chaotic dynamical systems essentially read initial conditions, they are exquisitely sensitive to the infinities and infinitesimals manifest in the continuum of real numbers which underlie almost all mathematical descriptions of natural phenomena. In contrast, regular systems, such as those studied in almost every textbook, are relatively insensitive to the mathematical pathologies of infinitely long digit-strings.

The difficulty with the continuum of real numbers lies in the fact that, although most real numbers can be proved to have random digit-strings, it is impossible to prove that a given digit-string is random. You simply can never exhaust all the possible tests for underlying order. This is a specific example of a class of true statements which cannot be proved, statements first shown to exist by Gödel in his celebrated incompleteness theorem (38). (For a clear discussion of the connection between random digit-strings and Gödel's incompleteness theorem, see ref. 31). Moreover, by definition these numbers cannot be computed by any algorithm shorter than the digit-string itself. As a consequence, most real numbers are uncomputable. Therefore, now that our understanding of chaotic dynamical systems has revealed that the root of the disease lies in these mathematical pathologies of the real numbers, Joe Ford has suggested that these uncomputable and undefinable objects should be excised from any meaningful physical theory (7). In addition to providing some logical consistency in the description of natural phenomena, this restriction might also provide the missing argument for the validity of the second law of thermodynamics. For example, if we assume that nature is a finite-state computer (or Turing machine), then the inevitable truncation of real numbers could provide the coarse-graining necessary to ensure the irreversibility of chaotic systems.

Such a conclusion would surely herald a revolution

in natural science, and a number of research groups have already begun to explore the possibilities of so-called cellular automata models for natural phenomena which are defined on discrete sets of numbers (39). However, it is possible that the scale at which the truncation of real numbers occurs may be so small that no practical consequences of the distinction between continuum and discrete theories can be deduced or verified. In that case the issue of the ultimate discretization of the real world will pass from the domain of physics to that of philosophy. Nevertheless, cellular automata are a fascinating subject in their own right and promise to play an important role in future studies of nonlinear dynamical systems.

The second fundamental question which arises from our improved understanding of classical nonlinear systems is whether chaos persists in microscopic physical systems, such as atoms and molecules, where the theory of quantum mechanics is expected to apply (40, 41). The difficulty here lies in the fact that the Schrödinger equation for the evolution of the quantum mechanical wave function is a linear equation which, strictly speaking, is incapable of exhibiting the chaotic behavior of nonlinear classical systems. Since quantum mechanics is presumed to be the fundamental theory for all physical systems, and since the predictions of quantum theory must agree with those of classical mechanics at the limit of the highest quantum numbers, according to Bohr's correspondence principle, one of these physical descriptions—classical chaos or quantum mechanics—threatens to undermine the other. Does this mean that the role of classical chaos in explaining the origins of turbulence and the foundations of statistical mechanics is merely an illusion? That Bohr's correspondence principle is invalid for systems that are classically chaotic? And that there isn't any problem with the continuum of real numbers after all?

The answers to these questions are naturally the goals of much current research. Preliminary results indicate that the evolution of the quantum mechanical wave function appears to mimic chaotic behavior for very long times (42, 43), in many cases longer than the age of the universe (44). Nevertheless, without chaos we have lost some of the necessary ingredients for the foundations of statistical mechanics. The validity of the correspondence principle, which guided the early development of quantum mechanics, also remains an outstanding problem, although recent experiments on the ionization of highly excited hydrogen atoms exposed to intense electromagnetic radiation (which study the behavior of a quantum system that is classically chaotic) suggest that the correspondence principle is remarkably robust (45, 46).

In conclusion, we have seen how deceptively simple mathematical models for nonlinear dynamical systems, like the logistic and standard maps, have provided new hope for the description of the complexity and chaos which surround us in the natural world. However, these and more recent studies have also opened a Pandora's box of new problems which ask profound and disturbing questions about the proper mathematical description of both macroscopic and microscopic natural phenomena and which promise to lie at the forefront of scientific research for many years to come.

# 104

## References

1. See articles in *Order in Chaos, Physica 7D*, North Holland, 1983.
2. R. M. May. 1976. Simple mathematical models with very complicated dynamics. *Nature* 261:459.
3. L. Glass, M. R. Guevara, and A. Shrier. 1983. Bifurcation and chaos in a periodically stimulated cardiac oscillator. *Physica 7D*, p. 89.
4. R. V. Jensen and R. Urban. 1984. Chaotic price behavior in a nonlinear cobweb model. *Econ. Lett.* 15:235.
5. A. M. Saperstein. 1984. Chaos—a model for the outbreak of war. *Nature* 309:303.
6. J. P. Crutchfield, J. D. Farmer, N. H. Packard, and R. S. Shaw. 1986. Chaos. *Sci. Am.* 255, Dec., p. 46.
7. J. Ford. 1983. How random is a coin toss? *Phys. Today* 36, Apr., p. 40.

    ———. 1986. Chaos: Solving the unsolvable, predicting the unpredictable! In *Chaotic Dynamics and Fractals*, Academic Press.

    ———. 1986. What is chaos, that we should be mindful of it? In *The New Physics*, ed. S. Capelin and P. C. W. Davies. Cambridge Univ. Press.
8. M. J. Feigenbaum. 1983. Universal behavior in nonlinear systems. *Physica 7D*, p. 16.
9. P. Cvitanovic. 1984. *Universality in Chaos*. Bristol, UK: Adam Hilger.
10. H. G. Schuster. 1984. *Deterministic Chaos*. Weinheim, FRG: Physik-Verlag.
11. P. Collet and J.-P. Eckmann. 1980. *Iterated Maps on the Interval as Dynamical Systems*. Birkhäuser.
12. R. V. Jensen and C. R. Meyers. 1985. Images of critical points of nonlinear maps. *Phys. Rev. A* 32:1222.
13. C. Grebogi, E. Ott, and J. Yorke. 1983. Crises, sudden changes in chaotic attractors, and transient chaos. *Physica 7D*, p. 181.
14. B. V. Chirikov. 1979. A universal instability of many-dimensional oscillator systems. *Phys. Rep.* 52:263.
15. V. I. Arnold. 1978. *Mathematical Methods of Classical Mechanics*. Springer-Verlag.
16. E. Ott. 1981. Strange attractors and chaotic motions of dynamical systems. *Rev. Mod. Phys.* 53:655.
17. R. V. Jensen and C. R. Oberman. 1982. Statistical properties of chaotic dynamical systems which exhibit strange attractors. *Physica 4D*, p. 183.
18. B. B. Mandelbrot. 1982. *The Fractal Geometry of Nature*. W. H. Freeman.
19. A. J. Lichtenberg and M. A. Lieberman. 1983. *Regular and Stochastic Processes*. Springer-Verlag.
20. R. S. MacKay and I. C. Percival. 1985. Converse KAM: Theory and practice. *Comm. Math. Phys.* 98:469.
21. J. M. Greene. 1979. A method for determining a stochastic transition. *J. Math. Phys.* 20:1183.
22. G. M. Zaslavskii and B. V. Chirikov. 1972. Stochastic instability of nonlinear oscillations. *Soviet Physics USPEKHI* 14:549.
23. C. F. F. Karney. 1983. Long-time correlations in the stochastic regime. *Physica 8D*, p. 360.
24. H. Poincaré. 1952. *Science and Method*. Dover.
25. J. L. Lebowitz and O. Penrose. 1973. Modern ergodic theory. *Phys. Today*, Feb., p. 23.
26. Ya. G. Sinai. 1977. *Introduction to Ergodic Theory*. Princeton Univ. Press.
27. N. S. Krylov. 1979. *Works on the Foundations of Statistical Physics*. Princeton Univ. Press.
28. V. I. Arnold and A. Avez. 1968. *Ergodic Problems of Classical Mechanics*. Benjamin.
29. Ya. B. Pesin. 1977. Characteristic Liapunov exponents and smooth ergodic theory. *Russ. Math. Surv.* 32:55.
30. G. Benettin, L. Galgani, and J.-M. Strelcyn. 1976. Kolmogorov entropy and numerical experiments. *Phys. Rev. A* 14:2338.
31. G. J. Chaitin. 1975. Randomness and mathematical proof. *Sci. Am.* 232, May, p. 47.

    ———. 1982. Gödel's theorem and information. *Int. J. Theor. Phys.* 21:941.
32. P. Martin-Löf. 1966. The definition of random sequences. *Info. Contr.* 9:602.
33. J. L. Borges. 1964. The Library of Babel. In *Labyrinths*, p. 51. New Directions.
34. S. Wolfram. 1985. Undecidability and intractability in theoretical physics. *Phys. Rev. Lett.* 54:735.
35. T. A. Bass. 1985. *The Eudemonic Pie: Or Why Would Anyone Play Roulette without a Computer in His Shoe?* Houghton Mifflin.
36. K. Sreenivasan. 1985. Transitions and turbulence in fluid flows and low dimensional chaos. In *Frontiers in Fluid Mechanics*, ed. S. H. Davis and J. L. Lumley, p. 41. Springer-Verlag.
37. I. Prigogine and I. Stengers. 1984. *Order out of Chaos*. Bantam.
38. D. Hofstadter. 1979. *Gödel, Escher, Bach: An Eternal Golden Braid*. Basic Books.
39. See articles in *Cellular Automata, Physica 10D*, North Holland, 1984.
40. See articles in *Chaotic Behavior in Quantum Systems*, ed. G. Casati, Plenum, 1985.
41. M. V. Berry. 1983. Semi-classical mechanics of regular and irregular motion. In *Chaotic Behavior of Deterministic Systems*, ed. G. Iooss, R. H. G. Helleman, and R. Stora, p. 171. North Holland.
42. R. V. Jensen and R. Shankar. 1985. Statistical behavior of deterministic quantum systems with few degrees of freedom. *Phys. Rev. Lett.* 54:1879.
43. G. Casati, B. V. Chirikov, I. Guarneri, and D. L. Shepelansky. 1986. Dynamical stability of quantum "chaotic" motion in a hydrogen atom. *Phys. Rev. Lett.* 56:2437.
44. A Peres. 1982. Recurrence phenomena in quantum mechanics. *Phys. Rev. Lett.* 49:1118.
45. K. A. H. van Leeuwen et al. 1985. Microwave ionization of hydrogen atoms: Experiment versus classical dynamics. *Phys. Rev. Lett.* 55:2231.
46. R. V. Jensen. In press. Chaos in atomic physics. In *Proceedings of the Xth International Conference on Atomic Physics (ICAP-X)*, ed. H. Narumi. North Holland.

## 2. KAM Theorem and Stochasticity in Classical Hamiltonian Systems

### PRESERVATION OF CONDITIONALLY PERIODIC MOVEMENTS
### WITH SMALL CHANGE IN THE HAMILTON FUNCTION*

Academician A. N. Kolmogorov
Department of Mathematics
Moscow State University
117234 Moscow, B-234
U.S.S.R.

#### ABSTRACT

This paper is a translation of Kolmogorov's original article announcing the theorem now known as the KAM theorem.

#### THEOREM AND DISCUSSION OF PROOF

Let us consider in the 2s-dimensional phase space of a dynamic system with s degrees of freedom the region G, represented as the product of an s-dimensional torus, T, and a region S, of a Euclidian s-dimensional space. We will designate the points of the torus, T, by the circular coordinates $q_1, \ldots, q_s$ (replacing $q_\alpha$ with $q'_\alpha = q_\alpha + 2\pi$ does not change points q), and the coordinates of the points, p, of S we will designate as $p_1, \ldots, p_s$. We will assume that in region G, in the coordinates $(q_1, \ldots, q_s, p_1, \ldots, p_s)$ the equations of motion have the canonical form

$$\frac{dq_\alpha}{dt} = \frac{\partial}{\partial p_\alpha} H(q,p), \quad \frac{dp_\alpha}{dt} = -\frac{\partial}{\partial q_\alpha} H(q,p). \tag{1}$$

The Hamilton function, H, is further assumed as dependent on the parameter $\theta$ and determined for all (q,p) $\epsilon G$, $\theta\epsilon(-c; +c)$, but not time-dependent. Moreover, further considerations require fairly significant restrictions on the smoothness of the function $H(q, p, \theta)$, stronger than infinite differentiability. For simplicity, in the following it is assumed that the function $H(q, p, \theta)$ is analytic over the set of variables $(q, p, \theta)$.

Summation over the Greek indices is further assumed to be from 1 to s. The usual vector designations $(x,y) = \sum_\alpha x_\alpha y_\alpha$, $|x| = + \sqrt{(x,x)}$ are used. A whole number vector indicates a vector for which all the components are whole numbers. The set of points (q, p) of G with

---

*Los Alamos Scientific Laboratory translation LA-TR-71-67 by Helen Dahlby of Akad. Nauk. S.S.S.R., Doklady <u>98</u>, 527 (1954).

p = c are designated by $T_c$. In theorem 1 it is assumed that S contains the point p = 0, i.e., $T_0 \subseteq G$.

Theorem 1. Let

$$H(q,p,0) = m + \sum_\alpha \lambda_\alpha P_\alpha + \frac{1}{2} \sum_{\alpha\beta} \Phi_{\alpha\beta}(q) P_\alpha P_\beta + O(|p|^3), \qquad (2)$$

where m and $\lambda_\alpha$ are constants and for a certain choice of constants c > 0 and $\eta$ > 0 for all whole-number vectors, n, the inequality

$$(n,\lambda) \geqslant \frac{c}{|n|^\eta} . \qquad (3)$$

is satisfied.

Let, moreover, the determinant composed of the average values

$$\varphi_{\alpha\beta}(0) = \frac{1}{(2\pi)^s} \int_0^{2\pi} \cdots \int_0^{2\pi} \Phi_{\alpha\beta}(q) \, dq_1 \cdots dq_s$$

of the functions

$$\Phi_{\alpha\beta}(q) = \frac{\partial^2}{\partial p_\alpha \partial p_\beta} H(q,0,0)$$

be different from zero:

$$|\varphi_{\alpha\beta}(0)| \neq 0. \qquad (4)$$

Then there exist analytical functions $F_\alpha(Q, P, \theta)$ and G (Q, P, θ) which are determined for all sufficiently small θ and for all points (Q, P) of some neighborhood, V, of the set $T_0$, which bring about a contact transformation

$$q_\alpha = Q_\alpha + \theta F_\alpha(Q,P,\theta), \quad p_\alpha = P_\alpha + \theta G (Q,P,\theta)$$

of V into $V' \subseteq G$, which reduced H to the form

$$H = M(\theta) + \sum_\alpha \lambda_\alpha P_\alpha + O(|P|^2) \qquad (5)$$

(M(θ) does not depend on Q and P).

It is easy to grasp the meaning of Theorem 1 for mechanics. It indicates that an s-parametric family of conditionally periodic motions

$$q_\alpha = \lambda_\alpha t + q_\alpha^{(0)}, \quad p_\alpha = 0,$$

53

which exists at $\theta = 0$ cannot, under conditions (3) and (4), disappear as a result of a small change in the Hamilton function H: there occurs only a displacement of the s-dimensional torus, $T_0$, around which the trajectories of these motions run, into the torus $P = 0$, which remains filled by the trajectories of conditionally periodic motions with the same frequencies $\lambda_1, \ldots, \lambda_s$.

The transformation

$$(Q, P) = K_\theta(q, p),$$

the existence of which is confirmed in Theorem 1, can be constructed in the form of the limit of the transformations

$$(Q^{(k)}, P^{(k)}) = K_\theta^{(k)}(q, p),$$

where the transformations

$$(Q^{(1)}, P^{(1)}) = L_\theta^{(1)}(q, p), \quad (Q^{(k+1)}, P^{(k+1)}) = L_\theta^{(k+1)}(Q^{(k)}, P^{(k)})$$

are found by the "generalized Newton method" (see Ref. 1). In this note we confine ourselves to the construction of the transformation: $K_\theta^{(1)} = L_\theta^{(1)}$, which itself permits grasping the role of conditions (3) and (4) of Theorem 1. Let us apply the transformation $L_\theta^{(1)}$ to the equations

$$Q_\alpha^{(1)} = q_\alpha + \theta Y_\alpha(q),$$

$$P_\alpha = P_\alpha^{(1)} = \theta \left\{ \sum_\beta P_\beta^{(1)} \frac{\partial Y_\beta}{\partial q_\alpha} + \xi_\alpha + \frac{\partial}{\partial q_\alpha} X(q) \right\} \tag{6}$$

(it is easy to verify that this is a contact transformation) and seek the constants $\xi_\alpha$ and $\zeta$ and the functions $X(q)$ and $Y_\beta(q)$, starting from the requirement that

$$H = m + \sum_\alpha \lambda_\alpha P_\alpha + \frac{1}{2} \sum_{\alpha\beta} \Phi_{\alpha\beta}(q) P_\alpha P_\beta +$$

$$+ \theta \left\{ A(q) + \sum_\alpha B_\alpha(q) P_\alpha \right\} + O(|P|^3 + \theta |P|^2 + \theta^2) \tag{7}$$

take the form

$$H = m + \theta\zeta + \sum_\alpha \lambda_\alpha P_\alpha^{(1)} + O(|P^{(1)}|^2 + \theta^2). \tag{8}$$

Substituting (6) into (7), we get

$$H = m + \sum_\alpha \lambda_\alpha P_\alpha^{(1)} + \theta \left\{ A + \sum_\alpha \lambda_\alpha \left( \xi_\alpha + \frac{\partial X}{\partial q_\alpha} \right) \right\} +$$
$$+ \theta \sum_\alpha P_\alpha^{(1)} \left\{ B_\alpha + \sum_\beta \phi_{\alpha\beta}(q) \left( \xi_\beta + \frac{\partial X}{\partial q_\beta} \right) + \sum_\beta \lambda_\beta \frac{\partial Y_\beta}{\partial q_\beta} \right\} + O(|P^{(1)}|^2 + \theta^2).$$

Thus, our requirement (8) reduces to the equations

$$A + \sum_\alpha \lambda_\alpha \left( \xi_\alpha + \frac{\partial X}{\partial q_\alpha} \right) = \zeta \tag{9}$$

$$B_\alpha + \sum_\beta \phi_{\alpha\beta} \left( \xi_\beta + \frac{\partial X}{\partial q_\beta} \right) + \sum_\beta \lambda_\beta \frac{\partial Y_\alpha}{\partial q_\beta} = 0. \tag{10}$$

being fulfilled.

Let us introduce the functions

$$Z_\alpha(q) = \sum_\beta \phi_{\alpha\beta}(q) \frac{\partial}{\partial q_\beta} X(q). \tag{11}$$

Expanding the functions $\phi_{\alpha\beta}$, $A$, $B_\alpha$, $X$, $Y_\alpha$, $Z_\alpha$ into a Fourier series of the type

$$X(q) = \sum x(n) e^{i(n,q)}$$

and assuming for definiteness that

$$x(0) = 0, \quad y(0) = 0, \tag{12}$$

we get for the remaining Fourier coefficients $x(n)$, $y_\alpha(n)$, and $z_\alpha(n)$ and constants $\xi_\alpha$ and $\zeta$ of the equation which are relevant to the determination

$$a(0) + \sum \lambda_\alpha \xi_\alpha = \zeta, \tag{13}$$

$$a(n) + (n,\lambda) x(n) = 0 \quad \text{for } n \neq 0, \tag{14}$$

$$b_\alpha(0) + \sum_\beta \boldsymbol{\varphi}_\alpha(0) \xi_\beta + z_\alpha(0) = 0, \tag{15}$$

$$b_\alpha(n) + \sum_\beta \boldsymbol{\varphi}_{\alpha\beta}(n) \xi_\beta + z_\alpha(n) + (n,\lambda) y_\alpha(n) = 0 \quad \text{for } n \neq 0. \tag{16}$$

It is easy to see that the system (11) - (16) is unambiguously solved under conditions (3) and (4). Condition (3) is important in the determination of x(n) from (14), and in the determination of $y_\alpha$(n) from (16). Condition (4) is important in the determination of $\xi_\beta$ from (15). Since, as |n| increases, the coefficients of the Fourier series of the analytical functions $\boldsymbol{\varphi}_{\alpha\beta}$, A, and $B_\alpha$ have an order of decrease not less than $\rho^{|h|}$, $\rho < 1$, then from condition (3) there results not only the formal solvability of equations (13) – (16) but also the convergence of the Fourier series for the functions X, $Y_\alpha$, and $Z_\alpha$ and the analyticity of these functions. The construction of further approximations is not associated with new difficulties. Only the use of condition (3) for proving the convergence of the recursions, $K_\theta^{(k)}$, to the analytical limit for the recursion $K_\theta$ is somewhat more subtle.

The condition of the absence of "small denominators" (3) should be considered, "generally speaking," as fulfilled, since for any $\eta > s - 1$ for all points of an s-dimensional space $\lambda = (\lambda_1, \ldots, \lambda_s)$ except the set of Lebesque measure zero it is possible to find $c(\lambda)$, for which

$$(n, \lambda) \geqslant \frac{c(\lambda)}{|n|^\eta}$$

whatever the integers $n_1$, $n_2$, $\ldots$, $n_s$ were[2]. It is also natural to consider condition (4) as, "generally speaking," fulfilled. Since

$$\boldsymbol{\varphi}_{\alpha\beta}(0) = \frac{\partial}{\partial p_\alpha} \lambda_\beta(0),$$

where

$$\lambda_\beta(p) = \frac{1}{(2\pi)s} \int_0^{2\pi} \cdots \int_0^{2\pi} \frac{dq_\beta}{dt} dq_1 \ldots dq_s$$

is the frequency averaged over the coordinate $q_\beta$ with fixed momenta $p_1, \ldots, p_s$, condition (3) means that the Jacobian of the average frequencies over the momenta is different from zero.

Let us turn now to a consideration of the special case where H(q, p, 0) depends only on p, i.e., H(q, p, 0) = W(p). In this case, for $\theta = 0$ each torus, $T_p$, consists of the complete trajectories of the conditionally periodic movements with frequencies

$$\lambda_\alpha(p) = \frac{\partial W}{\partial p_\alpha} .$$

56

If the Jacobian

$$J = \left| \frac{\partial \lambda_\alpha}{\partial p_\beta} \right| = \left| \frac{\partial^2 W}{\partial p_\alpha \partial p_\beta} \right| \tag{17}$$

is different from zero, then it is possible to apply Theorem 1 to almost all tori, $T_p$. There arises the natural hypothesis that at small $\theta$ the "displaced tori" obtained in accordance with Theorem 1 fill a larger part of region G. This is also confirmed by Theorem 2, pointed out later. In the formulation of this theorem we will consider the region S to be bounded and will introduce into the consideration the set, $M_\theta$, of those points $(q^{(0)}, p^{(0)})$ $\varepsilon G$ for which the solution

$$q_\alpha(t) = f_\alpha(t; q^{(0)}, p^{(0)}, \theta), \quad p_\alpha(t) = G_\alpha(t; q^{(0)}, p^{(0)}, \theta)$$

of the system of equations (1) with initial conditions

$$q_\alpha(0) = q_\alpha^{(0)}, \quad p_\alpha(0) = p_\alpha^{(0)}$$

leads to trajectories not moving out of region G with change in t from $-\infty$ to $+\infty$, and conditionally periodic with periods $\lambda_\alpha = \lambda_\alpha(q^{(0)}, p^{(0)}, \theta)$, i.e., it has the form

$$f_\alpha(t) = \varphi_\alpha(e^{i\lambda_1 t}, \ldots, e^{i\lambda_s t}), \quad g_\alpha(t) = \psi_\alpha(e^{i\lambda_1 t}, \ldots, e^{i\lambda_s t}).$$

Theorem 2. If $H(q, p, 0) = W(p)$ and determinant (17) is not equal to zero in region S, then for $\theta \to 0$ the Lebesque degree of the set $M_\theta$ converges to the complete degree of region S.

Apparently, in the usual sense of the phrase, "general case" is when the set $M_\theta$ at all positive $\theta$ is everywhere dense. In such a case the complications arising in the theory of analytical dynamic systems are indicated more specifically in my note.[3]

## REFERENCES

1. L. V. Kantorovich. Uspekhimatem. Nauk $\underline{3}$, 163 (1948).

2. J. F. Koksma. Diophantische Approximationen, Chelsea 1936. 157p.

3. A. N. Kolmogorov. Doklady Akad. Nauk $\underline{93}$, 763 (1953).

## 3. The Problem of Onset of Turbulence

4   The Problem of Onset of Turbulence

(1)  On the problem of turbulence
     by L.D. Landau, Doklad. Akad. Nauk (1944),
     English translation in Collected Papers of L.D. Landau,
     p. 387 (1965) edited by D. Ter Haar

(2)  On the nature of turbulence
     by D. Ruelle and F. Takens, Commun. Math. Phys.
     20, 167; 23, 343 (1971).

# 52. ON THE PROBLEM OF TURBULENCE

ALTHOUGH the turbulent motion has been extensively discussed in literature from different points of view, the very essence of this phenomenon is still lacking sufficient clearness. To the author's opinion, the problem may appear in a new light if the process of initiation of turbulence is examined thoroughly.

In the case of incompressible fluids the unsteadiness of the laminar motion is known to be determined as follows. Upon the principal motion with a velocity distribution $v_0(x, y, z)$ there is superimposed a small disturbance $v_1(x, y, z, t)$; the substitution of $v = v_0 + v_1$ in the equation of motion of a viscous fluid and the neglect of terms of the second order of smallness lead to a linear differential equation for the perturbation $v_1$. Further, $v_1$ is sought in the form

$$v_1 = A(t) f(x, y, z), \tag{1}$$

where the time function $A(t)$ may be represented as

$$A(t) = \text{const} \cdot e^{-i\Omega t}. \tag{2}$$

The problem of determining the possible values of the "frequencies" $\Omega$ where the boundary conditions of motion are given, is an "Eigenwert" problem. By solving it one will obtain a spectrum of proper frequencies $\Omega$ (which are complex values in the general case). This spectrum, generally speaking, contains separate, isolated, values ("discrete spectrum") and also contains frequencies continuously filling whole intervals of values ("continuous spectrum"). It may be supposed that the frequencies of the continuous spectrum correspond to such motions $v_1$ as are not damped at infinity, while the frequencies of the discrete spectrum correspond to motions which are damped at infinity rather rapidly (as is the case in many other Eigenwert problems).

For the problem of steadiness of the principal motion those of the frequencies $\Omega = \omega + i\gamma$ ($\omega$, $\gamma$ are real) are relevant in which the imaginary part is negative ($\gamma < 0$). The presence of such proper frequencies in the spectrum indicates the unsteadiness of the principal motion with respect to infinitely small perturbations. Such values of $\Omega$ are only possible among the frequencies of the discrete spectrum. In fact, the principal motion presents at infinity a plane-parallel homogeneous flow (we mean a flow past a body of finite dimensions. In so far as a plane–parallel flow is in no case steady, it will be evident that any perturbation that fails to disappear at infinity must necessarily be damped in time, or, in other words, correspond to frequencies $\Omega$ with $\gamma > 0$ Accordingly, only the $\Omega$ frequencies of the discrete spectrum can be considered below.

Л. Ландау, К проблеме турбулентности, *Доклады Академии Наук СССР*, **44**, 339 (1944).
L. Landau, On the problem of turbulence, *C. R. Acad. Sci. URSS*, **44**, 311 (1944).

In the case of sufficiently small velocities the principal motion is a steady one (inasmuch as a resting fluid is in any case steady). On the other hand, with sufficiently large Reynolds numbers the laminar flow past a body is unsteady at any rate. In fact, with large Reynolds numbers the motion far away from the body is not appreciably different from a plane–parallel flow unless in the region of the narrow "track". Now it follows from Lord Rayleigh's work that no motion with a two-dimensional velocity distribution of such a type is steady, and one may expect that the same is true of the three-dimensional track.

If the values of the proper frequencies $\Omega$ are taken to be functions of the Reynolds number of the principal motion, then the critical value $\text{Re}_{cr}$ is determined by the fact that for $\text{Re} = \text{Re}_{cr}$ the imaginary part of one of the frequencies $\Omega$ will vanish; suppose this frequency to be $\Omega_1 = \omega_1 + i\,\gamma_1$. For $\text{Re} > \text{Re}_{cr}$ we have $\gamma_1 > 0$; for such Reynolds numbers as are near to the critical value $\text{Re}_{cr}$, $\gamma_1$ is small in comparison with $\omega_1$. However, the expression (1–2) for the respective function $v_1(x, y, z, t)$ (with $\Omega = \Omega_1$) is only true for a very brief interval of time, as measured from the instant at which the stationary regime is broken. This is owing to the fact that the factor $e^{\gamma_1 t}$ grows rapidly with time. As a matter of fact, the modulus $|A|$ of the amplitude of nonstationary motion does not increase infinitely, but rather tends to a certain limit. With $\text{Re}$ near to $\text{Re}_{cr}$ ($\text{Re}$ is always supposed to be greater than $\text{Re}_{cr}$), this limit is yet very small, too, and for determining it one may proceed as follows.

For very small times, when (2) is still applicable, we have

$$\frac{d\,|A|^2}{dt} = 2\gamma_1 |A|^2.$$

In substance this expression is but the first term of a series of powers of $A$ and $A^*$. With the increase of the modulus $|A|$ the subsequent three terms of this series must be taken into account. The next terms are terms of the third order. We, however, are interested not in the exact value of the differential quotient $d\,|A|^2/dt$, but in its mean value with respect to time, the averaging being made over time intervals that are large in comparison with the period $2\pi/\omega_1$ of the periodic spectrum $e^{-i\omega_1 t}$ (as $\omega_1 \gg \gamma_1$, this period is small compared to the time $1/\gamma_1$ during which the modulus $|A|$ changes appreciably). But the terms of the third order involve a periodic spectrum, and so they are eliminated upon averaging. (Strictly speaking, they do not vanish altogether, but yield quantities of order four; these quantities are supposed to be included into the terms of the fourth order). The terms of the fourth order include a term proportional to $A^2 A^{*2} = |A|^4$; this term is not eliminated by averaging. Thus, up to terms of the fourth order we have

$$\overline{\frac{d\,|A|^2}{dt}} = 2\gamma_1 |A|^2 - \alpha |A|^4. \tag{3}$$

Here $\alpha$ is a positive constant (the case of negative $x$ is considered below).

There are no signs of averaging over $|A|^2$ and $|A|^4$, because this operation is carried out over such time intervals as are small in comparison with $1/\gamma_1$. For the same reason in solving this equation we must disregard the bar over the derivative in the left hand member. The solution of the equation (3) has the form

$$\frac{1}{|A|^2} = \frac{\alpha}{2\gamma_1} + \text{const} \cdot e^{-2\gamma_1 t},$$

i.e. $|A|^2$ tends asymptotically to a limit

$$|A|^2_{\max} = 2\gamma_1/\alpha; \tag{4}$$

$\gamma_1$ is a function of Reynolds' number; it vanishes with $\text{Re} = \text{Re}_{cr}$. Therefore for small $\text{Re} - \text{Re}_{cr}$ we have $\gamma_1 = \text{const} \cdot (\text{Re} - \text{Re}_{cr})$. Substituting this in (4) we shall see that

$$|A|_{\max} \sim \sqrt{\text{Re} - \text{Re}_{cr}}. \tag{5}$$

Thus, the unsteadiness of the laminar motion for $\text{Re} > \text{Re}_{cr}$ leads to the appearance of a non-stationary periodic motion. When Re is close to $\text{Re}_{cr}$, this motion can be represented as a superposition of a stationary motion $v_0(x, y, z)$ over a periodic motion $v_1(x, y, z, t)$, having a small but finite amplitude which varies with Re directly as $\sqrt{\text{Re} - \text{Re}_{cr}}$. The velocity distribution in this motion has therefore the form

$$v_1 = f(x, y\ z) e^{-i(\omega_1 t + \beta_1)}, \tag{6}$$

where $\beta_1$ is a constant initial phase. When the differences $\text{Re} - \text{Re}_{cr}$ are large, there is no longer any sense in separating the velocities into two parts $v_0$ and $v_1$. Here we have to deal simply with a periodic motion of frequency $\omega_1$. If instead of the time the phase $\varphi_1 \equiv \omega_1 t + \beta_1$ is used as the independent variable, the function $v(x, y, z, \varphi_1)$ may be said to be a periodic function of $\varphi_1$ with a period $2\pi$, but no simple trigonometric function. It can be represented as a Fourier series

$$v = \sum A_p(x, y, z) e^{-i\varphi_1 p} \tag{7}$$

(the summation is carried out over all positive and negative integers $p$).

The essential fact is that only the absolute value of the factor, but not its phase are determined by the equation (3). The phase $\varphi_1$ remains in substance indefinite and depends upon the initial conditions which are a matter of change and may cause $\beta_1$ to take any value. It will be obvious that the periodic motion under consideration is not determined uniquely by the given stationary boundary conditions of motion; one quantity, the phase, remains arbitrary. This motion may be said to have one degree of freedom, whereas stationary motion is completely determined by the given boundary conditions, and enjoys not a single degree of freedom.

As Re is further increased, this periodic motion, too, eventually becomes unsteady. The investigation of its unsteadiness should be conducted in a manner analogous to that described above. The role of the principal motion is now played by the periodic motion $v_0(x, y, z, t)$ of frequency $\omega_1$. Substituting $v = v_0 + v_2$ with small $v_2$ into the equation of motion, we shall again obtain for $v_2$ a linear equation, but this time the coefficients of this equation are not only functions of the co-ordinates, but of time also; with respect to time, they are periodic functions with a period $2\pi/\omega_1$. The solution of such an equation should be sought in the form $v_2 = \Pi(x, y, z, t) e^{-i\Omega t}$ where $\Pi(x, y, z, t)$ is a periodical function of time (with a period $2\pi/\omega_1$). Unsteadiness sets in again when the frequency $\Omega_2 = \omega_2 + i\gamma_2$ turns up whose imaginary part $\gamma_2$ is positive and the corresponding real part $\omega_2$ determines then the newly appearing frequency.

The result is a quasi-periodic motion characterised by two different periods. It involves two arbitrary quantities (phases), i.e. has two degrees of freedom.

In the course of a further increase of the Reynolds number more and more new periods appear in succession, and the motion assumes an involved character typical of a developed turbulence. For every value of Re the motion has a definite number of degrees of freedom; in the limit as Re tends to infinity, the number of degrees of freedom becomes likewise infinitely large. With $n$ degrees of freedom the velocity distribution is described by an expression of the type

$$v(x, y, z, t) = \sum_{p_1, p_2, \ldots, p_n} A_{p_1, \ldots, p_n}(x, y, z) e^{-i\sum_{i=1}^{n} p_i \varphi_i} \tag{8}$$

(summation over all integral numbers $p_1$, $p_2$, ..., $p_n$) where the phases are $\varphi_i = \omega_i t + \beta_i$; it contains $n$ arbitrary initial phases $\beta_i$. The frequencies $\omega_i$ being incommensurable, it will be apparent that during a sufficiently long interval of time the fluid will pass through the states which are as close as we will it to a state set beforehand by choosing freely a set of simultaneous values for the phases $\varphi_i$. It should, of course, be borne in mind that the states whose phases differ only by a multiple of $2\pi$ are identical physically. So a turbulent motion is to a certain extent a quasi-periodical motion.

The setting-up of a turbulent regime has a somewhat different character in those exceptional cases (the Poiseuille motion and others) where the laminar motion remains stable with respect to infinitesimal perturbations, no matter how large are the Reynolds numbers. If the latter are sufficiently small, no non-stationary motion is possible here at all; a steady non-stationary motion becomes possible only after a certain value of Re is reached, which is here in the nature of a critical value. With very large Reynolds numbers, the stationary motion may, notoriously, be materialised only if one is careful enough in eliminating the perturbations superimposed upon the motion. Contrary to this, if Re is close to $Re_{cr}$, the non-stationary motion is difficult to materialise. It may be thought therefore that the true value of $Re_{cr}$, say, in the case of Poiseuille motion, lies in any case below the value generally adopted at present. As for the properties of the turbulent motion that appears here with $Re > Re_{cr}$, it should, contrary to the preceding case, enjoy from the outset a large number of degrees of freedom.

Finally, in principle, there is one more possible type of the loss of steadiness by the laminar motion; this corresponds to the case where the coefficient before $|A|^4$ in (3) is positive, so that

$$\frac{d|A|^2}{dt} = 2\gamma |A|^2 + \alpha |A|^4$$

with positive $\alpha$. If Re is somewhat smaller than $\text{Re}_{cr}$, the term of the second order is negative (since $\gamma_1 < 0$ for $\text{Re} < \text{Re}_{cr}$). But, the term of the fourth order being positive, the derivative $\overline{d|A|^2/dt}$ will become positive when the amplitude of $|A|$ is sufficiently large. This means that the motion becomes steady with respect to sufficiently large perturbations even for $\text{Re} > \text{Re}_{cr}$. Thus, this type of unsteadiness is characterised by the fact that for a certain value, $\text{Re}_{cr}$, of the Reynolds number the motion becomes unsteady with respect to infinitesimal disturbances, but even with $\text{Re} > \text{Re}_{cr}$ there is unsteadiness in response to perturbations of a finite magnitude. In this case along with the above-mentioned critical Reynolds number there should exist another, "lower" number which determines the instant of appearance of stable non-stationary solutions of the equations of motion.

120

Commun. math. Phys. 20, 167—192 (1971)
© by Springer-Verlag 1971

# On the Nature of Turbulence

DAVID RUELLE and FLORIS TAKENS*
I.H.E.S., Bures-sur-Yvette, France

Received October 5, 1970

**Abstract.** A mechanism for the generation of turbulence and related phenomena in dissipative systems is proposed.

## § 1. Introduction

If a physical system consisting of a viscous fluid (and rigid bodies) is not subjected to any external action, it will tend to a state of rest (equilibrium). We submit now the system to a steady action (pumping, heating, etc.) measured by a parameter $\mu^1$. When $\mu = 0$ the fluid is at rest. For $\mu > 0$ we obtain first a *steady state*, i.e., the physical parameters describing the fluid at any point (velocity, temperature, etc.) are constant in time, but the fluid is no longer in equilibrium. This steady situation prevails for small values of $\mu$. When $\mu$ is increased various new phenomena occur; (a) the fluid motion may remain steady but change its symmetry pattern; (b) the fluid motion may become periodic in time; (c) for sufficiently large $\mu$, the fluid motion becomes very complicated, irregular and chaotic, we have *turbulence*.

The physical phenomenon of turbulent fluid motion has received various mathematical interpretations. It has been argued by Leray [9] that it leads to a breakdown of the validity of the equations (Navier-Stokes) used to describe the system. While such a breakdown may happen we think that it does not necessarily accompany turbulence. Landau and Lifschitz [8] propose that the physical parameters $x$ describing a fluid in turbulent motion are quasi-periodic functions of time:

$$x(t) = f(\omega_1 t, \ldots, \omega_k t)$$

where $f$ has period 1 in each of its arguments separately and the frequences $\omega_1, \ldots, \omega_k$ are not rationally related [2]. It is expected that $k$ becomes large for large $\mu$, and that this leads to the complicated and irregular behaviour

* The research was supported by the Netherlands Organisation for the Advancement of Pure Research (Z.W.O.).
[1] Depending upon the situation, $\mu$ will be the Reynolds number, Rayleigh number, etc.
[2] This behaviour is actually found and discussed by E. Hopf in a model of turbulence [A mathematical example displaying features of turbulence. Commun. Pure Appl. Math. 1, 303–322 (1948)].

characteristic of turbulent motion. We shall see however that a dissipative system like a viscous fluid will not in general have quasi-periodic motions[3]. The idea of Landau and Lifschitz must therefore be modified.

Consider for definiteness a viscous incompressible fluid occupying a region $D$ of $\mathbb{R}^3$. If thermal effects can be ignored, the fluid is described by its velocity at every point of $D$. Let $H$ be the space of velocity fields $v$ over $D$; $H$ is an infinite dimensional vector space. The time evolution of a velocity field is given by the Navier-Stokes equations

$$\frac{dv}{dt} = X_\mu(v) \tag{1}$$

where $X_\mu$ is a vector field over $H$. For our present purposes it is not necessary to specify further $H$ or $X_\mu$[4].

In what follows we shall investigate the nature of the solutions of (1), making only assumptions of a very general nature on $X_\mu$. It will turn out that the fluid motion is *expected* to become chaotic when $\mu$ increases. This gives a justification for turbulence and some insight into its meaning. To study (1) we shall replace $H$ by a finite-dimensional manifold[5] and use the qualitative theory of differential equations.

For $\mu = 0$, every solution $v(\cdot)$ of (1) tends to the solution $v_0 = 0$ as the time tends to $+\infty$. For $\mu > 0$ we know very little about the vector field $X_\mu$. Therefore it is reasonable to study *generic* deformations from the situation at $\mu = 0$. In other words we shall ignore possibilities of deformation which are in some sense exceptional. This point of view could lead to serious error if, by some law of nature which we have overlooked, $X_\mu$ happens to be in a special class with exceptional properties[6]. It appears however that a three-dimensional viscous fluid conforms to the pattern of generic behaviour which we discuss below. Our discussion should in fact apply to very general dissipative systems[7].

The present paper is divided into two chapters. Chapter I is oriented towards physics and is relatively untechnical. In Section 2 we review

---

[3] Quasi-periodic motions occur for other systems, see Moser [10].

[4] A general existence and uniqueness theorem has not been proved for solutions of the Navier-Stokes equations. We assume however that we have existence and uniqueness locally, i.e., in a neighbourhood of some $v_0 \in H$ and of some time $t_0$.

[5] This replacement can in several cases be justified, see § 5.

[6] For instance the differential equations describing a Hamiltonian (conservative) system, have very special properties. The properties of a conservative system are indeed very different from the properties of a dissipative system (like a viscous fluid). If a viscous fluid is observed in an experimental setup which has a certain symmetry, it is important to take into account the invariance of $X_\mu$ under the corresponding symmetry group. This problem will be considered elsewhere.

[7] In the discussion of more specific properties, the behaviour of a viscous fluid may turn out to be nongeneric, due for instance to the local nature of the differential operator in the Navier-Stokes equations.

some results on differential equations; in Section 3–4 we apply these results to the study of the solutions of (1). Chapter II contains the proofs of several theorems used in Chapter I. In Section 5, center-manifold theory is used to replace $H$ by a finite-dimensional manifold. In Sections 6–8 the theory of Hopf bifurcation is presented both for vector fields and for diffeomorphisms. In Section 9 an example of "turbulent" attractor is presented.

*Acknowledgements.* The authors take pleasure in thanking R. Thom for valuable discussion, in particular introducing one of us (F. T.) to the Hopf bifurcation. Some inspiration for the present paper was derived from Thom's forthcoming book [12].

## Chapter I

### § 2. Qualitative Theory of Differential Equations

Let $B = \{x : |x| < R\}$ be an open ball in the finite dimensional euclidean space $H^8$. Let $X$ be a vector field with continuous derivatives up to order $r$ on $\bar{B} = \{x : |x| \leq R\}$, $r$ fixed $\geq 1$. These vector fields form a Banach space $\mathscr{B}$ with the norm

$$\|X\| = \sup_{1 \leq i \leq \nu} \sup_{|\varrho| \leq r} \sup_{x \in B} \left| \frac{\partial^{|\varrho|}}{\partial x^\varrho} X^i(x) \right|$$

where

$$\frac{\partial^{|\varrho|}}{\partial x^\varrho} = \left( \frac{\partial}{\partial x^1} \right)^{\varrho_1} \cdots \left( \frac{\partial}{\partial x^\nu} \right)^{\varrho_\nu}$$

and $|\varrho| = \varrho_1 + \cdots + \varrho_\nu$. A subset $E$ of $\mathscr{B}$ is called *residual* if it contains a countable intersection of open sets which are dense in $\mathscr{B}$. Baire's theorem implies that a residual set is again dense in $\mathscr{B}$; therefore a residual set $E$ may be considered in some sense as a "large" subset of $\mathscr{B}$. A property of a vector field $X \in \mathscr{B}$ which holds on a residual set of $\mathscr{B}$ is called *generic.*

The *integral curve* $x(\cdot)$ through $x_0 \in B$ satisfies $x(0) = x_0$ and $dx(t)/dt = X(x(t))$; it is defined at least for sufficiently small $|t|$. The dependence of $x(\cdot)$ on $x_0$ is expressed by writing $x(t) = \mathscr{D}_{X,t}(x_0)$; $\mathscr{D}_{X,\cdot}$ is called *integral* of the vector field $X$; $\mathscr{D}_{X,1}$ is the time one integral. If $x(t) \equiv x_0$, i.e. $X(x_0) = 0$, we have a *fixed point* of $X$. If $x(\tau) = x_0$ and $x(t) \neq x_0$ for $0 < t < \tau$ we have a closed orbit of period $\tau$. A natural generalization of the idea of *closed orbit* is that of *quasi-periodic* motion:

$$x(t) = f(\omega_1 t, \ldots, \omega_k t)$$

where $f$ is periodic of period 1 in each of its arguments separately and the frequencies $\omega_1, \ldots, \omega_k$ are not rationally related. We assume that $f$ is

---

[8] More generally we could use a manifold $H$ of class $C^r$.

D. Ruelle and F. Takens:

a $C^k$-function and its image a $k$-dimensional torus $T^k$ imbedded in $B$. Then however we find that a quasi-periodic motion is non-generic. In particular for $k = 2$, Peixoto's theorem[9] shows that quasi-periodic motions on a torus are in the complement of a dense open subset $\Sigma$ of the Banach space of $C^r$ vector fields on the torus: $\Sigma$ consists of vector fields for which the non wandering set $\Omega$[10] is composed of a finite number of fixed points and closed orbits only.

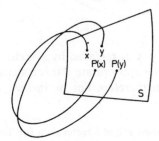

Fig. 1

As $t \to +\infty$, an integral curve $x(t)$ of the vector field $X$ may be attracted by a fixed point or a closed orbit of the vector field, or by a more general attractor[11]. It will probably not be attracted by a quasi-periodic motion because these are rare. It is however possible that the orbit be attracted by a set which is not a manifold. To visualize such a situation in $n$ dimensions, imagine that the integral curves of the vector field go roughly parallel and intersect transversally some piece of $n-1$-dimensional surface $S$ (Fig. 1). We let $P(x)$ be the first intersection of the integral curve through $x$ with $S$ ($P$ is a Poincaré map).

Take now $n - 1 = 3$, and assume that $P$ maps the solid torus $\Pi_0$ into itself as shown in Fig. 2,

$$P \Pi_0 = \Pi_1 \subset \Pi_0.$$

The set $\bigcap_{n>0} P^n \Pi_0$ is an attractor; it is locally the product of a Cantor set and a line interval (see Smale [11], Section I.9). Going back to the vector field $X$, we have thus a "strange" attractor which is locally the product of a Cantor set and a piece of two-dimensional manifold. Notice that we

---

[9] See Abraham [1].

[10] A point $x$ belongs to $\Omega$ (i.e. is non wandering) if for every neighbourhood $U$ of $x$ and every $T > 0$ one can find $t > T$ such that $\mathcal{P}_{x,t}(U) \cap U \neq \emptyset$. For a quasi-periodic motion on $T^k$ we have $\Omega = T^k$.

[11] A closed subset $A$ of the non wandering set $\Omega$ is an attractor if it has a neighbourhood $U$ such that $\bigcap_{t>0} \mathcal{P}_{x,t}(U) = A$. For more attractors than those described here see Williams [13].

124

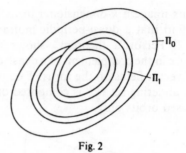

Fig. 2

keep the same picture if $X$ is replaced by a vector field $Y$ which is sufficiently close to $X$ in the appropriate Banach space. An attractor of the type just described can therefore not be thrown away as non-generic pathology.

## § 3. A Mathematical Mechanism for Turbulence

Let $X_\mu$ be a vector field depending on a parameter $\mu$[12]. The assumptions are the same as in Section 2, but the interpretation we have in mind is that $X_\mu$ is the right-hand side of the Navier-Stokes equations. When $\mu$ varies the vector field $X_\mu$ may change in a number of manners. Here we shall describe a pattern of changes which is physically acceptable, and show that it leads to something like turbulence.

For $\mu = 0$, the equation

$$\frac{dx}{dt} = X_\mu(x)$$

has the solution $x = 0$. We assume that the eigenvalues of the Jacobian matrix $A_k^j$ defined by

$$A_k^j = \frac{\partial X_0^j}{\partial x^k}(0)$$

have all strictly negative real parts; this corresponds to the fact that the fixed point 0 is attracting. The Jacobian determinant is not zero and therefore there exists (by the implicit function theorem) $\xi(\mu)$ depending continuously on $\mu$ and such that

$$X_\mu(\xi(\mu)) = 0 .$$

In the hydrodynamical picture, $\xi(\mu)$ describes a steady state.

We follow now $\xi(\mu)$ as $\mu$ increases. For sufficiently small $\mu$ the Jacobian matrix $A_k^j(\mu)$ defined by

$$A_k^j(\mu) = \frac{\partial X_\mu^j}{\partial x^k}(\xi(\mu)) \tag{2}$$

[12] To be definite, let $(x, \mu) \to X_\mu(x)$ be of class $C^r$.

172                           D. Ruelle and F. Takens:

has only eigenvalues with strictly negative real parts (by continuity).
We assume that, as $\mu$ increases, successive pairs of complex conjugate
eigenvalues of (2) cross the imaginary axis, for $\mu = \mu_1, \mu_2, \mu_3, \ldots$[13]. For
$\mu > \mu_1$, the fixed point $\xi(\mu)$ is no longer attracting. It has been shown by
Hopf[14] that when a pair of complex conjugate eigenvalues of (2) cross the
imaginary axis at $\mu_i$, there is a one-parameter family of closed orbits of the
vector field in a neighbourhood of $(\xi(\mu_i), \mu_i)$. More precisely there are
continuous functions $y(\omega)$, $\mu(\omega)$ defined for $0 \leqq \omega < 1$ such that

    (a) $y(0) = \xi(\mu_i)$, $\mu(0) = \mu_i$,

    (b) the integral curve of $X_{\mu(\omega)}$ through $y(\omega)$ is a closed orbit for $\omega > 0$.

Generically $\mu(\omega) > \mu_i$ or $\mu(\omega) < \mu_i$ for $\omega \neq 0$. To see how the closed
orbits are obtained we look at the two-dimensional situation in a
neighbourhood of $\xi(\mu_1)$ for $\mu < \mu_1$ (Fig. 3) and $\mu > \mu_1$ (Fig. 4). Suppose
that when $\mu$ crosses $\mu_1$ the vector field remains like that of Fig. 3 at large
distances of $\xi(\mu)$; we get a closed orbit as shown in Fig. 5. Notice that
Fig. 4 corresponds to $\mu > \mu_1$ and that the closed orbit is attracting.
Generally we shall assume that the closed orbits appear for $\mu > \mu_i$ so
that the vector field at large distances of $\xi(\mu)$ remains attracting in accordance with physics. As $\mu$ crosses we have then replacement of an
attracting fixed point by an attracting closed orbit. The closed orbit is
physically interpreted as a periodic motion, its amplitude increases
with $\mu$.

      Figs. 3 and 4                              Fig. 5

### § 3a) Study of a Nearly Split Situation

To see what happens when $\mu$ crosses the successive $\mu_i$, we let $E_i$ be
the two-dimensional linear space associated with the $i$-th pair of eigenvalues of the Jacobian matrix. In first approximation the vector field
$X_\mu$ is, near $\xi(\mu)$, of the form

$$\tilde{X}_\mu(x) = \tilde{X}_{\mu 1}(x_1) + \tilde{X}_{\mu 2}(x_2) + \cdots \tag{3}$$

---

[13] Another less interesting possibility is that a real eigenvalue vanishes. When this
happens the fixed point $\xi(\lambda)$ generically coalesces with another fixed point and disappears
(this generic behaviour is changed if some symmetry is imposed to the vector field $X_\mu$).

[14] Hopf [6] assumes that $X$ is real-analytic; the differentiable case is treated in Section 6
of the present paper.

where $\tilde{X}_{\mu i}$, $x_i$ are the components of $\tilde{X}_\mu$ and $x$ in $E_i$. If $\mu$ is in the interval $(\mu_k, \mu_{k+1})$, the vector field $\tilde{X}_\mu$ leaves invariant a set $\tilde{T}^k$ which is the cartesian product of $k$ attracting closed orbits $\Gamma_1, ..., \Gamma_k$ in the spaces $E_1, ..., E_k$. By suitable choice of coordinates on $\tilde{T}^k$ we find that the motion defined by the vector field on $\tilde{T}^k$ is quasi-periodic (the frequencies $\tilde{\omega}_1, ..., \tilde{\omega}_k$ of the closed orbits in $E_1, ..., E_k$ are in general not rationally related).

Replacing $\tilde{X}_\mu$ by $X_\mu$ is a perturbation. We assume that this perturbation is small, i.e. we assume that $X_\mu$ nearly splits according to (3). In this case there exists a $C^r$ manifold (torus) $T^k$ close to $\tilde{T}^k$ which is invariant for $X_\mu$ and attracting [15]. The condition that $X_\mu - \tilde{X}_\mu$ be small depends on how attracting the closed orbits $\Gamma_1, ..., \Gamma_k$ are for the vector field $\tilde{X}_{\mu 1}, ..., \tilde{X}_{\mu k}$; therefore the condition is violated if $\mu$ becomes too close to one of the $\mu_i$.

We consider now the vector field $X_\mu$ restricted to $T^k$. For reasons already discussed, we do not expect that the motion will remain quasi-periodic. If $k = 2$, Peixoto's theorem implies that generically the non-wandering set of $T^2$ consists of a finite number of fixed points and closed orbits. What will happen in the case which we consider is that there will be one (or a few) attracting closed orbits with frequencies $\omega_1$, $\omega_2$ such that $\omega_1/\omega_2$ goes continuously through rational values.

Let $k > 2$. In that case, the vector fields on $T^k$ for which the non-wandering set consists of a finite number of fixed points and closed orbits are no longer dense in the appropriate Banach space. Other possibilities are realized which correspond to a more complicated orbit structure; "strange" attractors appear like the one presented at the end of Section 2. Taking the case of $T^4$ and the $C^3$-topology we shall show in Section 9 that in any neighbourhood of a quasi-periodic $\tilde{X}$ there is an open set of vector fields with a strange attractor.

We propose to say that the motion of a fluid system is turbulent when this motion is described by an integral curve of the vector field $X_\mu$ which tends to a set $A$ [16], and $A$ is neither empty nor a fixed point nor a closed orbit. In this definition we disregard nongeneric possibilities (like $A$ having the shape of the figure 8, etc.). This proposal is based on two things:

(a) We have shown that, when $\mu$ increases, it is not unlikely that an attractor $A$ will appear which is neither a point nor a closed orbit.

---

[15] This follows from Kelley [7], Theorem 4 and Theorem 5, and also from recent work of Pugh (unpublished). That $T^k$ is attracting means that it has a neighbourhood $U$ such that $\bigcap_{t>0} \mathcal{D}_{X,t}(U) = T^k$. We cannot call $T^k$ an attractor because it need not consist of non-wandering points.

[16] More precisely $A$ is the $\omega^+$ limit set of the integral curve $x(\cdot)$, i.e., the set of points $\xi$ such that there exists a sequence $(t_n)$ and $t_n \to \infty$, $x(t_n) \to \xi$.

(b) In the known generic examples where $A$ is not a point or a closed orbit, the structure of the integral curves on or near $A$ is complicated and erratic (see Smale [11] and Williams [13]).

We shall further discuss the above definition of turbulent motion in Section 4.

### § 3 b) Bifurcations of a Closed Orbit

We have seen above how an attracting fixed point of $X_\mu$ may be replaced by an attracting closed orbit $\gamma_\mu$ when the parameter crosses the value $\mu_1$ (Hopf bifurcation). We consider now in some detail the next bifurcation; we assume that it occurs at the value $\mu'$ of the parameter [17] and that $\lim_{\mu \to \mu'} \gamma_\mu$ is a closed orbit $\gamma_{\mu'}$ of $X_{\mu'}$. [18]

Let $\Phi_\mu$ be the Poincaré map associated with a piece of hypersurface $S$ transversal to $\gamma_\mu$, for $\mu \in (\mu_1, \mu']$. Since $\gamma_\mu$ is attracting, $p_\mu = S \cap \gamma_\mu$ is an attracting fixed point of $\Phi_\mu$ for $\mu \in (\mu_1, \mu')$. The derivative $d\Phi_\mu(p_\mu)$ of $\Phi_\mu$ at the point $p_\mu$ is a linear map of the tangent hyperplane to $S$ at $p_\mu$ to itself.

We assume that the spectrum of $d\Phi_{\mu'}(p_{\mu'})$ consists of a finite number of isolated eigenvalues of absolute value 1, and a part which is contained in the open unit disc $\{z \in \mathbb{C} \mid |z| < 1\}$ [19]. According to § 5, Remark (5.6), we may assume that $S$ is finite dimensional. With this assumption one can say rather precisely what kind of generic bifurcations are possible for $\mu = \mu'$. We shall describe these bifurcations by indicating what kind of attracting subsets for $X_\mu$ (or $\Phi_\mu$) there are near $\gamma_{\mu'}$ (or $p_{\mu'}$) when $\mu > \mu'$.

Generically, the set $E$ of eigenvalues of $d\Phi_{\mu'}(p_{\mu'})$, with absolute value 1, is of one of the following types:
1. $E = \{+1\}$,
2. $E = \{-1\}$,
3. $E = \{\alpha, \bar{\alpha}\}$ where $\alpha, \bar{\alpha}$ are distinct.

For the cases 1 and 2 we can refer to Brunovsky [3]. In fact in case 1 the attracting closed orbit disappears (together with a hyperbolic closed orbit); for $\mu > \mu'$ there is no attractor of $X_\mu$ near $\gamma_{\mu'}$. In case 2 there is for $\mu > \mu'$ (or $\mu < \mu'$) an attracting (resp. hyperbolic) closed orbit near $\gamma_{\mu'}$, but the period is doubled.

If we have case 3 then $\Phi_\mu$ has also for $\mu$ slightly bigger than $\mu'$ a fixed point $p_\mu$; generically the conditions (a)', ..., (e) in Theorem (7.2) are

---

[17] In general $\mu'$ will differ from the value $\mu_2$ introduced in § 3a).

[18] There are also other possibilities: If $\gamma_\mu$ tends to a point we have a Hopf bifurcation with parameter reversed. The cases where $\lim_{\mu \to \mu'} \gamma_\mu$ is not compact or where the period of $\gamma_\mu$ tends to $\infty$ are not well understood; they may or may not give rise to turbulence.

[19] If the spectrum of $d\Phi_{\mu'}(p_{\mu'})$ is discrete, this is a reasonable assumption, because for $\mu_1 < \mu < \mu'$ the spectrum is contained in the open unit disc.

satisfied. One then concludes that when $\gamma_{\mu'}$ is a "vague attractor" (i.e. when the condition (f) is satisfied) then, for $\mu > \mu'$, there is an attracting circle for $\Phi_\mu$; this amounts to the existence of an invariant and attracting torus $T^2$ for $X_\mu$. If $\gamma_{\mu'}$ is not a "vague attractor" then, generically, $X_\mu$ has no attracting set near $\gamma_{\mu'}$ for $\mu > \mu'$.

## § 4. Some Remarks on the Definition of Turbulence

We conclude this discussion by a number of remarks:

1. The concept of genericity based on residual sets may not be the appropriate one from the physical view point. In fact the complement of a residual set of the $\mu$-axis need not have Lebesgue measure zero. In particular the quasi-periodic motions which we had eliminated may in fact occupy a part of the $\mu$-axis with non vanishing Lebesgue measure [20]. These quasi-periodic motions would be considered turbulent by our definition, but the "turbulence" would be weak for small $k$. There are arguments to define the quasi-periodic motions, along with the periodic ones, as non turbulent (see (4) below).

2. By our definition, a periodic motion (= closed orbit of $X_\mu$) is not turbulent. It may however be very complicated and appear turbulent (think of a periodic motion closely approximating a quasi-periodic one, see § 3 b) second footnote).

3. We have shown that, under suitable conditions, there is an attracting torus $T^k$ for $X_\mu$ if $\mu$ is between $\mu_k$ and $\mu_{k+1}$. We assumed in the proof that $\mu$ was not too close to $\mu_k$ or $\mu_{k+1}$. In fact the transition from $T^1$ to $T^2$ is described in Section 3b, but the transition from $T^k$ to $T^{k+1}$ appears to be a complicated affair when $k > 1$. In general, one gets the impression that the situations not covered by our description are more complicated, hard to describe, and probably turbulent.

4. An interesting situation arises when statistical properties of the motion can be obtained, via the pointwise ergodic theorem, from an ergodic measure $m$ supported by the attracting set $A$. An observable quantity for the physical system at a time $t$ is given by a function $x_t$ on $H$, and its expectation value is $m(x_t) = m(x_0)$. If $m$ is "mixing" the time correlation functions $m(x_t y_0) - m(x_0) m(y_0)$ tend to zero as $t \to \infty$. This situation appears to prevail in turbulence, and "pseudo random" variables with correlation functions tending to zero at infinity have been studied by Bass [21]. With respect to this property of time correlation functions the quasi-periodic motions should be classified as non turbulent.

---

[20] On the torus $T^2$, the rotation number $\omega$ is a continuous function of $\mu$. Suppose one could prove that, on some $\mu$-interval, $\omega$ is non constant and is absolutely continuous with respect to Lebesgue measure; then $\omega$ would take irrational values on a set of non zero Lebesgue measure.

[21] See for instance [2].

5. In the above analysis the detailed structure of the equations describing a viscous fluid has been totally disregarded. Of course something is known of this structure, and also of the experimental conditions under which turbulence develops, and a theory should be obtained in which these things are taken into account.

6. Besides viscous fluids, other dissipative systems may exhibit time-periodicity and possibly more complicated time dependence; this appears to be the case for some chemical systems [22].

## Chapter II

### § 5. Reduction to Two Dimensions

**Definition (5.1).** Let $\Phi : H \to H$ be a $C^1$ map with fixed point $p \in H$, where $H$ is a Hilbert space. The spectrum of $\Phi$ at $p$ is the spectrum of the induced map $(d\Phi)_p : T_p(H) \to T_p(H)$.

Let $X$ be a $C^1$ vectorfield on $H$ which is zero in $p \in H$. For each $t$ we then have $d(\mathscr{D}_{X,t})_p : T_p(H) \to T_p(H)$, induced by the time $t$ integral of $X$. Let $L(X) : T_p(H) \to T_p(H)$ be the unique continuous linear map such that $d(\mathscr{D}_{X,t})_p = e^{t \cdot L(X)}$.

We define the spectrum of $X$ at $p$ to be the spectrum of $L(X)$, (note that $L(X)$ also can be obtained by linearizing $X$).

**Proposition (5.2).** *Let $X_\mu$ be a one-parameter family of $C^k$ vectorfields on a Hilbert space $H$ such that also $X$, defined by $X(h, \mu) = (X_\mu(h), 0)$, on $H \times \mathbb{R}$ is $C^k$. Suppose:*

(a) *$X_\mu$ is zero in the origin of $H$.*

(b) *For $\mu < 0$ the spectrum of $X_\mu$ in the origin is contained in $\{z \in \mathbb{C} \mid \mathrm{Re}(z) < 0\}$.*

(c) *For $\mu = 0$, resp. $\mu > 0$, the spectrum of $X_\mu$ at the origin has two isolated eigenvalues $\lambda(\mu)$ and $\overline{\lambda(\mu)}$ with multiplicity one and $\mathrm{Re}(\lambda(\mu)) = 0$, resp. $\mathrm{Re}(\lambda(\mu)) > 0$. The remaining part of the spectrum is contained in $\{z \in \mathbb{C} \mid \mathrm{Re}(z) < 0\}$.*

*Then there is a (small) 3-dimensional $C^k$-manifold $\tilde{V}^c$ of $H \times \mathbb{R}$ containing $(0, 0)$ such that:*

1. *$\tilde{V}^c$ is locally invariant under the action of the vectorfield $X$ ($X$ is defined by $X(h, \mu) = (X_\mu(h), 0)$); locally invariant means that there is a neighbourhood $U$ of $(0, 0)$ such that for $|t| \leq 1$, $\tilde{V}^c \cap U = \mathscr{D}_{X,t}(\tilde{V}^c) \cap U$.*

2. *There is a neighbourhood $U'$ of $(0, 0)$ such that if $p \in U'$, is recurrent, and has the property that $\mathscr{D}_{X,t}(p) \in U'$ for all $t$, then $p \in \tilde{V}^c$*

3. *in $(0, 0)$ $\tilde{V}^c$ is tangent to the $\mu$ axis and to the eigenspace of $\lambda(0), \overline{\lambda(0)}$.*

---

[22] See Pye, K., Chance, B.: Sustained sinusoidal oscillations of reduced pyridine nucleotide in a cell-free extract of Saccharomyces carlbergensis. Proc. Nat. Acad. Sci. U.S.A. **55**, 888–894 (1966).

*Proof.* We construct the following splitting $T_{(0,0)}(H \times \mathbb{R}) = V^c \oplus V^s$: $V^c$ is tangent to the $\mu$ axis and contains the eigenspace of $\lambda(\mu)$, $\overline{\lambda(\mu)}$; $V^s$ is the eigenspace corresponding to the remaining (compact) part of the spectrum of $L(X)$. Because this remaining part is compact there is a $\delta > 0$ such that it is contained in $\{z \in \mathbb{C} \mid \mathrm{Re}(z) < -\delta\}$. We can now apply the centermanifold theorem [5], the proof of which generalizes to the case of a Hilbert space, to obtain $\tilde{V}^c$ as the centermanifold of $X$ at $(0,0)$ [by assumption $X$ is $C^k$, so $\tilde{V}^c$ is $C^k$; if we would assume only that, for each $\mu$, $X_\mu$ is $C^k$ (and $X$ only $C^1$), then $\tilde{V}^c$ would be $C^1$ but, for each $\mu_0$, $\tilde{V}^c \cap \{\mu = \mu_0\}$ would be $C^k$].

For positive $t$, $d(\mathcal{D}_{X,t})_{0,0}$ induces a contraction on $V^s$ (the spectrum is contained in $\{z \in \mathbb{C} \mid |z| < e^{-\delta t}\}$). Hence there is a neighbourhood $U'$ of $(0,0)$ such that

$$U' \cap \left[ \bigcap_{t=1}^{\infty} \mathcal{D}_{X,t}(U') \right] \subset (U' \cap \tilde{V}^c).$$

Now suppose that $p \in U'$ is recurrent and that $\mathcal{D}_{X,t}(p) \in U'$ for all $t$. Then given $\varepsilon > 0$ and $N > 0$ there is a $t > N$ such that the distance between $p$ and $\mathcal{D}_{X,t}(p)$ is $< \varepsilon$. It then follows that $p \in (U' \cap \tilde{V}^c) \subset \tilde{V}^c$ for $U'$ small enough. This proves the proposition.

*Remark (5.3).* The analogous proposition for a one parameter set of diffeomorphisms $\Phi_\mu$ is proved in the same way. The assumptions are then:

(a)' The origin is a fixed point of $\Phi_\mu$.

(b)' For $\mu < 0$ the spectrum of $\Phi_\mu$ at the origin is contained in $\{z \in \mathbb{C} \mid |z| < 1\}$.

(c)' For $\mu = 0$ resp. $\mu > 0$ the spectrum of $\Phi_\mu$ at the origin has two isolated eigenvalues $\lambda(\mu)$ and $\overline{\lambda(\mu)}$ with multiplicity one and $|\lambda(\mu)| = 1$ resp. $|\lambda(\mu)| > 1$. The remaining part of the spectrum is contained in $\{z \in \mathbb{C} \mid |z| < 1\}$.

One obtains just as in Proposition (5.2) a 3-dimensional center manifold which contains all the local recurrence.

*Remark (5.4).* If we restrict the vectorfield $X$, or the diffeomorphism $\Phi$ [defined by $\Phi(h, \mu) = (\Phi_\mu(h), \mu)$], to the 3-dimensional manifold $\tilde{V}^c$ we have locally the same as in the assumptions (a), (b), (c), or (a)', (b)', (c)' where now the Hilbert space has dimension 2. So if we want to prove a property of the local recurrent points for a one parameter family of vectorfield, or diffeomorphisms, satisfying (a) (b) and (c), or (a)', (b)' and (c)', it is enough to prove it for the case where $\dim(H) = 2$.

*Remark (5.5).* Everything in this section holds also if we replace our Hilbert space by a Banach space with $C^k$-norm; a Banach space $B$ has $C^k$-norm if the map $x \to \|x\|$, $x \in B$ is $C^k$ except at the origin. This $C^k$-norm is needed in the proof of the center manifold theorem.

*Remark (5.6).* The Propositions (5.2) and (5.3) remain true if
1. we drop the assumptions on the spectrum of $X_\mu$ resp. $\Phi_\mu$ for $\mu > 0$.
2. we allow the spectrum of $X_\varrho$ resp. $\Phi_\varrho$ to have an arbitrary but finite number of isolated eigenvalues on the real axis resp. the unit circle. The dimension of the invariant manifold $\tilde{V}^c$ is then equal to that number of eigenvalues plus one.

## § 6. The Hopf Bifurcation

We consider a one parameter family $X_\mu$ of $C^k$-vectorfields on $\mathbb{R}^2$, $k \geq 5$, as in the assumption of proposition (5.2) (with $\mathbb{R}^2$ instead of $H$); $\lambda(\mu)$ and $\overline{\lambda(\mu)}$ are the eigenvalues of $X_\mu$ in $(0, 0)$. Notice that with a suitable change of coordinates we can achieve $X_\mu = (\text{Re}\lambda(\mu)x_1 + \text{Im}\lambda(\mu)x_2)\dfrac{\partial}{\partial x_1}$
$+ (-\text{Im}\lambda(\mu)x_1 + \text{Re}\lambda(\mu)x_2)\dfrac{\partial}{\partial x_2}$ + terms of higher order.

**Theorem (6.1).**(Hopf [6]). *If* $\left(\dfrac{d(\lambda(\mu))}{d\mu}\right)_{\mu = 0}$ *has a positive real part, and if* $\lambda(0) \neq 0$, *then there is a one-parameter family of closed orbits of* $X(=(X_\mu, 0))$ *on* $\mathbb{R}^3 = \mathbb{R}^2 \times \mathbb{R}^1$ *near* $(0, 0, 0)$ *with period near* $\dfrac{2\pi}{|\lambda(0)|}$; *there is a neighbourhood* $U$ *of* $(0, 0, 0)$ *in* $\mathbb{R}^3$ *such that each closed orbit of* $X$, *which is contained in* $U$, *is a member of the above family.*

*If* $(0, 0)$ *is a "vague attractor" (to be defined later) for* $X_0$, *then this one-parameter family is contained in* $\{\mu > 0\}$ *and the orbits are of attracting type.*

*Proof.* We first have to state and prove a lemma on polar-coordinates:

**Lemma (6.2).** *Let* $X$ *be a* $C^k$ *vectorfield on* $\mathbb{R}^2$ *and let* $X(0, 0) = 0$. *Define polar coordinates by the map* $\Psi : \mathbb{R}^2 \to \mathbb{R}^2$, *with* $\Psi(r, \varphi) = (r\cos\varphi, r\sin\varphi)$. *Then there is a unique* $C^{k-2}$-*vectorfield* $\tilde{X}$ *on* $\mathbb{R}^2$, *such that* $\Psi_*(\tilde{X}) = X$ *(i.e. for each* $(r, \varphi)$ $d\Psi(\tilde{X}(r, \varphi)) = X(r\cos\varphi, r\sin\varphi))$.

*Proof of Lemma (6.2).* We can write

$$
\begin{aligned}
X &= X_1 \frac{\partial}{\partial x_1} + X_2 \frac{\partial}{\partial x_2} \\
&= \frac{x_1 X_1 + x_2 X_2}{\sqrt{x_1^2 + x_2^2}} \left( \frac{1}{\sqrt{x_1^2 + x_2^2}} \left( x_1 \frac{\partial}{\partial x_1} + x_2 \frac{\partial}{\partial x_2} \right) \right) \\
&\quad + \frac{(-x_2 X_1 + x_1 X_2)}{(x_1^2 + x_2^2)} \left( -x_2 \frac{\partial}{\partial x_1} + x_1 \frac{\partial}{\partial x_2} \right) \\
&= \frac{f_r(x_1, x_2)}{r} \cdot \Psi_*(\tilde{Z}_r) + \frac{f_\varphi(x_1, x_2)}{r^2} \Psi_*(\tilde{Z}_\varphi).
\end{aligned}
$$

Where $\tilde{Z}_r\left(=\dfrac{\partial}{\partial r}\right)$ and $\tilde{Z}_\varphi\left(=\dfrac{\partial}{\partial\varphi}\right)$ are the "coordinate vectorfields" with respect to $(r,\varphi)$ and $r=\pm\sqrt{x_1^2+x_2^2}$. (Note that $r$ and $\Psi_*(\tilde{Z}_r)$ are bi-valued.)

Now we consider the functions $\Psi^*(f_r)=f_r\circ\Psi$ and $\Psi^*(f_\varphi)$. They are zero along $\{r=0\}$; this also holds for $\dfrac{\partial}{\partial r}(\Psi^*(f_r))$ and $\dfrac{\partial}{\partial r}(\Psi^*(f_\varphi))$.

By the division theorem $\dfrac{\Psi^*(f_r)}{r}$, resp. $\dfrac{\Psi^*(f_\varphi)}{r^2}$, are $C^{k-1}$ resp. $C^{k-2}$.

We can now take $\tilde{X}=\dfrac{\Psi^*(f_r)}{r}\tilde{Z}_r+\dfrac{\Psi^*(f_\varphi)}{r^2}\tilde{Z}_\varphi$; the uniqueness is evident.

**Definition (6.3).** We define a Poincaré map $P_X$ for a vectorfield $X$ as in the assumptions of Theorem (6.1):

$P_X$ is a map from $\{(x_1,x_2,\mu)\mid |x_1|<\varepsilon,\ x_2=0,\ |\mu|\le\mu_0\}$ to the $(x_1,\mu)$ plane; $\mu_0$ is such that $\operatorname{Im}(\lambda(\mu))\neq 0$ for $|\mu|\le\mu_0$; $\varepsilon$ is sufficiently small. $P_X$ maps $(x_1,x_2,\mu)$ to the first intersection point of $\mathcal{D}_{X,t}(x_1,x_2,\mu)$, $t>0$, with the $(x_1,\mu)$ plane, for which the sign of $x_1$ and the $x_1$ coordinate of $\mathcal{D}_{X,t}(x_1,x_2,\mu)$ are the same.

*Remark (6.4).* $P_X$ preserves the $\mu$ coordinate. In a plane $\mu=$ constant the map $P_X$ is illustrated in the following figure $\operatorname{Im}(\lambda(u))\neq 0$ means that

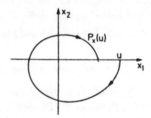

Fig. 6. Integral curve of $X$ at $\mu=$ constant

$X$ has a "non vanishing rotation"; it is then clear that $P_X$ is defined for $\varepsilon$ small enough.

*Remark (6.5).* It follows easily from Lemma (6.2) that $P_X$ is $C^{k-2}$. We define a *displacement function* $V(x_1,\mu)$ on the domain of $P_X$ as follows:

$$P_X(x_1,0,\mu)=(x_1+V(x_1,\mu),0,\mu);\qquad V\text{ is }C^{k-2}.$$

This displacement function has the following properties:

(i) $V$ is zero on $\{x_1=0\}$; the other zeroes of $V$ occur in pairs (of opposite sign), each pair corresponds to a closed orbit of $X$. If a closed orbit $\gamma$ of $X$ is contained in a sufficiently small neighbourhood of $(0,0)$, and intersects $\{x_1 = 0\}$ only twice then $V$ has a corresponding pair of zeroes (namely the two points $\gamma \cap (\text{domain of } P_x)$).

(ii) For $\mu < 0$ and $x_1 = 0$, $\dfrac{\partial V}{\partial x_1} < 0$; for $\mu > 0$ and $x_1 = 0$, $\dfrac{\partial V}{\partial x_1} > 0$

and for $\mu = 0$ and $x = 0$, $\dfrac{\partial^2 V}{\partial \mu \, \partial x_1} > 0$. This follows from the assumptions

on $\lambda(\mu)$. Hence, again by the division theorem, $\tilde{V} = \dfrac{V}{x_1}$ is $C^{k-3}$. $\tilde{V}(0,0)$

is zero, $\dfrac{\partial \tilde{V}}{\partial \mu} > 0$, so there is locally a unique $C^{k-3}$-curve $l$ of zeroes of $\tilde{V}$ passing through $(0,0)$. Locally the set of zeroes of $V$ is the union of $l$ and $\{x_1 = 0\}$. $l$ induces the one-parameter family of closed orbits.

(iii) Let us say that $(0,0)$ is a "vague attractor" for $X_0$ if $V(x_1, 0) = -Ax_1^3 + $ terms of order $> 3$ with $A > 0$. This means that the 3rd order terms of $X_0$ make the flow attract to $(0,0)$. In that case $\tilde{V} = \alpha_1 \mu - Ax_1^2 + $ terms of higher order, with $\alpha_1$ and $A > 0$, so $\tilde{V}(x_1, \mu)$ vanishes only if $x_1 = 0$ or $\mu > 0$. This proves that the one-parameter family is contained in $\{\mu > 0\}$.

(iv) The following holds in a neighbourhood of $(0,0,0)$ where $\dfrac{\partial V}{\partial x_1} > -1$.

If $V(x_1, \mu) = 0$ and $\left(\dfrac{\partial V}{\partial x_1}\right)_{(x_1, \mu)} < 0$, then the closed orbit which cuts the domain of $P_x$ in $(x_1, \mu)$ is an attractor of $X_\mu$. This follows from the fact that $(x_1, \mu)$ is a fixed point of $P_x$ and the fact that the derivative of $P_x$ in $(x_1, \mu)$, restricted to this $\mu$ level, is smaller than 1 (in absolute value).

Combining (iii) and (iv) it follows easily that, if $(0,0)$ is a vague attractor, the closed orbits of our one parameter family are, near $(0,0)$, of the attracting type.

Finally we have to show that, for some neighbourhood $U$ of $(0,0)$, every closed orbit of $X$, which is contained in $U$, is a member of our family of closed orbits. We can make $U$ so small that every closed orbit $\gamma$ of $X$, which is contained in $U$, intersects the domain of $P_x$.

Let $p = (x_1(\gamma), 0, \mu(\gamma))$ be an intersection point of a closed orbit $\gamma$ with the domain of $P_x$. We may also assume that $U$ is so small that $P_x[U \cap (\text{domain of } P_x)] \subset (\text{domain of } P_x)$. Then $P_x(p)$ is in the domain of $P_x$ but also $P_x(p) \subset U$ so $(P_x)^2(p)$ is defined etc.; so $P_x^i(p)$ is defined.

Restricted to $\{\mu = \mu(\gamma)\}$, $P_x$ is a local diffeomorphism of a segment of the half line $(x_1 \geq 0$ or $x_1 \leq 0$, $x_2 = 0$, $\mu = \mu(\gamma))$ into that half line.

If the $x_1$ coordinate of $P_x^i(p)$ is $<$ (resp. $>$) than $x_1(\gamma)$ then the $x_1$ coordinate of $P_x^{i+1}(p)$ is $<$ (resp. $>$) than the $x_1$ coordinate of $P_x^i(p)$, so $p$ does not lie on a closed orbit. Hence we must assume that the $x_1$ coordinate of $P_X(p)$ is $x_1(\gamma)$, hence $p$ is a fixed point of $P_X$, hence $p$ is a zero of $V$, so, by property (ii), $\gamma$ is a member of our one parameter family of closed orbits.

## § 7. Hopf Bifurcation for Diffeomorphisms[*]

We consider now a one parameter family $\Phi_\mu : \mathbb{R}^2 \to \mathbb{R}^2$ of diffeomorphisms satisfying (a)′, (b)′ and (c)′ (Remark (5.3)) and such that:

(d) $\dfrac{d}{d\mu}(|\lambda(\mu)|)_{\mu=0} > 0$.

Such a diffeomorphism can for example occur as the time one integral of a vectorfield $X_\mu$ as we studied in Section 2. In this diffeomorphism case we shall of course not find any closed (circular) orbit (the orbits are not continuous) but nevertheless we shall, under rather general conditions, find, near $(0,0)$ and for $\mu$ small, a one parameter family of invariant circles.

We first bring $\Phi_\mu$, by coordinate transformations, into a simple form:
We change the $\mu$ coordinate in order to obtain
(d)′ $|\lambda(\mu)| = 1 + \mu$.

After an appropriate ($\mu$ dependent) coordinate change of $\mathbb{R}^2$ we then have $\Phi(r, \varphi, \mu) = ((1 + \mu)r, \varphi + f(\mu), \mu) +$ terms of order $r^2$, where $x_1 = r \cos\varphi$ and $x_2 = r \sin\varphi$; "$\Phi = \Phi' +$ terms of order $r^l$" means that the derivatives of $\Phi$ and $\Phi'$ up to order $l - 1$ with respect to $(x_1, x_2)$ agree for $(x_1, x_2) = (0, 0)$.

We now put in one extra condition:

(e) $f(0) \neq \dfrac{k}{l} \cdot 2\pi$ for all $k, l \leq 5$.

**Proposition (7.1).** *Suppose $\Phi_\mu$ satisfies (a)′, (b)′, (c)′, (d)′ and (e) and is $C^k$, $k \geq 5$. Then for $\mu$ near 0, by a $\mu$ dependent coordinate change in $\mathbb{R}^2$, one can bring $\Phi_\mu$ in the following form:*

$$\Phi_\mu(r, \varphi) = ((1 + \mu)r - f_1(\mu) \cdot r^3, \ \varphi + f_2(\mu) + f_3(\mu) \cdot r^2) + \text{terms of order } r^5.$$

*For each $\mu$, the coordinate transformation of $\mathbb{R}^2$ is $C^\infty$; the induced coordinate transformation on $\mathbb{R}^2 \times \mathbb{R}$ is only $C^{k-4}$.*

The next paragraph is devoted to the proof of this proposition[**]. Our last condition on $\Phi_\mu$ is:

---

[*] *Note added in proof.* J. Moser kindly informed us that the Hopf bifurcation for diffeomorphisms had been worked out by Neumark (reference not available) and R. Sacker (Thesis, unpublished). An example of "decay" (loss of differentiability) of $T^2$ under perturbations has been studied by N. Levinson [a second order differentiable equation with singular solutions. Ann. of Math. **50**, 127–153 (1949)].

[**] *Note added in proof.* The desired normal form can also be obtained from § 21 of C. L. Siegel, Vorlesungen über Himmelsmechanik, Springer, Berlin, 1956 (we thank R. Jost for emphasizing this point).

182                                D. Ruelle and F. Takens:

(f) $f_1(0) \neq 0$. We assume even that $f_1(0) > 0$ (this corresponds to the case of a vague attractor for $\mu = 0$, see Section 6); the case $f_1(0) < 0$ can be treated in the same way (by considering $\Phi_{-\mu}^{-1}$ instead of $\Phi_\mu$).

*Notation.* We shall use $N\Phi_\mu$ to denote the map

$$(r, \varphi) \to ((1 + \mu)r - f_1(\mu) \cdot r^3, \ \varphi + f_2(\mu) + f_3(\mu) \cdot r^2)$$

and call this "the simplified $\Phi_\mu$".

**Theorem (7.2).** *Suppose $\Phi_\mu$ is at least $C^5$ and satisfies (a)', (b)', (c)', (d)' and (e) and $N\Phi_\mu$, the simplified $\Phi_\mu$, satisfies (f). Then there is a continuous one parameter family of invariant attracting circles of $\Phi_\mu$, one for each $\mu \in (0, \varepsilon)$, for $\varepsilon$ small enough.*

*Proof.* The idea of the proof is as follows: the set $\Sigma = \{\mu = f_1(\mu) \cdot r^2\}$ in $(r, q, \mu)$-space is invariant under $N\Phi$; $N\Phi$ even "attracts to this set". This attraction makes $\Sigma$ stable in the following sense: $\{\Phi^n(\Sigma)\}_{n=0}^\infty$ is a sequence of manifolds which converges (for $\mu$ small) to an invariant manifold (this is actually what we have to prove); the method of the proof is similar to the methods used in [4, 5].

First we define $U_\delta = \left\{(r, \varphi, \mu) \mid r \neq 0 \text{ and } \dfrac{\mu}{r^2} \in [f_1(\mu) - \delta, \ f_1(\mu) + \delta]\right\}$, $\delta \ll f_1(\mu)$, and show that $N\Phi(U_\delta) \subset U_\delta$ and also, in a neighbourhood of $(0, 0, 0)$, $\Phi(U_\delta) \subset U_\delta$. This goes as follows:

If $p \in \partial U_\delta$, and $r(p)$ is the $r$-coordinate of $p$, then the $r$-coordinate of $N\Phi(p)$ is $r(p) \pm \delta \cdot (r(p))^3$ and $p$ goes towards the interior of $U_\delta$. Because $\Phi$ equals $N\Phi$, modulo terms of order $r^5$, also, locally, $\Phi(U_\delta) \subset U_\delta$. From this it follows that, for $\varepsilon$ small enough and all $n \geq 0$ $\Phi^n(\Sigma_\varepsilon) \subset U_\delta$; $\Sigma_\varepsilon = \Sigma \cap \{0 < \mu < \varepsilon\}$.

Next we define, for vectors tangent to a $\mu$ level of $U_\delta$, the slope by the following formula: for $X$ tangent to $U_\delta \cap \{\mu = \mu_0\}$ and $X = X_r \dfrac{\partial}{\partial r}$ $+ X_\varphi \dfrac{\partial}{\partial \varphi}$ the slope of $X$ is $\left| \dfrac{X_r}{\mu_0 \cdot X_\varphi} \right|$; for $X_\varphi = 0$ the slope is not defined.

By direct calculations it follows that if $X$ is a tangent vector of $U_\delta \cap \{\mu = \mu_0\}$ with slope $\leq 1$, and $\mu_0$ is small enough, then the slope of $d(N\Phi)(X)$ is $\leq (1 - K\mu_0)$ for some positive $K$. Using this, the fact that $\dfrac{\mu}{r^2} \sim$ constant on $U_\delta$ and the fact that $\Phi$ and $N\Phi$ only differ by terms of order $r^5$ one can verify that for $\varepsilon$ small enough and $X$ a tangent vector of $U_\delta \cap \{\mu = \mu_0\}$, $\mu_0 \leq \varepsilon$, with slope $\leq 1$, $d\Phi(X)$ has slope $< 1$.

From this it follows that for $\varepsilon$ small enough and any $n \geq 0$,
1. $\Phi^n(\Sigma_\varepsilon) \subset U_\delta$ and
2. the tangent vectors of $\Phi^n(\Sigma_\varepsilon) \cap \{\mu = \mu_0\}$, for $\mu_0 \leq \varepsilon$ have slope $< 1$.

This means that for any $\mu_0 \leqq \varepsilon$ and $n \geqq 0$

$$\Phi^n(\Sigma_\varepsilon) \cap \{\mu = \mu_0\} = \{(f_{n,\mu_0}(\varphi), \varphi, \mu_0)\}$$

where $f_{n,\mu_0}$ is a unique smooth function satisfying:

1'. $f_{n,\mu_0}(\varphi) \in \left[\left|\sqrt{\dfrac{\mu_0}{f_1(\mu_0)+\delta}}, \sqrt{\dfrac{\mu_0}{f_1(\mu_0)-\delta}}\right.\right]$ for all $\varphi$

2'. $\dfrac{d}{d\varphi}(f_{n,\mu_0}(\varphi)) \leqq \mu_0$ for all $\varphi$.

We now have to show that, for $\mu_0$ small enough, $\{f_{n,\mu_0}\}_{n=0}^{\infty}$ converges. We first fix a $\varphi_0$ and define

$$p_1 = (f_n(\varphi_0), \varphi_0, \mu_0), \qquad p_1' = \Phi(p_1) = (r_1', \varphi_1', \mu),$$

$$p_2 = (f_{n+1}(\varphi_0), \varphi_0, \mu_0), \qquad p_2' = \Phi(p_2) = (r_2', \varphi_2', \mu).$$

Using again the fact that $(f_{n,\mu_0}(\varphi))^2/\mu_0 \sim$ constant (independent of $\mu_0$), one obtains:

$$|r_1' - r_2'| \leqq (1 - K_1\mu_0)\,|f_{n,\mu_0}(\varphi_0) - f_{n+1,\mu_0}(\varphi_0)|$$

and

$$|\varphi_1' - \varphi_2'| \leqq K_2\sqrt{\mu_0}\cdot|f_{n,\mu_0}(\varphi_0) - f_{n+1,\mu_0}(\varphi_0)| \quad \text{where} \quad K_1, K_2 > 0$$

and independent of $\mu_0$.

By definition we have $f_{n+1,\mu_0}(\varphi_1') = r_1'$ and $f_{n+2,\mu_0}(\varphi_2') = r_2'$. We want however to get an estimate for the difference between $f_{n+1,\mu_0}(\varphi_1')$ and $f_{n+2,\mu_0}(\varphi_1')$. Because

$$\frac{d}{d\varphi}(f_{n+2,\mu_0}(\varphi)) \leqq \mu_0,$$

$$|f_{n+2,\mu_0}(\varphi_2') - f_{n+2,\mu_0}(\varphi_1')| \leqq \mu_0|\varphi_2' - \varphi_1'| \leqq K_2\cdot\mu_0^{\frac{3}{2}}|f_{n,\mu_0}(\varphi_0) - f_{n+1,\mu_0}(\varphi_0)|.$$

We have seen that

$$|f_{n+1,\mu_0}(\varphi_1') - f_{n+2,\mu_0}(\varphi_2')| = |r_1' - r_2'|$$

$$\leqq (1 - K_1\mu_0)\,|f_{n,\mu_0}(\varphi_0) - f_{n+1,\mu_0}(\varphi_0)|.$$

So

$$|f_{n+1,\mu_0}(\varphi_1') - f_{n+2,\mu_0}(\varphi_1')| \leqq (1 + K_2\mu_0^{\frac{3}{2}} - K_1\mu_0)\,|f_{n,\mu_0}(\varphi_0) - f_{n+1,\mu_0}(\varphi_0)|.$$

We shall now assume that $\mu_0$ is so small that $(1 + K_2\mu_0^{\frac{3}{2}} - K_1\mu_0) = K_3(\mu_0) < 1$, and write $\varrho(f_{n,\mu_0}, f_{n+1,\mu_0}) = \max_{\varphi}(|f_{n,\mu_0}(\varphi) - f_{n+1,\mu_0}(\varphi)|)$.

It follows that

$$\varrho(f_{m,\mu_0}, f_{m+1,\mu_0}) \leqq (K_3(\mu_0))^m \cdot \varrho(f_{0,\mu_0}, f_{1,\mu_0}).$$

This proves convergence, and gives for each small $\mu_0 > 0$ an invariant and attracting circle. This family of circles is continuous because the limit functions $f_{\infty,\mu_0}$ depend continuously on $\mu_0$, because of uniform convergence.

*Remark (7.3).* For a given $\mu_0$, $f_{\infty,\mu_0}$ is not only continuous but even Lipschitz, because it is the limit of functions with derivative $\leqq \mu_0$. Now we can apply the results on invariant manifolds in [4, 5] and obtain the following:

If $\Phi_\mu$ is $C^r$ for each $\mu$ then there is an $\varepsilon_r > 0$ such that the circles of our family which are in $\{0 < \mu < \varepsilon_r\}$ are $C^r$. This comes from the fact that near $\mu = 0$ in $U_\delta$ the contraction in the $r$-direction dominates sufficiently the maximal possible contraction in the $\varphi$-direction.

## § 8. Normal Forms (the Proof of Proposition (7.1))

First we have to give some definitions. Let $\underline{V}_r$ be the vectorspace of $r$-jets of vectorfields on $\mathbb{R}^2$ in 0, whose $(r-1)$-jet is zero (i.e. the elements of $\underline{V}_r$ can be uniquely represented by a vectorfield whose component functions are homogeneous polynomials of degree $r$). $V_r$ is the set of $r$-jets of diffeomorphisms $(\mathbb{R}^2, 0) \to (\mathbb{R}^2, 0)$, whose $(r-1)$-jet is "the identity". Exp: $\underline{V}_r \to V_r$ is defined by: for $\alpha \in \underline{V}_r$, Exp$(\alpha)$ is the ($r$-jet of) the diffeomorphism obtained by integrating $\alpha$ over time 1.

*Remark (8.1).* For $r \geqq 2$, Exp is a diffeomorphism onto and Exp$(\alpha) \circ$ Exp$(\beta)$ = Exp$(\alpha + \beta)$. The proof is straightforward and left to the reader.

Let now $A: (\mathbb{R}^2, 0) \to (\mathbb{R}^2, 0)$ be a linear map. The induced transformations $A_r: \underline{V}_r \to \underline{V}_r$ are defined by $A_r(\alpha) = A_* \alpha$, or, equivalently, Exp$(A_r(\alpha)) = A \circ$ Exp$\alpha \circ A^{-1}$.

*Remark (8.2).* If $[\Psi]_r$ is the $r$-jet of $\Psi: (\mathbb{R}^2, 0) \to (\mathbb{R}^2, 0)$ and $d\Psi$ is $A$, then, for every $\alpha \in \underline{V}_r$, the $r$-jets $[\Psi]_r \circ$ Exp$(\alpha)$ and Exp$(A_r\alpha) \circ [\Psi]_r$ are equal. The proof is left to the reader.

A splitting $\underline{V}_r = \underline{V}_r' \oplus \underline{V}_r''$ of $\underline{V}_r$ is called an *A-splitting*, $A: (\mathbb{R}^2, 0) \to (\mathbb{R}^2, 0)$ linear, if

1. $\underline{V}_r'$ and $\underline{V}_r''$ are invariant under the action of $A_r$.
2. $A_r \mid \underline{V}_r''$ has no eigenvalue one.

*Example (8.3).* We take $A$ with eigenvalues $\lambda, \bar{\lambda}$ and such that $|\lambda| \neq 1$ or such that $|\lambda| = 1$ but $\lambda \neq e^{k/l \, 2\pi i}$ with $k, l \leqq 5$. We may assume that $A$

is of the form

$$|\lambda| \begin{pmatrix} \cos\alpha & \sin\alpha \\ -\sin\alpha & \cos\alpha \end{pmatrix}.$$

For $2 \leq i \leq 4$ we can obtain a $A$-splitting of $V_i$ as follows:

$V_i'$ is the set of those ($i$-jets of) vectorfields which are, in polar coordinates of the form $\alpha_1 r^i \dfrac{\partial}{\partial r} + \alpha_2 r^{i-1} \dfrac{\partial}{\partial \varphi}$. More precisely $V_2' = 0$, $V_3'$ is generated by $r^3 \dfrac{\partial}{\partial r}$ and $r^2 \dfrac{\partial}{\partial \varphi}$ and $V_4' = 0$ (the other cases give rise to vectorfields which are not differentiable, in ordinary coordinates).

$V_i''$ is the set of ($i$-jets of) vectorfields of the form

$$g_1(\varphi) r^i \frac{\partial}{\partial r} + g_2(\varphi) r^{i-1} \frac{\partial}{\partial \varphi} \quad \text{with} \quad \int\limits_0^{2\pi} g_1(\varphi) = \int\limits_0^{2\pi} g_2(\varphi) = 0.$$

$g_1(\varphi)$ and $g_2(\varphi)$ have to be linear combinations of $\sin(j \cdot \varphi)$ and $\cos(j \cdot \varphi)$, $j \leq 5$, because otherwise the vectorfield will not be differentiable in ordinary coordinates (not all these linear combinations are possible).

**Proposition (8.4).** *For a given diffeomorphism $\Phi : (\mathbb{R}^2, 0) \to (\mathbb{R}^2, 0)$ with $(d\Phi)_0 = A$ and a given $A$-splitting $V_i = V_i' \oplus V_i''$ for $2 \leq i \leq i_0$, there is a coordinate transformation $\varkappa : (\mathbb{R}^2, 0) \to (\mathbb{R}^2, 0)$ such that:*
1. *$(d\varkappa)_0 = $ identity.*
2. *For each $z \leq i \leq i_0$ the $i$-jet of $\Phi' = \varkappa \circ \Phi \circ \varkappa^{-1}$ is related to its $(i-1)$-jet as follows: Let $[\Phi']_{i-1}$ be the polynomial map of degree $\leq i-1$ which has the same $(i-1)$-jet. The $i$-jet of $\Phi'$ is related to its $(i-1)$-jet if there is an element $\alpha \in V_i'$ such that $\mathrm{Exp}\,\alpha \circ [\Phi']_{i-1}$ has the same $i$-jet as $\Phi'$.*

*Proof.* We use induction: Suppose we have a map $\varkappa$ such that 1 and 2 hold for $i < i_1 \leq i_0$. Consider the $i_1$ jet of $\varkappa \circ \Phi \circ \varkappa^{-1}$. We now replace $\varkappa$ by $\mathrm{Exp}\,\alpha \circ \varkappa$ for some $\alpha \in V_{i_1}''$. $\varkappa \circ \Phi \circ \varkappa^{-1}$ is then replaced by $\mathrm{Exp}(\alpha) \circ \varkappa \circ \Phi \circ \varkappa^{-1} \circ \mathrm{Exp}(-\alpha)$, according to remark (8.2) this equal to $\mathrm{Exp}(-A_{i_1}\alpha) \circ \mathrm{Exp}(\alpha) \circ \varkappa \circ \Phi \circ \varkappa^{-1} = \mathrm{Exp}(\alpha - A_{i_1}\alpha) \circ \varkappa \circ \Phi \circ \varkappa^{-1}$.

$A_{i_1} \mid V_{i_1}''$ has no eigenvalue one, so for each $\beta \in V_{i_1}''$ there is a unique $\alpha \in V_{i_1}''$ such that if we replace $\varkappa$ by $\mathrm{Exp}\,\alpha \circ \varkappa$, $\varkappa \circ \Phi \circ \varkappa^{-1}$ is replaced by $\mathrm{Exp}\,\beta \circ \varkappa \circ \Phi \circ \varkappa^{-1}$. It now follows easily that there is a unique $\alpha \in V_{i_1}''$ such that $\mathrm{Exp}\,\alpha \circ \varkappa$ satisfies condition 2 for $i \leq i_1$. This proves the proposition.

*Proof of Proposition (7.1).* For $\mu$ near 0, $d\Phi_\mu$ is a linear map of the type we considered in example (8.3). So the splitting given there is a $d\Phi_\mu$-splitting of $V_i$, $i = 2, 3, 4$, for $\mu$ near zero. We now apply Proposition (8.4) for each $\mu$ and obtain a coordinate transformation $\varkappa_\mu$ for each $\mu$ which brings $\Phi_\mu$ in the required form. The induction step then becomes:

14*

Given $\varkappa_\mu$, satisfying 1 and 2 for $i < i_1$ there is for each $\mu$ a unique $\alpha_\mu \in \underline{V}''_{i_1}$ such that $\mathrm{Exp}\alpha_\mu \circ \varkappa_\mu$ satisfies 1 and 2 for $i \leqq i_1$. $\alpha_\mu$ depends then $C^r$ on $\mu$ if the $i_1$-jet of $\Phi$ depends $C^r$ on $\mu$; this gives the loss of differentiability in the $\mu$ direction.

## § 9. Some Examples

In this section we show how a small perturbation of a quasi-periodic flow on a torus gives flows with strange attractors (Proposition (9.2)) and, more generally, flows which are not Morse-Smale (Proposition (9.1)).

**Proposition (9.1).** *Let $\omega$ be a constant vector field on $T^k = (\mathbb{R}/\mathbb{Z})^k$, $k \geqq 3$. In every $C^{k-1}$-small neighbourhood of $\omega$ there exists an open set of vector fields which are not Morse-Smale.*

We consider the case $k = 3$. We let $\omega = (\omega_1, \omega_2, \omega_3)$ and we may suppose $0 \leqq \omega_1 \leqq \omega_2 \leqq \omega_3$. Given $\varepsilon > 0$ we may choose a constant vector field $\omega'$ such that

$$\|\omega' - \omega\|_2 = \|\omega' - \omega\|_0 < \varepsilon/2\,,$$

$$\omega'_3 > 0, \quad 0 < \frac{\omega'_1}{\omega'_3} = \frac{p_1}{q_1} < 1, \quad 0 < \frac{\omega'_2}{\omega'_3} = \frac{p_2}{q_2} < 1\,,$$

where $p_1, p_2, q_1, q_2$ are integers, and $p_1 q_2$ and $p_2 q_1$ have no common divisor. We shall also need that $q_1, q_2$ are sufficiently large and satisfy

$$\tfrac{1}{2} < q_1/q_2 < 2\,.$$

All these properties can be satisfied with $q_1 = 2^{m_1}$, $q_2 = 3^{m_2}$.

Let $I = \{x \in \mathbb{R} : 0 \leqq x \leqq 1\}$ and define $g, h : I^3 \to T^3$ by

$$g(x_1, x_2, x_3) = (x_1 (\mathrm{mod}\, 1), x_2 (\mathrm{mod}\, 1), x_3 (\mathrm{mod}\, 1))$$

$$h(x_1, x_2, x_3) = (q_1^{-1} x_1 + p_1 q_2 x_3 (\mathrm{mod}\, 1), q_2^{-1} x_2 + p_2 q_1 x_3 (\mathrm{mod}\, 1),$$
$$q_1 q_2 x_3 (\mathrm{mod}\, 1))\,.$$

We have $gI^3 = hI^3 = T^3$ and $g$ (resp. $h$) has a unique inverse on points $gx$ (resp. $hx$) with $x \in \overset{\circ}{I}{}^3$.

We consider the map $f$ of a disc into itself (see [11] Section I.5, Fig. 7) used by Smale to define the horseshoe diffeomorphism. Imbedding $\Delta$ in $T^2$:

$$\Delta \subset \{(x_1, x_2) : \tfrac{1}{3} < x_1 < \tfrac{2}{3}, \tfrac{1}{3} < x_2 < \tfrac{2}{3}\} \subset T^2$$

we can arrange that $f$ appears as Poincaré map in $T^3 = T^2 \times T^1$. More precisely, it is easy to define a vector field $X = (\tilde{X}, 1)$ on $T^2 \times T^1$ such that if $\xi \in \Delta$, we have

$$(f(\xi), 0) = \mathcal{D}_{X,1}(\xi, 0)$$

where $\mathcal{D}_{X,1}$ is the time one integral of $X$ (see Fig. 7). Finally we choose the restriction of $X$ to a neighbourhood of $g(\partial I^2 \times I)$ to be $(0, 1)$ (i.e. $\tilde{X} = 0$).

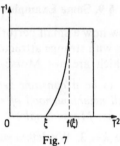

Fig. 7

If $x \in g\mathring{I}^3$, then $\Phi x = h \circ g^{-1}$ is uniquely defined and the tangent mapping to $\Phi$ applied to $X$ gives a vector field $Y$:

$$Y(\Phi(x)) = [d\Phi(x)] X(x)$$

where

$$[d\Phi(x)] = \begin{pmatrix} q_1^{-1} & & p_1 q_2 \\ & q_2^{-1} & p_2 q_1 \\ & & q_1 q_2 \end{pmatrix}.$$

$Y$ has a unique smooth extension to $T^3$, again called $Y$. Let now $Z = (q_1 q_2)^{-1} \omega_3' Y$. We want to estimate

$$\|Z - \omega'\|_r = \sup_{\varrho;\, |\varrho| \leq r} N^\varrho$$

where

$$N^\varrho = \sup_{y \in T^3} \sup_{i=1,2,3} |D^\varrho Z_i(y) - D^\varrho \omega_i'| \qquad (*)$$

and $D^\varrho$ denotes a partial differentiation of order $|\varrho|$. Notice that it suffices to take the first supremum in (*) over $y \in h\mathring{I}^3$, i.e. $y = \Phi x$ where $x \in g\mathring{I}^3$. We have

$$\frac{\partial}{\partial y} = \begin{pmatrix} q_1 & & \\ & q_2 & \\ -p_1 & -p_2 & (q_1 q_2)^{-1} \end{pmatrix} \frac{\partial}{\partial x}$$

so that

$$\sup_i \left| \frac{\partial}{\partial y_i} \right| < (q_1 + q_2) \sup_i \left| \frac{\partial}{\partial x_i} \right|.$$

Notice also that

$$Z_i(y) - \omega_i' = (q_1 q_2)^{-1} \omega_3' \begin{pmatrix} q_1^{-1} X_1 + p_1 q_2 \\ q_2^{-1} X_2 + p_2 q_1 \\ q_1 q_2 \end{pmatrix} - \omega_3' \begin{pmatrix} p_1 q_1^{-1} \\ p_2 q_2^{-1} \\ 1 \end{pmatrix}$$

$$= (q_1 q_2)^{-1} \omega_3' \begin{pmatrix} q_1^{-1} X_1 \\ q_2^{-1} X_2 \\ 0 \end{pmatrix}.$$

Therefore

$$N^\varrho \leq (q_1 q_2)^{-1} \omega_3' (q_1 + q_2)^{|\varrho|} \left( \sup_{i=1,2} q_i^{-1} \right) \sup_{i=1,2} \|X_i\|_{|\varrho|}$$

$$\|Z - \omega'\|_r \leq (q_1 q_2)^{-2} (q_1 + q_2)^{r+1} [\omega_3' \|\tilde{X}\|_r].$$

If we have chosen $q_1, q_2$ sufficiently large, we have

$$\|Z - \omega'\|_2 < \varepsilon/2$$

and therefore

$$\|Z - \omega\|_2 < \varepsilon.$$

Consider the Poincaré map $P: T^2 \to T^2$ defined by the vector field $Z$ on $T^3 = T^2 \times T^1$. By construction the non wandering set of $P$ contains a Cantor set, and the same is true if $Z$ is replaced by a sufficiently close vector field $Z'$. This concludes the proof for $k = 3$.

In the general case $k \geq 3$ we approximate again $\omega$ by $\omega'$ rational and let

$$0 < \frac{\omega_i'}{\omega_k'} = \frac{p_i}{q_i} < 1 \quad \text{for} \quad i = 1, ..., k-1.$$

We assume that the integers $p_1 \prod_{i \neq 1} q_i, ..., p_{k-1} \prod_{i \neq k-i} q_i$ have no common divisor. Furthermore $q_1, ..., q_{k-1}$ are chosen sufficiently large and such that

$$(\max_i q_i)/(\min_i q_i) < C$$

where $C$ is a constant depending on $k$ only.

The rest of the proof goes as for $k = 2$, with the horseshoe diffeomorphism replaced by a suitable $k-1$-diffeomorphism. In particular, using the diffeomorphism of Fig. 2 (end of § 2) we obtain the following result.

**Proposition (9.2).** Let $\omega$ be a constant vector field on $T^k$, $k \geq 4$. In every $C^{k-1}$-small neighbourhood of $\omega$ there exists an open set of vector fields with a strange attractor.

## Appendix

## Bifurcation of Stationary Solutions of Hydrodynamical Equations

In this appendix we present a bifurcation theorem for fixed points of a non linear map in a Banach space. Our result is of a known type [23], but has the special interest that the fixed points are shown to depend differentiably on the bifurcation parameter. The theorem may be used to study the bifurcation of stationary solutions in the Taylor and Bénard [24] problems for instance. By reference to Brunovský (cf. § 3 b) we see that the bifurcation discussed below is *nongeneric*. The bifurcation of stationary solutions in the Taylor and Bénard problems is indeed nongeneric, due to the presence of an invariance group.

**Theorem.** *Let $H$ be a Banach space with $C^k$ norm, $1 \leq k < \infty$, and $\Phi_\mu : H \to H$ a differentiable map such that $\Phi_\mu(0) = 0$ and $(x, \mu) \to \Phi_\mu x$ is $C^k$ from $H \times \mathbb{R}$ to $H$. Let*

$$L_\mu = [d\Phi_\mu]_0, \quad N_\mu = \Phi_\mu - L_\mu. \tag{1}$$

*We assume that $L_\mu$ has a real simple isolated eigenvalue $\lambda(\mu)$ depending continuously on $\mu$ such that $\lambda(0) = 1$ and $(d\lambda/d\mu)(0) > 0$; we assume that the rest of the spectrum is in $\{ \mathfrak{z} \in \mathbb{C} : |\mathfrak{z}| < 1 \}$.*

(a) *There is a one parameter family $(a\, C^{k-1}$ curve $l)$ of fixed points of $\Phi : (x, \mu) \to (\Phi_\mu x, \mu)$ near $(0, 0) \in H \times \mathbb{R}$. These points and the points $(0, \mu)$ are the only fixed points of $\Phi$ in some neighbourhood of $(0, 0)$.*

(b) *Let $\mathfrak{z}$ (resp. $\mathfrak{z}^*$) be an eigenvector of $L_0$ (resp. its adjoint $L_0^*$ in the dual $H^*$ of $H$) to the eigenvalue 1, such that $(\mathfrak{z}^*, \mathfrak{z}) = 1$. Suppose that for all $\alpha \in \mathbb{R}$*

$$(\mathfrak{z}^*, N_0 \alpha \mathfrak{z}) = 0. \tag{2}$$

*Then the curve $l$ of (a) is tangent to $(\mathfrak{z}, 0)$ at $(0, 0)$.*

From the center-manifold theorem of Hirsch, Pugh, and Shub [25] it follows that there is a 2-dimensional $C^k$-manifold $V^c$, tangent to the vectors $(\mathfrak{z}, 0)$ and $(0, 1)$ at $(0, 0) \in H \times \mathbb{R}$ and locally invariant under $\Phi$. Furthermore there is a neighbourhood $U$ of $(0, 0)$ such that every fixed point of $\Phi$ in $U$ is contained in $V^c \cap U$ [26].

We choose coordinates $(\alpha, \mu)$ on $V^c$ so that

$$\Phi(\alpha, \mu) = (f(\alpha, \mu), \mu)$$

---

[23] See for instance [16] and [14].

[24] The bifurcation of the Taylor problem has been studied by Velte [18] and Yudovich [19]. For the Benard problem see Rabinowitz [17], Fife and Joseph [15].

[25] Usually the center manifold theorem is only formulated for diffeomorphisms: C. C. Pugh pointed out to us that his methods in [5], giving the center manifold, also work for differentiable maps which are not diffeomorphisms.

[26] See § 5.

D. Ruelle and F. Takens:

with $f(0, \mu) = 0$ and $\dfrac{\partial f}{\partial \alpha}(0, \mu) = \lambda(\mu)$. The fixed points of $\Phi$ in $V^c$ are given by $\alpha = f(\alpha, \mu)$, they consist of points $(0, \mu)$ and of solutions of

$$g(\alpha, \mu) = 0$$

where, by the division theorem, $g(\alpha, \mu) = \dfrac{f(\alpha, \mu)}{\alpha} - 1$ is $C^{k-1}$. Since $\dfrac{\partial g}{\partial \mu}(0, 0) = \dfrac{d\lambda}{d\mu}(0) > 0$, the implicit function theorem gives (a).

Let $(x, 0) \in V^c$. We may write

$$x = \alpha_3 + Z \tag{3}$$

where $(3^*, Z) = 0$, $Z = 0(\alpha^2)$. Since $(3^*, 3) = 1$ we have

$$\begin{aligned}
f(\alpha, 0) &= (3^*, \Phi_0(\alpha_3 + Z)) \\
&= \alpha + (3^*, L_0 Z + N_0(\alpha_3 + Z)) \\
&= \alpha + (3^*, L_0 Z + N_0 \alpha_3) + 0(\alpha^3).
\end{aligned} \tag{4}$$

Notice that

$$(3^*, L_0 Z) = (L_0^* 3^*, Z) = (3^*, Z) = 0.$$

We assume also that (2) holds:

$$(3^*, N_0 \alpha_3) = 0.$$

Then

$$\begin{aligned}
f(\alpha, 0) &= \alpha + 0(\alpha^3), \\
f(\alpha, \mu) &= \alpha(\lambda(\mu) + 0^2),
\end{aligned} \tag{5}$$

where $0^2$ represents terms of order 2 and higher in $\alpha$ and $\mu$. The curve $l$ of fixed points of $\Phi$ introduced in (a) is given by

$$\lambda(\mu) - 1 + 0^2 = 0 \tag{6}$$

and (b) follows from $\lambda(0) = 0$. $\dfrac{d\lambda}{d\mu}(0) > 0$.

*Remark 1.* From (2) and the local invariance of $V^c$ we have

$$\Phi_0(\alpha_3 + Z) = \alpha_3 + Z + 0(\alpha^3)$$

hence

$$Z = L_0 Z + N_0 \alpha_3 + 0(\alpha^3),$$

$$Z = (1 - L_0)^{-1} N_0 \alpha_3 + 0(\alpha^3),$$

and (5) can be replaced by the more precise

$$f(\alpha, 0) = \alpha + (3^*, N_0[\alpha_3 + (1 - L_0)^{-1} N \alpha_3]) + 0(\alpha^4) \tag{7}$$

from which one can compute the coefficient $A$ of $\alpha^3$ in $f(\alpha, 0)$. Then (6) is, up to higher order terms

$$\left[\frac{d\lambda}{d\mu}(0)\right]\mu + A\alpha^2 = 0 .$$

Depending on whether $A < 0$ or $A > 0$, this curve lies in the region $\mu > 0$ or $\mu < 0$, and consists of attracting or non-attracting fixed points. This is seen by discussing the sign of $f(\alpha, \mu) - \alpha$ (see Fig. 8 b, c); the fixed points $(0, \mu)$ are always attracting for $\mu < 0$, non attracting for $\mu > 0$.

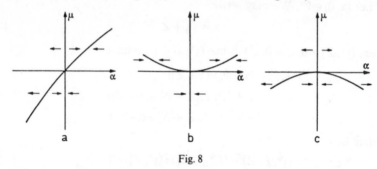

Fig. 8

*Remark 2.* If the curve $l$ of fixed points is not tangent to $(3, 0)$ at $(0, 0)$, then the points of $l$ are attracting for $\mu > 0$, non-attracting for $\mu < 0$ (see Fig. 8 a).

*Remark 3.* If it is assumed that $L_\mu$ has the real simple isolated eigenvalue $\lambda(\mu)$ as in the theorem and that the rest of the spectrum lies in $\{3 \in \mathbb{C} : |3| \neq 1\}$ (rather than $\{3 \in \mathbb{C} : |3| < 1\}$. the theorem continues to hold but the results on the attractive character of fixed points are lost.

## References

1. Abraham, R., Marsden. J.: Foundations of mechanics. New York: Benjamin 1967.
2. Bass. J.: Fonctions stationnaires. Fonctions de corrélation. Application à la représentation spatio-temporelle de la turbulence. Ann. Inst. Henri Poincaré. Section B 5, 135—193 (1969).
3. Brunovsky. P.: One-parameter families of diffeomorphisms. Symposium on Differential Equations and Dynamical Systems. Warwick 1968—69.
4. Hirsch, M., Pugh, C. C., Shub. M.: Invariant manifolds. Bull. A.M.S. **76**, 1015—1019 (1970).
5. — — — Invariant manifolds. To appear.
6. Hopf. E.: Abzweigung einer periodischen Lösung von einer stationären Lösung eines Differentialsystems. Ber. Math.-Phys. Kl. Sächs. Akad. Wiss. Leipzig **94**. 1—22 (1942).
7. Kelley, A.: The stable. center-stable. center, center-unstable, and unstable manifolds. Published as Appendix C of R. Abraham and J. Robbin: Transversal mappings and flows. New York: Benjamin 1967.

192         D. Ruelle and F. Takens: On the Nature of Turbulence

8. Landau, L. D., Lifshitz, E. M.: Fluid mechanics. Oxford: Pergamon 1959.
9. Leray, J.: Sur le mouvement d'un liquide visqueux emplissant l'espace. Acta Math. 63, 193—248 (1934).
10. Moser, J.: Perturbation theory of quasiperiodic solutions of differential equations. Published in J. B. Keller and S. Antman: Bifurcation theory and nonlinear eigenvalue problems. New York: Benjamin 1969.
11. Smale, S.: Differentiable dynamical systems. Bull. Am. Math. Soc. 73, 747—817 (1967).
12. Thom, R.: Stabilité structurelle et morphogénèse. New York: Benjamin 1967.
13. Williams, R. F.: One-dimensional non-wandering sets. Topology 6, 473—487 (1967).
14. Berger, M.: A bifurcation theory for nonlinear elliptic partial differential equations and related systems. In: Bifurcation theory and nonlinear eigenvalue problems. New York: Benjamin 1969.
15. Fife, P. C., Joseph, D. D.: Existence of convective solutions of the generalized Bénard problem which are analytic in their norm. Arch. Mech. Anal. 33, 116—138 (1969).
16. Krasnosel'skii, M.: Topological methods in the theory of nonlinear integral equations. New York: Pergamon 1964.
17. Rabinowitz, P. H.: Existence and nonuniqueness of rectangular solutions of the Bénard problem. Arch. Rat. Mech. Anal. 29, 32—57 (1968).
18. Velte, W.: Stabilität und Verzweigung stationärer Lösungen der Navier-Stokesschen Gleichungen beim Taylorproblem. Arch. Rat. Mech. Anal. 22, 1—14 (1966).
19. Yudovich, V.: The bifurcation of a rotating flow of fluid. Dokl. Akad. Nauk SSSR 169, 306—309 (1966).

D. Ruelle
The Institute for Advanced Study
Princeton, New Jersey 08540, USA

F. Takens
Universiteit van Amsterdam
Roetersstr. 15
Amsterdam, The Netherlands

Commun. math. Phys. 23, 343—344 (1971)
© by Springer-Verlag 1971

# Note Concerning our Paper
# "On the Nature of Turbulence"

D. RUELLE and F. TAKENS

Commun. math. Phys. **20**, 167—192 (1971)

Received November 2, 1971

After the paper referred to in the title was published, references to the relevant Russian literature were made available to us by Ya. G. Sinai and V. I. Arnold. From these it appears that much of the work on bifurcation described in our paper duplicates results already published by Russian authors. On the other hand the mathematical interpretation which we give of turbulence seems to remain our own responsability!

We thank Sinai and Arnold for informing us of the Russian references. The following list has been compiled from their indications.

Brušlinskaja, N. N.: Qualitative integration of a system of $n$ differential equations in a region containing a singular point and a limit cycle. Dokl. Akad. Nauk SSSR **139** N1, 9—12 (1961); — Soviet Math. Dokl. **2**, 845—848 (1961). MR 26, No. 5212; Errata MR 30, p. 1203.
— Limit cycles for equations of motion of a rigid body and Galerkin's equations for hydrodynamics. Dokl. Akad. Nauk SSSR **157**, N5, 1017—1020 (1964); — Soviet Math. Dokl. **5**, 1051—1054 (1964). MR 29, No. 6133.
— The behavior of solutions of the equations of hydrodynamics when the Reynolds number passes through a critical value. Dokl. Akad. Nauk SSSR **162** N4, 731—734 (1965);— Soviet Math. Dokl. **6**, 724—728 (1965). MR 31, No. 6460.
— On the generation of periodic flows and tori from laminar flows, p. 57—79 in *Nekotorje voprosy mehaniki gornych porod* (some problems of mechanics of minerals), Trudy gornogo instituta, Moscow, 1968.
Gurtovnik, A. S., Neǐmark, Ju. I.: On the question of the stability of quasi-periodic motions. Diff. Uravn. **5** N5, 824—832 (1969).
Neǐmark, Ju. I.: On some cases of dependence of periodic solutions on parameters. Dokl. Akad. Nauk SSSR **129** N4, 736—739 (1959).
— Motions close to doubly-asymptotic motion. Dokl. Akad. Nauk SSSR **172** N5, 1021—1024 (1967); — Soviet Math. Dokl. **8**, 228—231 (1967). MR 38, No. 669.
— Izv. Vysš. Učebn. Zav., Radiofizika **1** N5—6 (1958); **2** N3 (1959); **8** N3 (1965); **10** N3 (1967).
Judovič, V. I.: An example of the generation of a secondary or periodic flow due to the loss of stability of a laminar flow of a viscous incompressible fluid. Prikl. Mat. Meh. **29** N3, 453—467 (1965); — J. Appl. Math. Mech. **29**, 527—544 (1965).
— On the bifurcation of rotationary flows of fluids. Dokl. Akad. Nauk SSSR **169** N2, 306—309 (1966).

Judovič, V. I.: An example of loss of stability and generation of a secondary flow in a closed volume. Mat. Sbornik **74** (116) N 4, 565—579 (1967).
— Generation of secondary stationary and periodic solutions by destabilization of a stable stationary flow of fluid. Abstracts of short scientific communications, Sec. 12. Internat. Congress of Math. (Moscow, 1966). Mir, Moscow, 1968.
— Questions of the mathematical theory of stability of flows of fluids. Vsesojusnyi siesd po teoretičeskoi i prikladnoi mehanike. Moscow: Annotacii Dokladov 1968.
— On the stability of forced oscillations of fluid. Dokl. Akad. Nauk SSSR **195** N 2, 292—295 (1970); — Soviet Math. Dokl. **11**, 1473—1477 (1970).
— Appearance of self-oscillations in fluids, I. Prikl. Mat. Meh., 1971.

D. Ruelle                              F. Takens
Institut des Hautes Etudes             Dept. of Mathematics
Scientifiques                          University of Groningen
F-91 Bures-sur Yvette, France          Groningen. The Netherlands

*Nature Vol. 261 June 10 1976*

# Simple mathematical models with very complicated dynamics

Robert M. May*

*First-order difference equations arise in many contexts in the biological, economic and social sciences. Such equations, even though simple and deterministic, can exhibit a surprising array of dynamical behaviour, from stable points, to a bifurcating hierarchy of stable cycles, to apparently random fluctuations. There are consequently many fascinating problems, some concerned with delicate mathematical aspects of the fine structure of the trajectories, and some concerned with the practical implications and applications. This is an interpretive review of them.*

THERE are many situations, in many disciplines, which can be described, at least to a crude first approximation, by a simple first-order difference equation. Studies of the dynamical properties of such models usually consist of finding constant equilibrium solutions, and then conducting a linearised analysis to determine their stability with respect to small disturbances: explicitly nonlinear dynamical features are usually not considered.

Recent studies have, however, shown that the very simplest nonlinear difference equations can possess an extraordinarily rich spectrum of dynamical behaviour, from stable points, through cascades of stable cycles, to a regime in which the behaviour (although fully deterministic) is in many respects "chaotic", or indistinguishable from the sample function of a random process.

This review article has several aims.

First, although the main features of these nonlinear phenomena have been discovered and independently rediscovered by several people, I know of no source where all the main results are collected together. I have therefore tried to give such a synoptic account. This is done in a brief and descriptive way, and includes some new material: the detailed mathematical proofs are to be found in the technical literature, to which signposts are given.

Second, I indicate some of the interesting mathematical questions which do not seem to be fully resolved. Some of these problems are of a practical kind, to do with providing a probabilistic description for trajectories which seem random, even though their underlying structure is deterministic. Other problems are of intrinsic mathematical interest, and treat such things as the pathology of the bifurcation structure, or the truly random behaviour, that can arise when the nonlinear function $F(X)$ of equation (1) is not analytical. One aim here is to stimulate research on these questions, particularly on the empirical questions which relate to processing data.

Third, consideration is given to some fields where these notions may find practical application. Such applications range from the abstractly metaphorical (where, for example, the transition from a stable point to "chaos" serves as a metaphor for the onset of turbulence in a fluid), to models for the dynamic behaviour of biological populations (where one can seek to use field or laboratory data to estimate the values of the parameters in the difference equation).

*King's College Research Centre, Cambridge CB2 1ST; on leave from Biology Department, Princeton University, Princeton 08540.

Fourth, there is a very brief review of the literature pertaining to the way this spectrum of behaviour—stable points, stable cycles, chaos—can arise in second or higher order difference equations (that is, two or more dimensions; two or more interacting species), where the onset of chaos usually requires less severe nonlinearities. Differential equations are also surveyed in this light; it seems that a three-dimensional system of first-order ordinary differential equations is required for the manifestation of chaotic behaviour.

The review ends with an evangelical plea for the introduction of these difference equations into elementary mathematics courses, so that students' intuition may be enriched by seeing the wild things that simple nonlinear equations can do.

## First-order difference equations

One of the simplest systems an ecologist can study is a seasonally breeding population in which generations do not overlap[1-4]. Many natural populations, particularly among temperate zone insects (including many economically important crop and orchard pests), are of this kind. In this situation, the observational data will usually consist of information about the maximum, or the average, or the total population in each generation. The theoretician seeks to understand how the magnitude of the population in generation $t+1$, $X_{t+1}$, is related to the magnitude of the population in the preceding generation $t$, $X_t$: such a relationship may be expressed in the general form

$$X_{t+1} = F(X_t) \qquad (1)$$

The function $F(X)$ will usually be what a biologist calls "density dependent", and a mathematician calls nonlinear; equation (1) is then a first-order, nonlinear difference equation.

Although I shall henceforth adopt the habit of referring to the variable $X$ as "the population", there are countless situations outside population biology where the basic equation (1) applies. There are other examples in biology, as, for example in genetics[5,6] (where the equation describes the change in gene frequency in time) or in epidemiology[7] (with $X$ the fraction of the population infected at time $t$). Examples in economics include models for the relationship between commodity quantity and price[8], for the theory of business cycles[9], and for the temporal sequences generated by various other economic quantities[10]. The general equation (1) also is germane to the social sciences[11], where it arises, for example, in theories of

460

*Nature Vol. 261 June 10 1976*

learning (where $X$ may be the number of bits of information that can be remembered after an interval $t$), or in the propagation of rumours in variously structured societies (where $X$ is the number of people to have heard the rumour after time $t$). The imaginative reader will be able to invent other contexts for equation (1).

In many of these contexts, and for biological populations in particular, there is a tendency for the variable $X$ to increase from one generation to the next when it is small, and for it to decrease when it is large. That is, the nonlinear function $F(X)$ often has the following properties: $F(0)=0$; $F(X)$ increases monotonically as $X$ increases through the range $0 < X < A$ (with $F(X)$ attaining its maximum value at $X=A$); and $F(X)$ decreases monotonically as $X$ increases beyond $X=A$. Moreover, $F(X)$ will usually contain one or more parameters which "tune" the severity of this nonlinear behaviour; parameters which tune the steepness of the hump in the $F(X)$ curve. These parameters will typically have some biological or economic or sociological significance.

A specific example is afforded by the equation[1,4,12-23]

$$N_{t+1} = N_t(a - bN_t) \qquad (2)$$

This is sometimes called the "logistic" difference equation. In the limit $b=0$, it describes a population growing purely exponentially (for $a > 1$); for $b \neq 0$, the quadratic nonlinearity produces a growth curve with a hump, the steepness of which is tuned by the parameter $a$. By writing $X = bN/a$, the equation may be brought into canonical form[1,4,12-23]

$$X_{t+1} = aX_t(1 - X_t) \qquad (3)$$

In this form, which is illustrated in Fig. 1, it is arguably the simplest nonlinear difference equation. I shall use equation (3) for most of the numerical examples and illustrations in this article. Although attractive to mathematicians by virtue of its extreme simplicity, in practical applications equation (3) has the disadvantage that it requires $X$ to remain on the interval $0 < X < 1$; if $X$ ever exceeds unity, subsequent iterations diverge towards $-\infty$ (which means the population becomes extinct). Furthermore, $F(X)$ in equation (3) attains a maximum value of $a/4$ (at $X=\frac{1}{2}$); the equation therefore possesses non-trivial dynamical behaviour only if $a < 4$. On the other hand, all trajectories are attracted to $X=0$ if $a < 1$. Thus for non-trivial

Fig. 1 A typical form for the relationship between $X_{t+1}$ and $X_t$ described by equation (1). The curves are for equation (3), with $a = 2.707$ (*a*); and $a = 3.414$ (*b*). The dashed lines indicate the slope at the "fixed points" where $F(X)$ intersects the 45° line: for the case *a* this slope is less steep than $-45°$ and the fixed point is stable; for *b* the slope is steeper than $-45°$, and the point is unstable.

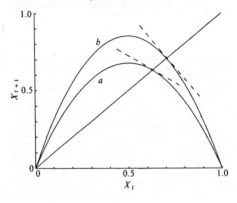

dynamical behaviour we require $1 < a < 4$; failing this, the population becomes extinct.

Another example, with a more secure provenance in the biological literature[1,23-27], is the equation

$$X_{t+1} = X_t \exp[r(1 - X_t)] \qquad (4)$$

This again describes a population with a propensity to simple exponential growth at low densities, and a tendency to decrease at high densities. The steepness of this nonlinear behaviour is tuned by the parameter $r$. The model is plausible for a single species population which is regulated by an epidemic disease at high density[28]. The function $F(X)$ of equation (4) is slightly more complicated than that of equation (3), but has the compensating advantage that local stability implies global stability[1] for all $X > 0$.

The forms (3) and (4) by no means exhaust the list of single-humped functions $F(X)$ for equation (1) which can be culled from the ecological literature. A fairly full such catalogue is given, complete with references, by May and Oster[1]. Other similar mathematical functions are given by Metropolis *et al.*[16]. Yet other forms for $F(X)$ are discussed under the heading of "mathematical curiosities" below.

## Dynamic properties of equation (1)

Possible constant, equilibrium values (or "fixed points") of $X$ in equation (1) may be found algebraically by putting $X_{t+1}=X_t=X^*$, and solving the resulting equation

$$X^* = F(X^*) \qquad (5)$$

An equivalent graphical method is to find the points where the curve $F(X)$ that maps $X_t$ into $X_{t+1}$ intersects the 45° line, $X_{t+1}=X_t$, which corresponds to the ideal nirvana of zero population growth; see Fig. 1. For the single-hump curves discussed above, and exemplified by equations (3) and (4), there are two such points: the trivial solution $X=0$, and a non-trivial solution $X^*$ (which for equation (3) is $X^*=1-[1/a]$).

The next question concerns the stability of the equilibrium point $X^*$. This can be seen[24,25,19-21,1,4] to depend on the slope of the $F(X)$ curve at $X^*$. This slope, which is illustrated by the dashed lines in Fig. 1, can be designated

$$\lambda^{(1)}(X^*) = [dF/dX]_{X=X^*} \qquad (6)$$

So long as this slope lies between 45° and $-45°$ (that is, $\lambda^{(1)}$ between $+1$ and $-1$), making an acute angle with the 45° ZPG line, the equilibrium point $X^*$ will be at least locally stable, attracting all trajectories in its neighbourhood. In equation (3), for example, this slope is $\lambda^{(1)}=2-a$: the equilibrium point is therefore stable, and attracts all trajectories originating in the interval $0 < X < 1$, if and only if $1 < a < 3$.

As the relevant parameters are tuned so that the curve $F(X)$ becomes more and more steeply humped, this stability-determining slope at $X^*$ may eventually steepen beyond $-45°$ (that is, $\lambda^{(1)} < -1$), whereupon the equilibrium point $X^*$ is no longer stable.

What happens next? What happens, for example, for $a > 3$ in equation (3)?

To answer this question, it is helpful to look at the map which relates the populations at successive intervals 2 generations apart; that is, to look at the function which relates $X_{t+2}$ to $X_t$. This second iterate of equation (1) can be written

$$X_{t+2} = F[F(X_t)] \qquad (7)$$

or, introducing an obvious piece of notation,

$$X_{t+2} = F^{(2)}(X_t) \qquad (8)$$

The map so derived from equation (3) is illustrated in Figs 2 and 3. Population values which recur every second generation (that

*Nature Vol. 261 June 10 1976*

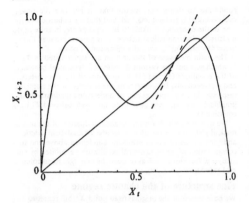

Fig. 2 The map relating $X_{t+2}$ to $X_t$, obtained by two iterations of equation (3). This figure is for the case (a) of Fig. 1, $a = 2.707$: the basic fixed point is stable, and it is the only point at which $F^{(2)}(X)$ intersects the 45° line (where its slope, shown by the dashed line, is less steep than 45°).

is, fixed points with period 2) may now be written as $X^*_2$, and found either algebraically from

$$X^*_2 = F^{(2)}(X^*_2) \qquad (9)$$

or graphically from the intersection between the map $F^{(2)}(X)$ and the 45° line, as shown in Figs 2 and 3. Clearly the equilibrium point $X^*$ of equation (5) is a solution of equation (9); the basic fixed point of period 1 is a degenerate case of a period 2 solution. We now make a simple, but crucial, observation[1]: the slope of the curve $F^{(2)}(X)$ at the point $X^*$, defined as $\lambda^{(2)}(X^*)$ and illustrated by the dashed lines in Figs 2 and 3, is the square of the corresponding slope of $F(X)$

$$\lambda^{(2)}(X^*) = [\lambda^{(1)}(X^*)]^2 \qquad (10)$$

This fact can now be used to make plain what happens when the fixed point $X^*$ becomes unstable. If the slope of $F(X)$ is less than $-45°$ (that is, $|\lambda^{(1)}| < 1$), as illustrated by curve a in Fig. 1, then $X^*$ is stable. Also, from equation (10), this implies $0 < \lambda^{(2)} < 1$ corresponding to the slope of $F^{(2)}$ at $X^*$ lying between 0° and 45°, as shown in Fig. 2. As long as the fixed point $X^*$ is stable, it provides the only non-trivial solution to equation (9). On the other hand, when $\lambda^{(1)}$ steepens beyond $-45°$ (that is, $|\lambda^{(1)}| > 1$), as illustrated by curve b in Fig 1, $X^*$ becomes unstable. At the same time, from equation (10) this implies $\lambda^{(2)} > 1$, corresponding to the slope of $F^{(2)}$ at $X^*$ steepening beyond 45°, as shown in Fig. 3. As this happens, the curve $F^{(2)}(X)$ must develop a "loop", and two new fixed points of period 2 appear, as illustrated in Fig. 3.

In short, as the nonlinear function $F(X)$ in equation (1) becomes more steeply humped, the basic fixed point $X^*$ may become unstable. At exactly the stage when this occurs, there are born two new and initially stable fixed points of period 2, between which the system alternates in a stable cycle of period 2. The sort of graphical analysis indicated by Figs 1, 2 and 3, along with the equation (10), is all that is needed to establish this generic result[1,4].

As before, the stability of this period 2 cycle depends on the slope of the curve $F^{(2)}(X)$ at the 2 points. (This slope is easily shown to be the same at both points[1,20], and more generally to be the same at all $k$ points on a period $k$ cycle.) Furthermore, as is clear by imagining the intermediate stages between Figs 2 and 3, this stability-determining slope has the value $\lambda = +1$ at the birth of the 2-point cycle, and then decreases through zero

towards $\lambda = -1$ as the hump in $F(X)$ continues to steepen. Beyond this point the period 2 points will in turn become unstable, and bifurcate to give an initially stable cycle of period 4. This in turn gives way to a cycle of period 8, and thence to a hierarchy of bifurcating stable cycles of periods 16, 32, 64, . . ., $2^n$. In each case, the way in which a stable cycle of period $k$ becomes unstable, simultaneously bifurcating to produce a new and initially stable cycle of period $2k$, is basically similar to the process just adumbrated for $k = 1$. A more full and rigorous account of the material covered so far is in ref. 1.

This "very beautiful bifurcation phenomenon"[22] is depicted in Fig. 4, for the example equation (3). It cannot be too strongly emphasised that the process is generic to most functions $F(X)$ with a hump of tunable steepness. Metropolis *et al.*[16] refer to this hierarchy of cycles of periods $2^n$ as the harmonics of the fixed point $X^*$.

Although this process produces an infinite sequence of cycles with periods $2^n$ ($n \to \infty$), the "window" of parameter values wherein any one cycle is stable progressively diminishes, so that the entire process is a convergent one, being bounded above by some critical parameter value. (This is true for most, but not all, functions $F(X)$: see equation (17) below.) This critical parameter value is a point of accumulation of period $2^n$ cycles. For equation (3) it is denoted $a_c$: $a_c = 3.5700 \ldots$

Beyond this point of accumulation (for example, for $a > a_c$ in equation (3)) there are an infinite number of fixed points with different periodicities, and an infinite number of different periodic cycles. There are also an uncountable number of initial points $X_0$ which give totally aperiodic (although bounded) trajectories; no matter how long the time series generated by $F(X)$ is run out, the pattern never repeats. These facts may be established by a variety of methods[1,4,20,22,29]. Such a situation, where an infinite number of different orbits can occur, has been christened "chaotic" by Li and Yorke[20].

As the parameter increases beyond the critical value, at first all these cycles have even periods, with $X_t$ alternating up and down between values above, and values below, the fixed point $X^*$. Although these cycles may in fact be very complicated (having a non-degenerate period of, say, 5,726 points before repeating), they will seem to the casual observer to be rather like a somewhat "noisy" cycle of period 2. As the parameter value continues to increase, there comes a stage (at $a = 3.6786$ . . for equation (3)) at which the first odd period cycle appears. At first these odd cycles have very long periods, but as the parameter value continues to increase cycles with smaller and smaller odd periods are picked up, until at last the three-point

Fig. 3 As for Fig. 2, except that here $a = 3.414$, as in Fig. 1b. The basic fixed point is now unstable: the slope of $F^{(2)}(X)$ at this point steepens beyond 45°, leading to the appearance of two new solutions of period 2.

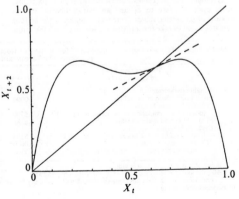

154

*Nature Vol. 261 June 10 1976*

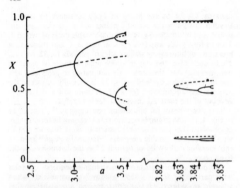

X

1.0

0.5

0

2.5   3.0   *a*   3.5   3.82   3.83   3.84   3.85

**Fig. 4** This figure illustrates some of the stable (———) and unstable (— — — —) fixed points of various periods that can arise by bifurcation processes in equation (1) in general, and equation (3) in particular. To the left, the basic stable fixed point becomes unstable and gives rise by a succession of pitchfork bifurcations to stable harmonics of period $2^n$; none of these cycles is stable beyond $a = 3.5700$. To the right, the two period 3 cycles appear by tangent bifurcation: one is initially unstable; the other is initially stable, but becomes unstable and gives way to stable harmonics of period $3 \times 2^n$, which have a point of accumulation at $a = 3.8495$. Note the change in scale on the $a$ axis, needed to put both examples on the same figure. There are infinitely many other such windows, based on cycles of higher periods.

cycle appears (at $a = 3.8284$ . . for equation (3)). Beyond this point, there are cycles with every integer period, as well as an uncountable number of asymptotically aperiodic trajectories: Li and Yorke[20] entitle their original proof of this result "Period Three Implies Chaos".

The term "chaos" evokes an image of dynamical trajectories which are indistinguishable from some stochastic process. Numerical simulations[12,15,21,23,25] of the dynamics of equation (3), (4) and other similar equations tend to confirm this impression. But, for smooth and "sensible" functions $F(X)$ such as in equations (3) and (4), the underlying mathematical fact is that for any specified parameter value there is one unique cycle that is stable, and that attracts essentially all initial points[22,29] (see ref. 4, appendix A, for a simple and lucid exposition). That is, there is one cycle that "owns" almost all initial points; the remaining infinite number of other cycles, along with the asymptotically aperiodic trajectories, own a set of points which, although uncountable, have measure zero.

As is made clear by Tables 3 and 4 below, any one particular stable cycle is likely to occupy an extraordinarily narrow window of parameter values. This fact, coupled with the long time it is likely to take for transients associated with the initial

conditions to damp out, means that in practice the unique cycle is unlikely to be unmasked, and that a stochastic description of the dynamics is likely to be appropriate, in spite of the underlying deterministic structure. This point is pursued further under the heading "practical applications", below.

The main messages of this section are summarised in Table 1, which sets out the various domains of dynamical behaviour of the equations (3) and (4) as functions of the parameters, $a$ and $r$ respectively, that determine the severity of the nonlinear response. These properties can be understood qualitatively in a graphical way, and are generic to any well behaved $F(X)$ in equation (1).

We now proceed to a more detailed discussion of the mathematical structure of the chaotic regime for analytical functions, and then to the practical problems alluded to above and to a consideration of the behavioural peculiarities exhibited by non-analytical functions (such as those in the two right hand columns of Table 1).

## Fine structure of the chaotic regime

We have seen how the original fixed point $X^*$ bifurcates to give harmonics of period $2^n$. But how do new cycles of period $k$ arise?

The general process is illustrated in Fig. 5, which shows how period 3 cycles originate. By an obvious extension of the notation introduced in equation (8), populations three generations apart are related by

$$X_{t+3} = F^{(3)}(X_t) \qquad (11)$$

If the hump in $F(X)$ is sufficiently steep, the threefold iteration will produce a function $F^{(3)}(X)$ with 4 humps, as shown in Fig. 5 for the $F(X)$ of equation (3). At first (for $a < 3.8284$ . . in equation 3) the 45° line intersects this curve only at the single point $X^*$ (and at $X = 0$), as shown by the solid curve in Fig. 5. As the hump in $F(X)$ steepens, the hills and valleys in $F^{(3)}(X)$ become more pronounced, until simultaneously the first two valleys sink and the final hill rises to touch the 45° line, and then to intercept it at 6 new points, as shown by the dashed curve in Fig. 5. These 6 points divide into two distinct three-point cycles. As can be made plausible by imagining the intermediate stages in Fig. 5, it can be shown that the stability-determining slope of $F^{(3)}(X)$ at three of these points has a common value, which is $\lambda^{(3)} = +1$ at their birth, and thereafter steepens beyond $+1$: this period 3 cycle is never stable. The slope of $F^{(3)}(X)$ at the other three points begins at $\lambda^{(3)} = +1$, and then decreases towards zero, resulting in a stable cycle of period 3. As $F(X)$ continues to steepen, the slope $\lambda^{(3)}$ for this initially stable three-point cycle decreases beyond $-1$; the cycle becomes unstable, and gives rise by the bifurcation process discussed in the previous section to stable cycles of period 6, 12, 24, . . ., $3 \times 2^n$. This birth of a stable and unstable pair of period 3 cycles, and the subsequent harmonics which arise as the initially stable cycle becomes unstable, are illustrated to the right of Fig. 4.

**Table 1** Summary of the way various "single-hump" functions $F(X)$, from equation (1), behave in the chaotic region, distinguishing the dynamical properties which are generic from those which are not

| The function $F(X)$ of equation (1) | $aX(1-X)$ | $X \exp[r(1-X)]$ | $aX$; if $X < \frac{1}{2}$ $a(1-X)$; if $X > \frac{1}{2}$ | $\lambda X$; if $X < 1$ $\lambda X^{1-b}$; if $X > 1$ |
|---|---|---|---|---|
| Tunable parameter | $a$ | $r$ | $a$ | $b$ |
| Fixed point becomes unstable | 3.0000 | 2.0000 | 1.0000* | 2.0000 |
| "Chaotic" region begins [point of accumulation of cycles of period $2^n$] | 3.5700 | 2.6924 | 1.0000 | 2.0000 |
| First odd-period cycle appears | 3.6786 | 2.8332 | 1.4142 | 2.6180 |
| Cycle with period 3 appears [and therefore every integer period present] | 3.8284 | 3.1024 | 1.6180 | 3.0000 |
| "Chaotic" region ends | 4.0000† | ∞‡ | 2.000† | ∞‡ |
| Are there stable cycles in the chaotic region? | Yes | Yes | No | No |

\* Below this $a$ value, $X = 0$ is stable.
† All solutions are attracted to $-\infty$ for $a$ values beyond this.
‡ In practice, as $r$ or $b$ becomes large enough, $X$ will eventually be carried so low as to be effectively zero, thus producing extinction in models of biological populations.

*Nature Vol. 261 June 10 1976*　**463**

Table 2 Catalogue of the number of periodic points, and of the various cycles (with periods $k = 1$ up to 12), arising from equation (1) with a single-humped function $F(X)$

| $k$ | 1 | 2 | 3 | 4 | 5 | 6 | 7 | 8 | 9 | 10 | 11 | 12 |
|---|---|---|---|---|---|---|---|---|---|---|---|---|
| Possible total number of points with period $k$ | 2 | 4 | 8 | 16 | 32 | 64 | 128 | 256 | 512 | 1,024 | 2,048 | 4,096 |
| Possible total number of points with non-degenerate period $k$ | 2 | 2 | 6 | 12 | 30 | 54 | 126 | 240 | 504 | 990 | 2,046 | 4,020 |
| Total number of cycles of period $k$, including those which are degenerate and/or harmonics and/or never locally stable | 2 | 3 | 4 | 6 | 8 | 14 | 20 | 36 | 60 | 108 | 188 | 352 |
| Total number of non-degenerate cycles (including harmonics and unstable cycles) | 2 | 1 | 2 | 3 | 6 | 9 | 18 | 30 | 56 | 99 | 186 | 335 |
| Total number of non-degenerate, stable cycles (including harmonics) | 1 | 1 | 1 | 2 | 3 | 5 | 9 | 16 | 28 | 51 | 93 | 170 |
| Total number of non-degenerate, stable cycles whose basic period is $k$ (that is, excluding harmonics) | 1 | – | 1 | 1 | 3 | 4 | 9 | 14 | 28 | 48 | 93 | 165 |

There are, therefore, two basic kinds of bifurcation processes[1,4] for first order difference equations. Truly new cycles of period $k$ arise in pairs (one stable, one unstable) as the hills and valleys of higher iterates of $F(X)$ move, respectively, up and down to intercept the 45° line, as typified by Fig. 5. Such cycles are born at the moment when the hills and valleys become tangent to the 45° line, and the initial slope of the curve $F^{(k)}$ at the points is thus $\lambda^{(k)} = +1$: this type of bifurcation may be called[1,4] a tangent bifurcation or a $\lambda = +1$ bifurcation. Conversely, an originally stable cycle of period $k$ may become unstable as $F(X)$ steepens. This happens when the slope of $F^{(k)}$ at these period $k$ points steepens beyond $\lambda^{(k)} = -1$, whereupon a new and initially stable cycle of period $2k$ is born in the way typified by Figs 2 and 3. This type of bifurcation may be called a pitchfork bifurcation (borrowing an image from the left hand side of Fig. 4) or a $\lambda = -1$ bifurcation[1,4].

Putting all this together, we conclude that as the parameters in $F(X)$ are varied the fundamental, stable dynamical units are cycles of basic period $k$, which arise by tangent bifurcation, along with their associated cascade of harmonics of periods $k2^n$, which arise by pitchfork bifurcation. On this basis, the constant equilibrium solution $X^*$ and the subsequent hierarchy of stable cycles of periods $2^n$ is merely a special case, albeit a conspicuously important one (namely $k=1$), of a general phenomenon. In addition, remember[1,4,22,29] that for sensible, analytical functions (such as, for example, those in equations (3) and (4)) there is a unique stable cycle for each value of the parameter in $F(X)$. The entire range of parameter values ($1 < a < 4$ in equation (3), $0 < r$ in equation (4)) may thus be regarded as made up of infinitely many windows of parameter

values—some large, some unimaginably small—each corresponding to a single one of these basic dynamical units. Tables 3 and 4, below, illustrate this notion. These windows are divided from each other by points (the points of accumulation of the harmonics of period $k2^n$) at which the system is truly chaotic, with no attractive cycle: although there are infinitely many such special parameter values, they have measure zero on the interval of all values.

How are these various cycles arranged along the interval of relevant parameter values? This question has to my knowledge been answered independently by at least 6 groups of people, who have seen the problem in the context of combinatorial theory[16,30], numerical analysis[13,14], population biology[1], and dynamical systems theory[22,31] (broadly defined).

A simple-minded approach (which has the advantage of requiring little technical apparatus, and the disadvantage of being rather clumsy) consists of first answering the question, how many period $k$ points can there be? That is, how many distinct solutions can there be to the equation

$$X^*_k = F^{(k)}(X^*_k)? \qquad (12)$$

If the function $F(X)$ is sufficiently steeply humped, as it will be once the parameter values are sufficiently large, each successive iteration doubles the number of humps, so that $F^{(k)}(X)$ has $2^{k-1}$ humps. For large enough parameter values, all these hills and valleys will intersect the 45° line, producing $2^k$ fixed points of period $k$. These are listed for $k \leqslant 12$ in the top row of Table 2. Such a list includes degenerate points of period $k$, whose period is a submultiple of $k$; in particular, the two period 1 points ($X=0$ and $X^*$) are degenerate solutions of equation (12) for all $k$. By working from left to right across Table 2, these degenerate points can be subtracted out, to leave the total number of non-degenerate points of basic period $k$, as listed in the second row of Table 2. More sophisticated ways of arriving at this result are given elsewhere[13,14,16,22,30,31].

For example, there eventually are $2^6=64$ points with period 6. These include the two points of period 1, the period 2 "harmonic" cycle, and the stable and unstable pair of triplets of points with period 3, for a total of 10 points whose basic period is a submultiple of 6; this leaves 54 points whose basic period is 6.

The $2^k$ period $k$ points are arranged into various cycles of period $k$, or submultiples thereof, which appear in succession by either tangent or pitchfork bifurcation as the parameters in $F(X)$ are varied. The third row in Table 2 catalogues the total number of distinct cycles of period $k$ which so appear. In the fourth row[14], the degenerate cycles are subtracted out, to give the total number of non-degenerate cycles of period $k$: these numbers must equal those of the second row divided by $k$. This fourth row includes the (stable) harmonics which arise by pitchfork bifurcation, and the pairs of stable-unstable cycles arising by tangent bifurcation. By subtracting out the cycles which are unstable from birth, the total number of possible stable cycles is given in row five; these figures can also be obtained by less pedestrian methods[13,16,30]. Finally we may subtract out the stable cycles which arise by pitchfork bifurcation, as harmonics of some simpler cycle, to arrive at the final

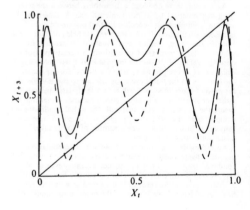

Fig. 5 The relationship between $X_{t+3}$ and $X_t$, obtained by three iterations of equation (3). The solid curve is for $a = 3.7$, and only intersects the 45° line once. As $a$ increases, the hills and valleys become more pronounced. The dashed curve is for $a = 3.9$, and six new period 3 points have appeared (arranged as two cycles, each of period 3).

464

*Nature Vol. 261 June 10 1976*

**Table 3** A catalogue of the stable cycles (with basic periods up to 6) for the equation $X_{t+1} = aX_t(1 - X_t)$

| Period of basic cycle | *a* value at which: Basic cycle first appears | Basic cycle becomes unstable | Subsequent cascade of "harmonics" with period $k2^n$ all become unstable | Width of the range of *a* values over which the basic cycle, or one of its harmonics, is attractive |
|---|---|---|---|---|
| 1 | 1.0000 | 3.0000 | 3.5700 | 2.5700 |
| 3 | 3.8284 | 3.8415 | 3.8495 | 0.0211 |
| 4 | 3.9601 | 3.9608 | 3.9612 | 0.0011 |
| 5(a) | 3.7382 | 3.7411 | 3.7430 | 0.0048 |
| 5(b) | 3.9056 | 3.9061 | 3.9065 | 0.0009 |
| 5(c) | 3.99026 | 3.99030 | 3.99032 | 0.00006 |
| 6(a) | 3.6265 | 3.6304 | 3.6327 | 0.0062 |
| 6(b) | 3.937516 | 3.937596 | 3.937649 | 0.000133 |
| 6(c) | 3.977760 | 3.977784 | 3.977800 | 0.000040 |
| 6(d) | 3.997583 | 3.997585 | 3.997586 | 0.000003 |

row in Table 2, which lists the number of stable cycles whose basic period is $k$.

Returning to the example of period 6, we have already noted the five degenerate cycles whose periods are submultiples of 6. The remaining 54 points are parcelled out into one cycle of period 6 which arises as the harmonic of the only stable three-point cycle, and four distinct pairs of period 6 cycles (that is, four initially stable ones and four unstable ones) which arise by successive tangent bifurcations. Thus, reading from the foot of the column for period 6 in Table 2, we get the numbers 4, 5, 9, 14.

Using various labelling tricks, or techniques from combinatorial theory, it is also possible to give a generic list of the order in which the various cycles appear[1,13,16,22]. For example, the basic stable cycles of periods 3, 5, 6 (of which there are respectively 1, 3, 4) must appear in the order 6, 5, 3, 5, 6, 6, 5, 6: compare Tables 3 and 4. Metropolis *et al.*[16] give the explicit such generic list for all cycles of period $k \leqslant 11$.

As a corollary it follows that, given the most recent cycle to appear, it is possible (at least in principle) to catalogue all the cycles which have appeared up to this point. An especially elegant way of doing this is given by Smale and Williams[22], who show, for example, that when the stable cycle of period 3 first originates, the total number of other points with periods $k$, $N_k$, which have appeared by this stage satisfy the Fibonacci series, $N_k = 2, 4, 5, 8, 12, 19, 30, 48, 77, 124, 200, 323$ for $k = 1, 2, \ldots, 12$: this is to be contrasted with the total number of points of period $k$ which will eventually appear (the top row of Table 2) as $F(X)$ continues to steepen.

Such catalogues of the total number of fixed points, and of their order of appearance, are relatively easy to construct. For any particular function $F(X)$, the numerical task of finding the windows of parameter values wherein any one cycle or its harmonics is stable is, in contrast, relatively tedious and inelegant. Before giving such results, two critical parameter values of special significance should be mentioned.

Hoppensteadt and Hyman[21] have given a simple graphical method for locating the parameter value in the chaotic regime at which the first odd period cycle appears. Their analytic recipe is as follows. Let $\alpha$ be the parameter which tunes the steepness of $F(X)$ (for example, $\alpha = a$ for equation (3), $\alpha = r$ for equation (4)), $X^*(\alpha)$ be the fixed point of period 1 (the nontrivial solution of equation (5)), and $X_{max}(\alpha)$ the maximum value attainable from iterations of equation (1) (that is, the value of $F(X)$ at its hump or stationary point). The first odd period cycle appears for that value of $\alpha$ which satisfies[21,31]

$$X^*(\alpha) = F^{(2)}(X_{max}(\alpha)) \qquad (13)$$

As mentioned above, another critical value is that where the period 3 cycle first appears. This parameter value may be found numerically from the solutions of the third iterate of equation (1): for equation (3) it is[14] $a = 1 + \sqrt{8}$.

Myrberg[13] (for all $k \leqslant 10$) and Metropolis *et al.*[16]. (for all $k \leqslant 7$) have given numerical information about the stable cycles in equation (3). They do not give the windows of parameter

values, but only the single value at which a given cycle is maximally stable; that is, the value of $a$ for which the stability-determining slope of $F^{(k)}(X)$ is zero, $\lambda^{(k)} = 0$. Since the slope of the $k$-times iterated map $F^{(k)}$ at any point on a period $k$ cycle is simply equal to the product of the slopes of $F(X)$ at each of the points $X^*_k$ on this cycle[1,8,20], the requirement $\lambda^{(k)} = 0$ implies that $X = A$ (the stationary point of $F(X)$, where $\lambda^{(1)} = 0$) is one of the periodic points in question, which considerably simplifies the numerical calculations.

For each basic cycle of period $k$ (as catalogued in the last row of Table 2), it is more interesting to know the parameter values at which: (1) the cycle first appears (by tangent bifurcation); (2) the basic cycle becomes unstable (giving rise by successive pitchfork bifurcations to a cascade of harmonics of periods $k2^n$); (3) all the harmonics become unstable (the point of accumulation of the period $k2^n$ cycles). Tables 3 and 4 extend the work of May and Oster[1], to give this numerical information for equations (3) and (4), respectively. (The points of accumulation are not ground out mindlessly, but are calculated by a rapidly convergent iterative procedure, see ref. 1, appendix A.) Some of these results have also been obtained by Gumowski and Mira[32].

## Practical problems

Referring to the paradigmatic example of equation (3), we can now see that the parameter interval $1 < a < 4$ is made up of a one-dimensional mosaic of infinitely many windows of $a$-values, in each of which a unique cycle of period $k$, or one of its harmonics, attracts essentially all initial points. Of these windows, that for $1 < a < 3.5700 \ldots$ corresponding to $k = 1$ and its harmonics is by far the widest and most conspicuous. Beyond the first point of accumulation, it can be seen from Table 3 that these windows are narrow, even for cycles of quite low periods, and the windows rapidly become very tiny as $k$ increases.

As a result, there develops a dichotomy between the underlying mathematical behaviour (which is exactly determinable) and the "commonsense" conclusions that one would draw from numerical simulations. If the parameter $a$ is held constant at one value in the chaotic region, and equation (3) iterated for an arbitrarily large number of generations, a density plot of the observed values of $X_t$ on the interval 0 to 1 will settle into $k$ equal spikes (more precisely, delta functions) corresponding to the $k$ points on the stable cycle appropriate to this $a$-value. But for most $a$-values this cycle will have a fairly large period, and moreover it will typically take many thousands of generations before the transients associated with the initial conditions are damped out: thus the density plot produced by numerical simulations usually looks like a sample of points taken from some continuous distribution.

An especially interesting set of numerical computations are due to Hoppensteadt (personal communication) who has combined many iterations to produce a density plot of $X_t$ for each one of a sequence of $a$-values, gradually increasing from $3.5700 \ldots$ to 4. These results are displayed as a movie. As can be expected from Table 3, some of the more conspicuous cycles

*Nature Vol. 261 June 10 1976*

do show up as sets of delta functions: the 3-cycle and its first few harmonics; the first 5-cycle; the first 6-cycle. But for most values of $a$ the density plot looks like the sample function of a random process. This is particularly true in the neighbourhood of the $a$-value where the first odd cycle appears ($a=3.6786$ . .), and again in the neighbourhood of $a=4$: this is not surprising, because each of these locations is a point of accumulation of points of accumulation. Despite the underlying discontinuous changes in the periodicities of the stable cycles, the observed density pattern tends to vary smoothly. For example, as $a$ increases toward the value at which the 3-cycle appears, the density plot tends to concentrate around three points, and it smoothly diffuses away from these three points after the 3-cycle and all its harmonics become unstable.

I think the most interesting mathematical problem lies in designing a way to construct some approximate and "effectively continuous" density spectrum, despite the fact that the exact density function is determinable and is always a set of delta functions. Perhaps such techniques have already been developed in ergodic theory[33] (which lies at the foundations of statistical mechanics), as for example in the use of "coarse-grained observers". I do not know.

Such an effectively stochastic description of the dynamical properties of equation (4) for large $r$ has been provided[28], albeit by tactical tricks peculiar to that equation rather than by any general method. As $r$ increases beyond about 3, the trajectories generated by this equation are, to an increasingly good approximation, almost periodic with period $(1/r)\exp(r-1)$.

The opinion I am airing in this section is that although the exquisite fine structure of the chaotic regime is mathematically fascinating, it is irrelevant for most practical purposes. What seems called for is some effectively stochastic description of the deterministic dynamics. Whereas the various statements about the different cycles and their order of appearance can be made in generic fashion, such stochastic description of the actual dynamics will be quite different for different $F(X)$: witness the difference between the behaviour of equation (4), which for large $r$ is almost periodic "outbreaks" spaced many generations apart, versus the behaviour of equation (3), which for $a \to 4$ is not very different from a series of Bernoulli coin flips.

## Mathematical curiosities

As discussed above, the essential reason for the existence of a succession of stable cycles throughout the "chaotic" regime is that as each new pair of cycles is born by tangent bifurcation (see Fig. 5), one of them is at first stable, by virtue of the way the smoothly rounded hills and valleys intercept the 45° line. For analytical functions $F(X)$, the only parameter values for which the density plot or "invariant measure" is continuous and truly ergodic are at the points of accumulation of harmonics, which divide one stable cycle from the next. Such exceptional parameter values have found applications, for example, in the use of equation (3) with $a=4$ as a random number generator[34,35]: it has a continuous density function proportional to $[X(1-X)]^{-\frac{1}{2}}$ in the interval $0 < X < 1$.

Non-analytical functions $F(X)$ in which the hump is in fact a spike provide an interesting special case. Here we may imagine spikey hills and valleys moving to intercept the 45° line in Fig. 5, and it may be that both the cycles born by tangent bifurcation are unstable from the outset (one having $\lambda^{(k)} > 1$, the other $\lambda^{(k)} < -1$), for all $k > 1$. There are then no stable cycles in the chaotic regime, which is therefore literally chaotic with a continuous and truly ergodic density distribution function.

One simple example is provided by

$$X_{t+1} = aX_t;\ \text{if } X_t < \tfrac{1}{2} \qquad (14)$$
$$X_{t+1} = a(1-X_t);\ \text{if } X_t > \tfrac{1}{2}$$

defined on the interval $0 < X < 1$. For $0 < a < 1$, all trajectories are attracted to $X=0$; for $1 < a < 2$, there are infinitely many periodic trajectories, along with an uncountable number of aperiodic trajectories, none of which are locally stable. The first odd period cycle appears at $a=\sqrt{2}$, and all integer periods are represented beyond $a=(1+\sqrt{5})/2$. Kac[36] has given a careful discussion of the case $a=2$. Another example, this time with an extensive biological pedigree[1-3], is the equation

$$X_{t+1} = \lambda X_t;\ \text{if } X_t < 1 \qquad (15)$$
$$X_{t+1} = \lambda X_t^{1-b};\ \text{if } X_t > 1$$

If $\lambda > 1$ this possesses a globally stable equilibrium point for $b < 2$. For $b > 2$ there is again true chaos, with no stable cycles: the first odd cycle appears at $b=(3+\sqrt{5})/2$, and all integer periods are present beyond $b=3$. The dynamical properties of equations (14) and (15) are summarised to the right of Table 2.

The absence of analyticity is a necessary, but not a sufficient, condition for truly random behaviour[31]. Consider, for example,

$$X_{t+1} = (a/2)X_t;\ \text{if } X_t < \tfrac{1}{2}$$
$$X_{t+1} = aX_t(1-X_t);\ \text{if } X_t > \tfrac{1}{2} \qquad (16)$$

This is the parabola of equation (3) and Fig. 1, but with the left hand half of $F(X)$ flattened into a straight line. This equation does possess windows of $a$ values, each with its own stable cycle, as described generically above. The stability-determining slopes $\lambda^{(k)}$ vary, however, discontinuously with the parameter $a$, and the widths of the simpler stable regions are narrower than for equation (3): the fixed point becomes unstable at $a=3$; the point of accumulation of the subsequent harmonics is at $a=3.27$ . .; the first odd cycle appears at $a=3.44$ . .; the 3-point cycle at $a=3.67$ . . (compare the first column in Table 1).

These eccentricities of behaviour manifested by non-analytical functions may be of interest for exploring formal questions in ergodic theory. I think, however, that they have no relevance to models in the biological and social sciences, where functions such as $F(X)$ should be analytical. This view is elaborated elsewhere[37].

As a final curiosity, consider the equation

$$X_{t+1} = \lambda X_t[1 + X_t]^{-b} \qquad (17)$$

---

**Table 4** Catalogue of the stable cycles (with basic periods up to 6) for the equation $X_{t+1} = X_t \exp[r(1-X_t)]$

| Period of basic cycle | *r* value at which: Basic cycle first appears | Basic cycle becomes unstable | Subsequent cascade of "harmonics" with period $k2^n$ all become unstable | Width of the range of *r* values over with the basic cycle, or one of its harmonics, is attractive |
|---|---|---|---|---|
| 1 | 0.0000 | 2.0000 | 2.6924 | 2.6924 |
| 3 | 3.1024 | 3.1596 | 3.1957 | 0.0933 |
| 4 | 3.5855 | 3.6043 | 3.6153 | 0.0298 |
| 5(a) | 2.9161 | 2.9222 | 2.9256 | 0.0095 |
| 5(b) | 3.3632 | 3.3664 | 3.3682 | 0.0050 |
| 5(c) | 3.9206 | 3.9255 | 3.9347 | 0.0141 |
| 6(a) | 2.7714 | 2.7761 | 2.7789 | 0.0075 |
| 6(b) | 3.4558 | 3.4563 | 3.4567 | 0.0009 |
| 6(c) | 3.7736 | 3.7745 | 3.7750 | 0.0014 |
| 6(d) | 4.1797 | 4.1848 | 4.1880 | 0.0083 |

*Nature Vol. 261 June 10 1976*

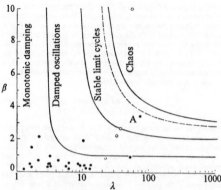

**Fig. 6** The solid lines demarcate the stability domains for the density dependence parameter, $\beta$, and the population growth rate, $\lambda$, in equation (17); the dashed line shows where 2-point cycles give way to higher cycles of period $2^n$. The solid circles come from analyses of life table data on field populations, and the open circles from laboratory populations (from ref. 3, after ref. 39).

This has been used to fit a considerable amount of data on insect populations[38,39]. Its stability behaviour, as a function of the two parameters $\lambda$ and $\beta$, is illustrated in Fig. 6. Notice that for $\lambda < 7.39$ . . there is a globally stable equilibrium point for all $\beta$; for $7.39 . . < \lambda < 12.50$ . . this fixed point becomes unstable for sufficiently large $\beta$, bifurcating to a stable 2-point cycle which is the solution for all larger $\beta$; as $\lambda$ increases through the range $12.50 . . < \lambda < 14.77$ . . various other harmonics of period $2^n$ appear in turn. The hierarchy of bifurcating cycles of period $2^n$ is thus truncated, and the point of accumulation and subsequent regime of chaos is not achieved (even for arbitrarily large $\beta$) until $\lambda > 14.77$ . . .

## Applications

The fact that the simple and deterministic equation (1) can possess dynamical trajectories which look like some sort of random noise has disturbing practical implications. It means, for example, that apparently erratic fluctuations in the census data for an animal population need not necessarily betoken either the vagaries of an unpredictable environment or sampling errors: they may simply derive from a rigidly deterministic population growth relationship such as equation (1). This point is discussed more fully and carefully elsewhere[1].

Alternatively, it may be observed that in the chaotic regime arbitrarily close initial conditions can lead to trajectories which, after a sufficiently long time, diverge widely. This means that, even if we have a simple model in which all the parameters are determined exactly, long term prediction is nevertheless impossible. In a meteorological context, Lorenz[15] has called this general phenomenon the "butterfly effect": even if the atmosphere could be described by a deterministic model in which all parameters were known, the fluttering of a butterfly's wings could alter the initial conditions, and thus (in the chaotic regime) alter the long term prediction.

Fluid turbulence provides a classic example where, as a parameter (the Reynolds number) is tuned in a set of deterministic equations (the Navier–Stokes equations), the motion can undergo an abrupt transition from some stable configuration (for example, laminar flow) into an apparently stochastic, chaotic regime. Various models, based on the Navier–Stokes differential equations, have been proposed as mathematical metaphors for this process[15,40,41]. In a recent review of the theory of turbulence, Martin[42] has observed that the one-

dimensional difference equation (1) may be useful in this context. Compared with the earlier models[15,40,41], it has the disadvantage of being even more abstractly metaphorical, and the advantage of having a spectrum of dynamical behaviour which is more richly complicated yet more amenable to analytical investigation.

A more down-to-earth application is possible in the use of equation (1) to fit data[1,2,3,38,39,43] on biological populations with discrete, non-overlapping generations, as is the case for many temperate zone arthropods. Figure 6 shows the parameter values $\lambda$ and $\beta$ that are estimated[39] for 24 natural populations and 4 laboratory populations when equation (17) is fitted to the available data. The figure also shows the theoretical stability domains: a stable point; its stable harmonics (stable cycles of period $2^n$); chaos. The natural populations tend to have stable equilibrium point behaviour. The laboratory populations tend to show oscillatory or chaotic behaviour; their behaviour may be exaggerated nonlinear because of the absence, in a laboratory setting, of many natural mortality factors. It is perhaps suggestive that the most oscillatory natural population (labelled $A$ in Fig. 6) is the Colorado potato beetle, whose present relationship with its host plant lacks an evolutionary pedigree. These remarks are only tentative, and must be treated with caution for several reasons. Two of the main caveats are that there are technical difficulties in selecting and reducing the data, and that there are no single species populations in the natural world: to obtain a one-dimensional difference equation by replacing a population's interactions with its biological and physical environment by passive parameters (such as $\lambda$ and $\beta$) may do great violence to the reality.

Some of the many other areas where these ideas have found applications were alluded to in the second section, above[8-11]. One aim of this review article is to provoke applications in yet other fields.

## Related phenomena in higher dimensions

Pairs of coupled, first-order difference equations (equivalent to a single second-order equation) have been investigated in several contexts[4,44-46], particularly in the study of temperate zone arthropod prey–predator systems[2-4,23,47]. In these two-dimensional systems, the complications in the dynamical behaviour are further compounded by such facts as: (1) even for analytical functions, there can be truly chaotic behaviour (as for equations (14) and (15)), corresponding to so-called "strange attractors"; and (2) two or more different stable states (for example, a stable point and a stable cycle of period 3) can occur together for the same parameter values[4]. In addition, the manifestation of these phenomena usually requires less severe nonlinearities (less steeply humped $F(X)$) than for the one-dimensional case.

Similar systems of first-order ordinary differential equations, or two coupled first-order differential equations, have much simpler dynamical behaviour, made up of stable and unstable points and limit cycles[48]. This is basically because in continuous two-dimensional systems the inside and outside of closed curves can be distinguished; dynamic trajectories cannot cross each other. The situation becomes qualitatively more complicated, and in many ways analogous to first-order difference equations, when one moves to systems of three or more coupled, first-order ordinary differential equations (that is, three-dimensional systems of ordinary differential equations). Scanlon (personal communication) has argued that chaotic behaviour and "strange attractors", that is solutions which are neither points nor periodic orbits[48], are typical of such systems. Some well studied examples arise in models for reaction–diffusion systems in chemistry and biology[49], and in the models of Lorenz[15] (three dimensions) and Ruelle and Takens[40] (four dimensions) referred to above. The analysis of these systems is, by virtue of their higher dimensionality, much less transparent than for equation (1).

An explicit and rather surprising example of a system which

*Nature Vol. 261 June 10 1976*

has recently been studied from this viewpoint is the ordinary differential equations used in ecology to describe competing species. For one or two species these systems are very tame: dynamic trajectories will converge on some stable equilibrium point (which may represent coexistence, or one or both species becoming extinct). As Smale[50] has recently shown, however, for 3 or more species these general equations can, in a certain reasonable and well-defined sense, be compatible with any dynamical behaviour. Smale's[50] discussion is generic and abstract: a specific study of the very peculiar dynamics which can be exhibited by the familiar Lotka-Volterra equations once there are 3 competitors is given by May and Leonard[51].

## Conclusion

In spite of the practical problems which remain to be solved, the ideas developed in this review have obvious applications in many areas.

The most important applications, however, may be pedagogical.

The elegant body of mathematical theory pertaining to linear systems (Fourier analysis, orthogonal functions, and so on), and its successful application to many fundamentally linear problems in the physical sciences, tends to dominate even moderately advanced University courses in mathematics and theoretical physics. The mathematical intuition so developed ill equips the student to confront the bizarre behaviour exhibited by the simplest of discrete nonlinear systems, such as equation (3). Yet such nonlinear systems are surely the rule, not the exception, outside the physical sciences.

I would therefore urge that people be introduced to, say, equation (3) early in their mathematical education. This equation can be studied phenomenologically by iterating it on a calculator, or even by hand. Its study does not involve as much conceptual sophistication as does elementary calculus. Such study would greatly enrich the student's intuition about nonlinear systems.

Not only in research, but also in the everyday world of politics and economics, we would all be better off if more people realised that simple nonlinear systems do not necessarily possess simple dynamical properties.

I have received much help from F. C. Hoppensteadt, H. E. Huppert, A. I. Mees, C. J. Preston, S. Smale, J. A. Yorke, and particularly from G. F. Oster. This work was supported in part by the NSF.

1 May, R. M., and Oster, G. F., *Am. Nat.*, **110** (in the press).
2 Varley, G. C., Gradwell, G. R., and Hassell, M. P., *Insect Population Ecology* (Blackwell, Oxford, 1973).
3 May, R. M. (ed.), *Theoretical Ecology: Principles and Applications* (Blackwell, Oxford, 1976).
4 Guckenheimer, J., Oster, G. F., and Ipaktchi, A., *Theor. Pop. Biol.* (in the press).
5 Oster, G. F., Ipaktchi, A., and Rocklin, I., *Theor. Pop. Biol.* (in the press).
6 Asmussen, M. A., and Feldman, M. W., *J. theor. Biol.* (in the press).
7 Hoppensteadt, F. C., *Mathematical Theories of Populations: Demographics, Genetics and Epidemics* (SIAM, Philadelphia, 1975).
8 Samuelson, P. A., *Foundations of Economic Analysis* (Harvard University Press, Cambridge, Massachusetts, 1947).
9 Goodwin, R. E., *Econometrica*, **19**, 1–17 (1951).
10 Baumol, W. J., *Economic Dynamics*, 3rd ed. (Macmillan, New York, 1970).
11 See, for example, Kemeny, J., and Snell, J. L., *Mathematical Models in the Social Sciences* (MIT Press, Cambridge, Massachusetts, 1972).
12 Chaundy, T. W., and Phillips, E., *Q. Jl Math. Oxford*, **7**, 74–80 (1936).
13 Myrberg, P. J., *Ann. Akad. Sc. Fennicae, A*, **I**, No. 336/3 (1963).
14 Myrberg, P. J., *Ann. Akad. Sc. Fennicae, A*, **I**, No. 259 (1958).
15 Lorenz, E. N., *J. Atmos. Sci.*, **20**, 130–141 (1963); *Tellus*, **16**, 1–11 (1964).
16 Metropolis, N., Stein, M. L., and Stein, P. R., *J. Combinatorial Theory*, **15(A)**, 25–44 (1973).
17 Maynard Smith, J., *Mathematical Ideas in Biology* (Cambridge University Press, Cambridge, 1968).
18 Krebs, C. J., *Ecology* (Harper and Row, New York, 1972).
19 May, R. M., *Am. Nat.*, **107**, 46–57 (1972).
20 Li, T-Y., and Yorke, J. A., *Am. Math. Monthly*, **82**, 985–992 (1975).
21 Hoppensteadt, F. C., and Hyman, J. M. (Courant Institute, New York University: preprint, 1975).
22 Smale, S., and Williams, R. (Department of Mathematics, Berkeley: preprint, 1976).
23 May, R. M., *Science*, **186**, 645–647 (1974).
24 Moran, P. A. P., *Biometrics*, **6**, 250–258 (1950).
25 Ricker, W. E., *J. Fish. Res. Bd. Can.*, **11**, 559–623 (1954).
26 Cook, L. M., *Nature*, **207**, 316 (1965).
27 Macfadyen, A., *Animal Ecology: Aims and Methods* (Pitman, London, 1963).
28 May, R. M., *J. theor. Biol.*, **51**, 511–524 (1975).
29 Guckenheimer, J., *Proc. AMS Symposia in Pure Math., XIV*, 95–124 (1970).
30 Gilbert, E. N., and Riordan, J., *Illinois J. Math.*, **5**, 657–667 (1961).
31 Preston, C. J. (King's College, Cambridge: preprint, 1976).
32 Gumowski, I., and Mira, C. *r, hebd. Séanc. Acad. Sci., Paris*, **281a**, 45–48 (1975); **282a**, 219–222 (1976).
33 Layzer, D., *Sci. Am.*, **233(6)**, 56–69 (1975).
34 Ulam, S. M., *Proc. Int. Congr. Math.1950, Cambridge, Mass.; Vol. II*, pp. 264–273 (AMS, Providence R.I., 1950).
35 Ulam, S. M., and von Neumann, J., *Bull. Am. math. Soc.* (abstr.), **53**, 1120 (1947).
36 Kac, M., *Ann. Math.*, **47**, 33–49 (1946).
37 May, R. M., *Science*, **181**, 1074 (1973).
38 Hassell, M. P., *J. Anim. Ecol.*, **44**, 283–296 (1974).
39 Hassell, M. P., Lawton, J. H. and May, R. M., *J. Anim. Ecol.* (in the press).
40 Ruelle, D., and Takens, F., *Comm. math. Phys.*, **20**, 167–192 (1971).
41 Landau, L. D., and Lifshitz, E. M., *Fluid Mechanics* (Pergamon, London, 1959).
42 Martin, P. C., *Proc. Int. Conf. on Statistical Physics, 1975, Budapest* (Hungarian Acad. Sci., Budapest, in the press).
43 Southwood, T. R. E., in *Insects, Science and Society* (edit. by Pimentel, D.), 151–199 (Academic, New York, 1975).
44 Metropolis, N., Stein, M. L., and Stein, P. R., *Numer. Math.*, **10**, 1–19 (1967).
45 Gumowski, I., and Mira, C. *Automatica*, **5**, 303–317 (1969).
46 Stein, P. R., and Ulam, S. M., *Rosprawy Mat.*, **39**, 1–66 (1964).
47 Beddington, J. R., Free, C. A., and Lawton, J. H., *Nature*, **255**, 58–60 (1975).
48 Hirsch, M. W., and Smale, S., *Differential Equations, Dynamical Systems and Linear Algebra* (Academic, New York, 1974).
49 Kolata, G. B., *Science*, **189**, 984–985 (1975).
50 Smale, S. (Department of Mathematics, Berkeley: preprint, 1976).
51 May, R. M., and Leonard, W. J., *SIAM J. Appl. Math.*, **29**, 243–253 (1975).

160

*Journal of Statistical Physics, Vol. 19, No. 1, 1978*

# Quantitative Universality for a Class of Nonlinear Transformations

**Mitchell J. Feigenbaum** [1]

*Received October 31, 1977*

A large class of recursion relations $x_{n+1} = \lambda f(x_n)$ exhibiting infinite bifurcation is shown to possess a rich quantitative structure essentially independent of the recursion function. The functions considered all have a unique differentiable maximum $\bar{x}$. With $f(\bar{x}) - f(x) \sim |x - \bar{x}|^z$ (for $|x - \bar{x}|$ sufficiently small), $z > 1$, the universal details depend only upon $z$. In particular, the local structure of high-order stability sets is shown to approach universality, rescaling in successive bifurcations, asymptotically by the ratio $\alpha$ ($\alpha = 2.5029078750957...$ for $z = 2$). This structure is determined by a universal function $g^*(x)$, where the $2^n$th iterate of $f$, $f^{(n)}$, converges locally to $\alpha^{-n} g^*(\alpha^n x)$ for large $n$. For the class of $f$'s considered, there exists a $\lambda_n$ such that a $2^n$-point stable limit cycle including $\bar{x}$ exists; $\lambda_\infty - \lambda_n \sim \delta^{-n}$ ($\delta = 4.669201609103...$ for $z = 2$). The numbers $\alpha$ and $\delta$ have been computationally determined for a range of $z$ through their definitions, for a variety of $f$'s for each $z$. We present a recursive mechanism that explains these results by determining $g^*$ as the fixed-point (function) of a transformation on the class of $f$'s. At present our treatment is heuristic. In a sequel, an exact theory is formulated and specific problems of rigor isolated.

**KEY WORDS:** Recurrence; bifurcation; limit cycles; attractor; universality; scaling; population dynamics.

## 1. INTRODUCTION

Recursion equations $x_{n+1} = f(x_n)$ provide a description for a variety of problems. For example, a numerical computation of a zero of $h(x)$ is obtained recursively according to

$$x_{n+1} = x_n + \frac{\epsilon h(x_n)}{h(x_n - \epsilon) - h(x_n)} \equiv f(x_n)$$

Research performed under the auspices of the U.S. Energy Research and Development Administration.

[1] Theoretical Division, Los Alamos Scientific Laboratory, Los Alamos, New Mexico.

**25**

If $\bar{x} = \lim_{n \to \infty} x_n$ exists, then $h(\bar{x}) = 0$. As $\bar{x}$ satisfies

$$\bar{x} = f(\bar{x})$$

the desired zero of $h$ is obtained as the "fixed point" of the transformation $f$. In a natural context, a (possibly fictitious) discrete population satisfies the formula $p_{n+1} = f(p_n)$, determining the population at one time in terms of its previous value. We mention these two examples purely for illustrative purposes. The results of this paper, of course, apply to any situation modeled by such a recursion equation. Nevertheless, we shall focus attention throughout this section on the population example, both for the intuitive appeal of so tangible a realization as well as for a definite viewpoint, rather different from the usual one toward this situation, that shall emerge in the discussion. It is to be emphasized, though, that our results are generally applicable.

If the population referred to is that of a dilute group of organisms, then

$$p_{n+1} = bp_n \tag{1}$$

accurately describes the population growth so long as it remains dilute, with the solution $p_n = p_0 b^n$. For a given species of organism in a fixed milieu, $b$ is a constant—the static birth rate for the configuration. As the population grows, the dilute approximation will ultimately fail: sufficient organisms are present and mutually interfere (e.g., competition for nutrient supply). At this point, the next value of the population will be determined by a *dynamic* or effective birth rate:

$$p_{n+1} = b_{\text{eff}} p_n$$

with $b_{\text{eff}} < b$. Clearly $b_{\text{eff}}$ is a function of $p$, with

$$\lim_{p \to 0} b_{\text{eff}}(p) = b$$

the only model-independent quantitative feature of $b_{\text{eff}}$. Since the volume and nutrient available to a population are limited, it is clear that $b_{\text{eff}} \simeq 0$ for $p$ sufficiently large. Accordingly, the simplest form of $b_{\text{eff}}(p)$ to reproduce the qualitative dynamics of such a population should resemble Fig. 1, where $b_{\text{eff}}(0) = b$ is an adjustable parameter [say, the nutrient level of the milieu held fixed independent of $p(t)$, and measurable by observing very dilute populations in that milieu]. A simple specific form of $b_{\text{eff}}$ is

$$b_{\text{eff}} = b - ap$$

so that

$$p_{n+1} = bp_n - ap_n{}^2$$

By defining $p_n \equiv (b/a)x_n$, we obtain the standard form

$$x_{n+1} = bx_n(1 - x_n) \tag{2}$$

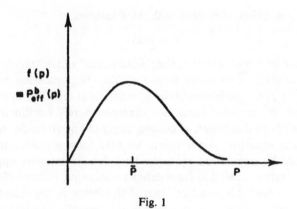

Fig. 1

In (2) the adjustable parameter $b$ is purely multiplicative. With a different choice of $b_{\text{eff}}$, $x_{n+1}$ would not in general depend upon $b$ in so simple a fashion. Nevertheless, the internal $b$ dependence may be (and often is) sufficiently mild in comparison to the multiplicative dependence that at least for qualitative purposes the internal dependence can be ignored. Thus, with $f(p) = pb_{\text{eff}}(p)$ any function like Fig. 1,

$$p_{n+1} = bf(p_n) \tag{3}$$

is compatible and representative of the population discussed. So long as $f'(0) = 1$ (so that the static birth rate is $b$ and the dilute regime is correctly modeled) and $f$ goes to zero for large $p$ with a single central maximum, relation (3) correctly (at least qualitatively) models the situation. However, $f_2(p) = \sin(ap)$ affords an (a priori) equally good modeling as $f_1(p) = p - ap^2$. Thus only detailed quantitative results of (3) could determine which (if either) is empirically correct. One should then ask what the dynamical behavior of (3) is with $f$ as in Fig. 1. It turns out that (3) enjoys a rich spectrum of excitations, with a universal behavior that would frustrate any attempt to discriminate among possible $f$'s qualitatively. That is, providing (3) affords an honest model of a population's dynamics, so far as qualitative aspects are concerned, $f$ is sufficiently specified by Fig. 1: the data could not qualitatively determine any more specific form [such as (2), say]. Conversely, *any* such choice of $f$—say Eq. (2)—is fully sufficient for study to comprehend all qualitative aspects of the dynamics. If the data should in any way disagree qualitatively with the predictions of (2), then (3) for any believable $f$ must be an incorrect model.

The qualitative information available pertaining to (3) for any $f$ of the form considered (see Appendix A for the exact requirements on $f$) is quite specific and detailed. In discussing the numerical solution to $h(x) = 0$ a fixed point was considered. In a population context, a fixed point

$$p^* = bf(p^*)$$

signifies zero population growth: $p_n = p^*$ for all $n$. However, $p^*$ is "interesting" only so long as it is stable: if $p$ fluctuates away from $p^*$, it should return to $p^*$ in successive generations. For example, if $g(\bar{x})$ is finite, then

$$x_{n+1} = x_n + h(x_n)g(x_n) \tag{4}$$

will possess $\bar{x}$ as a fixed point if $h(\bar{x}) = 0$. However, unless

$$x_n \to \bar{x}$$

(4) is of no value to obtain $\bar{x}$; indeed, $g$ is *chosen* so as to maximize the stability of $\bar{x}$. A stable fixed point is termed an "attractor," since points in its neighborhood approach it when iterated. An attractor is "global" if almost all points are eventually attracted to it. It is not necessary that an attractor be a unique isolated point. Thus, there might be $n$ points $\bar{x}_1, \bar{x}_2,..., \bar{x}_n$ such that

$$\bar{x}_{i+1} = f(\bar{x}_i), \quad i = 1,..., n - 1; \qquad \bar{x}_1 = f(\bar{x}_n)$$

Such a set is called an "$n$-point limit cycle." Every $n$ applications of $f$ return an $\bar{x}_i$ to itself: each $\bar{x}_i$ is a fixed point of the $n$th iterate of $f$, $f^{(n)}$:

$$f^{(n)}(\bar{x}_i) = \bar{x}_i, \qquad i = 1,..., n$$

Accordingly, $\{\bar{x}_1,..., \bar{x}_n\}$ is a stable $n$-point limit cycle if each $\bar{x}_i$ is a stable fixed point of $f^{(n)}$. If it is a global attractor, then for almost every $x_0$, the sequence $x_n = f^{(n)}(x_0)$, $n = 1, 2,...$, approaches the sequence

$$\bar{x}_1,..., \bar{x}_n, \bar{x}_1,..., \bar{x}_n,...$$

Finally, there can be infinite stability sets $\{\bar{x}_i\}$ with

$$\bar{x}_{i+1} = f(\bar{x}_i)$$

such that the sequence $x_n = f^{(n)}(x_0)$ eventually becomes the sequence $\{\bar{x}_i\}$.

With this terminology, some of the detailed qualitative features of (3) can be stated as follows. (See Appendix A for more precise statements.) Depending upon the parameter value $b$, (3) possesses stable attractors of every order, with one attractor present and global for each fixed choice of $b$. As $b$ is increased from a sufficiently small positive value, a fixed point $p^* > 0$ is stable until a value $B_0$ is reached when it becomes unstable. As $b$ increases above $B_0$, a two-point cycle is stable, until at $B_1$ it becomes unstable, giving rise to a stable four-point cycle. As $b$ is increased, this phenomenon recurs, with a $2^n$-point cycle stable for

$$B_{n-1} < b < B_n$$

giving rise to a $2^{n+1}$-point cycle above $B_n$ until $B_{n+1}$, etc. The sequence of $B_n$ is *bounded above* converging to a finite $B_\infty$. This set of cycles (of order $2^n$, $n = 1, 2,...$) is termed the set of "harmonics" of the two-point cycle. For any value of $b > B_\infty$ (but not too large) some particular stable $n$-point cycle will be present. As $b$ is increased, it becomes unstable, and is replaced with a stable $2n$-point cycle. Until the cycle has doubled ad infinitum, no new stability sets save for these appear. Moreover, the ordering (with respect to $b$) of the onset of new size stability sets (e.g., seven-point before five-point) is also independent of $f$. Thus, if $b$ is the unique parameter governing a population, any deviation of the ordering of stability sets upon increase of $b$ from that determined by (2), say, constitutes empirical proof that (3) for any believable $f$ incorrectly models the population. On the other hand, if (3) is appropriate for some $f$, then (2), for all qualitative purposes, comprises the full theory of the population's evolution. The exact quantitative theory reduces to the problem of determining the particular $f$. Unfortunately, even if (3) might be applicable, the data of biological populations are too crude at present to significantly discriminate among $f$'s.

With so much specific qualitative information about (3) independent of $f$ available, we may ask if the form of (3) might not also imply some *quantitative* information independent of $f$. It is the content of the following to answer this inquiry in the affirmative. Thus, the local structure of high-order stability sets (the quantitative locations of all elements of a stability set nearby one another) is independent of $f$. The role of a specific $f$ is to set a local scale size for each cluster of stability points and to set the spacing between them. If one plots the points of, say, a $2^8$-point limit cycle of (2) (or any cycle highly bifurcated from some low-order one), then by unevenly stretching the axis, the same $2^8$-point cycle of (3) for another $f$ is produced. The points are distributed unevenly in clusters sufficiently small that the stretching is essentially a pure magnification over the scale of a cluster. Moreover, for a fixed $f$, if $b$ is increased to produce a $2^9$-point cycle, that cluster about $\bar{x}$ (the maximum point) reproduces itself on a scale approximately $\alpha$ times smaller, where

$$\alpha = 2.5029078750957...$$

when $f$ has a normal (i.e., quadratic) maximum. (This shall be assumed unless specifically stated otherwise.) The presence of the number $\alpha$ is a binding test on whether or not (3) is a correct model. $\alpha$ is a reflection of the infinitely bifurcative structure of (3), independent of any particular $f$. That is, the great bulk of the detailed quantitative aspect of solutions to (3) is independent of a specific choice of $f$: Eq. (3) and Fig. 1 comprise the bulk of the quantitative theory of such a population. Indeed, it is very difficult to extract the exact form of $f$ from data, as so much quantitative information is determined purely by (3). In addition to $\alpha$, another universal number determined by (3) should leave its mark on the data of a system described by (3). Thus, let $b_0$ be the value of $b$ such that $\bar{x}$ (the abscissa of the maximum) is an element of a stable $r$-point cycle, and generally $b_n$ the value of $b$ such that $\bar{x}$ is an element of the stable $(r \times 2^n)$-point cycle $n$ times bifurcated from the original. Then

$$\delta = \lim_{n \to \infty} \frac{b_{n+1} - b_n}{b_{n+2} - b_{n+1}}$$

is universal, with

$$\delta = 4.6692016091029...$$

It must be stressed that the numbers $\alpha$ and $\delta$ are *not* determined by, say, the set of all derivatives of (an analytic) $f$ at same point. (Indeed, $f$ need not be analytic.) Rather, universal functions exist that describe the local structure of stability sets, and these functions obey functional equations [independent of the $f$ of (3)] implicating $\alpha$ and $\delta$ in a fundamental way.

## 2. QUALITATIVE ASPECTS OF BIFURCATION AND UNIVERSALITY

For definiteness (with no loss of generality), $f$ is taken to map $[0, 1]$ *onto* itself. At the unique differentiable maximum $\bar{x}$, $f(\bar{x}) = 1$,

$$x_{n+1} = \lambda f(x_n)$$

and $\lambda$ lies in the interval $[0, 1]$ to guarantee that if $x_0 \in [0, 1]$ then so, too, will all its iterates. When $\lambda = \bar{x}$,

$$\lambda f(\bar{x}) = \bar{x} f(\bar{x}) = \bar{x}$$

and $\bar{x}$ is a fixed point (Fig. 2). There is a simple graphical technique to determine the successive iterates of an initial point $x_0$:

(a) Draw a vertical segment along $x = x_0$ up to $\lambda f(x)$, intersecting at $P$.
(b) Draw a horizontal segment from $P$ to $y = x$. The abscissa of the point of intersection is $x_1$.
(c) Repeat (a) and (b) to obtain $x_{n+1}$ from $x_n$.

It is obvious from Fig. 2 that $\bar{x}$ is stable. Stability is locally analyzed by linear approximation about a fixed point. Setting

$$x_n = \bar{x} + \xi_n, \qquad \bar{x} f(x) \equiv g(x), \qquad g(\bar{x}) = \bar{x}$$
$$x_{n+1} = g(x_n) \Rightarrow \bar{x} + \xi_{n+1} = g(\bar{x} + \xi_n) = g(\bar{x}) + \xi_n g'(\bar{x}) + \cdots$$
$$\Rightarrow \xi_{n+1} = g'(\bar{x})\xi_n + O(\xi_n^2)$$

166

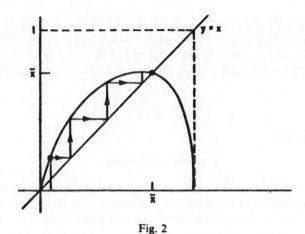

Fig. 2

Clearly $\xi_n \to 0$ if $|g'(\bar{x})| < 1$, the criterion for local stability. But $g'(\bar{x}) = \bar{x}f'(\bar{x}) = 0$, so that $\bar{x}$ is stable. With $r \equiv |g'(\bar{x})| < 1$,

$$\xi_n \propto r^n$$

so that convergence is geometric for $r \neq 0$. For $r = 0$, convergence is faster than geometric, and $\lambda = \bar{x}$ is that value of $\lambda$ determining the most stable fixed point. We denote this value of $\lambda$ by $\lambda_0$. Increasing $\lambda$ just above $\lambda_0$ causes the fixed point $x^*$ to move to the right with $g'(x^*) < 0$. At $\lambda = \Lambda_0$, $g'(x^*) = -1$ and $x^*$ is marginally stable; for $\lambda > \Lambda_0$ it is unstable. According to Metropolis et al.,[1] a two-point cycle should now become stable. Stability of either of these points, say $x_1^*$, is determined by $|g^{(2)'}(x_1^*)|$, where

$$g^{(2)}(x) = g(g(x)); \qquad g^{(n+1)}(x) = g(g^{(n)}(x)) = g^{(n)}(g(x))$$

Accordingly, consider $g^{(2)}(x)$ when $g'(x^*) < -1$ (Fig. 3). Several details of Fig. 3 are especially important. First, $g^{(2)}$ has two maxima: this because $\bar{x}$ has two inverses for $\lambda > \lambda_0$. Each maximum is of identical character to that of $g$: a neighborhood of $x_m^{(1)}$ is mapped into a neighborhood about $\bar{x}$ by $g$; $g$ has a nonvanishing derivative at $x_m^{(1)}$, so that the imaged neighborhood is the original simply translated and stretched; accordingly, $g$ applied to this new neighborhood is simply a magnification of $g$ about $\bar{x}$. Thus, if $g(x) \propto |x - \bar{x}|^z + g(\bar{x})$, $z > 1$ for $|x - \bar{x}|$ small, then $g^{(2)} \propto |x - x_m^{(1)}|^z + g(\bar{x})$ for $|x - x_m^{(1)}|$ small. Similarly, the minimum (located at $\bar{x}$) is of order $z$. This is, of course, the content of the chain rule: $g^{(n)'}(x_0) = \prod_{i=0}^{n-1} g'(x_i)$ with $x_i = g^{(i)}(x_0)$ [$g^{(0)}(x) \equiv x$]. In particular, observe that $\bar{x}$ is a point of extremum of $g^{(n)}$ for all $n$. Also, if $g(x^*) = x^*$, then $g^{(n)'}(x^*) = [g'(x^*)]^n$. With $g'(x^*) < -1$,

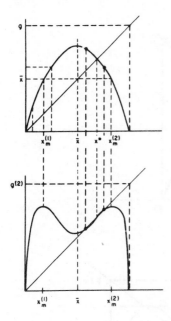

Fig. 3

$g^{(2)\prime}(x^*) > 1$, so that $g^{(2)}$ must develop two fixed points besides $x^*$: these two new fixed points are a two-point cycle of $g$ itself, and for $\lambda - \Lambda_0$ sufficiently small, $0 < g^{(2)\prime} < 1$ at these points. Moreover, since $g(x_1^*) = x_2^*$ and $g(x_2^*) = x_1^*$, the chain rule implies that $g^{(2)\prime}(x_1^*) = g^{(2)\prime}(x_2^*)$, so that each element of the cycle enjoys identical stability. As $\lambda$ is increased, the maxima of $g^{(2)}$ ($g^{(2)} = \lambda$ at maximum) also increase until a value $\lambda_1$ is reached when the abscissa of the rightmost maximum $x_m^{(2)} = \lambda_1$. By the chain rule, the other fixed point is now also at an extremum, and must be at $\bar{x}$ (Fig. 4).

As $\lambda$ increases above $\lambda_1$, $g^{(2)}(\bar{x})$ decreases below $\bar{x}$, so that $g^{(2)\prime} < 0$ for the leftmost fixed point, and so, for the rightmost one. At $\lambda = \Lambda_1$, $g^{(2)\prime} = -1$ for both: otherwise the two-point cycle would always remain stable, in violation of the results of Metropolis *et al.* Thus, $g^{(2)\prime} < -1$ for $\lambda > \Lambda_1$, the two-point cycle is unstable, and we are now motivated to consider $g^{(4)}$, as a four-point cycle should now be stable. Alternatively, the region "a" of $g^{(2)}$ of Fig. 4 bears a distinct resemblance to $g$ of Fig. 2 turned upside down and reduced in scale: the transition that led from Fig. 2 to Fig. 4 is now being reexperienced, with $g^{(2)}$ replacing $g$ and $g^{(4)}$ replacing $g^{(2)}$. In particular, at $\lambda = \lambda_2 > \Lambda_1$ the fixed points of $g^{(4)}$ beyond those of $g^{(2)}$ will occur at extrema (Fig. 5). The region "a" of $g^{(4)}$ is again an upside-down, reduced version of that of $g^{(2)}$ in Fig. 4; the square box construction including $\bar{x}$ for $g^{(2)}$ of Fig. 5 is an upside-down, reduced version of that of $g$ in Fig. 4. Since the boxes are *squares*, the Fig. 5 box is reduced by the *same* scale on both height

168

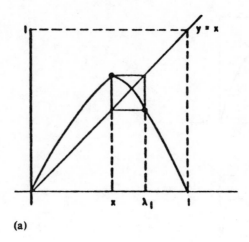

(a)

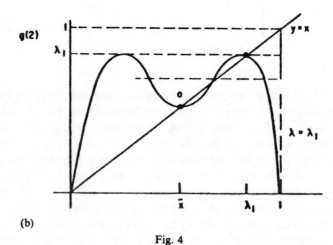

(b)

Fig. 4

and width from Fig. 4. Accordingly, the regions "a" are also rescaled identically on height and width.

It is very important to realize that in Fig. 5, $g$ itself was not drawn since it is unnecessary: $g^{(2)}$ is sufficient to determine $g^{(4)}$:

$$g^{(4)}(x) = g(g(g(g(x)))) = g(g(g^{(2)}(x))) = g^{(2)}(g^{(2)}(x))$$

[and similarly, $g^{(2n+1)}(x) = g^{(2n)}(g^{(2n)}(x))$]. At the level of discussion of Fig. 5, $g^{(2)}$ has effectively replaced $g$ as the fundamental function considered. $g^{(2)}$, though, is not simply proportional to $\lambda$, possessing internal $\lambda$ dependence: the underlying role of $g$ is exposed by $g^{(2)}$ in the simultaneous occurrence of the two box constructions. Similarly, by the $n$th bifurcation, only $g^{(2n-1)}$ and $g^{(2n)}$ are important. If at $\lambda_{n+1}$ (at $\lambda = \lambda_n$, $\bar{x}$ is an element of a $2^n$-point cycle)

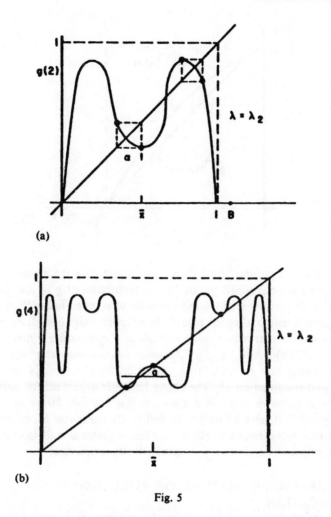

Fig. 5

we magnify the box containing $\bar{x}$ of $g^{(2^n)}$ and invert it to overlay that of $g^{(2^{n-1})}$ at $\lambda = \lambda_n$ (Fig. 6), we have two curves of identical order of maximum $z$, of identical height with identical zeros. Through a set of operations, $g^{(2^{n-1})}$ determines $g^{(2^n)}$, just as will $g^{(2^n)}$ determine $g^{(2^{n+1})}$. Referring back to Fig. 4, observe that the restriction of $g^{(2)}$ to the interval between maxima is determined entirely by the restriction of $g$ itself to this same interval. The region "a" of $g^{(2)}$ is determined by $g$ restricted a smaller interval plus essentially just the slope of $g$ at $\lambda_1$ if $g$ is sufficiently smooth. Analogously, the restriction of $g^{(2^n)}$ to its box part is determined through a similar restriction of $g^{(2^{n-1})}$. With the $n$ scale reductions that have taken place by this level of iteration, $g^{(2^n)}$ is determined by $g$ restricted to an increasingly small interval about $\bar{x}$ together with

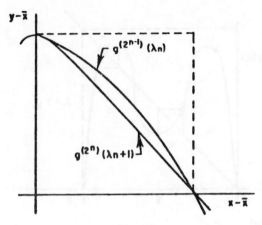

Fig. 6

the slope of $g$ at $n$ points. These slopes determine only the absolute scale of $g^{(2^n)}$: its shape is determined purely by the restriction of $g$ to the immediate vicinity of $\bar{x}$. If we now set by hand the scale of a magnified $g^{(2^n)}$ so that the square is of unit length, then the role of the $n$ slopes is eliminated. *Accordingly, we now conjecture that the rescaled $g^{(2^n)}$ about $\bar{x}$ approaches a function $g^*(x)$ independent of $f(x)$ for all $f$'s of a fixed order of maximum $z$: $g^*$ depends only on $z$.* It remains now to make this discussion formal, exactly defining the rescaling and the function $g^*$. The above heuristic argument for universality regrettably remains in want of a rigorous justification. However, we have carefully verified it, and all details to follow by computer experiment. In a sequel to this work we shall establish exact equations and isolate specific questions whose resolutions would establish the conjecture.

## 3. THE RECURSIVE NATURE OF SUCCESSIVE BIFURCATION

We have described a process that can be summarized as follows.

(0) We start at $\lambda = \lambda_n$, and look at $g^{(2^n)}$ near $x = \bar{x}$. Alternatively, we might look at $g^{(2^{n-1})}$ for the same $\lambda$ and range of $x$, as depicted in Fig. 7.

(i) Form $g^{(2^n)}(x) = g^{(2^{n-1})}(g^{(2^{n-1})}(x))$, depicted in Fig. 8.

(ii) Increase $\lambda$ from $\lambda_n$ to $\lambda_{n+1}$, depicted in Fig. 9.

(iii) Rescale: $g^{(2^n)}(x) \rightarrow \alpha_n g^{(2^n)}(x/\alpha_n)$, depicted in Fig. 10 ($|\alpha| > 1$).

Calling the operations (i)–(iii) $B_{n-1}$, we have

$$\tilde{g}_n(x) = B_{n-1}[\tilde{g}_{n-1}(x)], \qquad n = 2, 3,...$$

and are claiming $\tilde{g}_n(x) \rightarrow g^*(x)$ locally about $\bar{x}$.

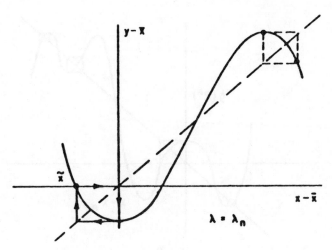

Fig. 7

Clearly (i) of $B_n$ is recursive and $n$-independent; we call this part of $B_n$ "doubling." We will motivate that (ii) becomes asymptotically $n$-independent; we term this part of $B_n$ "$\lambda$-shifting." Also, with $\alpha_n \to \alpha$ essentially by (i), part (iii) of $B_n$ becomes asymptotically $n$-independent; we term this part (obviously) "rescaling." Thus, $B_n \to B$. That is,

$$\lim_{r \to \infty} B^r[\tilde{g}_n(x)] = g^*(x)$$

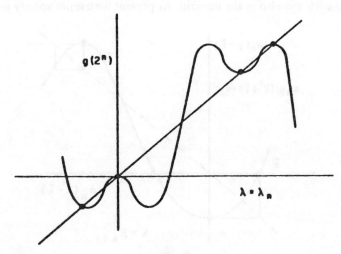

Fig. 8

172

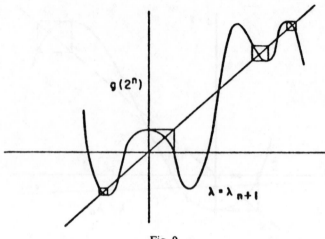

Fig. 9

Accordingly, $g^*$ satisfies the equation

$$g^* = B[g^*] \tag{5}$$

Universality, thus, is the consequence of a recursion on the class of functions $f(x)$ considered. Under high-order bifurcation, the fixed point of $B$ is approached—that fixed point being, within a certain domain, a property of $B$ itself and not of the starting $f(x)$. Evidently, domains of the various fixed points of $B$ are disjoint for different $z$. Also, each fixed-$z$ domain clearly exceeds the class of $f$'s specified by properties 1–4 of Appendix A, since $(f)^{(2^n)}$ for each $n$ is also in the domain. At present we cannot specify just how

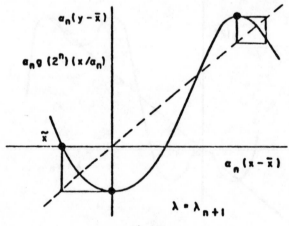

Fig. 10

large this domain is. The fixed-point equation (5) will certainly, for a given $z$, determine the rescaling ratio $\alpha$ as well as $g^*$. [For a variety of functions $f(x)$ with $z = 2$, we have determined $g^*$, with $\tilde{x}$ of Fig. 10 set to unit length.]

## 4. DETAILED FEATURES OF THE BIFURCATION RECURSION

We first indicate roughly how the parameters $\alpha$ and $\delta$ are interrelated and determined by $g^*$. At $\lambda = \lambda_n$, $\tilde{g}_{n-1}$ and $\tilde{g}_{n-1} \circ \tilde{g}_{n-1}$ appear as in Fig. 11. Increasing $\lambda$ has $\tilde{g}_{n-1}(0)$ increase above 1, producing Fig. 12, where $\tilde{g}_{n-1}$ and $\tilde{g}_{n-1} \circ \tilde{g}_{n-1}$ at $\lambda_n$ are shown dashed. By the definition of $\alpha_n$, $h_n$ of Fig. 12 satisfies

$$h_n = \alpha_n^{-1}$$

Clearly, though, in some rough sense

$$h_n \simeq (h_{n-1} - 1)|\tilde{g}'_{n-1}(1)| \equiv \delta h_{n-1}|\tilde{g}'_{n-1}(1)|$$

i.e.,

$$\delta h_{n-1} \simeq |\alpha_n \tilde{g}'_{n-1}(1)|^{-1} \tag{6}$$

Also,

$$\delta h_{n-2} \simeq |\tilde{g}'_{n-2}(1)|^{-1} \delta h_{n-1} \tag{7}$$

This is more nearly accurate than (6), since $\tilde{g}_{n-2}$ shifts less than $\tilde{g}_{n-1}$ for the same $\lambda$ increase. Thus,

$$\delta h_{n-1} \simeq \prod_{2}^{n} |\tilde{g}'_{n-i}(1)| \, \delta h_0 \tag{8}$$

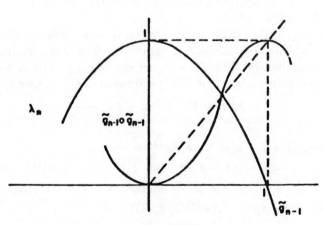

Fig. 11

174

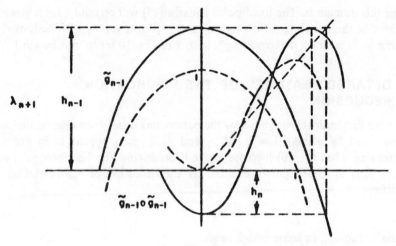

**Fig. 12**

However, $\delta h_0 = \delta\lambda_n = \lambda_{n+1} - \lambda_n$. Assuming $\tilde{g}_n \to g^*$ (this is not quite correct; see Section 5) one has, so far as $n$ dependence is concerned,

$$\delta h_{n-1} \simeq \mu|g^{*\prime}(1)|^{n-1}\,\delta\lambda_n \qquad (9)$$

with $\mu \sim 1$ an asymptotically $n$-independent factor. Substituting in (6),

$$\delta\lambda_n \simeq \mu^{-1}/|\alpha g^{*\prime}(1)|^n$$

with $\alpha = \lim \alpha_n$.

Accordingly, $\delta\lambda_n \propto \delta^{-n}$, with

$$\delta \simeq \alpha|g^{*\prime}(1)| \qquad (10)$$

For $z = 2$, the computer-experimental value for $|g^{*\prime}(1)|$ is $+1.89$, to be compared with $\delta/\alpha = 1.87$.

With $f(x)$ real-analytic in an arbitrarily small domain about $\bar{x}$, the manner in which the $\tilde{g}_n$ are formed ensures for them a systematically larger domain of analyticity. With $\tilde{g}_n \to g^*$, an equivalent procedure for defining the $\alpha_n$ is to require (at least one-sided) agreement in $\tilde{g}_n^{(2)}(0)$. One has

$$\tilde{g}_n(x) = 1 - |x|^z(a + bx^2 + \cdots) \qquad (11)$$

Then,

$$
\begin{aligned}
\tilde{g}_n \circ \tilde{g}_n(x) &= 1 - a|\tilde{g}_n|^z - b|\tilde{g}_n|^{z+2} + \cdots \\
&= 1 - a|1 - a|x|^z \cdots|^z - b|1 - a|x|^z + \cdots|^{z+2} + \cdots \\
&= 1 - a - b + \cdots + a[az|x|^z + \cdots + b(z + 2)|x|^z + \cdots] \\
&= \tilde{g}_n(1) + a|x|^z[1 - \tilde{g}_n{}'(1)] + \cdots \\
&= a|x|^z[1 - \tilde{g}_n{}'(1)] + \cdots
\end{aligned}
$$

Next, the $\lambda$ shift is performed:

$$\tilde{g}_n \to \tilde{g}_n \circ \tilde{g}_n \to -\nu + \mu a |x|^z [1 - \tilde{g}_n'(1)] + \cdots$$

and finally, $\alpha$ rescaling:

$$\tilde{g}_n \to -\{1 - \mu(a/\alpha^{z-1})[1 - \tilde{g}^{*\prime}(1)]|x|^z + \cdots\} \tag{12}$$

For (11) and (12) to agree, one has

$$\alpha^{z-1} \sim 1 - g^{*\prime}(1) \tag{13}$$

where $\mu \lesssim 1$ corresponds to $\lambda$-shifting being mostly a displacement in the immediate environs of $\bar{x}$. Again, for $z = 2$, one compares $\alpha = 2.50$ with $1 - g^{*\prime}(1) = 2.87$. Combining (10) and (13), one has

$$\delta \simeq |g^{*\prime}(1)|[1 - g^{*\prime}(1)]^{1/z - 1} \tag{14}$$

While (13) and (14) are crude, they are roughly correct for $z \gtrsim 2$, but more important, indicate that $g^*$ ultimately determines everything.

We now proceed to describe the situation more carefully, tacitly assuming convergence, and successively illustrating its details through consistency arguments.

By definition

$$g^*(x) = \lim(-1)^n \alpha^n g^{(2^n)}(x/\alpha^n, \lambda_{n+1}) \equiv \lim \tilde{g}_n(x) \tag{15}$$

where $\alpha^n$ is symbolic for $\alpha_n$ which becomes asymptotically a multiple of $\alpha^n$: the multiple has been absorbed in $g^{(2^n)}$. For all $n$, $\tilde{g}_n$ satisfies

$$\tilde{g}_n(1) = 0, \qquad \tilde{g}_n(0) = 1, \qquad \tilde{g}_n'(0) = 0 \tag{16}$$

and near $x = 0$, $1 - \tilde{g}_n(x) \sim |x|^z$.

We now furnish an approximate equation for $g^*$:

$$(-1)^n \alpha^{n-1} g^{(2^n)}(x, \lambda_n) = (-1)^n \alpha^{n-1} g^{(2^{n-1})}(g^{(2^{n-1})}(x, \lambda_n), \lambda_n)$$

$$= (-1)^n \alpha^{n-1} g^{(2^{n-1})}\left(\frac{1}{\alpha^{n-1}} \alpha^{n-1} g^{(2^{n-1})}(x, \lambda_n), \lambda_n\right)$$

$$= -\tilde{g}_n \circ \tilde{g}_n(x\alpha^{n-1}) \tag{17}$$

or

$$(-1)^n \alpha^n g^{(2^n)}(x/\alpha^n, \lambda_n) = -\alpha \tilde{g}_n \circ \tilde{g}_n(x/\alpha) \tag{18}$$

or

$$-\alpha \tilde{g}_n \circ \tilde{g}_n(x/\alpha) = \tilde{g}_{n+1}(x) - (-1)^n \alpha^n(g^{(2^n)}(x/\alpha_n, \lambda_{n+1}) - g^{(2^n)}(x/\alpha_n, \lambda_n))$$

or

$$-\alpha \tilde{g}_n \circ \tilde{g}_n(x/\alpha) \simeq \tilde{g}_{n+1}(x) - (-1)^n \alpha^n(\lambda_{n+1} - \lambda_n) \partial_\lambda g^{(2^n)}(x/\alpha_n, \lambda_n) \tag{19}$$

assuming a "mild" $\lambda$-shifting.

Clearly, $\alpha^n \, \partial_\lambda g^{(2^n)}(x/\alpha_n, \lambda_n)$ diverges with $n$ since $\lambda_{n+1} - \lambda_n \to 0$. Thus, a more careful analysis, like that used to treat Eq. (10), needs to be done. By (17),

$$\partial_\lambda g^{(2^n)}(x, \lambda_n) = \partial_\lambda g^{(2^{n-1})}(g^{(2^{n-1})}(x, \lambda_n), \lambda_n)$$
$$+ \partial_x g^{(2^{n-1})}(g^{(2^{n-1})}(x, \lambda_n), \lambda_n) \, \partial_\lambda g^{(2^{n-1})}(x, \lambda_n)$$
$$= \partial_\lambda g^{(2^{n-1})}(g^{(2^{n-1})}(x, \lambda_n), \lambda_n)$$
$$+ \partial_x \tilde{g}_{n-1}(\tilde{g}_{n-1}(x/\alpha^{n-1})) \partial_\lambda g^{(2^{n-1})}(x, \lambda_n)$$

So,

$$\alpha^n \, \partial_\lambda g^{(2^n)}\left(\frac{x}{\alpha^n}, \lambda_n\right) = \alpha^n \, \partial_\lambda g^{(2^{n-1})}\left(\frac{1}{\alpha^{n-1}} \, \tilde{g}_{n-1}\left(\frac{x}{\alpha}\right), \lambda_n\right)$$

$$+ \tilde{g}'_{n-1} \circ \tilde{g}_{n-1}\left(\frac{x}{\alpha}\right) \alpha^n \, \partial_\lambda g^{(2^{n-1})}\left(\frac{x}{\alpha^n}, \lambda_n\right) \tag{20}$$

At $x = 0$,

$$\alpha^n \, \partial_\lambda g^{(2^n)}(0, \lambda_n) = \alpha^n \, \partial_\lambda g^{(2^{n-1})}(1/\alpha^{n-1}, \lambda_n) + \tilde{g}'_{n-1}(1)\alpha^n \, \partial_\lambda g^{(2^{n-1})}(0, \lambda_n) \tag{21}$$

With

$$\alpha^n \, \partial_\lambda g^{(2^{n-1})}(1/\alpha^{n-1}, \lambda_n) = \mu\alpha^n \, \partial_\lambda g^{(2^{n-1})}(0, \lambda_n)$$

(such a $\mu$ exists if $\lambda$-shifting becomes $n$-independent), (21) becomes

$$\alpha^n \, \partial_\lambda g^{(2^n)}(0, \lambda_n) = [\mu + \tilde{g}'_{n-1}(1)]\alpha^n \, \partial_\lambda g^{(2^{n-1})}(0, \lambda_n)$$
$$\simeq [\mu + \tilde{g}'(1)]\alpha^n \, \partial_\lambda g^{(2^{n-1})}(0, \lambda_{n-1}) \tag{22}$$

($g^{(2^{n-1})}$ shifts more slowly than $g^{(2^n)}$: higher order $\lambda$ derivatives have been neglected). Iterating (22), one has

$$\partial_\lambda g^{(2^n)}(0, \lambda_n) \simeq \rho[\mu + \tilde{g}'(1)]^n \tag{23}$$

with $\rho \sim 1$, $n$-independent. So,

$$(-1)^n(\lambda_{n+1} - \lambda_n)\alpha^n \, \partial_\lambda g^{(2^n)}(0, \lambda_n) \simeq \rho[\alpha(-\tilde{g}'(1) - \mu)]^n(\lambda_{n+1} - \lambda_n)$$

By (19) this is $n$-independent, and so,

$$\lambda_{n+1} - \lambda_n \sim \delta^{-n}$$

with

$$\delta \simeq \alpha(-\tilde{g}'(1) - \mu) \tag{24}$$

Defining $\bar{h}_n(x) = (-1)^n\alpha^n(\lambda_{n+1} - \lambda_n) \, \partial_\lambda g^{(2^n)}(x/\alpha_n, \lambda_n)$, (19) reads

$$\tilde{g}_{n+1}(x) = \bar{h}_n(x) - \alpha\tilde{g}_n \circ \tilde{g}_n(x/\alpha) \tag{25}$$

or

$$g^*(x) = h^*(x) - \alpha g^* \circ g^*(x/\alpha) \tag{26}$$

Returning to (20), multiplied by $\lambda_{n+1} - \lambda_n$, neglecting higher order derivatives,

$$\tilde{h}_n(x) \simeq -\omega(\tilde{h}_{n-1} \circ \tilde{g}_{n-1}(x/\alpha) + \tilde{h}_{n-1}(x/\alpha)\tilde{g}'_{n-1} \circ \tilde{g}_{n-1}(x/\alpha)) \qquad (27)$$

with some $\omega \sim 1$, or, as $n \to \infty$, and repeating (26),

$$h^*(x) = -\omega(h^* \circ g^*(x/\alpha) + h^*(x/\alpha)g^{*'} \circ g^*(x/\alpha))$$

and

$$g^*(x) = h^*(x) - \alpha g^* \circ g^*(x/\alpha) \qquad (28)$$

These constitute first-order (approximate) fixed-point equations, satisfying the boundary conditions

$$g^*(0) = 1, \qquad g^{*'}(0) = 0, \qquad g^*(1) = 0, \qquad h^*(0) = 1 \qquad (29)$$

[We comment that (28) is recursively stable, and for $z = 2$ affords a 10% approximate solution.]

At this point, some remarks concerning convergence (say of $\tilde{g}_n \to g^*$) are in order. The function $g^*(x)$ describes the stability set for large $n$ in the vicinity of $\bar{x}$: those $x_i$ such that

$$g^* \circ g^*(x_i) = x_i$$

[and, of course, $g^{*'}(g^*(x_i)) \cdot g^{*'}(x_i) = 0$] are the stability set points near $\bar{x}$. Accordingly, all such $x_i$ scale with $\alpha$ upon bifurcation: $|x_i - x_j| \to (1/\alpha)|x_i - x_j|$. For example, the distance between $\bar{x}$ and the nearest element to it of the stability set of order $2^n$ is $\alpha$ times greater than that distance in the stability set of order $2^{n+1}$. (Also, if $x_1$ is the nearest point to $\bar{x}$ and $x_2$ the next nearest, then for all $n$ large enough, $|\bar{x} - x_1|/|\bar{x} - x_2| \equiv \gamma$ is fixed.) This immediately leads to a difficulty: distances near $\bar{x}$ and those near $\lambda_n$ (the furthest right element of a stability set) cannot possibly scale identically.

As is obvious from Fig. 13, with $\Delta_n$ the distance from $\bar{x}$ to $x_1$, and $d_n$ the distance from $\lambda_n$ to $\tilde{x}$ (the next to rightmost point), $d_n \sim \Delta_n^z$, so that with $\Delta_n \propto \alpha^{-n}$,

$$d_n \propto (\alpha^z)^{-n} \neq \alpha^{-n} \qquad (30)$$

Thus, convergence of $\tilde{g}_n(x)$ to $g^*(x)$ must be local in nature. The scale for which $g^*(0) = 1$ and $g^*(1) = 0$ is, of course, $\alpha^{-n}$ finer than usual measure on $[0,1]$: for large $n$, $\sup|\tilde{g}_n - g^*| < \epsilon_{N_n}$ for $|x| < N_n$ is uniform convergence in "real" $x$ of $|x| < N/\alpha^n$. To allow for a shifting rescaling of parts of $g^*$, $N_n \ll \alpha^n$. Thus, one anticipates that $\tilde{g}_n \to g^*$ (say in sup-norm) over *any* bounded part of $R$ but with the $g^*(1) = 0$ measure. In any (small) interval about a given point in the stability set of order $n$, one sets the origin of $g^{(2^n)}$ at the point in question and forms $\tilde{g}_n$ with an appropriate (local) scale factor. As $n$ increases, in the $\tilde{g}_n(1) = 0$ measure, any other point a finite distance away

178

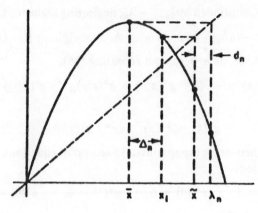

Fig. 13

in *usual* [0, 1] measure grows far remote from the chosen point: so far, that the local $\tilde{g}_n$ never converge to it. Thus, in effect, a class of $g^*$ exist, each determining the large-$n$ limiting stability set about a point.

For example, defining $\tilde{f}_n(x)$ about $\lambda_{n+1}$ by Fig. 14 [$x_n \equiv g^{(2^n)}(\lambda_{n+1})$, $\lambda_{n+1} = g^{(2^n)}(x_n)$], we have

$$\tilde{f}_n(x) = [g^{(2^n)}((\lambda_{n+1} - x_n)x + x_n) - x_n]/(\lambda_{n+1} - x_n) \rightarrow f^*(x) \quad (31)$$

[so that $\tilde{f}_n(0) = 1, \tilde{f}_n{}'(0) = 0, \tilde{f}_n(1) = 0$]. In the notation of Fig. 13,

$$\tilde{f}_n(x) = [g^{(2^n)}(xd_n + x_n) - x_n]/d_n$$

and so $\tilde{f}_n$ scales by $d^z$ rather than $d$. It is straightforward to relate $f^*$ to $g^*$:

$$g^{(2^n)}(\lambda_{n+1}f(x)) = \lambda_{n+1}f(g^{(2^n)}(x)) \quad (32)$$

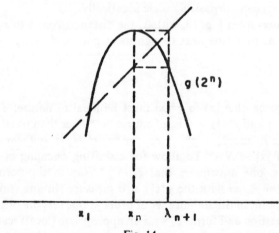

Fig. 14

so that for $x \sim 0$ (we have conveniently set $\bar{x} = 0$), $\lambda_{n+1} f(x)$ is near $\lambda_{n+1}$ and (32) relates $g^{(2^n)}$ about $\lambda_{n+1}$ to $g^{(2^n)}$ about 0. Thus

$$x \text{ small} \Rightarrow \lambda_{n+1} f(x) = \lambda_{n+1} - a\lambda_{n+1}|x|^z + O(|x|^{z+1})$$

$$g^{(2^n)}(\lambda_{n+1} f(x)) = g^{(2^n)}((\lambda_{n+1} - x_n)(1 - a|x|^z) - ax_n|x|^z + x_n)$$

$$= x_n + (\lambda_{n+1} - x_n)\tilde{f}_n\left(1 - a|x|^z - \frac{ax_n}{\lambda_{n+1} - x_n}|x|^z\right)$$

Also,

$$\lambda_{n+1} f(g^{(2^n)}(x)) = \lambda_{n+1} - a\lambda_{n+1}|g^{(2^n)}(x)|^z + \cdots$$

$$= \lambda_{n+1} - a\lambda_{n+1}\Delta_n^z|\tilde{g}_n(x/\Delta_n)|^z$$

Accordingly, (32) implies for small $x$ that

$$(\lambda_{n+1} - x_n)\left[1 - \tilde{f}_n\left(1 - a|x|^z - \frac{ax_n}{\lambda_{n+1} - x_n}|x|^z\right)\right] = a\lambda_{n+1}\Delta_n^z\left|\tilde{g}_n\left(\frac{x}{\Delta_n}\right)\right|^z$$

By Fig. 14, $\lambda_{n+1} - x_n = d_n = a\Delta_n^z\lambda_{n+1}$, so that

$$1 - \tilde{f}_n\left(1 - \frac{\lambda_{n+1}a|x|^z}{\lambda_{n+1} - x_n}\right) = \left|\tilde{g}_n\left(\frac{x}{\Delta_n}\right)\right|^z$$

or

$$1 - \tilde{f}_n(1 - |x|^z/\Delta_n^z) = |\tilde{g}_n(x/\Delta_n)|^z$$

or

$$\tilde{f}_n(1 - |\xi|^z) = -|\tilde{g}_n(\xi)|^z + 1$$

or

$$\tilde{f}_n(x) = -|\tilde{g}_n((1 - x)^{1/z})|^z + 1 + \cdots \tag{33}$$

For large $n$, neglected terms are powers of $\Delta_n \to 0$. So,

$$f^*(x) = -|g^*((1 - x)^{1/z})|^z + 1 \tag{34}$$

and $g^*$ determines $f^*$. (This has, of course, been computationally verified to full precision.) We are unsure of the size of the set of rescalings: clearly $\alpha$ and $\alpha^z$ belong to the set. However, about any point a fixed, finite number of iterates prior to $\bar{x}$, scaling goes by $\alpha$, whereas any region a finite number of iterates after $\lambda_{n+1}$ scales with $\alpha^z$. On the other hand, points $2^{n-1}$ iterates from $\bar{x}$, as well as $2^{n-2},..., 2^{n-N}$ from $\bar{x}$, are just the $N$ points nearest $\bar{x}$ that scale with $\alpha$. That is, we are uncertain how to define a region which possesses a scaling intermediate between $\alpha$ and $\alpha^z$; possibly the situation is simply an interspersal of regions scaling by either $\alpha$ or $\alpha^z$. What is missing is some notion of ordered measure along a stability set.

At this point it is perhaps illuminating to indicate just what $g^*$ looks like. Evidently Fig. 5 is a somewhat distorted version of $g^*$ very near $x = 0$.

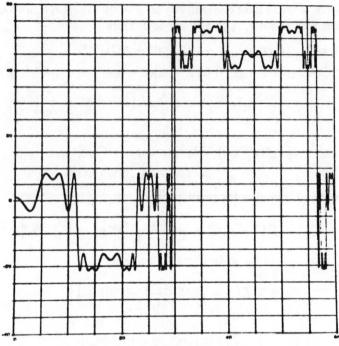

Fig. 15

Indeed, each large jump generates precursors about more mild features. At successive levels of $\tilde{g}_n$, more and more precursors are produced, whose oscillations grow narrower. Also, $g^*$ grows with $|x|$, with a long string of features of roughly the same height as the end height of the string. Figure 15 shows $g^*$ for $|x| < 50$ as computationally obtained. Evidently, convergence to such a function worsens with increasing $|x|$.

## 5. CONCLUSIONS AND A BRIEF SUMMARY OF THE EXACT THEORY

In this paper we have attempted to heuristically motivate our conjecture of universality, and indicate the form of an exact theory of highly bifurcated attractor sets. Our conclusion is that both of the numbers $\alpha$ and $\delta$ as well as the local structure of highly bifurcated attractors as determined by the universal function $g^*$ are determined by functional equations. We have provided approximate such equations, but failed in establishing exact equations for an inability to exactly reflect an increase in the parameter $\lambda$. As described, the local structure determined by $g^*$ pertains to values of $\lambda$ asymptotically near $\lambda_\infty$ at the specific values $\lambda_n$: we have not described that structure for values of

$\lambda$ between $\lambda_n$ and $\lambda_{n+1}$. However, at these choices of $\lambda$, the theory we describe holds regardless of the attractor from which the bifurcated attractors arise. (That is, $\lambda_n$ refers to that attractor of order $m \cdot 2^n$, which includes the critical point $\bar{x}$ for any $m$.) Or, the local structure of all *infinite* attractors of $x_{n+1} = \lambda f(x_n)$ is described by $g^*$. Also, the rescaling parameter $\alpha$ and the convergence rate $\delta$ are common to all highly bifurcated attractors.

At this point we should like to briefly summarize some results of the exact theory; a full treatment of these results will shortly appear in a sequel.[2]

Figure 5a shows, near $\bar{x}$, what shall evolve into $g^*$, and represents the local structure at a "two-point" level. Figure 5b represents the graph of a function that shall evolve into $g^* \circ g^*$, and accordingly must also be universal: it represents the identical local structure as does $g^*$, but now at a "one-point" level of description. Evidently, one can view this local structure at a "$2^r$-point" level from the function

$$g_r(x) = \lim_{n \to \infty} (-\alpha)^n \tilde{g}^{(2^n)}(\lambda_{n+r}, x/(-\alpha)^n), \qquad r = 0, 1, 2,... \qquad (35)$$

where $g_1(x)$ is exactly $g^*(x)$. All $g_r$ are of identical qualitative shape: each bump of $g_1$ aligned along $y = x$ contains two points of an attractor set, whereas each bump of $g_r$ similarly situated now contains $2^r$ points of that set. Evidently, the lower $r$, the more magnified the local structure. Following immediately from the definition,

$$g_{r-1}(x) = -\alpha g_r(g_r(x/\alpha)) \qquad (36)$$

(all the functions $g_r$ are symmetric). The content of this equation is essentially the Cantor-set-like nature of highly bifurcated attractors: at each bifurcation, the rough locations of attractor points are unchanged, with a "microscopic" splitting of each such point; the scale of splitting is $\alpha$ below the previous level, so that rescaling by $\alpha$ after a bifurcation reveals the set at a next level of magnification. In fact, (36) provides the entire exact description, as we now synoptically elucidate. The central bump of $g_r$ is effectively a $\lambda f(x)$ containing a $2^r$-point cycle, which, as $r$ increases, quickly approaches a $\lambda f$ containing the infinite attractor. That is, one expects

$$g(x) \equiv \lim_{r \to \infty} g_r(x) \qquad (37)$$

to exist; $g$ no longer affords the same description of attractor points as does $g_r$. Rather, $g$ is the description at the level of infinite clusters of points, which is again a universal property. But $g$ defined by (37) is simply a fixed point of (36). Accordingly,[2]

$$g(x) = -\alpha g(g(x/\alpha)) \qquad (38)$$

---

[2] This exact equation was discovered by P. Cvitanović during discussion and in collaboration with the author.

The great virtue of $g$ is that $\lambda$ has been set at $\lambda_\infty$ at the outset, and so the difficulty of modeling $\lambda$-shifting is totally bypassed. The price paid for this is that (38) defines no recursively stable equation like

$$\bar{g}_{n+1}(x) = -\alpha_n \bar{g}_n(\bar{g}_n(x/\alpha_n)) \tag{39}$$

[By (36), iterating produces $g_r$'s for smaller values of $r$ and hence diverging from $g$.] There are a variety of ways to solve (38). A method based on the fact that $\bar{g}_0(x) = f(\lambda_\infty(f)x)$ *must* cause (39) to converge [by (35), $g = \lim(-\alpha)^n \bar{g}^{(2^n)}(\lambda_\infty, x/(-\alpha)^n)$] together with the general recursive instability of (39) allows very fast, high-accuracy estimates of all $\lambda_\infty$ for any chosen $f$. Alternatively, one can simply solve (38) by a numerical functional-Newton's method. (The result of the latter method is a 20-place determination of both $\alpha$ and $g$ for $z = 2$.)

The $g$ of (38) is a fixed point of (36). By setting

$$g = g_r + y_r$$

in (38), employing (36), and expanding to first order in $y$, one obtains

$$y_{r-1}(x) = -\alpha[y_r(g(x/\alpha)) + g'(g(x/\alpha))y_r(x/\alpha)] \tag{40}$$

(40) simply separates with the substitution

$$y_r = \lambda^{-r}\psi(x) \tag{41}$$

where $\psi$ obeys

$$\mathscr{L}[\psi(x)] \equiv -\alpha[\psi(g(x/\alpha)) + g'(g(x/\alpha))\psi(x/\alpha)] = \lambda\psi(x) \tag{42}$$

The eigenvalue $\lambda$ can clearly attain the value $+1$ corresponding to

$$\psi = g - xg'$$

reflecting the dilatation invariance of (38). In addition to a spectrum $|\lambda| \leqslant 1$, computationally there exists a unique alternate value $\delta$ strictly greater than $+1$.

It is possible to show that the eigenvalue $\delta$ is exactly that convergence rate discussed in this paper. Heuristically, if $\lambda$ is held fixed at $\lambda_n$ for $n \gg 1$, and $\lambda_n f$ iterated, it is indistinguishable from the iterates of $\lambda_\infty f$ that approximate $g$ after an initial transient, until roughly $n$ iterations have been performed to magnify the deviation of $\lambda_n$ from $\lambda_\infty$. Thus the argument about Eq. (8) can be made exact where the function $\tilde{g}_{n-i}$ there is essentially $g$ for (logarithmically) all iterations.

One can next begin to investigate the nature of the $n$ limit of (35). Defining

$$g_{r,n}(x) \equiv (-\alpha)^n \tilde{g}^{(2^n)}(\lambda_{n+r}, x/(-\alpha)^n)$$

it is immediate to verify that

$$g_{r-1,n+1}(x) = -\alpha g_{r,n}(g_{r,n}(-x/\alpha)) \tag{43}$$

But (36) is the large-$n$ fixed point of (43), and so one can discuss in linear approximation the stability of (36). (We mention at this point that the anti-symmetric parts of $g_{r,n}$ vanish in the large-$n$ limit at the rate of $-\alpha$; this result is exactly observed computationally.)

With $\alpha$ and $g$ obtained from (38), (42) determines both $h$ and $\delta$. (Again, both have been obtained to 20-place accuracy.) We stress that (38) and (42) are totally free of any reference to (1) and do produce the same $\alpha$ and $\delta$ to the 14-place accuracy of our best recursion data. Next,

$$g_r \sim g - \delta^{-r}h \tag{44}$$

so that a $g_r$ for $r \gg 1$ is available. By successive application of (36) to this asymptotic $g_r$, $g_1$ can be obtained. (Regarding $r = 3$ as asymptotic produces a $g_1$ to six-place accuracy, to give an idea of the speed of onset of the asymptotic regime.) The approximate equations (28) are a high-$z$ approximation to (38) and (42).

Since $\delta^{-r} \sim \lambda_\infty - \lambda_r$, (44) has an immediate continuation:

$$g_\lambda(x) \sim g(x) - (\lambda_\infty - \lambda)h(x)$$

which allows the determination of local structure for $\lambda$ between $\lambda_n$ and $\lambda_{n+1}$ as well. Thus, the bifurcation points $\Lambda_n$ also geometrically converge to $\lambda_\infty$ at the rate $\delta$, and logarithmically the behavior of bifurcation is periodic with period $\log \delta$. A demonstration that (36) is in fact a stable fixed point of (43) would constitute a proof of our universality conjecture: with exact (functional) equations at hand, it is possible to focus on the exact details requiring proof.[3]

## APPENDIX A

In the formula $x_{n+1} = \lambda f(x_n)$ with $0 < \lambda < 1$, $f(x)$ satisfies the following conditions:

1. $f(x)$ is continuous, single-valued, piecewise $C^{(1)}$ on $[0, 1]$ possessing a unique, differentiable maximum at $\bar{x}$ with $f(\bar{x}) = 1$.
2. $f(x) > 0$ on $(0, 1)$, $f(0) = f(1) = 0$, and $f$ is strictly decreasing on $(\bar{x}, 1)$ and strictly increasing on $(0, \bar{x})$.
3. For $\Lambda_0 < \lambda < 1$, $\lambda f(x)$ has two fixed points [$x^* = 0$, and some other $x^* \in (x, 1)$] both of which are repellant (i.e., $|f'(x)| > 1/\lambda$).
4. In the interval $N$ about $\bar{x}$ such that $|f'(x)| < 1$, $f$ is concave downward.

[3] We are in possession of extensive high-precision data pertaining to all details discussed in this paper, as well as for the solutions to the functional equations discussed in the last sections. We will consider reasonable requests from individuals for copies of specific parts of this library.

Given these conditions, Metropolis *et al.*[1] have established among others the following universal, qualitative features:

(a) For $\Lambda_0 < \lambda < \Lambda_\infty$, there exists stability sets of order $2^n$, $n = 1, 2,...$ only, with $n$ increasing with $\lambda$.

(b) For $\Lambda_0 < \Lambda_{n-1} < \lambda < \Lambda_n < \Lambda_\infty$ only $2^n$-order stability sets exist. In particular, at $\lambda_n$, with $\Lambda_{n-1} < \lambda_n < \Lambda_n$, the $2^n$-order stability set contains $\bar{x}$ as an element.

(c1) For $\lambda = \lambda_1$, under repeated application of $\lambda f$, one has $\bar{x} \to x' \to \bar{x} \to \cdots$, where $x' > \bar{x}$. Calling an $x$ "*R*" if $x > \bar{x}$ and "*L*" if $x < \bar{x}$, the "pattern" of motion through the stability set is abbreviated as *R*—meaning $\bar{x} \to R \to \bar{x}$.

(c2) The "harmonic" of a pattern P is a stability set of twice the order of P, with pattern PLP if P contains an odd number of Rs and PRP otherwise. The $2^n$-order stability sets are exactly the successive harmonics of R (e.g., for $\lambda_2$, RLR; for $\lambda_3$, RLRRRLR; etc.).

(d) If P is a basic pattern (say, for an $r$-point cycle), then (a)–(c) hold with $2^n$ replaced by $r \cdot 2^n$.

## APPENDIX B. COMPUTATIONAL RESULTS

The parameter values $\lambda_n$ for a given recurrence function $f$ are obtained by definition from

$$(\lambda_n f)^{(2^n m)}(\bar{x}) - \bar{x} = 0 \tag{B1}$$

(B1) possesses in general many roots. Accordingly, $\lambda_0$ is first obtained for a given fundamental pattern. $\lambda$ is slowly increased to find the first new zero for $n = 1$; this $\lambda$ by definition is $\lambda_1$. Next, $\lambda_2$ is similarly found as the next largest zero of (B1) for $n = 2$. At this point $\delta_1$ is calculated

$$\delta_1 = (\lambda_1 - \lambda_0)/(\lambda_2 - \lambda_1)$$

and used to estimate–predict $\lambda_3$

$$\lambda_3 \simeq \lambda_2 + \delta_1^{-1}(\lambda_2 - \lambda_1) \tag{B2}$$

As $n$ increases, $\delta_n \to \delta$ and the predicted value increases in precision, so that for large $n$, several Newton's-method iterations suffice to locate $\lambda_n$ to full precision. Since the number of iterations increases geometrically and the number of zeros of (B1) similarly increases with collateral decrease of spacing between them, the prediction method is essential to locate high-$n$ $\lambda$'s. (For example, the set of all $\lambda_n$ up to $n = 20$ for $f = x - x^2$ to 29-place precision requires just a few minutes of CDC 6600 time.)

Analogous to $\delta_n$, one can also compute the rate of convergence of $\delta_n$ to $\delta$ through $\delta_n'$:

$$\delta_n' \equiv (\delta_{n+1} - \delta_n)/(\delta_{n+2} - \delta_{n+1})$$

With $\lambda_n$ of 29-place accuracy, $\delta_n$ converges to $\delta$ to 13 places by $n = 20$ and $\delta_n'$ converges to three or four places. We quote some typical results in Tables I–III. Observe that for an $f$ symmetric about its maximum, $\delta' \simeq \delta$ for $z \leqslant 2$, whereas $\delta' < \delta$ for $z > 2$. As related in the text, we shall explain these results in the sequel to this work.

With the $\lambda_n$ determined, the parameter $\alpha$ is next obtained. We transform to variables in which $\bar{x} = 0$. The element of the limit cycle nearest to $\bar{x}$ is obtained as the $2^{n-1}$ iterate of $\bar{x}$,

$$z_n \equiv (\lambda_n f)^{(2^{n-1})}(\bar{x})$$

and the $n$th rescaling $\alpha_n$, defined by

$$\alpha_n = -z_n/z_{n+1}$$

These $\alpha_n$ converge to $\alpha$, also typically to 13 places.

Table I.[a]  Two-Cycle Data for $f = 1 - 2x^2$

| $N$ | $\lambda$ | $\delta$ | $\delta'$ |
|---|---|---|---|
| 1 | 0.70710678118654752440084443621 | 4.74430946893705 | — |
| 2 | 0.80953772034934631684595410218 | 4.67444782765301 | 2.7504 |
| 3 | 0.83112799388303047024828338891 | 4.67079115022921 | 6.7888 |
| 4 | 0.83574677974388888508230093951 | 4.66946164833746 | 3.7990 |
| 5 | 0.83673564559387058460370949661 | 4.66926580979910 | 5.2553 |
| 6 | 0.83694741858280471080227218461 | 4.66921427043589 | 4.3595 |
| 7 | 0.83699277324830473230907131621 | 4.66920445137251 | 4.8560 |
| 8 | 0.83700248680244259434596829761 | 4.66920220132661 | 4.5641 |
| 9 | 0.83700456714701499933137326301 | 4.66920173797283 | 4.7307 |
| 10 | 0.83700501269305963494575502661 | 4.66920163645133 | 4.6340 |
| 11 | 0.83700510811537583348518876201 | 4.66920161499127 | 4.6896 |
| 12 | 0.83700512855191373187027246601 | 4.66920161036023 | 4.6575 |
| 13 | 0.83700513292879431735839903441 | 4.66920160937272 | 4.6759 |
| 14 | 0.83700513386618810557617655111 | 4.66920160916069 | 4.6684 |
| 15 | 0.83700513406694914924927446461 | 4.66920160911533 |  |
| 16 | 0.83700513410994601691059292491 | 4.66920160910564 |  |
| 17 | 0.837005134119154629273224400071 |  |  |
| 18 | 0.837005134121126832046536501512 |  |  |

[a] In this and the following tables, the cycle of size $2^N$ for two-cycle data and $3 \times 2^{N-1}$ for three-cycle data is referenced by $N$. The parameter is denoted by $\lambda$; $\delta_N = (\lambda_{N+1} - \lambda_N)/(\lambda_{N+2} - \lambda_{N-1})$ and $\delta_N' = (\delta_{N+1} - \delta_N)/(\delta_{N+2} - \delta_{N+1})$.

### Table II.$^a$ Three-Cycle Data for $f = 1 - 2x^2$

| $N$ | $\lambda$ | $\delta$ | $\delta'$ |
|---|---|---|---|
| 1 | 0.9367170507273508470331116496 | 3.36345171892599 | 5.6876 |
| 2 | 0.9415128526423905912356034646 | 4.42501749338226 | 4.0902 |
| 3 | 0.9429387098616436488660308130 | 4.61166338330503 | 4.9053 |
| 4 | 0.9432609362079542493235619775 | 4.65729621052512 | 4.5349 |
| 5 | 0.9433308082518413013515037559 | 4.66659897033153 | 4.7439 |
| 6 | 0.9433458109574302537416522425 | 4.66865036240409 | 4.6264 |
| 7 | 0.9433490258695502652989696376 | 4.66908278613357 | 4.6938 |
| 8 | 0.9433497144865745558168755447 | 4.66917625427465 | 4.6550 |
| 9 | 0.9433498619710069020581737454 | 4.66919616732989 | 4.6774 |
| 10 | 0.9433498935578274276740167548 | 4.66920044506702 | 4.6774 |
| 11 | 0.9433499003227649872965866091 | 4.66920135962593 | |
| 12 | 0.9433499017716078115042431354 | | |
| 13 | 0.9433499020819055832053356825 | | |

$^a$ See footnote to Table I.

### Table III.$^a$ Two-Cycle Data for $f = x(1 - x^2)$

| $N$ | $\lambda$ | $\delta$ | $\delta'$ |
|---|---|---|---|
| 1 | 2.1213203435596425732025330874 | 4.59165349403582 | 4.7073 |
| 2 | 2.2629896545363471897845547814 | 4.65266937815338 | 4.6198 |
| 3 | 2.2938433134983575307521580083 | 4.66563137254676 | 4.6704 |
| 4 | 2.3004747021602907471027097715 | 4.66843710204515 | 4.6686 |
| 5 | 2.3018960293360623306427444234 | 4.66903785250679 | 4.6680 |
| 6 | 2.3022004839414695502983032343 | 4.66916653116789 | 4.6700 |
| 7 | 2.3022656910815752484787734523 | 4.66919409734649 | 4.6687 |
| 8 | 2.3022796565590690203670882644 | 4.66920000018325 | 4.6695 |
| 9 | 2.3022826475414960966948147845 | 4.66920126453840 | 4.6690 |
| 10 | 2.3022832881185603889813142737 | 4.66920153530566 | 4.6693 |
| 11 | 2.3022834253105609079220759031 | 4.66920159329811 | 4.6691 |
| 12 | 2.3022834546928867363664720370 | 3.66920160571803 | 4.6692 |
| 13 | 2.3022834609856811769873007175 | 4.66920160837802 | 4.6831 |
| 14 | 2.3022834623334050432933708933 | 4.66920160894770 | 4.2652 |
| 15 | 2.3022834626220462242369870735 | 4.66920160906935 | |
| 16 | 2.3022834626838643261562712194 | 4.66920160909788 | |
| 17 | 2.3022834626971036705355950345 | 4.66920160909268 | |
| 18 | 2.3022834626999393755079069366 | | |
| 19 | 2.3022834627005466538414854464 | | |

$^a$ See footnote to Table I.

Finally, one computes the functions (where $\bar{x} = 0$)

$$g_n{}^*(x) \equiv (-1)^n \alpha_n^{-1} (\lambda_n f)^{(2^{n-1})} (\alpha_n x)$$

so normalized that

$$g_n{}^*(0) = 1, \qquad g_n{}^*(1) = 0$$

and observes convergence to $g^*$ (in the interval $[0, 1]$ also to 13 places).

## ACKNOWLEDGMENTS

Initial thoughts on this work occurred during the author's stay as Aspen; accordingly, he thanks the Aspen Institute for Physics for their hospitality, and especially E. Kerner for early discussions. A crucial hint at the existence of $\delta$ followed from conversation with P. Stein. The author's rapid acquaintance with computational technique was strongly abetted by M. Bolsterli and L. Carruthers. Discussion with E. Larsen and R. Haymaker was especially useful during initial attempts at a comprehension of the existence of $\delta$. To E. Lieb we owe strong thanks for his most profitable, critical remarks as well as for his high enthusiasm. Throughout this research, D. Campbell, F. Cooper, R. Menikoff, M. Nieto, D. Sharp, and P. Stein have offered continued critical interest. Finally, we thank P. Stein and P. Cvitanović for criticism of the manuscript. Beyond these acknowledgments, the author feels the strongest gratitude to Peter Carruthers, whose unflagging support of the author's research career has made this work possible.

## REFERENCES

1. N. Metropolis, M. L. Stein, and P. R. Stein, On Finite Limit Sets for Transformations on the Unit Interval, *J. Combinatorial Theory* **15**(1):25 (1973).
2. M. Feigenbaum, The Formal Development of Recursive Universality, Los Alamos preprint LA-UR-78-1155.
3. B. Derrida, A. Gervois, Y. Pomeau, Iterations of Endomorphisms on the Real Axis and Representations of Numbers, Saclay preprint (1977).
4. J. Guckenheimer, *Inventiones Math.* **39**:165 (1977).
5. R. H. May, *Nature* **261**:459 (1976).
6. J. Milnor, W. Thurston, Warwick Dynamical Systems Conference, *Lecture Notes in Mathematics*, Springer Verlag (1974).
7. P. Stefan, *Comm. Math. Phys.* **54**:237 (1977).

Reprinted from JOURNAL OF COMBINATORIAL THEORY
All Rights Reserved by Academic Press, New York and London

Vol. 15, No. 1, July 1973

# On Finite Limit Sets for Transformations
# on the Unit Interval

N. METROPOLIS, M. L. STEIN, AND P. R. STEIN*

*University of California, Los Alamos Scientific Laboratory,
Los Alamos, New Mexico 87544*

*Communicated by G.-C. Rota*

Received June 8, 1971

An infinite sequence of finite or denumerable limit sets is found for a class of many-to-one transformations of the unit interval into itself. Examples of four different types are studied in some detail; tables of numerical results are included. The limit sets are characterized by certain patterns; an algorithm for their generation is described and established. The structure and order of occurrence of these patterns is universal for the class.

1. Introduction. The iterative properties of 1-1 transformations of the unit interval into itself have received considerable study, and the general features are reasonably well understood. For many-to-one transformations, however. the situation is less satisfactory, only special and fragmentary results having been obtained to date [1, 2]. In the present paper we attempt to bring some coherence to the problem by exhibiting an infinite sequence of finite limit sets whose structure is common to a broad class of non 1-1 transformations of [0, 1] into itself. Generally speaking, the limit sets we shall construct are not the only possible ones belonging to an arbitrary transformation in the underlying class. Nevertheless, our sequence—which we shall call the "*U*-sequence"—constitutes perhaps the most interesting family of finite limit sets in virtue of the universality of their structure and of their order of occurrence. With regard to infinite limit sets we shall have little to say. There is reason to believe, however, that for a non-vacuous (in fact, infinite) subset of the class of transformations considered here, our construction—suitably extended to the limit of "periods of infinite length"—is exhaustive in the

* Work performed under the auspices of the U.S. Atomic Energy Commission under contract W-7405-ENG-36.

25

probabilistic sense, namely, that with "probability 1" every limit set belongs to the $U$-sequence.

**2.** We begin by describing the class of transformations to which our construction applies, but make no claim that the conditions imposed are strictly necessary. The description is complicated by our attempt to exclude, insofar as is possible, certain finite limit sets, not belonging to the $U$-sequence, whose existence and structure depend on detailed properties of the particular transformation in question.

All our transformations will be of the form

$$T_\lambda(x) : x' = \lambda f(x),$$

where $x'$ denotes the first iterate of $x$ (*not* the derivative!) and $\lambda$ varies in a certain open interval to be specified below. The fundamental properties of $f(x)$ will be:

A.1. $f(x)$ is continuous, single-valued, and piece-wise $C^{(1)}$ on $[0, 1]$, and strictly positive on the open interval, with $f(0) = f(1) = 0$.

A.2. $f(x)$ has a unique maximum, $f_{max} \leqslant 1$, assumed either at a point or in an interval. To the left or right of this point (or interval) $f(x)$ is strictly increasing or strictly decreasing, respectively.

A.3. At any $x$ such that $f(x) = f_{max}$, the derivative exists and is equal to zero.

We allow the possibility that $f(x)$ assumes its maximum in an interval so as to include certain broken-linear functions with a "flat top" (cf. example (3.4) below).

In addition to the properties (A) we need some further conditions which will serve to define the range of the parameter $\lambda$:

B. Let $\lambda_{max} = 1/f_{max}$. Then there exists a $\lambda_0$ such that, for $\lambda_0 < \lambda < \lambda_{max}$, $\lambda f(x)$ has only two fixed points, the origin and $x_F(\lambda)$, say, both of which are repellent. For functions of class $C^{(1)}$ this means simply that

$$\lambda \frac{df}{dx}\Big|_{x=0} > 1$$

$$\lambda \frac{df}{dx}\Big|_{x=x_F(\lambda)} < -1 \qquad (\lambda_0 < \lambda \leqslant \lambda_{max}).$$

For piece-wise $C^{(1)}$ functions the generalization of these conditions is obvious.

The above conditions are sufficient to guarantee the existence of the

$U$-sequence; that they are not necessary can be shown by various examples, but we shall not pursue this matter here.

$f(x)$ as defined above clearly has the property that its piecewise derivative is less than 1 in absolute value in some interval $N$ which includes that for which $f(x) = f_{\max}$. In order to exclude certain unwanted finite limit sets, we append the following condition:

C. $f(x)$ is convex in the interval $N$; at every point $x \notin N$, the piece-wise derivative of $f(x)$ is greater than 1 in absolute value.

Unfortunately, property (C) is not sufficient to exclude all finite limit sets not given by our construction; to achieve this end it might be necessary to restrict the underlying class of transformations rather drastically. We shall return to this point in Section 4 below.

It will simplify the subsequent discussion to make the non-essential assumption that $f(x)$ assumes its maximum at the point $x = \frac{1}{2}$ (or, if the function assumes its maximum in an interval, that the interval includes $x = \frac{1}{2}$). In the sequel we shall make this assumption without further comment. A particular iterate $x'$ will then be said to be of "type L" or of "type R" according as $x' < \frac{1}{2}$ or $x' > \frac{1}{2}$, respectively. Given an initial point $x_0$, the minimum distinguishing information about the sequence of iterates $T_\lambda^{(k)}(x_0)$, $k = 1, 2,...$, will consist in a "pattern" of R's and L's, the $k$-th letter giving the relative position of the $k$-th iterate of $x_0$ with respect to the point $x = \frac{1}{2}$. The patterns turn out to play a fundamental role in our construction; they will be discussed in detail in the following sections.

3. Let us give some simple examples of the class of transformations we are considering:

$$Q_\lambda(x): \quad x' = \lambda x(1 - x)$$

$$3 < \lambda < 4 \tag{3.1}$$

$$S_\lambda(x): \quad x' = \lambda \sin \pi x$$

$$\lambda_0 < \lambda < 1 \quad \text{(with .71} < \lambda_0 < .72) \tag{3.2}$$

$$C_\lambda(x): \quad x' = \lambda W(3 - 3W + W^2), \qquad W \equiv 3x(1 - x)$$

$$\lambda_0 < \lambda < \frac{64}{63} \quad \text{(with .872} < \lambda_0 < .873) \tag{3.3}$$

In the last two examples, more precise limits for $\lambda_0$ are available, but they are not important for our discussion. All these examples are convex functions of class $C^{(\infty)}$ which are, moreover, symmetric about $x = \frac{1}{2}$.

With regard to the existence of the $U$-sequence, these restrictions are in no way essential. As will be remarked in Section 4, however, these examples happen to belong to the subclass for which our construction does exhaust all finite limit sets.

As a further example, consider the broken-linear mapping:

$$
\begin{aligned}
L_\lambda(x; e): \quad x' &= \frac{\lambda}{e} x, & 0 \leqslant x \leqslant e, \\[2mm]
x' &= \lambda, & e \leqslant x \leqslant 1 - e, \\[2mm]
x' &= \frac{\lambda}{e}(1 - x), & 1 - e \leqslant x \leqslant 1,
\end{aligned}
\tag{3.4}
$$

$$
\text{with} \quad 1 - e < \lambda < 1.
$$

Here $e$ is a parameter characterizing the width 1-2e of the maximum, and may be chosen to have any value in the range $0 < e < \frac{1}{2}$. It remains fixed as $\lambda$ is varied, and different choices of $e$ yield distinct transformations.

4.  The finite limit sets of our class of transformations—and, in particular, of the four special transformations given above—are attractive periods of order $k = 2, 3, \ldots$ . (We exclude the case $k = 1$ by invoking property (B).) The reader will recall that an "attractive period of order $k$" is a set of $k$ periodic points $x_i$, $i = 1, 2, \ldots, k$, with $T_\lambda(x_i) = x_{i+1}$ (in some order). Each of these is a fixed point of the $k$-th power of $T_\lambda : T_\lambda^{(k)}(x_i) = x_i$, for which, moreover, the (piece-wise) derivative satisfies

$$
\left| \frac{dT_\lambda^{(k)}}{dx} \right|_{x=x_i} < 1.
$$

(By the chain rule, the slope is the same at all points in the period.) As a consequence of this slope condition, there exists for each $x_i$ an attractive neighborhood $n(x_i)$ such that for any $x^* \in n(x_i)$ the sequence of iterates $T_\lambda^{(jk)}(x^*)$, $j = 1, 2, \ldots$, will converge to $x_i$. Periodic points which do not satisfy this slope condition (more precisely, for which the absolute value of the derivative is greater than 1) have no attractive neighborhood; they are consequently termed repellent (or unstable). These points belong to what is sometimes called "the set of exceptional points," a set of measure zero in the interval which plays no role in a discussion of limit sets.

The sequence of finite periods which we shall exhibit will be characterized *inter alia* by the following property:

J. For every period belonging to the $U$-sequence there is a period point whose attractive neighborhood includes the point $x = \frac{1}{2}$.

Now it follows from a theorem of G. Julia [3] that, if $T_\lambda(x)$ is the restriction to [0, 1] of some function analytic in the complex plane whose derivative vanishes at a single point in the interval, then the only possible finite limit sets $(k > 1)$ are those with the property (J). The transformations (3.1) through (3.3) are clearly of this type, so that, with respect to finite limit sets, the $U$-sequence will exhaust all possibilities for them. That Julia's criterion is not necessary is shown by example (3.4); in this case there cannot be any attractive periods which do not have a period point lying in the region $e \leqslant x \leqslant 1 - e$. Such a period, however, clearly is of the type described by property (J), and hence belongs to the $U$-sequence.

Taking our clue from property (J), we now investigate the solutions $\lambda$ of the equation:

$$T_\lambda^{(k)}(\tfrac{1}{2}) = \tfrac{1}{2}. \tag{4.1}$$

The corresponding periodic limit sets will be attractive, since the slope of $\lambda f(x)$ at $x = \tfrac{1}{2}$ is zero by hypothesis (property (A.3)). By way of example, we choose $k = 5$. Then for each of the four transformations of Section 3 there are precisely three distinct solutions of equation (4.1). The three patterns—common to all four transformations—are:

$$\tfrac{1}{2} \to R \to L \to R \to R \to \tfrac{1}{2},$$
$$\tfrac{1}{2} \to R \to L \to L \to R \to \tfrac{1}{2},$$
$$\tfrac{1}{2} \to R \to L \to L \to L \to \tfrac{1}{2}.$$

Omitting the initial and final points as understood, we write these patterns in the simplified form:

$$RLR^2,$$
$$RL^2R, \tag{4.2}$$
$$RL^3.$$

In accordance with this convention, a pattern with $k - 1$ letters R or L will be said to be of "length $k$."

These solutions are clearly ordered on the parameter $\lambda$. In Table I we give the full set of solutions of (4.1), through $k = 7$, for all four special transformations; in the broken-linear case we choose $e = .45$. The numerical values of $\lambda$ were found by a simple iterative technique (the "binary chopping process"); although they are given to only seven decimal digits, they are actually known to approximately twice that precision. Of course, once a particular $\lambda$ has been found, the corresponding pattern can be generated by direct iteration.

30                    METROPOLIS, STEIN, AND STEIN

TABLE I

|   |       |          | Values of $\lambda_i$ | | | |
|---|-------|----------|-----------------------|-----------|-----------|------------|
| $i$ | $k_i$ | $P_i$ | $Q_\lambda(x)$ | $S_\lambda(x)$ | $C_\lambda(x)$ | $L_\lambda(x;.45)$ |
| 1 | 2 | R | 3.2360680 | .7777338 | .9325336 | .6581139 |
| 2 | 4 | RLR | 3.4985617 | .8463822 | .9764613 | .7457329 |
| 3 | 6 | RLR$^3$ | 3.6275575 | .8811406 | .9895107 | .7806832 |
| 4 | 7 | RLR$^4$ | 3.7017692 | .9004906 | .9955132 | .8031673 |
| 5 | 5 | RLR$^2$ | 3.7389149 | .9109230 | .9990381 | .8180892 |
| 6 | 7 | RLR$^2$LR | 3.7742142 | .9213346 | 1.0024311 | .8318799 |
| 7 | 3 | RL | 3.8318741 | .9390431 | 1.0073533 | .8645337 |
| 8 | 6 | RL$^2$RL | 3.8445688 | .9435875 | 1.0083134 | .8858150 |
| 9 | 7 | RL$^2$RLR | 3.8860459 | .9568445 | 1.0111617 | .8977794 |
| 10 | 5 | RL$^2$R | 3.9057065 | .9633656 | 1.0123766 | .9085993 |
| 11 | 7 | RL$^2$R$^3$ | 3.9221934 | .9687826 | 1.0132699 | .9187692 |
| 12 | 6 | RL$^2$R$^2$ | 3.9375364 | .9735656 | 1.0140237 | .9278274 |
| 13 | 7 | RL$^2$R$^2$L | 3.9510322 | .9782512 | 1.0146450 | .9361518 |
| 14 | 4 | RL$^2$ | 3.9602701 | .9820353 | 1.0149542 | .9462185 |
| 15 | 7 | RL$^3$RL | 3.9689769 | .9857811 | 1.0152122 | .9564172 |
| 16 | 6 | RL$^3$R | 3.9777664 | .9892022 | 1.0154974 | .9635343 |
| 17 | 7 | RL$^3$R$^2$ | 3.9847476 | .9919145 | 1.0156711 | .9702076 |
| 18 | 5 | RL$^3$ | 3.9902670 | .9944717 | 1.0157727 | .9775473 |
| 19 | 7 | RL$^4$R | 3.9945378 | .9966609 | 1.0158320 | .9846165 |
| 20 | 6 | RL$^4$ | 3.9975831 | .9982647 | 1.0158621 | .9903134 |
| 21 | 7 | RL$^5$ | 3.9993971 | .9994507 | 1.0158718 | .9957404 |

We note that the set of 21 patterns and its $\lambda$-ordering is common to all four transformations. This remains true when we extend our calculations through $k = 15$. As $k$ increases, the total number of solutions of (4.1) becomes large, as indicated in Table II. Thus for $k \leq 15$ there is a total of 2370 distinct solutions of equation (4.1). In the appendix we give a complete list of ordered patterns for $k \leq 11$.

The fact that these patterns and their $\lambda$-ordering are a common property of four apparently unrelated transformations (note that they are not connected by ordinary conjugacy, a relation which will be discussed in Section (6) suggests that the pattern sequence is a general property of a wide class of mappings. For this reason we have called this sequence of patterns the $U$-sequence where "$U$" stands (with some exaggeration) for

TABLE II

| $k$ ... | 2 | 3 | 4 | 5 | 6 | 7 | 8 | 9 | 10 | 11 | 12 | 13 | 14 | 15 |
|---|---|---|---|---|---|---|---|---|---|---|---|---|---|---|
| Number of solutions ... | 1 | 1 | 2 | 3 | 5 | 9 | 16 | 28 | 51 | 93 | 170 | 315 | 585 | 1091 |

"universal." In the next section we shall state and prove a logical algorithm which generates the $U$-sequence for any transformation having the properties (A) and (B) of Section 2. In the present section we confine ourselves to describing what might be called the "$\lambda$-structure" of the limit sets associated with the patterns of the $U$-sequence. No proofs are included, since the results given here will not be used in the proof of our main theorem.

As constructed, the patterns of the $U$-sequence correspond to distinct solutions of equation (4.1); they are attractive $k$-periods containing the point $x = \frac{1}{2}$ and possessing the property (J). It is clear by continuity that, given any solution $\lambda$ (with finite $k$) and its associated pattern $P_k(\lambda)$, then for sufficiently small $\epsilon > 0$ there will exist periodic limit sets with the same pattern for all $\bar{\lambda}$ in the interval $\lambda - \epsilon \leqslant \bar{\lambda} \leqslant \lambda + \epsilon$. In other words, each period has a finite "$\lambda$-width." It is also clear that there exist critical values $m_1(\lambda)$ and $m_2(\lambda)$ such that, for $\bar{\lambda} < \lambda - m_1$ and $\bar{\lambda} > \lambda + m_2$, the pattern $P_k(\lambda)$ does *not* correspond to an attractive period of $T_{\bar{\lambda}}(x)$.

Consider now for simplicity the case in which the transformation is $C^{(1)}$, and take $m_1$ and $m_2$ to be boundary values such that for $\lambda - m_1 < \bar{\lambda} < \lambda + m_2$ the periodic limit set with pattern $P_k(\lambda)$ is attractive. As $\bar{\lambda}$ varies in this interval, the slope $dT_{\bar{\lambda}}^{(k)}/dx$ at a period point varies continuously from $+1$ to $-1$, the values $\pm 1$ being assumed at the boundary points. It is natural to ask: what happens if $\bar{\lambda}$ lies just to the left or just to the right of the above interval? The question as to what the limit sets look like if $\bar{\lambda} = \lambda - m_1 - \delta$ ($\delta$ small) is a difficult one; the conjectured behavior will be described in Section 6, but rigorous proof is lacking. For $\bar{\lambda} = \lambda + m_2 + \delta$ we are in better case. As shown in Section 5, corresponding to any solution $\lambda$ of (4.1) and its associated pattern there exists an infinite sequence of solutions $\lambda < \lambda^{(1)} < \lambda^{(2)} < \cdots < \lambda^{(\infty)}$ with associated patterns $H^{(1)}(\lambda^{(1)})$, $H^{(2)}(\lambda^{(2)})$,..., called "harmonics," with the property that they exhaust all possible solutions $\lambda^*$ in the interval $\lambda \leqslant \lambda^* \leqslant \lambda^{(\infty)}$. The sequence of harmonics of a given solution is a set of periods of order $2^m k$, $m = 1, 2,...$, with contiguous $\lambda$-widths and well-defined pattern structure; no other periods of the $U$-sequence can exist for any $\lambda^*$ in the given interval (harmonics have been encountered before in a more restrictive context: cf. reference 4). From the construction of Section 5 it will be obvious that $\lambda^{(\infty)}$ exists as a right-hand limit; the question as to the nature of the limit sets for $\lambda^* = \lambda^{(\infty)} + \epsilon$ remains open (but cf. Conjecture 2 of Section 6).

In order to prepare the ground for the discussion in the next section, we give here the following formal

DEFINITION 1.    Let $P = RL^{\alpha_1}R^{\alpha_2}L^{\alpha_3} \cdots$ be a pattern corresponding to some solution of (4.1). Then the (first) harmonic of $P$ is the pattern $H = P\mu P$, where $\mu = L$ if $P$ contains an odd number of $R$'s, and $\mu = R$ otherwise.

For example, the pattern $RLR^2$ has the harmonic $H = RLR^2LRLR^2$, while for $RL^2R$ we have $H = RL^2R^3L^2R$.

Naturally, the construction of the harmonic can be iterated, so that one may speak of the second, third,..., $m$-th harmonic, etc. When necessary, we shall write $H^{(j)}$ to denote the $j$-th harmonic.

In addition to the harmonic $H$ of a pattern $P$, there is another formal construct which will be used in the sequel:

DEFINITION 2.    The "antiharmonic" $A$ of a pattern $P$ is constructed analogously to the harmonic $H$ except that $\mu = L$ when $P$ contains an even number of R's, while $\mu = R$ otherwise.

Thus in passing from a pattern to its harmonic the "R-parity" changes while for the antiharmonic the parity remains the same. The antiharmonic is a purely formal construct and never corresponds to any periodic limit set; the reason for this will become clear in the next section. Note that, like that of the harmonic, the antiharmonic construction can be iterated to any desired order.

5.    We begin by defining the "extension" of a pattern:

DEFINITION 3.    The $H$-extension of a pattern $P$ is the pattern generated by iterating the harmonic construction applied $j$ times to $P$, where $j$ increases indefinitely.

DEFINITION 4.    The $A$-extension of $P$ is the pattern $A^{(j)}(P)$, where $j$ increases indefinitely. Here $A^{(j)}(P)$ denotes the $j$-th iterate of the anti-harmonic.

In these definitions we avoid writing $j \to \infty$, in order to avoid raising questions concerning the structure of the limiting pattern. In practice, all that will be required is that $j$ is "sufficiently large."

We are now in a position to state

THEOREM 1. *Let $K$ be an integer. Consider the complete ordered sequence of solutions of (4.1) and their associated patterns for $2 \leqslant k \leqslant K$. Let $\lambda_1$ be any such solution with pattern $P_1$ and length $k_1$, and let $\lambda_2 > \lambda_1$ be the "adjacent" solution (i.e., the next in order) with pattern $P_2$ and length $k_2$.*

*Form the H-extension of $P_1$ and the A-extension of $P_2$. Reading from left to right, the two extensions $H(P_1)$ and $A(P_2)$ will have a maximal common leading subpattern $P^*$ of length $k^*$, so that we may write*

$$H(P_1) = P^*\mu_1 \ldots, \quad A(P_2) = P^*\mu_2 \ldots, \quad \mu_1 \neq \mu_2,$$

*where $\mu_i$ stands for one of the letters L or R.*

*Case 1. $k^* \geqslant 2k_1$. Then the solution $\lambda^*$ of lowest order such that $\lambda_1 < \lambda^* < \lambda_2$ is the harmonic of $P_1$.*

*Case 2. $k^* < 2k_1$. Then the solution $\lambda^*$ of lowest order such that $\lambda_1 < \lambda^* < \lambda_2$ corresponds to the pattern $P^*$ of length $k^*$ ($> K$ necessarily).*

A simple consequence of this theorem is the following:

COROLLARY. *Let $|k_1 - k_2| = 1$ in Theorem 1. Then the lowest order solution $\lambda^*$ with $\lambda_1 < \lambda^* < \lambda_2$ has length $k^* = 1 + \max(k_1, k_2)$.*

This follows from the theorem on noting that all patterns have the common leading subpatterns (not maximal!) RL; therefore, in forming the extensions, the first disagreement will indeed come at the indicated value of $k^*$.

We give some examples of the application of Theorem 1.

*Example 1.* Take $K = 9$. Reference to the table in the appendix shows that patterns #12 and #14 are adjacent. We have

$$P_1 = RLR^4, \ k_1 = 7, \ H(P_1) = RLR^4LRLR^4\ldots,$$
$$P_2 = RLR^4LR, \ k_2 = 9, \ A(P_2) = RLR^4LRLRL\ldots,$$

so $P^* = RLR^4LRLR$, with $k^* = 11$, as verified by the table.

*Example 2.* Again take $K = 9$. Patterns #16 and #19 are adjacent and $P_1 = RLR^2$ with $k_1 = 5$; here $k^* \geqslant 2k_1$. Therefore, by Case 1, the lowest order solution between the two patterns is the harmonic of $P_1$, namely, $RLR^2LRLR^2$, as given in the table (pattern #17).

To prove Theorem 1 we must first introduce some new concepts. Consider the transformation:

$$T_m(x): x' = \lambda_{max} f(x) \tag{5.1}$$

This transformation maps $[0, 1]$ *onto* itself; hence, for any point in the interval, the inverses of all orders exist. Let us restrict ourselves for the moment to the point $x = \frac{1}{2}$ and its set ($2^k$ in number) of $k$-th order inverses. At each step in constructing a $k$-th order inverse we have the free choice of taking a point on the right or on the left. For example, designating the point $x = \frac{1}{2}$ by the letter O, a possible inverse of order 5 would be represented by the sequence of letters

$$RLR^2O, \tag{5.2}$$

which is to be read from *right to left*. Let us call a sequence like (5.2), when read from right to left, a "5-th order inverse path of the point $x = \frac{1}{2}$." Note that (5.2) is precisely the pattern associated with the first solution of equation (4.1) for $k = 5$. Another possible inverse path of the same order would be $L^2R^2O$, but this clearly does not correspond to any solution of (4.1).

Choosing a particular $k$-th order inverse path of $x = \frac{1}{2}$, let us call the numerical value of the corresponding $k$-th inverse the "coordinate" of the path. Obviously, no path whose coordinate is less than $\frac{1}{2}$ can correspond to a pattern associated with a solution of (4.1). In order to achieve a 1-1 correspondence between a subclass of inverse paths and our periodic patterns we introduce the concept of a "legal inverse path," which we abbreviate as "l.i.p."

DEFINITION 5. For the transformation $T_m(x)$ (cf. (5.1)), an l.i.p. of order $k$ is a $k$-th order inverse path of $x = \frac{1}{2}$ whose coordinate $x_k$ has the greatest numerical value of any point on the path.

In other words, of all the inverses constituting the path, the coordinate (i.e., the $k$-th inverse) lies farthest to the right. Note that any inverse path of $x = \frac{1}{2}$ can be inversely extended to an l.i.p. by appending on the left some suitable sequence, e.g., the sequence $RL^\alpha$ with $\alpha$ sufficiently large. Now consider the transformation $T_\lambda(x)$ corresponding to $T_m(x)$ with $\lambda < \lambda_{max}$. As $\lambda$ decreases, the original l.i.p. is deformed into an inverse path with varying coordinate $x_k(\lambda)$, but with the same pattern. By continuity, there clearly exists a $\lambda^*$ for which

$$T_{\lambda^*}(\tfrac{1}{2}) = x_k(\lambda^*);$$

this in turn implies that for $\lambda = \lambda^*$ there exists a solution of equation (4.1) with the same pattern as that of the original l.i.p. On the other hand, for an inverse path (with, say, an $R$-type coordinate) which is *not* an l.i.p. the cycle will close on some intermediate point of the path (farther to the right than $x_k(\lambda)$), so that the path cannot be further inverted; this means that the original pattern cannot correspond to a solution of (4.1). Thus we have proved

LEMMA 1.  *There is a 1-1 correspondence between the set of l.i.p.'s and the patterns associated with the solutions of equation* (4.1).

We note that the l.i.p.'s are naturally ordered on the values of their coordinates. By Lemma 1, any true statement about the pattern structure and coordinate ordering of the set of l.i.p.'s corresponds to a true statement about the pattern structure and $\lambda$-ordering of the set of solutions of (4.1).

Given some l.i.p. of order $k$, we construct an *inverse extension* $I(P)$ of the path according to the prescription $I(P) = P\mu PO$, where $\mu$ is $R$ or $L$. Obviously, one choice corresponds to the harmonic, the other to the antiharmonic (Definitions 1 and 2). We can therefore speak of the harmonic or antiharmonic of an l.i.p. as well as of a pattern. Now, because of the monotonicity property (A.2) it follows that, given any two points, taking the left-hand inverse of both points preserves their relative order, while taking the right-hand inverse reverses it. A simple argument shows that $x_A < x < x_H$, where $x$ is the coordinate of some l.i.p. and $x_A$, $x_H$ are the coordinates of its antiharmonic and harmonic, respectively. This explains why the harmonic of an l.i.p. is again an l.i.p.. while the antiharmonic is not (and hence can never correspond to an attractive period of the $U$-sequence).

One final concept, the "projection" of an interval, will be of value in the subsequent discussion.

DEFINITION 6.  Choose any two points $x_1$, $x_2$ in $(0, 1)$; they define some interval $I$. Let $\bar{P}$ be an arbitrary sequence of R's and L's with $k - 1$ letters in all. Now, for some $T_m(x)$, construct the inverse paths $\bar{P}x_1$ and $\bar{P}x_2$. The coordinate $x_1^*$ and $x_2^*$ of these two paths define a new interval $I^*$, called the "projection under $\bar{P}$ of $I$." Because the defining inverse paths are of length $k$, we refer to it as a $k$-th order projection. (If we wish to exhibit explicitly the end-points $x_1$, $x_2$ of the interval I, we write $I(x_1, x_2)$; in contrast to the usual notation for an interval, no ordering is implied.)

*Proof of Theorem* 1.  It is clear that, if two intervals $I$, $I^*$ are related by a $k$-th order projection, then for any point $x \in I^*$ we have $T_m^{(k)}(x) \in I$.

Consider now any l.i.p. with pattern PO and coordinate $x_1$. Its harmonic is again an l.i.p., with pattern $P\mu PO$ and coordinate $x_H$, $\mu$ being either R or L depending on the R-parity of P. If $x_\mu$ is the point corresponding to the choice $\mu$, then this construction shows that the interval $I^*(x_1, x_H)$ is the ($k$-th order) projection under P of the interval $I(\frac{1}{2}, x_\mu)$. Now any point $x$ in the interior of $I^*$ must map into the interior of $I$, and the end-points must map into end-points. Thus no inverse path of $x = \frac{1}{2}$—which is one of the end-points of $I$—can have a coordinate $x^*$ satisfying $x_1 < x^* < x_H$. Precisely the same argument can be made for the antiharmonic. This proves

LEMMA 2. *Let* PO *be some l.i.p. with coordinate* $x_1$. *Form the* $H^{(j)}$-*extension of P, with coordinate* $x_H^{(j)}$, *and the* $A^{(j)}$-*extension of P with coordinate* $x_A^{(j)}$. *We then have* $x_A^{(j)} < x_1 < x_H^{(j)}$. *The intervals* $I^*(x_1, x_H^{(j)})$ *and* $I(x_1, x_A^{(j)})$ *do not contain the coordinate of any inverse path of* $x = \frac{1}{2}$.

The left-hand interval $I^*(x_1, x_A^{(j)})$ is of no significance for the limit sets of $T_\lambda(x)$; in fact, for $\lambda$ a solution of (4.1) this interval shrinks to zero (and for $\lambda^* < \lambda$, neither the harmonic nor the antiharmonic exists). The right-hand, interval, however, is important. Using Lemma 2 and Lemma 1 we immediately derive

LEMMA 3. *If* $\lambda_1$ *is a solution of equation* (4.1) *and* $\lambda_H$ *is the solution corresponding to its harmonic, then there does not exist any solution* $\lambda^*$ *of* (4.1) *with the property* $\lambda_1 < \lambda^* < \lambda_H$.

Iterating this argument, we verify the statement of Section 4 that the sequence of harmonics is contiguous, i.e., that harmonics are always adjacent.

The adjacency property of harmonics serves to prove Case 1 of Theorem 1. We now proceed to Case 2.

Given some $K$, let $(P_1, x_1, k_1)$ and $(P_2, x_2, k_2)$ be two adjacent l.i.p.'s with $x_1 < x_2$ and $K + 1 < 2k_1$. Form the $H$-extension of $P_1$ and the $A$-extension of $P_2$; these can be written in the form

$$H(P_1) = P^*\mu_1 \cdots$$
$$A(P_2) = P^*\mu_2 \cdots \qquad (\mu_1 \neq \mu_2).$$

The coordinates $x_H$ and $x_A$ define an interval $I^*$ which is a projection of $I(x_{\mu_1}, x_{\mu_2})$; clearly, $I^*$ is contained in the original interval $I(x_1, x_2)$. Since $I$ contains the point $x = \frac{1}{2}$, there must exist an inverse path of $\frac{1}{2}$, $P^*O$, with coordinate $x^*$ satisfying $x_1 < x^* < x_2$. But P*O must be an l.i.p. since it is a leading subpattern of the interated harmonic of $P_1$. Moreover, by the adjacency assumption, its length $k^*$ must necessarily be greater than $k_1$ or $k_2$. On the other hand, P*O is the shortest pattern for which an interval with non-zero content exists. Invoking Lemma 1, we see that the proof of the theorem is complete.

The formulation of a practical algorithm, using the results of Theorem 1, needs little comment. Given the complete $U$-sequence for $k \leqslant K$, one generates the sequence for $K + 1$ by inserting the appropriate pattern of length $K + 1$ between every two (non-harmonic) adjacent patterns whenever the theorem permits it. The pattern $R(k = 2)$ remains the lowest pattern; as is easily shown, for any $k$ the last pattern is always of the form $RL^{k-2}$, and this is simply appended to the list. As previously mentioned, the algorithm has been checked (to $k \leqslant 15$) for the four special transformations of Section 3 by actually finding the corresponding solutions of equation (4.1)—a simple process in which there are no serious accuracy limitations.

We remark here that the combinatorial problem of enumerating all l.i.p.'s of a given length $k$ has been solved [5]; the number of patterns turns out to be just the number of symmetry types of primitive periodic sequences (with two "values" or letters allowed) under the cyclic group $C_k$ (so that the full symmetry group is $C_k \times S_2$, where $S_2$ is the symmetric group on two letters). For $k$ a prime, this number is simple, and turns out to be given by the expression

$$\frac{1}{k} (2^{k-1} - 1).$$

We encountered this enumeration problem previously (cf. reference 4, Table 1); at that time we were not aware of the work of Gilbert and Riordan [5].

6. In this final section we collect some observations and conjectures concerning the nature of limit sets not belonging to the $U$-sequence, ending with a few remarks on the relation of conjugacy.

(1) *Other finite limit sets.* As remarked in the introduction, it does not seem possible to exclude "anomalous" limit sets without seriously restricting the underlying class of transformations. To convince the reader that such anomalous periods can in fact exist we give a simple example:

Let us alter the special transformation (3.4) in the following way (we take $e = .45$):

$$
\begin{aligned}
x' &= 4.5\lambda x, & 0 \leqslant x \leqslant .2 & \\
x' &= \lambda(.4x + .82), & .2 \leqslant x \leqslant .45 & \\
x' &= \lambda, & .45 \leqslant x \leqslant .55 & \quad (.55 < \lambda < 1). \quad (6.1)\\
x' &= \frac{\lambda}{.45}(1 - x), & .55 \leqslant x \leqslant 1 &
\end{aligned}
$$

Then, in addition to the $U$-sequence (with $\lambda$ values different from those of the original transformation), there exists an attractive 2-period in the range $\lambda_1 < \lambda \leqslant 1$ with

$$\lambda_1 = \tfrac{1}{2} + \tfrac{1}{2}\sqrt{.19}.$$

Note that the 2-period remains attractive even for $\lambda = 1$. While the anomalous periods do not affect the existence of the $U$-sequence, they do cause additional partitioning of the unit interval because their existence implies that there is a set of points (with non-zero measure) whose sequence of iterated images will converge to the periods in question.

These anomalous periods, however, differ radically from those belonging to the $U$-sequence in that they do not possess the property (J). This in turn means that the slope at a period point is strongly bounded away from zero. Thus, at least for transformations with the property (C), it seems reasonable to conjecture that such periods cannot have arbitrary length and still remain attractive. Hence we make

CONJECTURE 1. *For transformations with properties* (A), (B) *and* (C), *the anomalous attractive periods constitute at most a finite sequence.*

(2) *Infinite limit sets.* For simplicity we consider the case in which there are no anomalous periods, e.g., functions covered by Julia's theorem (or some valid extension thereof). We assign to each period of the $U$-sequence a $\lambda$-measure equal to its $\lambda$-width. The question is then: is the $\lambda$-measure of the full $U$-sequence equal to $\lambda_{max} - \lambda_0$? Or, put otherwise, is there a set of non-zero measure in the interval $(\lambda_0, \lambda_{max})$ such that the sequence of iterates of $x = \tfrac{1}{2}$ does *not* converge to a member of the $U$-sequence? Numerical experiments with the four special transformations of Section 3 together with some heuristic arguments based on the iteration of the algorithm of Theorem 1 leads us to make the modest

CONJECTURE 2. *For an infinite subclass of transformations with properties* (A) *and* (B), *the λ-measure of the U-sequence is the whole λ interval.*

(3) *A limiting case.* Take the transformation $L_\lambda(x; e)$ of Section 3 and set $e = \frac{1}{2}$. We then have

$$L_\lambda(x; \tfrac{1}{2}): x' = 2\lambda x, \quad 0 \leqslant x \leqslant \tfrac{1}{2}$$

$$x' = 2\lambda(1 - x), \quad \tfrac{1}{2} \leqslant x \leqslant 1 \qquad (\tfrac{1}{2} < \lambda < 1). \qquad (6.2)$$

Although we cannot speak of attractive periods in this case (since the slope of the function is nowhere less than 1 in absolute value), it is still of interest to investigate the corresponding solutions of equation (4.1). These turn out to be a subset of the $U$-sequence in which the 2-period, all harmonics, and all patterns algorithmically generated from the harmonics and adjacent nonharmonics, are absent. The count through $k \leqslant 15$ is given in Table III, which may be compared with Table II.

TABLE III

| $k \ldots$ | 3 | 4 | 5 | 6 | 7 | 8 | 9 | 10 | 11 | 12 | 13 | 14 | 15 |
|---|---|---|---|---|---|---|---|---|---|---|---|---|---|
| Number of solutions | 1 | 1 | 3 | 4 | 9 | 14 | 27 | 48 | 93 | 163 | 315 | 576 | 1085 |

One can explain this behavior by saying that, as the width 1-2e of the flat-top shrinks to zero, the harmonics and harmonic-generated periods "coalesce" in structure with their fundamentals. This provides another illustration of the nature of the harmonics outlined in Section 4.

(4) *Conjugacy.* Two transformations $f(x)$, $g(x)$ on [0, 1] are said to be conjugate to each other if there exists a continuous, 1-1 mapping $h(x)$ of [0, 1] onto itself such that

$$g(x) = hf[h^{-1}(x)], \quad x \in [0, 1]. \qquad (6.3)$$

If $f(x)$ and $g(x)$ are themselves 1-1, the question of the existence of an $h(x)$ satisfying (6.3) is settled by a theorem of Schreier and Ulam [6]. When $f(x)$, $g(x)$ are not homeomorphisms, very little is known about the existence or nonexistence of a conjugating function $h(x)$.

The importance of (6.3) for our purpose is that the attractive nature of limit sets is preserved under conjugacy; in particular, if $T_\lambda(x)$ possess the $U$-sequence, then so does every conjugate of it. Clearly, our class of transformations must be invariant under conjugation by the set of all continuous, 1-1 functions $h(x)$ on [0, 1]. (Incidentally, we now see why our special choice of the point $x = \frac{1}{2}$ is no restriction, since it can be shifted by conjugation with an appropriate $h(x)$.)

It has long been known [7] that the parabolic transformation (3.1) with $\lambda = \lambda_{max} = 4$ is conjugate to the broken-linear transformation (6.2) with $\lambda = 1$, the conjugating function being

$$h(x) = \frac{2}{\pi} \sin^{-1}(\sqrt{x}).$$

In general, no such pairwise conjugacy exists for the four special transformations of Section 3. For particular choices of the parameters this can be shown by making the following simple test. If $f(x_0) = x_0$ and $g(x_1) = x_1(x_0, x_1 \neq 0)$, then a short calculation shows that

$$\frac{df(x)}{dx}\bigg|_{x=x_0} = \frac{dg(x)}{dx}\bigg|_{x=x_1}. \tag{6.4}$$

It is easily established that (6.4) does not hold in general for any pair of our special transformations.

In view of the existence of the $U$-sequence, it is of interest to speculate whether there is not some well-defined but less restrictive equivalence relation that will serve to replace conjugacy (for one such suggestion-due to S. Ulam—see the remarks in reference 1, p. 49). Of course, Theorem 1 itself provides such an equivalence relation, albeit not a very useful one:

Let $T_{1\lambda}(x)$, $T_{2\mu}(x)$ be two transformations with properties (A) and (B). Then there exists a mapping function $M_{12}$ such that $M_{12}(\lambda) = \mu$, the domain of $M$ being the union of the $\lambda$-widths of the $U$-sequence for $T_1$ and the range being the union of the $\mu$-widths of the $U$-sequence for $T_2$.

Since at present nothing whatsoever is known about these mappings $M_{ij}$, the above correspondence amounts to nothing more than a restatement of the existence of the $U$-sequence itself.

## APPENDIX

The following table gives the complete ordered set of patterns associated with the $U$-sequence for $K \leqslant 11$; $i$ is a running index, $K$ gives the pattern length, and $I(K)$ indicates the relative order of periods of given length $K$. The ordering corresponds to the $\lambda$-ordering of solutions of equation (4.1).

| $i$ | $K$ | $I(K)$ | Pattern | $i$ | $K$ | $I(K)$ | Pattern |
|---|---|---|---|---|---|---|---|
| 1 | 2 | 1 | $R$ | 41 | 10 | 9 | $RL^2RLR^3L$ |
| 2 | 4 | 1 | $RLR$ | 42 | 7 | 3 | $RL^2RLR$ |
| 3 | 8 | 1 | $RLR^3LR$ | 43 | 10 | 10 | $RL^2RLRLRL$ |
| 4 | 10 | 1 | $RLR^3LRLR$ | 44 | 11 | 15 | $RL^2RLRLRLR$ |
| 5 | 6 | 1 | $RLR^3$ | 45 | 9 | 7 | $RL^2RLRLR$ |
| 6 | 10 | 2 | $RLR^5LR$ | 46 | 11 | 16 | $RL^2RLRLR^3$ |
| 7 | 8 | 2 | $RLR^5$ | 47 | 10 | 11 | $RL^2RLRLR^2$ |
| 8 | 10 | 3 | $RLR^7$ | 48 | 11 | 17 | $RL^2RLRLR^2L$ |
| 9 | 11 | 1 | $RLR^8$ | 49 | 8 | 5 | $RL^2RLRL$ |
| 10 | 9 | 1 | $RLR^6$ | 50 | 11 | 18 | $RL^2RLRL^2RL$ |
| 11 | 11 | 2 | $RLR^6LR$ | 51 | 5 | 2 | $RL^2R$ |
| 12 | 7 | 1 | $RLR^4$ | 52 | 10 | 12 | $RL^2R^3L^2R$ |
| 13 | 11 | 3 | $RLR^4LRLR$ | 53 | 11 | 19 | $RL^2R^3L^2RL$ |
| 14 | 9 | 2 | $RLR^4LR$ | 54 | 8 | 6 | $RL^2R^3L$ |
| 15 | 11 | 4 | $RLR^4LR^3$ | 55 | 11 | 20 | $RL^2R^3LR^2L$ |
| 16 | 5 | 1 | $RLR^2$ | 56 | 10 | 13 | $RL^2R^3LR^2$ |
| 17 | 10 | 4 | $RLR^2LRLR^2$ | 57 | 11 | 21 | $RL^2R^3LR^3$ |
| 18 | 11 | 5 | $RLR^2LRLR^3$ | 58 | 9 | 8 | $RL^2R^3LR$ |
| 19 | 9 | 3 | $RLR^2LRLR$ | 59 | 11 | 22 | $RL^2R^3LRLR$ |
| 20 | 11 | 6 | $RLR^2LRLRLR$ | 60 | 10 | 14 | $RL^2R^3LRL$ |
| 21 | 7 | 2 | $RLR^2LR$ | 61 | 7 | 4 | $RL^2R^3$ |
| 22 | 11 | 7 | $RLR^2LR^3LR$ | 62 | 10 | 15 | $RL^2R^5L$ |
| 23 | 9 | 4 | $RLR^2LR^3$ | 63 | 11 | 23 | $RL^2R^5LR$ |
| 24 | 11 | 8 | $RLR^2LR^5$ | 64 | 9 | 9 | $RL^2R^5$ |
| 25 | 10 | 5 | $RLR^2LR^4$ | 65 | 11 | 24 | $RL^2R^7$ |
| 26 | 8 | 3 | $RLR^2LR^2$ | 66 | 10 | 16 | $RL^2R^6$ |
| 27 | 10 | 6 | $RLR^2LR^2LR$ | 67 | 11 | 25 | $RL^2R^6L$ |
| 28 | 11 | 9 | $RLR^2LR^2LR^2$ | 68 | 8 | 7 | $RL^2R^4$ |
| 29 | 3 | 1 | $RL$ | 69 | 11 | 26 | $RL^2R^4LRL$ |
| 30 | 6 | 2 | $RL^2RL$ | 70 | 10 | 17 | $RL^2R^4LR$ |
| 31 | 9 | 5 | $RL^2RLR^2L$ | 71 | 11 | 27 | $RL^2R^4LR^2$ |
| 32 | 11 | 10 | $RL^2RLR^2LR^2$ | 72 | 9 | 10 | $RL^2R^4L$ |
| 33 | 10 | 7 | $RL^2RLR^2LR$ | 73 | 11 | 28 | $RL^2R^4L^2R$ |
| 34 | 11 | 11 | $RL^2RLR^2LRL$ | 74 | 6 | 3 | $RL^2R^2$ |
| 35 | 8 | 4 | $RL^2RLR^2$ | 75 | 11 | 29 | $RL^2R^2LRL^2R$ |
| 36 | 11 | 12 | $RL^2RLR^4L$ | 76 | 9 | 11 | $RL^2R^2LRL$ |
| 37 | 10 | 8 | $RL^2RLR^4$ | 77 | 11 | 30 | $RL^2R^2LRLR^2$ |
| 38 | 11 | 13 | $RL^2RLR^5$ | 78 | 10 | 18 | $RL^2R^2LRLR$ |
| 39 | 9 | 6 | $RL^2RLR^3$ | 79 | 11 | 31 | $RL^2R^2LRLRL$ |
| 40 | 11 | 14 | $RL^2RLR^3LR$ | 80 | 8 | 8 | $RL^2R^2LR$ |

42                    METROPOLIS, STEIN, AND STEIN

| i | K | I(K) | Pattern | i | K | I(K) | Pattern |
|---|---|---|---|---|---|---|---|
| 81 | 11 | 32 | $RL^2R^2LR^3L$ | 121 | 9 | 17 | $RL^3R^3L$ |
| 82 | 10 | 19 | $RL^2R^2LR^3$ | 122 | 11 | 50 | $RL^3R^3LR^2$ |
| 83 | 11 | 33 | $RL^2R^2LR^4$ | 123 | 10 | 30 | $RL^3R^3LR$ |
| 84 | 9 | 12 | $RL^2R^2LR^2$ | 124 | 11 | 51 | $RL^3R^3LRL$ |
| 85 | 11 | 34 | $RL^2R^2LR^2LR$ | 125 | 8 | 11 | $RL^3R^3$ |
| 86 | 10 | 20 | $RL^2R^2LR^2L$ | 126 | 11 | 52 | $RL^3R^5L$ |
| 87 | 7 | 5 | $RL^2R^2L$ | 127 | 10 | 31 | $RL^3R^5$ |
| 88 | 10 | 21 | $RL^2R^2L^2RL$ | 128 | 11 | 53 | $RL^3R^6$ |
| 89 | 11 | 35 | $RL^2R^2L^2RLR$ | 129 | 9 | 18 | $RL^3R^4$ |
| 90 | 9 | 13 | $RL^2R^2L^2R$ | 130 | 11 | 54 | $RL^3R^4LR$ |
| 91 | 11 | 36 | $RL^2R^2L^2R^3$ | 131 | 10 | 32 | $RL^3R^4L$ |
| 92 | 10 | 22 | $RL^2R^2L^2R^2$ | 132 | 11 | 55 | $RL^3R^4L^2$ |
| 93 | 11 | 37 | $RL^2R^2L^2R^2L$ | 133 | 7 | 7 | $RL^3R^2$ |
| 94 | 4 | 2 | $RL^2$ | 134 | 11 | 56 | $RL^3R^2LRL^2$ |
| 95 | 8 | 9 | $RL^3RL^2$ | 135 | 10 | 33 | $RL^3R^2LRL$ |
| 96 | 11 | 38 | $RL^3RL^2R^2L$ | 136 | 11 | 57 | $RL^3R^2LRLR$ |
| 97 | 10 | 23 | $RL^3RL^2R^2$ | 137 | 9 | 19 | $RL^3R^2LR$ |
| 98 | 11 | 39 | $RL^3RL^2R^3$ | 138 | 11 | 58 | $RL^3R^2LR^3$ |
| 99 | 9 | 14 | $RL^3RL^2R$ | 139 | 10 | 34 | $RL^3R^2LR^2$ |
| 100 | 11 | 40 | $RL^3RL^2RLR$ | 140 | 11 | 59 | $RL^3R^2LR^2L$ |
| 101 | 10 | 24 | $RL^3RL^2RL$ | 141 | 8 | 12 | $RL^3R^2L$ |
| 102 | 11 | 41 | $RL^3RL^2RL^2$ | 142 | 11 | 60 | $RL^3R^2L^2RL$ |
| 103 | 7 | 6 | $RL^3RL$ | 143 | 10 | 35 | $RL^3R^2L^2R$ |
| 104 | 11 | 42 | $RL^3RLR^2L^2$ | 144 | 11 | 61 | $RL^3R^2L^2R^2$ |
| 105 | 10 | 25 | $RL^3RLR^2L$ | 145 | 9 | 20 | $RL^3R^2L^2$ |
| 106 | 11 | 43 | $RL^3RLR^2LR$ | 146 | 11 | 62 | $RL^3R^2L^3R$ |
| 107 | 9 | 15 | $RL^3RLR^2$ | 147 | 5 | 3 | $RL^3$ |
| 108 | 11 | 44 | $RL^3RLR^4$ | 148 | 10 | 36 | $RL^4RL^3$ |
| 109 | 10 | 26 | $RL^3RLR^3$ | 149 | 11 | 63 | $RL^4RL^3R$ |
| 110 | 11 | 45 | $RL^3RLR^3L$ | 150 | 9 | 21 | $RL^4RL^2$ |
| 111 | 8 | 10 | $RL^3RLR$ | 151 | 11 | 64 | $RL^4RL^2R^2$ |
| 112 | 11 | 46 | $RL^3RLRLRL$ | 152 | 10 | 37 | $RL^4RL^2R$ |
| 113 | 10 | 27 | $RL^3RLRLR$ | 153 | 11 | 65 | $RL^4RL^2RL$ |
| 114 | 11 | 47 | $RL^3RLRLR^2$ | 154 | 8 | 13 | $RL^4RL$ |
| 115 | 9 | 16 | $RL^3RLRL$ | 155 | 11 | 66 | $RL^4RLR^2L$ |
| 116 | 11 | 48 | $RL^3RLRL^2R$ | 156 | 10 | 38 | $RL^4RLR^2$ |
| 117 | 10 | 28 | $RL^3RLRL^2$ | 157 | 11 | 67 | $RL^4RLR^3$ |
| 118 | 6 | 4 | $RL^3R$ | 158 | 9 | 22 | $RL^4RLR$ |
| 119 | 10 | 29 | $RL^3R^3L^2$ | 159 | 11 | 68 | $RL^4RLRLR$ |
| 120 | 11 | 49 | $RL^3R^3L^2R$ | 160 | 10 | 39 | $RL^4RLRL$ |

| i | K | I(K) | Pattern | i | K | I(K) | Pattern |
|---|---|---|---|---|---|---|---|
| 161 | 11 | 69 | $RL^4RLRL^2$ | 201 | 11 | 89 | $RL^6R^2L$ |
| 162 | 7 | 8 | $RL^4R$ | 202 | 8 | 16 | $RL^6$ |
| 163 | 11 | 70 | $RL^4R^3L^2$ | 203 | 11 | 90 | $RL^7RL$ |
| 164 | 10 | 40 | $RL^4R^3L$ | 204 | 10 | 50 | $RL^7R$ |
| 165 | 11 | 71 | $RL^4R^3LR$ | 205 | 11 | 91 | $RL^7R^2$ |
| 166 | 9 | 23 | $RL^4R^3$ | 206 | 9 | 28 | $RL^7$ |
| 167 | 11 | 72 | $RL^4R^5$ | 207 | 11 | 92 | $RL^8R$ |
| 168 | 10 | 41 | $RL^4R^4$ | 208 | 10 | 51 | $RL^8$ |
| 169 | 11 | 73 | $RL^4R^4L$ | 209 | 11 | 93 | $RL^9$ |
| 170 | 8 | 14 | $RL^4R^2$ | | | | |
| 171 | 11 | 74 | $RL^4R^2LRL$ | | | | |
| 172 | 10 | 42 | $RL^4R^2LR$ | | | | |
| 173 | 11 | 75 | $RL^4R^2LR^2$ | | | | |
| 174 | 9 | 24 | $RL^4R^2L$ | | | | |
| 175 | 11 | 76 | $RL^4R^2L^2R$ | | | | |
| 176 | 10 | 43 | $RL^4R^2L^2$ | | | | |
| 177 | 11 | 77 | $RL^4R^2L^3$ | | | | |
| 178 | 6 | 5 | $RL^4$ | | | | |
| 179 | 11 | 78 | $RL^5RL^3$ | | | | |
| 180 | 10 | 44 | $RL^5RL^2$ | | | | |
| 181 | 11 | 79 | $RL^5RL^2R$ | | | | |
| 182 | 9 | 25 | $RL^5RL$ | | | | |
| 183 | 11 | 80 | $RL^5RLR^2$ | | | | |
| 184 | 10 | 45 | $RL^5RLR$ | | | | |
| 185 | 11 | 81 | $RL^5RLRL$ | | | | |
| 186 | 8 | 15 | $RL^5R$ | | | | |
| 187 | 11 | 82 | $RL^5R^3L$ | | | | |
| 188 | 10 | 46 | $RL^5R^3$ | | | | |
| 189 | 11 | 83 | $RL^5R^4$ | | | | |
| 190 | 9 | 26 | $RL^5R^2$ | | | | |
| 191 | 11 | 84 | $RL^5R^2LR$ | | | | |
| 192 | 10 | 47 | $RL^5R^2L$ | | | | |
| 193 | 11 | 85 | $RL^5R^2L^2$ | | | | |
| 194 | 7 | 9 | $RL^5$ | | | | |
| 195 | 11 | 86 | $RL^6RL^2$ | | | | |
| 196 | 10 | 48 | $RL^6RL$ | | | | |
| 197 | 11 | 87 | $RL^6RLR$ | | | | |
| 198 | 9 | 27 | $RL^6R$ | | | | |
| 199 | 11 | 88 | $RL^6R^3$ | | | | |
| 200 | 10 | 49 | $RL^6R^2$ | | | | |

44                    METROPOLIS, STEIN, AND STEIN

REFERENCES

1. P. R. STEIN AND S. M. ULAM, Non-linear transformation studies on electronic computers, *Rozprawy Mat.* **39** (1964), 1–66.
2. O. W. RECHARD, Invariant measures for many-one transformations, *Duke Math. J.* **23** (1956), 477.
3. G. JULIA, Mémoire sur l'itération des functions rationelles, *J. de Math. Ser.* 7, **4** (1918), 47–245. The relevant theorem appears on p. 129 ff.
4. N. METROPOLIS, M. L. STEIN, AND P. R. STEIN, Stable states of a non-linear transformation, *Numer. Math.* **10** (1967), 1–19.
5. E. N. GILBERT AND J. RIORDAN, Symmetry types of periodic sequences, *Illinois J. Math.* **5** (1961), 657.
6. J. SCHREIER AND S. ULAM, Eine Bemerkung über die Gruppe der topologischen Abbildung der Kreislinie auf sich selbst, *Studia Math.* **5** (1935), 155–159.
7. See reference 1, p. 52. The result is due to S. Ulam and J. von Neumann.

*Commun. in Theor. Phys. (Beijing, China)*   *Vol.3, No.3 (1984)*   *283-295*

# SCALING PROPERTY OF PERIOD-n-TUPLING SEQUENCES
# IN ONE-DIMENSIONAL MAPPINGS

ZENG Wan-zhen ( 曾婉贞 ), HAO Bai-lin ( 郝柏林 )

*The Institute of Theoretical Physics, Academia Sinica, Beijing, China*

WANG Guang-rui ( 王光瑞 ) and CHEN Shi-gang ( 陈式刚 )

*The Institute of Physics, Academia Sinica, Beijing, China*

Received March 20, 1984

#### Abstract

*We calculated the universal scaling function g(x) and the scaling factor α as well as the convergence rate δ for period-tripling, -quadrapling and-quintupling sequences of RL, RL², RLR², RL²R and RL³ types. The superstable periods are closely connected to a set of polynomial $P_n$ defined recursively by the original mapping. Some notable properties of these polynomials are studied. Several approaches to solving the renormalization group equation and estimating the scaling factors are suggested.*

## I. Introduction

One-dimensional unimodal mappings have been serving as a paradigm in studying the bifurcation and chaos phenomena in dissipative systems. Dissipation plays a global stabilizing role against local orbital instability and causes the volume of phase space representing the initial states to contract in the process of evolution. This contraction makes the phase volume approach one-dimensional objects in some of its sections and thus enables higher dimensional systems to enjoy the universal properties of one-dimensional mappings.

Although there has been a large amount of numerical and analytical studies on one-dimensional mappings since the discovery of structural universality and U-sequence by Metropolis et al.[1] and Derrida et al.[2], and the renormalization-group analysis of metric universality by Feigenbaum[3,4] many questions remain to be clarified. In particular, the bifurcation and chaos "spectrum" of nonlinear differential equations have to be outlined by first identifying various periodic regimes in the parameter space. To understand the nature of more complicated periodic behaviour encountered in real experiments, one needs a detailed analysis of scaling properties of higher periods. In this paper we confine ourselves to the scaling properties of the n-tupling sequences in unimodal mappings.

To fix upon the notations we summarize briefly the known properties of unimodal mappings[5]. Consider a family of nonlinear mappings of the interval I, i.e., f: I→I, written in the form of iterations

$$x_{n+1} = f(\mu,x_n) \tag{1,1}$$

depending on a real parameter $\mu$. Without loss of generality we put the maximum of f at x=0 and require that

(1)  $\max_{x} f(\mu,x)=f(\mu,0)$, $f'(\mu,0)=0$.

(2)  $f(\mu,x)$ be monotonically increasing when x<0, and monotonically decreasing when x>0.

(3)  In the neighbourhood of x=0 f can be expanded as

$$f(\mu,x) = f(\mu,0)-\alpha x^z+\cdots \ , \tag{1.2}$$

where z=2,4,6,8..., etc.

All concrete numerical calculations will be carried out for the mapping

$$x_{n+1} = 1-\mu x_n^z \ , \tag{1.3}$$

which maps the interval (-1,+1) into itself when parameter $\mu$ varies in the range (0,2). The mapping  (1.1)  or (1.3)  has countable infinite period-doubling sequences.  Using the composition symbol * of Derrida et al.[2], these sequences correspond to sequences of words $P*R^{*n}$, n=0,1,2,..., where P may be any primitive word, i.e. such that cannot be decomposed further by using *, made of the two letters R and L.  The scaling property of these sequences (near the origin x=0) when n goes to infinity is given by the Feigenbaum renormalization-group equation

$$g(x) = -\alpha g(g(\tfrac{x}{\alpha})) \ ,$$
$$g(0)=1, \ g'(0)=0 \ . \tag{1.4}$$

Every period in a given sequence occupies a "window" of finite width on the $\mu$ axis.  The convergence rate $\delta$ of a period-doubling sequence described by R* is determined from the linearized functional equation

$$\delta h(x) = \alpha(h(g(\tfrac{x}{\alpha}))+g'(g(\tfrac{x}{\alpha}))h(\tfrac{x}{\alpha})) \equiv \mathscr{L}(h(x)) \tag{1.5}$$

near the fixed point function g(x).  Scaling properties of period-doubling sequences are described completely by $\alpha,\delta$, g(x) and h(x).

Actually in mapping (1.1) there are  infinitely many sequences of the type $P*Q^{*n}$, where both P and Q are primitive words.  Feigenbaum studied the simplest case P=Q=R.  We shall consider the cases $P=Q=RL,RL^2,RLR^2,RL^2R$ and $RL^3$, which correspond to sequences of periods $3^n,4^n$ and three different kinds of $5^n$. Such higher periods can be seen in real experiments e.g. Fig.10 in Ref.[6] corresponds to $3^2$.  We shall see that each primitive word P has its own scaling constant $\alpha_p$ and convergence rate $\delta_p$, but $\alpha_p$'s for the same period are determined by the same renormalization group equation.  We give also numerical methods for calculating $\alpha_p$ and the corresponding g(x).

We need also the notion of superstable period.  Each periodic window can

210

be characterized by a superstable orbit, i.e. orbit corresponding to the parameter value $\tilde{\mu}$ where the superstability condition

$$F'(n,\tilde{\mu},x) = \prod_{i=1}^{n} f'(\tilde{\mu},x_i) = 0 \qquad\qquad (1.6)$$

holds. In Eq.(1.6) $F(n,\mu,x)$ stands for the n-th iteration of $f$, i.e.,

$$F(n,\mu,x) = \underbrace{f(\mu,f(\mu,\ldots f(\mu,x)\ldots)}_{n\ \text{times}} \quad . \qquad\qquad (1.7)$$

According to our convention on the f-function a periodic orbit containing the
x=0 point must be superstable  because $f'(\mu,0)=0$ naturally leads to

$$F'(p,\mu,0)=0 \qquad \text{for all} \quad p\geqslant 1 \quad . \qquad\qquad (1.8)$$

We shall denote the parameter value corresponding to the n-th superstable
orbit in a given sequence by $\tilde{\mu}_n$. The convergence rate of the sequence is
defined by

$$\delta = \lim_{n\to\infty} \frac{\tilde{\mu}_n - \tilde{\mu}_{n-1}}{\tilde{\mu}_{n+1} - \tilde{\mu}_n} \quad . \qquad\qquad (1.9)$$

In practice one can estimate the value of $\delta$ by taking the ratio in Eq.(1.9)
with finite n.

## II. Calculation of Parameter Values for Superstable Periods

Many "lifting" relations have appeared in the research on one-dimensional
mappings. In Feigenbaum's renormalization-group analysis the fixed points of
specific functional iterations were lifted to a universal fixed point equation
(1.4) in functional space. We shall use two other liftings in this section.
First, let us turn the numerical iteration Eq.(1.1) into a recursive definition of a set of polynomials, namely, defining recursively

$$P_0(\mu) = 0,$$
$$P_n(\mu) = f(\mu,P_{n-1}(\mu)) \quad , \qquad\qquad (2.1)$$

we get a set of polynomials of $\mu$: $P_n(\mu)$, n=1,2,3,$\cdots$. Using any algebraic
manipulation language (REDUCE in our case), it is easy to get the first
$P_n(\mu)$'s. For example, for mapping Eq.(1.3) with z=2 we have

$$\begin{aligned}
P_1(\mu) &= 1 \ , \\
P_2(\mu) &= 1-\mu \ , \\
P_3(\mu) &= 1-\mu+2\mu^2-\mu^3 \ , \\
P_4(\mu) &= 1-\mu+2\mu^2-5\mu^3+6\mu^4-6\mu^5+4\mu^6-\mu^7 \ ,
\end{aligned} \qquad\qquad (2.2)$$

286
ZENG Wan-zhen, HAO Bai-lin
WANG Guang-rui and CHEN Shi-gang

$$P_5(\mu) = 1-\mu+2\mu^2-5\mu^3+14\mu^4-26\mu^5+44\mu^6-69\mu^7+94\mu^8-114\mu^9$$
$$+116\mu^{10}-94\mu^{11}+60\mu^{12}-28\mu^{13}+8\mu^{14}-\mu^{15} ,$$

...

In fact, these $P_n(\mu)$'s are nothing but the iteration (1.7) taken at x=0:

$$P_n(\mu) = F(n,\mu,0) \tag{2.3}$$

Since orbits including the point x=0 are superstable, real zeros of the polynomial $P_n(\mu)$ within the interval (0,2) determine the parameter values for superstable period n, e.g. the three roots of $P_5(\mu)=0$, 1.6254137, 1.8607825 and 1.9854242 correspond to three different periods of type $RLR^2$, $RL^2R$ and $RL^3$ respectively. However, one cannot go very far in this way owing to instability of numerical algorithms for solving high order algebraic equation when the roots get too close. Later in this section we shall use another "lifting" relation to calculate the parameter value for any given superstable period. Let us look at the properties of the polynomials $P_n(\mu)$ for the moment.

We take the case z=2, i.e. formulae (2.2), for example. For other even z in mapping (1.3) the discussion goes similarly. Skipping a few algebraic computation, we just list the results.

First, the iteration of x=0, i.e. $P_1(\mu)=1$ corresponds to the maximal value of the mapping. The iteration of x=1, i.e. $P_2(\mu)=1-\mu$, leads to the minimal value of the mapping at this $\mu$. Other $P_n(\mu)$'s also give upper or lower bounds of the mapping when it splits into segments, i.e., when there are chaotic bands of a certain period. Near these extreme values successive points in the mapping get denser. In other words, these $P_n(\mu)$'s are the dark lines seen in the usual x-$\mu$ bifurcation diagram, e.g. Fig.I. 19 of Ref.[5] or Figs.13-15 of Ref.[7], in the chaotic region. Therefore, they also determine the location of band-merging points. We shall return to this question from a different point of view.

Secondly, from the recursion formula

$$P_{n+1}(\mu) = 1-\mu P_n(\mu)^2 , \tag{2.4}$$

it follows that at those $\tilde{\mu}$ values where $P_n(\tilde{\mu})=0$ one must have $P_{n+1}(\tilde{\mu})=1$.

Thirdly, Eq.(2.4) can be further iterated to yield

$$P_n(\mu) = P_k(\mu)+(-1)^k 2^{k-1}\mu^k P_{n-k}^2(\mu)Q_{n-k}^{(k)}(\mu) , \tag{2.5}$$

where $Q_{n-k}^{(k)}(\mu)$ is a certain polynomial which can be calculated for given n and k. Therefore, if $P_{n-k}(\tilde{\mu})=0$ at a certain $\tilde{\mu}$ value, then it follows from Eq.(2.5) that $P_n(\tilde{\mu})=P_k(\tilde{\mu})$ for this $\tilde{\mu}$, or, denoting n-k by m, we have

$$P_m(\tilde{\mu})=0 \text{ leads to } P_{m+k}(\tilde{\mu})=P_k(\tilde{\mu}) \text{ for } m\geq2, k>1 . \tag{2.6}$$

212

A consequence of Eq.(2.6) is that

$$P_m(\bar{\mu})=0 \quad \text{leads to} \quad P_{n*m}(\bar{\mu})=0 \quad \text{for all } n>1 , \tag{2.7}$$

i.e., zeros of $P_m(\mu)$ must be zeros of $P_{n*m}(\mu)$. They correspond to harmonics of the original superstable orbit and do not represent new superstable orbit of period n*m.

Taking m=2 in Eq.(2.6) and recollecting that $P_2(1)=0$, we have

$$P_n(0)=1 \quad \text{for all } n ,$$
$$P_n(1)=\begin{cases} 0 & \text{for odd } n , \\ 1 & \text{for even } n . \end{cases} \tag{2.8}$$

Naturally, these relations can be obtained directly from the recursive definition. Similarly, at the zero $\bar{\mu}$ of $P_3(\mu)=0$ we have, besides $P_{3*n}(\bar{\mu})=0$,

$$P_1(\bar{\mu})=P_4(\bar{\mu})=P_7(\bar{\mu})=\cdots=1 ,$$
$$P_2(\bar{\mu})=P_5(\bar{\mu})=P_8(\bar{\mu})=\cdots=1-\bar{\mu} .$$

Analogous relations can be written for other m's. Together with what is said below formula (2.4) they determine various intersection or tangent points among different $P_n(\mu)$ curves.

On the other hand, from Eqs.(2.4) and (2.5) one can derive a recurrence relation for $Q_m^{(k)}(\mu)$:

$$Q_m^{(k)}(\mu)=1/2[P_{k+m-1}(\mu)+P_{k-1}(\mu)]Q_m^{(k-1)}(\mu) \tag{2.9}$$

with the obvious initial condition $Q_m^{(1)}(\mu)=1$. From Eq.(2.9) it follows, in particular, that if $Q_m^{(k)}(\mu)=0$ at a certain value $\tilde{\mu}$ then $Q_m^{(k+1)}(\tilde{\mu})$ must vanish at the same $\tilde{\mu}$ value. This means whenever for a certain k

$$Q_m^{(k)}(\tilde{\mu})=0 \quad \text{at} \quad \tilde{\mu} ,$$

and one is led to

$$P_{m+k}(\tilde{\mu})\mp P_k(\tilde{\mu}) \quad \text{at the same } \tilde{\mu} . \tag{2.10}$$

This kind of relation determines the merging point of period $2^{n+1}$ chaotic band into the period $2^n$ band. For example, a nontrivial root of

$$Q_1^{(3)}(\mu)=0$$

is given by the real root of

$$P_2(\mu)+P_3(\mu)=0 ,$$

i.e.,

*ZENG Wan-zhen, HAO Bai-lin*
*WANG Guang-rui and CHEN Shi-gang*

$$2-2\mu+2\mu^2-\mu^3 = 0$$

at $\tilde{\mu}=1.543689013$. At this point we have

$$P_3(\tilde{\mu})=P_4(\tilde{\mu})=P_5(\tilde{\mu})=P_6(\tilde{\mu})=\cdots \tag{2.11}$$

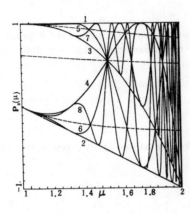

We shall not go into details of similar analysis for other merging points and illustrate all that has been said above in Fig.1, where curves of $P_1$ to $P_8$ are drawn. In addition, the unstable fixed point of Eq.(1.3) passes through the merging point (2.11) and the unstable period 2 orbits from the main period-doubling sequence pass through the next band-merging point where

$$P_5=P_7=P_9=\cdots,$$
$$P_6=P_8=P_{10}=\cdots,$$

*Fig.1  Polynomials $P_n(\mu)$ versus $\mu$, n=1 to 8.*
*Dash lines are period 1 and 2 unstable*
*orbits from the main sequence.*

and so on. These unstable orbits are shown in Fig.1.

Now let us return to the question of calculating the parameter value for any given superstable period. Take for example the period 5 orbit of type $RLR^2$. It proceeds from x=0 to x=0, i.e.,

$$0=F(5,\mu,0)=f_R(f_R(f_L(f_R(f(\mu,0))))) . \tag{2.12}$$

The subscript of f indicates whether the Right or the Left half of the unimodal mapping f is used at each iteration. Since the inverse mapping $f^{-1}$ is two-valued, we can define

$$R(x)=f_R^{-1}(\mu,x)=+[(1-x)/\mu]^{\frac{1}{2}} ,$$
$$L(x)=f_L^{-1}(\mu,x)=-[(1-x)/\mu]^{\frac{1}{2}} , \tag{2.13}$$

depending on which half of the mapping is used. By taking successive inverse of Eq.(2.12) and remembering that $f(\mu,0)=1$, the word RLRR in the U-sequence of MSS[1] is lifted to a functional relation

$$R(L(R(R(0))))=1 . \tag{2.14}$$

This is an equation for $\mu$. In our case of Eq.(1.3) it reads

$$\sqrt{\frac{1}{\mu}\left[1+\sqrt{\frac{1}{\mu}[1-\sqrt{\frac{1}{\mu}(1-\sqrt{1/\mu})}]}\right]} = 1 \ . \tag{2.15}$$

If we get rid of the radicals by squaring Eq.(2.15) several times, we go back to the equation

$$P_5(\mu)=0$$

and it gives also the parameter value for the other two types of period 5 orbits, namely, $RL^2R$ and $RL^3$. In order to separate the parameter value for $RLR^2$ only, multiply Eq.(2.15) by $\mu$ and get

$$\sqrt{\mu+\sqrt{\mu-\sqrt{\mu-\sqrt{\mu}}}} = \mu \ . \tag{2.16}$$

Kaplan[8] suggested solving this equation by iteration, i.e., replacing it by

$$\mu_{n+1} = \sqrt{\mu_n+\sqrt{\mu_n-\sqrt{\mu_n-\sqrt{\mu_n}}}} \quad , \tag{2.17}$$

and then iterating it for suitable $\mu_0$. Similarly, we have for the other two period 5 orbits $RL^2R$ and $RL^2$

$$\mu_{n+1} = \sqrt{\mu_n+\sqrt{\mu_n+\sqrt{\mu_n-\sqrt{\mu_n}}}}$$

and

$$\mu_{n+1} = \sqrt{\mu_n+\sqrt{\mu_n+\sqrt{\mu_n+\sqrt{\mu_n}}}}$$

respectively.

It is a curious fact that all the real roots of high order algebraic equation $P_n(\mu)=0$ in the interval $(0,2)$ can be approached separately by using a single irrational operation, i.e., taking the square root in our $z=2$ case. This procedure happens to be a very efficient method of determining the superstable orbits. Table 1 shows some results obtained in this way.

From Table 1 it is easy to estimate the convergence rate $\delta$ for each sequence using Eq.(1.9). We list the results in Table 2. For comparison we give also the values for the main period-doubling sequence $R*^n$ (See Refs.[3],[9] and [10]). The period-tripling sequence has been studied in Refs.[11] and [12]).

Table 1  Parameter values for superstable orbits in Eq.(1.3)

| Sequence | n | period | z=2 | z=4 | z=6 |
|---|---|---|---|---|---|
| $(RL)^{*n}$ | 1 | 3 | 1.75487766624669 | 1.85667488385450 | 1.89865387615624 |
| | 2 | 9 | 1.78586564641069 | 1.90869485541770 | 1.94845447243645 |
| | 3 | 27 | 1.78642985805576 | 1.90932805587416 | 1.94886316337701 |
| | 4 | 81 | 1.78644006735908 | 1.90933538425728 | 1.94886624516105 |
| | 5 | 243 | 1.78644025215705 | 1.90933546978620 | 1.94886626893769 |
| | 6 | 729 | 1.78644025550198 | 1.90933547078290 | 1.94886626912018 |
| $(RL^2)^{*n}$ | 1 | 4 | 1.94079980652949 | 1.98196443585403 | 1.99145720984726 |
| | 2 | 16 | 1.94270241385659 | 1.98550184517166 | 1.99420412925522 |
| | 3 | 64 | 1.94270435277818 | 1.98550465814397 | 1.99420541697422 |
| | 4 | 256 | 1.94270435475345 | 1.98550466034811 | 1.99420541755191 |
| $(RLR^2)^{*n}$ | 1 | 5 | 1.62541372512330 | 1.72570156646440 | 1.77674511880702 |
| | 2 | 25 | 1.63190116771613 | 1.74329046242399 | 1.79587583617159 |
| | 3 | 125 | 1.63192655475500 | 1.74335080762877 | 1.79591994253629 |
| | 4 | 625 | 1.63192665409982 | 1.74335101433455 | 1.79592004466609 |
| $(RL^2R)^{*n}$ | 1 | 5 | 1.86078252220486 | 1.94207538300024 | 1.96763991773560 |
| | 2 | 25 | 1.86222290265199 | 1.94585592158251 | 1.97097125924881 |
| | 3 | 125 | 1.86222402174850 | 1.94585858174374 | 1.97097261455148 |
| | 4 | 625 | 1.86222402261799 | 1.94585858361935 | 1.97097261510824 |
| $(RL^3)^{*n}$ | 1 | 5 | 1.98542425305421 | 1.99774624322048 | 1.99928785957603 |
| | 2 | 25 | 1.98553952325185 | 1.99797401060435 | 1.99943243778625 |
| | 3 | 125 | 1.98553953006001 | 1.99797402146537 | 1.99943244117849 |
| | 4 | 625 | 1.98553953006041 | 1.99797402146589 | 1.99943244117857 |

[+] *Parameters for other periods and for z=8 will be sent on request.*

Table 2  Convergence rate $\delta$ of period-n-tupling sequences

| Sequence | z=2 | z=4 | z=6 | z=8 |
|---|---|---|---|---|
| $(R)^{*n}$ | 4.669 | 7.285 | 9.298 | 10.9 |
| $(RL)^{*n}$ | 55.247 | 85.81 | 130.3 | 178.9 |
| $(RL^2)^{*n}$ | 981.6 | 1275 | 2220 | 3542 |
| $(RLR^2)^{*n}$ | 255.5 | 291.9 | 431.9 | 590.4 |
| $(RL^2R)^{*n}$ | 1287 | 1418 | 2434 | 3800 |
| $(RL^3)^{*n}$ | 17000(?) | 21000(?) | 43000(?) | 70000(?) |

## III. Estimate of the Scaling Factors

Using the polynomial $P_n(\mu)$ defined in the last section and the limiting parameter value $\mu_\infty$ for each n-tupling sequence (in practice one takes the last $\mu_n$ from Table 1 for $\mu_\infty$ and, if neccesary, improves it by using

$$\mu_\infty = \frac{\delta}{\delta-1} \mu_n - \frac{1}{\delta-1} \mu_{n-1} \tag{3.1}$$

with $\delta$ taken from Table 2), we can estimate the scaling factor $\alpha$ for each period-n-tupling sequence. To do this we replace the yet unknown universal function g(x) in the renormalization group equation

$$\alpha^{-1}g(\alpha x)=-g^{(n)}(x), \quad g(0)=1, \quad g'(0)=0 \tag{3.2}$$

by $f(\mu_\infty,x)$ and take x=0. Therefore, we are led to

$$\alpha = -\frac{1}{g^{(n)}(0)} \simeq -\frac{1}{P_n(\mu_\infty)} \; . \tag{3.3}$$

The $\alpha$ values obtained in this way are only approximate, but can be used as initial values in seeking more precise solutions to the renormalization group equation (3.2). We list these $\alpha$ values in Table 3. Estimates from numerical experiments are given in parentheses in the same Table.

Table 3  Estimated scaling factor $\alpha$ for period-n-tupling sequences

| Sequence | z=2 | z=4 | z=6 | z=8 |
|---|---|---|---|---|
| $(R)*^n$ | 2.50 (2.50) | 1.68 (1.69) | 1.46 (1.47) | 1.36 (1.36) |
| $(RL)*^n$ | 9.53 (9.09) | 3.27 (3.15) | 2.37 (2.28) | 2.00 (1.95) |
| $(RL^2)*^n$ | 39.7 (40.0) | 6.56 (6.18) | 3.90 (3.66) | 2.98 (2.80) |
| $(RLR^2)*^n$ | 20.1 (20.2) | 4.34 (4.30) | 2.84 (2.79) | 2.29 (2.24) |
| $(RL^2R)*^n$ | -46.0 (-46.1) | -6.51 (-6.4) | -3.82 (-3.73) | -2.91 (-2.83) |
| $(RL^3)*^n$ | 161. (161.3) | 13.11 (12.3) | 6.43 (6.0) | 4.42 (4.10) |

Equipped with an algebraic manipulation language such as REDUCE, we have another way to solve the renormalization group equation (3.2) approximately. The unknown function g(x) is expanded (for z=2)

$$g(x)=1+Ax^2+Bx^4+Cx^6+Dx^8+\cdots. \tag{3.4}$$

and substituted into Eq.(3.2). By equating the coefficients at equal powers of x one gets a system of algebraic equations, which is then solved numerically. To show how it works we give in Table 4 the $\alpha$ value versus number of terms retained in Eq.(3.4) for the case of period-doubling.

ZENG Wan-zhen, HAO Bai-lin
WANG Guang-rui and CHEN Shi-gang

Table 4  Convergence of the series solution (3.4) for period-doubling

| Number of terms retained in Eq.(3.4) | $\alpha$ | A | B | C | D |
|---|---|---|---|---|---|
| 2 | 2.73205 | -1.366025 | | | |
| 3 | 2.534030 | -1.522426 | 0.1276133 | | |
| 4 | 2.478897 | -1.521880 | 0.0729350 | 0.0455088 | |
| 5 | 2.50316 | -1.52779 | 0.10533 | 0.02631 | -0.00334 |
| "exact"[4] | 2.502907 | -1.527633 | 0.104815 | 0.0267056 | -0.00352 |

For the period-tripling sequence we get $\alpha$=9.0497 and 9.2764 by retaining in Eq.(3.4) 3 and 4 terms respectively. This is to be compared with the "exact" value 9.2773 computed in the following section.  Keeping only 2 terms in Eq.(3.4) we have $\alpha$=20.03, -46.2 and 161.9 for $RLR^2$, $RL^2R$ and $RL^3$ respectively.

We need more precise values of these $\alpha$ to study the fine structure in power spectra of different periodic orbits.  To this end the estimates for $\alpha$'s listed above are used as initial values in solving the renormalization group equation (3.2).

## IV.  Numerical Solution of the Renormalization Group Equation

As analysed by Feigenbaum[4] in the period-doubling case, the fixed point described by equation (3.2) corresponds to a saddle point in the functional space.  Therefore, if one tries to approach the universal scaling function g(x) by constructing an iteration

$$g_{k+1}(x) = -\alpha_k g_k^{(n)} \left(\frac{x}{\alpha_k}\right) , \qquad (4.1)$$

then $g_k$ will first approach the scaling function g due to the vanishing of deviations from the conjugacy transformation, but, on the other hand, owing to the existence of the eigenvalue $\delta$>1 of the linearized operator , the iteration will eventually  go away from the scaling function g as $g_k=g-\delta^k h$, unless one has set on the parameter value $\mu_\infty$ with high precision.  This situation resembles the renormalization group analysis of continuous phase transitions, where the rescaling transformation can be performed infinite times only when one settles in the critical surface $T=T_c$ from the outset, otherwise the "relevant" variables corresponding to eigenvalues greater than 1 will sooner or later lead the transformation away from the saddle point.

To solve the renormalization group equation Feigenbaum[4] mentioned an idea based on iteration and coordinate rescaling, but did not give any elaboration or application of the idea.  In fact, he adopted a direct parametrization  for g(x).  Inspired by this idea, we have devised an iteration scheme of solving Eq.(3.2) for n$\geq$3. The essentials are as follows.

For unimodal mapping with $f(0)=1$ one can take the rescaling factor for the coordinate as the control parameter. We use $f(\lambda_0 x)$ with $\lambda_0$ chosen as good as possible as the initial function

$$g_0(x) = f(\lambda_0 x) \tag{4.2}$$

for the iteration Eq.(4.1). Because $\lambda_0$ cannot be equal to $\lambda_\infty = \sqrt{\mu_\infty}$ exactly, we rescale x during iteration, choosing $\lambda_k$ in such a way that at $x=1$ the equality

$$\alpha_k(\lambda_k) g_k^{(n)}\left(\lambda_k, \frac{1}{\alpha_k(\lambda_k)}\right) = g_k(\lambda_k) \tag{4.3}$$

holds, where

$$g_k^{(n)}(\lambda, x) \equiv \underbrace{g_k\left(\lambda g_k\left(\lambda g_k \ldots \lambda g_k(\lambda x)\ldots\right)\right)}_{n \text{ times}} \tag{4.4}$$

and

$$\alpha_k(\lambda) = \frac{1}{g_k^{(n)}(\lambda,0)} \quad . \tag{4.5}$$

After $\lambda_k$ is determined let

$$g_{k+1}(x) = \alpha_k(\lambda_k) g_k^{(n)}\left(\lambda_k, \frac{x}{\alpha_k(\lambda_k)}\right) \tag{4.6}$$

as the result of the (k+1)-th iteration. In practice, g(x) is approximated by using the Lagrange interpolation formula

$$g(x) = \sum_{i=0}^{N} g(x_i) \frac{\prod_{j \neq i}^{N} (x^2 - x_j^2)}{\prod_{j \neq i}^{N} (x_i^2 - x_j^2)} \tag{4.7}$$

with $g(0)=1$ at $x=0$. The iterated $g_{k+1}$ is determined by $g_{k+1}(x_i)$ at N points $x_i$.

Our calculation shows that the iteration converges quickly to make the difference between two successive iterations less than $10^{-D}$, where D is the number of digits in the computer, D=10 in our case. The errors of approximation can be estimated by the order of magnitude of the coefficient at $x^{2(N+1)}$. Because of $\delta\alpha = \alpha^2 * \delta(g^{(n)}(0))$, the precision of $\alpha$ and the corresponding g(x) would be lower for larger $\alpha$. Skipping the computational technicalities, we put together the calculated scaling factors in Table 5.

Table 5  Scaling factors $\alpha$ for period-n-tupling sequences

| $(RL)^{*n}$ | $(RL^2)^{*n}$ | $(RLR^2)^{*n}$ | $(RL^2R)^{*n}$ | $(RL^3)^{*n}$ |
|---|---|---|---|---|
| 9.2773 | 38.8189 | 20.128 | -45.804 | 160. |

The first number in Table 5 may be compared with α=9.2774 in Ref.[11] and α=9.28 in Ref.[12].

Since the universal scaling functions are useful, we list them in the Appendix for future reference.

## Acknowledgements

We thank DING Ming-zhou and LI Jia-nan for many discussions. Dr. A.C. Hearn provided the REDUCE language and Ms. ZHANG Shu-yu implemented it at the Computing Centre of Academia Sinica. We express our gratitude to both of them.

## Appendix Universal Scaling Functions

We define the function g(x) either by their value at N equally spaced points (See Eq.(4.7)) or, equivalently, by N coefficients in expansion (3.4). In what follows, N=4

1. Period-tripling sequence of type RL:

   $g(0.5)=0.537283471,$  $A=-1.874308657,$
   $g(0.25)=0.883222201,$  $B=0.0938363652,$
   $g(0.75)=-0.0246581575,$  $C=-0.0002519955,$
   $g(1.0)=-.0780775237,$  $D=-0.0000509494.$

2. Period-quadrupling sequence of type $RL^2$:

   $g(0.25)=0.87690435,$  $A=-1.97140411,$
   $g(0.5)=0.5090210,$  $B=0.0299862827,$
   $g(0.75)=-0.09945102,$  $C=-0.0001354147,$
   $g(1.0)=-0.941552789,$  $D=0.0000004534.$

3. Period-quintupling sequence of type $RLR^2$:

   $g(0.25)=0.89846035,$  $A=-1.623915517,$
   $g(0.5)=0.5933389,$  $B=-0.0117005348,$
   $g(0.75)=0.08340150,$  $C=0.003152553,$
   $g(1.0)=-0.6325133,$  $D=-0.0000498001.$

4. Period-quintupling sequence of type $RL^2R$:

   $g(0.25)=0.88299627,$  $A=-1.872725665,$
   $g(0.5)=0.532490341,$  $B=0.0106228366,$
   $g(0.75)=-0.04995812,$  $C=0.0005024008,$
   $g(1.0)=-0.86160524,$  $D=-0.0000048115.$

5. Period-quintupling sequence of type $RL^3$:

   $g(0.25)=0.875456,$  $A=-1.993191921,$
   $g(0.5)=0.5021893,$  $B=0.007808485,$
   $g(0.75)=-0.1187070,$  $C=-0.0000539135,$
   $g(1.0)=-0.9854134,$  $D=0.0000239507.$

# References

*[1]*   *N. Metropolis, M.L. Stein, P.R. Stein, J. Combinatorial Theory(A), 15(1973)25.*

*[2]*   *B. Derrida, A. Gervois, Y. Pomeau, Ann. Inst. Henri Poincaré, 29A(1978)305.*

*[3]*   *M.J. Feigenbaum, J. Stat. Phys., 19(1978)25.*

*[4]*   *M.J. Feigenbaum, J. Stat. Phys., 21(1979)669.*

*[5]*   *P. Collet, J. -P. Eckmann, Iterated maps on the interval as dynamical systems,
Birkhäuser, Boston, 1980.*

*[6]*   *A. Libchaber, S. Fauve, C. Laroche, Physica, 7D(1983)73.*

*[7]*   *HAO Bai-lin, Progress in Physics (China),3(1983)329.*

*[8]*   *H. Kaplan, Phys. Lett., 97A(1983)365.*

*[9]*   *R. Vilela Mendes, Phys. Lett., 84A(1981)1.*

*[10]*  *J.B. McGuive, C.J. Thompson, Phys. Lett., 84A(1981)451.*

*[11]*  *B. Derrida, A. Gervois, Y. Pomeau, J. Phys., A12(1979)269.*

*[12]*  *B. Hu, I.I. Satija, Phys. Lett., 98A(1983)143.*

International Journal of Modern Physics B Vol. 3, No. 2 (1989) 235–246
© World Scientific Publishing Company

# SYMBOLIC DYNAMICS OF UNIMODAL MAPS REVISITED*

HAO BAI-LIN and ZHENG WEI-MOU

*Institute of Theoretical Physics, Academia Sinica, P.O. Box 2735, Beijing 100080, China*

Received 5 September 1988

Symbolic dynamics of unimodal maps has been recast into a more natural and down-to-numbers way. The median itineraries are built without such artificial constructions as "antiharmonics" and "harmonics" by making use of the newly established periodic window theorem. A generalized composition rule extends the *-composition introduced by Derrida, Gervois and Pomeau. Periodic as well as chaotic orbits are described systematically. The location of all superstable periodic orbits and band-merging points may be calculated by solving equations obtained directly from the corresponding symbolic sequences.

## 1. Introduction

Historically symbolic dynamics appeared in the theory of topological dynamical systems in the 1930's.[1] Being the only rigorous way to describe chaotic motion in dynamical systems, in its abstract mathematical form (see, e.g., Ref. 2) it is hard to gather how to make use of it in studying concrete models. However, in the case of unimodal mappings when the assignment of letters in the alphabet reflects the specific feature of the phase space, symbolic dynamics can be shaped into a more practical tool.[3,4] We have been extending it to more general one-dimensional mappings[5,6,7] and applying it to ordinary differential equations in an empirical way.[8-10] In particular, the systematics of periodic solutions in the celebrated Lorenz model[11] was shown to be described basically by that of the cubic map.[10] Recently, one of the authors[12,13,14] succeeded in recasting the symbolic dynamics of one-dimensional mappings into a more natural and workable form. Combined with our previous approach, we now use it in a down-to-numbers fashion. In this review we shall show how it works in practice without going into detailed proofs.

We would like to emphasize that although most of our examples are based on the logistic map, the power of symbolic dynamics goes beyond any particular map. It reflects those robust properties that are universal to a whole class of dynamical systems and provides a bridge to the renormalization group approach.

---

* Based on an invited talk at the First Asia-Pacific Conference on Condensed Matter Physics, 27 June–1 July, 1988, Singapore.

## 2. The Order of Orbits

To have a taste of symbolic dynamics let us look at the logistic map

$$x_{n+1} = f(\mu, x_n) = 1 - \mu x_n^2, \tag{1}$$

It maps the interval $(-1, 1)$ into itself when the parameter $\mu$ varies in between 0 and 2. The map (1) has a critical point $x_c = 0$ where the function $f$ reaches its maximal value 1. We shall denote $x_c$ by a single letter $C$, meaning *Center* or *Critical*. All points that fall to the *Right* of $C$ will be assigned the letter $R$ and those to the *Left* of $C$ — the letter $L$. We have the natural order

$$L < C < R. \tag{2}$$

In this way any iteration sequence (also called an orbit)

$$x_0, x_1 = f(x_0), x_2 = f(x_1), \ldots, x_n = f(x_{n-1}), \ldots$$

may be assigned a sequence of symbols. Sometimes this establishes only a many-to-one correspondence, since a great many different numerical sequences may correspond to one and the same symbolic sequence (see Sec. 3 below). However, it is just this multivalueness that opens up the possibility of classifying the orbits.

We first introduce an ordering for all possible orbits for a given map, i.e., at a fixed $\mu$, using a few simple properties of monotonic functions. The map (1) has two monotonic branches, a *Left* $f_L$ for $x < C$ and a *Right* $f_R$ for $x > C$. The $L$ branch is monotonically increasing and we assign an even parity to it in accordance with its positive slope. The $R$ branch, being monotonically decreasing, acquires an odd parity. It is well known that monotonic functions have the following nice properties:

**Property 1.** If a function $f$ is monotonic increasing on an interval, then for points $x_1 < x_2$ belonging to the interval we have the same order $f(x_1) < f(x_2)$. If $f$ is monotonic decreasing, then $x_1 < x_2$ implies the reverse order $f(x_1) > f(x_2)$.

**Property 2.** If $f$ is monotonic increasing (decreasing), so is the inverse function $f^{-1}$.

**Property 3.** If two functions $f$ and $g$ are both monotonic increasing or decreasing, then the composition $f \circ g$ is monotonic increasing. If one of the two is monotonic increasing whereas the other one is monotonic decreasing, then $f \circ g$ will be monotonic decreasing. Therefore, we can represent what has been stated by a multiplication table of parities. This observation generalizes naturally to compositions of more monotonic functions.

If we are given two numerical sequences

$$\Sigma_1 : x_1, x_2, x_3, \ldots$$
$$\Sigma_2 : x_1', x_2', x_3', \ldots$$

then we define their order by the order of the leading elements, i.e., $\Sigma_1 > \Sigma_2$ if $x_1 > x_1'$ and vice versa. (Two numerical sequences coincide if $x_1 = x_1'$ and there is no need to order them.)

This ordering carries over to symbolic sequences on the basis of their natural order (2). Namely, if we have two sequences

$$\Sigma_1 = \sigma_1 \sigma_2 \sigma_3 \ldots$$

$$\Sigma_2 = \tau_1 \tau_2 \tau_3 \ldots$$

(3)

then we say $\Sigma_1 > \Sigma_2$ if $\sigma_1 > \tau_1$ in the sense of (2) and vice versa. However, $\sigma_1$ and $\tau_1$ may happen to be the same letter, still the two sequences are different. In this case one has to compare the next letters. Now the parity comes into play. $\sigma_2 > \tau_2$ implies $\Sigma_1 > \Sigma_2$ only when the first numbers, i.e., $\sigma_1$ and $\tau_1$, are located on the $L$ side so they iterate to $\sigma_2$ and $\tau_2$ using the monotonically increasing, even parity branch $f_L$. Otherwise, $\sigma_2 > \tau_2$ leads to the opposite relation $\Sigma_1 < \Sigma_2$. In general, we compare two sequences

$$\Sigma_1 = \Sigma^* \sigma \ldots$$

$$\Sigma_2 = \Sigma^* \tau \ldots$$

(4)

where $\Sigma^*$ denotes their common leading string and $\sigma$ differs from $\tau$. It is clear now that the order of the two sequences is determined by the order of $\sigma$ and $\tau$ in the sense of (2) only when the common part $\Sigma^*$ is of even parity. When $\Sigma^*$ has odd parity, the order of the two sequences are determined by the anti-order of $\sigma$ and $\tau$.

For a given map each point $x$ in the interval may serve as a seed that starts an iteration to yield a symbolic sequence. The order of any two different sequences is essentially given by the location of the corresponding seeds.

## 3. Periodic Orbits

A periodic orbit is described by the repetition of a finite string of letters

$$(\sigma_1 \ldots \sigma_n)^\infty .$$

Among the cyclic permutations of a finite string there is always a maximal one. We take the convention to represent the orbit by this maximal string. At a fixed parameter $\mu$ there may be a great number of different periodic orbits, among which only the orbit that may be reached from the critical point $C$ is stable.[5] We start the iteration from the critical point $C$. When the point $C$ is reached again at a later iteration, we have a superstable periodic orbit, since the property

$$f'(\mu, C) = 0$$

leads to

$$f^{(n)'} = 0$$

for any finite $n$. Since $f(C)$ is the maximum of the map, the sequence corresponding to $f(C)$ must be the largest among all the sequences. The largest symbolic sequence containing the letter $C$ is sometimes called an *admissible* word in symbolic dynamics.

To look at the parameter dependence of superstable orbits, it is more convenient to normalize the map in a slightly different way, using, e.g., Eq. (14) below. Since the height of the map (1) varies monotonically with the parameter $\mu$, when we decrease $\mu$ continuously from $\mu = 2$, one of the maximal sequences may become admissible after another, giving rise to an ordering of stable orbits along the parameter axis. This is nothing but the ordering of windows seen in the bifurcation diagram.

Simple continuity consideration tells us that a slight shift of the initial point from $C$ would not destroy a stable periodic orbit. Due to the continuous dependence of the mapping function on the parameter, there must be a finite interval on the parameter axis where a given periodic orbit thrives. The parameter range cannot be determined without resorting to the map itself, but symbolic dynamics alone is capable of yielding the symbols for the whole window. The only thing we need is the fact that an orbit may lose stability by undergoing either a period-doubling bifurcation when the derivative in $f^{(n)'}$ reaches $-1$ (so the corresponding symbolic sequence must have odd parity) or a tangent bifurcation when the derivative is $+1$ (corresponding to even parity). When an orbit arises from the period-doubling of another orbit, the derivative must be $+1$, since

$$\left.\frac{\partial f^{(2n)}}{\partial x}\right|_* = \left(\left.\frac{\partial f^{(n)}}{\partial x}\right|_*\right)^2 = (-1)^2$$

where $*$ indicates that the derivatives are taken at the bifurcation point. In fact, we have the following

**Periodic Window Theorem.**[12] If $(\Sigma C)^\infty$ is a maximal sequence where $\Sigma$ does not contain $C$, then both $(\Sigma R)^\infty$ and $(\Sigma L)^\infty$ are maximal sequences.

Instead of giving the proof we look at a few examples. The simplest assertion is that all the three sequences $C^\infty$, $R^\infty$ and $L^\infty$ are maximal. This covers the period 1 (fixed point) window

$$(L, C, R) \tag{5}$$

(we drop the $\infty$ power whenever confusion does not occur). The superstable period 2 orbit is given by $RC$, so we have the window

$$(RR, RC, RL) \tag{6}$$

always putting the odd sequence to the right of the superstable sequence (that ends with $C$) in accordance with the direction of the period-doubling cascade. Notice that the "upper" sequence in the window (5) and the "lower" sequence in the window (6) coincide, both being $R^\infty$. The actual border between these periods follows from the concrete functional dependence of the map.

The periodic window theorem provides a way to construct the shortest admissible word included in between two given ones.[12] Suppose we are given two maximal sequences $\Sigma_1 C < \Sigma_2 C$; construct their windows by replacing the last $C$ by $R$ or $L$ to meet the parity requirement:

$$\text{even sequence } (\Sigma C)_-^\infty < (\Sigma C) < \text{odd sequence } (\Sigma C)_+^\infty .$$

Then the common leading part of the even sequence of $(\Sigma_2 C)_-^\infty$ and the odd sequence $(\Sigma_1 C)_+^\infty$ is, after adding to it a letter $C$, the word we are looking for. Consider an example: given $RC$ and $RL^2C$, construct $(RR, RC, RL)$ and $(RL^2R, RL^2C, RL^3)$; the common leading part is $RL$. We notice that all sequences which appear in this construction are maximal, thus avoiding such artificial objects, as, for example, so-called antiharmonics.[3]

The practical significance of the sequence-ordering and periodic window theorem is that if one has observed two sequences, then all sequences included in between these two are expected to be seen, provided the resolution is high enough. The assertion holds in both the phase space and the parameter space.

Furthermore, the relation between the two windows

$$(L, C, R)$$

and

$$(RR, RC, RL)$$

may be taken as a substitution rule, namely,

$$R \to RL$$

$$C \to RC \tag{7}$$

$$L \to RR .$$

Applying these rules to the period 2 window, we get the period 4 window

$$(RLRL, RLRC, RLRR).\qquad(8)$$

This process may be repeated to get the whole Feigenbaum period-doubling cascade. In fact, any of these windows may be taken as a substitution rule to generate selected members of the sequence. For example, taking Eq. (8)

$$R \rightarrow RLRR,$$

$$C \rightarrow RLRC,\qquad(9)$$

$$L \rightarrow RLRL$$

one would pick up the $4^n$ members from the Feigenbaum sequence.

A question arises naturally. Is it possible to define general substitution rules that would yield new maximal sequences from shorter ones? This question has been answered in the confirmative. We have the following **Generalized Composition Rule.**[13] Suppose

$$\Sigma = \sigma_1 \sigma_2 \sigma_3 \cdots$$

is a maximal symbolic sequence where the $\sigma_i$'s are either $R$ or $L$, excluding the extreme case $\Sigma = L^\infty$, then the substitutions

$$R \rightarrow \rho,$$

$$L \rightarrow \lambda\qquad(10)$$

transform $\Sigma$ into another maximal sequence, provided the symbolic strings $\rho$ and $\lambda$ satisfy the following conditions:

1. $\rho$ is odd, while $\lambda$ is even,
2. $\rho > \lambda$,
3. $\rho$ with its last letter replaced by $C$, denoted hereafter as $\rho|_C$, is maximal,
4. $\rho\lambda|_C$ is maximal,
5. $\rho\lambda^\infty$ is maximal.

Notice that no maximality of $\lambda$ is required. The rules (10) are a far-reaching generalization of the *-composition introduced by Derrida, Gervois and Pomeau.[4] Skipping again its detailed proof, we give a few examples:

We first note that if both $\lambda$ and $\rho$ are maximal sequences and $\lambda|_C = \rho|_C$, then

$$(\lambda, \rho|_C, \rho)$$

may represent a stable period. Its period-doubled regime will be described by

$$(\lambda_1 \equiv \rho\rho, \rho_1 \big|_C \equiv \rho\lambda \big|_C, \rho_1 \equiv \rho\lambda) \, .$$

Repetition of this procedure leads to the entire period-doubling cascade. One should be aware that these symbols say nothing about the stability range of the periods that depend on the particular function defining the map. However, we shall show how to locate the superstable parameter in Sec. 5.

**Example 1.** Take, for instance, one of the admissible period 5 words $\Sigma C = RL^2RC$. Replacing $C$ by $R$ or $L$ to ensure the required parity we get $\rho = RL^2RR$ and $\lambda = RL^2RL$. These strings satisfy the conditions of the generalized composition rule and may be used to generate an infinite period-quintupling sequence. This example can be equally well treated by the ordinary *-composition rule.[4]

**Example 2.** Take $\rho = RLL$ and $\lambda = RR$, then it follows from the maximality of $RL$ and $RLRR$ that $\rho\lambda = RLLRR$ and $\rho\lambda\rho\rho = RLLRRRLLRLL$ are also maximal. These results are beyond the reach of the *-composition rule.

In both of the examples above the composition rules hint on the form of the renormalization group equation. The substitutions in Example 1 imply a rescaling of the form

$$g_R \to g_R \circ g_R \circ g_L \circ g_L \circ g_R \, ,$$

$$g_L \to g_L \circ g_R \circ g_L \circ g_L \circ g_R \, .$$

(11)

(Notice the reverse order of $g$'s as compared to the original word. The use of $g^{-1}$'s would preserve the order.) For symmetric functions $g$ there is no need to distinguish $g_R$ from $g_L$. Apart from a scaling factor Eqs. (11) lead to the renormalization group equation for the period-quintupling sequence. In fact, any period $n$-tupling sequence[5] may be studied in this way.

## 4. A Class of Chaotic Orbits

In one-dimensional mappings any infinite orbit that is not periodic or quasiperiodic may be called chaotic. How many chaotic orbits are there and how to classify them? To the best of our knowledge this problem has not yet been completely solved. However, we do understand very well one class of chaotic orbits.

As early as in 1947, Ulam and von Neumann[16] studied the surjective logistic map, i.e., Eq. (1) at $\mu = 2$. They showed that the points $\{x_n \mid n = 0, 1, \cdots\}$ obey a continuous invariant density distribution

$$\rho(x) = \frac{1}{\pi\sqrt{1 - x^2}}$$

(12)

satisfying thus the requirement of a random variable. Now we know that the surjective map appears as the $n \to \infty$ limit of $RL^n$ orbits, i.e., it corresponds to the $RL^\infty$ orbit. In the vicinity of $\mu = 2$ there are infinitely many aperiodic orbits. It was conjectured and partially proved that in the interval $\mu \in (2 - \epsilon, 2)$

$$\frac{\text{the measure of aperiodic orbits}}{\epsilon} \to 1 \tag{13}$$

as $\epsilon \to 0$ (see e.g., Ref. 17). Aperiodic orbits here include quasiperiodic as well as chaotic orbits. (We mention in passing that the problem of quasiperiodic orbits in unimodal mappings may also be analysed by using the method of symbolic dynamics; this work is now under way.)

Therefore, the surjective logistic map may be taken as a prototype for chaotic motion. Whenever there is a $R \sim \rho, L \sim \lambda$ substitution the interval in between $\lambda^\infty$ and $\rho\lambda^\infty$ may be put into correspondence with that in between the $L^\infty$ and $RL^\infty$ and all possible symbolic sequences must be present. Their motion, smoothed over the scale of $\rho$ and $\lambda$, would look just like the corresponding orbits in the surjective logistic map. In fact, one can find a locally surjective map on the $\lambda^\infty - \rho\lambda^\infty$ interval by looking at the appropriate iterate of the mapping.

We can construct an infinite number of sequences of the form $\rho\lambda^\infty$. First of all, for maximal $\rho|_C = \lambda|_C$, there are the periodic windows generated by $(\lambda, \rho|_C, \rho)$, each giving rise to a $\rho\lambda^\infty$ orbit. It is easy to see that this orbit happens to be a band-merging point where a chaotic band of many pieces merges into a band of fewer pieces. Secondly, for each and every $\rho$ one can take other $\lambda$'s (not neccesarily the one in its own window) to form $\rho\lambda^\infty$ orbits. Lastly, we can give up the requirement of $\lambda$ being maximal and take any string that satisfies the conditions in the generalized composition rule to build a sequence $\rho\lambda^\infty$. The class of chaotic orbits comprises at least all these types of sequences. In principle, they are realistic and observable chaotic orbits as long as we can locate their parameter values with very high precision (see Sec. 5 below) and the conjecture (13) holds.

Furthermore, a $\rho\lambda^\infty$ sequence describes a homoclinic orbit. Homoclinic orbits in one-dimensional mappings are not so fully fledged as in higher dimensional mappings. The stable set degenerates into a finite number of points, represented by the symbolic string $\rho$. $\lambda^\infty$ always represents an unstable orbit and the $\rho$ string describes the finite number of steps that leads to the unstable orbit.

These $\rho\lambda^\infty$ sequences also determine the crisis points[18] as $\lambda^\infty$ describes the unstable object that collides with the chaotic attractor.

To summarize, we see that a $\rho\lambda^\infty$ sequence corresponds to:

1. A locally surjective map,
2. A homoclinic orbit,
3. A band-merging or band-ending point,
4. A crisis point, and
5. A set of all possible orbits as rich as the surjective logistic map.

## 5. Down to Numbers

Now we discuss a more practical problem of how to determine the parameter of a superstable orbit described by the sequence $\rho |_C$ or a chaotic orbit represented by $\rho \lambda^\infty$. To simplify the derivation, we use an equivalent form of the logistic map (1), replacing $x$ by $x/\mu$:

$$x_{n+1} = \mu - x_n^2. \tag{14}$$

The parameter values for all orbits remain the same as for the map (1). We shall need the two inverse branches of the map. We name these branches according to their subscripts, i.e.,

$$R(\mu, y) \equiv f_R^{-1}(\mu, y) = \sqrt{\mu - y},$$

$$L(\mu, y) \equiv f_L^{-1}(\mu, y) = -\sqrt{\mu - y}. \tag{15}$$

Consider a superstable period $n$ orbit and write it down explicitly as an $n$-cycle starting and ending at $x_c$:

$$\underbrace{f \circ f \circ \cdots f \circ f(x_c)}_{n \ times} = x_c. \tag{16}$$

Now shift all but one function $f$ from the left-hand side of Eq. (16) to the right-hand side. Since the inverse function is multivalued one must attach a subscript $\alpha_i = R$ or $L$ to the function to indicate which branch has been used in taking the inverse. Therefore, we have

$$f(x_c) = \underbrace{f_{\alpha_1}^{-1} \circ f_{\alpha_2}^{-1} \circ \cdots f_{\alpha_{n-1}}^{-1}}_{n-1\ times} (x_c). \tag{17}$$

Using the inverse functions $R$ and $L$ defined in Eqs. (15), Eq. (17) acquires the very simple form[5]

$$f(x_c) = \Sigma(x_c) \tag{18}$$

where $\Sigma(y)$ is the composite function

$$\Sigma(y) = \alpha_1 \circ \alpha_2 \circ \cdots \alpha_{n-1}(y)$$

obtained by "lifting" the word

replacing each letter by the function of the same name according to Eq. (15). The sequence $\Sigma$ is just the word describing the periodic orbit in question. Going from Eq. (16) to (17) a specific periodic orbit is selected from many $n$-cycles. In other words, the symbolic dynamics eliminates the "degeneracy".

Take for example the period 5 orbit $RL^2RC$. It lifts into the equation

$$f(x_c) = R \circ L \circ L \circ R(x_c)$$

which, by taking into account Eqs. (14) and (15), yields

$$\mu = \sqrt{\mu + \sqrt{\mu + \sqrt{\mu - \sqrt{\mu}}}}\ .$$

To solve this equation we transform it into an iteration

$$\mu_{n+1} = \sqrt{\mu_n + \sqrt{\mu_n + \sqrt{\mu_n - \sqrt{\mu_n}}}}$$

which converges quickly for any reasonable initial value, say, $\mu_0 = 2$ and gives $\mu = 1.860782522$.

Moreover, we can calculate the exact parameter values for any $\rho\lambda^\infty$ sequence by slightly extending the "word-lifting" technique described above. For example, the sequence $RL(RR)^\infty$, which is the $2 \to 1$ merging point, leads to an infinitely nested square-root equation which reduces to two finite equations by introducing one more variable

$$\mu = \sqrt{\mu + \sqrt{\mu - v}}$$

$$v = \sqrt{\mu - v}\ .$$

(19)

The $4 \to 2$ merging point $RLRR(RLRL)^\infty$ leads to

$$\mu = \sqrt{\mu + \sqrt{\mu - \sqrt{\mu - \sqrt{\mu - v}}}}$$

(20)

$$v = \sqrt{\mu + \sqrt{\mu - \sqrt{\mu + \sqrt{\mu - v}}}}\ .$$

The situation resembles that which we have encountered in calculating the superstable parameter values. By eliminating $v$ from, say, Eqs. (19), one ends up with a polynomial equation and one has to single out the required root which cannot always be done numerically. Our old trick of transforming these equations into iterations again comes to our help. For instance, Eqs. (20) yield the quickly converging iteration schemes

$$\mu_{n+1} = \sqrt{\mu_n + \sqrt{\mu_n - \sqrt{\mu_n - \sqrt{\mu_n - v_n}}}}$$

$$\tag{21}$$

$$v_{n+1} = \sqrt{\mu_n + \sqrt{\mu_n - \sqrt{\mu_n + \sqrt{\mu_n - v_n}}}}$$

which lead to $\mu = 1.4303576 \cdots v = 1.3248379 \cdots$ from any reasonable initial values, say, $\mu_0 = 2.0$ and $v_0 = 1.95$.

In fact, any infinite word of the form $\rho\lambda^\infty$ leads to a pair of iteration relations like Eqs. (21), the first corresponding to the $\rho$ part and the second to the $\lambda$ part. The only exception is $RL^\infty$ which reduces to the simple equation

$$\mu = \sqrt{\mu + \mu}$$

yielding $\mu = 2$ (apart from the trivial solution $\mu = 0$). Actually, there is nothing peculiar in $RL^\infty$. It may be considered as the zeroth band-merging point beyond which the chaotic band ceases to exist.

What has been said applies to all higher order band-merging points seen in the tails of the bifurcation sequences developed from tangent bifurcations. For example, the period 3 orbits $RLL - RLC - RLR$ develop into an inverse band-merging sequence which ends at $RLL(RLR)^\infty$. The iteration scheme generated by this infinite word gives $\mu = 1.79032749 \cdots$ which is the exact parameter of the crisis studied first by Grebogi, Ott and Yorke (denoted as $C_{+3}$ in their paper).[18]

Another numerical application of symbolic dynamics is to accurately locate *unstable* orbits. Once the symbolic sequence is known, we can recover the orbit by tracing the sequence backwards, hence the process is convergent. This shows the inadequacy of using the term "non-invertible" to one-dimensional mappings, since symbolic dynamics overcomes the multivalueness of taking the inverse.

We emphasize that the above described method is not restricted to the logistic map only. Whenever the inverse branches of a map are known explicitly, all the superstable periodic orbits and any chaotic orbit of the form $\rho\lambda^\infty$ including all the band-merging points can be located with as high precision as one wishes. In particular, a detailed symbolic dynamics analysis of symmetry breaking and restoration in antisymmetric mappings has been carried out.[19]

## Acknowledgements

We thank Lu Li-sha, Yang Wei-ming and Zou Chuan-ming for numerous discussions, as well as Prof. Rainer Radok for reading the manuscript. This work was partially supported by the Chinese Natural Science Foundation.

## References

1. M. Morse and G. A. Hedlund, *Am. J. Math.* **60** (1938) 815.
2. V. M. Alekseev and M. V. Yakobson, *Phys. Reports* **75** (1981) 287.
3. N. Metropolis, M. L. Stein and P. R. Stein, *J. Combinat. Theory* **A15** (1973) 25.
4. B. Derrida, A. Gervois and Y. Pomeau, *Ann. Inst. Henri Poincaré* **29A** (1978) 305.
5. W. Z. Zeng, B. L. Hao, G. R. Wang and S. G. Chen, *Commun. Theor. Phys.* **3** (1984) 283.
6. H. J. Zhang, J. H. Dai, P. Y. Wang, C. D. Jin and B. L. Hao, *Chin. Phys. Lett.* **2** (1985) 5; *Commun. Theor. Phys.* **8** (1987) 281.
7. W. Z. Zeng, M. Z. Ding and J. N. Li, *Chin. Phys. Lett.* **2** (1985) 293; *Commun. Theor. Phys.* **9** (1988) 141.
8. B. L. Hao, *Physica* **104A** (1986) 85.
9. B. L. Hao, Chapter 14 in *Order and Chaos in Nonlinear Physical Systems*, eds. S. Lundqvist, N. H. March and M. P. Tosi, (Plenum, 1988).
10. M. Z. Ding and B. L. Hao, *Commun. Theor. Phys.* **9** (1988) 374.
11. E. N. Lorenz, *J. Atmos. Sci.* **20** (1963) 130.
12. W. M. Zheng, "Construction of median itineraries without using the antiharmonic", Institute of Theoretical Physics, Academia Sinica, Preprint ASITP-88-006.
13. W. Z. Zheng, "Generalized composition law for symbolic sequences", ASITP-88-007.
14. W. M. Zheng, 'The *W*-sequence for circle maps", ASITP-88-010.
15. S. M. Ulam and J. von Neumann, *Bull. Amer. Math. Soc.* **53** (1947) 1120.
16. D. Singer, *SIAM J. Appl. Math.* **35** (1978) 260.
17. P. Collet and J. P. Eckmann, *Iterated Maps on the Interval as Dynamical Systems*, (Birkhaüser, 1980), p. 32.
18. C. Grebogi, E. Ott and J. A. Yorke, *Phys. Rev. Lett.* **48** (1982) 1507.
19. W. M. Zheng and B. L. Hao, "Symmetry breaking and restoration in antisymmetric mappings", preprint.

Commun. math. Phys. 50, 69—77 (1976)

Communications in
**Mathematical
Physics**
© by Springer-Verlag 1976

# A Two-dimensional Mapping with a Strange Attractor

M. Hénon

Observatoire de Nice, F-06300 Nice, France

**Abstract.** Lorenz (1963) has investigated a system of three first-order differential equations, whose solutions tend toward a "strange attractor". We show that the same properties can be observed in a simple mapping of the plane defined by: $x_{i+1} = y_i + 1 - ax_i^2$, $y_{i+1} = bx_i$. Numerical experiments are carried out for $a = 1.4$, $b = 0.3$. Depending on the initial point $(x_0, y_0)$, the sequence of points obtained by iteration of the mapping either diverges to infinity or tends to a strange attractor, which appears to be the product of a one-dimensional manifold by a Cantor set.

## 1. Introduction

Lorenz (1963) proposed and studied a remarkable system of three coupled first-order differential equations, representing a flow in three-dimensional space. The divergence of the flow has a constant negative value, so that any volume shrinks exponentially with time. Moreover, there exists a bounded region $R$ into which every trajectory becomes eventually trapped. Therefore, all trajectories tend to a set of measure zero, called *attractor*. In some cases the attractor is simply a point (which is then a stable equilibrium point) or a closed curve (known as a limit cycle). But in other cases the attractor has a much more complex structure; it appears to be locally the product of a two-dimensional manifold by a Cantor set. This is known as a *strange attractor*. Inside the attractor, trajectories wander in an apparently erratic manner. Moreover, they are highly sensitive to initial conditions. These phenomena are of interest for weather prediction (Lorenz, 1963) and more generally for turbulence theory (Ruelle and Takens, 1971; Ruelle, 1975). Further numerical explorations of the Lorenz system have been made by Lanford (1975) and Pomeau (1976).

We present her a "reductionist" approach in which we try to find a model problem which is as simple as possible, yet exhibits the same essential properties as the Lorenz system. Our aim is (i) to make the numerical exploration faster and more accurate, so that solutions can be followed for a longer time, more

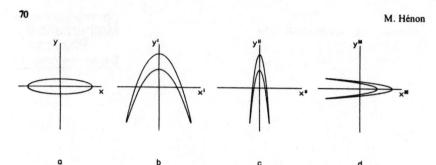

**Fig. 1.** The initial area $a$ is mapped by $T'$ into $b$, then by $T''$ into $c$, and finally by $T'''$ into $d$

detailed explorations can be conducted, etc.; (ii) to provide a model which might lend itself more easily to mathematical analysis.

## 2. The Model

Our first step is classical (Birkhoff, 1917) and consists in considering not the whole trajectories in the three-dimensional space, but only their successive intersections with a two-dimensional *surface of section S*. We define a mapping $T$ of $S$ into itself as follows: given a point $A$ of $S$, we follow the trajectory which originates from $A$ until it intersects $S$ again; this new point is $T(A)$. This mapping is sometimes called a *Poincaré map*. A trajectory is thus replaced by an infinite set of points in $S$, obtained by repeated application of the mapping $T$. The essential properties of the trajectory are reflected into corresponding properties of the set of points. We have thus formally reduced the problem to the study of a two-dimensional mapping.

At this point, however, the only advantage really gained is in clarity of presentation of the results; the actual computation of the mapping still requires the numerical integration of the differential equations. Now comes the second and decisive step: we forget about the differential system, and we define a mapping $T$ by explicit equations, giving directly $T(A)$ when $A$ is known. This of course simplifies the computation drastically. The new mapping $T$ does not any more correspond to the Lorenz system; however, by choosing it carefully we may hope to retain the essential properties which we wish to study. Past experience in the measure-preserving case (see Hénon, 1969, and references therein) has shown indeed that the same features are found in dynamical systems defined by differential equations and in mappings defined as such.

The third step consists in specifying $T$. Here we have been inspired by the numerical results of Pomeau (1976) on the Lorenz system, which show clearly how a volume is stretched in one direction, and at the same time folded over itself, in the course of one revolution. This folding effect has been also described by Ruelle (1975, Fig. 5 and 6). We simulate it by the following chain of three mappings of the $(x, y)$ plane onto itself. Consider a region elongated along the $x$ axis (Fig. 1a). We begin the folding by

$$T' : x' = x, \quad y' = y + 1 - ax^2 , \qquad (1)$$

which produces Figure 1b; $a$ is an adjustable parameter. We complete the folding by a contraction along the $x$ axis:

$$T'' : x'' = bx', \quad y'' = y', \tag{2}$$

which produces Figure 1c; $b$ is another parameter, which should be less than 1 in absolute value. Finally we come back to the orientation along the $x$ axis by

$$T''' : x''' = \gamma'', \quad y''' = x'', \tag{3}$$

which results in Figure 1d.

Our mapping will be defined as the product $T = T''' T'' T'$. We write now $(x_i, y_i)$ for $(x, y)$ and $(x_{i+1}, y_{i+1})$ for $(x''', y''')$ (as a reminder that the mapping will be iterated) and we have

$$T : x_{i+1} = y_i + 1 - ax_i^2, \quad y_{i+1} = bx_i. \tag{4}$$

This mapping has some interesting properties. Its Jacobian is a constant:

$$\frac{\partial(x_{i+1}, y_{i+1})}{\partial(x_i, y_i)} = -b. \tag{5}$$

The geometrical interpretation is quite simple: $T'$ preserves areas; $T'''$ also preserves areas but reverses the sign; and $T''$ contracts areas, multiplying them by the constant factor $b$. The property (5) is welcome because it is the natural counterpart of the constant negative divergence in the Lorenz system.

A polynomial mapping satisfying (5) is known as an *entire Cremona transformation*, and the inverse mapping is also given by polynomials (Engel, 1955, 1958). Indeed we have here

$$T^{-1} : x_i = b^{-1} y_{i+1}, \quad y_i = x_{i+1} - 1 + ab^{-2} y_{i+1}^2. \tag{6}$$

Thus $T$ is a one-to-one mapping of the plane onto itself. This is also a welcome property, because it is the natural counterpart of the fact that in the Lorenz system there is a unique trajectory through any given point.

The selection of $T$ could have been approached in a different way, by looking for the "simplest" non-trivial mapping. It is natural then to consider polynomial mappings of progressively increasing order. Linear mappings are trivial, so the polynomials must be at least of degree 2. The most general quadratic mapping is

$$x_{i+1} = f + ax_i + by_i + cx_i^2 + dx_i y_i + ey_i^2,$$
$$y_{i+1} = f' + a'x_i + b'y_i + c'x_i^2 + d'x_i y_i + e'y_i^2 \tag{7}$$

and depends on 12 parameters. But if we impose the condition that the Jacobian is a constant, some relations must be satisfied by these parameters. We can further reduce the number of parameters by an appropriate linear change of co-ordinates in the plane. In this way, by a slight extension of the results of Engel (1958), it can be shown that the general form (7) is reducible to a "canonical form" depending on two parameters only. This is a generalization of our earlier result (Hénon, 1969) that a quadratic *area-preserving* mapping can be brought into a form depending on one parameter only. The canonical form can be written in several different ways; and one of them turns out to be identical with (4), which is thus reached by an entirely different road! The mapping (4), which was initially constructed in empirical fashion, is in fact the most general quadratic mapping with constant Jacobian.

One difference with the Lorenz problem is that the successive points obtained by repeated application of $T$ do not always converge towards an attractor; sometimes they "escape" to infinity. This is because the quadratic term in (4) dominates when the distance from the origin becomes large. However, for particular values of $a$ and $b$ it is still possible to prove the existence of a bounded "trapping region" $R$, from which the points can never escape once they have entered it (see below Section 5).

$T$ has two invariant points, given by

$$x = (2a)^{-1}[-(1-b) \pm \sqrt{(1-b)^2 + 4a}], \qquad y = bx. \qquad (8)$$

These points are real for

$$a > a_0 = (1-b)^2/4. \qquad (9)$$

When this is the case, one of the points is always linearly unstable, while the other is unstable for

$$a > a_1 = 3(1-b)^2/4. \qquad (10)$$

## 3. Choice of Parameters

We select now particular values of $a$ and $b$ for a numerical study. $b$ should be small enough for the folding described by Figure 1 to occur really, yet not too small if one wishes to observe the fine structure of the attractor. The value $b = 0.3$ was found to be adequate. A good value of $a$ was found only after some experimenting. For $a < a_0$ or $a > a_3$, where $a_0$ is given by (9) and $a_3$ is of the order of 1.55 for $b = 0.3$, the points always escape to infinity: apparently there exists no attractor in these cases. For $a_0 < a < a_3$, depending on the initial values $(x_0, y_0)$, either the points escape to infinity or they converge towards an attractor, which appears to be unique for a given value of $a$. We concentrate now on this attractor. For $a_0 < a < a_1$, where $a_1$ is given by (10), the attractor is the stable invariant point. When $a$ is increased over $a_1$, at first the attractor is still simple and consists of a periodic set of $p$ points. (An equivalent attractor in the Lorenz problem would be a limit cycle intersecting the surface of section $p$ times). The value of $p$ increases through successive "bifurcations" as $a$ increases, and appears to tend to infinity as a approaches $a$ critical value $a_2$, of the order of 1.06 for $b = 0.3$. For $a_2 < a < a_3$, the attractor is no more simple, and the behaviour of the points becomes erratic. This is the case in which we are interested. We adopt the following values:

$$a = 1.4, \qquad b = 0.3. \qquad (11)$$

## 4. Numerical Results

Figure 2 shows the result of plotting 10000 successive points, obtained by iteration of $T$, starting from the arbitrarily chosen initial point $x_0 = 0$, $y_0 = 0$; the vertical scale is enlarged to give a better picture. Figure 3 shows the result of 10000

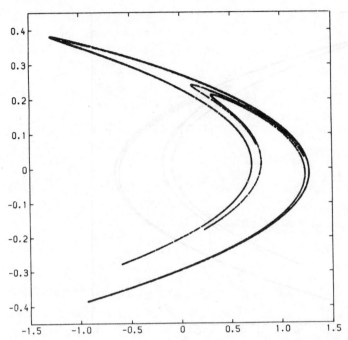

**Fig. 2.** 10000 successive points obtained by iteration of the mapping $T$ starting from $x_0 = 0$, $y_0 = 0$

iterations of $T$ again, starting from a different point: $x_0 = 0.63135448$, $y_0 = 0.18940634$ (this choice will be explained below). The two figures are seen to be almost identical. This suggests strongly that what we see in both figures is essentially the attractor itself: the successive points quickly approach the attractor and soon become undistinguishable from it at the scale of the figure. This is confirmed if one looks at the first few points on Figure 2. The initial point at $x_0 = 0$, $y_0 = 0$ and the first iterate at $x_1 = 1$, $y_1 = 0$ are clearly visible; the second iterate is still visible at $x_2 = -0.4$, $y_2 = 0.3$; the third iterate can barely be distinguished at $x_3 = 1.076$, $y_3 = -0.12$; and the fourth iterate at $x_4 = -0.7408864$, $y_4 = 0.3228$ is already lost inside the attractor at the resolution of Figure 2. The following points then wander over the attractor in an apparently erratic manner.

One of the two unstable invariant points has the coordinates, given by (8):

$$x = 0.63135448\ldots, \qquad y = 0.18940634\ldots . \tag{12}$$

This point appears to belong to the attractor. The two eigenvalues $\lambda_1$, $\lambda_2$ and the slopes $p_1$, $p_2$ of the corresponding eigenvectors are

$$\lambda_1 = 0.15594632\ldots, \qquad p_1 = 1.92373886\ldots,$$
$$\lambda_2 = -1.92373886\ldots, \qquad p_2 = -0.15594632\ldots . \tag{13}$$

The instability is due to $\lambda_2$. The corresponding slope $p_2$ appears to be tangent to the "curves" in Figure 2.

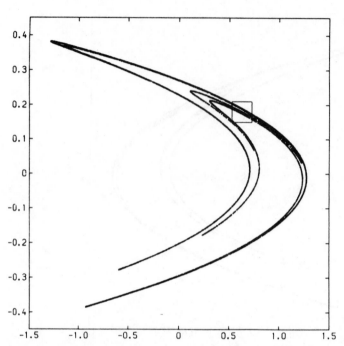

**Fig. 3.** Same as Figure 2, but starting from $x_0 = 0.63135448$, $y_0 = 0.18940634$

These properties allow us to eliminate the "transient regime" in which the points approach the attractor, and which is not of much interest: we simply start from the close vicinity of the unstable point (12), by rounding off its coordinates to 8 digits. This is done in Figure 3 and in the following figures. The points quickly move away along the line of slope $p_2$ since $|\lambda_2|$ is appreciably larger than 1.

The attractor appears to consist of a number of more or less parallel "curves"; the points tend to distribute themselves densely over these curves. The few gaps that can still be seen on Figures 2 and 3 have probably no particular significance. Their locations are not the same on the two figures. They are simply due to statistical fluctuations in the quasi-random distribution of points, and they would disappear if more moints were plotted. Thus, the *longitudinal structure* of the attractor (along the curves) appears to be simple, each curve being essentially a one-dimensional manifold.

The *transversal structure* (across the curves) appears to be entirely different, and much more complex. Already on Figures 2 and 3 a number of curves can be seen, and the visible thickness of some of them suggests that they have in fact an underlying structure. Figure 4 is a magnified view of the small square of Figure 3: some of the previous "curves" are indeed resolved now into two or more components. The number $n$ of iterations has been increased to $10^5$, in order to have a sufficient number of points in the small region examined. The small square in Figure 4 is again magnified to produce Figure 5, with $n$ increased to $10^6$: again the

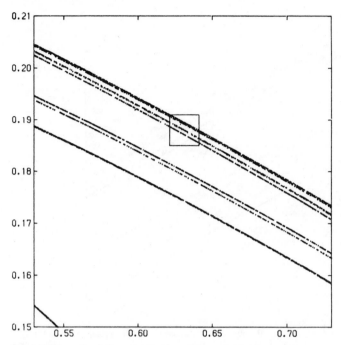

**Fig. 4.** Enlargement of the squared region of Figure 3. The number of computed points is increased to $n = 10^5$

number of visible "curves" increases. One more enlargement results in Fig. 6, with $n = 5 \times 10^6$: the points become sparse but new curves can still easily be traced.

These figures strongly suggest that the process of multiplication of "curves" will continue indefinitely, and that each apparent "curve" is in fact made of an infinity of quasi-parallel curves. Moreover, Figures 4 to 6 indicate the existence of a hierarchical sequence of "levels", the structure being practically identical at each level save for a scale factor. This is exactly the structure of a Cantor set.

The frames of Figures 4 to 6 have been chosen so as to contain the invariant point (12). This point appears to lie on the upper boundary of the attractor. Surprisingly, its presence is completely invisible on the figures; this contrasts with the area-preserving case, were stable and unstable invariant points play a very conspicuous role (see for instance Hénon, 1969). On the other hand, the presence of the invariant point explains, locally at least, the hierarchy of similar structures: at each application of the mapping, the scale of the transversal structure is multiplied by $\lambda_1$ given by (13). At the same time, the points spread out along the curves, as dictated by the value of $\lambda_2$.

## 5. A Trapping Region

The fact that even after $5 \times 10^6$ iterations the points have not diverged to infinity suggests that there is a region of the plane from which the points cannot escape.

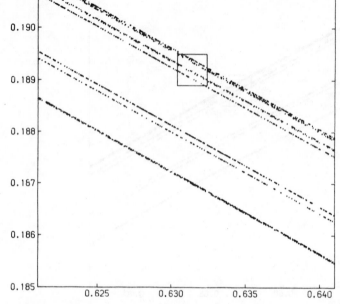

**Fig. 5.** Enlargement of the squared region of Figure 4; $n = 10^6$

This can be actually proved by finding a region $R$ which is mapped inside itself. An example of such a region is the quadrilateral $ABCD$ defined by

$$x_A = -1.33, \quad y_A = 0.42, \quad x_B = 1.32, \quad y_B = 0.133,$$

$$x_C = 1.245, \quad y_C = -0.14, \quad x_D = -1.06, \quad y_D = -0.5. \tag{14}$$

The image of $ABCD$ is a region bounded by four arcs of parabola, and it can be shown by elementary algebra that this image lies inside $ABCD$. Plotting the quadrilateral on Figure 2 or 3, one can verify that it encloses the observed attractor.

## 6. Conclusions

The simple mapping (4) appears to have the same basic properties as the Lorenz system. Its numerical exploration is much simpler: in fact most of the exploratory work for the present paper was carried out with a programmable pocket computer (HP-65). For the more extensive computations of Figures 2 to 6, we used a IBM 7040 computer, with 16-digit accuracy. The solutions can be followed over a much longer time than in the case of a system of differential equations. The accuracy is also increased since there are no integration errors.

Lorenz (1963) inferred the Cantor-set structure of the attractor from reasoning, but could not observe it directly because the contracting ratio after one "circuit"

Strange Attractor                                                              77

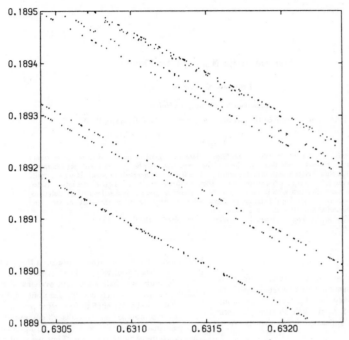

**Fig. 6.** Enlargement of the squared region of Figure 5; $n = 5 \times 10^6$

was too small: $7 \times 10^{-5}$. A similar experience was reported by Pomeau (1976). In the present mapping, the contracting ratio after one iteration is 0.3, and one can easily observe a number of successive levels in the hierarchy. This is also facilitated by the larger number of points.

Finally, for mathematical studies the mapping (4) might also be easier to handle than a system of differential equations.

### References

Birkhoff, G. D.: Trans. Amer. Math. Soc. **18**, 199 (1917)
Engel, W.: Math. Annalen **130**, 11 (1955)
Engel, W.: Math. Annalen **136**, 319 (1958)
Hénon, M.: Quart. Appl. Math. **27**, 291 (1969)
Lanford, O.: Work cited by Ruelle, 1975
Lorenz, E. N.: J. atmos. Sci. **20**, 130 (1963)
Pomeau, Y.: to appear (1976)
Ruelle, D., Takens, F.: Comm. math. Phys. **20**, 167; **23**, 343 (1971)
Ruelle, D.: Report at the Conference on "Quantum Dynamics Models and Mathematics" in Bielefeld, September 1975

Communicated by K. Hepp

Received March 25, 1976

Reprinted from Journal of the Atmospheric Sciences, Vol. 20, No. 2, March, 1963, pp. 130–141

# Deterministic Nonperiodic Flow[1]

Edward N. Lorenz

*Massachusetts Institute of Technology*

(Manuscript received 18 November 1962, in revised form 7 January 1963)

### Abstract

Finite systems of deterministic ordinary nonlinear differential equations may be designed to represent forced dissipative hydrodynamic flow. Solutions of these equations can be identified with trajectories in phase space. For those systems with bounded solutions, it is found that nonperiodic solutions are ordinarily unstable with respect to small modifications, so that slightly differing initial states can evolve into considerably different states. Systems with bounded solutions are shown to possess bounded numerical solutions.

A simple system representing cellular convection is solved numerically. All of the solutions are found to be unstable, and almost all of them are nonperiodic.

The feasibility of very-long-range weather prediction is examined in the light of these results.

## 1. Introduction

Certain hydrodynamical systems exhibit steady-state flow patterns, while others oscillate in a regular periodic fashion. Still others vary in an irregular, seemingly haphazard manner, and, even when observed for long periods of time, do not appear to repeat their previous history.

These modes of behavior may all be observed in the familiar rotating-basin experiments, described by Fultz, *et al.* (1959) and Hide (1958). In these experiments, a cylindrical vessel containing water is rotated about its axis, and is heated near its rim and cooled near its center in a steady symmetrical fashion. Under certain conditions the resulting flow is as symmetric and steady as the heating which gives rise to it. Under different conditions a system of regularly spaced waves develops, and progresses at a uniform speed without changing its shape. Under still different conditions an irregular flow pattern forms, and moves and changes its shape in an irregular nonperiodic manner.

Lack of periodicity is very common in natural systems, and is one of the distinguishing features of turbulent flow. Because instantaneous turbulent flow patterns are so irregular, attention is often confined to the statistics of turbulence, which, in contrast to the details of turbulence, often behave in a regular well-organized manner. The short-range weather forecaster, however, is forced willy-nilly to predict the details of the large-scale turbulent eddies—the cyclones and anticyclones—which continually arrange themselves into new patterns.

[1] The research reported in this work has been sponsored by the Geophysics Research Directorate of the Air Force Cambridge Research Center, under Contract No. AF 19(604)-4969.

Thus there are occasions when more than the statistics of irregular flow are of very real concern.

In this study we shall work with systems of deterministic equations which are idealizations of hydrodynamical systems. We shall be interested principally in nonperiodic solutions, i.e., solutions which never repeat their past history exactly, and where all approximate repetitions are of finite duration. Thus we shall be involved with the ultimate behavior of the solutions, as opposed to the transient behavior associated with arbitrary initial conditions.

A closed hydrodynamical system of finite mass may ostensibly be treated mathematically as a finite collection of molecules—usually a very large finite collection —in which case the governing laws are expressible as a finite set of ordinary differential equations. These equations are generally highly intractable, and the set of molecules is usually approximated by a continuous distribution of mass. The governing laws are then expressed as a set of partial differential equations, containing such quantities as velocity, density, and pressure as dependent variables.

It is sometimes possible to obtain particular solutions of these equations analytically, especially when the solutions are periodic or invariant with time, and, indeed, much work has been devoted to obtaining such solutions by one scheme or another. Ordinarily, however, nonperiodic solutions cannot readily be determined except by numerical procedures. Such procedures involve replacing the continuous variables by a new finite set of functions of time, which may perhaps be the values of the continuous variables at a chosen grid of points, or the coefficients in the expansions of these variables in series of orthogonal functions. The governing laws then become a finite set of ordinary differential equations again, although a far simpler set than the one which governs individual molecular motions.

In any real hydrodynamical system, viscous dissipation is always occurring, unless the system is moving as a solid, and thermal dissipation is always occurring, unless the system is at constant temperature. For certain purposes many systems may be treated as conservative systems, in which the total energy, or some other quantity, does not vary with time. In seeking the ultimate behavior of a system, the use of conservative equations is unsatisfactory, since the ultimate value of any conservative quantity would then have to equal the arbitrarily chosen initial value. This difficulty may be obviated by including the dissipative processes, thereby making the equations nonconservative, and also including external mechanical or thermal forcing, thus preventing the system from ultimately reaching a state of rest. If the system is to be deterministic, the forcing functions, if not constant with time, must themselves vary according to some deterministic rule.

In this work, then, we shall deal specifically with finite systems of deterministic ordinary differential equations, designed to represent forced dissipative hydrodynamical systems. We shall study the properties of nonperiodic solutions of these equations.

It is not obvious that such solutions can exist at all. Indeed, in dissipative systems governed by finite sets of *linear* equations, a constant forcing leads ultimately to a constant response, while a periodic forcing leads to a periodic response. Hence, nonperiodic flow has sometimes been regarded as the result of nonperiodic or random forcing.

The reasoning leading to these conclusions is not applicable when the governing equations are nonlinear. If the equations contain terms representing advection—the transport of some property of a fluid by the motion of the fluid itself—a constant forcing can lead to a variable response. In the rotating-basin experiments already mentioned, both periodic and nonperiodic flow result from thermal forcing which, within the limits of experimental control, is constant. Exact periodic solutions of simplified systems of equations, representing dissipative flow with constant thermal forcing, have been obtained analytically by the writer (1962a). The writer (1962b) has also found nonperiodic solutions of similar systems of equations by numerical means.

## 2. Phase space

Consider a system whose state may be described by $M$ variables $X_1, \cdots, X_M$. Let the system be governed by the set of equations

$$dX_i/dt = F_i(X_1, \cdots X_M), \quad i=1, \cdots, M, \qquad (1)$$

where time $t$ is the single independent variable, and the functions $F_i$ possess continuous first partial derivatives. Such a system may be studied by means of *phase space*—

an $M$-dimensional Euclidean space $\Gamma$ whose coordinates are $X_1, \cdots, X_M$. Each *point* in phase space represents a possible instantaneous state of the system. A state which is varying in accordance with (1) is represented by a moving *particle* in phase space, traveling along a *trajectory* in phase space. For completeness, the position of a stationary particle, representing a steady state, is included as a trajectory.

Phase space has been a useful concept in treating finite systems, and has been used by such mathematicians as Gibbs (1902) in his development of statistical mechanics, Poincaré (1881) in his treatment of the solutions of differential equations, and Birkhoff (1927) in his treatise on dynamical systems.

From the theory of differential equations (e.g., Ford 1933, ch. 6), it follows, since the partial derivatives $\partial F_i/\partial X_j$ are continuous, that if $t$ is any time, and if $X_{10}, \cdots X_{M0}$ is any point in $\Gamma$, equations (1) possess a unique solution

$$X_i = f_i(X_{10}, \cdots, X_{M0}, t), \quad i=1, \cdots, M, \qquad (2)$$

valid throughout some time interval containing $t_0$, and satisfying the condition

$$f_i(X_{10}, \cdots, X_{M0}, t_0) = X_{i0}, \quad i=1, \cdots, M. \qquad (3)$$

The functions $f_i$ are continuous in $X_{10}, \cdots, X_{M0}$ and $t$. Hence there is a unique trajectory through each point of $\Gamma$. Two or more trajectories may, however, approach the same point or the same curve asymptotically as $t \to \infty$ or as $t \to -\infty$. Moreover, since the functions $f_i$ are continuous, the passage of time defines a continuous deformation of any region of $\Gamma$ into another region.

In the familiar case of a conservative system, where some positive definite quantity $Q$, which may represent some form of energy, is invariant with time, each trajectory is confined to one or another of the surfaces of constant $Q$. These surfaces may take the form of closed concentric shells.

If, on the other hand, there is dissipation and forcing, and if, whenever $Q$ equals or exceeds some fixed value $Q_1$, the dissipation acts to diminish $Q$ more rapidly then the forcing can increase $Q$, then $(-dQ/dt)$ has a positive lower bound where $Q \geq Q_1$, and each trajectory must ultimately become trapped in the region where $Q < Q_1$. Trajectories representing forced dissipative flow may therefore differ considerably from those representing conservative flow.

Forced dissipative systems of this sort are typified by the system

$$dX_i/dt = \sum_{j,k} a_{ijk}X_jX_k - \sum_j b_{ij}X_j + c_i, \qquad (4)$$

where $\sum a_{ijk}X_iX_jX_k$ vanishes identically, $\sum b_{ij}X_iX_j$ is positive definite, and $c_1, \cdots, c_M$ are constants. If

$$Q = \tfrac{1}{2} \sum_i X_i^2, \qquad (5)$$

and if $e_1, \cdots, e_M$ are the roots of the equations

$$\sum_j (b_{ij}+b_{ji})e_j = c_i, \tag{6}$$

it follows from (4) that

$$dQ/dt = \sum_{i,j} b_{ij}e_i e_j - \sum_{i,j} b_{ij}(X_i-e_i)(X_j-e_j). \tag{7}$$

The right side of (7) vanishes only on the surface of an ellipsoid $E$, and is positive only in the interior of $E$. The surfaces of constant $Q$ are concentric spheres. If $S$ denotes a particular one of these spheres whose interior $R$ contains the ellipsoid $E$, it is evident that each trajectory eventually becomes trapped within $R$.

### 3. The instability of nonperiodic flow

In this section we shall establish one of the most important properties of deterministic nonperiodic flow, namely, its instability with respect to modifications of small amplitude. We shall find it convenient to do this by identifying the solutions of the governing equations with trajectories in phase space. We shall use such symbols as $P(t)$ (variable argument) to denote trajectories, and such symbols as $P$ or $P(t_0)$ (no argument or constant argument) to denote points, the latter symbol denoting the specific point through which $P(t)$ passes at time $t_0$.

We shall deal with a phase space $\Gamma$ in which a unique trajectory passes through each point, and where the passage of time defines a continuous deformation of any region of $\Gamma$ into another region, so that if the points $P_1(t_0)$, $P_2(t_0)$, $\cdots$ approach $P_0(t_0)$ as a limit, the points $P_1(t_0+\tau)$, $P_2(t_0+\tau)$, $\cdots$ must approach $P_0(t_0+\tau)$ as a limit. We shall furthermore require that the trajectories be uniformly bounded as $t \to \infty$; that is, there must be a bounded region $R$, such that every trajectory ultimately remains with $R$. Our procedure is influenced by the work of Birkhoff (1927) on dynamical systems, but differs in that Birkhoff was concerned mainly with conservative systems. A rather detailed treatment of dynamical systems has been given by Nemytskii and Stepanov (1960), and rigorous proofs of some of the theorems which we shall present are to be found in that source.

We shall first classify the trajectories in three different manners, namely, according to the absence or presence of transient properties, according to the stability or instability of the trajectories with respect to small modifications, and according to the presence or absence of periodic behavior.

Since any trajectory $P(t)$ is bounded, it must possess at least one *limit point* $P_0$, a point which it approaches arbitrarily closely arbitrarily often. More precisely, $P_0$ is a limit point of $P(t)$ if for any $\epsilon>0$ and any time $t_1$ there exists a time $t_2(\epsilon,t_1)>t_1$ such that $|P(t_2)-P_0|<\epsilon$. Here absolute-value signs denote distance in phase space. Because $\Gamma$ is continuously deformed as $t$ varies, every point on the trajectory through $P_0$ is also a limit point of $P(t)$, and the set of limit points of $P(t)$ forms a trajectory, or a set of trajectories, called the *limiting trajectories* of $P(t)$. A limiting trajectory is obviously contained within $R$ in its entirety.

If a trajectory is contained among its own limiting trajectories, it will be called *central*; otherwise it will be called *noncentral*. A central trajectory passes arbitrarily closely arbitrarily often to any point through which it has previously passed, and, in this sense at least, separate sufficiently long segments of a central trajectory are statistically similar. A noncentral trajectory remains a certain distance away from any point through which it has previously passed. It must approach its entire set of limit points asymptotically, although it need not approach any particular limiting trajectory asymptotically. Its instantaneous distance from its closest limit point is therefore a transient quantity, which becomes arbitrarily small as $t \to \infty$.

A trajectory $P(t)$ will be called *stable at a point* $P(t_1)$ if any other trajectory passing sufficiently close to $P(t_1)$ at time $t_1$ remains close to $P(t)$ as $t \to \infty$; i.e., $P(t)$ is stable at $P(t_1)$ if for any $\epsilon>0$ there exists a $\delta(\epsilon,t_1)>0$ such that if $|P_1(t_1)-P(t_1)|<\delta$ and $t_2>t_1$, $|P_1(t_2)-P(t_2)|<\epsilon$. Otherwise $P(t)$ will be called *unstable* at $P(t_1)$. Because $\Gamma$ is continuously deformed as $t$ varies, a trajectory which is stable at one point is stable at every point, and will be called a *stable* trajectory. A trajectory unstable at one point is unstable at every point, and will be called an *unstable* trajectory. In the special case that $P(t)$ is confined to one point, this definition of stability coincides with the familiar concept of stability of steady flow.

A stable trajectory $P(t)$ will be called uniformly stable if the distance within which a neighboring trajectory must approach a point $P(t_1)$, in order to be certain of remaining close to $P(t)$ as $t \to \infty$, itself possesses a positive lower bound as $t_1 \to \infty$; i.e., $P(t)$ is uniformly stable if for any $\epsilon>0$ there exists a $\delta(\epsilon)>0$ and a time $t_0(\epsilon)$ such that if $t_1>t_0$ and $|P_1(t_1)-P(t_1)|<\delta$ and $t_2>t_1$, $|P_1(t_2)-P(t_2)|<\epsilon$. A limiting trajectory $P_0(t)$ of a uniformly stable trajectory $P(t)$ must be uniformly stable itself, since all trajectories passing sufficiently close to $P_0(t)$ must pass arbitrarily close to some point of $P(t)$ and so must remain close to $P(t)$, and hence to $P_0(t)$, as $t \to \infty$.

Since each point lies on a unique trajectory, any trajectory passing through a point through which it has previously passed must continue to repeat its past behavior, and so must be *periodic*. A trajectory $P(t)$ will be called *quasi-periodic* if for some arbitrarily large time interval $\tau$, $P(t+\tau)$ ultimately remains arbitrarily close to $P(t)$, i.e., $P(t)$ is quasi-periodic if for any $\epsilon>0$ and for any time interval $\tau_0$, there exists a $\tau(\epsilon,\tau_0)>\tau_0$ and a time $t_1(\epsilon,\tau_0)$ such that if $t_2>t_1$, $|P(t_2+\tau)-P(t_2)|$

$<\epsilon$. Periodic trajectories are special .cases of quasi-periodic trajectories.

A trajectory which is not quasi-periodic will be called *nonperiodic*. If $P(t)$ is nonperiodic, $P(t_1+\tau)$ may be arbitrarily close to $P(t_1)$ for some time $t_1$ and some arbitrarily large time interval $\tau$, but, if this is so, $P(t+\tau)$ cannot remain arbitrarily close to $P(t)$ as $t \to \infty$. Nonperiodic trajectories are of course representations of deterministic nonperiodic flow, and form the principal subject of this paper.

Periodic trajectories are obviously central. Quasi-periodic central trajectories include multiple periodic trajectories with incommensurable periods, while quasi-periodic noncentral trajectories include those which approach periodic trajectories asymptotically. Non-periodic trajectories may be central or noncentral.

We can now establish the theorem that a trajectory with a stable limiting trajectory is quasi-periodic. For if $P_0(t)$ is a limiting trajectory of $P(t)$, two distinct points $P(t_1)$ and $P(t_1+\tau)$, with $\tau$ arbitrarily large, may be found arbitrary close to any point $P_0(t_0)$. Since $P_0(t)$ is stable, $P(t)$ and $P(t+\tau)$ must remain arbitrarily close to $P_0(t+t_0-t_1)$, and hence to each other, as $t \to \infty$, and $P(t)$ is quasi-periodic.

It follows immediately that a stable central trajectory is quasi-periodic, or, equivalently, that a nonperiodic central trajectory is unstable.

The result has far-reaching consequences when the system being considered is an observable nonperiodic system whose future state we may desire to predict. It implies that two states differing by imperceptible amounts may eventually evolve into two considerably different states. If, then, there is any error whatever in observing the present state—and in any real system such errors seem inevitable—an acceptable prediction of an instantaneous state in the distant future may well be impossible.

As for noncentral trajectories, it follows that a uniformly stable noncentral trajectory is quasi-periodic, or, equivalently, a nonperiodic noncentral trajectory is not uniformly stable. The possibility of a nonperiodic non-central trajectory which is stable but not uniformly stable still exists. To the writer, at least, such trajectories, although possible on paper, do not seem characteristic of real hydrodynamical phenomena. Any claim that atmospheric flow, for example, is represented by a trajectory of this sort would lead to the improbable conclusion that we ought to master long-range forecasting as soon as possible, because, the longer we wait, the more difficult our task will become.

In summary, we have shown that, subject to the conditions of uniqueness, continuity, and boundedness prescribed at the beginning of this section, a central trajectory, which in a certain sense is free of transient properties, is unstable if it is nonperiodic. A noncentral trajectory, which is characterized by transient properties, is not uniformly stable if it is nonperiodic, and,

if it is stable at all, its very stability is one of its transient properties, which tends to die out as time progresses. In view of the impossibility of measuring initial conditions precisely, and thereby distinguishing between a central trajectory and a nearby noncentral trajectory, all nonperiodic trajectories are effectively unstable from the point of view of practical prediction.

## 4. Numerical integration of nonconservative systems

The theorems of the last section can be of importance only if nonperiodic solutions of equations of the type considered actually exist. Since statistically stationary nonperiodic functions of time are not easily described analytically, particular nonperiodic solutions can probably be found most readily by numerical procedures. In this section we shall examine a numerical-integration procedure which is especially applicable to systems of equations of the form (4). In a later section we shall use this procedure to determine a nonperiodic solution of a simple set of equations.

To solve (1) numerically we may choose an initial time $t_0$ and a time increment $\Delta t$, and let

$$X_{i,n}=X_i(t_0+n\Delta t). \qquad (8)$$

We then introduce the auxiliary approximations

$$X_{i(n+1)}=X_{i,n}+F_i(P_n)\Delta t, \qquad (9)$$

$$X_{i((n+2))}=X_{i(n+1)}+F_i(P_{(n+1)})\Delta t, \qquad (10)$$

where $P_n$ and $P_{(n+1)}$ are the points whose coordinates are

$$(X_{1,n}, \cdots, X_{M,n}) \quad \text{and} \quad (X_{1(n+1)}, \cdots, X_{M(n+1)}).$$

The simplest numerical procedure for obtaining approximate solutions of (1) is the forward-difference procedure,

$$X_{i,n+1}=X_{i(n+1)}. \qquad (11)$$

In many instances better approximations to the solutions of (1) may be obtained by a centered-difference procedure

$$X_{i,n+1}=X_{i,n-1}+2F_i(P_n)\Delta t. \qquad (12)$$

This procedure is unsuitable, however, when the deterministic nature of (1) is a matter of concern, since the values of $X_{1,n}, \cdots, X_{M,n}$ do not uniquely determine the values of $X_{1,n+1}, \cdots, X_{M,n+1}$.

A procedure which largely overcomes the disadvantages of both the forward-difference and centered-difference procedures is the double-approximation procedure, defined by the relation

$$X_{i,n+1}=X_{i,n}+\tfrac{1}{2}[F_i(P_n)+F_i(P_{(n+1)})]\Delta t. \qquad (13)$$

Here the coefficient of $\Delta t$ is an approximation to the time derivative of $X_i$ at time $t_0+(n+\tfrac{1}{2})\Delta t$. From (9) and (10), it follows that (13) may be rewritten

$$X_{i,n+1}=\tfrac{1}{2}(X_{i,n}+X_{i((n+2))}). \qquad (14)$$

A convenient scheme for automatic computation is the successive evaluation of $X_{i(n+1)}$, $X_{i((n+2))}$, and $X_{i,n+1}$ according to (9), (10) and (14). We have used this procedure in all the computations described in this study.

In phase space a numerical solution of (1) must be represented by a jumping particle rather than a continuously moving particle. Moreover, if a digital computer is instructed to represent each number in its memory by a preassigned fixed number of bits, only certain discrete points in phase space will ever be occupied. If the numerical solution is bounded, repetitions must eventually occur, so that, strictly speaking, every numerical solution is periodic. In practice this consideration may be disregarded, if the number of different possible states is far greater than the number of iterations ever likely to be performed. The necessity for repetition could be avoided altogether by the somewhat uneconomical procedure of letting the precision of computation increase as $n$ increases.

Consider now numerical solutions of equations (4), obtained by the forward-difference procedure (11). For such solutions,

$$Q_{n+1}=Q_n+(dQ/dt)_n\Delta t+\tfrac{1}{2}\sum_i F_i^2(P_n)\Delta t^2. \qquad (15)$$

Let $S'$ be any surface of constant $Q$ whose interior $R'$ contains the ellipsoid $E$ where $dQ/dt$ vanishes, and let $S$ be any surface of constant $Q$ whose interior $R$ contains $S'$.

Since $\sum F_i^2$ and $dQ/dt$ both possess upper bounds in $R'$, we may choose $\Delta t$ so small that $P_{n+1}$ lies in $R$ if $P_n$ lies in $R'$. Likewise, since $\sum F_i^2$ possesses an upper bound and $dQ/dt$ possesses a *negative* upper bound in $R-R'$, we may choose $\Delta t$ so small that $Q_{n+1}<Q_n$ if $P_n$ lies in $R-R'$. Hence $\Delta t$ may be chosen so small that any jumping particle which has entered $R$ remains trapped within $R$, and the numerical solution does not blow up. A blow-up may still occur, however, if initially the particle is exterior to $R$.

Consider now the double-approximation procedure (14). The previous arguments imply not only that $P_{(n+1)}$ lies within $R$ if $P_n$ lies within $R$, but also that $P_{((n+2))}$ lies within $R$ if $P_{(n+1)}$ lies within $R$. Since the region $R$ is convex, it follows that $P_{n+1}$, as given by (14), lies within $R$ if $P_n$ lies within $R$. Hence if $\Delta t$ is chosen so small that the forward-difference procedure does not blow up, the double-approximation procedure also does not blow up.

We note in passing that if we apply the forward-difference procedure to a conservative system where $dQ/dt=0$ everywhere,

$$Q_{n+1}=Q_n+\tfrac{1}{2}\sum_i F_i^2(P_n)\Delta t^2. \qquad (16)$$

In this case, for any fixed choice of $\Delta t$ the numerical solution ultimately goes to infinity, unless it is asymp-totically approaching a steady state. A similar result holds when the double-approximation procedure (14) is applied to a conservative system.

## 5. The convection equations of Saltzman

In this section we shall introduce a system of three ordinary differential equations whose solutions afford the simplest example of deterministic nonperiodic flow of which the writer is aware. The system is a simplification of one derived by Saltzman (1962) to study finite-amplitude convection. Although our present interest is in the nonperiodic nature of its solutions, rather than in its contributions to the convection problem, we shall describe its physical background briefly.

Rayleigh (1916) studied the flow occurring in a layer of fluid of uniform depth $H$, when the temperature difference between the upper and lower surfaces is maintained at a constant value $\Delta T$. Such a system possesses a steady-state solution in which there is no motion, and the temperature varies linearly with depth, If this solution is unstable, convection should develop.

In the case where all motions are parallel to the $x$-$z$-plane, and no variations in the direction of the $y$-axis occur, the governing equations may be written (see Saltzman, 1962)

$$\frac{\partial}{\partial t}\nabla^2\psi=-\frac{\partial(\psi,\nabla^2\psi)}{\partial(x,z)}+\nu\nabla^4\psi+g\alpha\frac{\partial\theta}{\partial x}, \qquad (17)$$

$$\frac{\partial}{\partial t}\theta=-\frac{\partial(\psi,\theta)}{\partial(x,z)}+\frac{\Delta T}{H}\frac{\partial\psi}{\partial x}+\kappa\nabla^2\theta. \qquad (18)$$

Here $\psi$ is a stream function for the two-dimensional motion, $\theta$ is the departure of temperature from that occurring in the state of no convection, and the constants $g$, $\alpha$, $\nu$, and $\kappa$ denote, respectively, the acceleration of gravity, the coefficient of thermal expansion, the kinematic viscosity, and the thermal conductivity. The problem is most tractable when both the upper and lower boundaries are taken to be free, in which case $\psi$ and $\nabla^2\psi$ vanish at both boundaries.

Rayleigh found that fields of motion of the form

$$\psi=\psi_0\sin(\pi aH^{-1}x)\sin(\pi H^{-1}z), \qquad (19)$$

$$\theta=\theta_0\cos(\pi aH^{-1}x)\sin(\pi H^{-1}z), \qquad (20)$$

would develop if the quantity

$$R_a=g\alpha H^3\Delta T\nu^{-1}\kappa^{-1}, \qquad (21)$$

now called the *Rayleigh number*, exceeded a critical value

$$R_c=\pi^4a^{-2}(1+a^2)^3. \qquad (22)$$

The minimum value of $R_c$, namely $27\pi^4/4$, occurs when $a^2=\tfrac{1}{2}$.

Saltzman (1962) derived a set of ordinary differential equations by expanding $\psi$ and $\theta$ in double Fourier series in $x$ and $z$, with functions of $t$ alone for coefficients, and substituting these series into (17) and (18). He arranged the right-hand sides of the resulting equations in double-Fourier-series form, by replacing products of trigonometric functions of $x$ (or $z$) by sums of trigonometric functions, and then equated coefficients of similar functions of $x$ and $z$. He then reduced the resulting infinite system to a finite system by omitting reference to all but a specified finite set of functions of $t$, in the manner proposed by the writer (1960).

He then obtained time-dependent solutions by numerical integration. In certain cases all except three of the dependent variables eventually tended to zero, and these three variables underwent irregular, apparently nonperiodic fluctuations.

These same solutions would have been obtained if the series had at the start been truncated to include a total of three terms. Accordingly, in this study we shall let

$$a(1+a^2)^{-1}\kappa^{-1}\psi = X\sqrt{2}\,\sin\,(\pi a H^{-1}x)\,\sin\,(\pi H^{-1}z), \quad (23)$$

$$\pi R_c^{-1}R_a\Delta T^{-1}\theta = Y\sqrt{2}\,\cos\,(\pi a H^{-1}x)\,\sin\,(\pi H^{-1}z)$$
$$-Z\,\sin\,(2\pi H^{-1}z), \quad (24)$$

where $X$, $Y$, and $Z$ are functions of time alone. When expressions (23) and (24) are substituted into (17) and (18), and trigonometric terms other than those occurring in (23) and (24) are omitted, we obtain the equations

$$X^{\cdot} = \quad -\sigma X + \sigma Y, \quad (25)$$

$$Y^{\cdot} = -XZ + rX - Y, \quad (26)$$

$$Z^{\cdot} = \quad XY \qquad -bZ. \quad (27)$$

Here a dot denotes a derivative with respect to the dimensionless time $\tau = \pi^2 H^{-2}(1+a^2)\kappa t$, while $\sigma = \kappa^{-1}\nu$ is the *Prandtl number*, $r = R_c^{-1}R_a$, and $b = 4(1+a^2)^{-1}$. Except for multiplicative constants, our variables $X$, $Y$, and $Z$ are the same as Saltzman's variables $A$, $D$, and $G$. Equations (25), (26), and (27) are the convection equations whose solutions we shall study.

In these equations $X$ is proportional to the intensity of the convective motion, while $Y$ is proportional to the temperature difference between the ascending and descending currents, similar signs of $X$ and $Y$ denoting that warm fluid is rising and cold fluid is descending. The variable $Z$ is proportional to the distortion of the vertical temperature profile from linearity, a positive value indicating that the strongest gradients occur near the boundaries.

Equations (25)–(27) may give realistic results when the Rayleigh number is slightly supercritical, but their solutions cannot be expected to resemble those of (17) and (18) when strong convection occurs, in view of the extreme truncation.

## 6. Applications of linear theory

Although equations (25)–(27), as they stand, do not have the form of (4), a number of linear transformations

will convert them to this form. One of the simplest of these is the transformation

$$X' = X, \quad Y' = Y, \quad Z' = Z - r - \sigma. \quad (28)$$

Solutions of (25)–(27) therefore remain bounded within a region $R$ as $\tau \to \infty$, and the general results of Sections 2, 3 and 4 apply to these equations.

The stability of a solution $X(\tau)$, $Y(\tau)$, $Z(\tau)$ may be formally investigated by considering the behavior of small superposed perturbations $x_0(\tau)$, $y_0(\tau)$, $z_0(\tau)$. Such perturbations are temporarily governed by the linearized equations

$$\begin{bmatrix} x_0 \\ y_0 \\ z_0 \end{bmatrix}^{\cdot} = \begin{bmatrix} -\sigma & \sigma & 0 \\ (r-Z) & -1 & -X \\ Y & X & -b \end{bmatrix} \begin{bmatrix} x_0 \\ y_0 \\ z_0 \end{bmatrix}. \quad (29)$$

Since the coefficients in (29) vary with time, unless the basic state $X$, $Y$, $Z$ is a steady-state solution of (25)–(27), a general solution of (29) is not feasible. However, the variation of the volume $V_0$ of a small region in phase space, as each point in the region is displaced in accordance with (25)–(27), is determined by the diagonal sum of the matrix of coefficients; specifically

$$V_0^{\cdot} = -(\sigma + b + 1)V_0. \quad (30)$$

This is perhaps most readily seen by visualizing the motion in phase space as the flow of a fluid, whose divergence is

$$\frac{\partial X^{\cdot}}{\partial X} + \frac{\partial Y^{\cdot}}{\partial Y} + \frac{\partial Z^{\cdot}}{\partial Z} = -(\sigma + b + 1). \quad (31)$$

Hence each small volume shrinks to zero as $\tau \to \infty$, at a rate independent of $X$, $Y$, and $Z$. This does not imply that each small volume shrinks to a point; it may simply become flattened into a surface. It follows that the volume of the region initially enclosed by the surface $S$ shrinks to zero at this same rate, so that all trajectories ultimately become confined to a specific subspace having zero volume. This subspace contains all those trajectories which lie entirely within $R$, and so contains all central trajectories.

Equations (25)–(27) possess the steady-state solution $X = Y = Z = 0$, representing the state of no convection. With this basic solution, the characteristic equation of the matrix in (29) is

$$[\lambda + b][\lambda^2 + (\sigma + 1)\lambda + \sigma(1 - r)] = 0. \quad (32)$$

This equation has three real roots when $r > 0$; all are negative when $r < 1$, but one is positive when $r > 1$. The criterion for the onset of convection is therefore $r = 1$, or $R_a = R_c$, in agreement with Rayleigh's result.

When $r > 1$, equations (25)–(27) possess two additional steady-state solutions $X = Y = \pm\sqrt{b(r-1)}$, $Z = r - 1$.

For either of these solutions, the characteristic equation of the matrix in (29) is

$$\lambda^3 + (\sigma + b + 1)\lambda^2 + (r + \sigma)b\lambda + 2\sigma b(r-1) = 0. \quad (33)$$

This equation possesses one real negative root and two complex conjugate roots when $r > 1$; the complex conjugate roots are pure imaginary if the product of the coefficients of $\lambda^2$ and $\lambda$ equals the constant term, or

$$r = \sigma(\sigma + b + 3)(\sigma - b - 1)^{-1}. \quad (34)$$

This is the critical value of $r$ for the instability of steady convection. Thus if $\sigma < b + 1$, no positive value of $r$ satisfies (34), and steady convection is always stable, but if $\sigma > b + 1$, steady convection is unstable for sufficiently high Rayleigh numbers. This result of course applies only to idealized convection governed by (25)–(27), and not to the solutions of the partial differential equations (17) and (18).

The presence of complex roots of (34) shows that if unstable steady convection is disturbed, the motion will oscillate in intensity. What happens when the disturbances become large is not revealed by linear theory. To investigate finite-amplitude convection, and to study the subspace to which trajectories are ultimately confined, we turn to numerical integration.

## 7. Numerical integration of the convection equations

To obtain numerical solutions of the convection equations, we must choose numerical values for the constants. Following Saltzman (1962), we shall let $\sigma = 10$ and $a^2 = \frac{1}{2}$, so that $b = 8/3$. The critical Rayleigh number for instability of steady convection then occurs when $r = 470/19 = 24.74$.

We shall choose the slightly supercritical value $r = 28$. The states of steady convection are then represented by the points $(6\sqrt{2}, 6\sqrt{2}, 27)$ and $(-6\sqrt{2}, -6\sqrt{2}, 27)$ in phase space, while the state of no convection corresponds to the origin $(0,0,0)$.

We have used the double-approximation procedure for numerical integration, defined by (9), (10), and (14). The value $\Delta\tau = 0.01$ has been chosen for the dimensionless time increment. The computations have been performed on a Royal McBee LGP-30 electronic com-

TABLE 1. Numerical solution of the convection equations. Values of $X$, $Y$, $Z$ are given at every fifth iteration $N$, for the first 160 iterations.

| N | X | Y | Z |
|---|---|---|---|
| 0000 | 0000 | 0010 | 0000 |
| 0005 | 0004 | 0012 | 0000 |
| 0010 | 0009 | 0020 | 0000 |
| 0015 | 0016 | 0036 | 0002 |
| 0020 | 0030 | 0066 | 0007 |
| 0025 | 0054 | 0115 | 0024 |
| 0030 | 0093 | 0192 | 0074 |
| 0035 | 0150 | 0268 | 0201 |
| 0040 | 0195 | 0234 | 0397 |
| 0045 | 0174 | 0055 | 0483 |
| 0050 | 0097 | −0067 | 0415 |
| 0055 | 0025 | −0093 | 0340 |
| 0060 | −0020 | −0089 | 0298 |
| 0065 | −0046 | −0084 | 0275 |
| 0070 | −0061 | −0083 | 0262 |
| 0075 | −0070 | −0086 | 0256 |
| 0080 | −0077 | −0091 | 0255 |
| 0085 | −0084 | −0095 | 0258 |
| 0090 | −0089 | −0098 | 0266 |
| 0095 | −0093 | −0098 | 0275 |
| 0100 | −0094 | −0093 | 0283 |
| 0105 | −0092 | −0086 | 0297 |
| 0110 | −0088 | −0079 | 0286 |
| 0115 | −0083 | −0073 | 0281 |
| 0120 | −0078 | −0070 | 0273 |
| 0125 | −0075 | −0071 | 0264 |
| 0130 | −0074 | −0075 | 0257 |
| 0135 | −0076 | −0080 | 0252 |
| 0140 | −0079 | −0087 | 0251 |
| 0145 | −0083 | −0093 | 0254 |
| 0150 | −0088 | −0098 | 0262 |
| 0155 | −0092 | −0099 | 0271 |
| 0160 | −0094 | −0096 | 0281 |

TABLE 2. Numerical solution of the convection equations. Values of $X$, $Y$, $Z$ are given at every iteration $N$ for which $Z$ possesses a relative maximum, for the first 6000 iterations.

| N | X | Y | Z | N | X | Y | Z |
|---|---|---|---|---|---|---|---|
| 0045 | 0174 | 0055 | 0483 | 3029 | 0117 | 0075 | 0352 |
| 0107 | −0091 | −0083 | 0287 | 3098 | 0123 | 0076 | 0365 |
| 0168 | −0092 | −0084 | 0288 | 3171 | 0134 | 0082 | 0383 |
| 0230 | −0092 | −0084 | 0289 | 3268 | 0155 | 0069 | 0435 |
| 0292 | −0092 | −0083 | 0290 | 3333 | −0114 | −0079 | 0342 |
| 0354 | −0093 | −0083 | 0292 | 3400 | −0117 | −0077 | 0350 |
| 0416 | −0093 | −0083 | 0293 | 3468 | −0125 | −0083 | 0361 |
| 0478 | −0094 | −0082 | 0295 | 3541 | −0129 | −0073 | 0378 |
| 0540 | −0094 | −0082 | 0296 | 3625 | −0146 | −0074 | 0413 |
| 0602 | −0095 | −0082 | 0298 | 3695 | 0127 | 0079 | 0370 |
| 0664 | −0096 | −0083 | 0300 | 3772 | 0136 | 0072 | 0394 |
| 0726 | −0097 | −0083 | 0302 | 3853 | −0144 | −0077ʼ | 0407 |
| 0789 | −0097 | −0081 | 0304 | 3926 | 0129 | 0072 | 0380 |
| 0851 | −0099 | −0083 | 0307 | 4014 | 0148 | 0068 | 0421 |
| 0914 | −0100 | −0081 | 0309 | 4082 | −0120 | −0074 | 0359 |
| 0977 | −0100 | −0080 | 0312 | 4153 | −0129 | −0078 | 0375 |
| 1040 | −0102 | −0080 | 0315 | 4233 | −0144 | −0082 | 0404 |
| 1103 | −0104 | −0081 | 0319 | 4307 | 0135 | 0081 | 0385 |
| 1167 | −0105 | −0079 | 0323 | 4417 | −0162 | −0069 | 0450 |
| 1231 | −0107 | −0079 | 0328 | 4480 | 0106 | 0081 | 0324 |
| 1295 | −0111 | −0082 | 0333 | 4544 | 0109 | 0082 | 0329 |
| 1361 | −0111 | −0077 | 0339 | 4609 | 0110 | 0080 | 0334 |
| 1427 | −0116 | −0079 | 0347 | 4675 | 0112 | 0076 | 0341 |
| 1495 | −0120 | −0077 | 0357 | 4741 | 0118 | 0081 | 0349 |
| 1566 | −0125 | −0072 | 0371 | 4810 | 0120 | 0074 | 0360 |
| 1643 | −0139 | −0077 | 0396 | 4881 | 0130 | 0081 | 0376 |
| 1722 | 0140 | 0075 | 0401 | 4963 | 0141 | 0068 | 0406 |
| 1798 | −0135 | −0072 | 0391 | 5035 | −0133 | −0081 | 0381 |
| 1882 | 0146 | 0074 | 0413 | 5124 | −0151 | −0076 | 0422 |
| 1952 | −0127 | −0078 | 0370 | 5192 | 0119 | 0075 | 0358 |
| 2029 | −0135 | −0070 | 0393 | 5262 | 0129 | 0083 | 0372 |
| 2110 | 0146 | 0083 | 0408 | 5340 | 0140 | 0079 | 0397 |
| 2183 | −0128 | −0070 | 0379 | 5419 | −0137 | −0067 | 0399 |
| 2268 | −0144 | −0066 | 0415 | 5495 | 0140 | 0081 | 0394 |
| 2337 | 0126 | 0079 | 0368 | 5576 | −0141 | −0072 | 0405 |
| 2412 | 0137 | 0081 | 0389 | 5649 | 0135 | 0082 | 0384 |
| 2501 | −0153 | −0080 | 0423 | 5752 | 0160 | 0074 | 0443 |
| 2569 | 0119 | 0076 | 0357 | 5816 | −0110 | −0081 | 0332 |
| 2639 | 0129 | 0082 | 0371 | 5881 | −0113 | −0082 | 0339 |
| 2717 | 0136 | 0070 | 0395 | 5948 | −0114 | −0075 | 0346 |
| 2796 | −0143 | −0079 | 0402 | | | | |
| 2871 | 0134 | 0076 | 0388 | | | | |
| 2962 | −0152 | −0072 | 0426 | | | | |

puting machine. Approximately one second per iteration, aside from output time, is required.

For initial conditions we have chosen a slight departure from the state of no convection, namely (0,1,0).
Table 1 has been prepared by the computer. It gives the
values of $N$ (the number of iterations), $X$, $Y$, and $Z$ at
every fifth iteration for the first 160 iterations. In the
printed output (but not in the computations) the values
of $X$, $Y$, and $Z$ are multiplied by ten, and then only
those figures to the left of the decimal point are printed.
Thus the states of steady convection would appear as
0084, 0084, 0270 and $-0084$, $-0084$, 0270, while the
state of no convection would appear as 0000, 0000, 0000.

The initial instability of the state of rest is evident. All
three variables grow rapidly, as the sinking cold fluid
is replaced by even colder fluid from above, and the
rising warm fluid by warmer fluid from below, so that by
step 35 the strength of the convection far exceeds that
of steady convection. Then $Y$ diminishes as the warm
fluid is carried over the top of the convective cells, so
that by step 50, when $X$ and $Y$ have opposite signs,
warm fluid is descending and cold fluid is ascending. The
motion thereupon ceases and reverses its direction, as
indicated by the negative values of $X$ following step 60.
By step 85 the system has reached a state not far from
that of steady convection. Between steps 85 and 150 it
executes a complete oscillation in its intensity, the
slight amplification being almost indetectable.

The subsequent behavior of the system is illustrated
in Fig. 1, which shows the behavior of $Y$ for the first
3000 iterations. After reaching its early peak near step
35 and then approaching equilibrium near step 85, it
undergoes systematic amplified oscillations until near
step 1650. At this point a critical state is reached, and
thereafter $Y$ changes sign at seemingly irregular intervals, reaching sometimes one, sometimes two, and sometimes three or more extremes of one sign before changing
sign again.

Fig. 2 shows the projections on the $X$-$Y$- and $Y$-$Z$-
planes in phase space of the portion of the trajectory
corresponding to iterations 1400–1900. The states of
steady convection are denoted by $C$ and $C'$. The first
portion of the trajectory spirals outward from the
vicinity of $C'$, as the oscillations about the state of
steady convection, which have been occurring since step
85, continue to grow. Eventually, near step 1650, it
crosses the $X$-$Z$-plane, and is then deflected toward the
neighborhood of $C$. It temporarily spirals about $C$, but
crosses the $X$-$Z$-plane after one circuit, and returns to
the neighborhood of $C'$, where it soon joins the spiral
over which it has previously traveled. Thereafter it
crosses from one spiral to the other at irregular intervals.

Fig. 3, in which the coordinates are $Y$ and $Z$, is based
upon the printed values of $X$, $Y$, and $Z$ at every fifth
iteration for the first 6000 iterations. These values determine $X$ as a smooth single-valued function of $Y$ and $Z$
over much of the range of $Y$ and $Z$; they determine $X$

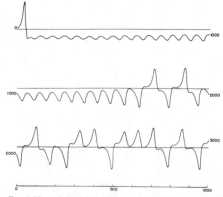

FIG. 1. Numerical solution of the convection equations. Graph
of $Y$ as a function of time for the first 1000 iterations (upper
curve), second 1000 iterations (middle curve), and third 1000
iterations (lower curve).

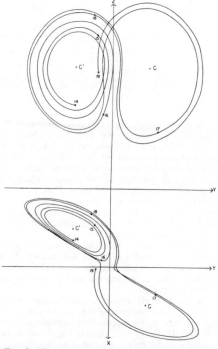

FIG. 2. Numerical solution of the convection equations.
Projections on the $X$-$Y$-plane and the $Y$-$Z$-plane in phase space
of the segment of the trajectory extending from iteration 1400 to
iteration 1900. Numerals "14," "15," etc., denote positions at
iterations 1400, 1500, etc. States of steady convection are denoted
by $C$ and $C'$.

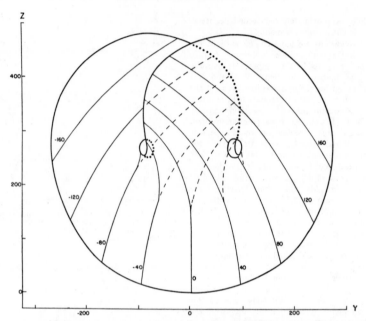

FIG. 3. Isopleths of $X$ as a function of $Y$ and $Z$ (thin solid curves), and isopleths of the lower of two values of $X$, where two values occur (dashed curves), for approximate surfaces formed by all points on limiting trajectories. Heavy solid curve, and extensions as dotted curves, indicate natural boundaries of surfaces.

as one of two smooth single-valued functions over the remainder of the range. In Fig. 3 the thin solid lines are isopleths of $X$, and where two values of $X$ exist, the dashed lines are isopleths of the lower value. Thus, within the limits of accuracy of the printed values, the trajectory is confined to a pair of surfaces which appear to merge in the lower portion of Fig. 3. The spiral about $C$ lies in the upper surface, while the spiral about $C'$ lies in the lower surface. Thus it is possible for the trajectory to pass back and forth from one spiral to the other without intersecting itself.

Additional numerical solutions indicate that other trajectories, originating at points well removed from these surfaces, soon meet these surfaces. The surfaces therefore appear to be composed of all points lying on limiting trajectories.

Because the origin represents a steady state, no trajectory can pass through it. However, two trajectories emanate from it, i.e., approach it asymptotically as $\tau \rightarrow -\infty$. The heavy solid curve in Fig. 3, and its extensions as dotted curves, are formed by these two trajectories. Trajectories passing close to the origin will tend to follow the heavy curve, but will not cross it, so that the heavy curve forms a natural boundary to the region which a trajectory can ultimately occupy. The

holes near $C$ and $C'$ also represent regions which cannot be occupied after they have once been abandoned.

Returning to Fig. 2, we find that the trajectory apparently leaves one spiral only after exceeding some critical distance from the center. Moreover, the extent to which this distance is exceeded appears to determine the point at which the next spiral is entered; this in turn seems to determine the number of circuits to be executed before changing spirals again.

It therefore seems that some single feature of a given circuit should predict the same feature of the following circuit. A suitable feature of this sort is the maximum value of $Z$, which occurs when a circuit is nearly completed. Table 2 has again been prepared by the computer, and shows the values of $X$, $Y$, and $Z$ at only those iterations $N$ for which $Z$ has a relative maximum. The succession of circuits about $C$ and $C'$ is indicated by the succession of positive and negative values of $X$ and $Y$. Evidently $X$ and $Y$ change signs following a maximum which exceeds some critical value printed as about 385.

Fig. 4 has been prepared from Table 2. The abscissa is $M_n$, the value of the $n$th maximum of $Z$, while the ordinate is $M_{n+1}$, the value of the following maximum. Each point represents a pair of successive values of $Z$ taken from Table 2. Within the limits of the round-off in tabulating $Z$, there is a precise two-to-one relation between $M_n$ and $M_{n+1}$. The initial maximum $M_1 = 483$ is shown as if it had followed a maximum $M_0 = 385$, since maxima near 385 are followed by close approaches to the origin, and then by exceptionally large maxima.

It follows that an investigator, unaware of the nature of the governing equations, could formulate an empirical prediction scheme from the "data" pictured in Figs. 2 and 4. From the value of the most recent maximum of $Z$, values at future maxima may be obtained by repeated applications of Fig. 4. Values of $X$, $Y$, and $Z$ between maxima of $Z$ may be found from Fig. 2, by interpolating between neighboring curves. Of course, the accuracy of predictions made by this method is limited by the exactness of Figs. 2 and 4, and, as we shall see, by the accuracy with which the initial values of $X$, $Y$, and $Z$ are observed.

Some of the implications of Fig. 4 are revealed by considering an idealized two-to-one correspondence between successive members of sequences $M_0$, $M_1$, $\cdots$, consisting of numbers between zero and one. These sequences satisfy the relations

$$
\begin{aligned}
M_{n+1} &= 2M_n & &\text{if } M_n < \tfrac{1}{2} \\
M_{n+1} &\text{ is undefined} & &\text{if } M_n = \tfrac{1}{2} \quad (35) \\
M_{n+1} &= 2 - 2M_n & &\text{if } M_n > \tfrac{1}{2}.
\end{aligned}
$$

The correspondence defined by (35) is shown in Fig. 5, which is an idealization of Fig. 4. It follows from repeated applications of (35) that in any particular sequence,

$$M_n = m_n \pm 2^n M_0, \qquad (36)$$

where $m_n$ is an even integer.

Consider first a sequence where $M_0 = u/2^p$, where $u$ is odd. In this case $M_{p-1} = \tfrac{1}{2}$, and the sequence terminates. These sequences form a denumerable set, and correspond to the trajectories which score direct hits upon the state of no convection.

Next consider a sequence where $M_0 = u/2^p v$, where $u$ and $v$ are relatively prime odd numbers. Then if $k > 0$, $M_{p+1+k} = u_k/v$, where $u_k$ and $v$ are relatively prime and $u_k$ is even. Since for any $v$ the number of proper fractions $u_k/v$ is finite, repetitions must occur, and the sequence is periodic. These sequences also form a denumerable set, and correspond to periodic trajectories.

The periodic sequences having a given number of distinct values, or phases, are readily tabulated. In particular there are a single one-phase, a single two-phase, and two three-phase sequences, namely,

$$2/3, \cdots,$$
$$2/5, 4/5, \cdots,$$
$$2/7, 4/7, 6/7, \cdots,$$
$$2/9, 4/9, 8/9, \cdots.$$

The two three-phase sequences differ qualitatively in that the former possesses two numbers, and the latter only one number, exceeding $\tfrac{1}{2}$. Thus the trajectory corresponding to the former makes two circuits about $C$, followed by one about $C'$ (or vice versa). The trajectory corresponding to the latter makes three circuits about $C$, followed by three about $C'$, so that actually only $Z$ varies in three phases, while $X$ and $Y$ vary in six.

Now consider a sequence where $M_0$ is not a rational fraction. In this case (36) shows that $M_{n+k}$ cannot equal

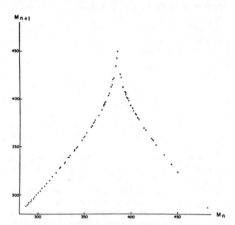

Fig. 4. Corresponding values of relative maximum of $Z$ (abscissa) and subsequent relative maximum of $Z$ (ordinate) occurring during the first 6000 iterations.

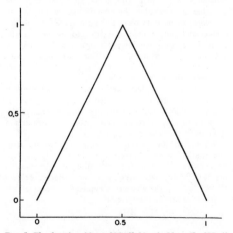

Fig. 5. The function $M_{n+1} = 2M_n$ if $M_n < \tfrac{1}{2}$, $M_{n+1} = 2 - 2M_n$ if $M_n > \tfrac{1}{2}$, serving as an idealization of the locus of points in Fig. 4.

$M_n$ if $k>0$, so that no repetitions occur. These sequences, which form a nondenumerable set, may conceivably approach periodic sequences asymptotically and be quasi-periodic, or they may be nonperiodic.

Finally, consider two sequences $M_0$, $M_1$, $\cdots$ and $M_0'$, $M_1'$, $\cdots$, where $M_0'=M_0+\epsilon$. Then for a given $k$, if $\epsilon$ is sufficiently small, $M_k'=M_k\pm2^k\epsilon$. All sequences are therefore unstable with respect to small modifications. In particular, all periodic sequences are unstable, and no other sequences can approach them asymptotically. All sequences except a set of measure zero are therefore nonperiodic, and correspond to nonperiodic trajectories.

Returning to Fig. 4, we see that periodic sequences analogous to those tabulated above can be found. They are given approximately by

$$398, \cdots,$$
$$377, 410, \cdots,$$
$$369, 391, 414, \cdots,$$
$$362, 380, 419, \cdots.$$

The trajectories possessing these or other periodic sequences of maxima are presumably periodic or quasi-periodic themselves.

The above sequences are temporarily approached in the numerical solution by sequences beginning at iterations 5340, 4881, 3625, and 3926. Since the numerical solution eventually departs from each of these sequences, each is presumably unstable.

More generally, if $M_n'=M_n+\epsilon$, and if $\epsilon$ is sufficiently small, $M_{n+k}'=M_{n+k}+\Lambda\epsilon$, where $\Lambda$ is the product of the slopes of the curve in Fig. 4 at the points whose abscissas are $M_n$, $\cdots$, $M_{n+k-1}$. Since the curve apparently has a slope whose magnitude exceeds unity everywhere, all sequences of maxima, and hence all trajectories, are unstable. In particular, the periodic trajectories, whose sequences of maxima form a denumerable set, are unstable, and only exceptional trajectories, having the same sequences of maxima, can approach them asymptotically. The remaining trajectories, whose sequences of maxima form a nondenumerable set, therefore represent deterministic nonperiodic flow.

These conclusions have been based upon a finite segment of a numerically determined solution. They cannot be regarded as mathematically proven, even though the evidence for them is strong. One apparent contradiction requires further examination.

It is difficult to reconcile the merging of two surfaces, one containing each spiral, with the inability of two trajectories to merge. It is not difficult, however, to explain the *apparent* merging of the surfaces. At two times $\tau_0$ and $\tau_1$, the volumes occupied by a specified set of particles satisfy the relation

$$V_0(\tau_1)=e^{-(\sigma+b+1)(\tau_1-\tau_0)}V_0(\tau_0), \tag{37}$$

according to (30). A typical circuit about $C$ or $C'$ requires about 70 iterations, so that, for such a circuit,

$\tau_2=\tau_1+0.7$, and, since $\sigma+b+1=41/3$,

$$V_0(\tau_1)=0.00007\,V_0(\tau_0). \tag{38}$$

Two particles separated from each other in a suitable direction can therefore come together very rapidly, and appear to merge.

It would seem, then, that the two surfaces merely appear to merge, and remain distinct surfaces. Following these surfaces along a path parallel to a trajectory, and circling $C$ or $C'$, we see that each surface is really a pair of surfaces, so that, where they appear to merge, there are really four surfaces. Continuing this process for another circuit, we see that there are really eight surfaces, etc., and we finally conclude that there is an infinite complex of surfaces, each extremely close to one or the other of two merging surfaces.

The infinite set of values at which a line parallel to the $X$-axis intersects these surfaces may be likened to the set of all numbers between zero and one whose decimal expansions (or some other expansions besides binary) contain only zeros and ones. This set is plainly nondenumerable, in view of its correspondence to the set of all numbers between zero and one, expressed in binary. Nevertheless it forms a set of measure zero. The sequence of ones and zeros corresponding to a particular surface contains a history of the trajectories lying in that surface, a one or zero immediately to the right of the decimal point indicating that the last circuit was about $C$ or $C'$, respectively, a one or zero in second place giving the same information about the next to the last circuit, etc. Repeating decimal expansions represent periodic or quasi-periodic trajectories, and, since they define rational fractions, they form a denumerable set.

If one first visualizes this infinite complex of surfaces, it should not be difficult to picture nonperiodic deterministic trajectories embedded in these surfaces.

## 8. Conclusion

Certain mechanically or thermally forced nonconservative hydrodynamical systems may exhibit either periodic or irregular behavior when there is no obviously related periodicity or irregularity in the forcing process. Both periodic and nonperiodic flow are observed in some experimental models when the forcing process is held constant, within the limits of experimental control. Some finite systems of ordinary differential equations designed to represent these hydrodynamical systems possess periodic analytic solutions when the forcing is strictly constant. Other such systems have yielded nonperiodic numerical solutions.

A finite system of ordinary differential equations representing forced dissipative flow often has the property that all of its solutions are ultimately confined within the same bounds. We have studied in detail the properties of solutions of systems of this sort. Our principal results concern the instability of nonperiodic solutions. A nonperiodic solution with no transient component must be unstable, in the sense that solutions temporarily approximating it do not continue to do so. A nonperiodic solution with a transient component is sometimes stable, but in this case its stability is one of its transient properties, which tends to die out.

To verify the existence of deterministic nonperiodic flow, we have obtained numerical solutions of a system of three ordinary differential equations designed to represent a convective process. These equations possess three steady-state solutions and a denumerably infinite set of periodic solutions. All solutions, and in particular the periodic solutions, are found to be unstable. The remaining solutions therefore cannot in general approach the periodic solutions asymptotically, and so are nonperiodic.

When our results concerning the instability of nonperiodic flow are applied to the atmosphere, which is ostensibly nonperiodic, they indicate that prediction of the sufficiently distant future is impossible by any method, unless the present conditions are known exactly. In view of the inevitable inaccuracy and incompleteness of weather observations, precise very-long-range forecasting would seem to be non-existent.

There remains the question as to whether our results really apply to the atmosphere. One does not usually regard the atmosphere as either deterministic or finite, and the lack of periodicity is not a mathematical certainty, since the atmosphere has not been observed forever.

The foundation of our principal result is the eventual necessity for any bounded system of finite dimensionality to come arbitrarily close to acquiring a state which it has previously assumed. If the system is stable, its future development will then remain arbitrarily close to its past history, and it will be quasi-periodic.

In the case of the atmosphere, the crucial point is then whether analogues must have occurred since the state of the atmosphere was first observed. By analogues, we mean specifically two or more states of the atmosphere, together with its environment, which resemble each other so closely that the differences may be ascribed to errors in observation. Thus, to be analogues, two states must be closely alike in regions where observations are accurate and plentiful, while they need not be at all alike in regions where there are no observations at all, whether these be regions of the atmosphere or the environment. If, however, some unobserved features are implicit in a succession of observed states, two successions of states must be nearly alike in order to be analogues.

If it is true that two analogues have occurred since atmospheric observation first began, it follows, since the atmosphere has not been observed to be periodic, that the successions of states following these analogues must eventually have differed, and no forecasting scheme could have given correct results both times. If, instead,

analogues have not occurred during this period, some accurate very-long-range prediction scheme, using observations at present available, may exist. But, if it does exist, the atmosphere will acquire a quasi-periodic behavior, never to be lost, once an analogue occurs. This quasi-periodic behavior need not be established, though, even if very-long-range forecasting is feasible, if the variety of possible atmospheric states is so immense that analogues need never occur. It should be noted that these conclusions do not depend upon whether or not the atmosphere is deterministic.

There remains the very important question as to how long is "very-long-range." Our results do not give the answer for the atmosphere; conceivably it could be a few days or a few centuries. In an idealized system, whether it be the simple convective model described here, or a complicated system designed to resemble the atmosphere as closely as possible, the answer may be obtained by comparing pairs of numerical solutions having nearly identical initial conditions. In the case of the real atmosphere, if all other methods fail, we can wait for an analogue.

*Acknowledgments.* The writer is indebted to Dr. Barry Saltzman for bringing to his attention the existence of nonperiodic solutions of the convection equations. Special thanks are due to Miss Ellen Fetter for handling the many numerical computations and preparing the graphical presentations of the numerical material.

#### REFERENCES

Birkhoff, G. O., 1927: *Dynamical systems.* New York, Amer. Math. Soc., Colloq. Publ., 295 pp.
Ford, L. R., 1933: *Differential equations.* New York, McGraw-Hill, 264 pp.
Fultz, D., R. R. Long, G. V. Owens, W. Bohan, R. Kaylor and J. Weil, 1959: Studies of thermal convection in a rotating cylinder with some implications for large-scale atmospheric motions. *Meteor. Monog,* 4(21), Amer. Meteor. Soc., 104 pp.
Gibbs, J. W., 1902: *Elementary principles in statistical mechanics.* New York, Scribner, 207 pp.
Hide, R., 1958: An experimental study of thermal convection in a rotating liquid. *Phil. Trans. Roy. Soc. London,* (A), 250, 441–478.
Lorenz, E. N., 1960: Maximum simplification of the dynamic equations. *Tellus,* 12, 243–254.
——, 1962a: Simplified dynamic equations applied to the rotating-basin experiments. *J. atmos. Sci.,* 19, 39–51.
——, 1962b: The statistical prediction of solutions of dynamic equations. *Proc. Internat. Symposium Numerical Weather Prediction,* Tokyo, 629–635.
Nemytskii, V. V., and V. V. Stepanov, 1960: *Qualitative theory of differential equations.* Princeton, Princeton Univ. Press, 523 pp.
Poincaré, H., 1881: Mémoire sur les courbes définies par une équation différentielle. *J. de Math.,* 7, 375–442.
Rayleigh, Lord, 1916: On convective currents in a horizontal layer of fluid when the higher temperature is on the under side. *Phil. Mag.,* 32, 529–546.
Saltzman, B., 1962: Finite amplitude free convection as an initial value problem—I. *J. atmos. Sci.,* 19, 329–341.

# Chaos, Strange Attractors, and Fractal Basin Boundaries in Nonlinear Dynamics

CELSO GREBOGI, EDWARD OTT, JAMES A. YORKE

Recently research has shown that many simple nonlinear deterministic systems can behave in an apparently unpredictable and chaotic manner. This realization has broad implications for many fields of science. Basic developments in the field of chaotic dynamics of dissipative systems are reviewed in this article. Topics covered include strange attractors, how chaos comes about with variation of a system parameter, universality, fractal basin boundaries and their effect on predictability, and applications to physical systems.

I N THIS ARTICLE WE PRESENT A REVIEW OF THE FIELD OF chaotic dynamics of dissipative systems including recent developments. The existence of chaotic dynamics has been discussed in the mathematical literature for many decades with important contributions by Poincaré, Birkhoff, Cartwright and Littlewood, Levinson, Smale, and Kolmogorov and his students, among others. Nevertheless, it is only recently that the wide-ranging impact of chaos has been recognized. Consequently, the field is now undergoing explosive growth, and many applications have been made across a broad spectrum of scientific disciplines—ecology, economics, physics, chemistry, engineering, fluid mechanics, to name several. Specific examples of chaotic time dependence include convection of a fluid heated from below, simple models for the yearly variation of insect populations, stirred chemical reactor systems, and the determination of limits on the length of reliable weather forecasting. It is our belief that the number of these applications will continue to grow.

We start with some basic definitions of terms used in the rest of the article.

*Dissipative system.* In Hamiltonian (conservative) systems such as arise in Newtonian mechanics of particles (without friction), phase space volumes are preserved by the time evolution. (The phase space is the space of variables that specify the state of the system.) Consider, for example, a two-dimensional phase space $(q, p)$, where $q$ denotes a position variable and $p$ a momentum variable. Hamilton's equations of motion take the set of initial conditions at time $t = t_0$ and evolve them in time to the set at time $t = t_1$. Although the shapes of the sets are different, their areas are the same. By a dissipative system we mean one that does not have this property (and cannot be made to have this property by a change of variables). Areas should typically decrease (dissipate) in time so that the area of

the final set would be less than the area of the initial set. As a consequence of this, dissipative systems typically are characterized by the presence of attractors.

*Attractor.* If one considers a system and its phase space, then the initial conditions may be attracted to some subset of the phase space (the attractor) as time $t \to \infty$. For example, for a damped harmonic oscillator (Fig. 1a) the attractor is the point at rest (in this case the origin). For a periodically driven oscillator in its limit cycle the limit set is a closed curve in the phase space (Fig. 1b).

*Strange attractor.* In the above two examples, the attractors were a point (Fig. 1a), which is a set of dimension zero, and a closed curve (Fig. 1b), which is a set of dimension one. For many other attractors the attracting set can be much more irregular (some would say pathological) and, in fact, can have a dimension that is not an integer. Such sets have been called "fractal" and, when they are attractors, they are called strange attractors. [For a more precise definition see (1).] The existence of a strange attractor in a physically interesting model was first demonstrated by Lorenz (2).

*Dimension.* There are many definitions of the dimension $d$ (3). The simplest is called the box-counting or capacity dimension and is defined as follows:

$$d = \lim_{\epsilon \to 0} \frac{\ln N(\epsilon)}{\ln(1/\epsilon)} \qquad (1)$$

where we imagine the attracting set in the phase space to be covered by small $D$-dimensional cubes of edge length $\epsilon$, with $D$ the dimension of the phase space. $N(\epsilon)$ is the minimum number of such cubes needed to cover the set. For example, for a point attractor (Fig. 1a), $N(\epsilon) = 1$ independent of $\epsilon$, and Eq. 1 yields $d = 0$ (as it should). For a limit cycle attractor, as in Fig. 1b, we have that $N(\epsilon) \sim \ell/\epsilon$, where $\ell$ is the length of the closed curve in the figure (dotted line); hence, for this case, $d = 1$, by Eq. 1. A less trivial example is illustrated in Fig. 2, in the form of a Cantor set. This set is

C. Grebogi is a research scientist at the Laboratory for Plasma and Fusion Energy Studies, E. Ott is a professor in the departments of electrical engineering and physics, and J. A. Yorke is a professor of mathematics and is the director of the Institute for Physical Science and Technology, University of Maryland, College Park, MD 20742.

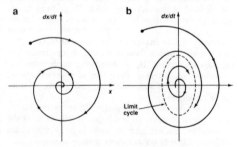

**Fig. 1.** (a) Phase-space diagram for a damped harmonic oscillator. (b) Phase-space diagram for a system that is approaching a limit cycle.

**Fig. 2.** Construction of a Cantor set.

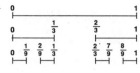

**Fig. 3.** Poincaré surface of section.

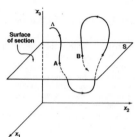

**Fig. 4.** The Hénon chaotic attractor. (**a**) Full set. (**b**) Enlargement of region defined by the rectangle in (a). (**c**) Enlargement of region defined by the rectangle in (b).

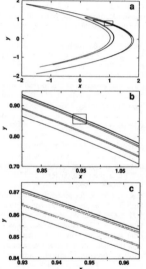

an arbitrarily fine-scaled interwoven structure of regions where orbit trajectories are dense and sparse. Such attractors have been called multifractals and can be characterized by subsidiary quantities that essentially give the dimensions of the dense and sparse regions of the attractor. In this review we shall not attempt to survey this work. Several papers provide an introduction to recent work on the dimension of chaotic attractors (3–5).

*Chaotic attractor.* By this term we mean that if we take two typical points on the attractor that are separated from each other by a small distance $\Delta(0)$ at $t = 0$, then for increasing $t$ they move apart exponentially fast. That is, in some average sense $\Delta(t) \sim \Delta(0)\exp(ht)$ with $h > 0$ (where $h$ is called the Lyapunov exponent). Thus a small uncertainty in the initial state of the system rapidly leads to inability to forecast its future. [It is not surprising, therefore, that the pioneering work of Lorenz (2) was in the context of meteorology.] It is typically the case that strange attractors are also chaotic [although this is not always so; see (1, 6)].

*Dynamical system.* This is a system of equations that allows one, in principle, to predict the future given the past. One example is a system of first-order ordinary differential equations in time, $dx(t)/dt = G(x,t)$, where $x(t)$ is a $D$-dimensional vector and $G$ is a $D$-dimensional vector function of $x$ and $t$. Another example is a map.

*Map.* A map is an equation of the form $x_{i+1} = F(x_i)$, where the "time" $t$ is discrete and integer valued. Thus, given $x_0$, the map gives $x_1$. Given $x_1$, the map gives $x_2$, and so on. Maps can arise in continuous time physical systems in the form of a Poincaré surface of section. Figure 3 illustrates this. The plane $x_3 = $ constant is the surface of section (S in the figure), and $\Lambda$ denotes a trajectory of the system. Every time $\Lambda$ pierces S going downward (as at points A and B in the figure), we record the coordinates $(x_1,x_2)$. Clearly the coordinates of A uniquely determine those of B. Thus there exists a map, $B = F(A)$, and this map (if we knew it) could be iterated to find all subsequent piercings of S.

## Chaotic Attractors

As an example of a strange attractor consider the map first studied by Hénon (7):

$$x_{n+1} = \alpha - x_n^2 + \beta y_n \quad (2)$$
$$y_{n+1} = x_n \quad (3)$$

Figure 4a shows the result of plotting $10^4$ successive points obtained by iterating Eqs. 2 and 3 with parameters $\alpha = 1.4$ and $\beta = 0.3$ (and the initial transient is deleted). The result is essentially a picture of the chaotic attractor. Figure 4, b and c, shows successive enlargements of the small square in the preceding figure. Scale invariant, Cantor set–like structure transverse to the linear structure is evident. This suggests that we may regard the attractor in Fig. 4c, for example, as being essentially a Cantor set of approximately straight parallel lines. In fact, the dimension $d$ in Eq. 1 can be estimated numerically (8) to be $d \cong 1.26$ so that the attractor is strange.

As another example consider a forced damped pendulum described by the equation

$$d^2\theta/dt^2 + vd\theta/dt + \omega_0^2\sin\theta = f\cos(\omega t) \quad (4)$$

where $\theta$ is the angle between the pendulum arm and the rest position, $v$ is the coefficient of friction, $\omega_0$ is the frequency of natural oscillation, and $f$ is the strength of the forcing. In Eq. 4, the first term represents the inertia of the pendulum, the second term represents friction at the pivot, the third represents the gravitational force, and the right side represents an external sinusoidally varying torque of strength $f$ and frequency $\omega$ applied to the pendulum at the pivot. In Fig. 5a, we plot the Poincaré surface of section of a strange

formed by taking the line interval from 0 to 1, dividing it in thirds, then discarding the middle third, then dividing the two remaining thirds into thirds and discarding their middle thirds, and so on ad infinitum. The Cantor set is the closed set of points that are left in the limit of this repeated process. If we take $\epsilon = 3^{-n}$ with $n$ an integer, then we see that $N(\epsilon) = 2^n$ and Eq. 1 (in which $\epsilon \to 0$ corresponds to $n \to \infty$) yields $d = (\ln 2)/(\ln 3)$, a number between 0 and 1, hence, a fractal. The topic of the dimension of strange attractors is a large subject on which much research has been done. One of the most interesting aspects concerning dimension arises from the fact that the distribution of points on a chaotic attractor can be nonuniform in a very singular way. In particular, there can be

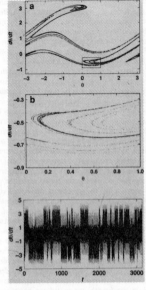

**Fig. 5.** (a) Poincaré surface of section of a pendulum strange attractor. (b) Enlargement of region defined by rectangle in (a).

**Fig. 6.** Chaotic time series for pendulum shown as a plot of angular velocity versus time.

attractor for the pendulum, where we choose $\nu = 0.22$, $\omega_0 = 1.0$, $\omega = 1.0$, and $f = 2.7$ in Eq. 4. This surface of section is obtained by plotting 50,000 dots, one dot for every cycle of the forcing term, that is, one dot at every time $t = t_n = 2\pi n$ (where $n$ is an integer). The strange attractor shown in Fig. 5a exhibits a Cantor set–like structure transverse to the linear structure. This is evident in Fig. 5b, which shows an enlargement of the square region in Fig. 5a. The dimension of this strange attractor in the surface of section is $d \cong 1.38$. Figure 6 shows the angular velocity $d\theta/dt$ as a function of $t$ for the parameters of Fig. 5. Note the apparently erratic nature of this plot.

In general, the form of chaotic attractors varies greatly from system to system and even within the same system. This is indicated by the sequence of chaotic attractors shown in Fig. 7. All of these attractors were generated from the same map $(9)$,

$$\psi_{n+1} = [\psi_n + \omega_1 + \epsilon P_1(\psi_n, \theta_n)] \bmod 1 \qquad (5)$$

$$\theta_{n+1} = [\theta_n + \omega_2 + \epsilon P_2(\psi_n, \theta_n)] \bmod 1 \qquad (6)$$

where $P_1$ and $P_2$ are periodic with period one in both their arguments. The $P_1$ and $P_2$ are the same in all of the cases shown in Fig. 7; only the parameters $\omega_1$, $\omega_2$, and $\epsilon$ have been varied. The results show the great variety of form and structure possible in chaotic attractors as well as their aesthetic appeal. Since $\psi$ and $\theta$ may be regarded as angles, Eqs. 5 and 6 are a map on a two-dimensional toroidal surface. [This map is used in $(9)$ to study the transition from quasiperiodicity to chaos.]

Because of the exponential divergence of nearby orbits on chaotic attractors, there is a question as to how much of the structure in these pictures of chaotic attractors (Figs. 4, 5, and 7) is an artifact due to chaos-amplified roundoff error. Although a numerical trajectory will diverge rapidly from the true trajectory with the same initial point, it has been demonstrated rigorously $(10)$ in important cases [including the Hénon map $(11)$] that there exists a true

trajectory with a slightly different initial point that stays near the noisy trajectory for a long time. [For example, for the Hénon map for a typical numerical trajectory computed with 14-digit precision there exists a true trajectory that stays within $10^{-7}$ of the numerical trajectory for $10^7$ iterates $(11)$.] Thus we believe that the apparently fractal structure seen in pictures such as Figs. 4, 5, and 7 is real.

## The Evolution of Chaotic Attractors

In dissipative dynamics it is common to find that for some value of a system parameter only a nonchaotic attracting orbit (a limit cycle, for example) occurs, whereas at some other value of the parameter a chaotic attractor occurs. It is therefore natural to ask how the one comes about from the other as the system parameter is varied continuously. This is a fundamental question that has elicited a great deal of attention $(9, 12–19)$.

To understand the nature of this question and some of the possible answers to it, we consider Fig. 8a, the so-called bifurcation diagram for the map.

$$x_{n+1} = C - x_n^2 \qquad (7)$$

where $C$ is a constant. Figure 8a can be constructed as follows: take $C = -0.4$, set $x_0 = -0.5$, iterate the map 100 times (to eliminate transients), then plot the next 1000 values of $x$; increase $C$ by a small amount, say 0.001, and repeat what was done for $C = -0.4$; increase again, and repeat; and so on, until $C = 2.1$ is reached. We see from Fig. 8a that below a certain value, $C = C_0 = -0.25$, there is no attractor in $-2 < x < 2$. In fact, in this case all orbits go to $x \to -\infty$, hence the absence of points on the plot. This is also true for $C$ above the "crisis value" $C_c = 2.0$. Between these two values there is an attractor. As $C$ is increased we have an attracting orbit of "period one," which, at $C = 0.75$, bifurcates to a period-two attracting orbit $(x_\alpha \to x_\beta \to x_\alpha \to x_\beta \to \cdots)$, which then bifurcates (at $C = 1.25$) to a period-four orbit $(x_a \to x_b \to x_c \to x_d \to x_a \to x_b \to x_c \to x_d \to x_a \to \cdots)$. In fact, there are an infinite number of such bifurcations of period $2^n$ to period $2^{n+1}$ orbits, and these accumulate as $n \to \infty$ at a finite value of $C$, which we denote $C_\infty$ (from Fig. 8a, $C_\infty \cong 1.4$). [The practical importance of this phenomenology was emphasized early on by May $(12)$.]

What is the situation for $C_\infty < C < C_c$? Numerically what one sees is that for many $C$ values in this range the orbits appear to be chaotic, whereas for others there are periodic orbits. For example, Fig. 8b shows an enlargement of Fig. 8a for $C$ in the range $1.72 < C < 1.82$. We see what appear to be chaotic orbits below $C = C_0^{(3)} = 1.75$. However, just above this value, a period-three orbit appears, supplanting the chaos. The period-three orbit then goes through a period-doubling cascade, becomes chaotic, widens into a three-piece chaotic attractor, and then finally at $C = C_c^{(3)} \cong 1.79$ widens back into a single chaotic band. We call the region $C_0^{(3)} < C < C_c^{(3)}$ a period-three window. (Such windows, but of higher period, appear throughout the region $C_\infty < C < C_c$, but are not as discernible in Fig. 8a because they are much narrower than the period-three window.)

An infinite period-doubling cascade is one way that a chaotic attractor can come about from a nonchaotic one $(13)$. There are also two other possible routes to chaos exemplified in Fig. 8, a and b. These are the intermittency route $(14)$ and the crisis route $(15)$.

*Intermittency.* Consider Fig. 8b. For $C$ just above $C_0^{(3)}$ there is a period-three orbit. For $C$ just below $C_0^{(3)}$ there appears to be a chaotic orbit. To understand the character of this transition it is useful to examine the chaotic orbit for $C$ just below $C_0^{(3)}$. The character of this orbit is as follows: The orbit appears to be a period-three orbit for long stretches of time after which there is a short

**Fig. 7.** Sequence of chaotic attractors for system represented by Eqs. 5 and 6. Plot shows iterated mapping on a torus for different values of $\omega_1$, $\omega_2$, and $\epsilon$. (**Top**) $\omega_1 = 0.54657$, $\omega_2 = 0.36736$, and $\epsilon = 0.75$. (**Center**) $\omega_1 = 0.45922$, $\omega_2 = 0.53968$, and $\epsilon = 0.50$. (**Bottom**) $\omega_1 = 0.41500$, $\omega_2 = 0.73500$, and $\epsilon = 0.60$.

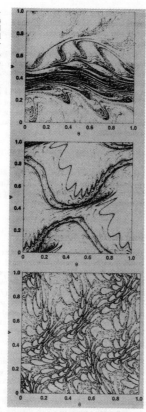

interval $-2 \leq x \leq 2$, and then rapidly begins to move to large negative $x$ values (that is, it begins to approach $x = -\infty$). This is called a chaotic transient (15). The length of a chaotic transient will depend on the particular initial condition chosen. One can define a mean transient duration by averaging over, for example, a uniform distribution of initial conditions in the interval $-2 < x < 2$. For the quadratic map, this average duration is

$$\tau \sim 1/(C - C_c)^\gamma \tag{8}$$

with the exponent $\gamma$ given by $\gamma = 1/2$. Thus as $C$ approaches $C_c$ from above, the lifetime of a chaotic transient goes to infinity and the transient is converted to a chaotic attractor for $C < C_c$. Again, this type of phenomenon occurs widely in chaotic systems. For example, the model of Lorenz (2) for the nonlinear evolution of the Rayleigh-Bénard instability of a fluid subjected to gravity and heated from below has a chaotic onset of the crisis type and an accompanying chaotic transient. In that case, $\gamma$ in Eq. 8 is $\gamma \sim 4$ (20). In addition, a theory for determining the exponent $\gamma$ for two-dimensional maps and systems such as the forced damped pendulum has recently been published (21). Thus we have seen that the period doubling, intermittency, and crisis routes to chaos are illustrated by the simple quadratic map (Eq. 7).

We emphasize that, although a map was used for illustrating these routes, all of these phenomena are present in continuous-time systems and have been observed in experiments. As an example of chaotic transitions in a continuous time system, we consider the set of three autonomous ordinary differential equations studied by Lorenz (2) as a model of the Rayleigh-Bénard instability,

$$dx/dt = Py - Px \tag{9}$$

$$dy/dt = -xz + rx - y \tag{10}$$

$$dz/dt = xy - bz \tag{11}$$

where $P$ and $b$ are adjustable parameters. Fixing $P = 10$ and $b = 8/3$ and varying the remaining parameter, $r$, we obtain numerical solutions that are clear examples of the intermittency and crisis types of chaotic transitions discussed above. We illustrate these in Fig. 9, a through d; the behavior of this system is as follows:

1) For $r$ between 166.0 and 166.2 there is an intermittency transition from a periodic attractor ($r = 166.0$, Fig. 9a) to a chaotic attractor ($r = 166.2$, Fig. 9b) with intermittent turbulent bursts. Between the bursts there are long stretches of time for which the orbit oscillates in nearly the same way as for the periodic attractor (14) (Fig. 9a).

2) For a range of $r$ values below $r = 24.06$ there are two periodic attractors, that represent clockwise and counterclockwise convections. For $r$ slightly above 24.06, however, there are three attractors, one that is chaotic (shown in the phase space trajectory in Fig. 9c), whereas the other two attractors are the previously mentioned periodic attractors. The chaotic attractor comes into existence as $r$ increases through $r = 24.06$ by conversion of a chaotic transient. Figure 9d shows an orbit in phase space executing a chaotic transient before settling down to its final resting place at one of the periodic attractors. Note the similarity of the chaotic transient trajectory in Fig. 9d with the chaotic trajectory in Fig. 9c.

The various routes to chaos have also received exhaustive experimental support. For instance, period-doubling cascades have been observed in the Rayleigh-Bénard convection (22, 23), in nonlinear circuits (24), and in lasers (25); intermittency has been observed in the Rayleigh-Bénard convection (26) and in the Belousov-Zhabotinsky reaction (27); and crises have been observed in nonlinear circuits (28–30), in the Josephson junction (31), and in lasers (32).

Finally, we note that period doubling, intermittency, and crises do not exhaust the possible list of routes to chaos. (Indeed, the

burst (the "intermittent burst") of chaotic-like behavior, followed by another long stretch of almost period-three behavior, followed by a chaotic burst, and so on. As $C$ approaches $C_0^{(3)}$ from below, the average duration of the long stretches between the intermittent bursts becomes longer and longer (14), approaching infinity and proportional to $(C_0^{(3)} - C)^{-1/2}$ as $C \to C_0^{(3)}$. Thus the pure period-three orbit appears at $C = C_0^{(3)}$. Alternatively we may say that the attracting periodic attractor of period three is converted to a chaotic attractor as the parameter $C$ decreases through the critical value $C_0^{(3)}$. It should be emphasized that, although our illustration of the transition to chaos by way of intermittency is within the context of the period-three window of the quadratic map given by Eq. 7, this phenomenon (as well as period-doubling cascades and crises) is very general; in other systems it occurs for other periods (period one, for example) in easily observable form.

*Crises.* From Fig. 8a we see that there is a chaotic attractor for $C < C_c = 2$, but no chaotic attractor for $C > C_c$. Thus, as $C$ is lowered through $C_c$, a chaotic attractor is born. How does this occur? Note that at $C = C_c$ the chaotic orbit occupies the interval $-2 \leq x \leq 2$. If $C$ is just slightly larger than $C_c$, an orbit with initial condition in the interval $-2 < x < 2$ will typically follow a chaotic-like path for a finite time, after which it finds its way out of the

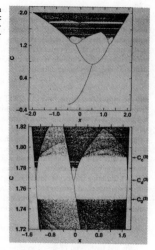

**Fig. 8.** (**Top**) Bifurcation diagram for the quadratic map. (**Bottom**) Period-three window for the quadratic map.

routes are not all known.) In particular, chaotic onsets involving quasiperiodicity have not been discussed here (9, 16, 18).

## Universality

Universality refers to the fact that systems behave in certain quantitative ways that depend not on the detailed physics or model description but rather only on some general properties of the system. Universality has been examined by renormalization group (33) techniques developed for the study of critical phenomena in condensed matter physics. Feigenbaum (13) was the first to apply these ideas, and he has extensively developed them, particularly for period doubling for dissipative systems. [See (17) for a collection of papers on universality in nonlinear dynamics.] For period doubling in dissipative systems, results have been obtained on the scaling behavior of power spectra for time series of the dynamical process (34), on the effect of noise on period doubling (35), and on the dependence of the Lyapunov exponent (36) on a system parameter. Applications of the renormalization group have also been made to intermittency (19, 37), and the breakdown of quasiperiodicity in dissipative (18) and conservative (38) systems.

As examples, two "universal" results can be stated within the context of the bifurcation diagrams (Fig. 8, a and b). Let $C_n$ denote the value of $C$ at which a period $2^n$ cycle period doubles to become a period $2^{n+1}$ cycle. Then, for the bifurcation diagram in Fig. 8a, one obtains

$$\lim_{n \to \infty} \frac{C_n - C_{n-1}}{C_{n+1} - C_n} = 4.669201\ldots \quad (12)$$

The result given in Eq. 12 is not restricted to the quadratic map. In fact, it applies to a broad class of systems that undergo period doubling cascades (13, 39). In practice such cascades are very common, and the associated universal numbers are observed to be well approximated by means of fairly low order bifurcations (for example, $n = 2,3,4$). This scaling behavior has been observed in

many experiments, including ones on fluids, nonlinear circuits, laser systems, and so forth. Although universality arguments do not explain why cascades must exist, such explanations are available from bifurcation theory (40).

Figure 8b shows the period-three window within the chaotic range of the quadratic map. As already mentioned, there are an infinite number of such periodic windows. [In fact, they are generally believed to be dense in the chaotic range. For example, if $k$ is prime, there are $(2^k - 2)/(2k)$ period-$k$ windows.] Let $C_0^{(k)}$ and $C_c^{(k)}$ denote the upper and lower values of $C$ bounding the period-$k$ window and let $C_d^{(k)}$ denote the value of $C$ at which the period-$k$ attractor bifurcates to period $2k$. Then we have that, for typical $k$ windows (41).

$$\lim_{k \to \infty} \frac{C_c^{(k)} - C_0^{(k)}}{C_d^{(k)} - C_0^{(k)}} \to 9/4 \quad (13)$$

In fact, even for the $k = 3$ window (Fig. 8b) the 9/4 value is closely approximated (it is $9/4 - 0.074\ldots$). This result is universal for one-dimensional maps (and possibly more generally for any chaotic dynamical process) with windows.

## Fractal Basin Boundaries

In addition to chaotic attractors, there can be sets in phase space on which orbits are chaotic but for which points near the set move away from the set. That is, they are repelled. Nevertheless, such chaotic repellers can still have important macroscopically observable effects, and we consider one such effect (42, 43) in this section.

Typical nonlinear dynamical systems may have more than one time-asymptotic final state (attractor), and it is important to consider the extent to which uncertainty in initial conditions leads to uncertainty in the final state. Consider the simple two-dimensional phase space diagram schematically depicted in Fig. 10. There are two attractors denoted A and B. Initial conditions on one side of the boundary, $\Sigma$, eventually asymptotically approach B; those on the other side of $\Sigma$ eventually go to A. The region to the left or right of $\Sigma$ is the basin of attraction for attractor A or B, respectively, and $\Sigma$ is the basin boundary. If the initial conditions are uncertain by an amount $\epsilon$, then for those initial conditions within $\epsilon$ of the boundary we cannot say a priori to which attractor the orbit eventually tends.

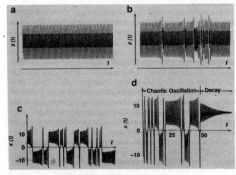

**Fig. 9.** Intermittency, crisis, and period doubling in continuous time systems. Intermittency in the Lorenz equations (**a**) $r = 166.0$; (**b**) $r = 166.2$. Crisis transition to a chaotic attractor in the Lorenz equations: (**c**) $r = 28$; (**d**) $r = 22$.

**Fig. 10.** A region of phase space divided by the basin boundary Σ into basins of attraction for the two attractors A and B. Points 1 and 2 are initial conditions with error ε.

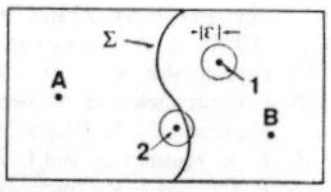

For example, in Fig. 10, points 1 and 2 are initial conditions with an uncertainty ε. The orbit generated by initial condition 1 is attracted to attractor B. Initial condition 2, however, is uncertain in the sense that the orbit generated by 2 may be attracted either to A or B. In particular, consider the fraction of the uncertain phase space volume within the rectangle shown and denote this fraction $f$. For the case shown in Fig. 10, we clearly have $f \sim \epsilon$. The main point we wish to make in what follows is that, from the point of view of prediction, much worse scalings of $f$ with ε frequently occur in nonlinear dynamics. Namely, the fraction can scale as

$$f \sim \epsilon^{\alpha} \tag{14}$$

with the "uncertainty exponent" α satisfying $\alpha < 1$ (42, 43). In fact, $\alpha \ll 1$ is fairly common. In such a case, a substantial reduction in the initial condition uncertainty, ε, yields only a relatively small decrease in the uncertainty of the final state as measured by $f$.

Although α is equal to unity for simple basin boundaries, such as that depicted in Fig. 10, boundaries with noninteger (fractal) dimension also occur. We use here the capacity definition of dimension, Eq. 1. In general, since the basin boundary divides the phase space, its dimension $d$ must satisfy $d \geq D - 1$, where $D$ is the dimension of the phase space. It can be proven that the following relation between the index α and the basin boundary dimension holds (42, 43)

$$\alpha = D - d \tag{15}$$

For a simple boundary, such as that depicted in Fig. 10, we have $d = D - 1$, and Eq. 15 then gives $\alpha = 1$, as expected. For a fractal basin boundary, $d > D - 1$, and Eq. 15 gives $\alpha < 1$.

We now illustrate the above with a concrete example. Consider the forced damped pendulum as given by Eq. 4. For parameter values $\nu = 0.2$, $\omega_0 = 1.0$, $\omega = 1.0$, and $f = 2.0$, we find numerically that the only attractors in the surface of section $(\theta, d\theta/dt)$ are the fixed points $(-0.477, -0.609)$ and $(-0.471, 2.037)$. They represent solutions with average counterclockwise and clockwise rotation at the period of the forcing. The cover shows a computer-generated picture of the basins of attraction for the two fixed point attractors. Each initial condition in a 1024 by 1024 point grid is integrated until it is close to one of the two attractors (typically 100 cycles). If an orbit goes to the attractor at $\theta = -0.477$, a blue dot is plotted at the corresponding initial condition. If the orbit goes to the other attractor, a red dot is plotted. Thus the blue and red regions are essentially pictures of the basins of attraction for the two attractors to the accuracy of the grid of the computer plotter. Fine-scale structure in the basins of attraction is evident. This is a consequence of the Cantor-set nature of the basin boundary. In fact, magnifications of the basin boundary show that, as we examine it on a smaller and smaller scale, it continues to have structure.

We now wish to explore the consequences for prediction of this infinitely fine-scaled structure. To do this, consider an initial condition $(\theta, d\theta/dt)$. What is the effect of a small change ε in the θ-coordinate? Thus we integrate the forced pendulum equation with the initial conditions $(\theta, d\theta/dt)$, $(\theta, d\theta/dt + \epsilon)$, and $(\theta, d\theta/dt - \epsilon)$ until they approach one of the attractors. If either or both of the perturbed initial conditions yield orbits that do not approach the same attractor as the unperturbed initial condition, we say that $(\theta, d\theta/dt)$ is uncertain. Now we randomly choose a large number of initial conditions and let $f$ denote the fraction of these that we find

to be uncertain. As a result of these calculations, we find that $f \sim \epsilon^{\alpha}$ where $\hat{\alpha} \cong 0.275 \pm 0.005$. If we assume that $f$, determined in the way stated above, is approximately proportional to $f$ [there is some support for this conjecture from theoretical work (44)], then $\alpha = \hat{\alpha}$. Thus, from Eq. 15, the dimension of the basin boundary is $d \cong 1.725 \pm 0.005$. We conclude, from Eq. 14, that in this case if we are to gain a factor of 2 in the ability to predict the asymptotic final state of the system, it is necessary to increase the accuracy in the measurement of the initial conditions by a factor substantially greater than 2 (namely by $2^{1/0.275} \cong 10$). Hence, fractal basin boundaries ($\alpha < 1$) represent an obstruction to predictability in nonlinear dynamics.

Some representative works on fractal basin boundaries, including applications, are listed in (42–47). Notable basic questions that have recently been answered are the following:

1) How does a nonfractal basin boundary become a fractal basin boundary as a parameter of the system is varied (45)? This question is similar, in spirit, to the question of how chaotic attractors come about.

2) Can fractal basin boundaries have different dimension values in different regions of the boundary, and what boundary structures lead to this situation? This question is addressed in (46) where it is shown that regions of different dimension can be intertwined on an arbitrarily fine scale.

3) What are the effects of a fractal basin boundary when the system is subject to noise? This has been addressed in the Josephson junction experiments of (31).

## Conclusion

Chaotic nonlinear dynamics is a vigorous, rapidly expanding field. Many important future applications are to be expected in a variety of areas. In addition to its practical aspects, the field also has fundamental implications. According to Laplace, determination of the future depends only on the present state. Chaos adds a basic new aspect to this rule: small errors in our knowledge can grow exponentially with time, thus making the long-term prediction of the future impossible.

Although the field has advanced at a great rate in recent years, there is still a wealth of challenging fundamental questions that have yet to be adequately dealt with. For example, most concepts developed so far have been discovered in what are effectively low-dimensional systems; what undiscovered important phenomena will appear only in higher dimensions? Why are transiently chaotic motions so prevalent in higher dimensions? In what ways is it possible to use the dimension of a chaotic attractor to determine the dimension of the phase space necessary to describe the dynamics? Can renormalization group techniques be extended past the borderline of chaos into the strongly chaotic regime? These are only a few questions. There are many more, and probably the most important questions are those that have not yet been asked.

REFERENCES AND NOTES

1. C. Grebogi, E. Ott, S. Pelikan, J. A. Yorke, *Physica* 13D, 261 (1984).
2. E. N. Lorenz, *J. Atmos. Sci.* 20, 130 (1963).
3. J. D. Farmer, E. Ott, J. A. Yorke, *Physica* 7D, 153 (1983).
4. J. Kaplan and J. A. Yorke, *Lecture Notes in Mathematics No. 730* (Springer-Verlag, Berlin, 1978), p. 228; L. S. Young, *Ergodic Theory Dyn. Syst.* 1, 381 (1981).
5. P. Grassberger and I. Procaccia, *Phys. Rev. Lett.* 50, 346 (1983); H. G. E. Hentschel and I. Procaccia, *Physica* 8D, 435 (1983); P. Grassberger, *Phys. Lett.* A97, 227 (1983); T. C. Halsey *et al.*, B. I. Shraiman, *Phys. Rev.* A 33, 1141 (1986); C. Grebogi, E. Ott, J. A. Yorke, *ibid.* 36, 3522 (1987).
6. A. Bondeson *et al.*, *Phys. Rev. Lett.* 55, 2103 (1985); F. J. Romeiras, A. Bondeson, E. Ott, T. M. Antonsen, C. Grebogi, *Physica* 26D, 277 (1987).
7. M. Hénon, *Commun. Math. Phys.* 50, 69 (1976).
8. D. A. Russell, J. D. Hanson, E. Ott, *Phys. Rev. Lett.* 45, 1175 (1980).
9. C. Grebogi, E. Ott, J. A. Yorke, *Physica* 15D, 354 (1985).

10. D. V. Anosov, *Proc. Steklov Ins. Math.* **90** (1967); R. Bowen, *J. Differ. Equations* **18**, 333 (1975).
11. S. M. Hammel, J. A. Yorke, C. Grebogi, *J. Complexity*, **3**, 136 (1987).
12. R. M. May, *Nature (London)* **261**, 459 (1976).
13. M. J. Feigenbaum, *J. Stat. Phys.* **19**, 25 (1978).
14. Y. Pomeau and P. Manneville, *Commun. Math. Phys.* **74**, 189 (1980).
15. C. Grebogi, E. Ott, J. A. Yorke, *Physica* 7D, 181 (1983).
16. D. Ruelle and F. Takens, *Commun. Math. Phys.* **20**, 167 (1971).
17. P. Cvitanovic, Ed. *Universality in Chaos* (Hilger, Bristol, 1984).
18. For example, S. J. Shenker, *Physica* 5D, 405 (1982); K. Kaneko, *Progr. Theor. Phys.* **71**, 282 (1984); M. J. Feigenbaum, L. P. Kadanoff, S. L. Shenker, *Physica* 5D, 370 (1982); D. Rand, S. Ostlund, J. Sethna, E. Siggia, *Physica* 8D, 303 (1983); S. Kim and S. Ostlund, *Phys. Rev. Lett.* **55**, 1165 (1985); D. K. Umberger, J. D. Farmer, I. I. Satija, *Phys. Lett. A* **114**, 341 (1986); P. Bak, T. Bohr, M. H. Jensen, *Phys. Scr.* T9, 50 (1985); P. Bak, *Phys. Today* **39** (No. 12), 38 (1987).
19. J. E. Hirsch, M. Nauenberg, D. J. Scalapino, *Phys. Lett. A* **87**, 391 (1982).
20. J. A. Yorke and E. D. Yorke, *J. Stat. Phys.* **21**, 263 (1979); in *Topics in Applied Physics* (Springer-Verlag, New York, 1981), vol. 45, p. 77.
21. C. Grebogi, E. Ott, J. A. Yorke, *Phys. Rev. Lett.* **57**, 1284 (1986).
22. A. Libchaber and J. Maurer, *J. Phys. (Paris)* **41**, C3-51 (1980); A. Libchaber, C. Laroche, S. Fauve, *J. Phys. (Paris) Lett.* **43**, L211 (1982).
23. J. P. Gollub, S. V. Benson, J. F. Steinman, *Ann. N.Y. Acad. Sci.* **357**, 22 (1980); M. Giglio, S. Musazzi, U. Perini, *Phys. Rev. Lett.* **47**, 243 (1981).
24. P. S. Linsay, *Phys. Rev. Lett.* **47**, 1349 (1981).
25. F. T. Arecchi, R. Meucci, G. Puccioni, J. Tredicce, *ibid.* **49**, 1217 (1982).
26. M. Dubois, M. A. Rubio, P. Bergé, *ibid.* **51**, 1446 (1983).
27. J. C. Roux, P. DeKepper, H. L. Swinney, *Physica* 7D, 57 (1983).
28. C. Jeffries and J. Perez, *Phys. Rev. A* **27**, 601 (1983); S. K. Brorson, D. Dewey, P. S. Linsay, *ibid.* **28**, 1201 (1983).
29. H. Ikezi, J. S. deGrasse, T. H. Jensen, *ibid.* **28**, 1207 (1983).
30. R. W. Rollins and E. R. Hunt, *ibid.* **29**, 3327 (1984).
31. M. Iansiti *et al.*, *Phys. Rev. Lett.* **55**, 746 (1985).
32. D. Dangoisse, P. Glorieux, D. Hannequin, *ibid.* **57**, 2657 (1986).
33. K. G. Wilson and J. Kogut, *Phys. Rep.* C **12**, 75 (1974); B. Hu, *ibid.* **91**, 233 (1982).
34. M. J. Feigenbaum, *Phys. Lett. A* **74**, 375 (1979); R. Brown, C. Grebogi, E. Ott,
*Phys. Rev. A* **34**, 2248 (1986); M. Nauenberg and J. Rudnick, *Phys. Rev. B* **24**, 493 (1981); B. A. Huberman and A. B. Zisook, *Phys. Rev. Lett.* **46**, 626 (1981); J. D. Farmer, *ibid.* **47**, 179 (1981).
35. J. Crutchfield, M. Nauenberg, J. Rudnick, *Phys. Rev. Lett.* **46**, 933 (1981); B. Shraiman, C. E. Wayne, P. C. Martin, *ibid.*, p. 935.
36. B. A. Huberman and J. Rudnick, *ibid.* **45**, 154 (1980).
37. B. Hu and J. Rudnick, *ibid.* **48**, 1645 (1982).
38. L. P. Kadanoff, *ibid.* **47**, 1641 (1981); D. F. Escande and F. Doveil, *J. Stat. Phys.* **26**, 257 (1981); R. S. MacKay, *Physica* 7D, 283 (1983).
39. P. Collet, J. P. Eckmann, O. E. Lanford III, *Commun. Math. Phys.* **76**, 211 (1980).
40. J. A. Yorke and K. A. Alligood, *ibid.* **100**, 1 (1985).
41. J. A. Yorke, C. Grebogi, E. Ott, L. Tedeschini-Lalli, *Phys. Rev. Lett.* **54**, 1095 (1985).
42. C. Grebogi, S. W. McDonald, E. Ott, J. A. Yorke, *Phys. Lett. A* **99**, 415 (1983).
43. S. W. McDonald, C. Grebogi, E. Ott, J. A. Yorke, *Physica* 17D, 125 (1985).
44. S. Pelikan, *Trans. Am. Math. Soc.* **292**, 695 (1985).
45. C. Grebogi, E. Ott, J. A. Yorke, *Phys. Rev. Lett.* **56**, 1011 (1986); *Physica* 24D, 243 (1987); F. C. Moon and G.-X. Li, *Phys. Rev. Lett.* **55**, 1439 (1985).
46. C. Grebogi, E. Kostelich, E. Ott, J. A. Yorke, *Phys. Lett.* **A118**, 448 (1986); *Physica* 25D, 347 (1987); C. Grebogi, E. Ott, J. A. Yorke, H. E. Nusse, *Ann. N.Y. Acad. Sci.* **497**, 117 (1987).
47. C. Mira, *C. R. Acad. Sci.* **288A**, 591 (1979); C. Grebogi, E. Ott, J. A. Yorke, *Phys. Rev. Lett.* **50**, 935 (1983); R. G. Holt and I. B. Schwartz, *Phys. Lett.* **A105**, 327 (1984); I. B. Schwartz, *ibid.* **106**, 339 (1984); I. B. Schwartz, *J. Math. Biol.* **21**, 347 (1985); S. Takesue and K. Kaneko, *Progr. Theor. Phys.* **71**, 35 (1984); O. Decroly and A. Goldbeter, *Phys. Lett.* **A105**, 259 (1984); E. G. Gwinn and R. M. Westervelt, *Phys. Rev. Lett.* **54**, 1613 (1985); *Phys. Rev. A* **33**, 4143 (1986); Y. Yamaguchi and N. Mishima, *Phys. Lett.* **A109**, 196 (1985); M. Napiorkowski, *ibid.* **113**, 111 (1985); F. T. Arecchi, R. Badii, A. Politi, *Phys. Rev. A* **32**, 402 (1985); S. W. McDonald, C. Grebogi, E. Ott, J. A. Yorke, *Phys. Lett.* **A107**, 51 (1985); J. S. Nicolis and I. Tsuda, in *Simulation, Communication, and Control*, S. G. Tzafestas, Ed. (North-Holland, Amsterdam, 1985); J. S. Nicolis, *Rep. Prog. Phys.* **49**, 1109 (1986); J. S. Nicolis, *Kybernetes* **14**, 167 (1985).
48. This work was supported by the Air Force Office of Scientific Research, the U.S. Department of Energy, the Defense Advanced Research Projects Agency, and the Office of Naval Research.

*Commun. in Theor. Phys. (Beijing, China)*          *Vol.9, No.4 (1988)*     375-389

# SYSTEMATICS OF THE PERIODIC WINDOWS IN THE LORENZ MODEL AND ITS RELATION WITH THE ANTISYMMETRIC CUBIC MAP[+]

DING Ming-zhou( 丁明洲 ) and HAO Bai-lin( 郝柏林 )

*Institute of Theoretical Physics, Academia Sinica*
*P.O.Box 2735, Beijing, China*

Received July 10, 1986

### Abstract

   *The Lorenz model and the antisymmetric cubic map enjoy the same discrete symmetry. A careful study of a one-dimensional bifurcation diagram obtained numerically from the Lorenz model reveals that the systematics of periodic windows is closely related to that of the cubic map. In addition, we determined the drift of the fundamental frequency and thus proposed a nomenclature for the periodic windows by associating an absolute period to each window, which in turn agrees with the corresponding word made of three letters for most of the observed periods.*

## I. Introduction

   The systematics of periodic windows, being a universal characteristics of the nonlinear systems, plays an important role in understanding the overall bifurcation and chaos "spectrum" in their parameter space.   In one-dimensional mappings it has been well-established that the periods embedded in chaotic region can be ordered and generated in a systematic way according to symbolic dynamics of two or more letters[1-3], whereas in the circle mapping the mode-locking frequencies embedded in the quasiperiodic sea are ordered in accordance with the Farey sequence(see, e.g., Ref.[4]).   However, for ODE's which are more realistic in describing many physical systems, this kind of global property is less understood comparing with other universal features, such as, period-doubling cascades, Feigenbaum convergent rate, etc., due to complexity of the system.   No means are readily accessible at present to systems other than one-

---

[+] *Projects Supported by the Science Fund of the Chinese Academy of Sciences.*

dimensional mappings. What we have been doing toward this direction
so far is to go round about, i.e., first try to reduce the high-dimensional trajectories of the ODE's to one-dimensional iterations in
properly chosen intersection plane and then relate the results to
the known properties of one-dimensional mappings. This aprroach has
been proved successful when applied to a driven system, i.e., the
periodically forced Brusselator[5-6], where a perfect MSS[1] sequence
made of two letters R and L has been discovered. In this paper we
report another numerical study on the well-known Lorenz model.
Taking into account the antisymmetry we have related its systematics
of periodic windows to that of the cubic map, whose periodic structure
has been investigated in Refs.[2-3].

## II. The Antisymmetric Cubic Map

We summarize briefly some notations and theorems on the cubic
map for later reference[2-3].

The cubic map

$$x_{n+1}=F(A,x_n)=Ax_n^3+(1-A)x_n \tag{2.1}$$

maps the interval $[-1,1]$ into itself
when A varies in $[1,4]$. The shape of
$F(A,x)$ is shown in Fig.1 where $x_c^-,x_c^+$
denote two critical points and L, M
and R mark the three monotonic segments. As usual, the symbolic description of periodic orbits will be made of
these three letters. Owing to the antisymmetric property of (2.1), there are
two kinds of periodic orbits, symmetric
if $F^{(n/2)}(A,x)=-x$, n being the period

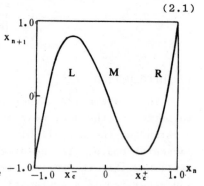

Fig.1 Function F(A,x) for A=3.5.

of the orbit and x being an element on the orbit, and asymmetric
which does not satisfy the above relation. The asymmetric orbits
always appear in pairs located symmetrically with respect to the
origin. It has been shown that[7] a stable symmetric periodic orbit
will first undergo symmetry-breaking bifurcation to a pair of asymmetric orbits of the same period as the original one, then enjoy period-
doubling and symmetry-restoration(we shall see that this conclusion
equally applies to the Lorenz system as well).

Symbolically, each point x in a cycle can be labelled by one of the three letters L, M or R according to whether it satisfies $x < x_c^-$, $x_c^- < x < x_c^+$ or $x > x_c^+$. In this way, a symmetric superstable orbit starting from $x_c$ will be of the form

$$x_c^- \sigma_1 \sigma_2 \ldots \sigma_n x_c^+ \bar{\sigma}_1 \bar{\sigma}_2 \ldots \bar{\sigma}_n x_c = x_c^- P x_c^+ \bar{P} x_c^- \, ,$$

where $\sigma_i$ is L, M or R and $\bar{\sigma}_i$ is the conjugate of $\sigma_i$ obtained by interchanging L's and R's but leaving M unchanged. Similarly, an asymmetric superstable cycle looks like

$$x_c^- \sigma_1 \sigma_2 \ldots \sigma_n x_c^- = x_c^- P x_c^- \, .$$

We omit $x_c^-$ hereafter as understood and simply call P a pattern or a word.

Between any two arbitrarily given patterns $P_1$ and $P_2$ an order can be defined referring to the natural order

$$L < x \quad < M < x \quad < R \, .$$

Expressing $P_1$ and $P_2$ as $P_1 = P^* \sigma_1 \ldots$ and $P_2 = P^* M_1 \ldots$, where $P^*$ is the common part and $\sigma_1 \neq \mu_1$, then if $P^*$ contains an even number of M's (we simply say $P^*$ even), then $P_2 > P_1$ if $\mu_1 > \sigma_1$ and $P_2 < P_1$ otherwise. If $P^*$ is odd, then $P_2 > P_1$ if $\mu_1 < \sigma_1$, and $P_2 < P_1$ otherwise.

We relate this order to the corresponding values on the parameter axis. Let $A_p$ denote the parameter value associated with a pattern P, then we have the following:

Theorem 1. If $P_1 < P_2$, then $A_{p1} < A_{p2}$, and the reverse is true.

This theorem can be partially verified by inspecting Table 1, where all the admissible patterns and the corresponding parameter values with periods less than 7 are listed.

In addition, a composition rule can be proven and it helps to generate all the legal words between any two given words $P_1$ and $P_2$.

Table 1 Periodic windows for the cubic map

| i | k | P | A |
|---|---|---|---|
| 1 | 2 | R | 3.1213203 |
| 2 | 4 | RMR | 3.2628786 |
| 3 | 6 | RMRLR | 3.3340241 |
| 4 | 6 | RMMLM | 3.4632834 |
| 5 | 6 | RMMLR | 3.5282272 |
| 6 | 4 | RMM | 3.5480858 |
| 7 | 6 | RMMMR | 3.5659880 |
| 8 | 6 | RMMMM | 3.5911819 |
| 9 | 5 | RMMM | 3.6150319 |
| 10 | 5 | RMMR | 3.6662070 |

| | | | |
|---|---|---|---|
| 11 | 3 | RM | 3.7003155 |
| 12 | 6 | RMLRM | 3.7029894 |
| 13 | 5 | RMLR | 3.7339407 |
| 14 | 5 | RMLM | 3.7753839 |
| 15 | 6 | RMLMM | 3.7909088 |
| 16 | 6 | RMLMR | 3.8073689 |
| 17 | 4* | RRL  ($x_C^- R x_C^+ L x_C^-$) | 3.8398944 |
| 18 | 6 | RRLMR | 3.8610860 |
| 19 | 6 | RRLMM | 3.8734615 |
| 20 | 5 | RRLM | 3.8835860 |
| 21 | 6 | RRLML | 3.8933550 |
| 22 | 6 | RRLRL | 3.8982992 |
| 23 | 5 | RRLR | 3.9069063 |
| 24 | 6 | RRLRM | 3.9144901 |
| 25 | 3 | RR | 3.9249907 |
| 26 | 6 | RRMRR | 3.9254576 |
| 27 | 6 | RRMRM | 3.9350271 |
| 28 | 5 | RRMR | 3.9409044 |
| 29 | 6 | RRMRL | 3.9464110 |
| 30 | 6 | RRMML | 3.9504721 |
| 31 | 5 | RRMM | 3.9553247 |
| 32 | 6 | RRMMM | 3.9597015 |
| 33 | 6 | RRMMR | 3.9637898 |
| 34 | 4 | RRM | 3.9675403 |
| 35 | 6 | RRMLR | 3.9710914 |
| 36 | 6 | RRMLM | 3.9745198 |
| 37 | 5 | RRML | 3.9777816 |
| 38 | 6* | RRRLL  ($x_C^- R R x_C^+ L L x_C^-$) | 3.9818990 |
| 39 | 5 | RRRL | 3.9854885 |
| 40 | 6 | RRRLM | 3.9878905 |
| 41 | 6 | RRRLR | 3.9900272 |
| 42 | 4 | RRR | 3.9919300 |
| 43 | 6 | RRRMR | 3.9936280 |
| 44 | 6 | RRRMM | 3.9951295 |
| 45 | 5 | RRRM | 3.9964269 |
| 46 | 6 | RRRML | 3.9975231 |
| 47 | 6 | RRRRL | 3.9984117 |
| 48 | 5 | RRRR | 3.9991078 |
| 49 | 6 | RRRRM | 3.9996037 |
| 50 | 6 | RRRRR | 3.9999009 |

*'*'   denotes the orbit after the symmetry-breaking bifurcation.*
*The symmetric patterns are included in the parentheses.*

It is important to mention that the periodic sequence constructed in this way is independent of the particular model. It is universal for all the mappings with two critical points and the antisymmetric property. Of course, the parameter values in Table 1 do depend on the mapping (2.1). Therefore, we shall call this ordering U-sequence too, following MSS[1].

As mentioned above, not every word made of L, M and R is admissible in the U-sequence. We present a criterion below which allows one to exclude all the illegal patterns. Similar argument was first given in Ref.[8] for unimodal mappings.

Let a pattern be $Px_c^- = \sigma_1\sigma_2\sigma_3\ldots\sigma_n x_c^-$. We associate a number
sequence $(a_1, a_2, a_3, \ldots, a_n, C)$ mode of 1, -1, 0, 1/2 and -1/2 to it
according to the following rule:

$a_j = 0$ if $\sigma_j = M$; $a_j = 1$ if $\sigma_j = R$ and the number of M's before $\sigma_j$ is
even or $\sigma_j = L$ otherwise; $a_j = -1$ if $\sigma_j = R$ and the number of the M's before $\sigma_j$ is odd or $\sigma_j = L$ otherwise.  C will take the value 1/2 or -1/2
depending on P being odd or even.  If the pattern is admissible,
then the following inequalities must be satisfied

$$(a_1, a_2, \ldots, a_n, C) > \pm(a_j, a_{j+1}, \ldots, a_n, C, 0, 0, \ldots) \qquad (2.2)$$

$$i = 1, 2, \ldots, n$$

Inequalities (2.2) should be understood in the sense that the
pairs $(a_m, a_{i+m-1})$ satisfy $a_m > \pm a_{i+m-1}$ for $i = 1, 2, \ldots, n+1$ while $m = 1, 2,$
$\ldots, n+2-i(a_{n+1} = C)$.  If any of them fails then the (2.2) fails too.
In this later case the corresponding word is illegal.

We give several examples to illustrate this criterion.

Example 1. P=RRL.  The associated number-sequence is $(1, 1, -1$
$-1/2)$.  It can be easily verified that the following inequalities
hold

$$(1, 1, -1, -1/2) > \pm(1, -1, -1/2) ,$$
$$(1, 1, -1, -1/2) > \pm(-1, -1/2) ,$$
$$(1, 1, -1, -1/2) > \pm(-1/2).$$

Therefore, the word RRL is admissible.

Example 2. P=RRLL.  This word does not belong to the U-sequence,
since at least one of the inequalities fails, namely,

$$(1, 1, -1, -1/2) \not> -(-1, -1, -1/2) ,$$

because one encounters $-1 < 1/2$ for the third pair of the elements.

Table 1 shows a clear regularity, schematically illustrated in
Fig.1(b).

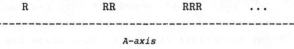

R          RR          RRR       ...

*A-axis*

*Fig.1(b)*

In the first region, labelled by R, every word begins with only
one letter R.  It will be called the R-region.  In the next region,
labelled by RR, every word has two consecutive R's in the leftmost

leading part followed by M or L.  It will be called the RR-region
later on.  In this way, we have RRR-region, RRRR-region,..., successively.  We shall see that similar regularity manifests in the Lorenz
model when suitably assigning letter to each point in the Poincare
maps of periodic orbits.

## III. The Lorenz Model and the Nomenclature of Its Periods

In this section we briefly review some known results on the
Lorenz model and suggest a nomenclature of periods by using power
spectra analysis.

The Lorenz model[9] is

$$\dot{x}=\sigma(y-x) ,$$
$$\dot{y}=rx-xz-y ,$$
$$\dot{z}=xy-bz ,$$

(3.1)

where $\sigma,r,b$ are positive parameters corresponding to the Prandtl
number, the Rayleigh number and a geometric ratio respectively.
It was first proposed to describe thermal convection in fluids by
B.  Saltzman[10] and E.N. Lorenz in early sixties then attracted
much attention as one of the first examples to show strange attractors.  Many interesting features, such as, preturbulence, chaos,
period-doubling, noisy-periodicity and intermittency have been revealed and interpreted (see, e.g., Ref.[11] and references therein).
Up to now, most of the numerical and analytical studies are restricted to the one-parameter subfamily given by $\sigma=10$, $b=8/3$ and $r>0$,
another subfamily given by $\sigma=16$, $b=4$ and $r>0$ was investigated mainly
by·Japanese authors[12] but did not lead to qualitatively new picture.  The situation has been extensively reviewed in monogragh [11].
We summarize some of the known results along this parameter line
below.

The origin is globally attracting for $r<1$, $r=1$ being a transition point.  The origin loses stability and another two fixed points
$(C_{\pm}=(\pm\sqrt{b*(r-1)}, \pm\sqrt{b*(r-1)}, (r-1))$ appear at this r-value.  For $1<r$
$<r_0=13.926$ all trajectories tend to one of the fixed points.  At
$r_0=13.926$ the first homoclinic explosion takes place and creates an
invariant set which includes a countable infinite number of periodic
orbits, an uncountable number of aperiodic orbits and an uncountable
number of initial values which go to the origin finally.  For $r>r_0$
the above invariant set keeps qualitatively unchanged.  At $r_A=24.06$

another transition takes place.  The invariant set becomes a strange
attractor.  In addition, an infinite sequence of homoclinic explosions begins at this r-value.  For $r_A<r<r_H=24.74$ there exist a
chaotic stable attractor and a pair of stable fixed points.  For
$r_H=24.74$ $C_{\pm}$ become unstable.  For $r>r_H$ there are no stable fixed
points.  For $r>r_1=30.1$ stable periodic orbits may appear.  The
systematics of their appearance is the major problem addressed in
this paper.

In the case of periodically driven systems the external forcing
frequency is at one's disposal and serves as the reference frequency
to measure all other periodic components occurring in the motion.
However, in an autonomous system like (3.1) the fundamental frequency
of the motion may drift as the parameter varies and this seems to
prevent us from associating an encountered period with an absolute
name like period 5.  However, every periodic component should show
up in the power spectrum.  Detailed Fourier analysis of all the available periodic orbits along the r-axis provides some information on
the drift of fundamental frequency.  Fig.2 is the resulted diagram[13].  Each point in the diagram corresponds  to a peak in the
spectrum at given r.  The dashed line joins the fundamental frequencies and shows a clear tendency of drifting.   The period of a stable
limit cycle measured as the longest return time in the Poincare map
corresponds  to the lowest peak in the spectrum and one may refer

to the curve in Fig.2 to tell
which subharmonic it represents.
Thus Fig.2 has served  as  a
calibration curve to determine
the period of observed windows.
All the known windows and newly
discovered by us have been assigned a period in this way,
e.g., the one near r=160 and
the one near r=100, which can
be recognized in Fig.2 and
has been classified as symmetric period 4 and period
3 respectively.

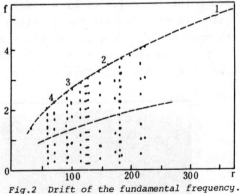

*Fig.2  Drift of the fundamental frequency.*

The above nomenclature enables us to identify the period of a
stable periodic orbit, but it cannot tell the relation between different periods.  To do this we need a global systematics of the

270

periodic windows what is the subject of the next section, where we
shall introduce a symbolic system relying on a one-dimensional bifurcation diagram extracted from the Lorenz model and relate it to
the well-studied one-dimensional cubic map. It is worthy to mention
that the period got from power spectra as described above agrees
perfectly with that measured from the one-dimensional bifurcation
diagram and agrees with the word in the symbolic dynamics of three
letters in an overwhelming majority of cases. The origin of this
agreement, besides the common discrete symmetry, still requires
further study.

## IV. The Systematics of Periodic Windows in the Lorenz Model

In order to understand the order of occurrence of different
periodic windows one has to devise some symbolic system, more precis-
ly, to label each periodic window with a suitable word made of a
certain number of letters and study the underlying symbolic dynamics.
This is the usual way adopted in studying one-dimensional mappings.
However, in high-dimensional systems the reasonable assignment
of words to windows does not seem so apparent. In the monogragh [11].
Sparrow introduced a two-symbol system using x and y. An orbit is
assigned a letter x every time it makes a revolution around $C_+$(see
Sec.II for the notation of $C_\pm$), while a letter y indicates a revolution around $C_-$. Consecutive revolutions around one and the same
fixed point is indicated by a power, e.g., $x^2y=xxy$. Later in this
section we will discuss the relation between this symbolic system
with what we are going to introduce below.

The symbolic system we use is based on the bifurcation diagram
shown in Figs.3(a-c). This diagram is obtained by plotting

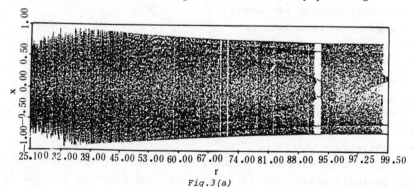

25.100 32.00 39.00 45.00 53.00 60.00 67.00 74.00 81.00 88.00 95.00 97.25 99.50

r

*Fig.3(a)*

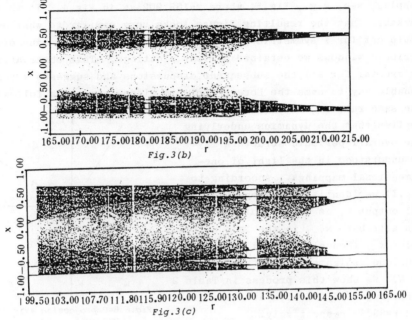

Fig.3(b)

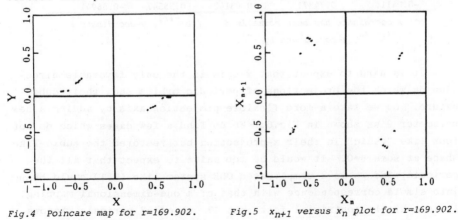

Fig.3(c)

*Fig.3 (a-c) Bifurcation diagram for the Lorenz model.*

the stationary x on the Poincare return plane z=r-1 versus the
parameter r. The z=r-1 plane has been used by many authors for computing the Poincare map since all the interesting trajectories intersect this plane and it contains both $C_+$ and $C_-$. We show in

*Fig.4  Poincare map for r=169.902.*      *Fig.5  $x_{n+1}$ versus $x_n$ plot for r=169.902.*

Fig.4 a typical Poincare map for r=169.902. For a given value of r
where a stable periodic orbit exists, plotting the successive $x_{n+1}$
versus $x_n$ will give an outline of the underlying one-dimensional

mapping, see, e.g., Fig.5, where r=169.902 as in Fig.4. It is remarkable that the resulting figure looks very similar to what one
would get for a symmetric period 16 from Fig.1.    For many of the
periodic windows we obtained similar plottings.  This observation
is crucial for all the subsequent discussions and suggests a reasonable way to name the Lorenz periodic windows by words following
the same rule of Sec.II.  This fact
implies that the dynamics underlying
the overall periodic structure may
be understood in the light of one-
dimensional mappings.  According to
Sec.II we identify the largest numerical output $x_i$ as the rightmost R point
and attribute $x_{i-1}$ to the left critical
point $x_c^-$, then all other $x_i$'s acquire
a unique assignment of letters L, M
or R.  We show this process in Table 2
for two period 5 orbits at r=114.00
and r=83.39 respectively.

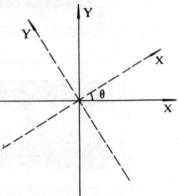

Fig.6 $\theta=0$ projection axis versus
$\theta\neq0$ axis.

Table 2   Letter-assignment of two period 5 orbits

| $x_c^-$ | R | R | L | R |
|---------|-------|-------|---------|-------|
| -0.26525 | 0.69690 | 0.53309 | -0.45523 | 0.58495 |
| $x_c^-$ | R | R | R | L |
| -0.04111 | 0.74371 | 0.63140 | 0.35347 | -0.56941 |

'*' x-coordinate has been normalized by $(rb)^{-1/2}$, y-coordinate by
$r^{-1}b^{-1/2}$, z-coordinate by $r^{-1}$.

It is hard to expect that x-axis is the only favorable direction on which the projections of periodic orbits show their cubic
nature.  So we take a more flexible projection axis by adding a new
parameter $\theta$ as shown in Fig.6. We do find a few cases which do not
look like "cubic" in their x-projection but restored the cubic-like
shape at some $\theta\neq0$.  It would be too naive to expect that all the
periodic windows in a complicated ODE system like (3.1) could be put
into simple correspondence with that of a one-dimensional mapping.
It is a highly nontrivial fact that 47 out of 53 periodic windows
(the period-doubling regimes did not count) fit into the cubic
scheme.  There are indeed a few stable orbits which do not have
one-dimensional correspondence, no matter how one adjusts the

projection angle θ.  In Table 3 all the periodic windows discovered
so far by us and by other authors are listed in descending r order
along with their periods, words and locations on the parameter axis.
The blanks in the third column indicate these windows which do not
fit into the "cubic" scheme.

Table 3 Periodic windows for the Lorenz model

| i | k | p | r |
|---|---|---|---|
| 1 | 24 | | 215.08-215.06 |
| 2 | 12 | | 214.06-213.95 |
| 3 | 6 | RMRLR($\theta$=5.0) | 209.45-209.31 |
| 4 | 6 | RMRLR | 208.81-208.78 |
| 5 | 10 | | 207.10-207.096 |
| 6 | 8 | RMMLRLR | 205.47-205.46 |
| 7 | 10 | RMMLRLRLR | 204.12-204.11 |
| 8 | 14* | RMMLRLRLMMRLR($x_C^-$RMMLRLx$_C^+$) | 200.66 |
| 9 | 10* | RMMLRLMMR($x_C^-$RMMLx$_C^+$) | 198.97-198.95 |
| 10 | 10* | RMRMRLMLM($x_C^-$RMRMx$_C^+$) | 191.99 |
| 11 | 5 | RMMR | 190.81-190.80 |
| 12 | 7 | RMMRLR | 189.549 |
| 13 | 16* | RMLRLRLMLMRLRLR($x_C^-$RMLRLRLx$_C^+$) | 187.25 |
| 14 | 12* | RMLRLMLMRLR($x_C^-$RMLRLx$_C^+$) | 185.80-185.74 |
| 15 | 8* | RMLMLMR($x_C^-$RML$x_C^+$) | 180.97-180.93 |
| 16 | 10 | RMLMLMRLR | 180.127 |
| 17 | 10 | RMLMLMLMR | 178.075 |
| 18 | 12* | RMLMLRLMRMR($x_C^-$RMLML$x_C^+$) | 177.81-177.78 |
| 19 | 6 | RMLMR | 172.712 |
| 20 | 16* | RMRMRMLMLMLMLMR($x_C^-$RMRMRMLx$_C^+$) | 169.902 |
| 21 | 10 | RMLMRMLMR | 168.58 |
| 22 | 4* | RRL($x_C^-$Rx$_C^+$) | 166.01-146.2 |
| 23 | 12 | | 145.99-145.93 |
| 24 | 20 | | 144.38-144.35 |
| 25 | 12* | RRLMRMLLRML($x_C^-$RRLMRx$_C^+$) | 143.35-143.32 |
| 26 | 6 | RRLML | 136.79 |
| 27 | 10 | RRLMLLRRL | 136.21 |
| 28 | 16* | RRLMLLRMLLRMRRL($x_C^-$RRLMLLRx$_C^+$) | 135.48 |
| 29 | 8* | RRLRLLR($x_C^-$RRLx$_C^+$) | 132.04-132.03 |
| 30 | 16* | RRLRLLRRLLRLRRL($x_C^-$RRLRLLRx$_C^+$) | 129.134 |
| 31 | 6 | RRLRL | 126.44-126.41 |
| 32 | 12* | RRLRLRLLRLR($x_C^-$RRLRLx$_C^+$) | 123.57-123.56 |
| 33 | 8 | RRLRLRL | 121.687 |
| 34 | 5 | RRLR | 114.00-113.91 |
| 35 | 10* | RRLRRLLRL($x_C^-$RRLRx$_C^+$) | 110.506 |
| 36 | 7 | RRLRRL | 107.613 |
| 37 | 14* | RRLRRLRLLRLLR($x_C^-$RRLRRLRx$_C^+$) | 106.74 |
| 38 | 8 | RRLRRLR | 104.19-104.18 |
| 39 | 16* | RRLRRLRRLLRLLRL($x_C^-$RRLRRLRx$_C^+$) | 103.637 |
| 40 | 3 | RR | 100.07-99.93 |
| 41 | 9 | | 99.28-99.26 |
| 42 | 12* | RRMLLMLLMRR($x_C^-$RRMLLx$_C^+$) | 94.55 |
| 43 | 6* | RRRLL($x_C^-$RRx$_C^+$) | 92.60-92.17 |
| 44 | 12* | RRRLLRLLLRR($x_C^-$RRRLLx$_C^+$) | 90.197 |
| 45 | 5 | RRRL | 83.39-83.36 |
| 46 | 10* | RRRLRLLLR($x_C^-$RRRLx$_C^+$) | 82.01 |

| 47 | 12* | RRRMMRLLLMM($x_c^-$RRRMM$x_c^+$) | 76.315 |
|----|-----|------|-------|
| 48 | 4 | RRR | 71.52-71.41 |
| 49 | 8* | RRRRLLL($x_c^-$RRR$x_c^+$) | 69.718 |
| 50 | 5 | RRRR | 59.25-59.24 |
| 51 | 10* | RRRRRLLLL($x_c^-$RRRR$x_c^+$) | 58.71-58.70 |
| 52 | 6 | RRRR`. | 52.456 |
| 53 | 12* | RRRRRRLLLLL($x_c^-$RRRRR$x_c^+$) | 52.246 |

'*' same as in Table 1.

It is easy to verify that all the words shown in Table 3 belong
to the U-sequence defined in Sec.Il.  The most remarkable observation from Table 3 is the order of all the words except the one at
r=76.315 below r=189.549 being exactly the same as that of their
one-dimensional counterparts along the increasing A direction.  Take
$P_1$=RRLR at $r_1$=114.00 and $P_2$=RRLRRL at $r_2$=107.613 for example.  According ·to the definition in Sec.II $P_2$>$P_1$, then it follows from
Theorem 1 that $r_1$>$r_2$ in accordance with the present situation.  We
emphasize that parameter r in the Lorenz model is in a sense opposite to the parameter A in the cubic map (2.1).  Therefore, descending r is the same as increasing A.  The perfect ordering below
r=189.549 reflects that the cubic dynamics underlying the Lorenz
periodic windows depend monotonically on the parameter r besides
one exceptional case.  On the other hand, the stable periods between
r=209.45 and r=191.99 are ordered just opposite to the cubic order
but the period-doubling cascades within each window still develop
along the descending r direction.  The reason for this anomalous
ordering is not clear yet. Comparing Table 3 with Table 1,another
striking similarity could be recognized.  Above r=166.01 we have R-
region according to the notation of section II.  Between r=92.6 and
r=166.01 RR-region could be identified, and RRR-, RRRR-, RRRRR-
regions are ordered accordingly on the r-axis, ending at r=52.246
with symmetric period 12 $x_c^-$RRRRR$x_c^-$.  We will come back to this regularity later.  Asymmetric orbits and their symmetry-restored par-
tners are frequently encountered in the table.  This is also a typical phenomenon in the cubic map (2.1) due to the presence of the
antisymmetry.  At r=209.45 and r=208.81 there are two period 6 orbits described by the same word RMRLR.  This can be understood by
looking at their trajectories in the phase space.  If we slightly
change the observation angle $\theta$ as denoted in Table 1 for r=209.45
orbit the resulting projection figure will look topologically the
same as the r=208.81 orbit.

Now we are in  a  position to say that apart from a few exceptions the skeleton of the systematics for the occurrence of periodic windows obeys the "cubic" law.  Or put it another way, the symbolic dynamics underlying the global periodic structure below $r=189.549$ appears to be the same as that in the cubic map.  Above $r=189.549$ most of the periods still can be described by words belonging to the U-sequence, but the parameter dependence of the periodic windows becomes quite complicated.

We make a brief comparison between our results and Sparrow's work[11].  As it can be seen from Sec.III the first homoclinic explosion takes place at $r_0=13.926$ and it creates an invariant set which includes an infinite number of periodic orbits.  We call it the original set in accordance with Ref.[11]. A lot of the periodic orbits will be liquidated by the subsequent homoclinic explosions beginning at $r_A=24.06$.  Each homoclinic explosion generates a pair of homoclinic orbits which will leave the origin as r increases. After $r_1=30.1$ some of the periodic orbits may become stable and form periodic windows and then annihilate with other orbits.  Those windows involving the orbits coming from the original set are called normal windows and others are called extra windows.  Using Sparrow's symbolic notation described at the beginning of this section the systematics of the normal windows is that the windows with more consecutive x's (or y's) occur at smaller r values.  For example, the symmetric $x^{n+1}y^{n+1}$ window comes earlier than $x^n y^n$ window as r goes up. In simple cases the coorespondence between the two symbolic systems is straightforward.  For instance, $x^n y^n$ symmetric orbit corresponds to $x_c^- R^{n-1} x_c^+ L^{n-1} x_c^-$ orbit in our language.  At the suggestion of the referee we list in Table 4 the words made of R,M and L for the 6 "noncubic" windows.  The numbering i is the same as in Table 3.

Table 4  "Noncubic" windows

| i | period | word |
|---|--------|------|
| 1 | 24 | $(RM)^{11}R$ |
| 2 | 12 | $(RM)^5 R$ |
| 5 | 10 | $(RM)^3 MLR$ |
| 23 | 12 | $(R^2 LM)^2 R^2 L$ |
| 24 | 20 | $x_c^- P\ x_c^+ P\ x_c^-$      $P=R^2 LMR^2 L^2 R$ |
| 41 | 9 | $R^2 MR^2 MR^2$ |

Therefore, the ordered $R^n$-regions in Table 3 agree with the systematics of normal windows.  The systematics of the extra windows is unclear yet. Our symbolic system has some advantages over Sparrow's. With one additional letter it helps to reveal more subtle structure underlying the global occurrence of the periodic windows and makes the comparison with one-dimensional mappings possible, which provides a way to understand the complicated dynamics in terms of simpler ones.

It is clear that a better understanding of the results calls for a deeper investigation on the origin of periodic orbits in the Lorenz model, i.e., on the nature of homoclinic explosions and on the drift of the resulting periodic orbits in phase space. Toward this end a great deal has been done in Ref.[11], but much more needs to be done.

In conclusion, we recall that one of the motivations for Henon to introduce the well-known two-dimensional map[14] was to mimic the Lorenz strange attractor. However, the Henon map did not share the same discrete symmetry as the Lorenz model. Consequently, the Henon map has become something in its own and has not brought about much understanding to the Lorenz system. A two-dimensional mapping incorporating the antisymmetric property would be closer to the Lorenz model. Some investigation along this direction has been under way.

## Acknowledgement

We thank ZENG Wan-zhen for many discussions to understand the symbolic description of the antisymmetric cubic map.

## References

[1]  N. Metropolis, M. Stein and P. Stein, J. Combinat. Theor. _A15_(1973)25.

[2]  Zeng Wanzhen, Ding Mingzhou and Li Jianan, Chinese Phys. Lett. _2_(1985)293.

[3]  ZENG Wan-zhen, DING Ming-zhou and LI Jianan, Commun. Theor. Phys. _9(1988)141._

[4]  P. Cvitanovic, M.H. Jensen, L?P. Kadanoff and I. Procaccia, Phys. Rev. Lett., _56(1985)343._

[5]  HAO Bai-lin, WANG Guang-rui and ZHANG Shu-yu, Commun. Theor. Phys. _2(1983)198._

[6]  HAO Bai-lin, in "Collected Papers Dedicated to Prof. Tomita on the occasion of his retirement", Kyoto University, 1987, 82.

[7]  J. Swift and K. Wiesenfield, Phys. Rev. Lett. _52(1984)705._

[8]  B. Derrida, A Gervois and Y. Pomeau, Ann. Henri. Poncare _29A(1978)305._

[9]   E.N. Lorenz, J. Atoms. Sci. 20(1963)130.

[10]  B. Saltzman, J. Atmos. Sci. 19(1962)329.

[11]  C. Sparrow, "The Lorenz Equations, Bifurcation, Chaos and Strange Attractors", Springer-Verlag, New York, 1982.

[12]  K. Tomita and I. Tsuda, Prog. Theor. Phys. Supp. 69(1980)185.

[13]  Ding Mingzhou, Hao Bai-lin and Hao Xin, Chinese Phys. Lett. 2(1984)1.

[14]  M. Henon, Commun. Math. Phys. 50(1976)69.

## 6. Intermittent Transitions

Commun. Math. Phys. 74, 189–197 (1980)

Communications in
**Mathematical
Physics**
© by Springer-Verlag 1980

# Intermittent Transition to Turbulence in Dissipative Dynamical Systems

Yves Pomeau and Paul Manneville*

Commissariat à l'Énergie Atomique, Division de la Physique. Service de Physique Théorique,
F-91190 Gif-sur-Yvette, France

**Abstract.** We study some simple dissipative dynamical systems exhibiting a transition from a stable periodic behavior to a chaotic one. At that transition, the inverse coherence time grows continuously from zero due to the random occurrence of widely separated bursts in the time record.

## Introduction

A number of investigators [1] have observed in convective fluids an intermittent transition to turbulence. In these experiments the external control parameter, say $r$, is the vertical temperature difference across a Rayleigh-Bénard cell. Below a critical value $r_T$ of this parameter, measurements show well behaved and regular periodic oscillations. As $r$ becomes slightly larger than $r_T$ the fluctuations remain apparently periodic during long time intervals (which we shall call "laminar phases") but this regular behavior seems to be randomly and abruptly disrupted by a "burst" on the time record. This "burst" has a finite duration, it stops and a new laminar phase starts and so on. Close to $r_T$, the time lag between two bursts is seemingly at random and much larger than – and not correlated to – the period of the underlying oscillations. As $r$ increases more and more beyond $r_T$ it becomes more and more difficult and finally quite impossible to recognize the regular oscillations (see Fig. 1).

This sort of transition to turbulence is also present in simple dissipative dynamical systems [2] such as the Lorenz model [2a]. We present here the results of some numerical experiments on this problem.

When a burst starts at the end of a laminar phase this denotes an instability of the periodic motion due to the fact that the modulus of at least one Floquet multiplier [3] is larger than one. This may occur in three different ways: a real Floquet multiplier crosses the unit circle at $(+1)$ or at $(-1)$ or two complex conjugate multipliers cross simultaneously. To each of these three typical crossings we may associate one type of intermittency that we shall call for convenience type

---

* DPh. G. PSRM, Cen Saclay, Boîte Postale 2, F-91190 Gif-sur-Yvette, France

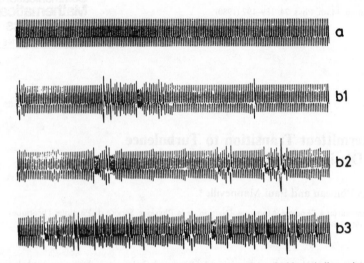

**Fig. 1a and b.** Time record of one coordinate ($z$) in the Lorenz model. **a** Stable periodic motion for $r = 166$. **b** Above the threshold the oscillations are interrupted by bursts which become more frequent as $r$ is increased

1: crossing at $(+1)$; type 2: complex crossing; and type 3: crossing at $(-1)$ respectively. In all these three cases our numerical studies show that the Lyapunov number grows continuously from zero beyond the onset of turbulence. In what follows we shall present some simple estimates for the "critical behavior" of this Lyapunov number in the vicinity of the turbulence threshold and compare them with the results of numerical experiments.

### Type 1. Intermittency in the Lorenz Model

The Lorenz system reads [4]:

$$\frac{dx}{dt} = \sigma(y-z); \quad \frac{dy}{dt} = -xz+rx-y; \quad \frac{dz}{dt} = xy-bz, \tag{1}$$

where $\sigma$, $b$, and $r$ are parameters. We have kept $b$ and $\sigma$ fixed at their original values ($\sigma = 10$, $b = 8/3$). Integrating system (1) around $r = 166$ one finds for $r$ slightly less than $r_T(\simeq 166.06)$ regular and stable oscillations for a random choice of initial condition (Fig. 1a). For $r$ slightly larger than $r_T$ these oscillations are interrupted by bursts (Fig. 1b). This can be explained quite simply by studying the Poincaré map (restricted here to be 1-dimensional without loss of significance). Let $f$ be the function such that $y_{n+1} = f(y_n, r)$ where $y_n$ is the $y$-coordinate of the $n^{th}$ crossing of the plane $x = 0$. Near $r = r_T$ the curve of equation $y' = f(y, r)$ is nearly tangent to the first bissectrix (Fig. 2). For $r$ slightly below $r_T$, this curve has two intersections with the bisectrix, they collapse into a single point at $r = r_T$ while for $r > r_T$ the curve is lifted up and no longer crosses the first bissectrix so that a "channel" appears between them (Fig. 3). Hence the successive iterates generated by the map

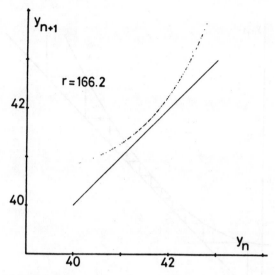

**Fig. 2.** A part of the Poincaré map along the y-coordinate for $r = 166.2$ slightly beyond the intermittency threshold ($r_T \simeq 166.06$)

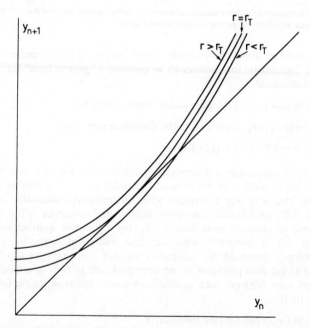

**Fig. 3.** Idealized picture of the deformation of $y_{n+1}(y_n)$ explaining the transition via intermittency. For $r < r_t$ two fixed points coexist one stable the other unstable. They collapse at $r = r_T$ and then disappear leaving a channel between the curve and the first bisectrix

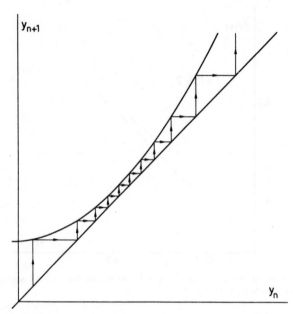

**Fig. 4.** The motion through the channel corresponds to the laminar phase of the movement. The slow drift is quite imperceptible on the time record of Fig. 1b

$y \to f(y, r)$ travel along this channel, which requires a large number of iterations (Fig. 4). To estimate this number let us consider a "generic form" for $f(y, r)$ in the region considered:

$$f(y, r) = y + \varepsilon + y^2 \text{ (+ higher order terms } - \text{H.O.T.)},$$

where $\varepsilon = (r - r_T)/r_T$. Near $\varepsilon = 0_+$ the difference equation

$$y_{n+1} = y_n + \varepsilon + y_n^2 \text{ (+ H.O.T.)}$$

can be approximated by a differential equation over $n$ and an elementary estimate shows that a number of iteration of the order of $\varepsilon^{-1/2}$ is needed to cross the channel. This is in nice agreement with our numerical simulation of system (1) (Fig. 5). After each transfer the burst destroys the coherence of the motion. This leads one to conclude that near $\varepsilon = 0_+$ the Lyapunov number varies as $\varepsilon^{1/2}$. Though this is consistent with our first numerical estimates, close to the intermittency threshold the Lyapunov number converges so slowly that it is difficult to get with precision, so we have preferred to turn to a modelling of the Poincaré map. We have got a qualitatively similar behavior for the following map of $S^1 = [0, 1[$

$$.\theta \to 2\theta + r \sin 2\pi\theta + 0.1 \sin 4\pi\theta \,(\text{mod } 1). \tag{2}$$

As shown in Fig. 6 this applies $S^1$ twice on itself and it is intermittent at $r_T \simeq -0.24706$. In this model, as well as those we shall consider later, the

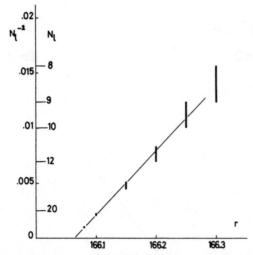

**Fig. 5.** The square of $N_l$ the largest number of cycles during a laminar period is inversely proportional to the distance from the threshold $r - r_T$. $N_l$ is given within 1 cycle to account for the uncertainty in the definition of the beginning/end of a laminar phase

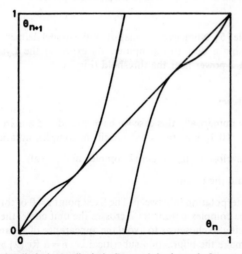

**Fig. 6.** Model mapping displaying qualitatively the same behavior as the Lorenz model around $r = 166$

possibility of starting a laminar phase after a burst comes from the fact that the map is not invertible. In diffeomorphisms the "relaminarization" cannot occur in this way due to the uniqueness of preimages. However dynamical systems for which the "reduced" Poincaré map takes a form similar to (2) [and later to (3) or (5)] can be constructed simply by adding other dimensions along which fluc-

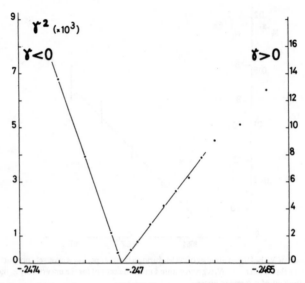

**Fig. 7.** For the model mapping $r_T \simeq -0.24706$. For $r < r_T$ the Lyapunov number $\gamma$ is negative and varies as $-\sqrt{r_T - r}$ while for $r > r_T$ it is positive and grows like $\sqrt{r - r_T}$

tuations are stable [5]. Numerical simulation of the model defined by (2) can easily be performed using a desk-top computer. As expected the Lyapunov number grows with the 1/2-power near the threshold (Fig. 7).

**Type 2. Intermittency**

In order to study numerically this case we have considered a map that applies the torus $T^2 = [0, 1[ \times [0, 1[$ four times onto itself: in complex notations $z = x + iy$

$$z' = \lambda z + \mu |z|^2 z \text{ close to the origin (} \lambda \text{ complex and } \mu \text{ real),} \tag{3a}$$

$$z' = 2z \text{ far from the origin} \tag{3b}$$

with a smooth interpolation inbetween. The fixed point $z = 0$ of this map looses its stability when the complex parameter $\lambda$ crosses the unit circle, the coefficient $\mu$ of the cubic term being so chosen as to avoid the appearance of a stable limit cycle or equivalently to make the bifurcation subcritical i.e. $\mu = \mu \operatorname{Re} \{\lambda\} > 0$. Iterations of the above map show intermittency when $|\lambda| = 1 + \varepsilon$ and $\varepsilon \to 0_+$. Once an iterate falls near $z = 0$, it enters a laminar phase and a large number of further iterations are needed to expell it towards the "bursting region" (where correlations are broken) defined by $|z| > \varrho^*$, $\varrho^*$ being fixed and $\varepsilon$-independent, roughly in the interpolating region. To find how the Lyapunov number grows near the intermittency threshold one may reason as follows: Let $\varrho_j = |z_j|$ be the distance of the $j^{\text{th}}$ iterate to the fixed point. The iterates rotate around the fixed point due to the complex nature of $\lambda$ but we shall neglect the angular variation and only consider the growth of the modulus

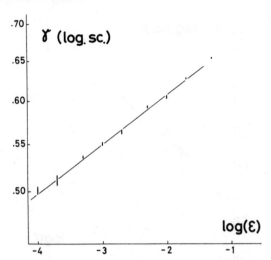

**Fig. 8.** On the torus $T_2$ for $\lambda=(1+\varepsilon)\exp i\varphi$, $\varphi=0.05$ rd and $\mu=20$ "mean field" theory predicts $\gamma \sim \ln(1/\varepsilon)$ while the numerical simulation gives $\gamma \sim \varepsilon^{\alpha}$ with $\alpha \sim 0.04$

$\varrho$. Near $\varrho=0$ it is approximatively given by

$$\varrho_{n+1}=(1+\varepsilon)\varrho_n+\bar{\mu}\varrho_n^3+\text{H.O.T.} \tag{4}$$

Now let us examine a laminar cycle with starting point at $\varrho=\tilde{\varrho}\ll\varrho^*$. If $\tilde{\varrho}\gg\varepsilon^{1/2}$ one easily sees that a number of iteration of order $1/\tilde{\varrho}^2$ are needed to reach $\varrho^*$ and enter a turbulent burst. On the other hand if $\tilde{\varrho}\ll\varepsilon^{1/2}$ the laminar cycle ends after $\varepsilon^{-1}\ln\varrho$ iterations approximately. Assuming then that $\tilde{\varrho}$ is at random with probability $\tilde{\varrho}\,d\tilde{\varrho}$ in the circle of center 0 and radius $\varrho^*$ the estimates given above yield $\ln(1/\varepsilon)$ as an order of magnitude for both the mean duration of a laminar period and the inverse Lyapunov number near $\varepsilon=0_+$. This is in slight disagreement with our computer experiments which seem to indicate rather a power-like growth of the Lyapunov number $\gamma\sim\varepsilon^{\alpha}$ $\alpha$ small and positive (Fig. 8). This descrepancy between the naive theory presented above and computer results may come from the neglect of fluctuations about the mean length of the laminar cycles, which makes the procedure used sound much like a "mean field theory" in the usual jargon of phase transitions (it may also come from the neglect of the rotation of iterates affecting the statistics in an unknown way).

## Type 3. Intermittency

The last type of intermittency we shall examine may occur when the Floquet multiplier is real and crosses the unit circle at $(-1)$. Although a differential system has been found which displays this kind of behavior [2b] we shall report here on the simulation of the following mapping of the circle $S_1$ onto itself:

$$\theta\to 1-2\theta-\frac{1}{2\pi}(1-\varepsilon)\cos\left[2\pi\left(\theta-\frac{1}{12}\right)\right](\text{mod}\,1). \tag{5}$$

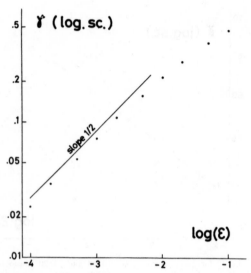

**Fig. 9.** On the torus $T_1$ for type 3 intermittency the Lyapunov number $\gamma$ grows as $\varepsilon^{1/2}$

This map applies $S^1$ twice onto itself and it reverse the orientation so that the eigenvalue of the map linearized near the fixed point $\theta_F = 1/3$ can easily be made negative. Near the fixed point the map expands as

$$\bar{\theta}_{n+1} = -(1+\varepsilon)\bar{\theta}_n - \frac{(2\pi)^2}{6}\bar{\theta}_n^3 + \text{H.O.T.} \ (\bar{\theta} = \theta - \theta_F). \tag{6}$$

The most general form would be

$$\bar{\theta}_{n+1} = -(1+\varepsilon)\bar{\theta}_n + a\theta_n^2 + b\theta_n^3, \tag{7}$$

$a$ and $b$ being constant. If the r.h.s. of (7) has a positive Schwarzian derivative that is here $b + a^2 < 0$ then the bifurcation at $\varepsilon = 0$ is subcritical and type 3 intermittency can occur. This is precisely the case with (5) since $a = 0$ and $b < 0$. To estimate the mean length of a laminar phase one considers instead of (6) or (7) the equation giving $\bar{\theta}_{n+2}$ in function of $\bar{\theta}_n$. This relation is basically of the same form as Eq. (4) (quadratic terms vanish at $\varepsilon = 0$ and are in inessential for $\varepsilon$ small enough). Thus one reasons as for type 2 intermittency with the difference that now the problem is strictly unidimensional so that the probability measure for for the starting point of a laminar cycle is now the usual Lebesgue measure instead of $\varrho d\varrho$ previously. An elementary calculation shows that the Lyapunov number should grow like $\varepsilon^{1/2}$ near threshold, this time in agreement with the computer experiment (Fig. 9).

### Conclusion

Intermittency is a quite common phenomenon in experimental turbulence. The theory sketched in this paper is more especially related with the case of convection in confined geometries [1] but intermittency is also well known in boundary layers and pipe flows [6] and even in $1/f$ – noise theory [7]. Despite the different meanings of the term "intermittency", the possibility remains that the kind of dynamics described by the models we have studied could afford a qualitative understanding of all these phenomena.

## References

1. Maurer, J., Libchaber, A., Bergé, P., Dubois, M.: Personal communications
2. (a) Manneville, P., Pomeau, Y.: Phys. Lett. **75**A, 1 (1979)
   (b) Arneodo, A., Coullet, P., Tresser, C.: Private communication
3. Iooss, G.: Bifurcation of maps and applications. In: North-Holland Math. Studies 36. Amsterdam, New York: North-Holland 1979
4. Lorenz, E.N.: J. Atmos. Science **20**, 130 (1963)
5. Pomeau, Y.: Intrinsic stochasticity in plasmas, Cargèse 1979. (eds. G. Laval, D. Gresillon). Orsay: Editions de Physique 1979
6. Tritton, D.J.: Physical fluids dynamics. New York: Van Nostrand-Reinhold 1977
7. Mandelbrot, B.: Fractals form chance and dimension. San Francisco: Freeman 1977

Communicated by D. Ruelle

Received January 17, 1980

Volume 87A, number 8          PHYSICS LETTERS          1 February 1982

# INTERMITTENCY IN THE PRESENCE OF NOISE: A RENORMALIZATION GROUP FORMULATION

J.E. HIRSCH and M. NAUENBERG [1]

*Institute for Theoretical Physics, University of California, Santa Barbara, CA 93106, USA*

and

D.J. SCALAPINO

*Institute for Theoretical Physics and Department of Physics, University of California, Santa Barbara, CA 93106, USA*

Received 14 November 1981

A renormalization group (RG) formulation of the transition to chaotic behavior via intermittency in one-dimensional maps is presented. The known scaling behavior of the length of the laminar regions in the presence of external noise is obtained from the leading relevant eigenvalues of the RG transformation. In addition, the complete spectrum of eigenvalues and corresponding eigenfunctions is found.

The renormalization group equations describing a phase transition are regular functions of the parameters such as temperature and external field which determine the state of the system. As these parameters are continuously varied through a phase transition, the singular behavior of the system arises from an infinite iteration of the regular renormalization group equations. Recent work suggests that a similar point of view provides a useful framework for understanding the onset of irregular or chaotic behavior of dynamical systems as a parameter is continuously varied. In particular, iterates of the one-dimensional logistic map

$$x_{n+1} = Rx_n(1 - x_n) , \qquad (1)$$

can change from a regular to an irregular pattern as $R$ is varied. The logistic map exhibits two types of such transitions. One of these involves an infinite cascade of period-doubling or pitchfork bifurcations, while the other arises from a saddle or tangent bifurcation leading to intermittency. Feigenbaum [1] and others [2] have developed a renormalization approach to

[1] Permanent address: Natural Science, University of California, Santa Cruz, CA 95060, USA.

describe the scaling and universal properties of the transition to chaos through period-doubling for the class of one-dimensional maps $x_{n+1} = f(x_n)$, with

$$f(x) = 1 - a|x|^z , \qquad (2)$$

where $z = 2$ for the logistic map. A scaling theory describing the effect of external noise on the period-doubling cascade has also been developed [3,4].

Here we are interested in the second type of transition exhibited by the logistic map. Following the initial ideas of Pomeau and Manneville [5], the onset of chaotic behavior characterized by the occurrence of regular or "laminar" sequences of $x_n$ values separated by intermittent bursts has also been shown to scale [6,7]. For the class of saddle point maps

$$f(x) = x + a|x|^z + \epsilon , \qquad (3)$$

with $z > 1$, the length of the laminar regions $l$ varies for small $\epsilon$ as $\epsilon^{-(1-1/z)}$. In the presence of a stochastic noise source of amplitude $g$, $l$ satisfies the scaling equation

$$l(\epsilon, g) = \epsilon^{-(1-1/z)} f(g/\epsilon^{(z+1)/2z}) . \qquad (4)$$

These relations were established by considering a Langevin equation describing the map near the saddle point, and using Fokker–Planck techniques to determine the time of passage in the presence of noise.

Here we develop a renormalization approach for saddle point maps which puts the known scaling results for intermittency in the same framework as Feigenbaum's treatment of the period-doubling cascade. We consider the class of maps given by eq. (3) in the presence of external noise,

$$x' = f(x) + g\xi , \tag{5}$$

where $\xi$ is a random variable of unit standard deviation. The idea of the renormalization approach is to evaluate the map $x \to x''$ associated with two consecutive iterations of eq. (5) and by rescaling cast it back into the original form. This requires new parameters, $\epsilon', g'$ which in the limit $\epsilon, g \to 0$ satisfy the relation

$$\epsilon' = \lambda_\epsilon \epsilon , \quad g' = \lambda_g g , \tag{6}$$

where $\lambda_\epsilon$ and $\lambda_g$ are the largest relevant eigenvalues of the linearized renormalization group transformation. Then the length $l(\epsilon, g)$ satisfies the homogeneity relation

$$l(\epsilon, g) = 2l(\epsilon', g') , \tag{7}$$

which leads in the usual way to the scaling relation

$$l(\epsilon, g) = \epsilon^{-\nu} f(g/\epsilon^\mu) , \tag{8}$$

with exponents

$$\nu = \log 2/\log \lambda_\epsilon , \quad \mu = \log \lambda_g/\log \lambda_\epsilon . \tag{9}$$

The functional recursion relation we use to define our renormalization procedure is the same as in Feigenbaum's case:

$$T\{f(x)\} = \alpha f(f(x/\alpha)) , \tag{10}$$

where $\alpha$ is a rescaling factor, but with boundary conditions

$$f(0) = 0 , \quad f'(0) = 1, \tag{11}$$

appropriate to a saddle point bifurcation at $x = 0$. It can be readily verified that

$$f^*(x) = x/(1 - ax) , \tag{12}$$

is a fixed point of the transformation (10) with $\alpha = 2$ and $a$ an arbitrary constant. For small $x$ this solution corresponds to eq. (3) for $\epsilon = 0$ and $z = 2$. For $z \neq 2$, we can find the fixed point of (10) by series expansion, and to third non-vanishing order obtain

$$f^*(x) = x + a|x|^z + \tfrac{1}{2} z a^2 |x|^{2z-1} + \dots , \tag{13}$$

with the scale factor $\alpha = 2^{1/(z-1)}$.

The next step is to consider the effect of small perturbations around the fixed point. We write

$$f(x) = f^*(x) + \epsilon h(x) , \tag{14}$$

where $\epsilon$ is a small parameter, and determine the eigenfunction from the usual condition of form invariance after rescaling:

$$f^{*\prime}(f^*(x))h_n(x) + h_n(f^*(x)) = (\lambda_n/\alpha)h_n(\alpha x) , \tag{15}$$

where $\lambda_n$ is the $n$th eigenvalue. For the case $z = 2$ we find $\lambda_n = 4/2^n$ and obtain the eigenfunctions $h_n$ by series expansions. The relevant eigenfunction with eigenvalue $\lambda_\epsilon = 4$ is, to second order in $x$,

$$h_\epsilon(x) = 1 + ax + \tfrac{4}{3}a^2 x^2 + \dots . \tag{16}$$

The other relevant eigenfunction, with eigenvalue $\lambda_1 = 2$, does not correspond to the physical situation of interest here and will not be considered [+1]. The marginal eigenfunction, with $\lambda_2 = 1$, is associated with the arbitrary constant $a$ in the map eq. (12) and can be found in closed form:

$$h_2(x) = x^2/(1 - ax)^2 . \tag{17}$$

Finally, the irrelevant eigenfunctions with eigenvalue $\lambda_n, n > 2$, have the leading behavior $h_n(x) = x^n + \dots$.

In the general case $z > 1$, the form of the eigenvalues is $\lambda_n = 2^{(z-n)/(z-1)}$, and the leading behavior of the eigenfunctions is $x^n$. The largest relevant eigenvalue is $\lambda_\epsilon = 2^{z/(z-1)}$. The case $n = 1$ is again not of interest here. For $1 < n < z$ the perturbation is still relevant and gives a crossover to a behavior described by the map eq. (3) with $z$ replaced by $n$. The marginal eigenvalue, for $n = z$, is again associated with the arbitrary constant $a$, and the eigenfunctions with $n > z$ are irrelevant.

We consider now the effect of adding a small amount of external noise to the fixed point function:

$$f(x) = f^*(x) + g(x)\xi . \tag{18}$$

Under iteration, this leads to the eigenvalue equation [4]

---

[+1] The eigenfunction corresponding to $\lambda = 2$ has the leading behavior $h_1(x) = x$ which changes the character of eq. (3), eliminating the intermittent behavior.

Volume 87A, number 8        PHYSICS LETTERS        1 February 1982

$$f^{*'}(f^*(x))g^2(x) + g^2(f^*(x)) = (\lambda_g^2/\alpha^2)g^2(\alpha x) . \quad (19)$$

For the leading eigenvalue $\lambda_g$ one obtains the exact result

$$\lambda_g = 2^{(z+1)/2(z-1)} , \quad (20)$$

and the corresponding eigenfunction is

$$g(x) \propto 1 + \tfrac{1}{2}za|x|^z/x + \dots . \quad (21)$$

Using the above results for $\lambda_\varepsilon$ and $\lambda_g$ we obtain for the exponents defined in eq. (9),

$$\nu = (z-1)/z , \quad \mu = (z+1)/2z , \quad (22)$$

and from eq. (8) the scaling behavior eq. (4) follows.

In conclusion, we have shown that the scaling behavior for the average length of the laminar regions in the transition to chaos via intermittency can be easily derived from a renormalization group formulation of the problem. We have obtained exact results for the complete spectrum of eigenvalues and the leading terms of the corresponding eigenfunctions of the renormalization group transformation.

Two of us (J.E.H. and M.N.) would like to acknowledge support by the NSF under PHY77-27084 and PHY78-22253. D.J.S. would like to acknowledge the support of the ONR under N0014-79-C-0707.

*References*

[1] M. Feigenbaum, J. Stat. Phys. 19 (1978) 25; 21 (1979) 669.
[2] P. Collet and J.-P. Eckmann, in: Iterated maps on the interval as dynamical systems (Birkhäuser, Boston, 1980).
[3] J.P. Crutchfield, M. Nauenberg and J. Rudnick, Phys. Rev. Lett. 46 (1981) 933.
[4] B. Shraiman, C.E. Wayne and P.C. Martin, Phys. Rev. Lett. 46 (1981) 935.
[5] P. Manneville and Y. Pomeau, Phys. Lett. 75A (1979) 1; Commun. Math. Phys. 74 (1980) 189.
[6] J.-P. Eckmann, L. Thomas and P. Wittwer, to be published.
[7] J.E. Hirsch, B.A. Huberman and D.J. Scalapino, to be published.

# Exact Solutions to the Feigenbaum Renormalization-Group Equations for Intermittency

Bambi Hu

*Department of Physics, University of Houston, Houston, Texas 77004*

and

Joseph Rudnick [a]

*Department of Physics, University of California, Davis, California 95616*

(Received 19 March 1982)

Exact solutions to the Feigenbaum renormalization-group recursion relation, and the associated eigenvalue equations describing deterministic as well as stochastic perturbations, are found for the case of intermittency. These solutions are generated by a reformulation of the one-dimensional iterated map that exploits its topological equivalence to a translation. Direct resummation of series expansions gives the same results.

PACS numbers: 05.40.+j, 02.50.+s

The study of bifurcation and the transition to chaos has attracted intense interest recently, and considerable progress has been made. The three most commonly discussed scenarios,[1] associated, respectively, with the works of Feigenbaum,[2] Manneville and Pomeau,[3] and Ruelle and Takens,[4] are based on three different types of bifurcations: the pitchfork, tangent, and Hopf bifurcations. The much discussed period-doubling route to chaos is based on the pitchfork bifurcation.

The tangent bifurcation, on the other hand, offers a different route to chaos via intermittency. In this scenario, intermittency is a precursor to periodic behavior. It consists of long-lived episodes of nearly periodic behavior, the duration of which becomes arbitrarily long as the transition, via a tangent bifurcation [Fig. 2(a)], is approached.

Recently, Hirsch, Huberman, and Scalapino,[5] following the initial ideas of Manneville and Pomeau, proposed a detailed theory of intermittency. Scaling relations for the length of laminarity in the presence of noise were established[5,6] by considering a Langevin equation describing the map near the saddle point, and using the Fokker-Planck techniques to determine the time of passage. Very remarkably, Hirsch, Nauenberg, and Scalapino[7] later found that the same results can be simply explained by using the same functional renormalization-group equations first proposed by Feigenbaum in his study of period doubling—with a mere change of boundary conditions appropriate to the tangency condition. Thus the renormalization group provides a unified and elegant approach to both period doubling and intermittency.

The renormalization-group approach as formulated by Feigenbaum postulates the existence of a universal map, obtained by repeated compositions and rescalings of the original map, at the onset of chaos. The rescaling factor needed to generate the universal map yields one universal exponent. Eigenvalues describing the rate at which perturbations of this map grow provide the others.

To find the spectrum of eigenvalues and corresponding eigenfunctions, Hirsch, Nauenberg, and Scalapino used series-expansion techniques. In the simplest $z = 2$ case they were able to sum the series and obtain a closed-form solution to the universal function. However, the universal function for arbitrary $z$ and all eigenfunctions were only computed to the first few orders.

We have found that it is possible to obtain not only all the exponents for intermittency, but also closed-form results for the universal functions and all eigenfunctions corresponding to deterministic as well as stochastic perturbations for arbitrary $z$. This was achieved by a simple transformation that recasts the map near a tangent bifurcation into a simple translational map $x_{i+1} = x_i + b$, with $b$ a constant. Direct resummation of series expansions corroborates our results. Whether this technique will prove to be of general utility remains to be seen, but the remarkable simplification it leads to in the renormalization-group study of intermittency induces us to believe that it may well prove a useful tool in the study of other dynamical transitions.

The tangent bifurcation as it occurs in iterated one-dimensional maps is illustrated in Fig. 1. Here the map $f(x) = rx(1-x)$ and its third iterate $f^{(3)}(x) = f(f(f(x)))$ are shown at $r = r_3 = 1 + \sqrt{8}$. For

294

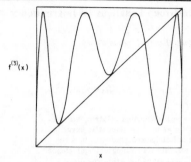

FIG. 1. The third-iterated map $f^{(3)}(x)$ at $r_3$.

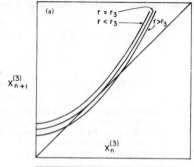

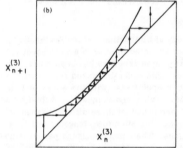

FIG. 2. (a) Tangent bifurcation near $r_3$. (b) Slow passage through the channel region.

$r \gtrsim r_3$, $f^{(3)}(x)$ has two unstable fixed points, at $x = 0$ and $x = (r - 1)/r$. These are the unstable fixed points of $f(x)$. As $r$ passes through $r_3$ [see Fig. 2(a)], $f^{(3)}(x)$ acquires six new fixed points, three stable and three unstable. The three stable fixed points are the three elements of a stable period-three limit cycle of $f(x)$. Even though the map has no stable period-three cycle when $r \gtrsim r_3$, it is evident [Fig. 2(b)] that under repeated iterations of $f^{(3)}(x)$, $x_i$'s spend several iterations in the immediate vicinity of the points of closest approach to the 45° line. This behavior corresponds to orbits under $f(x)$ that look nearly periodic for a sizable number of iterations, i.e., they almost repeat themselves every third iteration, but eventually slip out of this pattern, and, shortly thereafter, establish another pattern of near periodicity. This sequence of long-lived episodes is the phenomenon of intermittency.

To study the transition to periodicity of order $n$ we consider the $n$th iterated map in the immediate vicinity of one of the $n$ points at which it achieves tangency to the 45° line at the transition. Shifting the origin of coordinates to that point, we have for the map at tangency

$$f^n(x) = x(1 + ux^{z-1}) + O(x^z), \qquad (1)$$

where $u$ is the coefficient of expansion. The exponent $z$ determines the "universality classes." Most commonly $z$ will be equal to 2. Here we keep it general with the understanding that

$$x^{z-1} \equiv |x|^{z-1} \operatorname{sgn}(x). \qquad (2)$$

The universal map $f^*(x)$ has a power-series expansion in $x$ whose two lowest-order terms match the right-hand side of Eq. (1). The map furthermore satisfies

$$f^*(f^*(x)) = \alpha^{-1} f^*(\alpha x), \qquad (3)$$

where $\alpha$ is the rescaling factor mentioned earlier. If we add a small perturbation $\epsilon h_\lambda(x)$ to $f^*(x)$ then the composition of $f_\epsilon(x) = f^*(x) + \epsilon h_\lambda(x)$ satisfies

$$f_\epsilon(f_\epsilon(x)) = \alpha^{-1} f^*(\alpha x) + \epsilon(\lambda/\alpha) h_\lambda(\alpha x) + O(\epsilon^2) \quad (4)$$

when the eigenfunction $h_\lambda(x)$ satisfies

$$f^{*\prime}(f^*(x)) h_\lambda(x) + h_\lambda(f^*(x)) = (\lambda/\alpha) h_\lambda(\alpha x). \qquad (5)$$

There will actually prove to be a *spectrum* of eigenvalues $\lambda$ and corresponding eigenfunctions $h_\lambda(x)$. Stochastic exponents[8,9] are associated with the rate of growth of stochastic perturbations of the form $\xi g_{\lambda_g}(x)$, with $\xi$ a random variable controlled by a probability distribution of unit width. Here the eigenfunctions satisfy

$$f^{*\prime 2}(f^*(x)) g_{\lambda_g}{}^2(x) + g_{\lambda_g}{}^2(f^*(x))$$
$$= (\lambda_g/\alpha)^2 g_{\lambda_g}{}^2(\alpha x). \qquad (6)$$

Consider now the following recursion relation:

$$G(x') = G(x) - a \qquad (7)$$

with $G(x)$ a function to be determined shortly. Iterating this recursion relation we have

$$G(x'') = G(x') - a = G(x) - 2a.\qquad(8)$$

We can generate the universal map for intermittency by choosing a $G(x)$ for which a rescaling of $x$ yields the original recursion relation Eq. (7) from the iterated recursion relation Eq. (8). Such a function satisfies

$$G(x) = 2G(\alpha x).\qquad(9)$$

The form we need is $G(x) = x^{-(z-1)}$, with

$$\alpha = 2^{1/(z-1)}.\qquad(10)$$

Note that here the quantity $z$ is arbitrary. Thus, a function satisfying Eq. (3) is obtained by recasting the recursion relation Eq. (7) into explicit form. Using $G(x) = x^{-(z-1)}$ we obtain

$$x' = f^*(x) = (x^{-(z-1)} - a)^{-1/(z-1)}\qquad(11)$$

Replacing $a$ by $(z-1)u$ in Eq. (11) we have

$$f^*(x) = x[1 - (z-1)ux^{z-1}]^{-1/(z-1)}.\qquad(12)$$

It can be verified explicitly that $f^*(x)$ in Eq. (12) satisfies Eq. (3) and reproduces the correct power-series expansion. Furthermore the scale factor $\alpha$ given by Eq. (10) is correct.

We now consider the effect of a perturbation to $G(x)$ in Eq. (7). Our new implicit recursion relation is

$$x'^{-(z-1)} + \epsilon H(x') = x^{-(z-1)} + \epsilon H(x) - u(z-1).\qquad(13)$$

If $H(x) = x^{-p}$, then iterating the recursion relation Eq. (13) and rescaling by $\alpha$ as given by Eq. (10), we obtain our original recursion relation except that the coefficient $\epsilon$ has been increased by the factor $\lambda$, where

$$\lambda = 2^{p-z+1/(z-1)}.\qquad(14)$$

The associated eigenfunction is obtained by recasting Eq. (13) into an explicit recursion relation. Solving for $x'$ in terms of $x$ to order $\epsilon$ we obtain

$$x' = x[1 - u(z-1)x^{z-1}]^{-1/(z-1)} - \frac{\epsilon}{z-1}[x^{-(z-1)} - u(z-1)]^{-z/(z-1)}\{x^{-p} - [x^{-(z-1)} - u(z-1)]^{p/(z-1)}\} + O(\epsilon^2)$$

$$\equiv f^*(x) - \frac{\epsilon up}{z-1} h_\lambda(x) + O(\epsilon^2).\qquad(15)$$

The eigenfunction $h_\lambda(x)$ has been normalized so that its lowest-order term in $x$ is $x^{2z-1-p}$. If we want an eigenfunction corresponding to a shift from tangency, that lowest-order term must be 1, and so we must choose $p = 2z - 1$, which means than $\lambda$ in Eq. (14) is equal to $2^{z/(z-1)}$. This matches with the relevant eigenvalue of Hirsch, Nauenberg, and Scalapino.[7]

The stochastic eigenfunctions are variants of the nonstochastic ones. They are

$$g_{\lambda_g}{}^2(x) = (1/uq)[x^{-(z-1)} - u(z-1)]^{-2z/(z-1)}\{x^{-q} - [x^{-(z-1)} - u(z-1)]^{q/(z-1)}\}$$

$$= (x/uq)^{2z-q}\{[1 - u(z-1)x^{z-1}]^{-2z/(z-1)} - [1 - u(z-1)x^{z-1}]^{-(2z-q)/(z-1)}\}.\qquad(16)$$

with

$$\lambda_g = 2^{[q-2(z-1)]/2(z-1)}.\qquad(17)$$

The lowest-order term in $g_{\lambda_g}{}^2$ is $x^{3z-1-q}$. If we want that term to be a constant we must choose $q = 3z - 1$, in which case $\lambda_g$ in Eq. (17) is equal to $2^{(z+1)/2(z-1)}$. All these results can also be obtained directly by resumming the series expansions.

The fact that a reformulation of the recursion relation leads to an immediate and complete solution of the renormalization-group equations for the iterated map near a tangent bifurcation is highly intriguing. Whether or not this kind of reformulation proves useful in the study of other transitions in dynamical systems remains to be

seen. It certainly deserves to be considered as a viable approach.

This complete set of exact solutions provides a rare laboratory where ideas and theories can be experimented with and tested. The underlying mathematical structure, physical implications, and experimental consequences are still to be ruminated. However, since the method employed here depends crucially on the fact that the map for intermittency is topologically equivalent to a translation, most likely it will not prove to be fruitful for the study of period doubling.

One of us (B.H.) would like to thank Professor C. N. Yang for his interest in this work and encouragement. This work was supported in part by the National Science Foundation.

(a)Permanent address: Department of Physics, University of California, Santa Cruz, Cal. 95064.

[1]J.-P. Eckmann, Rev. Mod. Phys. 53, 643 (1981).

[2]M. J. Feigenbaum, J. Stat. Phys. 19, 25 (1978), and 21, 669 (1979).

[3]P. Manneville and Y. Pomeau, Phys. Lett. 75A, 1 (1979), and Commun. Math. Phys. 74, 189 (1980).

[4]D. Ruelle and F. Takens, Commun. Math. Phys. 20, 167 (1971).

[5]J. E. Hirsch, B. A. Huberman, and D. J. Scalapino, Phys. Rev. A 25, 519 (1982).

[6]J.-P. Eckmann, L. Thomas, and P. Wittwer, J. Phys. A 14, 3153 (1981).

[7]J. E. Hirsch, M. Nauenberg, and D. J. Scalapino, Phys. Lett. 87A, 391 (1982).

[8]J. P. Crutchfield, M. Nauenberg, and J. Rudnick, Phys. Rev. Lett. 46, 933 (1981).

[9]B. Shraiman, C. E. Wayne, and P. C. Martin, Phys. Rev. Lett. 46, 935 (1981).

## 7. Transition from Quasiperiodicity to Chaos

790

IEEE TRANSACTIONS ON CIRCUITS AND SYSTEMS, VOL. 35, NO. 7, JULY 1988

# Quasi-Periodicity and Dynamical Systems: An Experimentalist's View

JAMES A. GLAZIER AND ALBERT LIBCHABER

*Abstract* — A great variety of natural and artificial systems exhibit chaos and frequency locking associated with quasi-periodicity. In this tutorial paper we present an overview of current theoretical and experimental work on quasi-periodicity. In Section I, we discuss the concept of universality and its relevance to experiments on nonlinear multifrequency systems. In Section II, we describe the reduction of experimental data by means of Poincaré sections, and the mathematical properties of the one-dimensional circle map. In Section III, we present the various dynamical systems techniques for determining scaling and multifractal properties as well as other more traditional methods of analysis. We emphasize the experimental observations that would support or refute the one-dimensional circle map model. In Section IV, we summarize the experimental results, concentrating on forced Rayleigh–Bénard convection and solid state systems. In Section V, we conclude with a brief discussion of the accomplishments and open problems of the dynamical systems theory of quasi-periodicity.

## I. INTRODUCTION

THE phenomenon known today as *frequency locking* was discovered over three hundred years ago when the Dutch physicist Christian Huygens noted that the pendula of two clocks placed near each other tended to synchronize [74]. This effect, in which an oscillator adjusts its frequency in response to a periodic stimulus (either externally or internally generated), is used today in many electronic systems requiring precise control of frequencies. Examples including the phase locking circuitry of atomic clocks, in which a quartz oscillator is locked to a Cesium standard, radio receivers, stereo turntables, and disk drives.

In the natural world, systems which exhibit frequency locking behavior are almost bewildering common; the most visible example being the moon, whose orbital and rotational periods are locked in a one-to-one ratio because of dissipative tidal forces. In any system in which two or more frequencies couple nonlinearly, either because of external perturbations or internal generation, a rich variety of effects can occur, including frequency locking, quasi-periodicity, pattern formation, intermittency, period doubling and other subharmonic generation, and both temporal and spatial chaos. It is sobering even to attempt to list the systems which have been examined experimentally: in mechanics, the damped driven pendulum [3], [34], [60], [101]; in hydrodynamics, the vortices behind an obstacle in a wind tunnel or an airplane wing [122], the dripping of a faucet [116], the convective rolls in a heated pan of water,

and the oscillations of acoustically driven helium [82], [121]; in chemistry, the Belousov–Zhabotinsky reaction, the Chlorite–Thiosulphate reaction and many others [38], [111]; in solid state physics, charge density waves in niobium selenide [22], and other compounds [117], the conductivity of barium sodium niobate [94], oscillations in Josephson junctions [105], and in germanium [62], [72], in biology, cardiac cells [45], the brain [10], the slime mode *dictyostelium discoideum* [104], menstrual cycles in human females, and elsewhere [136]. This list could probably be extended almost indefinitely. We refer the reader interested in additional reading on experiments and theory to the bibliographies contained in the many surveys of specialized topics [11], [15], [33], [49], [102], [108], [128], [137].

One major characteristic of the above list is that, though all of its members exhibit complicated multifrequency behavior, they seem to share almost no other features. It is clearly impossible to produce a single theory which describes the detailed behavior of all of them. A theory which describes voltage oscillations in Josephson junctions can scarcely be expected to describe the life cycle of slime molds. A further problem is that full mathematical descriptions are not known for many of these systems; often, when the equations are known, they are effectively insoluble.

For a long time these various effects were seen as unrelated, if occasionally useful, curiosities. It required the development of a new branch of physics to allow us to appreciate them for what they are, the diverse results of a single elegant and simple theory. We find that if we examine all these systems at a high enough level, that is if we ignore detailed causes, they can be grouped into a few classes of generic behavior. This concept of *universality* — that seemingly unrelated systems can behave in essentially the same way — is central to many recent advances in physics. A Fortran programmer knows instinctively the lesson that physicists have had to learn with effort: it is the result of the program, not the particular machine language implementation that matters. In this case the "high level language" is, *dynamical systems theory*, the formalism which describes complicated behavior and chaos in terms of sequences produced by the repeated iteration of simple functions, and the relation of these iterated functions to the "machine code" of differential equations.

We may illustrate this point by considering an example from our own research [95]. The behavior of a fluid in a

Manuscript received September 25, 1987.
The authors are with The James Franck and Enrico Fermi Institutes, The University of Chicago, Chicago, IL 60637.
IEEE Log Number 88201084.

box, heated from below (Rayleigh–Bénard convection) has been studied for almost a hundred years [83]. Yet, 10 years ago, if you had asked a theoretical physicst to predict the flow pattern in such a box of fluid subject simultaneously to a magnetic field, heating and an alternating injected current, he would probably have said that it would be complicated and, because the result would apply only to one specialized system, uninteresting. If you had convinced him to try to calculate the flow he would have written down three coupled nonlinear partial differential equations (the Navier Stokes equation for the fluid flow and transport equations for the current and heat) and paused, for, with realistic boundary conditions, the equations would be completely intractable. To proceed further he would have assumed simplified boundary conditions, linearized the equations about a known solution [26], and with luck (for even the linearized equations are non-trivial to solve), produced an "approximate solution."

This answer would have had two fundamental problems: 1) changing the boundary conditions or the geometry even slightly would require resolving the whole problem from scratch; and 2) the answer would be both quantitatively and qualitatively wrong, because the linearization would be invalid in the range of parameters of interest. In particular, in the region of chaotic behavior, the "approximation solution" would be completely meaningless.

Today, despite extraordinary advances in computers and the techniques for solving partial differential equations, a physicist could still not solve the problem asked above. Except in the most trivial cases the Navier–Stokes equations remain insoluble. However, as first shown by Lorenz in his classic work on convection and weather prediction [87], in certain types of Rayleigh–Bénard convection we can predict many properties with excellent numerical accuracy without solving any differential equations at all. Indeed, for the small aspect ratio, forced Rayleigh–Bénard system, many of the calculations described in this article could be done with nothing more elaborate than a programmable calculator. Furthermore, the results we obtain for the Rayleigh–Bénard system apply, with only minor modifications, to many of the other systems we listed above. Using the techniques of dynamical systems theory, we can attain a universal result without a detailed solution to the underlying equations of motion.

We should add a note of caution. The dynamical systems approach is not a panacea. There are classes of questions, just as there are classes of systems, and we will discuss in this paper the sorts of questions our "high level" theory can answer. One thing we can definitely not predict is the detailed motion of a large volume of fluid in space and time. Though originally developed in the context of Hamiltonian systems, the iterated map approach discussed in this paper works best in heavily damped (or dissipative) systems in which most of the degrees of freedom are suppressed and only a few contribute to the behavior. In fact, the existing theory is only well developed for systems with one or two independent degrees of freedom. Fortunately, most of the systems on our list have this prop-

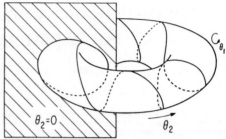

Fig. 1. Schematic diagram of a two frequency torus and a Poincaré section. The helical line on the torus traces out the system's trajectory in phase space. Angular coordinates $\theta_1$ and $\theta_2$ are indicated. The vertical plane indicates the stroboscopy at times $i\tau$. The Poincaré section is the intersection between the plane and the system trajectory. From [76].

erty. However, if we want to design an airplane wing, or a turbine, both of which depend on true many dimensional effects, we still need to solve the Navier–Stokes equations in detail. For the same reason we cannot address the problem of turbulence using existing dynamical systems techniques. With these caveats in mind we may turn to the theory of systems with two degrees of freedom.

## II. Theory of the Circle Map

### From Phase Space to the Iterated Map

Let us consider the simplest possible two-frequency system, two uncoupled harmonic oscillators with frequencies $f_1$ and $f_2$. We may characterize the state of the system by coordinates describing the amplitude of the oscillators $x_1$ and $x_2$ and their time derivatives $\dot{x}_1$ and $\dot{x}_2$. In this simple case, we can immediately reduce the number of variables to two by expressing both coordinate pairs in terms of angular coordinates $\theta_1 \equiv f_1 t$ and $\theta_2 \equiv f_2 t$, with $x_1 = \sin(\theta_1)$, $\dot{x}_1 = f_1 \cos(\theta_1)$, etc. $\cdots$. We can understand this system in a simple geometrical way. If we make a rotation by 360° correspond to $\theta = 1$ and identify $\theta \cong \theta + 1$, we may represent the time evolution of the system as a helical motion on a torus with the small diameter corresponding to $\theta_1$ and the large diameter corresponding to $\theta_2$, as shown in Fig. 1. Even this trivial system exhibits two qualitatively distinct behaviors depending on the ratio $f_1/f_2$, the number of rotations in the $\theta_1$ direction per rotation in the $\theta_2$ direction. If $f_1/f_2 = p/q$ is rational, then the motion is periodic, and the path will close after $q$ circuits around the big circle. We say that the system is *periodic with period q* and completes *p cycles per period*. If, on the other hand, $f_1/f_2$ is irrational, then the path never intersects itself and the trajectory will cover the torus densely, that is, the trajectory will come arbitrarily close to any point on the torus. A system containing two or more incommensurate frequencies is said to be *quasi-periodic*.

Visualizing a torus is inconvenient. We can simplify the picture by using the equivalent of a strobe light to freeze the motion in the $\theta_2$ direction and to eliminate the frequency $f_2$ from the problem. If we record $\theta_1$ at a fixed

792                                                        IEEE TRANSACTIONS ON CIRCUITS AND SYSTEMS, VOL. 35, NO. 7, JULY 1988

time interval $\tau \equiv 1/f_2$ and define $\theta_i \equiv \theta_1(i\tau)$, we flash the strobe at the frequency $f_2$, and take a slice through the torus at a fixed value of $\theta_2$ as indicated by the vertical plane shown in Fig. 1. It is a general theorem [36] that the structure of the stroboscopy we obtain will be the same for almost all (in a measure theoretic sense) choices of $\theta_2$. We have now reduced our four-dimensional problem to one dimension. We may therefore encode all of the dynamics of the problem in the form of a map from the circle onto itself, where the *return map*, $F(\theta)$ is defined by $\theta_{i+1} = F(\theta_i)$. In our example of a uniform rotation, $F(\theta)$ is the rotation map, $F(\theta) = \theta + \Omega$, where $\Omega = f_1/f_2$. For an arbitrary system we will obtain a return map of form, $F(\theta) = \theta + \Omega + f(\theta)$, $f_1$ will vary with $f_2$, and the period of the sequence $\{\theta_i\}$ will not equal the denominator of $\Omega$. In this case it is convenient to describe the frequency of the system using the *winding number* [17], [76]:

$$W \equiv \lim_{i \to \infty} \frac{\theta_i - \theta_0}{i}$$

which is, in fact, the measured frequency ratio, $f_1/f_2$. In the case of uniform rotation, $W = \Omega$. If $W$ is rational $\{\theta_i\}$ will be a finite periodic set of points, if irrational, $\{\theta_i\}$ will be quasi-periodic and cover the circle densely.

This formalism may seem elaborate for the problem in hand, but it does yield one immediately useful result. We have reduced a problem on the torus to the study of a map from the circle to itself.

In a real experiment, we measure the value of an oscillating variable $T$, at times $i\tau$ as described above, and plot $T_i$ versus $T_{i+1}$. This produces a tangled one-dimensional loop or a finite set of discrete points lying on a bumpy and folded surface, not a smooth circular doughnut. A theory of Takens [112], [131] assures that this attractor contains the same information as a plot of $T$ versus $\dot{T}$. For many purposes, e.g., the calculation of a local scaling or of a fractal dimension, this folded attractor is perfectly adequate. However, if we wish to calculate a return map, or an $f(\alpha)$ spectrum (to be discussed later) we must map the two-dimensional pairs $(T_i, T_{i+1})$ into the $\theta_i$'s, using a method developed by Thomae [77], [124] in which we measure the unknown winding number by plotting the time series versus a known rotation frequency (or the stroboscopy $T_i$ versus $iW$) and looking for a one-dimensional Lissajous pattern.

We pick a $W$ and plot $T$ versus $Wt$. If we have chosen $W$ correctly, the periods of the experimental data and $W$ will correspond and we will obtain a one-dimensional curve as seen in Fig. 11 column 1. If our guess is close but not exact, we will see a gradually drifting Lissajous pattern. We repeatedly guess values of $W$ and plot the results until we obtain a satisfactory agreement. An experienced operator with good data can calculate $W$ to one part in $10^5$ in four or five iterations. Using this $W$ we can define an unambiguous order on the experimental attractor and assign a value of $\theta$ to each point. It is then a simple matter to calculate the return map or the $f(\alpha)$ spectrum. Unfortunately this method only works efficiently for one-dimen-

sional sets and hence cannot be used in the strongly chaotic regime where many experiments show fundamentally two-dimensional behavior.

At this point it is helpful to introduce a few definitions. The reduction of a continuous time series to a discrete sequence $(\{T_i\})$ using stroboscopy is known as *taking a Poincaré section* [28], [37] and may be employed in an arbitrary number of dimensions. The plot of $T_i$ versus $T_{i+1}$ (versus $T_{i+2}, \cdots$ in higher dimensions) is the *Poincaré section*. It is also called an *attractor* because all points initially lying in some volume containing it, rapidly iterate towards it. This attraction is equivalent to the damping of an initial transient or perturbation. A given system may have more than one attractor for the same parameters, in which case it is said to be *multistable*. The reduction of the attractor to the $\{\theta_i\}$ form is known as *unwinding* and the resulting sequence $\{\theta_i\}$ is called an *orbit*. Because of the fundamental equivalence of these two representations, we shall use the terms orbit, section and attractor, interchangeably.

In an arbitrary two variable system the reduction procedure can break down at any point. It may not be possible to eliminate the time derivatives. If it is, the set of points produced by taking the Poincaré section may be a two-dimensional cloud, not a loop or finite set of points (this is the case for many strongly chaotic systems). It is the surprising experimental fact that many systems do produce one-dimensional Poincaré sections that makes the one dimensional theory discussed below useful.

*The Circle Map*

We next consider a slightly more complicated return map, which we will use as our model for the rest of this paper. We define the one-dimensional *standard circle map*, or *sine map* by

$$F(\theta) = \theta + \Omega - \frac{k}{2\pi} \sin(2\pi\theta).$$

In an experimental system, we define $\Omega$ to be the ratio $f_1^0/f_2$, where $f_1^0$ is the natural unperturbed oscillation frequency. The exact choice of the function $\sin(2\pi\theta)$ is not critical in this definition. Essentially any function with a single cubic inflection point will yield identical qualitative and similar quantitative behavior. The relative independence of the properties of the iterates on the exact form of the map makes the sine map model very general.

The big advance from the rotation map discussed in the previous section, is that we now have a nonlinear term with an adjustable strength and hence can examine what happens as we vary the nonlinearity. For $k = 0$ we are back to the linear situation described above, but for $0 < k \leqslant 1$ the situation is more interesting. $F(\theta)$ is still a simple invertible map of the circle onto itself, but the winding number $W$, no longer equals $\Omega$. Each irrational $W$ corresponds to a unique $\Omega$ as before. However, there is a finite interval $[\Omega_{W,1}, \Omega_{W,2}]$ over which the iterated map achieves each rational $W = p/q$ and $\{\theta_i\}$ (we should really write $\{\theta_i(k, \Omega)\}$) is periodic with period $q$. We say that the

302

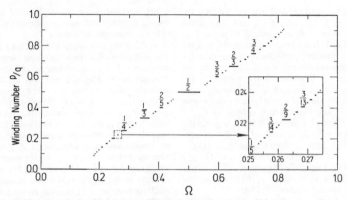

Fig. 2. Devil's staircase for the critical circle map. The steps indicate the regions in which $W$ is constant. Fractions indicate $W$ for a few of the wider steps. Inset shows an expanded view of the indicated section of the staircase. The structure of the sub-region is the same as that of the entire curve. From [76].

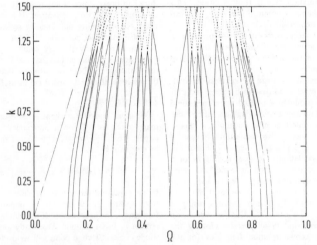

Fig. 3. Arnol'd tongue diagram. The pattern of locked tongues is shown in the $\Omega$ versus $k$ plane for the first few Farey orders. The fractions indicate the winding number achieved inside the tongues. The relative widths of the tongues decrease as the tongue denominator increases and the tongues bend away from each other. The tongues do not overlap below the $k = 1$ critical line. From [76].

equation *locks* to the winding number $W$ over the interval. Experimentally, when the frequency $f_2$ is changed, there is a transient during which $f_1$ gradually "pulls in" to the appropriate ratio to $f_2$. The function $W(\Omega)$ forms a *Devil's Staircase* [12], a monotonic increasing continuous function, with plateaus of finite width at every rational $W$. We show the Devil's staircase for the sine map at $k = 1$ in Fig. 2. The plateaus are *self-similar*; that is, if we enlarge any given segment of the staircase, its texture remains the same, as shown in the inset of Fig. 2.

As we increase the strength of the nonlinearity $k$, the width of each locked interval increases. If we plot these regions in the $k$ versus $\Omega$ plane they form a series of slightly distorted narrow triangles (known as *Arnol'd Tongues* [7]) with their apices on the $k = 0$ axis as shown in Fig. 3. Each tongue represents a region in parameter space associated with a particular rational winding number and we will refer to the tongue associated with a winding number $p/q$ as the *$p/q$-tongue*. Surprisingly, for $k \leqslant 1$ the tongues bend away from each other and do not overlap.

794                                          IEEE TRANSACTIONS ON CIRCUITS AND SYSTEMS, VOL. 35, NO. 7, JULY 1988

The area covered by the locked regions increases smoothly and monotonically from 0 at $k = 0$ to 1 at $k = 1$. At $k = 1$ (the *critical line*) almost any $\Omega$ yields a rational $W$ and the set of $\Omega$ corresponding to irrationals forms a fractal of measure 0. Above the critical line ($k > 1$), $F(\theta)$ is no longer invertible. The tongues begin to overlap, leading to hysteresis effects and chaos. Inside the tongues there are period doubling cascades leading gradually to chaotic motion (the "period doubling" route to chaos). Outside the tongues the remaining quasi-periodic orbits disappear abruptly, giving rise to further chaotic orbits (the "quasi-periodic" route to chaos).

The non-overlapping of tongues below $k = 1$ implies that the width of a tongue corresponding to a rational winding number with denominator $q$, ($w(q)$) must decrease rapidly as $q$ increases. We may make a quick estimate as follows: The number of tongues with a denominator $q$, $n(q)$ is of order $q$ (strictly $n(q) \underset{q \to \infty}{\to} n$). The total width is $w \sim \sum_{q=1}^{\infty} q \cdot w(q)$. For $w$ to remain finite requires $w(q) \sim q^{\beta}$ where $\beta < -2$. Detailed calculations by Bohr, Bak, and Jensen yield the result that at $k = 1$, $\beta = -2.29$ [17].

*Irrational Numbers*

Because the circle map distinguishes strongly between rational and irrational winding numbers, it is worth recalling a few facts about irrationals and methods of approximating them by sequences of rationals. Approximating an irrational by truncating its finite decimal expansion is universally familiar. We will discuss a different method here. Any irrational number, $\sigma \in [0,1]$ can be uniquely represented in *continued fraction* form [7], [28] as

$$\sigma = \cfrac{1}{n_1 + \cfrac{1}{n_2 + \cfrac{1}{n_3 + \cdots}}}$$

where the $n_i$ are positive integers. This formula may be written more conveniently as $\sigma \equiv \langle n_1, n_2, n_3, \cdots \rangle$. If we truncate the expansion after $i$ terms we may define $\sigma_i \equiv \langle n_1, \cdots, n_i \rangle = p_i/q_i$. This yields a sequence of rational approximants, known as the *truncation sequence*, converging to $\sigma$: $\{\sigma_i\} \underset{i \to \infty}{\to} \sigma$.

The truncation sequence is closely related to the Farey ordering of the rationals [2]. For any pair of rational numbers $p/q < p'/q'$ we define their *Farey sum*

$$\frac{p''}{q''} = \frac{p}{q} \oplus \frac{p'}{q'} \equiv \frac{p + p'}{q + q'}.$$

This sum has three properties: 1) $p/q < p''/q'' < p'/q'$, 2) $p''/q''$ is the rational with smallest denominator between $p/q$ and $p'/q'$, and 3) if $|pq' - p'q| = 1$ then $p''/q''$ is in lowest terms. If we construct a "Farey tree" by starting with 0 and 1 and Farey adding nearest neighbors at a given level, property 3) will always be satisfied. We can then construct an approximation sequence converging to $\sigma$ by successively bracketing $\sigma$ and taking appropriate Farey sums. We define $\sigma_0' = 0$ and $\sigma_1' = 1$ and

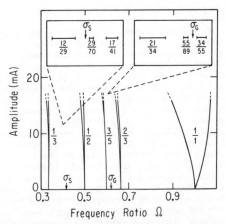

Fig. 4.   Experimental Arnol'd tongue diagram for small aspect ratio forced Rayleigh–Bénard convection in mercury. The tongues are shown in the $\Omega$ versus $A$ plane. The numbers indicate the winding number achieved inside the tongue. Insets show the relative widths of tongues near the golden ($\sigma_G$) and silver ($\sigma_S$) means and mark the position of the critical line. The widths and spacings of these tongues may be used to calculate the fractal dimension and scaling of the unlocked critical set. From [123].

let $\sigma_j' = \sigma_{i-1}' \oplus \sigma_j'$ where $j$ is chosen as large as possible such that $\sigma_j'$ lies on the opposite side of $\sigma$ from $\sigma_{i-1}'$. The sequence $\{\sigma_i'\}$ is in some sense the "best" approximation to the given irrational. It is the sequence of fractions with lowest monotonically increasing denominators which converge to $\sigma$. These lowest denominator tongues are the widest so the "best" sequence is the most significant to the experimentalist. In general the sequence $\{\sigma_i\}$ will be a subset of $\{\sigma_i'\}$ as the reader may easily verify by examining the sequence of fractions in Figs. 2 and 3.

These notions allow us to characterize the "degree of irrationality" of an irrational. We say that a number is strongly irrational if it is hard to approximate by rationals. In particular, numbers which have continued fractions of form $\langle n_1, n_2, \cdots, 1, 1, 1, \cdots \rangle$ are the most strongly irrational. The *golden mean*, $\sigma_G \equiv (\sqrt{5} - 1)/2 = \langle 1, 1, 1, \cdots \rangle$ is the simplest of these. Of all irrationals in the interval $[0,1]$, it is furthest from rationals of any given denominator. The golden mean has several other convenient properties. The sequence given by truncation of the continued fraction is the "best" sequence, and the terms are easily calculable: $\sigma_{i+1} = \sigma_i \oplus \sigma_{i-1} = F_i/F_{i+1}$ where $F_i$ is the $i$th Fibonacci number defined by $F_{i+1} = F_i + F_{i-1}$ for $i \geqslant 1$, $F_0 = 0$ and $F_1 = 1$. Because of its distance from rational approximants, the golden winding number is the easiest place to observe quasi-periodicity experimentally (other winding numbers are more likely to lock to low denominator tongues). Thus the majority of both experimental and theoretical work on quasi-periodicity has been done at the golden mean. A second, slightly less irrational, winding number often selected for study is the silver mean, $\sigma_S \equiv \sqrt{2} - 1 = \langle 2, 2, 2, \cdots \rangle$, for which $\sigma_{i+1} = \sigma_i \oplus \sigma_i \oplus \sigma_{i-1}$. The positions of $\sigma_G$ and $\sigma_S$ and a few of their approximants are indicated in Fig. 4.

We must be a bit careful when we consider the notation for periodic states. For rational winding numbers $f_1/f_2 = p/q$ the notation means that the system returns to its original state after $p$ cycles in $f_1$ or $q$ cycles in $f_2$. Thus we cannot in general divide out common factors between $p$ and $q$ (the use of the Farey construction guarantees that common factors will not appear accidentally). A period doubling represents the appearance of low frequency subharmonics at $f_1/2$ and $f_2/2$. The time the period doubled system takes to return to its original state is now twice as long, but the ratio $f_1/f_2$ is the same. We denote this state using the somewhat bizarre looking notation $2 \otimes p/q$ or $2p/2q$. It may help to think of $p$ and $q$ as elements in a matrix, rather than as a fraction. We denote a state with multiplicity $m$ by writing it out in terms of its prime factors, e.g., a period-18 state would be denoted $3^2 \otimes 2 \otimes p/q$ and call it a *period m* or *multiplicity m* state. There are also additional conventions for distinguishing the qualitative nature of such highly multiplied states which need not concern us here.

## III. EXPERIMENTALLY VERIFIABLE PREDICTIONS

With this basic mathematical formalism we can consider the ways in which an experimental system might behave like a circle map. We will pay particular attention to the feasibility of measurements and to experimental behaviors which are incompatible with the circle map model.

### Global Structure

We have already discussed the typical pattern of Arnol'd tongues produced by the circle map. The presence of a heirarchy of locked states with a unique locked tongue for each rational winding numbers is the most characteristic feature of this map. Other systems which exhibit frequency locking, like phase locked loops, will typically lock only one fixed frequency ratio [34]. Because we do not expect the experimental system to correspond exactly to the simple circle map, we can not hope for exact quantitative correspondence in all aspects of the tongue structure. Nevertheless, the ordering of the tongues and their relative widths as given by the Farey construction are robust, as is the presence of a well-defined critical line. No missing or duplicated tongues are possible. Below the critical line, tongues do not overlap. There is no hysteresis (each value of $k$ and $\Omega$ yields a unique winding number) and only periodic and quasi-periodic states exist. Above the critical line tongues overlap with hysteretic and multistable effects and only periodic and chaotic states exist. For higher dimensional iterated maps there is, in general, no single well-defined critical line and tongues can split and merge in complicated patterns [8], [90].

The sequence of states leading from periodic to chaotic motion within a tongue has been studied by MacKay and Tresser [90], Schell, Fraser, and Kapral [114], and by Glass and Perez [13], [45], [58], [107]. They find that above criticality, the locked states in the tongue undergo

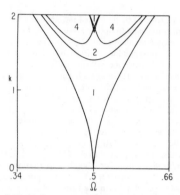

Fig. 5. Subharmonic structure of an Arnol'd tongue. A theoretical calculation of the period doubling structure of the 1/2-tongue for the circle map is shown in the $\Omega$ versus $k$ plane. The lines in the tongue indicate the borders between different periodic states. The areas labelled by $n$ correspond to $n \otimes 1/2$ states. Overlaps indicate regions of multistability. Note the symmetry breaking at periods 4 and the sequence of doubling. From [114].

Feigenbaum period doubling [28], [43]. That is, the length of time (number of iterations) the system takes to return to its initial condition successively doubles. This corresponds to a splitting of the torus into $2^n$ overlapping sheets, where $n$ is the order of the doubling. In the Poincaré section, each point of the locked state splits into $2^n$ distinct points. The separation between successive doublings decreases in accord with the Feigenbaum theory [33], and at a critical value of $k = k_c$, the *accumulation point of the cascade*, the period becomes infinite, i.e., the state becomes chaotic. For still higher $k$ the full bestiary of complicated multiplicities predicted by Metropolis, Stein, and Stein is expected [28], [97], [99], however, the overlapping of tongues makes it difficult to observe experimentally [18]. We present the first few doublings of the 1/2-tongue as calculated by Schell, Fraser, and Kapral, in Fig. 5. The first doubling is symmetric and occurs in a single U-shaped band across the tongue. Higher doublings are composed of at least two distinct doubled states centered in the right- and left-hand sides of the tongue. The orbits of these states are different and "break the symmetry" of the tongue. The order in which these periods appear is generic to all one-dimensional circle maps with cubic inflection points. The presence of a different sequence in experimental observations would rule out identification with the circle map [27].

### Scaling

We have noted that we do not expect quantitative correspondence between experimental and theoretical tongue widths. However, the sensitivity of tongue widths to the exact form of the circle map decreases for large denominators. At smaller length scales the locking sees only the narrow region around the inflection point of $F(\theta)$, much as a Taylor series sees only its lowest order terms for small arguments, so we expect that the ratios of tongue widths or tongue separations for large denominators will be universal quantities depending only on the order of the inflection point. There are many such ratios or *scalings* which can be calculated. We have already mentioned one in passing, the ratio between the $k$ intervals for period doubling bifurcations. However, we will discuss the Shenker $\delta$ [32], [118] (corresponding to the Feigenbaum $\delta$ for the logistic map [43]) which is the most easily calculated from experimental data. Choose an irrational $\sigma$ and let $w_i$ be the width of the tongue corresponding to the winding number $\sigma_i$ in the truncation series. We then define

$$\delta_\sigma \equiv \lim_{i \to \infty} \frac{w_{i-1} - w_i}{w_i - w_{i+1}}$$

or equivalently,

$$\delta_\sigma \equiv \lim_{i \to \infty} \frac{w_i}{w_{i+1}}.$$

Depending on the system, either the first or the second definition may converge more rapidly. It may be helpful to refer to Fig. 4 to see that this limit makes sense around $\sigma_G$ and $\sigma_S$. Renormalization group analysis by Shenker, Shraiman, Bohr, and others gives $\delta_{\sigma_G} = 2.833$ and $\delta_{\sigma_S} = 6.799$ [32], [76], [118]. Fortunately for the experimentalist, the limit converges rapidly and it is only necessary to measure tongues with denominators up to $\approx 100$ to obtain a value of $\delta$ to a few percent.

*Fractal Dimension*

We have mentioned that at $k = 1$ the set of $\Omega$ corresponding to irrational winding numbers is a fractal of measure 0. Fractals are objects whose apparent density or length changes depending on the length scale examined [92]. Such objects are ubiquitous in nature, classic examples being coastlines (which are short if measured in mile lengths but inconceivably long if measured at the scale of a grain of sand), the pattern of branches in a tree, the silhouette of a mountain, and cloud formations [113]. In mathematics the best known example (and a close analogue to the set under consideration) is the Cantor set consisting of all the numbers between 0 and 1 which have no 1's in their ternary expansion.

All objects are characterized by a dimension ($\delta$) which describes how their volume ($V$) changes with length scale ($l$). For an ordinary object: $V \sim l^\delta$, where $\delta$ is an integer. The dimension of a fractal is determined in exactly the same way except that, in the case of a set of points we use not volumes but an effective number of points. We may define the *fractal dimension*, $D$, by the method known as box counting [53], [54], which resembles measuring an area by counting squares on a piece of graph paper. Consider a volume containing the set to be measured and divide it into rectangular $n$-dimensional boxes of side $l$. Let $N(l)$ be the number of boxes containing one or more points. Then, in the limit $l \to 0$, $N(l) \sim l^{-D}$. For a normal object, the two methods yield identical integer results. For a fractal the dimension can be any positive real number. The only restrictions on this method are that the embedding dimension, $n$, must be larger than $D + 1$ and the minimum sample of the points goes like $10^D$. The latter means that box counting is an inefficient way to calculate dimensions, and there exist myriads of specialized tricks for calculating the dimensions of particular systems [56].

As an example we describe the calculation of the dimension of the set of irrational winding numbers at $k = 1$, first by box-counting and then using a trick. Let $w(l)$ be the total width of locked tongues on the interval $[0, 1]$ which have width greater than or equal to $l$. Then $1 - w(l)$ is the total width of the regions which are unlocked at this length scale (i.e., for which the denominator of the winding number is too large). Therefore the number of unlocked boxes at this length scale is $N(l) = (1 - w(l))/l$. We then calculate the fractal dimension $D$ by the box counting method, as $D = -\lim_{l \to 0}(\log(N(l))/\log(l))$. Numerical computations by Jensen, Bak, and Bohr [76] give $D = 0.87 \pm 3.7 \times 10^{-4}$.

For the experimentalist, measuring an arbitrarily large number of tongues to determine $w(l)$ is impractical. It is much more convenient to use a local method developed by Hentschel and Procaccia [73], which depends only on the scaling of the spacing between three Farey neighbors and yields a result within a few percent of the fractal dimension. If we pick an irrational $\sigma$ and look at the "best" sequence of rational approximants we can obtain a fair approximation to $D$ as follows: let $S_i$ be the length of the interval between the tongues corresponding to $\sigma'_{i-1}$ and $\sigma'_i$. Let $S'_i$ and $S''_i$ be the lengths of the intervals between these two tongues and the tongue corresponding to $\sigma'_{i+1}$. See Fig. 4. Then we may define $D'$ by

$$\lim_{i \to \infty} \left\{ \left( \frac{S'_i}{S_i} \right)^{D'} + \left( \frac{S''_i}{S_i} \right)^{D'} \right\} = 1.$$

The numerically computed value of $D'$ for the circle map is $D' = 0.868 \pm 0.002$. Again, as in the computation of scaling constants, one need only measure tongues with denominators up to about 100 to obtain experimental values of $D'$ to a few percent. Like $D$, $D'$ is the same for a wide variety of maps similar to the sine map. This method has the additional advantage that it establishes an implicit relationship between local dimension and scaling.

*The Multifractal Spectrum*

The frequency locking structure at the critical line is not the only fractal generated by the circle map. For irrational winding numbers, the Poincaré section itself is fractal at the critical line. The local density or scaling ($\alpha$) is nonuniform as seen in Fig. 12. That is, if we measure the density of the Poincaré section at different points we obtain different results. The simple fractal dimension is less useful for such sets, since it averages out much of the structure. We would like to be able to characterize inhomogeneities in scaling consistently. There are two different ways to view the problem, leading to equivalent results.

The first is the method of *generalized dimensions*, $D_q$ defined by Hentschel and Procaccia [73]. We examine the moments of the density distribution, much as we would the

306

multipole expansion of an electric field, and repeat the basic fractal dimension calculation keeping track of the number of points per box. We give each box a weight $p_i(l)$ associated with the number of points it contains by defining

$$p_i(l) = \lim_{N \to \infty} \frac{N_i}{N}$$

where $N_i$ is the number of points in the $i$th box when we restrict to a randomly chosen subset of $N$ points. We then define the $q$th moment of the probability distribution,

$$D_q \equiv \frac{1}{q-1} \lim_{l \to 0} \frac{\log \sum_i p_i^q}{\log(l)}$$

where $q$ is any real number. For $q = 0$, $D_0$ is the ordinary fractal dimension defined above. For $q$ large and positive, $D_q$ gives information about the most dense regions of the fractal. For $q$ large and negative, $D_q$ gives information about the least dense regions. Experimentally, we find that $D_q$ is more sensitive to high frequency noise for positive $q$ and to low frequency drifts and finite time series lengths for negative $q$.

Alternatively we may characterize the variation in density of a set by looking at the local scaling ($\alpha$) and calculating the dimension ($f$) of that subset of points which have a given value of the scaling. This function, the *multifractal spectrum*, $f(\alpha)$ encodes all the global scaling information of the set of a compact form [63], [130].

Following the method of Jensen *et al.* [77] we determine the $f(\alpha)$ spectrum as follows: we pick a point $x_i$ in the set and find the density of points, $p_i$, around it. As in our fractal dimension calculation we may do this by picking boxes of size $m$ and letting $p_i(m) = \lim_{N \to \infty} (N_m/N)$, where $N_m$ is the number of points in the box when we restrict to $N$ points. We may then define $\alpha_i$ by the relation $p_i(m) = m^{\alpha_i(l)}$ in the limit $m \to 0$. Letting $p_i = \lim_{m \to 0} p_i(m)$ we obtain the local scaling at each point. We can now measure the fractal dimension of the set of points with a given value of $\alpha$ by box counting a second time. If we set an acceptance interval $d\alpha$ and let $n(\alpha, l)$ be the number of boxes of side $l$ with $\alpha \in [\alpha_0, \alpha_0 + d\alpha]$, we obtain, in the limit $\alpha \to 0$ and $l \to 0$,

$$n(\alpha, l) = d\alpha \rho(\alpha) l^{-f(\alpha)}$$

where $\rho(\alpha)$ is a smooth function independent of $l$ that does not affect the value of $f(\alpha)$.

Once again, box counting is not the most convenient way to calculate $f(\alpha)$. However, it is the method which makes the meaning of the function most explicit (the $f(\alpha)$ spectrum can also be understood in the context of thermodynamics [42], [132]). In practice one calculates first the generalized dimension and defines a rescaled generalized dimension $\tau(q) \equiv (q-1)D_q$. Then, using the equivalence of dimension and scaling mentioned above, $f(\alpha)$ is the

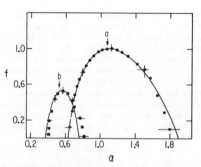

Fig. 6. Multifractals. $f(\alpha)$ curves for (a) the quasi-periodic transition to chaos at the golden mean and (b) the period doubling transition in the 8/13-tongue. Solid lines indicate theoretical calculations for the circle map. Dots and error bars indicate experimental results for small aspect ratio forced Rayleigh–Bénard convection in mercury. From [46] and [77].

Legendre transform of $\tau(q)$ given by

$$\alpha = \frac{d\tau(q)}{dq}$$

and

$$f(\alpha) = \alpha q - \tau(q).$$

Large scaling exponents correspond to low density, so the low $\alpha$ side of the $f(\alpha)$ curve corresponds to positive $q$ and the high $\alpha$ side to negative $q$. Depending on the system under study different tricks can be used to obtain $\tau(q)$ [46], [47], [62], [63].

For the circle map, $f(\alpha)$ characterizes unambiguously the transition to chaos for both locked and quasi-periodic states. It may either be calculated at $k = 1$ for irrational winding numbers or at accumulation points of period doubling cascades inside tongues. In both cases certain landmarks, e.g., the position and value of the maximum of $f$ ($f_{max} = D$), the smallest and largest $\alpha$'s, the upside-down paraboloid shape of the curve, etc. are extremely robust to minor changes in the map. For example, the period doubling cascades of the logistic map $F(\theta) = k\theta(1 - \theta)$ and the sine map produce identical $f(\alpha)$ curves. For theoretical curves see Fig. 6(a) for the quasi-periodic transition and Fig. 6(b) for the period doubling transition.

There is also a hidden bonus for the experimentalist. Below criticality the exact $f(\alpha)$ curve collapses immediately to a point located at $f = 1$, $\alpha = 1$ for quasi-periodic states and at $f = 0$, $\alpha = 0$ for periodic states. It does this at irrational winding numbers because, below $k = 1$, the map $F(\theta)$ is conjugate to a pure rotation. This means that at very small length scales the Poincaré section looks just like a circle and has uniform dimension equal to 1. There is no interesting scaling. Similarly, below the accumulation point of the period doubling cascade, the Poincaré section consists merely of a finite number of points. Hence at small length scales we see only a discrete set of points of uniform dimension 0. However, an experi-

798                                                                IEEE TRANSACTIONS ON CIRCUITS AND SYSTEMS, VOL. 35, NO. 7, JULY 1988

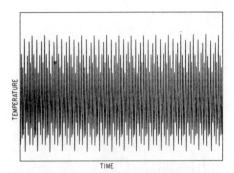

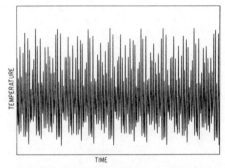

Fig. 7.   Experimental time series. An 8/13 locked state for small aspect ratio forced Rayleigh–Bénard convection in mercury. The basic oscillation is the low frequency ($f_1$) so the blocks contain 8 oscillations. The apparent phase drift is an artifact due to slow sampling rate of the digitization.

Fig. 8.   Experimental time series. A $3 \otimes 34/55$ locked state for small aspect ratio forced Rayleigh–Bénard convection in mercury. The basic period 34/55 and the period tripling envelope modulation are clearly visible. The basic oscillation is the low frequency ($f_1$) so the blocks contain 34 oscillations.

mentalist has access only to finite length scales because he cannot record an infinite number of points. Thus he will always observe density variations and obtain a nontrivial $f(\alpha)$ spectrum, even much below criticality, as seen in Fig. 10. This is useful for two reasons. 1) He need not worry if his data is taken a little away from the critical point. 2) Arneodo [6] has shown that away from criticality the calculated $f(\alpha)$ curves will narrow in a predictable way as either $k$ is decreased or the number of points used increased. By observing this *Arneodo narrowing* an experimentalist can derive a value of $k$ directly from an experimental time series. This is extremely helpful because the experimental control parameters do not in general correspond exactly to $k$ and $\Omega$ and the scaling the Fourier spectrum (discussed below), which is the only other technique for determining the amplitude of the nonlinearity $k$, is only quantitative at the critical line.

*Time Series, Poincaré Sections, and Spectra*

To bridge the gap between experiment and theory we must consider the general features of a signal generated by two nonlinear coupled oscillators. We may then ask what the time series of an experiment agreeing with the circle map should look like. We will assume throughout the following discussion that $f_2 > f_1$, however, the same arguments will hold in the opposite case. The time series will show a more or less sinusoidal oscillation at the natural frequency $f_1$. The second frequency $f_2$ will produce a beat pattern superimposed on this basic oscillation. In a locked state, $f_1/f_2 = p/q$, we will see a repeated unit block composed of $p$ fundamental oscillations. We show an experimentally observed 8/13 signal in Fig. 7. A period doubling will appear as a second modulation with period $2p$, making the amplitude of the blocks alternately large and small. As mentioned previously, the period doubling does not change the winding number. It takes $p/q \rightarrow 2p/2q$. A general period multiplication by $n$ of a state $p/q$, will result in a periodic time series $(n \otimes p/q)$ with a

period $np/f_1$. We show another experimentally observed state of form $3 \otimes 34/55$ in Fig. 8. The basic period 34 (the time series showing the numerator of the fraction) and the period tripling envelope modulation are clearly visible. A quasi-periodic state will look similar, except that the envelope will gradually drift in phase with respect to the fundamental oscillation. Chaotic states have additional irregular modulations. The attractors of weakly chaotic quasi-periodic states are nearly impossible to distinguish from ordinary quasi-periodic attractors. However, chaotic period doubled attractors are clearly distinguishable from ordinary locked states, showing smeared pointlike attractors which may drift to such an extent that they fill the entire circle.

In applying the above model to a real experiment, we must take into account the conflicting effects of noise. On the one hand, we can never achieve a true quasi-periodic state, since we cannot set the winding number exactly, and the external frequency inevitably varies slightly in time. The system will always tend to lock since the presence of even an arbitrarily small amount of noise will shift a quasi-periodic state to a nearby locked state. On the other hand, noise also smears out high denominator tongues. We may think of each tongue as a potential well whose depth varies inversely with the denominator of the tongue. In the presence of noise the state can tunnel between nearby tongues of high denominator. The system will not remain in a given large denominator tongue for an arbitrarily long time. Thus when we speak of an experimental quasi-periodic state, we mean only that we cannot measure any true periodicity over the duration of the experiment. Another limitation is that for a finite measurement time we cannot distinguish arbitrarily low frequencies. For a true quasi-periodic state with $W = \sigma$, all of the $\{\sigma_i\}$ will be approximate periods. For some value of $i$ our experiment will fail to distinguish the difference between $\sigma_i$ and $\sigma_{i+1}$. This is the limiting resolution of the experiment [46], [125]. Because the golden mean is furthest from rationals of any

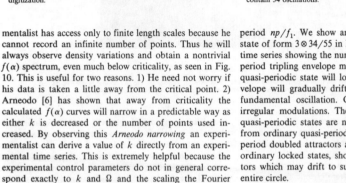

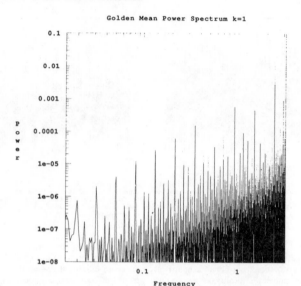

Fig. 9. Power spectrum for the critical golden mean circle map. The series of large peaks are at frequencies $\sigma_G^n$. The pattern of peaks between each pair of main peaks is the same and the envelope of the peak heights goes as $\omega^2$.

given denominator, it is the irrational winding number most resistant to noise. A golden mean state can be knocked further without locking to a low denominator tongue.

Using time series alone, it is rather difficult to distinguish quasi-periodic states from chaotic states arising from quasi-periodicity. The power spectrum:

$$P(\omega) = \left| \frac{1}{2\pi} \int dt f(\theta) e^{2\pi i \omega t} \right|^2$$

provides an immediate indication, however, and is not sensitive to random variations in winding number. If two oscillators of frequencies $f_1$ and $f_2$ are coupled nonlinearly, all frequencies of form $f_{n,m} \equiv nf_1 + mf_2$ (where $n$ and $m$ are integers) will be present in the power spectrum, with the amplitude of the peaks decreasing rapidly with increasing $m$ and $n$. Surprisingly, going from a continuous system to the circle map does not affect the global properties of the spectrum. It merely sets $f_2 = 1$. We may define the power spectrum for the discrete series as

$$P(\omega) = \lim_{i \to \infty} \left| \frac{1}{q_i} \sum_{j=0}^{q_i - 1} \theta_j e^{2\pi i \omega j} \right|^2$$

where $q_i$ is the denominator of $\sigma_i$. For periodic states the number of distinct $f_{n,m} \leqslant f_1$ is just $p$, the lowest frequency being $f_1/p$ and the low frequency spectrum will consist of a finite number of lines of form $jf_1/p$. If the system is quasi-periodic, however, the $f_{n,m}$ are distinct for all $n$ and

$m$ and the spectrum will consist of a countable infinity of lines. The combination frequencies are particularly well behaved at the golden and silver means. Because the golden mean has the property that its $n$th power $\sigma_G^n = F_{n-1} - \sigma_G F_n$, all powers of the golden mean are linear combinations of the fundamental frequencies and hence will be present. If we plot the low frequency part of a golden mean spectrum on a log scale, as shown in Fig. 9, we obtain a set of equally spaced peaks at frequencies $\sigma_G, \sigma_G^2, \sigma_G^3, \cdots$ [32], [106], [118]. There are also smaller amplitude sequences of peaks at frequencies, $\{\sigma_G^n(\sigma_G + m)\}$ which lie between the main peaks. The silver mean behaves identically, substituting $\sigma_S$ for $\sigma_G$. The pattern of the peaks is self similar, that is, the pattern between, for example, $\sigma_G^3$ and $\sigma_G^4$ is identical, up to scale factors in frequency and amplitude, to the pattern between $\sigma_G^6$ and $\sigma_G^7$ as can be seen in Fig. 9.

We may use the amplitude of the combination peaks to estimate $k$. For $k < 1$ the amplitude of these peaks drops off exponentially with $m$ and $n$ resulting in a clean spectrum with a finite number of measurable lines, as seen in Fig. 10 (a2) and (b2). However, at $k = 1$ the amplitude of each series drops off algebraically, $P(\sigma_G^n) \sim \sigma_G^{2n}$ as can be seen in Fig. 9 and Fig. 10 (c2). If we divide by $\omega^2$ to define the normalized power spectrum, $P(\omega)/\omega^2$, its envelope remains flat as $\omega \to 0$. Finding the power of $\omega$ that yields a flat envelope and counting the number of visible combination peaks, gives a qualitative estimate of $k$. The lower the power of $\omega$ and the more combination peaks are visible, the closer the state is to criticality.

800                                                    IEEE TRANSACTIONS ON CIRCUITS AND SYSTEMS, VOL. 35, NO. 7, JULY 1988

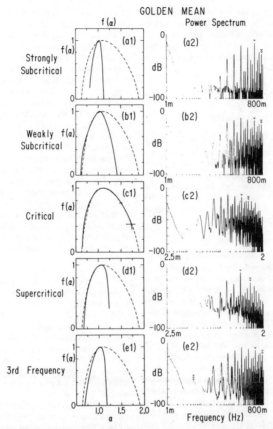

Fig. 10. Experimental Spectra and $f(\alpha)$ curves for small aspect ratio forced Rayleigh–Bénard convection in mercury: Column 1, golden mean $f(\alpha)$ curves (dashed lines show the numerically calculated critical curve); column 2, golden mean spectra. Row a, strongly subcritical; row b, slightly subcritical; row c, critical; row d, supercritical; row e, third frequency present. In column 2 single arrows mark the frequency of the oscillatory instability ($f_{int}$), double arrows mark the forcing frequency ($f_{ext}$). In (e2) triple arrows mark the third frequency. The number of points used to calculate the $f(\alpha)$ curves were: (a1), 89; (b1), 144; (c1), 377; (d1), 383; (e1), 123. The equivalent Arneodo couplings are: (a1), $k = 0.960$; (b1), $k = 0.977$; (c1), $k = 0.995$. Note the $1/\omega^2$ scaling in (c2). The error bars indicated in (c1) and (c3) are maximal deviations. For other figures the errors are typically a few percent. From [47].

Above $k = 1$ there are chaotic and period multiplied states. The $n$th period doubling appears as a subharmonic at $f_1/2^n$. In general any multiplication of period by $m$ will appear in the form of $m$th subharmonics, $m-1$ peaks between each of the main combination peaks. Chaotic states are qualitatively different from those discussed above. Their aperiodicity does not result from combinations of well defined incommensurate frequencies but from the presence of a $1/f$-like continuum of low frequencies. These frequencies produce broadband noise throughout the spectrum. The noise increases in amplitude with $k$, gradually swallowing the quasi-periodic combination peaks and, for large $k$, resulting in a nearly smooth spectrum.

We must also consider noncircle map effects. A complex experimental system can only resemble the simple circle map when all but one degree of freedom in its motion is suppressed. For small perturbations this suppression is not surprising, but for large external perturbations we expect that we may excite additional degrees of freedom, raising the system's effective dimension [17]. One symptom of an increase in dimensionality is wrinkling of the Poincaré section—that is, the Poincaré section folding up on itself to form a fractal with dimension between 1 and 2. As long as the wrinkling is small and the dimension is close to 1 our simple circle map will be a reasonable approximation. However, for very strong forcing, the Poincaré section tends to dissolve into a gnatlike sea of points and the dimension approaches two [47]. In this case we need to embed our system in a higher dimensional space and use a higher dimensional model like the Henon or Standard maps [5]. Two-dimensional extensions of the circle-map, like the dissipative Standard map:

$$
\begin{pmatrix} \theta_{i+1} \\ x_{i+1} \end{pmatrix} = \begin{pmatrix} \theta_i + \Omega - \dfrac{k}{2\pi}\sin(\theta_i) + \epsilon x_i \\ \epsilon x_i - \dfrac{k}{2\pi}\sin(\theta_i) \end{pmatrix}
$$

are particularly appealing since they can be reduced continuously to the one-dimensional case. Unfortunately, if it is difficult to determine $k$ experimentally, it is nearly impossible to determine $\epsilon$. Experimental and theoretical techniques exist to treat this case but they are much less well developed than those for the simple circle map. We will discuss two dimensional effects briefly in the experimental section of this paper.

## IV. EXPERIMENTAL RESULTS

Now that we know what to look for we may examine the experimental evidence. We will limit our discussion to frequency locking effects and neglect the related one-dimensional mapping problems of phase locking and simple period doubling. We have studied quasi-periodic effects in small aspect ratio forced Rayleigh–Bénard convection [46]–[48], [77], [123], [124], and we will describe this system in detail. A large range of effects have also been observed in oscillations in Germanium [61], [62], [69]–[71]. We will discuss other experimental results when relevant. Experiments on quasi-periodicity fall into two broad classes, those in which the frequency $f_2$ is externally controlled (which we will denote *type I*, and those in which it arises internally, which we will denote *type II*. Experiments of type I allow much greater control over the varieties of quasi-periodic behavior observed and we will concentrate on them.

### Forced Rayleigh–Bénard Convection

Our Rayleigh–Bénard convection experiment consists of a small mercury-filled rectangular cell ($1.4 \times 0.7 \times 0.7$ cm$^3$) with plexiglass walls and copper plates on the top and bottom. The cell has its temperature regulated to a few thousandths of a degree celsius and is placed in a horizontal magnetic field of $\approx 200$ G aligned perpendicular to its long axis. We begin with a motionless fluid and heat the cell from below. The heating causes the fluid to expand and lose density. At a few degrees temperature difference the inverted density gradient becomes unstable and the hot and cold fluids exchange places (convect) forming horizontal, time independent rolls. The magnetic field acts to damp motion perpendicular to its axis and aligns the convective rolls parallel to the short side of the cell. If we increase the heating further, to about 10°C, the rolls themselves become unstable to the *oscillatory instability* and begin to oscillate transverse to their

axes. Further increasing the heating results in the appearance of additional low frequencies which have been used to observe type II frequency locking effects [49], [84]. Other fluids with different thermal properties show different sequences of instabilities but the basic sequence: motionless, steady motion, periodic oscillation, and finally, multiperiod oscillation, is the same. The oscillatory instability produces a well defined frequency (defining $f_1$) which depends on the box size, magnetic field strength and temperature difference. In the experiments to be discussed, $f_1$ is typically between 0.2 and 0.4 Hz. We then inject an alternating pulsed current sheet (frequency $f_2$ and amplitude $A$) asymmetrically through the mercury. The current and magnetic field produce an alternating Lorentz force which couples nonlinearly to the oscillations of the rolls [124]. The whole procedure closely resembles stirring a pot of soup while heating it on a stove.

The experimental control parameters related to the nonlinearity are the amplitude of the current pulses ($A$) and their duty cycle ($x$). If the total forcing power ($xA^2$) is too large, it can drive the oscillatory instability off resonance and suppress it. Since the amplitude of the nonlinearity depends on the product of the internal and external oscillator amplitudes, and is more sensitive to peak height than power (it goes roughly as $xA$), we use narrow $\delta$-function pulses to reach the maximum possible nonlinearity without killing the internal oscillation.

We measure the temperature of the system at the center of the bottom of the cell using a semiconductor bolometer. A very useful theorem proposed by Poincaré and proved by Takens and Swinney [120], [128], [131] assures that as long as the flow in the cell is coherent (i.e., the flow is not turbulent) all information about the cell behavior can be reconstructed from any local measurement of any system variable.

In the first set of experiments we scanned the $A, f_2$ plane to map the locking behavior. We present the results in Fig. 4. We found excellent agreement with the standard Arnol'd structure with no duplicated or missing winding numbers and the correct qualitative tongue widths. We then located the position of the critical line at $\sigma_G$ and $\sigma_S$ (here defined as the $A$ value at which broadband noise first appears in the spectrum) and calculated the scaling exponent $\delta$ and the approximate fractal dimension of the quasi-periodic structure $D'$ by explicitly measuring the widths of tongues in the "best" approximant series with a denominator of 100 or less, as shown in Fig. 4. We obtain: $\delta_{\sigma_G} = 7.0 \pm 10\%$, $\delta_{\sigma_S} = 2.8 \pm 10\%$, and $D' = 0.86 \pm 3\%$ for both winding numbers. This agrees with the theoretical predictions discussed earlier and supports the hypothesis that the fractal dimension of the set of quasi-periodic winding numbers at criticality is uniform. However, the critical line is by no means straight, indicating that the correspondence between $A$ and $k$ is only approximate. Together these results establish the global similarity between forced Rayleigh–Bénard convection and the simple circle map. We have also, by means of a three-dimensional Poincaré section ($T_i, T_{i+1}, T_{i+2}$), been able to untangle the

IEEE TRANSACTIONS ON CIRCUITS AND SYSTEMS, VOL. 35, NO. 7, JULY 1988

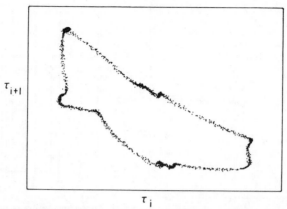

$\tau_{i+1}$

$\tau_i$

Fig. 11. Experimental critical attractor for small aspect ratio forced Rayleigh–Bénard convection in mercury. The attractor is wrinkled, indicating two dimensional effects, and varies in density. The width of the attractor results from temperature drift in the experiment. From [125].

attractors and calculate the return map $F(\theta)$ explicitly from the experimental data [124]. We present three-dimensional Poincaré sections and unwindings for weak, moderate and critical forcing in Fig. 11. In each case the resulting curve is clearly one dimensional, justifying the circle map model.

We next examined the nature of the transition to chaos at $\sigma_G$ and $\sigma_S$. The experimental procedure consisted of selecting a winding number $\sigma$ and then, for each change of the forcing amplitude, approximating it to the desired accuracy by adjusting the forcing frequency to lock successively to each of the best approximants $\{\sigma_i'\}$. By directly examining the periodicity of the locked time series we could rapidly tune the winding number to 5 parts in $10^6$, since tongues with denominators up to around 1000 are stable. One problem is the low basic frequency of the system, which results in long data acquisition times and sensitivity to long term temperature drift.

Fein, Heutmaker, and Gollub made the first experimental observation of the golden mean critical power spectrum in a hydrodynamic system. They studied forced Rayleigh–Bénard convection in water using a $2.1 \times 1.6 \times 0.8$-cm$^3$ cell driven by thermal pulses and detected density gradients in the fluid using optical techniques. Drifts and pattern competition instabilities prevented them from tuning their frequency ratio to better than 1 part in $10^3$. Nevertheless, they were able to observe a roughly self-similar spectrum with $\omega^4$ scaling of the peak heights [41]. Our work on spectra has confirmed and amplified those results. If we examine spectra at the golden and silver means we find the predicted pattern: well defined line spectra below criticality, power law ($\omega^2$) scaling at criticality and broadband noise above [47], [123]. We present subcritical spectra in Fig. 10 (a2) and (b2), critical spectra in Fig. 10 (c2), a rescaled critical spectrum in Fig. 12, and supercritical spectra in Fig. 10 (d2), all at the golden mean. We have

calculated the $f(\alpha)$ spectrum below, at, and above criticality for both $\sigma_G$ and $\sigma_S$ and find excellent agreement with theoretical predictions. We present golden mean results in Fig. 10 (a1)–(d1) and Fig. 6(a). We find good agreement with the predicted Arneodo narrowing for subcritical $f(\alpha)$ curves, and are able to use it to calculate $k$. We have also examined the effects of a third frequency on the spectrum and critical $f(\alpha)$ curve and find adequate agreement with theoretical predictions. We present these results in Fig. 10 (e1) and (e2).

The agreement between the experiment and circle map is surprising. Indeed it is so good one might wonder if the system were not simply a slow analogue computer simulating the circle map. Fortunately things are not that simple. We know that for very strong forcing we have turbulent (high-dimensional) behavior. These extra dimensions must begin to make themselves felt at some point. In practice, the first sign of the breakdown of sine map behavior is the wrinkling of the Poincaré section which begins near criticality as seen in Fig. 13. Just above criticality the dimension of the attractor creeps above 1, and 10% above, the attractor has exploded into a full two-dimensional set.

If we look for period doublings inside a tongue we see higher dimensional behavior of a different kind [46]. The doubled attractors themselves are unsurprising. In Fig. 14 we show Poincaré sections of a pure 8/13 state and a triply period doubled, $2^3 \otimes 8/13$ state. The periodic attractors have nearly the same shape as the quasi-periodic attractor shown in Fig. 13, indicating that the form of the torus is independent of the trajectory. This agreement is an example of universality that would be difficult to explain outside the context of dynamical systems theory. However, the sequence of multifurcations within the tongues is not that predicted by the one dimensional model. We have mapped the 8/13 tongue in detail and present it in Fig. 15. At the edge of the tongue and for small nonlinearities the

312

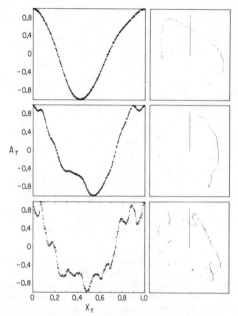

Fig. 12. Three-dimensional Poincaré sections and unwindings of experimental golden mean time series for small aspect ratio forced Rayleigh–Bénard convection in mercury. Column 1 shows the one dimensional unwinding described in the text. Column 2 shows the three dimensional Poincaré section. The top section shows weak forcing, the middle moderate and the bottom critical. From [124].

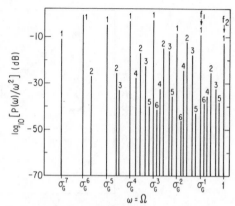

Fig. 13. Rescaled experimental critical spectrum at the golden mean for small aspect forced Rayleigh–Bénard convection in mercury. Only the first few series of peaks are shown. The envelope is flat for all series, indicating $\omega^2$ critical scaling. From [123].

structure is still largely one dimensional. The overlapping of neighboring tongues is that predicted by the one dimensional circle map, as are the various intermittency effects [126]. The first period doubling occurs slightly above criticality, and according to the one-dimensional scheme of Glass and Perez [13], [107]. We also observe the predicted symmetry breaking bifurcation (that is there are two distinct types of period 4 cycles), followed by period doubling cascades at the sides of the tongue. We have measured the $f(\alpha)$ curve for one of these cascades and find excellent agreement with that predicted for simple period doubling, as seen in Fig. 6(b). So far this is normal one dimensional behavior. However, when we move towards the axis of the tongue, the situation changes. Instead of the predicted pure period doubling cascade, we find paired "bells" of odd subharmonics, with the order of the subharmonic increasing toward the tongue axis. Inside each of these subharmonic bells we observe a subsidiary period doubling cascade. That is, proceeding from tongue edge to center, we find $2^n \otimes \frac{8}{13}, 2^n \otimes 3 \otimes \frac{8}{13}, 2^n \otimes 5 \otimes \frac{8}{13}, \cdots$. We are able to observe subharmonics up to 13 but have only succeeded in mapping up to 5. Furthermore these subharmonic bells overlap, leading to a strongly hysteretic multisheeted structure in which states of form $m \otimes \frac{8}{13}, n \otimes \frac{8}{13}$, and $m \otimes n \otimes \frac{8}{13}$

can all exist for the same values of $A$ and $f_2$. The simple circle map predicts no hysteresis in these regions. There are also various two-dimensional intermittency effects resulting from noise induced jumps between sheets within the tongue.

At first this structure seems completely incomprehensible in terms of the circle map model. However, similar, though not identical, structures have been predicted and more recently calculated for various two dimensional maps by MacKay and Tresser [89]–[91], [134]. We may understand this mixture of one- and two-dimensional behavior by considering the nature of a locked state. Near the edge of the tongue the system is detuned, resulting in a relatively large damping (effectively the nonlinearity is smaller near the edge of a tongue). As we move toward the center of the tongue and the resonance condition, the damping decreases, allowing the normally suppressed second dimension to appear. Taking the dissipative standard map as an example, we might say that $\epsilon = 0$ at the edge of the tongue and increases to a maximum on the tongue axis. In this case the sequence in which the subharmonics appear should allow calculation of $\epsilon$ as a function of the distance from the tongue axis.

Recently, Ecke and Haucke have succeeded in observing a similar range of phenomena, including well-defined Arnol'd tongues and period doubling, as well as a variety of three frequency states in binary convection in a $^3$He-superfluid $^4$He mixture [67], [68], [35], [66], [64]. Their system contains two internal frequencies whose ratio is precisely controlled by varying the Rayleigh number.

Thus in the simple convection experiments, it is possible to produce the full range of circle map behavior and to introduce higher dimensional effects in a controlled fashion. This holds out the hope that we can treat more complex hydrodynamic systems using extensions of the simple theory, rather than having to start from scratch.

IEEE TRANSACTIONS ON CIRCUITS AND SYSTEMS, VOL. 35, NO. 7, JULY 1988

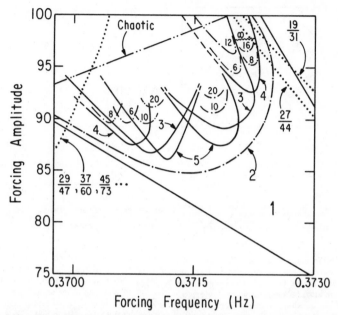

Fig. 14.  Subharmonic structure of a tongue in small aspect ratio forced Rayleigh–Bénard convection in mercury. The 8/13-tongue is shown in the $A$ versus $\Omega$ plane. The lines inside the tongue indicate the borders between different periodic states. Overlaps indicate regions of multistability. The areas labelled by $n$ correspond to $n \otimes 8/13$ states. The single dotted lines indicate smooth period doubling transitions. The solid lines indicate discontinuous transitions. The dotted lines on the right and left hand sides of the tongue indicate the position of overlapping tongues. The double dashed line indicates the region where the system is chaotic for all forcing frequencies. Note the symmetry breaking which follows the first period doubling and the sequence of subharmonics. Compare Fig. 5. From [46].

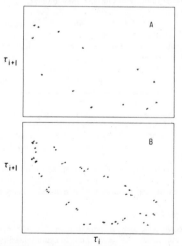

Fig. 15.  Experimental Poincaré sections in the 8/13-tongue for small aspect ratio forced Rayleigh–Bénard convection in mercury. (a) A pure 8/13 state. (b) A $2^3 \otimes 8/13$ state. The overall shape of the attractor is the same as in Fig. 11. From [46].

*Solid-State Systems*

Of the numerous solid state systems exhibiting quasi-periodic behavior, electrically forced germanium is the most studied. In the experiment of Held and Jeffries [70], [72] a single crystal (1 mm$^3$) of $n$-type germanium with a electron injecting contact made of diffused lithium and a hole injecting contact of diffused boron, is cooled to liquid nitrogen temperatures. When the crystal is subject to a dc electric field (15.02 V) and magnetic field (9.32 kG) it produces measurable oscillations in the form of traveling density waves in the electron-hole plasma. The typical frequency is 235 kHz which makes data acquisition some-what inconvenient. Depending on the angle between the magnetic and electric fields and their relative amplitudes, the system can oscillate at either one or many frequencies (resulting in type II frequency locking [69]). However, in the experiment, the angle and drive voltage were adjusted to produce a single well defined $f_1$, and an alternating voltage was applied between the contacts to define $f_2$. The measured variable was the total current through the sample $I(t)$.

They found the standard Arnol'd tongue structure with the fractal dimension for the unlocked set at the critical line being $D' = 0.90 \pm 0.03$ and the scaling $\delta_{\sigma_G} = 2.7 \pm 0.5$

314

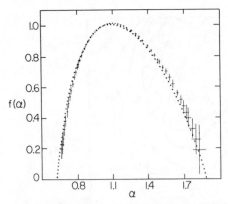

Fig. 16. Multifractal spectrum at the critical golden mean in electrically forced germanium. The solid line is the theoretical curve for the circle map. Dots and error bars are the experimental results. The high accuracy for large $\alpha$ (low densities) is remarkable. From [62].

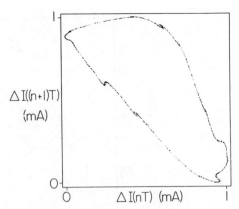

Fig. 17. Experimental attractor at the critical golden mean in electrically forced germanium. The attractor is wrinkled, indicating the presence of two-dimensional effects. From [62].

in agreement with the simple sine map. The power spectrum at the golden mean shows the expected behavior and critical ($\omega^2$) scaling. Unlike the Rayleigh–Bénard attractor, the germanium attractor does not wrinkle or explode until well above the transition to chaos [72]. They have also examined forcing at two incommensurate external frequencies and found three frequency quasi-periodicity with frequency locking to either forcing frequency. There is a well defined critical surface in agreement with the predictions of Ostlund, Kim and Siggia [55], [79], [80], [106]. Unfortunately their signal-to-noise ratio was not sufficient to allow them to map a three frequency tongue in detail.

Gwinn and Westervelt [61], [62] have performed a similar experiment on oscillating electric field domains, using cooled (4°C) germanium with boron implanted contacts. They applied a dc bias to induce oscillations at roughly 10 kHz and superimposed an ac driving voltage. The measured quantity was $I(t)$. They observe the expected Arnol'd diagram with $\omega^2$ power spectrum scaling at the golden mean. Their calculations of $f(\alpha)$ at the critical golden mean are in excellent agreement with those for the circle map. We reproduce their $f(\alpha)$ curve from [62] in Fig. 16. The high accuracy they obtain for large $\alpha$ is particularly impressive, since these are measurements made where the attractor density is lowest and hence require long time series and good stability. Their attractor (Fig. 17) shows clear signs of wrinkling at criticality but is otherwise extremely clean.

Martin and Martienssen [93], [94] have obtained a very clean circle-map-like return map and devil's staircase behavior in electrically driven Barium Sodium Niobate at high temperatures (500°C). Because of temperature drift problems their Poincaré sections are limited to about 500 points. They observe a standard Arnol'd diagram but are limited to denominators < 10. Zettl and Grüner have

obtained a nearly complete devil's staircase and incomplete period doubling cascades in forced charge density waves in NbSe$_3$ [22], [57]. They obtain a fractal dimension for the unlocked set at the critical line of $D = 0.91 \pm 0.03$. Other groups have obtained similar results in a variety of charge density wave systems [88].

Winful, Chen, and Liu [138] have observed frequency locking behavior in a periodically forced semiconductor laser. An internal instability in their AlGaAs/GaAs laser generates $f_1$ at a frequency between 0.5 and 3 GHz, and an externally applied RF modulation at a lower frequency defines $f_2$. For weak forcing they obtained a standard Arnol'd tongues diagram with appropriate tongue widths. They are able to follow the golden and silver mean quasi-periodic states over a wide range of forcing amplitudes, and to observe the expected transition from line spectra to broad band noise at criticality. Unfortunately the extremely high frequencies limit their signal-to-noise ratio to 40 dB, so they are unable to resolve more than the first few combination peaks or to tune their frequency ratio to better than 1 percent. For large forcing amplitudes (well above the onset of chaos) they observe a complicated pattern of disappearing tongues without period doubling. This behavior is typically two dimensional, and they provide a two dimensional differential equation model to explain it.

That such a variety of natural systems exhibit behavior quantitatively identical to the one-dimensional circle map is one of the great successes of dynamical systems theory.

*Analogue Simulators*

Another category of type I system is the nonlinear electronic circuit specifically built to exhibit quasi-periodic behavior. Bryant and Jeffries [23], [24] have studied a system incorporating a magnetic inductor with hysteresis. The system exhibits frequency locking and Arnol'd tongues,

806                                                                    IEEE TRANSACTIONS ON CIRCUITS AND SYSTEMS, VOL. 35, NO. 7, JULY 1988

however its behavior is complicated. The ordering of the tongues is not that of the simple circle map. There are missing and extra tongues with unexpected widths. There is no well defined critical line and subcritical tongues overlap. The period doubling behavior within the tongues is also anomalous, without any apparent symmetry breaking. The data recalls that of many two dimensional systems of differential equations [8], and Bryant and Jeffries provide a multidimensional model. Gollub et al. [51] have observed similar complicated locking behavior in tunnel diode circuits. We will not describe here the numerous other experiments in electronic systems built purposely to emulate circle map type behavior [25], [75], [86], [109], [133].

Recently, Su, Rollins, and Hunt [127], and Cumming and Linsay [31], have measured $f(\alpha)$ curves for analog simulators. Su, Rollins, and Hunt employed coupled diode-inductor resonators, Cumming and Linsay, a more complicated analog circuit. Both obtain reasonable agreement with the predictions of the circle map, though the results of Cummings and Linsay show anomalous behaviors apparently due to discontinuities in the return map describing their system.

How one judges analogue simulators is a philosophical problem. They have the advantage that they can be described by ordinary differential equations and thus simulated on a computer, but their relevance to dynamical systems theory is less clear. A dynamical systems approach is most valuable for systems like convection, which cannot be solved exactly. It is somewhat ironic that the simple theory works best in intractably complex continuous systems, and fails badly in simple simulators.

*Multiple Internal Frequency Quasi-Periodicity*

We can only mention a few of the many experiments done on type II systems. The problem with these systems is that it is difficult to predict whether changing a given parameter will affect both oscillators in the same way. One cannot tune the frequency ratio without affecting the strength of the nonlinearity. Trying to scan a two parameter phase space with one control parameter is a bit like moving in a maze. One can move further or backward but one cannot choose a direction. Nevertheless, in the cleanest of these systems varying a single control parameter results in a well defined sequence of periodic, quasi-periodic and chaotic states. In many cases the return map can also be determined. The typical signature of a circle-map-like system is a well behaved "Devil's Staircase," followed by period doubling cascades and windows of periodicity [12]. Full zoologies have been observed by Maurer and Libchaber in liquid helium [85], [95], [96] and Swinney et al. in both chemical (Belousov–Zhabotinsky reaction) and various hydrodynamic systems [52], [135]. In the latter they have obtained $D' = 0.87 \pm 0.02$ (Couette flow) [129] and attractor blowup above the onset of chaos (channel flow) [20]. Gollub and Benson have observed a similar variety of effects including three frequency quasi-periodic-

ity in unforced small aspect ratio Rayleigh–Bénard convection in water [49]. Three frequency quasi-periodic states and more or less complete Devil's staircases have also been observed in $CO_2$ lasers [16], [100], and Devil's staircases and other quasi-periodic effects in Fabry–Perot interferometers [4], [98], mechanical systems [34], [60], [101], Josephson junctions [105], and yttrium iron garnet oscillators [1], [44]. Braun et al. in plasmas [21] and Keolian et al. in the Faraday experiment [78] have found mixed period doubling-subharmonic cascades similar to those observed in our forced Rayleigh–Bénard convection experiment. In general the numerical accuracy of type II experiments is much lower than in the type I systems described above, so these results serve chiefly to confirm the large range of validity of the one-dimensional circle map model.

## IV. Conclusions and the Future

The great accomplishment of dynamical systems theory and experiment has been to establish that the same basic description of behavior can apply to a wide range of systems. This universal behavior is independent of the detailed structure which gives rise to it. The same language can describe both Josephson junctions and slime moulds. To return to our computer metaphor: We have indeed developed a machine independent code, and without having had to learn machine language! The recognition of universality in behavior is a tremendous advance in the way we think about complex systems. It tells us what sorts of behavior we can expect from unknown systems, and what questions we can ask to establish the class to which a system belongs. However, in its present state, dynamical systems theory is incomplete, both as a technical and as a theoretical tool.

Even in the study of one-dimensional systems the techniques of data analysis are cumbersome. We need more convenient and reliable ways to determine important quantities. The measurement of dimension is now a standard technique. We need to be able to determine winding numbers, $f(\alpha)$ curves, nonlinearity strengths and scaling parameters automatically. Only when making these calculations is as simple as taking a Fourier spectrum will the dynamical systems approach achieve its true potential as a tool for the study of systems. Recent work by Farmer and Sidorowich [39], [40], and Kosterlich and Yorke [81] on the extrapolation of chaotic time series is an excellent example of the type of tools that need to be developed.

Dynamical systems theory, though beautiful mathematically, is chiefly valuable to an experimentalist as an analytical technique. One thinks of basic quantum mechanics as an analogy. One does not, except in a few specialized cases, do experiments to check the basic mechanics of the quantum theory. Instead one uses it to analyze systems that are interesting in their own right. Similarly, with time, model dynamical systems will become less and less important. Once we know that many systems behave like a circle map, finding additional circle-map-like systems becomes less interesting.

We finish by listing, in approximate order of difficulty a few of the remaining problems which dynamical systems theory would need to solve to become a useful theory for explaining the real world: noise is present in all real systems, yet dynamical systems theory is essentially deterministic. We might expect that as systems become noisier the dynamical systems approach will be less relevant. This is not necessarily true, and there has been some progress on a theory of noisy dynamical systems [29], [30], [119]. However, the work of Stratonovich [126] which predates dynamical systems theory by twenty years, remains the most complete treatment of noisy systems. A complete theory of the significance of noise does not exist and experimental work is just beginning [19], [65], [40].

Most natural systems contain more than two frequencies. The behavior of systems with three frequencies is still poorly understood. It is not even known whether there are universal three frequency behaviors. The theoretical predictions concerning three frequency systems are vague and apparently contradictory [55], [103], [110], [115]. There has been some preliminary experimental research but the parameter space in a three frequency problem is huge, and without theoretical guidance it is easy for the experimentalist to lose his way [16], [47], [72].

The theory we have developed works only near the transition to chaos. Even in few dimensions, strongly chaotic attractors are still not well understood, though there has been some recent progress in pushing above the critical line (for example, in analyzing intermittency effects caused by the overlapping of locked tongues [48], [59]).

There is still no satisfactory general theory for many dimensional systems, though there has been significant recent progress toward extending the theory of symbolic dynamics, from the circle map to its two dimensional analogues [5], [9], [17], [59]. The hope has always been that dynamical systems theory could be extended in a consistent fashion dimension by dimension, but preliminary results indicate that the universality of two-dimensional systems is weaker than in the one-dimensional case.

Will it be possible to extend dynamical systems to higher dimensions? Will universality disappear completely in three or more dimensions? Barring such catastrophies, it is clear that any general theory will be significantly more complex than that for one-dimension. One thinks of the extension of simple quantum mechanics to field theory and beyond. For the experimentalist the problems are similarly daunting. We do not know what quantities are of interest, and the data sets required to analyze even two-dimensional systems are enormous. More significantly, most experiments lack a sufficient number of control parameters to explore a many dimensional parameter space fully. It seems that all many degree of freedom systems will be type II and hence hard to interpret. Large systems have a further problem, they tend to develop complex spatial as well as temporal behavior. The measurement of a single variable no longer fully describes the system, and instrumentation and data collection become significant problems. Spatio-temporal chaos is an almost untouched sub-

ject. Even Rayleigh–Bénard experiments in wide cells are not understood [14], [50], let alone more complicated natural systems.

The problems are considerable, but the potential payoffs is a deep understanding of a range of natural phenomena which would free physics from its usual restriction to artificially simplified systems. The dream of every hydrodynamicist is an effective and complete theory of turbulence. Early workers on chaos thought they had found this philosopher's stone when they discovered that simple systems could have complex aperiodic behavior. They were disappointed when they discovered that chaos and turbulence were fundamentally different. But this early failure has not prevented significant advances in understanding a large class of previously inexplicable phenomena. Dynamical systems still seems the best approach to a theory of complexity.

REFERENCES

[1] F. M. de Aguiar and S. M. Rezende, "Observation of subharmonic routes to chaos in parallel-pumped spin wave in yttrium iron garnet," *Phys. Rev. Lett.*, vol. 56, pp. 1070–1073, 1986.

[2] T. Allen, "On the arithmetic of phase locking: Coupled neurons as a lattice on $R^2$," *Physica*, vol. 6D, pp. 305–320, 1983.

[3] P. Alstrøm, B. Christiansen, P. Hyldgaard, M. T. Levinsen, and R. Rasmussen, "Scaling relations at the critical line and the period-doubling route for the sine map and the driven damped pendulum," *Phys. Rev. A*, vol. 34, pp. 2220–2233, 1986.

[4] F. T. Arecchi, W. Gadomski, and R. Menucci, "Generation of chaotic dynamics by feedback on a laser," *Phys. Rev. A*, vol. 34 (RC), pp. 1617–1620, 1986.

[5] A. Arneodo, G. Grasseau and E. J. Kosterlich, "Fractal dimensions and $f(\alpha)$ spectrum of the Hénon attractor," preprint, 1987.

[6] A. Arneodo and M. Holschneider, "Crossover effect in the $f(\alpha)$ spectrum for quasiperiodic trajectories at the onset of chaos," *Phys. Rev. Lett.*, vol. 58, pp. 2007–2010, 1987.

[7] V. I. Arnol'd, *Mathematical Methods of Classical Mechanics*. Berlin, Germany: Springer, 1974.

[8] D. G. Aronson, R. P. McGehee, I. G. Kevrekidis, and R. Aris, 'Entrainment regions for periodically forced oscillators," *Phys. Rev. A*, vol. 33 (RC), pp. 2190–2192, 1986.

[9] D. Auerbach, P. Cvitanovic, J. P. Eckmann, G. H. Gunaratne and I. Procaccia, "Exploring chaotic motion through periodic orbits," *Phys. Rev. Lett.*, vol. 58, pp. 2387–2390, 1987.

[10] A. Babloyantz, "Evidence of chaotic dynamics of brain activity during the sleep cycle," in *Dimensions and Entropies in Chaotic Systems*, (G. Mayer-Kress, Ed.) Berlin, Germany: Springer, 1986, pp. 241–245.

[11] H. Bai-Lin, *Chaos*. Singapore: World Scientific, 1984. (A collection of important papers with a useful introduction.)

[12] P. Bak, "The devil's staircase," *Physics Today*, vol. 39, pp. 38–45, Dec. 1986.

[13] J. Bélair and L. Glass, "Universality and self-similarity in the bifurcations of circle maps," *Physica*, vol. 16D, pp. 143–154, 1985.

[14] P. Bergé and M. Dubois, "Rayleigh–Bénard convection," *Contemp. Phys.*, vol. 25, pp. 535–582, 1984.

[15] P. Bergé, Y. Pomeau, and C. Vidal, *L'Ordre dans le Chaos: Vers une approche déterministe de la turbulence*. Paris, France: Hermann, 1984.

[16] D. J. Biswas and R. G. Harrison, "Experimental evidence of three-mode quasiperiodicity and chaos in a single longitudinal, multi-transverse-mode cw $CO_2$ laser," *Phys. Rev. A*, vol. 32, pp. 3835–3837, 1985.

[17] T. Bohr, P. Bak and M. H. Jensen, "Transition to chaos by interaction of resonances in dissipative systems. II. Josephson junctions, charge-density waves, and standard maps," *Phys. Rev. A*, vol. 30, pp. 1970–1981, 1984.

[18] T. Bohr and G. Gunaratne, "Scaling for supercritical circle-maps: Numerical investigation of the onset of bistability and period doubling," *Phys. Lett.*, vol. 113A, pp. 55–60, 1985.

[19] H. R. Brand, S. Kai, and S. Wakabayashi, "External noise can suppress the onset of spatial turbulence," preprint.

[20] A. Brandstäter, H. L. Swinney, and G. T. Chapman, "Characterizing turbulent channel flow," in *Dimensions and Entropies in Chaotic Systems*, (G. Mayer-Kress, Ed.), Berlin, Germany: Springer, pp. 150–157, 1986.

[21] T. Braun, J. A. Lisboa, R. E. Francke, and J. A. C. Gallas,

IEEE TRANSACTIONS ON CIRCUITS AND SYSTEMS, VOL. 35, NO. 7, JULY 1988

"Observation of deterministic chaos in electrical discharges in gases," *Phys. Rev. Lett.*, vol. 59, pp. 613–616, 1987.

[22] S. E. Brown, G. Mozurkewich, G. Grüner, "Subharmonic Shapiro steps and devil's staircase behavior in driven charge-density-wave systems," *Phys. Rev. Lett.*, vol. 52, pp. 2277–2280, 1984.

[23] P. Bryant and C. Jeffries, "The dynamics of phase locking and points of resonance in a forced magnetic oscillator," preprint, 1986.

[24] ——, "Bifurcations of a forced magnetic oscillator near points of resonance," *Phys. Rev. Lett.*, vol. 53, pp. 250–253, 1987.

[25] R. V. Buskirk and C. Jeffries, "Observation of chaotic dynamics of coupled nonlinear oscillators," *Phys. Rev. A*, vol. 31, pp. 3332–3357, 1985.

[26] S. Chandrasekhar, *Hydrodynamic and Hydromagnetic Stability*. New York: Dover, 1981.

[27] K. Coffman, W. D. McCormick, and H. L. Swinney, "Multiplicity in a chemical reaction with one-dimensional dynamics" *Phys. Rev. Lett.*, vol. 56, pp. 999–1002, 1986.

[28] P. Collet and J.-P. Eckmann, *Iterated Maps on the Interval as Dynamical Systems*. Boston, MA: Birkhäuser, 1980.

[29] J. P. Crutchfield and B. A. Huberman, "Fluctuations and the onset of chaos," *Phys. Lett.*, vol. 77A, pp. 407–410, 1980.

[30] J. P. Crutchfield and N. H. Packard, "Symbolic dynamics of noisy chaos," *Physica*, vol. 7D, pp. 201–223, 1983.

[31] A. Cumming and P. S. Linsay, "Deviations from universality in the transition from quasi-periodicity to chaos," *Phys. Rev. Lett.*, vol. 59, pp. 1633–1636, 1987.

[32] P. Cvitanović, B. Shraiman, and B. Söderberg, "Scaling laws for mode lockings in circle maps," *Phy. Scr.*, vol. 32, pp. 263–270, 1985.

[33] P. Cvitanović, *Universality in Chaos*. Bristol, England: Adam Hilger, 1984. (A collection of reprints of important articles on birfurcation theory and chaos.)

[34] D. D'Humieres, M. R. Beasley, B. A. Huberman, and A. Libchaber, "Chaotic states and routes to chaos in the forced pendulum," *Phys. Rev. A*, vol. 26, pp. 3483–3496, 1982.

[35] R. E. Ecke and I. G. Kevrekidis, "Interactions of resonances and global bifurcations in Rayleigh–Bénard convection," preprint, 1987.

[36] J.-P. Eckmann and D. Ruelle, "Ergodic theory of chaos and strange attractors," *Rev. Mod. Phys.*, vol. 57, pp. 617–656, 1985.

[37] J.-P. Eckmann, "Roads to turbulence in dissipative dynamical systems," *Rev. Mod. Phys.*, vol. 53, pp. 643–654, 1981.

[38] I. R. Epstein, "Oscillations and chaos in chemical systems," *Physica*, vol. 7D, pp. 47–56, 1983.

[39] J. D. Farmer and J. J. Sidorowich, "Predicting chaotic time series," *Phys. Rev. Lett.*, vol. 59, pp. 845–848, 1987.

[40] ——, "Exploiting chaos to predict the future and reduce noise," preprint, 1988.

[41] A. P. Fein, M. S. Heutmaker, and J. P. Gollub, "Scaling at the transition from quasiperiodicity to chaos in a hydrodynamic system," *Phys. Scr.*, vol. T9, pp. 79–84, 1985.

[42] M. J. Feigenbaum, M. H. Jensen, and I. Procaccia, "Time ordering and the thermodynamics of strange sets," *Phys. Rev. Lett.*, vol. 57, pp. 1503–1506, 1987.

[43] M. J. Feigenbaum, "Universal behavior in nonlinear systems," *Physica*, vol. 7D, pp. 16–39, 1983.

[44] G. Gibson and C. Jeffries, "Observation of period doubling and chaos in spin-wave instabilities in yttrium iron garnet," *Phys. Rev. A*, vol. 29, pp. 811–818, 1984.

[45] L. Glass, M. R. Guevara, A. Shrier, and R. Perez, "Bifurcation and chaos in a periodically stimulated cardiac oscillator," *Physica*, vol. 7D, pp. 89–101, 1983.

[46] J. A. Glazier, M. H. Jensen, A. Libchaber, and J. Stavans, "The structure of Arnold tongues and the $f(\alpha)$ spectrum for period-doubling: Experimental results," *Phys. Rev. A*, vol. 34, pp. 1621–1624, 1986.

[47] J. A. Glazier, G. Gunaratne, and A. Libchaber, "$f(\alpha)$ curves—Experimental results," *Phys. Rev. A*, vol. 37, pp. 523–530, 1988.

[48] J. A. Glazier, G. Gunaratne, A. Libchaber, and M. Vinson, "Tongue crossing intermittency in Rayleigh–Bénard convection," preprint, 1988.

[49] J. P. Gollub and V. Benson, "Many routes to turbulent convection," *J. F. Mech.*, vol. 100, pp. 449–470, 1980.

[50] J. P. Gollub and A. R. McCarriar, "Convection patterns in Fourier space," *Phys. Rev. A*, vol. 26, pp. 3470–3476, 1982.

[51] J. P. Gollub, E. J. Romer, and J. E. Socolar, "Trajectory divergence for coupled relaxation oscillators: Measurements and models," *J. Stat. Phys.*, vol. 23, pp. 321–333, 1980.

[52] J. P. Gollub and H. L. Swinney, "Onset of turbulence in a rotating fluid," *Phys. Rev. Lett.*, vol. 35, pp. 927–930, 1975.

[53] P. Grassberger and I. Procaccia, "Characterization of strange attractors," *Phys. Rev. Lett.*, vol. 50, pp. 346–349, 1983.

[54] ——, "Estimation of the Kolomogorov entropy from a chaotic signal," *Phys. Rev. A*, vol. 28 (RC), pp. 2591–2593, 1983.

[55] C. Grebogi, E. Ott, and J. A. Yorke, "Are three-frequency quasiperiodic orbits to be expected in typical nonlinear systems?," *Phys. Rev. Lett.*, vol. 51, pp. 339–342, 1983.

[56] H. S. Greenside, A. Wolf, J. Swift, and T. Pignataro, "Impracticality of a box-counting algorithm for calculating the dimensionality of strange attractors," *Phys. Rev. A*, vol. 25, pp. 3453–3456, 1982.

[57] G. Grüner and A. Zettl, "Charge density wave conduction: A novel collective transport phenomenon in solids," preprint, 1987.

[58] M. R. Guevara, L. Glass, and A. Shrier, "Phase locking, period-doubling bifurcations, and irregular dynamics in periodically stimulated cardiac cells," *Science*, vol. 214, pp. 1350–1353, 1981.

[59] G. Gunaratne, M. H. Jensen, and I. Procaccia, "Universal strange attractors on wrinkled tori," *Nonlinearity*, vol. 1, pp. 157–180, 1988.

[60] E. G. Gwinn and R. M. Westervelt, "Intermittent chaos and low-frequency noise in the driven damped pendulum," *Phys. Rev. Lett.*, vol. 54, pp. 1613–1616, 1985.

[61] ——, "Frequency locking, quasiperiodicity, and chaos in extrinsic Ge," *Phys. Rev. Lett.*, vol. 57, pp. 1060–1063, 1986.

[62] ——, "Scaling structure of attractors at the transition from quasiperiodicity to chaos in electronic transport in Ge," *Phys. Rev. Lett.*, vol. 59, pp. 157–160, 1987.

[63] T. Halsey, M. H. Jensen, L. P. Kadanoff, I. Procaccia, and B. I. Shraiman, "Fractal measures and their singularities: The characterization of strange sets," *Phys. Rev. A*, vol. 33, pp. 1141–1151, 1986.

[64] H. Haucke, "Time-dependent convection in a $^3$He–$^4$He Solution," Univ. California, San Diego, unpublished thesis.

[65] H. Haucke, R. E. Ecke, Y. Maeno, and J. C. Wheatley, "Noise-induced intermittency in a convecting dilute solution of $^3$He in Superfluid $^4$He," *Phys. Rev. Lett.*, vol. 53, pp. 2090–2093, 1984.

[64] H. Haucke and R. E. Ecke, "Mode locking and chaos in Rayleigh–Bénard convection," *Physica*, vol. 25D, pp. 307–329, 1987.

[67] H. Haucke and Y. Maeno, "Phase space analysis of convection in a $^3$He-superfluid $^4$He solution," *Physica*, vol. 7D, pp. 69–72, 1983.

[68] H. Haucke, Y. Maeno, and J. C. Wheatley, "Dimension and entropy for quasiperiodic and chaotic convection," in *Dimensions and Entropies in Chaotic Systems*, (G. Mayer-Kress, Ed.), Berlin, Germany: Springer, pp. 198–206, 1986.

[69] G. A. Held, C. Jeffries, and E. E. Haller, "Observation of chaotic behavior in an electron-hole plasma in Ge," *Phys. Rev. Lett.*, vol. 52, pp. 1037–1040, 1984.

[70] G. A. Held and C. Jeffries, "Spatial and temporal structure of chaotic instabilities in an electron-hole plasma in Ge," *Phys. Rev. Lett.*, vol. 55, pp. 887–890, 1985.

[71] ——, "Characterization of chaotic instabilities in an electron-hole plasma in germanium," in *Dimensions and Entropies in Chaotic Systems*, (G. Mayer-Kress, Ed.), Berlin, Germany: Springer, pp. 158–170, 1986.

[72] ——, "Quasiperiodic transition to chaos of instabilities in an electron-hole plasma excited by ac perturbations at one and two frequencies," *Phys. Rev. Lett.*, vol. 56, pp. 1183–1186, 1986.

[73] H. G. E. Hentschel and I. Procaccia, "The infinite number of generalized dimensions of fractals and strange attractors," *Physica*, vol. 8D, pp. 435–444, 1983.

[74] C. Huyghens, letter to his father, dated 26 Feb. 1665, *Oeuvres completes de Christian Huyghens*, (M. Nijhoff, Ed.), The Hague, The Netherlands: Société Hollandaise des Sciences, 1893, vol. 5, pp. 243–244. Cited in [12].

[75] H. Ikezi, J. S. deGrassie, and T. H. Jensen, "Observation of multiple-valued attractors and crises in a driven nonlinear circuit," *Phys. Rev. A*, vol. 28 (RC), pp. 1207–1209, 1983.

[76] M. H. Jensen, P. Bak, T. Bohr, "Transition to chaos by interaction of resonances in dissipative systems. I. Circle maps," *Phys. Rev. A*, vol. 30, pp. 1960–1969, 1984.

[77] M. H. Jensen, L. P. Kadanoff, A. Libchaber, I. Procaccia and J. Stavans, "Global universality at the onset of chaos: Results of a forced Rayleigh–Bénard experiment," *Phys. Rev. Lett.*, vol. 55, pp. 2798–2801, 1985.

[78] R. Keolian, L. A. Turkevich, S. J. Putterman, I. Rudnick, and J. A. Rudnick, "Subharmonic sequences in the Faraday experiment: Departures from period doubling," *Phys. Rev. Lett.*, vol. 47, pp. 1133–1136, 1981.

[79] S. Kim and S. Ostlund, "Simultaneous rational approximations in the study of hydrodynamical systems," *Phys. Rev. A*, vol. 34, pp. 3426–3434, 1986.

[80] ——, "Renormalization mappings of the two-torus," *Phys. Rev. Lett.*, vol. 55, pp. 1165–1168, 1985.

[81] E. J. Kosterlich and J. A. Yorke, "Noise reduction in dynamical systems," preprint, 1988.

[82] W. Lauterborn and E. Cramer, "Subharmonic route to chaos observed in acoustics," *Phys. Rev. Lett.*, vol. 47, pp. 1445–1448, 1981.

[83] A. Libchaber, "The onset of weak turbulence. An experimental introduction," in *Turbulence and Predictability in Geophysical Fluid Dynamics*, proc. Enrico Fermi Summer School, Corso LXXXVIII, Bologna, Italy: Società Italiana di Fisica, pp. 17–28, 1985.

[84] A. Libchaber, S. Fauve, and C. Laroche, "Two-parameter study of the routes to chaos," *Physica*, vol. 7D, pp. 73–84, 1983.

[85] A. Libchaber and J. Maurer, "Une expérience de Rayleigh–Bénard

de geometrie reduite; multiplication, accorchage et demultiplication de frequencies," *J. Phys. (Paris)*, vol. 41 C3, pp. C3-51–C3-56, 1980.

[86] P. S. Linsay, "Period doubling and chaotic behavior in a driven anharmonic oscillator," *Phys. Rev. Lett.*, vol. 47, pp. 1349–1352, 1981.

[87] E. Lorenz, "Deterministic nonperiodic flow," *J. Atmos Sci.*, vol. 20, pp. 130–144, 1963.

[88] *Low-Dimensional Conductors and Superconductors*, (D. Jérome and L. G. Caron, Eds.), London, England: Plenum, 1987, (contains an excellent sample of current work on charge density waves).

[89] R. S. Mackay and C. Tresser, "Transition to chaos for two-frequency systems," *J. Phys. Lett. (Paris)*, vol. 45, pp. L-741–L-746, 1984.

[90] ____, "Transition to topological chaos for circle maps," *Physica*, vol. 19D, pp. 206–237, 1986.

[91] ____, "Boundary of chaos for bimodal maps on the interval," and "Some flesh on the skeleton: The bifurcation structure of bimodal maps," (preprints).

[92] B. B. Mandelbrot, *Fractals, Form, Chanee, and Dimension*, San Francisco, CA: Freeman, 1977. To be used only with caution.

[93] S. Martin and W. Martienssen, "Transition from quasiperiodicity into chaos in the periodically driven conductivity of BSN crystals," in *Dimensions and Entropies in Chaotic Systems*, (G. Mayer-Kress, Ed.), Berlin, Germany: Springer, pp. 191–197, 1986.

[94] S. Martin and W. Martienssen, "Circle maps and mode locking in the driven electrical conductivity of barium sodium niobate crystals," *Phys. Rev. Lett.*, vol. 56, pp. 1522–1525, 1986.

[95] J. Maurer and A. Libchaber, "Rayleigh–Bénard experiment in liquid helium; frequency locking and the onset of turbulence," *J. Phys. Lett. (Paris)*, vol. 40, pp. L-419–L-423, 1979.

[96] ____, "Effects of Prandtl number on the onset of turbulence in liquid $^4$He," *J. Phys. Lett. (Paris)*, vol. 41, pp. L-515–L-518, 1980.

[97] R. M. May, "Simple mathematical models with very complicated dynamics," *Nature*, vol. 261, pp. 459–467, 1976.

[98] S. L. McCall, "Instability and regenerative pulsation phenomena in Fabry–Perot nonlinear optical media devices," *Appl. Phys. Lett.*, vol. 32, pp. 284–286, 1977.

[99] N. Metropolis, M. L. Stein, and P. R. Stein, "On finite limit sets for transformations of the unit interval," *J. Comb. Theor.*, vol. 15, pp. 25–44, 1973.

[100] T. Midavaine, D. Dangoisse, and P. Glorieux, "Observation of chaos in a frequency-modulated $CO_2$ laser," *Phys. Rev. Lett.*, vol. 55, pp. 1989–1992, 1985.

[101] F. C. Moon, J. Cusumano, and P. J. Holmes, "Evidence for homoclinic orbits as a precursor to chaos in a magnetic pendulum," *Physica*, vol. 24D, pp. 383–390, 1987.

[102] *Nonlinear Oscillations in Biology and Chemistry*, (S. Levin, Ed.), Berlin, Germany: Springer, 1986.

[103] S. Newhouse, D. Ruelle, and F. Takens, "Occurrence of strange axiom A attractors near quasi periodic flows on $T^m$, $m \geqslant 3$," *Comm. Math. Phys.*, vol. 64, pp. 35–40, 1978.

[104] P. C. Newell, "Attractor and adhesion in the slime mold dictyostelium," in *Fungal Differentiation*, (J. Smith, Ed.), New York: Marcel-Dekker, pp. 43–59, 1983.

[105] M. Octavio and C. R. Nasser, "Chaos in a dc-bias Josephson junction in the presence of microwave radiation," *Phys. Rev. B*, vol. 30 (RC), pp. 1586–1588, 1984.

[106] S. Ostlund, D. Rand, J. Sethna, and E. Siggia, "Universal properties of the transition from quasi-periodicity to chaos in dissipative systems," *Physica*, vol. 8D, pp. 303–342, 1983.

[107] R. Perez and L. Glass, "Bistability, period doubling bifurcations and chaos in a periodically forced oscillator," *Phys. Lett.*, vol. 90A, pp. 441–443, 1982.

[108] *Physica Scripta*, vol. T9, 1985, (contains a useful collection of experimental and theoretical papers on quasiperiodicity).

[109] R. W. Rollins and E. R. Hunt, "Exact solvable model of a physical system exhibiting universal chaotic behavior," *Phys. Rev. Lett.*, vol. 49, pp. 1295–1298, 1982.

[110] F. J. Romeiras, A. Bondeson, E. Ott, T. M. Antonsen, Jr., and C. Grebogi, "Quasiperiodically forced dynamical systems with strange nonchaotic attractors," *Physica*, vol. 26D, pp. 277–294, 1987.

[111] J.-C. Roux, R. H. Simoyi, and H. L. Swinney, "Observation of a strange attractor," *Physica*, vol. 8D, pp. 257–266, 1983.

[112] D. Ruelle and F. Takens, "On the nature of turbulence," *Comm. Math. Phys.*, vol. 20, pp. 167–192, 1971.

[113] F. S. Rys and A. Waldvogel, "Fractal shape of hail clouds," *Phys. Rev. Lett.*, vol. 56, pp. 784–787, 1986.

[114] M. Schell, S. Fraser, and R. Kapral, "Subharmonic bifurcation in the sine map: An infinite hierarchy of cusp bistabilities," *Phys. Rev. A*, vol. 28, pp. 373–378, 1983.

[115] J. P. Sethna, E. D. Siggia, "Universal transition in a dynamical system forced at two incommensurate frequencies," *Physica*, vol. 11D, pp. 193–211, 1984.

[116] R. S. Shaw, *The Dripping Faucet*. Santa Cruz, CA: Ariel, 1984.

[117] M. Sherwin, R. Hall, and A. Zettl, "Chaotic ac conductivity in the charge-density-wave state of $(TaSe_4)_2I$," *Phys. Rev. Lett.*, vol. 53,

pp. 1387–1390, 1984.

[118] S. J. Shenker, "Scaling behavior in a map of a circle onto itself: Empirical results," *Physica*, vol. 5D, pp. 405–411, 1982.

[119] B. Shraiman, C. E. Wayne, P. C. Martin, "Scaling theory for noisy period-doubling transitions to chaos," *Phys. Rev. Lett.*, vol. 46, pp. 935–939, 1981.

[120] R. H. Simoyi, A. Wolf, and H. L. Swinney, "One-dimensional dynamics in a multicomponent chemical reaction," *Phys. Rev. Lett.*, vol. 49, pp. 245–248, 1982.

[121] C. W. Smith and M. J. Tejwani, "Bifurcation and the universal sequence for first-sound subharmonic generation in superfluid helium-4," *Physica*, vol. 7D, pp. 85–88, 1983.

[122] K. R. Sreenivasan, "Chaos in open flow systems," in *Dimensions and Entropies in Chaotic Systems*, (G. Mayer-Kress, Ed.), Berlin, Germany: Springer, pp. 222–230, 1986.

[123] J. Stavans, F. Heslot, and A. Libchaber, "Fixed winding number and the quasiperiodic route to chaos in a convective fluid," *Phys. Rev. Lett.*, vol. 55, pp. 596–599, 1985.

[124] J. Stavans, S. Thomae, and A. Libchaber, "Experimental study of the attractor of a driven Rayleigh–Bénard system," in *Dimensions and Entropies in Chaotic Systems*, (G. Mayer-Kress, Ed.), Berlin, Germany: Springer, pp. 207–214, 1986.

[125] J. Stavans, "Experimental study of quasiperiodicity in a hydrodynamical system," *Phys. Rev. A*, vol. 35, pp. 4314–4328, 1987.

[126] R. L. Stratonovich, *Topics in the Theory of Random Noise* [2 vols.], trans. R. A. Silverman, New York: Gordon and Breach, 1967.

[127] Z. Su, R. W. Rollins, and E. R. Hunt, "Measurements of $f(\alpha)$ spectra of attractors at transitions to chaos in driven resonator systems," *Phys. Rev. A*, vol. 36, pp. 3515–3517, 1987.

[128] H. L. Swinney, "Observations of order and chaos in nonlinear systems," *Physica*, vol. 7D, pp. 3–15, 1983.

[129] H. L. Swinney and J. Maselko, "Comment on 'Renormalization, unstable manifolds, and the fractal structure of mode locking'," *Phys. Rev. Lett.*, vol. 55, p. 2366, 1985.

[130] P. Szépfalusy and T. Tél, "Dynamical fractal properties of one-dimensional maps," *Phys. Rev. A*, vol. 35 (RC), pp. 47–480, 1987.

[131] F. Takens, "Detecting strange attractors in turbulence," in *Dynamical Systems and Turbulence*, (D. A. Rand and L.-S. Young, Eds.) Berlin, Germany: Springer, 1981.

[132] T. Tél, "Dynamical spectrum and thermodynamic functions of strange sets from an eigenvalue problem," preprint.

[133] J. Testa, J. Pérez, and C. Jeffries, "Evidence for universal behavior in a driven nonlinear oscillator," *Phys. Rev. Lett.*, vol. 48, pp. 714–717, 1982.

[134] C. Tresser, private communication.

[135] J. S. Turner, J.-C. Roux, W. D. McCormick, and H. L. Swinney, "Alternating periodic and chaotic regimes in a chemical reaction —Experiment and theory," *Phys. Lett.*, vol. 85A, pp. 9–12, 1981.

[136] A. T. Winfree, *The Geometry of Biological Time*. Berlin, Germany: Springer, 1980.

[137] ____, *When Time Breaks Down*, Princeton, NJ: Princeton Univ., 1987.

[138] H. G. Winful, Y. C. Chen, and J. M. Liu, "Frequency locking, quasiperiodicity, and chaos in modulated self-pulsing semiconductor lasers," *Appl. Phys. Lett.*, vol. 48, pp. 616–618, 1986.

✷

**James A. Glazier** was born in Cambridge, MA, in 1962. He received the B.A. degree in physics and mathematics from Harvard University, Cambridge, MA, in 1984 and the M.S. degree in physics from the University of Chicago, Chicago, IL, in 1987.

His interests are dynamical systems and disorder.

✷

**Albert Libchaber** was born in Paris, France, in 1934. He received the Ph.D. degree from the Ecole Normale Supérieure in 1965.

He is presently professor of Physics at the University of Chicago. Previously he was directeur de recherche at the Ecole Normale Supérieure, Paris. His interests are all aspects of non-linear physics. He was co-recipient in 1986 of the Wolff prize in Physics. In 1986 he received a MacArthur fellowship.

Dr. Libchaber is a member of the AAAS and the French Academie des Sciencès.

PHYSICAL REVIEW A    VOLUME 30, NUMBER 4    OCTOBER 1984

# Transition to chaos by interaction of resonances in dissipative systems. I. Circle maps

Mogens Høgh Jensen

*H. C. Ørsted Institute, Universitetsparken 5, DK-2100 Copenhagen Ø, Denmark*

Per Bak

*Physics Department, Brookhaven National Laboratory, Upton, New York 11973*

Tomas Bohr

*Laboratory of Atomic and Solid State Physics, Cornell University, Ithaca, New York 14853*
(Received 9 May 1984)

Dissipative dynamical systems with two competing frequencies exhibit transitions to chaos. We have investigated the transition through a study of discrete maps of the circle onto itself. The transition is caused by interaction and overlap of mode-locked resonances and occurs at a critical line where the map loses invertibility. At this line the mode-locked intervals trace up a complete devil's staircase whose complementary set is a Cantor set with fractal dimension $D \sim 0.87$. Numerical results indicate that the dimension is universal for maps with cubic inflection points. Below criticality the staircase is incomplete, leaving room for quasiperiodic behavior. The Lebesgue measure of the quasiperiodic orbits seems to be given by an exponent $\beta \sim 0.35$ which can be related to $D$ through the scaling relation $D = 1 - \beta/\nu$. The exponent $\nu$ characterizes the cutoff of narrow plateaus near the transition. A variety of other exponents describing the transition to chaos is defined and estimated numerically.

## I. INTRODUCTION

Our quantitative knowledge about highly nonlinear dynamical systems is very meager. In a few cases exact solution of the dynamical equations exist, but their behavior is atypical—the very possibility of obtaining analytical solutions excludes the occurrence of chaotic motion which is of importance in any "truly" nonlinear system. A major breakthrough came—especially through the work of Feigenbaum[1]—with the realization that one-dimensional maps are an important laboratory for nonlinear studies. Not only do these maps qualitatively model the kinds of behavior found in dynamical systems, but, more astonishingly, behavior found in the maps carry quantitatively over to real systems.

In this paper and the following one (denoted II) we shall study scaling behavior for one-dimensional circle maps and show that the same scaling exponents can be found in dissipative dynamical systems that exhibit mode locking. Mode locking is a resonant response occurring in systems of coupled oscillators or oscillators coupled to periodic external forces. In general, resonances occur whenever the frequency of a harmonic, $P\omega_1$, of one oscillator approaches some harmonic, $Q\omega_2$, of another; and in the resonant region the frequencies of the two oscillators locks exactly into the rational ratio $P/Q$.

The mechanism, in these systems, leading eventually to chaotic behavior is interactions between the different resonances, caused by the nonlinear couplings, and overlap between the resonant regions when the couplings exceed a certain critical value. In some sense the mechanism is the analog, for dissipative systems, of Chirikov's instability of quasiperiodic orbits in Hamiltonian systems.[2]

In II some specific systems from condensed-matter physics (Josephson junctions in microwave fields, charge-density waves in periodic electric fields) and from classical mechanics (the "swing" or the damped driven pendulum) will be considered. The main result is that the behavior of these systems, including the transition to chaos, can be described by one-dimensional discrete maps of the circle onto itself, the so-called "circle maps," which is the subject of this paper. In general, circle maps are defined through

$$\theta_{n+1} = f_\Omega(\theta_n) = \theta_n + \Omega + g(\theta_n), \tag{1.1}$$

where

$$g(\theta_n) = g(\theta_n + 1) \pmod{1} \tag{1.2}$$

and can thus be thought of as "lifts" of mappings from the circle to itself. The advantage of studying simple maps on this form is obvious. It is much easier to identify periodic, quasiperiodic, and chaotic solutions by iterating the map than by a cumbersome numerical integration of the underlying differential equation. The variables $\theta_n$ represent the phase of the oscillating system measured stroboscopically at periodic time intervals $t_n = 2\pi n/\omega_2$, using the frequency of the external force, or one of the oscillating parts as a clock. A phase shift $\theta_n \to \theta_n + 1$ represents a full rotation; hence the periodic property (1.2) of $g$. The map has a linear term $\theta_n$ and a bias term $\Omega$ representing the frequency of the system in the absence of the nonlinear coupling $g$.

To study the mode locking in the circle map we consider iterations of the map, $\theta, f(\theta), f^2(\theta), \ldots$, or $\theta_1, \theta_2, \theta_3, \ldots$. The iteration of the map is conveniently described by the winding number

320

$$W = \lim_{n \to \infty} [(f_\Omega^n - \theta)/n] . \qquad (1.3)$$

The winding number is the mean number of rotations per iteration, i.e., the frequency of the underlying dynamical system, so $W = \Omega$ in the absence of the nonlinear coupling. Under iteration the variable $\theta_n$ may converge to a series which is either *periodic*, $\theta_{n+Q} = \theta_n + P$, with rational winding number $W = P/Q$; *quasiperiodic*, with irrational winding number $W = q$; or *chaotic* where the series behaves irregularly.

Although the question of the existence of smooth behavior in circle maps has very much the flavor of the general problem of the existence of smooth invariant tori in dynamical systems [the Kolmogorov-Arnol'd-Moser (KAM) problem], much stronger theorems[3] due to Arnol'd[4] and Herman[5] hold for the one-dimensional circle maps. As long as $f(\theta)$ is a diffeomorphism, i.e., smooth and invertible, these theorems guarantee that no chaotic motion can occur.

The nontrivial scaling behavior that we shall discuss occurs precisely at the point (subsequently denoted the critical point) in parameter space where $f(\theta)$ loses its invertibility. In that case the theorems mentioned above break down and not much is known in general. The first exposition of interesting scaling behavior at the critical point was given in a numerical investigation by Shenker[6] followed by renormalization-group treatments by Feigenbaum, Kadanoff, and Shenker[7] and by Rand, Ostlund, Sethna, and Siggia.[8] These studies concentrated on specific well-behaved winding numbers—mostly on the "golden mean," $(\sqrt{5}-1)/2$—and showed that nontrivial scaling behavior is found when the golden mean is approached through a sequence of rational winding numbers.

In our work we have generalized these ideas and asked for the global scaling properties of the mode-locking pattern. From the outset it was not clear whether any simple universal properties should exist globally since the renormalization-group treatments[7,8] are only valid for a measure-zero set of winding numbers. We do, however, find strong numerical evidence for nontrivial scaling behavior, and from this we can derive general universal "average" exponents distinctly different from their golden-mean values. A short account of these findings has already been published.[9]

Most of our results are obtained for the sine map

$$\theta_{n+1} = f_\Omega(\theta_n) = \theta_n + \Omega - (K/2\pi)\sin 2\pi\theta_n , \qquad (1.4)$$

but in order to check universality we have also investigated maps in which the sine function has been replaced by higher-order polynomials (Sec. IV).

The mapping (1.4) is sketched in Fig. 1(a) for $\Omega = 0.2$ and $K = 0.9$. Because of the periodicity of the map we have reduced it to the square $0 \le \theta_n < 1$, $0 \le \theta_{n+1} < 1$. We see two branches in the unit square. When $K < 1$ the map is strictly monotonic. At $K = 1$ [Fig. 1(b)] the map develops a cubic inflection point at $\theta = 0$, so the map is still invertible but the inverse map has a singularity. For $K > 1$ [Fig. 1(c)] the map develops a local maximum and a local minimum and is no longer invertible. The figure shows a chaotic trajectory.

We shall here be concentrating on the situation for $K$

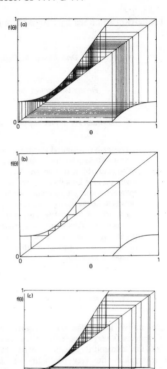

FIG. 1. Evolution of iterations of the circle map (1.3) for $\Omega = 0.2$ and (a) $K = 0.9$, (b) $K = 1.0$, and (c) $K = 1.1$. For $K > 1$ the map develops local maxima (and minima) and chaotic behavior may occur.

equal to or slightly below 1. For $0 < K < 1$ it has been shown[5] that the winding number locks-in at every single rational number $P/Q$ in a nonzero interval of $\Omega$, $\Delta\Omega(P/Q)$. For $K$ close to zero all intervals are quite small so the probability that the winding number for a random value of $\Omega$ is rational is almost zero, i.e., the probability of hitting an irrational winding number is almost one. However, with increasing $K$ the widths of all the phase-locked intervals increase (Fig. 2), so for $K = \frac{1}{2}$ the probabilities of observing rational and irrational winding numbers are almost equal. For $K \sim 1$ the probability of finding a rational winding number is close to 1. The regimes in $(\Omega, K)$ space where $W$ assumes rational values are called "Arnol'd tongues." Clearly, the widths of the resonances cannot grow indefinitely: at some point they will overlap. It will be shown numerically that at $K = 1$ the resonances will completely fill up the critical line, confining the quasiperiodic orbits to a Cantor set of zero measure.

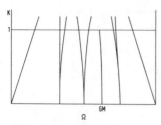

FIG. 2. Schematic phase diagram for circle map in $(\Omega, K)$ space. Note the Arnol'd tongues where the winding number assumes locked rational values. The winding number assumes irrational values (such as the golden mean) along one-dimensional curves ending at $K = 1$.

Let us briefly summarize our findings. Figure 3 shows the winding number as a function of $\Omega$ at $K = 1$. The plateaus in this function forms a complete "devil's staircase," a structure which has previously been found in quite different contexts, such as the one-dimensional Ising model with long-range interactions,[10] the Frenkel-Kontorowa model of atoms adsorbed on a periodic substrate,[11] and the three-dimensional (3D) Ising model with competing interactions.[12]

The complementary set (on the $\Omega$ axis) to a complete staircase is a Cantor set of fractal dimension $D \leq 1$. For the staircase of the circle map we find $D \sim 0.87$; this number is the universal index characterizing the transition to chaos by mode locking.

For $K < 1$ the staircase is no longer complete; there is room for quasiperiodic orbits. Near $K = 1$ we find that the integrated measure $M$ of the quasiperiodic orbits is described by an exponent $\beta$:

$$M \sim (1-K)^\beta , \tag{1.5}$$

where $\beta$ is related to the fractal dimension $D$ through the scaling law $D = 1 - \beta/\nu$. Here, $\nu$ is an exponent characterizing the cutoff of narrow plateaus versus $(1-K)$, i.e., $\nu$ is the exponent for the crossover between the critical behavior with $D \sim 0.87$ at the transition and the regular behavior with $D = 1$ below the transition. Hence, below criticality the resonances are separated by quasiperiodic orbits, at criticality the resonances fill up the critical line,

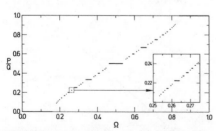

FIG. 3. Winding number $W$ vs $\Omega$ for the circle map at $K = 1$. Steps with $\Delta(P/Q) > 0.0015$ are shown. The inset shows intervals with $\Delta > 0.000 \, 15$. Note the self-similar nature of the staircase under magnification.

and above criticality the resonances overlap. The chaos occurring in the supercritical region is a "frustrated" response of the system due to the overlaps: the orbit jumps between resonances in an erratic way.

The remaining part of the paper is organized as follows. In Sec. II we present the numerical methods; in Sec. III the scaling of resonances at the critical line is investigated for the sine circle map. In Sec. IV the universality conjecture is tested by investigating circle maps with different periodic functions. In Sec. V additional critical exponents are defined and derived; in particular, we calculate the exponent $\beta$ characterizing the integrated Lebesgue measure of the quasiperiodic orbits below criticality. Finally, in Sec. VI the scaling near rational winding numbers is investigated.

## II. NUMERICAL METHODS

In order to find the widths of the various resonances $\Delta\Omega(P/Q)$ we consider the stability of an orbit with rational winding number. The cycle $\theta_1, \theta_2, \ldots, \theta_Q$ ($= \theta_1 + P$), with period $Q$ is stable as long as

$$\frac{df_\Omega^Q(\theta_i)}{d\theta_i} = \prod_{i=1}^{Q} f'_\Omega(\theta_i) < 1 . \tag{2.1}$$

Thus, the endpoints of the plateaus are determined by the condition $df_\Omega^Q(\theta_i)/d\theta_i = 1$, together with the condition $f_\Omega^Q(\theta_i) = \theta_i + P$. From the condition $f(\theta^*) = \theta^*$ and $f'(\theta^*) = 1$, we find analytically that

$$\Delta\Omega(0/1) = (-K/2\pi, K/2\pi)$$

for the map (1.7). The stability of a general $P/Q$ step is found by a two-dimensional Newton iteration method. We define the functions

$$g_1(\theta, \Omega) = f_\Omega^Q(\theta) - \theta - P ,$$

$$g_2(\theta, \Omega) = \frac{df_\Omega^Q(\theta)}{d\Omega} - 1 , \tag{2.2}$$

or

$$\vec{g}(\theta, \Omega) = \begin{bmatrix} g_1(\theta, \Omega) \\ g_2(\theta, \Omega) \end{bmatrix} .$$

The stability criteria can be expressed simply as $\vec{g}(\theta^*, \Omega^*) = \vec{g}^* = \vec{0}$. Expanding $\vec{g}^*$ around the initial point of the iteration, $\vec{g}(\theta^0, \Omega^0) = \vec{g}_0$,

$$\vec{g}^* \simeq \vec{g}_0 + \vec{\Delta} \underline{M} , \tag{2.3}$$

where

$$\vec{\Delta} = (\theta^*, \Omega^*) - (\theta^0, \Omega^0) \tag{2.4a}$$

and

$$\underline{M} = \begin{vmatrix} \dfrac{\partial g_1}{\partial \theta} & \dfrac{\partial g_1}{\partial \Omega} \\ \dfrac{\partial g_2}{\partial \theta} & \dfrac{\partial g_2}{\partial \Omega} \end{vmatrix} , \tag{2.4b}$$

we find (for $\vec{g}^* = \vec{0}$)

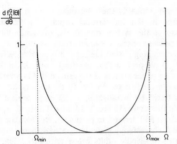

FIG. 4. $F'(\theta_i) = \prod_{i=1}^{Q} f'_{\Omega}(\theta_i)$ within the stability interval for the periodic orbit with $W = P/Q = \frac{2}{9}$.

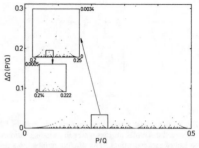

FIG. 5. $\Delta(P/Q)$ vs $P/Q$. Note the self-similarity of the diagram under scaling.

$$\vec{\Delta} \simeq -\underline{M}^{-1}\vec{g}_0 , \qquad (2.5)$$

and so, as the first approximation,

$$(\theta^*, \Omega^*) \simeq (\theta^1, \Omega^1) = -\underline{M}^{-1}\vec{g}_0 + (\theta^0, \Omega^0) . \qquad (2.6)$$

Iterating the equations (2.5) and (2.6), it is possible to locate the endpoints of a $P/Q$ interval even when $Q$ is very large ($Q \sim 4000$). Note that all derivatives can be derived recursively

$$\frac{\partial f_{\Omega}^{i+1}(\theta)}{\partial \theta} = [1 - K\cos(2\pi\theta_i)]\frac{\partial f_{\Omega}^{i}(\theta)}{\partial \theta} ,$$

$$\frac{\partial^2 f_{\Omega}^{i+1}(\theta)}{\partial\Omega\,\partial\theta} = 2\pi K\sin(2\pi\theta_i)\frac{\partial f_{\Omega}^{i}(\theta)}{\partial \theta}\frac{\partial_{\Omega}^{i}f(\theta)}{\partial \Omega} \qquad (2.7)$$

$$+ [1 - K\cos(2\pi\theta_i)]\frac{\partial^2 f_{\Omega}^{i}(\theta)}{\partial\Omega\,\partial\theta} .$$

To initiate the iteration is is always convenient to locate the superstable point $(\theta,\Omega) = (\theta_s,\Omega_s)$, where $df_{\Omega_s}^{Q}(\theta_s)/d\theta = 0$. At $K = 1$, $\theta_s = 0$ and $\Omega_s$ is determined by $f_{\Omega_s}^{Q}(0) = P$. This point is always close to the midpoint of the interval. Figure 4 shows the variation of $df_{\Omega_s}^{Q}(\theta)/d\theta$ within the stability interval. This function has infinite slope at the end points of the interval and is close to a half-ellipse.

With this numerical method we have found all intervals with $0 < P/Q < \frac{1}{2}$ and $Q \leq 95$ giving 1388 intervals $\Delta(P/Q)$ in the range between 0.3 and 0.000 002. All steps were found to an accuracy of $10^{-8}$. Due to the symmetry of the map (1.4) as $\theta \rightarrow -\theta$, the staircase is symmetric around $\Omega = \frac{1}{2}$, so $\Delta(P/Q) = \Delta(1 - P/Q)$.

## III. SCALING OF THE STAIRCASE AT THE TRANSITION TO CHAOS

All the intervals with $Q \leq 95$ were found to be stable in a nonzero interval for $K = 1$. Figure 5 shows the widths of the steps $\Delta(P/Q)$ versus $P/Q$. Note the self-similarity of the function under rescaling. We conjecture that eventually $\Delta(P/Q) > 0$ for all $P$ and all $Q$. By including more and more steps, with higher $Q$ and smaller widths the $\Omega$ axis becomes more and more "filled up." This is not different from the situation for $K < 1$.[4]

However, one might speculate that eventually the mode-locked intervals will cover the entire $\Omega$ axis. In this case the staircase is called *complete*. To investigate whether or not this is the case we have calculated the total width $S(r)$ of all steps which are larger than a given scale $r$. We are interested in the space between the steps, $1 - S(r)$, and have measured it on the scale $r$ to find the "number of holes," $N(r) = [1 - S(r)]/r$. Here the $\Omega$ interval is of length 1, in general, the interval may have any length $\Omega_0$ and the number of holes is $N(r) = [\Omega_0 - S(r)]/r$. In Fig. 6 $\log_{10}N(r)$ has been plotted versus $\log_{10}1/r$ for 40 values of $r$ in the interval (0.0009, 0.000017). The points fall excellently on a straight line indicating a power law

$$N(r) \sim \left[\frac{1}{r}\right]^{D} . \qquad (3.1)$$

From the slope of the straight line we find $D = 0.8700 \pm 3.7 \times 10^{-4}$. The uncertainty on $D$ was found from a standard linear regression analysis. The result (3.1) means that the space between the steps vanishes as

$$1 - S(r) \sim r^{1-D} \qquad (3.2)$$

as $r \rightarrow 0$. We, therefore, conjecture that *the staircase is complete*. The exponent $D$ is the fractal dimension[13] of

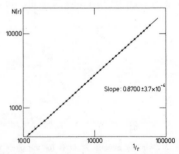

FIG. 6. Plot of $\log_{10}N(r)$ vs $\log_{10}(1/r)$ for the critical circle map. The slope of the straight line yields $D = 0.8700 \pm 3.7 \times 10^{-4}$.

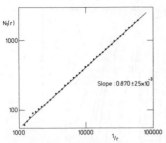

FIG. 7. Plot of $\log_{10}N_1(r)$ vs $\log_{10}(1/r)$ for the critical circle map. The slope of the line yields $D_1=0.870\pm2.5\times10^{-3}\simeq D$.

the staircase, or rather the fractal dimension of the Cantor set of zero Lebesgue measure which is the complementary set to the mode-locked intervals on the $\Omega$ axis.

The fractal dimension can be determined by an alternative method by simply counting the number of steps $N_1(r)$ which are larger than a given scale $r$.[10] This number is given by the equation

$$\frac{\partial N_1}{\partial r}=-\frac{1}{r}\frac{\partial S(r)}{\partial r}\sim\frac{1}{r}\frac{\partial r^{1-D}}{\partial r}\sim r^{-1-D} ,\qquad(3.3)$$

so the total number of steps wider than $r$ is

$$N_1(r)=\int_r^{r_0}\frac{\partial N_1}{\partial r}dr=r^{-D}+\text{const} .\qquad(3.4)$$

To investigate whether or not $N_1$ fulfills the condition (3.4), we have counted $N_1(r)$ for several values of $r$. Figure 7 shows $\log_{10}N_1(r)$ plotted versus $\log_{10}(1/r)$. Again

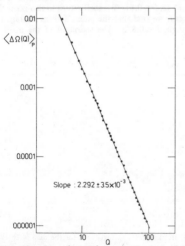

FIG. 8. Plot of $\log_{10}\langle\Delta\Omega(Q)\rangle_P$ vs $Q$. The slope of the line yields $\delta=2.292\pm3.4\times10^{-3}$.

the points fall on a straight line so

$$N_1(r)\sim\left[\frac{1}{r}\right]^{D_1}\qquad(3.5)$$

with $D_1=0.870\pm2.5\times10^{-3}\simeq D$ as it should be. The latter method seems easier to use when analyzing experiments since uncertainties in the determination of the stepwidth are not accumulated as when $S(r)$ is calculated. On the other hand, even if $N_1(r)$ obeys the simple power law (3.5) there is no guarantee that the staircase is complete. Integration of (3.5) leads to

$$S(r)\sim r^{1-D}+C ,$$

where $C$ is an integration constant. Thus, the power law (3.5) does not rule out a finite probability $\sim(1-C)$ of quasiperiodic orbits, so the two methods are equivalent only when the staircase is known to be complete.

Another dimension $D_2$ can be calculated as follows. The mean values of steps with a given denominator $Q$, $\langle\Delta\Omega(Q)\rangle_P$, can be found by averaging over the numerator $P$. Figure 8 shows $\log_{10}\langle\Delta\Omega(Q)\rangle_P$ versus $\log_{10}Q$, again indicating a power-law behavior,

$$r_Q=\langle\Delta\Omega(Q)\rangle_P\sim Q^{-\delta}\qquad(3.6)$$

with $\delta=2.292\pm3.4\times10^{-3}$. The exponent $\delta$ is related to a dimension $D_2$ in the following way. The number of rationals with denominator $Q_0$ is approximately $(3/\pi^2)Q_0$. The total number of rationals with denominator smaller than $Q_0$ is thus

$$N_0\sim\int_1^{Q_0}Q\,dQ\sim Q_0^2$$

and $D_2$ is defined as

$$D_2=\frac{\log_{10}N_0}{\log_{10}1/r_0}=\frac{2\log_{10}Q_0}{\delta\log_{10}Q_0}=\frac{2}{\delta}$$

$$=0.873\pm2.1\times10^{-3}.\qquad(3.7)$$

We stress that since $\Delta(P/Q)$ is a function of *both* $P$ and $Q$, not only of $Q$, it is not a mathematical necessity that $D_2=D$; it is not even a certainty that $D_2$ is well defined even if $D$ is. However, our numerical results are consistent with $D_2$ being identical to $D$.

For intervals $\Delta(P_i/P_{i+1})$, where $P_i$ are Fibonacci numbers, converging to the inverse golden-mean winding number, Shenker[6] found

$$\Delta(P/Q)\sim Q^{-\delta'} ,\quad\delta'\simeq2.16 ,$$

so the average exponent found from (3.6) is distinctly different.

When passing beyond the $K=1$ line the steps continue to increase. Since they fill up the whole $\Omega$ axis for $K=1$, they must necessarily overlap for $K>1$ (see Fig. 9). In an experimental situation, the transition to chaos is most easily identified by considering hysteresis involving the smallest steps. As soon as two steps overlap, an infinity of smaller steps in between are squeezed out. The overlap regimes correspond to chaotic or hysteretic solutions. The "in between" steps yield an infinity of metastable solu-

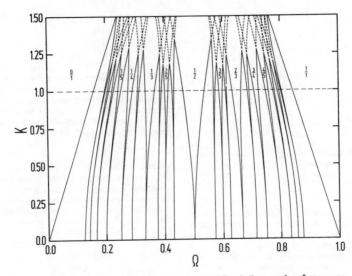

FIG. 9. Phase diagram for the sine circle map. The dotted lines indicate overlap of resonances.

tions which may all be observed in a numerical iteration process just by varying the initial point $\theta_0$. Chaotic behavior arises because the orbit jumps between the various overlapping resonances in an erratic way. *The transition to chaos is caused by overlap of resonances.* A transition of this type has been observed in a variety of physical systems as for instance Josephson junctions in microwave fields[14] and sliding charge-density waves;[15] we will discuss these systems in the following paper. Most nonlinear periodic systems perturbed by an external periodic field (sinusoidal or pulsed) will probably exhibit a transition to chaos caused by overlap of resonances as described here.

For $K > 1$ the map develops quadratic maxima and minima. It is well known from the work of Feigenbaum[1] that iterations of this type of mapping exhibit infinite series of period doubling leading to chaos. This type of chaos (associated with instabilities near the superstable points—not the edges of the steps) has been studied in detail by Glass and Perez[16] and by Kaneko.[17] Bifurcations of a $P/Q$ cycle lead to the cycles $2P/2Q$, $4P/4Q$, $8P/8Q,\ldots$, so the winding number is unaffected.

## IV. UNIVERSALITY

It is important to know whether or not the critical behavior at the transition to chaos is "universal," i.e., whether or not it depends on the specific function $f(\theta)$ in (1.1). In an experiment we do not know the function $f(\theta)$, and it is unlikely that it is a simple sine function (see the following paper). For the theory here to be *predictive* in such cases it is imperative that the critical behavior is universal. To check the universality of the scaling dimension $D \sim 0.87$, we have studied a class of mappings

$$f_{\Omega,a}(\theta) = \theta + \Omega - (K/2\pi)[\sin(2\pi\theta) + a\sin^3(2\pi\theta)] . \quad (4.1)$$

For $-\frac{4}{3} < a < \frac{1}{6}$ the function $f$ is monotonic and has a cubic inflection point at $\theta = 0$. Generally, the details of the staircases are different from the staircase shown in Fig. 3. Some steps become wider, some become narrower. The scaling behavior, however, remains the same, independently of $a$. Figure 10 shows $\log_{10}[1 - S(r)/r]$ versus $\log_{10}(1/r)$ for $a = -0.8$, $-0.25$, and 0.15. The points for $a = 0.15$ seemingly exhibit a crossover from $D \sim 0.81$ to $D \sim 0.87$. We shall return to this point shortly. Again we find that the points fall on a straight line with slope $D = 0.870$. The staircases of the maps (4.1)

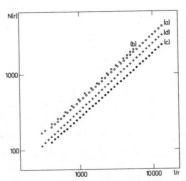

FIG. 10. $\log_{10}N(r)$ vs $\log_{10}(1/r)$ for the map (4.1) with (a) $a = -0.25$, (b) $a = -0.8$, and (c) $a = 0.15$, and for the map (4.2) with (d) $b = 0.2$.

also obey the symmetry $\Delta(P/Q)=\Delta(1-P/Q)$. In order to check that this symmetry does not influence the fractal dimension, we have studied the map

$$f_{\Omega,b}(\theta)=\theta+\Omega-(K/2\pi)[\sin(2\pi\theta)+b\sin^4(2\pi\theta)] \qquad (4.2)$$

for $b=0.2$ and $K=1$. Due to the even term the staircase is not symmetric but it is still complete with $D\sim0.87$ [see Fig. 10(d)].

From these investigations we conjecture that staircases constructed from maps with cubic inflection points are complete with a universal fractal dimension 0.87. This number is thus a universal index characterizing the transition to chaos.

From the map (4.1) with $a=\frac{1}{6}$ the lowest-order term in an expansion of $f(\theta)$ versus $\theta$ is of fifth order. This also leads to a complete staircase but with $D\simeq0.81$, so the fractal dimension depends upon the nature of the inflection point. This explains the behavior of curve (c) in Fig. 10 where $a$ is very close to $\frac{1}{6}$. We have not studied this crossover in a quantitative way. Of course, in an experimental situation one would not expect the first-order term and the third-order term to vanish simultaneously, so the generic critical exponent is $D\sim0.87$.

Clearly, it is a *local* property of the map, namely the behavior around the inflection point which determines the fractal dimension. The behavior of the map away from the inflection point does not affect the scaling properties associated with very-high-order iterates.

We would like to stress that although the dimension was calculated by considering steps in a large interval of $\Omega$, it is a well-defined number at any *point* on the transition line. The index $D$ expresses the self-similarity everywhere. In principle, we could choose any infinitesimal interval $\Delta\Omega$ around this particular point and derive the scaling properties. We have checked this by investigating the scaling properties of steps in different small intervals of $\Omega$ on the critical line. If the scaling index is universal a scaling law for an interval must necessarily apply to any part of the interval. Also, in principle, the locality of the scaling behavior implies that the same scaling behavior would apply if $\Delta(P/Q)$ is considered as a function of a variable $\Omega'$ which is a smooth function of $\Omega$. This is of importance when analyzing experiments since the effective $\Omega$ which enters the circle map is generally a complicated function of the variables in the experiment, such as currents and voltages in Josephson junctions and charge-density-wave (CDW) systems.

The transition to chaos is caused by the competition between two temporal periods. There is an analogous situation in condensed-matter physics where mode locking and chaos occur as a consequence of competition between spatial periods, namely the commensurate-incommensurate transition. In the Frenkel-Kontorowa model[11] and the axial next-nearest-neighbor Ising model[12] there is a competition between the lattice constant and the periods of structurally or magnetically ordered structures. The latter model has a phase diagram almost identical to Fig. 9: There is a regime with regular incommensurate ("quasiperiodic") structures between commensurate ("periodic") structures, and a regime with overlapping metastable commensurate and spatially chaotic structures, separated

(probably) by a line along which a staircase is complete. The critical properties are not universal, but depend on the actual interactions in the models,[10] probably because the discrete mappings constructed from these models are Hamiltonian and not dissipative.

## V. CROSSOVER BEHAVIOR FOR $K\lesssim1$

The steps do not fill up the entire $\Omega$ axis for $K<1$ and the slope $D$ in the $\log_{10}N(r)$ versus $\log_{10}(1/r)$ plot must then necessarily converge towards $D=1$. In fact, when $K$ is only slightly smaller than 1 it seems that the scaling follows $D\sim0.87$ down to a certain scale (depending on $1-K$) and then makes a smooth crossover to the trivial scaling characterized by $D=1$. In this section we shall define and estimate the exponents characterizing this crossover, and the measure of quasiperiodic orbits for $K<1$.

First, let us follow the consequences of treating $Q$ as a "scaling variable" as in (3.6) and (3.7). Thus we assume that the average widths at criticality have the scaling behavior

$$\langle\Delta\Omega(Q)\rangle_P\sim Q^{-(2/D_2)}, \quad D_2\sim D\sim0.87 \ .$$

A plausible scaling ansatz for $K<1$ would be

$$\langle\Delta\Omega(Q)\rangle_P Q^{(2/D_2)}\sim\exp[-a(1-K)^\phi Q] \ . \qquad (5.1)$$

We, therefore, plot the quantity on the left-hand side of (5.1) versus $Q$ (Fig. 11) for 10 different values of $(1-K)$ ranging from 0.0025 to 0.1, using staircases found by means of the numerical methods of Sec. II. The linear behavior indicates that

$$\langle\Delta\Omega(Q)\rangle_P Q^{(2/D_2)}\sim\exp[-A(K)Q] \ . \qquad (5.2)$$

Figure 12 shows $\log_{10}A(K)$ versus $(1-K)$. From the slope of the apparently straight line it seems plausible that the ansatz (5.1) indeed holds, with $\phi\sim1$, in agreement with the nonsingular behavior of the widths of the plateaus as $K$ approaches 1 (see Fig. 9). Somewhat surprisingly, this means that the functional form of $\Delta(P/Q)$ for

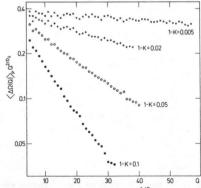

FIG. 11. Plot of $\log_{10}\langle\Delta\Omega(Q,K)\rangle_P Q^{2/D_2}$ vs $Q$ for various values of $(1-K)$.

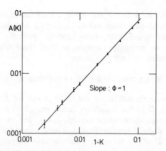

FIG. 12. Plot of $\log_{10}A(K)$ [where $A(K)$ are the slopes of the straight lines found in Fig. 11] vs $\log_{10}(1-K)$. The straight line is consistent with an exponent $\phi \sim 1$.

$K \ll 1$ is the same as for $K=1$: for small $K$, $\Delta(P/Q) \sim K^Q = e^{Q\ln K}$, which becomes $e^{Q(1-K)}$ for $K \sim 1$. The Arnol'd tongues grow in a uniform way from $K=0$ to $K=1$.

Equation (5.1) indicates that steps with $Q \gtrsim 1/(1-K)$ are effectively cut off for $K<1$, leaving room for quasi-periodic orbits. The integrated measure $M(K)$ of the support of these orbits becomes

$$M(K) = \int_{Q=1/(1-K)}^{\infty} dQ \, Q Q^{-2/D_2} \sim (1-K)^{\beta_2} , \quad (5.3)$$

where the exponent $\beta$ obeys the scaling law

$$\beta_2 = \frac{2}{D_2} - 2 = \delta - 2 \sim 0.29 . \quad (5.4)$$

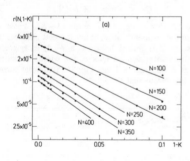

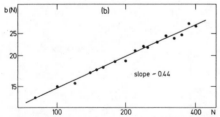

FIG. 13. (a) Scale $r$ for which there are $N$ intervals wider than $r$ plotted vs $(1-K)$. (b) Plot of $\log_{10}b(N_1)$, defined by Eq. (5.5) vs $\log_{10}N_1$.

This approach seems not entirely satisfactory: there is no *a priori* reason why $Q$ should be the natural scaling variable and thus for the scaling ansatz to make sense. The considerable spread around the straight line in Fig. 12 also points in that direction.

A more natural choice of scaling variable is the actual width of the resonances, the quantity that directly enters into the calculation of $M(K)$. Hence, for various values of $(1-K)$ we have calculated the scale $r(N_1,K)$ such that the number of resonances in the interval $[0,1]$ which are wider than $r$ is precisely $N_1$. Obviously, this function is a decreasing function of $(1-K)$ since the intervals become narrower. Figure 13(a) shows $\log_{10}r(N_1,K)$ versus $1-K$ for several values of $N_1$. The straight lines indicate exponential behavior:

$$r(N_1,K) = r(N_1,0)\exp[-b(N_1)(1-K)] . \quad (5.5)$$

Figure 13(b) shows $\log_{10}b(N_1)$ versus $\log_{10}N_1$. The linear behavior allows us to define an exponent $\nu$:

$$b(N_1) \sim N_1^{1/D\nu}, \quad 1/D\nu \simeq 0.44 \pm 0.02 . \quad (5.6)$$

Equations (5.5) and (5.6) give a cutoff of the number of resonances, $N_0(K)$, which give a contribution to the integrated measure below criticality:

$$N_0(K) \sim (1-K)^{-D\nu} . \quad (5.7)$$

These $N_0$ resonances which survive below the transition are precisely those which are wider than a scale $r_0$ at $(1-K)=0$, with $N_0$ related to $r_0$ through Eq. (3.5), so

$$r_0 \approx (1-K)^\nu, \quad \nu \sim 2.63 . \quad (5.8)$$

In other words, the plateaus which are narrower than $r_0$ at $K=1$ are effectively cut off at a value of $K<1$ given by (5.8).

In a sense, $(1-K)$ plays the role of the reduced temperature near a second-order phase transition, and $1/r_0$ is the "correlation length" which diverges at the transition. The measure of the quasiperiodic orbits is a valid order parameter for the transition since it is zero above the transition and nonzero below the transition. This measure is precisely the measure of the periodic orbits which are cut off below the transition:

$$M(K) = \int_0^{(1-K)^\nu} -\frac{\partial N_1}{\partial r} r \, dr$$

$$\sim (1-K)^{\nu(1-D)}$$

$$\equiv (1-K)^\beta, \quad \beta \sim 0.34 \pm 0.02 . \quad (5.9)$$

Equation (5.9) defines the scaling relation

$$D = 1 - \beta/\nu \quad (5.10)$$

which is very similar to the relation

$$D = d - \beta/\nu \quad (5.11)$$

which has been derived for second-order phase transitions. Here, $d$ is the Euclidean dimension. The exponent $\beta$ defined here seems to differ somewhat from the exponent $\beta_2$ derived from (5.4). We believe that the equation (5.11)

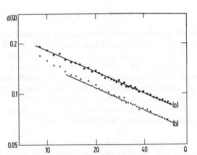

FIG. 14. Plot of $\log_{10}d(Q)$, defined by Eq. (5.5), vs $\log_{10}Q$. The slope yields an exponent $\alpha=0.421\pm2.5\times10^{-3}$ for $a=0$. The curve (b) is calculated for $a=-0.25$ in Eq. (4.1).

gives the proper asymptotic behavior. The expressions (5.4) and (5.11) are identical only for $\delta=\nu$. We do not believe that this relation holds; thus the result (5.4) seems to be spurious due to the use of the wrong scaling variable $Q$.

Besides the exponent $D$ which characterizes the scaling in the $\Omega$ variable, we can also define a scaling index for the $\theta$ variable. for the superstable cycle of the $P/Q$ step we find the point $\theta_i$ of the cycle $\theta_2, \dots, \theta_Q$ which is closest to zero, and define $d_1(P/Q)=\min(\theta_i,1-\theta_i)$. For constant $Q$, $d_1(P/Q)$ is averaged over the numerators $P$:

$$d(Q)=\langle d_1(P/Q)\rangle_P , \qquad (5.12)$$

and $d(Q)$ is plotted against $Q$ on a log-log scale (Fig. 14). The straight line indicates

$$d(Q)\sim Q^{-\alpha}, \quad \alpha=0.421\pm2.5\times10^{-3} . \qquad (5.13)$$

This number also seems to be universal as indicated by the line (b) in the figure which is based on the map (4.1) with $a=-0.25$. Our value for $\alpha$ is distinctly different from the corresponding value for limit cycles converging to the golden mean found by Shenker[6] ($\alpha_G=0.527$).

The smallest distance between any two points in the cycle also scales with a power law,

$$d_{\min}(Q)=\langle d_{\min}(P/Q)\rangle_P \sim Q^{-\alpha'} \qquad (5.14)$$

with $\alpha'\sim1.58$.

## VI. SCALING NEAR RATIONAL WINDING NUMBERS

Close to the instability point of the (0/1) plateau the increments in phase between two iterations, $\theta_i-\theta_{i-1}$, become infinitesimally small, and the map (1.3) may be studied in the continuum approximation, $\theta_i=\theta_{i-1}$ $\sim d\theta/dz$:

$$\frac{d\theta}{dz}=\Omega-\frac{K}{2\pi}\sin(2\pi\theta) . \qquad (6.1)$$

This equation can be integrated to yield

$$\theta=\frac{1}{\pi}\tan^{-1}\left[\frac{K}{2\pi\Omega}-\frac{\omega}{\Omega}\tan(\omega\pi z)\right] , \qquad (6.2)$$

where

$$\omega=[\Omega^2-(K/2\pi)^2]^{1/2} .$$

This equation shows that (6.1) has a transition from periodic behavior with $W=0$ to quasiperiodic behavior with $W>0$. The critical value of $\Omega$ is $\Omega_0=(K/2\pi)$. For $\Omega\geq\Omega_0$ the winding number is given by

$$W=\omega=[\Omega^2-(K/2\pi)^2]^{1/2}$$
$$\sim[\Omega-(K/2\pi)]^{1/2} . \qquad (6.3)$$

The square-root behavior can easily be identified in Fig. 3. For the series of rational numbers $1/Q$, which converges to zero, the distance between two consecutive midpoints of intervals, $\Omega_M(Q)$ and $\Omega_m(Q+1)$, therefore scales as

$$S(Q)=\Omega_m(Q)-\Omega_m(Q+1)$$
$$\sim\frac{1}{Q^2}-\frac{1}{(Q+1)^2}\sim Q^{-3} . \qquad (6.4)$$

By expanding around other steps such as $P/Q=\frac{1}{2}$, one finds similar square-root behavior; in fact, the square-root behavior must occur around every single step. The result (6.4) has been derived previously by Kaneko[17] using a phenomenological theory.

When the staircase is complete the widths $\Delta\Omega(1/Q)$ cannot decay with an exponent which is smaller than 3,

$$\Delta\Omega(1/Q)\sim Q^{-\delta'}, \quad \delta'>3 , \qquad (6.5)$$

since for $\delta'<3$ there would not be sufficient room for intervals on small scales [the widths must decay at least as the difference $S(Q)$ between steps]. Figure 15 shows $\log_{10}\Delta\Omega(1/Q)$ versus $\log_{10}Q$. The asymptotic slope yields $\delta'=3$, but the convergence is rather slow. The value of the exponent $\delta'$ when approaching rational numbers is thus much bigger then the value 2.16 for rationals approaching the golden mean, and the value 2.29 for the total staircase.

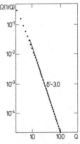

FIG. 15. Plot of $\log_{10}\Delta(1/Q)$ vs $\log_{10}Q$. The asymptotic straight line yields an exponent $\delta'=3$.

For the series $k/(2k+1)$ converging towards $\frac{1}{2}$ we find $\Delta\Omega(k/(2k+1))\sim(2k+1)^{-3}$, and similar behavior around several other numbers such as $K/(3k+1)$: $k/(4k+1)$ converging towards rationals. In fact, the convergence seems to be exponential,[18]

$$\Delta\Omega(P/Q)Q^3 \sim Ae^{bQ^{-c}},$$

for these rational series, with nonuniversal constants $A$, $b$, and $c$.

## ACKNOWLEDGMENT

We are grateful to Boris Shraiman, P. V. Christiansen, I. Satija, P. Cvitanovic, J. Doyne Farmer, M. J. Feigenbaum, J. Myrheim, and L. Glass, for stimulating discussions on circle maps. This work was supported by the Division of Materials Sciences of the U. S. Department of Energy under Contract No. DE-AC02-76CH00016 and by the Danish Natural Science Research Council. T. B. would also like to acknowledge support by National Science Foundation Grant No. DMR-83-14625.

[1]M. J. Feigenbaum, J. Stat. Phys. 19, 25 (1979); 21, 669 (1979).
[2]B. V. Chirikov, Phys. Rep. 52, 263 (1979).
[3]For an introduction see, e.g., V. I. Arnol'd, *Geometrical Methods in the Theory of Ordinary Differential Equations* (Springer, Berlin, 1982).
[4]V. I. Arnol'd, Am. Math. Soc. Trans., Ser. 2 46, 213 (1965).
[5]M. R. Herman, in *Geometry and Topology*, edited by J. Palis (Springer, Berlin, 1977), Vol. 597, p. 271.
[6]S. J. Shenker, Physica (Utrecht) 5D, 405 (1982).
[7]M. J. Feigenbaum, L. P. Kadanoff, and S. J. Shenker, Physica (Utrecht) 5D, 370 (1982).
[8]D. Rand, S. Ostlund, J. Sethna, and E. Siggia, Phys. Rev. Lett. 49, 132 (1982); Physica (Utrecht) 6D, 303 (1984).
[9]M. H. Jensen, P. Bak, and T. Bohr, Phys. Rev. Lett. 50, 1637 (1983).
[10]P. Bak and R. Bruinsma, Phys. Rev. Lett. 49, 249 (1982);

Phys. Rev. B 27, 5824 (1983).
[11]S. Aubry, in *Solitons and Condensed Matter Physics,* edited by A. R. Bishop and T. Schneider (Springer, Berlin, 1979), p. 264.
[12]P. Bak and J. von Boehm, Phys. Rev. B 21, 5297 (1980); M. H. Jensen and P. Bak, *ibid.* 27, 6853 (1983).
[13]B. B. Mandelbrot, *Fractals: Form, Change, and Dimension* (Freeman, San Francisco, 1977).
[14]V. N. Belykh, N. F. Pedersen, and O. H. Soerensen, Phys. Rev. B 16, 4860 (1977).
[15]S. E. Brown, G. Mozurkewich, and G. Grüner, Phys. Rev. Lett. 52, 2272 (1984).
[16]L. Glass and R. Perez, Phys. Rev. Lett. 48, 1772 (1982).
[17]K. Kaneko, Prog. Theor. Phys. 69, 669 (1982); 69, 403 (1983).
[18]B. Shraiman made us aware of this possibility.

PHYSICAL REVIEW A          VOLUME 30, NUMBER 4          OCTOBER 1984

# Transition to chaos by interaction of resonances in dissipative systems. II. Josephson junctions, charge-density waves, and standard maps

Tomas Bohr

*Laboratory of Atomic and Solid State Physics, Cornell University, Ithaca, New York 14853*

Per Bak

*Physics Department, Brookhaven National Laboratory, Upton, New York 11973*

Mogens Høgh Jensen

*H. C. Ørsted Institute, Universitetsparken 5, DK-2100 Copenhagen Ø, Denmark*

(Received 9 May 1984)

We have studied the transition to chaos caused by interaction and overlap of resonances in some condensed-matter systems by constructing and analyzing appropriate return maps. In particular, the resistively shunted Josephson junction in microwave fields and charge-density waves in rf electric fields may be described by the differential equation of the damped driven pendulum in a periodic force. The two-dimensional return map for this equation is shown to collapse to a one-dimensional circle map in a parameter regime including the transition to chaos. Phase locking, noise, and hysteresis in these systems can thus be understood in a simple and coherent way by taking over theoretical results for the circle map, some of which were derived in the preceding paper. In order to understand the contraction to one dimensionality we have studied the two-dimensional Chirikov standard map with dissipation. A well-defined transition line along which the system exhibits circle-map critical behavior was found. At this line the system is always phase locked. We conclude that recent theoretical results on universal behavior can readily be checked experimentally by studying systems in condensed-matter physics. The relation between theory and experiment is simple and direct.

## I. INTRODUCTION

The purpose of this paper is to demonstrate that there exist some simple condensed-matter systems which exhibit a transition to chaos caused by overlap of resonances with universal critical behavior as described in the preceding paper. Differential equations for the dynamics of these systems can be represented by one- and two-dimensional discrete maps, permitting direct confrontation of recent theories with experiment, and providing an understanding of phase locking, hysteresis, and noise phenomena which have been observed.

In the 17th century Christiaan Huyghens noted that two clocks hanging back to back on the wall tend to synchronize their motion.[1] More generally, strongly coupled damped oscillators tend to lock into commensurate motion where the ratio of their frequencies is a rational number. This phenomenon is known as phase locking and it is generally present in dissipative dynamical systems with competing frequencies. The two frequencies might arise dynamically within the system (as Huyghens's coupled clocks) or through the coupling of an oscillating or rotating motion to an external periodic force—as in a "swing." In many-dimensional systems the effective loss of degrees of freedom through dissipation may reduce the phase-locking phenomenon to basically two coupled oscillators.

If some parameter is varied, the system may pass through regimes which are phase locked and regimes which are not. For weakly nonlinear coupling the phase-locked intervals will have small measure. The motion is either (with small probability) periodic or, more likely, quasiperiodic, i.e., the ratio between the two frequencies $\omega_1/\omega_2$ is irrational. As the nonlinearity increases, the phase-locked portions increase and chaotic motion may occur in addition to the periodic and quasiperiodic (incommensurate) motion. The onset of chaos is basically caused by the growth of the phase-locked intervals until they eventually overlap, causing hysteretic response as well as truly chaotic behavior.

In the preceding paper (I) we gave strong numerical evidence for universal scaling behavior of the phase-locked steps on the critical line of the one-dimensional circle map. We shall here give evidence that the *same* scaling behavior should be found for the dissipative phase-locked systems described above. In particular, the theory applies to the resistively shunted Josephson junction in a microwave field, and sliding charge-density waves (CDW) in rf electric fields. The return maps for these systems are two-dimensional maps which collapse to 1D because of the dissipation. A short account of some of our results has already been published.[2]

For the circle maps studied in I the critical line is just $K=1$, but for higher-dimensional systems no such simple relation exists in general. We shall present evidence, however, derived from a two-dimensional dissipative map that a "critical line" does exist where the measure of the

330

phase-locked intervals is 1 and above which the intervals start to overlap. Most importantly, this line seems to be a continuous function of the parameters in the sense that the high-order mode-locked intervals determine a smooth critical curve (see Fig. 1). Along this line the complement to the locked portions is a measure-0 Cantor set, and we find the fractal (or Hausdorff) dimension of the set to be the same universal number $D \sim 0.87$ as found for the circle map.

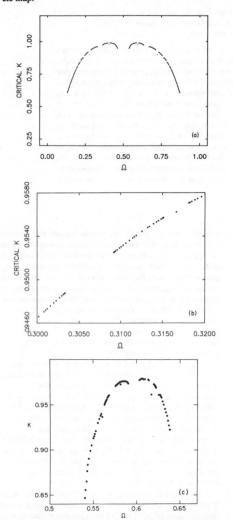

FIG. 1. Critical curves for 2D standard map with dissipation, calculated as discussed in Sec. V. (a) The curve was calculated for $b=0.25$; (b) the curve is a magnification $g$ of (a); (c) the curve was calculated for $b=0.5$, $\frac{1}{2} < W < \frac{2}{3}$.

Schematically, the connection between the physical system and the circle map can be represented as follows:

Physical system

(pendulum, CDW, Josephson junction)

$\updownarrow$

Differential equation

$\updownarrow$

2D discrete return map

$\updownarrow$

1D circle map (universal behavior)

The physical system is described by a differential equation; the return map for the differential equation is a two-dimensional (2D) discrete map; finally, the 2D return map collapses to a 1D circle map exhibiting universal behavior, i.e., there exists a 1D invariant curve.

The close resemblance between real coupled oscillator systems and the one-dimensional circle maps which we are suggesting might—from a mathematical viewpoint—seem to be a strange result. For the one-dimensional maps there exist strong theorems[3] guaranteeing that any monotonic, smooth map exhibits only periodic or quasi-periodic behavior. On the other hand, for higher-dimensional maps the analog [Komolgorov-Arnol'd-Moser (KAM) -like] theorems[4] only prove the existence of quasiperiodic motion (invariant tori) for maps that are "sufficiently close" to the trivial ones and no simple criterion for the breakdown of tori exists.

What we find here is the following: If, instead of looking at the original two-dimensional map, we project out the angular variable and consider the corresponding one-dimensional map (termed the "reduced map" in Sec. III), the criterion for having smooth invariant tori is—just as for the circle-map case—that this map be monotonic. Of course, we do not in general know where the monotonicity breaks down in terms of the parameters of the original map, but it does tell us that the situation, in terms of re-normalized parameters, is very close to the 1D case.

The layout of this paper is as follows. In Sec. II we shall discuss the Josephson junction in microwave fields and sliding CDW's (as found, for instance, in niobium triselenides, NbSe₃) in applied dc plus ac electric fields. These systems are described by the differential equation for a damped pendulum driven by a constant plus a periodic torque—so, in principle, our results can be tested on this simple mechanical device. We hope that experiments on Josephson junctions and CDW's be performed since they can be done with great precision.

In Sec. III we discuss general properties of return map for dissipative dynamical systems. We show numerically that the 2D return map of the equation for the Josephson junction reduces to a 1D map.

In Sec. IV we shall discuss in general the "criticality" for a 2D map which collapses to a 1D circle map, taking as the basic definition that the map becomes critical when the smooth invariant curve disappears.

As a concrete example we shall study a particular 2D

map, the "Chirikov standard map with dissipation" in Sec. V. We present numerical evidence that the critical line is smooth and that there is scaling behavior of the mode-locked intervals as for the circle map.

## II. PHASE LOCKING IN JOSEPHSON JUNCTIONS AND CDW SYSTEMS

Phase locking is basically a resonance effect between two oscillators. Even simpler, one can consider systems where one of the oscillators is replaced by an external perturbation; this also facilitates quantitative measurements of high-order locked motion since one of the frequencies is given as a parameter. The ubiquitous differential equation

$$\alpha\ddot{\theta}+\beta\dot{\theta}+\gamma\sin\theta=A+B\cos\omega t \qquad (2.1)$$

is precisely of this form. It describes a periodically forced, damped pendulum, with mass $\alpha$, damping coefficient $\beta$, and gravitational field $\gamma$. The equation is also—among other things—believed to give a fair description of a Josephson junction when the current has both a super and a normal part, and of sliding CDW's in electric fields.

It should be emphasized at this point that the precise form of (2.1) is immaterial for our investigation. Our aim is to find universal properties which should be independent of additional nonlinear terms in the equation, substitution of other periodic functions for $\sin\theta$, etc. We have chosen (2.1) as the simplest form to capture the essential physics. Figure 2 shows the equivalent electric diagram for the Josephson junction. Here, $\theta$ is the phase difference across the junction, and it is seen that $\alpha=\hbar C/2e$, $\beta=\hbar/2eR$, and $\gamma$ is the critical current $I_c$. Finally, $A$ and $B$ are the amplitudes of the dc and ac microwave components of the current through the junction. This model is usually referred to as the resistively shunted Josephson junction (RSJ) model, and a vast literature exists about it.[5] For certain values of the parameters the junction can be driven to a noisy state,[6,7] and indeed numerical simulations have indicated that the noise arises as chaotic solutions to the differential equation.[8-10] A sequence of bifurcations leading to chaotic behavior is known in a qualitative way from Refs. 7 and 11.

In the CDW systems, $\theta$ is the position of a sliding

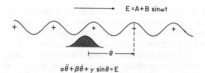

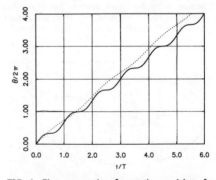

FIG. 3. Sliding charge density wave in ac plus dc electric field, and a quenched impurity pinning potential. For a single-domain sample the pinning potential could be a contact potential. The motion is that of a particle rolling down an oscillating washboard.

CDW relative to an "impurity" pinning potential (Fig. 3). For a single-domain sample, the pinning potential is probably a contact potential. The parameters $\alpha$, $\beta$, and $\gamma$ are phenomenological parameters representing the effective mass, damping, and periodic potential.[12,13] $A$ is a dc electric field which depins the CDW when it exceeds a critical value, and $B$ is the amplitude of an oscillating rf electric field.

In those systems the phase-locking phenomenon shows up as tendency of the average (angular) velocity $\langle\dot{\theta}\rangle$ to lock into rational multiples of the frequency of the external field,

$$\langle\dot{\theta}\rangle=\frac{P}{Q}\omega . \qquad (2.2)$$

Why does this mode locking occur? For small torque $A$ on the pendulum (or dc current in the Josephson junction, or dc voltage in the CDW system) the pendulum stays near its downward position. When $A$ exceeds a critical value, the pendulum enters a running "rotating" mode with average velocity $\sim A/\beta$ (for $\gamma$ and $B$ not too large).

Due to the periodicity of $\gamma\sin\theta$ and $B\cos\omega t$ the configuration space $(\theta,t)$ should really be thought of as a periodic lattice on which the motion takes place as shown in

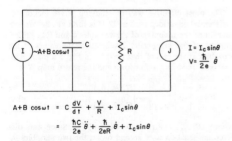

A+B cosωt $= C\dfrac{dV}{dt} + \dfrac{V}{R} + I_c\sin\theta$

$\qquad = \dfrac{\hbar C}{2e}\ddot{\theta} + \dfrac{\hbar}{2eR}\dot{\theta} + I_c\sin\theta$

FIG. 2. Diagram for the resistively shunted Josephson junction, driven by a constant current $A$ and a microwave current with amplitude $B$.

FIG. 4. Phase versus time for rotating pendulum, for a torque $A$ (or dc current $I$ in the Josephson junction or dc field $E$ in the CDW systems) exceeding the critical value. The time is measured in units of $2\pi/\omega$, i.e., the external force is used as a clock. The dotted curve indicates quasiperiodic motion, the solid curve periodic motion.

Fig. 4. Alternatively, one can pick out a rectangle of this lattice and identify opposite sides, thereby obtaining a two-dimensional torus which then forms the configuration space for the equation (Fig. 5). The dotted motion in Fig. 4 is quasiperiodic or incommensurate because its periodicity is unrelated to the underlying lattice, and on the torus the orbit would be dense and never close on itself. When the nonlinearities $\gamma$ and $B$ become large there will be a more pronounced tendency of the motion to fit into the lattice (as for the commensurate-incommensurate transition in adsorbed monolayers, etc.[15]) as shown by the solid curve. Here there exist integer numbers $N$ and $M$ such that

$$\theta(t_0 + MT) = \theta(t_0) + 2\pi N \tag{2.3}$$

or

$$\langle \dot{\theta} \rangle = \frac{N}{M}\frac{2\pi}{T} = \frac{N}{M}\omega , \tag{2.4}$$

where $T$ is the "clock" period of the external force.

If the situation (2.3) is realized for a certain parameter value (for instance a value of $A$), then there always exist an entire interval around this value where (2.4) is satisfied for the same $N$ and $M$; we have phase locking.

The above argument merely makes it plausible that locking might occur, and it would seemingly work just as well for $\alpha = 0$, the zero mass, or overdamped case, where the model reduces to the Stewart-McCumber[16] model. In fact, it has been shown rigorously that in this limit only the whole multiples ($M = 1$) of $\omega$ survive since (2.1) is then a linear equation (the Mathieu equation) in disguise.[17] The coefficient of the $\ddot{\theta}$ term is thus a measure of the non-linearity of the system.[18]

For the Josephson junction the voltage $V$ is given by the Josephson relation

$$V = \frac{\hbar}{2e}\dot{\theta} , \tag{2.5}$$

so a locking of $\langle \dot{\theta} \rangle$ implies a locking of $\langle V \rangle$ and steps will be seen in the $I$-$V$ characteristics. For $M = 1$ these are the Shapiro steps,[19] but in between them subharmonic steps (with $M > 1$) can often be seen.[20] Figure 6 shows the

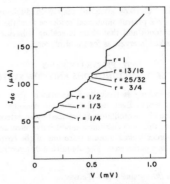

FIG. 6. $I$-$V$ characteristics of an $NB$-$Nb$ Josephson point junction in 295 GHz microwave field at $T = 4.2$ K. (Belykh, Pedersen, and Soerensen, Ref. 7.)

striking experimental observation by Belykh et al.[7] of a multitude of such substeps in the $I$-$V$ characteristics of a Nb-Nb Josephson junction.

In the CDW systems the current carried by the sliding charge-density wave is proportional to the velocity $\dot{\theta}$, so the average current is

$$I_{\text{CDW}} \sim \langle \dot{\theta} \rangle . \tag{2.6}$$

Hence, a locking of $\langle \dot{\theta} \rangle$ implies a locking of the current carried by the CDW (the current carried by the normal electrons behaves in a smooth way, $I_n \sim E$). The roles of currents and voltages are the reverse for Josephson junctions and CDW systems.

It has been suggested that the mass $\alpha$ of the CDW systems is essentially zero, so there would be essentially no subharmonic steps in apparent agreement with early measurements.[13,14] Recently, however, a multitude of such steps have been observed in a striking experiment by Brown, Mozurkewich, and Grüner;[21] the situations for CDW systems and Josephson junctions seem to be essentially the same.

### III. THE RETURN MAP

The most effective way of studying phase locking of differential equations such as (2.1) is through their return maps, i.e., the mapping of the variables $\theta$ and $\dot{\theta}$ at the beginning of the $n$th period $T = 2\pi/\omega$ to the values of the variables at the end of that period. In Fig. 5 a plane with $t = \text{const (mod } 2\pi/\omega)$ is shown and the return map $R$ is

$$\underline{R}\begin{bmatrix} \theta_n \\ \dot{\theta}_n \end{bmatrix} = \begin{bmatrix} \theta_{n+1} \\ \dot{\theta}_{n+1} \end{bmatrix} = \begin{bmatrix} G_1(\theta_n, \dot{\theta}_n) \\ G_2(\theta_n, \dot{\theta}_n) \end{bmatrix} , \tag{3.1}$$

where $(\theta_n, \dot{\theta}_n) = (\theta(t = nT), \dot{\theta}(t = nT))$. Since the differential equation is of second order, the two variables $\theta_n$ and $\dot{\theta}_n$ contain all information about the system at a given time; therefore, the return map is two dimensional.

The functions $G_1$ and $G_2$ must be periodic in $\theta$ with

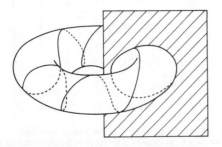

FIG. 5. The configuration space as the surface of a 2D torus. A possible orbit is indicated on the torus, and a plane with constant $t$ mod$2\pi/\omega$ is shown, on which the return map is generated.

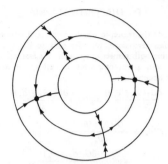

FIG. 7. Annular map (schematic) showing the invariant circle and the invariant manifolds. The simplest locked state is shown with a stable and unstable fixed point.

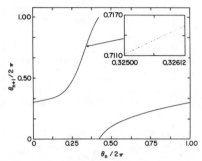

FIG. 8. Return map calculated for $B=\alpha=\gamma=1$, $\omega=1.76$, $\beta=1.576$, $A=1.4$. The function $f(\theta_n)=\theta_{n+1}$ is monotonically increasing indicating regular behavior. The inset is a magnification, emphasizing the one dimensionality of the map.

period $2\pi$ since that period is explicit in (2.1). Thus the map $R$ can be thought of either as mapping the plane into itself, or—identifying $\theta$ values differing only by multiples of $2\pi$—as a mapping of a cylinder (or an annulus) to itself, as shown in Fig. 7.

Since the points $(\theta_n,\dot\theta_n)$ are points on a solution curve for a differential equation, the map must be invertible and orientation preserving. This means that the Jacobian

$$J=J(\theta_n,\dot\theta_n)=\det\begin{vmatrix}\dfrac{\partial G_1}{\partial \theta_n} & \dfrac{\partial G_1}{\partial \dot\theta_n}\\[2ex] \dfrac{\partial G_2}{\partial \theta_n} & \dfrac{\partial G_2}{\partial \dot\theta_n}\end{vmatrix} \qquad (3.2)$$

*must always be greater than zero.*

The value of $J$ is simply related to the parameters of the differential equation, namely

$$J=e^{-(2\pi/\omega)/(\beta/\alpha)}, \qquad (3.3)$$

so that $J<1$ (the map is area contracting) and independent of $\theta$ and $\dot\theta$ because of the simple choice of damping term in (2.1).[22]

If the motion is phase locked with

$$\langle \dot\theta \rangle = \frac{P}{Q}\omega, $$

it means that there exist periodic fixed points $(\theta^*,\dot\theta^*)$ such that

$$\underline{R}^Q\begin{bmatrix}\theta^*\\\dot\theta^*\end{bmatrix}=\begin{bmatrix}\theta^*\\\dot\theta^*\end{bmatrix}+\begin{bmatrix}2\pi P\\0\end{bmatrix}. \qquad (3.4)$$

For small nonlinearity one expects that any motion will be either phase locked or quasiperiodic in which case the frequency, or winding number

$$W=\lim_{n\to\infty}(\theta_n/2\pi n), \qquad (3.5)$$

is irrational. The initial condition will soon be forgotten due to the damping of the motion. The mapping is area

contracting and hence destroys information on the initial configuration, so the motion might asymptotically be confined to a smooth invariant curve $\theta(t)$ on the torus (Fig. 5). This means, of course, also that asymptotically $\dot\theta$ is just a given function of $\theta$ so that in terms of the map $R$ there exists a smooth invariant curve

$$\dot\theta_n=g(\theta_n) \qquad (3.6)$$

on which the asymptotic behavior takes place. Referring back to (3.1) and inserting (3.6), we find a unique relation

$$\theta_{n+1}=f(\theta_n)=G_1(\theta_n,g(\theta_n)), \qquad (3.7)$$

where $f$ is a circle map since it is periodic in $\theta_n$. We shall denote this one-dimensional map the "reduced (return) map." In the following we shall assume that (3.6) and (3.7) are single-valued functions although all arguments below would hold just as well if they were just orientable curves.

Whether this dimensional reduction actually takes place depends crucially upon the assumption (3.6) for the ex-

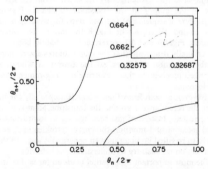

FIG. 9. Return map calculated for $\beta=1.253$, $A=1.2$; other parameters as in Fig. 8. The function develops a cubic inflection point indicating a transition to chaos. The inset shows an enlargement of the curve around $f^2(\theta_I)$ where $\theta_I$ $\mathbf{\not}$ the inflection point $[f'(\theta_I)\sim0, f^2(\theta_I)\to\infty]$.

334

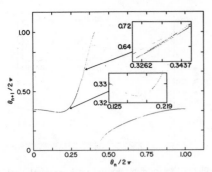

FIG. 10. Return map calculated for $\beta = 1.081$, $A = 1.094$. The map develops a local minimum and "wiggles" (insets) indicating chaotic behavior.

istence of an invariant curve, and for given values of the parameters we do not known whether or not it is satisfied (except in the limit $\alpha \ll 1$ where the connection with the circle map has been established analytically[18]). The best we can do is to generate (3.6) and (3.7) by solving the differential equation (2.1) numerically on the computer. Figure 8 shows the "reduced" map in a situation where this does indeed happen. The asymptotic form is quasiperiodic and the map is one dimensional (see inset). Changing some parameters (for instance, increasing $A$) we can generate plots that asymptotically display only a discrete set of points, namely the $Q$ points $\theta_1^*, \theta_2^*, \ldots, \theta_Q^*$ (mod$2\pi$) which are the stable periodic points of the map $R$ [or $f$ in (3.7)]. The scenario is precisely the same as for the circle map studied in I. Increasing $A$ still further leads again to quasiperiodic motion.

As described in I, chaotic motion in circle maps sets in when the map acquires zero slope somewhere in the interval, thus ceasing to be an invertible diffeomorphism. As long as the map is everywhere monotonic, erratic behavior can never occur.[3] Analogously, let us ask about the structure of the map $f$ at the transition to chaos. Figures 8–10 show a sequence of reduced maps for winding numbers around $W \simeq 0.38$. In Fig. 8 the motion is regular, whereas Figs. 9 and 10 correspond to chaotic behavior. Superficially it appears that the map acquires a zero slope at the transition to chaos, so the transition would be described precisely as that occurring in a one-dimensional circle map.

However, as seen from Figs. 9 and 10 (especially the insets), the behavior of $f$ around the transition point is more complicated: Instead of just turning over to form isolated local maxima and minima, the curve "crinkles up," and seems to be filled up in an uneven nonergodic way. In the next section we shall try to understand this behavior.

The most important observation made so far is that the return map is effectively one dimensional, up to and including the transition point. In Sec. V it will be argued that the critical behavior is indeed that of the circle map, despite the different features of the map near the transition.

## IV. THE "CRITICAL LINE"

To study the region in which the reduced return map $f$ approaches zero slope, we refer back to Eq. (3.1). Differentiating with respect to $\theta_n$, using the assumption (3.6) of a smooth asymptotic orbit as well as the definition (3.7), we obtain

$$f'(\theta_n) = \frac{\partial G_1(\theta_n, g(\theta_n))}{\partial \theta_n} + g'(\theta_n) \frac{\partial G_1(\theta_n, g(\theta_n))}{\partial \dot{\theta}_n}, \quad (4.1a)$$

$$g'(\theta_{n+1}) f'(\theta_n) = \frac{\partial G_2(\theta_n, g(\theta_n))}{\partial \theta_n} + g'(\theta_n) \frac{\partial G_2(\theta_n, g(\theta_n))}{\partial \dot{\theta}_n}. \quad (4.1b)$$

Now, if we focus our attention on a phase-locked state with winding number $P/Q$, the asymptotic motion really amounts to jumping between the $Q$ different stable periodic points $\theta_1^*, \theta_2^*, \ldots, \theta_Q^*$ (mod$2\pi$), so the question arises whether $g'(\theta_n)$ has any meaning.

Quite generally, the map (3.1) is a map of the annulus. Figure 7 shows schematically the fixed-point structure for a periodic (phase-locked) state. In addition to the $Q$ stable periodic points $(\theta_1^*, \dot{\theta}_1^*), \ldots, (\theta_Q^*, \dot{\theta}_Q^*)$, there are also $Q$ unstable periodic points $(\theta_1^u, \dot{\theta}_1^u), \ldots, (\theta_Q^u, \dot{\theta}_Q^u)$. The unstable manifold of these unstable periodic points [i.e., the point set for which $R^{-n}(\overline{X})$ approaches the unstable periodic cycle as $n \to \infty$] is by definition the invariant circle of the map. The coordinates of this invariant circle are precisely $(\theta, g(\theta))$.

Having now given meaning to $g'(\theta)$ [and therefore to $f'(\theta)$], we shall see that a zero slope for $f(\theta)$ leads to trouble. Thus assume that

$$f'(\theta_n) = 0. \quad (4.2)$$

Then it follows from (4.1a) and (4.1b) that

$$g'(\theta_{n+1}) f'(\theta_n) \neq 0, \quad (4.3)$$

since the Jacobian (3.2) is never zero.

Now (4.3) implies that

$$|g'(\theta_{n+1})| \to \infty \quad (4.4)$$

if $f'(\theta_n) \to 0$. Using (4.1a) with $n \to n+1$, we find that

$$|f'(\theta_{n+1})| \to \infty \quad (4.5)$$

unless

$$\frac{\partial G_1(\theta_{n+1}, g(\theta_{n+1}))}{\partial \dot{\theta}_{n+1}} = 0. \quad (4.6)$$

Assuming that $f'(\theta_{n+1})$ remains finite, we must accept (4.6), and because the Jacobian is nonzero we obtain

$$\frac{\partial G_2(\theta_{n+1}, g(\theta_{n+1}))}{\partial \dot{\theta}_{n+1}} \neq 0; \quad (4.7)$$

therefore, from (4.1b) with $n \to n+1$

$$|g'(\theta_{n+2}) f'(\theta_{n+1})| \to \infty \quad (4.8)$$

because of (4.4). Hence

$$|g'(\theta_{n+2})| \to \infty \ . \tag{4.9}$$

Iterating this argument we see that *in the generic case* $f'(\theta_n) \to 0$ *implies* $f(\theta_{n+1}) \to \infty$, as indicated numerically for the return map of the differential equation (2.1) (see Fig. 9). If the derivatives $\partial G_1(\theta_{n+j}, g(\theta_{n+j}))/\partial\dot\theta_{n+j}$ vanish for $j=1,\ldots,N$ exceptionally, we will find that $f'(\theta_{n+1})$, $f'(\theta_{n+2}),\ldots,f'(\theta_{n+N})$ remain finite whereas $|f'(\theta_{n+N+1})| \to \infty$.

More precisely, we can follow the large terms in (4.1a) and (4.1b) as $f'(\theta_n) \to 0$. Assuming for simplicity that $\partial G_1/\partial\dot\theta$ never vanishes on the orbit, we find

$$f'(\theta_{n+1}) \sim - \frac{\left.\dfrac{\partial G_1}{\partial\dot\theta}\right|_{(\theta_{n+1}, g(\theta_{n+1}))}}{\left.\dfrac{\partial G_1}{\partial\dot\theta}\right|_{(\theta_n, g(\theta_n))}} \frac{J(\theta_n)}{f'(\theta_n)} \ . \tag{4.10}$$

Now $J$ must always be positive. If, in addition, the derivatives of $G_1$ appearing in (4.10) have the same sign, the coefficient of $[f'(\theta_n)]^{-1}$ is negative. This is certainly true for a large class of two-dimensional maps showing phase locking and chaos, so we assume this to be "generically" true, i.e., only exceptionally violated.

If so, then (4.10) states that

$$f'(\theta_n) \to 0^+ \Longrightarrow f'(\theta_{n+1}) \to -\infty \ . \tag{4.11}$$

Consider now the graph of $f$ as a function of a parameter and assume to start with that $f$ is an increasing circle map. Varying the parameter decreases $f'(0)$ toward zero and by (4.11) $f'(f(0))$ must approach *minus infinity*. But then $f'(\theta)$ must be zero somewhere in the interval $[0, f(0)]$ giving another infinite slope, etc. The whole curve crinkles up as sketched in Figs. 11(a) and 11(b). Thus, the loss of monotonicity is tied to the loss of smoothness of the invariant circle: *If there exists a value of $\theta$ with $f'(\theta)=0$, then the invariant circle has already broken up and the initial assumption of a smooth orbital $g(\theta)$ is contradictory.*[23]

There is one trivial way in which the curve might avoid crinkling up, namely if it forms a closed circuit. This means, however, that we have chosen the wrong variables: $\theta$ does not act as an angular variable and even for small nonlinearity there does not exist a single-valued $g(\theta)$ in (3.6).

A more interesting possibility is that the curve spirals as shown in Fig. 11(c). Here the analyticity is lost only in a finite number of points: the stable periodic points of the map. In this case the breakdown can be located by linear analysis of the map. To see this we return to (4.1a) and (4.1b), dividing (4.1a) with (4.1b). We then obtain an equation containing explicitly only $g$:

$$g'(\theta_{n+1}) = \frac{\partial G_2/\partial\theta_n + g'(\theta_n)\partial G_2/\partial\dot\theta_n}{\partial G_1/\partial\theta_n + g'(\theta_n)\partial G_1/\partial\dot\theta_n} \ . \tag{4.12a}$$

We now introduce functions $m(\theta)$ and $n(\theta)$ through

$$g'(\theta) = m(\theta)/n(\theta) \ , \tag{4.12b}$$

and rewrite (4.11a) as

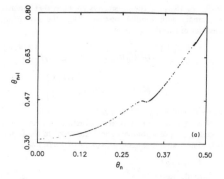

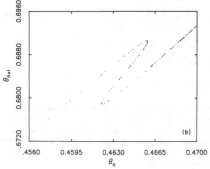

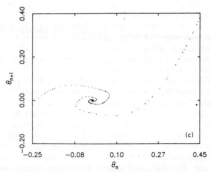

FIG. 11. Loss of smoothness of the invariant circle shown by different reduced return maps for the map (5.1) with $b=0.25$. (a) The curve crinkles up close to criticality ($K=1.1$, $\Omega=0.292$). (b) Magnification of (a). (c) The map loses smoothness at the periodic point ($K=1.25$, $\Omega=0$).

$$\begin{bmatrix} n(\theta_{n+1}) \\ m(\theta_{n+1}) \end{bmatrix} = \underline{D}(\vec{X}_n) \begin{bmatrix} n(\theta_n) \\ m(\theta_n) \end{bmatrix} \ , \tag{4.13}$$

where $\vec{X}_n = (\theta_n, \dot\theta_n)$ and $\underline{D}$ is the Jacobian of the map (3.1). This defines $n$ and $m$ completely once they are specified at some initial point [for instance by $n(\theta_0)=1$,

$m(\theta_0) = g'(\theta_0)]$. If the parameters are such that the map is periodic with winding number $P/Q$, the asymptotic behavior of $(n,m)$ is given by

$$\begin{bmatrix} n(\theta_{n+Q}) \\ m(\theta_{n+Q}) \end{bmatrix} = \underline{D}(\vec{X}_Q^*)\underline{D}(\vec{X}_{Q-1}^*) \cdots \underline{D}(\vec{X}_1^*) \begin{bmatrix} n(\theta_n) \\ m(\theta_n) \end{bmatrix}$$

$$= \underline{M}(\vec{X}_Q^*, \ldots, \vec{X}_1^*) \begin{bmatrix} n(\theta_n) \\ m(\theta_n) \end{bmatrix}, \qquad (4.14)$$

where $\theta_n$ and $\theta_{n+Q} = \theta_1^*$ (mod$2\pi$).

Now $\underline{M}$ is a real matrix and has either two real eigenvalues $\lambda_1 > \lambda_2$ or a complex conjugate pair. In the former case almost all initial choices $(n(\theta), m(\theta))$ will lead to

$$\begin{bmatrix} n(\theta_{n+NQ}) \\ m(\theta_{n+NQ}) \end{bmatrix} \to \lambda_1^N a \vec{e}_1 , \qquad (4.15)$$

where $\vec{e}_1 = (e_1^x, e_1^y)$ is the eigenvector of $\underline{M}$ corresponding to the largest eigenvalue and $a$ is the projection,

$$a = (m(\theta_n), n(\theta_n)) \cdot \vec{e}_1 . \qquad (4.16)$$

Hence $g'$ converges to the value

$$g'(\theta_{n+Q}) = g'(\theta_1^*) = \frac{m(\theta_{n+Q})}{n(\theta_{n+Q})} \to \frac{e_1^y}{e_1^x} . \qquad (4.17)$$

When the eigenvalues are complex conjugate pairs no simple asymptotic relation like (4.16) exists and $g'(\theta_1^*)$ becomes undefined. *The map loses its smoothness at the limit-cycle fixed point when the eigenvalues of the product of the "Jacobians" at the limit cycle points are identical,* $\lambda_1 = \lambda_2 = \sqrt{\det \underline{M}}$.[24]

In the next section we shall look at a particular two-dimensional dissipative map. Since we are looking for universal behavior we might as well choose a simple analytic map instead of doing cumbersome integration of the differential equation. For that particular map we shall try to locate the critical line, i.e., the line in parameter space where the smoothness breaks down.

The fundamental question is whether the critical line in any sense defines a smooth curve in parameter space. One might fear—since the condition of criticality at the periodic points relies upon properties of the matrix $\underline{M}$ defined in (4.13) and which explicitly depends upon $Q$, the denominator of the winding number—that the critical curve would be fundamentally fractal. Our finding—for the particular map which will be studied—is that it is not fractal: The dissipation present in the map secures that the critical line defined through the very high-order rational steps is smooth.

## V. SCALING BEHAVIOR
## OF THE "DISSIPATIVE STANDARD MAP"

In this section we show numerical results for a particular two-dimensional map which could in principle be the return map of a second-order differential equation such as (2.1). The map is defined by recursion relations in two variables $\theta$ and $r$:

$$\theta_{n+1} = \theta_n + \Omega - (K/2\pi)\sin(2\pi\theta_n) + br_n ,$$
$$\qquad (5.1)$$
$$r_{n+1} = br_n - (K/2\pi)\sin(2\pi\theta_n) ,$$

where $b$ is between 0 and 1. The equation has the required symmetry in $\theta$ and the variable $r_n$ plays a similar role as $\theta_n$ in (3.1). Note that we have inserted $2\pi$'s in the argument so that the map is periodic mod1.

The Jacobian matrix of the mapping is

$$\underline{D} = \begin{bmatrix} 1 - K\cos(2\pi\theta) & b \\ -K\cos(2\pi\theta) & b \end{bmatrix} , \qquad (5.2)$$

with $\det \underline{D} = b$, showing that $b < 1$ indeed defines an area contracting map corresponding to a dissipative dynamical system. When $b \to 0$ we recover the sine circle map (see I), and when $b \to 1$ we obtain the so-called standard area preserving map (Chirikov, Ref. 25).

In this case (4.10) reduces to

$$f'(\theta_{n+1}) \sim -\frac{b}{f'(\theta_n)} \qquad (5.3)$$

as $f'(\theta_n) \to 0$, so the conclusions below (4.10) are certainly valid in this case.

This map has been studied earlier both numerically[26] and by renormalization-group techniques[26,27] and it has been found that, for a series of winding numbers converging to the golden mean, $(\sqrt{5}-1)/2$, the scaling is the same as for one-dimensional circle maps. In the work by Rand et al.[27] the general structure of the locked steps was discussed on the basis of earlier findings (for a different map) by Aronson et al.[28] to which we shall return shortly.

As an illustrative example, let us consider the simplest phase-locked state, namely the one with winding number $W = 0/1$. Here the map has two fixed points, one stable and one unstable. To find these we must solve the equations

$$\theta^* = \theta^* + \Omega - \frac{K}{2\pi}\sin(2\pi\theta^*) + br^* ,$$
$$\qquad (5.4)$$
$$r^* = br^* - \frac{K}{2\pi}\sin(2\pi\theta^*) ,$$

giving

$$\sin(2\pi\theta^*) = \frac{2\pi\Omega(1-b)}{K} ,$$
$$\qquad (5.5)$$
$$r^* = -\Omega ,$$

Indeed, (5.5) has either 0 or two solutions, the latter being the case when

$$|\Omega| < \Omega_e \equiv \frac{K}{2\pi(1-b)} . \qquad (5.6)$$

The eigenvalues of the Jacobian matrix $\underline{D}$ are

$$\lambda = \tfrac{1}{2}(1 + b - K\cos(2\pi\theta)$$
$$\pm \{[1 + b - K\cos(2\pi\theta)]^2 - 4b\}^{1/2}) \qquad (5.7)$$

and the line confining the region where there exists a stable fixed point is found by solving

$$|\lambda| = 1 . \qquad (5.8)$$

Setting $\lambda = 1$ gives us

$$\Omega = \pm \Omega_e \, , \qquad (5.9)$$

but $\lambda = -1$ leads to a new hyperbolic constraint

$$\frac{K^2}{[2(1+b)]^2} - \frac{\Omega^2}{[2(1+b)/2\pi(1-b)]^2} = 1 \, . \qquad (5.10)$$

The criticality condition for the fixed point, $\lambda_1 = \lambda_2$, can be written

$$1 + b - K\cos(2\pi\theta^*) = \pm 2\sqrt{b} \, , \qquad (5.11)$$

which depends on $\theta^*$. Inserting (5.5) we can write it as

$$\frac{K^2}{(K_c^{\pm})^2} - \frac{\Omega^2}{[K_c^{\pm}/2\pi(1-b)]^2} = 1 \, , \qquad (5.12)$$

where

$$K_c^{\pm} = 1 + b \pm 2\sqrt{b} = (1 \pm \sqrt{b})^2 \, . \qquad (5.13)$$

The lower hyperbola (5.13) gives the maximum value of $K$ for which $f$ is smooth through the fixed point. Figure 12 shows one-half of the phase-locked region for $b=0.25$ (the other half is the mirror image). The curve through $O$ and $A'$ is the (lower) hyperbola and the straight line through $A$ is the edge $\Omega = \Omega_e$. In order to locate the critical line we must know when other parts of the map $f$ lose smoothness, i.e., at what parameters the more general crinkling up [Fig. 11(b)] takes place. As found by Aronson et al.[28] for a different map, and verified for our map by Rand et al.[27] this seems generically to happen on curves connecting the hyperbola with the edge of the Arnol'd tongue. The triangles in Fig. 12 are points on this curve intersecting the hyperbola at $A'$. The lowest part of the hyperbola ($OA'$) thus actually represents the critical line, but further away from the center the critical line follows $AA'$ (the triangles).

For higher-order locked steps the situation is similar, but we find that the point $A'$ moves down towards $O$ so that the critical line touches less of the hyperbola. Further, the whole line $OA'A$ becomes flatter: Due to the dissipation everything is squeezed together in a narrow $K$ interval, rapidly decreasing with the order of the step. For high-order steps, the whole picture looks very much like the "skeleton" diagram for circle maps,[29] with "hyperbolas" just touching a smooth line.

For the general phase-locked step $P/Q$ the analogs of (5.4), (5.8), and (5.11) cannot be solved analytically. If $(\theta_i, r_i)$, $i = 1, \ldots, Q$ denote the $Q$ stable periodic points, the stability condition for the step is that the eigenvalues of the matrix

$$\underline{M} = \prod_{i=1}^{Q} \underline{D}(\theta_i, r_i) \qquad (5.14)$$

should have absolute value less than unity and the edges of the steps are found by solving

$$\lambda_{\max} = 1 \qquad (5.15a)$$

or

$$\mathrm{Tr}\underline{M} = 1 + b^Q \, . \qquad (5.15b)$$

The $Q$-cycle fixed point becomes critical along a curve in $(K, \Omega)$ space where the eigenvalues of $\underline{M}$ are equal,

$$\lambda_1 = \lambda_2 = \sqrt{\det\underline{M}} = b^{Q/2} \qquad (5.16a)$$

or

$$\mathrm{Tr}\underline{M} = 2b^{Q/2} \, . \qquad (5.16b)$$

This curve corresponds to the lower hyperbola for the $0/1$ case (Fig. 12). In our numerical work we have taken the tip of the hyperbola (i.e., the analog of the point 0 in Fig. 12) simply as the point on the curve (5.16) with smallest $K$. Since the hyperbolas are generally tilted this is an approximation and we shall later explain how it can be improved. We have found these points numerically by a four-dimensional Newton iteration method (see Appendix) giving the values $\theta$, $r$, $\Omega$, and $K$ at the tips of the hyperbolas for different $P/Q$. The resulting critical lines are shown in Figs. 1(a) and 1(b) for $b=0.25$ and in Fig. 1(c) for $b=0.5$, respectively.

The important finding is that even though the "curves" in Fig. 1 are wildly discontinuous at the low-order locked steps, *they seem to approach smoothness very quickly as the order $Q$ of the resonance increases.* Thus on each edge of a given rational step there is a unique accumulation point for the tips of hyperbolas coming from very-high-order rationals converging to the given one, and if a smooth critical curve really exists, *these points are precisely the endpoints (denoted by A in Fig. 12) of the critical curve within the step.* This is shown clearly in Fig. 12 where the circles in the right-hand corner are the tips of high-order hyperbolas close to the locked region. They accumulate to the same point $A$ and seem to form a smooth curve together with the triangles. Thus one can imagine the critical line extending outside of the step in Fig. 12, i.e., beyond $A$, where the invariant circle (in some very-high-

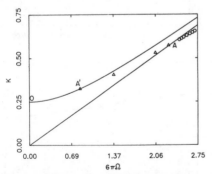

FIG. 12. Stability regime for fixed point with $W=0/1$ for $b=0.25$. Only one-half is shown since there is mirror symmetry around the $y$ axis. The straight line through $A$ is the edge $\Omega = \Omega_e$. The points marked by triangles indicate where the map crinkles up. The circles are tips of hyperbolas in the narrow locked regions just outside the edge. Note that the circles and triangles together seem to lie on a smooth curve cutting through the edge at $A$. This is strong evidence for the existence of a smooth critical line.

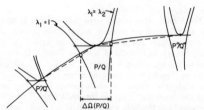

FIG. 13. Schematic diagram showing the hyperbolas where the invariant circle becomes critical through the stable periodic points, and the "true" critical curve. The squares, and the $+$'s represent the edges of the $P/Q$ steps at criticality as estimated using the numerical methods presented here. The superstable point where the hyperbola touches the stability curve is approximated by the minimum point ●.

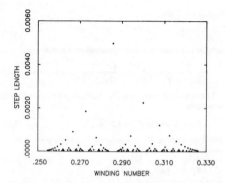

FIG. 15. The widths of phase-locked steps at criticality for the map (5.1) with $b=0.25$. Note the similarity with the circle map (see I).

order steps) crinkles up. By emphasizing only the behavior within a single Arnol'd tongue, earlier treatments[27,28] have failed to make this clear.

Our conclusion is then that *there exists a smooth limiting critical line on which the invariant circles with irrational winding numbers break down.* Since our primary aim is to study the metric properties of the parameter values corresponding to irrational winding numbers on the critical line (i.e., the complement of the locked steps), we can continue this line through rational steps in any convenient way for calculational purposes. On crossing the critical line the reduced map loses monotonicity as shown in Figs. 11(a) and 11(b) (for $b=0.25$). The crinkling shown here is very much like the behavior seen in Figs. 9 and 10 for the "Josephson map" (although the damping in the latter case was much larger).

We can now return to the approximation made in locating the tips of the hyperbolas. Since we now conjecture that the "limit curve" is smooth (Fig. 1), we should actually locate them as the touching points of the envelope for the hyperbolas as shown in Fig. 13. For more accurate calculations this should be taken into account.

Having argued for the existence of a smooth critical line we may address the question of possible scaling behavior along this line as found for one-dimensional

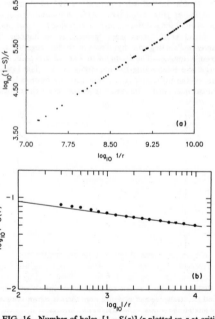

FIG. 16. Number of holes, $[1-S(r)]/r$ plotted vs $r$ at criticality for the standard map (5.1). The linear behavior indicates that the staircase is complete. The critical index $D$ (the fractal dimension of the staircase) was estimated from the slope. (a) $b=0.25$, $\frac{1}{4}<W<\frac{1}{3}$; $D=0.86\pm0.01$. (b) $b=0.5$, $\frac{1}{2}<W<\frac{2}{3}$; $D=0.87\pm0.01$.

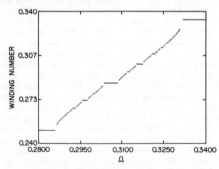

FIG. 14. Winding number $P/Q$ vs $\Omega$ at criticality for the map (5.1) with $b=0.25$. Note the self-similar structure of the staircase as for the circle map (see I).

maps (I). To find the endpoints of the $P/Q$ steps we use the critical line defined by the highest-order steps. This was done for $b=0.25$, $\frac{1}{4} \leq W \leq \frac{1}{3}$ and for $b=0.5$, $\frac{1}{2} \leq W \leq \frac{2}{3}$; in both cases approximately 100 steps were found. For $b=0.25$ the edge was found by moving along a critical line defined by linear interpolation between the approximate accumulation points on the edges (the points marked by squares in Fig. 13), whereas the edge points for $b=0.5$ simply have $K$ equal to the $K$ values of the corresponding accumulation points (the points marked by plus in Fig. 13). In the limit the two methods are, of course, identical. For a given rational step the projection of the endpoints to the $\Omega$ axis determines $\Delta\Omega(P/Q)$—the width of the step. Figure 14 shows the winding number versus $\Omega$ on the critical line (for $b=0.25$)—the "devil's staircase" similar to the diagram (I, Fig. 3) for the circle map. Figure 15 shows the widths of the steps $\Delta(P/Q)$ plotted versus $W = P/Q$; the self-similar structure is quite apparent and the similarity with (I, Fig. 5) for the circle map is striking. Figure 16(a) and 16(b) shows the "number of holes" $[1-S(r)]/r$ plotted versus $r$ as for the circle map, where $S(r)$ is the total width of steps which are wider than $r$. The plot in Fig. 16(a) is for $b=0.25$; the plot in Fig. 16(b) is for $b=0.5$. The linear behavior indicates scaling behavior at criticality. The slope of the straight line yields

$$[1-S(r)]/r \sim r^D \qquad (5.17)$$

with $D \sim 0.86 \pm 0.01$. Hence, the staircase is complete, and the fractal dimension of the staircase, or the complementary Cantor set is $D \sim 0.86$. The accuracy of this estimate is much less than the one for the circle map in I since (i) a much smaller number of steps were used to estimate $D$ and (ii) there is additional uncertainty related to the estimate of the critical curve, which was given simply by $K=1$ in the circle-map case. Within the uncertainty, the critical behavior for the standard map is the same as for the circle map.

## VI. CONCLUSION

In papers I and II we have investigated the transition to chaos caused by overlap of resonances by studying circle maps, differential equations representing actual physical systems, and 2D dissipative maps. Our conjecture is that the critical behavior of all these systems is the same, namely that of the circle map. We urge that experiments on Josephson junctions, charge-density-wave systems, and other systems with two competing periodicities be performed to check our predictions.

We are aware of three experiments which have been performed since we first announced our results: Kao et al.[30] and Alstrom et al.[31] have studied the differential equation (5.1) with a "Josephson junction simulator;" they find scaling behavior as predicted here, with $D \sim 0.91 \pm 0.04$ and $D \sim 0.87 \pm 0.02$, respectively. More importantly, Brown, Mozurkewich, and Grüner[21] have measured subharmonic steps in the CDW system $NbSe_3$ in ac plus dc electric fields; they also find scaling behavior in agreement with our conjecture, with $D \sim 0.91 \pm 0.03$. We feel that the accuracy of the experiments could be improved; in particular, more work should be done to locate

the critical line, following, for instance, the ideas presented here. It would be of interest to measure the return map directly. The most precise measurements can probably be performed on Josephson junctions in microwave fields, but no experiment on the scaling behavior near the critical point has been reported so far.

## ACKNOWLEDGMENTS

We have enjoyed many helpful discussions with our colleagues on the subject of this article. It is a pleasure to thank especially Mitchell Feigenbaum, Michael Fisher, George Grüner, Phil Holmes, Yittan Kao, Mogens Levinsen, David Rand, John Sampson, Scott Shenker, Eric Siggia, and Gertrud Zwicknagel. We are grateful to K. Fesser, A. Bishop, and P. Kumar for discussions on the possibility of having one-dimensional return maps for higher-dimensional systems. One of us (M.H.J.) is grateful to Brookhaven National Laboratory for kind hospitality. Another one of us (T.B.) would like to acknowledge support by National Science Foundation Grant No. DMR-83-14625. The work at Brookhaven National Laboratory was supported by The Division of Material Sciences of the U.S. Department of Energy under Contract No. DE-AC02.76CH00016.

## APPENDIX: NUMERICAL PROCEDURES FOR DETERMINING CRITICAL LINE AND STEP WIDTHS

The critical points, indicated by ● in Fig. 14, fulfill the fixed point conditions:

$$g_1 = \theta_n - \theta_0 - P = 0 , \qquad (A1)$$

$$g_2 = r_n - r_0 = 0 , \qquad (A2)$$

and the criticality condition,

$$g_3 = \text{Tr}\underline{M}(\theta_0, r_0) = \text{Tr} \prod_i \underline{D}(\theta_i, r_i) = 2b^{Q/2} . \qquad (A3)$$

The fourth condition is that $\text{Tr}\underline{M}$ be a minimum subject to the constraints (A1) and (A2) for a given $K$.

Using Lagrange-multiplier technique, we form the function

$$F = g_3 - L_1 g_1 - L_2 g_2 , \qquad (A4)$$

where $L_1$ and $L_2$ are Lagrange multipliers. The constrained minimum condition can be expressed as

$$\frac{\partial F}{\partial \theta_0} = \frac{\partial g_3}{\partial \theta_0} - L_1 \frac{\partial g_1}{\partial \theta_0} - L_2 \frac{\partial g_2}{\partial \theta_0} ,$$

$$\frac{\partial F}{\partial r_0} = \frac{\partial g_3}{\partial r_0} - L_1 \frac{\partial g_1}{\partial r_0} - L_2 \frac{\partial g_2}{\partial r_0} , \qquad (A5)$$

$$\frac{\partial F}{\partial \Omega} = \frac{\partial g_3}{\partial \Omega} - L_1 \frac{\partial g_1}{\partial \Omega} - L_2 \frac{\partial g_2}{\partial \Omega} .$$

This system of equations has solutions for $L_1$ and $L_2$ only if its determinant is zero:

$$H = \begin{vmatrix} \dfrac{\partial g_1}{\partial \theta_0} & \dfrac{\partial g_2}{\partial \theta_0} & \dfrac{\partial g_3}{\partial \theta_0} \\[2mm] \dfrac{\partial g_1}{\partial r_0} & \dfrac{\partial g_2}{\partial r_0} & \dfrac{\partial g_3}{\partial r_0} \\[2mm] \dfrac{\partial g_1}{\partial \Omega} & \dfrac{\partial g_2}{\partial \Omega} & \dfrac{\partial g_3}{\partial \Omega} \end{vmatrix} = 0 . \tag{A6}$$

Hence, the fourth condition is

$$g_4 = H = 0 .$$

The critical point is the point where the parameters $\theta_0$, $r_0$, $\Omega$, and $K$ fulfil the conditions (A1), (A2), (A3), and (A6). This point is found by means of a four-dimensional Newton iteration method as described in I.

The end points of the limit cycle steps are found by a similar Newton iteration method. The parameters $\theta_0$, $r_0$, and $\Omega$ must fulfil the fixed-point conditions (A1) and (A2) in addition to the stability condition

$$g_3 = \mathrm{Tr}\underline{M}(\theta_0, r_0) = 1 + b^Q . \tag{A7}$$

In all cases the quantities $\partial g_i / \partial \theta_0$, etc. (which involves derivatives up to third order) can be found recursively as described in I.

[1]Stated in a footnote in B. van der Pol, Philos. Mag. 3, 13 (1927). We thank David Rand for making us aware of this.

[2]P. Bak, T. Bohr, M. H. Jensen, and P. V. Christiansen, Solid State Commun. 51, 231 (1984).

[3]V. I. Arnol'd, Am. Math. Soc. Trans., Ser. 2, 46, 213 (1965); M. R. Herman, in Geometry and Topology, edited by J. Palis (Springer, Berlin, 1979), Vol. 579, p. 271.

[4]J. Moser, Stable and Random Motions in Dynamical Systems (Princeton University Press, Princeton, 1973).

[5]For reviews, see P. E. Lindelof, Rep. Prog. Phys. 44, 949 (1981); Y. Imry, in Statics and Dynamics of Nonlinear Systems, edited by G. Benedek, H. Bilz, and R. Zeyher (Springer, Berlin, 1983), p. 170.

[6]R. Y. Chiao, M. J. Feldman, D. W. Peterson, B. A. Tucker, and M. T. Levinsen, in Future Trends in Superconducting Electronics (Charlottesville, 1978), proceedings of the Conference on the Future Trends in Superconductive Electronics, edited by B. S. Deaver, C. M. Falco, J. H. Harris, and S. A. Wolf (AIP, New York, 1979). p. 259.

[7]V. N. Belykh, N. F. Pedersen, and O. H. Soerensen, Phys. Rev. B 16, 4860 (1978).

[8]Y. Braiman, E. Ben-Jacob, and Y. Imry, in SQUID 80, edited by H. D. Hahlbohm and H. Lubbig (de Gruyter, Berlin, 1980); E. Ben-Jacob, Y. Braiman, R. Shansky, and Y. Imry, Appl. Phys. Lett. 38, 822 (1981); E. Ben-Jacob, I. Goldhirsch, Y. Imry, and S. Fishman, Phys. Rev. Lett. 49, 1599 (1982); R. L. Kautz, J. Appl. Phys. 52, 6241 (1981).

[9]B. A. Huberman, J. P. Crutchfield, and N. H. Packard, Appl. Phys. Lett. 37, 751 (1980); D. D'Humieres, M. R. Beasley, B. A. Huberman, and A. Libchaber, Phys. Rev. A 26, 3483 (1982); N. F. Pedersen and A. Davidson, Appl. Phys. Lett. 39, 830 (1981); M. Cirillo and N. F. Pedersen, Phys. Lett. 90A, 150 (1982); W. J. Yeh and Y. H. Kao, Phys. Rev. Lett. 49, 1888 (1982); A. H. MacDonald and M. Plischke, Phys. Rev. B 27, 201 (1983). W. J. Yeh and Y. H. Kao, Appl. Phys. Lett. 42, 299 (1983). These papers deal with the case $A = 0$, which is less relevant here.

[10]M. T. Levinsen, J. Appl. Phys. 53, 4294 (1982).

[11]F. M. A. Salam and S. S. Sastry (unpublished).

[12]G. Grüner, A. Zawadowski, and P. M. Chaikin, Phys. Rev.

Lett. 46, 511 (1981); J. Bardeen, E. Ben-Jacob, A. Zettl, and G. Grüner, Phys. Rev. Lett. 49, 493 (1982).

[13]A. Zettl and G. Grüner, Solid State Commun. 46, 501 (1983); Phys. Rev. B 29, 755 (1984).

[14]P. Monceau, J. Richard, and M. Renard, Phys. Rev. Lett. 45, 43 (1980); Phys. Rev. B 25, 931 (1982).

[15]P. Bak, Rep. Prog. Theor. Phys. 45, 587 (1982).

[16]W. C. Stewart, Appl. Phys. Lett. 12, 277 (1981); D. E. McCumber, J. Appl. Phys. 39, 3113 (1968).

[17]M. J. Renne and D. Polder, Rev. Phys. Appl. 9, 25 (1974); J. R. Waldram and P. H. Wu, J. Low Temp. Phys. 47, 363 (1982).

[18]M. Ya. Azbel and P. Bak, Phys. Rev. B (to be published).

[19]S. Shapiro, Phys. Rev. Lett. 11, 80 (1963).

[20]Although workers in the field using Josephson junctions as measuring devices would prefer to get rid of the subharmonic steps as well as the noise.

[21]S. E. Brown, G. Mozurkewich, and G. Grüner, Phys. Rev. Lett. 54, 2272 (1984).

[22]For a general method of computing J see, e.g., M. L. Cartwright, in Contributions to the Theory of Nonlinear Oscillations, Vol. 20 of Annals of Mathematics Studies (Princeton University, Princeton, N.J., 1950), p. 149.

[23]A consequence of this is that the return map of a second-order differential equation cannot be a simple 1D logistic map with a local maximum. At best, a 1D map is a good approximation to the 2D map.

[24]For a different argument leading to this criterion, see Ref. 28.

[25]B. V. Chirikov, Phys. Rep. 52, 263 (1979).

[26]M. J. Feigenbaum, L. P. Kadanoff, and S. J. Shenker, Physica (Utrecht) 5D, 370 (1982).

[27]D. Rand, S. Ostlund, J. Sethna, and E. Siggia, Phys. Rev. Lett. 49, 132 (1982); Physica (Utrecht) 6D, 303 (1984).

[28]D. G. Aronson, M. A. Chory, G. R. Hall, and R. P. McGehee, Commun. Math. Phys. 83, 303 (1982).

[29]L. Glass and R. Perez, Phys. Rev. Lett. 48, 1772 (1982).

[30]W. J. Yeh, Da-Ren He, and Y. H. Kao, Phys. Rev. Lett. 52, 480 (1984).

[31]P. Alstrom, M. T. Levinsen, and M. H. Jensen, Phys. Lett. 103A, 171 (1984).

Physica 8D (1983) 435–444
North-Holland Publishing Company

# THE INFINITE NUMBER OF GENERALIZED DIMENSIONS OF FRACTALS AND STRANGE ATTRACTORS

H.G.E. HENTSCHEL and Itamar PROCACCIA
*Department of Chemical Physics, Weizmann Institute of Science, Rehovot 76100, Israel*

Received 23 December 1982
Revised 30 March 1983

We show that fractals in general and strange attractors in particular are characterized by an infinite number of generalized dimensions $D_q$, $q > 0$. To this aim we develop a rescaling transformation group which yields analytic expressions for all the quantities $D_q$. We prove that $\lim_{q \to 0} D_q$ = fractal dimension ($D$), $\lim_{q \to 1} D_q$ = information dimension ($\sigma$) and $D_{q=2}$ = correlation exponent ($v$). $D_q$ with other integer $q$'s correspond to exponents associated with ternary, quaternary and higher correlation functions. We prove that generally $D_q > D_{q'}$ for any $q' > q$. For homogeneous fractals $D_q = D_{q'}$. A particularly interesting dimension is $D_{q=\infty}$. For two examples (Feigenbaum attractor, generalized baker's transformation) we calculate the generalized dimensions and find that $D_\infty$ is a non-trivial number. All the other generalized dimensions are bounded between the fractal dimension and $D_\infty$.

## 1. Introduction

Fractal objects seem to appear in a variety of physical applications [1], and in particular seem to occur widely in the context of chaotic dynamical systems [2]. Here attractors are typically fractals and are termed "strange" [3]. The most basic property of an attractor is probably its dimension. The characterization of strange attractors on the basis of their dimensions has attracted a considerable effort recently [4–9]. It seems that so far only three truly different dimensions have been discussed: the similarity dimension $D$, the information dimension $\sigma$ and the correlation dimension $v$ [10–13]. These dimensions are defined as follows: Consider a strange attractor (or any fractal measure) which is embedded in a $d$-dimensional space. Let $\{X_i\}_{i=1}^N$ be the points of a long time series ($N$ very large but necessarily finite) on the attractor. Cover space with a mesh of $d$-dimensional cubes of size $b^d$. Let $M(b)$ be the number of cubes that contain points of the series $\{X_i\}_{i=1}^N$ and let $p_k \equiv N_k/N$ where $N_k$ is the number of points in the

$k$th cube. The similarity dimension is then defined by [1, 14]

$$D = -\lim_{b \to 0} \lim_{N \to \infty} \log M(b)/\log b . \tag{1.1}$$

In most cases of interest the similarity dimension coincides with the fractal dimension as defined by Mandelbrot. The information dimension is defined by [15, 16]

$$\sigma = -\lim_{b \to 0} \lim_{N \to \infty} S(b)/\log b , \tag{1.2}$$

where

$$S(b) = -\sum_{k=1}^{M(b)} p_k \log p_k . \tag{1.3}$$

The correlation dimension $v$ is defined by [10–13]

$$v = \lim_{b \to 0} \lim_{N \to \infty} \log C(b)/\log b , \tag{1.4}$$

where

$$C(b) = \frac{1}{N^2} \sum_{i \neq j} \theta(b - |X_i - X_j|) . \tag{1.5}$$

where $\theta$ is the Heaviside function. $C(b)$ is simply the correlation integral which counts how many pairs of points are there whose distance $|X_i - X_j|$ is less then $b$. A variety of other definitions of dimension has been described in ref. 16 with the conclusion that they are all equivalent to $D$ or $\sigma$ in all cases of physical interest. The authors of ref. 16 interpreted this fact as a signal that indeed there are only two relevant dimensions that characterize fractal attractors. We have shown however that $v$ is different from $D$ and $\sigma$ [12], and also argued that $v$ seems to be the most useful number in terms of characterization of strange attractors due to the ease of its experimental measurement [10–12]. (All other dimensions seem so far to be more difficult to measure [9].)

In this paper we want to show that in fact there is *an infinite number of different (and relevant) generalized dimensions* that characterize an attractor.

These exist as a hierarchy of generalized dimensions $D_q$ which are defined for any $q \geqslant 0$. When $q$ is an integer, $D_q$ has a physical meaning. We shall show that

$$\lim_{q \to 0} D_q = D , \tag{1.6}$$

$$\lim_{q \to 1} D_q = \sigma , \tag{1.7}$$

$$D_{q=2} = v \tag{1.8}$$

and for $q = 3, 4, \ldots, n$ we have generalized dimensions associated with correlation integrals of triplets, quadruplets ... and $n$-tuplets of points on the attractor. We shall be able to prove

$$D_q > D_{q'} \tag{1.9}$$

for any $q' > q$, and the inequality will be replaced by an equality if and only if the fractal is homogeneous [17] (the term "homogeneous" will be made precise below).

In order to prove all the above assertions we develop here a rescaling transformation group for fractals, which allows an analytical calculation of all the dimensions $D_q$. This rescaling transformation applies to a large class of fractals, but not to all. However, even when attractors do not fit exactly to our rescaling transformation, it remains useful. For example the attractor of the Feigenbaum map [18] fits in only approximately but the analytic calculation of its generalized dimensions yields

$$D = 0.537, \quad \sigma = 0.518, \quad v = 0.501, \ldots,$$
$$D_\infty = 0.394 , \tag{1.10}$$

which are values in excellent agreement (for $D$, $\sigma$, $v$) with previous estimates [8, 10, 12] which involved much more tedious work. For any fractal which falls into our rescaling class, the calculation of its dimensions becomes immediate.

The structure of this paper is as follows: In section 2 we develop the rescaling transformation group. For any fractal that falls in our transformation class we then give equations for $D$, $\sigma$ and $v$. We prove $v < \sigma < D$ with equalities when the fractal is uniform. In section 2.4 we present two examples: the Feigenbaum attractor and the generalized baker's transformation. In section 3 we introduce the infinity of generalized dimensions. We first define and calculate the countable infinite set $D_n$, $n = 2, 3, \ldots, n$ and later the uncountable infinite set $D_q$. We prove $D_q > D_{q'}$ for $q' > q$ with equalities in the case of homogeneous fractals. Examples from the Feigenbaum attractor and the generalized Baker's transformation are presented in section 3.3. In section 4 we summarize the paper and offer a discussion.

## 2. Rescaling transformation group

### 2.1. Definition of the rescaling hierarchy

Consider a fractal embedded in $d$-dimensional space and $N$ points on that fractal. In the context of strange attractors we think about a long time series $\{X_i\}_{i=1}^N$, $N \to \infty$. Let $l_0$ be the linear size of the fractal. Consider now the following rescaling hierarchy:

0th level: On the coarsest scale we have $N$ points in a volume $l_0^d$.

1st level: On a finer resolution we see that the $N$ points are partitioned as follows: There are $M_1$ boxes of size $(l_0/s_1)^d$ each of which contains $N_1$ points. $M_2$ boxes of size $(l_0/s_2)^d$ each of which contains $N_2$ points, ..., $M_R$ boxes of size $(l_0/s_R)^d$ each of which contains $N_R$ points.

Notice that boxes of the same size (i.e. same $s_\alpha$) which contain a different number of points get different indices. We define at this level a probability $p_\alpha \equiv N_\alpha/N$. This is the probability to fall in *one of the boxes* of type $\alpha$.

$(n+1)$th level: Given the $n$th level of the hierarchy we define the $(n+1)$th level by taking every box of the $n$th level and replacing it in a self similar fashion to the formation of the first level. A box of size $l^d$ on the $n$th level, to which a probability $p$ is attached, is replaced by the set

$$\{M_\alpha(l/s_\alpha)^d, \quad p_\alpha p\}_{\alpha=1}^R . \tag{2.1}$$

An example of such a hierarchy of shown in fig. 1. We note that not all fractals belong to this rescaling class. However, since all the generalized dimensions of interest are defined in the limit $b \to 0$ (cf. eqs. (1.1)–(1.4)), it is sufficient that there exists a length scale $l_0$ *below* which such self similarity

exists. We shall see below that this class is general enough to be of considerable use.

Some properties of this rescaling transformation are immediately apparent. The first is simply conservation of probability,

$$\sum_{\alpha=1}^R M_\alpha p_\alpha = 1 . \tag{2.2}$$

In addition, since we discuss objects whose $d$-dimensional volume is zero, we have

$$\sum_{\alpha=1}^R M_\alpha(l/s_\alpha)^d < l^d , \tag{2.3}$$

or

$$\sum_{\alpha=1}^R M_\alpha s_\alpha^{-d} < 1 . \tag{2.4}$$

The most important property for later calculation is a relation between probabilities on the various levels of the hierarchy. To see this property we consider a box of size $l^d$ on the $n$th level and cover it with bins of size $b^d (b \ll l)$. The bins can be indexed by $i = 1, 2, \ldots, (l/b)^d$. Denote the probability of the $i$th bin by $p_i(l, b)$. Now within this box of size $l$ there are boxes of the $(n+1)$th level which are of size $l/s_\alpha$. Cover these boxes by bins of size $b/s_\alpha$. Clearly these bins can be also indexed [19] by $i = 1, \ldots, (l/b)^d$. By construction there exists self similarity between the box of size $l^d$ on the $n$th level

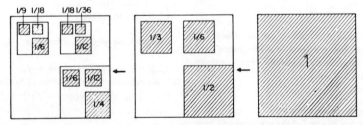

Fig. 1. An example of a fractal obeying the renormalization scheme. Here $M_1 = 1$, $s_1 = 2$, $p_1 = 1/2$, $M_2 = 1$, $s_2 = 16/5$, $p_2 = 1/6$, $M_3 = 1$, $s_3 = 16/5$, $p_3 = 1/3$.

of the hierarchy and the box of size $(l/s)_\alpha)^d$ on the $(n + 1)$th level. Because of this, when we cover these boxes by $(l/b)^d$ bins of sizes $b^d$ and $(b/s_\alpha)^d$, respectively, then in the identically situated bin $i$ we have

$$p_i(l/s_\alpha, b/s_\alpha) = p_\alpha p_i(l, b) . \tag{2.5}$$

This is the fundamental relation of the rescaling class and will be used below to calculate the dimensions associated with the natural measure.

### 2.2. Calculation of fractal, information and correlation dimensions

#### 2.2.1. The fractal dimension

Denote by $n_b(l, b)$ the number of bins of size $b^d$ that are needed to cover the fractal in a box of size $l^d$. It is apparent that

$$n_b(l, b) = \sum_{\alpha=1}^{R} M_\alpha n_b(l/s_\alpha, b) . \tag{2.6}$$

However, due to the self-similarity,

$$n_b(l/s_\alpha, b) = n_b(l, s_\alpha b) . \tag{2.7}$$

For self-similar fractals one can use the relation

$$n_b(l, b) \sim n_b(l) b^{-D} . \tag{2.8}$$

Substituting eqs. (2.7) and (2.8) in eq. (2.6) we find

$$\sum_{\alpha=1}^{R} M_\alpha s_\alpha^{-D} = 1 . \tag{2.9}$$

This is an implicit equation for $D$. Once the properties of the rescaling transformation are given, $D$ is immediately calculable. For examples see subsection 2.4 below.

#### 2.2.2. The information dimension $\sigma$

Consider a box of size $l$ of the $n$th level and bins of size $b$. The information theoretic entropy for this box is

$$S(l, b) = - \sum_{i=1}^{(l/b)^d} p_i(l, b) \ln p_i(l, b) . \tag{2.10}$$

Consider now all the boxes of the next level of the hierarchy which belong to this box. Evidently, we can write $S(l, b)$ also as

$$S(l, b) = - \sum_{\alpha=1}^{R} M_\alpha \sum_{j_\alpha=1}^{(l/bs_\alpha)^d} p_{j_\alpha}(l/s_\alpha, b) \ln p_{j_\alpha}(l/s_\alpha, b) . \tag{2.11}$$

This is simply due to the fact that $p_i$ of eq. (2.10) is zero wherever there is no box of the $(n + 1)$th level.

Using eq. (2.5) we can rewrite eq. (2.11) as

$$S(l, b) = - \sum_{\alpha=1}^{R} M_\alpha \sum_{j_\alpha=1}^{(l/s_\alpha b)^d} p_\alpha p_{j_\alpha}(l, s_\alpha b) \ln [p_\alpha p_{j_\alpha}(l, s_\alpha b)] . \tag{2.12}$$

Comparing eq. (2.12) with eq. (2.10) we find

$$S(l, b) = - \sum_{\alpha=1}^{R} M_\alpha p_\alpha \ln p_\alpha + \sum_{\alpha=1}^{R} M_\alpha p_\alpha S(l, s_\alpha b). \tag{2.13}$$

This equation has a solution of the form

$$S(l, b) = A(l) - \sigma \ln b , \tag{2.14}$$

where

$$\sigma = \frac{- \sum_{\alpha=1}^{R} M_\alpha p_\alpha \ln p_\alpha}{\sum_{\alpha=1}^{R} M_\alpha p_\alpha \ln s_\alpha} . \tag{2.15}$$

Examples are given in subsection 2.4.

#### 2.2.3. The correlation exponent $v$

We first note that $C(b)$ of eq. (1.5) can be written [10, 12] (up to corrections of $\mathcal{O}(1)$) as

$$C(b) = \sum_{k=1}^{M(b)} p_k^2 . \tag{2.16}$$

Consider then again a box of the $n$th level and the quantity

$$C(l, b) = \sum_{i=1}^{(l/b)^d} p_i^2(l, b) = \sum_{\alpha=1}^{R} M_\alpha \sum_{j_\alpha=1}^{(l/bs_\alpha)^d} p_{j_\alpha}^2(l/s_\alpha, b).$$

(2.17)

Using eq. (2.5) we have

$$C(l, b) = \sum_{\alpha=1}^{R} M_\alpha \sum_{j_\alpha=1}^{(l/bs_\alpha)^d} [p_\alpha p_{j_\alpha}(l, s_\alpha b)]^2$$

$$= \sum_{\alpha=1}^{R} M_\alpha p_\alpha^2 C(l, s_\alpha b).$$

(2.18)

This equation has a solution of the form $C(l, b) = C(l) b^\nu$ where $\nu$ is given by the implicit equation

$$\sum_{\alpha=1}^{R} M_\alpha p_\alpha^2 s_\alpha^\nu = 1.$$

(2.19)

Examples of the calculation of $\nu$ are given below in subsection 2.4.

### 2.3. Relations between $D$, $\sigma$ and $\nu$

In this subsection we prove the set of inequalities

$$\nu < \sigma < D,$$

(2.20)

with the inequalities replaced by equalities in this model only when the fractal has the special property $p_\alpha = s_\alpha^{-D}$. The proof is based on the theorem on arithmetic and geometric means [20] which reads

$$\sum_{\alpha=1}^{R} q_\alpha a_\alpha > \prod_{i=1}^{R} a_\alpha^{q_\alpha}$$

(2.21)

for any set of numbers $a_\alpha > 0$ and $q_\alpha$ which satisfy $\Sigma q_\alpha = 1$. The inequality is replaced by an equality *only when* all the $a_\alpha$ are equal.

Denote now $M_\alpha p_\alpha$ by $q_\alpha$. From eq. (2.2) we see that the condition $\Sigma_\alpha q_\alpha = 1$ is met. Next write

the equations for $D$, $\sigma$ and $\nu$ as

$$\sum_\alpha q_\alpha (p_\alpha s_\alpha^D)^{-1} = 1,$$

(2.22)

$$\sigma = -\sum_\alpha q_\alpha \ln p_\alpha \bigg/ \sum_\alpha q_\alpha \ln s_\alpha,$$

(2.23)

$$\sum_\alpha q_\alpha (p_\alpha s_\alpha^\nu) = 1.$$

(2.24)

Eq. (2.22) and the inequality (2.21) can be now used to write

$$1 = \sum_\alpha q_\alpha (p_\alpha s_\alpha^D)^{-1} > \prod_\alpha (p_\alpha s_\alpha^D)^{-q_\alpha},$$

(2.25)

or by taking the logarithm

$$0 > -\sum_\alpha q_\alpha \ln (p_\alpha s_\alpha^D),$$

(2.26)

which is rewritten as

$$D > -\sum_\alpha q_\alpha \ln p_\alpha \bigg/ \sum_\alpha q_\alpha \ln s_\alpha = \sigma.$$

(2.27)

Next use eq. (2.24) and the inequality (2.21) to write

$$1 = \sum_\alpha q_\alpha (p_\alpha s_\alpha^\nu) > \prod_\alpha (p_\alpha s_\alpha^\nu)^{q_\alpha},$$

(2.28)

or

$$0 > \sum_\alpha q_\alpha \ln (p_\alpha s_\alpha^\nu),$$

(2.29)

which is rewritten as

$$\nu < -\sum_\alpha q_\alpha \ln p_\alpha \bigg/ \sum_\alpha q_\alpha \ln s_\alpha = \sigma.$$

(2.30)

Collecting eqs. (2.27) and (2.30) we have

$$D > \sigma > \nu,$$

(2.31)

348

440          *H.G.E. Hentschel and I. Procaccia / Characterization of fractals by dimension*

with an equality $D = \sigma = \nu$ obtaining if and only if all $p_\alpha s_\alpha^D$ are constant. This condition, which translates to

$$p_\alpha = s_\alpha^{-D} \tag{2.32}$$

is simply the condition, for this class of fractals, that the points are distributed uniformly over the fractal. For non-uniform fractals of this class the inequalities (2.31) are obtained. Notice that if only one scale exists (i.e. $R = 1$), then (2.32) is obeyed trivially and

$$D = \sigma = \nu = \ln M / \ln s. \tag{2.33}$$

One never finds $D > \sigma = \nu$.

### 2.4. Examples: The Feigenbaum attractor and the generalized Baker's transformation

#### 2.4.1. The Feigenbaum attractor

As is well known [18] one-dimensional maps of the interval of the type

$$X_{n+1} = F(X_n) \tag{2.34}$$

which have a unique quadratic maximum, possess universal scaling features. The $n$th iterate converges towards a universal even function $g(x)$ which satisfies the exact scaling relation [18]

$$g(g(x)) = -\frac{1}{\alpha} g(\alpha x), \tag{2.35}$$

with $\alpha = 2.50290\ldots$ and normalization condition $g(0) = 1$. The attractor of $g(x)$ consists of the sequence $\zeta_n$, $n = 0, 1, 2\ldots$ with $\zeta_0 = 0$ and $\zeta_{n+1} = g(\zeta_n)$ (see fig. 2). Noticing that $\zeta_1 = g(0) = 1$,

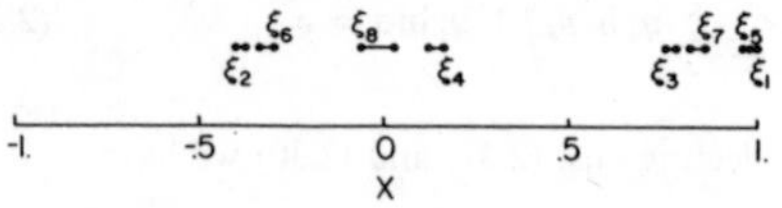

Fig. 2. First 16 points $\zeta_1, \ldots, \zeta_{16}$ of the Feigenbaum equation.

$\zeta_2 = g(g(0)) = -1/\alpha$, $\zeta_3 = g(1/\alpha)$, $\zeta_4 = 1/\alpha^2$, and noticing further that the points $\zeta_5, \zeta_6 \ldots$ fall evenly into the intervals $[\zeta_4, \zeta_2]$ and $[\zeta_3, \zeta_1]$, we come to the conclusion that this attractor falls into our rescaling class if we choose

$$M_1 = 1, \quad p_1 = 1/2, \quad s_1 = \alpha,$$

$$M_2 = 1, \quad p_2 = 1/2, \quad s_2 = \frac{1 + 1/\alpha}{1 - g(1/\alpha)}. \tag{2.36}$$

Notice that the attractor is not *exactly* of this class. If it were, the interval $[\zeta_2, \zeta_6]$ for example would have been precisely of the same size as the interval $[\zeta_3, \zeta_7]$. Calculating the iterates one sees that they are not exactly the same. The error is however small. If we disregard it we calculate [18] now $s_2 = 5.8053648$ and solve

$$1 = s_1^{-D} + s_2^{-D}, \tag{2.37}$$

$$\sigma = \frac{2 \ln 2}{\ln s_1 s_2}, \tag{2.38}$$

$$4 = s_1^\nu + s_2^\nu, \tag{2.39}$$

to get

$$D = 0.537, \tag{2.40}$$

$$\sigma = 0.518, \tag{2.41}$$

$$\nu = 0.501, \tag{2.42}$$

in excellent agreement with previous (rather tedious) numerical [10–12] and analytical [8, 10–12] calculations.

#### 2.4.2. The generalized baker's transformation

The generalized baker's transformation is defined [15] by the recursion relations on the unit square

$$X_{n+1} = \begin{cases} \lambda_a X_n, & \text{if } Y_n < \alpha, \\ \frac{1}{2} + \lambda_b X_n, & \text{if } Y_n > \alpha; \end{cases} \tag{2.43}$$

$$Y_{n+1} = \begin{cases} \frac{1}{\alpha} Y_n, & \text{if } Y_n < \alpha, \\ \frac{1}{1-\alpha}(Y_n - \alpha), & \text{if } Y_n > \alpha. \end{cases} \tag{2.44}$$

Two iterations of the map for $\alpha$, $\lambda_a$, $\lambda_b < 1/2$, $\lambda_b > \lambda_a$ are shown in fig. 3. We see that the fractal generated here falls in our rescaling class if we choose

$$M_1 = \lambda_a^{-1}, \quad s_1 = \lambda_a^{-1}, \quad P_1 = \alpha\lambda_a , \tag{2.45}$$
$$M_2 = \lambda_b^{-1}, \quad s_2 = \lambda_b^{-1}, \quad P_2 = (1-\alpha)\lambda_b ,$$

remember that $p_\alpha$ is the probability to fall in one of the boxes $M_\alpha$! From eqs. (2.22)–(2.24) we now get

$$\lambda_a^{D-1} + \lambda_b^{D-1} = 1 , \tag{2.46}$$

$$\sigma = 1 + \frac{\alpha \ln \alpha + (1-\alpha)\ln(1-\alpha)}{\alpha \ln \lambda_a^{-1} + (1-\alpha)\ln \lambda_b^{-1}} \tag{2.47}$$

and

$$\alpha^2\lambda_a^{1-\nu} + (1-\alpha)^2\lambda_b^{1-\nu} = 1 . \tag{2.48}$$

Eqs. (2.46) and (2.47) are in agreement with the calculation of ref. 15.

Clearly, any other example for which $M_\alpha$, $s_\alpha$ and $p_\alpha$ are calculable would give the wanted information with the same ease.

## 3. The infinite set of generalized dimensions

### 3.1. The countable infinity of generalized dimensions

The exponent $\nu$ has been related to correlations between pairs of point on the fractal. In fact there is no reason not to consider higher order correlation functions, or correlation integrals. We thus define $C_n(l)$ by

$$C_n(l) = \lim_{N\to\infty} \frac{1}{N^n} \text{ [number of } n\text{-tuplets of points}$$
$$(i_1, i_2, \ldots, i_n) \text{ whose distances } |X_{i_\alpha}$$
$$- X_{i_\beta}| \text{ are less then } l \text{ for all } i_\alpha, i_\beta]. \tag{3.1}$$

As in the case of the binary correlation integral we note that up to a factor of $\mathcal{O}(1)$

$$C_n(b) \approx C_n(l, b) = \sum_{i\in I} p_i^n(l, b) . \tag{3.2}$$

We shall argue that $C_n(l, b)$ scales like

$$C_n(l, b) = C_n(l)b^{\nu_n} . \tag{3.3}$$

We shall also show that for the fractals of our rescaling class

$$\frac{n}{n-1}\nu_n > \nu_{n+1} . \tag{3.4}$$

Unless the fractal is uniform, where the inequality is replaced by an equality. Clearly $\nu_2 = \nu$.

To see these points we note that for fractals of our rescaling class $C_n(l, b)$ obeys the following scaling relation:

$$C_n(l, b) = \sum_{\alpha=1}^{R} M_\alpha p_\alpha^n C_n(l, s_\alpha b) . \tag{3.5}$$

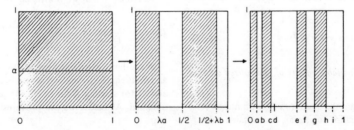

Fig. 3. Two iterations of the generalized baker's transformation. In the third panel $a = \lambda a^2$, $b = \lambda_a/2$, $c = \lambda_a(1/2 + \lambda_b)$, $d = \lambda_a$, $e = 1/2$, $f = 1/2 + \lambda_a\lambda_b$ $g = 1/2 + \lambda_b/2$, $h = 1/2 + \lambda_b (1/2 + \lambda_b)$, $i = 1/2 + \lambda_b$.

This relation indeed has a solution of the form (3.3) where $v_n$ are the solutions of the implicit equation

$$1 = \sum_{\alpha=1}^{R} M_\alpha p_\alpha^n s_\alpha^{v_n} .$$ (3.6)

To prove the inequalities (3.4) we use the following argument: Denote by $M_r(a)$ the quantity

$$m_r(a) = \left[ \sum_\alpha q_\alpha a'_\alpha \right]^{1/r} .$$ (3.7)

The quantities $m_r(a)$ obey an inequality which is an extension of Hölder's inequality [20], that if $r' > r$,

$$m_{r'}(a) > m_r(a) ,$$ (3.8)

unless all $a_\alpha$ are the same. We can now rewrite eq. (3.6) in the form

$$m_{n-1}(ps^{v_n/(n-1)}) = 1 .$$ (3.9)

With this equation in mind we can write

$$m_{n-1}(ps^{v_n/(n-1)}) = m_{n'-1}(ps^{v_{n'}/(n'-1)}) .$$ (3.10)

Remembering that $s_\alpha > 1$ we see that the inequality (3.8) implies

$$\frac{v_n}{n-1} > \frac{v_{n'}}{n'-1} .$$ (3.11)

In the case of homogeneous fractals we obtain for $n > 2$

$$D_n = v_n/(n-1) = v = \sigma = D .$$ (3.12)

3.2. *The uncountable infinity of generalized dimensions*

So far we got the quantities $D_n$ with integer $n$'s. We can generalize now to an uncountable infinity of quantities $D_q$ with any $q > 0$, not necessarily integer. To do so we consider the quantities $D_q$ defined by

$$D_q = \frac{1}{(q-1)} \lim_{l \to 0} \frac{\log \sum_{i \in I} p_i^q}{\log l} ,$$ (3.13)

where as before $p_i$ is the probability of a bin in the $l$th box. In a measure theoretic language we can get an analogous definition by letting $\mu$ denote the natural probability measure on the fractal and $B_l(x)$ denote a ball of radius $l$ centered about a point $x$ on the fractal. Then

$$D_q = \frac{1}{(q-1)} \lim_{l \to 0} \frac{1}{\log l} \log \int d\mu(x)\mu(B_l(x))^{q-1} .$$ (3.14)

We can see that eqs. (3.13) or (3.14) define a generalized dimension by considering the homogeneous fractal. Then $\mu(B_l(x)) \approx l^D$ and $D_q = D$ for any $q$. In general however $D_q \neq D$. It is easy to see that for integer $q$, we have (cf. eqs. (3.2), (3.13)) $v_n = (n-1)D_n$, $n > 2$. We now show that also

$$\lim_{q \to 0} D_q = D ,$$ (3.15)

$$\lim_{q \to 1} D_q = \sigma .$$ (3.16)

Eq. (3.15) is most clearly seen from eq. (3.13). Here when $q \to 0$, $p_i^q$ is 1 for $p_i \neq 0$ and 0 for $p_i = 0$. Thus the sum $\sum p_i^q$ = number of bins that contain a piece of the fractal. Eq. (3.13) then simply becomes the definition of the fractal dimension.

To see eq. (3.16) we rewrite eq. (3.13) as

$$D_q = \frac{1}{(q-1)} \lim_{l \to 0} \frac{\log \sum p_i \exp(q-1) \ln p_i}{\log l} .$$ (3.17)

In the limit $q \to 1$ one expands the exponential and finds eq. (1.2). Analogous proofs to eqs. (3.15) and (3.16) starting from eq. (3.14) can be easily obtained. As before we can prove using eq. (3.8) that

$$D_q > D_{q'}$$ (3.18)

351

for $q' > q$. Thus we see that the inequalities $D > \sigma > v$ are special cases of eq. (3.18).

The calculation of all the dimensions $D_q$ is easy when the properties of the rescaling group are given. One simply solves the implicit equation

$$m_{(q-1)}(ps^{D_q}) = 1. \tag{3.19}$$

Apart from the dimensions, $D$, $\sigma$, $v$, a very interesting generalized dimension is $D_\infty$. This quantity is given by

$$D_\infty = \inf \ln p_\alpha^{-1} / \ln s_\alpha. \tag{3.20}$$

### 3.3. Examples

For examples we turn again to the Feigenbaum attractor and the baker's transformation.

#### 3.3.1. The Feigenbaum attractor

Here the most interesting number is $D_\infty$. From eqs. (2.38) and (3.20) we find

$$D_\infty = 0.3941061 \ldots. \tag{3.21}$$

Clearly, any other dimension $D_q$ can be calculated. However, we see that the uncountable infinity of generalized dimensions is bounded from above and below by 0.538 and 0.394, respectively.

#### 3.3.2. The baker's transformation

By using the values of eq. (2.45) we find

$$D_\infty = \inf\left\{1 + \frac{\ln \alpha}{\ln \lambda_a}; \quad 1 + \frac{\ln(1-\alpha)}{\ln \lambda_b}\right\}. \tag{3.22}$$

Again we see that $D_\infty$ is a non-trivial number. All other generalized dimensions of the baker's transformation are sandwiched between $D$ and $D_\infty$.

## 4. Discussion

We have shown that in general fractals are characterized by an infinite number of generalized

dimensions that are all different as long as the fractal is non-uniform. We developed a rescaling transformation group which resulted in closed form expressions for these dimensions. We gave two examples for the usefulness of the approach. We now turn to some further discussion of the results.

A complete knowledge of the set of dimensions $D_q$ is equivalent to a complete physical characterization of the fractal. All correlation functions (doublet, triplet etc.) are determined uniquely. For strange attractors the set $D_q$ contains information on the appearance and statistical properties of the chaotic time series.

If the properties of the rescaling transformation (i.e. the numbers $M_\alpha$, $p_\alpha$, $s_\alpha$) are not known, one has to resort to numerical estimates of the generalized dimensions. Of all the uncountable infinity of dimensions $D_q$, the only one which is easy to calculate on the basis of a time series is $v$. It has been demonstrated previously [10–12] that algorithms for calculating $v$ are much more efficient than the box counting algorithms needed to calculate $D$, $\sigma$, etc. If one *is* running a box counting algorithm, however, one can get all the dimensions $D_q$ at one go, simply by calculating eq. (3.13).

It seems to us that in the case of strange attractors it might be worthwhile to direct numerical procedures towards an estimate of the numbers $M_\alpha$, $p_\alpha$, $s_\alpha$. We have seen that once these are given, the calculation of all dimensions $D_q$ becomes immediate. Any self similar attractor might be a candidate for such an approach.

A good example is the attractor of the Hénon map about the fixed point. It has been demonstrated that this attractor is a self-similar fractal [21, 22] and it seems easy to estimate the parameters of the rescaling transformation group.

We have seen that all the quantities $D_q$ are bounded between $D$ and $D_\infty$. The quantity $D_\infty$ has been calculated explicitly for the two examples and found to be a non-trivial number. Further applications of the above ideas would be presented in future publications.

## Acknowledgements

This work has been supported in part by the Israel Commission for Basic Research. IP wishes to acknowledge the fact that seemingly the first time when a quéstion on higher order correlation exponents was raised, was in a discussion with S. Grossmann during the Minerva meeting in Deidesheim.

## References

[1] B.B. Mandelbrot, Fractal–Form, Chance and Dimension (Freeman, San Francisco, 1977).

[2] E.N. Lorenz, J. Atm. Sci. 20 (1963) 130.

[3] D. Ruelle and F. Takens, Comm. Math. Phys. 20 (1971) 167. In fact it is only true for differmorphisms or con-'tinuous flows that the attractors are typically fractals.

[4] H. Mori, Prog. Theor. Phys. 63 (1980) 1044.

[5] J.L. Kaplan and J.A. Yorke, in: Functional Differential Equations and Approximations of Fixed Points, H.-O. Peitgen and H.-O. Walther, eds, Lecture Notes in Math. 730 (Springer, Berlin, 1979), p. 204.

[6] D.A. Russel, J.D. Hanson and E. Ott, Phys. Rev. Lett. 45 (1980) 1175.

[7] H. Froehling, J.D. Crutchfield, D. Farmer, N.H. Packard and R. Shaw, Physica 30 (1981) 605.

[8] P. Grassberger, J. Stat. Phys. 26 (1981) 173.

[9] H.S. Greenside, A. Wolf, J. Swift and T. Pignataro, Phys. Rev. A25 (1982) 3453.

[10] P. Grassberger and I. Procaccia, Phys. Rev. Lett. 50 (1983) 346.

[11] I. Procaccia, P. Grassberger and H.G.E. Hentschel, in: Dynamical Systems and Chaos, L. Garrido, ed. (Springer, Berlin, 1982).

[12] P. Grassberger and I. Procaccia, Physica D, in press.

[13] F. Takens, in: Proc. of the Warwick Symposium 1980, D. Rand and B.S. Young, eds., Lecture Notes in Math. 898 (Springer, Berlin, 1981).

[14] The limit $b \to 0$, $N \to \infty$ means in practical applications going to a cube size which is extremely small on the global scale but which is still large enough to ensure that cubes contain a large number of points.

[15] J. Balatoni and A. Renyi, Pub. Math. Inst. Hungarian Acad. Sci. 1 (1956) 9.

[16] J.D. Farmer, E. Ott and J.A. Yorke, to appear in Physica D.

[17] The fact that generally speaking fractal measures lead to inequalities between exponents has been noted by Mandelbrot, see: J. Fluid Mech. 62 (1974) 381. Mandelbrot (private communication) has also warned against the use of the term "dimension" for quantities like $D_q$, $q > 0$, as they have some but not all the properties of dimensions. For this reason we refer to these as generalized dimensions.

[18] M. Feigenbaum, J. Stat. Phys. 19 (1978) 25; 21 (1979) 669.

[19] Notice that $(l/b)$ is not necessarily an integer. However, one can either choose $b$ such that $l/b$ is an integer or remember that $b \ll l$, and therefore only slight inaccuracies are caused by indexing $i$ between 1 and $(l/b)^d$.

[20] G.H. Hardy, J.E. Littlewood and G. Pólya, Inequalities (Cambridge Univ. Press, Cambridge, 1952).

[21] M. Hénon, Comm. Math. Phys. 50 (1976) 69.

[22] J.D. Farmer, preprint.

Physica 9D (1983) 189–208
North-Holland Publishing Company

# MEASURING THE STRANGENESS OF STRANGE ATTRACTORS

Peter GRASSBERGER† and Itamar PROCACCIA

*Department of Chemical Physics, Weizmann Institute of Science, Rehovot 76100, Israel*

Received 16 November 1982
Revised 26 May 1983

We study the correlation exponent $v$ introduced recently as a characteristic measure of strange attractors which allows one to distinguish between deterministic chaos and random noise. The exponent $v$ is closely related to the fractal dimension and the information dimension, but its computation is considerably easier. Its usefulness in characterizing experimental data which stem from very high dimensional systems is stressed. Algorithms for extracting $v$ from the time series of a single variable are proposed. The relations between the various measures of strange attractors and between them and the Lyapunov exponents are discussed. It is shown that the conjecture of Kaplan and Yorke for the dimension gives an upper bound for $v$. Various examples of finite and infinite dimensional systems are treated, both numerically and analytically.

## 1. Introduction

It is already an accepted notion that many nonlinear dissipative dynamical systems do not approach stationary or periodic states asymptotically. Instead, with appropriate values of their parameters, they tend towards strange attractors on which the motion is chaotic, i.e. not (multiply) periodic and unpredictable over long times, being extremely sensitive on the initial conditions [1–4].

A natural question is by which observables this situation is most efficiently characterized. Even more basically, when observing a seemingly strange behaviour, one would like to have clear-cut procedures which could exclude that the attractor is indeed multiply periodic, or that the irregularities are e.g. caused by external noise [5].

The first possibility can be ruled out by making a Fourier analysis, but for the second one has to turn to some other measures. These measures should be sensitive to the *local* structure, in order to distinguish the blurred tori of a noisy (multi-) periodic motion from the strictly deterministic

† Permanent address: Department of Physics, University of Wuppertal, W. Germany.

motion on a fractal. Also, they should be able to distinguish between different strange attractors.

In this paper we shall propose such a measure. Before doing so we shall discuss however the existing approaches to the subject.

In a system with $F$ degrees of freedom, an attractor is a subset of $F$-dimensional phase space towards which almost all sufficiently close trajectories get "attracted" asymptotically. Since volume is contracted in dissipative flows, the volume of an attractor is always zero, but this leaves still room for extremely complex structures.

Typically, a strange attractor arises when the flow does not contract a volume element in *all* directions, but stretches it in some. In order to remain confined to a bounded domain, the volume element gets folded at the same time, so that it has after some time a multisheeted structure. A closer study shows that it finally becomes (locally) Cantor-set like in some directions, and is accordingly a fractal in the sense of Mandelbrot [6].

Ever since the notion of strange attractors has been introduced, it has been clear that the Lyapunov exponents [7, 8] might be employed in studying them. Consider an infinitesimally small $F$-dimensional ball in phase space. During its evolution it will become distorted, but being infinitesimal, it will remain an ellipsoid. Denote the principal axes of this ellipsoid by $\epsilon_i(t)$ $(i = 1, \ldots, F)$. The Lyapunov exponents $\lambda_i$ are then determined by

$$\epsilon_i(t) \approx \epsilon_i(0)\, e^{\lambda_i t}. \tag{1.1}$$

The sum of the $\lambda_i$, describing the contraction of volume, has of course to be negative. But since a strange attractor results from a stretching and folding process, it requires at least one of the $\lambda_i$ to be positive. Inversely, a positive Lyapunov exponent implies sensitive dependence on initial conditions and therefore chaotic behaviour.

One drawback of the $\lambda_i$'s is that they are not easily measured in experimental situations. Another limitation is that while they describe the *stretching* needed to generate a strange attractor, they don't say much about the *folding*.

That these two are at least partially independent is best seen by looking at a horshoe-like map† embedded in 3-dimensional space (fig. 1). Assume that each step of the evolution consists of (i) stretching in the $x$-direction by a factor of 2, (ii) squeezing in the $y$- and $z$-direction by different factors $\mu_z < \mu_y < \frac{1}{2}$, and (iii) folding in the $(x, y)$ plane (fig. 1a) or in the $(x, z)$ plane (fig. 1b). From fig. 1 one realizes already that the attractor will in both cases be a Cantorian set of lines, being more "plane-filling" in the first case than in the second case. Indeed, using the results of Section 7, one finds easily that the fractal dimensions are $D_a = 1 + \ln 2/|\ln \mu_y|$ and $D_b = 1 + \ln 2/|\ln \mu_z|$, respectively.

It is this fractal (or Hausdorff–Besikovich) dimension which has until now attracted most attention [9–14] as a measure of the local structure of fractal attractors. In order to define it [5], one first covers the attractor by $F$-dimensional hypercubes of side length $l$ and considers the limit $l \to 0$. If the

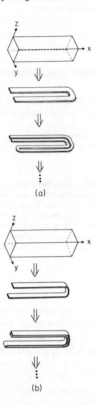

Fig. 1. Shape of an originally rectangular volume element after two iterations, each consisting of stretching, squeezing and folding. In fig. 1a (1b), the folding is in the $y(z)$-direction, which is the direction of lesser (stronger) squeezing.

minimal number of cubes needed for the covering grows like

$$M(l) \underset{l \to 0}{\simeq} l^{-D}, \tag{1.2}$$

the exponent $D$ is called the Hansdorff dimension of the attractor [5].

Being a purely geometric measure, $D$ is independent of the frequency with which a typical trajectory visits the various parts of the attractor.

† Notice that this is not a Smale's horseshoe. We also neglect in the following the bent parts of the horseshoe, in comparison to the parallel parts (i.e. we assume $L_x \gg L_y, L_z$; see fig. 1).

Even if these frequencies are very inequal, developing maybe even singularities somewhere, all parts contribute to $D$ equally. It has been documented [12, 14] that the calculation of $D$ is exceedingly hard and in fact impractical for higher dimensional systems.

Another measure which has been considered and which is sensitive to the frequency of visiting, is the information entropy of the attractor. By "information entropy" here we understand the information gained by an observer who measures the actual state $X(t)$ of the system with accuracy $l$, and who knows all properties of the system but not the initial condition $X(0)$. This is very similar to the entropy in statistical mechanics if we relate $X(t)$ to the microstate ($F \approx 10^{23}$), and the "system" to the macrostate. It is *not* the Kolmogorov entropy which is essentially the sum of all positive Lyapunov exponents.

Using the above partition of phase space into cells with length $l$, the information entropy can be written as

$$S(l) = - \sum_{i=1}^{M(l)} p_i \ln p_i , \qquad (1.3)$$

where $p_i$ is the probability for $X(t)$ to fall into the $i$th cell. For all attractors studied so far, $S(l)$ increases logarithmically with $1/l$ as $l \to 0$, and we shall accordingly make the ansatz

$$S(l) \simeq S_0 - \sigma \ln l . \qquad (1.4)$$

The constant $\sigma$ will be called, following ref. 8, the information dimension. It is always a lower bound to the Hausdorff dimension, and in most cases they are almost the same within numerical errors.

The measure on which we shall concentrate mostly in this paper, has been recently introduced by the present authors [15]. It is obtained from the correlations between random points on the attractor. Consider the set $\{X_i, i = 1 \cdots N\}$ of points on the attractor, obtained e.g. from a time series, i.e. $X_i \equiv X(t + i\tau)$ with a fixed time increment $\tau$ between successive measurements. Due to the ex-

ponential divergence of trajectories, most pairs $(X_i, X_j)$ with $i \neq j$ will be *dynamically* uncorrelated pairs of essentially random points. The points lie however on the attractor. Therefore they will be spatially correlated. We measure this spatial correlation with the correlation integral $C(l)$, defined according to

$$C(l) = \lim_{N \to \infty} \frac{1}{N^2} \times \{\text{number of pairs } (i, j) \text{ whose}$$

$$\text{distance } |X_i - X_j| \text{ is less than } l\} . \qquad (1.5)$$

The correlation integral is related to the standard correlation function

$$c(r) = \lim_{N \to \infty} \frac{1}{N^2} \sum_{\substack{i,j=1 \\ i \neq j}}^{N} \delta^F(X_i - X_j - r) \qquad (1.6)$$

by

$$C(l) = \int_0^l d^F r c(r) . \qquad (1.7)$$

One of the central aims of this paper is to establish that for small $l$'s $C(l)$ grows like a power

$$C(l) \sim l^\nu , \qquad (1.8)$$

and that this "correlation exponent" can be taken as a most useful measure of the local structure of a strange attractor. It seems that $\nu$ is more relevant, in this respect, than $D$. In any case, its calculation yields also an estimate of $\sigma$ and $D$, since we shall argue that in general one has

$$\nu \leq \sigma \leq D . \qquad (1.9)$$

We found that the inequalities are rather tight in most cases, but not in all. Given an experimental signal, if one finds eq. (1.8) with $\nu < F$, one knows that the signal stems from deterministic chaos rather than random noise, since random noise will always result in $C(l) \sim l^F$. Explicit algorithms will be proposed below.

One of the main advantages of $v$ is that it can easily be measured, at least more easily than either $\sigma$ or $D$. This is particularly true for cases where the fractal dimension is large ($\gtrsim 3$) and a covering by small cells becomes virtually impossible. We thus expect that the measure $v$ will be used in experimental situations, where typically high dimensional systems exist.

In theoretical cases, when the evolution law is known analytically, the easiest quantities to evaluate are the Lyapunov exponents. General formulae expressing $D$ in terms of the $\lambda_i$ have been proposed by Mori [9] and by Kaplan and Yorke [10]. If they were correct, they would obviously be very useful. They have been verified in simple cases [11, 14]. But Mori's formula was shown to be wrong in one case by Farmer [8], and the above example shown in fig. 1 shows that also the Kaplan–Yorke formula

$$D = D_{KY} \equiv j + \frac{\lambda_1 + \lambda_2 + \cdots + \lambda_j}{|\lambda_{j+1}|} \qquad (1.10)$$

does not hold even in all those cases where $v = \sigma = D$. Here, the exponents are ordered in descending order $\lambda_1 \geq \lambda_2 \geq \cdots \geq \lambda_F$, and $j$ is the largest integer for which $\lambda_1 + \lambda_1 + \ldots + \lambda_j \geq 0$.

In section 7 we shall take up this question again. We shall show that the counterexample in fig. 1b is not generic. We shall however claim that eq. (1.10) cannot generally be expected to be correct, and that in fact $D_{KY}$ is an upper bound, if $v = \sigma = D$.

In the next section, we shall present numerical results for several simple models, for which the fractal dimensions are known from the literature. This will serve to illustrate the scaling law (1.8), and to verify the inequality $v \leq D$. This inequality and its stronger version, eq. (1.9), will be derived in section 3. The case of one-dimensional maps at infinite bifurcation (Feigenbaum [16]) points is special in that there the information dimension $\sigma$ and the exponent $v$ can be calculated exactly, with the result $v \neq \sigma \neq D$. It is treated in section 4. Section 5 is dedicated to an important modification

which allows to extract $v$ from a time series of one single variable, instead of from the series $\{X_i\}$. This is of course most important for infinite-dimensional systems, but it is also very useful in low-dimensional cases where it diminishes systematic errors. Among others, we shall apply this method in section 6 to the Mackey–Glass [17] delay equation studied in great detail in ref. 8.

In section 7 we discuss the relation of $v$ to the Lyapunov exponents, and establish the result

$$v \leq D_{KY}. \qquad (1.11)$$

A summary and a discussion of the actual method of treating experimental signals is offered in section 8.

## 2. Case studies of low-dimensional systems

In this section we shall establish that $C(l)$ can be very well represented by a power law $l^v$, by exhibiting numerical results for a number of low dimensional systems. These results are summarized in table I. In section 5 we shall show that this is the case also in high (and infinite) dimensional systems. Details of the numerical algorithms are discussed in appendix A.

### 2.1. One-dimensional maps

The simplest cases of chaotic system are represented by maps of some interval into itself, as e.g. the logistic map [2]

$$x_{n+1} = ax_n(1 - x_n). \qquad (2.1)$$

We shall study this map both at the point of onset of chaos via period doubling bifurcations, i.e. when $a = a_\infty = 3.5699456\ldots$ and for the case $a = 4.0$. In fig. 2 we show the result for the first case. It is well known [2, 16] that for this map the attractor* is

---

\* Note that the term "attractor" would not be universally accepted here due to the fact that in any neighbourhood there exist trajectories which do not tend towards it asymptotically.

Table I

| | $\nu$ | No. of iterations, time increment $\tau$ | $D$ | $\sigma$ |
|---|---|---|---|---|
| Hénon map | $1.21 \pm 0.01^{d}$) | 15000 | 1.26 (ref. 11) | - |
| $a = 1.4,\ b = 0.3$ | $1.25 \pm 0.02^{e}$) | | | |
| Kaplan–Yorke map | $1.42 \pm 0.02$ | 15000 | 1.431(ref. 11) | - |
| $\alpha = 0.2$ | | | | |
| Logistic eq., | $0.500 \pm 0.005$ | 25000 | 0.538(ref. 13) | 0.5170976 |
| $b = 3.5699456 \cdots$ | $0.4926 < \nu < 0.5024^{f}$) | | | |
| Lorenz eq.[a]) | $2.05 \pm 0.01$ | 15000; $\tau = 0.25$ | $2.06 \pm 0.01$ | - |
| Rabinovich–[b]) | $2.19 \pm 0.01$ | 15000; $\tau = 0.25$ | - | - |
| Fabrikant eq. | | | | |
| Zaslavskii map[c]) | $(\approx 1.5)$ | 25000 | 1.39(ref. 11) | - |

[a])Parameters as in refs. 7 and 11.
[b])Parameters as in section 3 of ref. 20.
[c])Parameters as in ref. 11.
[d])From eqs. (1.5) and (1.8).
[e])From single variable time series, with $f = 3$.
[f])Exact analytic bound.

Cantor-like with a fractal dimension satisfying the exact bound [13] $0.5376 < D < 0.5386$. In section 4 we shall prove exactly that $\sigma = 0.517097\ldots$, and that $0.4926 < \nu < 0.5024$ while from Fig. 2 we find $\nu = 0.500 \pm 0.005$. For very small distances, the data for $C(l)$ deviate from a power law, but that was to be expected: the behaviour at $a = a_{\infty}$ is not yet chaotic, and therefore the values $x_{n}$ are strongly

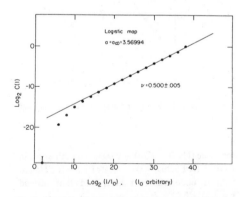

Fig. 2. Correlation integral for the logistic map (2.1) at the infinite bifurcation point $a = a_{\infty} = 3.699\ldots$ The starting point was $x_{0} = \frac{1}{2}$, the number of points was $N = 30.000$.

correlated. We verified that indeed the powerlaw holds down to smaller values of $l$ if we increase $N$ or use only values $x_{i}, x_{i+p}, x_{i+2p}, x_{i+2p}, \ldots$ with $p$ being a large odd number.

The same map can be used also to introduce the important issue of corrections to scaling. These are found for the parameter value $a = 4$. It is well known that in this the attractor* consists of the interval [0, 1], and that the invariant probability density is equal to

$$p(x) \equiv \lim_{N \to \infty} \frac{1}{N} \sum_{i=1}^{N} \delta(x_{i} - x) \tag{2.2}$$

$$= \frac{1}{\pi} [x(1-x)]^{-1/2}. \tag{2.3}$$

From this, one finds easily

$$\nu = \sigma = D = 1. \tag{2.4}$$

Notice, however, that while the scaling laws (1.2) and (1.4) are exact, the scaling law (1.8) for $C(l)$

---
* Again, the term is questionable, as no point outside the interval [0, 1] gets attracted towards it. We shall ignore this irrelevant point, which could be avoided by using $a = 4 - \epsilon$.

requires logarithmic corrections, due to the singular behaviour of $p(x)$:

$$C(l) = \int_0^1 \int dx \, dy \, p(x) p(y) \theta(|x - y| - l)$$

$$\underset{l \to 0}{\simeq} \frac{4}{\pi^2} l \ln 1/l. \tag{2.5}$$

Thus, a numerical calculation of $\nu$ is expected to converge very slowly. This problem and a remedy for it are discussed further in section 5.

### 2.2. Maps of the plane

Here we examined the Hénon [18] map

$$x_{n+1} = y_n + 1 - ax_n^2,$$
$$y_{n+1} = bx_n, \tag{2.6}$$

with $a = 1.4$ and $b = 0.3$, the Kaplan–Yorke [10] map

$$x_{n+1} = 2x_n \pmod 1,$$
$$y_{n+1} = \alpha y_n + \cos 4\pi x_n \tag{2.7}$$

with $\alpha = 0.2$, and the Zaslavskii [19] map

$$x_{n+1} = [x_n + \nu(1 + \mu y_n) + \epsilon \nu \mu \cos 2\pi x_n] \pmod 1,$$
$$y_{n+1} = e^{-\Gamma}(y_n + \epsilon \cos 2\pi x_n), \tag{2.8}$$

with the parameters

$$\mu = \frac{1 - e^{-\Gamma}}{\Gamma} \tag{2.9}$$

and $\Gamma = 3.0$, $\nu = 400/3$, and $\epsilon = 0.3$ taken from ref. 11.

Figs. 3–5 exhibit the results for the correlation integrals. In the first two cases, we find excellent agreement with a power law; while for the Kaplan–Yorke map we find $\nu = 1.42 \pm .02$ in agreement with the published [11] value of $D$, a fit to the Hénon map yields $\nu_{eff} = 1.21$, smaller than

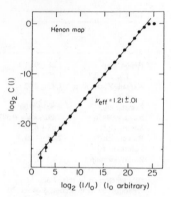

Fig. 3. Correlation integral for the Hénon map (2.6) with $a = 1.4$, $b = 0.03$ and $N = 15.000$.

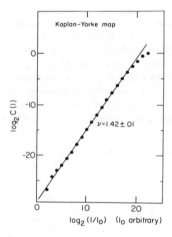

Fig. 4. Same as fig. 3, but for Kaplan–Yorke map (2.7) with $\alpha = 0.2$.

the value [11] $D = 1.261 \pm 0.003$. We shall argue in sectin 5 that actually the value of $\nu$ for the Hénon map is underestimated here, and that instead $\nu = 1.25 \pm 0.02 \approx D$.

The case of the Zaslavskii map is exceptional as it was the only system for which we did not find

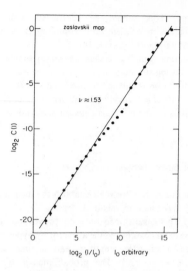

Fig. 5. Correlation integral for Zaslavskii map (eqs. (2.8), (2.9)); $N = 25.000$, parameters as in the text. For faster scaling, the $y$-coordinate was blown up by a factor of 25, rendering the attractor square-like at low resolution (see fig. 6; without this, the attractor would have looked effectively 1-dimensional for $l \gtrsim l_{max}/25$).

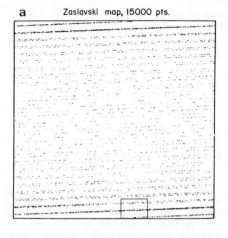

Fig. 6. Attractor of the Zaslavskii map. a) entire attractor (15.000 points plotted; y-scale blown up by factor 25); b) Blown up view of part indicated in part a (10.000 points plotted).

clear-cut power behaviour. Also, an (admittedly poor) fit would yield $v \approx 1.5$, in clear violation of the bound $v < D$. The reasons why our method has to fail for this map – with the parameters as quoted above – becomes clear when looking at fig. 6. Call $l_0$ the outer length scale. From fig. 6a one sees that the attractor looks 2-dimensional for $l \gtrsim l_0 \times 2^{-5}$ and $\approx$ 1-dimensional for $l_0 \times 2^{-5} \gtrsim l \gtrsim l_0 \times 2^{-9}$. From fig. 6b one sees that it looks $\approx$ 2-dimensional again down to $\approx l_0 \times 2^{-14}$, scaling behaviour setting in only at about that scale (which is beyond our resolution). It seems to us that the box-counting algorithm of ref. 11 in which $D$ is evaluated, should confront the same problem†.

† Note added: Dr. Russel kindly provided us with the original data of $M(\epsilon)$ versus $\epsilon$. From these, it seems that indeed a similar phenomenon occurs and that accordingly a value $D \approx 1.5$ cannot be excluded.

2.3. *Differential equations*

We have studied the Lorenz [1] model

$$\dot{x} = \sigma(y - x),$$
$$\dot{y} = -y - xz + Rx, \qquad (2.10)$$
$$\dot{z} = xy - bz,$$

with $R = 28$, $\sigma = 10$, and $b = 8/3$, and the Rabinovich–Fabrikant [20] equations

$$\dot{x} = y(z - 1 + x^2) + \gamma x,$$
$$\dot{y} = x(3z + 1 - x^2) + \gamma y, \qquad (2.11)$$
$$\dot{z} = -2z(\alpha + xy),$$

with $\gamma = 0.87$ and $\alpha = 1.1$.

As seen in fig. 7 we get adequate power laws for $C(l)$, and in the case of the Lorenz model, where $D$ is known [11], we obtain $v \simeq D$.

Further examples will be studied in section 6, in the context of higher dimensional systems.

It should be stressed that the algorithm used to calculate $v$ converged quite rapidly. Although each entry in table I and figs. 2–7 were based on $\approx 15.000$–$25.000$ points each, reasonable results (i.e. results for $v$ within $\pm 5\%$) were obtained in most cases already with only a few thousand

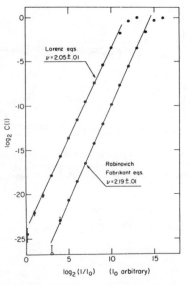

Fig. 7. Correlation integrals for the Lorenz equations (eq. (2.10); dots) and for the Rabinovich–Fabrikant equation (eq. (2.11); open circles). In both cases, $N = 15.000$ and $\tau = 0.25$.

points. This should be contrasted with the difficulties associated with estimating $D$ in box-counting algorithms [11, 14].

Summarizing this section, we can say that except for the logistic map at $a = a_\infty$ ("Feigenbaum attractor") we found in all cases that $v \approx D$ within the limits of accuracy. We now turn to a theoretical analysis of the relations between $v$, $\sigma$ and $D$.

## 3. Relations between $v$, $\sigma$ and $D$

In this section we shall establish the inequalities (1.9). We shall do this in 3 steps.

a) The easiest inequality to prove is $\sigma \leqslant D$. Consider a covering of the attractor by hypercubes ("cells") of edge length $l$, and a time series $\{X_k; k = 1, \ldots, N\}$. The probabilities $p_i$ for an arbitrary $X_k$ to fall into cell $i$ are simply

$$p_i = \lim_{N \to \infty} \frac{1}{N} \mu_i. \qquad (3.1)$$

where $\mu_i$ is the number of points $X_k$ which fall into cell $i$.

If the coverage of the attractor is uniform, one has,

$$p_i = \frac{1}{M(l)}, \qquad (3.2)$$

where $M(l)$ is the number of cells needed to cover the attractor, and one finds from eqs. (1.3) and (1.2)

$$S(l) = S^{(0)}(l) = \ln M(l) = \text{const} - D \ln l. \qquad (3.3)$$

In the general case, one uses the convexity of $x \ln x$ in the usual way to prove that $S(l) \leqslant S^{(0)}(l)$. Invoking the ansatz $S(l) = \text{const} - \sigma \ln l$, we find $\sigma \leqslant D$.

b) Instead of showing immediately $v \leqslant \sigma$, let us proceed slowly and show first that $v \leqslant D$.

From the definition of $C(l)$, we get up to a factor

of order unity

$$C(l) \simeq \lim_{N \to \infty} \frac{1}{N^2} \sum_{i=1}^{M(l)} \mu_i^2 = \sum_{i=1}^{M(l)} p_i^2 . \qquad (3.4)$$

Here, we have replaced the number of pairs with distance $< l$ by the number of pairs which fall into the same cell of length $l$. The error committed should be independent on $l$, and thus should not affect the estimation of $\nu$. Using the Schwartz inequality we get

$$C(l) = M(l)\langle p_i^2 \rangle \geqslant M(l)\langle p_i \rangle^2 = \frac{1}{M(l)} \sim l^D . \qquad (3.5)$$

In this equation square brackets denote average over all cells. Comparing eqs. (3.5) and (1.8) we find immediately $\nu \leqslant D$.

c) In order to derive $\nu \leqslant \sigma$, consider two nested coverings with cubes of lengths $l$ and $2l$. The numbers of cubes that contain a piece of the attractor are then related by

$$M(l) = 2^D M(2l) . \qquad (3.6)$$

Denote by $p_i$ the probability to fall in cube $i$ of the finer coverage, and by $P_j$ the probability to fall in cube $j$ of the coarser. Define $\omega_i (i = 1, \ldots M(l))$ by

$$p_i = \omega_i P_j \quad (i \in j) . \qquad (3.7)$$

Evidently we have

$$P_j = \sum_{i \in j} p_i, \quad \sum_{i \in j} \omega_i = 1 . \qquad (3.8)$$

We can then write the correlation integral as

$$C(l) \simeq \sum_{i=1}^{M(l)} p_i^2 = \sum_{j=1}^{M(2l)} P_j^2 \sum_{i \in j} \omega_i^2 . \qquad (3.9)$$

Consider now the ratio

$$\frac{C(l)}{C(2l)} = \frac{\sum_j P_j^2 \sum_{i \in j} \omega_i^2}{\sum_j P_j^2} , \qquad (3.10)$$

and compare it to the entropy difference

$$\begin{aligned} S(2l) - S(l) &= \sum_{i=1}^{M(l)} p_i \ln p_i - \sum_{j=1}^{M(2l)} P_j \ln P_j \\ &= \sum_{j=1}^{M(2l)} P_j \sum_{i \in j} \omega_i \ln \omega_i . \end{aligned} \qquad (3.11)$$

In order to estimate eq. (3.10) in terms of eq. (3.11), we have to introduce a new assumption. We assume that the $\omega_i$'s are distributed independently of the $P_j$. This means essentially that locally the attractor looks the same in regions where it is rather dense ($P_j$ large) as in regions where $P_j$ is small. Although we cannot further justify this assumption, it seems to us very natural. It leads immediately to

$$\frac{C(l)}{C(2l)} = \frac{\langle \omega^2 \rangle}{\langle \omega \rangle} = 2^D \langle \omega^2 \rangle , \qquad (3.12)$$

and to

$$S(2l) - S(l) = 2^D \langle \omega \ln \omega \rangle . \qquad (3.13)$$

Define now a normalized variable $W$ by

$$W = \frac{\omega}{\langle \omega \rangle} = 2^D \omega . \qquad (3.14)$$

Using the inequality [21]

$$\langle W^2 \rangle > \exp \langle W \ln W \rangle , \qquad (3.15)$$

we establish

$$\frac{C(l)}{C(2l)} \geqslant \exp[S(2l) - S(l)] \qquad (3.16)$$

and thus

$$\nu \leqslant \sigma . \qquad (3.17)$$

Remarks. From the proofs it is clear that if the attractor is uniformly covered, one has equalities

$$\nu = \sigma = D . \qquad (3.18)$$

It is an interesting question how non-uniform the coverage must be in order to break them. With the exception of the Feigenbaum map (logistic map with $a = a_\infty$), which is however not generic, all examples of the last section were compatible with eq. (3.18).

In cases where $v \neq D$, we claim that indeed $v$ is the more relevant observable. In these cases, the neighbourhoods of certain points have higher "seniority" in the sense that they are visited more often than others. The fractal dimension is ignorant of seniority, being a purely geometric concept. But both the correlation integral and the entropy dimension weight regions according to their seniority.

Eqs. (1.9) and (3.18) have been used previously in the context of fully developed homogeneous turbulence [22]. The connection

$$c(l) \propto l^{D-F}, \quad l \in R^f$$

following from $v = D$ has been used previously also in percolation theory [23] and in a model for dendritic growth [24].

## 4. Information entropy and $v$ of the Feigenbaum attractor

In this section we shall compute exactly the information dimension and $v$ of one-dimensional maps

$$x_{n+1} = F(x_n) \tag{4.1}$$

at the onset of chaos. The method follows closely the one of ref. 13.

It is well known that such maps – provided they have a unique quadratic maximum – have universal scaling features, studied in most detail by Feigenbaum [16]. This behaviour is most easily described by observing that the iterations

$$F^{(2^n)}(x) = \underbrace{F(F(\dots F(x) \dots))}_{2^n \text{ times}} \tag{4.2}$$

tend after a suitable rescaling towards a universal function

$$g(x) = \lim_{n \to \infty} \frac{1}{F^{(2^n)}(0)} F^{(2^n)}(xF^{(2^n)}(0)). \tag{4.3}$$

This "Feigenbaum function" $g(x)$ satisfies the exact scaling relation

$$g(g(x)) = -\frac{1}{\alpha} g(\alpha x), \tag{4.4}$$

with $\alpha = 2.50290\dots$, and the normalization condition $g(0) = 1$. We have here assumed that the maximum of $F(x)$ is at $x = 0$, which can always be achieved by a change of variables. In order to obtain the information dimension of the logistic map at $a = a_\infty = 3.5699345\dots$, it is thus sufficient to compute $\sigma$ for the Feigenbaum map.

The "attractor" (see the reservations in section 2) of $g(x)$ consists of the sequence $\{\xi_n, n = 0, 1, 2, \dots\}$ with

$$\xi_0 = 0 \tag{4.5}$$

and

$$\xi_{n+1} = g(\xi_n). \tag{4.6}$$

The first few $\xi_k$'s are shown in fig. 8. There, it is also indicated how they build up the Cantorian structure of the attractor: the points $\xi_1$, $\xi_2$, $\xi_3, \dots, \xi_{2^k+1}$ form the end-points of $2^k$ intervals, and the following $\xi_k$'s fall all into these intervals. Furthermore, any sequence $\{\xi_n, \xi_{n+1} \dots \xi_{n+2^k-1}\}$ of $2^k$ successive points visits each of these intervals exactly once. Thus, the a priori probabilities $p_i (i = 1, \dots, 2^k)$ for an arbitrary $x_n$ to fall into the $i$th interval are all equal to $p_i = 2^{-k}$.

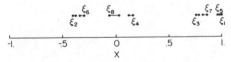

Fig. 8. First 16 points $\xi_1, \dots, \xi_{16}$ of the attractor of the Feigenbaum equation describing the onset of chaos in 1-dimensional systems.

By the grouping axiom, we can first write the information entropy as

$$S(l) = \frac{1}{2}[S_{[2,4]}(l) + S_{[3,1]}(l)] + \ln 2, \qquad (4.7)$$

where we denote by $S_{[i,j]}$ the information needed to specify the point on the interval $[\xi_i, \xi_j]$, and where we have used the fact that an arbitrary $x_n$ has equal probability to be on $[\xi_2, \xi_4]$ or on $[\xi_3, \xi_1]$. From eq. (4.4) we find, however, that

$$\xi_{2.} = -\frac{1}{\alpha}\xi_n. \qquad (4.8)$$

Thus, the interval $[\xi_2, \xi_4]$ is a down-scaled image of the whole attractor, and we have

$$S_{[2,4]}(l) = S(\alpha l) \approx S(l) - \sigma \ln \alpha, \qquad (4.9)$$

where we have used the scaling ansatz (1.4).

In order to estimate $S_{[3,1]}(l)$, we decompose the interval [3, 1] into the $2^{k-1}$ subintervals discussed above, defined by the $\xi_n$ with odd $n$'s:

$$S_{[3,1]}(l) = (k-1)\ln 2 + 2^{-k+1}\sum_{i=1}^{2^{k-1}} S_i(l). $$

Again, we have applied the grouping axiom, using that $p_i = 2^{-k}$. The $S_i(l)$ are the informations needed to pin down $x_n$ provided one knows that it falls into the $i$th subinterval. Since each subinterval maps onto one on the left-hand piece $[\xi_2, \xi_4]$, each $S_i(l)$ is equal to the information $\tilde{S}_i(|g_i'|l)$ needed to pin $x_{n+1}$ on the corresponding interval on the left-hand side. Here, $g_i'$ is some average derivative of $g(x)$ in the $i$th subinterval. Using that $\tilde{S}_i(|g_i'|l) \simeq \tilde{S}_i(l) - \sigma \ln|g_i'|$, we obtain

$$S_{[3,1]}(l) = (k-1)\ln 2 + 2^{-k+1}\sum_{i=1}^{2^{k-1}} \tilde{S}_i(l)$$

$$- \sigma \sum_{i=1}^{2^{k-1}} \ln|g_i'|$$

$$= S_{[2,4]}(l) - \sigma \sum_{i=1}^{2^{k-1}} \ln|g_i'|. \qquad (4.10)$$

Inserting this and eq. (4.9) into eq. (4.7), we find

after a few manipulations and after taking the limit $k \to \infty$

$$\sigma = \lim_{k\to\infty} \frac{\ln 2}{\ln\alpha + \frac{1}{2^{k+1}}\sum_{i=1}^{2^k} |g'(\xi_{2i-1})|}. \qquad (4.11)$$

The limit converges very quickly, leading (for $k > 7$) to

$$\sigma = 0.5170976. \qquad (4.12)$$

The calculation of the correlation exponent, or rather of the exponent of the Renyi entropy (see eq. (3.4))

$$R(l) = \sum_{i=1}^{M(l)} p_i^2 \qquad (4.13)$$

follows even more closely the one in ref. 13.

As in that paper, we obtain a nested set of bounds. The first (and least stringent) is obtained by writing

$$R(l) = \frac{1}{4}\{R_{[2,4]}(l) + R_{[3,1]}(l)\} \qquad (4.14)$$

and using $R_{[2,4]}(l) = R(\alpha.l)$ and $R_{[3,1]}(l) = R(\alpha g' l)$ with

$$|g'(\xi_3)| < g' < |g'(\xi_1)|. \qquad (4.15)$$

Assuming $R(l) \sim l^\nu$, we obtain

$$1 + |g'(\xi_3)|^\nu < \frac{4}{\alpha^\nu} < 1 + |g'(\xi_1)|^\nu, \qquad (4.16)$$

leading to $0.4857 < \nu < 0.5235$.

For the next more stringent bounds, we write further

$$R_{[3,1]}(l) = \frac{1}{4}\{R_{[3,7]}(l) + R_{[5,1]}(l)\}, \qquad (4.17)$$

with

$$R_{[3,7]}(l) + R(\alpha^2 g^{(1)}.l), \quad |g'(\xi_3)| < g^{(1)} < |g'(\xi_7)| \qquad (4.18)$$

and

$$R_{[5,1]}(l) + R_{[3,1]}(\alpha g^{(2)} l), \quad |g'(\xi_5)| < g^{(2)} < |g'(\xi_1)|. \tag{4.19}$$

Some algebra leads then to

$$|g'(\xi_5)|^\nu + \frac{\alpha^\nu}{4 - \alpha^\nu} |g'(\xi_3)|^\nu < \frac{4}{\alpha^\nu}$$

$$< |g'(\xi_1)|^\nu + \frac{\alpha^\nu}{4 - \alpha^\nu} |g'(\xi_7)|^\nu, \tag{4.20}$$

with the result

$$0.4926 < \nu < 0.5024, \tag{4.21}$$

in agreement with the numerical value $\nu = 0.500 \pm 0.005$.

## 5. Using a single-variable time series

Very often one does not have access to a time series $\{X_n\}$ of $F$-dimensional vectors. Instead one follows only one or at most a few components of $X_n$. This is particularly relevant for real (as opposed to computer) experiments where the number of degrees of freedom often is very high if not infinite. Such systems nevertheless can have low-dimensional attractors. It would be very desirable to have a reliable method which allows a characterization of this attractor from a single-variable time series. $\{x_i, i = 1, \ldots, N; x_i \in R\}$.

The essential idea [25, 26] consists in constructing d-dimensional vectors

$$\xi_i = (x_i, x_{i+1}, \ldots, x_{i+d-1}) \tag{5.1}$$

and using $\xi$-space instead of $X$-space. The correlation integral would e.g. be

$$C(l) = \lim_{N \to \infty} \frac{1}{N^2} \sum_{i,j=1}^N \theta(l - |\xi_i - \xi_j|). \tag{5.2}$$

More generally, one can use

$$\xi_i = (x(t_i), x(t_i + \tau) \ldots x(t_i + (d-1)\tau)), \tag{5.3}$$

with $\tau$ some fixed interval. The magnitude of $\tau$ should not be chosen too small since otherwise $x_i \approx x_{i+\tau} \approx x_{i+2\tau} \approx \cdots$ so that the attractor in $\xi$-space would be stretched along the diagonal and thus difficult to disentangle. On the other hand, $\tau$ should not be chosen too large since distant values in the time series are not strongly correlated (due to the exponential divergence of trajectories and unavoidable small errors).

A similar compromise must be chosen for the dimension $d$. Clearly, $d$ must be larger than the Hausdorff dimension $D$ of the attractor (otherwise, $C(l) \sim l^d$). If the attractor is Cantorian in more than one dimension, this might however not be sufficient. Also, it might be that, when looked at in $d$ dimension, the density

$$p(\xi) = \lim_{N \to \infty} \frac{1}{N} \sum_{i=1}^N \delta(\xi_i - \xi) \tag{5.4}$$

develops singularities which are absent in more than $d$ dimensions (such singularities occur e.g. when one projects a sphere with constant density, $p(\xi) = p\delta(x^2 + y^2 + z^2 - R^2)$, onto the $x$–$y$ plane: the new density $\tilde{p}(x, y)$ is infinite at $x^2 + y^2 = R^2$).

On the other hand, one cannot make $d$ too large without getting lost in experimental errors and lack of statistics.

In the next section, we shall study an infinite-dimensional system from this point of view. In the remainder of the present section, we shall apply these considerations to the logistic map with $a = 4$, and to the Hénon map.

In the logistic map, we have seen that there are logarithmic corrections to the power law $C(l) \sim l^\nu$. They result precisely from singularities of $p(x)$, at $x = 0$ and $x = 1$. While embedding the attractor in a higher dimensional space does not completely remove these singularities, it substantially reduces their influence. The reason is that embedding in higher dimensional space always results in stretching the attractor. However, the portions which are most strongly stretched are those which are most densely populated at the lower dimension. For example in the logistic map with $a = 4$ the "attractor" is the interval [0, 1] in 1$d$ but is the parabola in 2$d$. The parabola has highest slopes at the end points, exhibiting the stronger stretching associated with regions of singular distributions at a lower dimension. A similar effect appears when going from $d = 2$ to $d = 3$. We thus expect that the importance of the singularities in the distribution would be reduced in higher dimensions.

In order to check this, we have calculated for the logistic map at $a = 4$ the original correlation integral and the modified integral obtained by embedding in a 2- and 3-dimensional space. The results are shown in fig. 9. We observe indeed the expected decrease of systematic error when increasing $d$, accompanied by an increase of the statistical error.

Analogous results for the Hénon map are shown in fig. 10. There, we used as time series the series $\{x_n, x_{n+2}, x_{n+4}, \ldots\}$. While the 2-dimensional correlation integral gives an effective $\nu$ in agreement with the result of section 2, the 3-dimensional embedding gives a larger values $\nu = 1.25 \pm 0.02$ which agrees with the value of $D$ found in refs. 11 and 14.

No such effects were observed in the Lorenz model, where both the originally defined $C(l)$ and

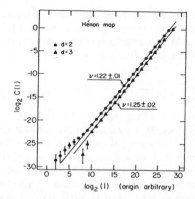

Fig. 10. Modified correlation integrals for the Hénon map (2.6). The time series consisted of coordinates $x_n$, $x_{n+2}$, $x_{n+4} \ldots$, and $\zeta = (x_n, x_{n+2}, \ldots, x_{n+2(d-1)})$ for each $d$. For $d = 2$, we took $N = 30.000$; for $d = 3$, we took $N = 20.000$.

the modified correlation integral using only a single coordinate time series gave values of $\nu$ which agreed with $D$ [15].

The conclusion drawn from these examples is that it is often useful to represent the attractor in a higher dimensional space than absolutely necessary, in order to reduce systematic errors. These errors result from a strongly non-uniform coverage of the attractor, provided this non-uniformity is not so strong as to make $\nu \neq D$.

## 6. Infinite-dimensional systems: an example

An extremely convenient way of generating very high dimensional systems is to consider delay differential equations of the type

$$\frac{dx(t)}{dt} = F(x(t), x(t - \tau)), \tag{6.1}$$

where $\tau$ is a given time delay. Such a delay equation is in fact infinite dimensional, as is most easily seen from the initial conditions necessary to solve eq. (6.1): they consist of the function $x(t)$ over a whole interval of length $\tau$.

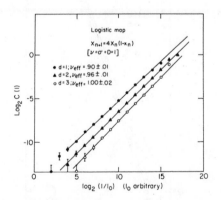

Fig. 9. Modified correlation integrals for the logistic map (2.1) with $a = 4$. The distance $l$ between 2 points $\zeta_n$ and $\zeta_m$ on the attractor is defined as $l^2 = (\zeta_n - \zeta_m)^2 = (x_n - x_m)^2 + \ldots + (x_{n+d-1} - x_{m+d-1})^2$. For each value of $d$, we took $N = 15.000$.

Following ref. 8, we shall study a particular example, introduced by Mackey and Glass [17] as a model for regeneration of blood cells in patients with leukemia. It is

$$\dot{x}(t) = \frac{ax(t-\tau)}{1 + [x(t-\tau)]^{10}} - bx(t).\qquad(6.2)$$

As in ref. 8, we shall keep $a = 0.2$ and $b = 0.1$ fixed, and study the dependence on the delay time $\tau$.

For the numerical investigation, eq. (6.2) is turned into an $n$-dimensional set of difference equations, with $n = 600$–$1200$. Details are described in the appendix. The time series was always chosen as $\{x(t),\ x(t+\tau),\ x(t+2\tau),\dots\}$ except for some runs with $\tau = 100$, where we took points at times $t,\ t+\tau/2,\ t+2\tau/2,\dots$.

The results for the correlation integral are shown in figs. 11–14. Estimated values of $\nu$ are given in table II, together with values of $D$ obtained in ref. 8 by applying the defining eq. (1.2) to a Poincaré return map. Also shown in table II are the

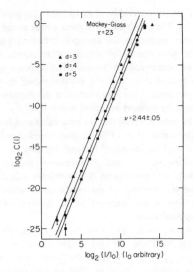

Fig. 12. Same as fig. 11, but for $\tau = 23$.

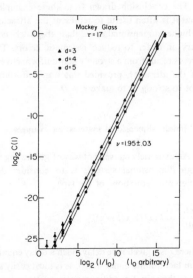

Fig. 11. Modified correlation integrals for the Mackey–Glass delay equation (6.2), with delay $\tau = 17$. The time series consisted of $\{X(t + i\tau);\ i = 1,\dots, 25.000\}$.

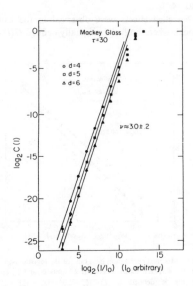

Fig. 13. Same as fig. 11, but for $\tau = 30$.

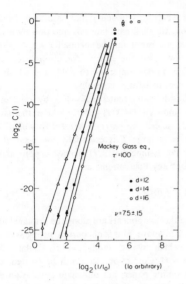

Fig. 14. Same as fig. 11, but for $\tau = 100$. For $d = 16$, the time series consisted of points $\{X(t + i\tau/2); i = 1, \ldots, 25.000\}$.

Kaplan–Yorke dimension $D_{KY}$ (see eq. (1.10)) which will in the next section be shown to be an upper bound to $\nu$, and the number of positive Lyapunov exponents, both taken from ref. 8. It is obvious that this latter number, called $D_{LB}$, is a lower bound to $D$. If the density of trajectories on the attractor is not too non-uniform, we expect that $D_{LB}$ yields also a lower bound to $\nu$.

From table II we see that indeed in all cases

$$D_{LB} \leqslant \nu \leqslant D \leqslant D_{KY}, \qquad (6.3)$$

except for $\tau = 17$ where $\nu$ is slightly less than $D_{LB}$. However, for those small values of $\tau$ for which box-counting according to the definition of $D$ had been feasible, our values of $\nu$ are considerably smaller than the values of $D$ found in ref. 8, while the values of $D$ were fairly close to $D_{KY}$.

In all cases, the linearity of the plot of $\log C(l)$ versus $\log l$ improved substantially when increasing $d$ above its minimal required value. For increasing values of $d$, the effective exponent at first also

Table II
Estimates of the correlation exponent $\nu$ for the Mackey–Glass equation (6.2) with $a = 0.2$, $b = 0.1$. Values for $D_{LB}$, $D$ and $D_{KY}$ are from ref. 8. For $\tau = 100$ the value of $\nu$ saturated at $d = 16$

| $\tau$ | $D_{LB}$ | $\nu$ | $D$ | $D_{KY}$ |
|---|---|---|---|---|
| 17.0 | 2 | $1.95 \pm 0.03$ ($d = 3$)<br>$1.35 \pm 0.03$ ($d = 4$)<br>$1.95 \pm 0.03$ ($d = 5$) | $2.13 \pm 0.03$ | $2.10 \pm 0.02$ |
| 23.0 | 2 | $2.38 \pm 0.15$ ($d = 3$)<br>$2.43 \pm 0.05$ ($d = 4$)<br>$2.44 \pm 0.05$ ($d = 5$)<br>$2.42 \pm 0.1$ ($d = 6$) | $2.76 \pm 0.06$ | $2.92 \pm 0.03$ |
| 30.0 | 3 | $2.87 \pm 0.3$ ($d = 4$)<br>$3.0 \pm 0.2$ ($d = 5$)<br>$3.0 \pm 0.2$ ($d = 6$)<br>$2.8 \pm 0.3$ ($d = 7$) | $> 2.94$ | $3.58 \pm 0.04$ |
| 100.0 | 6 | $5.8 \pm 0.3$ ($d = 10$)<br>$6.6 \pm 0.2$ ($d = 12$)<br>$7.2 \pm 0.2$ ($d = 14$)<br>$7.5 \pm 0.15$ ($d = 16$) | - | $\approx 10.0$ |

increases, but settles at a value which we assume to be the true value of $v$. We must stress that we have no *proof* that the values of $v$ obtained with the highest chosen $d$ represent the "true" exponent. We feel however that they surely represent reasonable estimates even for attractors with dimensions as high as $\approx 7$.

In real experiments, where Lyapunov exponents are not available and thus $D_{LB}$ and $D_{KY}$ not easily obtained, our method seems the only one which could distinguish such an attractor from a system where the stochasticity is due to random noise. In that case, one would expect $C(l) \sim l^d$ as the trajectory is space-filling, in clear distinction from what we observe.

## 7. Relation to Lyapunov exponents and the Kaplan–Yorke conjecture

As we already mentioned in the introduction, the Lyapunov exponents are related to the evolution of the shape of an infinitesimal $F$-dimensional ball in phase space: being infinitesimal, it depends only on the linearized part of the flow, and thus becomes an ellipsoid with exponentially shrinking or growing axes. Denoting the principal axes by $\epsilon_i(t)$, the Lyapunov exponents are given by

$$\lambda_i = \lim_{t \to \infty} \lim_{\epsilon_i(0) \to 0} \frac{1}{t} \ln \frac{\epsilon_i(t)}{\epsilon_i(0)}. \tag{7.1}$$

Directions associated with positive Lyapunov exponents are called "unstable", those associated with negative exponents are called "stable".

Originally [10], Kaplan and Yorke had conjectured that $D_{KY}$ is equal to $D$. In a recent preprint [27], they claim that $D_{KY}$ is generically equal to a "probabilistic dimension", which seems to be the same as $\sigma$.

This latter claim has been partially supported in ref. 28, where essentially $D_{KY}$ is proven to be an upper bound to the probabilistic dimension.

As shown by the counter example mentioned in the introduction, there are (possibly exceptional)

cases where this bound is not saturated. In this section, we shall elucidate this question by giving a heuristic proof for the inequality $v \leqslant D$. From this, we see necessary conditions for the Kaplan–Yorke conjecture to hold, and which do not seem to be met generally.

Consider two infinitesimally close-by trajectories $X(t)$ and $X'(t) = X(t) + \Delta(t)$, where the latter could indeed be $X'(t) = X(t + T)$, which for sufficiently large $T$ is essentially independent of $X(t)$. We assume that $\Delta_i(t)$ increase exponentially, without any fluctuations, as

$$\Delta_i(t) = \Delta_i(0)\, e^{\lambda_i t}, \tag{7.2}$$

where the components are along the principal axes discussed above. This is of course a strong assumption which would imply, in particular, that $v = \sigma = D$. Corrections to it will be treated in a forthcoming paper, but our main conclusion will remain unchanged. Conservation of the number of trajectories implies that the correlation function increases like

$$c(\Delta(t)) = \left| \frac{\partial(\Delta(0))}{\partial(\Delta(t))} \right| c(\Delta(0)) = e^{-t\sum_{f-1}^{f} \lambda_i} c(\Delta(0)). \tag{7.3}$$

To proceed further, we need a scaling assumption which generalizes the scaling ansatz

$$c(|\Delta|) \sim |\Delta|^{v-F}. \tag{7.4}$$

Observing that the attractor is locally a topological product of an $R^n$ with Cantor sets, and that the relevant axes are the principal axes, we associate with each axis an exponent $v_i$, $0 < v_i < 1$, and make the ansatz

$$c(\Delta) \approx \prod_{i=1}^{F} c_i(\Delta_i), \tag{7.5}$$

with

$$c_i(x) \propto \begin{cases} x^{v_i - 1}, & \text{if } 0 < v_i \leqslant 1, \\ \delta(x), & \text{if } v_i = 0. \end{cases} \tag{7.6}$$

If $v_i = 0$, this means that the motion along this axis dies asymptotically (example: directions normal to a limit cycle). Directions with $v_i = 1$ are the unstable directions, with the continuous density. Directions with $0 < v_i < 1$, finally, are either Cantorian or, in exceptional cases, directions along which the distribution is continuous but singular at $\Delta_i = 0$. Notice that $v_i > 1$ is impossible.

Substituting eq. (7.5) into (7.3), we find

$$\prod_i (\Delta_i(0)^{v_i - 1} e^{t\lambda_i(v_i - 1)}) = e^{-\alpha \Sigma_i \lambda_i} \prod_i \Delta_i(0)^{v_i - 1}, \quad (7.7)$$

or

$$\sum_{i=1}^{F} \lambda_i v_i = 0. \quad (7.8)$$

In addition we have, from eqs. (7.5) and (7.4),

$$\sum_{i=1}^{F} v_i = v, \quad (7.9)$$

and

$$0 \leqslant v_i \leqslant 1. \quad (7.10)$$

It is now easy to find the maximum of $v$ subject to the constraints (7.8)–(7.10). It is obtained when

$$v_i = \begin{cases} 1, & \text{for } i \leqslant j, \\ 0, & \text{for } i \geqslant j + 2, \end{cases} \quad (7.11a)$$

and

$$v_{j+1} = \frac{1}{|\lambda_{j+1}|} \sum_{i < j} \lambda_i. \quad (7.11b)$$

Here, we have used that $\lambda_1 \geqslant \lambda_2 \geqslant \ldots$, and that $\Sigma_j \lambda_i < 0$. Expressed in words, the distribution (7.11) means that the attractor is the most extended along the most unstable directions. Inserting it into eq. (7.9), we obtain

$$v \leqslant j + \frac{\sum_{i \leqslant j} \lambda_i}{|\lambda_{j+1}|} \equiv D_{KY}. \quad (7.12)$$

as we had claimed.

From the derivation it is clear that the Kaplan–Yorke conjectures $\sigma = D_{KY}$ or $D = D_{KY}$ cannot be expected to hold when *either the attractor is Cantorian in more than one dimension, or if the folding occurs in a direction which is not the minimally contracting one*. The latter was indeed the case for example b in fig. 1. But example b of fig. 1 is not generic, the generic case being the one where the folding is in a plane which encloses an arbitrary angle $\phi$ with the $z$-axis (see fig. 15). It is easy to convince oneself that $D = D_{KY}$ whenever $\phi \neq 0$, i.e. nearly always. A still more general case is obtained if we fold in each $(2n)$th iteration in a plane characterized by $\phi_1$, and each $(2n + 1)$st iteration in a different plane. Again, it seems that $D = D_{KY}$ is generic.

The examples might suggest that indeed $D = D_{KY}$ in all those generic cases in which $v = D$, but we consider it as not very likely in high-dimensional cases. For invertible two-dimensional maps, the above conditions are of course satisfied, and thus $\sigma = D_{KY}$ if $v = \sigma = D$ (see ref. 29).

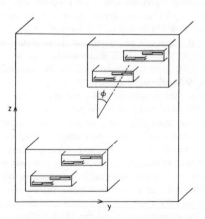

Fig. 15. Cross section through a rectangular volume element and its first 4 iterations under a map which stretches in $x$-direction, contracts in $y$- and $z$-directions (factors $\frac{1}{2}$ and $\frac{1}{4}$, respectively), and folds back under an angle $\phi$ with respect to the $z$-direction.

370

## 8. Conclusions

The theoretical arguments of section 3 and 6 of this paper have shown (though not with mathematical rigour) that the correlation exponent $v$ introduced in this paper is closely related to other quantities measuring the local structure of strange attractors.

The numerical results presented in section 2, 5 and 7 have yielded proof that $v$ can indeed be calculated with reasonable efforts. While all results presented in this paper were based on time series of 10.000–30.000 points, reasonable estimates of $v$ can already be obtained with series of a few thousand points, in most cases. Surely, for higher dimensional attractors one needs longer time series. However, rather than taking longer time series, we found it in general more important to embed the attractor in higher dimensional spaces, and to choose this embedding dimension judiciously. Compared to box-counting algorithms used previously by other authors, our method has two advantages: First, our storage requirements are drastically reduced. Secondly, in a box-counting algorithm one should iterate until *all* non-empty boxes of a given size $l$ have been visited. This is clearly impractical, in particular if $l$ is very small. Thus, one has systematic errors even if the number of iterations $N$ is excessively large. In our method, there is no such problem. In particular, the finiteness of $N$ induces no systematic errors beyond the corrections to the scaling law $C(l) \sim l^v$.

We found that in most cases $v$ was very close to the Hausdorff dimensions $D$ and to the information dimension $\sigma$, with two notable exceptions. One was the Feigenbaum map, corresponding to the onset of chaos in 1 dimension. In that case, we were able to compute $\sigma$ exactly in an analytic way, with the result $\sigma \neq D$, supporting the numerical evidence for $v < \sigma$.

The other exception was the Mackey–Glass delay equation, where we found numerically $v < D$. The information dimension has not been calculated directly in this case. Accepting the claim made in ref. 8 that the Kaplan–Yorke formula

(1.10) predicts correctly $\sigma$, we would have $v < \sigma = D = D_{KY}$. This seems somewhat surprising, since we argued in section 7 that a rather direct connection (as an inequality $v \leqslant D_{KY}$) exists between $v$ and $D_{KY}$, while a connection between $\sigma$ and $D_{KY}$ seems less evident to us.

The main conclusion of this paper, as far as experiments are concerned, is that one can distinguish deterministic chaos from random noise. By analyzing the signal as explained in section 5, and embedding the attractor in an increasingly high dimensional space, one finds whether $C(l)$ scales like $l^v$ or $l^d$. With a random noise the slope of $\log C(l)$ vs. $\log l$ will increase indefinitely as $d$ is increased. For a signal that comes from a strange attractor the slope will reach a value of $v$ and will then become $d$ independent.

An issue of experimental importance is the effect of random noise *on top* of the deterministic chaos. The treatment of this question is beyond the scope of this paper and is treated elsewhere [30]. Here we just remark that when there is an external noise of a given mean square magnitude, a plot of $\log C(l)$ vs. $\log l$ has two regions. For length scales above those on which the random component blurs the fractal structure, $C(l)$ continues to scale like $l^v$. On length scales below those that are affected by the random jitter of the trajectory, $C(l)$ scales like $l^d$. The analysis of experimental signals along these lines can therefore yield simultaneously a characterization of the strange attractor *and* and estimate of the size of the random component. For more details see ref. 30.

It is thus our hope that the correlation exponent will indeed be measured in experiments whose dynamics is governed by strange attractors.

### Acknowledgements

This work has been supported in part by the Israel Comission for Basic Research. P.G. thanks the Minerva Foundation for financial support. We thanks Drs. H.G.E. Hentschel and R.M. Mazo for a number of useful discussions.

## Appendix A

All numerical calculations were performed in double precision arithmetic on an IBM 370/165 at the Weizmann Institute.

The integrations of the Lorenz and Rabinovich–Fabrikant equations were done using a standard Merson–Runge–Kutta subroutine of the NAG library.

In order to integrate the Mackey–Glass delay equation we approximated it by a $N$-dimensional set of difference equations by introducing a time step

$$\Delta t = \tau/n, \tag{A.1}$$

with $n$ being some large integer, and writing

$$x(t + \Delta t) \approx x(t) + \frac{\Delta t}{2}(\dot{x}(t) + \dot{x}(t + \Delta t)). \tag{A.2}$$

Notice that this, being the optimal second-order approximation, is a very efficient algorithm – provided we can compute $\dot{x}(t + \Delta t)$. In the present case we can, due to the special form

$$\dot{x}(t) = f(x(t - \tau)) - bx(t). \tag{A.3}$$

Inserting this in eq. (A.2) and rearranging terms, we arrive at

$$x(t + \Delta t) = \frac{2 - b\Delta t}{2 + b\Delta t} x(t) + \frac{\Delta t}{2 + b\Delta t}$$
$$\times \{f(x(t - \tau)) + f(x(t - \tau + \Delta t))\}. \tag{A.4}$$

In all runs shown in this paper, we used $n = 600$ (corresponding to $0.03 \lesssim \Delta t \lesssim 0.15$), except for the runs with $\tau = 100$, where we used $n = 1200$ and with $n = 600$, finding no appreciable differences.

We also performed control runs with a fourth-order approximation instead of eq. (A.2). The correlation integral was unchanged within statistical errors, and the stability of the solutions did not seem to improve much. This could result from the very large higher derivatives of $x$, resulting from the tenth power in eq. (6.2).

In order to ensure that all $x_i$ are on the attractor, the first 100–200 iterations were discarded.

Generating the time series $\{X_i\}_{i=1}^N$ was indeed the less time-consuming part of our computation, the more important part consisting of calculating the $N(N-1)/2 \gtrsim 10^8$ pairs of distances $r_{ij} = |X_i - X_j|$ and summing them up to get the correlation integral.

In particular, we found that an efficient algorithm for the latter was instrumental in applying the method advocated in this paper.

Such a fast algorithm was found using the fact that floating-point numbers are stored in a computer in the form

$$r = \pm \text{ mantissa} \cdot \text{base}^{+\text{exp}}. \tag{A.5}$$

with base $= 16$ in our case $1/\text{base} < \text{mantissa} < 1$, and exp being an integer. If one can extract the exponent, one can bin the $r_{ij}$'s in bins of widths increasing geometrically. By extracting the exponent of an arbitrary power $r^p$ of $r$, one can furthermore choose the width of this binning arbitrarily. Access to the exponent is made very easy and fast by using the shifting and masking operations available e.g. in extended IBM and in CDC Fortran. After having computed the numbers $N_K$ of pairs $(i, j)$ in the interval $2^{k-1} < r_{ij} < 2^k$, the correlation integrals are obtained by

$$c(r = 2^k) = \frac{1}{N^2} \sum_{k'=-\infty}^{k} N_{k'}. \tag{A.6}$$

We found this method to be nearly an order of magnitude faster than computing e.g. the logarithmics of $r_{ij}$ directly, and binning by taking their integer parts. A typical run with 20.000 points took – depending on the model studied – between 15 and 30 minutes CPU time.

## References

[1] E.N. Lorenz, J. Atmos. Sci. 20 (1963) 130.
[2] R.M. May, Nature 261 (1976) 459.
[3] D. Ruelle and F. Takens, Commun. Math. Phys. 20 (1971) 167.

[4] E. Ott, Rev. Mod. Phys. 53 (1981) 655.

[5] J. Guckenheimer, Nature 298 (1982) 358.

[6] B. Mandelbrot, *Fractals – Form, Chance and Dimension* (Freeman, San Francisco, 1977).

[7] V.I. Oseledec, Trans. Moscow Math. Soc. 19 (1968) 197. D. Ruelle, Proc. N.Y. Acad. Sci. 357 (1980) 1 (R.H.G. Helleman, ed.).

[8] J.D. Farmer, Physica 4D (1982) 366.

[9] H. Mori, Progr. Theor. Phys. 63 (1980) 1044.

[10] J.L. Kaplan and J.A. Yorke, in: Functional Differential Equations and Approximations of Fixed Points, H.-O. Peitgen and H.-O. Walther, eds. Lecture Notes in Math. 730 (Springer, Berlin, 1979) p. 204.

[11] D.A. Russel, J.D. Hanson and E. Ott, Phys. Rev. Lett. 45 (1980) 1175.

[12] H. Froehling, J.P. Crutchfield, D. Farmer, N.H. Packard and R. Shaw, Physica 3D (1981) 605.

[13] P. Grassberger, J. Stat. Phys. 26 (1981) 173.

[14] H.S. Greenside, A. Wolf, J. Swift and T. Pignataro, Phys. Rev. A25 (1982) 3453.

[15] P. Grassberger and I. Procaccia, Phys. Rev. Lett. 50 (1983) 346. Related discussions can be found in a preprint by F. Takens "Invariants Related to Dimensions and Entropy".

[16] M. Feigenbaum, J. Stat. Phys. 19 (1978) 25; 21 (1979) 669.

[17] M.C. Mackey and L. Glass, Science 197 (1977) 287.

[18] M. Hénon, Commun. Math. Phys. 50 (1976) 69.

[19] G.M. Zaslavskii, Phys. Lett. 69A (1978) 145.

[20] M.I. Rabinovich and A.L. Fabrikant, Sov. Phys. JETP 50 (1979) 311. (Zh. Exp. Theor. Fiz. 77 (1979) 617).

[21] W. Feller, An Introduction to Probability Theory and its Applications, vol. 2, 2nd ed. (Wiley, New York, 1971) p. 155.

[22] B.B. Mandelbrot, in: *Turbulence and the Navier–Stokes Equations*, R. Teman, ed., Lecture Notes in Math. 565 (Springer, Berlin, 1975). H.G.E. Hentschel and I. Procaccia, Phys. Rev. A., in press.

[23] D. Stauffer, Phys. Rep. 54C (1979) 1.

[24] T.A. Witten, Jr., and L.M. Sander, Phys. Rev. Lett. 47 (1981) 1400.

[25] N.H. Packard, J.P. Crutchfield, J.D. Farmer and R.S. Shaw, Phys. Rev. Lett. 45 (1980) 712.

[26] F. Takens, in: Proc. Warwick Symp. 1980, D. Rand and B.S. Young, eds, Lectures Notes in Math. 898 (Springer, Berlin, 1981).

[27] P. Frederickson, J.L. Kaplan, E.D. Yorke and J.A. Yorke, "The Lyapunov Dimension of Strange Attractors" (revised), to appear in J. Diff. Eq.

[28] F. Ledrappier, Commun. Math. Phys. 81 (1981) 229.

[29] L.S. Young, "Dimension, Entropy, and Lyapunov Exponents" preprint.

[30] A. Ben-Mizrachi, I. Procaccia and P. Grassberger, Phys. Rev. A, submitted.

# Ergodic theory of chaos and strange attractors

## J.-P. Eckmann

*Université de Genève, 1211 Genève 4, Switzerland*

## D. Ruelle

*Institut des Hautes Etudes Scientifiques, 91440 Bures-sur-Yvette, France*

Physical and numerical experiments show that deterministic noise, or chaos, is ubiquitous. While a good understanding of the onset of chaos has been achieved, using as a mathematical tool the geometric theory of differentiable dynamical systems, moderately excited chaotic systems require new tools, which are provided by the *ergodic* theory of dynamical systems. This theory has reached a stage where fruitful contact and exchange with physical experiments has become widespread. The present review is an account of the main mathematical ideas and their concrete implementation in analyzing experiments. The main subjects are the theory of *dimensions* (number of excited degrees of freedom), *entropy* (production of information), and *characteristic exponents* (describing sensitivity to initial conditions). The relations between these quantities, as well as their experimental determination, are discussed. The systematic investigation of these quantities provides us for the first time with a reasonable understanding of dynamical systems, excited well beyond the quasiperiodic regimes. This is another step towards understanding highly turbulent fluids.

## CONTENTS

---

*Sections marked with * contain supplementary material which can be omitted at first reading.

## I. INTRODUCTION

In recent years, the ideas of differentiable dynamics have considerably improved our understanding of irregular behavior of physical, chemical, and other natural phenomena. In particular, these ideas have helped us to understand the onset of turbulence in fluid mechanics. There is now ample experimental and theoretical evidence that the qualitative features of the time evolution of many physical systems are the same as those of the solution of a typical evolution equation of the form

$$\dot{x}(t) = F_\mu(x(t)), \quad x \in \mathbb{R}^m \tag{1.1}$$

in a space of small dimension $m$. Here, $x$ is a set of coordinates describing the system (typically, mode amplitudes, concentrations, etc.), and $F_\mu$ determines the nonlinear time evolution of these modes. The subscript $\mu$ corresponds to an experimental *control parameter*, which is kept constant in each run of the experiment. (Typically, $\mu$ is the intensity of the force driving the system.) We write

$$x(t) = f_\mu^t(x(0)) . \tag{1.2}$$

We usually *assume* that there is a parameter value, say $\mu = 0$, for which the equation is well understood and leads to a motion in phase space which, after some transients, settles down to be stationary or periodic.

As the parameter $\mu$ is varied, the nature of the asymptotic motion may change.[1] The values $\mu$ for which this change of asymptotic regime happens are called bifurcation points. As the parameter increases through successive bifurcations, the asymptotic motion of the system typically gets more complicated. For special sequences of these bifurcations a lot is known, and even quantitative features are predicted, as in the case of the period-doubling cascades ("Feigenbaum scenario"). We do not, however, possess a complete classification of the possible transitions to more complicated behavior, leading eventually to turbulence. *Geometrically,* the asymptotic motion follows an *attractor* in phase space, which will become more and more complicated as $\mu$ increases.

The aim of the present review is to describe the current state of the theory of *statistical* properties of dynamical systems. This theory becomes relevant as soon as the system is "excited" beyond the simplest bifurcations, so that precise geometrical information about the shape of the attractor or the motion on it is no longer available. See Eckmann (1981) for a review of the *geometrical* aspects of dynamical systems. The statistical theory is still capable of distinguishing different degrees of complexity of attractors and motions, and presents thus a further step in bridging the gap between simple systems and fully developed turbulence. In particular, the present treatment does not exclude the description of space-time patterns.

After introducing precise dynamical concepts in Sec. II, we address the theory of characteristic exponents in Sec. III and the theory of entropy and information dimension in Sec. IV. In Sec. V we discuss the extraction of dynamical quantities from experimental time series.

It is necessary at this point to clarify the role of the physical concept of *mode,* which appears naturally in simple theories (for instance, Hamiltonian theories with quadratic Hamiltonians), but which loses its importance in nonlinear dynamical systems. The usual idea is to represent a physical system by an appropriate change of variables as a collection of independent oscillators or

---

[1]It is to be understood that the experiment is performed with a *fixed* value of the parameter.

modes. Each mode is periodic, and its state is represented by an angular variable. The global system is *quasiperiodic* (i.e., a superposition of periodic motions). From this perspective, a dissipative system becomes more and more turbulent as the number of *excited modes* grows, that is, as the number of independent oscillators needed to describe the system progressively increases. This point of view is very widespread; it has been extremely useful in physics and can be formulated quite coherently (see, for example, Haken, 1983). However, this philosophy and the corresponding intuition about the use of Fourier modes have to be completely modified when nonlinearities are important: *even a finite-dimensional motion need not be quasiperiodic in general.* In particular, the concept of "number of excited modes" will have to be replaced by new concepts, such as "number of non-negative characteristic exponents" or "information dimension." These new concepts come from a statistical analysis of the motion and will be discussed in detail below.

In order to talk about a statistical theory, one needs to say what is being averaged and in which sample space the measurements are being made. The theory we are about to describe treats *time averages.* This implies and has the advantage that *transients* become irrelevant. (Of course, there may be formidable experimental problems if the transients become too long.) Once transients are over, the motion of the solution $x$ of Eq. (1.1) settles typically near a subset of $\mathbb{R}^m$, called an *attractor* (mathematical definitions will be given later). In particular, in the case of dissipative systems, on which we focus our attention, the volume occupied by the attractor is in general very small relative to the volume of phase space. We shall not talk about attractors for conservative systems, where the volume in phase space is conserved. For dissipative systems we may assume that phase-space volumes are contracted by the time evolution (if phase space is finite dimensional). Even if a system contracts volumes, this does not mean that it contracts lengths in *all* directions. Intuitively, some directions may be stretched, provided some others are so much contracted that the final volume is smaller than the initial volume (Fig. 1). This seemingly trivial remark has profound consequences. It implies that, even in a dissipative system, the final motion may be unstable *within the attractor.* This instability usually manifests itself by an exponential separation of orbits (as time goes on) of points which initially are very close to each other (on the attractor). The exponential separation takes place in the direction of stretching, and an attractor having this stretching property will be called a *strange attractor.* We shall also say that a system with a strange attractor is *chaotic* or has *sensitive dependence on initial conditions.* Of course, since the attractor is in general bounded, exponential separation can only hold as long as distances are small.

Fourier analysis of the motion on a strange attractor (say, of one of its coordinate components) in general reveals a *continuous power spectrum.* We are used to interpreting this as corresponding to an infinite number of modes. However, as we have indicated before, this

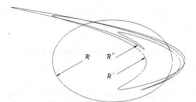

FIG. 1. The Hénon map $x'_1 = 1 - 1.4x_1^2 + x_2$, $x'_2 = 0.3x_1$ contracts volumes but stretches distances. Shown are a region $R$, and its first and second images $R'$ and $R''$ under the Hénon map.

reasoning is only valid in a "linear" theory, which then has to take place in an infinite-dimensional phase space. Thus, if we are confronted experimentally with a continuous power spectrum, there are two possibilities: We are either in the presence of a system that "explores" an infinite number of dimensions in phase space, or we have a system that evolves nonlinearly on a finite-dimensional attractor. Both alternatives are possible, and the second appears frequently in practice. We shall give below an algorithm which, starting from measurements, gives information on the effective dimension. This algorithm has been successfully used in several experiments, e.g., Malraison et al. (1983), Abraham et al. (1984), Grassberger and Procaccia (1983b); it has indicated finite dimensions in hydrodynamic systems, even though the phase space is infinite dimensional and the system therefore could potentially excite an infinite number of degrees of freedom.

The tool with which we want to measure the dimension and other dynamical quantities of the system is ergodic theory. Ergodic theory says that a time average equals a space average. The weight with which the space average has to be taken is an invariant measure. An invariant measure $\rho$ satisfies the equation

$$\rho[f^{-t}(E)] = \rho(E), \quad t > 0 , \qquad (1.3)$$

where $E$ is a subset of points of $\mathbf{R}^m$ and $f^{-t}(E)$ is the set obtained by evolving each of the points in $E$ backwards during time $t$. There are in general many invariant measures in a dynamical system, but not all of them are physically relevant. For example, if $x$ is an unstable fixed point of the evolution, then the $\delta$ function at $x$ is an invariant measure, but it is not observed. From an experimental point of view, a reasonable measure is obtained according to the following idea of Kolmogorov (see Sec. II.F). Consider Eq. (1.1) with an external noise term added,

$$\dot{x}(t) = F_\mu(x(t)) + \varepsilon\omega(t) , \qquad (1.4)$$

where $\omega$ is some noise and $\varepsilon > 0$ is a parameter. For suitable noise and $\varepsilon > 0$, the stochastic time evolution (1.3) has a unique stationary measure $\rho_\varepsilon$, and the measure we propose as "reasonable" is $\rho = \lim_{\varepsilon \to 0} \rho_\varepsilon$. We shall come back later to the problem of choosing a reasonable measure $\rho$.

For the moment we assume that such a measure exists, and call it the *physical measure*. Physically, we assume that it represents experimental time averages. Mathematically, we only require (for the moment) that it be invariant under time evolution.

A basic virtue of the ergodic theory of dynamical systems is that it allows us to consider only the long-term behavior of a system and not to worry about transients. In this way, the problems are at least somewhat simplified. The physical long-term behavior is on attractors, as we have already noted, but the geometric study of attractors presents great mathematical difficulties. Shifting attention from attractors to invariant measures turns out to make life much simpler.

An invariant probability measure $\rho$ may be *decomposable* into several different pieces, each of which is again invariant. If not, $\rho$ is said to be *indecomposable* or *ergodic*. In general, an invariant measure can be uniquely represented as a superposition of ergodic measures. In view of this, it is natural to assume that the physical measure is not only invariant, but also ergodic. If $\rho$ is ergodic, then the *ergodic theorem* asserts that for every continuous function $\varphi$,

$$\lim_{T \to \infty} \frac{1}{T} \int_0^T \varphi[f^t x(0)] dt = \int \rho(dx)\varphi(x) \qquad (1.5)$$

for almost all initial conditions $x(0)$ with respect to the measure $\rho$. Since the measure $\rho$ might be singular, for instance concentrated on a fractal set, it would be better if we could say something for almost all $x(0)$ with respect to the ordinary (Lebesgue) measure on some set $S \subset \mathbf{R}^m$. We shall see below that this is sometimes possible.

One crucial decision in our study of dynamics is to concentrate on the analysis of the separation in time of two infinitely close initial points. Let us illustrate the basic idea with an example in which time is discrete [rather than continuous as in (1.1)]. Consider the evolution equation

$$x(n+1) = f(x(n)), \quad x(i) \in \mathbf{R} , \qquad (1.6)$$

where $n$ is the discrete time. The separation of two initial points $x(0)$ and $x(0)'$ after time $N$ is then

$$x(N) - x(N)' = f^N(x(0)) - f^N(x(0)')$$

$$\approx \left[ \frac{d}{dx}(f^N)(x(0)) \right][x(0) - x(0)'] , \qquad (1.7)$$

where $f^N(x) = f(f(\cdots f(x) \cdots))$, $N$ times. By the chain rule of differentiation,

$$\frac{d}{dx}(f^N)(x(0)) = \frac{d}{dx}f(x(N-1))$$
$$\times \frac{d}{dx}f(x(N-2)) \cdots \frac{d}{dx}f(x(0)) . \qquad (1.8)$$

[In the case of $m$ variables, i.e., $x \in \mathbf{R}^m$, we replace the derivative $(d/dx)f$ by the Jacobian matrix, evaluated at $x$: $D_x f = (\partial f_i / \partial x_j)$.] Assuming that all factors in the above expression are of comparable size, it seems plausible that $df^N/dx$ grows (or decays) exponentially with $N$.

The same is true for $x(N)-x(N)'$, and we can define the average rate of growth as

$$\lambda = \lim_{N\to\infty} \frac{1}{N}\log|D_{x(0)}f^N\delta x(0)| \ . \qquad (1.9)$$

By the theorem of Oseledec (1968), this limit exists for almost all $x(0)$ (with respect to the invariant measure $\rho$). The average expansion value depends on the direction of the initial perturbation $\delta x(0)$, as well as on $x(0)$. However, if $\rho$ is ergodic, the largest $\lambda$ [with respect to changes of $\delta x(0)$] is independent of $x(0)$, $\rho$-almost everywhere. This number $\lambda_1$ is called the *largest Liapunov exponent* of the map $f$ with respect to the measure $\rho$. Most choices of $\delta x(0)$ will produce the largest Liapunov exponent $\lambda_1$. However, certain directions will produce smaller exponents $\lambda_2,\lambda_3,\ldots$ with $\lambda_1\geq\lambda_2\geq\lambda_3\geq\cdots$ (see Sec. III.A for details).

In the continuous-time case, one can similarly define

$$\lambda(x,\delta x) = \lim_{T\to\infty} \frac{1}{T}\log|(D_xf^T)\delta x| \ . \qquad (1.10)$$

We shall see that the Liapunov exponents (i.e., characteristic exponents) and quantities derived from them give useful bounds on the *dimensions of attractors*, and on the *production of information* by the system (i.e., *entropy* or Kolmogorov-Sinai invariant). It is thus very fortunate that $\lambda$ and related quantities are experimentally accessible. [We shall see below how they can be estimated. See also the paper by Grassberger and Procaccia (1983a).]

The Liapunov exponents, the entropy, and the Hausdorff dimension associated with an attractor or an ergodic measure $\rho$ all are related to how excited and how chaotic a system is (how many degrees of freedom play a role, and how much sensitivity to initial conditions is present). Let us see by an example how entropy (information production) is related to sensitive dependence on initial conditions.

We consider the dynamical system given by $f(x)=2x \bmod 1$ for $x\in[0,1)$. (This is "left-shift with leading digit truncation" in binary notation.) This map has sensitivity to initial conditions, and $\lambda=\log 2$. Assume now that our measuring apparatus can only distinguish between $x<\frac{1}{2}$ and $x>\frac{1}{2}$. Repeated measurements in time will nevertheless yield eventually all binary digits of the initial point, and it is in this sense that information is produced as "time" (i.e., the number of iterations) goes on. Thus changes of initial condition may be unobservable at time zero, but become observable at some later time. If we denote by $\rho$ the Lebesgue measure on $[0,1)$, then $\rho$ is an invariant measure, and the corresponding mean information produced per unit time is exactly one bit. More generally, the average rate $h(\rho)$ of information production in an ergodic state $\rho$ is related to sensitive dependence on initial conditions. [The quantity $h(\rho)$ is called the entropy of the measure $\rho$; see Sec. IV.] It may be bounded in terms of the characteristic exponents, and one finds

$$h(\rho) \leq \sum \text{positive characteristic exponents} \ . \qquad (1.11)$$

In fact, in many cases (but not all), when a physical measure $\rho$ may be identified, we have Pesin's formula (Pesin, 1977):

$$h(\rho) = \sum \text{positive characteristic exponents} \ . \qquad (1.12)$$

Another quantity of interest is the *Hausdorff dimension*. [This quantity has been brought very much to the attention of physicists by Mandelbrot (1982), who uses the term *fractal dimension*. This is also used as a sort of generic name for different mathematical definitions of dimension for "fractal" sets.] The dimension of a set is roughly the amount of information needed to specify points on it accurately. For instance, let $S$ be a compact set and assume that $N(\varepsilon)$ balls of radius $\varepsilon$ are needed to cover $S$. Then a dimension $\dim_K S$, the "capacity" of $S$, is defined by

$$\dim_K S = \limsup_{\varepsilon\to 0} \log N(\varepsilon)/|\log\varepsilon| \ .$$

[This is a little less than requiring $N(\varepsilon)\varepsilon^d\to$ finite, which means that the "volume" of the set $S$ is finite in dimension $d$.] Mañé (1981) has shown that the points of $S$ can be parametrized by $m$ real coordinates as soon as $m\geq 2\dim_K S + 1$.

The definition of the Hausdorff dimension $\dim_H S$ is slightly more complicated than that of $\dim_K S$; it does not assume that $S$ is compact (see Sec. II.J). We next define the *information dimension* $\dim_H \rho$ of a probability measure $\rho$ as the minimum of the Hausdorff dimensions of the sets $S$ for which $\rho(S)=1$. It is not *a priori* clear that sets defined by dynamical systems have *locally* the *same* Hausdorff dimension everywhere, but this follows from the ergodicity of $\rho$ in the case of $\dim_H\rho$. A result of Young [see Eq. (1.13) below] permits in many cases the evaluation of $\dim_H\rho$. Starting from different ideas, Grassberger and Procaccia (1983a,1983b) have arrived at a very similar way of computing the information dimension $\dim_H\rho$ of the measure $\rho$. Their proposal has been extremely successful, and has been used to measure reproducibly dimensions of the order of $3-10$ in hydrodynamical experiments (see, for example, Malraison *et al.*, 1983).

We present some details of the method. Let $\rho[B_x(r)]$ be the mass of the measure $\rho$ contained in a ball of radius $r$ centered at $x$, and assume that the limit

$$\lim_{r\to 0}\frac{\log\rho[B_x(r)]}{\log r} = \alpha \qquad (1.13)$$

exists for $\rho$-almost all $x$. The existence of the limit implies that it is constant, by the ergodicity of $\rho$. Under these conditions, $\alpha$ is equal to the information dimension $\dim_H\rho$, as noted by Young. In an experimental situation, one takes $N$ points $x(i)$, regularly spaced in time, on an orbit of the dynamical system, and estimates $\rho[B_{x(i)}(r)]$ by

$$\frac{1}{N}\sum_{j=1}^{N}\Theta[r-|x(j)-x(i)|] \quad (N \text{ large}) \ , \qquad (1.14)$$

where $\Theta(u)=(1+\operatorname{sgn}u)/2$. This permits us in principle to test the existence of the limit. In practice (Grassberger and Procaccia, 1983a,1983b) one defines

$$C(r) = \frac{1}{N^2} \sum_{i,j} \Theta[r - |x(j) - x(i)|] \quad (N \text{ large}), \qquad (1.15)$$

$$\text{information dimension} = \lim_{r \to 0} \frac{\log C(r)}{|\log r|}. \qquad (1.16)$$

The problem of associating an orbit in $\mathbf{R}^m$ with experimental results will be discussed later. We also postpone discussion of relations between the Hausdorff dimension and characteristic exponents [such relations are described in the work of Frederickson, Kaplan, Yorke, and Yorke (1983); Douady and Oesterlé (1980); and Ledrappier (1981a)].

One may ask to what extent the definition of the above quantities is more than wishful thinking: is there any chance that the dimensions, exponents, and entropies about which we have been talking are finite numbers? For the case of the Navier-Stokes equation,

$$\frac{\partial v_i}{\partial t} = -\sum_j v_j \partial_j v_i + \nu \Delta v_i - \frac{1}{d}\partial_i p + g_i, \qquad (1.17)$$

with the incompressibility conditions $\sum_j \partial_j v_j = 0$, one has some comforting results given below. [Note that, in the case of two-dimensional hydrodynamics, one has good existence and uniqueness results for the solutions to Eq. (1.17). Assuming the same to be true in three dimensions (for reasonable physical situations), the conclusions given below for the two-dimensional case will carry over.]

Consider the Navier-Stokes equation in a bounded domain $\Omega \subset \mathbf{R}^d$, where $d = 2$ or 3 is the spatial dimension. For *every* invariant measure $\rho$ one has the following relations between the energy dissipation $\varepsilon$ (per unit volume and time) and the ergodic quantities described earlier:

$$h(\rho) \leq \sum_{\lambda_i \geq 0} \lambda_i \leq \frac{B_d}{\nu^{1+d}} \left\langle \int_\Omega \varepsilon^{(2+d)/4} \right\rangle, \qquad (1.18)$$

$$\dim_H \rho \leq B_d' \frac{|\Omega|^{2/(d+2)}}{\nu^{d/2}} \left\langle \int_\Omega \varepsilon^{(2+d)/4} \right\rangle^{d/(d+2)}, \qquad (1.19)$$

where $B_d, B_d'$ are universal constants (see Ruelle, 1982b,1984, and Lieb, 1984, for a detailed discussion of these inequalities). Thus, if some average dissipation is finite, then all of these quantities are finite. In two dimensions, if the average dissipation is finite, i.e., if the *power* pumped into the system is finite, then $h(\rho)$ and $\dim_H \rho$ are also finite. In three dimensions, the situation is less clear because the average of $\int \varepsilon^{5/4}$ occurs instead of the average of $\int \varepsilon$. The lack of an existence and uniqueness theorem is in fact related to this difficulty. Experimentally, however, one finds that $\dim_H \rho$ is finite (implying that there are only finitely many $\lambda_i > 0$).

To conclude, let us remark that the dynamical theory of physical systems is a rather mathematical subject, in the sense that it appeals to difficult mathematical theories and results. On the other hand, these mathematical theories still have many loose ends. One might thus be tempted either to disregard rigorous mathematics and go ahead with the physics, or on the contrary to wait until

the mathematical situation is sufficiently clarified before going ahead with the physics. Both attitudes would be unfortunate. We believe in the value of the interplay between mathematics and physics, although either discipline offers only incomplete results. A mathematical theorem can prevent us from making "intuitive" assumptions that are already proved to be invalid. On the other hand, the relation between the two disciplines can help us to formulate mathematical conjectures which are made plausible on the basis of our experience as physicists. We are fortunate that the theory of dynamical systems has reached a stage where this kind of attitude seems especially fruitful.

The following are a few general references which are of interest in relation to the topics discussed in the present paper. (These references include books, conference proceedings, and reviews.)

Abraham, Gollub, and Swinney (1984): An overview of the experimental situation.

Bergé, Pomeau, and Vidal (1984): A very nice physics-oriented introduction, to be translated into English.

Bowen (1975): A more advanced introduction, stressing the ergodic theory of hyperbolic systems.

Campbell and Rose (1983): Los Alamos conference.

Collet and Eckmann (1980): A monograph, mostly on maps of the interval.

Cvitanović (1984): A very useful reprint collection.

Eckmann (1981): Review article on the geometric aspects of dynamical systems theory.

Ghil, Benzi, and Parisi (1985): Summer school proceedings on turbulence and predictability in geophysics.

Guckenheimer and Holmes (1983): An easy introduction to differential dynamical systems, oriented towards chaos.

Gurel and Rössler (1979): N.Y. Academy Conference.

Helleman (1980): N.Y. Academy Conference. These two conferences played an important historical role.

Iooss, Helleman, and Stora (1981): Proceedings of a summer school in Les Houches, 1981, with many interesting lectures.

Nobel symposium on chaos (1985).

Shaw (1981): A nice intuitive introduction to the information aspects of chaos.

Vidal and Pacault (1981): Conference proceedings on chemical turbulence.

Young (1984): A brief, but excellent, exposition of the inequalities for entropy and dimension.

## II. DIFFERENTIABLE DYNAMICS AND THE RECONSTRUCTION OF DYNAMICS FROM AN EXPERIMENTAL SIGNAL

### A. What is a differentiable dynamical system?

A differentiable dynamical system is simply a time evolution defined by an evolution equation

$$\frac{dx}{dt}=F(x) \tag{2.1}$$

(continuous-time case) or by a map

$$x(n+1)=f(x(n)) \tag{2.2}$$

(discrete-time case), where $f$ or $F$ are *differentiable* functions. In other words, $f$ or $F$ have continuous first-order derivatives. We may require $f$ or $F$ to be twice differentiable or more, i.e., to have continuous derivatives of second or higher order. Differentiability (possibly of higher order) is also referred to as *smoothness*. The physical justification for the assumed continuity of the derivatives of $f$ or $F$ is simply that physical quantities are usually continuous (small causes produce small effects). This philosophy, however, should, not be adhered to blindly (see Sec. III.D.2).

One introduces the nonlinear time-evolution operators $f^t$, $t$ real or integer, requiring sometimes $t \geq 0$. They have the property

$$f^0=\text{identity}, \quad f^s f^t=f^{s+t} .$$

The variable $x$ varies over the *phase space* $M$, which is $\mathbf{R}^m$, or a manifold like a sphere or a torus, or infinite dimensional (Banach spaces, in particular Hilbert spaces, are important in hydrodynamics). If $M$ is a linear space, we define the linear operator $D_x f^t$ (matrix of partial derivatives of $f$ at $x$, or a bounded operator if $M$ is a Banach space). Writing $f^1=f$, we have

$$D_x f^n=D_{f^{n-1}x}f \cdots D_{fx}fD_x f \tag{2.3}$$

by the chain rule.

*Example.*

A viscous fluid in a bounded container $\Omega \subset \mathbf{R}^2$ or $\mathbf{R}^3$ is described by the Navier-Stokes equation

$$\frac{\partial v_i}{\partial t}=-\sum_j v_j \partial_j v_i+\nu\Delta v_i-\frac{1}{d}\partial_i p+g_i , \tag{2.4}$$

where $(v_i)$ is the velocity field in $\Omega$, $\nu$ a constant (the kinematic viscosity), $d$ the (constant) density, $p$ the pressure, and $g$ an external force field. We add to Eq. (2.4) the incompressiblity condition

$$\sum_j \partial_j v_j=0 , \tag{2.5}$$

which expresses that $v_i$ is divergence free, and we impose $v_i=0$ on $\partial\Omega$ (the fluid sticks to the boundary). Note that the divergence-free vector fields are orthogonal to gradients, so that one can eliminate the pressure from Eq. (2.4) by orthogonal projection of the equation on the divergence-free fields. One obtains thus an equation of the type (2.1) where $M$ is the Hilbert space of square-integrable vector fields which are orthogonal to gradients. In two dimensions (i.e., for $\Omega \subset \mathbf{R}^2$), one has a good existence and uniqueness theorem for solutions of Eqs. (2.4) and (2.5), so that $f^t$ is defined for $t$ real $\geq 0$ (Ladyzhenskaya, 1969; Foias and Temam, 1979; Temam, 1979). In three dimensions one has only partial results (Caffarelli, Kohn, and Nirenberg, 1982).

## B. Dissipation and attracting sets

For a *conservative system* (Hamiltonian time evolution), Liouville's theorem says that the volume in phase space $M$ is conserved by the time evolution. We shall be mainly interested in *dissipative systems*, for which this is not the case and for which the volume is usually contracted. Let us therefore assume that there is an open set $U$ in $M$ which is contracted by time evolution asymptotically to a compact set $A$. To be precise, we say that $A$ is an *attracting set* with *fundamental neighborhood* $U$ if (a) for every open set $V \supset A$ we have $f^t U \subset V$ when $t$ is large enough, and (b) $f^t A=A$ for all $t$. (See Fig. 2.) The open set $\cup_{t>0}(f^t)^{-1}U$ is the *basin of attraction* of $A$. If the basin of attraction of $A$ is the whole of $M$, we say that $A$ is the *universal attracting set*.

*Examples.*

(a) If $U$ is an open set in $M$, and the closure of $f^t U$ is compact and contained in $U$ for all sufficiently large $t$, then the set $A=\cap_{t\geq 0}f^t U$ is a (compact) attracting set with fundamental neighborhood $U$ (see Ruelle, 1981).

(b) The Lorenz time evolution in $\mathbf{R}^3$ is defined by the equation

$$\frac{d}{dt}\begin{bmatrix}x_1\\x_2\\x_2\end{bmatrix}=\begin{bmatrix}-\sigma x_1+\sigma x_2\\-x_1x_3+rx_1-x_2\\x_1x_2-bx_3\end{bmatrix}, \tag{2.6}$$

with $\sigma=10$, $b=\frac{8}{3}$, $r=28$ (see Lorenz, 1963). If $U$ is a sufficiently large ball, [i.e., $U=\{(x_1,x_2,x_3):\sum x_i^2 \leq R^2\}$ with large $R$], then $U$ is mapped into itself by time evolution. It contains thus an attracting set $A$, and $A$ is universal (see Fig. 3).

(c) The Navier-Stokes time evolution in two dimensions also gives rise to a universal attracting set $A$, because one can again apply (a) to a sufficiently large ball (in a suitable Hilbert space). It can be shown that $A$ has finite dimension (see Mallet-Paret, 1976).

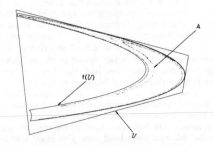

FIG. 2. Example of an attracting set $A$ with fundamental neighborhood $U$. (The map is the Hénon map.)

379

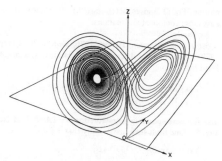

FIG. 3. The Lorenz attractor. From Lanford (1977).

## C. Attractors

Physical experiments and computer experiments with dynamical systems usually exhibit transient behavior followed by what seems to be an asymptotic regime. Therefore the point $f^t x$ representing the system should eventually lie on an attracting set (or near it). However, in practice *smaller* sets, which we call *attractors*, will be obtained (they should be carefully distinguished from attracting sets). This is because some parts of an attracting set may not be attracting (Fig. 4).

We should also like to include in the mathematical definition of an attractor $A$ the requirement of irreducibility (i.e., the union of two disjoint attractors is not considered to be an attractor). This (unfortunately) implies that one can no longer impose the requirement that there be an *open* fundamental neighborhood $U$ of $A$ such that $f^t U \to A$ when $t \to \infty$. Instead of trying to give a precise mathematical definition of an attractor, we shall use here the *operational definition*, that it is a set on which *experimental points* $f^t x$ accumulate for large $t$. We shall come back later to the significance of this operational definition and its relation to more mathematical concepts.

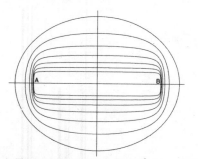

FIG. 4. The dynamical system is $\dot{x}_1 = x_1 - x_1^3$, $\dot{x}_2 = -x_2$. The segment $A, B$ is the universal attracting set, but only the points $A, B$ are attractors. In other words, the whole space is attracted to the segment $A, B$ but only $A$ and $B$ are attractors.

*Examples.*

(a) *Attracting fixed point.* Let $P$ be a *fixed point* for our dynamical system, i.e., $f^t P = P$ for all $t$. The derivative $D_P f^1$ of $f^1$ (time-one map) at the fixed point is an $m \times m$ matrix or an operator in Hilbert space. If its spectrum is in a disk $\{z : |z| < \alpha\}$ with $\alpha < 1$, then $P$ is an attracting fixed point. It is an attracting set (and an attractor). When the time evolution is defined by the differential equation (1.1) in $\mathbf{R}^m$, the attractiveness condition is that the eigenvalues of $D_P F_\mu$ all have a negative real part. For a discrete-time dynamical system, we say that $(P_1, \ldots, P_n)$ is an attracting periodic orbit, of period $n$, if $f P_1 = P_2, \ldots, f P_n = P_1$, and $P_i$ is an attracting fixed point for $f^n$.

(b) *Attracting periodic orbit for continuous time.* For a continuous-time dynamical system, suppose that there are a point $a$ and a $T > 0$, such that $f^T a = a$ but $f^t a \neq a$ when $0 < t < T$. Then $a$ is a periodic point of period $T$, and $\Gamma = \{f^t a : 0 \le t < T\}$ is the corresponding *periodic orbit* (or closed orbit). The derivative $D_a f^T$ has an eigenvalue 1 corresponding to the direction tangent to $\Gamma$ at $a$. If the rest of the spectrum is in $\{z : |z| < \alpha\}$ with $\alpha < 1$, then $\Gamma$ is an attracting periodic orbit. It is again an attracting set and an attractor. The attracting character of a periodic orbit may also be studied with the help of a Poincaré section (see Sec. II.H).

(c) *Quasiperiodic attractor.* A periodic orbit for a continuous system is really a circle, and the motion on it (by proper choice of coordinate $\varphi$) may be written

$$\varphi(t) = \varphi(0) + \omega t \pmod{2\pi}, \qquad (2.7)$$

where $\omega = 2\pi / T$. This may be thought of as representing the time evolution of a simple oscillator. Consider now a collection of $k$ oscillators with frequencies $\omega_1, \ldots, \omega_k$ (without rational relations between the $\omega_i$: no linear combination with nonzero integer coefficients vanishes). The motion of the oscillators is described by

$$\varphi_i(t) = \varphi_i(0) + \omega_i t \pmod{2\pi}, \ i = 1, \ldots, k, \qquad (2.8)$$

and this motion takes place on the product of $k$ circles, ($k > 1$), which is a $k$-dimensional torus $T^k$. Suppose that the torus $T^k$ is embedded in $\mathbf{R}^m$, $m \ge k$ (or in Hilbert space), as the periodic orbit $\Gamma$ was in the previous example; suppose, furthermore, that this torus is an attracting set. Then we say that $T^k$ is a quasiperiodic attractor. Asymptotically, the dynamical system will thus be described by

$$x(t) = f^t x = \Phi[\varphi_1(t), \ldots, \varphi_k(t)] \qquad (2.9)$$

$$= \Psi(\omega_1 t, \ldots, \omega_k t), \qquad (2.10)$$

where $\Psi$ is periodic, of period $2\pi$, in each argument. A function of the form $t \to \Psi(\omega_1 t, \ldots, \omega_k t)$ is known as a *quasiperiodic function* (with $k$ different periods). Quasiperiodic attractors are a natural generalization of periodic orbits, and they occur fairly frequently in the description of moderately excited physical systems.

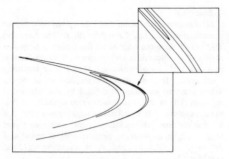

FIG. 5. The Hénon attractor for $a = 1.4$, $b = 0.3$. The successive iterates $f^k$ of $f$ have been applied to the point $(0,0)$, producing a sequence asymptotic to the attractor. Here, 30 000 points of this sequence are plotted, starting with $f^{20}(0,0)$.

### D. Strange attractors

The attractors discussed under (a), (b), and (c) above are also attracting sets. They are nice manifolds (point, circle, torus). Notice also that, if a small change $\delta x(0)$ is made to the initial conditions, then $\delta x(t) = (D_x f^t)\delta x(0)$ remains small when $t \to \infty$. [In fact, for a quasiperiodic

motion, Eq. (2.7) gives $\delta\varphi(t) = \delta\varphi(0)$.] We shall now discuss more complicated situations.

*Examples.*

*(a) Hénon attractor* (Hénon, 1976; Feit, 1978; Curry, 1979). Consider the discrete-time dynamical system defined by

$$f\begin{bmatrix} x_1 \\ x_2 \end{bmatrix} = \begin{bmatrix} 1 + x_2 - ax_1^2 \\ bx_1 \end{bmatrix} \qquad (2.11)$$

and the corresponding attractor, for $a = 1.4$, $b = 0.3$ (see Fig. 5). One finds here numerically that

$$\delta x(t) \approx \delta x(0)e^{\lambda t}, \quad \lambda = 0.42 ,$$

i.e., the errors grow exponentially. This is the phenomenon of *sensitive dependence on initial conditions*. In fact (Curry, 1979), computing the successive points $f^n x$ for $n = 1, 2, \ldots$, with 14 digits' accuracy, one finds that the error of the sixtieth point is of order 1. Sensitive dependence on initial conditions is also expressed by saying that the system is *chaotic* [this is now the accepted use of the word *chaos*, even though the original use by Li and Yorke (1975) was somewhat different].

*(b) Feigenbaum attractor* (Feigenbaum, 1978,1979,1980; Misiurewicz, 1981; Collet, Eckmann, and Lanford, 1980). A map of the interval [0,1] to itself is defined by

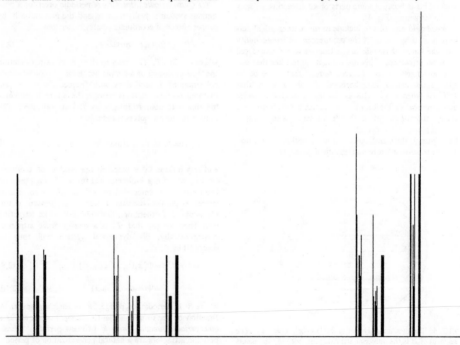

FIG. 6. The Feigenbaum attractor. Histogram of 50 000 points in 1024 bins. This histogram shows the unique ergodic measure, which is clearly singular.

$$f_\mu(x) = \mu x(1-x) \qquad (2.12)$$

when $\mu \in [0,4]$. It has attracting periodic orbits of period $2^n$, with $n$ tending to infinity as $\mu$ tends to 3.57... through lower values. For the limiting value $\mu = 3.57...$, there is a very special attractor $A$ shown in Fig. 6. We shall call it the Feigenbaum attractor (although it was known earlier to many authors). Note that interspersed with this attractor, and arbitrarily close to it, there are repelling periodic orbits of period $2^n$, for all $n$. Therefore the attractor $A$ cannot be an attracting set. One can show, moreover, that, for this very special attractor, there is no sensitive dependence on initial conditions (no exponential growth of errors): the Feigenbaum attractor is not chaotic.

The Hénon and Feigenbaum attractors, as depicted in Figs. 5 and 6, have a complicated aspect typical of *fractal* objects. In general, a fractal set is a set for which the Hausdorff dimension is different from the topological dimension, and usually not an integer. (The exact definition of the Hausdorff dimension is given in Sec. II.J.) The name *fractal* was coined by Mandelbrot. For the rich lore of fractal objects, see Mandelbrot (1982). While many attractors are fractals, and therefore complicated objects, they are by no means featureless. They are unions of *unstable manifolds* (to be defined in Sec. III.E) and often have a Cantor-set structure in the direction transversal to the unstable manifolds. (For the Feigenbaum attractor the unstable manifolds have dimension 0, and only a Cantor set is visible; for the Hénon attractor the unstable manifolds have dimension 1.) An attractor is by definition invariant under a dynamical evolution, and this creates a self-similarity that is often strikingly visible.

In view of both its chaotic and fractal characters, the Hénon attractor deserves to be called a *strange attractor* (this name was introduced by Ruelle and Takens, 1971). The property of being chaotic is actually a more important dynamical concept than that of being fractal, and we shall therefore say that the Feigenbaum attractor is *not* a strange attractor (this differs somewhat from the point of view in Ruelle and Takens). We therefore define a strange attractor to be an attractor with *sensitive dependence on initial conditions*. The notion of strangeness refers thus to the *dynamics* on the attractor, and *not just to its geometry*; it applies whether the time is discrete or continuous. This is again an operational definition rather than a mathematical one. We shall see in Sec. III what should be clarified mathematically. For physics, however, the above operational concept of strange attractors has served well and deserves to be kept.

*Example.*

(c) *Thom's toral automorphisms and Arnold's cat map.* Let $x_1 \pmod 1$ and $x_2 \pmod 1$ be coordinates on the 2-torus $T^2$; a map $f : T^2 \to T^2$ is defined by

$$f \begin{bmatrix} x_1 \\ x_2 \end{bmatrix} = \begin{bmatrix} x_1 + x_2 \\ x_1 + 2x_2 \end{bmatrix} \pmod 1 . \qquad (2.13)$$

[Because $\det \binom{1\,1}{1\,2} = 1$, the map $\mathbf{R}^2 \to \mathbf{R}^2$ defined by the matrix $\binom{1\,1}{1\,2}$ maps $Z^2$ to $Z^2$ and therefore, going to the quo-

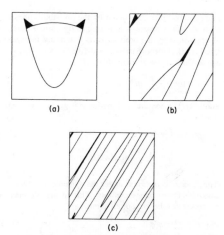

FIG. 7. Arnold's "cat map": (a) the cat; (b) its image under the first iterate; (c) image under the second iterate of the cat map.

tient $T^2 = \mathbf{R}^2/Z^2$, a map $f : T^2 \to T^2$ of the 2-torus to itself is defined. The system is area preserving, and the whole torus is an "attractor."] This is Arnold's celebrated cat map (Arnold and Avez, 1967), well known to be chaotic (see Fig. 7). In fact we have here

$$\delta x(t) \approx \delta x(0) e^{\lambda t} , \qquad (2.14)$$

with

$$\lambda = \log \frac{3+\sqrt{5}}{2} ,$$

$(3+\sqrt{5})/2$ being the eigenvalue larger than 1 of the matrix $\binom{1\,1}{1\,2}$.

More generally, if $V$ is an $m \times m$ matrix with integer entries and determinant $\pm 1$, it defines a toral automorphism $T^m \to T^m$, and Thom noted that these automorphisms have sensitive dependence on initial conditions if $V$ has an eigenvalue $\alpha$ with $|\alpha| > 1$.

Returning to Arnold's cat map, we may imbed $T^2$ as an attractor $A$ in a higher-dimensional Euclidean space. In this case $A$ is chaotic, but not fractal.

Our examples clearly show that the notions of fractal attractor and strange (i.e., chaotic) attractor are independent. A periodic orbit is neither strange nor chaotic, Arnold's map is strange but not fractal, Feigenbaum's attractor is fractal but not strange, the Lorenz and Hénon attractors are both strange and fractal. [Another strange and fractal attractor with a simple equation has been introduced by Rössler (1976).]

## E. Invariant probability measures

An attractor $A$, be it strange or not, gives a global picture of the long-term behavior of a dynamical system. A more detailed picture is given by the probability measure $\rho$ on $A$, which describes how frequently various parts of $A$ are visited by the orbit $t \rightarrow x(t)$ describing the system (see Fig. 8). Operationally, $\rho$ is defined as the time average of Dirac deltas at the points $x(t)$,

$$\rho = \lim_{t \to \infty} \frac{1}{T} \int_0^T dt\, \delta_{x(t)} . \qquad (2.15)$$

Similarly, if a continuous function $\varphi$ is given, then we define

$$\rho(\varphi) \equiv \int \rho(dx)\varphi(x)$$
$$= \lim_{T \to \infty} \frac{1}{T} \int_0^T dt\, \varphi[x(t)] . \qquad (2.16)$$

The measure is *invariant* under the dynamical system, i.e., invariant under time evolution. This invariance may be expressed as follows: For all $\varphi$ one has

$$\rho(\varphi \circ f^t) = \rho(\varphi) . \qquad (2.17)$$

Suppose that the invariant probability measure $\rho$ cannot be written as $\frac{1}{2}\rho_1 + \frac{1}{2}\rho_2$ where $\rho_1, \rho_2$ are again invariant probability measures and $\rho_1 \neq \rho_2$. Then $\rho$ is called *indecomposable*, or equivalently, *ergodic*.

*Theorem.* If the compact set $A$ is invariant under the dynamical system $(f^t)$, then there is a probability measure $\rho$ invariant under $(f^t)$ and with support contained in $A$. One may choose $\rho$ to be ergodic.

[The important assumptions are that the $f^t$ commute and are continuous $A \rightarrow A$ ($A$ compact). The theorem results from the Markov-Kakutani fixed-point theorem (see Dunford and Schwartz, 1958, Vol. I).] This is not a very detailed result; it is more in the class referred to as "general nonsense" by mathematicians. But since we shall talk a lot about ergodic measures in what follows, it is good to know that such measures are indeed present.

*Theorem (Ergodic theorem).* If $\rho$ is ergodic, then for $\rho$ almost all initial $x(0)$ the time averages (2.15) and (2.16) reproduce $\rho$.

The above theorems show that there are invariant (ergodic) measures defined by time averages. Unfortunately,

a strange attractor typically carries *uncountably many* distinct ergodic measures. Which one do we choose? We shall propose natural definitions in the next section.

*Example.*

The points of the circle $T^1$ may be parametrized by numbers in $[0,1)$, and each such number has a binary expansion $0.a_1 a_2 a_3 \cdots$, where, for each $i$, $a_i = 0$ or 1 (this coding introduces a little ambiguity, of no importance for what follows). We define a map $f : T^1 \rightarrow T^1$ by

$$f(x) = 2(x) \pmod{1} . \qquad (2.18)$$

Clearly, $f$ replaces $0.a_1 a_2 a_3 \cdots$ by $0.a_2 a_3 \cdots$ (an operation called a *shift*). We now choose $p$ between 0 and 1. A probability distribution $\rho_p$ on binary expansions $0.a_1 a_2 a_3 \cdots$ is then defined by requiring that $a_i$ be 0 with probability $p$, and 1 with probability $1 - p$ (independently for each $i$). One can check that $\rho_p$ is invariant under the shift, and in fact ergodic. It thus defines an ergodic measure for the differentiable dynamical system (2.18), $f : T^1 \rightarrow T^1$, and there are uncountably many such measures, corresponding to the different values of $p$ in $(0,1)$.

### F. Physical measures

Operationally, it appears that (in many cases, at least) the time evolution of physical systems produces well-defined time averages. The same applies to computer-generated time evolutions. There is thus a selection process of a particular measure $\rho$ which we shall call *physical measure* (another operational definition).

One selection process was discussed by Kolmogorov (we are not aware of a published reference) a long time ago. A physical system will normally have a small level $\varepsilon$ of random noise, so that it can be considered as a stochastic process rather than a deterministic one. In a computer study, roundoff errors should play the role of the random noise. Due to sensitive dependence on initial conditions, even a very small level $\varepsilon$ of noise has important effects, as we saw in Sec. II.D for the Hénon attractor. On the other hand, a stochastic process such as the one described above normally has only one stationary measure $\rho_\varepsilon$, and we may hope that $\rho_\varepsilon$ tends to a specific measure (the *Kolmogorov measure*) when $\varepsilon \rightarrow 0$. As we shall see below, this hope is substantiated in the case of *Axiom-A* dynamical systems. However, this approach may have difficulties in general, because an attractor $A$ does not always have an open basin of attraction, and thus the added noise may force the system to jump around on several attractors.

Another possibility is the following: Suppose that $M$ is finite dimensional, and that there is a set $S \subset M$ with Lebesgue measure $\mu(S) > 0$ such that $\rho$ is given by the time averages (2.15) and (2.16) when $x(0) \in S$. This property holds if $\rho$ is an *SRB measure* (to be defined and studied in Sec. IV.C; Sinai, 1972; Bowen and Ruelle, 1975; Ruelle, 1976). For Axiom-A systems, the Kolmogorov and SRB measures coincide, but in general SRB measures are easier to study.

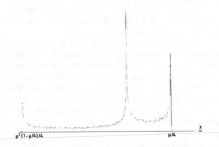

FIG. 8. Histogram of 50 000 iterates of the map $x \rightarrow \mu x (1-x)$, in 400 bins. The parameter $\mu = 3.678\,57 \ldots$ is the real solution of the equation $(\mu - 2)^2(\mu + 2) = 16$. It is known that the invariant density is smooth with square-root singularities.

Clearly, Kolmogorov measures and SRB measures are candidates for the description of physical time averages, but they are not always easy to define. Fortunately, many important results hold for an arbitrary invariant measure $\rho$. Results of this type, which constitute a large part of the ergodic theory of differentiable dynamical systems, will be discussed in Secs. III and IV of this paper.

### G. Reconstruction of the dynamics from an experimental signal

In a computer study of a dynamical system in $m$ dimensions, we have an $m$-dimensional signal $x(t)$, which can be submitted to analysis. By contrast, in a physical experiment one monitors typically only one scalar variable, say $u(t)$, for a system that usually has an infinite-dimensional phase space $M$. How can we hope to understand the system by analyzing the single scalar signal $u(t)$? The enterprise seems at first impossible, but turns out to be quite doable. This is basically because (a) we restrict our attention to the dynamics on a finite-dimensional attractor $A$ in $M$, and (b) we can generate several different scalar signals $x_i(t)$ from the original $u(t)$. We have already mentioned that the universal attracting set (which contains all attractors) has finite dimension in two-dimensional hydrodynamics, and we shall come back later to this question of finite dimensionality.

The easiest, and probably the best way of obtaining several signals from a single one is to use *time delays*. One chooses different delays $T_1=0, T_2, \ldots, T_N$ and writes $x_k(t)=u(t+T_k)$. In this manner an $N$-dimensional signal is generated. The experimental points in Fig. 9 below have been obtained by this method. Successive time derivatives of the signal have also been used: $x_{k+1}(t)=d^k x_1(t)/dt^k$, but the numerical differentiations tend to produce high levels of noise. Of course one should measure several experimental signals instead of only one whenever possible.

The reconstruction just outlined will provide an $N$-dimensional image (or projection) $\pi A$ of an attractor $A$ which has finite Hausdorff dimension, but lives in a usually infinite-dimensional space $M$. Depending on the choice of variables (in particular on the time delays), the projection will look different. In particular, if we use fewer variables than the dimension of $A$, the projection $\pi A$ will be bad, with trajectories crossing each other. There are some theorems (Takens, 1981; Mañé, 1981) which state that if we use enough variables, typically about twice the Hausdorff dimension, we shall generally get a good projection.

*Theorem* (Mañé). Let $A$ be a compact set in a Banach space $B$, and $E$ a subspace of finite dimension such that

$$\dim E > \dim_H(A \times A) + 1 \, ,$$

or let $A$ be compact and

$$\dim E > 2 \dim_K(A) + 1 \, ,$$

where $\dim_H$ is the Hausdorff dimension and $\dim_K$ is the

capacity. Then the set of projections $\pi: B \to E$ such that $\pi$ restricted to $A$ is injective (i.e., one to one into $E$) is dense among all projections $B \to E$ with respect to the norm operator topology.

[More precisely, the injective projections are "residual," i.e., contain a countable intersection of dense sets. As noticed by Mañé, his original statement of the theorem needs a slight correction, which is made in the above formulation.]

The choice of variables for the reconstruction of a dynamical system has to be made carefully (by trial and error). This is discussed in Roux, Simoyi, and Swinney (1983).

### H. Poincaré sections

The reconstruction process described above yields a line $(f^t x)_0^T$ that may look like a heap of spaghetti and may be difficult to interpret. It is often possible and useful to make a transverse cut through this mess, so that instead of a long curve in $N$ dimensions one now has a set of points $S$ in $N-1$ dimensions (Poincaré section). Figure 9 gives an experimental example corresponding to the Beloussov-Zhabotinski chemical reaction. Given a point $x$ of the Poincaré section, the *first return map* will bring it to $Px$, which is again in the Poincaré section. When a good model of $S$ and $P$ can be deduced from the experiment, one has essentially understood the dynamical system. This is, however, possible only for low-dimensional attractors.

Notice that the use of a Poincaré section is different from a *stroboscopic* study, where one looks at the system at integer multiples of a *fixed* time interval. By contrast, the time of first return to the Poincaré section is variable

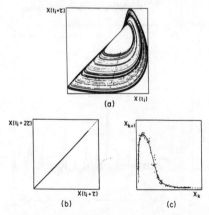

FIG. 9. Experimental plot of a Poincaré section in the Beloussov-Zhabotinski reaction, after Roux and Swinney (1981): (a) the attractor and the plane of Poincaré section; (b) the Poincaré section; (c) the corresponding first return map.

(and has to be determined numerically by interpolation). Sometimes, as for quasiperiodic motions, there is a *natural frequency* (or several) that will stabilize the stroboscopic image. But in general this is not the case, and therefore the stroboscopic study is useless.

### I. Power spectra

The power spectrum $S(\omega)$ of a scalar signal $u(t)$ is defined as the square of its Fourier amplitude per unit time. Typically, it measures the amount of energy per unit time (i.e., the power) contained in the signal as a function of the frequency $\omega$. One can also define $S(\omega)$ as the Fourier transform of the time correlation function $\langle u(0)u(t)\rangle$ equal to the average over $\tau$ of $u(\tau)u(t+\tau)$. If the correlations of $u$ decay sufficiently rapidly in time, the two definitions coincide, and one has (Wiener-Khinchin theorem; see Feller, 1966)

$$S(\omega)=(\text{const})\lim_{T\to\infty}\frac{1}{T}\left|\int_0^T dt\, e^{i\omega t}u(t)\right|^2$$

$$=(\text{const})\int_{-\infty}^{\infty} dt\, e^{i\omega t}\lim_{\tau\to\infty}\frac{1}{\tau}\int_0^\tau d\tau'u(\tau')u(t+\tau')\ .$$

$$(2.19)$$

Note that the above limit (2.19) makes sense only after averaging over small intervals of $\omega$. Without this averaging, the quantity

$$\frac{1}{T}\left|\int_0^T dt\, e^{i\omega t}u(t)\right|^2$$

fluctuates considerably, i.e., it is very noisy. (Instead of averaging over intervals of $\omega$, one may average over many runs).

The power spectrum indicates whether the system is periodic or quasiperiodic. The power spectrum of a periodic system with frequency $\omega$ has Dirac $\delta$'s at $\omega$ and its harmonics $2\omega,3\omega,\dots$. A quasiperiodic system with basic frequencies $\omega_1,\dots,\omega_k$ has $\delta$'s at these positions and also at all linear combinations with integer coefficients. (The choice of basic frequencies is somewhat arbitrary, but the number $k$ of independent frequencies is well defined.) In experimental power spectra, the Dirac $\delta$'s are not infinitely sharp; they have at least an "instrumental width" $2\pi/T$, where $T$ is the length of the time series used. The linear combinations of the basic frequencies $\omega_1,\dots,\omega_k$ are dense in the reals if $k>1$, but the amplitudes corresponding to complicated linear combinations are experimentally found to be small. (A mathematical theory for this does not seem to exist.) A careful experi-

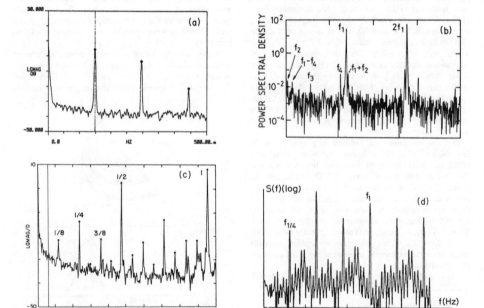

FIG. 10. Some spectra: (a) The power spectrum of a periodic signal shows the fundamental frequency and a few harmonics. Fauve and Libchaber (1982). (b) A quasiperiodic spectrum with four fundamental frequencies. Walden, Kolodner, Passner, and Surko (1984). (c) A spectrum after four period doublings. Libchaber and Maurer (1979). (d) Broadband spectrum invades the subharmonic cascade. The fundamental frequency and the first two subharmonics are still visible. Croquette (1982).

ment may show very convincing examples of quasiperiodic systems with two, three, or more basic frequencies. In fact, $k=2$ is common, and higher $k$'s are increasingly rare, because the nonlinear couplings between the modes corresponding to the different frequencies tend to destroy quasiperiodicity and replace it by chaos (see Ruelle and Takens, 1971; Newhouse, Ruelle, and Takens, 1978). However, for weakly coupled modes, corresponding for instance to oscillators localized in different regions of space, the number of observable frequencies may become large (see, for instance, Grebogi, Ott, and Yorke, 1983a; Walden, Kolodner, Passner, and Surko, 1984). Nonquasiperiodic systems are usually chaotic. Although their power spectra still may contain peaks, those are more or less broadened (they are no longer instrumentally sharp). Furthermore, a noisy background of *broadband spectrum* is present. For this it is *not* necessary that the system be infinite dimensional [Figs. 10(a)—10(d)].

In general, power spectra are very good for the visualization of periodic and quasiperiodic phenomena and their separation from chaotic time evolutions. However, the analysis of the chaotic motions themselves does not benefit much from the power spectra, because (being squares of absolute values) they lose phase information, which is essential for the understanding of what happens on a strange attractor. In the latter case, as already remarked, the dimension of the attractor is no longer related to the number of independent frequencies in the power spectrum, and the notion of "number of modes" has to be replaced by other concepts, which we shall develop below.

## J. Hausdorff dimension and related concepts[*]

Most concepts of dimension make use of a metric. Our applications are to subsets of $\mathbb{R}^m$ or Banach spaces, and the natural metric to use is the one defined by the norm.

Let $A$ be a compact metric space and $N(r,A)$ the minimum number of open balls of radius $r$ needed to cover $A$. Then we define

$$\dim_K A = \limsup_{r \to 0} \frac{\log N(r,A)}{\log(1/r)} .$$

This is the *capacity* of $A$ (this concept is related to Kolmogorov's $\varepsilon$ *entropy* and has nothing to do with *Newtonian capacity*). If $A$ and $B$ are compact metric spaces, their product $A \times B$ satisfies

$$\dim_K(A \times B) \le \dim_K A + \dim_K B . \qquad (2.20)$$

Given a nonempty set $A$, with a metric, and $r > 0$, we denote by $\sigma$ a covering of $A$ by a (countable) family of sets $\sigma_k$ with diameter $d_k = \text{diam}\,\sigma_k \le r$. Given $\alpha \ge 0$, we write

$$m_r^{\alpha}(A) = \inf_{\sigma} \sum_k (d_k)^{\alpha} .$$

When $r \downarrow 0$, $m_r^{\alpha}(A)$ increases to a (possibly infinite) limit $m^{\alpha}(A)$ called the *Hausdorff measure of A in dimension* $\alpha$. We write

$$\dim_H A = \sup\{\alpha : m^{\alpha}(A) > 0\}$$

and call this quantity the *Hausdorff dimension* of $A$. Note that $m^{\alpha}(A) = +\infty$ for $\alpha < \dim_H A$, and $m^{\alpha}(A) = 0$ for $\alpha > \dim_H A$. The Hausdorff dimension of a set $A$ is, in general, strictly smaller than the Hausdorff dimension of its closure. Furthermore, the inequality (2.20) on the dimension of a product does not extend to Hausdorff dimensions. It is easily seen that for every compact set $A$, one has $\dim_H A \le \dim_K A$.

If $A$ and $B$ are compact sets satisfying $\dim_H A = \dim_K A$, $\dim_H B = \dim_K B$, then

$$\dim_H(A \times B) = \dim_K(A \times B) = \dim_H A + \dim_H B .$$

We finally introduce a *topological* dimension $\dim_L A$. It is defined as the smallest integer $n$ (or $+\infty$) for which the following is true: For every finite covering of $A$ by open sets $\sigma_1, \ldots, \sigma_N$ one can find another covering $\sigma_1', \ldots, \sigma_N'$ such that $\sigma_i' \subset \sigma_i$ for $i = 1, \ldots, N$ and any $n+2$ of the $\sigma_i'$ will have an empty intersection:

$$\sigma_{i_0}' \cap \sigma_{i_1}' \cap \cdots \cap \sigma_{i_{n+1}}' = \varnothing .$$

The quantity $\dim_L A$ is also called the *Lebesgue* or *covering dimension* of $A$.

For more details on dimension theory, see Hurewicz and Wallman (1948) and Billingsley (1965).

## III. CHARACTERISTIC EXPONENTS

In this section we review the ergodic theory of differentiable dynamical systems. This means that we study invariant probability measures (corresponding to time averages). Let $\rho$ be such a measure, and assume that it is ergodic (indecomposable). The present section is devoted to the *characteristic exponents of* $\rho$ (also called *Liapunov exponents*) and related questions. We postpone until Sec. V the discussion of how these characteristic exponents can be measured in physical or computer experiments.

### A. The multiplicative ergodic theorem of Oseledec

If the initial state of a time evolution is slightly perturbed, the exponential rate at which the perturbation $\delta x(t)$ increases (or decreases) with time is called a *characteristic exponent*. Before defining characteristic exponents for differentiable dynamics, we introduce them in an abstract setting. Therefore, we speak of measurable maps $f$ and $T$, but the application intended is to continuous maps.

*Theorem* (multiplicative ergodic theorem of Oseledec). Let $\rho$ be a probability measure on a space $M$, and $f : M \to M$ a measure preserving map such that $\rho$ is ergodic. Let also $T : M \to$ the $m \times m$ matrices be a measurable map such that

$$\int \rho(dx) \log^+ ||T(x)|| < \infty ,$$

where $\log^+ u = \max(0, \log u)$. Define the matrix

386

$T_x^n = T(f^{n-1}x) \cdots T(fx)T(x)$. Then, for $\rho$-almost all $x$, the following limit exists:

$$\lim_{n \to \infty} (T_x^{n*} T_x^n)^{1/2n} = \Lambda_x . \qquad (3.1)$$

(We have denoted by $T_x^{n*}$ the adjoint of $T_x^n$, and taken the $2n$th root of the positive matrix $T_x^{n*} T_x^n$.)

The logarithms of the eigenvalues of $\Lambda_x$ are called *characteristic exponents*. We denote them by $\lambda_1 \geq \lambda_2 \geq \cdots$. They are $\rho$-almost everywhere constant. (This is because we have assumed $\rho$ ergodic. Of course, the $\lambda_i$ depend on $\rho$.) Let $\lambda^{(1)} > \lambda^{(2)} > \cdots$ be the characteristic exponents again, but no longer repeated by multiplicity; we call $m^{(i)}$ the multiplicity of $\lambda^{(i)}$. Let $E_x^{(i)}$ be the subspace of $\mathbf{R}^m$ corresponding to the eigenvalues $\leq \exp\lambda^{(i)}$ of $\Lambda_x$. Then $\mathbf{R}^m = E_x^{(1)} \supset E_x^{(2)} \supset \cdots$ and the following holds

*Theorem.* For $\rho$-almost all $x$,

$$\lim_{n \to \infty} \frac{1}{n} \log || T_x^n u || = \lambda^{(i)} \qquad (3.2)$$

if $u \in E_x^{(i)} \setminus E_x^{(i+1)}$. In particular, for all vectors $u$ that are not in the subspace $E_x^{(2)}$ (viz., almost all $u$), the limit is the largest characteristic exponent $\lambda^{(1)}$.

The above remarkable theorem dates back only to 1968, when the proof of a somewhat different version was published by Oseledec (1968). For different proofs see Raghunathan (1979), Ruelle (1979), Johnson, Palmer, and Sell (1984). What does the theorem say for $m = 1$? The $1 \times 1$ matrices are just ordinary numbers. Assuming them to be positive and taking the log, the reader will verify that the multiplicative ergodic theorem reduces to the ordinary ergodic theorem of Sec. II.E. The novelty and difficulty of the multiplicative ergodic theorem is that for $m > 1$ it deals with *noncommuting* matrices.

In some applications we shall need an extension, where $\mathbf{R}^m$ is replaced by an infinite-dimensional Banach or Hilbert space $E$ and the $T(x)$ are bounded operators. Such an extension has been proved under the condition that the $T(x)$ are *compact* operators. In the Hilbert case this means that the spectrum of $T(x)^* T(x)$ is discrete, that the eigenvalues have finite multiplicities, and that they accumulate only at 0.

*Theorem* (multiplicative ergodic theorem—compact operators in Hilbert space). All the assertions of the multiplicative ergodic theorem remain true if $\mathbf{R}^m$ is replaced by a separable Hilbert space $E$, and $T$ maps $M$ to compact operators in $E$. The characteristic exponents form a sequence tending to $-\infty$ (it may happen that only finitely many characteristic exponents are finite).

See Ruelle (1982a) for a proof. For compact operators on a Banach space, Eq. (3.1) no longer makes sense, but there are subspaces $E_x^{(1)} \supset E_x^{(2)} \supset \cdots$ such that (3.2) holds. This was shown first by Mañé (1983), with an unnecessary injectivity assumption, and then by Thieullen (1985) in full generality. (Thieullen's result applies in fact also to noncompact situations.)

## B. Characteristic exponents for differentiable dynamical systems

### 1. Discrete-time dynamical systems on $\mathbf{R}^m$

We consider the time evolution

$$x(n+1) = f(x(n)) , \qquad (3.3)$$

where $f : \mathbf{R}^m \to \mathbf{R}^m$ is a differentiable vector function. We denote by $T(x)$ the matrix $(\partial f_i / \partial x_j)$ of partial derivatives of the components $f_i$ at $x$. For the $n$th iterate $f^n$ of $f$, the corresponding matrix of partial derivatives is given by the chain rule:

$$\partial (f^n)_i / \partial x_j = T(f^{n-1}x) \cdots T(fx)T(x) . \qquad (3.4)$$

Now, if $\rho$ is an ergodic measure for $f$, with compact support, the conditions of the multiplicative ergodic theorem are all satisfied and the characteristic exponents are thus defined.

In particular, if $\delta x(0)$ is a small change in initial condition (considered as infinitesimally small), the change at time $n$ is given by

$$\delta x(n) = T_x^n \delta x(0)$$

$$= T(f^{n-1}x) \cdots T(x)\delta x(0) . \qquad (3.5)$$

For most $\delta x(0)$ [i.e., for $\delta x(0) \notin E_{x(0)}^{(2)}$] we have $\delta x(n) \approx \delta x(0) e^{n\lambda_1}$, and sensitive dependence on initial conditions corresponds to $\lambda_1 > 0$. Note that if $\delta x(0)$ is finite rather than infinitely small, the growth of $\delta x(n)$ may not go on indefinitely: if $x(0)$ is in a bounded attractor, $\delta x(n)$ cannot be larger than the diameter of the attractor.

### 2. Continuous-time dynamical systems on $\mathbf{R}^m$

If the time is continuous, we apply the multiplicative ergodic theorem to the time-one map $f = f^1$. The limits defining the characteristic exponents hold again, with $t \to \infty$ replacing $n \to \infty$ (because of continuity it is not necessary to restrict $t$ to integer values). To be specific, we define

$$T_x^t = \text{matrix } (\partial f_i^t / \partial x_j) . \qquad (3.6)$$

If $\rho$ is an ergodic measure with compact support for the time evolution, then, for $\rho$-almost all $x$, the following limits exist:

$$\lim_{t \to \infty} (T_x^{t*} T_x^t)^{1/2t} = \Lambda_x , \qquad (3.7)$$

$$\lim_{t \to \infty} \frac{1}{t} \log || T_x^t u || = \lambda^{(i)} \text{ if } u \in E_x^{(i)} \setminus E_x^{(i+1)} , \qquad (3.8)$$

where $\lambda^{(1)} > \lambda^{(2)} > \cdots$ are the logarithms of the eigenvalues of $\Lambda_x$, and $E_x^{(i)}$ is the subspace of $\mathbf{R}^m$ corresponding to the eigenvalues $\leq \exp\lambda^{(i)}$. Notice, incidentally, that if the Euclidean norm $|| \; ||$ is replaced by some other norm on $\mathbf{R}^m$, *the characteristic exponents and the* $E_x^{(i)}$ *do not change.*

### 3. Dynamical systems in Hilbert space

We assume that $E$ is a (real) Hilbert space, $\rho$ a probability measure with compact support in $E$, and $f^t$ a time evolution such that the linear operators $T_x^t = D_x f^t$ (derivative of $f^t$ at $x$) are compact linear operators for $t > 0$. This situation prevails, for instance, for the Navier-Stokes time evolution in two dimensions (as well as in three dimensions, so long as the solution has no singularities). The definition of characteristic exponents is the same here as for dynamical systems in $\mathbf{R}^m$.

### 4. Dynamical systems on a manifold M

For definiteness, let $M$ be a compact manifold like a sphere or a torus; $\rho$ is a probability measure on $M$, invariant under the dynamical system. If $M$ is $m$ dimensional, we may cut $M$ into a finite number of pieces which are smoothly parametrized by subsets of $\mathbf{R}^m$ (see Fig. 11). In terms of this new parametrization, the map $f$ is continuous except at the cuts, and so is the matrix of partial derivatives. Since only measurability is needed for the abstract multiplicative ergodic theorem, we can again define characteristic exponents. This definition is independent of the partition of the manifold $M$ that has been used, and of the choice of parametrization for the pieces. The reason is that, for any other choice, the norm used would differ from the original norm by a bounded factor, which disappears in the limit. One could alternatively use a Riemann metric on the manifold and define the characteristic exponents in terms of this metric. If $\mathcal{T}_x M$ denotes the tangent space at $x$, we now have $\mathcal{T}_x M = E_x^{(1)} \supset E_x^{(2)} \supset \cdots$.

### C. Steady, periodic, and quasiperiodic motions

#### 1. Examples and parameter dependence

Before proceeding with the general theory, we pause to discuss illustrations of the preceding results.

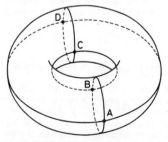

FIG. 11. A two-dimensional torus cut into four rectangular pieces by two horizontal and two vertical circles.

A *steady state* of a physical time evolution is associated with a fixed point $P$ of the corresponding dynamical system. The steady state is thus described by the probability measure $\rho = \delta_P$ (Dirac's delta at $P$), which is of course invariant and ergodic. We denote by $\alpha_1, \alpha_2, \ldots$, the eigenvalues of the operator $D_P f^1$ (derivative of the time-one map $f^1$ at $P$), in decreasing order of absolute values, and repeated according to multiplicity. Then the characteristic exponents are

$$\lambda_1 = \log|\alpha_1|, \quad \lambda_2 = \log|\alpha_2|, \quad \ldots. \tag{3.9}$$

In particular, a stable steady state associated with an attracting fixed point (see Sec. III.C.2) has negative characteristic exponents. If the dynamical system depends continuously on a bifurcation parameter $\mu$, the $\lambda_i = \log|\alpha_i|$ depend continuously on $\mu$, but we shall see in Sec. III.D that this situation is rather exceptional.

A *periodic state* of a physical time evolution is associated with a periodic orbit $\Gamma = \{f^t a : 0 \le t < T\}$ of the corresponding continuous-time dynamical system. It is thus described by the ergodic probability measure

$$\rho = \delta_\Gamma = \frac{1}{T} \int_0^T dt \, \delta_{f^t a}. \tag{3.10}$$

We denote by $\alpha_i^T$ the eigenvalues of $D_a f^T$; then one of these eigenvalues is 1 (corresponding to the direction tangent to $\Gamma$ at $a$). The characteristic exponents are the numbers

$$\lambda_i = \frac{1}{T} \log|\alpha_i^T|,$$

and one of them is thus 0. In particular, a stable periodic state, associated with an attracting periodic orbit (see Sec. II.C.2), has one characteristic exponent equal to zero and the others negative. Here again, if there is a bifurcation parameter $\mu$, the $\lambda_i$ depend continuously on $\mu$.

Consider now a *quasiperiodic state with $k$ frequencies*, stable for simplicity. This is represented by a quasiperiodic attractor (Sec. II.C.3), i.e., an attracting invariant torus $T^k$ on which the time evolution is described by translations (2.8) in terms of suitable angular variables $\varphi_1, \ldots, \varphi_k$. There is only one invariant probability measure here: the Haar measure $\rho$ on $T^k$, defined in terms of the angular variables by

$$(2\pi)^{-k} d\varphi_1 \cdots d\varphi_k.$$

Here, $k$ characteristic exponents are equal to zero, and the others are negative. If the dynamical system depends continuously on a bifurcation parameter $\mu$ and has a quasiperiodic attractor for $\mu = \mu_0$, it will still have an attracting $k$ torus for $\mu$ close to $\mu_0$, but the motion on this $k$ torus may no longer be quasiperiodic. For $k \ge 2$, frequency locking may lead to attracting periodic orbits (and negative characteristic exponents). For $k \ge 3$, strange attractors and positive characteristic exponents may be present for $\mu$ arbitrarily close to $\mu_0$ (see Ruelle and Takens, 1971; Newhouse, Ruelle, and Takens, 1978). Nevertheless, we have continuity at $\mu = \mu_0$: the characteristic exponents for $\mu$ close to $\mu_0$ are close to their values at $\mu_0$.

## 2. Characteristic exponents as indicators of periodic motion

The examples of the preceding section are typical for the case of negative characteristic exponents. We now point out that, conversely, it is possible to deduce from the negativity of the characteristic exponents that the ergodic measure $\rho$ describes a steady or a period state.

*Theorem* (continuous-time fixed point). Consider a continuous-time dynamical system and assume that all the characteristic exponents are different from zero. Then $\rho = \delta_P$, where $P$ is a fixed point. (In particular, if all characteristic exponents are negative, $P$ is an attracting fixed point.)

Another formulation: If the support of $\rho$ does not reduce to a fixed point, then one of the characteristic exponents vanishes.

*Sketch of proof.* One considers the vector function $F$,

$$F(x) = \frac{d}{d\tau} f^\tau x \bigg|_{\tau=0} . \tag{3.11}$$

If the support of $\rho$ is not reduced to a fixed point, we have $F(x) \neq 0$ for $\rho$-almost all $x$. Furthermore, Eq. (3.11) yields

$$T_x^t F(x) = (D_x f^t) F(x) = F(f^t x) .$$

Since $\rho$ is ergodic, $f^t x$ comes close to $x$ again and again, and we find for the limit (3.8)

$$\lim_{t \to \infty} \frac{1}{t} \log ||T_x^t F(x)|| = 0 .$$

Thus there is a characteristic exponent equal to 0.

In the next two theorems we assume that the dynamical system is defined by functions that have continuous second-order derivatives. (The proofs use the stable manifolds of Sec. III.E.)

*Theorem* (discrete-time periodic orbit). Consider a discrete-time dynamical system and assume that all the characteristic exponents of $\rho$ are negative. Then

$$\rho = \frac{1}{N} \sum_1^N \delta_{f^k a} ,$$

where $\{a, fa, \ldots, f^{N-1}a\}$ is an attracting periodic orbit, of period $N$.

*Proof.* See Ruelle (1979).

*Theorem* (continuous-time periodic orbit). Consider a continuous-time dynamical system and assume that all the characteristic exponents of $\rho$ are negative, except $\lambda_1$. There are then two possibilities: (a) $\rho = \delta_P$, where $P$ is a fixed point, (b) $\rho$ is the measure (3.9) on an attracting periodic orbit (and $\lambda_1 = 0$).

*Proof.* See Campanino (1980).

As an application of these results, consider the time evolution given by a differential equation (2.1) in two dimensions. We have the following possibilities for an ergodic measure $\rho$:

$$\lambda_1 = \lambda_2 = 0,$$

$\lambda_1 = 0$, $\lambda_2 < 0$: $\rho$ is associated with a fixed point or an attracting period orbit,

$\lambda_1 > 0$, $\lambda_2 = 0$: this reduces to the previous case by changing the direction of time, and therefore $\rho$ is associated with a fixed point or a repelling periodic orbit,

$\lambda_1$ and $\lambda_2$ are nonvanishing: $\rho$ is associated with a fixed point.

None of these possibilities corresponds to an *attractor* with a positive characteristic exponent. Therefore, an evolution (2.1) can be chaotic only in three or more dimensions.

### D. General remarks on characteristic exponents

We now fix an ergodic measure $\rho$, and the characteristic exponents that occur in what follows are with respect to this measure.

#### 1. The growth of volume elements

The rate of exponential growth of an infinitesimal vector $\delta x(t)$ is given in general by the largest characteristic exponent $\lambda_1$. The rate of growth of a surface element $\delta \sigma(t) = \delta_1 x(t) \wedge \delta_2 x(t)$ is similarly given in general by the sum of the largest two characteristic exponents $\lambda_1 + \lambda_2$. In general for a $k$-volume element $\delta_1 x(t) \wedge \cdots \wedge \delta_k x(t)$ the rate of growth is $\lambda_1 + \cdots + \lambda_k$. (Of course, if this sum is negative, the volume is contracted.) The construction above gives computational access to the lower characteristic exponents (and is used in the proof of the multiplicative ergodic theorem). For instance, for a dynamical system in $\mathbb{R}^m$, the rate of growth of the $m$-volume element is the rate of growth of the *Jacobian determinant* $|J_x^t| = |\det(\partial f_i^t / \partial x_j)|$, and is given by $\lambda_1 + \cdots + \lambda_m$. For a volume-preserving transformation we have thus $\lambda_1 + \cdots + \lambda_m = 0$. For a map $f$ with constant Jacobian $J$, we have $\lambda_1 + \cdots + \lambda_m = \log |J|$.

*Examples.*

In the case of the Hénon map [example (a) of Sec. II.D] we have $J = -b = -0.3$, hence $\lambda_2 = \log |J| - \lambda_1$ $\approx -1.20 - 0.42 = -1.62$.

In the case of the Lorenz equation [example (b) of Sec. II.B] we have $dJ^t/dt = -(\sigma + 1 + b)$. Therefore, if we know $\lambda_1 > 0$ we know all characteristic exponents, since $\lambda_2 = 0$ and $\lambda_3 = -(\sigma + 1 + b) - \lambda_1$.

#### 2. Lack of explicit expressions, lack of continuity

The ordinary ergodic theorem states that the time average of a function $\varphi$ tends to a limit ($\rho$-almost everywhere) and asserts that this limit is $\int \varphi(x) \rho(dx)$. By contrast, the multiplicative ergodic theorem gives no explicit expression for the characteristic exponents. It is true that in the proof of the theorem as given by Johnson *et al.* (1984) there is an integral representation of characteristic exponents in terms of a measure on the space of points $(x,Q)$, where $x$ is a point of our $m$-dimensional manifold, and $Q$ is an $m \times m$ orthogonal matrix. However, this measure is not constructively given. This situation is similar to that in statistical mechanics where, for example, there is in general no explicit expression for the pressure in terms of the interparticle forces.

For a dynamical problem depending on a bifurcation parameter $\mu$, one would like at least to know some continuity properties of the $\lambda_i$ as functions of $\mu$. The situation there is unfortunately quite bad (with some exceptions—see Sec. III.C). For each $\mu$ there may be several attractors $A_\alpha^\mu$, each having at least one physical measure $\rho_\alpha^\mu$. The dependence of the attractors on $\mu$ need not be continuous, because of captures and "explosions," and we do not know that $\rho_\alpha^\mu$ depends continuously on $A_\alpha^\mu$. Finally, even if $\rho_\alpha^\mu$ depends continuously on $\mu$, it is not true in general that the characteristic exponents do the same. To summarize: the characteristic exponents are in general discontinuous functions of the bifurcation parameter $\mu$.

*Example.*

The interval [0,1] is mapped into itself by $x \to \mu x (1-x)$ when $0 \leq \mu \leq 4$, and Fig. 12 shows $\lambda_1$ as a function of $\mu$. There are intervals of values of $\mu$ where $\lambda_1$ is negative, corresponding to an attracting periodic orbit. It is believed that these intervals are dense in [0,4]. If this is so, $\lambda_1$ is necessarily a discontinuous function of $\mu$ wherever it is positive. It is believed that $\{\mu \in [0,4]:\lambda_1 > 0\}$ has positive Lebesgue measure (this result has been announced by Jakobson, but no complete proof has appeared). For some positive results on these difficult problems see Jakobson (1981), Collet and Eckmann (1980a,1983), and Benedicks and Carleson (1984).

The wild discontinuity of characteristic exponents raises a philosophical question: should there not be at least a piecewise continuous dependence of physical quantities on parameters such as one sees, for example, in the solution of the Ising model? Yet we obtain here discontinuous predictions. Part of the resolution of this paradox lies in the fact that our mathematical predictions are *measurable functions* if not continuous, and that measurable functions have much more controllable discontinuities (cf. Luzin's theorem, for instance) than those one could construct with help of the axiom of choice. Another fact is that physical measurements are smoothed by the instrumental procedure. In particular, the definition of characteristic exponents involves a limit $t \to \infty$ [see Eqs. (3.7) and (3.8)], and the great complexity of a curve $\mu \to \lambda_1(\mu)$ will only appear progressively as $t$ is made larger and larger. The presence of noise also smooths out experimental results. At a given level of precision one may find, for instance, that there is one positive characteristic exponent $\lambda_1(\mu)$ in the interval $[\mu_1,\mu_2]$. *This is a meaningful statement,* even though it probably will have to be revised when higher-precision measurements are made; those may introduce *small* subintervals of $[\mu_1,\mu_2]$ where all characteristic exponents are negative. Let us also mention the possibility that for a *large* chaotic sys-

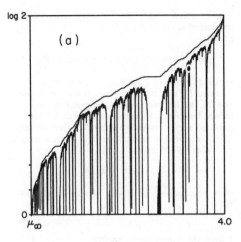

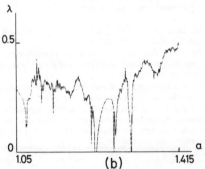

FIG. 12. (a) Topological entropy (upper curve) and characteristic exponent (lower curve) as a function of $\mu$ for the family $x \to \mu x (1-x)$. (Graph by J. Crutchfield.) Note the discontinuity of the lower curve. (b) Similar figure for the Hénon map, with $b = 0.3$, after Feit (1978).

tem (like a fully turbulent fluid) the distribution of characteristic exponents could again be a smooth function of bifurcation parameters.

### 3. Time reflection

Let us assume that the time-evolution maps $f^t$ are defined for $t$ negative as well as positive. In the discrete-time case this means that $f$ has an inverse $f^{-1}$ which is a smooth map (i.e., $f$ is a *diffeomorphism*). We may consider the time-reversed dynamical system, with time-evolution map $\bar{f}^t = f^{-t}$. If $\rho$ is an invariant (or ergodic) probability measure for the original system, it is also invariant (or ergodic) for the time-reversed system. Furthermore, *the characteristic exponents of an ergodic measure $\rho$ for the time-reversed system are those of the original* *system, but with opposite sign.* We have correspondingly a sequence of subspaces $\bar{E}_x^{(1)} \subset \bar{E}_x^{(2)} \subset \cdots$ for almost all $x$, such that

$$\lim_{t \to -\infty} \frac{1}{|t|} \log||T_x^t u|| = -\lambda^{(i)} \quad \text{if } u \in \bar{E}_x^{(i)} \setminus \bar{E}_x^{(i-1)} .$$

Define $F_x^{(i)} = E_x^{(i)} \cap \bar{E}_x^{(i)}$. Then, for $\rho$-almost all $x$, the subspaces $F_x^{(i)}$ span $\mathbf{R}^m$ (or the tangent space to the manifold $M$, as the case may be; compact operators in infinite-dimensional Hilbert space are excluded here because they are not compatible with $t < 0$). Furthermore, if $T_x^t$ is the derivative matrix or operator corresponding to $\bar{f}^{|t|}$ when $t < 0$, we have

$$\lim_{|t| \to \infty} \frac{1}{t} \log||T_x^t u|| = \lambda^{(i)} \quad \text{if } u \in F_x^{(i)} ,$$

where $t$ may go to $+\infty$ or $-\infty$. (For details see Ruelle, 1979.)

### 4. Relations between continuous-time and discrete-time dynamical systems

We have defined the characteristic exponents for a continuous-time dynamical system [see Eqs. (3.7) and (3.8)] so that they are the same as the characteristic exponents for the discrete-time dynamical system generated by the time-one map $f = f^1$.

Given a Poincaré section (see Sec. II.H), we want to relate the characteristic exponents $\lambda_i$ for a continuous-time dynamical system with the characteristic exponents $\tilde{\lambda}_i$ corresponding to the first return map $P$. Note that one of the $\lambda_i$ is zero (first theorem in Sec. III.C); we claim that the other $\lambda_i$ are given by

$$\lambda_i = \tilde{\lambda}_i / \langle \tau \rangle_\sigma , \tag{3.12}$$

where $\langle \tau \rangle_\sigma$ is the average time between two crossings of the Poincaré section $\Sigma$, computed with respect to the probability measure $\sigma$ on $\Sigma$ naturally associated with $\rho$. (The measure $\sigma$ gives the density of intersections of orbits with $\Sigma$.) The proof is not hard and is left to the reader.

### 5. Hamiltonian systems

Consider a Hamiltonian (i.e., conservative) system with $m$ degrees of freedom. This is a continuous-time dynamical system in $2m$ dimensions. We claim that the set of $\lambda_i$'s is symmetric with respect to 0. This is readily checked from Eq. (3.7) and the fact that $T_x^t$ is a symplectic matrix. Actually, two of the $\lambda_i$ vanish; we get rid of one by going to a $(2m-1)$-dimensional energy surface, and one zero characteristic exponent survives in accordance with the first theorem of Sec. III.C.2.

### E. Stable and unstable manifolds

The multiplicative ergodic theorem asserts the existence of *linear* spaces $E_x^{(1)} \supset E_x^{(2)} \supset \cdots$ such that

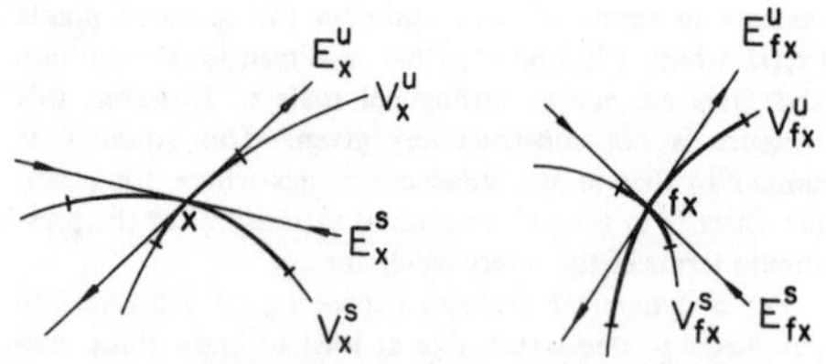

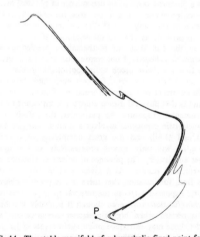

FIG. 13. Stable and unstable manifolds can be defined for points that are neither fixed nor periodic. The stable and unstable directions $E_x^s$ and $E_x^u$ are tangent to the stable and unstable manifolds $V_x^s$ and $V_x^u$, respectively. They are mapped by $f$ onto the corresponding objects at $fx$.

$$\lim_{t \to \infty} \frac{1}{t} \log||T_x^t u|| \leq \lambda^{(i)} \quad \text{if } u \in E_x^{(i)} .$$

This means that there exist subspaces $E_x^{(i)}$ such that the vectors in $E_x^{(i)} \setminus E_x^{(i+1)}$ are expanded exponentially by time evolution with the rate $\lambda^{(i)}$. (This expansion is of course a contraction if $\lambda^{(i)} < 0$.) See Fig. 13.

One can define a nonlinear analog of those $E_x^{(i)}$ which correspond to negative characteristic exponents. Let $\lambda < 0$, $\varepsilon > 0$, and write

$$V_x^s(\lambda, \varepsilon) = \{ y : d(f^t x, f^t y) \leq \varepsilon e^{\lambda t} \text{ for all } t \geq 0 \} ,$$

where $d(x, y)$ is the distance of $x$ and $y$ (Euclidean distance in $\mathbf{R}^m$, norm distance in Hilbert space, or Riemann distance on a manifold). We shall assume from now on that the time-one map $f^1$ has continuous derivatives of second as well as first order. If $\lambda^{(i-1)} > \lambda > \lambda^{(i)}$, the set $V_x^s(\lambda, \varepsilon)$ is in fact, for $\rho$-almost all $x$ and small $\varepsilon$, a piece of *differentiable* manifold, called a *local stable manifold* at $x$; it is tangent at $x$ to the linear space $E_x^{(i)}$ (and has the

FIG. 14. The stable manifold of a hyperbolic fixed point folds up on itself. (The map is after Hénon and Heiles, 1964.)

same dimension). One shows that $V_x^s(\lambda,\varepsilon)$ is differentiable as many times as $f^1$.

If we assume that our dynamical system is defined for negative as well as positive times, we can define *global stable manifolds* such that

$$V_x^{(i)s} = \left\{ y : \lim_{t\to\infty} \frac{1}{t}\log d\,(f^t x, f^t y) \le \lambda^{(i)} \right\}$$

$$= \bigcup_{t>0} f^{-t} V_x^s(\lambda,\varepsilon) \ ,$$

with negative $\lambda$ between $\lambda^{(i-1)}$ and $\lambda^{(i)}$ as above.

These global manifolds have the somewhat annoying feature that, while they are locally smooth, they tend to fold and accumulate in a very complicated manner, as suggested by Fig. 14. We can also define *the* stable manifold of $x$ by

$$V_x^s = \left\{ y : \lim_{t\to\infty} \frac{1}{t}\log d\,(f^t x, f^t y) < 0 \right\}$$

(it is the largest of the stable manifolds, equal to $V_x^{(i)s}$ where $\lambda^{(i)}$ is the largest negative characteristic exponent).

For a dynamical system where negative times are allowed, we obtain *unstable manifolds* $V^u$ instead of stable manifolds simply through replacement of $t$ by $-t$ in the definitions. Instead of assuming that $f^t$ is defined for $t<0$, we find it desirable to make the weaker assumption that $f^t$ and $Df^t$ (defined for $t\ge 0$) are *injective*. This means that $f^t x = f^t y$ implies $x=y$ and $D_x f^t u = D_x f^t v$ implies $u=v$. This injectivity assumption is satisfied when the dynamical system is defined for negative as well as positive times, but also in the case of the Navier-Stokes time evolution. The *global unstable manifold* $V_x^u$ is then defined, provided that for every $t>0$ there is $x_{-t}$ such that $f^t x_{-t} = x$; the definition is

$$V_x^u = \left\{ y : \text{there exists } y_{-t} \text{ such that } f^t y_{-t} = y \text{ and } \lim_{t\to\infty} \frac{1}{t}\log d\,(x_{-t}, y_{-t}) < 0 \right\} \ .$$

If $\lambda > 0$ we define similarly

$$V_x^u(\lambda,\varepsilon) = \{ y : \text{there exists } y_{-t} \text{ such that } f^t y_{-t} = y \text{ and } d\,(x_{-t}, y_{-t}) \le \varepsilon e^{-\lambda t} \text{ for all } t \ge 0 \} \ .$$

and if $\lambda > 0$ and $\lambda^{(i+1)} < \lambda < \lambda^{(i)}$, we write

$$V_x^{(i)u} = \left\{ y : \text{there exists } y_{-t} \text{ such that } f^t y_{-t} = y \text{ and } \lim_{t\to\infty} \frac{1}{t} d\,(x_{-t}, y_{-t}) \le -\lambda^{(i)} \right\}$$

$$= \bigcup_{t>0} f^t V_x^u(\lambda,\varepsilon)$$

The global unstable manifold $V_x^u$ is the largest of the $V_x^{(i)u}$, corresponding to the smallest positive characteristic exponent $\lambda^{(i)}$. Here again one shows that the local unstable manifolds $V_x^u(\lambda,\varepsilon)$ are differentiable (as many times, in fact, as $f^1$), while the global unstable manifolds $V_x^{(i)u}$ and $V_x^u$ are locally differentiable, but may accumulate on themselves in a complicated manner globally.

The theory of stable and unstable manifolds is part of Pesin theory (for some details, see Sec. III.G).

*Examples.*

(a) *Fixed points.* If $P$ is a fixed point for a dynamical system (with discrete or continuous time), the characteristic exponents of the $\delta$-measure $\delta_P$ at $P$ are called characteristic exponents of the fixed point. They are given explicitly by Eq. (3.9). The fixed point $P$ is said to be *hyperbolic* if all characteristic exponents $\lambda_i$ are nonzero. When all $\lambda_i < 0$, $P$ is *attracting*. When all $\lambda_i > 0$, $P$ is *repelling*. When some $\lambda_i$ are $>0$ and some $<0$, $P$ is of *saddle type*. The *stable* and *unstable manifolds* of the hyperbolic fixed point $P$ are defined to be the stable and unstable manifolds of $\delta_P$. One has

$$V_x^s = \left\{ y : \lim_{t\to+\infty} f^t y = x \right\} \ ,$$

$$V_x^u = \left\{ y : \lim_{t\to-\infty} f^t y = x \right\} \ .$$

(b) *Periodic orbits.* Let $\Gamma$ be a closed orbit for a continuous-time dynamical system. There is only one invariant measure with support $\Gamma$, namely $\delta_\Gamma$ given by Eq. (3.10); it is ergodic. If $u$ is a vector tangent to $\Gamma$ at $x$, the corresponding characteristic exponent is zero as one may easily check. If all other characteristic exponents are nonzero, $\Gamma$ is a *hyperbolic periodic orbit*. The *attracting, repelling,* and *saddle-type* periodic orbits are similarly defined. If $x \in \Gamma$ we have

$$V_x^s = \left\{ y : \lim_{t\to+\infty} d\,(f^t x, f^t y) = 0 \right\}$$

This is also called the *strong stable manifold* of $x$, and a stable manifold of $\Gamma$ is defined by

$$V_\Gamma^{cs} = \bigcup_{x\in\Gamma} V_x^s$$

$$= \bigcup_{t>0} f^{-t} V_\Gamma^{cs}(\varepsilon) \ ,$$

where the local stable manifold $V_\Gamma^{cs}(\varepsilon)$ is defined for small $\varepsilon$ by

$V_\Gamma^{ct}(\varepsilon) = \{ y : d(f^t x, f^t y) < \varepsilon \text{ for all } t \geq 0 \}$ .

*Theorem.* If $A$ is an attracting set, and $x \in A$, then $V_x^u \subset A$ i.e., the unstable manifold of $x$ is contained in $A$.

*Proof.* If $U$ is a fundamental neighborhood of $A$, and $y \in V_x^u$, then $f^{-\tau} y \in U$ for sufficiently large $\tau$ (because $f^{-\tau} y$ is close to $f^{-\tau} x \in A$). Therefore $y \in \bigcap_{\tau > T} f^\tau U = A$.

*Corollary.* Let $A$ be an attracting set. The number of characteristic exponents $\lambda_i > 0$ for any ergodic measure with support in $A$ is a lower bound to the dimension of $A$.

*Proof.* The dimension of $A$ is at least that of $V_x^u$, which is equal to the dimension of $\overline{E}_x^{(k)}$, where $\lambda^{(k)}$ is the smallest positive characteristic exponent. But $\dim \overline{E}_x^{(k)}$ is the sum of the multiplicities of the positive $\lambda^{(l)}$, i.e., the number of positive characteristic exponents $\lambda_l$.

(c) *Visualization of the unstable manifolds.* The Hénon attractor has a characteristic appearance of a line folded over many times (see Fig. 5). A similar picture appears for attractors of other two-dimensional dynamical systems generated by a diffeomorphism (differentiable map with differentiable inverse). The theorem stated above suggests that the convoluted lines seen in such attractors are in fact *unstable manifolds.* This suggestion is confirmed by the fact that in many cases the *physical measure* on an attractor is *absolutely continuous on unstable manifolds,* as we shall discuss below.

In higher dimensions, the unstable manifolds forming an attractor may be lines (one dimension), veils (two dimensions), etc. Attractors corresponding to noninvertible maps in two dimensions often have the characteristic appearance of folded veils or drapes, and it is thus immediately apparent that they do not come from a diffeomorphism (Fig. 15).

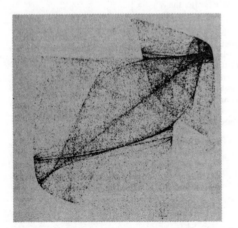

FIG. 15. The map $x' = (A - x - B_1 y)x$, $y' = (A - B_2 x - y)y$ with $A = 3.7$, $B_1 = 0.1$, $B_2 = 0.15$ is not invertible. Shown are 50 000 iterates. The map was described in Ushiki *et al.* (1980).

### F. Axiom-$A$ dynamical systems*

We discuss here some concepts of *hyperbolicity* which will be referred to in Sec. IV. The hurried reader may skip this discussion without too much disadvantage. In this section, $M$ will always be a compact manifold of dimension $m$. We shall denote by $T_x M$ the tangent space to $M$ at $x$. If $f : M \to M$ is a differentiable map, we shall denote by $T_x f : T_x M \to T_{fx} M$ the corresponding tangent map. (We refer the reader to standard texts on differential geometry for the definitions.) If a Riemann metric is given on $M$, the vector spaces $T_x M$ acquire norms $|| \; ||_x$.

#### 1. Diffeomorphisms

Let $f : M \to M$ be a diffeomorphism, i.e., a differentiable map with differentiable inverse $f^{-1}$.

We say that a point $a$ of $M$ is *wandering* if there is an open set $B$ containing $a$ [say a ball $B_a(\varepsilon)$] such that $B \cap f^k B = \emptyset$ for all $k > 0$ (or we might equivalently require this only for all $k$ large enough). The set of points that are not wandering is the *nonwandering set* $\Omega$. It is a closed, $f$-invariant subset of $M$.

Let $\Lambda$ be a closed $f$-invariant subset of $M$, and assume that we have linear subspaces $E_x^-, E_x^+$ of $T_x M$ for each $x \in \Lambda$, depending continuously on $x$ and such that

$$T_x M = E_x^+ + E_x^-, \quad \dim E_x^+ + \dim E_x^- = m .$$

Assume also that $T_x f E_x^- = E_{fx}^-$ and $T_x f E_x^+ = E_{fx}^+$ (i.e., $E^-, E^+$ form a continuous invariant splitting of $TM$ over $\Lambda$). One says that $\Lambda$ is a *hyperbolic set* if one may choose $E^-$ and $E^+$ as above, and constants $C > 0$, $\Theta > 1$ such that, for all $n \geq 0$,

$$|| T_x f^n u ||_{f^n x} \leq C \Theta^{-n} || u ||_x \quad \text{if } u \in E_x^- ,$$

$$|| T_x f^{-n} v ||_{f^{-n} x} \leq C \Theta^{-n} || v ||_x \quad \text{if } v \in E_x^+ .$$

[Note that, as a consequence, no ergodic measure with support in $\Lambda$ has characteristic exponents in the interval $(-\Theta^{-1}, \Theta^{-1})$.]

If the whole manifold $M$ is hyperbolic, $f$ is called an *Anosov diffeomorphism.* [Arnold's cat map, Sec. II.D, example (c), is an Anosov diffeomorphism.]

If the nonwandering set $\Omega$ is hyperbolic, and if the periodic points are dense in $\Omega$, $f$ is called an *Axiom-$A$ diffeomorphism.* (Every Anosov diffeomorphism is an Axiom-$A$ diffeomorphism.)

#### 2. Flows

Consider a continuous-time dynamical system $(f^t)$ on $M$, where $f^t$ is defined for all $t \in R$; $(f^t)$ is then also called a *flow.*

We say that a point $a$ of $M$ is *wandering* if there is an open set $B$ containing $a$ [say a ball $B_a(\varepsilon)$] such that $B \cap f^t B = \emptyset$ for all sufficiently large $t$. The set of points that are not wandering is the *nonwandering set* $\Omega$. It is a closed, $(f^t)$-invariant subset of $M$.

Let $\Lambda$ be a closed invariant subset of $M$ containing no fixed point. Assume that we have linear subspaces $E_x^-, E_x^0, E_x^+$ of $T_xM$ for each $x \in \Lambda$, depending continuously on $x$, and such that

$$T_xM = E_x^- + E_x^0 + E_x^+ \ ,$$

$$\dim E_x^0 = 1, \quad \dim E_x^- + \dim E_x^+ = m - 1 \ .$$

Assume also that $E_x^0$ is spanned by

$$\left. \frac{d}{dt} f^t x \right|_{t=0} \ ,$$

i.e., $E_x^0$ is in the direction of the flow, and that

$$T_x f^t E_x^- = E_{f^t x}^- \ ,$$

$$T_x f^t E_x^+ = E_{f^t x}^+ \ .$$

One says that $\Lambda$ is a *hyperbolic set* if one may choose $E^-$, $E^0$, and $E^+$ as above, and constants $C > 0$, $\Theta > 1$ such that, for all $t \geq 0$

$$||T_x f^t u||_{f^t x} \leq C\Theta^{-t}||u||_x \quad \text{if } u \in E_x^- \ ,$$

$$||T_x f^{-t} v||_{f^{-t} x} \leq C\Theta^{-t}||v||_x \quad \text{if } v \in E_x^+ \ .$$

More generally, we shall also say that $\Lambda^*$ is a *hyperbolic set* if $\Lambda^*$ is the union of $\Lambda$ as above and of a finite number of hyperbolic fixed points [Sec. III.E, example (a)].

If the whole manifold $M$ is a hyperbolic, $(f^t)$ is called an *Anosov flow*.

If the nonwandering set $\Omega$ is hyperbolic, and if the periodic orbits and fixed points are dense in the $\Omega$, then $(f^t)$ is called an *Axiom-A flow*.

### 3. Properties of Axiom-$A$ dynamical systems

Axiom-$A$ dynamical systems were introduced by Smale [for reviews, see Smale's original paper (1967) and Bowen (1978)]. Smale proved the following "*spectral theorem*" valid both for diffeomorphism and flows.

*Theorem.* $\Omega$ is the union of finitely many disjoint closed invariant sets $\Omega_1, \ldots, \Omega_s$, and for each $\Omega_i$ there is $x \in \Omega_i$ such that the orbit $\{f^t x\}$ is dense in $\Omega_i$. The decomposition $\Omega = \Omega_i \cup \cdots \cup \Omega_s$ is unique with these properties.

The sets $\Omega_i$ are called *basic sets*, while those which are attracting sets are called *attractors* (there is always at least one attractor among the basic sets).

Some of the ergodic properties of Axiom-$A$ attractors will be discussed in Sec. IV. The great virtue of these systems is that they can be analyzed mathematically in detail, while many properties of a map apparently as simple as the Hénon diffeomorphism [Sec. II.D, example (a)] remain conjectural.

It should be pointed out that there is a vast literature on the Axiom-$A$ systems, concerned in particular with *structural stability*.

### G. Pesin theory[*]

We have seen above that the stable and unstable manifolds (defined almost everywhere with respect to an ergodic measure $\rho$) are differentiable. This is part of a theory developed by Pesin (1976,1977).[2] Pesin assumes that $\rho$ has differentiable density with respect to Lebesgue measure, but this assumption is not necessary for the study of stable and unstable manifolds [see Ruelle (1979), and for the infinite-dimensional case Ruelle (1982a) and Mañé (1983)].

The earlier results on differentiable dynamical systems had been mostly geometric and restricted to *hyperbolic* (Anosov, 1967) or *Axiom-A* systems (Smale, 1967). Pesin's theory extends a good part of these geometric results to arbitrary differentiable dynamical systems, but working now *almost everywhere* with respect to some ergodic measure $\rho$. (The results are most complete when all characteristic exponents are different from zero.) The original contribution of Pesin has been extended by many workers, notably Katok (1980) and Ledrappier and Young (1984). Many of the results quoted in Sec. IV below depend on Pesin theory, and we shall give an idea of the present aspect of the theory in that section. Here we mention only one of Pesin's original contributions, a striking result concerning area-preserving diffeomorphisms (in two dimensions).

*Theorem* (Pesin). Let $f$ be an area-preserving diffeomorphism, and $f$ be twice differentiable. Suppose $fS = S$ for some bounded region $S$, and let $S'$ consist of the points of $S$ which have nonzero characteristic exponents. Then (up to a set of measure 0) $S'$ is a countable union of ergodic components.

In this theorem the area defines an invariant measure on $S$, which is not ergodic in general, and $S$ can therefore be decomposed into further invariant sets. This may be a continuous decomposition (like that of a disk into circles). The theorem states that *where the characteristic exponents are nonzero, the decomposition is discrete.*

## IV. ENTROPY AND INFORMATION DIMENSION

In this section we introduce two more ergodic quantities: the *entropy* (or *Kolmogorov-Sinai invariant*) and the *information dimension*. We discuss how these quantities are related to the characteristic exponents. The measurement of the entropy and information dimension in physical and computer experiments will be discussed in Sec. V.

### A. Entropy

As we have noted already in the Introduction, a system with sensitive dependence on initial conditions produces

---

[2]For a systematic exposition see Fathi, Herman, and Yoccoz (1983).

information. This is because two initial conditions that are different but indistinguishable at a certain experimental precision will evolve into indistinguishable states after a finite time. If $\rho$ is an ergodic probability measure for a dynamical system, we introduce the concept of *mean rate of creation of information* $h(\rho)$, also known as *measure-theoretic entropy* or the *Kolmogorov-Sinai invariant* or simply *entropy*. When we study the dynamics of a dissipative physicochemical system, it should be noted that the Kolmogorov-Sinai entropy is not the same thing as the thermodynamic entropy of the system. To define $h(\rho)$ we shall assume that the support of $\rho$ is a compact set with a given metric. (More general cases can be dealt with, but in our applications supp $\rho$ is indeed a compact metric space.) Let $\mathscr{A} = (\mathscr{A}_1, \ldots, \mathscr{A}_\alpha)$ be a finite ($\rho$-measurable) partition of the support of $\rho$. For every piece $\mathscr{A}_j$ we write $f^{-k}\mathscr{A}_j$ for the set of points mapped by $f^k$ to $\mathscr{A}_j$. We then denote by $f^{-k}\mathscr{A}$ the partition $(f^{-k}\mathscr{A}_1, \ldots, f^{-k}\mathscr{A}_\alpha)$. Finally, $\mathscr{A}^{(n)}$ is defined as

$$\mathscr{A}^{(n)} = \mathscr{A} \vee f^{-1}\mathscr{A} \vee \cdots \vee f^{-n+1}\mathscr{A},$$

which is the partition whose pieces are

$$\mathscr{A}_{i_1} \cap f^{-1}\mathscr{A}_{i_2} \cap \cdots \cap f^{-n+1}\mathscr{A}_{i_n}$$

with $i_j \in \{1, 2, \ldots, \alpha\}$. What is the significance of these partitions? The partition $f^{-k}\mathscr{A}$ is deduced from $\mathscr{A}$ by time evolution (note that $f^k\mathscr{A}$ need not be a partition, since $f$ might be many-to-one; this is why we use $f^{-k}\mathscr{A}$). The partition $\mathscr{A}^{(n)}$ is the partition generated by $\mathscr{A}$ in a time interval of length $n$. We write

$$H(\mathscr{A}) = -\sum_{i=1}^{\alpha} \rho(\mathscr{A}_i)\log\rho(\mathscr{A}_i), \qquad (4.1)$$

with the understanding that $u\log u = 0$ when $u = 0$. (We strongly advise using natural logarithms, but $\log_{10}$ and $\log_2$ have their enthusiasts.) Thus $H(\mathscr{A})$ is the information content of the partition $\mathscr{A}$ with respect to the state $\rho$, and $H(\mathscr{A}^{(n)})$ is the same, over an interval of time of length $n$. The following limits are asserted to exist, defining $h(\rho, \mathscr{A})$ and $h(\rho)$:

$$h(\rho, \mathscr{A}) = \lim_{n \to \infty} [H(\mathscr{A}^{(n+1)}) - H(\mathscr{A}^{(n)})]$$

$$= \lim_{n \to \infty} \frac{1}{n}H(\mathscr{A}^{(n)}), \qquad (4.2)$$

$$h(\rho) = \lim_{\text{diam}\mathscr{A} \to 0} h(\rho, \mathscr{A}), \qquad (4.3)$$

where $\text{diam}\mathscr{A} = \max_i\{$diameter of $\mathscr{A}_i\}$. Clearly, $h(\rho, \mathscr{A})$ is the rate of information creation with respect to the partition $\mathscr{A}$, and $h(\rho)$ its limit for finer and finer partitions. This last limit may sometimes be avoided [i.e., $h(\rho, \mathscr{A}) = h(\rho)$]; this is the case when $\mathscr{A}$ is a *generating partition*. This holds in particular if $\text{diam}\mathscr{A}^{(n)} \to 0$ when $n \to \infty$, or if $f$ is invertible and $\text{diam}f^n\mathscr{A}^{(2n)} \to 0$ when $n \to \infty$. For example, for the map of Fig. 8, a generating partition is obtained by dividing the interval at the singularity in the middle. For more details we must refer the reader to the literature, for instance the excellent book by

Billingsley (1965).

The above definition of the entropy applies to continuous as well as discrete-time systems. In fact, the entropy in the continuous-time case is just the entropy $h(\rho, f^1)$ corresponding to the time-one map. We also have the formula

$$h(\rho, f^T) = |T| h(\rho, f^1).$$

Note that the definition of the entropy in the continuous-time case *does not involve a time step t tending to zero*, contrary to what is sometimes found in the literature. Note also that the entropy does not change if $f$ is replaced by $f^{-1}$.

If $(f^t)$ has a Poincaré section $\Sigma$, we let $\sigma$ be the probability measure on $\Sigma$, invariant under the Poincaré map $P$ and corresponding to $\rho$ (i.e., $\sigma$ is the density of intersection of orbits of the continuous dynamical system with $\Sigma$). If we also let $\tau$ be the first return time, then we have *Abramov's formula*,

$$h(\rho) = \frac{h(\sigma)}{\langle \tau \rangle_\sigma}$$

which is analogous to Eq. (3.12) for the characteristic exponents.

The relationship of entropy to characteristic exponents is very interesting. First we have a general inequality.

*Theorem* (Ruelle, 1978). Let $f$ be a differentiable map of a finite-dimensional manifold and $\rho$ an ergodic measure with compact support. Then

$$h(\rho) \leq \Sigma \text{ positive } \lambda_i. \qquad (4.4)$$

The result is believed to hold in infinite dimensions as well, but no proof has been published yet.

It is of considerable interest that the *equality* corresponding to Eq. (4.4) seems to hold often (but not always) for the *physical measures* (Sec. II.F) in which we are mainly interested. This equality is called the *Pesin identity*:

$$h(\rho) = \Sigma \text{ positive } \lambda_i.$$

Pesin proved that it holds if $\rho$ is invariant under the diffeomorphism $f$, and $\rho$ has smooth density with respect to Lebesgue measure. More generally, the Pesin identity holds for the *SRB measures* to be studied in Sec. IV.B.

In Sec. V we shall use in addition an entropy concept different from that of Eqs. (4.1)–(4.3). It is given by

$$H_2(\mathscr{A}) = -\log \sum_{i=1}^{\alpha} \rho(\mathscr{A}_i)^2, \qquad (4.5)$$

$$K_2(\rho) = \lim_{\text{diam}\mathscr{A} \to 0} \lim_{n \to \infty} \frac{1}{n}H_2(\mathscr{A}^{(n)}),$$

if these limits exist (see Grassberger and Procaccia, 1983a). It can be shown that *the $K_2$ entropy is a lower bound to the entropy $h(\rho)$*:

$$K_2(\rho) \leq h(\rho). \qquad (4.6)$$

## B. SRB measures

We have seen in Sec. III.E that attracting sets are unions of unstable manifolds. Transversally to these, one often finds a discontinuous structure corresponding to the complicated piling up of the unstable manifolds upon themselves. This suggests that invariant measures may have very rough densities in the directions transversal to the foliations of the unstable manifolds. On the other hand, we may expect that—due to stretching in the unstable direction—the measure is smooth when viewed along these directions. We shall call SRB measures (for Sinai, Ruelle, Bowen) those measures that are smooth along unstable directions. They turn out to be a natural and useful tool in the study of physical dynamical systems.

Much of this section is concerned with consequences of the existence of SRB measures. These are mostly relations between entropy, dimensions, and characteristic exponents. To prove the existence of SRB measures for a given system is a hard task, and whether they exist is not known in general. Sometimes no SRB measures exist, but it is unclear how frequently this happens. On the other hand, we do not have much of physical relevance to say about systems without SRB measures.

To repeat, we should like to define, intuitively, SRB measures as measures with smooth density in the stretching, or *unstable*, directions of the dynamical system defined by $f$. The geometric complexities described above make a rather technical definition necessary. Before going into these technicalities, we discuss the framework in which we shall work.

(a) In the ergodic theory of differentiable dynamical systems, there is no essential difference between discrete-time and continuous-time dynamical systems. In fact, if we discretize a continuous-time dynamical system by restricting $t$ to integer values (i.e., use the time-one map $f = f^1$ as a generator), then the characteristic exponents, the stable and unstable manifolds, and the entropy are unchanged. (The information dimension to be defined in Sec. IV.C also remains the same.) We may thus, for simplicity, *consider only discrete-time systems*.

(b) If $f$ is a diffeomorphism (i.e., a differentiable map with differential inverse), then our dynamical system is defined for negative as well as positive times. If, in addition, $f$ is twice differentiable, then the inverse map is also twice differentiable. We shall assume a little less, namely, that $f$ is *twice differentiable* and either a *diffeomorphism* or at least such that $f$ and $Df$ are *injective* (i.e., $fx = fy$ implies $x = y$, and $D_x fu = D_x fv$ implies $u = v$; these conditions hold for the Navier-Stokes time evolution).

Given an ergodic measure $\rho$ (with compact support as usual), unstable manifolds $V_x^u$ are defined for almost all $x$ according to Eq. (3.13). Notice that $y \in V_x^u$ is the same thing as $x \in V_y^u$, so that the unstable manifolds $V^u$ partition the space into equivalence classes. It might seem natural to define SRB measures by using this partition for a decomposition of $\rho$ into pieces $\rho_\alpha$, carried by different unstable manifolds:

$$\rho = \int \rho_\alpha m(d\alpha) , \qquad (4.7)$$

where $\alpha$ parametrizes the $V^u$'s, and $m$ is a measure on the "space of equivalence classes." In reality, this space of equivalence classes does not exist in general (as a measurable space) because of the folding and accumulation of the global unstable manifolds [and the existence of a nontrivial decomposition (4.7) would contradict ergodicity].

The correct approach is as follows. Let $S$ be a $\rho$-measurable set of the form $S = \cup_{\alpha \in A} S_\alpha$, where the $S_\alpha$ are disjoint small open pieces of the $V^u$'s (say each $S_\alpha$ is contained in a *local* unstable manifold). If this decomposition is $\rho$ measurable, then one has

$$\rho \text{ restricted to } S = \int \rho_\alpha m(d\alpha) ,$$

where $m$ is a measure on $A$, and $\rho_\alpha$ is a probability measure on $S_\alpha$ called the *conditional probability measure* associated with the decomposition $S = \cup_{\alpha \in A} S_\alpha$. The $\rho_\alpha$ are defined $m$-almost everywhere. See Fig. 16. The situation of interest for the definition of SRB measures occurs when *the conditional probabilities $\rho_\alpha$ are absolutely continuous with respect to Lebesgue measure* on the $V^u$'s. This means that

$$\rho_\alpha(d\xi) = \varphi_\alpha(\xi) d\xi \text{ on } S_\alpha , \qquad (4.8)$$

where $d\xi$ denotes the volume element when $S_\alpha$ is smoothly parametrized by a piece of $\mathbf{R}^{m_+}$ and $\varphi_\alpha$ is an integrable function. The *unstable dimension* $m_+$ of $S_\alpha$ or $V^u$ is the sum of the multiplicities of the positive characteristic exponents. It is finite even for the case of the Navier-Stokes equation discussed earlier (because $\lambda_i \to -\infty$ when $i \to \infty$, as we have noted).

We say that the ergodic measure $\rho$ is an *SRB measure* if its conditional probabilities $\rho_\alpha$ are absolutely continuous with respect to Lebesgue measure for some choice of $S$ with $\rho(S) > 0$, and a decomposition $S = \cup_\alpha S_\alpha$ as above. The definition is independent of the choice of $S$ and its decomposition (this is an easy exercise in ergodic theory). We shall also say that *$\rho$ is absolutely continuous along unstable manifolds*.

*Theorem* (Ledrappier and Young, 1984). Let $f$ be a twice differentiable diffeomorphism of an $m$-dimensional manifold $M$ and $\rho$ an ergodic measure with compact support. The following conditions are then equivalent: (a) The measure $\rho$ is an SRB measure, i.e., $\rho$ is absolutely

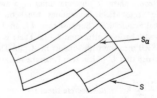

FIG. 16. A decomposition of the set $S$ into smooth leaves $S_\alpha$, each of which is contained in the unstable manifold.

continuous along unstable manifolds. (b) The measure $\rho$ satisfies Pesin's identity,

$$h(\rho) = \Sigma \text{ positive characteristic exponents} .$$

Furthermore, if these conditions are satisfied, the density functions $\varphi_\alpha$ in Eq. (4.8) are differentiable.

The theorem says that if $\rho$ is absolutely continuous along unstable manifolds, then the rate of creation of information is the mean rate of expansion of $m_+$-dimensional volume elements. If, however, $\rho$ is singular along unstable manifolds, then this rate is strictly less than the rate of expansion. These assertions are intuitively quite reasonable, but in fact quite hard to prove. The first proofs have been given for Axiom-$A$ systems (see Sec. III.F) by Sinai (1972; Anosov systems), Ruelle (1976; Axiom-$A$ diffeomorphisms), and Bowen and Ruelle (1975; Axiom-$A$ flows). The general importance of (a) and (b) was stressed by Ruelle (1980).

One hopes that there is an infinite-dimensional extension applying to Navier-Stokes, but such an extension has not yet been proved. Ledrappier (1981b) has obtained a version of the above theorem that is valid for noninvertible maps in one dimension.

The SRB measures are of particular interest for physics because one can show—in a number of cases—that the ergodic averages

$$\frac{1}{n} \sum_{k=0}^{n-1} \delta_{f^k x}$$

tend to the SRB measure $\rho$ when $n \to \infty$, not just for $\rho$-almost all $x$, but for $x$ in a set of positive *Lebesgue* measure. Lebesgue measure corresponds to a more *natural notion of sampling* than the measure $\rho$ (which is carried by an attractor and usually singular). The above property is thus both strong and natural.

To formulate this result as a theorem, we need the notion of a *subset of Lebesgue measure zero* on an $m$-dimensional manifold. We say that a set $S \subset M$ is Lebesgue measurable (has zero Lebesgue measure or positive Lebesgue measure) if for a smooth parametrization of $M$ by patches of $\mathbb{R}^m$ one finds that $S$ is Lebesgue measurable (has zero Lebesgue measure or positive Lebesgue measure). These definitions are independent of the choice of parametrization (in contrast to the *value* of the measure).

*Theorem* (SRB measures for Axiom-$A$ systems). Consider a dynamical system determined by a twice differentiable diffeomorphism $f$ (discrete time) or a twice differentiable vector field (continuous time) on an $m$-dimensional manifold $M$. Suppose that $A$ is an Axiom-$A$ attractor, with basin of attraction $U$. (a) There is one and only one SRB measure with support in $A$. (b) There is a set $S \subset U$ such that $U \setminus S$ has zero Lebesgue measure, and

$$\lim_{n \to \infty} \frac{1}{n} \sum_{k=0}^{n-1} \delta_{f^k x} = \rho \quad \text{(discrete time)} ,$$

or

$$\lim_{T \to \infty} \frac{1}{T} \int_0^T dt\, \delta_{f^t x} = \rho \quad \text{(continuous time)} ,$$

whenever $x \in S$.

For a proof, see Sinai (1972; Anosov systems), Ruelle (1976; Axiom-$A$ diffeomorphisms), or Bowen and Ruelle (1975; Axiom-$A$ flows). The "geometric Lorenz attractor" can be treated similarly.

The following theorem shows that the requirement of Axiom $A$ can be replaced by weaker information about the characteristic exponents.

*Theorem* (Pugh and Shub, 1984). Let $f$ be a twice differentiable diffeomorphism of an $m$-dimensional manifold $M$ and $\rho$ an SRB measure such that all characteristic exponents are different from zero. Then there is a set $S \subset M$ with positive Lebesgue measure such that

$$\lim_{n \to \infty} \frac{1}{n} \sum_{k=0}^{n-1} \delta_{f^k x} = \rho \quad (4.9)$$

for all $x \in S$.

This theorem is in the spirit of the "absolute continuity" results of Pesin. An infinite-dimensional generalization has been promised by Brin and Nitecki (1985). The theorem fails if 0 is a characteristic exponent, as the following example shows.

*Counterexample.* A dynamical system is defined by the differential equation

$$\frac{dx}{dt} = x^3$$

on $\mathbb{R}$. Its time-one map has $\delta_0$ as an ergodic measure, with $\lambda_1 = 0$. However, 0 is (weakly) repelling, so that Eq. (4.9) cannot hold for $x \neq 0$. In fact, if $x \neq 0$, $f^t x$ goes to infinity in a finite time.

We give now an example showing that there is not always an SRB measure lying around, and that there are physical measures that are not SRB.

*Counterexample* (Bowen, and also Katok, 1980). Consider a continuous-time dynamical system (flow) in $\mathbb{R}^2$ with three fixed points $A,B,C$ where $A,C$ are repelling and $B$ of saddle type, as shown in Fig. 17. The system has an invariant curve in the shape of a "figure 8" (or rather, figure $\infty$), which is attracting. It can be seen that any point different from $A$ or $C$ yields an ergodic average corresponding to a Dirac $\delta$ at $B$. Therefore $\delta_B$ is the physical measure for our system. Clearly it has zero entropy, one strictly positive characteristic exponent and the other strictly negative (and thus, in particular, not zero),

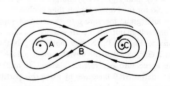

FIG. 17. The figure $\infty$ counterexample of Bowen (1975).

and is not absolutely continuous with respect to Lebesgue measure on the unstable manifold. (Note that the unstable manifold at $B$ consists of the "figure 8.") On the other hand, it is not hard to see that $\delta_B$ is a Kolmogorov measure; i.e., a system perturbed with a little noise $\varepsilon$ will spend most of its time near $B$, and as $\varepsilon \to 0$ the fraction of time spent near $B$ goes to 1.

## C. Information dimension

Given a probability measure $\rho$, we know that its *information dimension* $\dim_H \rho$ is the smallest Hausdorff dimension of a set $S$ of $\rho$ measure 1. Note that the set $S$ is not closed in general, and therefore the Hausdorff dimension $\dim_H(\mathrm{supp}\rho)$ of the support of $\rho$ may be strictly larger than $\dim_H \rho$.

*Example.* The rational numbers of the interval $[0,1]$, i.e., the fractions $p/q$ with $p,q$ integers, form a countable set. This means that they can be ordered in a sequence $(a_n)_1^\infty$. Consider the probability measure

$$\rho = \frac{1}{e} \sum_{n=1}^{\infty} \frac{1}{n!} \delta_{a_n} ,$$

where $\delta_x$ is the $\delta$ measure at $x$. Then $\rho$ is carried by the set $S$ of rational numbers of $[0,1]$, and since this is a countable set we have $\dim_H \rho = 0$. On the other hand, $\mathrm{supp}\rho = [0,1]$, so that $\dim_H(\mathrm{supp}\rho) = 1$.

It turns out that the information dimension of a physical measure $\rho$ is a more interesting quantity than the Hausdorff dimension of the attractor or attracting set $A$ which carries $\rho$. This is both because $\dim_H \rho$ is more accessible experimentally and because it has simple mathematical relations with the characteristic exponents. In any case, we have $\mathrm{supp}\rho \subset A$ and therefore

$$\dim_H \rho \le \dim_H(\mathrm{supp}\rho) \le \dim_H A .$$

The next theorem shows that the information dimension is naturally related to the measure of small balls in phase space.

*Theorem* (Young, 1982). Let $\rho$ be a probability measure on a finite-dimensional manifold $M$. Assume that

$$\lim_{r \to 0} \frac{\log\rho[B_x(r)]}{\log r} = \alpha \tag{4.10}$$

for $\rho$-almost all $x$. Then $\dim_H \rho = \alpha$.

Young shows that $\alpha$ is also equal to several other "fractal dimensions" (in particular, the "Rényi dimension").

We are of course mostly interested in the case when $\rho$ is ergodic for a differentiable dynamical system. In that situation, the requirement that

$$\lim_{r \to 0} \frac{\log\rho[B_x(r)]}{\log r}$$

exists $\rho$-almost everywhere already implies that the limit is almost everywhere constant, and therefore equal to $\dim_H \rho$. [The above limit does not always exist, as Ledrappier and Misiurewicz (1984) have shown for certain maps of the interval.]

An interesting relation between $\dim_H \rho$ and the characteristic exponents $\lambda_i$ has been conjectured by Yorke and others (see references below). We denote by

$$c_\rho(k) = \sum_{i=1}^{k} \lambda_i$$

the sum of the $k$ largest characteristic exponents, and extend this definition by linearity between integers (see Fig. 18):

$$c_\rho(s) = \sum_{i=1}^{k} \lambda_i + (s-k)\lambda_{k+1} \quad \text{if } k \le s < k+1 .$$

The function $c_\rho$ is defined on the interval $[0, +\infty)$ for a dynamical system on a Hilbert space, and on the interval $[0,m]$ for a system on $\mathbf{R}^m$ or an $m$-dimensional manifold. In the latter cases we write $c_\rho(s) = -\infty$ for $s > m$, so that $c_\rho$ is now in all cases a *concave* function on $[0, +\infty)$, as in Fig. 18. Notice that $c_\rho(0) = 0$, that the maximum of $c_\rho(s)$ is the sum of the positive characteristic exponents, and that $c_\rho(s)$ becomes negative for sufficiently large $s$. (This is because, in the Hilbert case, the $\lambda_i$ tend to $-\infty$.)

The *Liapunov dimension* of $\rho$ is now defined as

$$\dim_\Lambda \rho = \max\{s : c_\rho(s) \ge 0\} .$$

Notice that when $c_\rho(k) \ge 0$ and $c_\rho(k+1) < 0$ we have

$$\dim_\Lambda \rho = k + \frac{c_\rho(k)}{|\lambda_{k+1}|} .$$

The $(k+1)$-volume elements are thus contracted by time evolution, and this suggests that the dimension of $\rho$ must be less than $k+1$, a result made rigorous by Ilyashenko (1983). Yorke and collaborators have gone further and made the following guess.

*Conjecture* (Kaplan and Yorke, 1979; Frederickson, Kaplan, Yorke, and Yorke, 1983; Alexander and Yorke, 1984). If $\rho$ is an SRB measure, then generically

$$\dim_H \rho = \dim_\Lambda \rho . \tag{4.11}$$

The SRB measures have been defined in Sec. IV.B and "genericity" means here "in general." What concept of

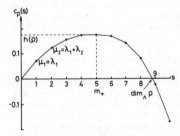

FIG. 18. Determination of the Liapunov dimension $\dim_\Lambda(\rho)$. The number of positive Liapunov exponents (unstable dimension) is $m_+$. The graph is from Manneville (1985), for the Kuramoto-Sivashinsky model.

genericity is adequate is a very difficult question; we do not know—among other things—how frequently a dynamical system has an SRB measure. An *inequality* is, however, available in full generality.

*Theorem.* Let $f$ be a twice continuously differentiable map and let $\rho$ be an ergodic measure with compact support. Then

$$\dim_H \rho \leq \dim_\Lambda \rho \ . \tag{4.12}$$

The basic result was proved by Douady and Oesterlé (1980), and from this Ledrappier (1981a) derived the theorem as stated. (It holds for Hilbert spaces as well as in finite dimensions.)

An *equality* is also known in special cases, notably the following.

*Theorem* (Young, 1982). Let $f$ be a twice differentiable diffeomorphism of a two-dimensional manifold, and let $\rho$ be an ergodic measure with compact support. Then the limit (4.10) exists $\rho$-almost everywhere, and we have

$$\dim_H \rho = h(\rho) \left[ \frac{1}{\lambda_1} + \frac{1}{|\lambda_2|} \right] , \tag{4.13}$$

where $\lambda_1 > 0$ and $\lambda_2 < 0$ are the characteristic exponents of $\rho$.

Note, incidentally, that if $f$ is replaced by $f^{-1}$, the characteristic exponents change sign, the entropy remains the same, and the formula remains correct, as it should. Note also that the cases where $\lambda_1$ and $\lambda_2$ are not of opposite sign are relatively trivial. From the inequality (4.4) applied for $f$ or $f^{-1}$ we see that $\lambda_1 = 0$ or $\lambda_2 = 0$ implies $h(\rho) = 0$, so that the right-hand side of Eq. (4.13) becomes indeterminate. If $\lambda_1 \geq \lambda_2 > 0$ or $0 > \lambda_1 \geq \lambda_2$, a theorem of Sec. III.C.2 applied to $f$ or $f^{-1}$ shows that $\rho$ is carried by a periodic orbit, so that Eq. (4.13) holds with $\dim_H \rho = h(\rho) = 0$.

Variants of the above theorem that do not assume the invertibility of $f$ are known (for one dimension see Ledrappier, 1981b, Proposition 4; for holomorphic functions see Manning, 1984).

If $\rho$ is an SRB measure, then Eq. (4.13) becomes $\dim_H \rho = 1 + \lambda_1 / |\lambda_2|$, which is just the conjecture (4.11). The next example shows that the conjecture does not always hold.

*Counterexample.* Notice first that if a measure $\rho$ has no positive characteristic exponent, then $h(\rho) = 0$ by a theorem in Sec. IV.A, and therefore $\rho$ is not an SRB measure. If $\dim_H \rho$ is strictly between 0 and 1, then Eq. (4.11) cannot hold (because $\dim_\Lambda \rho$ can only have the value 0 or a value $\geq 1$). In particular, the Feigenbaum attractor [example (b) of Sec. II.D] carries a unique probability measure $\rho$ with $\lambda_1 = 0$ and $\dim_H \rho = 0.538\ldots$ so that Eq. (4.11) is violated here. [For the Hausdorff dimension of the Feigenbaum measure see Grassberger (1981); Vul, Sinai, and Khanin (1984); Ledrappier and Misiurewicz (1984).]

Finally, let us mention *lower* bounds on $\dim_H \rho$.

*Theorem.* If $\rho$ is an SRB measure, then $\dim_H \rho \geq m_+$, where $m_+$ is the sum of the multiplicities of the positive characteristic exponents (unstable dimension).

This follows readily from the definitions.

## D. Partial dimensions

Given an ergodic measure $\rho$, we can associate with each characteristic exponent $\lambda^{(i)}$ a *partial dimension* $D^{(i)}$. Roughly speaking, $D^{(i)}$ is the Hausdorff dimension in the direction of $\lambda^{(i)}$. The entropy inequality (4.4) and the dimension inequality (4.12) will be natural consequences of the existence of the $D^{(i)}$.

In order to give a precise definition, we assume that $f$ is a twice differentiable diffeomorphism of a compact manifold $M$ (in the case of a continuous-time dynamical system we take for $f$ the time-one map). If $\lambda^{(i)}$ is a positive characteristic exponent and $\lambda^{(i)} > \lambda > \max(0, \lambda^{(i+1)})$, we have defined in Sec. III.E the local *unstable manifolds* $V_x^u(\lambda, \varepsilon)$, which we simply denote here by $V_{\text{loc}}^{(i)}$. Suppose $S = \cup_{\alpha \in A} S_\alpha$, where the $S_\alpha$ are open pieces of the $V_{\text{loc}}^{(i)}$, and define *conditional probability measures* $\rho_\alpha^{(i)}$ on $S_\alpha$ such that

$$\rho \text{ restricted to } S = \int \rho_\alpha^{(i)} m(d\alpha) \ ,$$

where $m$ is some measure on $A$. This definition is a bit more general than that given in Sec. IV.B. There $\lambda^{(i)}$ was the smallest positive characteristic exponent. We define

$$\delta^{(i)} = \dim_H \rho_\alpha^{(i)}$$

(this is a constant almost everywhere) and write

$$D^{(1)} = \delta^{(1)} \text{ if } \lambda^{(1)} > 0 \ ,$$

and

$$D^{(i)} = \delta^{(i)} - \delta^{(i-1)} \text{ if } i > 1 \text{ and } \lambda^{(i)} > 0 \ .$$

Similarly, if $\lambda^{(i)} < 0$, we define conditional probabilities $\rho_\alpha^{(j)}$ on pieces of *stable manifolds* and let

$$\delta^{(j)} = \dim_H \rho_\alpha^{(j)} \ .$$

We then write

$$D^{(r)} = \delta^{(r)}$$

if the smallest characteristic exponent $\lambda^{(r)}$ is negative, and

$$D^{(j)} = \delta^{(j)} - \delta^{(j+1)} \text{ if } j < r, \ \lambda^{(j)} < 0 \ .$$

The definition of $D^{(k)}$ for $\lambda^{(k)} = 0$ is somewhat arbitrary [between 0 and the multiplicity $m^{(k)}$ of $\lambda^{(k)} = 0$]; we take $D^{(k)} = m^{(k)}$.

*Theorem.* The partial dimensions $D^{(1)}, \ldots, D^{(r)}$ satisfy

$$0 \leq D^{(i)} \leq m^{(i)} \text{ for } i = 1, \ldots, r \ ,$$

where $m^{(i)}$ is the multiplicity of $\lambda^{(i)}$. The entropy is given by

$$h(\rho) = \sum_i {}^+ \lambda(i) D^{(i)} = -\sum_i {}^- \lambda^{(i)} D^{(i)} \ , \tag{4.14}$$

where $\sum^+$ ($\sum^-$) is the sum over positive (negative) characteristic exponents, in particular

$$\sum_i \lambda^{(i)} D^{(i)} = 0 \ . \tag{4.15}$$

The Hausdorff dimension satisfies

$$\dim_H \rho \leq \sum_i D^{(i)} . \tag{4.16}$$

(It is not known if there is equality when no characteristic exponent vanishes.)

The proof of this theorem by Ledrappier and Young (1984) is not easy, but brings further dividends, in particular an interpretation of the numbers $|\lambda^{(i)}| D^{(i)}$ as *partial entropies*.

Some earlier theorems on entropy and Hausdorff dimension are recovered as corollaries of the above theorem, as we now indicate. [We follow Ledrappier and Young (1984); Grassberger (1984); Procaccia (1984).]

(a) First we recover from

$$h(\rho) = \sum_i {}^+ \lambda^{(i)} D^{(i)}$$

the entropy inequality (4.4),

$$h(\rho) \leq \sum_i {}^+ \lambda^{(i)} m^{(i)} .$$

(b) The above inequality is in fact an equality (i.e., $\rho$ is an SRB measure) if and only if $D^{(i)} = m^{(i)}$ for all positive $\lambda^{(i)}$.

Next we check that we recover the dimension inequality (4.12):

$$\dim_H \rho \leq \dim_\Lambda \rho \tag{4.17}$$

from

$$\dim_H \rho \leq \sum_i D^{(i)} \leq \max \left\{ \sum_i d^{(i)} : 0 \leq d^{(i)} \leq m^{(i)} \text{ and } \sum_i d^{(i)} \lambda^{(i)} = 0 \right\} .$$

*Proof.* Let $k$ be such that

$$\sum_1^k \lambda^{(i)} \geq 0 > \sum_i^{k+1} \lambda^{(i)} .$$

We then have $\lambda^{(k+1)} < 0$ and

$$\dim_H \rho \leq \sum_i D^{(i)} = \frac{-1}{\lambda^{(k+1)}} \sum_i (-\lambda^{(k+1)}) D^{(i)}$$

$$= \frac{-1}{\lambda^{(k+1)}} \sum_i (\lambda^{(i)} - \lambda^{(k+1)}) D^{(i)} \leq \frac{-1}{\lambda^{(k+1)}} \sum_{i=1}^k (\lambda^{(i)} - \lambda^{(k+1)}) D^{(i)}$$

$$\leq \frac{-1}{\lambda^{(k+1)}} \sum_{i=1}^k (\lambda^{(i)} - \lambda^{(k+1)}) m^{(i)} = \sum_{i=1}^k m^{(i)} + \frac{\sum_{i=1}^k \lambda^{(i)} m^{(i)}}{|\lambda^{(k+1)}|} .$$

It is easily seen that the right-hand side is just the Liapunov dimension $\dim_\Lambda \rho$, and Eq. (4.17) follows.

(d) Suppose that we have equalities in the proof above, i.e., that the Kaplan-Yorke conjecture holds for $\rho$. Then we must have $D^{(i)} = m^{(i)}$ for $i = 1, \ldots, k$ and $D^{(i)} = 0$ for $i = k+2, \ldots, r$ (conversely these properties imply $\sum D^{(i)} = \dim_\Lambda \rho$). In particular, *if the Kaplan-Yorke conjecture holds for $\rho$ then $\rho$ is an SRB measure.*

*Remarks.*

(a) If $\sum \lambda^{(i)} m^{(i)} > 0$ we have $\dim_\Lambda \rho = m$ (the dimension of the manifold), which provides a trivial bound for $\dim_H \rho$. However, if one replaces $f$ by $f^{-1}$, changing the sign of the $\lambda^{(i)}$, one gets a new Liapunov dimension, which is $< m$ and provides a nontrivial bound on the dimension of $\rho$.

(b) If there are only two *distinct* characteristic exponents, then $D^{(1)}$ and $D^{(2)}$ can be computed from Eq. (4.14).

(c) Let $\rho$ be an SRB measure with $r$ characteristic exponents such that $\lambda^{(1)} > \cdots > \lambda^{(r-1)} > 0 > \lambda^{(r)}$ and $\sum_1^r \lambda^{(i)} m^{(i)} < 0$. Then what we have said shows that $\sum D^{(i)} = \dim_\Lambda \rho$.

## E. Escape from almost attractors

Before asymptotic behavior is reached by a dynamical system, transients of considerable duration are often observed experimentally. This is the case, for instance, for the Lorenz system [Sec. II.B, example (b)] as observed by Kaplan and Yorke (1979): for some values of the parameters *preturbulence* occurs in the form of long chaotic transients, even though the system does not yet have a strange attractor. One may say that the system has an *almost attractor* and try to estimate the escape rate from this set. More generally, one would like to have a precise description of *transient chaos* (see Grebogi, Ott, and Yorke, 1983b).

The situation, as usual, is best understood for the Axiom-$A$ systems, where the *basic sets* (see Sec. III.F) are the natural candidates to describe almost attractors. Let $\Omega_i$ be a basic set, $U$ a small neighborhood of $\Omega_i$, and $\mu$ a measure with positive continuous density with respect to Lebesgue measure on $U$. Let

$$p(T) = \mu \left[ \bigcap_{0 \leq t \leq T} f^t U \right]$$

be the amount of mass that has not left $U$ by time $T$. One finds that $p(T) \approx e^{Pt}$, where

$$P = \max \left\{ h(\rho) - \sum \text{ positive } \lambda_i(\rho) \right.$$

$$\left. \rho \text{ ergodic with support in } \Omega_i \right\} . \quad (4.18)$$

Note that $P$ vanishes, as it should, if $\Omega_i$ is an attractor; the maximum is given in that case by the SRB measure. If $\Omega_i$ is not an attractor, then $P < 0$; there is again a unique measure $\rho_i$ realizing the maximum of Eq. (4.18), but it is no longer SRB (see Bowen and Ruelle, 1975).

If our dynamical system is not necessarily Axiom $A$, the following is a natural guess.

*Conjecture.* Write

$$P = h(\rho) - \sum \text{ positive } \lambda_i(\rho) . \quad (4.19)$$

Then $|P|$ is the rate of escape from the support $K$ of $\rho$, provided

$$P \geq h(\sigma) - \sum \text{ positive } \lambda_i(\sigma)$$

for all ergodic $\sigma$ with support in $K$. If $P > h(\sigma) - \sum$ positive $\lambda_i(\sigma)$ when $\sigma \neq \rho$ then $\rho$ describes the time averages over transients near $K$.

A heuristic argument following the Axiom-$A$ case makes this plausible, but it is unknown how generally the conjecture holds. Some satisfactory experimental verifications have been given by Kantz and Grassberger (1984). They write Eq. (4.19) as follows in terms of the *partial dimensions* $D^{(i)}$ discussed in Sec. IV.D:

$$|P| = -P = \sum_{i: \lambda^{(i)} > 0} \lambda^{(i)} (m^{(i)} - D^{(i)}) .$$

### F. Topological entropy*

The measure-theoretic entropy of Sec. IV.A gave the rate of information creation with respect to an ergodic measure. A related concept, involving the *topology* rather than a measure, will be discussed here.

Let $K$ be a compact set and $f : K \to K$ a continuous map. If $\mathscr{A} = (\mathscr{A}_1, \ldots, \mathscr{A}_a)$ is a finite open cover of $K$ (i.e., $\cup_i \mathscr{A}_i \supset K$), we write

$$f^{-k} \mathscr{A} = (f^{-k} \mathscr{A}_1, \ldots, f^{-k} \mathscr{A}_k) ,$$

$$\mathscr{A}^{(n)} = \mathscr{A} \vee f^{-1} \mathscr{A} \vee \cdots \vee f^{-n+1} \mathscr{A}$$

$$= (\mathscr{A}_{i_1} \cap f^{-1} \mathscr{A}_{i_2} \cap \cdots \cap f^{-n+1} \mathscr{A}_{i_n}) .$$

Now let $N(\mathscr{A}, n)$ be the smallest number of sets in $\mathscr{A}^{(n)}$ that still covers $K$. The following limit is asserted to exist:

$$h_{\text{top}}(K, \mathscr{A}) = \lim_{n \to \infty} \frac{1}{n} \log N(\mathscr{A}, n) ,$$

and one defines the *topological entropy* of $K$ by

$$h_{\text{top}}(K) = \sup_{\mathscr{A}} h_{\text{top}}(K, \mathscr{A}) .$$

If we have a metric on $K$ we may write more conveniently

$$h_{\text{top}}(K) = \lim_{\text{diam} \mathscr{A} \to 0} h_{\text{top}}(K, \mathscr{A}) .$$

The following important theorem relates the topological entropy and the measure-theoretic entropies.

*Theorem.* If $K$ is compact and $f : K \to L$ continuous, then $h_{\text{top}}(K) = \sup\{h(\rho) : \rho$ is an ergodic measure with respect to $f\}$.

[This was conjectured by Adler, Konheim, and McAndrew (1965), and proved by Goodwyn, Dinaburg, and Goodman.] For references and more details on topological entropy we must refer the reader to Walters (1975) and Denker, Grillenberger, and Sigmund (1976).

### G. Dimension of attractors*

The estimates of $\dim_H \rho$ in Sec. IV.C can be completed by estimates of the dimension of compact invariant sets (like the support of $\rho$, or attractors and attracting sets).

*Theorem.* Let $A$ be a compact invariant set for a differentiable map $f$. Then

$$\dim_H A \leq \sup\{\dim_A \rho : \rho \text{ is ergodic with support in } A\} . \quad (4.20)$$

This result is due to Ledrappier (1981a), based on Douady and Oesterlé (1980); it is not known whether one can write $\dim_K A$ instead of $\dim_H A$ in Eq. (4.20). Note that, contrary to what Eq. (4.20) might suggest, there are cases where $\dim_H A > \sup\{\dim_H \rho : \rho$ is ergodic with support in $A\}$ (see McCluskey and Manning, 1983).

Lower bounds on $\dim_K A$ are also known. For instance, if a dynamical system has an attracting set $A$ and a fixed point $P$ with unstable dimension $m_+(P)$, then $\dim_K A \geq m_+(P)$ (see the corollary in Sec. III.E). For better estimates see Young (1981).

### H. Attractors and small stochastic perturbations*

In this section we discuss how physical measures and attractors are selected by their stability under small stochastic perturbations.

#### 1. Small stochastic perturbations

In Sec. II.F we discussed how the introduction of a small amount of noise in a deterministic system could select a particular invariant measure, the Kolmogorov measure. We can now be more precise. Consider first a discrete-time dynamical system generated by the map $f : M \to M$, where $M$ has finite dimension $m$. Let $\varepsilon > 0$, and for each $x \in M$, let $\mu_x^\varepsilon$ be a probability measure with support in the ball $\bar{B}_x(\varepsilon) = \{y : d(x, y) \leq \varepsilon\}$. More specifically, we assume that $\mu_x^\varepsilon(dy)$

$=\varepsilon^{-m}\varphi[x,\varepsilon^{-1}(y-x)]dy$, where $\varphi$ is continuous, $\varphi\geq0$, $\varphi(x,0)>0$, $\varphi(x,x-y)=0$ if $d(x,y)\geq1$, and $dy$ denotes the Lebesgue volume element if $M=\mathbf{R}^m$ (if $M$ is not $\mathbf{R}^m$ this prescription is modified by using a Riemann metric on $M$; see Kifer, 1974). A *stochastic dynamical system* is a time evolution defined not on $M$, but at the level of probability measures on $M$. In our case we replace $f:M\rightarrow M$ by the stochastic perturbation

$$\nu\rightarrow\int\mu_{fx}^{\varepsilon}\nu(dx)$$
$$=\left[\int\varepsilon^{-m}\varphi[fx,\varepsilon^{-1}(y-fx)]\nu(dx)\right]dy . \quad (4.21)$$

The limit of a small stochastic perturbation corresponds to taking $\varepsilon\rightarrow0$.

In the continuous-time case, the dynamical system defined on $\mathbf{R}^m$ by the equation

$$\frac{dx_i}{dt}=F_i(x)$$

is replaced by an evolution equation for the density $\Phi$ of $\nu$. We write $\nu(dx)=\Phi(x)dx$, and

$$\frac{d\Phi(x)}{dt}=\sum_i F_i\partial_i\Phi(x)+\varepsilon\Delta\Phi(x) . \quad (4.22)$$

If we have a Riemann manifold, the Laplacian $\Delta$ should be replaced by the Laplace-Beltrami operator. The limit of a small stochastic perturbation corresponds again to taking $\varepsilon\rightarrow0$.

*Theorem.* Let an Axiom-$A$ dynamical system on the compact manifold $M$ be defined by a twice differentiable diffeomorphism $f$ or a twice differentiable vector field $F$. Let $A_1,\ldots,A_r$ be the attractors, and $\rho_1,\ldots,\rho_r$ the corresponding SRB measures.

(a) In the discrete-time case, for $\varepsilon$ small enough, let $\rho_i^{\varepsilon}$ be a stationary measure for the process (4.21) with support near $A_i$. Then $\rho_i^{\varepsilon}\rightarrow\rho_i$ when $\varepsilon\rightarrow0$.

(b) In the continuous-time case, there is a unique stationary measure $\rho^{\varepsilon}$ for the process (4.22), and any limit of $\rho^{\varepsilon}$ when $\varepsilon\rightarrow0$ is a convex combination $\sum\alpha_i\rho_i$ where $\alpha_i\geq0$, $\sum\alpha_i=1$.

These results have been established by Kifer (1974), following Sinai's work on Anosov systems (1972). Another proof has been announced by Young. The idea behind the theorem is as follows. The noisiness of the stochastic time evolution yields measures which have continuous densities on $M$. The deterministic part of the time evolution will improve this continuity in the unstable directions by stretching, and roughen it in other directions due to contraction. In the limit one gets measures that are continuous along unstable directions, i.e., SRB measures.

Note the difference between the discrete-time and the continuous-time cases, which is due to the fact that in the discrete-time case, for small $\varepsilon$, a point near one attractor cannot jump out of its basin of attraction.

If we have a general dynamical system (not Axiom $A$), the stationary states for small stochastic perturbations will again tend to be continuous along unstable directions, but the limit when $\varepsilon\rightarrow0$ need not be SRB (see the coun-

terexample of Sec. IV.C). Moreover, there need not be an open basin of attraction associated with each SRB measure, so that, even in the discrete-time case, the stochastic perturbation may switch from one measure to another, and the limit may be a convex combination of many SRB measures (in particular be nonergodic). That basins of attraction may indeed be a mess is shown by the following result.

*Theorem* (Newhouse, 1974,1979). There is an open set $S\neq\varnothing$ in the space of twice differentiable diffeomorphisms of a compact two-dimensional manifold, and a dense subset $R$ of $S$ such that each $f\in R$ has an infinite number of attracting periodic orbits.

A variation of this result implies that for the Hénon map [see Sec. II.D, example (a)] the presence of infinitely many attracting periodic orbits is assured for some $b$ and a dense set of values of $a$ in some interval $(a_0,a_1)$. The basins of such attracting periodic orbits are mostly very small and interlock in a ghastly manner.

The study of stochastic perturbations of differentiable dynamical systems is at present quite active; see in particular Carverhill (1984a,1984b) and Kifer (1984).

### 2. A mathematical definition of attractors

We have defined attractors operationally in Sec. II.C. Here, finally, we discuss a mathematical definition.

If $a,b\in M$, let us write $a\rightarrow b$ ($a$ goes to $b$) provided for arbitrarily small $\varepsilon>0$ there is a chain $a=x_0,x_1,\ldots,x_n=b$ such that $d(x_k,f^{\Theta_k}x_{k-1})<\varepsilon$ with $\Theta_k\geq1$ for $k=1,\ldots,n$. We accept $a\rightarrow a$ (corresponding to a chain of length 0), and it is clear that $a\rightarrow b,b\rightarrow a$ imply $a\rightarrow c$. If for every $\varepsilon>0$ there is a chain $a\rightarrow a$ of length $\geq1$ we say that $a$ is *chain recurrent*. If $a$ is chain recurrent, we define its *basic class* $[a]=\{b:a\rightarrow b\rightarrow a\}$. If $[a]$ consists only of $a$, then $a$ is a fixed point. Otherwise if $b\in[a]$ then $b$ is chain recurrent and $[b]=[a]$.

We shall say that a basic class $[a]$ is an *attractor* if $a\rightarrow x$ implies $x\rightarrow a$ (i.e., $x\in[a]$). This definition ensures that $[a]$ is attracting, but in a weaker sense than the definition of attracting sets in Sec. II.B. Here, however, we have irreducibility: an attractor cannot be decomposed into two distinct smaller attractors. (More generally, the set of chain-recurrent points decomposes in a unique way into the union of basic classes.)

It can be shown that any limit when $\varepsilon\rightarrow0$ of a measure stable under small stochastic perturbations of a discrete-time dynamical system is "carried by attractors," at least in a weak sense. More precisely, this can be stated as the following theorem.

*Theorem.* Let $\Lambda$ be a compact attracting set for a discrete-time dynamical system, $m$ a probability measure with support close to $\Lambda$, and $\varepsilon$ sufficiently small. Let also $m_{\varepsilon}^k$ be obtained at time $k$ from the stochastic evolution (4.22). If $[a]$ is not an attractor, then

$$\lim_{k\rightarrow\infty}m_{\varepsilon}^k[B_a(\delta)]=0 ,$$

when $\delta$ is sufficiently small. [$B_a(\delta)$ denotes the ball of radius $\delta$ centered at $a$.] In particular, if $m_\varepsilon^\infty$ is a limit of $m_\varepsilon^k$ when $k \to \infty$, then $a$ does not belong to the support of $m_\varepsilon^\infty$. If $\rho$ is a limit when $\varepsilon \to 0$ of $m_\varepsilon^\infty$, and if $m$ is ergodic, then its support is contained in an attractor.

*Proof.* See Ruelle (1981).

The topological definition of an attractor given in this section follows the ideas of Conley (1978) and Ruelle (1981). A rather different definition has been proposed recently by Milnor (1985), based on the privileged role which the Lebesgue measure should play for physical dynamical systems.

## I. Systems with singularities and systems depending on time*

The theory of differentiable dynamical systems may to some extent be generalized to differentiable dynamical system *with singularities*. This is of interest, for instance, in Hamiltonian systems with collisions (billiards, hard-sphere problems). On this problem we refer the reader to the considerable work by Katok and Strelcyn (1985).

Another conceptually important extension of differentiable dynamical systems is to systems with *time-dependent forces*. One does not allow here for arbitrary non-autonomous systems, but assumes that

$$x(t+1) = f(x(t), \omega(t)) \quad \text{or} \quad \frac{dx}{dt} = F(x, \omega(t)),$$

where $\omega$ has a stationary distribution. (For instance, $\omega$ is defined by a continuous dynamical system.) It is surprising how many results extend to this more general situation; the extension is without pain, but the formalism more cumbersome. Here again we can only refer to the literature. See Ruelle (1984) for a general discussion, and Carverhill (1984a,1984b) and Kifer (1984) for problems involving stochastic differential equations and random diffeomorphisms.

## V. EXPERIMENTAL ASPECTS

Now that we have developed a theoretical background and a language in which to formulate our questions, it is time to discuss their experimental aspects. A basic conceptual problem is that of confronting the limited information that can be obtained in a real experiment with the various limits encountered in the mathematical theory. A similar situation occurs, for instance, in the application of statistical mechanics to the study of phase transitions. Other important problems in the relation between theory and experiment concern numerical efficiency and accuracy. The present section will address those problems.

We shall describe two different fields of experimentation—computer experiments and experiments with real physical systems. There is a quantitative difference between the two fields, since one can study dynamical evolution equations with *fixed* experimental condi-

tions more accurately on a computer than in reality. However, there is also a more important qualitative difference: Since the evolution equations are explicitly known in a computer experiment, it is generally easy to compute directly the "tangent map" $Df^t$. In a physical experiment, by contrast, only points on a trajectory are directly measurable, and the derivatives (tangents) have to be obtained by a delicate interpolation, to be discussed below.

It must be understood that the information currently being extracted from experiments goes a long way beyond the solid mathematical foundations that we have described in the previous sections. It is a challenge for the mathematical physicist to clarify the relations between the various quantities measured on dynamical systems. Most of them seem indeed very interesting, and very promising, but a lot of work is still necessary to prove the existence of these quantities and establish their relations. Our selection below reflects to some extent our personal taste for measurements based on sound ideas and for which a mathematical foundation can be expected.

### A. Dimension

The measurement of dimensions is discussed first because it is most straightforward. We concentrate on the determination of the *information dimension*, using the method advocated by Young (1982), and Grassberger and Procaccia (1983b). The idea is described in the first theorem (by Young) in Sec. IV.C. The method, developed independently by Grassberger and Procaccia, has gained wide acceptance through their work.

We start with an experimental time series $u(1), u(2), \ldots$, corresponding to measurements *regularly spaced in time*. We assume that $u(i) \in \mathbb{R}^\nu$, where $\nu = 1$ in the (usual) case of scalar measurements. From the $u(i)$, a sequence of points $x(1), \ldots$, in $\mathbb{R}^{m\nu}$ is obtained by taking $x(i) = [u(i+1), \ldots, u(i+m-1)]$. This construction associates with points $X(i)$ in the phase space of the system (which is, in general, infinite dimensional) their projections $x(i) = \pi_m X(i)$ in $\mathbb{R}^{m\nu}$. In fact, if $\rho$ is the physical measure describing our system ($\rho$ is carried by an attractor in phase space), then the points $x(i)$ are equidistributed with respect to the projected measure $\pi_m \rho$ in $\mathbb{R}^{m\nu}$. [Actually this is not always true: if the time spacing $\Delta t$ between consecutive measurements $u(i), u(i+1)$ is a "natural period" of the system—for instance, when the system is quasiperiodic—one does not have equidistribution. This exception is easily recognized and handled.] We wish to deduce $\dim_H \rho$ from this information (with the possibility of varying $m$ in the above construction). Before seeing how this is done, a general word of caution is in order. In any given experiment we have only a finite time series, and therefore there are natural limits on what can be extracted from it: some questions are too detailed (or the statistical fluctuations too large) for a reasonable answer to come out. See, for instance, Guckenheimer (1982) for a discussion of such matters.

A serious difficulty seems to arise here from the fact that $\dim_H \tau_m \rho$ need not be equal to the desired $\dim_H \rho$. We remove this objection with the observation that, if $\dim_H \rho \leq M$, then, for most $M$-dimensional projections $p$, $\dim_H p\rho = \dim_H \rho$. More precisely, we have the following result.

*Theorem.* Let $0 < M < n$. If $E$ is a Suslin set in $R^n$, and $\dim E \leq M$, then there is a Borel set $G$ in the space of orthogonal projections $p: R^n \to R^M$ such that its complement has measure zero with respect to the natural rotation-invariant measure on projections, and $\dim pE = \dim E$ for all $p$ in $G$.

*Proof.* See Lemma 5.3 in Mattila (1975). We are indebted to C. McCullen for this reference. It is interesting to compare this result with that of Mañé in Sec. II.G, where one obtains (with stronger restrictions) the injectivity of $p$.

Of course we have not proved that our projection $\tau_m$ belongs to the good set $G$ of the above theorem (with $M = m\nu$), but this appears to be a reasonable guess, and we shall proceed with the assumption that $\dim_H \tau_m \rho = \dim_H \rho$ for large enough $m$.

We now use our sequence $x(1), x(2), \ldots, x(N)$ in $R^{m\nu}$ to construct functions $C_i^m$ and $C^m$ as follows:

$$C_i^m(r) = N^{-1} \{\text{number of } x(j) \text{ such that } d[x(i), x(j)] \leq r\} , \tag{5.1}$$

$$C^m(r) = N^{-1} \sum_i C_i^m(r)$$

$$= N^{-2} \{\text{number of ordered pairs } [x(i), x(j)] \text{ such that } d[x(i), x(j)] \leq r\} . \tag{5.2}$$

[$C_i^m$ is obtained by sorting the $x(j)$ according to their distances to $x(i)$; $C^m$ is obtained more efficiently directly by sorting pairs than as an average of the $C_i^m$.] We may use $d(x, x') = $ Euclidean norm of $x' - x$, or any other norm, such as

$$|x' - x| = \max_\alpha |u'(\alpha) - u(\alpha)| ,$$

where the $u(\alpha)$ are the $m$ components of $x$, and $|u'(\alpha) - u(\alpha)|$ is for instance the Euclidean norm in $R^\nu$ (this will be used in Sec. V.B). Note that when $N \to \infty$, we have

$$\lim C_i^m = (\tau_m \rho)[B_{x(i)}(r)] \tag{5.3}$$

(except perhaps at discontinuity points of the right-hand side). Suppose now that

$$\lim_{r \to 0} \lim_{N \to \infty} \frac{\log C_i^m(r)}{\log r} = \lim_{r \to 0} \frac{\log(\tau_m \rho)[B_{x(i)}(r)]}{\log r} = \alpha_m . \tag{5.4}$$

Then $\dim_H \tau_m \rho = \alpha_m$ (first theorem of Sec. IV.C). Provided the projection $\tau_m$ is in the "good set" $G$, we have thus $\alpha_m = m\nu$ if $\dim_H \rho \geq m\nu$ and $\alpha_m = \dim_H \rho$ if $\dim_H \rho \leq m\nu$. Experimentally, $\alpha_m$ may be obtained by plotting $\log C_i^m(r)$ vs $\log r$ and determining the slope of the curve (see below). With a little bit of luck [existence of the limit (5.4), and $\tau_m$ in the good set] we may thus obtain $\dim_H \rho$ experimentally: *we choose $m$ such that $\alpha_m < m\nu$; then we have $\dim_H \rho = \alpha_m$.* Although we cannot completely verify that $\tau_m$ is in the good set, we can in principle check (within experimental accuracy) the existence of the limits $\alpha_m$, and the fact that $\alpha_m$ becomes independent of $m$ when $m$ increases beyond a value such that $\alpha_m \leq m\nu$. Note that $\alpha_m$ should also be independent of the index $i$ in Eq. (5.4).

The information dimension $\dim_H \rho$ may also be obtained by a modification of Eq. (5.4). We describe the method of Grassberger and Procaccia, which has been tested experimentally in a number of cases. This consists

in writing

$$\lim_{r \to 0} \lim_{N \to \infty} \frac{\log C^m(r)}{\log r} = \beta_m , \tag{5.5}$$

and asserting that for $m$ sufficiently large, $\beta_m$ is the information dimension. The only relation that can easily be established rigorously between Eqs. (5.4) and (5.5) is that if both limits exist, then $\alpha_m \geq \beta_m$. However, it seems quite reasonable to assume that in general $\alpha_m = \beta_m$ (i.e., if the $C_i^m$ behave like $r^\alpha$, then their linear superposition $C^m$ also behaves like $r^\alpha$). The $\beta_m$ obtained experimentally do become independent of $m$ for $m$ large enough, as expected. See Fig. 19.

To summarize: the method of Grassberger and Procaccia is a highly successful way of determining the information dimension experimentally. Values between 3 and 10 are obtained reproducibly. The method is not entirely justified mathematically, but nevertheless quite sound. The study of the limit (5.4) is also desirable, even though the statistics there is poorer.

### 1. Remarks on physical interpretation

#### a. The meaningful range for $C^m(r)$

Suppose we plot $\log C(r)/\log r$ as a function of $\log r$ (we suppress the superscript $m$ and possibly the subscript $i$ of $C$). First, for small $r$, we have a large scatter of points due to poor statistics; then there is a range $(r_0, r_1)$ of near constancy (the constant is the information dimension if $m$ is suitably large). For $r$ larger than $r_1$ we have deviation from constancy due to nonlinear effects. The "meaningful range" $(r_0, r_1)$ is that in which the distribution of distances between pairs of points is statistically useful.

#### b. Curves with "knees"

It is not uncommon that the $\log C(r)$ vs $\log r$ plot shows a "knee" (see Fig. 20), so that it has slope $\alpha$ in the range

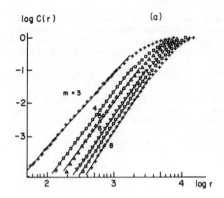

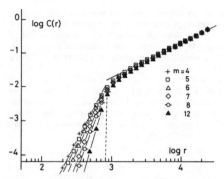

FIG. 20. A $\log C(r)$ vs $\log r$ plot may show a "knee," so that a dimension $\alpha$ appears in the range $(\log r_0, \log r_1)$, and a dimension $\alpha'$ in the range $(\log r_1, \log r_2)$. Here the scalar signal of a deterministic system with information dimension $\alpha'$ is perturbed by the addition of random noise, which yields a dimension $\alpha$ equal to the embedding dimension $m$ (see Atten et al., 1984).

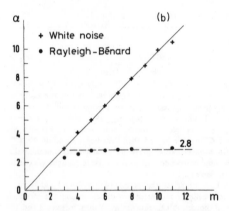

FIG. 19. Experimental results from Malraison et al. (1983) and Atten et al. (1984); see also Dubois (1982): (a) The plots show $\log C$ vs $\log r$ for different values of the embedding dimension, for the Rayleigh-Bénard experiment. (b) The measured dimension $\alpha$ as a function of the embedding dimension $m$, both for the Rayleigh-Bénard experiment and for numerical white noise. Note that $\alpha$ becomes nearly constant (but not quite) at $m = 3$. The $\alpha$ for white noise is nearly equal to $m$ (but not quite).

$(\log r_0, \log r_1)$ and a smaller slope $\alpha'$ in the range $(\log r_1, \log r_2)$. The dimension, or "number of degrees of freedom" is thus different for $r$ above and below $r_1$ (see, for instance, Riste and co-workers, 1985). To see how this situation can arise, let us consider a product dynamical system $I \times II$ formed of two noninteracting subsystems I and II. Take an observable $u = u_I + u_{II}$, where $u_I$ and $u_{II}$ depend only on the subsystems I and II, respectively, and let the amplitude $r_1$ of the signal $u_I$ be much smaller than that of $u_{II}$. In the range $r < r_1$ we have statistical information on the complete system $I \times II$, giving an information dimension $\alpha$. In the range $r \gg r_1$ we have statistical information only on the subsystem II, giving an information dimension $\alpha'$. More generally, suppose that

system II evolves independently of I, but that I has a time evolution that may depend on II; then the same conclusions persist for this "semidirect product." (The small-amplitude modes of I are driven here by system II, an example of Haken's "slaving principle.") The above argument makes clear, for instance, how the information dimension found by analysis of a turbulent hydrodynamic system does not take into account small ripples of amplitude less than the discrimination level $r_0$ of the analysis. (We thank P. C. Martin for useful discussions on this point.) A knee will also appear if the signal from a deterministic chaotic system is perturbed by adding random noise of smaller amplitude (see Fig. 20).

### c. Spatially localized degrees of freedom

We have just discussed dynamical systems that have a product structure $I \times II$, or where a subsystem II evolves independently and drives other degrees of freedom. Strictly speaking, such decoupling does not seem to occur in realistic situations like that of a turbulent viscous fluid (except for the trivial case where the fluid is in two different uncoupled containers). Normally, in a nonlinear system one may say that "every mode is coupled with all other modes," and exact factorization is impossible. An apparent exception is constituted by quasiperiodic motions where factorization is present, but the independent frequencies do not correspond to independent physical subsystems. In other words, if a physical variable $u(t)$ of the system is monitored (for instance a component of the velocity of a viscous fluid at one point), the whole dynamics of the system (on the appropriate attractor) can in principle be reconstructed from the time series $[u(t):t$ varying from 0 to $\infty]$. In particular, the information dimension of the system can be obtained indifferently from monitoring $u$ or any other physical variable.

As noted in Secs. V.A.1.a and V.A.1.b, the experimental uncertainties change this situation. At the level of accuracy of an experiment, some degrees of freedom may effectively be driven by others and, having small amplitude, pass unnoticed. (A case in point would be that of eddies of small size in three-dimensional turbulence.) Another frequent and important case occurs when some "oscillators" (possibly complex oscillators) are strongly localized in some region of space.[3] Consider, for instance, the flow between coaxial rotating cylinders in a regime where there is some turbulence superposed with Taylor cells. Some features of the flow are global (like the very existence of the Taylor cells), others seem to be restricted to one Taylor cell, having very little interaction with neighboring cells.

The information dimension $d_c$, obtained from moderate-precision measurements of one cell, is then likely to be different from the global information dimension $d_v$ of a column of $v$ cells (and one expects formulas like $d_c = a + b$, $d_v = a + vb$). Note that $d_v$ could be obtained by monitoring a vector signal with $v$ components, each corresponding to a scalar signal from one cell.

2. Other dimension measurements

The most straightforward way to find the "fractal" dimension of a set $A$ is to cover it with a grid of size $r$, to count the number $N(r)$ of occupied cells, and to compute

$$\lim_{r \to 0} \frac{\log N(r)}{|\log r|} .$$

This box-counting is computationally ineffective (Farmer,

1984). It gives access to the dimension of attractors rather than to the information dimension (the latter seems for the moment to have greater theoretical interest). Another problem with box-counting is that usually the population of boxes is very uneven, so that it may take a considerable amount of time before some "occupied" boxes really become occupied. For all these reasons, the box-counting approach is not used currently.

B. Entropy

The entropy (or Kolmogorov-Sinai invariant) $h(\rho)$ of a physical measure $\rho$ is an important quantity, as we have seen in Sec. IV. Early attempts to measure $h(\rho)$ were based directly on the definitions and used a partition $\mathscr{A}$ (see Sec. IV.A). These attempts (Shimada, 1979; Curry, 1981; Crutchfield, 1981) were interesting but not entirely successful. We describe here another approach due to Grassberger and Procaccia (1983a); see also Cohen and Procaccia (1984). [Similar ideas were developed independently by Takens (1983).] This approach has far greater potential for implementation in experimental situations.

The idea of Grassberger and Procaccia is to exploit the $m$ dependence of the functions $C_i^m(r)$ and $C^m(r)$ defined in Eqs. (5.1) and (5.2). As before, they use $C^m(r)$, which has better statistics, but it is easier to argue with the $C_i^m(r)$, which satisfy

$$C_i^m(r) \approx (\pi_m \rho)[B_{x(i)}(r)] \tag{5.6}$$

for large $N$ [see Eq. (5.3)]. In view of Eq. (5.6), $C_i^m(r)$ is the probability that $x(j)$ satisfies $d[x(j), x(i)] \le r$. Grassberger and Procaccia use the Euclidean norm, but we prefer to follow Takens and to take

$$d[x(j), x(i)] = \max\{|u(j) - u(i)|, \ldots, |u(j+m-1) - u(i+m-1)|\} .$$

Usually $v = 1$ (scalar signal), but the general case is not harder to handle [with $|u(i) - u(j)|$ being the Euclidean norm of $u(i) - u(j)$ in $\mathbf{R}^v$]. We may thus interpret $C_i^m(r)$ as the probability that the signal $u(j+k)$ remains in the ball $B_{u(i+k)}^v(r)$ for $m$ consecutive units of time [$B_{u(i)}^v(r)$ is the ball of radius $r$ in $\mathbf{R}^v$ centered at $u(i)$].

With this interpretation, and the fact that $C^m(r)$ is the average of the $C_i^m(r)$, it can be argued that

$$\lim_{r \to 0} \lim_{m \to \infty} \frac{1}{m} \lim_{N \to \infty} [-\log C^m(r)] = \Delta t K_2(\rho) , \tag{5.7}$$

where $K_2$ has been defined at the end of Sec. IV.A, and $\Delta t$ is the spacing between measurements of the signal $u$. Since $K_2(\rho)$ is a lower bound to $h(\rho)$, we see that if one obtains $K_2(\rho) > 0$ from Eq. (5.7) then one can conclude that $h(\rho) > 0$, i.e., that the system is chaotic.

It is, however, also possible to obtain $h(\rho)$ directly as follows. Define

$$\Phi^m(r) = \frac{1}{N} \sum_i \log C_i^m(r) .$$

Then

$$\Phi^{m+1}(r) - \Phi^m(r) = \text{average over } i \text{ of } \log[\text{probability that } u(j+m) \in B_{u(i+m)}^v(r)$$

$$\text{given that } u(j+k) \in B_{u(i+k)}^v(r) \text{ for } k = 0, \ldots, m-1] .$$

Therefore,

---

[3]For some discussion of the difficult problem of localization in hydrodynamics, see Ruelle (1982b).

$$\lim_{r\to 0}\lim_{m\to\infty}\lim_{N\to\infty}[\Phi^{m+1}(r)-\Phi^m(r)]=\Delta t h(\rho)\ . \qquad (5.8)$$

*Remarks.*

(a) Like the expressions of Sec. V.A, the identities (5.7) and (5.8) hold "if all goes well." Basically, the condition is that the monitored signal should reveal enough of what is going on in the system.

(b) While the information dimension could be obtained from $C^m(r)$ for one single $m$ (sufficiently large), we have a limit $m\to\infty$ in Eqs. (5.7) and (5.8). In this respect, (5.7) is not optimal (it will contain errors of order $1/m$); it is better to write

$$\lim_{r\to 0}\lim_{m\to\infty}\lim_{N\to\infty}\log\frac{C^m(r)}{C^{m+1}(r)}=\Delta t K_2(\rho)\ ,$$

or to use Eq. (5.8).

## C. Characteristic exponents: computer experiments

We recall that the characteristic exponents measure the exponential separation of trajectories in time and are computed from the *derivative* $D_x f^t$. In computer experiments, the derivative is often directly calculable, whereas in physical experiments it has to be obtained indirectly from the experimental signal. Therefore the methods for evaluating characteristic exponents are somewhat different in the two cases and will be treated separately. In this section, we discuss computed experiments, which have served and still serve an important purpose in the exploration of dynamical systems.

Let us mention here some interesting open problems. What is the distribution of characteristic exponents for a large or a highly excited system? Can one define a density of exponents per unit volume for a spatially extended system? What is the behavior near zero exponent? For a theoretical study in the case of turbulence, see Ruelle (1982b,1984). For an experimental study in the case of the Kuramoto-Sivashinsky model, see Manneville (1985).

In the case of a discrete-time dynamical system defined by a map $f:\mathbf{R}^m\to\mathbf{R}^m$, let

$$T(x)=D_x f\ .$$

This is the matrix of partial derivatives of the $m$ components of $f(x)$ with respect to the $m$ components of $x$. Write

$$T_x^n=T(f^{n-1}x)\cdots T(fx)T(x) \qquad (5.9)$$

(matrix multiplication on the right-hand side). Then the largest characteristic exponent is given by

$$\lambda_1=\lim_{n\to\infty}\frac{1}{n}\log||T_x^n u|| \qquad (5.10)$$

for almost any vector $u$, and this is a very efficient way to obtain $\lambda_1$. The other characteristic exponents can in principle be obtained by diagonalizing the positive matrices $(T_x^n)^* T_x^n$ and using the fact that their eigenvalues behave like $e^{2n\lambda_1},e^{2n\lambda_2},\ldots$. Obviously, for large $n$, the dif-

ferent eigenvalues have very different orders of magnitude, and this creates a problem if $T_x^n$ is computed without precaution. When $(T_x^n)^* T_x^n$ is diagonalized, the small relative errors on the large eigenvalues might indeed contaminate the smaller ones, causing intolerable inaccuracy. We shall see below how to avoid this difficulty.

Consider next a continuous-time dynamical system defined by a differential equation

$$\frac{dx(t)}{dt}=F(x(t))\ , \qquad (5.11)$$

in $\mathbf{R}^m$. An early proposal to estimate $\lambda_1$ (Benettin, Galgani, and Strelcyn, 1976) used solutions $x(t),x'(t),x''(t),\ldots$, chosen as follows. The initial condition $x'(0)$ is chosen very close to $x(0)$, and $x(t)$ remains close to $x'(t)$ up to some time $T_1$; one then replaces the solution $x'$ by a solution $x''$ such that $x''(T_1)-x(T_1)=\alpha[x'(T_1)-x(T_1)]$ with $\alpha$ small. Thus $x''(T_1)$ is again very close to $x(T_1)$, and $x''(t)$ remains close to $x(t)$ up to some time $T_2>T_1$, and so on. The rate of deviation of nearby trajectories from $x(\ )$ can thus be determined, yielding $\lambda_1$. This simple method has also been applied to physical experiments (Wolf *et al.*, 1984); we shall return to this topic in Sec. V.D.

In the case of (5.11) one can, however, do much better. Namely, one differentiates to obtain

$$\frac{d}{dt}u(t)=(D_{x(t)}F)[u(t)]\ , \qquad (5.12)$$

which is *linear* in $u$, but with nonconstant coefficients. The solution of (5.11) yields $x(t)=f^t(x(0))$, and the solution of (5.12) yields

$$u(t)=(D_{x(0)}f^t)u(0)\ .$$

Therefore one can readily compute the matrices $T_x^t=D_x f^t$ by integrating Eq. (5.12) with $m$ different initial vectors $u$. Better yet, one can use the matrix differential equation

$$\frac{d}{dt}T_{x(0)}^t=(D_{x(t)}F)T_{x(0)}^t\ ,$$

with $T_{x(0)}^0$ the identity matrix.

As in the discrete case, it is not advisable to compute $T_x^t$ for large $t$. We choose a reasonable unit of time $\tau$: not too large, so that the $e^{\lambda_i\tau}$ do not differ too much in their orders of magnitude, but not too small either, because we have to multiply a number of matrices proportional to $\tau^{-1}$. Having chosen $\tau$, we discretize the time (setting $\tilde{f}=f^\tau$) and proceed as in the discrete-time case. If the characteristic exponents for $\tilde{f}$ are $\tilde{\lambda}_i$, then the characteristic exponents for the continuous-time system are $\lambda_i=\tau^{-1}\tilde{\lambda}_i$.

Before discussing the accurate calculation of the $\lambda_i$ for $i>1$, let us mention that the knowledge of $\lambda_1$ [obtained from Eq. (5.10)] is sometimes sufficient to determine *all* characteristic exponents. This is certainly the case for one-dimensional systems, as well as for the Hénon map and the Lorenz equation, as we have seen in the examples of Sec. III.D.1. It is also possible to estimate successively

$\lambda_1$ by (5.10), then $\lambda_1 + \lambda_2$ as the rate of growth of surface elements, $\lambda_1 + \lambda_2 + \lambda_3$ as the rate of growth of three-volume elements, etc. This approach was first proposed by Benettin *et al.* (1978). In what follows, we discuss a somewhat different method.

The algorithm we propose for the calculation of the $\lambda_i$ is very close to the method presented by Johnson *et al.* (1984) for proving the multiplicative ergodic theorem of Oseledec. Remember that we are interested in the product (5.9):

$$T_x^n = T(f^{n-1}x) \cdots T(fx)T(x) .$$

To start the procedure, we write $T(x)$ as

$$T(x) = Q_1 R_1 , \qquad (5.13)$$

where $Q_1$ is an orthogonal matrix and $R_1$ is upper triangular with non-negative diagonal elements. [If $T(x)$ is invertible, this decomposition is unique.] Then for $k = 2, 3, \ldots$, we successively define

$$T_k' = T(f^{k-1}x)Q_{k-1}$$

and decompose

$$T_k' = Q_k R_k ,$$

where $Q_k$ is orthogonal and $R_k$ upper triangular with non-negative diagonal elements. Clearly, we find

$$T_x^n = Q_n R_n \cdots R_1 .$$

To exploit this decomposition, we shall make use of the results of Johnson *et al.*, but note that those are only proved in the "invertible case" of a dynamical system defined for negative as well as positive times. In the paper referred to, an orthogonal matrix $Q$ is chosen at random (i.e., $Q$ is equidistributed with respect to the *Haar* measure on the orthogonal group), and the initial $T(x)$ is replaced by $T(x)Q$ in Eq. (5.13), the matrices $T(f^{k-1}x)$ for $k > 1$ being left unchanged. It is then shown that the diagonal elements $\lambda_{ii}^{(n)}$ of the upper triangular matrix product $R_n \cdots R_1$ obtained from this modified algorithm satisfy

$$\lim_{n \to \infty} \frac{1}{n} \log \lambda_{ii}^{(n)} = \lambda_i \qquad (5.14)$$

almost surely with respect to the product of the invariant measure $\rho$ and the Haar measure (corresponding to the choice of $Q$). *On the right-hand side of Eq. (5.14) we have the characteristic exponents arranged in decreasing order.* For practical purposes, it is clearly legitimate to take $Q = $ identity.

In the case of constant $T(f^k x)$, i.e., $T(f^k x) = A$ for all $k$, the above algorithm is known as the "Analog of the treppen-iteration using orthogonalization."[4] See Wilkinson (1965, Sec. 9.38, p. 607). The multiplicative ergodic theorem can thus be viewed as the generalization of this

algorithm to the case when the $T(f^k x)$ are randomly chosen.

Let us again call $\lambda^{(1)}, \ldots, \lambda^{(r)}$ the *distinct* characteristic exponents, and $m^{(1)}, \ldots, m^{(r)}$ their multiplicities. The space $E^{(i)}$ associated with the characteristic exponent $\leq \lambda^{(i)}$ (see Sec. III.A) is obtained as follows. Consider the last $m_-^{(i)} = m^{(i)} + \cdots + m^{(r)}$ columns of the matrix

$$R_1^{-1} \cdots R_n^{-1}\Delta = (R_n \cdots R_1)^{-1}\Delta ,$$

where $\Delta$ is the diagonal matrix equal to the diagonal part of $R_n \cdots R_1$. Let $E^{(i)}(n)$ be the space generated by these $m_-^{(i)}$ column vectors. Then $E^{(i)} = \lim_{n \to \infty} E^{(i)}(n)$.

Note that if we are only interested in the largest $s$ characteristic exponents, $\lambda_1 \geq \cdots \geq \lambda_s$, then it suffices to do the decomposition to triangular form only in the upper left $s \times s$ submatrix, leaving the matrices untouched in the lower right $(m-s) \times (m-s)$ corner.

The practical task of decomposing a matrix $T_k'$ as $Q_k R_k$, as discussed above, is abundantly treated in the literature, and library routines exist for it. According to Wilkinson (1965, Secs. 4.47–4.56), the Householder triangularization is preferable to Schmidt orthogonalization, since it leads to more precisely orthogonal matrices. This algorithm is available in Wilkinson and Reinsch (1971, Algorithm I/8, procedure "decompose"). It exists as part of the packages EISPACK and NAG. This algorithm is numerically very stable, and in fact the size of the eigenvalues should not matter.

## D. Characteristic exponents: physical experiments

By contrast with computer experiments, experiments in the laboratory do not normally give direct access to the derivatives $D_x f^t$. These derivatives must thus be estimated by a detailed analysis of the data. Once the derivatives $D_x f^t$ are known, the problem is analogous to that encountered in computer experiments. The same algorithms can be applied to obtain either the largest characteristic exponent $\lambda_1$ or other exponents. Only the positive exponents will be determined, however, or part of them. We have seen above how to restrict the computation to the largest $s$ characteristic exponents, and we shall see below why one can only hope to determine the *positive* $\lambda_i$ in general.

As in Sec. V.A we start with a time series $u(1), u(2), \ldots$, in $\mathbb{R}^\nu$, and from this we construct a sequence $x(1), x(2), \ldots$, in $\mathbb{R}^{m\nu}$, with $x(i) = [u(i), \ldots, u(i+m-1)]$. We shall discuss in remark (c) below how large $m$ should be taken. We shall now try to estimate the derivatives $T_{x(i)}^\tau = D_{x(i)} f^\tau$. As in Sec. V.C, $\tau$ should be such that the $e^{\lambda_k \tau}$ are not too large (we are only interested in positive $\lambda_k$); this means that $\tau$ should not be larger (and rather smaller) than the "characteristic time" of the system. Also, $\tau$ should not be too small, since we have to multiply later a number of matrices $T_{x(i)}^\tau$ proportional to $\tau^{-1}$. Of course, $\tau$ will be a multiple $p\Delta t$ of the time interval $\Delta t$ between measurements, so that

---

[4]We thank G. Wanner for helpful discussions in relation to this problem.

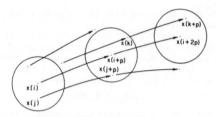

FIG. 21. The balls of radius $\bar{r}$ centered at $x(i)$, $x(i+p)$, and $x(i+2p)$.

$f_\tau x(i) = x(i+p)$.

The derivatives $T_{x(i)}^\tau$ will be obtained by a best linear fit of the map which, for $x(j)$ close to $x(i)$, sends $x(j) - x(i)$ to $f^\tau(x(j)) - f^\tau(x(i)) = x(j+p) - x(i+p)$ (see Fig. 21). Close here means that the map should be approximately linear. This will be ensured by choosing $\bar{r}$ sufficiently small and taking only those $x(j)$ for which

$$d[x(i),x(j)] \leq \bar{r} \quad \text{and} \quad d[x(i+p),x(j+p)] \leq \bar{r}$$

(these conditions imply that $x(j+k)$ is close to $x(i+k)$ for $0 < k < p$; one could also require $d[x(i+k), x(j+k)] \leq \bar{r}$ for each $k$ separately). The choice of $\bar{r}$ should probably be made by trial and error, monitoring how good a linear fit is obtained. Certainly $\bar{r}$ should be less than the upper bound $r_1$ of the "meaningful range for $C^m(r)$," discussed in remark (a) of Sec. V.A. Having chosen $\bar{r}$, we have to assume that the length $N$ of the original time series is sufficiently long so that there is a fair number of points $x(j)$ in the ball of radius $\bar{r}$ around $x(i)$, and such that $x(j+p)$ is also in the ball of radius $\bar{r}$ around $x(i+p)$. In principle, $m$ points are enough to determine a linear map, but we want many points: (a) to overcome the statistical scatter of the $x(j)$, (b) because a symmetric distribution of the $x(j)$ will yield a linear best fit from which the quadratic nonlinear terms have been eliminated. We repeat how $T_{x(i)}^\tau = D_{x(i)} f^\tau$ is obtained, with $\tau = p\Delta t$. Take all $x(j)$ such that $d[x(i),x(j)] \leq \bar{r}$ and $d[x(i+p),x(j+p)] \leq \bar{r}$, and determine the $mv \times mv$ matrix $T_{x(i)}^\tau$ by a least-squares fit[5] such that

$$T_{x(i)}^\tau[x(j) - x(i)] \approx x(j+p) - x(i+p) . \tag{5.15}$$

Note that when we estimate $T_{x(i+p)}^\tau$ we have to start looking again for *all* $x(k)$ such that $d[x(i+p),x(k)] \leq \bar{r}$ and $d[x(i+2p),x(k+p)] \leq \bar{r}$, and not just for the $x(k)$ of the form $x(j+p)$!

In general, the points $x(j)$ will not be uniformly distributed in all directions from $x(i)$. In other words, the vectors $x(j) - x(i)$ may not span $\mathbf{R}^{mv}$, and therefore the ma-

trix $T_{x(i)}^\tau$ may not be well defined by our prescription. Even if $T_{x(i)}^\tau$ is defined, there will in general be directions in which there are many fewer points than in others, so that the uncertainty on the elements of $T_{x(i)}^\tau$ corresponding to those directions is large. In fact, we can only expect with confidence that the vectors $x(j) - x(i)$ span the *expanding directions* around $x(i)$, i.e., the linear space tangent to the unstable manifold at $x(i)$ (because an SRB measure is absolutely continuous along the unstable manifold or because an attracting set contains the unstable manifolds of the points on it). The fact that the matrix $T_{x(i)}^\tau$ is only known with confidence in the unstable directions need not distress us: It means that we can determine with confidence only the *positive characteristic exponents.* This is done with the method of Sec. V.C, constructing triangular matrices from the $T_x^\tau$ for $x = x(i_0)$, $x(i_0+p), x(i_0+2p), \ldots$, computing the characteristic exponents, and discarding those which are $\leq 0$. The latter will usually (although not necessarily) be meaningless.

*Remarks.*

(a) The detailed method presented above for deriving characteristic exponents from experiments seems new. Up to now, attention has been concentrated on obtaining the largest exponent $\lambda_1$, using basically the method discussed in Sec. V.C after Eq. (5.11). For a different approach, see Wolf et al. (1984).

(b) The example of a time series $u(i) = \text{const}$, corresponding to an attracting fixed point, shows that it is not possible in general to obtain the negative characteristic exponents from the long-term behavior of a dynamical system. It is conceivable that our method works up to that $k$ after which the sum of the largest $k$ characteristic exponents becomes negative. If one has access to transients, then negative characteristic exponents are in principle accessible.

(c) We have discussed in this section the determination of the characteristic exponents of a measure $\rho$ from its projection $\pi_m \rho$. How large should one choose $m$ to be? For the determination of the information dimension, it was sufficient to take $mv \geq \dim_H \rho$. Here, however, this will usually be insufficient, because we have to reconstruct the *dynamics* in the support of $\rho$ from its projection in $\mathbf{R}^{mv}$. We want $\pi_m$ therefore to be *injective* on the support of $\rho$. According to Mañé's theorem in Sec. II.G. this may require $mv > 2 \dim_K(\text{support } \rho) + 1$. Probably the best evidence that $\pi_m$ is a "good" projection for the present purposes would be a reasonably good linear fit for Eq. (5.15).

### E. Spectrum, rotation numbers

The ergodic quantities that we have discussed in this paper—characteristic exponents, entropy, information dimension—are those which appear at this moment most important and most easily accessible. They are, however, not the only quantities one might consider. For a quasi-periodic system, the *generating frequencies* are of course important. More generally, Frisch and Morf (1981) have drawn attention to the *complex singularities* of the signal

---

[5]The most convenient algorithms are given in Wilkinson and Reinsch (1971), contribution I/8. (All algorithms in this book are available in the large libraries such as EISPACK and NAG.)

$u(t)$ and their relation to the high-frequency behavior of the power spectrum. One may also look for poles at complex values of the frequency in the power spectrum (*resonances*). Finally one should mention *rotation numbers*, which are not always defined but are interesting quantities when they make sense (see Ruelle, 1985).

## VI. OUTLOOK

Our review has led from the definitions of dynamical systems theory to a discussion of those quantities accessible today through the *statistical* analysis of time series for deterministic nonlinear systems. Together with the more geometrical aspects of bifurcation theory, this represents the main body of theoretically and experimentally successful ideas concerning nonlinear dynamics at this time. The purpose of this review is to make this knowledge accessible to a large number of scientists. The results presented here are the combined achievement of many investigators, only incompletely cited. We believe that the next step in the study of dynamical systems should lead to a better understanding of space-time patterns, for which only timid beginnings are now seen. We hope that the present review serves as an encouragement for the undertaking of this difficult problem.

*Note added in proof.* Another useful reprint collection to be added to the list of Sec. I is Hao Bai-Lin, 1984, *Chaos* (World Scientific, Singapore).

## ACKNOWLEDGMENTS

We wish to thank E. Lieb, A. S. Wightman, and especially P. Bergé and F. Ledrappier for their critical reading of the manuscript. This work was partially supported by the Fonds National Suisse.

## REFERENCES

Abraham, N. B., J. P. Gollub, and H. L. Swinney, 1984, "Testing nonlinear dynamics," Physica D **11**, 252.
Adler, R. L., A. G. Konheim, and M. H. McAndrew, 1965, "Topological entropy," Trans. Am. Math. Soc. **114**, 309.
Alexander, J. C., and J. A. Yorke, 1984, "Fat baker's transformations," Ergod. Theory Dynam. Syst. **4**, 1.
Anosov, D. V., 1967, "Geodesic flows on a compact Riemann manifold of negative curvature," Trudy Mat. Inst. Steklov **90**, [Proc. Steklov Math. Inst. **90** (1967)].
Arnold, V. I., and A. Avez, 1967, *Problèmes Ergodiques de la Mécanique Classique* (Gauthier-Villars, Paris) [*Ergodic Problems of Classical Mechanics* (Benjamin, New York, 1968)].
Atten, P., J. G. Caputo, B. Malraison, and Y. Gagne, 1984, "Détermination de dimension d'attracteurs pour différents écoulements," preprint.
Benedicks, M., and L. Carleson, 1984, "On iterations of $1 - ax^2$ on $(-1,1)$," Ann. Math., in press.
Benettin, G., L. Galgani, J. M. Strelcyn, 1976, "Kolmogorov entropy and numerical experiments," Phys. Rev. A **14**, 2338.
Benettin, G., L. Galgani, A. Giorgilli, and J. M. Strelcyn, 1978, "Tous les nombres de Liapounov sont effectivement calcul-

ables," C. R. Acad. Sci. Paris **286A**, 431.
Bergé, P., Y. Pomeau, and C. Vidal, 1984, *L'ordre dans le Chaos* (Herman, Paris).
Billingsley, P., 1965, *Ergodic Theory and Information* (Wiley, New York).
Bowen, R., 1975, *Equilibrium States and the Ergodic Theory of Anosov Diffeomorphisms*, Lecture Notes in Mathematics 470 (Springer, Berlin).
Bowen, R., 1978, *On Axiom A Diffeomorphisms*, Regional conference series in mathematics, No. 35 (Amer. Math. Soc., Providence, R.I.)
Bowen, R., and D. Ruelle, 1975, "The ergodic theory of Axiom A flows," Inventiones Math. **29**, 181.
Brin, H., and Z. Nitecki, 1985, private communication.
Caffarelli, R., R. Kohn, and L. Nirenberg, 1982, "Partial regularity of suitable weak solutions of the Navier-Stokes equations," Commun. Pure Appl. Math. **35**, 771.
Campanino, M., 1980, "Two remarks on the computer study of differentiable dynamical systems," Commun. Math. Phys. **74**, 15.
Campbell, D., and H. Rose, 1983, Eds., *Order in Chaos: Proceedings of the International Conference . . . Los Alamos, . . . 1982* (Physica D **7**, 1).
Carverhill, A., 1984a, "Flows of stochastic systems: ergodic theory," Stochastics, in press.
Carverhill, A., 1984b, "A 'Markovian' approach to the multiplicative ergodic (Oseledec) theorem for nonlinear dynamical systems," preprint.
Cohen, A., and I. Procaccia, 1984, "On computing the Kolmogorov entropy from the time signals of dissipative and conservative dynamical systems," Phys. Rev. A **31**, 1872.
Collet, P., J.-P. Eckmann, and O. E. Lanford, 1980, "Universal properties of maps of an interval," Commun. Math. Phys. **76**, 211.
Collet, P., and J.-P. Eckmann, 1980a, "On the abundance of aperiodic behavior for maps on the interval," Commun. Math. Phys. **73**, 115.
Collet, P., and J.-P. Eckmann, 1980b, *Iterated Maps on the Interval as Dynamical Systems* (Birkhauser, Cambridge, MA).
Collet, P., and J.-P. Eckmann, 1983, "Positive Liapunov exponents and absolute continuity for maps of the interval," Ergod. Theory Dynam. Syst. **3**, 13.
Conley, C., 1978, *Isolated Invariant Sets and the Morse Index*, Regional conference series in mathematics, No. 38 (Amer. Math. Soc., Providence, R.I.)
Croquette, V., 1982, "Déterminisme et chaos," Pour la Science **62**, 62.
Crutchfield, J., 1981, private communication.
Curry, J. H., 1979, "On the Hénon transformation," Commun. Math. Phys. **68**, 129.
Curry, J. H., 1981, "On computing the entropy of the Hénon attractor," J. Stat. Phys. **26**, 683.
Cvitanović, P., 1984, Ed., *Universality in Chaos* (Adam Hilger, Bristol).
Denker, M., C. Grillenberger, and K. Sigmund, 1976, *Ergodic Theory on Compact Spaces*, Lecture Notes in Mathematics 527 (Springer, Berlin).
Douady, A., and J. Oesterlé, 1980, "Dimension de Hausdorff des attracteurs," C. R. Acad. Sci. Paris **290A**, 1135.
Dubois, M., 1982, "Experimental aspects of the transition to turbulence in Rayleigh-Bénard convection," in *Stability of Thermodynamic Systems*, Lecture Notes in Physics 164, (Springer, Berlin), pp. 177–191.
Dunford, M., and J. T. Schwartz, 1958, *Linear Operators* (Inter-

science, New York).

Eckmann, J.-P., 1981, "Roads to turbulence in dissipative dynamical systems," Rev. Mod. Phys. 53, 643.

Farmer, D., 1984, private communication.

Fathi, A., M. R. Herman, and J.-C. Yoccoz, 1983, "A proof of Pesin's stable manifold theorem," in *Geometric Dynamics*, Lecture Notes in Mathematics 1007 (Springer, Berlin), pp. 177–215.

Fauve, S., and A. Libchaber, 1982, "Rayleigh-Bénard experiment in a low Prandtl number fluid, mercury," in *Chaos and Order in Nature: Proceedings of the International Symposium on Synergetics at Schloss Elmau, 1981*, edited by H. Haken (Springer, Berlin), pp. 25–35.

Feigenbaum, M. J., 1978, "Quantitative universality for a class of nonlinear transformations," J. Stat. Phys. 19, 25.

Feigenbaum, M. J., 1979, "The universal metric properties of nonlinear transformations," J. Stat. Phys. 21, 669.

Feigenbaum, M. J., 1980, "The transition to aperiodic behavior in turbulent systems," Commun. Math. Phys. 77, 65.

Feit, S. D., 1978, "Characteristic exponents and strange attractors," Commun. Math. Phys. 61, 249.

Feller, W., 1966, *An Introduction to Probability Theory and its Applications*, Vol. II (Wiley, New York).

Foias, C., and R. Temam, 1979, "Some analytic and geometric properties of the solutions of the evolution Navier-Stokes equations," J. Math. Pures Appl. 58, 339.

Frederickson, P., J. L. Kaplan, E. D. Yorke, and J. A. Yorke, 1983, "The Lyapunov dimension of strange attractors," J. Diff. Equ. 49, 185.

Frisch, U., and R. Morf, 1981, "Intermittency in nonlinear dynamics and singularities at complex times," Phys. Rev. A 23, 2673.

Ghil, M., R. Benzi, and G. Parisi, 1985, *Turbulence and Predictability in Geophysical Fluid Dynamics and Climate Dynamics* (North-Holland, Amsterdam).

Grassberger, P., 1981, "On the Hausdorff dimension of fractal attractors," J. Stat. Phys. 26, 173.

Grassberger, P., 1984, "Information aspects of strange attractors," preprint, Wuppertal.

Grassberger, P., and I. Procaccia, 1983a, "Estimating the Kolmogorov entropy from a chaotic signal," Phys. Rev. A 28, 2591.

Grassberger, P., and I. Procaccia, 1983b, "Measuring the strangeness of strange attractors," Physica D 9, 189.

Grebogi, C., E. Ott, and J. A. Yorke, 1983a, "Are three-frequency quasiperiodic orbits to be expected in typical nonlinear dynamical systems?" Phys. Rev. Lett. 53, 339.

Grebogi, C., E. Ott, and J. A. Yorke, 1983b, "Crises, sudden changes in chaotic attractors and transient chaos," Physica 7D, 181.

Guckenheimer, J., 1982, "Noise in chaotic systems," Nature 298, 358.

Guckenheimer, J., and P. Holmes, 1983, *Nonlinear Oscillations, Dynamical Systems, and Bifurcations of Vector Fields* (Springer, New York).

Gurel, O., and O. E. Rössler, 1979, Eds., *Bifurcation Theory and Applications in Scientific Disciplines* (Ann. N. Y. Acad. Sci. 316).

Haken, H., 1983, *Advanced Synergetics* (Springer, Berlin).

Helleman, R. H. G., 1980, Ed., *International Conference on Nonlinear Dynamics* (Ann. N. Y. Acad. Sci. 357).

Hénon, M., 1976, "A two-dimensional mapping with a strange attractor," Commun. Math. Phys. 50, 69.

Hénon, M., and C. Heiles, 1964, "The applicability of the third

integral of motion, some numerical experiments," Astron. J. 69, 73.

Hurewicz, W., and H. Wallman, 1948, *Dimension Theory* (Princeton University, Princeton).

Ilyashenko, Yu. S., 1983, "On the dimension of attractors of k-contracting systems in an infinite dimensional space," Vestn. Mosk. Univ. Ser. 1 Mat. Mekh. 1983 No. 3, 52–58.

Iooss, G., R. Helleman, and R. Stora, 1983, Eds., *Chaotic Behavior of Deterministic Systems* (North-Holland, Amsterdam).

Jakobson, M., 1981, "Absolutely continuous invariant measures for one-parameter families of one-dimensional maps," Commun. Math. Phys. 81, 39.

Johnson, R. A., K. J. Palmer, and G. Sell, 1984, "Ergodic properties of linear dynamical systems," preprint, Minneapolis.

Kantz, H., and P. Grassberger, 1984, "Repellers, semi-attractors, and long-lived chaotic transients," preprint, Wuppertal.

Kaplan, J. L., and J. A. Yorke, 1979, "Preturbulence: a regime observed in a fluid flow model of Lorenz," Commun. Math. Phys. 67, 93.

Katok, A., 1980, "Liapunov exponents, entropy and periodic orbits for diffeomorphisms," Publ. Math. IHES 51, 137.

Katok, A., and J.-M. Strelcyn, 1985, "Smooth maps with singularities, invariant manifolds, entropy and billiards" (in preparation).

Kifer, Yu. I., 1974, "On small random perturbations of some smooth dynamical systems," Izv. Akad. Nauk SSSR Ser. Mat. 38, No. 5, 1091 [Math. USSR Izv. 8, 1083 (1974)].

Kifer, Yu. I., 1984, "A multiplicative ergodic theorem for random transformations," preprint.

Ladyzhenskaya, O. A., 1969, *The Mathematical Theory of Viscous Incompressible Flow*, 2nd ed. (Nauka, Moscow, 1970). [2nd English edition Gordon and Breach, New York (1969)].

Lanford, O. E., 1977, in Turbulence Seminar, Proceedings 1976/77, edited by P. Bernard and T. Ratiu, Lecture Notes in Mathematics 615 (Springer, Berlin), pp. 113–116.

Ledrappier, F., 1981a, "Some relations between dimension and Liapunov exponents," Commun. Math. Phys. 81, 229.

Ledrappier, F., 1981b, "Some properties of absolutely continuous invariant measures on an interval," Ergod. Theory Dynam. Syst. 1, 77.

Ledrappier, F., and M. Misiurewicz, 1984, "Dimension of invariant measures," preprint.

Ledrappier, F., and L.-S. Young, 1984, "The metric entropy of diffeomorphisms. Part I. Characterization of measures satisfying Pesin's formula. Part II. Relations between entropy, exponents and dimension," preprints, University of California, Berkeley, 1984. "The metric entropy of diffeomorphisms," Bull. Am. Math. Soc. (New Ser.) 11, 343.

Li, T., and J. A. Yorke, 1975, "Period three implies chaos," Am. Math. Monthly 82, 985.

Libchaber, A., and J. Maurer, 1979, "Une expérience de Rayleigh-Bénard de géométrie réduite; multiplication, accrochage, et démultiplication de fréquences," J. Phys. (Paris) Colloq. 41, C3-51.

Lieb, E. H., 1984, "On characteristic exponents in turbulence," Commun. Math. Phys. 92, 473.

Lorenz, E. N., 1963, "Deterministic nonperiodic flow," J. Atmos. Sci. 20, 130.

Lundqvist, Stig, 1985, Ed., *The Physics of Chaos and Related Problems: Proceedings of the 59th Nobel Symposium, 1984* (Phys. Scr. T9, 1).

Mallet-Paret, J., 1976, "Negatively invariant sets of compact maps and an extension of a theorem of Cartwright," J. Diff. Equ. 22, 331.

Malraison, B., P. Atten, P. Bergé, and M. Dubois, 1983, "Dimension of strange attractors: an experimental determination for the chaotic regime of two convective systems," J. Phys. (Paris) Lett. 44, L-897.

Mandelbrot, B., 1982, The Fractal Geometry of Nature (Freeman, San Francisco).

Mañé, R., 1981, "On the dimension of the compact invariant sets of certain nonlinear maps," in Dynamical Systems and Turbulence, Warwick 1980, Lecture Notes in Mathematics 898 (Springer, New York), pp. 230–242.

Mañé, R., 1983, "Liapunov exponents and stable manifolds for compact transformations," in Geometrical Dynamics, Lecture Notes in Mathematics 1007 (Springer, Berlin), pp. 522–577.

Manneville, P., 1985, "Liapunov exponents for the Kuramoto-Sivashinsky model," preprint.

Manning, A., 1984, "The dimension of the maximal measure for a polynomial map," Ann. Math. 119, 425.

Mattila, P., 1975, "Hausdorff dimension, orthogonal projections and intersections with planes," Ann. Acad. Sci. Fenn. Ser. A1, Math., 1, 227.

McCluskey, H., and A. Manning, 1983, "Hausdorff dimension for horseshoes," Ergod. Theory Dynam. Syst. 3, 251.

Milnor, J., 1985, "On the concept of attractor," Commun. Math. Phys., in press.

Misiurewicz, M., 1981, "Structure of mappings of an interval with zero entropy," Publ. Math. IHES 53, 5.

Newhouse, S., 1974, "Diffeomorphisms with infinitely many sinks," Topology 13, 9.

Newhouse, S., 1979, "The abundance of wild hyperbolic sets and non-smooth stable sets for diffeomorphisms," Publ. Math. IHES 50, 102.

Newhouse, S., D. Ruelle, and F. Takens, 1978, "Occurrence of strange axiom A attractors near quasiperiodic flows on $T^m$, $m \geq 3$," Commun. Math. Phys. 64, 35.

Nobel Symposium, 1984, see Lundqvist, 1985.

Oseledec, V. I., 1968, "A multiplicative ergodic theorem. Lyapunov characteristic numbers for dynamical systems," Trudy Mosk. Mat. Obsc. 19, 179 [Moscow Math. Soc. 19, 197 (1968)].

Pesin, Ya. B., 1976, "Invariant manifold families which correspond to nonvanishing characteristic exponents," Izv. Akad. Nauk SSSR Ser. Mat. 40, No. 6, 1332 [Math. USSR Izv. 10, No. 6, 1261 (1976)].

Pesin, Ya. B., 1977, "Lyapunov characteristic exponents and smooth ergodic theory," Usp. Mat. Nauk 32, No. 4 (196), 55 [Russian Math. Survey 32, No. 4, 55 (1977)].

Procaccia, I., 1984, The static and dynamic invariants that characterize chaos and the relations between them in theory and experiments, proceedings of the 59th Nobel Symposium (Phys. Scr. T9, 40).

Pugh, C. C., and M. Shub, 1984, "Ergodic attractors" (in preparation).

Raghunathan, M. S., 1979, "A proof of Oseledec's multiplicative ergodic theorem," Israel J. Math. 32, 356.

Riste, T., and K. Otnes, 1984, Neutron scattering from a convecting nematic: multicriticality, multistability and chaos, proceedings of the 59th Nobel Symposium (Phys. Scr. T9, 76).

Rössler, O. E., 1976, "An equation for continuous chaos," Phys. Lett. A 57, 397.

Roux, J.-C., R. H. Simoyi, and H. L. Swinney, 1983, "Observation of a strange attractor," Physica D 8, 257.

Roux, J.-C., and H. L. Swinney, 1981, "Topology of chaos in a

chemical reaction," in Nonlinear Phenomena in Chemical Dynamics, edited by C. Vidal and A. Pacault (Springer, Berlin), pp. 38–43.

Ruelle, D., 1976, "A measure associated with Axiom A attractors," Am. J. Math. 98, 619.

Ruelle, D., 1978, "An inequality for the entropy of differentiable maps," Bol. Soc. Bras. Mat. 9, 83.

Ruelle, D., 1979, "Ergodic theory of differentiable dynamical systems," Phys. Math. IHES 50, 275.

Ruelle, D., 1980, "Measures describing a turbulent flow," Ann. N. Y. Acad. Sci. 357, 1.

Ruelle, D., 1981, "Small random perturbations of dynamical systems and the definition of attractors," Commun. Math. Phys. 82, 137.

Ruelle, D., 1982a, "Characteristic exponents and invariant manifolds in Hilbert space," Ann. Math. 115, 243.

Ruelle, D., 1982b, "Large volume limit of the distribution of characteristic exponents in turbulence," Commun. Math. Phys. 87, 287.

Ruelle, D., 1984, "Characteristic exponents for a viscous fluid subjected to time dependent forces," Commun. Math. Phys. 93, 285.

Ruelle, D., 1985, "Rotation numbers for diffeomorphisms and flows," Ann. Inst. Henri Poincaré, 42, 109.

Ruelle, D., and F. Takens, 1971, "On the nature of turbulence," Commun. Math. Phys. 20, 167. Note concerning our paper "On the nature of turbulence," Commun. Math. Phys. 21, 21.

Shaw, R. S., 1981, "Strange attractors, chaotic behavior and information flow," Z. Naturforsch. 36a, 80.

Shimada I., 1979, "Gibbsian distribution on the Lorenz attractor," Prog. Theor. Phys. 62, 61.

Sinai, Ya. G., 1972, "Gibbs measures in ergodic theory," Usp. Mat. Nauk 27, No. 4, 21 [Russian Math. Surveys 27, No. 4, 21 (1972)].

Smale, S., 1967, "Differentiable dynamical systems," Bull. Am. Math. Soc. 73, 747.

Takens, F., 1981, "Detecting strange attractors in turbulence," in Dynamical Systems and Turbulence, Warwick 1980, Lecture Notes in Mathematics 898 (Springer, Berlin), pp. 366–381.

Takens, F., 1983, "Invariants related to dimension and entropy," in Atas do 13. Col. brasiliero de Matematicas, Rio de Janerio, 1983.

Temam, R., 1979, Navier-Stokes Equations, 2nd ed. (North-Holland, Amsterdam).

Thieullen, P., 1985, "Exposant de Liapounov des fibrés dynamiques pseudocompacts," in preparation.

Ushiki, S., M. Yamaguti, and H. Matano, 1980, "Discrete population models and chaos," Lecture Notes Num. Appl. Anal. 2, 1.

Vidal, C., and A. Pacault, 1981, Eds., Nonlinear Phenomena in Chemical Dynamics: Proceedings (Springer, Berlin).

Vul, E. B., Ya. G. Sinai, and K. M. Khanin, 1984, "Universality of Feigenbaum and thermodynamic formalism," Usp. Mat. Nauk 39, No. 3 (237), 3.

Walden, R. W., P. Kolodner, A. Passner, and C. M. Surko, 1984, "Nonchaotic Rayleigh-Bénard convection with four and five incommensurate frequencies," Phys. Rev. Lett. 53, 242.

Walters, P., 1975, Ergodic Theory—Introductory Lectures, Lecture Notes in Mathematics 458 (Springer, Berlin).

Wilkinson, J. H., 1965, The Algebraic Eigenvalue Problem (Clarendon, Oxford).

Wilkinson, J. H., and C. Reinsch, 1971, Linear Algebra (Springer, Berlin).

Wolf, A., J. B. Swift, H. L. Swinney, and J. A. Vastano, 1984,

"Determining Liapunov exponents from a time series," Physica D, in press.

Young, L.-S., 1981, "Capacity of attractors," Ergod. Theory Dynam. Syst. 1, 381.

Young, L.-S., 1982, "Dimension, entropy and Liapunov exponents," Ergod. Theory Dynam. Syst. 2, 109.

Young, L.-S., 1984, "Dimension, entropy and Liapunov exponents in differentiable dynamical systems," Physica A 124, 639.

PHYSICAL REVIEW A          VOLUME 33, NUMBER 2          FEBRUARY 1986

# Fractal measures and their singularities: The characterization of strange sets

Thomas C. Halsey, Mogens H. Jensen, Leo P. Kadanoff, Itamar Procaccia,[*] and Boris I. Shraiman[†]

*The James Franck Institute, The Enrico Fermi Institute for Nuclear Studies, and Department of Chemistry,*
*The University of Chicago, 5640 South Ellis Avenue, Chicago, Illinois 60637*

(Received 26 August 1985)

We propose a description of normalized distributions (measures) lying upon possibly fractal sets; for example those arising in dynamical systems theory. We focus upon the scaling properties of such measures, by considering their singularities, which are characterized by two indices: $\alpha$, which determines the strength of their singularities; and $f$, which describes how densely they are distributed. The spectrum of singularities is described by giving the possible range of $\alpha$ values and the function $f(\alpha)$. We apply this formalism to the $2^\infty$ cycle of period doubling, to the devil's staircase of mode locking, and to trajectories on 2-tori with golden-mean winding numbers. In all cases the new formalism allows an introduction of smooth functions to characterize the measures. We believe that this formalism is readily applicable to experiments and should result in new tests of global universality.

## I. INTRODUCTION

Nonlinear physics presents us with a perplexing variety of complicated fractal objects and strange sets. Notable examples include strange attractors for chaotic dynamical systems,[1,2] configurations of Ising spins at critical points,[3] the region of high vorticity in fully developed turbulence,[4,5] percolating clusters and their backbones,[6] and diffusion-limited aggregates.[7,8] Naturally one wishes to characterize the objects and describe the events occurring on them. For example, in dynamical systems theory one is often interested in a strange attractor (the object) and how often a given region of the attractor is visited (the event). In diffusion-limited aggregation, one is interested in the probability of a random walker landing next to a given site on the aggregate.[8] In percolation, one may be interested in the distribution of voltages across the different elements in a random-resistor network.[6]

In general, one can describe such events by dividing the object into pieces labeled by an index $i$ which runs from 1 up to $N$. The size of the $i$th piece is $l_i$ and the event occurring upon it is described by a number $M_i$. For example, in critical phenomena, we can let $M_i$ be the magnetization of the region labeled by $i$. Such a picture is natural in the droplet theory of the Ising model, where one argues that if the region $i$ has a size of order $l$, the magnetization has a value of the order of

$$M_i \sim l^y , \tag{1.1}$$

where $y$ (or $y_\sigma$) is one of the standard critical indices.[9] Since these droplets are imagined to fill the entire space, the density of such droplets is simply

$$\rho(l) \sim \frac{1}{l^d} , \tag{1.2}$$

where $d$ is the Euclidean dimension of space. In fact, in critical phenomena we define a whole sequence of $y_q$'s by saying that the typical values of $(M_i)^q$ vary with $q$ and have the form[10]

$$(M_i)^q \sim l^{y_q}, \quad q = 1, 2, 3, \ldots . \tag{1.3}$$

Typically, our attention focuses upon the values of $y_q$ that are greater than zero and we have only a few distinct values of these.[11]

In this paper we are interested not in critical phenomena but instead in a broad class of strange objects. However, we specialize our treatment to the case in which $M_i$ has the meaning of a probability that some event will occur upon the $i$th piece. For example, in experiments on chaotic systems one measures a time series $\{x_i\}_{i=1}^N$. These points belong to a trajectory in some $d$-dimensional phase space. Typically, the trajectory does not fill the $d$-dimensional space even when $N \to \infty$, because the trajectory lies on a strange attractor of dimension $D$, $D < d$. One can ask now how many times, $N_i$, the time series visits the $i$th box. Defining $p_i = \lim_{N \to \infty} (N_i / N)$, we generate the measure on the attractor $d\mu(x)$, because

$$p_i = \int_{i\text{th box}} d\mu(x) .$$

In many nonlinear problems, the possible scaling behavior is richer and more complex than is the case in critical phenomena. If a scaling exponent $\alpha$ is defined by saying that

$$p_i^q \sim l_i^{\alpha q} , \tag{1.4}$$

then $\alpha$ [roughly equivalent to $y_q / q$ in Eq. (1.3)] can take on a range of values, corresponding to different regions of the measure. In particular, if the system is divided into pieces of size $l$, we suggest that the number of times $\alpha$ takes on a value between $\alpha'$ and $\alpha' + d\alpha'$ will be of the form

$$d\alpha' \rho(\alpha') l^{-f(\alpha')} , \tag{1.5}$$

where $f(\alpha')$ is a continuous function. The exponent $f(\alpha')$ reflects the differing dimensions of the sets upon which the singularities of strength $\alpha'$ may lie. This expression is roughly equivalent to Eq. (1.2), except that now, instead of the dimension $d$, we have a fractal dimension $f(\alpha)$

which varies with $\alpha$. Thus, we model fractal measures by interwoven sets of singularities of strength $\alpha$, each characterized by its own dimension $f(\alpha)$. The rest of our formalism attempts to unravel this complexity in a workable fashion.

The concept of a singularity strength $\alpha$ was stressed in the context of diffusion-limited aggregation in independent work of Turkevich and Scher[12] and of Halsey, Meakin, and Procaccia.[8] The latter group pointed out the significance of the density of singularities and expressed it in terms of $f$.

In order to determine the function $f(\alpha)$ for a given measure, we must relate it to observable properties of the measure. We relate $f(\alpha)$ to a set of dimensions which have been introduced by Hentschel and Procaccia, the set $D_q$ defined by[13]

$$D_q = \lim_{l \to 0} \left[ \frac{1}{q-1} \frac{\ln \chi(q)}{\ln l} \right] , \qquad (1.6)$$

where

$$\chi(q) = \sum_i p_i^q . \qquad (1.7)$$

$D_0$ is just the fractal dimension of the support of the measure, while $D_1$ is the information dimension and $D_2$ is the correlation dimension.[14]

As $q$ is varied in Eq. (1.7), different subsets, which are associated with different scaling indices, become dominant. Substituting Eqs. (1.4) and (1.5) into Eq. (1.7), we obtain

$$\chi(q) = \int d\alpha' \rho(\alpha') l^{-f(\alpha')} l^{q\alpha'} . \qquad (1.8)$$

Since $l$ is very small, the integral in Eq. (1.8) will be dominated by the value of $\alpha'$ which makes $q\alpha' - f(\alpha')$ smallest, provided that $\rho(\alpha')$ is nonzero. Thus, we replace $\alpha'$ by $\alpha(q)$, which is defined by the extremal condition

$$\frac{d}{d\alpha'} [q\alpha' - f(\alpha')] \Big|_{\alpha' = \alpha(q)} = 0 .$$

We also have

$$\frac{d^2}{d(\alpha')^2} [q\alpha' - f(\alpha')] \Big|_{\alpha' = \alpha(q)} > 0 ,$$

so that

$$f'(\alpha(q)) = q , \qquad (1.9a)$$

$$f''(\alpha(q)) < 0 . \qquad (1.9b)$$

It then follows from Eq. (1.6) that[8]

$$D_q = \frac{1}{q-1} [q\alpha(q) - f(\alpha(q))] . \qquad (1.10)$$

Thus, if we know $f(\alpha)$, and the spectrum of $\alpha$ values, we can find $D_q$. Alternatively, given $D_q$, we can find $\alpha(q)$ since

$$\alpha(q) = \frac{d}{dq} [(q-1)D_q] , \qquad (1.11)$$

and, knowing $\alpha(q)$, $f(q)$ can be obtained from Eq. (1.10). Equations (1.9)–(1.11) are the main formal results used

in this paper. In the next section we develop the formalism outlined here in somewhat more detail and apply it to systems with strong self-similarity properties. In Sec. III we apply the formalism to some important examples of measures arising in dynamical systems. We examine the $2^\infty$ cycle of period doubling,[15] the devil's staircase of mode locking in circle maps,[16,17] and the elements of the critical cycle at the onset of chaos in circle maps with golden-mean winding number.[18–20] Although all of these cases have been examined previously, we are able to find a *smooth* function with which to characterize them. Furthermore, these characterizations are universal. Other attempts to study these measures have led to nowhere smooth scaling functions.[15,21] Since the characterizations are functions rather than numbers, they offer much more information than fractal dimensions. Unlike power spectra, these functions possess an immediate connection to the metric properties of the measures involved, and do not call for cumbersome interpretation. Therefore, we believe that experimental measurements of $D_q$, and thus of $f(\alpha)$, should replace more common tests of universality in the transition to chaos. We give many examples of the procedures employed, and we hope to encourage experiments to follow these lines.

## II. EXACTLY SOLUBLE STRANGE SETS

### A. Preliminaries

We begin by introducing a more general definition of the dimensions $D_q$. Consider a strange set $S$ embedded in a finite portion of $d$-dimensional Euclidean space. Imagine partitioning the set into some number of disjoint pieces, $S_1, S_2, \ldots, S_N$, in which each piece has a measure $p_i$ and lies within a ball of radius $l_i$, where each $l_i$ is restricted by $l_i < l$. Then define a partition function

$$\Gamma(q, \tau, \{S_i\}, l) = \sum_{i=1}^{N} \frac{p_i^q}{l_i^\tau} . \qquad (2.1)$$

Eventually we shall argue that, for large $N$, this partition function is of the order unity only when

$$\tau = (q-1)D_q . \qquad (2.2)$$

To make this argument, consider now two regions:

region $A$: $q \geq 1$, $\tau \geq 0$, $\qquad (2.3a)$

region $B$: $q \leq 1$, $\tau \leq 0$. $\qquad (2.3b)$

In region $A$, adjust the partition $\{S_i\}$ so as to maximize $\Gamma$. In region $B$, adjust it so that $\Gamma$ is as small as possible. Then define

$$\Gamma(q, \tau, l) = \text{Sup } \Gamma(q, \tau, \{S_i\}, l) \quad (\text{region } A) , \qquad (2.4a)$$

$$\Gamma(q, \tau, l) = \text{Inf } \Gamma(q, \tau, \{S_i\}, l) \quad (\text{region } B) . \qquad (2.4b)$$

The supremum in region $A$ will exist as long as there are constants $a > 0$ and $\alpha_0 > 0$, so that for any possible subset of $S$, $\{S_i\}$, we have

$$p_i \leq a(l_i)^{\alpha_0} . \qquad (2.5)$$

Then $\Gamma(q,\tau,l)$ will exist and be less than infinity whenever

$$\alpha_0(q-1)>\tau . \tag{2.6}$$

Next define

$$\Gamma(q,\tau)=\lim_{l\to 0}[\Gamma(q,\tau,l)] . \tag{2.7}$$

Notice that $\Gamma(q,\tau)$ is a monotone nondecreasing function of $\tau$ and a monotone nonincreasing function of $q$. One can argue that there is a unique function $\tau(q)$ such that

$$\Gamma(q,\tau)=\begin{cases}\infty & \text{for } \tau>\tau(q),\\ 0 & \text{for } \tau<\tau(q) .\end{cases} \tag{2.8}$$

Equation (2.8) permits us to define $D_q$ as

$$(q-1)D_q=\tau(q) . \tag{2.9}$$

Once $D_q$ is known, Eqs. (1.10) and (1.11) will then give $\alpha(q)$ and $f(q)$. Notice that our definition of $D_q$ is precisely the one which makes $D_0$ the Hausdorff dimension.

### B. Connection to previously defined $D_q$

Hentschel and Procaccia[13] also defined a $D_q$, which we now denote as $D_q^{\mathrm{HP}}$. To relate the two quantities, recall that the authors of Ref. 13 defined a partition in which all the diameters $l_i$ had the same value $l$. We know that

$$\Gamma(q,\tau,l)\begin{cases}>l^{-\tau}\sum_{i=1}^{N}p_i^q & \text{(region } A)\\ \\ <l^{-\tau}\sum_{i=1}^{N}p_i^q & \text{(region } B) .\end{cases} \tag{2.10}$$

If $\tau$ is chosen correctly, i.e., $\tau=\tau(q)$, the left-hand side of Eq. (2.10) will neither go to zero nor diverge very strongly as $l\to 0$. In particular, we guess that $\Gamma(l)$ is no worse than logarithmically dependent upon these quantities. Then

$$\lim_{l\to 0}[\ln\Gamma(q,\tau(q),l)/\ln l]\to 0 .$$

We have now

$$\frac{\tau}{q-1}\le\lim_{l\to 0}\left[\frac{\ln\left[\sum_{i=1}^{N}p_i^q\right]}{(\ln l)(q-1)}\right] . \tag{2.11}$$

The right-hand side of (2.11) is $D_q^{\mathrm{HP}}$. We thus find

$$D_q\le D_q^{\mathrm{HP}} . \tag{2.12}$$

Since we believe that Eq. (2.10) will often be an order of magnitude equality when $\tau=\tau(q)$, we think that Eq. (2.12) will be an equality in most cases of interest.

At this point we turn to some simple examples to illustrate the quantities $\tau(q)$. These examples will enable us to gain intuition about the quantities $\alpha(q)$ and $f(\alpha)$.

### C. Exactly soluble examples

#### 1. Power-law singularity

One of the simplest possible applications of this formalism is to a probability measure with only one power-law singularity. Imagine a probability density $\rho(x)=\tilde{\alpha}x^{\tilde{\alpha}-1}$ on $x\in[0,1]$, where $0<\tilde{\alpha}<1$. Let us partition the interval into $N$ segments $[x_i,x_i+\Delta x]$, with $\Delta x=N^{-1}$. The total probability measure on all of these intervals except for that adjoining zero is well approximated by $\rho(x_i)\Delta x$. The probability upon the segment adjoining zero possesses a probability $\rho_0=(\Delta x)^{\tilde{\alpha}}$. The partition function is therefore

$$\Gamma(q,\tau,\Delta x)\approx\frac{(\Delta x)^{\tilde{\alpha}q}}{(\Delta x)^{\tau}}+\sum_{i\neq 0}\frac{\tilde{\alpha}x_i^{\tilde{\alpha}-1}(\Delta x)^q}{(\Delta x)^{\tau}} . \tag{2.13}$$

There are $(\Delta x)^{-1}$ terms in the sum, so that

$$\Gamma(q,\tau,\Delta x)\sim(\Delta x)^{\tilde{\alpha}q-\tau}+(\Delta x)^{q-1-\tau} . \tag{2.14}$$

Thus, since we require that $\Gamma$ neither go to zero nor infinity, we have that

$$\tau=\min\{q-1,\tilde{\alpha}q\} , \tag{2.15a}$$

or

$$D_q=\frac{1}{q-1}\min\{q-1,\tilde{\alpha}q\} . \tag{2.15b}$$

Thus for $q>q^*=1/(1-\tilde{\alpha})$, the dimensions correspond to a value of $\alpha=\tilde{\alpha}$ and of $f=0$, while for $q<q^*$ the dimensions correspond to $\alpha=1$ and $f=1$. Thus, in this example the $f$-$\alpha$ spectrum consists of two points, corresponding to the two types of behavior in the measure.

#### 2. Cantor sets and generators

If a measure possesses an exact recursive structure, one can find its $D_q$. Suppose that the measure can be generated by the following process. Start with the original region which has measure 1 and size 1. Divide the region into pieces $S_i$, $i=1,2,\ldots,N$, with measure $p_i$ and size $l_i$. Suppose that the maximum of $l_i$ is given by $l$. Then at the first stage we can construct a partition function,

$$\Gamma(q,\tau,l)=\sum_i\frac{p_i^q}{l_i^{\tau}} . \tag{2.16}$$

Continue the Cantor construction. At the next stage each piece of the set is further divided into $N$ pieces, each with a measure reduced by a factor $p_j$ and size by a factor $l_j$. At this level the partition function will be

$$\Gamma(q,\tau,l^2)=[\Gamma(q,\tau,l)]^2 . \tag{2.17}$$

We see at once that, for this kind of measure, the first partition function $\Gamma(l)$ will generate all the others, and that $\tau(q)$ is defined by

$$\Gamma(q,\tau(q),l)=1 . \tag{2.18}$$

If a partition with finite $N$ yields a $\Gamma$ which obeys (2.17), that partition is called a generator.[22]

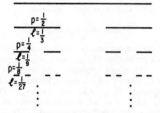

FIG. 1. The construction of the uniform Cantor set. At each stage of the construction the central third of each segment is removed from the set. Each segment has measure $p_0=(\frac{1}{2})^n$ and scale $l_0=(\frac{1}{3})^n$, where $n$ is the number of generations.

### 3. Uniform Cantor set

A simple example is the classical Cantor set obtained by dividing the interval [0,1] as shown in Fig. 1. We initially replace the unit interval with two intervals, each of length $l=\frac{1}{3}$. Each of these intervals receives the same measure $p=\frac{1}{2}$. At the next stage of the construction of the measure this same process is repeated on each of these two intervals. Thus, for this measure we require

$$2\left[\frac{(\frac{1}{2})^q}{(\frac{1}{3})^\tau}\right]=1 , \tag{2.19}$$

which yields

$$\tau=(q-1)[\ln(2)/\ln(3)] \quad [\text{or } D_q=\ln(2)/\ln(3)] . \tag{2.20}$$

If $l_0$ is the length scale of the intervals at a particular level of the partitioning, and $p_0$ is the measure for such an interval, then

$$p_0=l_0^{\ln(2)/\ln(3)} . \tag{2.21}$$

Calling the index of the singularity $\alpha$, i.e., $p_0\sim l_0^\alpha$, we have here $\alpha=\ln(2)/\ln(3)$. If we further ask what is the density of these singularities, we find immediately that it is simply the density of the set,

$$\rho(l_0)=\frac{1}{l_0^{\ln(2)/\ln(3)}} , \tag{2.22}$$

and Eq. (1.5) leads to $f=\ln(2)/\ln(3)$. Thus in this example, $\alpha=f$, and also

$$\tau(q)=q\alpha-f . \tag{2.23}$$

Although Eq. (2.23) is trivial here, we shall see that its analog, Eq. (1.10), also holds in the most general cases.

### 4. Two-scale Cantor set

A somewhat less trivial example is obtained by constructing a Cantor set as in Fig. 2. Here we use two rescaling parameters $l_1$ and $l_2$ and two measures $p_1$ and $p_2$, and then continue to subdivide self-similarly. We assume that $l_2>l_1$. It is apparent that this example also has a generator, since the condition

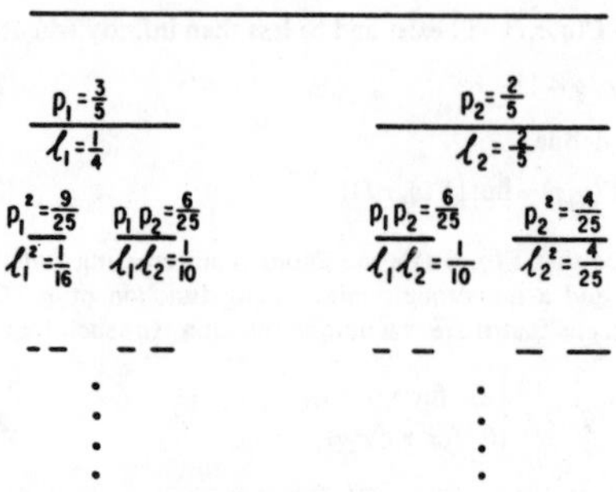

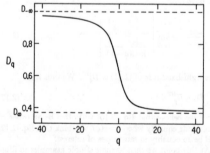

FIG. 2. A Cantor-set construction with two rescalings $l_1=0.25$ and $l_2=0.4$ and respective measure rescalings $p_1=0.6$ and $p_2=0.4$. The division of the set continues self-similarly.

$$\Gamma(q,\tau,l_2^n)=\left[\frac{p_1^q}{l_1^\tau}+\frac{p_2^q}{l_2^\tau}\right]^n=1 \tag{2.24}$$

results in a $\tau$ that does not depend on $n$. The value of $\tau$ depends, however, on $q$. In Fig. 3 we show $D_q=\tau(q)/(q-1)$ as a function of $q$, as obtained numerically by solving Eq. (2.24). To further understand this curve, we can examine the quantity $\Gamma(l_2^n)$ for this case explicitly:

$$\Gamma(q,\tau,l_2^n)=\sum_m \binom{n}{m} p_1^{mq}p_2^{(n-m)q}(l_1^m l_2^{n-m})^{-\tau}=1 . \tag{2.25}$$

We expect that in the limit $n\to\infty$ the largest term in this sum should dominate. To find the largest term we compute

$$\frac{\partial\ln\Gamma(l_2^n)}{\partial m}=0 . \tag{2.26}$$

Using the Stirling approximation, we find that Eq. (2.26) is equivalent to

$$\tau=\frac{\ln(n/m-1)+q\ln(p_1/p_2)}{\ln(l_1/l_2)} . \tag{2.27}$$

Since we expect that the maximal term dominates the sum, we have a second equation,

FIG. 3. $D_q$ plotted vs $q$ for the two-scale Cantor set of Fig. 2.

$$\begin{bmatrix} n \\ m \end{bmatrix} p_1^{mq} p_2^{(n-m)} \left[ \frac{1}{l_1^m l_2^{(n-m)}} \right]^\tau = 1 . \qquad (2.28)$$

Inserting Eq. (2.27) into Eq. (2.28) leads to an equation for $n/m$. After some algebraic manipulation, one finds

$$\ln(n/m)\ln(l_1/l_2) - \ln(n/m - 1)\ln l_1$$
$$= q(\ln p_1 \ln l_2 - \ln p_2 \ln l_1) . \qquad (2.29)$$

We thus see that for any given $q$ there will be a value of $n/m$ which solves Eq. (2.29) and, in turn, determines $\tau$ from Eq. (2.27). This maximal term which determines $\tau$ actually comes from a set of $\binom{n}{m}$ segments, all of which have the same size $l_1^m l_2^{(n-m)}$. Their density exponent $f$ is determined by

$$\begin{bmatrix} n \\ m \end{bmatrix} (l_1^m l_2^{(n-m)})^f = 1 , \qquad (2.30)$$

or

$$f = \frac{(n/m - 1)\ln(n/m - 1) - (n/m)\ln(n/m)}{\ln l_1 + (n/m - 1)\ln l_2} . \qquad (2.31)$$

The exponent determining the singularity in the measure, $\alpha$, is determined by

$$p_1^m p_2^{(n-m)} = (l_1^m l_2^{(n-m)})^\alpha , \qquad (2.32)$$

or

$$\alpha = \frac{\ln p_1 + (n/m - 1)\ln p_2}{\ln l_1 + (n/m - 1)\ln l_2} . \qquad (2.33)$$

Thus, for any chosen $q$, the measure scales as $\alpha(q)$ on a set of segments which converge to a set of dimension $f(q)$. As $q$ is varied, different regions of the set determine $D_q$. It can be shown that Eqs. (2.27), (2.29), (2.31), and (2.33) again lead to

$$\tau = (q-1)D_q = q\alpha(q) - f(q) . \qquad (2.34)$$

We can also understand the spectrum of scaling indices $\alpha$ by considering the "kneading sequences" for the segments. In the first level of the construction there are two segments of sizes $l_1$ and $l_2$ and measures $p_1$ and $p_2$ which we can label $L$ (left) and $R$ (right). At the next level we have four segments, which we can reach by going left or right: $LL$, $LR$, $RL$, and $RR$. Thus the measure and the size of any segment are determined by its address, the kneading sequence of $L$'s and $R$'s. For example, the size of a segment is $l_1^m l_2^{(n-m)}$, where $m$ and $n-m$ are, respectively, the numbers of $L$'s and $R$'s in the kneading sequence. Clearly, the sequence $LLL...LLL...$ is associated with $\alpha = \ln(p_1)/\ln(l_1) = D_\infty$, which lies on the edge of the spectrum, while the sequence $RRR...RRR...$ is associated with the singularity lying on the other edge of the spectrum. Other, less trivial kneading sequences lead to values of $\alpha$ between these two extremes. We note, however, that it is only the infinite "tail" of the sequence that determines the asymptotic scaling behavior. The number of sequences leading to the same singularity $\alpha$ may be simply found, and leads via Eq. (2.30) to exactly the same results for $f(\alpha)$ as the partition-function analysis above.

Finally, in Fig. 4 we display the curve $f(\alpha)$. The curve has been obtained for $l_1 = 0.25$, $l_2 = 0.4$ and $p_1 = 0.6$,

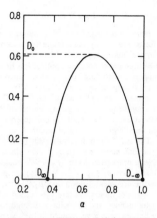

FIG. 4. The plot of $f$ vs $\alpha$ for the set in Fig. 2. Note that $f = 0$ corresponds to $\alpha$ values $D_{-\infty} = \ln(0.4)/\ln(0.4) = 1.0$ and $D_\infty = \ln(0.6)/\ln(0.25) = 0.3684$.

$p_2 = 0.4$. The leftmost point on the curve is $f = 0$, $\alpha = \ln(0.6)/\ln(0.25)$. This is the value that in Eqs. (2.31) and (2.33) obtains for $n = m$. At any level of the construction there is exactly one such segment ($f = 0$) and the singularity is

$$\ln p_1 / \ln l_1 = \text{Inf}\{\ln p_1 / \ln l_1, \ln p_2 / \ln l_2\} .$$

This value of $\alpha$ is also $D_\infty$. The rightmost point on the graph again corresponds to $f = 0$, but now

$$\alpha = \ln p_2 / \ln l_2 = \text{Sup}\{\ln p_1 / \ln l_1, \ln p_2 / \ln l_2\} .$$

This is also $D_{-\infty}$. Whereas $D_\infty$ corresponds to the region in the set where the measure is most concentrated, $D_{-\infty}$ corresponds to that where the measure is most rarefied. For $q = 0$ we simply obtain $f = D_0$, where $D_0$ is the Hausdorff dimension of the set. This is the maximum of the graph $f(\alpha)$.

Certain features of this curve are quite general, and follow from Eqs. (1.9)–(1.12). From Eq. (1.9) we find immediately that

$$\frac{\partial f}{\partial \alpha} = q , \qquad (2.35a)$$

$$\frac{\partial^2 f}{\partial \alpha^2} < 0 . \qquad (2.35b)$$

Thus, for any measure the curve $f(\alpha)$ will be convex, with a single maximum at $q = 0$, and with infinite slope at $q = \pm\infty$. Also from Eq. (1.10) with $q = 1$, we find that $\alpha(1) = f(1)$. The slope $\partial f / \partial \alpha$ there is unity. This general behavior of the curve $f(\alpha)$ will be seen in all cases where the measure possesses a continuous spectrum.

Although this example is rather simple, it contains many of the properties of the richer sets considered in Sec. III. In particular, we will not lose this intuitive view of the meaning of $\alpha$ and $f$.

### 5. Other types of spectra

We can obtain more insight into the meaning of the $f$-$\alpha$ spectrum for a measure by considering two examples of measures on continuous supports. Many of the most interesting measures encountered in applications lie on continuous supports, including the growth measure for diffusion-limited aggregates and strange attractors for systems of ordinary differential equations.

The first example is a simple generalization of the two-scale Cantor set defined by (2.24). A unit interval is subdivided into three segments, two of length $l_2$ and one of length $l_1$. The two former intervals each receive a proportion of the total measure given by $p_2$, and the latter interval receives a proportion given by $p_1$. We imagine that $l_1 + 2l_2 = 1$ and that $p_1 + 2p_2 = 1$. We also imagine, for the sake of the argument below, that $p_2/l_2 > p_1/l_1$ and that $l_2 > l_1$. Each of these three intervals is then subdivided in the same manner, and so forth. Although the measure on the line segment is rearranged at each step of the recursive process, the support for the measure remains at each step the original line segment. Thus we expect that $D_0$ for this measure will be 1. Furthermore, the densest intervals on the line segment contract not to one point (as was the case in the two-scale Cantor set), but to a set of points of finite dimension. Thus, we expect the lowest value of $\alpha$, and hence the value of $D_\infty$, to correspond to a nonzero value of $f$. Note that there is always only one segment at the lowest value of the density, so that we still expect $D_{-\infty}$ to correspond to a value of $f = 0$. The condition (2.18) above on $\Gamma$ requires that

$$\Gamma(q,\tau,l_2) = \frac{p_1^q}{l_1^\tau} + 2\frac{p_2^q}{l_2^\tau} = 1 . \tag{2.36}$$

The solution is simple and is displayed in Fig. 5. As predicted above, $f(q \to \infty) \neq 0$, so that the leftmost part of the $f$-$\alpha$ curve resembles a hook.

The second example is a set generated according to a different rule than the Cantor sets. The method is displayed in Fig. 6. At each stage, only the regions which

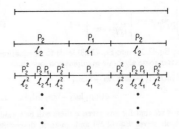

FIG. 6. The partitioning process for the measure yielding the partition function (2.37). Only those segments receiving a measure multiplied by $p_2$ at any stage of the construction are further subdivided. This measure is far less self-similar than that generated by the Cantor process.

have had their measure multiplied by a factor $p_2$ in the preceding stage are subdivided further, while the regions which have had their measure multiplied by a factor $p_1$ are not subdivided further. Thus the expression for the measure density of any region, at any stage of the iterative construction, will have, at most, one factor of $p_1$. The measure generated by this construction is much less self-similar than that considered in Sec. III C 4. For this measure the partition function is given for large $n$ by

$$\Gamma(q,\tau,l_2^n) = (p_1^q/l_1^\tau)\Gamma^U(l_2^{n-1}) + 2(p_2^q/l_2^\tau)\Gamma(l_2^{n-1}) , \tag{2.37}$$

where $\Gamma^U$ is the partition function for a uniform measure on a line segment. It is easy to show that

$$\tau(q) = \min\{q - 1, q\tilde{\alpha} - \tilde{f}\} , \tag{2.38}$$

with $\tilde{\alpha} = \ln(p_2)/\ln(l_2)$, and $\tilde{f} = \ln(\frac{1}{2})/\ln(l_2)$. This example corresponds to a discrete, rather than a continuous, $f$-$\alpha$ curve, consisting of a point at $(\tilde{\alpha}, \tilde{f})$ and a point at $(1,1)$. This result should not surprise us, as this measure is properly described as a nonsingular background interrupted by singularities upon a Cantor set of dimension $\tilde{f}$.

### III. EXAMPLES FROM DYNAMICAL SYSTEMS

In this section we examine the implications of the formalism of Sec. II for three examples: (i) the $2^\infty$ cycle at the accumulation point of period doubling, (ii) the set of irrational winding numbers at the onset of chaos via quasiperiodicity, and (iii) the critical cycle elements at the golden-mean winding number for the same problem. In all cases we calculate numerically the $D_q$, and use Eqs. (1.10) and (1.11) to extract $\alpha(q)$, $f(q)$, and a plot of $f(\alpha)$. In all three cases we can find theoretically $D_\infty$, $D_{-\infty}$, and thus $\alpha(q = \pm\infty)$.

### A. The $2^\infty$ cycle of period doubling

Dynamical systems that period double on their way to chaos can be represented by one-parameter families of maps $M_\lambda(\mathbf{x})$, where $M_\lambda$: $R^F \to R^F$, and $F$ is the number of degrees of freedom. At values of $\lambda = \lambda_n$ the system

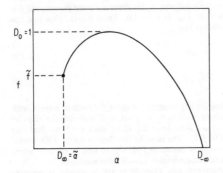

FIG. 5. The function $f(\alpha)$ for the measure defined by Eq. (2.36). Note that $D_\infty$ corresponds to a nonzero value of $f$. Also, $D_0 = 1$.

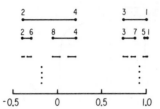

FIG. 7. The construction of the period-doubling attractor; the indices refer to the number of the iterate of $x=0$. The lines represent the scales $l_i$. Note the similarity with Fig. 2.

gains a stable $2^n$-periodic orbit. This period-doubling cascade accumulates at $\lambda_\infty$, where the system possesses a $2^\infty$ orbit. We generated numerically the set of elements of this orbit for the map $x'=\lambda(1-2x^2)$, with $\lambda_\infty \approx 0.837\,005\,134\ldots$[15] The points making up the cycle are displayed in Fig. 7. The iterates of $x=0$ form a Cantor set, with half the iterates falling between $f(0)$ and $f^3(0)$ and the other half between $f^2(0)$ and $f^4(0)$. The most natural partition, $\{S_i\}$, for this case simply follows the natural construction of the Cantor set as shown in Fig. 7. At each level of the construction of this set, each $l_i$ is the distance between a point and the iterate which is closest to it. The measures $p_i$ of these intervals are all equal.

With $2^{11}$-cycle elements we solved numerically $\Gamma=1$, thereby generating the $D_q$-versus-$q$ curve shown in Fig. 8. From these results we calculated $\alpha(q)$ from Eq. (1.11) and $f(\alpha)$ from (1.10). The curve $f(\alpha)$ is displayed in Fig. 9.

To understand the shape of the curve in Fig. 9 we first consider the end points of the curve (for which $f=0$). As with the example solved in Sec. II C 4, we expect these two points to be determined by the most rarefied and the most concentrated intervals in the set. As has been shown by Feigenbaum,[15] these have scales $l_{-\infty} \sim \alpha_{PD}^{-n}$ and $l_{+\infty} \sim \alpha_{PD}^{-2n}$, respectively, where $\alpha_{PD}=2.502\,907\,875\ldots$ is the universal scaling factor.[15] Since the measures there

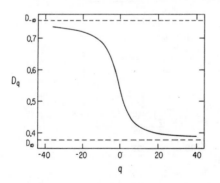

FIG. 8. $D_q$ vs $q$ calculated for the period-doubling attractor of Fig. 7.

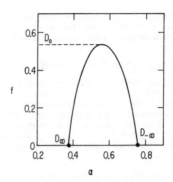

FIG. 9. The function $f(\alpha)$ for the period-doubling attractor of Fig. 7.

are simply $p_{-\infty} \sim 2^{-n}$, we expect these end points to be $\ln p_{-\infty}/\ln l_{-\infty}$ and $\ln p_\infty/\ln l_\infty$, respectively. These values are also $D_{-\infty}$ and $D_\infty$, so that we find

$$D_{-\infty}=\frac{\ln 2}{\ln \alpha_{PD}}=0.755\,51\ldots , \qquad (3.1a)$$

$$D_\infty=\frac{\ln 2}{\ln \alpha_{PD}^2}=0.377\,75\ldots . \qquad (3.1b)$$

These values are in extremely good agreement with the numerically determined endpoints of the graph. The curve $f(\alpha)$ is perfectly smooth. The maximum is at $D_0=0.537\ldots$, in agreement with previous calculations of the Hausdorff dimension for this set. Since the slope of the curve $f(\alpha)$ is $q$, $\alpha(q)$ will be very close to $D_{\pm\infty}$ even for $|q| \sim 10$. However, Fig. 8 indicates that $D_q$ is far from converged to $D_{\pm\infty}$ even for $q \sim \pm 40$. Thus, the transformations (1.10) and (1.11) lead more easily to good estimations of $D_{\pm\infty}$ than do direct calculations of the $D_q$'s.

### B. Mode-locking structure

Dynamical systems possessing a natural frequency $\omega_1$ display very rich behavior when driven by an external frequency $\omega_2$. When the "bare" winding number $\Omega=\omega_1/\omega_2$ is close to a rational number, the system tends to mode lock. The resulting "dressed" winding number, i.e., the ratio of the response frequency to the driving frequency, is constant and rational for a small range of the parameter $\Omega$. At the onset of chaos the set of irrational dressed winding numbers is a set of measure zero, which is a strange set of the type discussed above. The structure of the mode locking is best understood in terms of the "devil's staircase" representing the dressed winding number as a function of the bare one. Such a staircase is shown in Fig. 10 as obtained for the map[16,17]

$$\theta_{n+1}=\theta_n+\Omega-\frac{K}{2\pi}\sin(2\pi\theta_n) , \qquad (3.2)$$

with $K=1$, which is the onset value above which chaotic orbits exist.

To calculate $D_{\pm\infty}$ analytically we make use of previous findings that the most extremal behaviors of this staircase are found at the golden-mean sequence of dressed winding numbers

$$F_n/F_{n+1} \to w^* = (\sqrt{5}-1)/2 \approx 0.6108\ldots ,$$

where $F_n$ are the Fibonacci numbers, ($F_0=0$, $F_1=1$, and $F_n = F_{n-1}+F_{n-2}$ for $n \geq 2$) and at the harmonic sequence $1/Q \to 0$.[16,17] The most rarefied region of the staircase is located around the golden mean. Shenker found that the length scales $l_i$ vary in that neighborhood as $l_{-\infty} \sim F_n^{-\delta} \sim (w^*)^{n\delta}$, where $\delta = 2.1644\ldots$ is a universal number.[18] The corresponding changes in dressed winding number are

$$p_{-\infty} \sim F_n/F_{n+1} - F_{n+1}/F_{n+2} \sim (w^*)^{2n} .$$

We thus conclude that

$$D_{-\infty} = \frac{\ln p_{-\infty}}{\ln l_{-\infty}} = \frac{2}{\delta} = 0.9240\ldots . \tag{3.3a}$$

For the $1/Q$ series it has been shown that changes in dressed winding number go as the square root of changes in bare winding number, i.e., that $p_i \sim l_i^{1/2}$.[16] This series determines the most concentrated portion of the staircase (Fig. 10), which means that $p_\infty \sim l_\infty^{1/2}$, leading to

$$D_\infty = \ln p_\infty / \ln l_\infty = \tfrac{1}{2} . \tag{3.3b}$$

To construct the curve $f(\alpha)$ we generated 1024 mode-locked intervals following the Farey construction, which also defines the partition $\{S_i\}$.[17] For each two neighboring intervals (see Fig. 10) we measured the change both in bare and in dressed winding numbers. The changes in bare winding numbers determined the scales $l_i$ of the partition $\{S_i\}$, whereas the changes in dressed winding numbers were defined to be the measures $p_i$. Solving then the equation $\Gamma = 1$ we generated $D_q$ as shown in Fig. 11 (for $q > 0$ we accelerated the convergence as will be described shortly). Figure 12 shows $f(\alpha)$ for this case. Again the curve is smooth, in contrast to scaling functions found for

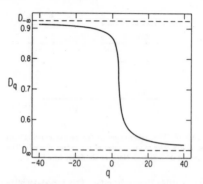

FIG. 11. $D_q$ vs $q$ for the staircase of Fig. 10.

the same problem by other authors.[15,21] Note that the maximum on Fig. 12 gives the fractal dimension $D_0$ of the mode-locking structure as $D_0 \approx 0.87\ldots$, in agreement with the predictions of Refs. 16 and 17. The rightmost branch of the curve $f(\alpha)$ in Fig. 12 (i.e., for $q < 0$) converges vary rapidly within the Farey partition. This is, however, not the case for the leftmost branch (i.e., for $q > 0$). To improve the convergence of this portion of the curve substantially, we made use of the following trick. In general, the partition function (2.1) will be of the form

$$\Gamma(l) = a e^{\gamma \ln l} , \tag{3.4}$$

where $a$ and $\gamma$ are constants. The convergence is often slowed down by the prefactor $a$ and by the logarithmic dependence on $l$. However, by considering instead the ratio

$$\frac{\Gamma(l)}{\Gamma(2l)} = e^{-\gamma \ln 2} , \tag{3.5}$$

we find that $a$ and $l$ do not appear in the equation. We thus determine $\tau(q)$ by requiring that

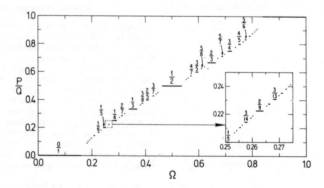

FIG. 10. The "devil's staircase" for the critical circle map of Eq. (3.2). The "dressed" winding number is plotted vs the "bare" winding number (Ref. 16).

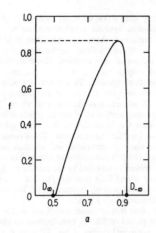

FIG. 12. A plot of $f$ vs $\alpha$ for the mode-locking structure of the circle map. The left portion of the curve is found by accelerated convergence as described in Sec. III B.

$$\frac{\Gamma(l)}{\Gamma(2l)}(\tau,q)=1 \; .$$

In general, the denominator can be chosen to be of the form $\Gamma(bl)$, where $b$ is a constant. The leftmost portion of the curve was generated with this method by calculating $\Gamma(l(1452))/\Gamma(l(886))=1$ [where $l(1452)$ and $l(886)$ are the maximal scales for partitions with 1452 and 886 intervals, respectively], and we observe that it passes through the point $(D_\infty,0)$. We found empirically that this method usually did not give reliable results for large values of $|q|$. Still, this method did successfully generate the entire curve in Fig. 12. We emphasize the ease of this measurement. The rightmost branch of the $f$-$\alpha$ curve of Fig. 12 converges very rapidly, even when only 8–16 mode-locked intervals are available.

### C. Quasiperiodic trajectories for circle maps

Circle maps of the type (3.1) exhibit a transition to chaos via quasiperiodicity. A well-studied transition takes place at $K=1$ with dressed winding number equal to the golden mean, $w^*$. [18–20] We have at this point studied the structure of the trajectory $\theta_1,\theta_2,\ldots,\theta_I,\ldots$. To perform the numerical calculation we chose $\theta_1=f(0)$ and truncated the series $\theta_I$ at $i=2584=F_{17}$. The distances $l_i=\theta_{i+F_{16}}-\theta_i$ (calculated mod 1) define natural scales for the partition with measures $p_i=1/2584$ attributed to each scale. Figure 13 shows $D_q$ versus $q$ calculated for this set and Fig. 14 shows the corresponding function $f(\alpha)$. Again the curve is smooth. Shenker found for this problem that the distances around $\theta \sim 0$ scale down by a universal factor $\alpha_{GM}=1.2885\ldots$ when the trajectory $\theta_i$ is truncated at two consecutive Fibonacci numbers,

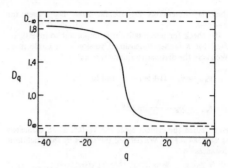

FIG. 13. $D_q$ plotted vs $q$ for the critical trajectory of a circle map with golden-mean winding number.

$F_n,F_{n+1}$.[18] This corresponds to the most rarefied region so that $l_{-\infty} \sim \alpha_{GM}^{-n}$. The corresponding measure scales as $p_{-\infty} \sim 1/F_n \sim (\omega^*)^n$, leading to

$$D_{-\infty}=\frac{\ln w^*}{\ln \alpha_{GM}^{-1}}=1.8980\ldots . \tag{3.6a}$$

The map (3.2) for $K=1$ has at $\theta=0$ a zero slope with a cubic inflection and is otherwise monotonic. The neighborhood around $\theta=0$, which is the most rarefied region of the set, will therefore be mapped onto the most concentrated region of the set. As the neighborhood around $\theta=0$ scales as $\alpha_{GM}$ when the Fibonacci index is varied, the most concentrated regime will scale as $\alpha_{GM}^3$ due to the cubic inflection. This means that $l_\infty=\alpha_{GM}^3$ and $p_\infty=(w^*)^n$, so that we obtain

$$D_\infty=\frac{\ln w^*}{\ln \alpha_{GM}^{-3}}=0.6326\ldots . \tag{3.6b}$$

Figure 14 shows that the curve passes very close to the points $(D_\infty,0)$ and $(D_{-\infty},0)$. Again, however, we find

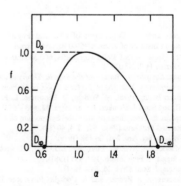

FIG. 14. A plot of $f$ vs $\alpha$ for the golden-mean trajectory for the circle map.

that the dimensions $D_q$ are far from $D_{\pm\infty}$ even for $q \sim \pm 40$.

To check for universality it is important to investigate $f(\alpha)$ for a higher-dimensional version of a circle map. We chose the dissipative standard map,

$$\theta_{n+1} = \theta_n + \Omega + br_n - \frac{K}{2\pi}\sin(2\pi\theta_n) ,$$
$$r_{n+1} = br_n - \frac{K}{2\pi}\sin(2\pi\theta_n) ,$$
(3.7)

and studied the critical cycle for $b=0.5$, again truncated at $i=2584=F_{17}$. We defined the scales by the Euclidean distances

$$l_i = [(\theta_{i+F_{16}} - \theta_i)^2 + (r_{i+F_{16}} - r_i)^2]^{1/2} .$$
(3.8)

We found that the convergence for the two-dimensional (2D) case was slightly slower than for the one-dimensional (1D) case. This is, however, to be expected since it was found by Feigenbaum, Kadanoff, and Shenker that the convergence of the scaling number $\alpha_{GM}$ is slower for the 2D case than for the 1D case.[19] To improve the convergence we again made use of the ratio trick as embodied in Eqs. (3.4) and (3.5). For this case we calculated the partition function for two consecutive Fibonacci numbers, $F_{16} = 1597$ and $F_{17} = 2584$, and found $\tau$ from the requirement

$$\frac{\Gamma(l(F_{17}))}{\Gamma(l(F_{16}))}(\tau, q) = 1$$

$[l(F_i)$ are the maximal scales for the partitions]. This improves the convergence significantly, and the $f$-$\alpha$ curve for this 2D case coincides almost completely with the curve found for the 1D case and displayed in Fig. 14.

## IV. CONCLUSION

Most previous characterizations of strange sets arising in physics have followed the example of critical phenomena in relying upon a few universal numbers to characterize the physical systems generating these sets. Thus, strange attractors are characterized by their Hausdorff dimensions, or by the scaling exponents of particularly divergent regions of their measure. However, these numbers reflect only a small part of the universal scaling structure of these systems. Feigenbaum introduced scaling functions in order to describe the complex scaling properties of attractors at the onset of chaos.[15] These scaling functions contain all of the geometric information about the attractor, in contrast to the partial information furnished by local scaling exponents. These functions are, however, nowhere differentiable, and are thus very difficult to use. The full complexity of this scaling structure is more conveniently reflected by the continuous spectrum of exponents $\alpha$ and their densities $f(\alpha)$, of which previously investigated scaling exponents and Hausdorff dimensions comprise only a part.

Not only does this spectrum enrich our conceptual vocabulary, it should enrich our experimental vocabulary as well. The numerical studies of Sec. III were straightforward and did not require large investments of computer time in order to obtain extremely accurate results. Furthermore, this spectrum can be measured, and has been measured, in experiments upon physical realizations of dynamical systems.[23] The measurement of this spectrum should result in new tests of scaling theories of nonlinear systems.

## ACKNOWLEDGMENTS

We would like to thank P. Jones and J. Rudnick for stimulating discussions. This work was supported by the U.S. Office of Naval Research, by the National Science Foundation through Grant No. DMR-83-16626, and through the Materials Research Laboratory of the University of Chicago. One of us (I.P.) wishes to thank Professor S. Berry for his warm hospitality and also acknowledge the support of the Minerva Foundation (Munich, West Germany).

*Permanent address: Department of Chemical Physics, The Weizmann Institute of Science, 76100 Rehovot, Israel.

†Present address: AT&T Bell Laboratories, 600 Mountain Avenue, Murray Hill, NJ 07974.

[1]R. Bowen, *Equilibrium States and the Ergodic Theory of Anisov Diffeomorphisms*, Vol. 470 of *Lectures Notes in Mathematics* (Springer, Berlin, 1975); M. Widom, D. Bensimon, L. P. Kadanoff, and Scott J. Shenker, J. Stat. Phys. 32, 443 (1983).

[2]I. Procaccia, in Proceedings of the Nobel Symposium on Chaos and Related Problems [Phys. Scr. T 9, 40 (1985)].

[3]K. G. Wilson, Sci. Am. 241, (2) 158 (1979).

[4]B. B. Mandelbrot, Ann. Isr. Phys. Soc. 225 (1977); H. Aref and E. D. Siggia, J. Fluid Mech. 109, 435 (1981).

[5]I. Procaccia, J. Stat. Phys. 36, 649 (1984).

[6]L. de Arcangelis, S. Redner, and A. Coniglio, Phys. Rev. B 31, 4725 (1985).

[7]T. A. Witten, Jr. and L. M. Sander, Phys. Rev. Lett. 47, 1400 (1981).

[8]T. C. Halsey, P. Meakin, and I. Procaccia (unpublished).

[9]B. Widom, J. Chem. Phys. 43, 3892 (1965); M. E. Fisher, in Proceedings of the University of Kentucky Centennial Conference on Phase Transitions (March 1965) (unpublished); A. Z. Patashinskii and V. L. Prokovskii, Zh. Eksp. Teor. Fiz. 50, 439 (1966) [Sov. Phys.—JETP 23, 292 (1966)]; L. P. Kadanoff, W. Gotze, D. Hamblen, R. Hecht, E. A. S. Lewis, V. V. Palciauskas, M. Rayl, J. Swift, D. Aspnes, and J. Kane, Rev. Mod. Phys. 39, 395 (1967).

[10]L. P. Kadanoff, Physics 2, 263 (1966).

[11]In fact, for many two-dimensional critical phenomena we know [see, for instance, D. Friedan, Z. Qiu, and Stephen Shenker, Phys. Rev. Lett. 52, 1575 (1984)] that there is only a finite set of dist $t$ $y$'s.

[12]L. Turkevich and H. Scher, Phys. Rev. Lett. 55, 1024 (1985).

[13]H. G. E. Hentschel and I. Procaccia, Physica 8D, 435 (1983).

[14]P. Grassberger and I. Procaccia, Physica 13D, 34 (1984).

[15]M. J. Feigenbaum, J. Stat. Phys. 19, 25 (1978); 21, 669 (1979).

See also the two reprint compilations: P. Cvitanović, *Universality in Chaos* (Hilger, Bristol, 1984); Hao Bai-lin, *Chaos* (World Scientific, Singapore, 1984).

[16]M. H. Jensen, P. Bak, and T. Bohr, Phys. Rev. Lett. **50**, 1637 (1983); Phys. Rev. A **30**, 1960 (1984); **30**, 1970 (1984).

[17]P. Cvitanović, M. H. Jensen, L. P. Kadanoff, and I. Procaccia, Phys. Rev. Lett. **55**, 343 (1985).

[18]Scott J. Shenker, Physica **5D**, 405 (1982).

[19]M. J. Feigenbaum, L. P. Kadanoff, and Scott J. Shenker, Physica **5D**, 370 (1982).

[20]S. Ostlund, D. Rand, J. P. Sethna, and E. D. Siggia, Phys. Rev. Lett. **49**, 132 (1982); Physica **8D**, 303 (1983).

[21]P. Cvitanović, B. Shraiman, and B. Soderberg (unpublished).

[22]A. Cohen and I. Procaccia, Phys. Rev. A **31**, 1872 (1985), and references therein.

[23]M. H. Jensen, L. P. Kadanoff, A. Libchaber, I. Procaccia, and J. Stavans, Phys. Rev. Lett. **55**, 2798 (1985).

424

# Global Universality at the Onset of Chaos: Results of a Forced Rayleigh-Bénard Experiment

Mogens H. Jensen, Leo P. Kadanoff, and Albert Libchaber

*The James Franck Institute, The University of Chicago, Chicago, Illinois 60637*

Itamar Procaccia

*Department of Chemical Physics, The Weizmann Institute of Science, Rehovot 76100, Israel*

and

Joel Stavans

*The James Franck Institute, The University of Chicago, Chicago, Illinois 60637*
(Received 15 October 1985)

We study an experimental orbit on a two-torus with a golden-mean winding number obtained from a forced Rayleigh-Bénard system at the onset of chaos. This experimental orbit is compared with the orbit generated by a simple theoretical model, the circle map, at its golden-mean winding number at the onset of chaos. The "spectrum of singularities" of the two orbits are compared. Within error, these are identical. Since the spectrum characterizes the metric properties of the entire orbit, this result confirms theoretical speculations that these orbits, taken as a whole, enjoy a kind of universality.

PACS numbers: 47.20.+m, 05.45.+b, 47.25.−c

In the study of the transition to chaos most theoretical attention has been paid to the behavior near special points in phase space. Thus Feigenbaum[1] concentrated upon the region of the maximum of the period-doubling map, Shenker[2] looked near the inflection point of the circle map, etc. In experimental situations, such distinguished points in phase space are not readily discernible. If one is to look experimentally for universality, one would do well to seek more global,[3] but still universal, features of the phase-space orbits.

In this Letter we report experimental results which, together with theoretical analysis, show that critical orbits in phase space at the onset of chaos exhibit global universal properties. The example discussed here is the cycle with golden-mean winding number at the point of breakdown of a 2-torus. The experiment is a periodically forced Rayleigh-Bénard system with mercury as a fluid. Recent measurements on this system revealed two scaling indices[4]: the index for the ratio between two successive Fibonacci resonances[2] and the dimension of the structure of mode locking.[5] Both were in agreement with the indices found for circle maps. We therefore compare the experimentally observed critical orbit with the corresponding orbit in the circle map.

In order to examine global scaling properties it is not sufficient to measure the dimension of the attracting set; the set certainly contains more topological information than can be characterized by a single number.[1b] To achieve a characterization that more fully describes those properties of such sets which remain unchanged under smooth changes of coordinates, it has been proposed to use a continuous spectrum of scaling indices.[6]

These spectra display the range of scaling indices and their density in the set. To clarify what we mean, consider the experimental cycle displayed in Fig. 1. One sees with bare eyes that the time series is concentrated with various intensities in different regions. The spectrum that we use quantifies this variation in density on the attractor, and allows us to show the similarity of

FIG. 1. The experimental attractor in two dimensions. 2500 points are plotted. Note the variation in the density of points on the attractor. Part of this variation is, however, due to the projection of the attractor onto the plane. The attractor is nonintersecting in three dimensions, in which it was embedded for the numerical analysis. In the absence of experimental noise the points should fall on a single curve. The smearing of the observed data set is mostly due to the slow drift in the experimental system during the run over about 2 h. Our method of analyzing the data to secure $f$ vs $\alpha$ (see Fig. 2) is intended to minimize the effect of the slow drift. This is realized by estimating the recurrence times which experimentally are matters of minutes rather than hours.

this cycle to sets produced by model equations that describe the onset of chaos via quasiperiodicity.[2,7,8] In fact, the approach proposed here constitutes a rare opportunity for an extensive quantitative comparison of experiments with universal results obtained from theoretical models.

The experiment which yielded the critical golden-mean trajectory has been described previously.[4] The experiment studies a small-aspect-ratio Rayleigh-Bénard system of size $0.7 \times 0.7 \times 1.4$ cm$^3$ with two convective rolls present. For a low-Prandl-number fluid like mercury, as the heat flux increases beyond the convection threshold $R_c$, the system undergoes a Hopf bifurcation, called the oscillatory instability, into a time-dependent periodic mode. This mode is characterized by an ac vertical vorticity otherwise absent in the static roll pattern. This oscillation is one of our two oscillators (frequency $\approx 230$ mHz). The second oscillator is introduced electromagnetically, mercury being an electrical conductor. An ac current sheet is passed through the mercury and the system is immersed in an horizontal magnetic field ($H \approx 200$ G) parallel to the rolls' axes. The geometry of electrode and field is such that the Lorentz force on the fluid produce ac vertical vorticity. In this way the oscillators are dynamically coupled. During the experiment the Rayleigh number is kept fixed at $R = 4.09R_c$ giving a large amplitude to the first oscillator.

The nonlinear interaction between the oscillators is controlled by the amplitude of the injected ac current. A signal is obtained from the experiment by means of a thermal probe located in the bottom plate of the cell. The winding number, which is the ratio between the two frequencies, is kept close to the golden mean, i.e., within $10^{-4}$. Time series are obtained by observation of the temperature signal at discrete times separated by the period of the forcing.

The theoretical work is aimed at estimating in a quantitative fashion how "bunched" the density on the orbit might be. In technical terms this bunching is a description of singularities in the probabilities of the orbit points.[9] Less technically, one can view a particular orbit point $x_i$, in a phase space like that of Fig. 1, and ask what is the probability for other points falling v thin the small distance, $l$, of this one. Call this probability $p_i(l)$. One can describe this probability by defining an index $\alpha_i(l)$ via

$$p_i(l) = l^{\alpha_i(l)} \tag{1}$$

In typical sets the scaling index $\alpha_i$ takes, for small $l$, a range of values between $\alpha_{min}$ and $\alpha_{max}$. We refer to this situation as a spectrum of singularities.

To analyze the experimental time series, we make use of a key theoretical idea that fractal sets in general, and critical orbits in particular, can be described as interwoven sets of singularities.[9] The density of singularities of type $\alpha$, $\alpha_{min} < \alpha < \alpha_{max}$, is determined by an index $f$ that can be interpreted as the dimension of the set of singularities of this type. In other words, if the system is divided into pieces of size $l$, then the number of times, $n(\alpha,l)$, that $\alpha$ takes on a value between $\alpha$ and $\alpha + d\alpha$ is of the form

$$n(\alpha,l) = d\alpha\rho(\alpha)l^{-f(\alpha)}, \tag{2}$$

where $\rho(\alpha)$ is nonsingular with respect to $l$. The intuitive meaning of $\alpha_{max}$ is that it is associated with the most rarefied regions of the measure, whereas $\alpha_{min}$ with the most concentrated. Typically, $f(\alpha_{max}) = f(\alpha_{min}) = 0$. Other types of singularities between $\alpha_{max}$ and $\alpha_{min}$ live on subsets of dimension $f$, $0 < f < D_0$ ($D_0$ being the dimension of the set). The functions $f(\alpha)$ are universal functions for critical cycles like the trajectory with golden-mean winding number at the onset of chaos via quasiperiodicity.[2] Another key point is that these functions are smooth, in contrast to the universal scaling functions of the type suggested by Feigenbaum,[1] which are nowhere differentiable (see for example Fig. 10 of Ref. 1b). The reason for this important difference is that the latter functions are constructed by following the *local* changes in scaling everywhere,[1] whereas the former are based on finding the *global density* of scaling indices of each type.

The $\alpha_i(l)$ of Eq. (1) are estimated in a very simple fashion: Start from the point $x_i$ on the trajectory. Count the number of steps along the time series required before a point returns to within $l$ of the starting point. We call this number the recurrence time and denote it $m_i$. We shall now make use of the fact that the orbit is conjugate to a pure rotation[2,7,8] with an irrational winding number, and is therefore ergodic. Thus, we simply estimate $p_i(l)$ as the inverse recurrence time, $(m_i)^{-1}$, and find

$$\alpha_i(l) = -\ln m_i/\ln l. \tag{3}$$

In principle, the remainder of the analysis is very simple. One estimates how many $\alpha_i(l)$ values live in a given range, substitutes that estimate into Eq. (2), and then chooses some very small value of $l$ to find $f(\alpha)$. In practice, given only a moderate amount of data, one cannot obtain a good estimate of $f(\alpha)$ by this direct method. Instead, we employ[6] an indirect method which smooths the data and gives an efficient calculation of $f(\alpha)$. To obtain this smoothness, we use the data to calculate the auxilary quantity

$$\Gamma(q,l) = \langle p_i(l)^{q-1} \rangle = \langle m_i^{1-q} \rangle, \tag{4}$$

where the brackets represent an average over all the trajectory elements $i$. The whole point of using the "partition function" Eq. (4) is that it is a smooth function of $l$ and $q$ and, for $l \ll 1$, is given by a power

2799

of $l$,

$$\Gamma(q,l) \sim l^{\tau(q)}. \qquad (5)$$

This $\tau(q)$ is related to the generalized dimensions, $D_q$, of Hentschel and Procaccia[10] by $D_q = (q-1)^{-1}\tau(q)$.

From the point of view of this paper $\tau(q)$ is not important in itself. Instead it is a kind of generating function which can be used to determine the function $f(\alpha)$ via the pair of formulas (derived in Ref. 6),

$$\alpha(q) = d\tau(q)/dq,$$
$$f(q) = \tau(q) - q \, d\tau/dq. \qquad (6)$$

This is essentially a Legendre transformation, as used in statistical thermodynamics. Once $q$ is eliminated from the pair of Eqs. (6), we discover that we have $f(\alpha)$ defined in a range of $\alpha$ values $\alpha_{min} < \alpha < \alpha_{max}$.

An example of a theoretical $f(\alpha)$ curve calculated in this manner is shown as the curve in Fig. 2. The attractor is the critical cycle of the circle map[2,7,8] $\theta_{n+1} = \theta_n + \Omega - (K/2\pi) \sin 2\pi\theta_n$ at the critical value $K=1$ and $\Omega = \Omega_{gm}$, where the orbit has a golden-mean winding number. Here, $\tau(q)$ was evaluated by calculation of the average Eq. (4) for several $l$ values and then finding $\tau$ as the slope of a straight line fit of a plot of $\ln\Gamma$ vs $\ln l$. The value of $\alpha_{max}$, which is also $D_{-\infty}$ agrees with the theoretical expectation[6]

$$\alpha_{max} = \ln\omega^*/\ln\alpha_{gm}^{-1} = 1.8980\ldots, \qquad (7)$$

where $\omega^*$ is the golden mean $\omega^* = (\sqrt{5}-1)/2$ and $\alpha_{gm}$ is the universal local scale factor[2] in the vicinity of the critical point $\theta = 0$, $\alpha_{gm} = 1.2885\ldots$. This is the most rarefied region in the trajectory. This region is mapped onto the most concentrated region in the set[6] which is characterized by the $\alpha$ value $\alpha_{min} = \ln\omega^*/\ln\alpha_{gm}^{-3} = 0.6326\ldots$. The curve turns around at the value of $f$ which is $f = D_0 = 1$. This is also to be expected since the support of the measure is the circle, which is one dimensional.

The experimental data were similarly analyzed. On the basis of a time series of 2500 points embedded in a three-dimensional space we first calculated $\Gamma$ as required by Eq. (4). Plotting again $\ln\Gamma$ vs $\ln l$, we typically fitted the $\tau$'s with over fifty different values of $l$ ranging over two decades. The $f(\alpha)$ values were computed via Eqs. (6) with the result shown as the dots in Fig. 2. For small $q$ ($|q| < 1$) the scaling was in general best for the largest values of $l$. As $|q|$ was increased, the best scaling regime gradually moved towards lower values of $l$. This is expected since high $|q|$ values correspond to isolated regimes on the attractor. The accuracy of the fits was always very good for positive $q$'s (corresponding to the leftmost branch of the curve), and we estimate the error bar on the point $(D_\infty, 0)$ to be a few percent. The accuracy was less for negative values of $q$ (corresponding to the rightmost branch) and the error bar on the point

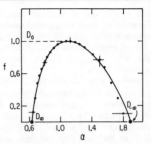

FIG. 2. The $f(\alpha)$ spectrum calculated for a critical circle map with golden-mean winding number is shown by the curve (Ref. 6). The curve ends in the points $(D_\infty, 0)$ and $(D_{-\infty}, 0)$, which are shown by the two large dots. The $f(\alpha)$ estimates for the experimental time series are marked by the smaller dots. The error bars are estimated by varying the range of $l$ used to fit the data.

$(D_{-\infty}, 0)$ is around $(10–12)\%$. The accuracy of the maximum point of the curve (i.e., $D_0$) is indicated by a vertical error bar. Theory and experiment agree. This agreement supports our conjecture that this Rayleigh-Bénard system at the onset of chaos and the critical circle map belong to the same universality class. We note in passing that from the value of $\alpha_{max}$ (and also of $\alpha_{min}$) one can read immediately $\alpha_{gm}$, cf. Eq. (7). To the best of our knowledge this is the first direct measurement of this universal scaling number.

To conclude we note that the raw experimental orbit in its reconstructed phase space looks nothing like the orbit of the circle map. To the eye, it does not appear to lie on a circle. It is twisted and contorted in a complicated way. Our results demonstrate, however, that from the metric point of view these two sets *are the same* within experimental accuracy. To date we are not aware of any other approach that can lead to such a strong conclusion.

We thank Thomas C. Halsey for many stimulating discussions. This work has been partially supported by the Materials Research Laboratory of the University of Chicago, the National Science Foundation through Grants No. DMR 83-16626 and No. DMR 83-16204, the Office of Naval Research, and the Minerva Foundation, Munich, Germany.

[1a]M. J. Feigenbaum, J. Stat. Phys. **19**, 25 (1978), and J. Stat. Phys. **21**, 669 (1979).

[1b]M. J. Feigenbaum, Los Alamos Sci. **1**, 4 (1980).

[2]Scott J. Shenker, Physica (Amsterdam) **5D**, 405 (1982).

[3]In parameter space however, global universal features have been proposed [M. H. Jensen, P. Bak, and T. Bohr,

Phys. Rev. Lett. **50**, 1637 (1983), and Phys. Rev. A **30**, 1960 (1984); P. Cvitanović, M. H. Jensen, L. P. Kadanoff, and I. Procaccia, Phys. Rev. Lett. **55**, 343 (1985)] and measured experimentally [J. Stavans, F. Heslot, and A. Libchaber, Phys. Rev. Lett. **55**, 596 (1985); see also A. Libchaber, C. Laroche, and S. Fauve, Physica (Amsterdam) **7D**, 73 (1983), and J. Phys. (Paris), Lett. **43**, L211 (1982)] for the structure of modelocking.

[4]Stavans, Heslot, and Libchaber, and Libchaber, Laroche, and Fauve, Ref. 3.

[5]Jensen, Bak, and Bohr, and Cvitanović *et al.*, Ref. 3.

[6]T. C. Halsey, M. H. Jensen, L. P. Kadanoff, I. Procaccia, and B. I. Shraiman, Phys. Rev. A (to be published).

[7]M. J. Feigenbaum, L. P. Kadanoff, and Scott J. Shenker, Physica (Amsterdam) **5D**, 370 (1982).

[8]S. Ostlund, D. Rand, J. P. Sethna, and E. D. Siggia, Phys. Rev. Lett. **49**, 132 (1982), and Physica (Amsterdam) **8D**, 303 (1983).

[9]T. C. Halsey, P. Meakin, and I. Procaccia, to be published.

[10]H. G. E. Hentschel and I. Procaccia, Physica (Amsterdam) **8D**, 435 (1983); I. Procaccia, Phys. Scr. **T9**, 40 (1985).

428

PHYSICAL REVIEW A       VOLUME 34, NUMBER 6       DECEMBER 1986

# Liapunov exponents from time series

J. -P. Eckmann and S. Oliffson Kamphorst

*Département de Physique Théorique, Université de Genève, CH-1211 Genève 4, Switzerland*

D. Ruelle

*Institut des Hautes Etudes Scientifiques (IHES), F-91440 Bures-sur-Yvette, France*

S. Ciliberto

*Istituto Nazionale di Ottica, I-50125 Arcetri (Firenze), Italy*

(Received 24 April 1986)

We analyze in detail an algorithm for computing Liapunov exponents from an experimental time series. As an application, a hydrodynamic experiment is investigated.

## I. INTRODUCTION

In Ref. 1 two of us proposed a method to compute Liapunov exponents from an experimental time series. Here we report on a detailed analysis of this algorithm for numerical and laboratory experiments. Note that a very similar proposal has been made independently by Sano and Sawada.[2] In the course of the discussion, we shall also point out some divergences between Refs. 1 and 2.

Before discussing our algorithm, we briefly state what we are trying to do. A time evolution is realized, in Nature, in the laboratory, or on the computer, and it is assumed that this time evolution can be described by a differentiable dynamical system in a phase space of possibly infinite dimensions. We want to obtain Liapunov exponents corresponding to the large-time behavior of the system. On a more mathematical level, the large-time behavior defines an ergodic measure in phase space for the time evolution, and we are interested in the corresponding Liapunov exponents. For a discussion of these concepts and precise definitions see, for instance, Ref. 1. What we know is a time series $(x_i)_{1 \leq i \leq N}$ obtained by monitoring a scalar signal for a finite time $T$ and with finite precision. Clearly, thus, there are limitations on how much we can say about the characteristic exponents—it is the aim of this paper to discuss some of these limitations. Certainly, we have to assume that the recording time $T$ is long, that the noise level is low, and that the measurements are made with good precision (viz., $10^{-3}$ or $10^{-4}$ if we want to determine one or two positive characteristic exponents). From a sufficiently good time series, one can in principle obtain all non-negative characteristic exponents, and it may or may not be possible to obtain also some negative ones (cf. Ref. 1).

A complete list of other methods for computing Liapunov exponents is given in Ref. 1. To our knowledge, the proposals in Refs. 1 and 2 are the only ones which allow a systematic computation of several Liapunov exponents.

## II. THE ALGORITHM

It is convenient to present the measured time series in the form of a sequence of integers $x_1, x_2, \ldots, x_N$, with $0 \leq x_i \leq 10\,000$. (The choice of integer values speeds up the computation without sacrificing experimental precision.) The upper bound $10\,000$ is in accordance with a precision of $10^{-4}$ and can easily be modified, if required. We assume that the time interval $\tau$ between measurements is fixed, so that $x_i = x(i\tau)$. Note that the recording time is $T = N\tau$. The present paper deals specifically with the case of a scalar signal, but the method can easily be extended to multidimensional signals.

Conceptually, the algorithm (a copy of the computer program implementing this algorithm can be obtained from the authors) to be discussed involves the following steps: (a) reconstructing the dynamics in a finite dimensional space, (b) obtaining the tangent maps to this reconstructed dynamics by a least-squares fit, (c) deducing the Liapunov exponents from the tangent maps. We now consider these different steps in detail.

(a) We choose an embedding dimension $d_E$ and construct a $d_E$-dimensional orbit representing the time evolution of the system by the time-delay method.[3] This means that we define

$$\vec{x}_i = (x_i, x_{i+1}, \ldots, x_{i+d_E-1}) \tag{1}$$

for $i = 1, 2, \ldots, N - d_E + 1$. In view of step (b), we have to determine the neighbors of $\vec{x}_i$, i.e., the points $\vec{x}_j$ of the orbit which are contained in a ball of suitable radius $r$ centered at $\vec{x}_i$,

$$\|\vec{x}_j - \vec{x}_i\| \leq r \tag{2}$$

with

$$\|\vec{x}_j - \vec{x}_i\| = \max_{0 \leq \alpha \leq d_E - 1} \{ |x_{j+\alpha} - x_{i+\alpha}| \} . \tag{3}$$

The use of (3) rather than the Euclidean norm allows a fast search for the $\vec{x}_j$ which satisfy (2). We first sort the $x_i$ (using "Quicksort," see, e.g., Knuth[4]) so that

$$x_{\Pi(1)} \leq x_{\Pi(2)} \leq \cdots \leq x_{\Pi(N)}$$

and store the permutation $\Pi$ and its inverse $\Pi^{-1}$. Then, to find the neighbors of $x_i$ in dimension 1, we look at $k = \Pi^{-1}(i)$ and scan the $x_{\Pi(s)}$ for $s = k+1, k+2, \ldots$ until $x_{\Pi(s)} - x_i > r$, and similarly for $s = k-1, k-2, \ldots$.

For an embedding dimension $d_E > 1$, we first select the values of $s$ for which $|x_{\Pi(s)} - x_i| \leq r$, as above, and then impose the further conditions

$$|x_{\Pi(s)+\alpha} - x_{i+\alpha}| \leq r ,$$

for $\alpha = 1, 2, \ldots, d_E - 1$.

(b) Having embedded our dynamical system in $d_E$ dimensions (it would be more correct to say that we have projected our dynamical system to $R^{d_E}$), we want to determine the $d_E \times d_E$ matrix $T_i$ which describes how the time evolution sends small vectors around $\vec{x}_i$ to small vectors around $\vec{x}_{i+1}$. The matrix $T_i$ is obtained by looking for neighbors $\vec{x}_j$ of $\vec{x}_i$ and imposing

$$T_i(\vec{x}_j - \vec{x}_i) \approx \vec{x}_{j+1} - \vec{x}_{i+1} . \tag{4}$$

The vectors $\vec{x}_j - \vec{x}_i$ may not span $R^{d_E}$ (think, for instance, of an embedding of the three-dimensional Lorentz system in four dimensions). Therefore, the matrix $T_i$ may only be partially determined. This indeterminancy does not spoil the calculation of the positive Liapunov exponents, but is nevertheless a nuisance because it introduces parasitic exponents which confuse the analysis, in particular with respect to zero or negative exponents which otherwise might be recoverable from the data. The way out of this difficulty is to allow $T_i$ to be a $d_M \times d_M$ matrix with a matrix dimension $d_M \leq d_E$, corresponding to the time evolution from $\vec{x}_i$ to $\vec{x}_{i+m}$.

Specifically, we assume that there is an integer $m \geq 1$ such that

$$d_E = (d_M - 1)m + 1 , \tag{5}$$

and associate with $\vec{x}_i$ a $d_M$-dimensional vector

$$\mathbf{x}_i = (x_i, x_{i+m}, \ldots, x_{i+(d_M-1)m})$$
$$= (x_i, x_{i+m}, \ldots, x_{i+d_E-1}) , \tag{6}$$

in which some of the intermediate components of (1) have been dropped. When $m > 1$ we replace (4) by the condition

$$T_i(\mathbf{x}_j - \mathbf{x}_i) \approx \mathbf{x}_{j+m} - \mathbf{x}_{i+m} . \tag{7}$$

Taking $m > 1$ does not mean that we delete points from the data file, i.e., all points are acceptable as $\mathbf{x}_j$, and the distance measurements are still based on $d_E$, not on $d_M$. Note that, in view of (6) and (7), the matrix $T_i$ has the form

$$T_i = \begin{bmatrix} 0 & 1 & 0 & \cdots & 0 \\ 0 & 0 & 1 & \cdots & 0 \\ \vdots & \vdots & \vdots & & \vdots \\ 0 & 0 & 0 & \cdots & 1 \\ a_1 & a_2 & a_3 & \cdots & a_{d_M} \end{bmatrix} .$$

If we define by $S_i^E(r)$ the set of indices $j$ of neighbors $\vec{x}_j$ of $\vec{x}_i$ within distance $r$, as determined by (2), then we obtain the $a_k$ by a least-squares fit

$$\sum_{j \in S_i^E(r)} \left[ \sum_{k=0}^{d_M-1} a_{k+1}(x_{j+km} - x_{i+km}) - (x_{j+d_M m} - x_{i+d_M m}) \right]^2 = \text{minimum} .$$

The least-squares fit is the most time-consuming part of our algorithm when $S_i^E(r)$ is large. We limit ourselves therefore typically to the first 30–45 neighbors of the a point. We use the least-squares algorithm by Householder.[6] This algorithm may fail for several reasons, the most prominent being that card $S_i^E(r) < d_M$. We therefore choose $r$ sufficiently large so that $S_i^E(r)$ contains at least $d_M$ elements.

In fact, we make a new choice of $r = r_i$ for every $i$. This choice is a compromise between two conflicting requirements: take $r$ sufficiently small so that the effect of nonlinearities can be neglected, take $r$ sufficiently large so that there are at least $d_M$ neighbors of $\vec{x}_i$, and in fact somewhat more than $d_M$ to improve statistical accuracy.

For the specific examples discussed in Sec. IV we have selected $r$ as follows. Count the number of neighbors of $x_i$ corresponding to increasing values of $r$ from a preselected sequence of possible values, and stop when the number of neighbors exceeds for the first time $\min(2d_M, d_M + 4)$. If with this choice the matrix $T_i$ is singular, or, more generally, does not have a previously fixed minimal rank, we again increase $r_i$. It should be noted that this last criterion only seems to come into operation for time series obtained for low-dimensional computer experiments (such as maps of the interval). We stress that the singularity of $T_i$ in itself is not catastrophic for the algorithm and the first $p$ positive Liapunov exponents are not affected provided the rank of the $T_i$ is at least $p$ (which may be a lot less than $d_M$). One should thus not stop the calculation, as suggested in Ref. 2 when the map is singular, since information about the expanding direction(s) will be lost.

(c) Step (b) gives a sequence of matrices $T_i, T_{i+m}, T_{i+2m}, \ldots$. One determines successively orthogonal matrices $Q_{(j)}$ and upper triangular matrices $R_{(j)}$ with positive diagonal elements such that $Q_{(0)}$ is the unit matrix and

$$T_1 Q_{(0)} = Q_{(1)} R_{(1)} ,$$

$$T_{1+m} Q_{(1)} = Q_{(2)} R_{(2)} ,$$

$$\ldots , \tag{8}$$

$$T_{1+jm} Q_{(j)} = Q_{(j+1)} R_{(j+1)} ,$$

$$\ldots .$$

This decomposition is unique except in the case of zero diagonal elements. Then the Liapunov exponents $\lambda_k$ are given by

$$\lambda_k m = \frac{1}{\tau K} \sum_{j=0}^{K-1} \ln R_{(j)kk} ,$$

where $K \leq (N - d_M m - 1)/m$ is the available number of matrices, and $\tau$ is sampling time step. Obviously, fewer matrices can be taken to shorten the computing time. [See Ref. 1 for a justification of the algorithm of Eq. (8).]

## III. REMARKS ON THE ALGORITHM

(a) Let us comment again on the usefulness of taking the matrix dimension $d_M$ different from the embedding dimension $d_E$. As we have said, if $d_M$ is not sufficiently low, there is some numerical indeterminacy in the coefficients on the $T_i$ which, combined with noise, produces undesirable parasitic Liapunov exponents (examples of this phenomenon will be shown in Sec. IV). It is thus natural to take $d_M$ relatively low (a little bigger than the expected number of positive Liapunov exponents). But if one takes $d_E$ too small, the embedding (or rather projection) of the dynamics in $R^{d_E}$ would not be well defined; orbits with different directions might go through the same point. The cure is to take $d_E > d_M$. This is, admittedly, a nonrigorous prescription, and leaves some "intuitive" freedom. We try to overcome this by examining the result for several $d_M$ and $d_E$. Note that for disentangling the dynamics, the important thing is the embedding time $d_E\tau$ rather than $d_E$; this is a first indication that it is not wise to take $\tau$ very small.

(b) As already discussed, the choice of the radius $r$ at $\vec{x}_i$ is a compromise between limitations due to nonlinearities and limitations due to noise. In fact, we have chosen the smallest ball around $\vec{x}_i$ which contains enough neighbors for an unambiguous determination of $T_i$ (note that the algorithm becomes impractically slow when there are more than about 45 neighbors). In principle, i.e., with very good experimental data one can do a little better.

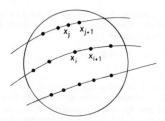

FIG. 1. Neighbors are picked up mostly on the orbit itself.

Since the effect of errors is the worst for the short vectors $\vec{x}_j - \vec{x}_i$ one could replace the ball

$$\{ \vec{x}_j : \ ||\vec{x}_j - \vec{x}_i|| \leq r \}$$

by a shell

$$\{ \vec{x}_j : \ r_{\min} \leq ||\vec{x}_j - \vec{x}_i|| \leq r \} \ .$$

(c) To obtain good statistics it is, of course, desirable to have a time series with a large number $N$ of measurements. However, the really important thing is the total recording time $T = N\tau$, and increasing $N$ at fixed $T$ by making $\tau$ very small would be useless. Actually, the experimental studies of Sec. IV show that for large embedding dimension $d_E$ (hence large $r$), and small $\tau$, many of the neighbors of $\vec{x}_i$ are in factor of the form $\vec{x}_{i\pm 1}, \vec{x}_{i\pm 2}, \ldots$ (see Fig. 1). Attempts at numerical projection onto the supplement of this line have given bad results.

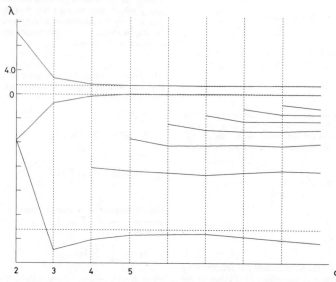

FIG. 2. The Liapunov exponents for the case $d_M = d_E$, connected in a "natural" way.

ECKMANN, KAMPHORST, RUELLE, AND CILIBERTO

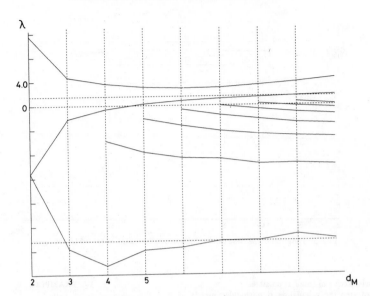

FIG. 3. Typical behavior when too few neighbors are chosen: card $S_E^i(r) \geq d_M$.

(d) Summary of advice.

(1) Use long recording time $T$, but not very small time step $\tau$.

(2) Use large embedding dimension $d_E$.

(3) Use a matrix dimension $d_M$ somewhat larger than the expected number of positive Liapunov exponents.

(4) Choose $r$ such that the number of neighbors is greater than $\min(2d_M, d_M + 4)$.

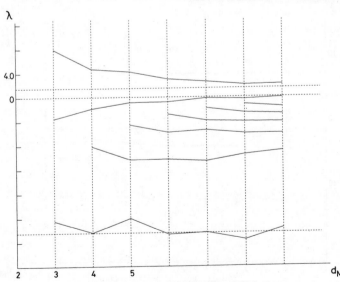

FIG. 4. A noise level of 0.4% has been added.

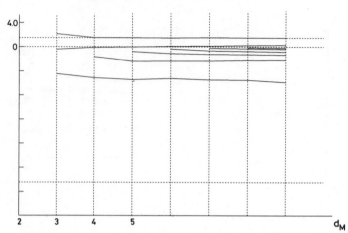

FIG. 5. The effect of the noise can be eliminated by increasing $m$ ( $m = 5$ for the figure).

(5) Otherwise keep $r$ as small as possible.

(6) Do not step the calculation if a singular matrix arises.

(7) Take a product of as many matrices as possible to determine the Liapunov exponents.

In particular this procedure eliminates the difficulties encountered by Vastano and Kostelich.[6]

### IV. EXAMPLES

We begin with the Lorentz equations

$$\frac{d}{dt} \begin{bmatrix} x \\ y \\ z \end{bmatrix} = \begin{bmatrix} -\sigma x + \sigma y \\ -xz + rx - y \\ xy - bz \end{bmatrix} ,$$

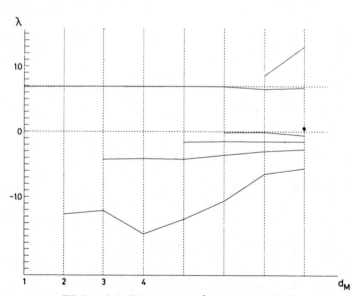

FIG. 6. Analysis of the map $x \to 1 - 2x^2$ with a resolution of $10^{-4}$.

4976         ECKMANN, KAMPHORST, RUELLE, AND CILIBERTO         <u>34</u>

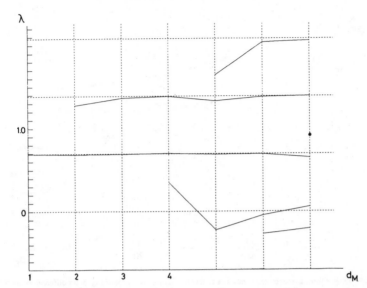

FIG. 7. Analysis of the map $x \to 1 - 2x^2$ with a resolution of $2^{-31}$. A spurious positive Liapunov exponent appears at $\approx 2 \ln 2$.

which we study for the parameter values $\sigma = 16$, $b = 4$, and $r = 45.92$. (These parameter values give the usual picture.) We take $\tau = 0.03$ and 64 000 data points. In Fig. 2, we take $m = 1$, i.e., $d_M = d_E$ and we require card $S_E^i(r) \geq 2 d_M$. In this case the agreement with the numerically known non-negative Liapunov exponents (dashed lines) is very good for $d_E \geq 5$. Note that there is a large deviation at $d_E = 2$. This serves as an indication that the system lives in a space with more than two dimensions. Also, as observed in Sec. III, $d_E = 3$ is not a sufficiently

large embedding dimension for precise values of the Liapunov exponents. Finally, it should be noted that increasing the minimal number of neighbors does not change the above observations. In Fig. 3 we illustrate the effect of taking too few neighbors. With the same parameters as in Fig. 2, we have required only card $S_E^i(r) \geq d_M$. The increase of the curves is a typical signature of a lack of sufficiently many neighbors. In Fig. 4 we analyze the influence of (artificially added) noise. We have added 0.4% noise (in terms of the total data latitude) and we observe that the prediction of the Liapunov exponent is wiped out. A typical signature of noise is the decrease of the Liapunov exponent with $d_M$. Note also that the effect of the noise can be essentially eliminated if we increase $m$, that is by increasing $d_E$ while keeping $d_M$ fixed. This is shown in Fig. 5, where we have chosen $m = 5$ and card $S_E^i(r) \geq d_M + 4$. The results are usually good for the positive Liapunov exponents, but the zero exponent tends to increase with $d_M$. The collection of Figs. 2–5 is a clear illustration of the summary of advice of Sec. III.

We next illustrate in more detail the effect of having too few data points. For this we shall deal with the very simple system defined by the map $x \to 1 - 2x^2$ of the interval $(-1,1)$. It has a Liapunov exponent $\ln 2$. Figure 6 shows the results analogous to Fig. 2, with a resolution of the data of $10^{-4}$ (obtained by multiplying each $x$ by 5000 and truncating). To make the statistical errors smaller, we have insisted on a minimum of 20 neighbors per point. In Fig. 7 we have applied the same algorithm, but with a precision of $2^{-31} \approx 0.5 \times 10^{-9}$. While this situation is unlikely to occur in laboratory experiments, it is typical for the appearance of spurious Liapunov exponents which are

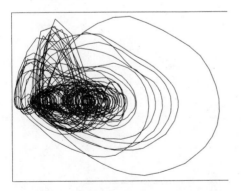

FIG. 8. An experimental orbit. Horizontal axis is $x_i$, vertical axis $x_i - x_{i-1}$.

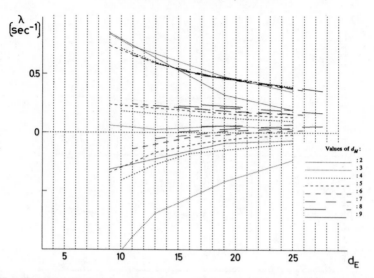

FIG. 9. Largest three Liapunov exponents as a function of $d_E$ for the signal of Fig. 8, for different values of $d_M$.

about twice (in principle even thrice) the real ones. This phenomenon is generated by the finiteness of the data set. This means that we cannot achieve the limit $r \rightarrow 0$ to determine the matrices $T_i$. Therefore one expects the nonlinearities to be important. A simple calculation shows that if one carries along second-order effects in the

equations leading to the $T_i$, for the map $x \rightarrow 1 - 2x^2$ in dimension $d_M = 2$, one obtains two Liapunov exponents, one at ln2 and one at 2 ln2. In the absence of noise, the nonlinear terms desingularize the equations for the $T_i$ when $d_M$ is larger than the true dimension of the system, and they tend to generate multiples of the "true" Liapunov ex-

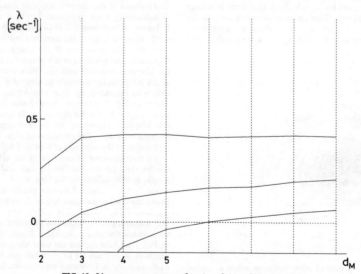

FIG. 10. Liapunov exponents as a function of $d_M$ at fixed $d_E = 22$.

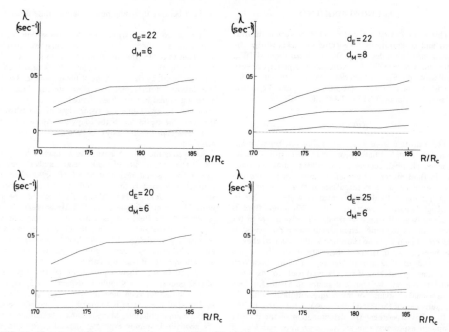

FIG. 11. Three largest Liapunov exponents as a function of Rayleigh number, for different $d_M$ and $d_E$. From theoretical arguments we know that one Liapunov exponent is equal to zero.

ponents. Having gone to very high precision in the calculations leading to Fig. 7, we have produced explicitly the generation of a "double" Liapunov exponent. (We suspect that Table I in Ref. 5 has the same origin.) In third-order terms are above the numerical imprecision one will in principle see a "triple" Liapunov exponent, and so on. For data coming from laboratory experiments, the equations for the $T_i$ are in general rather desingularized by the noise of the data. In that case a model calculation can be done, based on some independence assumptions of the noise (which may not be justified in general). For the map $x \rightarrow 1 - 2x^2$ and $d_M = 2$, this calculation predicts that the two Liapunov exponents will be $\ln 2$ and $\approx -1.2$.

Let us summarize. All of the above effects only concern the spurious Liapunov exponents, because they are caused by the desingularization of the equations for the $T_i$. If the noise is not too small relative to the precision and the density of the data, one will see the true Liapunov exponents and the spurious ones will all be negative. This seems to be the usual situation in laboratory experiments.

We now discuss the more difficult problem of analyzing an experimental time series. In particular, we have made measurements coming very close to the required desiderata of Sec. III. In this section we only analyze one of these runs, from the same point of view as the numerical experiments. In the Appendix, the experiments and

the complete results are described, as function of a varying parameter. To give a certain feeling of what is involved, we draw a piece of the experimental orbit (Fig. 8).

In Fig. 9 we summarize the results of varying the lags $m$ and matrix dimension $d_M$ in such a way that $d_E$ varies between 9 and 26. We limit $d_M$ between 2 and 9. One observes a relatively flat section in the region above $d_E = 20$ for $d_M > 3$. This limit is more visible in Fig. 10, where we plot, by interpolation, a section of Fig. 8 at $d_E = 22$. In view of the preceding discussions, we conclude that there are two positive Liapunov exponents. In the Appendix, we show the complete results for a series of experiments with varying parameters. (The main body of this paper was done by the first three authors, and the experiment, as well as the Appendix, have been provided by the fourth author. Our collaboration has made an optimization of the experiment and the analysis possible.)

*Noted added in proof.* An interesting recent paper by A. M. Fraser and H. L. Swinney [Phys. Rev. A **33**, 1134 (1986)] indicates how to choose time delays optimally in the reconstruction of dynamics from a time series. For other general literature see Ref. 1, which contains in particular a reference to a paper by Wolf *et al.* We remain unconvinced that the method described in that paper allows the systematic computation of several Liapunov exponents from noisy experimental data.

## ACKNOWLEDGMENTS

This work has been made possible by various institutions and organizations. We gratefully acknowledge the support of Institut des Hautes Etudes Scientifique (IHES), Bures-sur-Yvette (JPE), Coordenação de Aperfeiçoamento de Pessoal do Ensino Superior (CAPES), Brazil (SOK), Fonds National Suisse (DR,SC), European Economic Community Contract STI-082-J-C(CD), (SC).

## APPENDIX

The transition from a regular to a chaotic behavior has been widely studied both theoretically and experimentally in many physical, chemical, and other natural phenomena. In fluid mechanics one of the most used systems to investigate the onset of turbulence is thermal convection in a horizontal fluid layer heated from below, that is, Rayleigh-Benard convection (RB).[7] When the fluid is confined in a cell whose horizontal dimensions are of the same order of the fluid height (small aspect ratio cells), it has been found that the transition to the chaotic behavior can be explained in terms of the nonlinear interaction of a small number of degrees of freedom. This has been verified either by checking if the observed route to chaos was equal to one of the standard routes to chaos for low-dimensional systems[7] or more quantitatively with the measurement of the fractal dimension of the attractor.[8,9]

In a RB experiment the determination of the positive Liapunov exponents, that are indeed an important sign of the existence of a strange attractors, has been done only in Ref. 2 with a method similar to that outlined in the previous pages. Nevertheless, as pointed out in this text, there are some differences between the two methods and so we have applied the algorithm here proposed to evaluate the Liapunov exponents from a series of data recorded in the chaotic regime of a RB cell.

The fluid layer has horizontal sizes $l_x = 4$ cm, $l_y = 1$ cm, and height $d = 1$ cm (aspect ratios $\Gamma_x = 4, \Gamma_y = 1$). The fluid is silicon oil with Prandtl number 30. The bottom and top plates of the cell are made of copper and the temperature stability is about 1 mK. The lateral walls are made of glass to allow optical inspection. The detection system allows a semilocal measurement to be made. In fact, it consists of a laser beam, with a diameter of 1 mm, that crosses the fluid layer parallel to the rolls axis and is deflected by the thermal gradients inside the fluid. By measuring the deflection of the laser beam outside the cell we can measure the thermal gradient averaged along the optical path. We record the horizontal component of the gradient because it usually has the largest time-dependent amplitude.

In order to have a signal-to-noise ratio within the requirements specified in this paper, particular attention has been paid to reduce the environmental noise produced, for example, by the air convection along the optical path of the laser beam, by the vibrations of the mirrors, and of the laser cavity. Furthermore, to eliminate high-frequency noise, the signal has also been filtered at a suitable cutoff frequency to avoid that the rising time of the filter influencing the evaluation of the Liapunov exponents. This way a signal-to-noise ratio of about $10^{-4}$ has been achieved.

Analyzing the fluid behavior as a function of $R/R_c$ ($R_c$ is the critical value of the Rayleigh number), we find, except for a small region at $81 < R/R_c < 90$ where the convective motion is time dependent, a stable four-rolls structure for $R/R_c < 141$. Above this threshold a periodic oscillation at a frequency of 75 mHz is observed. Increasing $R$, the fluid crosses many different periodic and biperiodic states and it goes into the chaotic region via intermittency at $R/R_c > 170$. We have characterized this chaotic behavior by measuring the Liapunov exponents as a function of the Rayleigh number.

To satisfy all other requirements of the algorithm, 40 000 points, with a sampling frequency of 5 Hz, have been recorded for each measurement. This way, the time evolution of the system is followed for about 600 periods of the main oscillation. Many tests have been done to verify how the Liapunov exponents depend on $d_E$ and $d_M$. It has been found that the value of $\lambda$ is sufficiently stable in the interval $20 < d_E < 25$ and $5 < d_M < 8$. The results are reported in Fig. 11, where the values of the positive Liapunov exponents are shown as a function of $R$ for different $d_E$ and $d_M$. We see that the qualitative behavior of the curves is similar and the difference between them is about 10%. The measurements were done for $R/R_c = 171.41$, 174.08, 176.75, 182.10, 183.44, and 184.79. Figures 8–10 show details for $R/R_c = 182.10$.

By moving the detection point inside the cell by about 1 cm and keeping $R$ at the last value shown in Fig. 11 we find that the Liapunov exponents change by less than 5%. As a conclusion, the positive Liapunov exponents in the chaotic regime of a RB convection experiment have been determined using the method proposed in (Ref. 1). Even though the error of the measurement is not small (about 10%) it is still possible to follow how the number of the Liapunov exponents and their values change as a function either of the control parameter $R$ or of the position where the measurement has been taken inside the fluid.

[1]J.-P. Eckmann and D. Ruelle, Rev. Mod. Phys. 57, 617 (1985).

[2]M. Sano and Y. Sawada, Phys. Rev. Lett. 55, 1082 (1985).

[3]N. H. Packard, J. P. Crutchfield, J. D. Farmer, and R. S. Shaw, Phys. Rev. Lett. 45, 712 (1980).

[4]D. E. Knuth, The Art of Computer Programming (Addison-Wesley, New York, 1973), Vol. 3.

[5]J. A. Vastano and E. J. Kostelich, in Entropies and Dimensions, edited by G. Mayer-Kress (Springer-Verlag, Berlin, in press).

[6]J. H. Wilkinson, and C. Reinsch, Linear Algebra (Springer-Verlag, Berlin, 1971).

[7]For a recent review, see, e.g., R. P. Behringer, Rev. Mod. Phys. 57, 657 (1985).

[8]B. Malraison, P. Atten, P. Bergé, and M. Dubois, J. Phys. Lett. 44, L897, (1983).

[9]M. Giglio, S. Musazzi, and V. Perini, Phys. Rev. Lett. 53, 240 (1984).

PHYSICAL REVIEW A       VOLUME 38, NUMBER 3       AUGUST 1, 1988

# Noise reduction in dynamical systems

Eric J. Kostelich

*Center for Nonlinear Dynamics and Department of Physics, University of Texas, Austin, Texas 78712*

James A. Yorke

*Institute for Physical Science and Technology and Department of Mathematics, University of Maryland,
College Park, Maryland 20742*
(Received 29 October 1987)

A method is described for reducing noise levels in certain experimental time series. An attractor is reconstructed from the data using the time-delay embedding method. The method produces a new, slightly altered time series which is more consistent with the dynamics on the corresponding phase-space attractor. Numerical experiments with the two-dimensional Ikeda laser map and power spectra from weakly turbulent Couette-Taylor flow suggest that the method can reduce noise levels up to a factor of 10.

The ability to extract information from time-varying signals is limited by the presence of noise. Methods of noise reduction are a subject of widespread interest in communication,[1] physical systems,[2] and experimental measurements.[3] Recent experiments to study the transition to turbulence in systems far from equilibrium, like those by Fenstermacher et al.,[4] Behringer and Ahlers,[5] and Libchaber et al.,[6] succeeded largely because of instrumentation that enabled them to quantify and reduce the noise.

In recent years, traditional methods of time series analysis like power spectra have been augmented by new methods. In many cases, the time series can be viewed as a dynamical system with a low-dimensional attractor that can be reconstructed from the time series using time delays.[7] Because the dynamics of the phase-space attractor are not localized in a time or frequency domain, traditional noise-reduction methods like Wiener[8] and Kalman[9] filters are not applicable. In this paper we describe a noise-reduction procedure that works by taking many nearby points in phase space (corresponding to widely varying times in the original signal) to find a local approximation of the dynamics. These approximations can be used collectively to produce a new time series whose dynamics are more consistent with those on the phase-space attractor. We demonstrate the efficacy of the method using chaotic attractors obtained from the Ikeda laser map[10] and a Couette-Taylor fluid flow experiment.[11]

The discrete sampling of the original signal means that the points on the reconstructed attractor can be treated as iterates of a nonlinear map $f$ whose exact form is unknown. However, we assume that $f$ is nearly linear in a small neighborhood about each attractor point $x_i$ and write

$$x_{i+1} = f(x_i) \approx A_i x_i + b_i \equiv L(x_i)$$

for some matrix $A_i$ and vector $b_i$. (The matrix $A_i$ is the Jacobian of $f$ at $x_i$.) This can be done with least-squares

procedures similar to those described in Ref. 12. Let $\{x_j\}_{j=1}^n$ be a collection of points in a small ball around the $i$th reference point, and let $y_i = f(x_j)$ denote the observed image of $x_j$. The $k$th row $a_i^{(k)}$ of $A_i$ and $k$th component $b_i^{(k)}$ of $b_i$ are given by the least-squares solution of the equation

$$y_j^{(k)} = b_i^{(k)} + (a_i^{(k)} \mid x_j) \,, \tag{1}$$

where $y_j^{(k)}$ is the $k$th component of $y_j$ and ( $\mid$ ) denotes the dot product. (Farmer and Sidorowich[13] have generated similar approximations for the different purpose of forecasting chaotic time series.)

We remark that Eq. (1) can be ill-conditioned, for example, when the unstable manifold at $x_i$ is nearly one dimensional and $A_i$ is $2 \times 2$. We detect this situation by computing the singular values and right singular vectors[14] of the matrix $X$ whose $j$th row is $x_j$ to find the condition number of $X$, which is defined as the ratio of the largest to the smallest singular value. When the condition number is sufficiently large, we solve Eq. (1) using the components of $x_j$ contained in the subspace spanned by the singular vectors corresponding to the largest singular values. (For instance, we find a one-dimensional linear approximation of $f$ wherever the points $x_j$ fall nearly along a single line.) Moreover, because error exists both in the points $x_j$ and their observed images $y_j$, a modified least-squares procedure as described in Ref. 15 often gives better estimates of $A_i$ and $b_i$.

In the second stage of the method, we use the linear (more precisely, linear + constant) maps $L$ to correct errors in the observed trajectories as follows. Given a "window" of consecutive points $\{x_i, x_{i+1}, \ldots, x_{i+p}\}$ on the observed trajectory, we find the collection of points $\{\hat{x}_i, \hat{x}_{i+1}, \ldots, \hat{x}_{i+p}\}$ closest to the observed ones which also best satisfy the corresponding linear maps. More precisely, the new trajectory $\{\hat{x}_{i+k}\}_{k=0}^p$ minimizes the sum of squares

$$\sum_{k=0}^{p} \|\hat{x}_{i+k} - L(\hat{x}_{i+k-1})\|^2 + \|\hat{x}_{i+k} - x_{i+k}\|^2$$

$$+ \|\hat{x}_{i+k+1} - L(\hat{x}_{i+k})\|^2 \quad (2)$$

(terms with subscripts outside $[i, i+p]$ are omitted). This procedure can be iterated by replacing the original trajectory $\{x_i\}$ with the most recent least-squares trajectory $\{\hat{x}_i\}$, then finding a new solution to Eq. (2).[16] Moreover, the windows can overlap; for instance, the second window can begin in the middle of the first.

We have conducted numerical experiments using the attractor produced by the Ikeda map $f(z) = \rho + c_2 z \exp\{i[c_1 - c_3/(1 + |z|^2)]\}$, which models the dynamics of a bistable laser cavity.[10] We consider the attractor for the mapping $z_{j+1} = f(z_j)$, where $\rho = 1$, $c_1 = 0.4$, $c_2 = 0.9$, $c_3 = 6$. (The complex number $z_j$ is identified with the 2-vector $x_j$.) Numerical evidence suggests that initial conditions in $[0.5, 1.8] \times [-2, 1]$ are in the basin of a chaotic attractor whose numerically calculated Lyapunov exponents are $0.7296, -1.034$ (logarithms base 2) and whose Lyapunov dimension is 1.71.

We measure the noise level in terms of the *pointwise error* $e_j = \|x_{j+1} - L(x_j)\|$, i.e., the distance between the observed image and the predicted one [using the linear maps from Eq. (1)]. The *mean error* is $E = (\sum_j e_j^2/N)^{1/2}$, the root-mean-square value of the pointwise error over all

$N$ points on the attractor. We define the *noise reduction* $R = 1 - E_{\text{fitted}}/E_{\text{noisy}}$, where the mean errors are computed for the adjusted and original noisy attractor, respectively. ($R$ is a measure of the self-consistency of the time series, assuming that the linear maps are accurate approximations of the true dynamics.)

The numerical experiments on the Ikeda attractor use 65 536 iterates, to which 0.1% uniformly distributed random noise is added. The noise is independent of the dynamics, i.e., the input to the computer program is the series $\{z_j + \eta_j : \eta_j \text{ random}\}$, for which $E_{\text{noisy}} = 7.588 \times 10^{-4}$. The linear maps $L$ are computed using at least 50 points about each attractor point. Points are collected until the condition number of Eq. (1) is less than ten.[17] Trajectory adjustment is done in windows of 24 points, and the windows overlap by two points. After noise reduction, $E_{\text{fitted}} = 1.178 \times 10^{-4}$, so that the total noise reduction $R$ is 84%. When 1% noise is added, we find $R = 83\%$.

We have performed similar numerical trials with the Hénon attractor,[18] for which the $(j+1)$st time series' value is given by

$$x_{j+1} = f(x_j, x_{j-1}) = 1 - 1.4x_j^2 + 0.3x_{j-1}.$$

In this case the pointwise error can be measured exactly by replacing $L$ with $f$ (the mean error $E$ then becomes a "correctness index"). When 1% noise is added to the in-

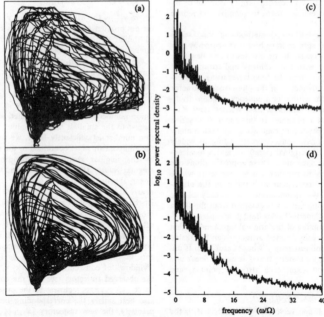

FIG. 1. Chaotic attractors from the Couette-Taylor fluid flow experiment described in Ref. 11 at $R/R_c = 12.9$. (a) Raw data. (b) Attractor after the noise reduction procedure described in the text. (c) and (d) Power spectra corresponding to (a) and (b), respectively. The units of frequency and power spectral density are as described in Ref. 11.

put as described above, the noise reduction (measured with the actual map) is 79%.[19] In addition, noise levels can be reduced almost as much in cases where the noise is added to the dynamics, i.e., where the input is of the form

$$\{x_{j+1}:x_{j+1}=f(x_j+\eta_j,x_{j-1}+\eta_{j-1}),\eta_j,\eta_{j-1}\text{ random}\}\ .$$

Next we consider the application of the method to data from the Couette-Taylor fluid flow experiment described in Ref. 11. Figure 1(a) shows a two-dimensional phase portrait of the raw time series at a Reynolds number $R/R_c=12.9$, which corresponds to weakly chaotic flow.[11] The corresponding phase portrait from the filtered time series[20] is shown in Fig. 1(b). The noise reduction, using the above criterion with the linear maps, is 63%.

Figure 1(c) and 1(d) show the power spectra for the corresponding time series. We emphasize that the dynamical information used to adjust the trajectories (viz., the motion of ensembles of points which are close together in phase space) corresponds to portions of the original signal that are widely and irregularly spaced in time. One question therefore is whether reducing the high-frequency noise corresponds to discovering the true dynamics which have been masked by noise. We believe that the answer is yes, based on those cases where there is an underlying low-dimensional dynamical system. However, in chaotic process some high-frequency components remain, because they are appropriate to the dynamics.

The method is particularly useful in calculations of dynamical quantities such as metric entropy and attractor dimension from experimental data. As an example, we consider the correlation dimension.[21] Let $C(x_i,\epsilon)$ denote the fraction of points on the attractor that fall within a distance $\epsilon$ of a randomly chosen (with respect to the natural measure) reference point $x_i$. Let $C(\epsilon)$ be the average values of $C(x_i,\epsilon)$ over the reference points $x_i$. Then $C(\epsilon)\sim\epsilon^d$ for small $\epsilon$, where $d$ is the correlation dimension.[21]

The dimension calculation illustrated in Fig. 2 is for the Ikeda attractor described above. The value of $C(\epsilon)$ is estimated from 1000 reference points using 48 values of $\epsilon$, equally spaced on a logarithmic scale from $2^{-10}$ to $2^{-2}$ (only the range $2^{-6}\leq\epsilon\leq2^{-3}$ is shown). Distances are normalized so that the total attractor extent is 1. The dimension, which is estimated as the derivative of $\log C(\epsilon)$ with respect to $\log\epsilon$, is taken as the slope of the regression line through six consecutive ($\log\epsilon,\log C(\epsilon)$) pairs. Although the noise level in the input is only 1%, noise inflates the dimension estimate even at ball sizes which are 3% of the attractor extent (top curve). However, the

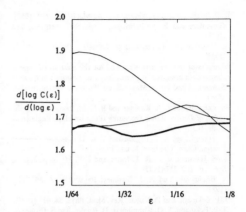

FIG. 2. Grassberger-Procaccia correlation dimension $d$ for the Ikeda attractor, using a data set with 1% uniformly distributed random noise (top curve), the same data set after the noise-reduction procedure (middle curve), and the original noiseless attractor (bold curve). The Lyapunov dimension of the attractor is 1.71.

dimension estimate for the fitted attractor (middle thin curve) compares favorably to that obtained from the noiseless attractor (bold line).

Since accurate linear approximations are essential for the success of the method, there must be an ample number of points in a small neighborhood about each point on the attractor. Thus, the data requirements depend on the dimension of the attractor. This method is best suited to situations where large amounts of data can be collected but the measurement precision is limited. This method promises to be of considerable value in the analysis of experimental data when the time series can be viewed as arising from a dynamical system with a low-dimensional attractor.

We thank Anke Brandstäter for providing data from her Couette-Taylor experiment. E. K. thanks Bill Jefferys, Andrew Fraser, Randy Tagg, and Harry Swinney for helpful discussions. This research is supported by the U.S. Department of Energy Office of Basic Energy Sciences, U.S. Air Force Office of Scientific Research, U.S. Defense Advanced Research Projects Agency–Applied and Computational Mathematics Program.

[1]For instance, see R. G. Gallager, *Information Theory and Reliable Communication* (Wiley, New York 1968).

[2]For instance, see *Sixth International Conference on Noise in Physical Systems,* edited by P.H.E. Meijer, R. D. Mountain, and R. J. Soulen (U.S. GPO, Washington, D.C., 1981), and similar volumes.

[3]For instance, see D. R. Brillinger, *Time Series: Data Analysis and Theory,* (Holden-Day, San Francisco, 1980); M. Priestly, *Spectral Analysis and Time Series* (Academic, London, 1981), two volumes.

[4]P. R. Fenstermacher, H. L. Swinney, and J. P. Gollub, J. Fluid Mech. **94**, 103 (1979).

440

[5]R. P. Behringer and G. Ahlers, J. Fluid Mech. **125**, 219 (1982); G. Ahlers and R. P. Behringer, Phys. Rev. Lett. **40**, 712 (1978).

[6]A. Libchaber, S. Fauve, and C. Laroche, Physica D **7**, 73 (1983).

[7]Examples of the use of this method for the analysis of experimental data are given in *Dimensions and Entropies in Chaotic Systems*, edited by G. Mayer-Kress (Springer-Verlag, Berlin, 1986).

[8]For example, see L. R. Rabiner and B. Gold, *Theory and Application of Digital Signal Processing* (Prentice-Hall, Englewood Cliffs, N.J., 1975).

[9]For example, see A. E. Bryson and Y. C. Ho, *Applied Optimal Control* (Ginn, Waltham, Mass., 1969).

[10]S. M. Hammel, C. K. R. T. Jones, and J. V. Moloney, J. Opt. Soc. Am. B **2**, 552 (1985).

[11]A. Brandstäter and H. L. Swinney, Phys. Rev. A **35**, 2207 (1987).

[12]J.-P. Eckmann and D. Ruelle, Rev. Mod. Phys. **54**, 617 (1985); J.-P. Eckmann, S. O. Kamphorst, D. Ruelle, and S. Ciliberto, Phys. Rev. A **34**, 4971 (1986); M. Sano and Y. Sawada, Phys. Rev. Lett. **55**, 1082 (1985).

[13]J. D. Farmer and J. J. Sidorowich, Phys. Rev. Lett. **59**, 845 (1987).

[14]J. J. Dongarra, C. B. Moler, J. R. Bunch, and G. W. Stewart, *LINPACK User's Guide* (Society of Industrial and Applied Mathematics, Philadelphia, 1979).

[15]W. H. Jefferys, Astron. J. **85**, 177 (1980); *ibid.* **86**, 149 (1981); M. G. Kendall and A. Stuart, *The Advanced Theory of Statistics* (Griffin, London, 1961), Vol. 2, p. 375.

[16]Because the input is a scalar time series, we constrain the trajectory adjustment to yield a scalar time as output. For instance, if $x_1 = (\xi_i, \xi_2)$ consists of the first two values in the time series, then the output begins with values $\hat{\xi}_1, \hat{\xi}_2, \hat{\xi}_3$ such that $\hat{x}_1 = (\hat{\xi}_1, \hat{\xi}_2), \hat{x}_2 = (\hat{\xi}_2, \hat{\xi}_3)$.

[17]To save CPU time, a maximum of 200 points is used to compute the linear maps. Despite these constraints, it is never necessary to compute a one-dimensional map approximation for this attractor as described in the text.

[18]M. Hénon, Commun. Math. Phys. **50**, 69 (1976).

[19]When the program is run on noiseless input, the mean error in the output is 0.025% of the attractor extent, which suggests that errors arising from small nonlinearities are negligible when the input contains enough points.

[20]In the noise reduction procedure, the attractor is reconstructed in four dimensions from a time series containing 32 768 values. Linear maps are computed using at least 50 points in each ball. Trajectories are fitted using windows of 24 attractor points which overlap by 6 points.

[21]P. Grassberger and I. Procaccia, Physica D **9**, 189 (1983).

LYAPUNOV EXPONENTS OF EXPERIMENTAL SYSTEMS

by Robert Conte and Monique Dubois,
Service de physique du solide et de résonance magnétique
(DPhG/PSRM)
Centre d'études nucléaires de Saclay
F-91191 Gif-sur-Yvette Cedex
France

(Text processing using Mathor)

## INTRODUCTION.

Recent developments in the study of dynamical systems have shown the importance of the calculation of quantitative values characteristic of the observed chaotic régimes. First, a calculation of the correlation dimension of the chaotic attractors has been proposed (Grassberger, Procaccia 1983) and then applied extensively, but this dimension reflects only the geometrical aspects, not the dynamical ones of the attractor. Moreover, the correlation dimension gives only a lower bound to the number of effective degrees of freedom. On the contrary, the knowledge of the Lyapunov exponents entirely determines the dynamics of the trajectories in the phase space. It is therefore of primary importance to correctly obtain their value from experimental data.

## DEFINITION OF LYAPUNOV EXPONENTS.

Let us recall it briefly. The dynamical system may be defined by a first order differential system (continuous case)

$$\frac{dx}{dt} = F(x)$$

or by an iterative map $x_n = f(x_{n-1})$ (discrete case). If we consider a small vector $\delta x(0)$ at initial time, then a short time later it will have evolved into $\delta x(t) = \delta x(0) \, e^{\lambda t}$. The scalar $\lambda$ is by definition called a characteristic (or Lyapunov) exponent of the dynamical system.

This intuitive definition can of course be expressed more rigorously:

assume there exists a probability measure $\rho$ invariant under f. If $\rho$ is ergodic (i.e. if $\rho$ is not the sum $\rho_1 + \rho_2$ of separately invariant measures), then **characteristic exponents** are by definition (multiplicative ergodic theorem of Oseledec,1968) the logarithm of the eigenvalues of the operator

$$\Lambda_x \;=\; \lim_{n \to +\infty} \left( \left( adj\; T_n^x \right) T_n^x \right)^{\frac{1}{2n}} .$$

which is the $2n^{th}$ root of a positive operator. In that formula, $T_n^x$ is the product of the tangent maps of the successive iterates

$$T_n^x = T(f^{n \cdot 1}(x)) \; \ldots \; T(f(x))\; T(x)$$

= tangent map of the $n^{th}$ iterate.

The ergodicity assumption implies that eigenvalues of the limit operator $\Lambda_x$ do not depend on x (on the contrary of eigenvectors).

For a recent review on that subject, with a good balance between mathematics and physics, see Eckmann, Ruelle 1985.


## METHODS FOR COMPUTING LYAPUNOV EXPONENTS.

### STATEMENT OF THE PROBLEM.

We only have a sequence of N measured values of a physical variable at regular times 0, $\Delta t$, 2 $\Delta t$,..., (N-1) $\Delta t$.

The problem is to compute Lyapunov exponents from those data only.

There are 3 main methods, the first one being valid only for analytically defined systems.

1) (Shimada,Nagashima 1979 ; Benettin, Galgani, Strelcyn 1980).

Their method can be summarized by the following algorithm:

- given an orthonormal set of vectors, compute its image one time step later; this image is a set of non orthogonal, unnormalized vectors
- store the lengths ratios of all vectors
- build an orthonormalized set by Gram-Schmidt method
- iterate until final time

Then characteristic exponents are given by

$$\lambda_m := \frac{1}{k} \sum_{i=1}^{k} \text{Log ratio}_{i,m} \quad \text{(i index of step=time; m index of dimension)}$$

2) (Wolf,Swift,Swinney,Vastano 1985).
- given a pair of points made of the current point and a second neighboring point, get its image one time step later (this image is just another pair of points from the data file)
- store the lengths ratio
- choose a new neighbor of the image of the current point such that the orientation of the new vector is close to the previous one
- iterate until exhaustion of the data file
Then $\lambda_1$ is given by the same formula than in method 1.
Afterwards, in order to get $\lambda_2$ (resp. $\lambda_3$ ,...), one must consider a triplet (resp. quadruplet,...) of points, which yields the sum $\lambda_1 + \lambda_2$ (resp. $\lambda_1 + \lambda_2 + \lambda_3$,...), from which $\lambda_2$ (resp. $\lambda_3$,...) is computed by difference.
3) (Eckmann,Ruelle 1985).
- given a set of points in the neighborhood of current point, get their image one time step later (this image is another set of points from the data file)
- compute the tangent map T by a numerical fit
- build an upper triangular matrix R from T, using the QR algorithm (Francis, Householder)
- store diagonal elements of R
- choose next set of points as new neighbors of the image of the current point
- iterate until exhaustion of the data file
Then

$$\lambda_m := \frac{1}{k} \sum_{i=1}^{k} \text{Log } |\text{diagonal element } (R_i)_{mm} | \text{ (see details below)}.$$

This is the method we are going to develop below.

## RECONSTRUCTION OF A PHASE SPACE FROM AN EXPERIMENTAL SIGNAL.

Data consists in the values of a single variable $u(i)$, $i=1,...,N$, at regularly spaced times $i \Delta t$.
The method is that of time delays (Takens, Young, Grassberger, Procaccia). The main idea of this method is that a phase space is made of positions and impulsions and therefore that the
couple (position, its time derivative) is a good candidate for defining a phase space. For practical purposes, delays are preferred to derivatives and a point in the reconstructed phase space is defined by

770

$x(i) := (u(i), u(i+\tau), \ldots, u(i+(d-1)\tau)) = (X_{i\alpha}, \alpha=1,\ldots,d)$

This method introduces 2 parameters, which up to now are arbitrary:

d the embedding dimension

$\tau$ the time delay $= p\,\Delta t$ (p integer)

It has been very successful for computing for instance the correlation dimension $\nu$ which is found to reach a limit when the embedding dimension d is increased (Malraison et al. 1983, Atten et al. 1984, Ciliberto and Gollub 1985). In fact, it has been shown that a better parameter than d is rather the underline embedding time defined as the length of that portion of the time series which is really used in the reconstruction process, i.e. $(d-1)\,p\,\Delta t$. When there is a characteristic orbital time in the system (Rössler chaos, temporal chaos related to oscillators,...), then this embedding time should be around once or twice this characteristic time. Eckmann et al. (1987) already found that criterium sound.

Note that Shaw, Fraser and Swinney (1986) have devised a criterium for getting rid of one of the two arbitrary parameters of the time delay method: given 2 coordinates of the reconstructed phase space, the best $\tau$ is the first time these 2 coordinates are functionnally independent (which corresponds to a local minimum of the mutual information), and not only linearly independent (which would select the first zero of the autocorrelation function).

For instance, in the Rössler chaos with parameters a=0.15, b=0.20, c=10.0, this leads to $\tau=1.40$ (recall that in this dynamical system the main characteristic time is T=6.07 )

**DETAILED IMPLEMENTATION OF THE ECKMANN-RUELLE ALGORITHM.**

The successive steps of the calculation are the following( parameters are underlined):

- **wait** for the transient régime to die out
- collect data points (number **N**, sampling time **$\Delta t$**), supposed to lay on the attractor
- (optionally) convert them to short integers in a **range** suited to the **precision** of data
- (optionally) sort them by increasing values to speed up following steps
- choose reconstruction parameters **d**, $\tau=\underline{p}\,\Delta t$
- perform the following loop (in time) over the set of points, until data file is exhausted:
-- $X_0$ denoting the current reconstructed point, define a small ball B( $X_0$ , $\epsilon$) of radius $\underline{\epsilon}$ centered at $X_0$ ($\epsilon$ is small as compared to the diameter of the attractor)

-- select at least $\underline{n}$ neighbors $X_i$ of $X_0$ inside that ball. If the norm for the calculation of distances is chosen as the max norm, then the sorted file immediately yields the candidate neighbors; with the choice of the euclidian norm, neighbors search is much longer since the sorted file is of no direct help

-- let them evolve in time :

$t \to t + e \, \Delta t$, where $\underline{e}$ is an integer

$\qquad X_i \to Y_i = X_{i+e}$ , $i=0,\ldots,n$

ball $B(X_0, \epsilon) \to$ ellipsoid centered at $Y_0$

-- compute the best linear application (also called tangent map, Jacobian matrix) T which applies vectors $X_i - X_0$ on vectors $Y_i - Y_0$, namely:

T:= the unique minimal length solution of the overdetermined linear least squares system defined by

$\qquad \forall i=1,\ldots,n \ \forall \alpha=1,\ldots,d \ : \ (Y_i - Y_0)_\alpha = T_{\alpha\beta} \ (X_i - X_0)_\beta$

(one least squares matrix $(X_i - X_0)_\beta$ with several second members $\alpha$)

-- multiply successive matrices T using QR factorization (see below)

- iterate $t \leftarrow t + e \, \Delta t$, i.e. go e points farther in data file


Let us now describe in detail the multiplication of tangent maps using the QR factorization.

Let $T_k \ T_{k-1} \ldots T_2 \ T_1$ be the non-commutative product to perform. We first recall the following property (QR factorization without column interchange):

$\forall$ square matrix T, $\exists$ orthogonal matrix Q

$\qquad\qquad\qquad \exists$ upper (right) triangular matrix R: T=QR.

Let us define:

$(Q_1, R_1) :=$ the QR factorization of $T_1$

$(Q_2, R_2) := $ " " " $\quad T_2 Q_2$

etc.

Then:

$$T_k T_{k-1} \ldots T_2 T_1 = Q_k R_k R_{k-1} \ldots R_2 R_1$$

and

$$^t(T_k \ldots T_1)(T_k \ldots T_1) = \ ^t(R_k \ldots R_1)(R_k \ldots R_1)$$

which eliminates $Q_k$.

We thus have replaced the multiplication of general matrices T by the multiplication of upper triangular matrices R. This new multiplication symbolically reads

$$(RR)_{ij} = \sum_{k=i}^{j} R_{ik} R_{kj}$$

772

instead of

$$(TT)_{ij} = \sum_{k=1}^{d} T_{ik}T_{kj}$$

i.e. it involves much less multiplications and additions (for instance 1 multiplication and 0 addition when i=j). Moreover, the matrix Q is exactly orthogonal (Householder, not Schmidt, algorithm).

There is a strong analogy with the QR algorithm of Francis used to diagonalize matrices: if we make "enough" turns on the attractor, the Q's will define a random sequence and therefore (Johnson, Palmer and Sell 1984) the diagonal of $R_k \ldots R_1$ will lead to Lyapunov exponents

$$\lambda_i = \lim_{n \to +\infty} \frac{1}{n} \text{Log } |(R_n \ldots R_1)_{ii}|$$

Moreover, like in QR algorithm, the $\lambda_i$'s are arranged in decreasing order.

We can also compute the eigenspaces (see Eckmann, Ruelle page 651): the eigenspace associated to the eigenvalues less than or equal to $\lambda_i$ is generated by the vectors defined by the last d-i+1 columns of matrix

$$(R_n \ldots R_1)^{-1} \text{diag}(R_n \ldots R_1)$$

Let us recall that eigenspaces do depend on starting point, on the contrary of eigenvalues.

Remarks:

What is the effect of time reversal on the computation?

There are 2 answers to that question (asked at the end of the talk).

1) If we collect the points after the end of the transient régime, and if we reverse time when going through the file of collected points, then one should get $-\lambda_i$ as new spectrum.

2) If, before collecting data points, we reverse time, then the attractor will be different and the spectrum will be that of the new attractor, which has nothing to do with the previous spectrum.

SOME ADVICE FOR OVERCOMING DIFFICULTIES OF THE METHOD.

The algorithm parameters are numerous:
- those related to the collection process of the experimental data: number N of points, their precision, sampling time $\Delta t$
- the parameters d and p of the phase space reconstruction
- the parameters of the calculation itself, i.e. mainly the diameter $\epsilon$ of the ball, which governs the amount of nonlinearity; when $\epsilon$ is too

small, the influence of the experimental noise is dominant, but if $\epsilon$ is too large the vectors $(X_i - X_0)$ are no more in the tangent space.

We have already discussed the choice of the phase space reconstruction.

The main questions to be solved in the algorithm implementation are the following:
- How many **neighbors n** should we require inside the ball of radius $\epsilon$? The condition that the linear least squares system be overdetermined is $n \geqslant d$, and we experimentally found the following behavior according to the value of $\frac{n}{d}$ :

$1 \leqslant \frac{n}{d} \leqslant 1.5$ : computation is imprecise

$1.5 \leqslant \frac{n}{d} \leqslant 2$ : acceptable

$2 \leqslant \frac{n}{d} \leqslant 2.5$ : good

$2.5 \leqslant \frac{n}{d}$ : not better
- What should be the **size $\epsilon$ of the ball** as compared to the attractor diameter?
The smaller the better, but this may conflict with the requirement for $\frac{n}{d}$ . Typically, $\frac{\epsilon}{\text{diameter}}$ should stay between 0.05 and 0.15, a condition which implies a minimal value for the number N of data points in order to keep a good statistics during the whole computation.
- What should we do in case there are not enough neighbors?
Then we should increase $\epsilon$ and retry. This can be done automatically for instance by choosing $\epsilon$ in a predetermined increasing sequence of fractions of the diameter. In no case should the step be skipped, as Sano and Sawada proposed it, for this amounts to choosing for T the identity, which is certainly wrong.
- What to do if the linear **least squares system** is **singular**?
It may be useful to briefly recall the mathematical results
concerning the solutions X of a linear system A X = B with arbitrary matrices orders by the linear least squares method (for a good reference book, see Lawson, Hanson).
When the rank of A is maximal, then the solution X is unique. When it's not, the solution X is the sum of the "minimal length solution" and of an arbitrary solution in the kernel, the minimal length solution being uniquely defined by the requirement that the euclidian

774

norm of X be minimal.

Consequently, if we choose a solution different from the minimal length solution, this encourages numerical noise propagation (Conte 1985 unpublished, on Rössler chaos)

The choice of the minimal length solution gives their smallest possible values to the elements of T, and moreover these values are found to be of the same order of magnitude than in the non singular cases.

- The number $d^2$ of unknown elements $T_{\alpha\beta}$ to be determined by solving the least squares system can be lowered by carefully choosing the ratio $\dfrac{e}{p}$ . If we choose $\dfrac{e}{p}$ = integer k ⩽ d-1, then the number of unknown elements goes down to k d. Indeed, coordinates of $X_i$ and $Y_i$ are both subsequences of a same sequence u(j), and the above mentioned choice corresponds to a non empty intersection of these 2 subsequences; the consequence on the structure of matrix T is the occurence of d-k lines made of d-1 null elements plus one unity element.

Ex. : the choice e=p leaves only d unknown elements, the others being 0 or 1.

This feature greatly increases the computation precision, for it avoids the spreading of round-off errors; moreover, it lowers the cost.

## PREVIOUS ATTEMPTS ON ECKMANN-RUELLE ALGORITHM.

- Sano, Sawada 1985. They computed the tangent map by solving a linear least squares system, but they incorrectly handled the case when this system becomes singular, and they did not use the QR factorization, preferring the Gram-Schmidt method like in Benettin et al.

- Vastano, Kostelich 1986. They performed a comparison of the methods of Wolf et al. and Eckmann-Ruelle, but, because they wanted to prove the superiority of the Wolf method, they made a wrong description of the Eckmann-Ruelle algorithm, namely:

. they forgot the use of the QR algorithm, which is an essential part of the method, and used the Gram-Schmidt method instead, which is known to be less precise concerning orthogonality,

. in case of degeneracy of the least squares system, they did not make choice of the minimal length solution,

. they did not adjust $\epsilon$ to n, but n to $\epsilon$ .

As a consequence, their experiments on Rössler chaos using their "Eckmann-Ruelle" algorithm disagree with ours.

- Eckmann, Oliffson Kamphorst, Ruelle, Ciliberto 1986. They found a good behavior of the algorithm, notably on Lorenz model and on experimental Rayleigh-Bénard data. Moreover, they found that requiring too much precision could lead to second order effects; for instance, with the map $x_n = 1 - 2 x_{n-1}^2$ , for which the only Lyapunov exponent is $\lambda_1 = \text{Log } 2$ , with a precision $10^{-4}$ one finds $\lambda_1$ as expected, but with a precision $10^{-9}$ one finds 2 positive exponents $\lambda_1$ and $2 \lambda_1$.

### OUR COMPUTATIONS.

- Hénon 2d map $x_{n+1} = 1 - a x_n^2$ , $y_{n+1} = b x_n$ . For a=1.4, b=0.3, the spectrum to be found is $\lambda_1 = 0.42$, $\lambda_2 = -1.6$ .
We tested the QR part alone by choosing the embedding dimension d equal to the true dimension 2 and the tangent map T equal to the analytically known one. Then, using only 128 data points, we recovered $\lambda_1$ with a precision 0.01 and the negative exponent $\lambda_2$ with a precision 0.05 .
- Logistic 1d map $x_{n+1} = a x_n (1-x_n)$ . For a=4, exponent $\lambda_1$ is Log 2.
We chose d=2 and $N=10^4$ points with a precision $10^{-4}$, and we required the minimal length solution for T. Then, at step number 10, matrix element $R_{22}$ was zero, meaning that the non physical exponent $\lambda_2$ had been detected as $-\infty$. This example is an illustration of the general fact that the additional exponents are meaningless and have no influence on the computation of the true exponents ($\lambda_1$ has the expected value).
- Rössler chaos, defined by the three dimensional differential system

$$\frac{d}{dt} \begin{pmatrix} x \\ y \\ z \end{pmatrix} = \begin{pmatrix} -(y + z) \\ x + a y \\ b + z (x - c) \end{pmatrix}$$

For a=0.15, b=0.20, c=10.0, the Lyapunov spectrum to be found is $\lambda_1$ =0.090, $\lambda_2$=0, $\lambda_3$=-9.7 , the mean orbital period time being T=6.07 .
We collected 40000 data points $y_i$ with a precision $10^{-4}$ represented by integers in interval [0,10000[, with a sampling time $\frac{T}{60}$ (60 points per mean orbital period).
The result of computation is that $\lambda_1$ and $\lambda_2$ are good over a wide range of values of (d, $\tau$= p $\Delta t$) provided d⩾5 and $\frac{n}{d}$ ⩾ 2. The exponents between $\lambda_3$ and $\lambda_{d-1}$ which have no physical meaning are all found to be negative. As to the least negative one $\lambda_d$, we find it closer to its true value when the time delay p $\Delta t$ is small:
$\lambda_d$= -6.5 with p=1 and $\Delta t$=0.1

**776**

$\lambda_d$= -2.6 with p=1 and $\Delta t$=0.3 .
- Rössler chaos with 2% flat noise added.
We added noise to the sampled variable u(i) by the formula
$$u(i) = u(i) \ (1 + random(-1,1) \ . \ noise\_level)$$
We could also have added noise by a term
$$(u_{max}-u_{min}) \ random(-1,1) \ . \ noise\_level$$
or by a combination of both types.
The result is that the absolute values of all $\lambda_i$'s except $\lambda_2$ are a little bit smaller but their signs are unaffected; $\lambda_2$ remains around 0, although very slightly below 0.
- Rayleigh-Bénard experimental data.
In the same way, experimental data have been processed. They came from a Rayleigh-Bénard experiment with confined geometry, the Prandtl number of the fluid being Pr = 38. The corresponding chaotic régimes had been previously studied in detail (Dubois, Bergé 1986) and were related to the interaction of two coupled oscillators: the expected Lyapunov spectrum is therefore (+ + 0 -), i.e. two positive, one zero and one negative exponents.
One particular behavior was retained here for the calculation; the correlation dimension $\nu$ of the attractor was 3.15 ± 0.15, as calculated by the algorithm proposed by Grassberger and Procaccia (data file: N=15000 points, sampling time $\Delta t = \dfrac{T}{16}$ ; T is the shorter period of the dynamics) .
Figure 1 shows the dependence of the computed exponents $\lambda_1$, $\lambda_2$, $\lambda_3$, $\lambda_4$ and $\lambda_d$ versus the embedding time (d-1) p $\Delta t$. The first two are clearly seen to be positive, while $\lambda_4$ and $\lambda_d$ remain negative. For the third one, there is some ambiguity: its value remains positive but small; since its behavior is different from that of other exponents for small and intermediate values of the embedding time, we think that $\lambda_3$ is the null exponent.
In agreement with the physical situation, this system has then 2 positive Lyapunov exponents and the physically meaningful exponents are therefore $\lambda_1$, $\lambda_2$, $\lambda_3$ and $\lambda_d$. Their numerical values can be estimated by taking the limit when increasing the embedding time:
$$\lambda_1 = 6 \pm 0.5 \ 10^{-3}$$
$$\lambda_2 = 3 \pm 0.3 \ 10^{-3}$$
$$\lambda_d = -0.05$$
We can check these numerical values by looking at the inequalities that must be satisfied between different estimates for entropies and dimensions.

A first inequality is

$$\kappa_2 \leqslant \text{Kolmogorov entropy} \leqslant \sum (\text{positive Lyapunov exponents})$$

where $\kappa_2$ is the metric entropy, which is a lower bound of the Kolmogorov entropy. The metric entropy $\kappa_2$ can be computed from a set of data points using an algorithm proposed by Grassberger and Procaccia (1984), which takes advantage of the calculation of the correlation dimension. This calculation has given $\kappa_2 = 5 \ 10^{-3} \ s^{-1}$ (Dubois 1987). There is therefore a relatively good agreement between this value and the sum of positive Lyapunov exponents:

$$5 \ 10^{-3} \ s^{-1} \leqslant \text{Kolmogorov entropy} \leqslant 9 \ 10^{-3} \ s^{-1} .$$

Another inequality links the correlation dimension $\nu$ and the Kaplan-Yorke dimension $D_{KY}$:

$$\nu \leqslant D_{KY} = j + \frac{\sum\limits_{i=1}^{j} \lambda_i}{|\lambda_{j+1}|} ,$$

where j is the smallest index which makes the sum $\sum\limits_{i=1}^{j+1} \lambda_i$ negative.

We find:

$$3.15 \leqslant 3 + \frac{0.01}{0.05} = 3.2$$

Therefore the agreement is correct, even although the value of the negative exponent is not precise.

## CONCLUSION

The Eckmann-Ruelle algorithm has now been sufficiently tested to be found reliable, at least in the case when the number of positive Lyapunov exponents is small. It has the great advantage of directly giving the number of positive Lyapunov exponents. Moreover, it is much more natural and easier to implement than the method of Wolf et al. Nevertheless, the influence of the experimental noise is not well understood yet.

A diskette is available on request (360 Kb, format IBM PC DOS).

Open questions.
1) When there is a zero exponent, how to recognize it?

778

We are currently testing a criterium to recognize among the numerically computed values the one which is exactly zero when the dynamical system is a flow. This is an important point since it determines also the number of positive exponents. This criterium has been suggested by Ruelle (private communication) and is the following: If we consider a flow with Lyapunov spectrum $\lambda_i$ , i=1,...,d with $\lambda_{i_0}$=0 and a Poincaré section of the flow, then the Poincaré section spectrum is $\frac{\lambda_i}{T}, i \neq i_0$ ,i=1,...,d , where T is the mean orbital period, i.e. the spectra are in a one-to-one correspondence except for the null exponent, which does not exist any more in the section.
2) Is there a criterium to have confidence in the computed negative exponents?

References
P. Atten, J. G. Caputo, B. Malraison, Y. Gagne (1984) Détermination de dimensions d'attracteurs pour différents écoulements, Journal de mécanique, volume spécial "Bifurcations et comportements chaotiques"
G. Benettin, L. Galgani, J. M. Strelcyn (1980) Lyapunov characteristic exponents for smooth dynamical systems and for hamiltonian systems; a method for computing all of them, Meccanica 15, 9-
S. Ciliberto, J. P. Gollub (1985), Chaotic mode competition in parametrically forced surface waves, J. fluid mech. 158, 381-
M. Dubois (1987) Dynamics of two coupled oscillators near the critical line in Rayleigh-Bénard convection, Proceedings of "The physics of chaos and systems far from equilibrium (CHAOS '87)", Monterey, California, January 11-14, 1987
M. Dubois, P. Bergé (1986) Rotation number dependence at onset of chaos in free Rayleigh-Bénard convection, Physica scripta 33, 159-162
J.-P. Eckmann, S. Oliffson Kamphorst, D. Ruelle (1987) Recurrence plots of dynamical systems, preprint Genève UGVA-DPT 1987/03-532
J.-P. Eckmann, S. Oliffson Kamphorst, D. Ruelle, S. Ciliberto (1986) Liapunov exponents from time series, Phys. rev. A34, 4971-4979
J.-P. Eckmann, D. Ruelle (1985) Ergodic theory of strange attractors, Reviews of modern physics 57, 617-656, addendum 57, 1115
A. M. Fraser, H. L. Swinney (1986) Independent coordinates for strange attractors from mutual information, Phys. rev. A33, 1134-1140
P. Grassberger, I. Procaccia (1983), Measuring the strangeness of strange attractors, Physica 9D, 189-208
P. Grassberger, I. Procaccia (1984) Dimensions and entropies of strange attractors from a fluctuating dynamics approach, Physica 13D,

34-54

R. A. Johnson, K. J. Palmer, G. R. Sell (1984) Ergodic properties of linear dynamical systems, preprint IMA 65, University of Minnesota, Minneapolis

B. Malraison, P. Atten, P. Bergé, M. Dubois (1983), Dimension of strange attractors: an experimental determination for the chaotic régime of two convective systems, J. physique lettres 44, L897-902

M. Sano, Y. Sawada (1985) Measurement of the Lyapunov spectrum from a chaotic time series, Phys. rev. lett. 55, 1082-1085

I. Shimada, T. Nagashima (1979) A numerical approach to ergodic problem of dissipative dynamical systems, Progress of theoretical physics 61, 1605-1616

C. W. Simm, M. L. Sawley, F. Skiff, A. Pochelon (1986) On the analysis of experimental signals for evidence of deterministic chaos, Helvetica physica acta, 60, 510-551

J. A. Vastano, E. J. Kostelich (1986) Comparison of algorithms for determining Lyapunov exponents from experimental data, in Dimensions and entropies in chaotic systems, edited by G. Mayer-Kress, Springer Verlag 1986

A. Wolf, J. B. Swift, H.L. Swinney, J.A. Vastano (1985) Determining Lyapunov exponents from a time series, Physica 16D, 285-317

L.-S. Young (1982) Dimension, entropy and Lyapunov exponents, Ergod. th. & dynam. sys. 2, 109-124

Figure caption.

Fig. 1. Dependence of the calculated Lyapunov exponents on the embedding time (D-1) p $\Delta$t. On the lower part, the most negative exponent $\lambda_D$ is plotted on a different scale. Above the highest exponent, the D and p values of the reconstructed phase space are reported. Sampling time $\Delta$t is 1s.

780

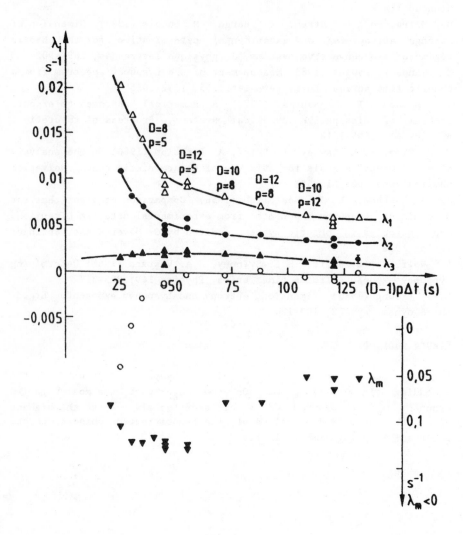

## 9. Scaling for External Noise

Volume 77A, number 6        PHYSICS LETTERS        23 June 1980

# FLUCTUATIONS AND THE ONSET OF CHAOS

J.P. CRUTCHFIELD [1] and B.A. HUBERMAN
*Xerox Palo Alto Research Center, Palo Alto, CA 94304, USA*

Received 3 April 1980

We consider the role of fluctuations on the onset and characteristics of chaotic behavior associated with period doubling subharmonic bifurcations. By studying the problem of forced dissipative motion of an anharmonic oscillator we show that the effect of noise is to produce a bifurcation gap in the set of available states. We discuss the possible experimental observation of this gap in many systems which display turbulent behavior.

It has been recently shown that the deterministic motion of a particle in a one-dimensional anharmonic potential, in the presence of damping and a periodic driving force, can become chaotic [1]. This behavior, which appears after an infinite sequence of subharmonic bifurcations as the driving frequency is lowered, is characterized by the existence of a strange attractor in phase space and broad band noise in the power spectral density. Furthermore, it was predicted that under suitable conditions such turbulent behavior may be found in strongly anharmonic solids [2]. Since condensed matter is characterized by many-body interactions, one may ask about the effects that random fluctuating forces have on both the nature of the chaotic regime and the sequence of states that lead to it. This problem is also of relevance to the behavior of stressed fluids, where it has been suggested that strange attractors play an essential role in the onset of the turbulent regime [3]. Although there are experimental results supporting this conjecture [4–6], other investigations have emphasized the possible role of thermodynamic fluctuations directly determining the chaotic behavior [7].

With these questions in mind, we study the role of fluctuations on the onset and characteristics of chaotic behavior associated with period doubling subharmonic bifurcations. We do so by solving the problem of

forced dissipative motion in an anharmonic potential with the aid of an analog computer and a white-noise generator. As we show, although the structure of the strange attractor is very stable even under the influence of large fluctuating forces, their effect on the set of available states is to produce a symmetric gap in the deterministic bifurcation sequence. The magnitude of this bifurcation gap is shown to increase with noise level. By keeping the driving frequency fixed we are also able to determine that increasing the random fluctuations induces further bifurcations, thereby lowering the threshold value for the onset of chaos. Finally, the universality of these results is tested by observing the effect of random errors on a one-dimensional map, and suggestions are made concerning the possible role of temperature in experiments that study the onset of turbulence.

Consider a particle of mass $m$, moving in a one-dimensional potential $V = a\eta^2/2 - b\eta^4/4$, with $\eta$ the displacement from equilibrium and $a$ and $b$ positive constants. If the particle is acted upon by a periodic force of frequency $\omega_d$ and amplitude $F$, and a fluctuating force $f(t)$, with its coupling to all other degrees of freedom represented by a damping coefficient $\gamma$, its equation of motion in dimensionless units reads

$$\frac{d^2\psi}{dt^2} + \alpha\frac{d\psi}{dt} + \psi - 4\psi^3 = \Gamma\cos\left(\frac{\omega_d}{\omega_0}\right)t + f(t) \quad (1)$$

with $\psi = \eta/2\eta_0$, the particle displacement normalized to the distance between maxima in the potential ($\eta_0$

---

[1] Permanent address: Physics Department, University of California, Santa Cruz, CA 95064, USA.

Volume 77A, number 6        PHYSICS LETTERS        23 June 1980

$= (a/b)^{1/2}$, $\alpha = \gamma/(ma)^{1/2}$, $\Gamma = Fb^{1/2}/2a^{3/2}$, $\omega_0$ $= (a/m)^{1/2}$ and $f(t)$ a random fluctuating force such that

$$\langle f(t) \rangle = 0 \tag{2a}$$

and

$$\langle f(0)f(t) \rangle = 2A\delta(t) \tag{2b}$$

with $A$ a constant proportional to the noise temperature of the system.

The range of solutions of eq. (1), in the case where $f(t) = 0$ (the deterministic limit) has been investigated earlier [1]. For values of $\Gamma$ and $\omega_d$ such that the particle can go over the potential maxima, as the driving frequency is lowered, a set of bifurcations takes place in which orbits in phase space acquire periods of $2^n$ times the driving period, $T_d$. At a threshold frequency $\omega_{th}$, a chaotic regime sets in, characterized by a strange attractor with "periodic" bands. Within this chaotic regime, as the frequency is decreased even further, another set of bifurcations takes place whereby $2^m$ bands of the attractor successively merge in a mirror sequence of the $2^n$ periodic sequence that one finds for $\omega \to \omega_{th}^+$. The final chaotic state corresponds to a single band strange attractor, beyond which there occurs an irreversible jump into a periodic regime of lower amplitude.

In order to study the effects of random fluctuations on the solutions we have just described, we solved eq. (1) using an analog computer in conjunction with a white-noise generator having a constant power spectral density over a dynamical range two orders of magnitude larger than that of the computer. Time series and power spectral densities were then obtained for different values of $\Gamma$, $A$ and $\omega_d$. While we found that the folding structure of the strange attractor is very stable under the effect of random forces, the bifurcation sequence that is obtained in the presence of noise differs from the one encountered in the deterministic limit.

Our results can be best summarized in the phase diagram of fig. 1, where we plot the observed set of bifurcations (or limiting set) as a function of the noise level, $N$, normalized to the rms amplitude of the driving term, $\Gamma$. The vertical axis denotes the possible states of the system, labeled by their periodicity $P = 2^n$, which is defined as the observed period normalized to the driving force, $T_d$. As can be seen, with increasing noise level a symmetric bifurcation gap appears, deplet-

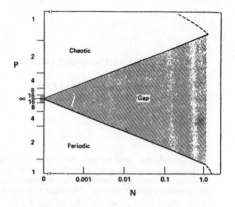

Fig. 1. The set of available states of a forced dissipative anharmonic oscillator as a function of the noise level. The vertical axis denotes the periodicity of a given state with $P = T/T_d$. The noise level is given by $N = A/\Gamma_{rms}$. The shaded area corresponds to inaccessible states.

ing states both in the chaotic and periodic phases. This set of inaccessible states is characterized by the fact that the longest periodicity which is observed before a strange attractor appears is a decreasing function of $N$, with the maximum number of bands which appear in the strange attractor behaving in exactly the same fashion. This gap extends over a large range of noise levels (up to $N = 1.5$), beyond which the motion either becomes unstable (i.e., $|\psi| \to \infty$; lower dashed line) or an amplitude jump takes place from the chaotic regime to a limit cycle of period 1 (upper dashed line).

We can illustrate this behavior by looking at the power spectral densities, $S(\omega)$ at fixed values of the driving frequency while increasing noise $N$. Fig. 2 shows such a sequence for $\omega_d/\omega_0 = 0.6339$, $\Gamma = 0.1175$, and $\alpha = 0.4$. Fig. 2(a) corresponds to $S(\omega)$ near the deterministic limit which, for the parameter values used, displays a limit cycle of period four. As $N$ is increased, a transition takes place into a chaotic regime characterized by broad band noise with subharmonic content of periodicity $P = 4$ (fig. 2(b)) [‡1]. As the noise is increased even further, a new bifurcation occurs from which a new

---

[‡1] We should mention that the Poincare map corresponding to this state clearly shows a four-band strange attractor with a single fold.

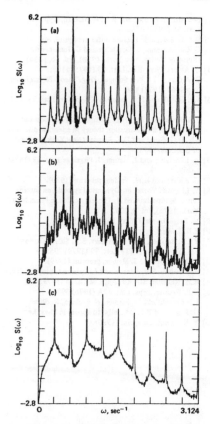

Fig. 2. Power spectral densities at increasing values of the effective noise temperature, for $\Gamma = 0.1175$, $\alpha = 0.4$, and $\omega_d = 0.6339\,\omega_0$. Fig. 2(a): $N = 10^{-4}$. Fig. 2(b): $N = 0.005$. Fig. 2(c): $N = 0.357$.

chaotic state with $P = 2$ emerges. Physically, this sequence reflects the fact that a larger effective noise temperature (and hence a larger fluctuating force) makes the particle gain enough energy so as to sample increasing nonlinearities of the potential, with a resulting motion which in the absence of noise could only occur for longer driving periods [2].

A different set of states appears if the noise level is kept fixed while changing the driving frequency. In this case the observed states of the system correspond to vertical transitions in the phase diagram of fig. 1, with the threshold value of the driving period, $T_{th}^n$, at which one can no longer observe periodicities $P \geqslant 2^n$, behaving like

$$T_{th}^n = T_{th}^\infty (1 - N_n^\gamma) \tag{3}$$

for $0 \leqslant N_n \leqslant 1$, with $N_n$ the corresponding noise level, $T_{th}^\infty$ the value of the driving period for which the deterministic equation undergoes a transition into the chaotic state, and $\gamma$ a constant which we determined to be $\gamma = \sim 1$ for $P \geqslant 2$ [3].

In order to test the universality of the bifurcation gap we have just described, we have also studied the bifurcation structure of the one-dimensional map described by

$$x_{L+1} = \lambda x_L (1 - x_L) + n_L(0, \sigma^2) \tag{4}$$

where $0 \leqslant x_L \leqslant 1$, $0 \leqslant \lambda \leqslant 4$, and $n_L$ is a gaussian random number of zero mean and standard deviation $\sigma$. For $n_L = 0$, eq. (4) displays a set of $2^n$ periodic states universal to all single hump maps [8–9], with a chaotic regime characterized by $2^m$ bands that merge pairwise with increasing $\lambda$ [10]. For $n_L \neq 0$ and a given value of $\sigma$, the effect of random errors on the stability of the limiting set is to produce a bifurcation gap analogous to the one shown in fig. 1.

The above results are of relevance to experimental studies of turbulence in condensed matter, for they show that temperature plays an important role in the observed behavior of systems belonging to this same universality class. In particular, Belyaev et al. [11], Libchaber and Maurer [12] and Gollub et al. [13] have reported that under certain conditions the transition to turbulence is preceded and followed by different finite sets of $2^n$ subharmonic bifurcations. It would therefore be interesting to see if temperature changes or external sources of noise in the fluids can either reduce or increase the set of observed frequencies, thus providing for a test of these ideas. In the case of solids such as superionic conductors, the expo-

---

[2] In the regime of subharmonic bifurcation the dependence of response amplitude on driving frequency is almost linear.

[3] Using the scaling relation $(T_{th} - T_n)/(T_{th} - T_{n+1}) = \delta$ [1] this implies that the threshold noise level scales like $N_n/N_{n+1} = \delta$, with $\delta = 4.669201609\ldots$ .

460

Volume 77A, number 6            PHYSICS LETTERS            23 June 1980

nential dependence on temperature of their large diffusion coefficients might provide for an easily tunable system with which to study the existence of bifurcation gaps. Last, but not least, these studies can serve as useful calibrations on the relative noise temperature of digital and analog simulations.

In concluding we would like to emphasize the wide applicability of the effects that we have reported. Beyond the experimental studies of turbulence, there exist other systems which belong to the same universality class as the anharmonic oscillator and one-dimensional maps. These systems range from the ordinary differential equations studied by Lorenz [10], Robbins [14], and Rossler [15] to partial differential equations describing chemical instabilities [16]. Since period doubling subharmonic bifurcation is a universal feature of all these models, our results provide a quantitative measure of the effect of noise on their non-linear solutions.

The authors wish to thank D. Farmer, N. Packard, and R. Shaw for helpful discussions and the use of their simulation system.

*References*

[1] B.A. Huberman and J.P. Crutchfield, Phys. Rev. Lett. 43 (1979) 1743.

[2] See also, C. Herring and B.A. Huberman, Appl. Phys. Lett. 36 (1980) 976.

[3] See, D. Ruelle, in Lecture notes in physics, eds. G. Dell'Antonio, S. Doplicher and G. Jona-Lasinio (Springer-Verlag, New York, 1978), Vol. 80, p. 341.

[4] G. Ahlers, Phys. Rev. Lett. 33 (1975) 1185;. G. Ahlers and R.P. Behringer, Prog. Theor. Phys. (Japan) Suppl. 64 (1978) 186.

[5] J.P. Gollub and H.L. Swinney, Phys. Rev. Lett. 35 (1975) 927; P.R. Fenstermacher, H.L. Swinney, S.V. Benson and J.P. Gollub, in: Bifurcation theory in scientific disciplines, eds. D.G. Gorel and D.E. Rossler (New York Academy of Sciences, 1978).

[6] A. Libchaber and I. Maurer, J. Physique Lett. 39 (1978) L-369.

[7] G. Ahlers and R.W. Walden, preprint (1980).

[8] T. Li and J. Yorke, in: Dynamical systems, an International Symposium, ed. L. Cesari (Academic Press, New York, 1972), Vol. 2, 203.

[9] M. Feigenbaum, J. Stat. Phys. 19 (1978) 25.

[10] E.N. Lorenz, preprint (1980).

[11] Yu.N. Belyaev, A.A. Monakhov, S.A. Scherbakov and I.M. Yavorshaya, JETP Lett. 29 (1979) 295.

[12] A. Libchaber and J. Maurer, preprint (1979).

[13] J.P. Gollub, S.V. Benson and J. Steinman, preprint (1980).

[14] K.A. Robbins, SIAM J. Appl. Math. 36 (1979) 451.

[15] O.E. Rossler, Phys. Lett. 57A (1976) 397; J.P. Crutchfield, D. Farmer, N. Packard, R. Shaw, G. Jones and R.J. Donnelly, to appear in Phys. Lett.

[16] Y. Kuramoto, preprint (1980).

461

# Scaling for External Noise at the Onset of Chaos

J. Crutchfield, M. Nauenberg, and J. Rudnick

*Physics Department, University of California, Santa Cruz, California 95064*

(Received 8 December 1980)

The effect of external noise on the transition to chaos for maps of the interval which exhibit period-doubling bifurcations are considered. It is shown that the Liapunov characteristic exponent satisfies scaling in the vicinity of the transition. The critical exponent for noise is calculated with the use of Feigenbaum's renormalization group approach, and the scaling function for the Liapunov characteristic exponent is obtained numerically by iterating a map with additive noise.

PACS numbers: 64.60.Fr, 02.90.+p, 47.25.Mr

The notion that the transition to turbulence in fluids has universality properties similar to those of critical phenomena has been suggested by Feigenbaum[1] on the basis of the scaling behavior of mathematical models near the onset of chaos.[2] A further impetus for an analogy between the transition to chaos and critical point phase transitions was given[3] by the observation that as a control parameter $r$ in these models increases past a critical value $r_c$ into the chaotic regime the measure-theoretic entropy—the Liapunov characteristic exponent $\bar{\lambda}$—has an envelope curve of the form $(r - r_c)^\tau$. The universal exponent $\tau$ is given by $\tau = \ln 2 / \ln \delta = 0.449\,806\,9\ldots$, where $\delta$ is the maximum eigenvalue associated with perturbations about the invariant map[1] of the interval. The transition to chaos in these models is heralded by a cascade of period-doubling bifurcations,[2] which is also of interest to an understanding of the onset of turbulence in physical systems.[4]

Motivated by the interpretation of experiments in fluids[5] and solids and by some recent numerical calculations,[6,7] we have considered theoretically the effect of added external noise on the transition to chaos in maps of the interval. The main result to be reported here is that the noise amplitude behaves as a *scaling variable* and that the dependence of the Liapunov characteristic exponent $\bar{\lambda}$ on the noise amplitude $\sigma$ and $\bar{r} = (r - r_c)/r_c$ is of the scaling form

$$\bar{\lambda}(r, \sigma) = \sigma^\theta L(\bar{r}/\sigma^\gamma) \qquad (1)$$

with $L(y)$ a universal function, and $\theta$ and $\gamma$ universal exponents. In the limit of vanishing noise $\sigma \to 0$ we have $\bar{\lambda} \propto \bar{r}^\tau$ which implies that as $y \to \infty$, $L(y) \propto y^\tau$, and leads to the exponent relation $\theta = \gamma\tau$.

The idea that the noise plays a role parallel to

that of the ordering field in a ferromagnetic transition was conjectured previously in Ref. 7. The noise exponent $\theta$ is a new critical exponent which we evaluate from an extension of Feigenbaum's scaling theory. Our result agrees with the recently observed value[7] of $\theta$ to within the limits of accuracy of the measurement. We also report on the measured form of the scaling function $L(y)$.

We start out by specifying the form of the one-dimensional map with additive noise. It is defined by the stochastic recursion relation

$$x_{k+1} = f(x_k; r) + \xi_k \sigma \qquad (2)$$

with $f(x; r)$ a continuous function of $x$ in a finite interval having a parabolic maximum, and $r$ a parameter that controls the shape of the function.[2] A common example is the function $rx(1-x)$ with $0 \le r \le 4$, and $0 \le x \le 1$. The quantity $\xi_k$ is a random variable controlled by an even distribution of unit width, and $\sigma$ is a variable that controls the width (or amplitude) of the noise. Note that when $\sigma = 0$ the map is perfectly deterministic.

We consider successive iterations of the stochastic map, Eq. (2) with $r$ at the critical value $r_c$, following techniques introduced by Feigenbaum. Setting the origin of coordinates to the $x$ for which the function $f(x; r)$ is a maximum and rescaling this maximum to 1, the $2^n$th iterate of $f(x; r_c)$ converges to $(-\alpha)^{-n} g(\alpha^n x)$, where $g(x)$ is a universal map satisfying the equation

$$g(g(x)) = -\alpha^{-1} g(\alpha x) \qquad (3)$$

with $\alpha = -1/g(1)$. Adding a small amount of noise $\xi\sigma$, we assume that the corresponding $2^n$th iterate of the map converges to $(-\alpha)^{-n}[g(\alpha^n x) + \xi\sigma\kappa^n D(\alpha^n x)]$ with $D(x)$ a universal $x$-dependent noise amplitude function and $\kappa$ a constant. When $\sigma$ is small enough, we have

$$g(g(x) + \xi\sigma D(x)) + \xi'\sigma D(g(x) + \xi\sigma D(x)) \approx g(g(x)) + \xi\sigma g'(g(x))D(x) + \xi'\sigma D(g(x)) + O(\sigma^2)$$

$$= g(g(x)) + \xi''\sigma\{[g'(g(x))D(x)]^2 + [D(g(x))]^2\}^{1/2}. \qquad (4)$$

933

In going to the last line we used the fact that $\xi$ and $\xi'$ are independent random variables, and that $\xi''$ is also a random variable. This and our above assumption implies that $D(x)$ must satisfy the eigenvalue equation

$$KD(\alpha x) = \alpha \{[g'(g(x))D(x)]^2 + [D(g(x))]^2\}^{1/2}. \quad (5)$$

We have solved Eq. (5) for the eigenvalue $\kappa$ and the corresponding eigenfunction $D(x)$ using the known results[1] for $\alpha$ and $g(x)$. Carrying out a calculation involving a polynomial interpolation for $D(x)$ we have found $\kappa = 6.61903....$

In the immediate vicinity above the transition to chaos the invariant probability distribution associated with the stochastic map will consist of $2^n$ bands, where $n$ is an integer that grows in the case of the deterministic map by unit steps to infinity as the transition is approached.[8,9] In the case of the stochastic map, $n$ grows to a finite value—and then decreases by unit steps as one passes to the other side of the transition. This modification of the deterministic bifurcation sequence is called a bifurcation gap.[6]

We now extend to the present case the previous discussion in Ref. 2 of the scaling behavior of the Liapunov characteristic exponent $\bar{\lambda}$, given by

$$\bar{\lambda} = \lim_{N \to \infty} \frac{1}{N} \sum_{k=1}^{N} \ln|f'(x_k; r)|, \quad (6)$$

or alternatively

$$\bar{\lambda} = \int p(x) \ln|f'(x; r)| dx, \quad (7)$$

where $p(x)$ is the invariant probability distribution associated with the map. Applying the above-mentioned considerations we obtain[10]

$$\bar{\lambda} = 2^{-n} L(\delta^n \bar{r}, \kappa^n \sigma). \quad (8)$$

Now, we assume that there will be $2^n$ bands in the chaotic regime when $\kappa^n \sigma$ is of order unity so that $n = -\ln\sigma/\ln\kappa$. Substituting this result into Eq. (9) we obtain Eq. (1) for $\bar{\lambda}$ with the two exponents $\theta$ and $\gamma$ given in terms of Feigenbaum's eigenvalue $\delta$ and the new eigenvalue $\kappa$ by $\theta = \ln2/\ln\kappa = 0.366754...$ and $\gamma = \ln\delta/\ln\kappa = 0.815359....$ The appearance of a bifurcation gap implies that $L(y)$ vanish at some $y = y_0$ which in turn implies that the maximum number $n$ of bifurcations is determined by the relation $\bar{r}_{n\,max} = y_0\sigma^\gamma$. This behavior has been observed numerically.[6]

Measurements of the behavior of $\bar{\lambda}$ as a function of $\sigma$ at $\bar{r} = 0$ have already been made by numerically calculating $\bar{\lambda}$ according to Eq. (6) with varying amounts of noise.[7] The measured value for $\theta$ is $0.37 \pm 0.01$. This agrees with our theoretical value for $\theta$ to within the experimental error.

To verify the existence of the scaling function $L(y)$ of Eq. (1) we used our values of $\theta$ and $\gamma$ to plot $\bar{\lambda}\sigma^{-\theta}$, with $\bar{\lambda}$ the result of numerical calculations of Eq. (6), as a function of $\bar{r}\sigma^{-\gamma}$. The results are shown in Figs. 1 and 2 for three different noise levels: $\sigma = 10^{-6}$, $10^{-8}$, and $10^{-10}$. The results for those three different noise levels all fall on a universal curve in the chaotic regime,

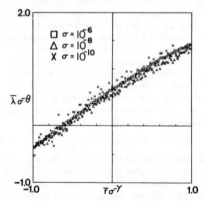

FIG. 1. Numerical determination of the scaling function $L(y)$, Eq. (1). The quantity $\bar{\lambda}\sigma^{-\theta}$ is plotted against 100 values of $y = \bar{r}\sigma^{-\gamma}$ at each of three noise levels: $\sigma = 10^{-6}$, $10^{-8}$, and $10^{-10}$. $\bar{\lambda}$ was calculated with use of Eq. (6), with $N = 10^6$ and with $\xi_\kappa$ a uniformly distributed random number of standard deviation $\sigma$.

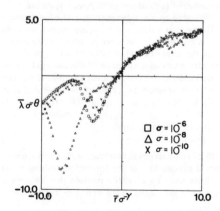

FIG. 2. $\bar{\lambda}\sigma^{-\theta}$ is plotted again, but over a wider range of $y = \bar{r}\sigma^{-\gamma}$ to illustrate the scaling regime. See text for discussion of various features. The details are the same as in Fig. 1, expect that $\bar{\lambda}$ was calculated with $N = 10^5$ in Eq. (6).

VOLUME 46, NUMBER 14     PHYSICAL REVIEW LETTERS     6 APRIL 1981

463

and in its immediate vicinity, Fig. 1, and fit the asymptotic behavior $L(y) \sim y^\tau$ for large $y$. The results do *not* coincide in the periodic regime, Fig. 2, but they could have been made to agree if we had chosen noise amplitudes differing by factors of $\kappa$, instead of factors of 100. This more restricted scaling follows from considerations of the type enunciated above.

These results appear to us to be both exciting and highly provocative. A theoretical picture of the transition to turbulence is just beginning to emerge; the analogy to critical phenomena should lead to new and important insights into the nature and characteristics of this transition.

The authors have benefited from conversations with B. A. Huberman and wish to thank him for the use of computing facilities at Xerox Palo Alto Research Center. One of us (J.C.) would also like to acknowledge useful discussions with N. Packard and the receipt of a University of California Regents Fellowship. This work is supported by the National Science Foundation.

[1]M. J. Feigenbaum, J. Statist. Phys. <u>19</u>, 25 (1978).

[2]For a recent monograph on this subject, see P. Collet and J. P. Eckmann, *Iterated Maps of the Interval as Dynamical Systems* (Birkhäuser, Boston, 1980).

[3]B. A. Huberman and J. Rudnick, Phys. Rev. Lett. <u>45</u>, 154 (1980). The exponent $\tau$ appears as $t$ in this reference. We have replaced the latin by a greek letter for consistency with other critical exponents.

[4]A. Libchaber and J. Maurer, J. Phys. (Paris), Colloq. <u>41</u>, C3-51 (1980); M. J. Feigenbaum, Phys. Lett. <u>74A</u>, 375 (1979); J. P. Gollub, S. V. Benson, and J. Steinman, Ann. N.Y. Acad. Sci. (to be published); B. A. Huberman and J. P. Crutchfield, Phys. Rev. Lett. <u>43</u>, 1743 (1979).

[5]G. Ahlers, private communication.

[6]J. P. Crutchfield and B. A. Huberman, Phys. Lett. <u>77A</u>, 407 (1980).

[7]J. P. Crutchfield, J. D. Farmer, and B. A. Huberman, to be published.

[8]J. P. Crutchfield, J. D. Farmer, N. Packard, R. Shaw, G. Jones, and R. J. Donnelly, Phys. Lett. <u>76A</u>, 1 (1980).

[9]E. N. Lorentz, Ann. N.Y. Acad. Sci. (to be published).

[10]The details of the derivation of the result (8), which involves a careful consideration of the structure of the bands, will be presented in a future paper.

PHYSICAL REVIEW A VOLUME 31, NUMBER 5 MAY 1985

# Observation of chaotic dynamics of coupled nonlinear oscillators

Robert Van Buskirk* and Carson Jeffries

*Department of Physics, University of California, Berkeley, California 94720*
*and Materials and Molecular Research Division, Lawrence Berkeley Laboratory, Berkeley, California 94720*
(Received 26 September 1984)

The nonlinear charge storage property of driven Si $p$-$n$ junction passive resonators gives rise to chaotic dynamics: period doubling, chaos, periodic windows, and an extended period-adding sequence corresponding to entrainment of the resonator by successive subharmonics of the driving frequency. The physical system is described; equations of motion and iterative maps are reviewed. Computed behavior is compared to data, with reasonable agreement for Poincaré sections, bifurcation diagrams, and phase diagrams in parameter space (drive voltage, drive frequency). $N=2$ symmetrically coupled resonators are found to display period doubling, Hopf bifurcations, entrainment horns ("Arnol'd tongues"), breakup of the torus, and chaos. This behavior is in reasonable agreement with theoretical models based on the characteristics of single-junction resonators. The breakup of the torus is studied in detail, by Poincaré sections and by power spectra. Also studied are oscillations of the torus and cyclic crises. A phase diagram of the coupled resonators can be understood from the model. Poincaré sections show self-similarity and fractal structure, with measured values of fractal dimension $d=2.03$ and $d=2.23$ for $N=1$ and $N=2$ resonators, respectively. Two line-coupled resonators display first a Hopf bifurcation as the drive parameter is increased, in agreement with the model. For $N=4$ and $N=12$ line-coupled resonators complex quasiperiodic behavior is observed with up to 3 and 4 incommensurate frequencies, respectively.

## I. INTRODUCTION

Many physical systems can be viewed as a collection of coupled oscillators or modes. In this paper we report the behavior of $N$ driven nonlinear oscillators, coupled in several ways, for $N = 1,2,4,12$. The oscillator is a passive resonator comprised of a silicon $p$-$n$ junction used as a nonlinear charge-storage element, together with an external inductance. This physical system can be approximately modeled as a driven damped oscillator with a very nonlinear asymmetric restoring force, and has been used previously, first by Linsay[1] who found that it exhibited a period-doubling sequence with convergence ratio $\delta$ and power spectra as predicted by Feigenbaum.[2] It was shown to display other universal behavior patterns[3] and has been much studied;[4] in particular, intermittency,[5] effects of added noise,[6,7] and crises[8,9] have been reported. For two or more coupled resonators (which we also refer to as "oscillators") the system displays a much richer dynamical structure:[10,11] period doubling, Hopf bifurcations to quasiperiodicity, entrainment horns, and breakup of the invariant torus. This is the main subject of this paper. We view the junction oscillator as an interesting *physical system* from the viewpoint of contemporary nonlinear dynamics theory.[12] It is not an analog computer and is to be clearly distinguished from the numerical solutions of mathematical models that approximately represent it.

To understand coupled junction oscillators we first attempt to understand a single-junction oscillator in detail, in Sec. II: we review the relevant physics of the system and differential equations that, *a priori*, might approximate its behavior. The observed basic oscillator response

function is discussed as well as elementary maps and differential equation models. In Sec. III we show the detailed behavior of a single oscillator by real-time signals, bifurcation diagrams, return maps, phase portraits, Poincaré sections, fractal dimension measurements, and phase diagrams in parameter space. These data are compared to predictions from theoretical models. In Sec. IV we give models for $N=2$ coupled oscillators, present our experimental results, and compare to theory. Section V gives some results for $N=4$ and $N=12$, where quasiperiodicity with up to four frequencies is observed.

## II. PHYSICAL SYSTEM AND MODELS

*The system.* In Fig. 1, the basic nonlinear element is the $p$-$n$ junction:[13] a single crystal of Si containing fixed donor ions and electrons to the right and acceptors and holes to the left of an interface in a region $\sim 10^{-4}$ cm wide. One solves the transport equation including drift and mobility terms in an electric field arising from an applied potential difference $V$. The establishment of electron-hole diffusive equilibrium at the interface results in a built-in potential difference $\Phi$, and parallel layers of fixed donor and acceptor ions, yielding an effective junction differential capacitance $C_j(V) = C_{j0}(1 - V/\Phi)^{-1/2}$ for negative applied voltage. If $V$ is positive, forward injection of holes (electrons) into the $n$ ($p$) regions creates a much larger stored charge limited, however, by the recombination and back diffusion of electrons and holes in minority carrier lifetime $\tau$. For times $t \leq \tau$ the system is approximated by an effective storage differential capacitance $C_s(V) = C_{s0}\exp(V/\phi)$, with $\phi \equiv kT/e$. Figure 2 shows typical data for the total differential capacitance

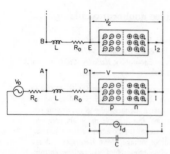

FIG. 1. Experimental system, showing a $p$-$n$ junction in a Si crystal with fixed donor ions ($+$) and electrons ($\cdot$) in the $n$ region, and acceptors ($-$) and holes ($\circ$) in the $p$ region. A single $p$-$n$ junction is driven by a sinusoidal voltage $V_0(t)$ through an inductor $L$ and resistances $R_C, R_0$. Connecting $A$ to $B$ makes two resistively coupled resonators. Connecting $D$ to $B$ makes two line-coupled resonators. This can be extended to $N$ coupled resonators by connecting $E$ to the next inductor, etc. A junction is modeled by a nonlinear capacitance $C(V)$ in shunt with an ideal Shockley diode $I_d(V)$.

$C(V) = dq/dV = C_j + C_s$ versus $V$ for junctions used here. The junction is modeled by $C(V)$ shunted by an ideal Shockley diode $I_d(V) = I_0[\exp(V/\phi) - 1]$, and is usually driven resonantly through an inductance $L$ by a driving voltage $V_0 = V_{os}\sin(\omega t)$ with $\omega \approx \omega_{res} = [LC(V=0)]^{-1/2}$.

*Equations of motion.* For a single junction and dynamical variables $(I, V, \theta)$, Kirchoff's laws for Fig. 1 yield three coupled first-order autonomous differential equations:

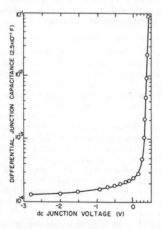

FIG. 2. Measured total differential capacitance $C(V)$ for $p$-$n$ junction vs junction voltage $V$. The steep rise is due to charge-storage capacitance $C_s(V)$, the negative voltage region to junction capacitance $C_j(V)$.

$$\dot{I} = \frac{V_0(\theta) - RI - V}{L} , \tag{1a}$$

$$\dot{V} = \frac{I - I_d(V)}{C(V)} , \tag{1b}$$

$$\dot{\theta} = \omega , \tag{1c}$$

where $R = R_C + R_0$ (Fig. 1), and $\theta = \omega t$. The motion can be presented in $(I, V, \theta)$ polar coordinates, so chosen that the orbits traverse the $(I, V)$ plane at, say, $\theta = 0, 2\pi, \ldots$ at consecutive times determined by the period $T = 2\pi/\omega$ of the driving voltage. This Poincaré section of the attractor can be observed directly by displaying $(I, V)$ on an oscilloscope and strobing the beam intensity at $t = nT$, $n = 1, 2, \ldots$ . Alternatively, one displays $(\dot{I}, I)$ or $(I_{n+1}, I_n)$, which are conjectured to be topologically equivalent.[14] Equations (1) are stiff, display a slow and a fast manifold, but can be numerically integrated by an explicit fourth-order Runge-Kutta algorithm.[15]

Equations (1) have a form discussed by Ott,[12]

$$dx_i(t)/dt = f_i(x_1(t), x_2(t), x_3(t)), \quad i = 1, 2, 3 \tag{2}$$

and a negative divergence of phase-space flow, $\sum_i \partial f_i / \partial x_i$. For the simpler Eq. (3) the divergence has the value $-|a|$. Phase-space volumes decrease roughly exponentially in time: $\Lambda(t) = \Lambda(0)\exp(-at)$. Since the system is observed to display chaotic motion, one can conclude that it has a strange attractor, characterized by a fractal dimension.

Physical insight comes from calculating the effective junction charge, $q(V) = \int_0^V C(V)dV = \phi C_{s0}[\exp(V/\phi) - 1] + 2\Phi C_{j0}(1 - \sqrt{1 - V/\Phi})$, and rewriting Eqs. (1) in the form of a driven damped oscillator

$$\ddot{q} + a(q)\dot{q} - f(q) = A_0 \sin(\omega t) , \tag{3}$$

with nonlinear damping coefficient

$$a(q) = \frac{R}{L} + \frac{1}{C(q)} \frac{\partial I_d}{\partial V} \tag{4a}$$

and nonlinear restoring force

$$f(q) = -\frac{1}{L}[V(q) + RI_d(V(q))] . \tag{4b}$$

Figure 3 is a plot of the force function for typical parameter values used and shows a weak, almost constant negative force $f(q) \propto -\ln(q+1)$ for positive $q$ (forward injection). For negative $q$ there is a strong positive force, $f(q) \sim q^2$. For $q$ small, the expansion of Eq. (4b), $f(q) = -Aq + Bq^2 - Cq^3 + Dq^4 + \ldots$, shows no symmetry; the system may show period doubling without first a symmetry-breaking bifurcation,[16] in contrast to the driven pendulum and to Duffing's equation with $f(q) \sim -q \pm q^3$. The junction oscillator is so nonlinear it can be driven hard enough to shift its resonant frequency $\omega_{res}$ down by an order of magnitude—an ultrasoft spring.

*Oscillator response.* Figure 4 shows data for the response voltage $V$ of a $p$-$n$ junction oscillator as a function of frequency, for various values of the driving voltage amplitude. We note the following. (i) A shift down of $\omega_{res}$ with increasing $V_{os}$: a soft spring. (ii) Hysteresis:

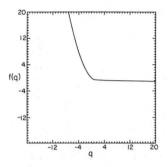

FIG. 3. Nonlinear restoring force $f(q)$ vs $q$ computed from Eq. (4b) using the values $C_{i0}=6\times10^{-13}$ F, $C_{j0}=6\times10^{-10}$ F, $I_0=4.8\times10^{-9}$ A, $\phi=0.04$ V, $\Phi=0.6$ V, $R=90$ $\Omega$, $\omega=4.08\times10^5$ sec$^{-1}$, $L=0.01$ H. The units of $f$ and $q$ have been rescaled. Note that $f(0)=0$.

this is the well-known "jump" phenomena for driven nonlinear oscillators;[17] it is a transcritical bifurcation.[18] (iii) Subharmonic response at $\omega_{res}/n$, $n=2,3,4\ldots$, also expected.[17] (iv) Increased resonance width due to increased value of damping coefficient $a(q)$ in Eq. (4a): the junction conducts when driven harder; $a(q)$ switches from $R/L$ at $V<0.3$ V to $I_0/\phi C_{so}$ at $V>0.3$ V, corresponding to a quality factor jump from 120 to $\approx1$ as the junction becomes conducting.

*Models.* Equations (1) or (3) may be numerically integrated and bifurcation diagrams, Poincaré sections, and

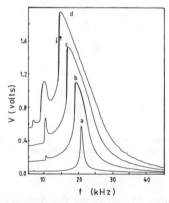

FIG. 4. Junction voltage $V$ vs drive frequency $f$ for a junction resonator for drive voltage $V_{os}$ in mV rms: $a$, 3; $b$, 41; $c$, 103; $d$, 179. The system is not yet chaotic and responds like a soft spring oscillator with subharmonic response; it also displays jump phenomena, with hysteresis. Type-1N4723 junction, $L=100$ mH, $R=53$ $\Omega$.

return maps computed; this is done below for some cases. The return map may be generally described by a two-dimensional map of the form

$$x_{n+1}=g(x_n,y_n,\Omega),\ \ y_{n+1}=h(x_n,y_n,\Omega),\ \ (5)$$

where $\Omega$ is the set of experimentally adjustable parameters; typically, $\Omega(R,V_{os},\omega)$. If the system is sufficiently dissipative, the map may reduce to one dimensional, e.g.,

$$x_{n+1}=1-Cx_n^2,\ \ (6)$$

the logistic map with one parameter $C$. In higher order it may reduce to the two-dimensional invertible map of Hénon,

$$x_{n+1}=1-Cx_n^2-y_n,\ \ y_{n+1}=Jx_n\ \ (7)$$

with an additional parameter $J$, the Jacobian determinant corresponding to the fractional area contraction per iteration, and thus to the system dissipation; furthermore, with $J\neq0$ there is hysteresis. As discussed in Sec. III, the driven junction oscillator is only very roughly modeled by Eq. (6) and somewhat better by Eq. (7). It turns out that the behavior can be better modeled by a generalization of Eq. (7),

$$x_{n+1}=f(x_n,\Omega)-y_n,\ \ y_{n+1}=Jx_n\ \ ,\ \ (8)$$

where the form of the function $f$ is not simply parabolic but is a unimodal or bimodal function chosen to model the junction oscillator characteristic behavior, e.g., Eq. (10).[10]

From the physical fact that the minority carrier density recovery after forward injection is a diffusion process, the motion may be more properly described by differential delay equations rather than Eqs. (1) and (3). *In principle* the system is rather high dimensional, and Eqs. (5) should be generalized to the form $x_{n+1}=g(x_n,x_{n-1},x_{n-2}\ldots,$ $y_n,y_{n-1}\ldots,\Omega)$, although present data do not seem to require this, owing to the dissipation. The question can be rephrased: how many previous cycles can the system remember in the steady state, i.e., what is the dimension of the phase space?

*Simple ODE model.* Returning to Eqs. (3) and (4) we make the simplifying assumptions $a(q)\rightarrow$const; $\omega\rightarrow1$, driving at resonance; $-f(q)\rightarrow-1+\exp q$, an exponential force function, to get a simple ordinary differential equation (ODE) model,

$$\ddot{x}+a\dot{x}+e^x-1=A\sin t\ \ (9)$$

which we numerically integrate to get a rough idea of expected chaotic behavior of driven junctions. Figure 5(a) shows a sequence of computed Poincaré sections, $x$ versus $\dot{x}$, for consecutive times $t=2\pi(n+\Delta/5)$ for $n=0,1,2,\ldots$, and strobe phase $\Delta=0,1,2,3,4,5,\ldots$. This shows that under the action of the Poincaré map the attractor, initially at $\Delta=0$, is stretched upward ($\Delta=1,2$), then stretched to the left ($\Delta=3$), then folded down ($\Delta=4$) to its final shape ($\Delta=5,0$). The stretching ratio measured from this figure is approximately $l_f/l_i\approx1.6$. Figures 5(b)–5(f) show the attractor computed from Eq. (9) for some sets of parameter values $(a,A)$, strobed at $\Delta=0$. It is clear that for small dissipation [Fig. 5(f),

470

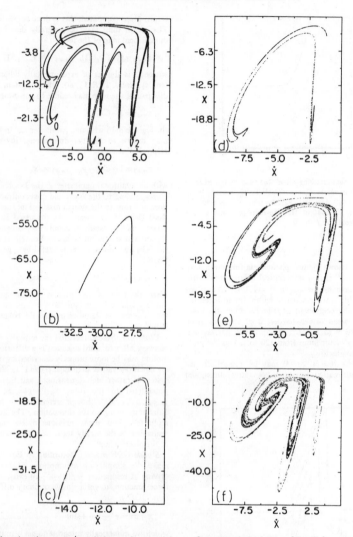

FIG. 5. Poincaré sections, $x$ vs $\dot{x}$, strobed at $t = 2\pi n$, $n = 1, 2, \ldots$, from numerical solutions of Eq. (9) for various parameter sets $(a, A)$: (a) (0.25,5), (b) (0.75,44), (c) (0.5,5), (d) (0.25,5), (e) (0.1,2.2), (f) (0.05,2). In (a) the five sections are strobed at $t = 2\pi(n + \Delta/5)$ with $\Delta$ shown on the figure.

$a = 0.05$] the attractor displays self-similarity and a complex fractal structure,[12] characteristic of chaotic dynamics. However, as the dissipation is increased, the fractal structure is damped out, and for Fig. 5(b) $(a = 0.75)$ the attractor is essentially one dimensional and could be modeled by a one-dimensional map. Under higher resolution the attractor appears ropelike. Figure 5 sequence

demonstrates the rapid decrease of dimension of a system as dissipation is increased; this is the essence of the present belief that high-dimensional dissipative systems may be usefully represented by low-dimensional maps.

It is straightforward to make semiquantitative calculations of the fractal dimension $d$ of the attractors in Fig. 5 using the conjecture of Kaplan and Yorke[19]

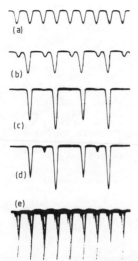

FIG. 6. Junction voltage $V$ (vertical, arbitrary units) vs time $t$ for driven resonator, Fig. 1, with $f = 20.7$ kHz, $L = 100$ mH, $R = R_C + R_0 = 53$ $\Omega$, 1N4723 junction. Drive voltage (volts rms): (a) 0.318; (b) 0.601; (c) 1.332; (d) 1.575, onset of chaos; (e) 1.978, one-band chaos.

$$d = j + \frac{\sum_{i=1}^{j} \lambda_i}{-\lambda_{j+1}} \, , \qquad (10a)$$

where $\lambda_i$ are the characteristic Lyapunov exponents and $j$ is the largest integer for which $(\lambda_1 + \lambda_2 + \cdots + \lambda_j) > 0$. For Eq. (9), with three degrees of freedom we have $\lambda_1 > 0$, $\lambda_2 = 0$, $\lambda_3 < 0$, $j = 2$, and

$$d = 2 - \frac{\lambda_1}{\lambda_3} \, .$$

Since nearby orbits on the attractor diverge at the rate $r(\tau) = r(0) \exp(\lambda_1 \tau)$, we estimate $\lambda_1 \approx \ln(l_f / l_i)$ from the measured stretching ratio per cycle of the map ($\tau = 1$). Since the total phase-space volume contraction ratio is $\exp(-at) = \exp(-2\pi a)$ after one cycle of the ODE ($t = 2\pi$), we set $\lambda_1 + \lambda_3 = -2\pi a$ to find

$$d = 2 + \frac{\lambda_1}{\lambda_1 + 2\pi a} \qquad (10b)$$

for the dimension of the whole attractor; this is reduced by 1 for a Poincaré section. For the sections of Figs. 5(b)—5(f) the dimension is found to be 1.1, 1.15, 1.3, 1.5, and 1.7, respectively. For map contraction ratios $\exp(-2\pi a) \approx 0.1$ these attractors bear a qualitative resemblance to those observed (cf. Fig. 15), and the dimension $d$ is comparable to that directly measured in Sec. III for driven junctions.

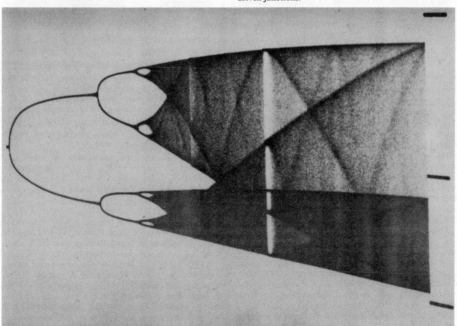

FIG. 7. Observed bifurcation diagram $[I_n]$ vertical (arbitrary units) vs drive voltage amplitude for driven $p$-$n$ junction resonator, $L = 470$ mH, $f = 3.87$ kHz, $R = 244$ $\Omega$, 300 A $p$-$n$ junction.

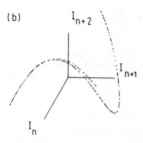

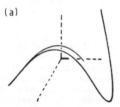

FIG. 8. (a) Observed 3D return map (arbitrary units) for driven $p$-$n$ junction. (b) Computed 3D return map (arbitrary units) from Eq. (7) with $C=1.5$, $J=-0.08$.

## III. EXPERIMENTAL RESULTS AND INTERPRETATION: SINGLE-JUNCTION OSCILLATOR

*Bifurcation diagram.* A recording of the junction voltage $V(t)$ for increasing values of the drive voltage $V_{os}$, Figs. 6(a)–6(c), shows a waveform of period 1,2,4, and then a nonperiodic form, 6(d), corresponding to onset of chaos. We sample and plot the set of consecutive current values $[I_n]$ separated by the driving period (the sampling phase is fixed at the current peaks in the periodic region) to obtain the bifurcation diagram of Fig. 7, showing period doubling, chaos, band merging, windows of periods 5 and 3, and veils. This is the *simplest* type of bifurcation diagram observed and is displayed by all junctions studied (approximately ten types) provided that $\omega \approx \omega_{res}$, $(2\pi/\omega) \sim \tau$, $a(q)$ is large enough, and $V_{os}$ is not too large.

*Return map.* Although Fig. 7 is similar to the bifurcation diagram of the logistic map, Eq. (6), the observed return map, Fig. 8(a), is not one-dimensional and shows a structure that can be reasonably fit by Fig. 8(b), computed from the two-dimensional Hénon map, Eq. (7), with $C=1.5$, $J=-0.08$. This value of $J$, the phase volume contraction per cycle, is typical for junction oscillators with moderate driving.

*Phase portrait and Poincaré section.* Figure 9(a) shows the projection of the attractor onto the $(\dot{I},I)$ plane at a drive voltage for the period-3 window; the black dots are a strobed Poincaré section. Figure 9(b) is the portrait for

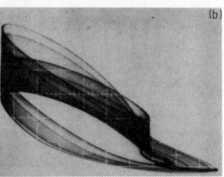

FIG. 9. (a) Observed phase portrait, $I$ vs $I$, for driven $p$-$n$ junction at period-3 window: $L=100$ mH, $f=19.64$ kHz, $R=53$ $\Omega$, 1N4723 junction, $V_{os}=3.82$ V rms; the three dark dots are a strobed Poincaré section. (b) Phase portrait and Poincaré section (dark "bent hairpin") for same system at one-band chaos, $V_{os}=3.48$ V rms.

one-band chaos just below this window. The dark rings correspond to the veils of Fig. 7, which are successive iterates of the critical point of the map. The dark line is a Poincaré section of the attractor showing structure topologically like Fig. 8.

*Period adding.* If driven hard enough, junction oscillators display the bifurcation diagram of Fig. 10(a), and an average (over one cycle) junction current $\overline{I(t)} = \overline{I}$ as in Fig. 10(b). In addition to period-doubling cascades to chaos, there is a larger sequence of periodic regions of period $\ldots, 3, 4, 5, \ldots, N, \ldots$ which we refer to as period adding, which can be physically understood as follows. We observe that $\overline{I}$ is constant in a region of period $N$ and furthermore that $N^{-1} \propto (\overline{I})^{-1/2}$; see data, Fig. 11. For strongly driven junctions the capacitance $C \approx C_s \propto \exp(V/\phi)$ is just proportional to the junction current $I_d$ for large forward injection (see Sec. II), so that the average (over one cycle) capacitance $\overline{C} \propto \overline{I}_d \sim \overline{I}$. Thus the oscillator resonant frequency $\omega_{res} \propto (\overline{C})^{-1/2} \propto (\overline{I})^{-1/2} \propto N^{-1}$.

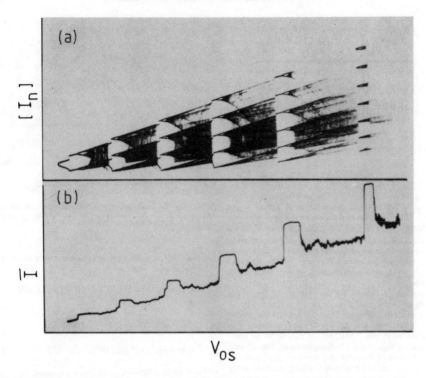

FIG. 10. (a) Bifurcation diagram $[I_n]$ vs drive voltage $V_{os}$ (arbitrary units) for $p$-$n$ junction showing period doubling and period adding (frequency locking). (b) Average junction current $\bar{I}$ vs $V_{os}$, showing peaks at locked regions. $f = 28$ kHz, $L = 10$ mH, $R = 8$ $\Omega$, 300 A $p$-$n$ junction.

That is, as this very soft spring oscillator is driven harder, it shifts its frequency down and becomes entrained or locked at successive subharmonics $\omega/N$ of the drive frequency $\omega$. Figure 12 is a two-parameter phase diagram of the observed entrainment regions, $N$:1. The waveforms of the junction voltage $V$, current $I$, and $\dot{I}$ for period $N = 7$ are shown in Fig. 13 and can be understood from numeri-

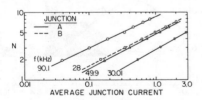

FIG. 11. Period $N$ of locked region vs average junction current $\bar{I}$ (in amperes) showing $N \propto (\bar{I})^{1/2}$. $L = 10$ mH, $R = 8$ $\Omega$. Junction $A$: 1N4721. Junction $B$: 300 A.

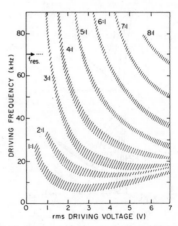

FIG. 12. Frequency of drive vs voltage at which locking of period $N$:1 occurs for driven $p$-$n$ junction, $L = 10$ mH, $R = 8$ $\Omega$, 1N4721 junction.

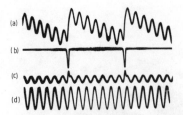

FIG. 13. Waveforms (a) $I(t)$, (b) $V(t)$, (c) $\dot{I}(t)$, and (d) drive voltage $V_0(t)$ for system of Fig. 12 at the $N=7$ locking.

cal integration of Eqs. (1).[3] The observed return map, Fig. 14, for the chaotic region between period 4 and period 5 [Fig. 10(a)] shows a structure which adds one more branch as $N \rightarrow N+1$. A high-resolution Poincaré section $(\dot{I}, I)$, Fig. 15, shows well-resolved self-similarity: four overall branches with the upper branch further divided into four sub-branches. The ratio of the tip-to-tip spreads is $\sim 0.1$, which corresponds to the area contraction ratio.

*Comparison to theory.* As an example of theoretical modeling of behavior of a single driven junction, Fig. 16 shows a bifurcation diagram from a numerical integration of Eq. (3) with $\omega = 1$ and $a = 0.45$, corresponding to a contraction ratio $\exp(-2\pi a) = 0.06$; this is to be compared to data, Figs. 7 and 17. Integration of Eqs. (1) with measured values of the junction parameters gives comparable results and also models period adding at large driving

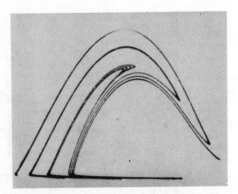

FIG. 15. Poincaré section, $I$ vs $I$, showing self-similarity and fractal structure. $L = 10$ mH, $f = 76$ kHz, $R = 43$ $\Omega$, 1N4721 junction.

voltage. Period adding can also be reasonably modeled by this two-dimensional map of the form of Eq. (8):

$$x_{n+1} = [x_n + 1 - S(x_n)] - S(A\{1 - [S(-x_n)]^2\}) - y_n ,$$

$$(11a)$$

$$y_{n+1} = Bx_n ,$$

$$(11b)$$

$$S(x) = \tfrac{1}{2}[x + (x^2 + 0.1)^{1/2}] .$$

$$(11c)$$

Figure 18 is a bifurcation diagram computed from Eqs. (11) with $B = 0.1$ and using $A$ as the control parameter; it is a reasonable fit to the data of Fig. 10(a), except for the

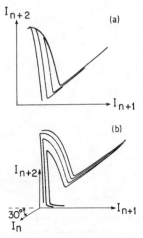

FIG. 14. (a) Two-dimensional and (b) three-dimensional return maps for the system of Fig. 10 in the chaotic region between $N = 4$ and 5.

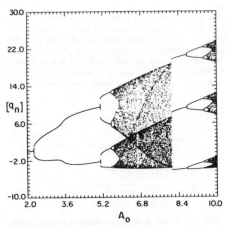

FIG. 16. Computed bifurcation diagram $[q_n]$ vs $A_0$ from Eq. (3) with $\omega = 1$, $a = 0.45$; compare to data, Fig. 17.

ROBERT VAN BUSKIRK AND CARSON JEFFRIES

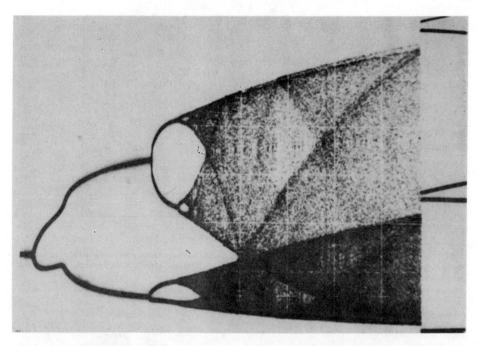

FIG. 17. Measured bifurcation diagram, $[I_n]$ vs $V_{os}$ (horizontal, arbitrary units) for driven junction. $L = 100$ mH, $R = 53$ $\Omega$, $f = 20.3$ kHz, 1N4723 junction.

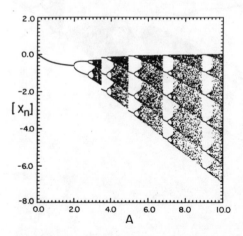

FIG. 18. Bifurcation diagram: $[X_n]$ vs $A$ computed from Eq. (11) with $B = 0.1$ showing period doubling, period adding, and hysteresis, with overall behavior similar to the data of Fig. 10.

$2 \rightarrow 2$ jump at low voltages, a subtle phenomena sensitive to system parameters and better explained by Fig. 16.

*Phase diagram.* An overview of junction oscillator behavior for the system used for Fig. 17 is provided by the two-parameter phase diagram of Fig. 19: a plot of the boundaries between various periodic and chaotic regions as a function of driving voltage $V_{os}$ and frequency $f = \omega/2\pi$. The junction resonance occurs at $f_{res} = 20$ kHz. Increasing $V_{os}$ upward along a line of constant frequency $f_{res}$ yields the simplest bifurcation sequence: periods $1, 2, 4, 8, \ldots$, chaos, $\ldots$, two-band chaos $C2$, one-band chaos $C1$, period-3 window (with hysteresis $\uparrow, \downarrow$), period 6, three-band chaos $C3$, and one-band chaos $C1$ (interior crisis). At higher drive voltage there begins a period-adding sequence—see the phase diagram of Fig. 12. Moving upward along $f = 34$ kHz in Fig. 19 gives a sequence $1, 2, 4, 8, 4, 2, 1$ without chaos. Increasing frequency at $V_{os} = 2$ V gives a sequence $1, 2, 4, 8, \ldots$, chaos, $\ldots$, $8, 4, 2, 1$. This makes clear why such a wide variety of bifurcation diagrams are observed, e.g., Fig. 20 showing reverse bifurcation.

To compare the observed phase diagram of Fig. 19 with our model we numerically integrated Eq. (3) for $a = 0.45$ for the two control parameters $0 \le A_0 < 20$, $0.4 \le \omega \le 3$, with the resulting theoretical phase diagram of Fig. 21,

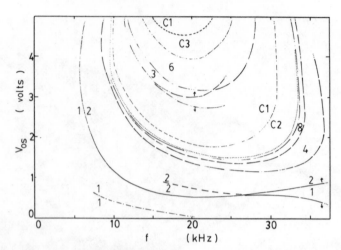

FIG. 19. Phase diagram: Drive voltage $V_{os}$ vs frequency $f$ of junction oscillator showing boundaries between periods 1,2,4,8; threshold for chaos; two-band chaos $C2$; one-band chaos $C1$; periods 3,6; three-band chaos $C3$; $C1$; hysteresis: $(\uparrow,\downarrow)$; 1:1 and 2:2 jump bifurcations. $L = 100$ mH, $R = 53 \, \Omega$, 1N4723 junction.

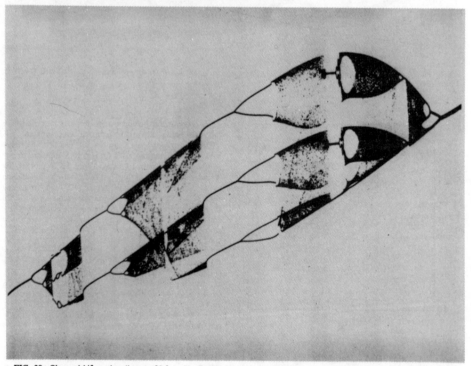

FIG. 20. Observed bifurcation diagram $[I_n]$ vs $V_{os}$ (horizontal, arbitrary units) for driven junction, $f = 11.79$ kHz, $L = 10$ mH, $R = 8 \, \Omega$, 300 A junction.

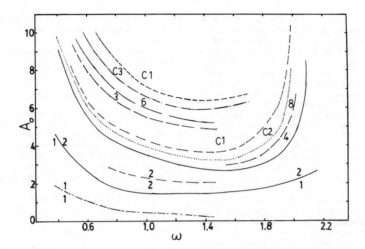

FIG. 21. Computed phase diagram, $A_0$ vs $\omega$, from integration of Eq. (3) with $a = 0.45$, showing boundaries between periods 1,2,4,8; two-band chaos $C2$; one-band chaos $C1$; periods 3,6; three-band chaos $C3$; $C1$; and 1:1 and 2:2 jump bifurcations.

which shows a satisfactory agreement with the data by making the usual correspondences $(A_0, \omega) \rightarrow$ (voltage, frequency of drive oscillator). That the theoretical diagram does not show sharply increasing values of drive voltage to cross boundaries on either side of resonance is probably due to use of a constant value $a(q) \rightarrow a$ in Eq. (3). Figures

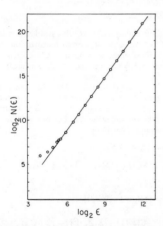

FIG. 22. Plot of $\log_2 \overline{N(\epsilon)}$ vs $\log_2 \epsilon$ for embedding dimension $D = 6$ giving line of slope $d = 2.04 \pm 0.03$, the fractal dimension of the attractor for driven $p$-$n$ junction in chaotic region just below period-3 window. $L = 100$ mH, $R = 77$ $\Omega$, $f = 19.881$ kHz, $V_{os} = 3.105$ V rms, 1N4723 junction.

19 and 21 both show a period-1$\rightarrow$period-1 bifurcation, and a period-2$\rightarrow$period-2 bifurcation. These are observed (Fig. 17) and predicted (Fig. 16) on bifurcation diagrams and are examples of the jump phenomena for driven nonlinear oscillators; cf. Fig. 4.

*Fractal dimension of attractor.*[20] The assumed equations of motion, Eqs. (1), contain three dynamical variables.[21] We suppose the motion can be described in a three-dimensional phase space, neglecting the possibility (Sec. II) that the system has a higher-dimensional memory from diffusive motion of injected charge. We test this supposition below. Since the system has negative divergence of phase-space flow, the attractor must have zero volume and thus must have dimension $d$ less than three; furthermore, to be chaotic the dimension must be greater than two.[12] So we expect the dimension to be a fractal, $2 < d < 3$. It is of interest to measure $d$, and we do so as follows.[22,23] We record a data set of $q = 96\,000$ values of the junction current $I(t)$ using a fast 12-bit (binary digit) analog-to-digital converter and a Digital Equipment Corporation LSI-11/23 minicomputer. By strobing the converter asynchronously with respect to the driving period, we collect data from the whole attractor rather than from a fixed Poincaré section. From the data set $\{A_0, A_1, \ldots, A_n, \ldots, A_q\}$ we construct $q$ vectors $\mathbf{B}_n = (A_n, \ldots, A_{n+D-1})$ in a $D$-dimensional phase space, the "embedding" space. We measure the number of points on the attractor $N(\epsilon)$ which are contained in a $D$-dimensional hypersphere of radius $\epsilon$ centered on a particular $\mathbf{B}_n$. One expects scaling of the form

$$N(\epsilon) \propto \epsilon^d ,$$

where $d$ is the attractor dimension. Thus a plot of

$\log_{10}\overline{N(\epsilon)}$ versus $\log_{10}\epsilon$ is expected to have a slope $d$, where $\overline{N(\epsilon)}$ is the average for hyperspheres centered on many different $\mathbf{B}_n$. This procedure can be carried out for consecutive values of $D = 2, 3, 4, \ldots$, to ensure that the embedding dimension is chosen sufficiently large (important if dimension of phase space is not known) and to discriminate against high-dimensional stochastic noise, not of deterministic origin. The resulting plot for $D = 6$ is shown in Fig. 22 for the system of Fig. 17 with $V_{os}$ set for a one-band chaotic attractor just below the period-3 window. From the slope we find $d = 2.04 \pm 0.03$. The same slope was found for $D = 4$, i.e., the slope converged for $D \geq 4$. Data were also taken for a Poincaré section by strobing the converter synchronously with the drive; these data gave a value $d_{PS} = 1.06 \pm 0.02$, less by unity than $d$, as expected. In summary, these fractal dimension measurements show that $2 < d < 3$ and are consistent with Eqs. (1) and (3). However, they do not exclude the possibility that the system has a higher-dimensional phase space, since sufficient dissipation could reduce $d$ to the value measured.

To summarize this section on a single-junction resonator we compare various theoretical models in their ability to predict observed behavior. The simplest, the logistic map, Eq. (6), while yielding a bifurcation diagram superficially like those observed, cannot explain hysteresis, period adding, nor the fractal structure. The Henon map, Eq. (7), with a Jacobian of $J \sim 0.1$, can model the return map at moderate drive, the bifurcation diagram with hysteresis, but not the period adding. The two-dimensional (2D) map of Eq. (11), of form tailored to the junction characteristics, can model period adding. Differential equations, Eqs. (1) and (3), with more parameters, seem to be the best models and can even model the phase diagram in parameter space. Brorson et al.[4] have accurately modeled behavior using an equation similar to Eq. (3) with ten measured input parameters.

## IV. TWO COUPLED JUNCTION OSCILLATORS

### A. Introduction

Having well characterized theoretically and experimentally the single driven $p$-$n$ junction resonator, one is prepared to predict and to observe the behavior when two or more are coupled together and driven. Coupled nonlinear oscillators or modes are central to the understanding of extended systems, e.g., a line of lattice oscillators or coupled plasma-wave modes. A system of two oscillators has the possibility to make a Hopf bifurcation to a second incommensurate frequency and follow a quasiperiodic route to chaos, in addition to period doubling and intermittency. Such routes, first discussed by Ruelle and Takens,[24] are not yet really understood.

*Equations of motion; ODE model.* It is possible to couple two identical junction resonators in various ways and we discuss only these two. (i) Resistive coupling: in Fig. 1 connect $A$ to $B$; the two resonators are coupled through their currents which flow in a common coupling resistance $R_C$ to the driving oscillator. (ii) Line coupling: connect $B$ to $D$ and set $R_C = 0$; the driving oscillator

drives the lower resonator, which excites the upper resonator in a configuration modeling a nonlinear transmission line.

For resistive coupling, the coupled equations of motion obtained from Kirchoff's laws and Eqs. (3) and (4) are

$$\ddot{q}_1 + \beta[b\dot{q}_1 + r(\dot{q}_1 + \dot{q}_2)] - f(q_1) = A_0\sin(\omega t) , \quad (12a)$$

$$\ddot{q}_2 + b\dot{q}_2 + r(\dot{q}_1 + \dot{q}_2) - f(q_2) = A_0\sin(\omega t) , \quad (12b)$$

where $r \equiv R_C/L$ is the coupling coefficient, $b(q) = (1/C(q))(\partial I_D/\partial V) \approx b$, $b + r = a$ in Eq. (3), $f(q)$ is given by Eq. (4b), and a small term $rI_d(q_2)$ is neglected in Eq. (12a) and a term $rI_d(q_1)$ neglected in Eq. (12b). The factor $\beta \lesssim 1$ is introduced to take into account small differences in the two $p$-$n$ junctions. From the form of Eq. (12) we see that the coupling is through the currents via the common resistance $R_C$.

For line coupling the coupled equations are

$$\ddot{q}_1 + a\dot{q}_1 - 2f(q_1) + f(q_2) = A_0\sin(\omega t) , \quad (13a)$$

$$\ddot{q}_2 + a\dot{q}_2 - f(q_2) + f(q_1) = 0 , \quad (13b)$$

where $a$ and $f(q)$ are given by Eq. (4). The coupling is through the $f(q)$ term which is essentially the potential across the junction, corresponding to the restoring force.

*Map models.* In the same way that a two-dimensional map [Eq. (8) and Eqs. (11)] was used to model a single junction, including its specific characteristics, we now model two coupled junctions by taking two two-dimensional maps and adding a simple linear coupling term to each:

$$z_{n+1} = f(\lambda, z_n) - y_n + CZ_n , \quad (14a)$$

$$y_{n+1} = Jz_n , \quad (14b)$$

$$Z_{n+1} = f(\lambda', Z_n) - Y_n + C'z_n , \quad (14c)$$

$$Y_{n+1} = JZ_n . \quad (14d)$$

For simplicity, choose $|C'| = C$; this still leaves open two choices, $C' = C$ or $C' = -C$. From linearization about a bifurcating fixed point we find, for $C' = C$, real eigenvalues of the matrix and a period-doubling bifurcation initially. For $C' = -C$, the eigenvalues are complex, leading to a Hopf bifurcation initially. The experiments below find that resistive coupling gives first a period-doubling bifurcation, while line coupling gives a Hopf bifurcation first. We take the following specific form of Eq. (14) to model two coupled junctions:

$$z'_{n+1} = \gamma[z_n + 1 - S(z_n)]$$
$$-S(A\{1 - [S(z_n)]^2\}) - y_n + CZ_n , \quad (15a)$$

$$y_{n+1} = Jz_n , \quad (15b)$$

$$Z_{n+1} = \gamma[Z_n + 1 - S(Z_n)]$$
$$-S(bA\{1 - [S(Z_n)]^2\}) - Y_n + C'z_n , \quad (15c)$$

$$Y_{n+1} = JZ_n , \quad (15d)$$

$$S(z) = 0.5[z + (z^2 + 0.1)^{1/2}] \quad (15e)$$

with $b \approx 1$ an asymmetry parameter; $C' = +C$ for resistive coupling; $C' = -C$ for line coupling; $A$ is control parameter; and $\gamma < 1$ so that map is globally stable and attracting (typically $\gamma = 0.5$ for resistive coupling and $\gamma = 0.85$ for line coupling).

*Coupled logistic maps.* A simpler map model is obtained from Eqs. (15) for very dissipative systems by taking $J = 0$ and a simple quadratic form for the nonlinearity; this yields two linearly coupled logistic maps

$$z_{n+1} = \lambda z_n (1 - z_n) + \epsilon y_n , \tag{16a}$$

$$y_{n+1} = \lambda y_n (1 - y_n) + \epsilon' z_n , \tag{16b}$$

where $\lambda$ is the drive parameter and $\epsilon$ is the coupling parameter. This two-dimensional map has been studied in detail by several authors,[25–29] particularly for the case $\epsilon = \epsilon'$. A phase diagram in $(\lambda, \epsilon)$ parameter space has been computed[25,28] showing the domains of period-doubling and symmetry-breaking bifurcations, Hopf bifurcation to quasiperiodicity, regions of entrainment or locking to a rational ratio of the two frequencies, and chaos. For all values of $\epsilon = \epsilon'$, one first finds period doubling to period 2, then a Hopf bifurcation as $\lambda$ is increased. Other generic behavior studied includes oscillation of the torus,[30] crises,[29] and intermittency between locked and quasiperiodic states. The chaotic attractors have a somewhat characteristic appearance of folded rugs or strange animals.

*Sine circle map.* This map has been used to model the

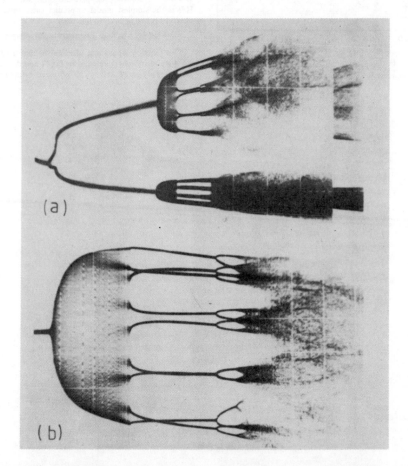

FIG. 23. (a) Bifurcation diagram $[I_n]$ vs $V_{os}$ for two identical resistively coupled junctions, showing period doubling, Hopf bifurcations, and chaos. (b) Enlarged bifurcation diagram $[V_n]$ vs $V_{os}$ of upper center section of (a), showing Hopf bifurcation, entrainment, doubling, chaos. $L = 100$ mH, $R_C = 1200$ Ω, $f = 27.127$ kHz, 1N4723 junction.

480

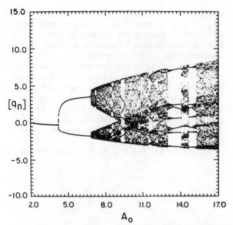

FIG. 24. Computed bifurcation $[q_n]$ vs $A_0$ from Eq. (12) with $\beta = \frac{19}{20}$, $b = 0.45$, $r = 0.6$, $\omega = 1.5$.

phase motion of two coupled oscillators:

$$\theta_{n+1} = \theta_n + \Omega - (\kappa/2\pi)\sin(2\pi\theta_n) , \qquad (17)$$

where the parameter $\Omega$ is the frequency ratio in the absence of the last term and $\kappa$ is the drive (or coupling) parameter. This equation has been abundantly studied.[31,32] In the phase diagram in $\kappa,\Omega$ parameter space, for $\kappa < 1$, there are entrainment horns at all rational values of $\Omega$, and quasiperiodic orbits in between. These "Arnol'd tongues"[33] merge at $\kappa = 1$ at onset of chaos, accompanied by hysteresis and intermittency; period doubling can occur at $\kappa > 1$. Universal behavior at $\kappa = 1$ includes scaling of the power spectra for rotation numbers equal to the reciprocal of the golden mean;[31] and a fractal dimension $D = 0.87\ldots$ for the quasiperiodic orbit set.[32] Equation (17) is the simplest model to predict entrainment horns, observed below in coupled junctions.

### B. Resistive coupling: Experiments and interpretation

Two junction resonators, identical to that used in Fig. 17, were resistively coupled as in Fig. 1 and driven at frequency $f = f_1 = 27$ kHz. The observed bifurcation dia-

FIG. 25. Measured bifurcation diagram $[I_n]$ vs $V_{os}$ for two identical resistively coupled junctions, $L = 8.2$ mH, $R_C = 100$ $\Omega$, $f = 120$ kHz, 1N4723 junction.

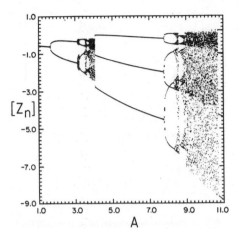

FIG. 26. Bifurcation diagram $[Z_n]$ vs A computed from Eq. (15) for two resistively coupled junctions, with $\gamma = 0.5$, $b = 0.95$, $C' = C = 0.05$, $J = 0.1$.

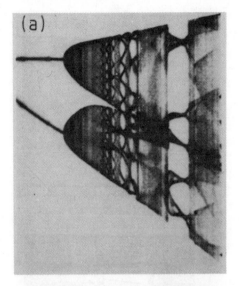

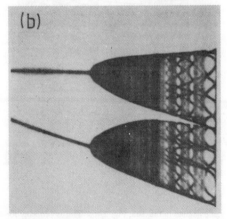

FIG. 27. (a) Observed bifurcation diagram $[I_n]$ vs $V_{os}$ for system of Fig. 25 with coupling increased to $R_C = 1200$ $\Omega$. Compare to model, Fig. 28(a). (b) Expanded view of center section of (a).

gram is shown in Fig. 23(a). After period doubling to $f_1/2$ there occurs a Hopf bifurcation to a second, incommensurate frequency, $f_2 \approx 0.22 f_1$, followed by narrow locked regions [see expanded diagram, Fig. 23(b)] and then a wide locked region with winding number $\rho = f_1/f_2 = \frac{9}{2}$. Then follows period doubling to chaos, an abrupt jump in attractor size, further locking, etc. (not shown). This figure shows the clear distinction in a bifurcation diagram between period doubling and Hopf bifurcations. To compare this data to a model, we use the coupled ODE's, Eqs. (12), to compute the bifurcation diagram of Fig. 24, using $\beta = \frac{19}{20}$, damping constant $b = 0.45$, coupling constant $r = 0.6$, and relative drive frequency $\omega = 1.5$. The model agrees with the data in showing first a period-doubling bifurcation, then a Hopf bifurcation with many narrow lockings. However, this model does not then show a wide locking and period doubling to chaos but rather more lockings, becoming wider at larger $A_0$. It would appear that to find ODE models that give detailed agreement is more difficult for $N = 2$ than $N = 1$ oscillators.

Figure 25 is another experimental bifurcation diagram taken under different experimental parameters corresponding to weaker coupling, $R_C = 100$ $\Omega$. The data show period doubling, Hopf bifurcation, locking, jump to period 4, chaos, period 3, Hopf bifurcation, rough period 6, chaos, crisis (jump), etc. We compare this data to a different model: Fig. 26 shows a bifurcation diagram computed from the iterative map model, Eq. (15), with $\gamma = 0.5$, asymmetry parameter $b = 0.95$, and coupling $C' = +C = 0.05$. The qualitative agreement with Fig. 25 is surprisingly good: The sequence of events is close to that of the data. Both the data and the model show first a period-doubling bifurcation, a Hopf bifurcation, period 4, chaos , and then period 3. The data then show a Hopf bi-

furcation and a period-6 band, whereas the model does not show this Hopf bifurcation but rather a clear period 6.

Figure 27 is a diagram for the same system as in Fig. 25 but with the coupling resistance increased to $R_C = 1200$ $\Omega$. Figure 28(a) is a diagram computed from Eq. (15) with coupling $C' = C = 0.6$; it compares well with the data of Fig. 27(a). Figure 28(b) is a high-resolution expansion of the diagram at $2 \rightarrow 1$ band merge. There are 21 lines resolved, to be compared to approximately the

482

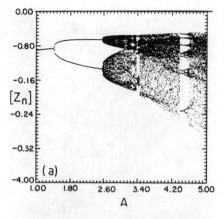

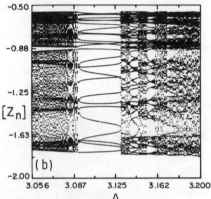

FIG. 28. (a) Bifurcation diagram $[Z_n]$ vs $A$ computed from Eq. (15) with $\gamma=0.5$, $b=0.95$, $J=0.1$, $C=C'=0.6$. (b) Blow-up of $1\rightarrow 2$ band merge region of (a).

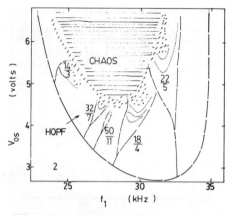

FIG. 29. Observed phase diagram for two resistively coupled junction oscillators: in parameter space, drive voltage $V_{os}$ and drive frequency $f_1$, the heavy line is the boundary of a Hopf bifurcation from $f_1/2$ to quasiperiodicity, with new frequency $f_2$. Entrainment horns are labeled by $P/Q=f_1/f_2$. Period doubling (dotted lines) occurs within horn.

the Hopf bifurcation the frequency ratio $\rho=f_1/f_2$ varies continuously from 4.732 18 (upper left of Fig. 29) to 4.344 42 (upper right). Where $\rho$ tends to a rational number $P/Q$, a point of resonance, there emerges from the boundary an entrainment horn or Arnol'd tongue.[33] There are many other narrow horns not shown. Period doubling to chaos occurs within the horns,[34] as shown. Where two horns overlap there is hysteresis and intermittency between the two attractors, leading to chaos: one can say that this "confusion leads to chaos."[35] The re-

same number in the data, Fig. 27(b). This is probably fortuitous since we do not expect such detailed agreement with the model.

In summary, for two resistively coupled junction resonators we find qualitative agreement with bifurcation diagram data using the ODE model, Eq. (12), and somewhat better agreement with the four-dimensional map model, Eq. (15). We note that bifurcation diagrams computed from coupled logistic maps (Ref. 28, Fig. 2 for coupling $\epsilon=0.06$; Ref. 25, Fig. 4 for coupling $d=0.1$) also bear a qualitative resemblance to our data.

*Phase diagram.* An overview of the behavior of two coupled junction resonators is given by the phase diagram in Fig. 29 in $(V_{os},f_1)$ parameter space (not shown is a bifurcation from period 1 to period 2 along a line similar to that in Fig. 19 for one resonator). Along the boundary of

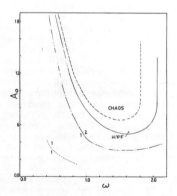

FIG. 30. Phase diagram, $A_0$ vs $\omega$, for two resistively coupled junctions computed from Eq. (12) with $\gamma=19/20$, $b=0.45$, $r=0.6$, showing boundary of 1:1 jump bifurcation, 1:2 period doubling, Hopf bifurcation, and approximate region of chaos.

gions of chaos (however reached: by period doubling, by overlap of horns, by following a "true" quasiperiodic route along an irrational rotation number) are widespread but fall roughly in the shaded region shown. We note that this phase diagram for two driven passive resonators is qualitatively similar to a much more detailed phase diagram for a driven active nonlinear oscillator.[36,37] In both cases the entrainment horns are very roughly modeled by Eq. (17).

To test the ODE model we have used Eqs. (12) to compute the phase diagram of Fig. 30 which shows reasonably well the principal features of the data, Fig. 29, including the boundaries of the period doubling and Hopf bifurcations and the region of chaos.

*Breakup of the torus.* From the five-dimensional phase space of two driven coupled resonators (e.g., $I_1, V_1, I_2, V_2, \theta$) we select the space $(I_2, V_1, \theta)$ to examine experimentally and look at the $(I_2(t), V_1(t))$ phase portrait.

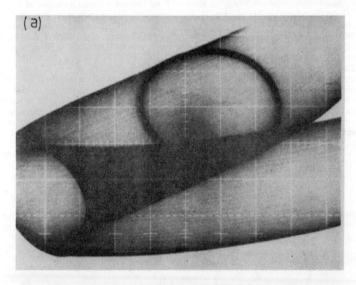

FIG. 31. (a) Phase portrait, $I_2(t)$ vs $V_1(t)$, for two resistively coupled junctions, showing piece of two-loop torus and strobed circular Poincaré section; $V_{os} = 3.2$ V rms. (b) At $V_{os} = 4.8$ V rms torus is broken up, the Poincaré section is the dark "rabbit"-like object. $L = 100$ mH, $R_C = 1200$ $\Omega$, $f = 27.1$ kHz, 1N4723 junction.

484

There is first a single loop which bifurcates to a double loop as the drive voltage is increased; then there occurs a Hopf bifurcation to a double-loop torus. The projection of one loop of this torus onto the $(I_2, V_1)$ plane is shown in Fig. 31(a) for $V_{os} = 3.2$ V rms; the dark circle is a Poincaré section strobed at $t = nT$. At $V_{os} = 4.2$ V rms the torus has broken up: the Poincaré section, Fig. 31(b), resembles a "strange rabbit." We next show the details of the breakup of this circle, a simple graphic view of events on the road to chaos in this system (for the exact same system, Fig. 23 shows the bifurcation diagram, Fig. 34 the power spectra, and Fig. 29 the phase diagram). As $V_{os}$ is increased, we see in the Poincaré sections of Fig. 32 (a) the invariant circle just after the Hopf bifurcation, (b) wrinkling of the circle, (c) more wrinkling, with small folds, (d) frequency locking, $f_1/f_2 = \frac{18}{4}$. In Fig. 33 we see (a) period doubling, (b) nine-band chaos, strange ("rabbit") attractor, (c) folding, (d) more folding, a "folded rug" attractor. There is another similar attractor corresponding to the lower branch of Fig. 23(a).

The models, Eqs. (12) and (15), yield computed Poincaré sections similar to those observed. We also point out the good correspondence between our data and sections computed for two coupled logistic maps. For example, the folded rug of Fig. 33(d) is visually quite similar to the attractor computed by Froyland (Ref. 25, Fig. 5, lower),

by Kaneko [Ref. 26, Fig. 2(f)], and by Hogg and Huberman [Ref. 28, Fig. 7(a)]. The rabbit of Fig. 33(b) is similar to Ref. 25, Fig. 5, upper. The general sequence, Figs. 32 and 33, is also qualitatively represented by Poincaré sections computed by Curry and Yorke[38] for a map of the plane. The sequence is perhaps even more similar to those computed by Kaneko[30] for a two-dimensional delayed logistic map.

*Fractal dimension.* Using the method described in Sec. II we measure a fractal dimension of the attractor under conditions similar to those for Fig. 33(d). For $R_C = 1200$ $\Omega$, $V_{os} = 7.191$ V rms, $f_1 = 29.671$ kHz, we sampled $q = 96\,000$ consecutive values of $I_1(t)$ by strobing asynchronously with reference to the drive period. This data yielded $d = 2.23 \pm 0.04$ in a plot similar to Fig. 22, with embedding dimension $D = 6$. It is not yet clear if anything significant can be said about this value of $d$. If the two oscillators are very strongly coupled, one expects the temporal behavior of $I_1(t)$ to be representative of the whole system, operationally represented by $I_s(t) = I_1(t) + I_2(t)$, so that a measurement of dimension $d_1$ from the time series $I_1(t)$ should yield essentially the same value as $d_s$ from $I_s(t)$. However, if the coupling is reduced to zero, we have complete localization and $I_1(t)$ and $I_2(t)$ have no temporal correlation; one then expects $d_s \approx 2d_1$. Similar ideas apply to a line of $N$ identical os-

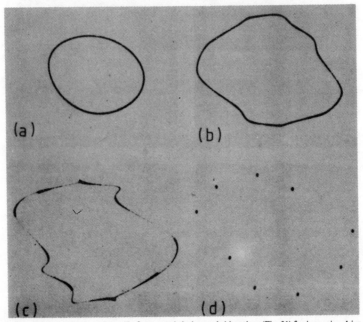

FIG. 32. Sequence of Poincaré sections, $I_2$ vs $V_1$, for two resistively coupled junctions (Fig. 31) for increasing drive voltage $V_{os}$ (V rms): (a) 3.165, smooth circle just after Hopf bifurcation; (b) 3.681, wrinkled circle; (c) 4.028, more wrinkled; (d) 4.190, entrainment (locking) at $f_1/f_2 = 18/4$.

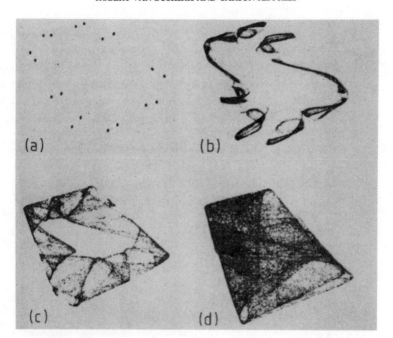

FIG. 33. Sequence of Poincaré sections, $I_2$ vs $V_1$, continued from Fig. 32. (a) $V_{os} = 4.409$, period doubling of locked state; (b) 4.882, strange rabbit attractor; (c) 5.132, folding; (d) 8.958, more folding.

cillators where some localization may occur even if the coupling is nonzero.[39]

*Power spectra.* For each of the values of the drive voltage in the sequence of Figs. 32 and 33 we recorded the power spectrum, shown in Fig. 34. For $V_{os} < 3.1$ V rms the spectrum is a set of sharp lines at $f_1/2, f_1, 3f_1/2$, etc. The new frequency $f_2$ appears after the Hopf bifurcation in Fig. 34(a), together with the combination frequencies $f_1/2 - f_2$, $f_1/2 + f_2$ (not shown), etc., all given by $f_{nm} = nf_1/2 + mf_2$ with $m, n$ positive and negative integers. In Fig. 34(f) the rabbit attractor has appeared and the spectrum has broadband character: onset of chaos, in this instance, by period doubling. In Fig. 34(h) the spectrum is very broadband with sharp peaks at $f_1/2$ and $f_1$ and their harmonics (not shown).

*Oscillations of the torus.* Figure 35 shows two Poincaré sections for increasing drive voltage following Hopf bifurcation just prior to locking at $P/Q = 4/1$. The counter clockwise orbit rapidly approaches the upper right-hand corner, bends left, slows down, and develops damped transverse oscillations. The orbit lingers near points $A$ and $B$, Fig. 35(b), which become stable fixed points. Similar-looking orbits have been computed for two coupled logistic maps [see Ref. 28, Fig. 6(a)]. Insight into the details for a similar case is given by Kaneko,[30] who studied the oscillations for a two-dimensional delayed logistic

map. He attributes the effect to damped oscillation of an unstable manifold of a periodic saddle. Figure 36 is a schematic showing two stable fixed points with manifolds $M_{sa}$ and $M_{sp}$ along the amplitude and phase directions, respectively, and two (unstable) saddle points with manifolds $M_{ua}$ and $M_{up}$. If $M_{up}$ crosses $M_{sp}$ once, it must cross an infinite number of times, hence the oscillations of $M_{up}$. The damping is determined by the eigenvalue $\lambda_{sa}$ of the Jacobian matrix, which is close to $-1$ near the bifurcation point, where the oscillations have maximum amplitude. This model gives a good qualitative explanation of our observations. It is related to heteroclinic crossings in area-preserving maps, but the oscillations in our case are damped.

*Crises of the attractor.*[8] Another example of characteristic behavior of coupled oscillators is shown in Fig. 37. After period doubling and a Hopf bifurcation, the system is entrained at $P/Q = 14/3$, Fig. 37(a); by increasing the drive voltage there is another Hopf bifurcation to 14 "island" attractors; the seven upper islands are shown in Fig. 37(b). As the drive voltage is further increased, these begin to break up, and a crisis ensues: a cyclic collision of the seven attractors with the boundaries that separate the basins of attraction, resulting in a sudden merging into one attractor, Fig. 37(c). This behavior is expected theoretically and has been noted in computations for two

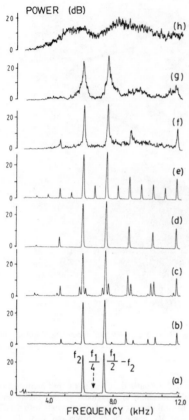

FIG. 34. Power spectra, $P$ in dB vs frequency for two resistively coupled junctions, for same sequences of drive voltage as in Figs. 32 and 33, $V_{os}$ (V rms): (a) 3.165, (b) 3.681, (c) 4.028, (d) 4.190, (e) 4.409, (f) 4.882, (g) 5.132, (h) 8.958.

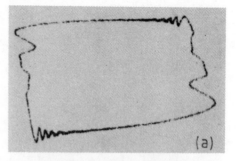

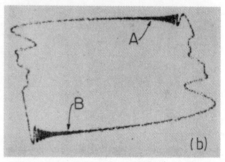

FIG. 35. Poincaré section, $I_2$ vs $V_1$, for two resistively coupled junctions showing oscillation of the torus near period-4 locking. (a) $V_{os} = 1.976$ V rms, (b) $V_{os} = 2.003$. At $V_{os} = 2.045$ points $A$ and $B$ become stable fixed points. $L = 100$ mH, $R_C = 510$ $\Omega$, $f = 27.164$ kHz, 1N4723 junction.

Hopf bifurcation [Fig. 38(h)] is reached; no period doubling occurs. These two types of symmetry, in phase and out of phase, correspond crudely to the two modes of two line-coupled oscillators. Generally, we observe that a Hopf bifurcation can occur only from an out-of-phase state. This is consistent with Kaneko's phase diagram[26] for two coupled logistic maps; see also Refs. 25 and 28 for a similar treatment of the effects of symmetry.

coupled logistic maps (Ref. 29, Fig. 5; Ref. 25, Fig. 6).

*Symmetry.* For two resistively coupled junctions with very weak coupling ($R_C = 83$ $\Omega$) Fig. 38 shows that the two junction waveforms $V_1(t)$ and $V_2(t)$ are in time phase (a) before and (b) after a period-doubling bifurcation, leading to chaos, (c). No Hopf bifurcation is observed. When the coupling is slightly increased ($R_C = 107$ $\Omega$), the waveforms are initially in time phase, Fig. 38(d), then become out of phase just at the period-doubling bifurcation, (e); there follows a Hopf bifurcation, (f); chaos is reached at much higher drive voltages (not shown). For two line-coupled junctions, where the coupling cannot be made very weak, Fig. 38(g) shows that even for low drive voltages the waveforms are out of phase, and remain so as a

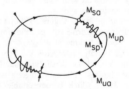

FIG. 36. Two periodic stable points (○) and manifolds $M_{sa}, M_{sp}$; two unstable saddle points (×) and manifolds $M_{ua}, M_{up}$. Damped radial oscillations occur where $M_{up}$ intersects $M_{sp}$ [after Ref. 30 (1984), Figs. 2 and 4].

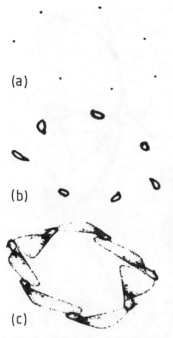

(a)

(b)

(c)

FIG. 37. Poincaré sections, $I_2$ vs $V_1$, for two resistively coupled junctions showing (a) frequency locking ($P/Q = 14/3$) at $V_{os} = 6.152$; there is a second set of seven dots corresponding to the lower branch of the attractor (not shown); (b) $V_{os} = 6.298$, second Hopf bifurcation; (c) $V_{os} = 6.359$, cyclic crisis. $L = 100$ mH, $R_C = 1200$ $\Omega$, $f = 24.46$ kHz, 1N4723 junction.

### C. Line coupling: Experiments and interpretation

Figure 39 is a bifurcation diagram observed for two line-coupled resonators, connected as explained in the caption of Fig. 1. This is analogous to a nonlinear transmission line with inductors in series and $p$-$n$ junctions in shunt.[40] This system displayed first a Hopf bifurcation, then locking, period doubling, chaos, locking, etc., in a quite complex diagram. For comparison, Fig. 40 is a bifurcation diagram computed from the map model, Eq. (15) with $\gamma = 0.8$, $b = 0.95$, and $-C' = C = 0.5$. It shows an overall resemblance to the data including the first Hopf bifurcation and the bifurcation to period 3 and period 6.

Figure 41 shows the breakup of the circle in the $(I_2, V_2)$ Poincaré section as the drive voltage is increased. Figure 42 shows corresponding sections computed from the map model, Eq. (15), with $\gamma = 0.8$, $b = 0.95$, and $-C' = C = 0.5$, $J = 0.1$. The overall agreement is good if one compares the structural features.

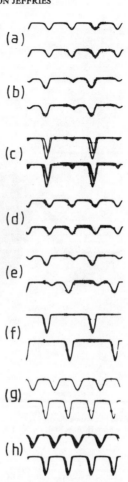

(a)

(b)

(c)

(d)

(e)

(f)

(g)

(h)

FIG. 38. Junction voltage waveforms, $V_1(t)$ and $V_2(t)$, for two coupled junctions for $R_C = 83$ $\Omega$: (a) $V_{os} = 0.513$ V rms; (b) $V_{os} = 0.645$, period doubling; (c) $V_{os} = 1.913$, onset of chaos. For $R_C = 106$ $\Omega$: (d) $V_{os} = 0.519$; (e) $V_{os} = 0.645$, jump to out of phase and period doubling; (f) $V_{os} = 1.641$, Hopf bifurcation from out-of-phase state. For line coupling, (g) $V_{os} = 1.134$; (h) $V_{os} = 1.158$, Hopf bifurcation. $L = 100$ mH, $f = 20$ kHz.

### V. COUPLED OSCILLATORS WITH $N > 2$

For $N = 4$ line-coupled junction resonators we observed the power spectra of Fig. 43 at increasing drive voltage at frequency $f_1$. In Fig. 43(a) the system has made a Hopf bifurcation to a second frequency $f_2$. In Fig. 43(b) a second bifurcation to a third frequency $f_3$ has occurred. In Fig. 43(c) the intensity of $f_3$ is more fully developed.

FIG. 39. Observed bifurcation diagram $[I_n]$ vs $V_{os}$ for two identical line-coupled resonators showing Hopf bifurcation and frequency locking. $L = 8.2$ mH, $R_0 = 70$ $\Omega$, $f = 90$ kHz, 1N4723 junction.

All lines in this quasiperiodic spectrum are fit by the expression for the combination frequency

$$f = m_1 f_1 + m_2 f_2 + \cdots + m_i f_i \qquad (18)$$

with the set of integers $(m_1, \ldots, m_i)$ shown in the figure. Figure 43(a) is fit by two frequencies, and (b)–(d) by three frequencies: $f_1 = 167$ kHz, $f_2 \cong 63.6$ kHz, $f_3 \cong 11.53$ kHz. $f_1$ is set by the drive oscillator, whereas $f_2$ and $f_3$ are determined by the system dynamics and depend on the drive voltage, but this dependence has not been measured. We believe that if any two frequencies $f_i$ and $f_j$ are locked, then the locking ratio $f_i/f_j$ must have integers larger than at least 30 since with our apparatus we could have observed such ratios. Within this error we believe that the three frequencies are incommensurate. We note that as the drive voltage is increased, the spectral intensity at $f_3$ and its combination frequencies is increased, e.g., the line $f_1$-$f_3$ in Fig. 43(c). Figure 43(d) shows onset of chaos: there is the beginning of a broadband line centered

at $f_1$-$f_3$. If there is a fourth frequency, its intensity must be at least 10 dB below that of $f_3$.

For $N = 12$ junction resonators with line coupling, the observed power spectra are shown in Fig. 44, for increasing values of drive voltage. As in all line-coupled systems there is a first Hopf bifurcation to a second frequency $f_2$, then to a third frequency $f_3$, etc. All spectral lines in Fig. 44 can be fit by Eq. (18) extended to four frequencies. Figure 44(a)–(d) require 2, 2, 3, and 4 frequencies, respectively. This conclusion is supported by the direct observation of the following Poincaré sections: Fig. 44(a), a single loop (a section of a 2-torus); Fig. 44(b), a complicated loop (but still a section of a 2-torus); Fig. 44(c), a complicated 2-torus (a section of a 3-torus); Fig. 44(d), an object suggestive of a 2D projection of a 3-torus (itself a section of a 4-torus). Chaos is just beginning to set in for Fig. 44(d). On the whole it was difficult to experimentally find the parameter values $(V_{os}, f_1)$ at which quasiperiodicity with four frequencies was observed.

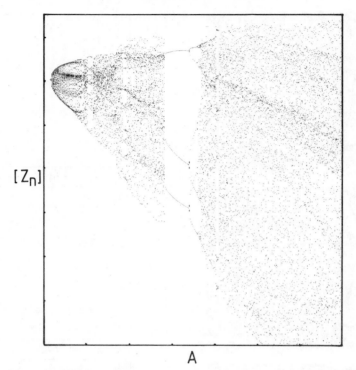

$[Z_n]$

A

FIG. 40. Bifurcation diagram $[Z_n]$ vs $A$ for line-coupled iterative map model, Eq. (15), with $\gamma=0.8$, $b=0.95$, $-C'=C=0.5$, $J=0.1$.

## VI. SUMMARY AND CONCLUSIONS

A driven $p$-$n$ junction resonator is highly nonlinear with an asymmetric weak-strong restoring force, owing to charge storage in forward injection. The system displays a period-doubling cascade to chaos, which is part of a larger period-adding sequence in which the resonator is entrained to successive subharmonics of the drive frequency. To model the effects of dissipation, attractors are computed for various values of $a$ in an exponential force model, Eq. (9); the results, Fig. 5, show a marked dependence of the fractal dimension on $a$. Measured bifurcation diagrams are reasonably similar to those computed from a three-dimensional ODE model [Eq. (3)] and a two-dimensional iterative map model [Eq. (11)] with a form chosen to represent the junction resonator characteristics. The measured phase diagram in parameter space (drive voltage, drive frequency) is similar to that computed from Eq. (3). At low drive voltage the observed return map is similar to that computed from Hénon's map, Eq. (7), with contraction ratio $J \approx -0.1$. Poincaré sections show self-similarity and fractal structure; a fractal dimension $d=2.04\pm0.03$ is measured for the one-band chaotic

attractor just before the period-3 window for a particular set of system parameters.

For two resistively coupled junction resonators we find two-frequency quasiperiodicity. As the drive voltage is increased we observe: period $1\rightarrow2$ doubling, Hopf bifurcation to a second incommensurate frequency, entrainment, additional Hopf bifurcations and/or period doubling, chaos. Bifurcation diagrams are compared to those computed from a coupled ODE model, Eq. (12); and also to an iterative map model, fashioned from coupling two two-dimensional maps, Eq. (15). Qualitative agreement is found; two coupled logistic maps are also found to be a reasonable model. The phase diagram in parameter space (drive voltage, drive frequency) is found to display entrainment horns emanating from the Hopf bifurcation boundary, with period doubling within a horn. The major boundaries in the phase diagram can be understood by computations using Eq. (12). The breakup of the torus is observed in detail, simultaneously in Poincaré sections and in power spectra. The strange attractor is found to be quite similar to that from maps of the plane. A fractal dimension $d=2.23\pm0.04$ was measured for a "fully folded" strange attractor. Other generic behavior reported in-

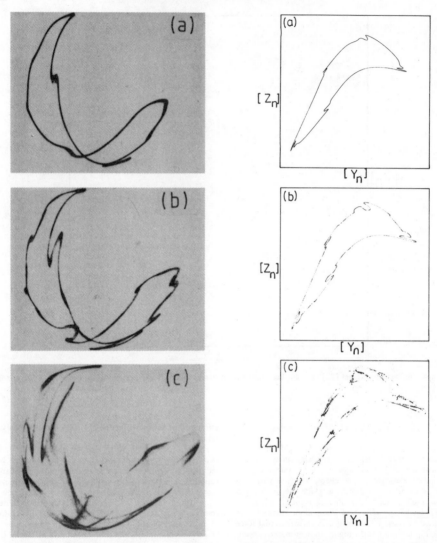

FIG. 41. Observed Poincaré sections, $V_2$ vs $I_2$, for two line-coupled junctions for increased drive voltage, from (a) to (c), showing breakup of the torus; system same as in Fig. 39.

FIG. 42. Computed Poincaré sections of $[Z_n]$ vs $[Y_n]$ computed from map coupled model, Eq. (15), with $\gamma=0.8$, $b=0.95$, $-C'=C=0.5$, $J=0.1$. (a) $A=2.25$, (b) $A=2.29$, (c) $A=2.357$.

cludes oscillators of the torus, cyclic crises of the attractor; and effects of coupling on the symmetry.

For two line-coupled junction resonators we find first a Hopf bifurcation as the drive voltage is increased in contrast to the resistively coupled case. This is found to be in

agreement with the model, Eq. (15), which also explains reasonably the Poincaré sections.

For a line of $N=4$ coupled resonators we find quite complex behavior in bifurcation diagrams; almost any sequence of patterns can occur. Power spectra are fit to

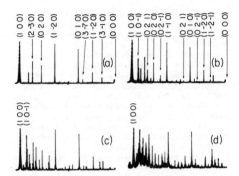

FIG. 43. Quasiperiodic power spectrum (relative dB vertical) observed for $N=4$ identical line-coupled junctions for increased drive voltage, from (a) to (d), with integers $(m_1, m_2, m_3)$ in Eq. (16), which classify the spectrum lines. This system is quasiperiodic with three incommensurate frequencies. $L = 8.2$ mH, $R = 70$ $\Omega$, drive frequency $f_1 = 167$ kHz, 1N4723 junctions.

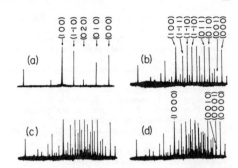

FIG. 44. Quasiperiodic power spectrum (relative dB vertical) observed for $N=12$ identical line-coupled junctions for increased drive voltage, from (a) to (d). The spectrum lines in (a) and (b) can be fit by Eq. (16) with $(m_1, m_2, m_3)$ shown. (c) requires three frequencies and (d) requires four frequencies. $L = 8.2$ mH, $R = 70$ $\Omega$, drive frequency is $f_1 = 220$ kHz, 1N4723 junction.

three-frequency quasiperiodicity. For $N=12$ resonators we find four-frequency quasiperiodicity. No attempts were made to model the detailed behavior for these cases.

In conclusion, we find that the chaotic dynamics of $N=1$ and $N=2$ coupled $p$-$n$ junction resonators can be reasonably understood by tractable models. The driven $p$-$n$ junction is a simple but very useful physical system for further study of quasiperiodicity in high-dimensional systems, e.g., the question of localization. Such studies are now in progress.

## ACKNOWLEDGMENTS

We thank Jose Perez for his help on some aspects of this work, Glenn Held for collaboration on the fractal measurements, and James Crutchfield, Paul Bryant, and James Testa for helpful discussions. One of us (C.D.J.) thanks the Miller Institute for Basic Research in Science for support. This work was supported by the Director, Office of Energy Research, Office of Basic Energy Sciences, Materials Sciences Division of the U.S. Department of Energy under Contract No. DE-AC03-76SF00098.

*Present address: Department of Physics, Harvard University, Cambridge, MA 02138.

[1] P. S. Linsay, Phys. Rev. Lett. **47**, 1349 (1981).

[2] M. J. Feigenbaum, J. Stat. Phys. **19**, 25 (1978).

[3] J. Testa, J. Perez, and C. Jeffries, Phys. Rev. Lett. **48**, 714 (1982); J. Perez, Thesis, 1983, University of California, Berkeley (Lawrence Berkeley Laboratory Report No. LBL-16898).

[4] R. W. Rollins and E. R. Hunt, Phys. Rev. Lett. **49**, 1295 (1982); E. R. Hunt and R. W. Rollins, Phys. Rev. A **29**, 1000 (1984); H. Ikezi, J. S. deGrassie, and T. H. Jensen, ibid. **28**, 1207 (1983); S. D. Brorson, D. Dewey, and P. S. Linsay, ibid. **28**, 1201 (1983); J. Perez and C. Jeffries, Phys. Lett. **92A**, 82 (1982); P. Klinker, W. Meyer-Ilse, and W. Lauterborn, ibid. **101A**, 371 (1984).

[5] C. Jeffries and J. Perez, Phys. Rev. A **26**, 2117 (1982).

[6] J. Perez and C. Jeffries, Phys. Rev. B **26**, 3460 (1982).

[7] C. Jeffries and K. Wiesenfeld, Phys. Rev. A **31**, 1077 (1985).

[8] C. Grebogi, E. Ott, and J. A. Yorke, Phys. Rev. Lett. **48**, 1507 (1982).

[9] C. Jeffries and J. Perez, Phys. Rev. A **27**, 601 (1983); R. W. Rollins and E. R. Hunt, ibid. **29**, 3327 (1984).

[10] R. Van Buskirk, Thesis, 1984, University of California, Berkeley (Lawrence Berkeley Laboratory Report No. LBL-17868).

[11] C. D. Jeffries, Phys. Scr. (to be published).

[12] For reviews of theory see J.-P. Eckmann, Rev. Mod. Phys. **53**, 643 (1981); E. Ott, ibid. **53**, 655 (1981); J. Guckenheimer and P. Holmes, Nonlinear Oscillators, Dynamical Systems and Bifurcations of Vector Fields (Springer, New York, 1983).

[13] See, e.g., S. Wang, Solid State Electronics (McGraw-Hill, New York, 1966), Chap. 6.

[14] N. H. Packard, J. P. Crutchfield, J. D. Farmer, and R. S. Shaw, Phys. Rev. Lett. **45**, 712 (1980).

[15] S. D. Conte and C. de Boor, Elementary Numerical Analysis (McGraw-Hill, New York, 1980).

[16] See, e.g., J. Swift and K. Wiesenfeld, Phys. Rev. Lett. **52**, 705 (1984).

[17] See, e.g., D. W. Jordan and P. Smith, Nonlinear Ordinary Differential Equations (Clarendon, Oxford, 1977), Chap. 7.

[18] See, e.g., J. Guckenheimer and P. Holmes, in Ref. 12, p. 149.

[19] J. L. Kaplan and J. A. Yorke, in Functional Differential Equations and Approximations of Fixed Points, Vol. 730 of Lecture Notes in Math, edited by Heinz-Otto Peitten and Heinz-Otto Walter, (Springer, New York, 1979) p. 228.

[20] See, e.g., J. D. Farmer, E. Ott, and J. A. Yorke, Physica **7D**,

153 (1983); P. Grassberger and I. Procaccia, Phys. Rev. Lett. **50**, 346 (1983).

[21]Although an oscillator has only two dynamical variables, the driving term introduces a third variable, the phase of the driver.

[22]We acknowledge with thanks the collaboration of G. Held in these experiments.

[23]The experimental procedure follows ideas put forth by P. Grassberger and I. Procaccia, in Ref. 20; A. Brandstäter *et al.*, Phys. Rev. Lett. **51**, 1442 (1983); A. Ben-Mizrachi, I. Procaccia, and P. Grassberger, Phys. Rev. A **29**, 975 (1984).

[24]D. Ruelle and F. Takens, Commun. Math. Phys. **20**, 167 (1971).

[25]J. Froyland, Physica **8D**, 423 (1983).

[26]K. Kaneko, Prog. Theor. Phys. **69**, 1427 (1983).

[27]J.-M. Yuan, M. Tung, D. H. Feng, and L. M. Narducci, Phys. Rev. A **28**, 1662 (1983).

[28]T. Hogg and B. A. Huberman, Phys. Rev. A **29**, 274 (1984).

[29]Y. Gu *et al.*, Phys. Rev. Lett. **52**, 701 (1984).

[30]K. Kaneko, Thesis, 1983, University of Tokyo; see also K. Kaneko, Prog. Theor. Phys. **72**, 202 (1984).

[31]S. J. Shenker, Physica 5D, 405 (1982); M. J. Feigenbaum, L. P. Kadanoff, and S. J. Shenker, *ibid.* **5D**, 370 (1982); S. Ostlund, D. Rand, J. Sethna, and E. Siggia, *ibid.* **8D**, 303 (1983).

[32]M. H. Jensen, P. Bak, and T. Bohr, Phys. Rev. Lett. **50**, 1637 (1983); (unpublished).

[33]V. I. Arnold, Trans. Am. Math. Soc., 2nd Ser. **46**, 213 (1965).

[34]L. Glass and R. Perez, Phys. Rev. Lett. **48**, 1772 (1982); M. Schell, S. Fraser, and R. Kapral, Phys. Rev. A **28**, 373 (1983).

[35]Similar qualitative expressions have been voiced. S. Ciliberto and J. Gollub, Phys. Rev. Lett. **52**, 922 (1984): ". . . pattern competition leads to chaos;" and M. H. Jensen, P. Bak, and T. Bohr, in Ref. 32, ". . .chaos is a frustrated response. . . ."

[36]P. Bryant and C. Jeffries, Phys. Rev. Lett. **53**, 250 (1984).

[37]P. Bryant and C. Jeffries, Lawrence Berkeley Laboratory Report No. LBL-16949 (unpublished).

[38]J. H. Curry and J. A. Yorke, Lect. Notes Math. **688**, 48 (1978).

[39]See, e.g., R. W. Walden, P. Kolodner, A. Passner, and C. M. Surko, Phys. Rev. Lett. **53**, 242 (1984).

[40]For early experiments on similar nonlinear transmission lines operated in the nonchaotic regime, see R. Hirota and K. Suzuki, J. Phys. Soc. Jpn. **28**, 1366 (1970), Proc. IEEE **61**, 1483 (1973).

# Nonlinear Dynamics of the Wake of an Oscillating Cylinder

D. J. Olinger and K. R. Sreenivasan

*Mason Laboratory, Yale University, New Haven, Connecticut 06520*
(Received 30 November 1987)

The wake of an oscillating cylinder at low Reynolds numbers is a nonlinear system in which a limit cycle due to natural vortex shedding is modulated, generating in phase space a flow on a torus. We experimentally show that the system displays Arnol'd tongues for rational frequency ratios, and approximates the devil's staircase along the critical line. The "singularity spectrum" as well as spectral peaks at various Fibonacci sequences accompanying quasiperiodic transition to chaos show's that the system belongs to the same universality class as the sine circle map.

PACS numbers: 47.15.Gf, 47.25.Gk

In low-dimensional dynamical systems, detailed predictions have been made for the "universal" features of transition to chaos by period-doubling[1,2] and quasiperiodic[3-6] routes. Experiments in small–aspect-ratio closed-flow systems[7-9] have gone a long way in establishing the validity of these predictions to fluid flows. However, experiments in open-flow systems (those with imposed unidirectional main flow) with little or no confinement have paid heed to these predictions only rarely.[10-13] The best case for showing some conformity with features of nonlinear dynamics is the flow behind circular cylinders.[10,13] Here, we study at low Reynolds numbers the flow behind a circular cylinder oscillating transverse to an oncoming stream, and show that it exhibits some quantitative features of universality.

Briefly, with increasing Reynolds number, the flow behind a stationary cylinder first undergoes a Hopf bifurcation[14] from the steady state to a periodic state characterized by the vortex-shedding mode at a frequency $f_0$, say. We have shown in Ref. 14 that the post-critical state can be modeled by the Landau equation, and determined the Landau constants. For cylinder aspect ratio (that is, the length to diameter ratio) exceeding about 60, details of this bifurcation are independent of the aspect ratio, and the critical Reynolds number (based on the cylinder diameter $D$ and the oncoming velocity) is about 46. For the present measurements, the working fluid was air, and the Reynolds number about 55. A modulation was imposed by our causing the cylinder to oscillate transverse to the main flow at a frequency $f_e$, the amplitude of oscillation being then a measure of the nonlinear coupling between the two modes. The system has two competing frequencies ($f_0$ and $f_e$) yielding two control parameters, $f_e/f_0$ and the nondimensional amplitude of oscillation, $a/D$. Once the external modulation is imposed, we expect $f_0$ to shift to $f_0'$, say. This is similar in spirit to the convection experiments of Refs. 7 and 8, and the well-studied sine circle map.

$$\theta_{n+1} = \theta_n + \Omega - (K/2\pi)\sin(2\pi\theta_n),$$

for which $\Omega$ is the bare winding number (equal to the average shift per iteration in the absence of nonlinear coupling—analogous to $f_e/f_0$), and $K$ is the nonlinearity parameter, comparable to $a/D$. The average shift per iteration in the presence of nonlinear coupling is the dressed winding number, $\omega$, comparable to $f_e/f_0'$. The sine circle map has been studied in recent years as a standard model for the transition from quasiperiodicity to chaos in dynamical systems, and its properties are believed to be universal for any map with a cubic inflection point. For $K < 1$ (subcritical behavior), iterates of the map lock on to rational $\omega$ values (in general different from $\Omega$ for nonzero amplitudes of oscillation) in the Arnol'd tongues[15] which increase in width as $K$ increases. At $K = 1$, the critical line, a universal transition to chaos occurs at a special value of the dressed winding number, $\sigma_G = (\sqrt{5}-1)/2$, the inverse of the golden mean. To observe the universal transition to chaos, it is best to move without phase locking along the line $\omega = \sigma_G$ up to the critical point, this choice being relevant because the irrational number $\sigma_G$ is least well approximated by rationals (containing only ones in the continued-fraction representation). The universal behavior at the critical golden-mean point is observed in the scaled power spectrum and in the self-similar devil's staircase structure along the critical line,[3-7] and the so-called $f(\alpha)$ curve.[16]

Our motivation for studying the oscillating cylinder within this framework is twofold. First, there is evidence[17] of lock-in when $f_e$ is near $f_0$ ($\Omega$ near 1). Our search for additional phase-locked tongues at other rational $\omega$'s constitutes a generalization of this work within the framework of dynamical systems. Second, analogies can be drawn between our flow and other fluid systems (most notably the forced Rayleigh-Bénard system) which have exhibited quantitative experimental results reminiscent of sine circle maps. Establishing this fact in a different, and by all accounts more complex, fluid system has broad implications transcending the immediate measurements.

The cylinder was placed in a wind tunnel of the suction type with double contractions, honeycomb, and

494

VOLUME 60, NUMBER 9          PHYSICAL REVIEW LETTERS          29 FEBRUARY 1988

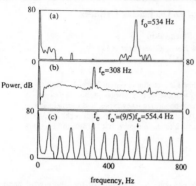

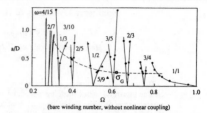

FIG. 1. (a) Power spectral density for the case of natural vortex shedding at $f_0 = 534$ Hz. (b) The corresponding data for the excitation source, measured with a photodiode; $f_e = 308$ Hz. (c) Frequency locking occurring as a result of excitation. The natural shedding frequency disappears in favor of the new peak at $\frac{9}{5} f_e$; peaks appear at other fractions of $f_e$.

FIG. 2. Arnol'd tongues (that is, the locked-in regions) in the wake of the oscillating cylinder. The ordinate is the amplitude of oscillation normalized by the cylinder diameter. About 30 such tongues were noted, but only those with reasonable width are shown. In each tongue, the natural shedding frequency disappears in favor of a rational multiple of the excitation frequency, and the appropriate multiplication factor is shown in each tongue. The critical line (corresponding to the $K = 1$ line in the circle map), as determined by the expected fractal dimension, is shown dashed.

control, no more than 30 such tongues have been identified. (To avoid cluttering, not all of them are shown.) In accordance with predictions for the circle map, these tongues increase in width as $a/D$ increases. The 1/1 tongue is in close agreement with the previously mentioned lock-in region near $f_0$. The dashed line represents the experimentally determined "best fit" critical line found by our determining, for various $\Omega$, the $a/D$ level at which the fractal dimension $D_0$ of the critical line was equal to 0.87 appropriate to the circle map. The dimension $D_0$ was computed with

$$\sum (S_i/S)^{D_0} = 1,$$

where $S$ is the distance between two parent tongues around an irrational winding number and the $S_i$'s ($i = 1, 2$) are the distances between a daughter tongue, constructed according to Farey arithmetic, and each of its parents. All possible parent-daughter combinations of tongues shown in Figs. 2 and 3 were used. A few measurements on the onset of chaos at different $\Omega$ yield-

several damping screens, and was made to oscillate at the desired frequency by the passage of a sinusoidally alternating current through it in the presence of a properly aligned magnetic field. Both the vortex-shedding and modulation frequencies were steady to $\pm 2$ parts in $10^4$. A hot wire, 5 $\mu$m in diameter and 0.6 mm in length, placed approximately $15D$ downstream of the cylinder, $0.5D$ to one side of its mean position, monitored the flow velocity. The hot-wire signal was amplified, digitized by a twelve-bit analog-to-digital converter, and stored in a computer (MASSCOMP 5500) for later analysis. A HP3561A spectrum analyzer was used for real-time analysis. The cylinder oscillation frequency and amplitude were varied over a range of (26–100)% of $f_0$ and (0–200)% of the cylinder diameter. The cylinder diameter varied from 0.03 to 0.09 cm, and its active length was 15 cm; the actual length of the cylinder, stretching outside the wind tunnel, was about 3 times as long. The cylinder always oscillated in its first mode.

Figure 1 highlights the effect of cylinder oscillations on the wake-velocity power spectrum. We note the shift of $f_0$ to $f_0'$ between Figs. 1(a) and 1(c), and the complete suppression of $f_0$ in favor of $f_0'$ in Fig. 1(c). All the principal peaks in Fig. 1(c) (linear combinations of $f_e$ and $f_0'$) are more than 5 orders of magnitude above background noise levels. From several such spectra, one can plot a "phase" diagram showing Arnol'd tongues (Fig. 2). All symbols represent boundaries of the larger tongues shown. The exception is the triangle symbol, which represents the $\frac{5}{9}$ lock-in shown in detail in Fig. 1. The dressed winding numbers in these tongues correspond to rationals constructed according to Farey arithmetic although, because of limitations in experimental

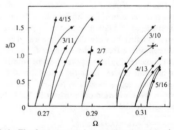

FIG. 3. The fine structure in a small region of the $(a/D, \Omega)$ plane of Fig. 2. Typical experimental uncertainties are shown.

VOLUME 60, NUMBER 9     PHYSICAL REVIEW LETTERS     29 FEBRUARY 1988

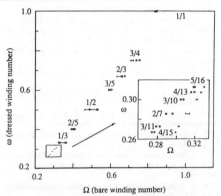

FIG. 4. The devil's-staircase construction with the data of Figs. 2 and 3 along the critical line. Although general pattern is the same as for the circle map, there are some noticeable departures.

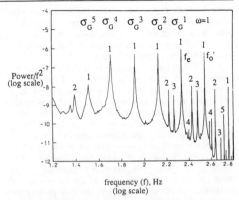

FIG. 5. Frequency-scaled power spectrum for the excited wake at the critical golden-mean point (to within 0.1%).

ed essentially the same critical line. The nonconstant level of $a/D$ along the critical line for small $\Omega$ is not understood, but not surprising considering similar findings in Ref. 7. In Fig. 3 we highlight the fine structure of the phase diagram in a region just below the $\frac{1}{3}$ tongue.

The experimentally determined devil's staircase along the critical line is shown and compared to the predictions for the circle map at $K = 1$ in Fig. 4. The symbols represent the limits of the experimental steps, while the solid lines with vertical limiting bars represent predictions. Although a staircase structure is definitely obtained in experiments, the limitation of the agreement between the measured fine structure and the devil's staircase is obvious especially from the inset enlarging the boxed region.

In Fig. 5 we show a typical scaled power spectrum of the wake velocity at the critical golden-mean point shown by the square symbol in Fig. 2. The dressed winding number $\omega$ is within 0.1% of $\sigma_G$, this being the best control possible in our experiments. The spectrum is averaged over approximately 65000 cycles of the cylinder oscillation frequency. The circle map predicts a self-similar power spectrum (when power is scaled with $f^2$) divided into bands by the principal sum and difference frequencies located at all powers of $\sigma_G$. These peaks are commonly designated as generation-1 peaks. Other generations are created by positive-integer mixing coefficients of various Fibonacci sequences, $f = |j\sigma_G - k|$ — generation 2 by the sequence $(2,2,4,6,\ldots)$, generation 3 by the sequence $(1,3,4,7,\ldots)$, etc; peaks within each generation are of constant amplitude for the circle map. In Fig. 5 we see that the principal peaks fall at powers of $\sigma_G$ down to $\sigma_G^5$. They are nearly of constant amplitude except for $\sigma_G^5$ which falls off. We note that the generation-2 and -3 peaks fall as predicted by the

mixing coefficients within the resolution of our power spectrum. We also see that generation-2 peaks show the constant-amplitude trend, but generation-3 peaks and beyond degrade considerably. Generation-2 and -3 peaks are not present at lower frequencies and higher generations are observed rather rarely.

Finally, from the time series of velocity obtained at the critical golden-mean point, we constructed a pseudo attractor by the usual time-delay methods and obtained Poincaré sections by sampling data at intervals separated by the period of forcing. The resulting Poincaré section was embedded in three dimensions (in which it was nonintersecting in all three views), and a smoothed attractor was obtained by performing averages locally. The data were then used to compute the so-called generalized dimensions[18] by using the standard box-counting methods; in each of the appropriate log-log plots, the

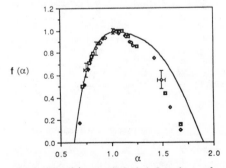

FIG. 6. The $f(\alpha)$ curve obtained via Legendre transform (Ref. 16) of the measured generalized dimensions. Levels of uncertainty are shown by error bars.

VOLUME 60, NUMBER 9     PHYSICAL REVIEW LETTERS     29 FEBRUARY 1988

scale similarity regime extended typically over two decades. The multifractal spectrum [or the $f(\alpha)$ curve[16]] was then obtained via a Legendre transform discussed in Ref. 16. The result is compared in Fig. 6 with the theoretical curve for the circle map.

Even though some departures from the circle-map behavior do exist, we think that the extent of the observed similarity is remarkable. It is not obvious whether these departures are real, or whether they occur because the control of experimental parameters was not as fine tuned as desired. It is known, however, that very small departures from criticality can produce similar behavior.[19,20] As already noted, inherent difficulties in the establishment of the flow made it impossible to control $\sigma_G$ to better than 0.1%. The departures observed in the devil's-staircase construction are of the same order of magnitude as the uncertainties in the flow parameters. Further, the largest departures in the $f(\alpha)$ occur for large $\alpha$, consistent with the relatively large influence of noise on the most sparsely populated (that is, large $\alpha$) regions of the attractor. Apropos of this somewhat unsatisfactory state of affairs, we reiterate that we exercised enormous care in the experiments, and believe that the residual problems of fine control cannot be eliminated without our resorting to unconventional ways of generating such flows; some thoughts on this are currently being investigated.

We acknowledge useful discussions with C. Meneveau and A. Chhabra. The research was supported by a grant from the U.S. Air Force Office of Scientific Research and a University Research Initiative Grant from the Defense Advanced Research Projects Agency.

[1]M. J. Feigenbaum, J. Stat. Phys. **19**, 25 (1979).

[2]M. J. Feigenbaum, Phys. Lett. **74A**, 375 (1979).

[3]S. J. Shenker, Physica (Amsterdam) **5D**, 405 (1982).

[4]M. J. Feigenbaum, L. P. Kadanoff, and S. J. Shenker, Physica (Amsterdam) **5D**, 370 (1982).

[5]S. Ostlund, D. Rand, J. Sethna, and E. Siggia, Physica (Amsterdam) **8D**, 303 (1983).

[6]D. Rand, S. Ostlund, J. Sethna, and E. Siggia, Phys. Rev. Lett. **49**, 132 (1982).

[7]J. Stavans, F. Heslot, and A. Libchaber, Phys. Rev. Lett. **55**, 596 (1985).

[8]M. H. Jensen, L. P. Kadanoff, A. Libchaber, I. Procaccia, and J. Stavans, Phys. Rev. Lett. **55**, 2798 (1985).

[9]A. P. Fein, M. S. Heutmaker, and J. P. Gollub, Phys. Scr. **T9**, 79 (1985).

[10]K. R. Sreenivasan, in *Frontiers of Fluid Mechanics,* edited by S. H. Davis and J. L. Lumley (Springer-Verlag, Berlin, 1985), p. 41.

[11]K. R. Sreenivasan and R. Ramshankar, Physica (Amsterdam) **23D**, 246 (1986).

[12]K. R. Sreenivasan and P. J. Strykowski, in *Turbulence and Chaotic Phenomena in Fluids,* edited by T. Tatsumi (North-Holland, Amsterdam, 1984), p. 191.

[13]G. Schewe, Phys. Lett. **109A**, 47 (1985).

[14]K. R. Sreenivasan, P. J. Strykowski, and D. J. Olinger, in *American Society for Mechanical Engineers Forum on Unsteady Flow Separation,* edited by K. N. Ghia, Fluids Engineering Division Vol. 52 (American Society for Mechanical Engineers, New York, 1987), p. 1.

[15]M. H. Jensen, P. Bak, and T. Bohr, Phys. Rev. A **30**, 1960 (1984).

[16]T. C. Halsey, M. H. Jensen, L. P. Kadanoff, I. Procaccia, and B. I. Shraiman, Phys. Rev. A **33**, 1141 (1986).

[17]G. H. Koopman, J. Fluid Mech. **28**, 501 (1967).

[18]H. G. E. Hentshel and I. Procaccia, Physica (Amsterdam) **8D**, 435 (1983).

[19]A. Arneodo and M. Holschneider, Phys. Rev. Lett. **58**, 2007 (1987).

[20]J. A. Glazier, G. Gunaratne, and A. Libchaber, Phys. Rev. A **37**, 504 (1988).

Reprinted from Accounts of Chemical Research, Vol. *20*, Page 436, December, 1987
Copyright © 1987 by the American Chemical Society and reprinted by permission of the copyright owner.

# Chemical Chaos: From Hints to Confirmation

F. ARGOUL, A. ARNEODO,[†] P. RICHETTI, and J. C. ROUX

*Centre de Recherche Paul Pascal, Université de Bordeaux I, Domaine Universitaire, 33405 Talence Cedex, France*

HARRY L. SWINNEY*

*Department of Physics and the Center for Nonlinear Dynamics, The University of Texas, Austin, Texas 78712*

*Received March 16, 1987 (Revised Manuscript Received August 25, 1987)*

## 1. Introduction

The term *chaotic*, as it is now widely used, describes nonperiodic behavior that arises from the nonlinear nature of *deterministic* systems, not noisy behavior arising from random driving forces.[1,2] Recent experiments on diverse nonlinear systems, including fluid flows and nonlinear electrical circuits, have revealed chaotic dynamics similar to that found in theoretical analyses. The intrinsically nonlinear properties of chemical kinetics suggest the possibility of chaos in chemical systems,[3] but there is a healthy skepticism regarding the actual existence of chaos in real well-controlled chemical reactions; for example:

"There certainly are experimental systems which exhibit 'chaotic' behavior in spite of heroic measures to control all recognized parameters. Such behavior does not prove that the chaos is inherent in the mechanism itself rather than due to unavoidable stochastic fluctuations."[4]

"In realistic models of the B. Z. [Belousov–Zhabotinskii] reaction, numerical computations, however, reveal only periodic patterns in both the continuous and discrete models."[5]

"...it is still an open question whether chaos arising from an homogeneous chemical mechanism has been obtained experimentally or whether it comes from the imperfect control of external features."[6]

Thus, it seems worthwhile to examine carefully the experimental and theoretical evidence for the existence of chemical chaos—that is the goal of this paper.

There is no question about the existence of *oscillations* in chemical reactions—many oscillating chemical reactions have been discovered in recent years.[7] Studies of oscillating chemical reactions have focused primarily on the Belousov–Zhabotinskii (BZ) reaction,[8] in which bromate ions are reduced in an acidic medium by an organic compound (usually malonic acid) with or without a catalyst (usually cerous and/or ferrous ions). The mechanism of the BZ reaction was elucidated[9] in 1972 and elaborated later,[10] and it is generally accepted in spite of some recent discussions. Research on rate-

Argoul, Richetti, and Roux are chemists at the Centre de Recherche Paul Pascal, which is a chemical laboratory that has for more than a decade pioneered in the study of chemical oscillations and chaos. Argoul and Richetti, who earned doctoral degrees from the Université de Bordeaux in 1986 and 1987, respectively, performed many of the recent experiments and analyses that are described in this Account. Arneodo, a theoretical physicist at the Université de Nice, switched fields from particle physics to dynamical systems in the early 1980s, and Swinney, an experimental physicist at the University of Texas, began to study instabilities in fluid dynamics in the mid-1970s. The French–Texas collaboration was initiated in 1980–1981, when Roux spent a year at Austin as a visiting scientist in the nonlinear dynamics research program in the physics department. During that visit Roux discovered the now much-studied Texattractor shown in Figure 2.

determining steps indicated that many of the more than 20 species identified in the reaction were slaved in their time dependence to that of a few species. Hence, it was possible to develop a skeletal model, the "Oregonator", which had only three species.[11,12] This model (and another version with seven species[13]) was shown to be sufficient to reproduce qualitatively the main features of the BZ reaction known a decade ago, namely, bistability, excitability, and oscillations.

Thus, the theoretical understanding as well as experimental knowledge of the BZ reaction was far ahead of that for other oscillating chemical reactions. Therefore, when Ruelle[14] suggested in 1973 that chemical reactions might exhibit nonperiodic behavior, chemists turned naturally to the BZ reaction. The results dealing with nonperiodic behavior come primarily from studies of this reaction, and this Account will deal only with them.

## 2. First Qualitative Findings

Schmitz, Graziani, and Hudson[15] were the first to report observations of chaos in a chemical reaction. They conducted an experiment on the BZ reaction in a continuous flow stirred tank reactor, where the flow

[†]Permanent address: Laboratoire de Physique Théorique, Université de Nice, Parc Valrose, 06034 Nice Cedex, France.

(1) See, for example, the following monographs: Bergé, P.; Pomeau, Y.; Vidal, C. *Order within Chaos*; Wiley: New York, 1986; Hermann: Paris, 1984. Thompson, J. M. T.; Stewart, H. B. *Nonlinear Dynamics and Chaos*; Wiley, New York, 1986. Schuster, H. G. *Deterministic Chaos*; Physik-Verlag: Weinheim, 1984.

(2) See, for example, the following collections of articles: Cvitanovic, P., Ed. *Universality in Chaos*; Hilger: Bristol, 1984. Holden, A. V., Ed. *Chaos*; Manchester University Press: Manchester, 1986. Bai-Lin, Hao, Ed. *Chaos*; World Scientific: Singapore, 1984. Shlesinger, M. F.; Cawley, R.; Saenz, A. W.; Zachary, W., Eds. *Perspectives in Nonlinear Dynamics*; World Scientific: Singapore, 1986.

(3) For an earlier discussion of chemical chaos, see: Swinney, H. L.; Roux, J. C. In *Nonequilibrium Dynamics in Chemical Systems*; Vidal, C., Pacault, A., Eds.; Springer: Berlin, 1984; p 124.

(4) Noyes, R. M. In *Stochastic Phenomena and Chaotic Behavior in Complex Systems*; Schuster, P., Ed.; Springer: Berlin, 1984; p 107.

(5) Schwartz, I. B. *Phys. Lett. A* 1984, *102*, 25.

(6) Gray, P.; Scott, S. K. *J. Phys. Chem.* 1985, *89*, 22.

(7) Field, R. J., Burger, M., Eds. *Oscillations and Travelling Waves in Chemical Systems*; Wiley: New York, 1985.

(8) See, for example: Field, R. J.; Noyes, R. M. *Acc. Chem. Res.* 1977, *10*, 214.

(9) Field, R. J.; Körös, E.; Noyes, R. M. *J. Am. Chem. Soc.* 1972, *94*, 8649.

(10) Edelson, D.; Field, R. J.; Noyes, R. M. *Int. J. Chem. Kinet.* 1975, *7*, 417.

(11) Field, R. J.; Noyes, R. M. *J. Chem. Phys.* 1974, *60*, 1877.

(12) Field, R. J. *J. Chem. Phys.* 1975, *63*, 2284.

(13) Showalter, K.; Noyes, R. M.; Bar-Eli, K. *J. Chem. Phys.* 1978, *69*, 2514.

(14) Ruelle, D. *Trans. N.Y. Acad. Sci.* 1973, *35*, 66.

(15) Schmitz, R. A.; Graziani, K. R.; Hudson, J. L. *J. Chem. Phys.* 1977, *67*, 3040.

498

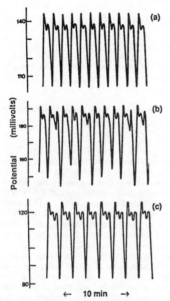

**Figure 1.** Time series records of the bromide ion electrode potential observed for periodic states that have (a) one large and one small oscillation per period and (c) one large and two small oscillations per period. (b) A chaotic state observed for conditions between those that yielded the periodic states (a) and (b). The residence time $\tau$ (the reactor volume divided by the total flow rate) was 0.104, 0.098, and 0.097 h in (a)–(c), respectively. From ref 16.

rate of the feed chemicals was maintained constant and the reaction behavior was monitored with bromide ion specific and platinum wire electrodes. For some range in flow rate, periodic oscillations in the concentrations were observed, as illustrated in Figure 1a. (This figure is from one of their later experiments.[16]) However, for another range in flow rate, chaotic (irregular) oscillations were observed, as illustrated in Figure 1b. The chaotic behavior persisted over about a 10% range in flow rate before there was a transition back to periodic oscillations.

Subsequent experiments on the BZ reaction[16-26] have shown that *periodic–chaotic* sequences are common. In these sequences two periodic states lie to either side (in

(16) Hudson, J. L.; Hart, M.; Marinko, D. *J. Chem. Phys.* **1979**, *71*, 1601.
(17) Rössler, O. E.; Wegmann, K. *Nature (London)* **1978**, *271*, 89.
(18) Wegmann, K.; Rössler, O. E. *Z. Naturforsch., A* **1978**, *33*, 1179.
(19) Bachelart, S. Bordeaux Thesis, 1981.
(20) Vidal, C.; Roux, J. C.; Rossi, A.; Bachelart, S. *C.R. Seances Acad. Sci., Ser. C* **1979**, *289*, 73.
(21) Vidal, C.; Roux, J. C.; Bachelart, S.; Rossi, A. *Ann. N.Y. Acad. Sci.* **1980**, *357*, 377.
(22) Vidal, C.; Bachelart, S.; Rossi, A. *J. Phys. (Les Ulis, Fr.)* **1982**, *43*, 7.
(23) Roux, J. C.; Turner, J. S.; McCormick, W. D.; Swinney, H. L. In *Nonlinear Problems: Present and Future*; Bishop, A. R., Campbell, D. K., Nicolaenko, B., Eds.; North-Holland: Amsterdam, 1982; p 409.
(24) Roux, J. C.; Swinney, H. L. In *Nonlinear Phenomena in Chemical Kinetics*; Pacault, A., Vidal, C., Eds.; Springer: Berlin, 1981; p 33.
(25) Turner, J. S.; Roux, J. C.; McCormick, W. D.; Swinney, H. L. *Phys. Lett. A* **1981**, *85*, 9.
(26) Simoyi, R. H.; Wolf, A.; Swinney, H. L. *Phys. Rev. Lett.* **1982**, *49*, 245.

flow rate) of a chaotic regime. Typically, one of the periodic states has $m$ oscillations per period and the other has $m + 1$ oscillations per period, and the chaotic state that is bracketed by these two periodic states appears to be a random mixture of the two periodic states. Figure 1 illustrates this behavior: the chaotic state in (b) appears to be a mixture of periodic states with two oscillations per period (Figure 1a) and three oscillations per period (Figure 1c).

In the early experiments the term chaotic was used when (1) the time series looked irregular,[15,17,18] (2) the power spectra of the concentration time series contained broad-band noise that was well above the instrumental noise level found for the periodic states,[21,22] and (3) the autocorrelation function of the concentration decayed to zero for large times.[19,21] However, these measures could characterize random noise as well as chaos. In fact, the observed irregular behavior could plausibly be interpreted as arising from random fluctuations in the flow rate or other control parameters, which would result in a random switching of the system from one to another of the adjacent periodic states. Even though the range in flow rate in which the irregular behavior was observed appeared to be large compared to fluctuations in the pumping rate, this argument cast reasonable doubt on the existence of deterministic nonperiodic behavior in a well-controlled chemical system.

These experimental findings motivated numerical studies of models of the BZ reaction. However, these simulations yielded sequences involving only periodic states, but these states were found to be very similar to the ones observed in the periodic–chaotic sequence (see, e.g., ref 13). No chaos was observed in these simulations, at least through a visual inspection of the time series. (In section 4.2 we shall see the full significance of this remark.) Thus, a reasonable conclusion, reached in 1978, was that[13] "the difference between experiments and simulations suggests that the chaotic behavior observed experimentally may result from fluctuations too small to measure in any other way." Later work appeared to support this conclusion.[4-6,27,28]

### 3. Dynamical Systems Theory: Hints for Identifying Low-Dimensional Deterministic Chaos

Theoretical and numerical studies of deterministic chaos were a rapidly growing area of nonlinear physics by 1980, and the knowledge gained there provided new tools for experimentalists to use in the analysis and understanding of nonperiodic data.[1-3] We will describe two ideas that have proved to be especially fruitful.

**3.1. Analysis of Experimental Data: Construction of Phase Space Portraits.** Any dynamical system at an instant of time can be described by a single point in an appropriate multidimensional phase space. The temporal evolution of the system is then given by the trajectory of that point in phase space. Periodic behavior is given by a closed curve called a limit cycle. A chaotic state is described by an irregular trajectory, called a *strange attractor*.

For a chemical reaction with $n$ species the $n$-dimensional phase space coordinates could be the concentrations of the $n$ species. The measurement of the time

(27) Ganapathisubramanian, N.; Noyes, R. M. *J. Chem. Phys.* **1982**, *76*, 1770.
(28) Rinzel, J.; Schwartz, I. B. *J. Chem. Phys.* **1984**, *80*, 5610.

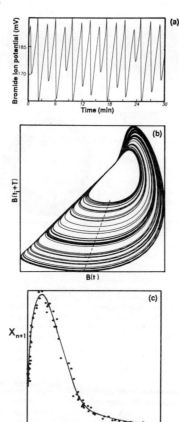

**Figure 2.** Graphs showing the analysis of chaotic time series data. (a) Bromide ion electrode potential time series $B(t)$. (b) A two-dimensional projection of the phase space attractor constructed from the time series $B(t+T)$ vs $B(t)$, where $T = 8.8$ s. (c) A one-dimensional map for the strange attractor shown in (b), constructed by plotting as ordered pairs $[X_n, X_{n+1}]$ the successive values of the ordinate $B(t+T)$ in the portrait when the orbit crosses the dashed line. From ref 33.

dependence of the concentrations of all $n$ chemical species would be an extremely difficult task, but fortunately the application of a 1936 theorem[29] makes it possible to construct a multidimensional phase space portrait from time series measurements of a *single* variable, $B(t)$. The coordinates of a point in phase space are obtained from time delay values of the original time series:[30,31] $[B(t), B(t+T), B(t+2T), ...]$, where the time delay $T$ is arbitrary for noiseless data. For real (i.e., noisy) data there is an optimum choice of $T$, as discussed by Fraser and Swinney;[32] the optimum delay is typically one-tenth to one-half the mean orbital period.

The time delay method was used to construct phase portraits for data obtained in Texas by Roux et al.[23-25] for another periodic–chaotic sequence, observed for similar chemical concentrations but much lower flow rates than in the experiments of Hudson et al.[16] Time series data for a chaotic state are shown in Figure 2a, and a two-dimensional projection of a phase space attractor constructed from these data is shown in Figure 2b. If the attractor is considered in three rather than two dimensions, the intersection of the orbits with a plane approximately normal to the orbits yields a set of points that, within the experimental resolution, lie along a smooth curve. Such a *Poincaré section* demonstrates the low-dimensional nature of this chaotic state—the orbits lie approximately on a two-dimensional sheet.

Further insight into the dynamics can be achieved by constructing a one-dimensional map: let the successive intersections of the ordinate $B(t+T)$ of the orbits with the dashed line in Figure 2b have values called $X_1$, $X_2$, ..., $X_n$, $X_{n+1}$, .... A plot of $X_{n+1}$ vs $X_n$ is shown in Figure 2c: the points fall on a smooth curve, a one-dimensional map. Thus, even though the behavior is nonperiodic and has a power spectrum with broad-band noise, the system is nevertheless completely (within the experimental resolution) *deterministic!*—for any $X_n$, the map gives the next value, $X_{n+1}$.

A hallmark of a strange attractor is exponential separation of nearby points on the attractor. Since a point on an attractor represents the entire physical system, exponential separation of nearby points on chaotic attractors means that systems that are initially nearly identical will inevitably evolve differently at long times. Hence, even though chaotic behavior is deterministic, long-term prediction of the state of the system is impossible! The quantity that characterizes the long-term separation rate of nearby points is the largest *Lyapunov exponent*; it is negative for a time-independent state of a system, zero for a periodic or multiperiodic state, and positive for a chaotic state.[1-3] A method has been developed for computing the largest Lyapunov exponent for time series data, and by use of this method the value of the exponent for the data in Figure 2 was found to be positive.[33,34] The exponential separation of nearby points cannot continue indefinitely (for example, none of the chemical concentrations can become infinite); therefore, the attracting set for strange attractors always has many folds. This folding, which results in a fractal (noninteger) dimension of the attractors, was directly observed for the data in Figure 2.[33]

**3.2. Universal Dynamics.** According to dynamical systems theory there are certain routes from regular to chaotic dynamics that are common in diverse systems. Some of these *universal* routes have been observed in recent experiments[1,2]—studies of convection, semiconductors, lasers, etc.—and these routes have also been found in chemical experiments.[3]

The best known and best understood route to chaos is through period doubling: the period of oscillation successively doubles in an infinite sequence of transitions that occurs as a control parameter is varied. The distance in control parameter between successive

(29) Whitney, H. *Ann. Math.* **1936**, *37*, 645.
(30) Packard, N. H.; Crutchfield, J. P.; Farmer, J. D.; Shaw, R. S. *Phys. Rev. Lett.* **1980**, *45*, 712.
(31) Takens, F.; *Lect. Notes Math.* **1981**, *No. 898*, 366.
(32) Fraser, A. M.; Swinney, H. L. *Phys. Rev. A* **1986**, *33*, 1134.

(33) Roux, J. C.; Simoyi, R. H.; Swinney, H. L. *Physica D (Amsterdam)* **1983**, *8*, 257.
(34) Wolf, A.; Swift, J. B.; Swinney, H. L.; Vastano, J. A. *Physica D (Amsterdam)* **1985**, *16*, 285.

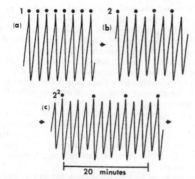

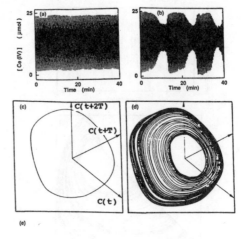

**Figure 3.** Bromide ion electrode potential time series illustrating a period doubling sequence. (a) The original periodic state ($\tau$ = 0.725 h). (b) The period has doubled; the waveform now repeats after every two oscillations ($\tau$ = 0.773 h). (c) The has doubled again; the waveform now repeats after every four oscillations ($\tau$ = 0.803 h). The dots above the time series are separated by one period. From ref 40.

**Figure 4.** These laboratory data illustrate the transition from singly periodic behavior, where the phase space attractor is a limit cycle, to doubly periodic (quasi-periodic) behavior, where the attractor is a torus with frequency ratio $f_2/f_1 \simeq 1/30$. Time series, phase portraits, and Poincaré sections are shown for two values of the residence time: $\tau$ = 0.44 h in (a), (c), and (e), and $\tau$ = 0.42 h in (b), (d), and (f). From ref 46.

transitions decreases geometrically at a universal rate; hence the sequence converges.[35-37] The accumulation point for the sequence marks the onset of chaos.

Period doubling is a generic property of systems described by a one-dimensional map with a single extremum.[38] The map in Figure 2c is of this type; therefore, it is not surprising (at least not now) that the chaotic state in Figure 2 is reached through a period doubling sequence.[26,39,40] Data illustrating this sequence are shown in Figure 3.

Another aspect of universality is the behavior found in the chaotic region beyond the end of a period doubling sequence. For a large class of mathematical models this chaotic regime contains an infinite number of periodic states (windows) that occur in a certain order—this is known as the *universal sequence*.[38] A large number of the distinct periodic states have been found in experiments on the regime that contains the chaotic state shown in Figure 2, and these periodic states have been found to have the same properties and occur in the same order as the states of the universal sequence.[33,40] The striking correspondence between the very complicated dynamics of the universal sequence and a sequence observed in experiments on the BZ reaction is alone strong evidence that deterministic chaos occurs in nonequilibrium chemical reactions.

Another kind of behavior often observed in nonlinear systems is intermittency, where regular oscillations are interrupted by occasional bursts of noise. At the onset of chaos these bursts are infinitely far apart in time, but as a control parameter is varied beyond onset, the bursts become more and more frequent. Pomeau and

Manneville[41] have identified three distinct types of intermittency, and two types have been observed in experiments on the BZ reaction.[42,43]

Theoretical studies have shown that chaotic dynamics may also emerge from a regime with two incommensurate frequencies (frequencies whose ratio is an irrational number).[44] Recently, such two-frequency *quasi-periodicity* has been discovered in the BZ reaction.[45] It was found that, following a transition to a periodic state,[46] the resulting limit cycle underwent a secondary transition to a quasi-periodic state where the second frequency corresponded to a slow modulation of the amplitude of the initial periodic oscillations, as illustrated in Figure 4. With further change in the control parameter, the inner part of the toroidal phase space attractor shrank to a thin tube, and finally the torus disappeared, leaving a fixed point attractor on its axis.[46] For other experimental conditions, a transition to a chaotic state was observed to occur through a wrinkling of the torus.[47]

(35) Grossmann, S.; Thomae, S. *Z. Naturforsch., A* **1977**, *32*, 1353.

(36) Feigenbaum, M. J. *J. Stat. Phys.* **1978**, *19*, 25.

(37) Coullet, P.; Tresser, C. *J. Phys. (Les Ulis, Fr.)* **1978**, *39*, C5. Tresser, C.; Coullet, P. *C. R. Seances Acad. Sci.* **1978**, *287*, 577.

(38) Collet, J.; Eckmann, J. P. *Iterated Maps of the Interval as Dynamical Systems*; Birkhauser: Boston, 1980.

(39) Coffman, K.; McCormick, W. D.; Swinney, H. L.; Roux, J. C. In *Nonequilibrium Dynamics in Chemical Systems*; Vidal, C., Pacault, A., Eds.; Springer: Berlin, 1984; p 251.

(40) Coffman, K. G.; McCormick, W. D.; Noszticzius, Z.; Simoyi, R. H.; Swinney, H. L. *J. Chem. Phys.* **1987**, *86*, 119. Noszticzius, Z.; McCormick, W. D.; Swinney, H. L. *J. Phys. Chem.* **1987**, *91*, 5129.

(41) Pomeau, Y.; Manneville, P. *Commun. Math. Phys.* **1980**, *24*, 189.

(42) Pomeau, Y.; Roux, J. C.; Rossi, A.; Bachelart, S.; Vidal, C. *J. Phys., Lett.* **1981**, *42*, L271.

(43) Arneodo, A.; Kreisberg, N.; McCormick, W. D.; Richetti, P.; Swinney, H. L., in preparation.

(44) See, e.g.: the discussions in ref 1 and 2; Aronson, D. G.; Chory, M. A.; Hall, G. R.; McGehee, R. P. *Commun. Math. Phys.* **1982**, *83*, 303; MacKay, R. S.; Tresser, C. *Physica D (Amsterdam)* **1986**, *19*, 206.

(45) Argoul, F.; Roux, J. C. *Phys. Lett. A* **1985**, *108*, 426.

(46) Argoul, F.; Arneodo, A.; Richetti, P.; Roux, J. C. *J. Chem. Phys.* **1987**, *86*, 3325.

(47) Roux, J. C.; Rossi, A. In *Nonequilibrium Dynamics in Chemical Systems*; Vidal, C., Pacault, A., Eds.; Springer: Berlin, 1984; p 141.

(48) Hudson, J. L.; Mankin, J. C. *J. Chem. Phys.* **1981**, *74*, 6171.

(49) Nagashima, H. *J. Phys. Soc. Jpn.* **1980**, *49*, 2427.

(50) Hudson, J. L. In *Nonlinear Phenomena in Chemical Kinetics*; Pacault, A., Vidal, C., Eds.; Springer: Berlin, 1981; p 33.

(51) Turner, J. S. In *Self Organization in Dissipative Structures*; Schieve, W. D., Allen, P., Eds.; University of Texas Press: Austin, 1982; p 41.

*Argoul et al.*

The observations in experiments on the BZ reaction of routes to chaos that have been well-established in dynamical systems theory provide very strong evidence for the existence of low-dimensional deterministic nonperiodic behavior—*chaos*—in nonequilibrium chemical reactions. Despite this evidence, doubts about the existence of chaos persist, largely because chaotic states have not been found[61] in most simulations.[4-6,13,27,28] We will now describe some recent theoretical results that reconcile these legitimate doubts with the experimental observations.

## 4. Insights from Simulations

**4.1. Simulations with Oregonator-Type Models.** One reduction of the original 20-variable Field–Körös–Noyes model[9] of the BZ reaction leads to the skeletal scheme given by reactions R1–R9 with nine intermediate species,[53,54] where R· is an oxidized derivative of malonic acid (MA), P is an inert organic product, and BrMA is bromomalonic acid. The concentrations of both bromate and cerous ions are assumed to be constant in the reactor; only the input flow of Br⁻ is taken into account in the calculations.

$$BrO_3^- + Br^- + 2H^+ \xrightarrow{k_1} HBrO_2 + HOBr \quad (R1)$$

$$HBrO_2 + Br^- + H^+ \xrightarrow{k_2} 2HOBr \quad (R2)$$

$$HOBr + Br^- + H^+ \xrightarrow{k_3} Br_2 + H_2O \quad (R3)$$

$$BrO_3^- + HBrO_2 + H^+ \underset{k_{-4}}{\overset{k_4}{\rightleftarrows}} 2BrO_2\cdot + H_2O \quad (R4)$$

$$2HBrO_2 \xrightarrow{k_5} HOBr + BrO_3^- + H^+ \quad (R5)$$

$$BrO_2\cdot + Ce(III) + H^+ \xrightarrow{k_6} Ce(IV) + HBrO_2 \quad (R6)$$

$$HOBr + MA \xrightarrow{k_7} BrMA + H_2O \quad (R7)$$

$$BrMA + Ce(IV) \xrightarrow{k_8} Br^- + R\cdot + Ce(III) + H^+ \quad (R8)$$

$$R\cdot + Ce(IV) \xrightarrow{k_9} Ce(III) + P \quad (R9)$$

Although the numerical results presented in this paper have been obtained with this particular reaction scheme, we should emphasize that this particular model is only one among a number of reasonably realistic models that have been proposed to describe the dynamics of the BZ reaction.

Reactions R1–R9 translate into a system of seven ordinary differential equations, which when simulated on a computer[54] yield a periodic–chaotic sequence similar to the one observed in the Texas experiments; for example, the time series, attractor, and one-dimensional map shown in Figure 5 compare fairly well with the corresponding experimental ones shown in Figure 2. Furthermore, for the model as well as in the experiments the transitions from the periodic to the chaotic regimes all occur by period doubling,[39,54] and the successive periodic regimes[23,24] differ by the addition of one more small-amplitude oscillation per period of the time

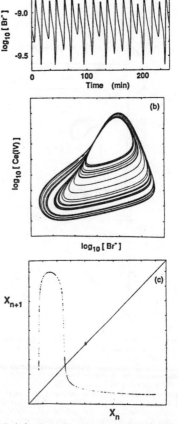

**Figure 5.** A chaotic state found in a numerical simulation of a 7-variable model (R1)–(R9) of the kinetics of the BZ reaction: (a) time series, (b) strange attractor, and (c) one-dimensional map with $X = \log[Ce(IV)]$. From ref 54.

series. Simulations for a similar model, one with seven species,[13] have yielded a transition from periodic to quasi-periodic behavior,[55] just as was found in the experiments described in section 3.2 (see Figure 4).

**4.2. Small-Scale Chaos.** Experiments and simulations on the BZ reaction often yield, in addition to periodic–chaotic sequences and quasi-periodicity, sequences in which there appear to be abrupt transitions from one multipeaked periodic state to another. No chaos has been evident in experiments[46,54,56] on the latter sequences, and previous simulations[4-6,13,27,28] have also yielded transitions that appear to be directly from one periodic state to another without any intervening chaos. However, a new study[54] of the 7-variable model (R1)–(R9) shows that chaos *can* indeed occur in the

(52) Ringland, J.; Turner, J. S. *Phys. Lett. A* **1984**, *105*, 93.
(53) Richetti, P.; Arneodo, A. *Phys. Lett. A* **1985**, *109*, 359.
(54) Richetti, P.; Roux, J. C.; Argoul, F.; Arneodo, A. *J. Chem. Phys.* **1987**, *86*, 3339.

(55) Barkley, D.; Ringland, J.; Turner, J. S. *J. Chem. Phys.* **1987**, *87*, 3812.
(56) Maselko, J.; Swinney, H. L. *Phys. Scr.* **1984**, *52*, 269; *J. Chem. Phys.* **1986**, *85*, 6430; *Phys. Lett. A* **1987**, *119*, 403. Swinney, H. L.; Maselko, J. *Phys. Rev. Lett.* **1985**, *55*, 2366.

502

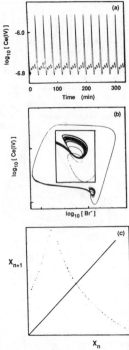

**Figure 6.** The possibility of chaos on a small scale, a scale too small to be observable in laboratory experiments, is illustrated by these results from a numerical study of a 7-variable model (R1)–(R9). The time series in (a), although apparently periodic, is actually chaotic: the one-dimensional map in (c), deduced from a Poincaré section for the attractor in (b), reveals the presence of a very small fluctuation in the value s of the ammplitude of the small-amplitude oscillation that immediately precedes the large-amplitude oscillation. The map demonstrates that these nonperiodic fluctuations in amplitude are *deterministic*: for any $X_i$, the map gives $X_{i+1}$, where $X_i$ is the amplitude of $i$th occurrence of the third of three small-amplitude oscillations. From ref 54.

neighborhood of the transition between different periodic states, but this chaos occurs on a very small scale, as Figure 6 illustrates: the time series in Figure 6a appears on first inspection to be periodic with four oscillations per period, but on closer inspection it can be seen that the amplitude of the small peaks that precede each large-amplitude oscillation varies irregularly by a very small amount. The one-dimensional map in Figure 6c, constructed from the phase portrait in Figure 6b, demonstates that the small irregularities are *deterministic*; as we have discussed, such a map is a hallmark of chaos. The *small-scale chaos* in Figure 6, in contrast to the large scale chaos in Figure 2, would be extremely difficult to observe directly in experiments, and even in simulations a definitive identification of the chaos is possible only by exploiting tools from dynamical systems theory.

Only one frequency is associated with the dynamics of the large-scale chaos shown in Figures 2 and 5,[53,57]

(57) Argoul, F.; Arneodo, A.; Richetti, P. *Phys. Lett. A* 1987, *120*, 269.

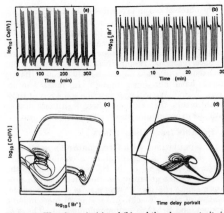

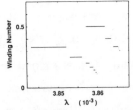

**Figure 7.** Waveforms in (a) and (b) and the phase portraits in (c) and (d) exhibit the spiraling in and spiraling out often observed for chaotic as well as periodic states in the BZ reaction: (a) and (c) illustrate a periodic state found in a numerical study of a 7-variable model (R1)–(R9); (b) and (d) illustrate respectively a periodic time series and a chaotic phase portrait obtained in experiments. (a) and (c) from ref 54, (b) from ref 56, and (d) from ref 46.

**Figure 8.** Devil's staircase obtained in a simulation of the 7-variable model (R1)–(R9). From ref 54.

but the states that have small-scale chaos are reminiscent of dynamics with two characteristic frequencies;[58] this is suggested by the convergent and divergent spiraling[46,48,54] that can be seen in time series and phase portraits such as those in Figures 6a,b and 7. However, this behavior, although characterized by two frequencies, is *periodic* rather than *quasi-periodic* (as in Figure 4a) because the two frequencies are locked together in rational ratios for finite ranges in control para.... 'er; these frequency-locked states are readily apparent in the experiments and simulations. The frequency-locked states can be labeled by a *winding number*, which can be computed as follows:[54,58] in a periodic state with a total of $N_t$ oscillations per period, and with $N_i$ changes per period from a large-amplitude to small-amplitude oscillation, the winding number is $N_i/N_t$. Thus, for example, the periodic state in Figure 7b has a total of 26 oscillations per period and there are 6 changes per period from large to small amplitude, hence the winding number of 6/26.

A succession of frequency-locked periodic states was observed in experiments by Maselko and Swinney[56] and

(58) Argoul, F. Bordeaux Doctoral Thesis, 1986.
(59) This winding number is defined differently from the firing number discussed in ref 56.

by Argoul et al.[46,54] and in simulations of the 7-variable model (R1)–(R9).[54] The winding numbers for sequences of frequency-locked states form a "devil's staircase"[44] when plotted as a function of a control parameter, as illustrated in Figure 8. In the simulations of the model[54] chaotic states were unambiguously observed, but only for very narrow control parameter ranges, and the chaotic nature of the state was localized on a very small part of the trajectory. However, observation of a devil's staircase with overlapping steps can be considered from dynamical systems theory to provide strong evidence for the existence of chaos in the experiments and simulations.[46,54,58]

In summary, it is now clear why chaos could have been present yet undetected in past experiments and numerical simulations that showed what appeared to be direct transitions from one periodic state to another: the domain of existence of chaos could have been too small to be seen, and moreover, small fluctuations in the amplitude of one (or more) of the many oscillations per period in a multipeaked waveform would naturally be interpreted as noise. For example, for a complex state like the one in Figure 7b, many full cycles of 26 oscillations per period would be required to determine the possible chaotic nature of the dynamics.

## 5. Conclusions

We have presented evidence for the existence of low-dimensional chaotic dynamics in the BZ reaction. Chemical kinetics is well-suited for studies of chaos because the behavior is often clearly nonperiodic. In fact, the first experimental strange attractor and one-dimensional map were extracted from laboratory data obtained in chemical experiments. Tools from dynamical systems theory and the observation of routes to chaos that are well-established theoretically have provided evidence for chaos in many chemical experiments.

Thus, the evidence for chaos when it occurs on a large scale is unequivocal. Surprisingly, our simulations have also revealed chaotic behavior even in situations that at first glance appear to be periodic: the chaotic dynamics occurs on a very small scale, a scale that could be very difficult to resolve in the laboratory. Although such small scale chaos was definitely proved for a time series obtained from a simulation, even in a simulation such chaos might easily go unnoticed or dismissed as

round-off error; such nonperiodic behavior, if it were noticed, could not be understood without the recently developed tools of dynamical system theory. Thus, behavior identified as periodic in some past simulations may have in fact been chaotic.

All of the strange attractors that we have discussed, obtained from both simulations and experiments, can be considered to be embedded in a three-dimensional space. Furthermore, although space has not permitted a discussion here, we should at least mention the subject of *normal forms*,[60] which are the simplest nonlinear equations that describe the interaction of a few instabilities (for example, an oscillatory instability and the hysteresis instability that is associated with bistability). A recent analysis of normal forms has shown that a system of coupled differential equations with only three variables can describe most of the dynamics found in the BZ reaction, including quasi-periodicity and large- and small-scale chaos.[54] Thus, the reduction of the original Field–Körös–Noyes scheme involving some 20 species to some 3-variable skeletal mechanism (like the Oregonato) appears to be justified. But we have to be very careful at this point. The three variables in the normal form, unlike the three variables in the Oregonator, do *not* represent three species involved in the reaction. Rather, the normal form variables are nonlinear combinations of the original $N$ species ($N \approx 20$). Thus, the normal form analysis specifies the appropriate three-dimensional subspace (in the $N$-dimensional phase space) on which the dynamics is confined asymptotically.

In conclusion, experimental, numerical, and theoretical evidence demonstrates the existence of chaos and that the chaos can be understood in terms of the chemical kinetics and the interaction of a few basic instabilities known to occur in the BZ reaction.

*The research in Bordeaux is partially supported by a CNRS ATP Grant, and the research in Texas is partially supported by the Department of Energy, Office of Basic Energy Sciences.*

(60) Arnold, V. I. *Geometrical Methods in the Theory of Ordinary Differential Equations*; Nauka: Moscow, 1978 (in English, Springer-Verlag: New York, 1983).
(61) A simulation of a 4-variable Oregonator-type model did exhibit chaos,[25,51] but this chaos was reached after a period doubling sequence from nearly sinusoidal oscillations,[52] while the chaotic states found in the experiments (e.g., see Figure 2) were reached through period doubling of relaxaton oscillations.

Reprinted from Nature, Vol. 321, No. 6068, pp. 394-401, 22 May 1986
© *Macmillan Journals Ltd., 1986*

# Chaos in light

## Robert G. Harrison

Physics Department, Heriot-Watt University, Edinburgh EH14 4AS, UK

## Dhruba J. Biswas

MDRS (Physics Group), Bhabha Atomic Research Centre, Bombay 400 085, India

*Chaos is an inherent feature of many nonlinear systems. In particular, the transition from a steady to chaotic state occurs independently of the physical properties of the system. Such behaviour occurs in optics, both in lasers and in nonlinear optical devices. Such devices, which are fundamentally simple both in construction and in the mathematics that describes them, provide excellent opportunities for investigating nonlinear phenomena as well as for technological innovation.*

MATHEMATICAL discoveries have revolutionized our understanding of nonlinear science. The preconception that physical systems, in general, behave in a predictable manner is seriously in question. Rather than yielding regular and repeatable behaviour, many nonlinear systems also exhibit unstable, even chaotic, solutions. Furthermore, the transition from stable to chaotic behaviour, which may occur when varying a control parameter of the system, follows specific, well-defined routes which are universal in the sense that they are independent of the physical properties of the system they describe. It is these signatures which have been a major impetus to experimentalists in the subsequent search for physical systems that exhibit these phenomena. Such behaviour has now been observed in many branches of science and the number is rapidly growing.

The most recent exciting development is the discovery that such phenomena exist in optics. Here we are concerned, in general, with the nonlinear interaction of light with media contained in optical resonators; the physical properties of the medium, such as absorption and refractive index, being modified by the intensity of the incident radiation.

There are two major areas. The first are lasers or active systems, in which the optical signal is derived from stimulated emission generated within an optical cavity containing a gain medium. The second are passive systems for which the optical signal is but the transmission of an input light signal through an optical cavity containing, for example, an absorptive medium.

Instabilities in laser emission, notably in the form of spontaneously coherent pulsations, have been observed almost since the first demonstration of laser action. However, subsequent theoretical efforts towards understanding these phenomena have been at a modest level, due in part to the wide variety of alternative areas of investigation provided by lasers. It is only with the new mathematical discoveries that these instability phenomena have been investigated to give a deeper insight into the mechanisms of laser action and its deterministic chaotic behaviour. Re-examination of many of these systems show that such effects are quite abundant; the operating window for conventional stable emission in some systems often proving surprisingly small while in some cases, the instabilities are found to prevail just where the lasing emission is optimum.

On the other hand, passive systems which are being increasingly recognized for their potential application as bistable all-optical logic elements, may give rise to similar phenomena. It may, nevertheless, be possible to take advantage of the periodic instabilities that precede chaos in the development of ultra-high frequency all-optical modulators.

Of the variety of physical systems that exhibit deterministic instability phenomena, optical systems, both lasers and passive devices, provide nearly ideal systems for quantitative investigation due to their simplicity both in construction and in the mathematics that describe them, enriched by the possibility of a quantum description. Notable also is the very short timescale (nanosecond to microsecond) over which optical instabilities occur which, in contrast to many other systems, ensures essentially constant environmental conditions during data acquisition. This is particularly important since even small extraneous perturbations, such as noise, may dramatically alter the form of the subsequent temporal evolution of the instability process. These features are fundamental to the rapid establishment of optical systems in this multidisciplinary field.

## Universality in chaos

In considering deterministic behaviour, one is tempted into the misconception that such behaviour must be regular since successive states evolve continuously from each other. However, as early as 1892 Poincaré showed that particular mechanical systems, where time evolution is governed by hamiltonian equations, could display chaotic behaviour. The subsequent discovery by Lorentz[1] in 1963 that even a simple set of three coupled first-order, nonlinear differential equations can lead to completely chaotic trajectories is recognized as a landmark. This work is fundamental to our understanding of laser instabilities.

At first sight, such behaviour appears alien to our conception of many problems in physical science. This prejudice stems from the dominance of mathematical theory pertaining to linear systems and its subsequent successful application to many fundamental linear problems in the physical sciences. Unfortunately, however, this has led to a narrow vision of the physical world where nonlinear behaviour is the rule and linear behaviour the exception.

Unlike linear systems, nonlinear systems must be treated in their full complexity, and so there is no general analytical approach for solving them. The advances made in understanding many previously intractable nonlinear problems can largely be attributed to the power of contemporary computers, where simulated solutions of nonlinear equations have provided insights into their behaviour and suggested directions for future research.

The temporal evolution in the behaviour of a system can be characterized when presented as a trajectory of a point in the phase space of its dynamical variables. In this representation, consider a familiar dynamical system such as a periodically-forced pendulum in a frictional environment. Such a dynamical system is characterized by the fact that the rate of change of its variable is given as a function of the value of the variable at that time. The space defined by the variables is called the phase space. The pendulum's behaviour can be described by the motion of a point in a two-dimensional phase space whose coordinates are the position and velocity of the pendulum. In

2

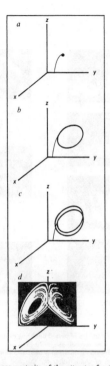

**Fig. 1** Phase space portraits of the attractor for the dynamical variables $x$, $y$ and $z$. $a$, Stable point corresponding to a steady state in time after initial transients have died out shown as the thin continuous line. $b$, Period-one limit cycle corresponding to a periodic solution in time of single frequency. $c$, Bifurcation to period-two limit cycle; a periodic solution in time with double the period of that in $b$. Successive period doubling bifurcations lead to an eventual chaotic colution. $d$ Shows the strange or chaotic attractor for a Lorenz–Haken system describing a single mode laser with a homogeneously broadened two-level gain medium.

more complicated systems, involving many variables, the dimension of the phase space will, however, be considerably larger. If an initial condition of a dissipative dynamical system, such as the pendulum, is allowed to evolve for a long time, the system, after all the transients have died out, will eventually approach a restricted region of the phase space called an attractor. A dynamical system can have more than one attractor in which case different initial conditions lead to different types of long-time behaviour.

The simplest attractor in phase space is a fixed point. The system is attracted towards this point and stays there. This is the case for a simple pendulum in the presence of friction; regardless of its initial position, the pendulum will eventually come to rest in a vertical position. When the pendulum is under the influence of an external periodic driving force, the system is then fully nonlinear and leads to strikingly different behaviour. Irrespectively of the initial conditions the pendulum always ends up making a periodic motion. The limit or attractor of the motion is a periodic cycle called a limit cycle. However, when the driving force exceeds a certain critical value, the periodic motion of the pendulum breaks down into a more complex chaotic pattern

which never repeats itself. This motion represents a third kind of attractor in phase space called a chaotic or strange attractor (see Fig. 1).

A trajectory on a chaotic attractor exhibits most of the properties intuitively associated with random functions, although no randomness is ever explicitly added. The equations of motion are purely deterministic; the random behaviour emerges spontaneously from the nonlinear system. Over short times, the trajectory of each point can be followed, but over longer periods small differences in position are greatly amplified making the predictions of long-term behaviour impossible. As such arbitrarily close initial conditions can lead to trajectories which after a sufficiently long time diverge widely; even for the simplest of systems in which all the parameters are determined exactly, long-term prediction is, therefore, impossible. This behaviour is in marked contrast to that of the fixed point and limits cycle attractors for which, irrespective of starting conditions, the system always settles down to the same solutions.

Irratic and aperiodic temporal behaviour of any of the systems' variables implies a corresponding continuous spectrum for its Fourier transform which is, therefore, also a further signature of chaotic motion. However, other factors including noise, can lead to continuous spectra, and distinguishing chaos from noise is one of the major problems of the field. Hence, although time series, power spectra and routes to chaos collectively provide strong evidence of deterministic behaviour further signatures are desirable for its full characterization and in discriminating it from stochastic behaviour. Here analysis of trajectories of a point in the phase space of its dynamical variables is required. However, for a system with, say, $N$ degrees of freedom it seemed that it would be necessary to measure $N$ independent variables; an awesome if not impossible task for complex system. Consequently mathematicians have long tried to develop practical techniques for extracting specific finite dimensional information from the limited output provided by experiment; typically the time record of a specific physical observable; that is, one variable of the system. Here embedding theorems[2] have been recently used to reconstruct phase portraits from which Lyapunov exponents may be determined that measure the average rate of exponential separation or contraction of nearby points on the attractor. These measure intrinsically dynamical properties, unlike power spectra, and provide quantitative measures by which chaotic motion may be distinguished from stochastic behaviour.

The discoveries that deterministic chaos proceeds through a limited number of specific routes when a control parameter of the nonlinear system is varied is profoundly significant as such behaviour is not restricted to a particular model description of a particular physical system. Rather, nonlinear physical systems in all branches of science which may be formally described by the same set of mathematical equations will give solutions that evolve identically in time through one or other routes to chaotic motion. The unique effect of such unification between many separate scientific diciplines forms the basis for the foundation of synergetics[3,4]. There are at least three common routes by which a nonlinear system may become chaotic. These are referred to as period doubling, intermittency and two-frequency scenarios.

**Period doubling.** From considering various difference equations, many of which can be reduced to simple one-dimensional maps, solutions have been found to oscillate between stable values, the period of which successively doubles at distinct values of the external control parameter[5,6,7]. This continues until the number of fixed points becomes infinite at a finite parameter value, where the variation in time of the solutions becomes irregular (see Fig. 1). One example showing such behaviour is the simple logistic map

$$X_{n+1} = rX_n(1 - X_n)$$

Perhaps the most popular application of this map is in describing

3

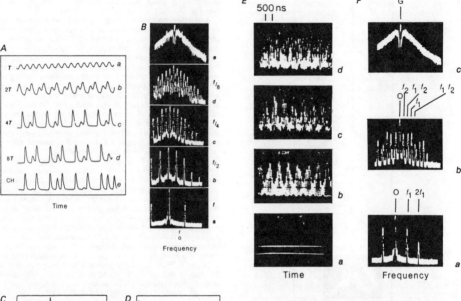

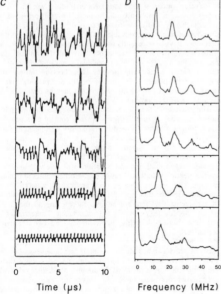

**Fig. 2** *A, B* Examples of period doubling route to chaos for laser systems. The time recordings shown in *A* are for a $CO_2$ laser in which cavity length modulation is used as a control parameter for the generation of chaos (after ref. 20) and *B* shows a spectral recording of the lasing emission for an optically pumped far-infrared ammonia laser with cavity length tuning as a control parameter (after ref. 49). The steady-state solution as in Fig. 1*a* are not shown in these recordings. As seen initial oscillation (*Aa* and *Ba*) undergoes successive period doublings (*Ab–d*) or frequency halving (*Bb–d*) before culminating in chaos where in time the output is aperiodic (*Ae*) and in frequency is a broad continuum (*Be*). *C, D,* Examples of intermittent route to chaos in lasers. The time recordings shown in *C* are for a He–Ne laser in which cavity mirror tilt is used as the control parameter (after ref. 50, and *D* shows a spectral recording of the lasing emission for a He–Xe laser with laser gain as a control parameter (after ref. 25). Steady-state solutions are not shown in the recordings. The initial instability here is characterized by a narrow spectral peak in frequency and periodic pulsation in time. On varying the control parameter the intermittent bursts of chaos become increasingly frequent in time with a corresponding steady broadening of the peaks in the frequency followed by fully chaotic motion. *E, F,* Examples of Ruelle-Takens route to chaos in lasers. The time recordings shown in *E* are for a multimode $CO_2$ laser in which cavity length tuning is used as a control parameter (after ref. 51) and *F* shows a spectral recording of the lasing emission for a He–Ne laser with cavity mirror tilt as a control parameter (after ref. 50. Steady-state solutions are not shown in the recordings. Initial instability (periodic pulsation, *Ea*; one frequency and its harmonics, *Fa*) develops into quasiperiodic motion with two incommensurate frequencies (*Eb* and *Fb*) as the control parameter is varied before going into eventual chaos (*Ed* and *Fc*).

4

changes in population $(X_n \to X_{n+1})$ of an organism or species from year to year $(n \to n+1)$ where there is no overlap between successive generations. The first term in the brackets describes population growth by birth and the second term, population decline due, for example, to preditorial or other environmental conditions. Dependent on the increase in the value of the control parameter $r$ the population may be steady, oscillatory or chaotic the transition following a period doubling scenario (see refs 8, 9). More generally many complex physical systems, often described by large numbers of coupled differential equations, may be reduced in some conditions to the form of this or similar maps. Period doubling bifurcation to chaos has been experimentally observed in numerous systems (see Fig. 2A, B).

**Intermittency.** Intermittency[10] means that a signal which behaves regularly in time becomes interrupted by statistically-distributed periods of irregular motion. The average number of these intermittent bursts increases with the external control parameter until the condition becomes completely chaotic (see Fig. 2C, D).

**Two frequency.** Turbulence in time was originally considered as a limit of an infinite sequence of instabilities (Hopf bifurcation) evolving from an initial stable solution each of which creates a new basic frequencies[11,12]. However, it has been recently shown[13,14] that after only two or perhaps three instabilities in the third step the trajectory becomes attracted to a bounded region of phase space in which initially closed trajectories separates exponentially; as such the motion becomes chaotic (see 2E, F).

## Chaos in lasers

Chaotic behaviour in lasers may exist in even the simplest of systems: one in which population inversion is established between two discrete energy levels of the medium and where the lasing transition between these two levels is homogeneously broadened. A variety of practical lasers may be controlled to operate in these conditions. A further simplification is that the laser cavity, a Fabry–Perot or ring resonator system surrounding the gain medium, be sufficiently short so that only one resonant frequency of the cavity lies within the bandwidth of the gain medium and that this mode be resonantly tuned to the gain centre frequency. The frequency spacing $(\Delta\nu)$ between cavity modes for a Fabry–Perot cavity is given by:

$$\Delta\nu = cn/2L$$

where $c$ is velocity of light, $n$ the refractive index of the lasing medium and $L$ the cavity length. Typical examples of single mode and many mode operation, for short and long cavity lengths respectively, are shown in Fig. 3. In conditions in which the gain or population inversion is maintained at a constant level by, for example, constant electrical or optical excitation, and for the single mode system, lasing occurs with a constant output power at the frequency of the single cavity mode. In this condition, the gain is reduced to a threshold level equal to the cavity losses. Note that even for a multimode system and in the absence of spatial hole burning, lasing is still restricted to a single mode, because the mode with highest gain, here at nor near gain centre, grows at the expense of all other modes. Such behaviour is the accepted operating characteristics of these systems.

However, the discovery[15] that for certain operating conditions emission could be periodic or even chaotic implies that the signal comprises more than one frequency, contrary to the accepted understanding of single mode operation. Prediction of such behaviour were initially identified by Haken through the mathematical equivalence of the equations describing laser action, the Maxwell–Bloch equations, and those derived earlier by Lorenz to describe chaotic motion in fluids. If we consider the trajectory of the Lorenz strange attractor (see Fig. 1d) where in the equivalent laser system the dynamic variables $x$, $y$ and $z$ are the field amplitude $(E)$, polarization of the medium $(P)$,

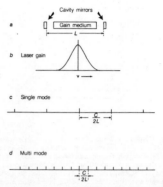

**Fig. 3** a, A Fabry–Perot laser cavity system with a partially transmitting mirror for coupling out the laser emission; b, lorentzian gain profile for a simple two-level homogeneously broadened lasing medium for which each and every atom/molecule emits identically. c, Relative position of the cavity modes for a short optical cavity. Here only one mode lies within the gain bandwidth resulting in a single mode emission, d, Corresponding position of the modes for a long cavity for which several modes lie within the gain bandwidth.

and the population inversion $(D)$, a point $(x, y, z)$ circles in one region for a while, but then suddenly jumps into another region, where it moves for a while until it jumps, seemingly randomly, back into the first region, and so on; the trajectory never intersects. For the laser, such behaviour not only requires a cavity with high transmission but also a gain of at least nine times that required to produce lasing, making the experimental realization of such operation rather impracticable for most lasers of this simple type. A notable exception are optically-pumped far-infrared molecular lasers[16] which will probably become an important research area.

General prerequisites for the onset of deterministic chaos include: that apart from nonlinear interaction there is a sufficiently large phase space; the minimum requirement being that the system possesses at least three degrees of freedom. For example, the simple pendulum possesses only two degrees of freedom and exhibits at most periodic behaviour. In a two-dimensional phase space, described by the variables, velocity and displacement, it is impossible for the trajectory to be restrained within a basin of attraction without intersecting itself; repetition of trajectory points resulting in only periodic or quasiperiodic solutions. The addition of one further degree of freedom, through the imposition of a modulated external force, leads to the possibility of a non-overlapping trajectory still in a limited phase space but now described by three variables; the so-called strange attractor (see ref. 17).

The Maxwell–Bloch equations described above for the special case of a single-mode laser with field tuned to the centre of the gain line such that both field and polarization are real quantities, satisfy the minimum condition of three independent variable equations, each of which has its own relaxation.

$$\frac{\mathrm{d}E}{\mathrm{d}t} = -\kappa E + \kappa P$$

$$\frac{\mathrm{d}P}{\mathrm{d}t} = \gamma_\perp ED - \gamma_\perp P$$

$$\frac{\mathrm{d}D}{\mathrm{d}t} = \gamma_\parallel(\lambda+1) - \gamma_\parallel D - \gamma_\parallel \lambda EP$$

where $\kappa$ is the cavity decay rate, $\gamma_\perp$ is the decay rate of atomic

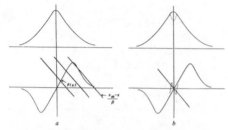

**Fig. 4** Graphical solution of the equation of active cavity mode: $F(x) = (x - x_m)/\beta$. Top traces represent gain profile and the corresponding dispersion features associated with these is shown in the bottom traces. *a*, Graphical solution is illustrated for different detunings of the cavity mode. As seen only when the mode is tuned sufficiently away from the line centre, intersection on more than one point is possible resulting in the splitting of the mode into different frequencies all of which fill the same number of half-wavelengths between the cavity mirrors; spontaneous or passive mode splitting. *b*, When the oscillating cavity mode burns a spectral hole into the gain, the associated dispersion takes just the shape required for giving rise to more than one intersection point. As seen, the gain at the split frequencies are indeed higher than that at the original frequency; induced or active mode splitting.

polarization, $\gamma_\parallel$ is the decay rate of population inversion, $\lambda$ is the pumping parameter, $E$ is the field inside the cavity, $D$ is the population inversion, and $P$ is the atomic polarization.

However, if one variable relaxes faster than the others the stationary solution for that variable may be taken, so resulting in a reduced number of coupled differential equations; commonly termed adiabatic elimination of the fast variables[3,4]. In many systems, polarization and population inversion have relaxation times much shorter than the cavity lifetime and both variables can be adiabatically eliminated. With just one variable describing the dynamics, the laser must show a stable behaviour (fixed point in phase space; see Fig. 1*a*). This group of lasers, comprises many common systems such as He–Ne, Ar+, Dye and lasers. In some cases, only polarization is fast and hence two variables describe the dynamics. In this class, we find ruby, Nd and $CO_2$ lasers which exhibit oscillating behaviour in some conditions, although ringing is always damped.

Since many lasers are not described by the full set of Maxwell–Bloch equations normally chaotic behaviour from these systems cannot be obtained. For these systems with less than three variables the addition of independent external control parameters to the system, as for the pendulum considered above, have been extensively considered[18] as a means to provide the extra degrees of freedom. Active modulation of a parameter such as population inversion, field, or cavity length as well as injection of a constant field detuned from the cavity resonance and also the use of intracavity saturable absorbers have all been considered[19-21]. For multimode rather than single-mode lasers intrinsic modulation of inversion (or photon flux) by multimode parametric interaction ensures additional degrees of freedom[18]. When the field is detuned from gain centre the field amplitude, polarization and population inversion are complex, providing (in the absence of adiabatic elimination) five rather than three nonlinear equations for single mode systems which is more than sufficient to yield deterministic chaos for suitable parameter values[22]. Also of significance is the remarkably low threshold found for the generation of instabilities and chaos in single mode inhomogeneously broadened laser systems[23-25]. Compared with homogeneously-broadened systems this is attributed to the increased number of independent gain packets available in inhomogeneous systems. Pulsating instabilities and routes to

chaos have also been reported for Raman lasers[26] where the instability threshold is again found to be reduced. Significantly, it is also found that instabilities are greatest in conditions for which the laser produces maximum output.

## Physical mechanism

A transition from a steady laser output to an oscillatory and subsequently chaotic emission implies the generation of further oscillating frequencies to that of the original stable emission. That this should occur even for a so-called single frequency (single mode) laser seems to be a contradiction in terms. The explanation is in the phenomena of mode splitting[27,28]. This occurs in a region of rapidly varying dispersion when the oscillating cavity mode splits into more than one frequency all of which fill the same number of half-wavelengths between the cavity mirrors. Coupling between these several frequencies having common mode index is a prerequisite for single-mode pulsating instabilities. Dispersion changes its value rapidly near the wings of the gain curve and the splitting that occurs in this region is known as passive mode splitting. An oscillating cavity mode can also induce large variation in dispersion if it can locally saturate the gain curve, hence this splitting is termed induced mode splitting.

To understand spontaneous mode splitting we consider the equation of the oscillating active cavity mode

$$\frac{mc}{2L} = \nu n(\nu) \tag{1}$$

where, $m$ is the mode index and $n(\nu)$ is the index of refraction at the laser oscillation frequency $\nu$. In terms of the empty resonator mode frequencies, $\nu_m = mc/2L$ obtained by putting $n(\nu) = 1$ the above equation may be re-written as,

$$\nu_m - \nu = \nu[n(\nu) - 1] \tag{2}$$

When the value of $n(\nu)$, either for a lorentzian or gaussian gain profile which are common to most lasers, is substituted this equation may be re-expressed as,

$$x_m - x = \beta F(x) \tag{3}$$

where $x$ and $x_m$ are respectively the laser oscillation frequency and empty cavity resonant frequency normalized as a detuning from the atomic resonance. $F(x)$ has the same functional dependence with $x$ as $n(\nu)$ has with $\nu$ and it depends on whether the gain is gaussian or lorentzian. The mode splitting factor, $\beta$, is dimensionless and is given by

$$\beta = K\frac{cg}{(\Delta\nu)} \tag{4}$$

where $g$ is the peak small-signal incremental gain, $\Delta\nu$ is the FWHM (full width at half maximum) value of gain-width, $c$ is the velocity of light in vacuum and $K$ is a numerical factor equal to $\pi^{-3/2}\sqrt{\ln 2}$ for gaussian gain profile and $\pi^{-1}$ for lorentzian gain profile. Equation (3) has been graphically solved for different detunings for a gaussian gain profile in Fig. 4*a*. Under certain detuning (near the wings of the gain) the equation is simultaneously satisfied by more than one value of $x$. Physically this means that the cavity mode is split into more than one frequency and all of which correspond to the same number of half-wavelengths within the resonator cavity. Such effects occur spontaneously at the wings of the gain curve and hence the name spontaneous mode splitting.

Induced mode splitting may be explained in terms of distortions in the dispersion caused by hole burning in the gain profile by a single mode operating above lasing threshold (see Fig. 4*b*). The dispersion may be so distorted that several new frequencies satisfy the boundary conditions. The onset of these sideband frequencies (the gain at which can be more than that at the parent oscillating frequency; see Fig. 4*b*) gives rise to pulsations

6

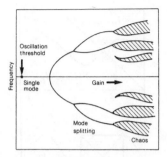

**Fig. 5** Mode splitting sequence in which an initially single mode (single frequency lasing oscillation) successively bifurcates on varying gain to generate emission with increasing complex frequency context culminating in a broad band (chaotic) spectrum (after ref. 52).

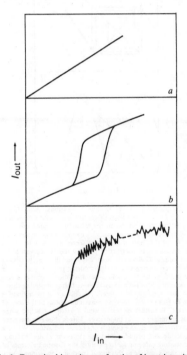

**Fig. 6** Transmitted intensity as a function of input intensity for a simple optical resonator, for example, Fabry-Perot or ring cavity. *a,* An empty cavity showing characteristic linear behaviour; *b,* a characteristic optical bistable action and associated hysteresis; *c,* appearance of periodic and chaotic behaviour in the transmitted signal, on increasing the input signal intensity, that may occur in the form of trace *b* for different parameter conditions of the nonlinear cavity.

in the intensity output of the laser. Because the dispersive effects are caused by the oscillating mode itself, this splitting is termed 'induced mode splitting'. This effect should be strongest at line centre and probably exists over much of the lasing tuning range. In view of these effects we may now consider two line broadening situations common to most lasers—homogeneous and inhomogeneous broadening.

The saturating characteristics of an inhomogeneously broadened gain medium is ideal for the realization of mode splitting. In such a medium, unlike the homogeneously broadened case more than one frequency may naturally oscillate since independent sets of atoms/molecules are responsible for lasing at different frequencies across the gain-bandwidth and thus passive mode splitting may be readily observed. The oscillating frequency may also burn a spectral hole into the gain medium thus favouring induced mode splitting. Furthermore, the side bands split again as they again burn spectral holes. This process will continue and may eventually lead to chaos (see Fig. 5). On the other hand, when a homogeneously-broadened gain medium saturates, all the atoms or molecules actually contribute at the oscillating frequency thereby making the possibility of spectral hole burning almost impossible. In such a system, therefore, only the frequency with highest gain would eventually grow. However, for the Haken-Lorentz system, in an extremely high-gain medium survival of more than one frequency is possible.

## Chaos in nonlinear optical devices

In parallel with work on laser instabilities, Ikeda predicted oscillation and chaos in passive optical systems[29]. In these systems, the cavity contains a medium whose refractive index is modified by the intensity of the input light signal. This may be expressed as

$$n(I) = n_0 + n_2 I \qquad (5)$$

where $n_0$ and $n_2$ are respectively the ordinary and nonlinear contribution towards the refractive index and $I$ is the light intensity.

To understand this phenomena we must consider how the light field within the cavity changes, according to the intensity dependent nonlinear phase shift produced by this refractive index, with successive round trips. The simplest case is one in which the intensity-dependent refractive index immediately responds to changes in intensity and where the cavity is of low finesse. The intensity of the signal inside the cavity, here a ring resonator, for successive round trips then obey the one-dimensional mapping rule[30]

$$I_{n+1} = I_{in}\left\{1 + C \cos\left[\frac{2\pi}{\lambda}(n_2 I_n)L + \phi_0\right]\right\}$$

$$= F(I_{in}, C, I_n) \qquad (6)$$

where $\phi_0 = (2\pi/\lambda)n_0 L$, is the normal round trip optical phase shift experienced by the input signal in the cavity in the absence of nonlinearity, $C$ is a constant for the particular cavity and we assume that the nonlinear medium fills the whole cavity of length $L$. The stationary solution of this mapping, denoted by $I_s$, occurs when $I_{n+1} = I_n$ and yields the relation

$$I_s = I_{in}\left\{1 + C \cos\left[\frac{2\pi L}{\lambda}(n_2 I_s) + \phi_0\right]\right\} \qquad (7)$$

where the transmitted signal ($I_T$) is then simply given by $I_T = I_s T$, where $T$ is the transmission of one of the cavity mirrors.

In the limit of low intensity, the nonlinear term is negligible and the transmitted signal of the ring cavity reduces to the standard expression

$$I_T = I_{in}\{1 + C \cos \phi_0\}T \qquad (8)$$

for this optical cavity. Here the intensity of transmitted signal scales linearly with that of the input signal (see Fig. 6a) its

510

magnitude depending on the degree to which the input signal frequency is resonant with the cavity. Tuning the length of the cavity, thereby varying the phase $\phi_0$, leads to the familiar resonant transmission peaks characteristic of optical cavities. Returning to the nonlinear equation a corresponding plot of transmitted intensity as a function of input intensity is shown in Fig. 6b. The nonlinear form of this curve is the signature of optical bistability, of interest because of its application to all optical logic elements. For example, this particular curve shows switching action and memory in which at a critical input intensity the output signal switches to a higher value but on the return cycle switches down for a reduced input intensity due to optical hysteresis. A stable and reproducible operation is required for such application. However, there are parameter conditions, for the input intensity, and cavity finesse for these systems for which the transmitted signal exhibits a behaviour far from this ideal, manifesting not only pulsating instabilities but also full chaotic characteristics. Such regions are schematically indicated on the bistability curve (Fig. 6c).

To understand this behaviour we need to examine more closely the mapping behaviour of equation (6) as shown in Fig. 7a. Intersection of this curve with the line $I_{n+1} = I_n$ gives the stationary solution to the mapping. Starting from an arbitrary value of $I_n$, the corresponding value of $I_{n+1}$ (vertical intersection with $F(I_{in}; C; I_n)$) is then the new value of $I_n$ (horizontal intersection with line $I_{n+1} = I_n$). Successive iteration generates a spiral of successively decreasing steps which converge to a single point intersection shown. This is then a stable single point solution. In time, the iteration steps are then transients which die out to this steady-state solution. Similar constructions are shown in Fig. 7b and c for two higher values of $C$; the input intensity could, of course, be used as an alternative control parameter. As seen in equation (6) this is manifested as an increase in amplitude of the map. In Fig. 7b, successive iterations, do not converge to a single point but stabilize to a two-point solution; shown by the intersection of the iterations with the mapping function $F(I_{in}; C; I_n)$. The corresponding signal in time, therefore, shows that after transients have died out the transmitted signal exhibits periodic oscillation or limit cycle behaviour. For further increase in $C$, the oscillatory signal successively bifurcates to yield oscillations which are each double the period of the preceeding ones; the so-called period doubling sequence, as shown in Fig. 2 for laser systems. Ultimately for sufficiently high $C$ a chaotic solution is obtained for which, as seen in Fig. 7d, successive bifurcations never converge to fixed points. This figure also shows that a small change in the starting conditions, value of $I_n$, will soon result in wildly different iterative steps; unlike the stable and periodic solutions such behaviour which characterizes deterministic chaos is often referred to as being sensitive to initial conditions. As such a small difference in the initial conditions is amplified by many operations. Mathematically, this is quantified by the Lyapunov exponent $\lambda$ which for this sytem is

$$\lambda = \log \frac{I_{in} C}{2} \qquad (9)$$

In general, a positive Lyapunov exponent implies exponential separation of nearby trajectories, the signature of deterministic chaos, while a negative exponent implies exponential contraction, the signature of an attracting fixed point. For chaotic behaviour one, therefore, requires $I_{in} C/2 > 1$, whereas bistable action without the presence of instability requires $I_{in} C/2 < 1$. Evidently a transition to chaos may then occur on increasing the input intensity or the value of the parameter $C$. For convenience, we have assumed that the medium responds instantaneously to the field or more precisely the relaxation time of the medium is much shorter than the cavity round trip time of the field. For systems in which the converse situation applies bistable operation is favoured although even here sideband instabilities may arise.

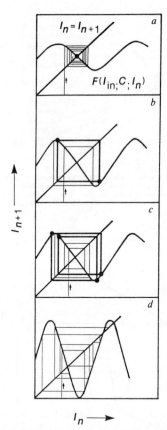

**Fig. 7** Graphical solutions of one dimensional map $I_{n+1} = F(I_{in}, C, I_n)$ for various values of control parameter $C$. The line $I_n = I_{n+1}$ is the condition when the transmitted signal does not change with successive iterations. a, A fixed point or steady-state solution $I_s$ for the field in the cavity since successive iterations converge to the single intersection point of the two curves. b, An increased value of $C$, shows the iterations to stabilize on a two-point solution corresponding to a periodic output in time. For further increase in the value of $C$, a four-point solution is obtained (c); that is iterations here have to cycle twice before repeating themselves so corresponding to a periodic oscillation of twice that for the trace b. This period doubling behaviour repeats as the control parameter $C$ is further increased until a value is reached when iterations no longer repeat themselves. This is shown in the trace d which characterizes an example of chaotic behaviour. We note that the stable and periodic solutions (traces a–c) are invariant with the starting value of $I_n$. In contrast, the chaotic trajectory of trace d is very sensitive to this initial condition; following a very different path for even the smallest change. The time series of Fig. 2A are representative of the behaviour described here.

Observations of these phenomena, although still limited, have been made in various optical systems such as hybrid bistable devices, and all optical bistable systems. In the hybrid system, nonlinearity in a transparent dielectric medium is caused by an externally-applied voltage rather than by the laser signal itself. The impressed voltage is coupled to the transmitted light signal

8

through a controllable electronic delay with variable gain which serves as the control parameter of the system. On increasing the feedback gain the transmitted optical signal undergoes a period doubling bifurcation though only up to period eight before becoming chaotic. Here[31] as in many other systems, both passive and active, the effect of external noise[32] interrupts the otherwise infinite series of period doublings to create what is termed a bifurcation gap.

The first observation of bifurcation to periodic and chaotic states in an all optical bistable system was made with a single mode optical fibre as the transparent nonlinear medium in a ring cavity[33], the refractive index here being directly modified by the input intensity through the Kerr effect. Here, the high intensities required to induce sufficient nonlinearity precluded the use of a continuous wave (cw) input source and instead a mode-locked train of intense short pulses, separated in time by an amount equal to the round trip time of the ring cavity, was used. Each pulse is, in effect, an iterative step in which the signal of one pulse circulating the ring cavity adds to the signal of the next incoming pulse, the process repeating for the whole train. Since bifurcations are caused by the interference between the incident field and cavity field which, suffers a nonlinear phase shift in each round trip of the cavity (see equation (6)), the transmitted signal is evidently modified by the intensity of the pulses. With this system period doubling followed by chaos was observed.

The most fundamental of all nonlinear optical systems is nevertheless, one containing a two-level medium as originally analysed by Ikeda for a ring cavity[33]. Near resonant excitation of the medium will, for a sufficiently intense incident signal, cause saturation of the absorption which thereby modifies the associated anomalous dispersion feature of the two-level transition. The mechanism is, in many respects, equivalent to that which gives rise to mode splitting in lasers described earlier for a two-level gain rather than absorptive medium. As for the laser, the intensity dependent change in refraction (dispersion) is resonantly enhanced resulting in considerably reduced power requirements to those used in the fibre experiments. Both atomic and molecular gases are potentially useful media for investigation here, providing discrete levels many of which are in near resonance with available laser lines and approximate reasonably closely to two-level systems. The first corroboration of period doubling routes to chaos in such systems were made in both ring and Fabry–Perot cavities containing ammonia gas near

resonantly-excited by pulsed $CO_2$ radiation[34,35] Recently, experiments have been extended to cw pump laser conditions[36] in which sodium vapour was used as the nonlinear medium. Preliminary results show evidence of period doubling with additional instabilities yet to be fully characterized.

Recognizing the problems that chaotic behaviour may pose for fast all-optical bistable elements, yet, operated in the period doubling window and, in particular, period two oscillation, for which the operational parameter window is largest, they offer the attractive prospect of high-frequency all-optical modulators. But such an operation requires that the relaxation time of the nonlinear medium be much faster than the cavity round trip time, then for small devices (<1 mm in length) the frequency of modulation may, in principle, be in the THz range provided that nonlinear materials may be found with sufficiently short relaxation times. The search for fast large nonlinearities is one of the main efforts in this area.

## Conclusions

Nonlinear optics is proving valuable to the fields of nonlinear dynamics and deterministic chaos. On one hand, basically simple optical systems can be constructed exhibiting the most interesting classes of chaotic behaviour enriched by the possibility of a quantum description. On the other hand, lasers and related nonlinear otical devices have a large and growing technical application, and the understanding, control and possible exploitation of sources of instability in these systems has considerable practical importance.

Experimental findings are, in general, not yet sufficiently comprehensive to permit the quantitative analysis necessary to fully test the theoretical models. More carefully controlled experiments are forthcoming from which, along with time series, power spectra and identification of routes to chaos, embedding procedures may be implemented to the attractors describing the dynamical behaviour of these systems.

More detailed discussions of points in this review can be found elsewhere. For laser instabilities see refs 37–39 and also refs 40, 41 which cover both active and passive systems. For passive systems see ref. 42. Comprehensive treatments on the more general principles of deterministic chaos can be found in refs 3, 4. See the recent text on deterministic chaos by Schuster[43]. See refs 8, 9, 44–46 for articles on the general aspects of deterministic chaos, and refs 47 and 48 for discussions on the device application of optical bistable systems.

1. Lorenz, E. N. J. atoms. Sci. 20, 130 (1963).
2. Whitney, H. Annls Math. 37, 645 (1936).
3. Haken, H. Synergetics—An Introduction (Springer, Berlin, 1983).
4. Haken, H. Advanced Synergetics (Springer, Berlin, 1983).
5. Grossman S & Thomae, S. Z. Naturforschung. 32A, 1353 (1977).
6. Feigenbaum, F. J. J. Stat. Phys. 19, 25 (1978).
7. Coullet, P. & Tresser, J. J. Phys. Paris C5, 25 (1978).
8. May, R. M. Nature 261, 459 (1976).
9. Feigenbaum M. J. Physica 7D, 16 (1983).
10. Manneville, P. & Pomeau, Y. Phys. Lett. 75A, 1 (1979).
11. Landau, L. D. Acad. Sci. USSR 44, 311 (1944).
12. Landau, L. D. & Lifshitz, E. M. Fluid Mechanics (Pergamon, Oxford, 1959).
13. Ruelle, D. & Takens, F. Commun. Math. Phys. 20, 167 (1971).
14. Newhouse, S., Ruelle, D. & Takens, F. Commun. math. Phys. 64, 35 (1978).
15. Haken, H. Phys. Lett. 53A, 77 (1975).
16. Hogenboom, E. H. M. Klische, W., Weiss, C. O. & Godone, A., Phys. Lett. 55, 2571 (1985).
17. Arecchi, F. T., Lippi, G. L., Puccioni, G. P. & Tredicce, J. R. Opt. Commun. 51, 308 (1984).
18. Scholz, H. J., Yamada, T., Brant, H. & Graham, R. Phys. Lett. 82A, 321 (1981).
19. Arecchi, F. T., Meucci, R., Puccioni, G. P. & Tredicce, J. R. Phys. Rev. Lett. 49, 1217 (1982).
20. Midavaine, T., Dangisse, D. & Glorieux, P. Phys. Rev. Lett. 55 1989 (1985).
21. Lugiato, L. A., Narducci, L., Bandy, D. K. & Pennise, C. A. Opt. Commun. 46, 64 (1983).
22. Mandel, P. & Zeglache, H. Opt. Commun. 47, 146 (1983).
23. Casperson, L. W. Phys. Rev. A 21, 911 (1980).
24. Abraham, N. B. et al. Lecture notes in Physics, Vol. 182, 107 (Springer, Berlin, 1983).
25. Gioggia, R. S. & Abraham, N. B. Phys. Rev. Lett. 51, 650 (1983).
26. Harrison, R. G. & Biswas, D. J. Phys. Rev. Lett. 55, 63 (1985).
27. Casperson, L. W. & Yariv, A. Appl. Phys. Lett. 17, 259 (1970).

28. Hendow, S. & Sargent, M. III, J. opt. soc. Am. B 2, 84 (1985).
29. Ikeda, K. Opt. Commun 30, 257 (1979).
30. Ikeda, K. & Daido, H. Phys. Rev. Lett 45, 709 (1980).
31. Gibbs, H. M., Hopf, F. A., Kaplan, D. L. & Shoemaker, R. L. Phys. Rev. Lett. 46, 474 (1981).
32. Crutchfield, J. P. & Huberman, B. A. Phys. Lett. 77A, 407 (1980).
33. Nakatsuka, H., Asaka, S., Itoh, H., Ikeda, K. & Matsuoka, M. Phys. Rev. Lett. 50, 109 (1983).
34. Harrison, R. G., Firth, W. J., Emshary, C. A. & Al-Saidi, I. A. Phys. Rev. Lett. 51, 562 (1983).
35. Harrison, R. G., Firth, W. J. & Al-Saidi, I. A. Phys. Rev. Lett. 53, 258 (1984).
36. Gibbs, H. M. et al. Proc. OSA Topical Meet. on Instabilities and Dynamics of Lasers and Nonlinear Optical Systems, Rochester (1985).
37. Abraham, N. B. Laser Focus 73 (May 1983).
38. Harrison, R. G. & Biswas, D. J. Progress in Quantum Electronics 10, 147 (Pergamon, Oxford, 1985).
39. J. Opt. Soc. Am. B, Spec. Iss. Instabilities in Active Optical Media 2, 1–272 (1985).
40. Ackerhalt, J. R., Milonni, P. W. & Shih, M. L. Phys. Rep. 128, 205–300 (1985).
41. Arecchi, F. T. & Harrison, R. G. (eds) Instabilities and Chaos in Quantum Optics (Springer, Berlin, in the press).
42. Gibbs, H. M. Optical Bistability: Controlling Light with Light (Academic, New York, 1985).
43. Schuster, H. G. Deterministic Chaos (Physik, Berlin, 1984).
44. Eckmann, J. P. & Ruelle, D. Rev. Mod. Phys. 57, 617 (1985).
45. Swinney, H. L. Physica 7D, 3 (1985).
46. Farmer, J. D., Ott, E. & Yorke, J. A. Physica 7D, 153 (1985).
47. Miller, D. A. B. Laser Focus 18, 79 (1982).
48. Smith, S. D. et al. Opt. Engn 24, 569 (1985).
49. Weiss, C. O., Klische, W., Ering, P. S. & Cooper, M. Opt. Commun. 52, 405 (1985).
50. Weiss, C. O., Godone, A. & Olafsson, A. Phys. Rev. A 28, 892 (1983).
51. Biswas, D. J. & Harrison, R. G. Phys. Rev. (Rap. comoun.) A 32, 3835 (1985).
52. Minden, M. L. & Casperson, L. W. J. opt. Soc. Am. B 2, 120 (1985).

*Quarterly Review of Biophysics* **18**, 2 (1985), pp. 165–225     165

*Printed in Great Britain*

# Chaos in biological systems

## LARS FOLKE OLSEN AND HANS DEGN

*Institute of Biochemistry, Odense University, Denmark*

166   L. F. OLSEN AND H. DEGN

1. INTRODUCTION

Chaos is a widespread and easily recognizable phenomenon that hardly anybody took notice of until recently. The reason may be that chaos has something profoundly counterintuitive about it. It will not fit easily into any familiar cause–effect frame. The best introduction to chaos is by the way of an example. Consider a leaking faucet (Shaw, 1984). When the weight of the accumulating drop exceeds the surface tension the drop falls and a new drop begins to form. If the leak is small and the pressure in the faucet is constant, the time taken for the drop to reach the critical weight is constant. The dripping is perfectly periodic, the period depending on the leak rate. If the leak is slightly increased, the period of dripping will decrease slightly and vice versa. However, somewhere beyond this point the leaking faucet becomes a nuisance. When the leak is increased beyond a certain point the dripping looses its regularity. The time interval between the drops will first alternate periodically between a short and a long time interval. After a further increase of the leak this double periodic pattern will become unstable and change into a new pattern where four different time intervals between the drops alternate periodically. As the leak is further increased the period will double again and again and finally the dripping becomes completely irregular without any repeating pattern. When this occurs we are observing chaos. At the same time we are posed with the problem of understanding how such a ridiculously simple system can show random behaviour.

It seems that chaos could not be recognized by natural science before it had been shown mathematically and by computer simulations that very simple models can indeed exhibit random behaviour. Among the first equations known to do so were one-dimensional difference equations. Such equations are used in population biology as models of populations without generation overlap. The occurrence of chaos in difference equations was described in the mid-seventies by May in some very influential reviews (May, 1976; May & Oster, 1976). At the same time a paper by Lorenz with the first description of chaos in a simple non-linear differential equation model emerged from a dozen years of dormancy (Lorenz, 1963). In the late seventies followed an avalanche of papers on chaos in mathematical and physical journals signalling a major breakthrough in the field of non-linear dynamics. Excellent reviews on chaos in physical systems have been published (Ott, 1981; Eckmann, 1981; Shaw, 1981).

Biology has non-linear systems in abundance and there is now an increasing stream of reports on chaos in biological systems ranging from enzyme reactions to population biology. Actually, the first chemical reaction reported to be chaotic was an enzyme reaction (Olsen & Degn, 1977). It is the aim of the present paper to review the literature on chaos in biology except for population dynamics that is well covered by existing reviews (May, 1976; May & Oster, 1976; Rogers, 1981). Since some of the mathematical methods used in the present field are not yet to be found in the textbooks accessible to most biologists we find it necessary to dedicate a few pages to the introduction of some unfamiliar mathematical concepts. Models of chaotic systems generally are non-linear equations that cannot be solved analytically and they tend to loose all non-trivial properties at linearization. As computer solutions are easily obtained one may be tempted to skip the mathematics altogether. However, in order to be able to utilize the power of the computer efficiently one must know the general properties of the equations.

The study of chaos is an endeavour that has brought together researchers of widely different backgrounds. Predictably, the language of the field resembles that of Babel. Until a clear terminology has crystallized one must be prepared to hear many different words for the same thing. Take for example the following expressions: Poincaré map, return relation, next amplitude plot, transfer function, transition function and transformation. They are all commonly used to designate essentially the same mathematical entity. We consider it our duty to familiarize the reader with the current synonyms rather than exercising our hands as language purists. We shall take exception though when the word hysteresis is used as a synonym for bistability.

## 2. DIFFERENCE EQUATIONS

Some biological systems lend themselves more easily to description by difference equations than by differential equations. This is the case with populations without generation overlap where the number of individuals one year can be assumed to depend only on the number of individuals the previous year. This is expressed with the difference equation

$$x_{t+1} = F(x_t) \qquad (2.1)$$

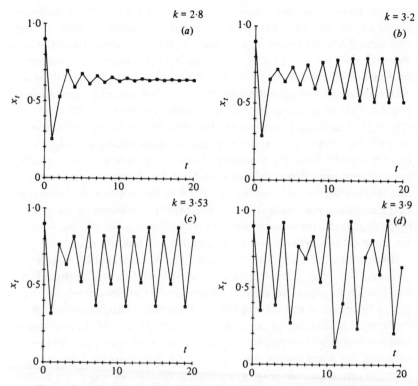

Fig. 1. Samples of the iterate of (2.2) at four different values of $k$ indicated in the graphs.

where $t$ denotes the generation time in discrete units and $x$ is the number of individuals in a generation. Despite their modest appearance difference equations are the most powerful vehicles to the understanding of chaos. They will take you right to the heart of the matter in a few paragraphs. Difference equations are not reserved for the few with interest in population dynamics. They are indispensable for the understanding of non-linear dynamical systems in general (May, 1976; Collet & Eckmann, 1980).

Consider for example the difference equation

$$x_{t+1} = kx_t(1 - x_t) \qquad (2.2)$$

This is said to be a one-dimensional equation since it has only one variable, $x$. Graphically it can be represented in two dimensions by plotting $x_{t+1}$ as a function of $x_t$. The graph is a parabola with its maximum at $x_t = \frac{1}{2}$. Provided $0 < k \leqslant 4$ this equation maps the unit interval two-to-one into itself. If we choose a value for $x_1$ between zero and one and insert it in (2.2) we will obtain an $x_2$ between zero and one. When we insert $x_2$ we obtain $x_3$ which is also within the unit interval and so on. The study of difference equations consists in finding out where $x_t$ goes in such iteration processes. Generally it will converge on something called an attractor. The attractor can be a fixed point a cycle or neither. In the latter case we have a so-called strange attractor. The sequence of numbers produced by the iteration process we call the iterate. In certain branches of mathematics an object like (2.2) is called a map or a mapping. In other branches it may be called a transformation. Since the iterate is bounded within the unit interval (2.2) is also said to be a map of the interval. Since the right hand term of the equation is a quadratic it is called the quadratic map. Since (2.2) maps or transforms two-to-one it is said to be non-invertible. Finally, for no good reason it is sometimes called the logistic map.

The iterate of (2.2) depends on $k$ in a remarkable manner as illustrated in Fig. 1. At a small value of $k$, the iterate approaches a fixed value as $t$ goes to infinity (Fig. 1*a*). At an intermediate value of $k$, a bifurcation has taken place and the iterate approaches a stable cycle, with $x$ alternating between two values (Fig. 1*b*). In population biology this might correspond to, say, a year with many potato beetles followed by a year with few beetles. As $k$ is increased the 2-cycle changes into a 4-cycle (Fig. 1*c*) which in turn gives way to an 8-cycle and so on. Each period doubling bifurcation appears at a smaller increment of $k$ than the previous one and a limit value of $k$ exists beyond which the iterate is of infinite period (Fig. 1*d*). At this point it is worthwhile pausing and reflecting over the meaning of the period being infinite. Clearly, it means that there is no pattern that repeats itself in a finite number of iterations. This is what is expected from a stochastic process. There is striking similarity between the behaviour of (2.2) and that of the leaking faucet described in the introduction. The phenomenon that links the two systems, the cascade of period doubling bifurcations, is all-pervasive in non-linear dynamics. When it is observed chaos is usually not far away.

The nature of the iterate of the quadratic map (2.2) is shown as a bifurcation diagram in Fig. 2. The diagram was obtained by iterating the equation at a slowly changing value of the parameter $k$, sampling on the average every fiftieth $x$ and plotting it as a function of $k$. We observe that for $k < 3$ the iterate converges on a fixed point (c.f. Fig. 1*a*). At $k = 3$ the iterate bifurcates into a 2-cycle, at $k = 3.444$.. the 2-cycle bifurcates into a 4-cycle, etc. This sequence of pitchfork bifurcations continues with decreasing intervals of $k$ until at the critical value $k_c = 3.570$ .. the period length becomes infinite. The length of the interval between the consecutive points of period doubling, $\varDelta_i$, decreases at an almost exponential rate for increasing $i$ (Grossmann & Thomae, 1977). Beyond the point of accumulation the sample points of the iterate of (2.2) spread randomly within four different intervals which expand and merge to cover the unit interval as $k$ approaches 4. There are however some visible exceptions to the random distribution of the points. A prominent one is observed at $k = 3.82$ .. where a 3-cycle appears and bifurcates into a 6-cycle, then a 12-cycle and so on as $k$ is increased. Intervals of $k$ where the iterate is periodic are called windows. There are windows for all uneven cycles from 3 to infinity. However, windows for cycles higher than 5 are so narrow that they cannot be distinguished in a graph with the resolution of Fig. 2.

Instead of localizing the 3-cycle by sampling the iterate as in Fig. 2 one could have solved the equation

$$x = F(F(F(x)))  \tag{2.3}$$

which equates the third iterate of $x$ with $x$ itself. Doing this we would find that (2.2) has two different 3-cycles. Only one of these cycles show up in the bifurcation diagram. In fact only one cycle is an attractor or stable cycle. The other one is a repellor or an unstable cycle. There are unstable cycles for all integers. In order to illustrate this point we have added the unstable 2-cycles and 3-cycles to the section of the bifurcation diagram of the quadratic map shown in Fig. 2. We observe that the curves corresponding to stable cycles in the window continue as unstable cycles outside the windows.

The above observations made on the quadratic map (2.2) are general for continuous one-dimensional difference equations that transform two-to-one. Some of the features are also found in difference equations of higher dimensions. Interestingly, the study of period doubling bifurcations in one-dimensional difference equations has revealed a new universal constant, Feigenbaums number, $\delta$, defined as the rate of convergence of the parameter value to the point where period doublings accumulate (Feigenbaum, 1978).

$$\delta = \lim_{i \to \infty} \frac{\varDelta_i}{\varDelta_{i+1}}.  \tag{2.4}$$

The meaning of $\varDelta$ is shown in Fig. 2. Some but not all real systems with period doubling bifurcations agree with (2.4).

518

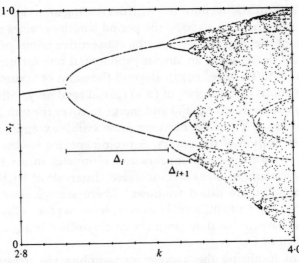

Fig. 2. Bifurcation diagram for (2.2). The diagram is drawn by increasing $k$ in very small steps and for each $k$ iterating the equation on the average 50 times and plotting the resulting $x_t$ against $k$. $\Delta$ indicates the distance between consecutive period doublings. The dashed curves indicate unstable cycles of period 2 and 3. The unstable cycles were calculated by reverse iteration as explained in Section 4.

In order to discover the origin of the period doubling bifurcations let us again study the quadratic map. Fig. 3 shows the graph representing (2.2) for different values of $k$. Included in the figure is also the identity line

$$x_{t+1} = x_t. \tag{2.5}$$

The points of intersection of the two curves are the fixed points

$$x = 0 \quad \text{and} \quad x^* = 1 - 1/k.$$

The slope of the tangent to the curve at the fixed point $x^* = 1 - 1/k$ determines the stability of that point. When the slope is between 0 and $-1$ the fixed point is stable. At any initial $x$ the iterate is attracted to the fixed point. When the slope is less than $-1$ the fixed point becomes unstable and the iterate diverges from the fixed point irrespective of how close the initial $x$ is to this point. Differentiating (2.2) and inserting the value $1 - 1/k$ we find that $x^*$ becomes unstable when $k > 3$. At the same value of $k$ a 2-cycle is born. This can be seen by solving the expression for the second iterate of (2.2)

$$x_t = k^2(x_t - (k+1)x_t^2 + 2kx_t^3 - kx_t^4). \tag{2.6}$$

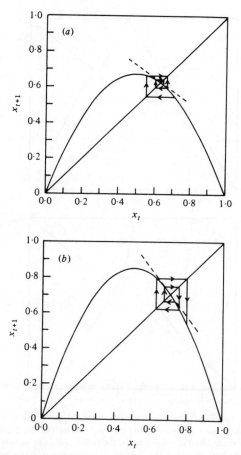

Fig. 3. Graphical representation of (2.2) for (a) $k = 2\cdot7$ and (b) $k = 3\cdot4$. The dashed line is the tangent of the curve at the fixed point where the curve intersects the identity line $x_{t+1} = x_t$. The lines marked with arrows represent graphic iteration. An horizontal arrow transforms the previous $x_{t+1}$ into $x_t$ for the next iteration and a vertical arrow determines $x_{t+1}$ from $x_t$. The graphical iterations indicate that the iterate (a) converges on or (b) diverges from the fixed point depending on whether the slope of the tangent at the fixed point is larger than $-1$ or less than $-1$, respectively.

We already know two solutions to this equation, namely $x = 0$ and $x = 1 - 1/k$. When these are eliminated (2.6) reduces to

$$kx^2 - (k+1)x + 1 + 1/k = 0 \qquad (2.7)$$

which has two positive real roots for $k > 3$.

The same result may be obtained graphically by drawing $x_{t+2}$ versus $x_t$ for different values of $k$ and inspecting the intersections of

520

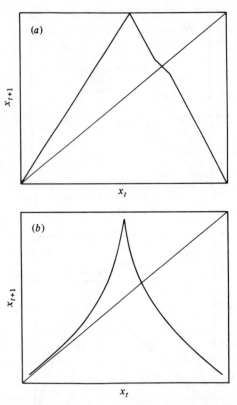

Fig. 4. (*a*) Piecewise linear transfer function whose slope at the fixed point can be adjusted without the maximum being affected. When the fixed point is stable and the initial *x* is far from the fixed point a chaotic transient occurs. When the fixed point is slightly unstable the iterate is intermittent chaos as shown in Fig. 5. (*b*) Transfer function with its left limb almost tangential to the identity line. Maps of this type may also produce intermittency.

the curve with its identity line $x_{t+2} = x_t$. We now apply the same reasoning to the stability of these points as we did to the stability of the fixed point of the first iterate. The analytical work becomes tedious but we may verify that at $k = 3.444..$ the absolute values of the slopes of the tangents to the fixed points of the second iterate exceed 1 and a 4-cycle is born which itself becomes unstable at $k = 3.544..$, etc. The sudden appearance of a window with a stable cycle in a chaotic region is due to a tangent bifurcation that has a different origin than a pitchfork bifurcation. A tangent bifurcation occurs whenever the curve $x_{t+n} = F(x_t)$ happens to have its identity line $x_t = x_{t+n}$ as a tangent.

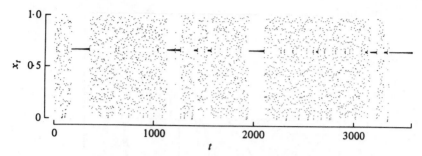

Fig. 5. Intermittent chaos produced by the iteration of piecewise linear transfer function of the type shown in Fig. 4a.

The prominent road to chaos in the quadratic map and in all similar one-humped maps is a sequence of period doubling bifurcations beginning when the parameter has been increased to a point where the slope at the fixed point is less than − 1 and the fixed point becomes unstable. However, the fixed point does not have to be unstable for chaos to occur in a one-humped map. In some maps where only a narrow range of attraction exists around the fixed point one may observe very long chaotic transients which perpetuate until by chance the iterate hits within the range of attraction and settles down in the fixed point. A system which has two different states, one stable state and one labile state is called monostable. The case discussed above would then be called monostable chaos. Another interesting situation arises if the slope at the fixed point in a map of the type shown in Fig. 4a is locally slightly less than − 1. The fixed point will repel very weakly so that it may take many iterations to leave the quasi-stable neighbourhood of the fixed point. Once the iterate has left this neighbourhood it suddenly becomes fully chaotic, only to get caught in the quasi-stable neighbourhood again sooner or later. This phenomenon which will repeat itself with a random period is called intermittency, a sample of which is shown in Fig. 5. Intermittency can also be found in connection with tangent bifurcations (Manneville & Pomeau, 1979, 1980) as illustrated by the map in Fig. 4b. Here the curve is very close to the identity line near the origin. This part of the curve is quasi-stable because it takes many iterations to get away from it. Once the iterate has escaped it becomes chaotic and remains so until by chance it hits the quasi-stable region again etc. Since the dynamics of the maps such as shown in Figs. 4a and 4b can be changed from stable states to intermittent states by the variation of parameters, these maps allow transition to chaos through intermittency. This is observed in many real systems.

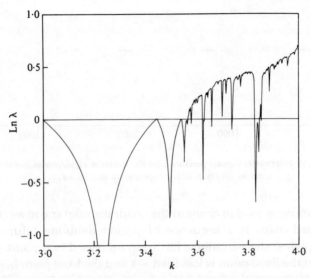

Fig. 6. Plot of the Lyapunov exponent against $k$ for (2.2). The plot was obtained by iterating the equation 10000 times at each $k$ value at a slowly increasing $k$. At each $x_t$ the numerical value of the slope is calculated and the logarithm of the slope is added to the sum of logarithms of the previous slopes. The Lyapunov exponent is then obtained by dividing the sum by 10000.

## 3. THE LYAPUNOV NUMBER

If (2.2) is iterated twice at $k = 4$ with two initial $x$ values very close to each other the iterates are observed to diverge rapidly from each other. Sensitivity to the initial conditions is a general property of systems which exhibit chaotic behaviour. This property can be quantified and made to serve as a measure of how chaotic a certain process is (Farmer, Ott & Yorke, 1983). The quantity used for this purpose is the Lyapunov number, $\lambda$, defined as

$$\lambda = \lim_{t \to \infty} J_t^{1/t} \tag{3.1}$$

where $J_t$ is defined as

$$J_t = \left| \frac{dF(x_t)}{dx} \right| \cdot \left| \frac{dF(x_{t-1})}{dx} \right| \cdot \ldots \cdot \left| \frac{dF(x_2)}{dx} \right| \cdot \left| \frac{dF(x_1)}{dx} \right|. \tag{3.2}$$

The Lyapunov number is the average absolute value of the slope of the map $x_{n+1} = F(x_n)$ at the points of the iterate. When the iterate converges on a fixed point or on a cycle the Lyapunov number is smaller than one. When the Lyapunov number is larger than one no stable cycles exist and for two close initial conditions the iterates will

7

diverge exponentially. The maximum possible value of the Lyapunov number is two for a two-to-one transformation, it is three for a three-to-one transformation and so on.

In practice the Lyapunov number is often evaluated numerically as $\ln \lambda$ which we shall call the Lyapunov exponent. Fig. 6 shows a plot of the Lyapunov exponent for (2.2) as a function of $k$. The method of calculation is explained in the figure legend. A comparison of this spectrum of Lyapunov exponents with the bifurcation diagram in Fig. 2 is instructive. For $k < 3.570$. the Lyapunov exponent is negative throughout, approaching zero at the bifurcation points. At $k = 3.570$. . the exponent becomes positive for the first time and the spectrum looses its regularity. Beyond this value of $k$ there are certain narrow intervals of $k$ at which the exponent is negative. These intervals coincide with the windows corresponding to the stable cycles in the bifurcation diagram.

So far we have used the term chaotic behaviour without giving a precise definition of this term. It is evident from the previous sections that chaotic behaviour has something to do with cyclic behaviour. We may define chaos as a cyclic phenomenon without a fixed period. Since the absence of a fixed period is linked with the absence of stability, the Lyapunov number, which is an average measure of the stability of the equation, may be used to give a quantitative definition of chaos. For a one-dimensional difference equation chaos can be defined as the state of the iterate when the Lyapunov number is greater than one (Farmer *et al.* 1983). As we shall see later, this definition may be expanded to two or higher dimensional difference equations and also to systems of coupled first order differential equations. The Lyapunov number has proved useful as a test for chaos in experimental data that could be approximated by a one-dimensional difference equation (Hudson & Mankin, 1981).

Another way to define chaos in one-dimensional difference equations is by the help of a theorem by Li & Yorke (1975) which can be stated as: Period three implies chaos. According to this definition chaos occurs in (2.2) when the 3-cycle has appeared, that is for $k > 3.8284$. . . . However, when chaos is defined by the help of the Lyapunov number chaos occurs when the Lyapunov number is greater than one, that is when $k > 3.570$. . . This discrepancy is solved if we abandon the Li and Yorke theorem and make use of a more general theorem by Sharkovsky (Kloeden, Deakin & Tirkel, 1976). According to Sharkovskys theorem there is an uncountably large number of initial conditions that do not lead to periodic

behaviour when all even cycles occur. For (2.2) this happens at
$k > 3\cdot570$.. Sharkovskys condition for chaos is therefore in agreement
with the one based on the Lyapunov number at least for the quadratic
map.

## 4. CHAOS AND DETERMINISM

Even after having got acquainted with a transparent case of chaos as
provided by (2.2) one may still feel uneasy about the whole business
because there appears to be a paradox. A stochastic process originates
from a deterministic system. The way to dismantle this paradox is
to sharpen the definition of a 'deterministic system'. Let us define
a deterministic system as one whose future and past are both unique
functions of the present. Clearly the quadratic map (2.2) is not
deterministic according to this definition because it is non-invertible.
The future is a unique function of the present but there are two
different pasts of equal probability.

We shall elaborate a little on the invertibility of difference equations,
again using (2.2) as an example. The inverse of (2.2) is

$$x_{t-1} = \tfrac{1}{2} \pm \tfrac{1}{2}\sqrt{(1 - 4x_t/k)} \qquad (4.2)$$

The fact that (4.2) is double valued does not preclude its iteration as
long as one first defines a rule how to choose one of the two values
(Degn, 1982). If the rule is that the high value is chosen every time
then for $k = 4$ the iteration of (4.2) converges on $x = \tfrac{3}{4}$ which is the
unstable fixed point $x$ of (2.2). If we iterate (4.2) choosing the low
value every time we obtain the other unstable fixed point $x = 0$ of
(2.2). Generally, what is unstable in the iteration of the forward
process is stable in the iteration of the inverse process and vice versa.
This also is true for cycles. If for example we choose high–high–low
repeatedly the iterate of (4.2) converges on the 3-cycle of (2.2)
$0\cdot950484$, $0\cdot611260$, $0\cdot188255$. If we choose high–low–low the iterate
converges on the 3-cycle $0\cdot969846$, $0\cdot413176$, $0\cdot116978$. Since the
convergence is fast this provides a convenient way to determine
unstable cycles of the forward process. Since all cycles of (2.2) are
unstable for $k = 4$ they are all stable in (4.2) for $k = 4$. Therefore the
iteration of (4.2) with the parameter slowly decreasing from 4 will
produce the unstable cycles of (2.2) for all values of $k$ down to the
window of the particular cycle calculated. This method which works
for any one-dimensional map was used to produce the unstable cycles

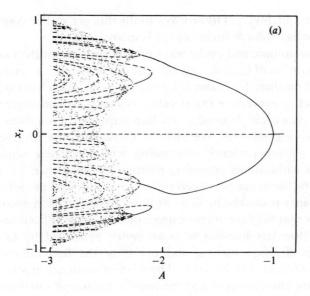

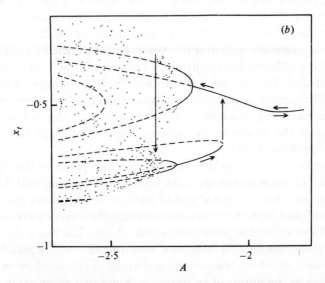

Fig. 7. (a) Bifurcation diagram of (5.1) for $B = 0.01$. The diagram was constructed in the same way as that in Fig. 2 except that the iterate of the equation was plotted for increasing as well as decreasing values of $A$. The dashed curves are unstable cycles of periods 1, 2 and 4. (b) Enlarged section of (a). The horizontal arrows indicate the path of the iterate for increasing and decreasing $A$. The vertical arrows indicate the crisis points at which the iterate jumps from one attractor to the other.

in Fig. 2 and Fig. 7. Other ways to do this are rather complicated (Metropolis, Stein & Stein, 1973; Kaplan, 1983).

The monotonic and cyclic ways of choosing one of the two values in the iteration of (4.2) do not exhaust the possibilities. One can also choose at random, for example by tossing a coin at each iteration step. With such a procedure the iterate obviously cannot converge on a point or on a cycle. Instead it will be a sequence of numbers without any repeating pattern. We already know that the inverse of the process we are presently discussing will produce a sequence of numbers without any repeating pattern. We may then ask whether the chaotic sequence produced by (2.2) has different properties from the sequence produced by (4.2) with stochastic decision making. The answer is that they are mirror images of each other. This is seen when we take the last number of a sequence produced by (4.2) with stochastic decision making and use it as the first number of an iteration of (2.2). The iterate will be the first sequence in reverse. The interesting conclusion is that the same sequence of numbers can be produced by a stochastic as well as a deterministic process if we consider the iteration of (4.2) supplemented with coin tossing a stochastic process and we consider the iteration of (2.2) a deterministic process. Since there can be no doubt about the stochasticity of the former process there is something suspicious about the determinism of the latter. Actually the latter process is deterministic in a peculiar sense as it is predetermined to produce random numbers, namely the ones produced by the reverse process with stochastic decision making. Thus chaos in a one-dimensional difference equation can be viewed as reversed random walk (Degn, 1982). An interesting discussion of the stochasticity of one-dimensional maps has been published by Ford (1983).

The shape of the curve did not enter at any point in the above discussion. The arguments are valid for any one-dimensional difference equation whatever the shape of the curve. It follows that any one-dimensional difference equation that maps the interval two-to-one onto itself can exhibit at least monostable chaos. The same must be true for transformations three-to-one, four-to-one, etc. We recall that the Lyapunov number represents the average of the absolute values of the slopes at the points of an iteration. For a two-to-one transformation that maps the interval onto itself the Lyapunov number is less than or equal to 2, for a three-to-one transformation it is less than or equal to 3, etc. If we integrate over the interval the absolute values of the slopes of a two-to-one transformation that maps the interval onto itself we obtain the number 2 irrespective of the shape of the curve. If we make the same integration of a three-to-one transformation that maps the interval onto itself we obtain the number 3, etc. If the Lyapunov number of such a map is equal to the integral we may conclude that all numbers of the interval have the same probability ($P = 0$) of being present in the iterate, and the process is ergodic. This is for example the case with (2.2) when $k = 4$.

## 5. BISTABLE CHAOS

In the preceding sections we have seen that difference equations with one extremum may exhibit period doubling bifurcations, stable chaos and monostability. Here we shall briefly discuss properties of difference equations with two extrema. Consider for example the cubic difference equation

$$x_{t+1} = Ax_t + B(1 - x_t^2) + (1 - A)x_t^3 \tag{5.1}$$

where $A$ and $B$ are parameters. For appropriate values of $A$ and $B$ (5.1) maps the interval $(-1, 1)$ into itself. For the special case $B = 0$ the map has two separate antisymmetric chaotic attractors (May, 1980; Testa & Held, 1983). Raising the parameter $A$ causes a so-called split bifurcation doubling the number of stable attractors. In the bifurcation diagram a split bifurcation looks like a pitchfork bifurcation with one arm missing. Both arms exist but only one can be realized at a time. Which one is realized depends on the initial value of $x$. The set of initial values which lead to the same attractor is called the basin of that attractor. The basins of the two attractors of (5.1) are not continuous. They consist of infinitely many intervals of fractal structure (see the following section).

In the case where $B \neq 0$ the cubic map (5.1) does not carry any symmetry. However, it may still have two distinct chaotic attractors for certain values of $A$ (Skjolding et al. 1983). This opens the possibility for bistable chaos as illustrated by the bifurcation diagrams of (5.1) shown in Fig. 7. The diagrams are produced in the same way as the one in Fig. 2. In order to reveal the bistability, the bifurcation diagram is plotted at increasing as well as decreasing values of the parameter $A$. Within a certain interval of $A$ the iterate will settle in one of two possible attractors depending on whether $A$ is increasing or decreasing. For all values of $A$ outside this interval there is only one attractor. Fig. 7b is enlarged section of Fig. 7a with arrows indicating the direction of change of the parameter.

At the boundaries of the interval where two different stable attractors are seen in Fig. 7 the iterate undergoes abrupt changes following small changes in $A$. There is an abrupt change from periodic motion to chaotic motion or vice versa. This phenomenon has been named crisis (Grebogi, Ott & Yorke, 1982, 1983). It has been shown to coincide with a collision between the chaotic attractor and an unstable cycle. In order to illustrate this the unstable $x$-cycles have been added to the bifurcation diagrams in Fig. 7. In the case of decreasing $A$ this collision results in a sudden change in the size and location of the chaotic attractor. In the case of increasing $A$ a collision between a periodic attractor and an unstable cycle gives rise to a transition from one periodic cycle to another. The bistability and crisis phenomena are accompanied by the existence of two different Lyapunov numbers at the same parameter value. The Lyapunov numbers change abruptly at the crisis points. Such phenomena can be found in many two-parameter maps. Bistability and the associated crisis probably does not exist in a one-parameter map.

## 6. Difference equations of two and higher dimensions

To our knowledge there are as yet no examples in the biological literature of the use of difference equations of two or higher dimensions. However we shall discuss some properties of such equations because they can be derived from differential equation models of relevance in biology.

The most thoroughly studied two-dimensional map is the Hénon map (Hénon, 1976)

$$\left.\begin{aligned} x_{t+1} &= 1 - ax_t^2 + y_t, \\ y_{t+1} &= bx_t. \end{aligned}\right\} \tag{6.1}$$

For $b = 0$ this map reduces to a quadratic one-dimensional map with identical properties to those of (2.2). Since it is possible to solve (6.1) to obtain explicit expressions for $x_t$ and $y_t$ (6.1) is said to be invertible. It is not possible to draw the graph of (6.1) since this would require a four dimensional representation. However, we may visualize its chaotic attractor by plotting successive points of the iteration of (6.1) as shown in Fig. 8a. This reveals that the map does not define a single curve but a bundle of curves. This bundle has a fine-structure that is seen when a small region of the bundle is enlarged. The enlarged

182   L. F. OLSEN AND H. DEGN

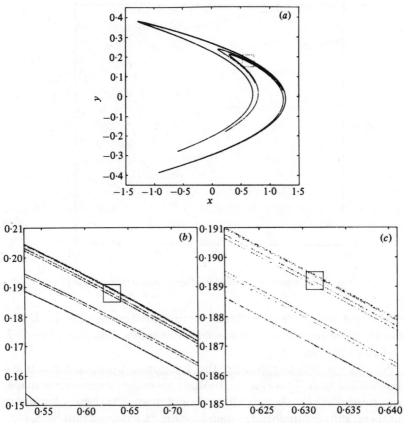

Fig. 8. (a) The chaotic attractor of the Hénon map (6.1) for $a = 1·4$ and $b = 0·3$. (b) Enlarged section indicated by box in (a). (c) Enlarged section indicated by box in (b). Reproduced with permission from Hénon (1976).

detail is similar to the original overall structure. Further enlargement gives the same result. This phenomenon is illustrated in Fig. 8b and c. When the overall structure of an object is found in successive enlargements of details, the object is said to be self-similar or fractal (Mandelbrot, 1977). Self-similarity is often found in two-dimensional difference equations with chaotic attractors. A cut normal to the bundle of curves produces a set of points that has features similar to the Cantor set, one of the simplest fractal objects in mathematics.

In the Hénon map there are two parameters. Therefore the dynamics of the map can be described by a bifurcation diagram at a constant value of $a$ and varying $b$ or the other way round. Fig. 9 shows the bifurcation diagram of the Hénon map for $0·2 < a < 1·4$

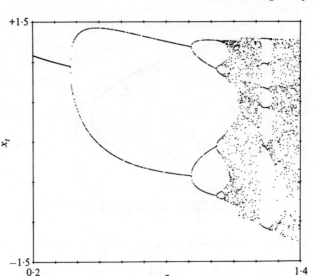

Fig. 9. Bifurcation diagram of the Hénon map (6.1) for $b = 0.3$.

with $b$ fixed at $0.3$. We note that as for the one-dimensional quadratic map (2.2) the transition to chaos goes through a sequence of period doubling bifurcations.

We have previously defined the Lyapunov number for a one-dimensional map. This can also be done for higher dimensional maps (see Appendix I). A two dimensional map has two Lyapunov numbers, a three dimensional map has three Lyapunov numbers, etc. Also in maps of higher dimensions the Lyapunov numbers can be used to diagnose chaos. Fortunately it is not necessary to evaluate all the Lyapunov numbers for this purpose because chaos exists in an $n$-dimensional map when the largest of the Lyapunov numbers is greater than one (Grebogi *et al.* 1983), and there are simple ways of calculating the largest Lyapunov number (see Appendix I). Fig. 10 shows a plot of the largest Lyapunov exponent of the Henon map against $a$ for $0.9 < a < 1.4$.

For certain values of $a$ and $b$ the Hénon map has two different chaotic attractors (Simo, 1979) with their own basins of attraction and with different Lyapunov numbers. A detailed account of this and an extensive discussion of crisis in the Henon map is given in a review by Gregobi *et al.* (1983).

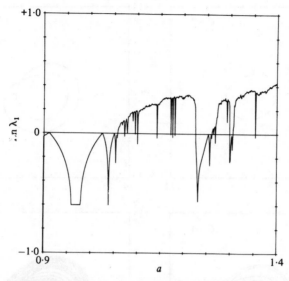

Fig. 10. Plot of the largest Lyapunov exponent against the parameter $a$ for the Hénon map (6.1). The method of calculation is described in Appendix I.

## 7. NON-LINEAR DIFFERENTIAL EQUATIONS

The most common way to simulate dynamical systems is by the help of first order differential equations. Thus a system with $n$ independent variables is described by $n$ coupled differential equations. There are two conditions to be fulfilled for chaos to occur in such equations:

1. The right hand side of the equations must contain at least one non-linear term.

2. There must be at least three independent variables.

Let us examine a simple chaotic differential equation system, the Rossler (1976$b$) equations

$$\left.\begin{aligned}
\dot{x} &= -(y+z), \\
\dot{y} &= x+ay, \\
\dot{z} &= b+z(x-c)
\end{aligned}\right\} \tag{7.1}$$

which contain only one non-linear term, $zx$. This is one of the simplest differential equation models capable of exhibiting chaotic behaviour. Fig. 11 shows the projection of the trajectories of (7.1) on the $(x, y)$ plane for different values of $c$. We observe that as $c$ is increased the trajectories change from a simple limit cycle with one

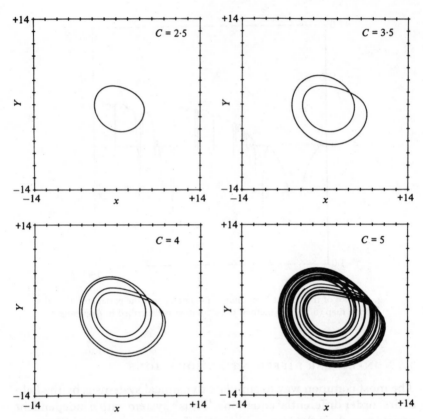

Fig. 11. $X-Y$ phase plot of the Rossler equation (7.1) at four different values of $c$ indicated in the graphs and $a = 0.2$ and $b = 0.2$.

maximum over a limit cycle with two maxima to a limit cycle with four maxima and so on until at $c = 5$ there is no longer any visible periodicity.

There is an obvious resemblance between the behaviour of the Rossler equations and that of the quadratic map (2.2). This becomes striking if we plot the maxima of the Rossler attractor as a function of the parameter $c$ as shown in Fig. 12. The graph is a bifurcation diagram showing a route to chaos through period doubling bifurcations similar to the bifurcation diagram of (2.2) shown in Fig. 2. One might then ask whether there exists a difference equation that would produce the same bifurcation diagram as the Rossler equations. If so this difference equation would be a model of the phenomenon the Rossler equations are a model of. If we plot each maximum of the

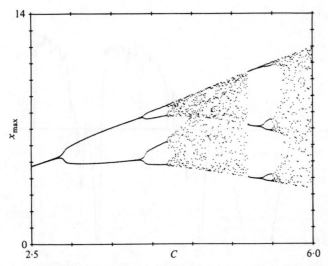

Fig. 12. Bifurcation diagram for the Rossler equation (7.1). The diagram is constructed essentially as those in Figs. 2, 7 and 9. Note the broad window of period 3.

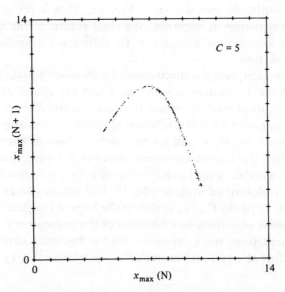

Fig. 13. Next amplitude plot of the Rossler equation (7.1) for $c = 5$, $a = 0.2$ and $b = 0.2$. Each amplitude of the oscillation of $x$ was plotted against the preceding amplitude.

534

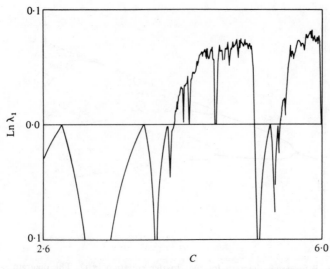

Fig. 14. Plot of the largest Lyapunov exponent of the Rossler equation (7.1) against $c$ for $a = 0\cdot2$ and $b = 0\cdot2$. Reproduced with permission from Nicolis, Mayer-Kress & Haubs (1984).

Rossler attractor as a function of the preceding maximum we obtain the next amplitude plot shown in Fig. 13. This is the graph of the difference equation in question. We shall return to the question of modelling continuous systems with difference equations in the following section.

In connection with the discussion of difference equations we have presented the Lyapunov number as a test for chaos. We shall do likewise for differential equations. In analogy with difference equations a set of $n$ coupled first order differential equations has associated with it $n$ Lyapunov numbers and we may define chaos as the state of the system when the largest Lyapunov number is larger than one. One method of calculating the Lyapunov number for a system of differential equations is described in Appendix II. It involves a large number of integration steps. In Fig. 14 is shown the largest Lyapunov exponent of the Rossler equations as a function of the parameter $c$. According to our calculations the Lyapunov number becomes larger than one and therefore the system becomes chaotic when $c > 4\cdot233$.

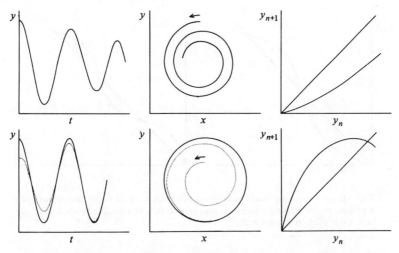

Fig. 15. Three different graphical representations of damped oscillations (upper panel) and limit cycle oscillations (lower panel). From left to right: Time series, phase trajectory and next amplitude plot.

## 8. DISCRETE REPRESENTATION OF CONTINUOUS PROCESSES

It was hinted above that the so called next amplitude plot can bridge the gap between differential equations and difference equations, i.e. between continuous and discrete representations. Let us therefore examine the physical meaning of next amplitude plots. Fig. 15 a shows three different representations of a damped oscillation, namely a three dimensional time series, a two dimensional phase trajectory and a one-dimensional next amplitude plot taken in the order of decreasing information content. The phase trajectory cannot be reconstructed from the next amplitude plot as the time series cannot be reconstructed from the phase trajectory. Actually the next amplitude plot retains only the stability properties of the original time series. The next amplitude plot of a damped oscillation is always below the identity line, intersecting it at the origin. The point of intersection corresponds to the stable focus in the phase plane. The next amplitude plot of a limit cycle oscillation intersects the identity line twice, at the point corresponding to the unstable focus in the phase plane and at the point corresponding to the limit cycle (Fig. 15 b). With these observations in mind one can easily translate the shape of a next amplitude plot into statements about the stability of the system studied. For example the next amplitude plot in Fig. 16 a which intersects the identity line at three points depicts a bistable system with a stable focus, an

536

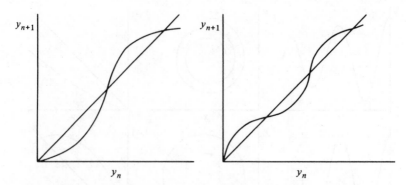

Fig. 16. (*a*) Next amplitude plot of system with a stable focus, anti-limit cycle and limit cycle. (*b*) Next amplitude plot of system with unstable focus, limit cycle, anti-limit cycle and limit cycle.

anti-limit cycle and a limit cycle. Likewise Fig. 16*b* depicts a system with an unstable focus, a limit cycle, an anti-limit cycle and a limit cycle.

Beside variations in the amplitude, non-linear oscillators can also show variations in the period. A plot of each period as a function of the preceding period has similar properties as the next amplitude plot. The slope of the next period plot at the fixed point determines the stability of the fixed point. The next period plot offers one advantage over the next amplitude plot, namely that real time is involved. Along with the iteration of the next period plot one can sum up the periods to account for the total time.

$$\left.\begin{aligned} p_{n+1} &= f(p_n), \\ t_{n+1} &= t_n + p_{n+1} \end{aligned}\right\} \tag{8.2}$$

where $t_n$ is the total time at the end of period $n$. This makes it possible to study periodic perturbations of difference equations as described in the next section.

The next amplitude plot is a special case of maps known as Poincaré maps (Shaw, 1981). A Poincaré map is constructed from the cross section of the phase trajectories normal to the flow in phase space. When the trajectories form a very thin sheet the cross section is almost one-dimensional. A plot of each point of the Poincaré section as a function of the preceding point may then define an apparently one-dimensional map as shown in Fig. 13. In many cases this does not hold and Poincaré cross sections look more like the attractor of the Henon map shown in Fig. 8.

Another type of discrete representation of a continuous system is the stroboscopic phase plot (Tomita, 1982). This is based on a similar idea as the Poincare map. Instead of determining the intersections of the phase trajectories with a plane the points are collected at a constant time interval. One advantage of the stroboscopic phase plot is that the points are easier to calculate than the Poincare sections except when the latter have been constructed from successive maxima or minima. Both types of maps have been used successfully in the analysis of both numerical and experimental data from biological systems. Examples will be discussed in more detail later.

There is no guarantee that a next amplitude plot of a real system is a well defined curve on the plane as it is for a two-dimensional system. However, it is not unusual to find such well defined curves from systems of higher dimensions. This is for example the case with the next amplitude plot of the Rossler equations at least for some values of the parameters as shown in Fig. 13. It is a one-humped map reminiscent of the quadratic map. For other values of the parameters it looks more like the cubic map (Fraser & Kapral, 1982). If one fits a curve to the next amplitude plot one can iterate the map and calculate its Lyapunov number which is then identical to the largest Lyapunov number of the differential equations under provisions discussed in Appendix II. The Lyapunov number cannot be obtained this way when the next amplitude plot is not a well defined curve.

## 9. PERIODIC PERTURBATIONS

It seems to be the rule rather than the exception that exposing a non-linear oscillator to periodic perturbations causes the oscillation to become chaotic for some parameter values. This effect has been studied in differential equation models (Tomita, 1982) and in difference equations (Degn, 1983). In the physical laboratory chaotic responses to periodic perturbations have been found in simple non-linear electronic circuits and in lasers. The effects of periodic perturbations on non-linear systems are of great interest in biology because many biological control systems are exposed to periodic perturbations from the environment or from other systems within the same organism.

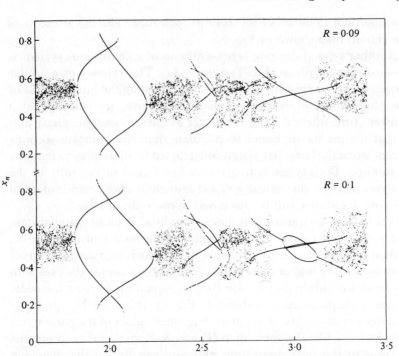

Fig. 17. Bifurcation diagrams of periodically perturbed period transfer function at two different values of the strength factor $R$ indicated in the graphs and the slope factor $A = -0.8$. On the average every fiftieth number was plotted when (9.1) was iterated at slowly increasing and decreasing value of the frequency factor $F$. The borders of the windows depend on the direction of change of $F$ revealing the existence of bistability.

A simple example of a perturbed difference equation is the periodically perturbed linear map

$$\left.\begin{aligned} p_{n+1} &= A(p_n - \tfrac{1}{2}) + \tfrac{1}{2} + R \sin (2\pi F t_n), \\ t_{n+1} &= t_n + p_{n+1} \end{aligned}\right\} \tag{9.1}$$

where $t$ denotes time and $A$, $R$ and $F$ are parameters (Degn, 1983). In the absence of perturbations, e.g. when the strength parameter $R$ is zero, (9.1) reduces to a linear next period map with its fixed point at $p^* = \tfrac{1}{2}$. For $-1 < A < 0$ the reduced equation corresponds to a limit cycle oscillation. However, when $R > 0$ and $F > 0$ (9.1) may be chaotic. Fig. 17 shows samples of the iterate of (9.1) at two different constant values of $R$ and a slowly changing value of $F$. Chaotic solutions are seen to alternate with periodic windows and bistability is revealed when the direction of change of $F$ is reversed. An example of remerging bifurcations is seen in the lower panel of Fig. 17. Equation (9.1) is closely related to an equation known as Chirikov's standard mapping (Chirikov, 1979), the difference being that (9.1) describes dissipative systems whereas Chirikov's equation describes Hamiltonian or energy conserving systems. Chaos in Hamiltonian systems is extensively discussed in the physical literature. It does not seem to be relevant in biology.

The addition of a periodic perturbation to a set of differential equations is equivalent to the introduction of a new variable, time. As a consequence it is not necessary to have at least three coupled differential equations to obtain chaos (Tomita, 1982). Two are sufficient. The stroboscopic phase plot is particularly useful in connection with periodically perturbed differential equations because the sampling period can be made identical to the perturbation period.

## 10. DIFFERENTIAL EQUATIONS WITH TIME DELAY

In some biological systems with continuous dynamics the change in state at time $t$ depends not only on the present state of the system but also on the state of the system at some fixed time $t - T$ in the past. If we denote the state by $x$, and for simplicity let $x$ be a real number, we may put the above into the form

$$dx(t)/dt = F(x(t), x(t-T)). \qquad (10.1)$$

Such an equation is called a differential-delay equation. Very few tools for the investigation of differential-delay equations are available. We shall restrict ourselves to mentioning that like difference equations and differential equations they may have stable points, stable cycles, period doubling bifurcations and chaos. This is illustrated in Fig. 18 for the equation

$$\frac{dx(t)}{dt} = \frac{AB^n x(t-T)}{B^n + x(t-T)^n} - Cx(t) \qquad (10.2)$$

where $A$, $B$, $C$ and $n$ are parameters. In order to obtain chaos only one differential-delay equation is needed. We observe that (10.1) and (10.2) have two variables $x(t)$ and $x(t-T)$. In addition the time $t$ participates as a third variable in a less obvious way. Therefore one differential-delay equation may be equivalent to a set of three coupled differential equations.

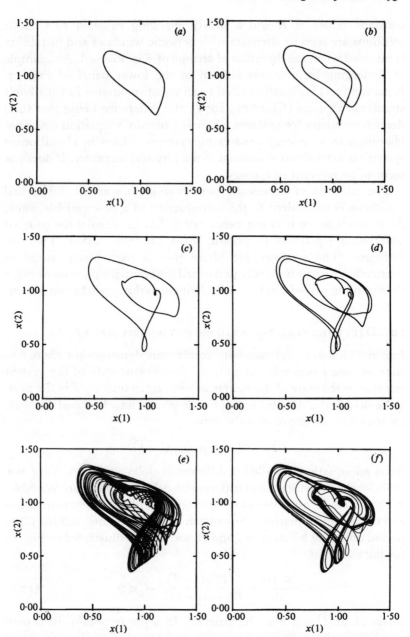

Fig. 18. Bifurcating solutions of (10.2) in the form of phase plots of $x(t-T)$ against $x(t)$. $A = 2$, $B = 1$, $C = 1$ and $T = 2$. (a) $n = 7$; (b) $n = 7.75$; (c) $n = 8.5$; (d) $n = 8.79$; (e) $n = 9.65$ and (f) $n = 9.69715$ $x(1)$ and $x(2)$ correspond to $x(t)$ and $x(t-T)$ respectively. Reproduced with permission from Glass & Mackey (1979).

194    L. F. OLSEN AND H. DEGN

Differential-delay equations are particularly useful in the description of population dynamics with continuous overlapping generations, in the dynamics of epidemics (May, 1980), and in simulations of physiological feed-back systems. In the latter case differential-delay equations may find clinical applications in explaining certain respirational and haematological disorders (Mackey & Glass, 1977; Glass & Mackey, 1979) and epileptic seizures (Kaczmarek & Babloyantz, 1977).

11. DIAGNOSIS OF CHAOS

In the previous sections we have described chaos from the viewpoint of mathematics and modelling. This may not be very useful to the experimentalist who observes some wild fluctuations in his measured data but who has no mathematical model of his system. This is often the situation of the experimental biologist. He needs some rules to tell chaos and noise apart. A commonly used method to analyse irregular oscillations is to construct a power spectrum in which the number of occurrences of oscillations of a given frequency is plotted against the frequency. For a periodic oscillation such a spectrum will show sharp peaks corresponding to the fundamental frequency and its higher harmonics whereas the spectrum of aperiodic oscillations will be a broad band with no discernible structure. A power spectrum has little diagnostic value with regard to chaos since it cannot distinguish between chaos and noise.

In order to diagnose chaos one important sign to look for is period doublings. These can usually be seen by direct inspection of a graph of a variable as a function of time. One may scrutinize either the amplitudes or the periods. In the following we assume the case of amplitude but the statements are also valid for period. If there are alternating high and low amplitudes or some repetitive pattern of amplitudes of different heights then it is advisable to change slightly some parameter and see if the periodicity is doubled or cut in half. If this is found to be the case, there is a good chance that changing the parameter in the direction of period doubling will lead to chaos. However, this is not invariably the case, since it is possible for a system to have period doubling bifurcations that do not go all the way to the point of accumulation where chaos begins (Bier & Bountis, 1984). On the other hand the absence of period doubling bifurcations does not necessarily imply that chaos is absent. The transition to chaos may also take place abruptly due to a crisis or through intermittency. A possible reason for not finding period doubling bifurcation when chaos is present may be that it is not practically possible to vary a parameter into the appropriate range.

When period doubling bifurcations are found one can try to determine the value of $\Delta$ of (2.4) and see whether Feigenbaum's number comes out. This is not as easy as it sounds. The distance between the parameter values at the bifurcation points decreases so rapidly that in practice one cannot determine more than the first three values of the sequence. This is not sufficient in order to estimate the limit value of (2.4). Even if a determination of $\Delta$ could be made it would not have much diagnostic value because also non-Feigenbaumian bifurcations usually lead to chaos.

If the measured data are a sequence of peaks without any obvious regularity then the first thing to do is to see if there is a functional relationship between successive amplitudes. This is done by the construction of a next amplitude plot. Each amplitude is plotted as a function of the preceding amplitude. The result may be anything from a well defined curve to a set of points without any discernible structure. If the next amplitude plot is a well defined curve there is the possibility that it maps two or more to one. In such case the system is chaotic. If the next amplitude plot shows some structure without being a well defined curve then the system may be chaotic and belong to a class exemplified by the Hénon map (6.1). If there is no discernible structure in the next amplitude plot it is an open question whether the fluctuations in the data are a result of noise or of chaos.

The experimentalist is often faced with the situation that only one variable can be measured in a system of many independent variables. Ideally one should be able to measure the time series of all the variables in order to describe the behaviour of the system. However, in non-linear dynamics methods have been developed to reconstruct a multidimensional attractor from the time series of a single observable. One way is by computing successive derivatives $\dot{x}(t)$, $\ddot{x}(t)$ . . ., etc. of the original time series $x(t)$ and plotting these in the resulting $n$-dimensional phase space (Packard *et al.* 1980). Often a three-dimensional representation $x(t)$, $\dot{x}(t)$, $\ddot{x}(t)$ will suffice. The use of this method requires smooth data which are not always obtainable. An alternative method is to choose a fixed time lag, $T$, and plotting values of $x(t), x(t+T), x(t+2T), \ldots, x(t+(m-1)T)$ in $m$-dimensional space. The value chosen for $T$ is not critical. If the system has $N$ independent variables the plot will have exactly the same properties as the plot of $x_1(t), x_2(t), \ldots, x_N(t)$ if $m > 2N+1$ (Takens, 1981). In

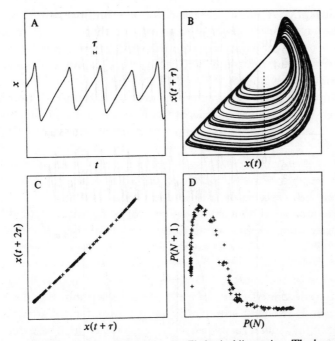

Fig. 19. Chaotic attractor of the Belousov–Zhabotinskii reaction. The bromide concentration, $x$, was recorded with a bromide sensitive electrode. ($A$) Bromide concentration as a function of time. $\tau$ is a time interval selected to construct a three dimensional phase trajectory $x(t)$, $x(t+\tau)$, $x(t+2\tau)$. ($B$) Two-dimensional projection of the phase trajectory. The dotted line indicates the position of the cut. ($C$) Cross section (Poincaré section) of attractor along cut. The Poincaré section is very thin indicating that the flow is confined to an almost two dimensional surface. ($D$) Almost one-dimensional map derived from the Poincaré section. Reproduced with permission from Roux, Turner, McCormick & Swinney (1982).

practice a much lower value of $m$ will do. The value of $m$ is selected so that increasing the value by one does not result in any additional structure. Fig. 19 illustrates the use of this method on a chaotic chemical reaction, the Belousov–Zhabotinskii reaction (Roux *et al.* 1982). Here $m$ could be set equal to 3. From the phase plot $x(t)$, $x(t+T)$, $x(t+2T)$ we obtain the Poincaré section by intersecting the trajectories in three-dimensional space with a plane normal to the paper as indicated by the dotted line. This Poincaré section is a thin sheet and the resulting discrete mapping, produced by plotting each point of the section against the preceding one is an almost one-dimensional map from which chaos can be diagnosed.

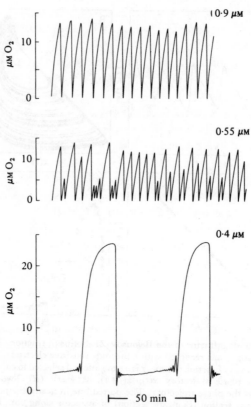

Fig. 20. Periodic and chaotic oscillations of the concentration of $O_2$ in the peroxidase–oxidase reaction in open system at three different enzyme concentrations indicated in the graphs. Reproduced with permission from Olsen (1983).

## 12. CHAOS IN PEROXIDASE REACTION

Shortly after its existence in chemical reaction systems was predicted (Rossler, 1976 *a*) chaos was observed in the oscillating peroxidase–oxidase reaction in an open system (Olsen & Degn, 1977). The reaction is the peroxidase catalysed reduction of molecular oxygen to water with NADH as the electron donor.

$$O_2 + 2NADH + 2H^+ \rightarrow 2H_2O + 2NAD^+.$$

In the open reaction system both substrates are fed continuously to a stirred, buffered medium containing a fixed amount of the enzyme horse radish peroxidase. Oxygen is entering by diffusion through the surface from a head-space gas of controlled oxygen partial pressure.

A concentrated solution of NADH is being infused at a constant rate. The peroxidase–oxidase reaction in the open system had previously yielded the first observation of bistability in a homogeneous chemical system (Degn, 1968) and it was the second enzyme system shown to oscillate (Yamazaki, Yokota & Nakajima, 1965), the first one being the glycolytic system (Ghosh & Chance, 1964).

The peroxidase–oxidase reaction is associated with spectral changes in the heam prosthetic group of the enzyme, spectral changes due to the oxidation of NADH and changes in dissolved oxygen concentration, all of which are easily measured. Fig. 20 shows an experiment where the oxygen concentration was recorded as a function of time at different enzyme concentrations. We observe that at the high enzyme concentration there is a stable oscillation. As the enzyme concentration is decreased the oscillation develops period multiplicity and finally it becomes completely irregular. Further decreases in the enzyme concentration result in the reappearance of periodic oscillations. The problem is to prove that these irregular spikes are what we call chaos and not a result of the reaction system being influenced by some source of noise. At the time of the discovery of these phenomena in the peroxidase–oxidase reaction there did not seem to be any other acceptable evidence than that of the next amplitude plot. Actually this turned out to be a set of points distributed quite narrowly around a bell-shaped curve and a geometrical construction showed that the Li and Yorke condition, namely the existence of period three, was fulfilled for the average curve. The rather stronger evidence of period doubling bifurcations was not taken into account.

The peroxidase–oxidase reaction is known to be a branched chain reaction with free radical intermediates. (Degn & Mayer, 1969; Olsen & Degn, 1978; Olsen, 1978, 1979). As is usually the case with simple oscillating reactions in homogeneous phase the reaction mechanism of the peroxidase oxidase reaction cannot be identified unambiguously. There are too many plausible intermediates that cannot be monitored in the concentration range and the time scale of the oscillating reaction. The problem with the peroxidase–oxidase reaction is mainly that the enzyme has five distinct active forms namely ferroperoxidase, ferriperoxidase, compound I, compound II and compound III. If an enzyme mechanism involving two distinct active forms of the enzyme is a ping-pong mechanism then the peroxidase–oxidase reaction is a pang-peng-ping-pong-pung mechanism. Fig. 21 shows two widely different examples of plausible reaction steps in the peroxidase–oxidase

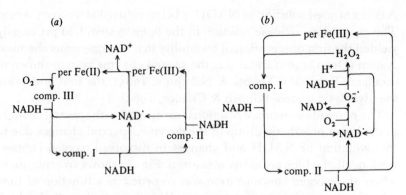

Fig. 21. Two autocatalytic schemes derived from known or proposed reaction steps in the peroxidase–oxidase reaction. Compounds I to III, per. Fe(III) and per. Fe(II) are enzyme intermediates. Reproduced with permission from Olsen (1984).

reaction joined together to form branched chain mechanisms. The truth probably is that several mechanisms of this type act together in a concerted mechanism (Olsen & Degn, 1978).

Because the mechanism of the peroxidase–oxidase reaction is not known in detail it cannot be expressed by a mathematical model where each intermediate is represented by a variable. However, it is possible to formulate a minimal model that simulates the experimentally established properties of the reaction system. There is strong evidence that the branched chain reaction in the peroxidase–oxidase reaction does not have simple linear branching but quadratic branching due to a positive interaction between chain propagating species (Olsen & Degn, 1978). The minimal chemical expression for quadratic branching is

$$
\left.
\begin{aligned}
A + X &\rightarrow 2X \\
2X &\rightarrow 2Y \\
A + Y &\rightarrow 2X
\end{aligned}
\right\} \tag{12.1}
$$

where $A$ is a reactant and $X$ and $Y$ are two free radical species. A scrutiny of (12.1) reveals that it includes two distinct feed back loops, one involving $X$ and the other one involving $Y$ and $X$. Since they have $X$ in common they are closely coupled.

Using the expression for quadratic branching (12.1) as a nucleus a minimal chemical model was constructed including all the important features of the peroxidase–oxidase reaction. The model is

$$
\left.
\begin{aligned}
B + X &\rightarrow 2X \\
2X &\rightarrow 2Y \\
A + B + \overset{\cdot}{Y} &\rightarrow 3X \\
X &\rightarrow P \\
Y &\rightarrow Q \\
X_0 &\rightarrow X \\
A_0 &\rightleftharpoons A \\
B_0 &\rightarrow B'
\end{aligned}
\right\} \qquad (12.2)
$$

where $A$ and $B$ denote $O_2$ and NADH respectively, $X$ and $Y$ are unidentified free radical intermediates and $P$ and $Q$ are products of the linear termination of $X$ and $Y$ respectively. The two last reaction steps simulate the inputs of $O_2$ and NADH from their respective sources. The model was translated into four coupled differential equations and studied on a computer. The results were periodic and chaotic oscillations closely resembling the experimental results (Olsen, 1983; 1984). A next amplitude plot of the chaotic solutions were in remarkably good agreement with the experimental data as shown in Fig. 22. Studies of similar models revealed that chaos could not be produced unless the model included two autonomous feed back loops.

As was pointed out above the peroxidase oxidase reaction has been a particularly rewarding object of study for those interested in non-linear phenomena in homogeneous systems (Degn, Olsen & Perram, 1979). One may wonder what is the biological significance of an enzyme residing in the roots of horseradish being capable of such wild things as bistability, oscillations and chaos. We do not have any answer to this question. There is no evidence that peroxidase behaves this way *in vivo*. Until evidence to the contrary has been collected we must assume that chaos in the peroxidase–oxidase reaction is an accidental result of the complicated branched chain nature of the reaction mechanism.

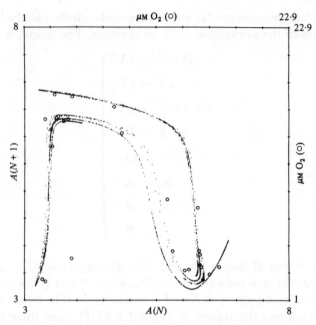

Fig. 22. Next amplitude plots of the chaotic oscillations in the peroxidase–oxidase reaction constructed from experimental data (O) and computer simulation (·). Reproduced with permission from Olsen (1984).

## 13. CHAOS IN GLYCOLYSIS

Unlike the peroxidase system the glycolytic system is known to produce oscillations *in vivo*. In fact glycolytic oscillations were first discovered in suspensions of actively metabolizing yeast cells (Ghosh & Chance, 1964). Later the oscillations were found in cell free yeast extracts as well (Chance, Hess & Betz, 1964). The minimal system required for glycolytic oscillations to occur involves several enzymes and reconstituting a system capable of oscillations from purified enzymes has proved difficult (Eschrich, Schellenberger & Hofmann, 1983). Accordingly studies of glycolytic oscillations are usually done with cell free extracts. So far there have been no reports of the occurrence of autonomous chaos in the glycolytic system, but recently chaos was found in a cell free extract of *Saccharomyces cerevisiae* exposed to a periodic glucose supply (Markus, Kuschmitz & Hess, 1984). An almost one-dimensional discrete map was constructed from the experimental data using the stroboscopic sampling technique. From this map chaos was diagnosed using the Li and Yorke theorem.

Many different mechanisms have been proposed to account for the

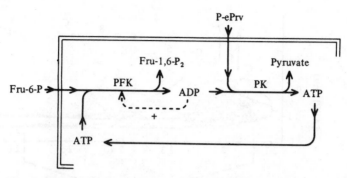

Fig. 23. Model to explain oscillations in glycolysis. The model involves only the enzymes phosphofructokinase (PFK) and pyruvate kinase (PK). The ADP formed by phosphorylation of fructose-6-phosphate (Fru-6-P) into fructose-1,6-bisphosphate (Fru-1,6-P$_2$) is reconverted to ATP by transfer of a phosphate group from phosphoenolpyruvate (P-ePrv). ADP also acts as an allosteric activator of PFK as indicated in the figure. Reproduced with permission from Markus & Hess (1984).

oscillations in glycolysis. They all include non-linear effects arising from allosteric control by glycolytic intermediates on glycolytic enzymes. Since there are many enzymes and many intermediates involved in glycolysis a model including everything would be too large to be workable. One must therefore work with an abbreviated model. In the actual case the experimental data on the periodically perturbed glycolytic system were compared with numerical solutions of a simplified glycolytic model involving only two enzymes, phosphofructokinase and pyruvate kinase, and having fructose-6-phosphate and phosphoenolpyruvate as substrates (Markus & Hess, 1984). The model is shown in Fig. 23. It is noted that the model has one feed back loop that can give rise to non-linear behaviour. With a constant influx of substrate only simple limit cycle oscillations were obtained with this model, whereas with a periodic influx of substrate complex oscillations with period doubling bifurcations and chaos were observed. Using the ratio of the frequency of the perturbation to the natural frequency of the oscillations as a parameter, bifurcation diagrams were constructed for the model and for the experimental system (Fig. 24). Both bifurcation diagrams contain chaotic regions interrupted by small integer periodic windows. However, the chaotic regions in the experimental system occur at somewhat higher frequencies of perturbation than those in the model and only two of the five numerically predicted periodic windows were observed experimentally. The transition to chaos occurs through period doubling

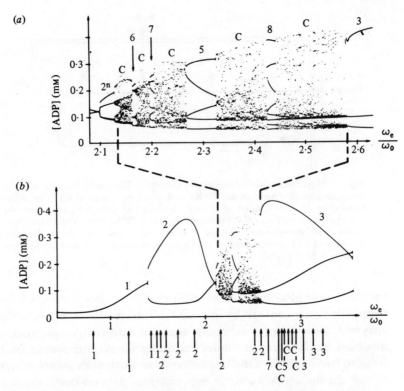

Fig. 24. Bifurcation diagram of the sinusoidally perturbed glycolytic system. Comparison of experimental results and computer results. The ADP concentration was monitored stroboscopically as the frequency $\omega_e$ of the oscillating substrate input flux was changed slowly. $\omega_0$ is the frequency of the autonomous oscillation of the system at constant substrate input flux. The graphs are the theoretical results obtained from the model in Fig. 23. The arrows indicate experimental results. Numbers indicate period multiplicity and $C$ means chaos. Reproduced with permission from Markus, Kuschmitz & Hess (1985).

bifurcations as the chaotic region is approached from one side. From the other side the transition to chaos is of a different nature. A similar observation was made for the peroxidase–oxidase reaction (Olsen, 1983; 1984).

A model that is considered to be a prototype of a metabolic control mechanism has been proposed by Decroly & Goldbeter (1983; 1984). It consists of a sequence of two enzymes that are allosterically activated by their own products. The product of the first enzyme is the substrate of the second one. The model shown in Fig. 25 is perhaps less faithful to established details of the glycolytic control

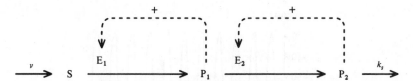

Fig. 25. Hypothetical enzyme model capable of displaying autonomous chaos. Substrate $S$ is supplied at a constant rate, $v$. The conversion $S \to P_1$ is catalysed by the enzyme $E_1$ that is activated allosterically by $P_1$. The conversion $P_1 \to P_2$ is catalysed by the enzyme $E_2$ that is activated allosterically by $P_2$. $P_2$ is removed by a first order reaction with the rate constant $k_s$. Reproduced with permission from Decroly & Goldbeter (1982).

mechanism than the model of Markus & Hess (1984). However it embodies one significant aspect of glycolytic control that is absent in the latter, namely that it has more than one feed back loop. As discussed in relation to the peroxidase system it seems that more than one feed back loop is required in a chemical model for chaos to occur. Accordingly the model of Decroly & Goldbeter was found to be capable of autonomous chaos. A bifurcation diagram was constructed using the rate constant of removal of the product of the second autocatalytic reaction as a parameter. This diagram showed that the transition to chaos in the model occurred through period doubling bifurcations. In addition the model was bistable with two different limit cycles or a limit cycle and a chaotic attractor. So far there are no experimental observations that can substantiate the predictions of the model of Decroly & Goldbeter. A possible obstacle is that the parameter interval where chaos occurs in the model is very narrow.

The literature contains a considerable amount of speculation on the purpose of glycolytic oscillations. The most modest view is that the oscillations have no purpose of their own. They are a harmless byproduct of the feed back regulation that has evolved to balance the glycolytic production of ATP and intermediates against the demand. According to this view autonomous chaos in glycolysis, if it exists, and chaos resulting from periodic perturbation of glycolysis by another oscillatory process within the same organism, would hardly have any purpose or consequence. According to another school of thought a limit cycle may be an optimal time pattern (Selkov, 1980) and chaos may allow trial and error switching between different time patterns (Markus, Kuschmitz & Hess, 1985; Hess & Markus, 1985 a; b).

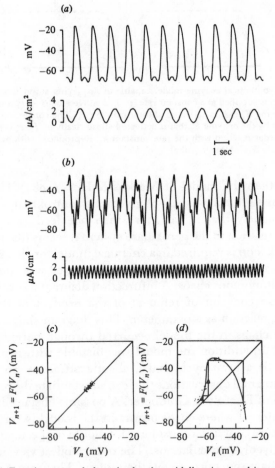

Fig. 26. Entrainment and chaos in the sinusoidally stimulated internodal cell of Nitella. (a) Repetitive firing (upper curve) synchronized with the periodic current stimulation (lower curve). (b) Non-periodic response to periodic stimulation. (c) and (d) Stroboscopic transfer functions obtained from (a) and (b) respectively. The membrane potential at each peak of the periodic stimulation was plotted against the preceding one. Period three is indicated graphically by arrows in (d). Reproduced with permission from Hayashi, Nakao & Hirakawa (1982).

## 14. CHAOS IN EXCITABLE CELLS

In plants as well as in animals are found giant cells that have been the favourites of electrophysiologists for several decades. These cells are excitable and they are large enough to allow the insertion of microelectrodes to study their electrical properties. Depending on the conditions spontaneous or provoked action potentials may appear.

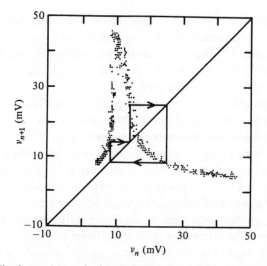

Fig. 27. Stroboscopic transfer function of the chaotic response to periodic current stimulation in the Onchidium giant neuron. The plot was obtained in the same way as those of Fig. 26 c and 26 d. The arrows indicate period three. Reproduced with permission from Hayashi et al. (1982).

Periodic perturbations were used to probe such cells a long time before chaotic dynamics was a defined concept. It is likely that chaotic responses have been observed many times and discarded because of a lack of theoretical background.

Sinusoidal electrical stimulation of the giant internodal cell of the freshwater algae *Nitella flexilis* caused entrainment, quasiperiodic behaviour and chaos as seen in Fig. 26 a and b (Hayashi, Nakao & Hirakawa, 1982; 1983). Stroboscopic phase portraits of the apparently aperiodic oscillations were used to construct a one-dimensional map from which chaos was diagnosed by the help of the Li and Yorke theorem (Fig. 26 d). Evidence of a period doubling road to chaos was not presented but the data indicated that the transition to chaos took place through intermittency.

Similar studies were made with the giant neurons and the pacemaker neurons from the marine pulmonate mollusc *Onchidium verruculatum* (Hayashi et al. 1982; Hayashi, Ishizuka & Hirakawa, 1983). In both cases an apparently one-dimensional map could be constructed from the stroboscopic phase portrait of the aperiodic oscillations (Fig. 27). The map which is very similar to the one shown in Fig. 4 b allowed for period three, demonstrating that the dynamics were chaotic. Due to the form of the map it seems likely that the transition to chaos takes place through intermittency. Apparently aperiodic behaviour was also observed experimentally in the sinusoidally stimulated squid axon (Guttman, Feldman & Jakobsson, 1980). However tests for chaotic behaviour were not made and an explanation based on membrane current noise was preferred.

Experimental observations of non-periodic oscillations were made in a periodically forced squid axon (Matsumoto *et al.* 1984) and were compared with a numerical study of the periodically forced Hodgkin–Huxley equations (Aihara, Matsumoto & Ikegaya, 1984). There was a good agreement between the time series of the experimental oscillations in membrane potential and those obtained numerically. The numerical study furthermore showed that transitions to chaos could either go through period doubling bifurcations or through intermittency. The attractors were visualized by stroboscopic phase plots. These plots did not define simple, single valued curves but had a fractal substructure like that of the Hénon map (Fig. 8). Consequently no simple one-dimensional transfer function could be made from the plots. Chaotic dynamics were independently demonstrated for the sinusoidally forced Hodgkin–Huxley equations by Holden & Muhamad (1984) and by Jensen *et al.* (1983). The latter authors calculated the maximal Lyapunov exponents for the different stimulation frequencies of the Hodgkin–Huxley equations using the method of Benettin, Galgani & Strelcyn (1976) (see Appendix II). For the frequency leading to chaotic dynamics the the Lyapunov exponent was $0\cdot11$ ms$^{-1}$ indicating a low rate of divergence.

Irregular bursting electrical activity has been reported for snail neurons following exposure to cocaine, strychnine or benzodiazepines (Labos & Lang, 1978; Klee, Faber & Hoyer, 1978; Hoyer, Park & Klee, 1978). The transition from periodic to aperiodic activity often occurred through doublet and triplet discharges. However, these abnormal discharges were not tested for chaos. Similar aperiodic oscillations in membrane potential were observed in a neuron from the pond snail *Lymnea stagnalis* following prolonged exposure to the K$^+$-channel blockers and convulsants tetraethyl ammonium ion and 4-amino-pyridine (Holden & Winlow, 1982; Holden, Winlow & Haydon, 1982). Also here the transition to chaos was beginning with the development of double discharges. Although it was recognized that these irregular oscillations might be a result of the non-linear properties of the nerve membrane, no quantitative tests were made to demonstrate that the membrane potential activity was chaotic. However, such irregular discharge patterns have been simulated by a neuronal model somewhat similar to the Hodgkin–Huxley equation (Chay, 1984). It was predicted from the model that changing the efficacy of various channels and electrogenic pumps by the addition of drugs or by changing the temperature could result in a change from periodic to chaotic firing in agreement with the experimental observations mentioned above. The transition to chaos as shown by the model of Chay sometimes followed the orthodox period doubling route whereas at other times the first period doubling was followed by a period tripling. A next amplitude plot was made for the chaotically oscillating intracellular Ca²⁺-concentration. Although this map was a well defined curve it was not always single valued.

The $\beta$-cells of pancreatic islets exhibit electrical activity when exposed to glucose or other insulin releasing agents. Autonomous chaotic behaviour was recently reported for a Hodgkin–Huxley type model of the electrical activity in $\beta$-pancreatic cells (Chay & Rinzel, 1985). Following changes in a parameter corresponding to the glucose dependent uptake of intracellular calcium, the self-sustained oscillations of the intracellular calcium and the membrane potential underwent period doubling bifurcations resulting in chaotic oscillations. Successive Ca²⁺-concentrations, $C$, at the upstroke in membrane potential were computed and used to construct a one-dimensional map

$$C_{n+1} = F(C_n)$$

from which period doubling and chaos could be verified. However, there is limited experimental evidence for the existence of chaos in pancreatic $\beta$-cells. Meissner (1976) has reported irregular electrical activity in these cells but his observations were done before the concept of chaotic dynamics was widely recognized.

## 15. CHAOS IN HEART BEAT

From a medical point of view there is a growing interest in chaos. It seems plausible that certain neurological disorders and cardiac arrhythmia can originate from chaotic dynamics (Mackey & Glass, 1977). Systematic research is beginning to substantiate this idea. In a recent work induced cardiac arrhythmia was studied in dogs (Ritzenberg, Adam & Cohen, 1984). A number of dogs were anaesthetized, noradrenaline was injected intravenously and the electrocardiograms and arterial blood pressure were recorded. Fourier transform analysis was performed on both types of measurements and indicated period multiplications following increased venticular rates.

556

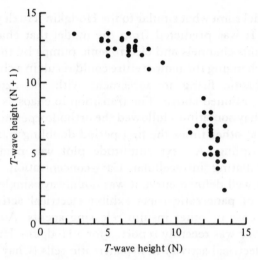

Fig. 28. Next amplitude plot of the $T$-waves during 48 heart beats of a noradrenalin treated dog. Reproduced with permission from Ritzenberg, Adam & Cohen (1984).

A next amplitude map was made for the $T$-wave. The map (Fig. 28) demonstrates some coupling between adjacent amplitudes but it does not seem to be a clear cut one-dimensional map. If it is viewed as a one dimensional map it is invertible and therefore it does not provide evidence for the occurrence of chaos. Nevertheless the results suggest that the heart beat has properties related to chaos and further *in vivo* investigations should be done.

Also *in vitro* experiments on the heart beat have been reported (Guevara, Glass & Shrier, 1981; Glass *et al.* 1984). Aggregates of embryonic cells of chick heart were exposed to brief single and periodic current pulses. The spontaneous periodic contraction of the aggregate was reset by a single pulse and the perturbed cycle length was measured as a function of the phase of the contraction at the time of the perturbation. From these data a Poincaré map was determined (Fig. 29). The resetting data were used to predict the effect of periodic perturbations. The observed responses to periodic perturbations were phase locking, period doubling and irregular dynamics as the perturbation frequency was increased. Iterations of the Poincaré map determined by the single pulse perturbations were in good agreement with the periodic perturbation results as shown in Fig. 30. In some cases the iteration of the Poincaré map produced period doublings which did not lead to chaos. Instead the bifurcation sequence

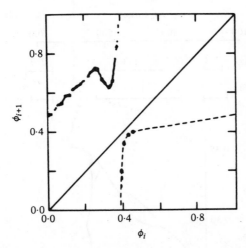

Fig. 29. Poincaré map constructed from single pulse perturbations of embryonic chick heart cells. $\phi$ is the phase of the spontaneous contraction cycle where the perturbation takes place. Reproduced with permission from Guevara, Glass & Shrier (1981).

reversed and a simple limit cycle reappeared as the perturbation frequency was increased.

A more detailed model of the electrical activity of the heart is the Beeler–Reuter differential equation system which is capable of reproducing the experimentally observed response of heart muscle under constant stimulation (Beeler & Reuter, 1977). The Beeler–Reuter equations are a set of eight non-linear differential equations and like the Hodgkin–Huxley equations they are based on membrane ionic currents measured in voltage clamp experiments. The response of the Beeler–Reuter equations to sinusoidal stimulation was investigated by Jensen, Christiansen & Scott (1984). They found that for some stimulation frequencies the system exhibits phase locking whereas for other frequencies the response seems to be aperiodic. In the latter case the power spectrum is a broad band as expected for chaotic dynamics. The complete set of Lyapunov exponents corresponding to the number of differential equations were calculated for one of the aperiodic solutions. The largest Lyapunov exponent was 0·004 ms$^{-1}$. This demonstrates that the aperiodic dynamics are indeed chaotic. The transition from periodic to chaotic behaviour was not investigated in detail, but there was some indication of period doubling bifurcations following changes in the stimulation frequency.

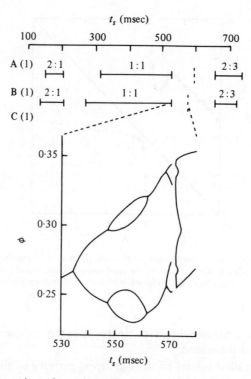

Fig. 30. Comparison of experimentally determined and theoretically computed response of embryonic chick heart cells to periodic perturbations. (*A*) Experimental results. Ratios of numbers indicate phase locking. Empty intervals indicate complicated dynamics. (*B*) Theoretical results from iteration of Poincaré map (Fig. 29). (*C*) Bifurcation diagram of Poincaré map in zone of complicated dynamics. Reproduced with permission from Guevara, Glass & Shrier (1981).

## 16. CHAOS IN EPIDEMICS

Infectious diseases may be divided into two classes: Those caused by microparasites such as viruses, bacteria and protozoa and those caused by macroparasites such as parasitic helminths and arthropods (Anderson & May, 1979; May & Anderson, 1979). The dynamics of some microparasitic infections show almost periodic yearly outbreaks. Examples are mumps and chicken pox epidemics as recorded in New York (Fig. 31). Other microparasitic infections have epidemic outbreaks every 2 or 3 years with very few cases occurring the following 1 or 2 years (London & Yorke, 1973; Yorke & London, 1973). Examples of the latter type are measles epidemics as recorded in New York and Baltimore (Fig. 31). Usually such irregular

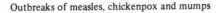

Outbreaks of measles, chickenpox and mumps

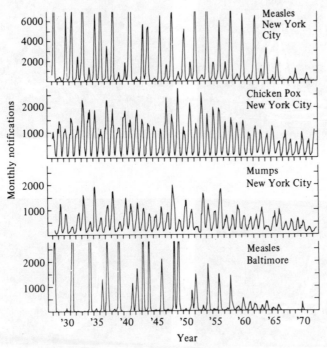

Fig. 31. The monthly reported cases of measles, chicken pox and mumps in New York and measles in Baltimore in the period 1928–72. Reproduced with permission from London & Yorke (1973).

dynamics in epidemiology are explained in terms of stochastic models (Bartlett, 1960; Anderson, 1982). However, recently attempts have been made to relate them to chaotic dynamics.

Schaffer & Kot (1985) have analysed the monthly records of chicken pox, mumps and measles in New York and Baltimore for chaotic behaviour using the method discussed in section 11. The data were plotted as phase plots of $N(t)$, $N(t+h)$, $N(t+2h)$ where $N$ is the number of cases per month and $h$ is a fixed time lag of 2–3 months (Figs. 32 and 33). The Poincaré cross sections were constructed for all three phase plots and shown to define V-shaped half lines, demonstrating that the flow is confined to a nearly two-dimensional conical surface. The points from one branch of the cross section were plotted against the preceding points resulting in an almost one-dimensional humped map for the New York and Baltimore measles data whereas for the chicken pox and mumps data consecutive points

560

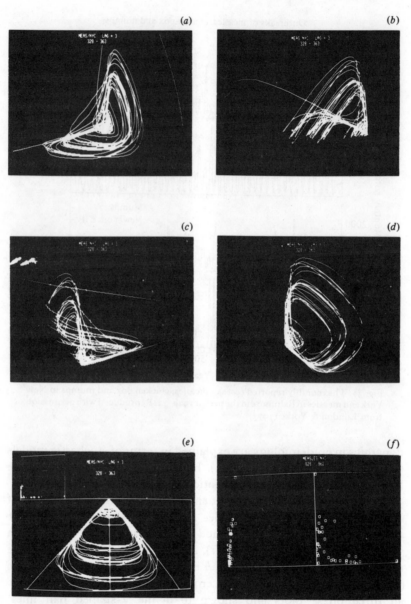

Fig. 32. Reconstructed trajectories of the New York measles data from Fig. 31 (smoothed and interpolated). (*a*)–(*d*) The same data embedded in three dimensions and viewed from different directions using a time lag of 3 months. (*e*) The orbit viewed from above and sliced twice with a plane (vertical line). The Poincaré sections are shown in the small box at the upper left. (*f*) One of the Poincaré sections magnified (left) together with the resulting one-dimensional map (right). Reproduced with permission from Schaffer & Kot (1985).

**214** L. F. OLSEN AND H. DEGN

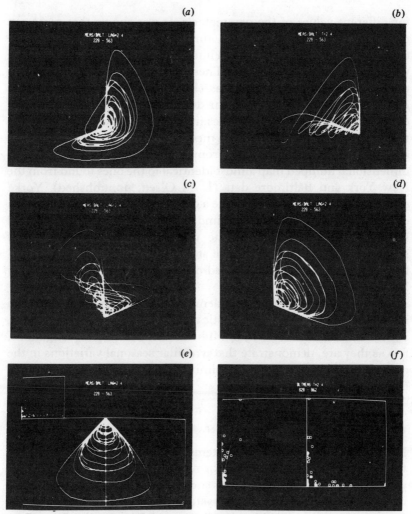

Fig. 33. Reconstructed trajectories from the Baltimore measles data using a time lag of 2·4 months. The order of displays are as in Fig. 32. Reproduced with permission from Schaffer and Kot (1985).

from the Poincaré cross sections were apparently uncorrelated. The Lyapunov numbers were calculated for the one-dimensional maps, in both cases yielding positive values.

Schaffer & Kot (1985) also studied the effects of imposing stochastic fluctuations on one-dimensional maps of the type

$$X_{t+1} = X_t \exp r(1 - X_t)$$

for different values of the parameter *r*. It was shown that noise can totally obscure the underlying functional relationship when the dynamics are periodic with small integer periods. This is not the case when the dynamics are chaotic. These results support the interpretation of the measles epidemics as chaotic and the other epidemics included in the study as irregular due to stochastic influences. The conclusion that measles epidemics in large cities may be chaotic due to a well defined, albeit unknown mechanism is also supported by the analysis of the measles data from Copenhagen yielding a one-dimensional humped map almost identical to the ones found from the New York and Baltimore data (L. F. Olsen, unpublished). Yorke *et al.* (1979) have previously made a detailed analysis of the outbreaks of measles in New York and Baltimore. They concluded that seasonal waves in transmissibility of the disease and in the number of susceptible individuals minimize stochastic effects in large populations. Opposite conclusions based on the same data were drawn by Bartlett (1960).

Recently simulations of epidemics using differential equation models entirely free from stochastic terms have been published (May, 1980; Aron & Schwartz, 1984). The results of these investigations, few as they are, demonstrate that irregular seasonal variations in the outbreaks of epidemics can be simulated by non-linear differential equations.

## 17. CHAOS IN CELLULAR SIGNAL TRANSMISSION

Feed-back in physiological systems is often mediated via signal transmission by hormones or hormone-like substances. It is to be anticipated that oscillations in such systems may sometimes be chaotic. A comparatively simple system for the study of cellular signal transmission is provided by the slime mold *Dictyostelium discoideum* where cyclic AMP takes part in the cellular aggregation process as a signal transmitter. Normally the signalling by cyclic AMP is periodic but aperiodic waves of aggregation in a mutant (Durston, 1974) may indicate an aperiodic signalling by cyclic AMP in this mutant.

Martiel & Goldbeter (1985) have studied a model of the cyclic AMP signalling in the slime mold and found very complicated dynamics. The model consists of seven differential equations and includes two coupled oscillatory mechanisms similar to the models proposed for chaos in glycolysis (Fig. 25). For some parameter values the model predicts that the intracellular cyclic AMP will oscillate in a simple periodic fashion whereas for other values birythmicity with two coexisting limit cycles, bursting activity and chaotic dynamics will exist. While birythmicity and bursting activity have not been observed in the slime mold the chaotic dynamics predicted by the model may correspond to the aperiodic aggregation waves observed by Durston (1974).

King, Barchas & Huberman (1984) investigated a one-dimensional model of the central dopaminergic neuronal system. They showed that for certain values of the parameter which monitors the efficacy of dopamine at the postsynaptic receptor and the parameter corresponding to the external depolarizing input to the dopamine cells, the firing rate of the dopamine neuron will be chaotic. The authors claimed that their result may find clinical applications in explaining certain symptoms found in schizophrenic psychoses.

## 18. Conclusion

It is only a few years ago that chaos was established as a well-defined phenomenon in the dynamics of non-linear systems. Many physical examples have been reported and chaos has rapidly progressed from the exotic to the commonplace, at least in physics. At the present time the concept of chaos is penetrating the biological sciences. Biologists are looking for it, provoking it and gauging its implications for the living organism. The immediate thought is that chaos may be behind some pathological conditions. Afterthought admits that living organisms may also derive something useful from chaos. In all events the literature on chaos in biology seems to be entering its exponential growth phase and future reviews may have answers where the present one has questions.

## 19. Acknowledgement

This research was supported by a grant from the Danish Natural Science Research Council to Lars Folke Olsen (11-4595).

## 20. Appendix I

The general definition of Lyapunov numbers for difference equations in two and higher dimensions is: Let $F$ be a discrete mapping $R^n \to R^n$ and let

$$J_t = J(X_t) J(X_{t-1}) \ldots J(X_2) J(X_1)$$

where $J(X_i)$ is the Jacobian matrix $J(X) = \delta F/\delta X$ at the $i$th iteration and let

$$j_1(t) > j_2(t) > \ldots > j_n(t)$$

be the eigenvalues of $J_t$. The Lyapunov numbers are then defined as

$$\lambda_i = \lim_{t \to \infty} (j_i(t))^{1/t} \quad (i = 1, 2, \ldots, n)$$

where the positive real $t$th root is taken. This implies that a $n$-dimensional map has $n$ Lyapunov numbers $\lambda_1 > \lambda_2 > \ldots > \lambda_n$ (Farmer, Ott & Yorke, 1983).

The definition of Lyapunov numbers leads to a general definition of chaos in $n$-dimensional discrete maps, namely that chaos exists when $\lambda_1 > 1$. From this definition follows that in order to diagnose chaos in such maps it is sufficient to calculate $\lambda_1$. There are various ways of doing this numerically. Here we shall describe the method used by Nicolis, Meyer-Kress & Haubs (1983) that utilizes the fact that the maximal Lyapunov number is a measure of the divergence rate of neighbouring trajectories.

Let $X$ be an $n$-dimensional vector and let $F$ be a discrete mapping

$$X_{t+1} = F(X_t).$$

We choose an initial condition $X_0^0$. The $t$th iterate is $X_t^0$. We also choose a neighbouring initial condition $x_0$ such that $X_0 = X_0^0 + \delta X_0$. Applying $F$ to $X_i$ and to $X_i^0$ yields $X_{i+1}$ where

$$X_{i+1} = X_{i+1}^0 + \delta X_{i+1}$$

and $$\delta X_{i+1} = J(X_i^0)\,\delta X_i$$

and $J(X)$ is the Jacobian matrix $\delta F(X)/\delta X$. For the initial conditions $X_0^0$ and $X_0$ we may compute the divergence of the two trajectories after $t$ iterations

$$\delta X_t = J(X_{t-1}^0)\,J(X_{t-2}^0) \ldots J(X_1^0)\,J(X_0^0)\,\delta X_0$$

The maximal Lyapunov number, $\lambda_1$, is then computed as

$$\ln \lambda_1 = \lim_{t \to \infty} \sup\, (1/t) \ln |\delta X_t|.$$

We may express $\delta X_0$ in terms of its length, $d_0$, and a vector, $\mathbf{e}(\text{o})$, of unit length

$$\delta X_0 = d_0\,\mathbf{e}(\text{o}).$$

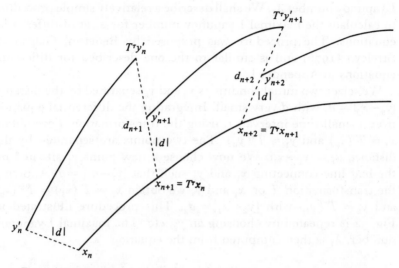

Fig. 34. Graphic illustration of the method of Bennettin, Galgani & Strelcyn (1976) for the numerical computation of the largest Lyapunov number of a differential equation system.

Multiplying $\delta X_0$ by $J(X_0^0)$ we obtain a new vector with direction $\mathbf{e}(1)$ and length $d_1 d_0$. Multiplying this vector by $J(X_1^0)$ we obtain a third vector with the length $d_2 d_1 d_0$ and the direction $\mathbf{e}(2)$. Repeating this procedure $t$ times we end up with the equation

$$\delta X_t = d_t d_{t-1} \ldots d_1 d_0 \mathbf{e}(t)$$

where $|\mathbf{e}(t)| = 1$. We obtain $\lambda_1$ as

$$\ln \lambda_1 = \lim_{t \to \infty}(1/t) \ln |d_t d_{t-1} \ldots d_1 d_0 e(t)|$$

$$= \lim_{t \to \infty}(1/t) \sum_{i=0}^{t} \ln d_1.$$

We have used this method to calculate $\ln \lambda_1$ for the Henon map (Fig. 10).

## 21. APPENDIX II

Like difference equations a set of $n$ first order differential equations has associated with it $n$ Lyapunov numbers. Computation of all the Lyapunov numbers offers considerable problems compared to the computation of all the Lyapunov numbers of difference equations. However, for diagnostic purposes we need only know the maximal

Lyapunov number $\lambda_1$. We shall describe a relatively simple procedure to calculate the maximal Lyapunov number for a set of differential equations. The procedure was proposed by Benettin, Galgani & Strelcyn (1976) and is similar to the one described for difference equations in Appendix I.

We select two initial conditions $x_0$ and $y_0$ separated by the distance $|y_0 - x_0| = d$, with $d$ very small. Integrating the differential equation over a small time interval, $\tau$, using the transformation $T$ we obtain $x_1 = T^\tau(x_0)$ and $y_1 = T^\tau(y_0)$. The two points are separated by the distance $d_1 = |y_1 - x_1|$. We now choose a new point $y_1'$ situated on the half line connecting $x_1$ and $y_1$ such that $|y_1' - x_1| = d$. Applying the transformation $T$ on $x_1$ and $y_1$ we obtain $x_2 = T^\tau(x_1) = T^{2\tau}(x_0)$ and $y_2 = T^\tau(y_1')$ with $|y_2 - x_2| = d_2$. This procedure illustrated in Fig. 34 is repeated by choosing an $y_2'$, etc. The maximal Lyapunov number, $\lambda_1$ is then computed from the equation

$$\ln \lambda_1 = \lim_{m \to \infty} (1/m\tau) \sum_{i=1}^{m} \ln (d_i/d)$$

where $m$ is the number of integration steps. The procedure is rather time consuming since $m$ has to be large ($10^5$) and $d$ must be small. The value of $\tau$ is not critical as long as it is larger than the step length in the integration procedure.

Unlike Lyapunov exponents calculated from difference equations the Lyapunov exponents calculated from differential equations involve real time. If one wants to compare Lyapunov exponents from differential equations with Lyapunov exponents from difference equations one must introduce a characteristic time scale (Shaw, 1981) which normally will be the average period, $P_{av}$, of the oscillation. The relationship between the maximal Lyapunov exponent, $\ln \lambda_1$, calculated as above from differential equations, and the Lyapunov exponent $\ln \lambda$, calculated from any corresponding one-dimensional map is

$$\ln \lambda_1 = \ln \lambda / P_{av}.$$

## 22. References

AIHARA, K., MATSUMOTO, G. & IKEGAYA, Y. (1984). Periodic and non-periodic responses of a periodically forced Hodgkin–Huxley oscillator. *J. theor. Biol.* **109**, 249–269.

ANDERSON, R. M. (1982). Directly transmitted viral and bacterial infections of man. In *The Population Dynamics of Infectious Diseases: Theory and*

220   L. F. OLSEN AND H. DEGN

*Applications* (ed. R. M. Anderson), pp. 1–37. London: Chapman & Hall.

ANDERSON, R. M. & MAY, R. M. (1979). Population biology of infectious diseases: I. *Nature* **280**, 361–367.

ARON, J. L. & SCHWARTZ, I. B. (1984). Seasonality and period-doubling bifurcations in an epidemic model. *J. theor. Biol.* **110**, 665–679.

BARTLETT, M. S. (1960). *Stochastic Population Models in Ecology and Epidemiology.* London: Methuen.

BEELER, G. W. & REUTER, H. (1977). Reconstruction of the action potential of ventricular myocardial fibres. *J. Physiol.* **268**, 177–210.

BENETTIN, G., GALGANI, L. & STRELCYN, J.-M. (1976). Kolmogorov entropy and numerical experiments. *Phys. Rev.* A **14**, 2338–2345.

BIER, M. & BOUNTIS, T. C. (1984). Remerging Feigenbaum Trees in Dynamical Systems. *Physics. Lett.* **104**A, 239–244.

CHANCE, B., HESS, B. & BETZ, A. (1964). DPNH oscillations in a cell-free extract of *S. carlsbergensis*. *Biochem. biophys. Res. Commun.* **16**, 182–187.

CHAY, T. R. (1984). Abnormal discharges and chaos in a neuronal model system. *Biol. Cybernet.* **50**, 301–311.

CHAY, T. R. & RINZEL, J. (1985). Bursting, beating and chaos in an excitable membrane model. *Biophys. J.* **47**, 357–366.

CHIRIKOV, B. V. (1979). A universal instability of many-dimensional oscillator systems. *Physics. Rep.* **52**, 263–379.

COLLET, P. & ECKMANN, J.-P. (1980). *Iterated maps on the interval as dynamical systems.* Basel: Birkhauser.

DECROLY, O. & GOLDBETER, A. (1982). Birhythmicity, chaos and other patterns of temporal self-organization in a multiply regulated biochemical system, *Proc. natn Acad. Sci. U.S.A.* **79**, 6917–6921.

DECROLY, O. & GOLDBETER, A. (1984). Multiple periodic regimes and final state sensitivity in a biochemical system. *Phys. Lett.* **105**A, 259–262.

DEGN, H. (1968). Bistability caused by substrate inhibition of peroxidase in an open reaction system. *Nature* **217**, 1047–1050.

DEGN, H. (1982). Discrete chaos is reversed random walk. *Phys. Rev.* A **26**, 711–712.

DEGN, H. (1983). Strange attractors in linear period transfer functions with periodic perturbations. In *Chemical Applications of Topology and Graph Theory* (ed. R. B. King), pp. 364–370. Amsterdam: Elsevier.

DEGN, H. & MAYER, D. (1969). Theory of oscillations in peroxidase catalyzed oxidation reactions in open system. *Biochim. biophys. Acta* **180**, 291–301.

DEGN, H., OLSEN, L. F. & PERRAM, J. W. (1979). Bistability, Oscillations and chaos in an enzyme reaction. *Ann. N.Y. Acad. Sci.* **316**, 623–637.

DURSTON, A. J. (1974). Pacemaker mutants of *Dictyostelium discoideum*. *Devl. Biol.* **38**, 308–319.

ECKMANN, J.-P. (1981). Roads to turbulence in dissipative dynamical systems. *Rev. Mod. Phys.* **53**, 643–654.

ESCHRICH, K., SCHELLENBERGER, W. & HOFMANN, E. (1983). Sustained oscillations in a reconstituted enzyme system containing phospho-fructokinase and fructose 1,6-bisphosphate. *Archs. Biochem. Biophys.* **222**, 657–660.

FARMER, J. D., OTT, E. & YORKE, J. A. (1983). The dimension of chaotic attractors. *Physica* **7**D, 153–180.

FEIGENBAUM, M. J. (1978). Quantitative universality for a class of non-linear transformations. *J. Statist. Phys.* **19**, 25–52.

FORD, J. (1983). How random is a coin toss? In *Physics Today*, April 40–47.

FRASER, S. & KAPRAL, R. (1982). Analysis of flow hysteresis by a one-dimensional map. *Phys. Rev. A* **25**, 3223–3233.

GHOSH, A. & CHANCE, B. (1964). Oscillations of glycolytic intermediates in yeast cells. *Biochem. biophys. Res. Commun.* **16**, 174–181.

GLASS, L., GRAVES, C., PETRILLO, G. A. & MACKEY, M. C. (1980). Unstable dynamics of a periodically driven oscillator in the presence of noise. *J. theor. Biol.* **86**, 455–475.

GLASS, L., GUEVARA, M. R., BELAIR, J. & SHRIER, A. (1984). Global bifurcations of a periodically forced biological oscillator. *Phys. Rev. A* **29**, 1348–1357.

GLASS, L., GUEVARA, M. R., SHRIER, A. & PEREZ, R. (1983). Bifurcation and chaos in a periodically stimulated cardiac oscillator. *Physica* **7**D, 89–101.

GLASS, L. & MACKEY, M. C. (1979). Pathological conditions resulting from instabilities in physiological control systems. *Ann. N.Y. Acad. Sci.* **316**, 214–235.

GREBOGI, C., OTT, E. & YORKE, J. A. (1982). Chaotic attractors in crisis. *Phys. Rev. Lett.* **48**, 1507–1510.

GREBOGI, C., OTT, E. & YORKE, J. A. (1983). Crises, sudden changes in chaotic attractors, and transient chaos. *Physica* **7**D, 181–200.

GROSSMANN, S. & THOMAE, S. (1977). Invariant distributions and stationary correlation functions of one-dimensional discrete processes. *Z. Naturf.* **32**a, 1353–1363.

GUEVARA, M. R. & GLASS, L. (1982). Phase locking, period doubling bifurcations and chaos in a mathematical model of a periodically driven oscillator: A theory for the entrainment of biological oscillators and the generation of cardiac dysrythmias. *J. Math. Biol.* **14**, 1–23.

GUEVARA, M. R., GLASS, L., MACKEY, M. C. & SHRIER, A. (1983). Chaos in neurobiology. *IEEE Trans. Sys. Man Cyber.* **13**, 790–798.

GUEVARA, M. R., GLASS, L. & SHRIER, A. (1981). Phase locking period-doubling bifurcations, and irregular dynamics in periodically stimulated cardiac cells. *Science* **214**, 1350–1353.

GUTTMAN, R., FELDMAN, L. & JAKOBSSON, E. (1980). Frequency entrainment of squid axon membrane. *J. Membrane Biol.* **56**, 9–18.

HAYASHI, H., ISHIZUKA, S. & HIRAKAWA, K. (1983). Transition to chaos via intermittency in the onchidium pacemaker neuron. *Phys. Lett.* **98**A, 474–476.

HAYASHI, H., ISHIZUKA, S., OHTA, M. & HIRAKAWA, K. (1982). Chaotic behavior in the onchidium giant neuron under sinusoidal stimulation. *Phys. Lett.* **88**A, 435–438.

HAYASHI, H., NAKAO, M. & HIRAKAWA, K. (1982). Chaos in the self-sustained oscillation of an excitable membrane under sinusoidal stimulation. *Phys. Lett.* **88**A, 265–266.

HAYASHI, H., NAKAO, M. & HIRAKAWA, K. (1983). Entrained, harmonic, quasiperiodic and chaotic responses of the self-sustained oscillation of nitella to sinusoidal stimulation. *J. phys. Soc. Japan* **52**, 344–351.

HÉNON, M. (1976). A two-dimensional mapping with strange attractor. *Commun. Math. Phys.* **50**, 69–77.

HESS, B. & MARKUS, M. (1985*a*). The diversity of biochemical time patterns. *Ber. Bunsenges. Phys. Chem.* **89**, 642–651.

HESS, B. & MARKUS, M. (1985*b*). Dynamic coupling and time patterns in biochemical processes. In *Temporal Order* (ed. L. Rensing and N. I. Jager). Berlin: Springer-Verlag. (In the Press.)

HOLDEN, A. V. & MUHAMAD, M. A. (1984). Chaotic activity in neural systems. In *Cybernetics and Systems Research* **2** (ed. R. Trappl), pp. 245–250. Amsterdam: Elsevier.

HOLDEN, A. V. & WINLOW, W. (1982). Bifurcation of periodic activity from periodic activity in a molluscan neurone. *Biol. Cybern.* **42**, 189–194.

HOLDEN, A. V., WINLOW, W. & HAYDON, P. G. (1982). The induction of periodic and chaotic activity in a molluscan neurone. *Biol. Cybern.* **43**, 169–173.

HOYER, J., PARK, M. R. & KLEE, M. R. (1978). Changes in ionic currents associated with flurazepam-induced abnormal discharges in Aplysia neurons. In: *Abnormal Neuronal Discharges* (ed. N. Chalazonitis and M. Boisson), pp. 301–310. New York: Raven Press.

HUDSON, J. L. & MANKIN, J. C. (1981). Chaos in the Belousov–Zhabotinskii reaction. *J. Chem. Phys.* **74**, 6171–6177.

JENSEN, J. H., CHRISTIANSEN, P. L. & SCOTT, A. C. (1984). Chaos in the Beeler–Reuter system for the action potential of ventricular myocardial fibres. *Physica* **13**D, 269–277.

JENSEN, J. H., CHRISTIANSEN, P. L., SCOTT, A. C. & SKOVGAARD, O. (1983). *Chaos in Nerve*, Proceedings of the Iasted Symposium, Copenhagen, pp. 15/6–15/9.

KACZMAREK, L. K. & BABLOYANTZ, A. (1977). Spatiotemporal patterns in epileptic seizures. *Biol. Cybern.* **26**, 199–208.

KAPLAN, H. (1983). New method for calculating stable and unstable periodic orbits of one-dimensional maps. *Phys. Lett.* **97**A, 365–367.

KING, R., BARCHAS, J. D. & HUBERMAN, B. A. (1984). Chaotic behaviour in dopamine neurodynamics. *Proc. natn Acad. Sci. U.S.A.* **81**, 1244–1247.

KLEE, M. R., FABER, D. S. & HOYER, J. (1978). Doublet discharges and bistable states induced by strychnine in a neuronal soma membrane. In *Abnormal Neuronal Discharges* (ed. N. Chalazonitis and M. Boisson), pp. 287–300. New York: Raven Press.

KLOEDEN, P., DEAKIN, M. A. B. & TIRKEL, A. Z. (1976). A precise definition of chaos. *Nature* **264**, 295.

LABOS, E. & LANG, E. (1978). On the behavior of snail neurons in the presence of cocaine. In *Abnormal Neuronal Discharges* (ed. N. Chalazonitis and M. Boisson), pp. 177–188. New York: Raven Press.

LI, T.-Y. & YORKE, J. A. (1975). Period three implies chaos. *Am. Math. Mon.* **82**, 985–992.

LONDON, W. P. & YORKE, J. A. (1973). Recurrent outbreaks of measles, chickenpox and mumps I: seasonal variations in contact rates. *Am. J. Epidemiol.* **98**, 453–468.

LORENZ, E. N. (1963). Deterministic nonperiodic flow. *J. atmos Sci.* **20**, 130–141.

MACKEY, M. C. & GLASS, L. (1977). Oscillation and chaos in physiological control systems. *Science* **197**, 287–289.

MANDELBROT, B. (1977). *Fractals: Form, Chance and Dimension.* San Francisco: Freeman.

MANNEVILLE, P. & POMEAU, Y. (1979). Intermittency and the Lorenz Model. *Phys. Lett.* **75**A, 1–2.

MANNEVILLE, P. & POMEAU, Y. (1980). Different ways to turbulence in dissipative dynamical systems. *Physica* 1D, 219–226.

MARKUS, M. & HESS, B. (1984). Transition between oscillatory modes in a glycolytic model system. *Proc. natn Acad. Sci. U.S.A.* **81**, 4394–4398.

MARKUS, M., KUSCHMITZ, D. & HESS, B. (1984). Chaotic dynamics in yeast glycolysis under periodic substrate input flux. *FEBS Lett.* **172**, 235–238.

MARKUS, M., KUSCHMITZ, D. & HESS, B. (1985). Properties of strange attractors in yeast glycolysis. *Biophys. Chem.* (In the Press.)

MARTIEL, J. L. & GOLDBETER, A. (1985). Autonomous chaotic behaviour of the slime mold *Dictyostelium discoideum* predicted by a model for cyclic AMP signalling. *Nature* **313**, 590–592.

MATSUMOTO, G., AIHARA, K., ICHIKAWA, M. & TASAKI, A. (1984). Periodic and nonperiodic responses of membrane potentials in squid giant axons during sinusoidal current stimulation. *J. theoret. Neurobiol.* **3**, 1–14.

MAY, R. M. (1976). Simple mathematical models with very complicated dynamics. *Nature* **261**, 459–467.

MAY, R. M. (1980). Non-linear phenomena in ecology and epidemiology. *Ann. N.Y. Acad. Sci.* **357**, 267–281.

MAY, R. M. & ANDERSON, R. M. (1979). Population biology of infectious diseases: II. *Nature* **280**, 455–461.

MAY, R. M. & OSTER, G. F. (1976). Bifurcations and dynamic complexity in simple ecological models. *Am. Nat.* **110**, 573–599.

MEISSNER, H. P. (1976). Electrical characteristics of the $\beta$-cells in pancreatic islets. *J. Physiol.* **72**, 757–767.

METROPOLIS, N., STEIN, M. L. & STEIN, P. R. (1973). On finite limit sets for transformations on the unit interval. *J. Combinat. Theor. Ser. A* **15**, 25–44.

224   L. F. OLSEN AND H. DEGN

NICOLIS, J. S., MEYER-KRESS, G. & HAUBS, G. (1983). Non-uniform chaotic dynamics with implications to information processing. *Z. Naturf.* **38**a, 1157–1169.

NICOLIS, J. S., MEYER-KRESS, G. & HAUBS, G. (1984). Non-uniform information processing by strange attractors of chaotic maps and flows. In *Stochastic Phenomena and Chaotic Behaviour in Complex Systems* (ed. P. Schuster), pp. 124–139. Berlin: Springer-Verlag.

OLSEN, L. F. (1978). The oscillating peroxidase–oxidase reaction in an open system: Analysis of the reaction mechanism. *Biochim. biophys. Acta* **527**, 212–220.

OLSEN, L. F. (1979). Studies of the chaotic behaviour in the peroxidase–oxidase reaction. *Z. Naturf.* **34**a, 1544–1546.

OLSEN, L. F. (1983). An enzyme reaction with a strange attractor. *Phys. Lett.* **94**A, 454–457.

OLSEN, L. F. (1984). The enyme and the strange attractor – comparisons of experimental and numerical data for an enzyme reaction with chaotic motion. In *Stochastic Phenomena and Chaotic Behaviour in Complex Systems* (ed. P. Schuster), pp. 116–123. Berlin: Springer-Verlag.

OLSEN, L. F. & DEGN, H. (1977). Chaos in an enzyme reaction. *Nature* **267**, 177–178.

OLSEN, L. F. & DEGN, H. (1978). Oscillatory kinetics of the peroxidase-oxidase reaction in an open system. *Experimental and Theoretical Studies, Biochim. biophys. Acta* **523**, 321–334.

OTT, E. (1981). Strange attractors and chaotic motion of dynamical systems. *Rev. Mod. Phys.* **53**, 655–671.

PACKARD, N. H., CRUTCHFIELD, J. P., FARMER, J. D. & SHAW, R. S. (1980). Geometry from a time series. *Phys. Rev. Lett.* **45**, 712–716.

RITZENBERG, A. L., ADAM, D. R. & COHEN, R. J. (1984). Period multupling – evidence for non-linear behaviour of the canine heart. *Nature* **307**, 159–161.

ROGERS, T. D. (1981). Chaos in systems in population biology. *Prog. theor. Biol.* **6**, 91–146.

ROSSLER, O. E. (1976a). Chaotic behaviour in simple reaction systems. *Z. Naturf.* **31**a, 259–264.

ROSSLER, O. E. (1976b). An equation for continuous chaos. *Phys. Lett.* **57**A, 397–398.

ROUX, J. C., TURNER, J. S., McCORMICK, W. D. & SWINNEY, H. L. (1982). Experimental observations of complex dynamics in a chemical reaction. In *Non-linear Problems: Present and Future* (ed. A. R. Bishop, D. K. Campbell and B. Nicolaenko), pp. 409–422. Amsterdam: North-Holland.

SCHAFFER, W. M. & KOT, M. (1985). Nearly one dimensional dynamics in an epidemic. *J. theor. Biol.* **112**, 403–427.

SELKOV, E. E. (1980). Instability and self-oscillation in the cell energy metabolism. *Ber. Buns.Ges. phys. chem.* **84**, 399–402.

SHAW, R. (1981). Strange attractors, chaotic behavior, and information flow. *Z. Naturf.* **36**a, 80–112.

SHAW, R. (1984). *The Dripping Faucet as a Model Chaotic System*. Santa Cruz, CA: Aerial Press.

SIMO, C. (1979). On the Henon–Pomeau attractor. *J. Statist. Phys.* **21**, 465–494.

SKJOLDING, H., BRANNER-JORGENSEN, B., CHRISTIANSEN, P. L. & JENSEN, H. E. (1983). Bifurcations in discrete dynamical systems with cubic maps. *SIAM Jl appl. Math.* **43**, 520–534.

TAKENS, F. (1981). Detecting strange attractors in turbulence. *Lect. Notes in Math.* **898**, 366–381.

TESTA, J. & HELD, G. A. (1983). Study of a one-dimensional map with multiple basins. *Phys. Rev. A* **28**, 3085–3089.

TOMITA, K. (1982). Chaotic response of non-linear oscillators. *Phys. Rep.* **86**, 113–167.

YAMAZAKI, I., YOKOTA, K. & NAKAJIMA, R. (1965). Oscillatory oxidations of reduced pyridine nucleotide. *Biochem. biophys. Res. Commun.* **21**, 582–586.

YORKE, J. A. & LONDON, W. P. (1973). Recurrent outbreaks of measles chickenpox and mumps: systematic differences in contact rates and stochastic effects. *Am. J. Epidemiol.* **98**, 469–482.

YORKE, J. A., NATHANSON, N., PIANIGIANI, G. & MARTIN, J. (1979). Seasonality and the requirements for perpetuation and eradication of viruses in populations. *Am. J. Epidemiol.* **109**, 103–123.

*IMA Journal of Mathematics Applied in Medicine & Biology* (1985) **2**, 221–252

# Can Nonlinear Dynamics Elucidate Mechanisms in Ecology and Epidemiology?

W. M. SCHAFFER

*Department of Ecology and Evolutionary Biology, The University of Arizona,*
*and*
*Program in Applied Mathematics, The University of Arizona, Tucson, Arizona*
*85721*

[Received 8 July 1985 and in revised form 14 November 1985]

## 1. Introduction

DYNAMICS is the study of motion, for example, fluctuations in a population's density or in the incidence of diseases that afflict it. For linear systems, the possibilities are restricted—stable and unstable fixed points, saddles, and periodic orbits of the sort observed in conservative systems such as the frictionless pendulum. The latter are of no interest to ecologists (e.g. May, 1973), since ecological systems are nothing if not dissipative. Moreover, there is every reason to believe that they are nonlinear. Thus, the simplest model of population growth for a single species in a finite environment,

$$dx/dt = rx(1-x) \tag{1}$$

is a nonlinear equation. Nonlinear systems exhibit a diversity of dynamics. Included in the 'dynamical zoo' are point attractors, stable limit cycles, toroidal flow, including quasiperiodic and phase-locked trajectories, and various sorts of chaotic attractors. Recent advances in nonlinear dynamics (e.g. Guckenheimer & Holmes, 1983) have focused on chaos, both because it is new and interesting, and because the concept of chaos may substantially change our view of the natural world. Specifically, chaotic systems are deterministic; yet to varying degrees, their properties appear stochastic. One way of quantifying this apparent stochasticity is by means of the concept of algorithmic complexity (e.g. Ford, 1983). Essentially, the algorithmic complexity of a time series increases with the amount of information necessary to specify it. For some chaotic systems, the most economical description of the solution turns out to be the time series itself. Such systems have maximal algorithmic complexity and in this sense are indistinguishable from a set of random numbers.

Chaotic motion exhibits a property called 'sensitivity to initial conditions' (Ruelle, 1979). By this it is meant that nearby trajectories diverge, on average, exponentially. Topologically, this results from the fact that trajectories on a chaotic attractor are successively stretched and folded. In effect, the attractor acts as a waring blender, homogenizing initial conditions (Rössler, 1976), and, as a consequence, errors in estimating the system's initial state undergo continuing amplification. Since there will always be some uncertainty, even in numerical simulations, it follows that chaotic motion is effectively indeterminate. From knowledge of the initial conditions, one cannot predict the system's long term behaviour.

How does one go about attempting to detect chaos in real-world data? To begin with, the traditional method of looking for order in a time series, i.e. computing its power spectrum, may be of little help. Some chaotic systems have spectra which lack prominent peaks. Other, so-called "phase-coherent," chaotic attractors exhibit very sharp spectral peaks superimposed on a noisy background (Crutchfield et al., 1980; Farmer et al., 1980). In the first case, an experimentalist might conclude that he was observing a random process; in the second, a periodic process in the presence of small amounts of noise. In both cases, he would be wrong.

Especially for systems in which the fractal dimension (e.g. Farmer et al., 1983), is between 2 and 3, it is often more revealing to plot the orbit in phase space. Phase portraits allow one to determine whether or not a particular system conforms to one of the known kinds of chaos. If it does, one can rule out the stochastic alternative. Of course, in experimental systems, there are often a great many state variables, not all of which can even be enumerated, much less measured systematically. For example, in the much studied (e.g. Roux et al., 1983) Belousov–Zhabotinskii (B–Z) reaction, there are upwards of 25–35 chemical species in the reaction vessel. Constructing a phase portrait would therefore seem an impossibility. Takens (1981) proposed a solution to this problem. Essentially, he showed that one can reconstruct a phase portrait for a system with an arbitrary number of state variables from a univariate time series. The application of Takens' method has led to major advances in physics and chemistry, in particular, in the case of the B–Z reaction and in studies of the road to turbulence in fluid dynamics (e.g. Brandstäter et al., 1983).

Can these same techniques to be put to use in ecology and epidemiology? Certainly, ecological data sets exhibit erratic fluctuations—potential candidates for chaos. On the other hand, ecological time series are generally short. If one considers, for example, that Roux et al. (1983) collected data files of 64K points for the B–Z reaction and compares this with the 200 years (one point per year) of data available for ecology's most famous oscillation, the Canadian lynx cycle (Elton & Nicholson, 1942), the magnitude of the problem becomes apparent. Ecological data sets are also also far noisier than their physical counterparts, although it is hard to say just how noisy, since we do not know what the data would look like in the absence of perturbations.

Despite these difficulties, it is interesting to consider the possible consequences to ecology and epidemiology were it the case that nonlinear methods applied, and further, if these methods revealed evidence for chaos.

The obvious consequence is that fluctuations previously believed random would turn out to have a deterministic basis. This would send the theorists back to their equations.

At the same time, the dynamical reference point, i.e. the baseline dynamics assumed for the deterministic part of the motion, would change. In the specific case of childhood epidemics, the emphasis on cyclic patterns of infection, e.g. biennial outbreaks in measles, three year cycles in pertussis, etc. (e.g. Anderson *et al.*, 1984), would prove misplaced. In ecology, the equilibrium view of the world (e.g. Lewontin, 1969) central to two decades of theorizing would go forever out of the window. In particular, it would no longer make sense to think of such systems in terms of a balance between intrinsic forces, forever searching out some mythical attracting point, and environmental vagaries perturbing the system away from it. Similar thinking pervades population genetics (e.g., Wright, 1968–78), most notably in the theory of adaptive topographies. Note the problem—in a chaotic universe the fixed points are repelling. Then, for the genetic problem, the action takes place not on adaptive peaks, but on ridges surrounding the valleys. These ridges, like Mandelbrot's (1977) mountains, would almost certainly prove to have a fractal structure.

Finally, there is the problem of ecological experiments. Currently, it is fashionable to perform some manipulation, for example, the removal of a species, in the field and then to interpret the results, often in terms of species' interactions (e.g. Brown *et al.*, 1979). But if ecological systems are on complex orbits, then both the sign and the magnitude of the interactions can depend on just where the system happens to be. Chaos in ecology would force experimentalists to determine the location of their systems in phase space *before* performing the manipulation.

In this paper, we concern ourselves with two types of low-dimensional chaos. By dimension, I refer to the fractal or Hausdorff (e.g. Farmer *et al.*, 1983) dimension of the orbit as opposed to the dimension of the space in which it lives. I distinguish between the following:

1. Motion that is essentially two-dimensional.
2. Motion for which the dimension is more or less squarely between two and three.

In both cases, the flow is contracting in all but two directions. Thus, to a greater or lesser extent, the motion takes place in two dimensions.

## 2. Chaos in nearly two dimensions

As an example of chaos that is essentially two-dimensional, consider the case of three Lotka–Volterra equations for a predator and two prey species. Gilpin (1979) was the first to show that these equations could be 'rigged', by the appropriate choice of parameter values, to produce essentially two-dimensional chaos. Figure 1 shows Gilpin's attractor viewed from different perspectives. Apparently, the orbit is confined to a sheet-like surface. We can show this explicitly, by slicing the orbit with a plane. This is called taking a Poincaré section. In Fig. 1(e), we display the intersections of the orbit with the slicing plane. As expected, the points lie along what appears to be a one-dimensional curve.

In Gilpin's equations, there are three directions and the flow is evidently *strongly* contracting in one of them. One way of quantifying rates of expansion and contraction of flows is to compute so-called Lyapunov numbers (e.g. Shimada

224                                    W. M. SCHAFFER

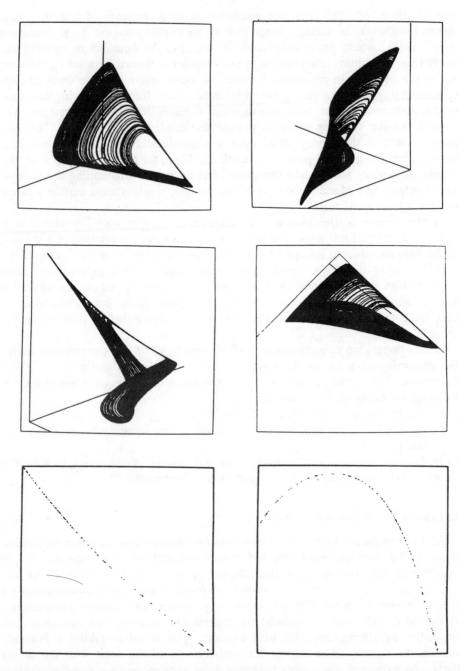

FIG. 1. Gilpin's attractor for three Lotka–Volterra Equations (one predator, two prey). Parameters as given by Schaffer, Ellner, and Kot (Ms.) with $a = 0.0039$. The top four photographs show the orbit viewed from different perspectives. Note the position of the slicing plane (normal to the page) in centre right photo. Poincaré section shown at bottom left and resulting 1-D map at bottom right. 100 000 post-transient points ($\delta t = 0.1$); transient = 10 000 points. From Schaffer & Kot (1985c).

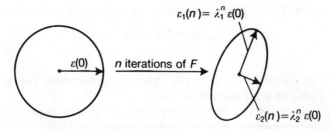

FIG. 2. Lyapunov numbers (exponents) define the average stretching or contraction of trajectories in characteristic directions. Here we show the effects of applying a two-dimensional mapping to a circle of initial conditions. From Farmer *et al.* (1983).

and Nagashima, 1979; Bennetin *et al.*, 1980; Wolf *et al.*, 1984). Figure 2, modified from Farmer *et al.* (1983), illustrates this computation for a two-dimensional mapping, i.e. a difference equation of the form

$$(X', Y') = F(X, Y) \tag{2}$$

Consider a set of initial conditions distributed on a circle of radius $\varepsilon(0)$. Now for each initial condition, we iterate (2) $n$ times. We can define the Lyapunov exponents $\lambda_1$ and $\lambda_2$ as follows:

$$\lambda_1 = [\varepsilon_1(n)/\varepsilon(0)]^{1/n}, \qquad \lambda_2 = [\varepsilon_2(n)/\varepsilon(0)]^{1/n} \tag{3}$$

where $\lambda_1^n \varepsilon(0)$ is the major axis of the resulting ellipsoid and $\lambda_2^n \varepsilon(0)$ is the minor axis. Clearly, Lyapunov numbers greater than 1 correspond to expansion, whereas Lyapunov numbers less than one correspond to contraction.

For flows (e.g. differential equations) we use a slightly different definition

$$\lambda_i = \lim_{t \to \infty} \lim_{\varepsilon(0) \to 0} \left( \frac{1}{t} \log \frac{\varepsilon_i(t)}{\varepsilon(0)} \right) \tag{4}$$

In this case, we speak of Lyapunov exponents. Positive exponents correspond to expansion; negative exponents, to contraction. Summarizing, Lyapunov exponents can be regarded as generalized eigenvalues averaged over the entire attractor.†

For Gilpin's attractor, the Lyapunov exponents are $\lambda_1 = 0.008$, $\lambda_2 = 0$, $\lambda_3 = -0.57$. Thus, there is a positive exponent corresponding to the fact that there is a direction in which, on average, trajectories diverge, a zero exponent corresponding to motion along the orbit, and a negative exponent corresponding to the fact that trajectories based at points off the attractor get pulled down onto it. Notice that the negative exponent is large in magnitude relative to the positive exponent. It is this fact which makes the orbit essentially two-dimensional. In this regard, there is a conjecture (Kaplan & Yorke, 1979), verified numerically for many models (e.g. Russell *et al.*, 1981), which relates the Lyapunov exponents to the

† Explicit computational instructions are given by Wolf *et al.* (1985).

fractal dimension. For Gilpin's equations, the conjecture states that

$$D_L = 2 + \lambda_1/|\lambda_3|$$
$$= 2 \cdot 01 \tag{5}$$

Here, we use the notation $D_L$ to signify Lyapunov dimension. This agrees with the observed thinness of the Poincaré section.

Since the points on the section approximate a 1-D curve, we look for a one-dimensional mapping, i.e. a difference equation of form

$$X' = F(X) \tag{6}$$

describing the temporal sequence of the points. To do this, first parametrize the points on some interval, typically $[0, 1]$, by fitting them to a regression of appropriate order. Then plot $X_{i+1}$ against $X_i$ for all pairs of points $i$ and $i + 1$. The resulting map for Gilpin's equation (Fig. 1(f)) is a unimodal curve. It gives a simple one-dimensional rule by which the sequence of excursions on the attractor may be summarized. Of interest, is the fact that a great deal (e.g. May, 1976; Collet & Eckmann, 1980) is known about the dynamics of such maps. Consequently, we also know a great deal about the dynamics of the flow from which the map was abstracted.

The classic application of these techniques, including Takens' reconstruction

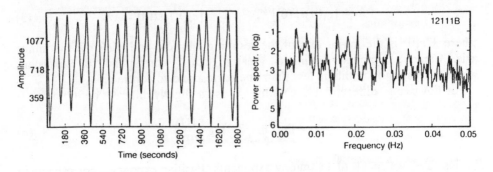

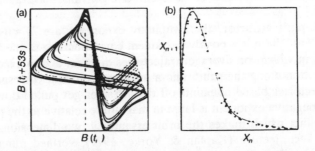

FIG. 3. Chaos in the Belousov–Zhabotinskii reaction. Top left: the time series for one of the products, Br⁻. Top right: the power spectrum. Bottom left: the strange attractor reconstructed using Takens' method. Bottom right: the resulting 1-D map. Modified from Simoyi *et al.* (1982).

algorithm, to an experimental system is the case of the B–Z reaction (Roux *et al.*, 1983; Simoyi *et al.*, 1982). Here, depending on the experimental circumstances, one obtains periodic or chaotic dynamics. Figure 3 summarizes the situation for a chaotic flow. Here we see irregular fluctuations in the concentration of one of the products, $Br^-$, a noisy power spectrum, the exquisite strange attractor obtained by application of Takens' method to the time series, and a 1-D map extracted therefrom.

### 3. Chaos in 2–3 dimensions

Even for Gilpin's equations, the attractor in section 2 is not a one-dimensional curve. If it were, the trajectory would be planar and exhibit self-crossing, thereby violating uniqueness. Thus, had we computed enough points and magnified the section sufficiently, we would have seen an elaborately folded structure (e.g., Shaw, 1981; Abraham & Shaw, 1983). Most (but see Grebogi *et al.*, 1984) chaotic attractors exhibit this Cantor-set-like structure in at least one dimension. Indeed, for orbits to undergo continuous stretching and remain bounded, it necessarily follows that they are also folded. Hence, the dimension of a chaotic attractor is generally non-integer. Now consider the case of an attractor with a single positive Lyapunov exponent, but for which the negative exponents are not all large in relative magnitude. In this case, the fractal structure is more easily discerned. As an example, we consider the seasonally-forced SEIR model that arises in the theory of epidemics. For parameters appropriate to measles in large cities, one computes the following Lyapunov exponents: $\lambda_1 = 0.41$, $\lambda_2 = 0.0$, $\lambda_3 = -0.87$, $\lambda_4 = -31.0$. Note that $\lambda_3$ is approximately the same order of magnitude as $\lambda_1$, and we compute a Lyapunov dimension, $D_L = 2.47$. Figure 4 shows a so-called 'time-one' map computed by plotting the state variables† every $T$ time units. Here $T$ is the period of the forcing—in this case, one year. What one observes is structure within structure within structure, . . . , the hallmark of a fractal set.

### 4. Application to ecological systems

Can one reasonably expect to be able to apply Takens' reconstruction scheme to ecological data, polluted—as it is—with relatively large amounts of noise? In some cases, the answer is yes, provided that the motion is chaotic (Schaffer *et al.*, Ms.). To see this, consider again Gilpin's equations. As pointed out by Schaffer & Kot (1986), Gilpin's equations exhibit a period-doubling route to chaos of the 'spiral' type. Figure 5 shows the consequences of integrating the equations in the presence of noise. In Figure 5(a), we choose parameter values corresponding to a stable limit cycle. Although the resulting Poincaré section is reasonably thin, attempting to construct a 1-D map produces an apparently random jumble of points. The situation improves somewhat, if one chooses parameter values

---

† As discussed below, there are four state variables, the numbers of susceptible, $S(t)$, exposed, $E(t)$, infectious, $I(t)$, and recovered individuals, $R(t)$. Aron & Schwartz (1984a) observe that to order $\varepsilon$, the ratio $E(t)/I(t)$ is constant. In cases where immunity is permanent, $R(t)$ can also be eliminated, and the $S$–$I$ plane is appropriate for a time-one map.

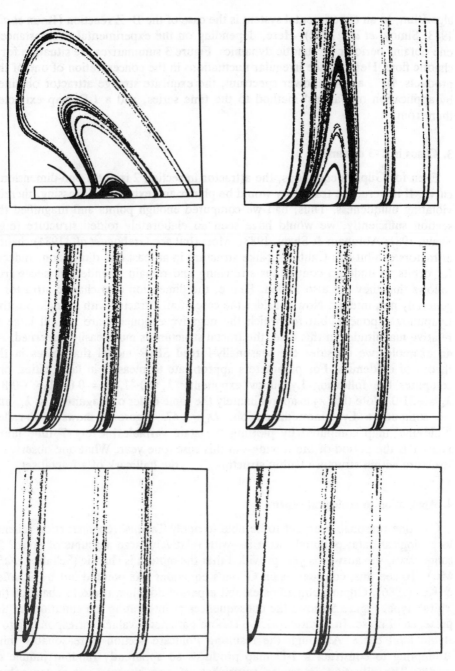

FIG. 4. Time-one map for the seasonally forced SEIR model. The state variables $S(t)$ and $I(t)$ are plotted every $T$ units of time where $T = 1$ year, the period of forcing. The portions of the map in the small boxes are successively magnified, i.e. everything in the top right photo is contained in the small box in the picture at the top left. Order of magnifications is left to right and top to bottom. Parameter values as follows: $m = 0 \cdot 02$, $a = 35 \cdot 84$, $g = 100$; $b_0 = 1800$; $b_1 = 0 \cdot 28$ [See Eqs (8).]. 100 000 post-transient points; transient = 10 000 points.

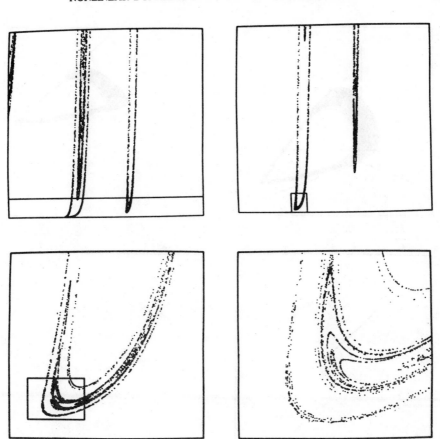

FIG. 4. Contd.

corresponding to a twice-periodic orbit (Fig. 5(b)). Finally, for the chaotic flow, one can extract a very tolerable map, indeed (Fig. 5c). Similar results hold for 1-D maps and for circle maps, discrete counterparts to quasiperiodic flow on a two-torus. We conclude that the search for essentially two-dimensional chaos in ecological data is not a pointless undertaking. If the chaos is there, we should be able to detect it.

Table 1 summarizes the results of the investigations undertaken so far. Apparently two-dimensional chaos crops up in a variety of cases. But these results need to be qualified. In all cases, save the diseases, the time series are woefully short. The 'yes' entries should therefore be read 'yes?'.

The foregoing approach, sifting through existing data sets, has been criticized on the grounds that categorizing the dynamics says nothing about the nature of the underlying mechanisms. Strictly speaking, this is not necessarily the case. The observation of toroidal flow, for example, suggests that one should look for factors that vary periodically. Beyond this, one can observe that reconstructing

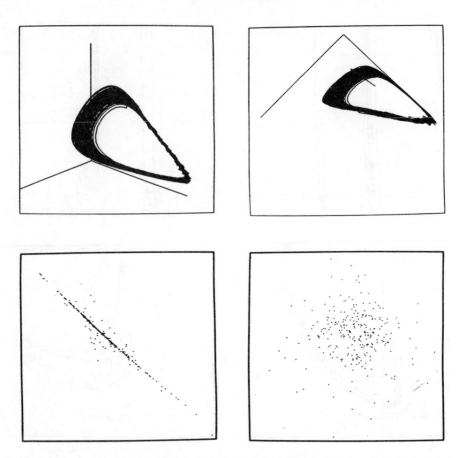

FIG. 5. Effects of adding noise to Gilpin's Equations. (a) ($a = 0.0028$). The deterministic dynamics are a stable limit cycle: Top left and right: the orbit in the presence of noise. Bottom left: Poincaré section. Bottom right: '1-D' map suggests a random scatter. (b) ($a = 0.035$). The deterministic dynamics are a twice-periodic orbit. The 1-D map suggests a unimodal curve. (c) ($a = 0.004$). The deterministic dynamics are a chaotic orbit (Fig. 1). The 1-D map is suggested quite clearly. For each experiment, white noise ($s = 0.02$) was added multiplicatively to each of the three state variables every 10 time units (4–7 times per orbit). For additional details, see Schaffer, Ellner, & Kot (Ms.).

the dynamics gives the theorist concerned with mechanism something to shoot for. If his model is a good one, it should reproduce the dynamics actually observed. Describing real-world dynamics might thus help confirm or reject models that otherwise tend to have a life of their own. The purpose of the present paper is to consider whether or not this is, in fact, feasible. Accordingly, we turn to a specific case—childhood diseases, in particular, epidemics of measles.

## 5. Measles

Prior to the introduction of the vaccine that led to its effective eradication, recurrent major outbreaks of measles occurred in most large American and

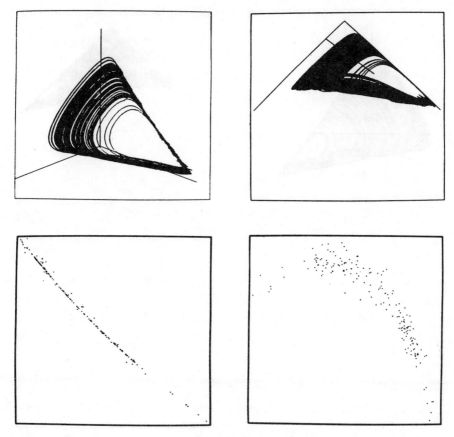

FIG. 5b. Contd.

European cities at intervals of two to five years (e.g. Bartlett, 1960). Even today, measles remains a major source of childhood mortality in poor countries such as the Sudan (Callum, 1983). Figure 6 summarizes the situation for New York and Baltimore for the years 1928–1963. Here, we show the numbers of cases reported monthly (Yorke & London, 1973) in each city (top) and the power spectra (bottom). In passing, we note the following.

1. In both cities, one observes an annual cycle with case rates peaking during the winter. This intra-annual variation has been ascribed to seasonal variation in contact rates (e.g. Yorke & London, 1973; Fine & Clarkson, 1982) and may also reflect the effects of temperature and humidity on the pathogen's ability to disperse (Anderson, 1982).

2. Superimposed upon the annual cycle, is enormous variation among years. In New York, for example, the yearly totals range from a low of 1870 cases to a high of 79 646.

3. In New York, the interval between major outbreaks is two to three years, with an essentially biennial pattern holding after 1945. In Baltimore, the inter-epidemic interval is less nearly constant, being two to five years.

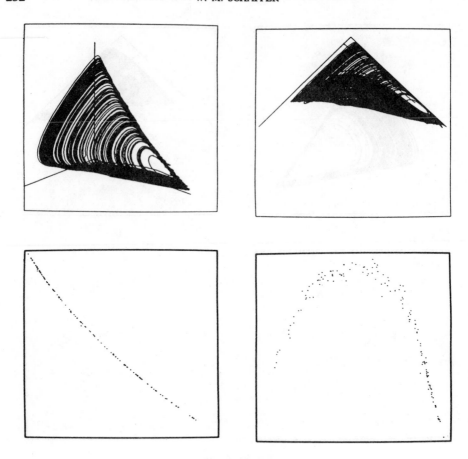

FIG. 5c. Contd.

4. Fluctuations in the two cities are *not* in synchrony. Thus one is dealing with two separate cases.

Conventionally (e.g. Yorke & London, 1973; Anderson *et al.*, 1984), these and similar data from other cities are interpreted as reflecting a two year cycle, in the presence of noise. In fact, for both cities, the prominent spectral peak corresponds to the yearly cycle. Evidence for a biennial cycle (Anderson *et al.*, 1984) emerges only upon the application of spectral smoothing techniques (e.g. Jenkins & Watts, 1968).

What happens if we apply Takens' (1981) reconstruction scheme to the reported case rates? The answer (Schaffer & Kot, 1985a) is given in Fig. 7 (left and centre columns). At the top, we show the time series embedded in three dimensions. The middle row shows the orbits viewed from above and sliced with a plane (vertical line) normal to the paper. The Poincare sections are displayed in the small box at the upper left in each photograph. [Note that there are two sections for each orbit, one corresponding to case numbers increasing (points along the left side of the box); the other to case numbers declining (points alo

TABLE 1

*Evidence for low dimensional chaos in ecology and epidemiology*

| System | References* | 2-D Orbit | 1-D Map |
|--------|-----------|-----------|---------|
| (Mammals) | | | |
| Canadian lynx | Elton & Nicholson (1942a); Schaffer (1984) | yes | ? |
| Muskrat | Elton & Nicholson (1942b); Boyce & Schaffer (unpubl.) | no | no |
| (Insects) | | | |
| *Thrips* | Davidson & Andrewartha (1948); Schaffer & Kot (1985b) | yes | yes |
| *Leucoptera caffeina* | Bigger & Tapley (1969); Schaffer (unpubl.) | yes | ? |
| *Leucoptera meyricki* | Bigger & Tapley, (1969); Schaffer (unpubl.) | yes | ? |
| Blowflies | Oster (1981); Schaffer (unpubl.) | no | yes |
| (Human diseases) | | | |
| Chickenpox–NYC | Yorke & London (1973); Schaffer & Kot (1985a) | no | no |
| Chickenpox– Copenhagen | Olsen (unpubl.) | no | no |
| Measles–NYC | Yorke & London (1973); Schaffer & Kot (1985a) | yes | yes |
| Measles–Baltimore | Yorke & London (1973); Schaffer & Kot (1985a) | yes | yes |
| Measles–Copenhagen | Olsen (unpubl.) | yes | yes |
| Mumps–NYC | Yorke & London (1973); Schaffer & Kot (1985a) | no | no |
| Mumps–Copenhagen | Olsen (unpubl.) | yes | yes? |
| Rubella–Copenhagen | Olsen (unpubl.) | yes | yes |
| Scarlet fever– Copenhagen | Olsen (unpubl.) | no | no |
| Whooping cough– Copenhagen | Olsen (unpubl.) | no | no |

* Where two reference are given, the first is data source; second, analysis

the bottom).] The bottom row shows one of the sections magnified and the associated 1-D maps. Although strongly compressed, the latter suggest unimodal curves. Similar results have been obtained by L. F. Olsen (personal communication) for measles statistics in Copenhagen and possibly also for mumps and rubella.

Fitting the maps in Fig. 7 to the equation

$$X' = aX \exp(bX) \tag{7}$$

yields estimated Lyapunov exponents of 0·5 to 0·6 (Schaffer and Kot, 1985a). This places the data squarely in the chaotic region for 1-D maps. It is to be emphasized that this is a risky procedure (Wolf & Swift, 1984). In general, such computations are unstable with respect to small changes in the data and fitting procedure. However, given the brevity of the data set, it is the only game in

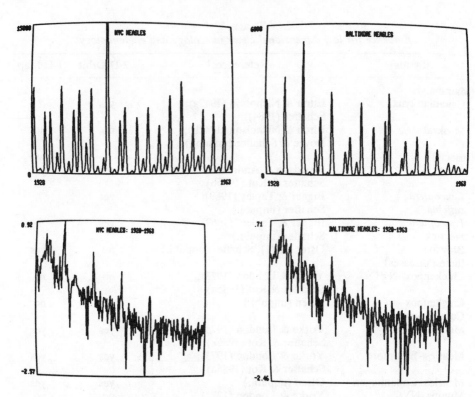

FIG. 6. Epidemics of measles in New York and Baltimore. Top. The numbers of cases reported monthly by physicians from 1928 to 1963. Bottom. Power spectra (From Schaffer & Kot, 1985a).

town. In particular, the exponents cannot be estimated directly from the reconstructed trajectory as described by Wolf *et al.* (1985).

One can also estimate the fractal dimension of the trajectory using a method devised by Grassberger & Procaccia (1983). Here the data are embedded in successively higher dimensions using the method of Takens. For each embedding, one computes the correlation integral $C(g)$, where $g$ is the appropriate length scale. Then for small $g$, the quantity $C(g)$ should scale according to a power law and that the resulting exponent $n$ is a lower bound on the fractal dimension. When dealing with experimental data, one hopes that the exponent converges with increasing embedding dimension. Figure 8 summarizes the result of applying this procedure to the New York data. Here, we see an apparent asymptote of 2·55. In the case of the Baltimore data, however, the exponent fails to converge. Again, we emphasize that given the brevity of the data set, 432 points, this is a risky procedure. Specifically, when one plots $\log C(g)$ against $\log g$, the region of linearity is poorly defined. Unfortunately, better definition would require substantially more data. Here it should be pointed out that increasing the size of the data set by interpolation (e.g. Nicolis & Nicolis, 1984) is a sure prescription for error; in particular, interpolating points increases the degree of local

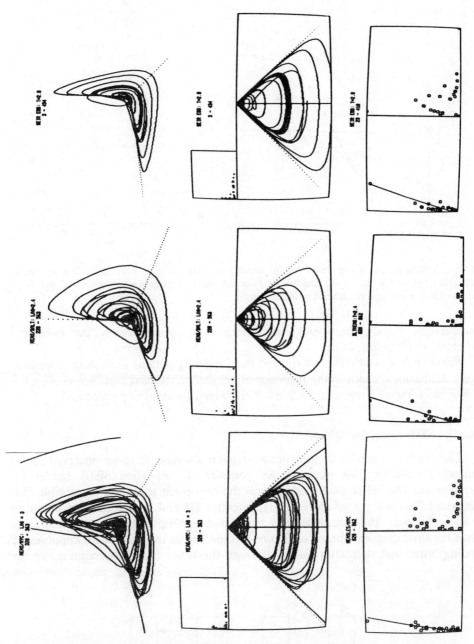

FIG. 7. Measles epidemics real and imagined. Top row. Orbits reconstructed from the numbers of infective individuals reported monthly with three-point smoothing and interpolation with cubic splines (Schaffer & Kot, 1985a). Time lag for reconstructions indicated in photos. Middle row. Orbits viewed from above (main part of the figures) and sliced with a plane (vertical line) normal to the paper. Poincaré sections shown in the small boxes at upper left. Bottom row. One of the Poincaré sections magnified (left) and resulting 1-D map (right). In each case, 36 years of data are shown. Left column: data from New York City. Middle column: data from Baltimore. Right column: SEIR equations with parameters as in Fig. 4, save $b_1 = 0.28$.

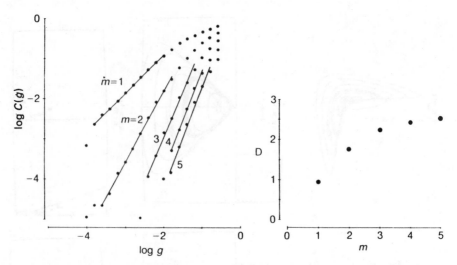

FIG. 8. Estimating the fractal dimension for measles epidemics in New York. Left. The correlation integral $C(g)$ plotted against the length scale $g$ for different embeddings $m$ of the data. Right: Slope of the log–log plot against embedding dimension.

correlation among them and can reduce in a dramatic way the estimated dimension (Schaffer, Ms.).

Schaffer & Kot (1985a) interpreted the foregoing data as indicating essentially two-dimensional chaos in the presence of noise. A disturbing fact, however, is the estimated fractal dimension for New York which substantially exceeds 2.

## 6. The SEIR model for epidemics

Can we find a model that generates dynamics similar to those observed in the data? To answer this question we consider the so-called SEIR model for epidemics. The SEIR model categorizes the host population into susceptible ($S$), exposed (but not yet infectious) ($E$), infective ($I$) and recovered (and immune) ($R$) individuals. Because, immunity in measles is almost always permanent, one has the flow diagram shown in Figure 9. One enters the class of susceptibles by being born, and thereafter moves through the other classes by contracting the

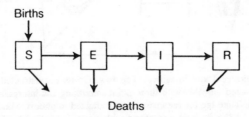

FIG. 9. Flow diagram for the SEIR model. Individuals enter the susceptible class, $S$, at birth, and thereafter the exposed, $E$, infective, $I$, and recovered, $R$, classes by contracting the disease, becoming infectious, and recovering. Since immunity is permanent, there is no feedback from $R$ to $S$.

disease, becoming contagious, and finally recovering. Individuals exit all four classes by dying.

Clearly, the SEIR scheme can be implemented in various ways, the simplest being to write coupled differential equations as shown below:

$$\left.\begin{array}{l} dS(t)/dt = m[1 - S(t)] - bS(t), \\ dE(t)/dt = bS(t)I(t) - (m + a)E(t), \\ dI(t)/dt = aE(t) - (m + g)I(t). \end{array}\right\} \tag{8}$$

Here, $1/m$ is the average life expectancy, $1/a$ is the average latency period, and $1/g$ is the average infectious period. For measles in large cities of rich countries, $m^{-1} \sim 10^2$, $a^{-1} \sim 10^{-1}$, and $g^{-1} \sim 10^{-2}$ (e.g. Anderson, 1982). The contact rate $b$ is the average number of susceptibles contacted yearly per infective and may be estimated from the other parameters and the mean age of infection (Dietz, 1976). Equations (8) assume a constant population—births balance deaths at rate $m$, and also that the various rate processes are independent of age. Relaxing this assumption leads to an analogous set of partial differential equations (e.g. Dietz, 1982; Anderson & May, 1983) or integral equations with delays (e.g. Hethcote & Tudor, 1980).

As written above, the SEIR model admits to two possibilities depending on the value of the so-called basic reproductive rate of infection

$$Q = ba/[(m + a)(m + g)]. \tag{9}$$

Specifically, if $Q < 1$, the disease dies out; if $Q > 1$, it persists at a constant level and is said to be endemic. In this case, what one sees dynamically are weakly damped oscillations about a stable fixed point. May & Anderson (1979) give a formula for the period which, for measles, yields a value of 2–4 years, in good agreement with observation. However, in the absence of perturbations, the oscillations eventually decay.

## 7. Two variants of the basic model

The observation of recurrent epidemics is at variance with the SEIR model as formulated above. As reviewed by May & Anderson (1979; 1982), essentially two hypotheses have been proposed to account for the discrepancy. The first, due originally to Bartlett (1956), suggests that the oscillations are sustained by random perturbations. The second (e.g. Dietz, 1976; Grossman, 1980; Aron & Schwartz, 1984a) holds that seasonal variation in transmission rates gives rise to resonance, in which case, cycles with periods equal to an integer number of years result. Does either of these hypotheses square with the data?

Figure 10 illustrates the effects of adding noise to the basic model with constant transmission. Although the orbit (top) is vaguely suggestive of actual data and the Poincaré section reasonably thin, attempts to extract a 1-D map result in an essentially random jumble of points (bottom, right). The addition of noise thus fails to reproduce the roughly one-dimensional rule specifying the observed sequence of excursions.

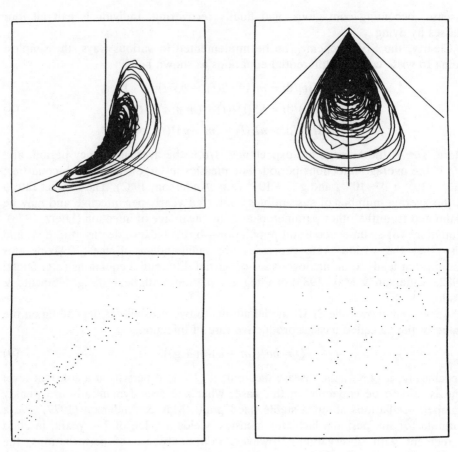

FIG. 10. Effect of adding noise to the SEIR model with constant contact rates. Attempts to recover a 1-D map produce a random scatter. Noise added as in Fig. 5, on average, 5 times per orbit.

To study the effects of seasonality, we replace the contact rate $b$ in (8) with a periodic function

$$b(t) = b_1(1 + b_0 \cos 2\pi t). \tag{10}$$

In this case, Aron & Schwartz (1984a) and Schwartz & Smith (1983) have observed period-doubling bifurcations (Fig. 11) leading to chaos. In the region of period doubling, the oscillations are roughly biennial—high and low years alternate. Hence Aron & Schwartz conclude that the model is consistent with the noisy two-cycle interpretation favoured by other workers. However, in the chaotic region, any suggestion of two year periodicity is lost (Fig. 11, bottom). Instead, one observes very erratic dynamics reminiscent of the forced Duffing equation (e.g. Guckenheimer & Holmes, 1983). This would seem both good and bad. On the one hand, the Baltimore data, and to a lesser extent those from New York, depart markedly from a two-year cycle. The fact that the SEIR model can generate similar irregularities is a point in its favour. At the same time, the resulting

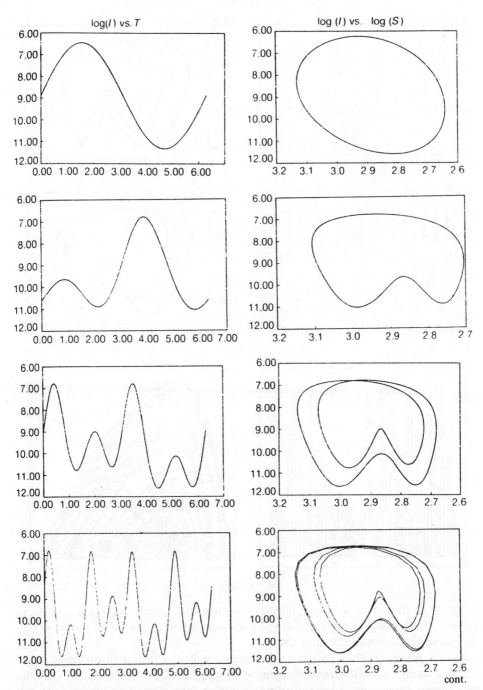

FIG. 11. Period-doubling route to chaos in the seasonally forced SEIR model. $b_1 = 0.05$ to $0.285$. Left column, the time series of infectives. Right column. Motion projected into the $S$–$I$ plane. Pre-chaotic orbits (from Aron & Schwartz, 1984a) depicted logarithmically.

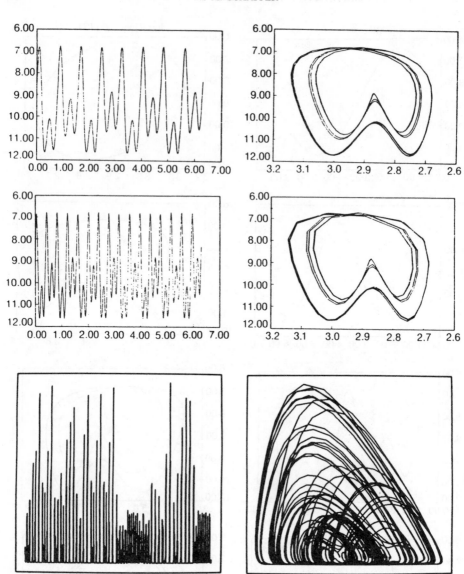

FIG. 11. Contd.

projections in the $S–I$ plane hardly resemble the orbits reconstructed from nature. Thus, it would appear that the seasonal variant of the SEIR model is also incapable of accounting for the data.

Recall, however, that Takens' result states only that the reconstructed phase portrait should have the same dynamical properties as the original, not that the two should necessarily look the same. Suppose that instead of viewing the full system, we reconstruct a phase portrait from the numbers of infectives, proceeding as we must in the case of real data. What do we get? The answer is shown in Fig. 7 (right column), and it is hoped that the reader will agree that the result bears striking similarity to the epidemics observed in New York and Baltimore. In particular, note that the orbit (top) shows deviations from an essentially two-dimensional flow similar to those seen in the data. Note also, the similarity of the Poincaré sections (middle), and finally that the '1-D' map (bottom), though fatter than what one sees in the data, exhibits similar departures from a true unimodal curve. (The Copenhagen data (L. F. Olsen personal communication) show a broader map more nearly resembling the model.)

The reconstructions in Fig. 7 correspond to 36 years of data, the number available for New York and Baltimore. Figure 12 shows the 1-D maps obtained for longer runs. What, for a limited number of points, resembles a noisy 1-D map, now clearly exhibits the same intricately folded geometry observed in the time-one map shown in Fig. 4.

In Table 2, we compute Lyapunov exponents for the SEIR model at several

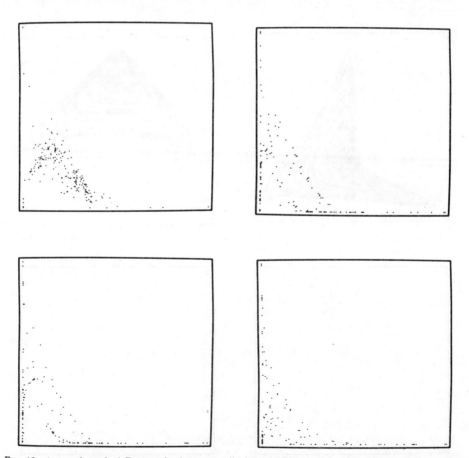

FIG. 12. Approximately 1-D maps in the seasonally forced SEIR model obtained by Poincaré section of reconstructed trajectories. Top left. $b_1 = 0.20$. Top right. $b_1 = 0.26$; Bottom left. $b_1 = 0.275$; Bottom right. $b_1 = 0.285$. In the first two cases, a small amount of noise was added to prevent the orbit from settling down to a stable limit cycle.

TABLE 2

*Lyapunov exponents[1] and dimension for the SEIR equations[2]*

| $b_1$ | $\lambda_1$ | $\lambda_2$ | $\lambda_3$ | $\lambda_4$ | $D_L$ |
|-------|-------------|-------------|-------------|-------------|-------|
| 0·270 | 0·23 | 0 | −0·67 | −32 | 2·35 |
| 0·275 | 0·41 | 0 | −0·86 | −32 | 2·48 |
| 0·280 | 0·41 | 0 | −0·87 | −31 | 2·47 |
| 0·285 | 0·43 | 0 | −0·89 | −31 | 2·48 |
| 0·290 | 0·51 | 0 | −0·97 | −30 | 2·53 |
| 0·295 | 0·46 | 0 | −0·92 | −30 | 2·50 |

[1] 200 post-transient orbits (transient = 800 orbits). Renormalized once per orbit (see Wolf *et al.*, 1985).
[2] $m = 0·02$; $a = 35·84$; $g = 100$; $b_0 = 1800$.

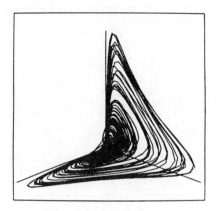
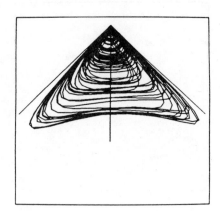

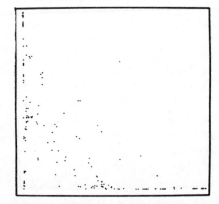

FIG. 13. The apparent nature of the reconstructed trajectory for the seasonally forced SEIR depends on the variable chosen. (a) The orbit reconstructed from the numbers of infectives. (b) The orbit reconstructed from the numbers of susceptibles. Parameters as in Fig. 4.

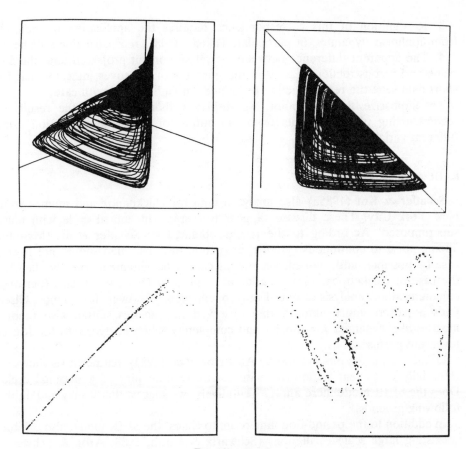

FIG. 13b. Contd.

points in the chaotic region. For most, we obtain a positive exponent ranging from 0·4 to 0·5. This is in reasonable agreement with the estimate of 0·5 to 0·6 obtained from the data using the 1-D map approximation. Table 2 also gives values for the Lyapunov dimension approximately ranging from 2·4 to 2·5. This compares with the fractal dimension of 2·55 obtained from the actual data for New York.

To conclude this section, we offer the following summary observations:

1. Bartlett's stochastic hypothesis fails to reproduce an essential feature of the data, namely the approximately one-dimensional relationship specifying the sequence of excursions when the attractor is reconstructed from the numbers of infectives.

2. There appears to be excellent agreement between the pattern of measles epidemics actually observed and the dynamics of the forced SEIR model in the chaotic region that follows the period-doubling sequence described by Aron & Schwartz (1984a). The necessity of claiming that the observed dynamics correspond to random departures from what is essentially a two year cycle is thereby obviated.

3. Recognition of this correspondence requires the application of methods from nonlinear dynamics, in particular, Takens[7] (1981) reconstruction scheme.

4. The apparent difference between the phase portrait projected onto the *S–I* plane and reconstructions based on the numbers of infectives indicates that for short data sets, the reconstruction techniques should be used with care.

To emphasize this last point, we display without comment the results of reconstructing phase portraits from the output of the same simulation using different variables (Fig. 13).

## 8. Other childhood diseases

Schaffer & Kot (1985a) also analyzed data for chickenpox and mumps from New York City. These diseases appear to display an annual cycle with noise superimposed. According to the results obtained by Schaffer *et al.* (Ms.) for adding noise to continuous systems, extraction of a 1-D map should not in this case be possible, and, indeed, such maps cannot be obtained from the data. In the case of chickenpox, these results are confirmed by L. F. Olsen's (personal communication) analysis of data from Copenhagen. However, for mumps, Olsen finds a pattern more akin to that observed for measles. Olsen also reports measles-like dynamics for rubella and completely different dynamics for scarlet fever and pertussis.

Of interest, is the observation by Anderson *et al.* (1984) that pertussis shows an essentially three-year periodicity with relatively little hint of a seasonal cycle. Does the SEIR model here apply? Tentatively, we suggest that it may and on the following grounds.

In addition to the period-doubling route to chaos, the SEIR model also exhibits coexisting large scale oscillations (Schwartz & Smith, 1983; Aron & Schwartz, 1984a, 1984b; Schwartz, Ms.). These appear to be continuous analogues of resonance structures called 'Arnol'd tongues' (e.g. Arnol'd, 1977; Aronson *et al.* 1980) observed in mappings of the plane. Basically, a cycle of base period *k* emerges and then undergoes its own period-doubling route to chaos. Tuning of the system beyond the strange attractor leads to what Grebogi *et al.* (1983) term a 'crisis'. Apparently, the attractor runs in to its basin boundary At this point, the attractor 'dies', so that solutions based thereon, eventually wind up on the coexisting stable solution.

Figures 14–16 illustrate the period-three resonance structure which coexists with the two-cycle discussed previously. Figure 17 shows the strange attractor with magnifications of one of the pieces. In Fig. 18, we show the crisis. For the first 800 years or so, the trajectory is on what was formerly the strange attractor. Thereafter it gets sucked into the two-cycle. Whether or not such large-amplitude cycles explain the three-year cycle of whooping cough remains to be investigated.

## 9. Other implementations of the SEIR model

As indicated above, equations (8) simplify nature to a considerable degree. For one thing, they ignore the fact that exposure varies with age. For another, the

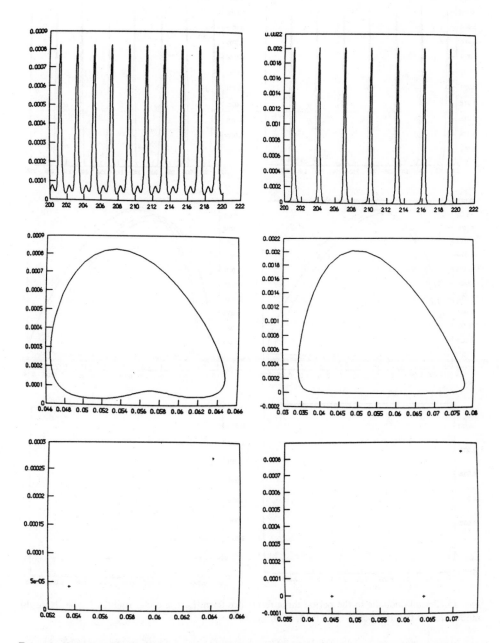

FIG. 14. Large amplitude period three oscillations (right) coexist with the two-cycle (left) in the seasonally forced SEIR model. Top: infectives plotted against time. Middle: orbits projected to the $S–I$ plane. Bottom: Time-one maps. $b_1 = 0 \cdot 13$.

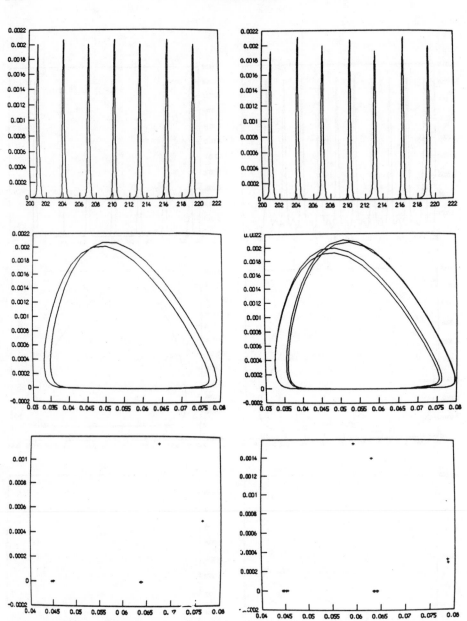

FIG. 15. Period-doubling in the period three resonance horn. Left: 6-cycle. Right: 12-cycle. Note the modulation of the time series. $b_1 = 0 \cdot 14$, $0 \cdot 15$.

transitions from one category to the next are more realistically modelled with delays (e.g. Grossman, 1980). Finally, to the extent that seasonal variations in contact rates depend on whether or not children are in school, equation (10) is a very bad model indeed, since the true state of affairs would more nearly resemble a step function. Recently, Schenzle (1984) and Dietz & Schenzle (1985)

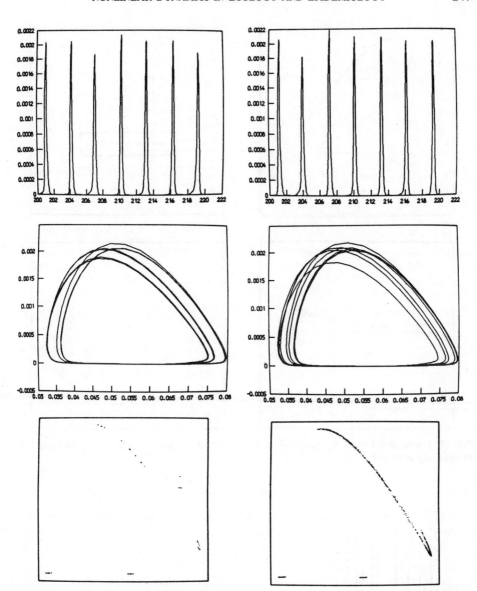

FIG. 16. Emergence of the strange attractor at the end of the period-doubling cascade of the 3-cycle. $b_1 = 0.153, 0.1535$.

have proposed a model which takes age structure into explicit account as well as the fact that children 'are promoted grade-wise into and out of school'. Essentially, Schenzle reduces a set of integro-differential equations (Dietz, 1975) to ordinary differential equations by dividing children into cohorts consisting of individuals born during the same school year. Thus Fig. 9 is replaced by a diagram in which there are $4n$ boxes with $n = n_c + n_a$, that is, the number of

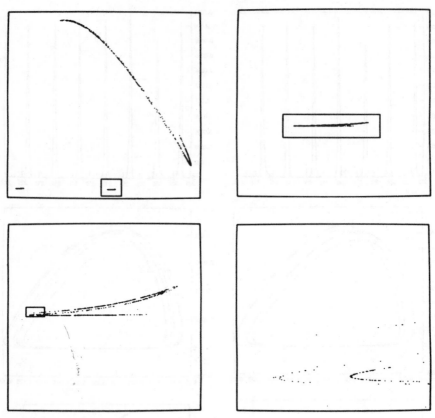

FIG. 17. Fractal structure of the period three strange attractor. The sequence of magnifications is left to right, top to bottom.

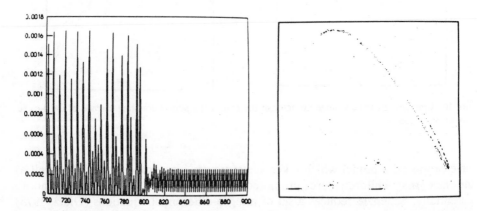

FIG. 18. Death of the period three strange attractor. Left: the time series of infectives. Right: time-one map. Note the abrupt transition from chaotic motion to the coexisting two cycle after about 800 years.

school age cohorts plus the number of 'adult' age classes. Contact and mortality rates are presumed to vary with age and exposure is turned on and off according to the details of the school year (including Sundays and vacations).

Analytic results for such a detailed model are, of course, impossible to come by, nor is it presently known if the global bifurcation pattern resembles that obtained for simpler implementations of the SEIR model. Of interest is the claim (Schenzle, 1984; Dietz & Schenzle, 1985) that this 'school-oriented' model better squares with observed patterns of incidence than simpler models. Unfortunately, it is not made clear whether the improved fit results from the incorporation of age structure, or the more precise representation of seasonality. Disentangling these factors would be an interesting undertaking as would characterizing the asymptotic dynamics of the Schenzle model in terms of the concepts discussed here.

## 10. Summary

Using recently developed methods in nonlinear dynamics, two hypotheses often advanced to account for recurrent outbreaks of childhood diseases such as measles are investigated. The first, maintenance of otherwise damped oscillations by noise, appears incapable of reproducing essential features of the data. The second, cycles and chaos sustained by seasonal variation in contact rates gives qualitative and quantitative agreement between model and observation. It is concluded that nonlinear dynamics offers a methodology which may allow students of ecology and epidemiology to distinguish between competing mechanistic hypotheses.

## Acknowledgements

I am grateful to Mark Kot for continuing discussions of nonlinear topics. The work reported herein was supported by the National Science Foundation, the John Simon Guggenheim Memorial Foundation, and the University of Arizona.

REFERENCES

ABRAHAM, R. H., & SHAW, C. D. 1983 *Dynamics: The geometry of behavior. Part 2. Chaotic behavior*. Santa Cruz, Ca.: Aerial Press.
ANDERSON, R. M. 1982 Directly transmitted viral and bacterial infections of man. In Anderson, R. M. (Ed.) *Population Dynamics of Infectious Diseases. Theory and Applications*. New York: Chapman and Hall. Pp. 1–37.
ANDERSON, R. M., GRENFELL, B. T. & MAY, R. M. 1984 Oscillatory fluctuations in the incidence of infectious disease and the impact of vaccination: time series analysis. *J. Hyg. Camb.* **93**, 587–608.
ANDERSON, R. M. & MAY, R. M. 1982 Directly transmitted infectious diseases: control by vaccination. *Science* **215**, 1053–1060.
ANDERSON, R. M. 1983 Vaccination against rubella and measles: quantitative investigations of different policies. *J. Hyg. Camb.* **90**, 259–325.
ARON, J. L. & SCHWARTZ, I. B. 1984a. Seasonality and period-doubling bifurcations in an epidemic model. *J. Theor. Biol.* **110**, 665–679.
ARON, J. L. & SCHWARTZ, I. B. 1984b. Some new directions for research in epidemic models. *IMA J. Math. Appl. Med. Biol.* **1**, 267–276.

ARONSON, D. G., CHORY, M. A., HALL, M. A. & McGEHEE, R. P. 1980 A discrete dynamical system with subtly wild behavior. In Holmes, P. (Ed.) *New Approaches to Nonlinear Problems in Dynamics.* SIAM. Pp. 339–359.

ARNOL'D. V. I. 1977 Loss of stability of self-oscillations close to resonance and versal deformations of èquivariant vector fields. *Funct. Anal. Appl.* **11**, 1–10.

BARTLETT, M. S. 1956 Deterministic and stochastic models for recurrent epidemics. *Proc. Third Berkeley Symp. Math. Stat. Prob.* **4**, 81–109.

BARTLETT, M. S. 1960 *Stochastic population models in ecology and epidemiology.* London: Methuen.

BENNETIN, G., GALGANI, L., GIORGILLI, A., & STRELCYN, J.-M. 1980 Lyapunov characteristic exponents for smooth dynamical systems and for Hamiltonian systems; a method for computing all of them. Part 1: Theory. *Meccanica* **15**, 9–20.

BIGGER, M., & TAPLEY, R. E. 1969 Prediction of outbreaks of coffee leaf-miners on Kilimanjaro. *Bull. Entomol. Res.* **58**, 601–617.

BRANDSTÄTER, A., SWIFT, J., SWINNEY, H. L., WOLF, A., FARMER, J. D., & JEN, E. 1983 Low dimensional chaos in a system with Avogadro's number of degrees of freedom. *Phys. Rev. Lett.* **51**, 1442–1445.

BROWN, J. H., DAVIDSON, D. W., & REICHMAN, O. J. 1979 An experimental study of competition between seed-eating desert rodents and ants. *Amer. Zool.* **19**, 1129–1143.

CALLUM, C. 1983 Results of an ad hoc survey on infant and child mortality in Sudan *World Health Stat. Quart.* **36**, 80–99.

COLLET, P., & ECKMANN, J.-P. 1980 *Iterated Maps on the Interval as Dynamical Systems.* Boston: Birkhauser.

CRUTCHFIELD, J., FARMER, D., PACKARD, N., & SHAW, R. 1980 Power spectral analysis of a dynamical system. *Phys. Lett.* **76a**, 1–4.

DAVIDSON, J., & ANDREWARTHA, H. G. 1948 Annual trends in a natural population of *Thrips imaginis* (Thysanoptera). *J. Anim. Ecol.* **17**, 193–199.

DIETZ, K. 1975 Transmission and control of arboviruses. In Ludwig, D. and Cooke, K. L. (Eds.) *Epidemiology.* Philadelphia: SIAM. Pp. 104–121.

DIETZ, K. 1976 The incidence of infectious diseases under the influence of seasonal fluctuations. *Lect. Notes Biomath.* **11**, 1–15.

DIETZ, K. 1982 Overall population patterns in the transmission cycle of infectious disease agents. In Anderson, R. M. and May R. M. (Eds). *Population biology of infectious diseases.* New York: Springer-Verlag. Pp. 87–102.

DIETZ, K., & SCHENZLE, D. 1985 Mathematical models for infectious disease statistics. In Atkinson, A. C. and Fienberg, S. E. (Eds) *A Celebration of Statistics.* The ISI centenary volume. New York & Berlin: Springer-Verlag. Pp. 167–204.

ELTON, C., & NICHOLSON, M. J. 1942a The ten-year cycle in numbers of the lynx in Canada. *J. Anim. Ecol.* **11**, 215–244.

ELTON, C., & NICHOLSON, M. J. 1942b Fluctuations in the numbers of muskrat (*Ondatra zibethica*) in Canada. *J. Anim. Ecol.* **11**, 96–126.

FARMER, D., CRUTCHFIELD, J., FROEHLING, H., PACKARD, N., & SHAW, R. 1980 Power spectra and mixing properties of strange attractors. *Ann. N. Y. Acad. Sci.* **357**, 453–472.

FARMER, J. D., OTT, E., & YORKE, J. A. 1983 The dimension of chaotic attractors. *Physica* **7D**, 153–180.

FINE, P. E. M., & CLARKSON, J. A. 1982 Measles in England and Wales. I. An analysis of factors underlying seasonal patterns. *Int. J. Epidem.* **11**, 5–14.

FORD, J. 1983 How random is a coin toss? In Horton, C. W., Reichl, L. E., & Szebehely, V. G. *Long-time Prediction in Dynamics.* New York: J. Wiley. Pp. 79–92.

GILPIN, M. E. 1979 Spiral chaos in a predator-prey model. *Amer. Natur.* **107**, 306–308.

GRASSBERGER, P., & PROCACCIA, I. 1983 Measuring the strangeness of strange attractors. *Physica* **9D**, 189–208.

GREBOGI, C., OTT, E., & YORKE, J. A. 1983 Cries, sudden changes in chaotic attractors and transient chaos. *Physica* **7D**, 181–200.

GREBOGI, C., OTT, E., PELIKAN, S., & YORKE, J. A. 1984 Strange attractors that are not chaotic. *Physica* **13D,** 261–268.

GROSSMAN, Z. 1980 Oscillatory phenomena in a model of infectious diseases. *Theor. Pop. Biol.* **18,** 204–243.

GUCKENHEIMER, J. & HOLMES, P. 1983 *Nonlinear Oscillations, Dynamical Systems and Bifurcations of Vector Fields.* New York: Springer-Verlag.

HETHCOTE, H. W., & TUDOR, D. W. 1980 Integral equations for endemic infectious diseases. *J. Math. Biol.* **9,** 37–47.

JENKINS, G. M. & WATTS, D. G. 1968 *Spectral analysis and its applications.* San Francisco: Holden-Day.

KAPLAN, J. & YORKE, J. A. 1979. Chaotic behavior of multidimensional difference equations. In Peitgen, O. and Walther, H. O. (Eds.) *Functional Differential Equations and the Approximation of Fixed Points.* New York: Springer-Verlag. Pp. 228–237.

LEWONTIN, R. C. 1969 The meaning of stability. *Brookhaven Symp. Biol.* **22,** 13–24.

MANDELBROT, B. 1977 *Fractals: Form, Chance and Dimension.* San Francisco: Freeman.

MAY, R. M. 1973 *Stability and complexity in model ecosystems.* Princeton, N.J.: Princeton Univ. Press.

MAY, R. M. 1976 Simple mathematical models with very complicated dynamics. *Nature* **261,** 459–467.

MAY, R. M., & ANDERSON, R. M. 1979 Population biology of infectious diseases: II. *Nature* **28,** 455–461.

NICOLIS, C., & NICOLIS, G. 1984. Is there a global climatic attractor? *Nature.* **311,** 529–532.

OSTER, G. 1981 Predicting populations. *Amer. Zool.* **21,** 831–844.

RÖSSLER, O. 1976 Chaotic behavior in simple reaction systems. *Zeit. f. Natürforsch.* **31a,** 259–264.

ROUX, J.-C., SIMOYI, H., & SWINNEY, H. L. 1983 Observation of a strange attractor. *Physica* **8D,** 257–266.

RUELLE, D. 1979. Sensitive dependence on initial conditions and turbulent behavior in dynamical systems. *Ann. N. Y. Acad. Sci.* **316,** 408–416.

RUSSELL, D. A., HANSON, J. D., & OTT, E. 1981 Dimension of strange attractors. *Phys. Rev. Lett.* **15,** 1175–1178.

SCHAFFER, W. M. 1984 Stretching and folding in lynx fur returns: evidence for a strange attractor in nature? *Amer. Natur.* **124,** 798–820.

SCHAFFER, W. M. 1985 Order and chaos in ecological systems. *Ecology* **66,** 93–106.

SCHAFFER, W. M. Ms. On the evidence for a Global Climatic Attractor.

SCHAFFER, W. M., ELLNER, S. E., & KOT, M. Ms. Effects of Noise on Dynamical Models in Ecological Systems.

SCHAFFER, W. M. & KOT, M. 1985a Nearly one dimensional dynamics in an epidemic. *J. Theory. Biol.* **112,** 403–427.

SCHAFFER, W. M. & KOT, M. 1985b Do strange attractors govern ecological systems? *Bioscience* **35,** 342–350.

SCHAFFER, W. M. & KOT, M. 1986 Differential systems in ecology and epidemiology. In Holden, A. V. (Ed.) *Chaos: an Introduction.* Univ. Manchester Press, UK. Pp. 158–178.

SCHENZLE, D. 1984 An age-structured model of pre- and post-vaccination measles transmission. *IMA J. Math. Appl. Med. Biol.* **1,** 169–191.

SCHWARTZ, I. B., & SMITH, H. L. 1983. Infinite subharmonic bifurcations in an SEIR model. *J. Math. Biol.* **18,** 233–253.

SCHWARTZ, I. B., & SMITH, H. L. 1983 Multiple recurrent outbreaks and predictability in seasonally forced nonlinear epidemic models. *J. Math. Biol.* **18,** 233–253.

SHAW, R. 1981 Strange attractors, chaotic behavior and information flow. *Zeit. f. Natürforsch.* **36a,** 80–112.

252 W. M. SCHAFFER

SHIMADA, I. & NAGASHIMA, T. 1979 A numerical approach to ergodic problem of dissipative dynamical systems. *Prog. Theory. Phys.* **61,** 1606–1616.

SIMOYI, R. H., WOLF, A., & SWINNEY, H. L. 1982 One dimensional dynamics in a multi-component chemical reaction. *Phys. Rev. Lett.* **49,** 245–248.

TAKENS, F. 1981. Detecting strange attractors in turbulence. In Rand, D. A. and Young, L. S. (Eds.) *Dynamical Systems and Turbulence, Warwick, 1980.* New York: Springer-Verlag. Pp. 366–381.

WOLF, A., SWIFT, J. B., SWINNEY, H. L. & VASTANO, J. A. 1985. Determining Lyapunov Exponents from a Time Series. *Physica* **16D,** 285–317.

WOLF, A. & SWIFT, J. 1984 Progress in computing Lyapunov exponents from experimental data. In Horton, C. W. and Reichl, L. E. *Statistical Physics and Chaos in Fusion Plasmas.* New York: J. Wiley and Sons. Pp. 111–126.

WRIGHT, S. 1968–1978 *Evolution and the Genetics of Populations.* Vols. 1–4. Chicago: Univ. Chicago Press.

YORKE, J. A. & LONDON, W. P. 1973 Recurrent outbreaks of measles, chickenpox and mumps. II. *Amer. J. Epidem.* **98,** 469–482.

# Empirical and theoretical evidence of economic chaos

Ping Chen

Empirical and theoretical investigations of chaotic phenomena in macroeconomic systems are presented. Basic issues and techniques in testing economic aggregate movements are discussed. Evidence of low-dimensional strange attractors is found in several empirical monetary aggregates. A continuous-time deterministic model with delayed feedback is proposed to describe the monetary growth. Phase transition from periodic to chaotic motion occurs in the model. The model offers an explanation of the multiperiodicity and irregularity in business cycles and of the low dimensionality of chaotic monetary attractors. Implications for monetary control policy and a new approach to forecasting business cycles are suggested.

In recent years, there has been rapid progress in the study of deterministic chaos, random behavior generated by deterministic systems with low dimensionality. This progress has been made not only in theoretical modeling but also in experimental testing (Abraham, Gollub, and Swinney 1984). Chaotic models have been applied to a variety of dynamic phenomena in the areas of fluid dynamics, optics, chemistry, climate, and neurobiology. Applications to economic theory have also been developed, especially in business cycle theory (see review article: Grandmont and Malgrange 1986).

Over the past century, the nature of business cycles has been one of the most important issues in economic theory (Zarnowitz 1985). Business cycles have several puzzling features. They have elements of a continuous wavelike movement; they are partly erratic and at the same time serially correlated. More than one periodicity has been identified in business cycles in addition to long growth trends. Most simplified models in macroeconomics address one of these features (Rau 1974), while system dynamics models describe economic movements in terms of a large number of variables (Forrester 1977).

Two basic questions arise in studies of business cycles: Are endogenous mechanisms or exogenous stochastics the main cause of economic fluctuations? and Can complex phenomena be characterized by mathematical models as simple as, say, those for planetary motion and electricity?

The early deterministic approach to business cycles with well-defined periodicity mainly discussed the endogenous mechanisms of economic movements. A linear deterministic model was first proposed by Samuelson (1939), which generated damped or explosive cycles. Nonlinearities were introduced in terms of limit cycles to explain the self-sustained wavelike movement in economics (Goodwin 1951).

A stochastic approach seems to be convenient for describing the fluctuating behavior in economic systems (Osborne 1959; Lucas 1981). The problem with stochastic models, however, lies in the fact that random noise with finite delay terms (usually less than ten lags, in practice) only explains the short-term fluctuating behavior. Most aggregate economic data are serially correlated not only in the short term but also over long periods. Two methods dealing with long correlations are often used: longer lags in regression studies and multiple differencing time series in ARIMA models. Longer lags require estimating more "free parameters," while ARIMA models are essentially whitening processes that wipe out useful information about deterministic mechanisms.

Actually, fluctuations may be caused by both intrinsic mechanisms and external shocks. An alternative to the stochastic approach, with a large number of variables

The author is greatly indebted to Professor Ilya Prigogine for inspiring the research of economic chaos and to Professor W. A. Barnett for suggesting the test of monetary aggregates. The author is grateful to Professors B. L. Hao, H. L. Swinney, G. Nicolis, W. W. Rostow, and P. Allen, and to Drs. A. Arneodo, Y. Yamaguchi, A. Wolf, A. Brandstater, and W. M. Zheng for their valuable discussions. He also appreciates the stimulating comments from Professors J. D. Sterman, E. Mosekilde, P. A. Samuelson, and R. Solow. This research is supported by the IC² Institute at Austin.

*System Dynamics Review* 4 (nos. 1–2, 1988): 81–108. ISSN 0883–7066. © 1988 by the System Dynamics Society.

**82** System Dynamics Review Volume 4 Numbers 1–2 1988

Ping Chen is a research fellow at the I. Prigogine Center for Studies in Statistical Mechanics and a research associate at the $IC^2$ Institute, University of Texas at Austin. He received a Ph.D. degree from that university through work on nonlinear dynamics and business cycle theory. *Address:* I. Prigogine Center for Studies in Statistical Mechanics, University of Texas, Austin, TX 78712.

and parameters, is deterministic chaos, with few variables or low-dimensional strange attractors (Schuster 1984). This is the approach adopted in the present article. Newly developed numerical techniques of nonlinear dynamics also shed light on a reasonable choice of the number of variables needed in characterizing a complex system.

An increasing number of works examine economic chaos. Most theoretical models are based on discrete time (Benhabib 1980; Stutzer 1980; Day 1982; Grandmont 1985; Deneckere and Pelikan 1986; Samuelson 1986); only one long-wave model is based on continuous time (Rasmussen et al. 1985). Ongoing empirical studies are being conducted by a few economists (Sayers 1986; Brock 1986; Scheinkman and Le Baron 1987; Ramsey and Yuan 1987; Frank and Stengers 1987). Some clues of nonlinearities have been reported, but no solid evidence of chaos has yet been found by these authors. Two efforts were made to fit nonlinear discrete models with empirical data (Dana and Malgrange 1984; Candela and Gardini 1986), but the parameters were found outside the chaotic regions.

We started searching for empirical evidence of chaos in economic time series in 1984. The main features of deterministic chaos, such as complex patterns of phase portraits and positive Lyapunov exponents, have been found in many economic aggregate data like GNP and IPP, but most of our studies have failed to identify the dimensionality of attractors because of limited data. Then we tested monetary aggregates at the suggestion of W. A. Barnett. Low-dimensional strange attractors from weekly data were found in 1985, and a theoretical model of low-dimensional monetary attractors was developed in 1986 (Chen 1987). A brief description of comprehensive studies of economic chaos is presented here for general readers.

In this article, a short comparison between stochastic and deterministic models is introduced. Positive evidence of low-dimensional strange attractors found in monetary aggregates is shown by a variety of techniques. A continuous-time model is suggested to describe the delayed feedback system in monetary growth. The period-doubling route to chaos occurs in the model (Feigenbaum 1978). The model offers an explanation for the low dimensionality of chaotic monetary time series and for the random periodic nature of business cycles and long waves. Finally, the implications of deterministic chaos in economics and econometrics are discussed.

## Simple pictures of deterministic and stochastic processes

To what extent economic fluctuations around trends should be attributed to endogenous mechanisms (described by deterministic chaos) or exogenous shocks (described by stochastic noise) is a question that can be addressed by empirical tests.

There are at least four possible candidates in describing fluctuating time series: linear stochastic processes, discrete deterministic chaos, continuous deterministic chaos, and nonlinear deterministic chaos plus noise. The test of the last one is only in its infancy, because a high level of noise will easily destroy the subtle signal of deterministic chaos. We mainly discuss the first three candidates here and give numerical examples of white noise and deterministic chaos as the background further discussions. The linear autoregressive AR(2) model adopted in explaining

Fig. 1. Comparison of the time series of model solutions. The time units are arbitrary. (a) AR(2) linear stochastic model. (b) Discrete logistic chaos generated by mapping $X(t + 1) = 4X(t)[1 - X(t)]$. (c) Rössler model of spiral chaos with time interval $dt = 0.05$.

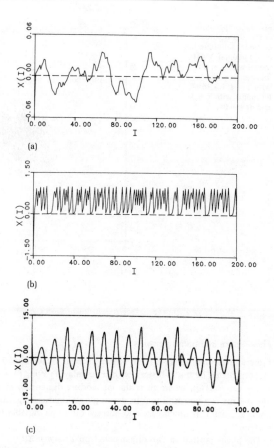

(a)

(b)

(c)

fluctuations of log linear detrended GNP time series (Brock 1986) is demonstrated as an example of a linear stochastic process. For deterministic chaos, two models are chosen: the discrete logistic model (May 1976), which is widely used in population studies and economics, and the continuous spiral chaos model (Rössler 1976).

The time sequences of these models are shown in Figure 1. They seem to be equally capable of describing economic fluctuations when appropriate scales are used to match real time series. But closer examination reveals the differences among them.

Fig. 2. Comparison of the phase portraits of model solutions. $N = 1,000$. (a) AR(2) model with $T = 20$. (b) Logistic chaos with $T = 1$. (c) Rössler model with $T = 1$ and $dt = 0.05$.

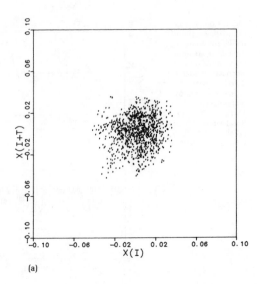

(a)

*Phase space and phase portraits*

From a given time series $X(t)$, an m-dimensional vector $\mathbf{V}(m, T)$ in phase space can be constructed by the m-history with time delay $T$: $\mathbf{V}(m, T) = \{X(t), X(t + T), \ldots, X[t + (m - 1)T]\}$, where $m$ is the embedding dimension of phase space (Takens 1981). This is a powerful tool in developing numerical algorithms of nonlinear dynamics, since it is much easier to observe only one variable to analyze a complex system.

The phase portrait in two-dimensional phase space $X(t + T)$ versus $X(t)$ gives a clear picture of the underlying dynamics of a time series. With the fixed-point solution (the so-called zero-dimensional attractor), the dynamic system is represented by only one point in the phase portrait. For a periodic solution (the one-dimensional attractor), the portrait is a closed loop. Figure 2 displays the phase portrait of the three models. The nearly uniform cloud of points in Figure 2a closely resembles the phase portrait of random noise (with infinite degrees of freedom). The curved image in Figure 2b is characteristic of one-dimensional unimodal discrete chaos. The spiral pattern in Figure 2c is typical of a strange attractor whose dimensionality is not an integer. Its wandering orbit differs from periodic cycles.

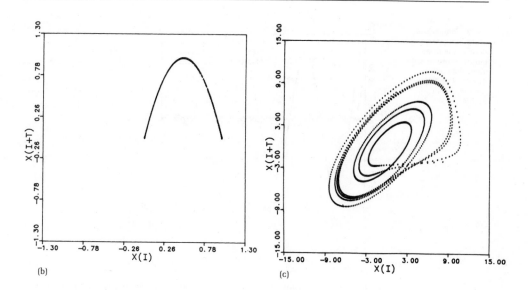

(b)                                      (c)

*Long-term autocorrelations*

The autocorrelation function is another useful concept in analyzing time series. The autocorrelation function $AC(I)$ is defined by

$$AC(I) = AC(t' - t) = \frac{\text{cov}[X(t'), X(t)]}{E[(X(t) - M)^2]} \tag{1}$$

where $M$ is the mean of $X(t)$ and $\text{cov}[X(t'), X(t)]$ is the covariance between $X(t')$ and $X(t)$. They are given by

$$M = E[X(t)] = \frac{\left\{ \sum_{t=1}^{N} X(t) \right\}}{N} \tag{2}$$

$$\text{cov}(X(t'), X(t)) = E[(X(t') - M)(X(t) - M)] \tag{3}$$

It is known that the autocorrelation function of the periodic motion is periodic and that of white noise is a delta function. Figure 3 shows autocorrelation functions for three models. The autocorrelations of the AR(2) process quickly decay to small disturbances. The autocorrelations of logistic chaos look the same as those of white noise. The Rössler attractor displays some resemblance to periodic cycles. Its autocorrelations have initial exponential decay after a characteristic decorrelation time

610

Fig. 3. Comparison
of the autocorrela-
tions of the three
model solutions with
1,000 data points.
The time units are
the same as in
Figure 1.

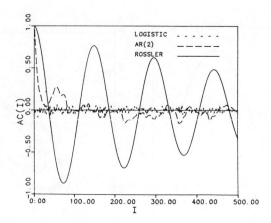

$T_d$, followed by wavelike fat tails. $T_d$ is determined by the first vanishing autocorrelations.

### Testing economic chaos in monetary aggregates

Testing economic aggregate time series is a complex process, since they contain growth trends. Not all the techniques in nonlinear dynamics developed by mathematicians and physicists (Mayer-Kress 1986) are applicable to economic time series. For example, Poincaré sections and spectral analysis used in physics require more than several thousand data points. Data quantity and quality are crucial in applying the currently existing techniques. After introducing the monetary indexes, we focus on the testing and modeling of monetary aggregates. The data source is Fayyad (1986).

*Monetary aggregates*

Observable indicators are essential to empirical investigations. In simple physical systems, some macroscopic quantities (such as mass and energy) can be simple summations of microscopic quantities. The right choice of aggregate indexes for economic systems remains an issue on which there is no consensus. For example, there are 27 component monetary assets—currency, travelers checks, demand deposit, Eurodollars, money market deposit, savings deposit, Treasury securities, commercial paper, and so on—according to the Federal Reserve's latest classification of the American monetary system. Four levels of simple-sum aggregate indexes, M1, M2, M3, and L, consisting of 6 to 27 monetary assets, are used by the Federal Reserve. There are

Fig. 4. The exponential growth trends in time series of monetary aggregates: official simple-sum index SSM2 and Divisia index DDM2 (January 1969–July 1984). The time unit is one week.

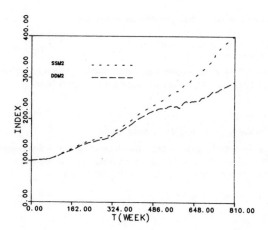

also parallel theoretical indexes, such as Divisia monetary aggregates, initiated by W. A. Barnett. Better aggregate indexes are needed to describe macroeconomic movements by simple mathematical methods.

We tested 12 types of monetary index time series, including official simple-sum monetary aggregates (denoted by SSM), Divisia monetary demand aggregates (DDM), and Divisia monetary supply aggregates (DSM); each yielded about 800 weekly data points between 1969 and 1984. Five of them were successful in testing strangeness: simple-sum SSM2, Divisia demand DDM2, DDM3, DDL, and Divisia supply DSM2 monetary aggregates. The behaviors of Divisia aggregates are very similar. We only discuss SSM2 and DDM2 here, for brevity. The exponential growth trends of these time series are shown in Figure 4.

*Observation reference and first-difference detrending*

Mathematical models with attractor solutions can greatly simplify descriptions of complex movements without obvious growth trends. The choice of detrending methods is basically a choice of reference system or transformation theory. Detrending is a solved problem for physicists when observations of physical systems are obtained in appropriate inertial reference systems. However, it is an unsolved issue in testing economic time series. How to choose a reference system to observe the global features of economic movements is a critical question for identifying the deterministic mechanisms of economic activities. We attempt to answer this question through numerical experiments on empirical data.

The percentage rate of change and its equivalent form, the logarithmic first differences, are widely used in fitting stochastic econometric models (Osborne 1959;

**88** System Dynamics Review Volume 4 Numbers 1–2 1988

Friedman 1969). It can be defined as follows:

$$Z(t) = \ln S(t + 1) - \ln S(t) = \ln \left\{ \frac{S(t + 1)}{S(t)} \right\} \tag{4}$$

where $S(t)$ is the original time series, and $Z(t)$ is the logarithmic first difference. Its ineffectiveness for observing chaos will be shown later.

### Log linear detrending and growth cycles

We detrended data using log linear detrending, which was suggested by W. A. Barnett. The same detrending has also been used by other economists (Dana and Malgrange 1984; Brock 1986). In log linear detrending, we have

$$X(t) = \ln S(t) - (k_0 + k_1 t) \tag{5}$$

or

$$S(t) = S_0 \exp(k_1 t) \exp(X(t)) \tag{6}$$

where $S(t)$ is the original time series, $X(t)$ is the resulting log linear detrended time series, $k_0$ is the intercept, $k_1$ is the constant growth rate, and $S_0 = \exp(k_0)$.

After numerical experiments on a variety of detrending methods and economic time series, we finally found that the percentage rate of change and its equivalent methods are whitening processes based on short-time scaling. Log linear detrending, on the other hand, retains the long-term correlations in economic fluctuations, since its time scale represents the whole period of the available time series. Findings of evidence of deterministic chaos mainly from log linear detrended economic aggregates lead to this conclusion. Figure 5a shows the time sequences of the log linear detrended (denoted by LD) monetary aggregates SSM2. Its almost symmetric pattern of nearly equal lengths of expansion and contraction is a typical feature of growth cycles in economic systems. The usual business cycles are not symmetric; their longer expansions and shorter contractions can be obtained by superimposing a trend with constant growth rate adding to the symmetric growth cycles. The logarithmic first-difference time series (denoted by FD) SSM2 is given in Figure 5b as a comparison. The latter is asymmetric and more erratic.

### Empirical evidence of deterministic and stochastic processes

Based on the phase portrait and autocorrelation analysis, we can easily distinguish qualitatively a stochastic process from a deterministic one. A comparison between IBM daily stock returns and monetary aggregates follows.

Figure 6a presents the phase portrait of detrended monetary aggregates LD SSM2. It rotates clockwise like the spiral chaos in Figure 2c. The complex pattern is a potential indicator of nonlinear deterministic movements and eliminates the possibilities of white noise or simple periodic motions. The phase portrait of IBM daily stock returns is shown in Figure 6b. It closely resembles Gaussian white noise. It is

Fig. 5. Comparison of the detrended weekly time series SSM2. (a) Symmetric LD SSM2: the log linear detrended SSM2 with a natural growth rate of 4 percent per year. (b) Asymmetric FD SSM2: the logarithmic first differences of SSM2.

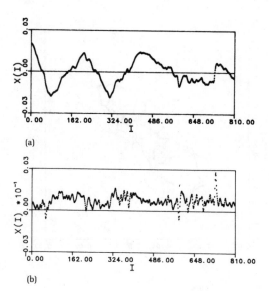

(a)

(b)

consistent with previous findings in economics (Osborne 1959; Fama 1970). The autocorrelations of the detrended time series are shown in Figure 7. Readers may compare these with the autocorrelations in Figure 3.

If we approximate the fundamental period $T_1$ by 4 times the decorrelation time $T_d$, as in the case of periodic motion, then $T_1$ is about 4.7 years for LD SSM2, which is very close to the common experience of business cycles. We will return to this point later.

### The numerical maximum Lyapunov exponent

Chaotic motion is sensitive to initial conditions. Its measure is the Lyapunov exponents, which are the average exponential rates of divergence or convergence of nearby orbits in phase space. Consider a very small ball with a radius $\epsilon(0)$ at time $t = 0$ in the phase space. The ball may distort into an ellipsoid as the dynamic system evolves. Let the length of the $i$th principal axis of this ellipsoid at time $t$ be $\epsilon_i(t)$. The spectrum of Lyapunov exponents $\lambda_i$ from an initial point can be obtained theoretically (Farmer 1982) by

$$\lambda_i = \lim_{t \to \infty} \lim_{\epsilon(0) \to 0} \left[ \frac{1}{t} \ln \frac{\epsilon_i(t)}{\epsilon_i(0)} \right] \tag{7}$$

Fig. 6. Comparison of the phase portraits of empirical time series. Time delay $T = 20$. (a) LD SSM2 time series. The time unit is one week. $N = 807$ points. (b) IBM daily common stock returns. The time interval is one day. $N = 1,000$ points, beginning on July 2, 1962.

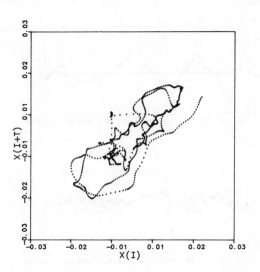

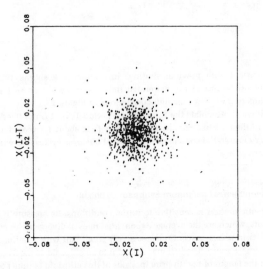

Fig. 7. Comparison of autocorrelation functions; AC($I$) plotted against $I$. There are three time series: LD SSM2, LD DDM2, and IBM daily stock returns, each in their original time units. $N = 807$.

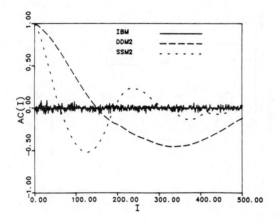

Fig. 8. An artist's sketch of the Wolf algorithm. The lower line, $y(t)$, is the reference orbit. The upper broken line, $z(t)$, starting in the neighborhood of $y(t_0)$, is traced to calculate the divergence of the lines. The points $z_0(t_1)$, $z_1(t_2)$, . . . are replaced by new nearest neighboring points $z_1(t_1)$, $z_2(t_2)$, . . . after an evolution time EVOLV for numerical calculation.

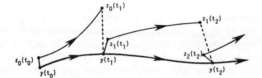

The maximum Lyapunov exponent $\lambda$ (the largest among $\lambda_i$) can be calculated numerically by the Wolf algorithm (Wolf et al. 1985), where the limiting procedure is approximated by an averaging process over the evolution time EVOLV. This algorithm is applicable when the noise level is small. A sketch of the algorithm is shown in Figure 8. The maximum Lyapunov exponent $\lambda$ is negative for stable systems with fixed points, zero for periodic or quasiperiodic motion, and positive for chaos.

In theory, the maximum Lyapunov exponent is independent of the choice of evolution time EVOLV, embedding dimension $m$, and time delay $T$. In practice, the value of the Lyapunov exponent does relate to these numerical parameters. The range of evolution time EVOLV must be chosen by numerical experiments. The positive maximum Lyapunov exponents of the investigated monetary aggregates are stable over some region in evolution time shown in Figure 9. The numerical Lyapunov exponent is less sensitive to the choice of embedding dimension $m$. In our tests, we fixed $m$ at 5 and time delay $T$ at 5 weeks, based on the numerical experiments. For example, the stable region of EVOLV is 45–105 weeks for SSM2 and 45–150 weeks for DDM2.

**92** System Dynamics Review Volume 4 Numbers 1–2 1988

Fig. 9. The maximum Lyapunov exponents of log linear detrended monetary aggregates LD SSM2 and LD DDM2. The maximum Lyapunov exponents of monetary aggregates plotted against the evolution time EVOLV, calculated in phase space with time delay $T$ = 5 weeks and embedding dimension $m$ = 5. The unit is bit per week. The evolution time EVOLV in calculating numerical Lyapunov exponents varies from 15 to 180 weeks at 15-week intervals.

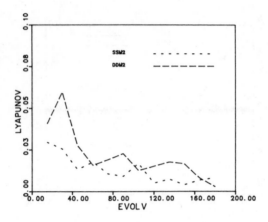

Their average maximum Lyapunov exponents $\lambda$ over this region are 0.0135 and 0.0184 (bit per week), respectively.

The characteristic decorrelation time $T_d$ of the LD SSM2 is 61 weeks. The reciprocal of the maximum Lyapunov exponent $\lambda^{-1}$ (= 74.1) for LD SSM2 is roughly of the same order of magnitude as the decorrelation time $T_d$ (Nicolis and Nicolis 1986). This relation does not hold for pure white noise.

### The correlation dimension

The most important characteristic of chaos is its fractal dimension (Mandelbrot 1977), which provides a lower bound to the degrees of freedom for the system (Grassberger and Procaccia 1983; 1984). The popular Grassberger-Procaccia algorithm estimates the fractal dimension by means of the correlation dimension $D$. The correlation integral $C_m(R)$ is the number of pairs of points in $m$-dimensional phase space whose distances between each other are less than $R$. For random or chaotic motion, the correlation integral $C_m(R)$ may distribute uniformly in some region of the phase space and has a scaling relation of $R^D$. Therefore, we have

$$\ln_2 C_m(R) = D \ln_2 R + \text{constant} \tag{8}$$

For white noise, $D$ is an integer equal to the embedding dimension $m$. For deterministic chaos, $D$ is less than or equal to the fractal dimension. The Grassberger-Procaccia plots of $\ln C_m(R)$ versus $\ln R$ and slope versus $\ln R$ for LD SSM2 and LD DDM2 are shown in Figures 10 and 11. For $R$ too large, $C_m(R)$ becomes too saturated at the total number of data points (see the right-hand regions of Figures 10 and 11). For $R$ too small, the algorithm detects the noise level of the data (see the left-hand regions of

Fig. 10. The Grass-
berger-Procaccia
plots for calculating
the correlation di-
mension of LD SSM2
time series with time
delay $T = 5$. The
embedding dimen-
sion $m = 2, \ldots, 6$ is
taken as a parameter.
(a) Plots of $\ln_2 C_m(R)$
versus $\ln_2 R$. The
plots rotate down-
wards and to the
right as $m$ increases.
(b) Plot of the slopes
of the curves in (a)
against $\ln_2 R$. The
linear region of the
curves in (a) can be
identified from the
plateau region in (b).
The correlation di-
mension is equal to
the saturated slope
1.5 for LD SSM2
measured from the
plateau region with
$m = 5$.

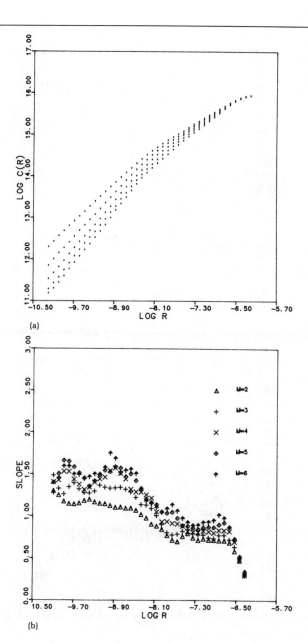

(a)

(b)

Fig. 11. (a) and (b) The Grassberger-Procaccia plots for calculating the correlation dimension of LD DDM2. The correlation dimension is 1.3.

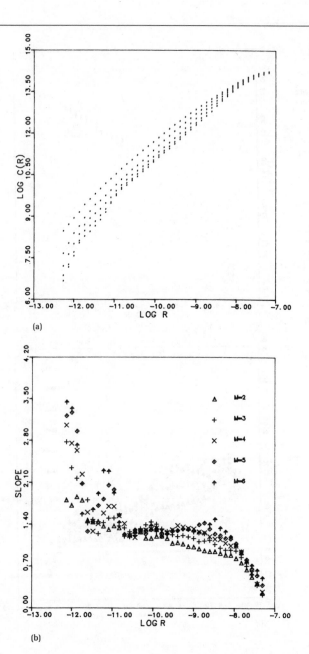

(a)

(b)

Figures 10 and 11). The existence of linear regions of intermediate $R$, which reflect the fractal structure of the attractors, is shown in Figures 10a and 11a. The correlation dimension can be determined from the saturated slope of the plateau region in Figures 10b and 11b.

We found that the correlation dimensions of the investigated five monetary aggregates, including four Divisia monetary indexes and one official simple-sum monetary index, were between 1.3 and 1.5. For other monetary aggregates, no correlation dimension could be determined. These findings are consistent with previous studies in economic aggregation theory and index number theory, which indicate that, except for SSM2, Divisia monetary aggregates are better indexes than simple-sum monetary aggregates (Barnett, Hinich, and Weber 1986; Barnett and Chen 1988).

*Some remarks about numerical algorithms*

Given a deterministic attractor whose correlation dimension is $D$, we first ask how many data points are needed to determine the dimensionality $D$ (Greenside et al. 1982). The minimum data points $N_D$ with a $D$-dimensional attractor can be estimated by the scaling relation $h^D$, where the constant $h$ varies with attractors. Practically, we can only identify low-dimensional attractors with finite data sets, since $N_D$ increases exponentially with $D$. For the Mackey-Glass model (1977), 500 points are needed for $D = 2$, and more than 10,000 points are needed for $D = 3$. In the Couette-Taylor experiment, $N_D$ is about 800 points for $D = 2.4$, 40,000 points for $D = 3$, and 50 billion points for $D = 7$ (Brandstater and Swinney 1987). This issue seems to be ignored by some economists. For example, in Brock (1986), the correlation dimension of GNP with 143 quarterly data points was calculated under the extremely high embedding dimension ($m = 20$) without showing the linear region of the Grassberger-Procaccia plots. In our experience, the width of the linear region shrinks rapidly to zero when m increases beyond 6, as seen in Figures 10 and 11. Practically, m is large enough when m reaches $2D + 1$.

There is another concern about the time expansion covered by the time series. In physics experiments, the sampling rate is typically 10–100 points per orbit. Therefore, 100–1,000 periods are needed for $D = 3$, and 5–50 periods for $D = 2$. We tested this estimation in terms of the Mackey-Glass attractor. When the time delay $\tau$ is 17, its correlation dimension $D$ is 1.95, calculated with 25,000 points (Grassberger and Procaccia 1983). To compare this result, we estimated the correlation dimension under a variety of sampling rates and time periods. We found the error to be within 1 percent with 100 periods, 3 percent with 30 periods, 8 percent with 10 periods, and 18 percent with 5 periods when using 1,000–3,000 data points. Similar results are obtained for the model we develop later.

It should be noted that there is no unique approach to identifying deterministic chaos with certainty. Several algorithms that may be complementary were used in our tests. At present, with only hundreds of data points, the discovery of economic strange attractors whose dimensionality is higher than 3 is unlikely.

We can only speculate why we were unable to identify correlation dimensions for other types of economic time series, such as GNP, IPP, and the Dow-Jones indexes, in our numerical tests. Either their dimensions are too high to be estimated for limited data, or their noise levels are too large to recover the subtle information of deterministic chaos.

## A delayed feedback model of economic growth

Let us consider modeling the low-dimensional monetary strange attractors as growth cycles. There are several problems to be solved: time scale, dynamic mechanism, and system stability.

### Continuous versus discrete time

Current economic studies are dominated by discrete models. Economists favor discrete models because economic data are often reported discretely in years, quarters, or months, and because discrete models are easier for numerical regression. However, continuous-time models are needed when the serial correlation of disturbances can no longer be neglected (Koopmans 1950). The decorrelation time $T_d$ of the autocorrelations of time series sets a lower bound to the time unit of the discrete model. For a typical discrete model, $T_d$ is in approximately the same length as the discrete time unit. The decorrelation time $T_d$ for monetary attractors is more than 60 weeks. The time scales of discrete models of deterministic chaos with one or two variables in business cycle theory (Benhabib 1980; Day 1982; Grandmont 1985) are usually larger than the time scale of real business cycles (Sims 1986; Sargent 1987). Clearly, the simple discrete model is not appropriate for describing monetary growth cycles. A continuous model is needed for the monetary time series.

The observed low correlation dimension of monetary aggregates sets additional constraints on the modeling of growth cycles. The minimum number of degrees of freedom required for chaotic behavior in autonomous differential equations is 3 (Ott 1981), so the fractal dimension will be larger than 2. Therefore, the driven oscillator in the long-wave model (Rasmussen, Mosekilde, and Sterman 1985) is not applicable in our case.

After comparing the correlation dimensions and the phase portraits of existing models, we believe that the differential-delay equation is a good candidate for modeling monetary growth. For simplicity, we consider only one variable here. The low dimensionality of monetary attractors leads to the belief of the separability of the monetary deviations from other macroeconomic movements that are integrated in the natural trends of monetary growth rate.

### Deviations from trend and feedback behavior

The apparent monetary strange attractors are mainly found in log linear detrended data. This is an important finding for studying control behavior in monetary policy. We believe that the human ability to manage information is limited even if decision makers have "perfect information." Economic behavior is more likely to follow some simple rule or procedure than to provide global optima (Simon 1979). We assume that the general trends of economic development, the natural growth rate, are perceived by people in economic activities as a common psychological reference or as an anchor in observing and reacting (Tversky and Kahneman 1974). Administrative activities are basically reactions to deviations from the trend. We choose the deviation from the natural growth rate as the main variable in the dynamic model of monetary growth.

There are a number of differential-delay models in theoretical biology and population dynamics (Mackey and Glass 1977; May 1980; Blythe, Nisbet, and Gurney 1982; Chow and Green 1985). Our model has a new feature that differs from previous models of population dynamics: its wave pattern should be symmetric, because we are dealing with detrended growth cycles. The wave form of business cycles is not symmetric, since they are observed in terms of the first difference of logarithmic macroeconomic indexes or annual percent rate of growth. In an economic system moving with a constant growth rate, we define the reference equilibrium state as zero. The proposed equation is

$$\frac{dX(t)}{dt} = aX(t) + F(X(t - \tau)) \tag{9}$$

$$F(X) = X \cdot G(X) \tag{10}$$

$X$ is here the relative growth index, which measures the deviation from the trend. $\tau$ is the time delay, $a$ is the expansion speed, $F$ is the control function, and $G$ is the feedback function.

There are two competing mechanisms in the growth system. The first is the stimulative growth that is an instantaneous response to market demand. It is described by the first term on the right side of Eq. 9. A linear term for exponential growth is used for mathematical convenience. The second term represents the endogenous system control described by the control function $F$. This consists of feedback signal $X(t - \tau)$ and feedback function $G$. The time delay $\tau$ exists in the feedback loop because of information and regulation lags.

*The flow diagram and the symmetric control function*

Figure 12 shows a flow diagram to describe our model. There are several considerations in specifying $F$ and $G$. We assume the control function $F(X)$ has two extrema at $\pm X_m$ for the control target floor and ceiling (Solomon 1981). $G(X)$ should be nonlinear and symmetric, $G(-X) = G(X)$, in order to describe the overshooting in economic management and the symmetry in growth cycles. These features are essential to generate complex behavior in the economic growth model.

In choosing the form of $G$, we do not use the    polynomial    function adopted in previous models with relaxation oscillations. Here, we suggest a simple exponential function to describe negative feedback reactions:

$$G(X) = -b \exp\left(\frac{-X^2}{\sigma^2}\right) \tag{11}$$

Fig. 12. The flow diagram for a de-layed feedback system of economic growth.

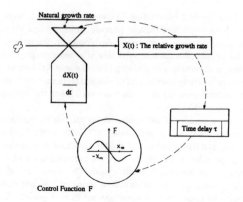

where $b$ is the control parameter, $\sigma$ is the scaling parameter, and the extrema of $F(X)$ are located at $X_m = \pm\sigma/\sqrt{2}$. Substituting Eqs. 11 and 10 into Eq. 9 gives the following differential-delay equation:

$$\frac{dX(t)}{dt} = aX(t) - bX(t - \tau) \exp\left(\frac{-X(t - \tau)^2}{\sigma^2}\right)$$  (12)

We may change the scale by $X = X'\sigma$ and $t = t'\tau$, then drop the prime for convenience:

$$\frac{dX(t)}{dt} = a\tau X(t) - b\tau X(t - 1) \exp(-X(t - 1)^2)$$  (13)

The rough behavior of the time delay Eq. 13 can be discussed in terms of linear stability analysis in determining the boundaries of damped and divergent oscillations in the parameter space.

*The period-doubling route to chaos*

We solved Eq. 13 numerically by the predictor-corrector approach. Time sequences and phase portraits of solutions with different $b$ for fixed $a$ and $\tau$ are shown in Figures 13 and 14. In order to identify the route to chaos, the power spectra are shown in Figure 15. The period-doubling route to chaos is observed when parameter changes induce bifurcations (Feigenbaum 1978). One observes the fundamental frequency $f_1$ and its subharmonic frequency $f_2$ before and after transition to chaos in Figure 15c. In addition to period-1 orbit P1 (limit cycle) in Figure 14a, period-2 orbit P2 in Figure 14b, and period-3 orbit P3 in Figure 14d, we also observe P4, P8, and P6 in the regions close to P2 and P3, respectively. The period-doubling route to chaos has also been found in other differential-delay models with asymmetric solutions (May 1980).

Fig. 13. The time sequences of the numerical solutions of Eq. 13. The parameters were fixed at $a = 0.1$ and $\tau = 1$ while changing the parameter $b$. (a) Period-1 solution P1 (limit cycle) with $b = 5.7$. Its cycle number is C1. (b) Period-2 solution P2 (C2) with $b = 5.8$. (c) Chaotic solution CH with $b = 6.0$. (d) Period-3 solution P3 (C3) with $b = 6.3$.

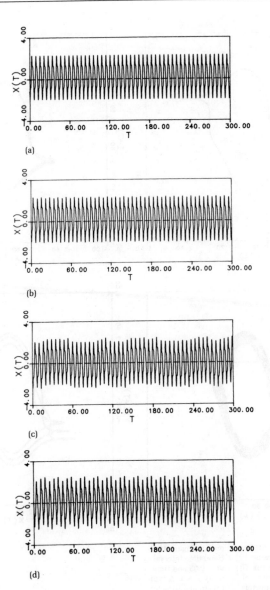

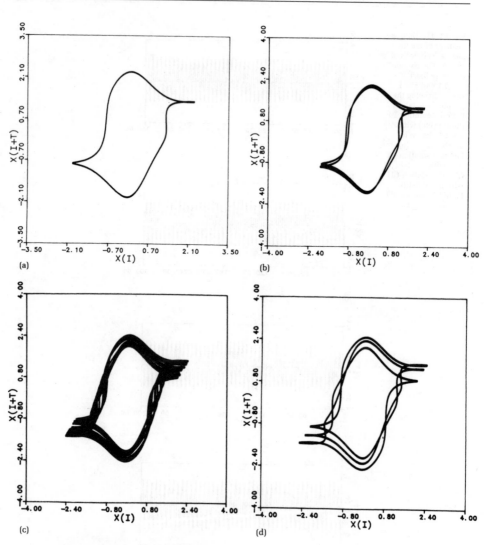

(a)

(b)

(c)

(d)

Fig. 14. (a)–(d) The phase portraits $X(t + T)$ versus $X(t)$ of the solutions of Eq. 13. Parameters are the same as the ones in Figure 13. Here, time interval $dt = 0.05$ and time delay $T = 1$. A typical strange attractor can be seen in (c).

Fig. 15. (a)–(d) The power spectra of the solutions of Eq. 13. The number of sampling points is 4,096. Only the lower quarter of the spectrum is displayed. The parameters are the same as in Figure 13. The highest peak in all plots is the fundamental frequency $f_1$. The second highest peak is the subharmonic frequency $f_2$, which can be seen in (b) and (d). A typical chaotic spectrum is shown in (c).

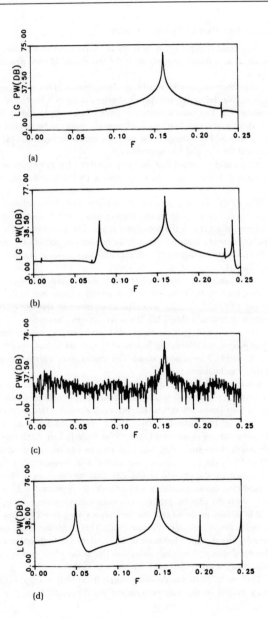

(a)

(b)

(c)

(d)

*Phase transition and pattern stability*

A more useful approach is to study the wave form of business cycles, since spectral analysis is difficult to apply with the few cycles of data available in economic time series.

The observed periodic repetition often consists of basic patterns with several shorter cycles. We define the number of shorter cycles in a basic wave pattern as the cycle number $Ck$. The basic pattern may have $L$ large amplitude oscillations followed by $S$ small amplitude oscillations. Each periodic state can be labeled with the cycle number $Ck = L + S$. For example, the periodic states in Figures 14a, 14b, and 14d can be labeled $C1$, $C2$, and $C3$, respectively.

We should point out that the cycle number $Ck$ is not necessarily equal to the period number $P$. For example, the wave form of P6 is $C3$, and those of P4 and P8 belong to $C2$.

The phase diagram in terms of cycle number of the solutions is useful in characterizing economic long waves. Figures 16a and 16b display qualitatively the phase diagram of Eq. 13 in the parameter space. The broad diversity of dynamic behavior includes steady state ST, limit cycle or periodic motion $C1$, and explosive solution EP. The complex regime CP includes alternate periodic states ($C1$, $C2$, $C3$) and chaotic regime CH.

When parameter values change within each region, the dynamic behavior is pattern-stable, because the dynamic mode occupies a finite area in the parameter space. The phase transition occurs when parameters cross the boundary between different phases. It is observable when the wave pattern changes.

The notation of cycle number $Ck$ is introduced for possible application in analyzing long waves. An interesting feature of the model is that only three periodic patterns $C1$, $C2$, and $C3$ have been found. The model gives a simple explanation of multiperiodicity in business cycles.

It is speculated that no unique periodicity is involved in business cycles. In addition to seasonal changes, several types of business cycles have been identified by economists (van Duijn 1983). The Kitchin cycles usually last 3–5 years; the Juglar cycles, 7–11 years; the Kuznets cycles, 15–25 years; and Kondratieff cycles, 45–60 years. Schumpeter suggested that these cycles were linked. Each longer wave may consist of two or three shorter cycles. This picture can be described by the periodic phase $C2$ or $C3$ in the CP regime of our model. The irregularity of long waves can also be explained by the chaotic regime CH. Our model gives a variety of possibilities of periodicity, multiperiodicity, and irregularity in economic history, although our data only show the chaotic pattern in monetary movements.

It is widely assumed that the long waves are caused by long lags, a belief coming from the linear paradigm (Rostow 1980). This condition is not necessary in our model, because the dynamic behavior of Eq. 13 depends both on $a\tau$ and $b\tau$. A strong overshooting plus a short time delay has the same effect as a weak control plus a long time delay, a point also made by Sterman (1985).

This model is so simple and general, it could have applications beyond the monetary system in the market economy we discussed here. For example, the growth

Fig. 16. The phase
diagram of the nu-
merical solutions of
Eq. 13 in parameter
space $a\tau$ and $b\tau$. The
dashed square area
in (a) is enlarged in
(b). Here, EP, ST,
and CP represent ex-
plosive regime,
steady state (after
damped oscillation),
and complex regime,
respectively. CH is
chaotic regime. C1,
C2, and C3 are peri-
odic patterns, whose
longer waves consist
of one, two, or three
shorter waves in
turn.

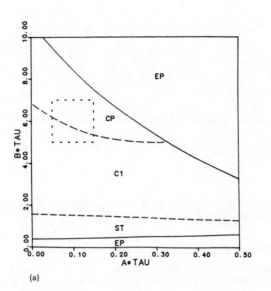

(a)

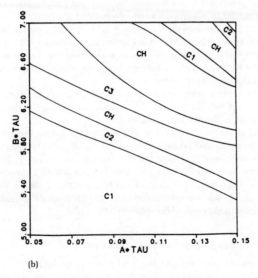

(b)

Fig. 17. The time path of medium-term growth cycles simulating the LD SSM2 time series in Figure 5a by means of Eq. 12. Its long-term picture is the same as that in Figure 13c with a changed time scale.

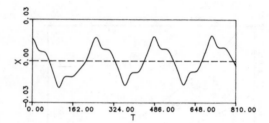

cycles and long waves caused by overshooting and time delay may also happen in centrally planned economies.

### Simulating empirical cycles and forecasting basic trends

In comparing model-generated patterns with empirical data, we may confine our experiments to certain regions of parameter space. For example, we can estimate the average period $T$ from 4 times the decorrelation time $T_d$. The time delay $\tau$ in monetary control due to regulation lag and information lag is between 20 and 56 weeks (Gordon 1978). If we estimate the time delay $\tau$ to be 39 weeks, we can simulate LD SSM2 time series by the solution shown in Figure 13c, by setting $\tau = 39$, $a = 0.00256$, $b = 0.154$, and $\sigma = 0.0125$. The model results match well the average amplitude $A_m$, decorrelation time $T_d$, positive maximum Lyapunov exponent $\lambda$, and correlation dimension $D$ of the empirical time series.

The medium-term picture of simulated LD SSM2 in Figure 17 has well-behaved peaks and troughs with a stable period. We can hardly imagine that its long-term behavior is chaotic (see Figure 13c).

We tested the theoretical models with power spectra and autocorrelation analysis. The approximated period $T$ of the chaotic solution can be estimated from $T_d$ measured by 3–5 cycles. It is close to the fundamental period $T_1$ ($= f_1^{-1}$), determined by power spectra measured by 100 cycles with an error within 3 percent. For LD SSM2 time series, the difference of $T_d$ measured between 10–15 years is less than 5 percent. We can obtain valuable information about the fundamental period $T_1$ without knowing the exact parameters of the deterministic model.

### Implications for forecasting and control policy

We should point out that the word *chaos* is misleading. Chaotic motion has both regular and irregular characteristics. We prefer to refer to continuous deterministic chaos as *imperfect periodic motion*, which has a stable fundamental period but an irregular wave shape and a changing amplitude. Actually, we may often recover more information from chaotic motion than from random movements. For example, econometric models based on linear stochastic processes mainly explain the variance of the residuals. They offer little information about the trends and periods of business cycles beyond the short term. We suggest a new forecasting approach based on detecting strangeness of growth cycles. Although the long-term prediction of the chaotic orbit is impossible from the point of view of nonlinear dynamics, a medium-term prediction of approximate period $T$ can be made if we can identify strange attractors from the time series.

Let us discuss the meaning of the control parameters in Eq. 12. When $b = 0$, the monetary deviation from the natural rate will grow at a speed $e^{at}$. We define a characteristic doubling time $t_a$, which measures the time needed to double the autonomous monetary expansion $X(t)$ without control. Similarly, we define a characteristic half time $t_b$, which measures the time needed to reduce the money supply to half its level when $a = 0$ and $X(t - \tau)$ reaches the control target $X_m = \sigma/\sqrt{2} = 1.4$ percent per year. The same is true for the contraction movements, since the feedback function $G(x)$ is symmetric. Here $t_a = 5.2$ years, and $t_b = 7.4$ weeks, for SSM2 in our simulation. We see that even modest time delay and overshooting may generate cycles and chaos.

For policy considerations, we suggest that the fluctuations in money supply can be moderated by reducing the time delay $\tau$ or the control parameter $b$. We can set $7.3 < \tau < 29.3$ weeks while fixing $a$ and $b$; or let $29.5 < t_b < 108.7$ weeks (when $1.51 > b > 0.41$). These figures give a qualitative picture of monetary target policy that seems reasonable for the real economy.

## Summary and discussion

Empirical evidence of low-dimensional strange attractors is found in log linear detrended monetary aggregate data for the United States. These results are very encouraging, since new information is revealed about macroeconomic movements.

A differential-delay equation with only two parameters is suggested to describe monetary growth cycles. Self-generated periodic, multiperiodic, and chaotic behaviors are observed in the deterministic model. This model sheds light on the mechanism of business cycles and long waves: the nonlinearity and time delay in feedback control may cause complex behavior. Although our model is simple and exploratory, it has enabled us to simulate the wave pattern and the low dimensionality of monetary growth cycles.

We do not deny the complexity of social phenomena and the usefulness of disaggregated approaches in econometrics and system dynamics. Low-dimensional economic chaos is not only useful but also testable in economic studies. It can be understood through the experience of physicists. It is often convenient to introduce projection operators that decompose the system into one low-dimensional space, whose movements can be effectively simplified, and one orthogonal to it (Prigogine 1980). In practice, the right choice of projection operator can only be made by empirical tests. Our work, together with previous efforts in the study of complex systems, strongly supports the hope that social phenomena can be quantitatively described by simple mathematical models in some aspects. The key issues are which pertinent variable to observe and what can be learned from the model.

Three problems remain to be solved for future studies of economic chaos.

- The main obstacle in empirical analysis arises from limited data sources in economics. In order to facilitate the testing of deterministic chaos and to improve our understanding of modern economies, it is worthwhile to develop numerical algorithms that work with moderate data sets, as well as to expand the data base of economic statistics.
- The second question is how to determine the reference system. In our numerical experiment, the starting and ending periods of the observations were arbitrarily dictated by the available data. We do not know if the natural rate of growth is a constant or changing over time. Perhaps this problem can be solved by future testing on longer periods combined with the efforts of a historian helping to identify turning points of economic history. It is advantageous for nonlinear dynamics to introduce a time arrow or a historical perspective in analyzing complex systems (Prigogine 1980).
- The third issue is how to estimate parameters from empirical data. We should point out that the solutions of a nonlinear differential-delay equation may not be approximated by one- or two-dimensional discrete models in fitting the empirical data. We should be cautious in applying conventional techniques of econometrics to chaos models.

Exploring economic chaos opens a new way to understand human behavior and social evolution. Studies of nonequilibrium and nonlinear phenomena have not only changed the techniques we use but also the ways in which we think (Prigogine and Stengers 1984).

### References

Abraham, N. B., J. P. Gollub, and H. L. Swinney. 1984. Testing Nonlinear Dynamics. *Physica D* 11:252–264.

Barnett, W. A., and P. Chen. 1988. The Aggregation-Theoretic Monetary Aggregates Are Chaotic and Have Strange Attractors: An Econometric Application of Mathematical Chaos. In *Dynamic Econometric Modelling*, ed. W. A. Barnett, E. Berndt, and H. White. Cambridge: Cambridge University Press. Forthcoming. See also W. A. Barnett and P. Chen. 1987. Economic Theory as a Generator of Measurable Attractors. In *Laws of Nature and Human Conduct*, ed. I. Prigogine and M. Sanglier. Brussels: Task Force of Research, Information, and Study on Science.

Barnett, W. A., M. J. Hinich, and W. E. Weber. 1986. The Regulatory Wedge Between the Demand-Side and Supply-Side Aggregation Theoretic Monetary Aggregates. *Journal of Econometrics* 33:165–185.

Benhabib, J. 1980. Adaptive Monetary Policy and Rational Expectations. *Journal of Economic Theory* 23:261–266.

Blythe, S. P., R. M. Nisbet, and W.S.C. Gurney. 1982. Instability and Complex Dynamic Behavior in Population Models With Long Time Delays. *Theoretical Population Biology* 22:147–176.

Brandstater, A., and H. L. Swinney. 1987. Strange Attractors in Weakly Turbulent Couette-Taylor Flow. *Physical Review A* 35:2207–2220.

Brock, W. A. 1986. Distinguishing Random and Deterministic Systems. *Journal of Economic Theory* 40:168–195.

Candela, G., and A. Gardini. 1986. Estimation of a Nonlinear Discrete Time Macro Model. *Journal of Economic Dynamics and Control* 10:249–254.

Chen, P. 1987. Nonlinear Dynamics and Business Cycles. Ph.D. Dissertation, University of Texas, Austin, TX 78712.

Chow, S. N., and D. Green, Jr. 1985. Some Results on Singular Delay-Differential Equations. In *Chaos, Fractals, and Dynamics,* ed. P. Fischer and W. R. Smith. New York: Dekker.

Dana, R. A., and P. Malgrange. 1984. The Dynamics of a Discrete Version of a Growth Cycle Model. In *Analyzing the Structure of Econometric Models,* ed. J. P. Ancot. The Hague: Nijhoff.

Day, R. H. 1982. Irregular Growth Cycles. *American Economic Review* 72:406–414.

Deneckere, R., and S. Pelikan. 1986. Competitive Chaos. *Journal of Economic Theory* 40:13–25.

Fama, E. F. 1970. Efficient Capital Markets: A Review of Theory and Empirical Work. *Journal of Finance* 25:383–417.

Farmer, J. D. 1982. Chaotic Attractors of an Infinite-Dimensional Dynamical System. *Physica D* 4:366–393.

Fayyad, S. 1986. Monetary Asset Component Grouping and Aggregation: An Inquiry Into the Definition of Money. Ph.D. Dissertation, University of Texas, Austin, TX 78712.

Feigenbaum, M. J. 1978. Quantitative Universality for a Class of Nonlinear Transformations. *Journal of Statistical Physics* 19:25–52.

Forrester, J. W. 1977. Growth Cycles. *Economist* (Leiden) 125:525–543.

Frank, M., and T. Stengers. 1987. Some Evidence Concerning Macroeconomic Chaos. Mimeographed notes, University of Guelph, Ontario, Canada.

Friedman, M. 1969. *The Optimum Quantity of Money.* Chicago: Aldine.

Goodwin, R. M. 1951. The Nonlinear Accelerator and the Persistence of Business Cycles. *Econometrica* 19:1–17.

Gordon, R. J. 1978. *Macroeconomics.* Boston: Little, Brown.

Grandmont, J. M. 1985. On Endogenous Competitive Business Cycles. *Econometrica* 53:995–1045.

Grandmont, J. M., and P. Malgrange. 1986. Nonlinear Economic Dynamics. *Journal of Economic Theory* 40:3–12.

Grassberger, P., and I. Procaccia. 1983. Measuring the Strangeness of Strange Attractors. *Physica D* 9:189–192.

———. 1984. Dimensions and Entropies of Strange Attractors From a Fluctuating Dynamic Approach. *Physica D* 13:34–54.

Greenside, H. S., A. Wolf, J. Swift, and T. Pignataro. 1982. Impracticality of a Box-Counting Algorithm for Calculating the Dimensionality of Strange Attractors. *Physical Review A* 25:3453–3456.

Koopmans, T. C. 1950. Models Involving a Continuous Time Variable. In *Statistical Inference in Dynamic Economic Models.* New York: Wiley.

Lucas, R. E., Jr. 1981. *Studies in Business Cycle Theory.* Cambridge, Mass.: MIT Press.

Mackey, M. C., and L. Glass. 1977. Oscillation and Chaos in Physiological Systems. *Science* 197:287–289.

Mandelbrot, B. 1977. *Fractals, Forms, Chances, and Dimension.* San Francisco: Freeman.

May, R. M. 1976. Simple Mathematical Models With Very Complicated Dynamics. *Nature* 261:459–467.

———. 1980. Nonlinear Phenomena in Ecology and Epidemiology. *Annals of the New York Academy of Science* 357:267–281.

Mayer-Kress, G. 1986. *Dimensions and Entropies in Chaotic Systems.* Berlin: Springer.

Nicolis, C., and G. Nicolis. 1986. Reconstructing the Dynamics of the Climatic System From Time Series Data. *Proceedings of the National Academy of Sciences USA* 83:536–540.

Osborne, M.F.M. 1959. Brownian Motion in the Stock Market. *Operations Research* 7:145–173.

Ott, E. 1981. Strange Attractors and Chaotic Motions of Dynamical Systems. *Reviews of Modern Physics* 53:655–671.

Prigogine, I. 1980. *From Being to Becoming.* San Francisco: Freeman.

Prigogine, I., and I. Stengers. 1984. *Order Out of Chaos.* New York: Bantam.

Ramsey, J. B., and H. J. Yuan. 1987. The Statistical Properties of Dimension Calculations Using Small Data Sets. Mimeographed notes, New York University, New York, NY 10003.

Rasmussen, S., E. Mosekilde, and J. D. Sterman. 1985. Bifurcations and Chaotic Behavior in a Simple Model of the Economic Long Wave. *System Dynamics Review* 1:92–110.

Rau, N. 1974. *Trade Cycles: Theory and Evidence.* London: Macmillan.

Rössler, O. E. 1976. An Equation for Continuous Chaos. *Physics Letters A* 57:397–398.

Rostow, W. W. 1980. *Why the Poor Get Richer and the Rich Slow Down.* Austin: University of Texas Press.

Samuelson, P. A. 1939. Interactions Between the Multiplier Analysis and the Principle of Acceleration. *Review of Economic Statistics* 21:75–78.

———. 1986. Deterministic Chaos in Economics: An Occurrence in Axiomatic Utility Theory. Mimeographed notes, M.I.T., Cambridge, MA 02139.

Sargent, T. J. 1987. *Dynamic Macroeconomic Theory.* Cambridge, Mass.: Harvard University Press.

Sayers, C. 1986. Workstoppages: Exploring the Nonlinear Dynamics. Mimeographed notes, University of Wisconsin, Madison, WI 53706.

Scheinkman, J., and B. Le Baron. 1987. Nonlinear Dynamics and GNP Data. Mimeographed notes, University of Chicago, Chicago, IL 60637.

Schuster, H. G. 1984. *Deterministic Chaos: An Introduction.* Weinheim, F.R.G.: Physik-Verlag.

Simon, H. A. 1979. Rational Decision Making in Business Organizations. (Nobel Lecture) *American Economic Review* 69:493–513.

Sims, C. 1986. Commentaries on the Grandmont paper "On Endogenous Competitive Business Cycles." In *Models of Economic Dynamics* (Lecture Notes in Economics and Mathematical Systems. Vol. 264), 37–39, ed. H. F. Sonnenschein. Berlin: Springer.

Solomon, A. M. 1981. Financial Innovation and Monetary Policy. In *Sixty-seventh Annual Report.* Federal Reserve Bank of New York.

Sterman, J. D. 1985. A Behavioral Model of the Economic Long Wave. *Journal of Economic Behavior and Organization* 6:17–53.

Stutzer, M. J. 1980. Chaotic Dynamics and Bifurcation in a Macro Model. *Journal of Economic Dynamics and Control* 2:353–376.

Takens, F. 1981. Detecting Strange Attractors in Turbulence. In *Dynamical Systems and Turbulence* (Lecture Notes in Mathematics. No. 898), 366–381, ed. D. A. Rand and L. S. Young. Berlin: Springer.

Tversky, A., and D. Kahneman. 1974. Judgment Under Uncertainty: Heuristics and Biases. *Science* 185 (September 27):1124–1131.

van Duijn, J. J. 1983. *The Long Wave in Economic Life.* London: Allen & Unwin.

Wolf, A., J. Swift, H. L. Swinney, and J. Vastano. 1985. Determining Lyapunov Exponents From a Time Series. *Physica D* 16:285–317.

Zarnowitz, V. 1985. Recent Work on Business Cycles in Historical Perspective: A Review of Theories and Evidence. *Journal of Economic Literature* 23:523–580.

PART THREE: BIBLIOGRAPHY ON CHAOS

# Bibliography on Chaos

## I. Books, Conference Proceedings, and Collection of Papers

Entries in this part are ordered by their year of publication and are referred to as, for example, Poincaré (B1899).

1. Poincaré H (1899), *Les Médothes Nouvelles de la Mécanique Céleste*, tom 3, Gauthier-Villars.
2. Arnold V I, and Avez A (1968), *Ergodic Problems of Classical Mechanics*, Addison-Wesley.
3. Moser J (1973), *Stable and Random Motions in Dynamical Systems*, Princeton University Press.
4. Marsden E J, and McCracken M (1976), *The Hopf Bifurcation and its Applications*, Applied Math. Sci. vol. 19, Springer-Verlag.
5. Casati G, and Ford J (1977), eds., *Stochastic Behaviour in Classical and Quantum Hamiltonian Systems*, Lect. Notes in Phys. **93**, Springer-Verlag.
6. Mandelbrot B B (1977) *Fractals, Form, Chance and Dimension*, W. H. Freeman and Co.
7. Abraham, and Marsden J E (1978), *Foundations of Mechanics*, 2nd ed., Addison-Wesley.
8. Arnold V I (1978), *Mathematical Methods of Classical Mechanics*, Springer-Verlag.
9. Jorna S (1978), ed., *Topics in Nonlinear Dynamics*, AIP Conference Proceedings **46**.
10. Markley N G, Martin J C, and Perrizo W (1978), eds., *The Structure of Attractors in Dynamical Systems*, Lect. Notes in Math. **668**, Springer-Verlag.
11. Mori H (1978), ed., *Nonlinear Nonequilibrium Statistical Mechanics*, Proceedings of the Oji Seminar, *Prog. Theor. Phys. Suppl.*, **64**.
12. Iooss G (1979), *Bifurcations of Maps and Applications*, North-Holland.
13. Szebehely V G (1979), ed., *Instabilities in Dynamical Systems with Applications to Celestial Mechanics*, Riedel.
14. Gurel O, and Rössler O E (1979), eds., *Bifurcation Theory and its Applications to Scientific Disciplines*, Annals of the New York Academy of Sciences, **316**.
15. Month M, and Herra J C (1979), eds., *Nonlinear Dynamics and the Beam-beam Interaction*, AIP Conference Proceedings **57**.
16. Collet P, and Eckmann J-P (1980), *Iterated Maps on the Interval as Dynamical Systems*, Birkhäuser.
17. Helleman H G (1980), ed., *Nonlinear Dynamics*, Annals of the New York Academy of Sciences, **357**.
18. Guckenheimer J, Moser J, and Newhouse S E (1980), *Dynamical Systems*, Birkhäuser.

19. Bardos C, and Besis D (1980), eds., *Bifurcation Phenomena in Mathematical Physics and Related Topics*, Riedel.

20. Laval G, and Gresillon D (1980), eds., *Intrinsic Stochasticity in Plasmas*, Les Editions de Physique, Courtboeuf, Orsay.

21. Holmes P (1980), ed., *New Approaches to Nonlinear Problems in Dynamics*, SIAM Publications, Philadelphia.

22. Nitecki Z, and Robinson C (1980), eds., *Global Theory of Dynamical Systems*, Lect. Notes in Math., **819**, Springer-Verlag.

23. Swinney H L, and Gollub J P (1981), eds., *Hydrodynamical Instabilities and the Transitions to Turbulence*, Springer-Verlag.

24. Haken H (1981), ed., *Chaos and Order in Nature*, Springer Series in Synergetics **11**, Springer-Verlag.

25. Rand D A, and Young L S (1981), eds., *Dynamical Systems and Turbulence*, Lect. Notes in Math. **898**, Springer-Verlag.

26. Vidal C, and Pacault A (1981), eds., *Nonlinear Phenomena in Chemical Dynamics*, Springer-Verlag.

27. Sparrow C (1982), *The Lorenz Equations: Bifurcations, Chaos, and Strange Attractors*, Springer-Verlag.

28. Kalia R K, and Vashishta P (1982), eds., *Melting, Localization and Chaos*, Elsevier Science Publishing Co.

29. Barrenblatt G I, Iooss G, and Joseph D D (1982), eds., *Nonlinear Dynamics and Turbulence*, Pittman.

30. Mandelbrot B B (1982) *The Fractal Geometry of Nature*, W. H. Freeman and Co.

31. Haken H (1982), ed., *Evolution of Order and Chaos*, Springer Series in Synergetics **17**, Springer-Verlag.

32. Bishop A R, Campbell D K, and Nicolaenko B (1982), eds., *Nonlinear Problems: Present and Future*, North-Holland.

33. Lichtenberg A J, and Liberman M A (1983), *Regular and Stochastic Motion*, Springer-Verlag.

34. Guckenheimer J, and Holmes P (1983), *Nonlinear Oscillations, Dynamical Systems and Bifurcations of Vector Fields*, Springer-Verlag.

35. Preston C (1983), *Iterates of Maps on an Interval*, Lect. Notes in Math., **999**, Springer-Verlag.

36. Abraham R H, and Shaw C D (1983-1985), *Dynamics: the Geometry of Behavior*. Part I. *Periodic Behavior*; Part II. *Chaotic Behavior*; Part III. *Global Behavior*; Part IV. *Bifurcation Behavior*, Aerial Press.

37. Horton C W, Reichl L, and Szebehely V (1983), eds., *Long-time Prediction in Nonlinear Dynamics*, Wiley.

38. Garrido L (1983), ed., *Dynamical Systems and Chaos*, Lect. Notes in Phys., **179**, Springer-Verlag.

39. Iooss G, Helleman R H G, and Stora R (1983), eds., *Chaotic Behaviour of Deterministic Systems*, Les Houches 1981, North-Holland.

40. Campbell D K, and Rosen H (1983), eds., *Order in Chaos*, Proceedings of the International Conference at Los Alamos, 1982, *Physica*, **7D**, No. 1-3, and separate book, North-Holland.

41. Schuster H G (1984), *Deterministic Chaos. An Introduction*, 2nd Edition, 1988, Physik-Verlag.

42. Shaw R (1984), *The Dripping Faucet as a Model Chaotic System*, Aerial Press.

43. Bergé P, Pomeau Y, and Vidal Ch (1984), *L'ordre dans le Chaos, vers une approache deterministe de la turbulence*, Hermann Editions des Sciences. English translation (1986): *Order within Chaos, towards a deterministic approach to turbulence*, Wiley.

44. Hao Bai-lin (1984), ed., *Chaos*, An introduction and reprints volume, World Scientific.

45. Cvitanovic P (1984), ed., *Universality in Chaos*, An introduction and reprints volume, Adam Hilger.

46. Schuster P (1984), ed., *Stochastic Phenomena and Chaotic Behaviour in Complex Systems*, Springer Series in Synergetics 21, Springer-Verlag.

47. Kuramoto Y (1984), ed., *Chaos and Statistical Methods*, Springer Series in Synergetics 24, Springer-Verlag.

48. Chandra J (1984), ed., *Chaos in Nonlinear Dynamical Systems*, SIAM.

49. Horton C W, and Reichl L (1984), eds., *Statistical Physics and Chaos in Fusion Plasmas*, Wiley.

50. Tatsumi T (1984), ed., *Turbulence and Chaotic Phenomena in Fluids*, North-Holland.

51. Vidal C, Pacault A (1984), eds., *Nonequilibrium Dynamics in Chemical Systems*, Springer-Verlag.

52. Arnold L, and Wihstutz V (1984), eds., *Lyapunov Exponents*, Proceedings of a Workshop, Lect. Notes in Math., 1186, Springer-Verlag.

53. Kuramoto Y (1984), ed., *Topics on Nonlinear Dynamics in Dissipative Systems*, *Prog. Theor. Phys. Suppl.*, 79.

54. Zaslavsky G M (1985), *Chaos in Dynamical Systems*, Harwood Academic Publishers.

55. Buchler J R, Perdang J M, and Spiegel E A (1985), eds., *Chaos in Astrophysics*, Riedel.

56. Casati G (1985), ed., *Chaotic Behaviour in Quantum Systems. Theory and Applications*, Plenum.

57. Velo G, and Wightman A S (1985), eds., *Regular and Chaotic Motion in Dynamical Systems*, Plenum.

58. Rensing L, and Jaeger N I (1985), eds., *Temporal Order*, Springer Series in Synergetics 29, Springer-Verlag.

59. Ghil M, Benzi R, and Parisi G (1985), eds., *Turbulence and Predictability in Geophysical Fluid Dynamics and Climate Dynamics*, North-Holland.

60. Shiraiwa K (1985), *Bibliography for Dynamical Systems*, Preprint Series No.1, 1985, Department of Mathematics, Nagoya University, Japan.

61. Fischer D, and Smith W R (1985), eds., *Chaos, Fractals, and Dynamics*, Dekka.

62. Lundqvist S (1985), ed., *Physics of Chaos and Related Problems*, Proceedings of the 59th Nobel Symposium, *Phys. Scripta*, T9.

63. Devaney R L (1986), *An Introduction to Chaotic Dynamical Systems*, Benjamin/Cummings.

64. Thompson J M T, and Stewart H B (1986), *Nonlinear Dynamics and Chaos. Geometrical Methods for Engineers and Scientists*, Wiley.

65. Kaneko K (1986), *Collapse of Tori and Genesis of Chaos in Dissipative Systems*, World Scientific.

66. Steeb W-H, and Louw J A (1986), *Chaos and Quantum Chaos*, World Scientific.

67. Peitgen H-O, and Richter P H (1986), *The Beauty of Fractals. Images of Complex Dynamical Systems*, Springer-Verlag.

68. Holden A V (1986), ed., *Chaos*, Manchester University Press.

69. Mayer-Kress G (1986), ed., *Dimensions and Entropies in Chaotic Systems. Quantification of complex behaviour*, Springer Series in Synergetics **32**, Springer-Verlag.

70. Shlesinger M F, Cawley R, Saenz A W, Zachary W (1986), eds., *Perspectives in Nonlinear Dynamics*, World Scientific.

71. Sarkar S (1986), ed., *Nonlinear Phenomena and Chaos*, Adam Hilger.

72. Campbell D K, Newell A C, Schrieffer R J, and Sugar H (1986), eds., *Solitons and Coherent Structures*, Chapter 6: Instabilities and Chaos, *Physica*, **18D**, No. 1-3.

73. Bishop A R, Gruener A R, and Nicolaenko B (1986), eds., *Spatio-Temporal Coherence and Chaos in Physical Systems*, Physica, **23D**, North-Holland.

74. Barnsley M F (1986), ed., *Chaotic Dynamics and Fractals*, Academic Press.

75. Seligman T H, and Nishioka (1986), *Quantum Chaos and Statistical Nuclear Physics*, Lect. Notes in Phys., **263**, Springer-Verlag.

76. Chrostowski W B, and Abraham N B (1986), eds., *Optical Chaos*, Proceedings of SPIE, No. 667.

77. Boyd R W, Raymer M G, and Narducci L M (1986), eds., *Optical Instabilities*, Cambridge University Press.

78. Mira C (1987), *Chaotic dynamics*, World Scientific.

79. Bhattacharjee J K (1987), *Convection and Chaos in Fluids*, World Scientific.

80. Moon F C (1987), *Chaotic Vibration. An Introduction for Applied Scientists and Engineers*, John Wiley & Sons.

81. Milonni P W, Shih M-L, and Ackerhalt J R (1987), *Chaos in Laser-Matter Interactions*, World Scientific.

82. Gleik J (1987), *Chaos: Making a New Science*, Viking.

83. Hao Bai-lin (1987), ed., *Directions in Chaos*, vol.1, World Scientific.

84. Buchler J R, and Eichhorn H (1987), eds., *Chaotic Phenomena in Astrophysics*, *Annals of the New York Academy of Sciences*, **497**.

85. MacKay R S, Meiss J D (1987), eds., *Hamiltonian Dynamical Systems*, A reprint volume, Adam Hilger.

86. Pike E R, Lugiato L (1987), eds., *Chaos, Noise and Fractals*, Adam Hilger.

87. Pike E R, and Sarkar S (1987), eds., *Quantum Measurements and Chaos*, Plenum.

88. Degn H, Holden A V, and Olsen L F (1987), eds., *Chaos in Biological Systems*, Plenum.

89. Perelson A, Goldstein B, Dembo M, and Jacquez J A (1987), eds., *Nonlinearity in Biology and Medicine*, reprinted from *Math. Bioscience*, **90**, No. 1-2, Elisevier.

90. Casati G, Gutzwiller M C, Arecchi T, Eriksson K-E, and Haken H (1987), eds., *Quantum Chaos and Physics of Structure and Complexity*, *Phys. Scripta*, **RS9**, 1-199.

91. Garrido L (1987), ed., *Fluctuation and Stochastic Phenomena in Condensed Matter*, Springer-Verlag.

92. Harrison R G, and Arecchi F T (1987), eds., *Instabilities and Chaos in Quantum Optics*, Springer-Verlag.

93. Duong-Van M (1987), ed., *Chaos'87. International Conference on the Physics of Chaos and Systems Far from Equilibrium*, *Nucl. Phys.* B, Proc. Suppl. **2**.

94. Othmer H G (1987), ed., *Nonlinear Oscillations in Chemistry and Biology*, Springer-Verlag.

95. Zweifel P, Gallavotti G, and Anile M (1987), eds., *Nonlinear Evolution and Chaotic Phenomena*, Plenum, ix + 337.

96. Kapitaniak T (1988), *Chaos in Systems with Noise*, World Scientific.

97. Glass L, Mackey M C (1988), *From Clocks to Chaos. The Rhythms of Life*, Princeton University Press.

98. Hao Bai-lin (1988), ed., *Directions in Chaos*, vol. 2, World Scientific.

99. Berry M V (1988), ed., *Dynamical Chaos*, Cambridge University Press.

100. Lundqvist S, March N H, and Tosi M P (1988), eds., *Order and Chaos in Nonlinear Physical Systems*, Plenum.

101. Velarde M G (1988), ed., *Synergetics, Order and Chaos*, World Scientific.

102. Livi R, Ruffo S, Ciliberto S, and Buiatti M (1988), eds., *Chaos and Complexity*, World Scientific.

103. Nicolaenko B (1988), ed., *The Connection Between Infinite Dimensional and Finite Dimensional Systems*, American Mathematical Society, 1988.

104. Peitgen H O, and Saupe P (1988), eds., *The Science of Fractal Images*, Springer-Verlag.

105. Flaschka H, Chirikov B (1988), eds., *Progress in Chaotic Dynamics. Essays in honour of Joseph Ford's 60th Birthday*, Physica, **33D**, No. 1-3.

106. Bandy A N, Oraevsky A N, and Tredicce T R (1988), eds., *Nonlinear Dynamics of Lasers*, J. Opt. Soc. Am., B5, No. 5.

107. Leon J J P (1988), ed., *Nonlinear Evolutions*, World Scientific.

108. Abraham N B, Arecchi F T, and Lugiato L A (1988), eds., *Instabilities and Chaos in Quantum Optics II*, Springer-Verlag.

109. Saenz A W (1988), ed., *Methods and Applications of Nonlinear Dynamics*, World Scientific.

110. Hao Bai-lin (1989), *Elementary Symbolic Dynamics and Chaos in Dissipative Systems*, World Scientific.

111. Parker T S, and Chua L O (1989), *Practical Numerical Algorithms for Chaotic Systems*, Springer-Verlag.

112. Ruelle D (1989), *Chaotic Evolution and Strange Attractors*, Cambridge University Press.

113. Tabor M (1989), *Chaos and Integrability in Nonlinear Dynamics An Introduction*, Wiley.

114. Rasband S N (1989) *Chaotic Dynamics of Nonlinear Systems*, Wiley.

115. Stewart I (1989) *Does God Play Dice? The Mathematics of Chaos*, Basil Blackwell.

116. Christiansen P L, and Parmentier R D (1989), *Structure, Coherence and Chaos in Dynamical Systems*, Manchester University Press.

117. Ottino J M (1989), *The Kinematics of Mixing: Stretching, Chaos, and Transport*, Cambridge Treatise in Applied Mathematics, Cambridge University Press.

## II. Papers including Reviews

There were 4405 entries in Shiraiwa (B1985), covering a wider field of the mathematical theory of dynamical systems. The titles listed below have been selected from a database maintained by Zhang Shu-yu. They are more or less restricted to *chaos* and to recent publications. Most of the papers, which have been included in contents of Part I, are not listed.

1. Abarbanel H D I, and Latham P (1982) Finite resolution approximation to the asymptotic distribution for dynamical systems, Phys. Lett. 89A, 55.

2. Abdulaev F Kh (1989) Dynamical chaos of solitons and nonlinear periodic waves, Phys. Reports 179, 1.

3. Abdulaev F Kh, Darmanyan S A, and Umarov B A (1985a) Chaos in the parametrically driven sine-Gordon system, Phys. Lett. 108A, 7.

4. Abdulaev F Kh, Darmanyan S A, and Umarov B A (1985b) Dynamic chaos of magnetization in ferromagnets in high-frequency fields, Sov. Phys. Solid State, 27, 976; Russ. Orig. 27,1620.

5. Abe S, and Mukamel S (1983) Anharmonic molecular spectra, self-consistent mode coupling, nonlinear maps, and quantum chaos, J. Chem. Phys. 79, 5457.

6. Abraham E, Firth W J, and Carr J (1982) Self-oscillation and chaos in nonlinear Fabry Perot resonators with finite response time, Phys. Lett. 91A, 47.

7. Abraham N B (1983) A new focus on laser instabilities and chaos, Laser Focus, 19, 73.

8. Abraham N B, Gollub J P, and Swinney H L (1983) Testing nonlinear dynamics, Physica 11D, 252.

9. Abraham N B, Albano A M, Das B, deGuzman G, Yong S, Gioggia R S, Puccioni G P, and Tredicce J P (1986) Calculating the dimension of attractors from small data sets, Phys. Lett. 114A, 217.

10. Abraham R H, and Stewart H B (1986) A chaotic blue sky catastrophe in forced relaxation oscillations, Physica 21D, 394.

11. Abul-Magd A Y, and Weidenmüller H A (1985) Regular versus chaotic dynamics in nuclear spectra near the ground state, Phys. Lett. 162B, 223.

12. Ackerhalt J R, Galbraith H W, and Milonni P W (1983) Chaos in multiple-photon excitation of molecules, Phys. Rev. Lett. 51, 1259.

13. Ackerhalt J R, Milonni P W, and Shih M-L (1985) Chaos in quantum optics, Phys. Reports, 128, 205.

14. Adachi S, Toda M, and Ikeda K (1988a) Potential for mixing in quantum chaos, Phys. Rev. Lett. 61, 635.

15. Adachi S, Toda M, and Ikeda K (1988b) Quantum-classical correspondence in many-dimensional quantum chaos, Phys. Rev. Lett. 61, 659.

16. Afraimovich V S, Rabinovich M I, and Sbitnev V I (1985) Dimensionality of attractors in a class of coupled oscillators, Sov. Tech. Phys. Lett. 11, 139; Russ. Orig. 11,338.

17. Afraimovich V S, Rabinovich M I, and Ugodnikov A D (1983) Critical points and 'phase transitions' in the stochastic behaviour of a nonautonomous unharmonic oscillator, JETP Lett. 38, 72; Russ. Orig. 38, 64.

18. Agarwal A K, Banerjee K, and Bhattacharjee J K (1986) Universality of fractal dimension at the onset of one-dimensional chaos, Phys. Lett. 119A, 290.

19. Agarwal A, Bhattacharjee J K, and Banerjee K (1985) The modulated double diffusion system: chaotic behavior near the convection threshold, Phys. Lett. 111A, 329.

20. Aguirregabiria J M, and Etxebarria J R (1987) Fractal basin boundaries of a delay-differential equation, Phys. Lett. 122A, 241.

21. Aizawa Y (1982) Global aspects of the dissipative dynamical system: I. Statistical identification and fractal properties of the Lorenz chaos, Prog. Theor. Phys. 68, 64.

22. Aizawa Y (1983) Symbolic dynamics approach to intermittent chaos, Prog. Theor. Phys. 70, 1249.

23. Aizawa Y (1984a) Symbolic dynamics approach to the two-dimensional chaos in area-preserving maps: a fractal geometrical model, Prog. Theor. Phys. 71, 1419.

24. Aizawa Y (1984b) On the $f^{-1}$ spectral chaos, Prog. Theor. Phys. 72, 659.

25. Aizawa Y (1984c) Kinetic theory of chaos-systems, Prog. Theor. Phys. 72, 662.

26. Aizawa Y, and Kohyama T (1984) Asymptotically non-stationary chaos, Prog. Theor. Phys. 71, 847.

27. Aizawa Y, and Murakami C (1983) Generalization of baker's transformation — chaos and stochastic process on a Smale's horseshoe, Prog. Theor. Phys. 69, 1416.

28. Aizawa Y, and Uezu T (1982a) Topological aspects in chaos and in $2^k$ period-doubling cascade, Prog. Theor. Phys. 67, 982.

29. Aizawa Y, and Uezu T (1982b) Global aspects of dissipative dynamical systems: II. Periodic and chaotic responses in the forced Lorenz system, Prog. Theor. Phys. 68, 1864.

30. Akhromeeva T S, Kurdyumov S P, Malinetskii G G, and Samarskii (1984) On diffusion-induced chaos in nonlinear dissipative systems, Sov. Phys. Dokl. 29, 991; Russ. Orig. 279, 191.

31. Albahadily F N, Ringland J, and Schell M (1989) Mixed-mode oscillations in an electrochemical system. I. A Farey sequence which does not occur on a torus, J. Chem. Phys. 90, 813.

32. Albano A M, Muench J, Schwartz C, Mees A I, and Rapp P E (1988) Singular value decomposition and the Grassberger-Procaccia algorithm, Phys. Rev. A38, 3017.

33. Alekseev K N, and Berman G P (1985a) A strange attractor in stationary coherent states, Phys. Lett. 111A, 326.

34. Alekseev K N, and Berman G P (1985b) Quantum chaos in stationary coherent states, Sov. Phys. JETP, 61, 569; Russ. Orig. 88, 968.

35. Alekseev K N, and Berman G P (1987) Dynamic chaos in the interaction between external monochromatic radiation and a two-level medium with allowance for cooperative effects, Sov. Phys. JETP, 65, 1115; Russ. Orig. 92, 1985.

36. Alekseev V M (1968-1969) Quasirandom dynamical systems: I, II and III, Math. USSR Sbornik, 5, 73; 6, 505; 7, 1.

37. Alekseev V M, and Yakobson M V (1981) Symbolic dynamics and hyperbolic dynamic systems, Phys. Reports, 75, 287.

38. Alekseev V V, and Loskutov A Yu (1987) Control of a system with a strange attractor through periodic parametric action, Sov. Phys. Dokl. 32, 270; Russ. Orig. 293, 1346.

39. Alfsen K H, and Frøyland J (1985) Systematics of the Lerenz model at $\sigma =10$, Phys. Scripta 31, 15.

40. Alhassid Y, and Levine R D (1986) Transition-strength fluctuations and the onset of chaotic motion, Phys. Rev. Lett. 57, 2879.

41. Ali M K, and Somorjai R L (1982) Reappearance of ordered motion in nonintegrable Hamiltonian systems — the strong coupling case, Prog. Theor. Phys. 68, 1854.

42. Allen T (1983a) Complicated, but rational, phase-locking responses of a simple time-based oscillator, IEEE Trans. Cir. & Sys. CAS-30, 627.

43. Allen T (1983b) On the arithmetic of phase locking: coupled neurons on a lattice on $R^2$, Physica 6D, 305.

44. Alligood K T, Yorke E D, and Yorke J A (1987) Why period-doubling cascades occur: periodic orbit creation followed by stability shedding, Physica 28D, 197.

45. Alligood K T, and Sauer T (1988) Rotation numbers of periodic orbits in the Hénon's map, Commun. Math. Phys. 120, 105.

46. Alstrøm P (1986) Map dependence of the fractal dimension deduced from iterations of cirle maps, Commun. Math. Phys. 104, 581.

47. Alstrøm P (1988) Phase-locking structure of 'integrate-and-fire' models with threshold modulation, Phys. Lett. 128A, 187.

48. Alstrøm P, Christiansen B, Hyldgaard P, Levinsen M T, and Rasmussen R (1986) Scaling relations at the critical line and the period-doubling route for the sine map and the driven damped pendulum, Phys. Rev. A34, 2220.

49. Alstrøm P, Christiansen B, and Levinsen M T (1988) Nonchaotic transitions from quasiperiodicity to complete phase locking, Phys. Rev. Lett. 61, 1679.

50. Alstrøm P, Hansen L K, and Rasmussen D R (1987) Crossover scaling for moments in multifractal systems, Phys. Rev. A36, 827.

51. Alstrøm P, Levinsen M T, and Rasmussen D R (1987) Scaling exponents, relations, and order dependence for circle maps, Physica 26D, 336.

52. Alstrøm P, and Levinson M T (1985) Fractal structure of the complete devil's staircase in dissipative systems described by a driven damped-pendulum equation with a distorted potential, Phys. Rev. B32, 1503.

53. Alstrøm P, and Ritala R K (1987) Mode locking in an infinite set of coupled circle maps, Phys. Rev. A35, 300.

54. Amritkar R E, and Gupte N (1987) Dimensional characterization of sets with partial scaling symmetry, Phys. Rev. A36, 2850.

55. Amritkar R E, and Gupte N (1988) Inverse problem for multifractals: extension to logarithmic and additive correlations, Phys. Rev. Lett. 60, 245.

56. Anania G, and Politi A (1988) Dynamical behavior at the onset of chaos, Europhys. Lett. 7, 119.

57. Ananthakrishna G, Balakrishnan R, and Hao B-L (1987) Spatially chaotic spin patterns in a field-perturbed Heisenberg Chain, Phys. Lett. 121A, 407.

58. Ananthakrishna G, and Valsakumar M C (1983) Chaotic flow in a model for repeated yielding, Phys. Lett. 95A, 69.

59. Andereck C D, Liu S S, and Swinney H L (1986) Flow regimes in a circular Couette system with independently rotating cylinders, J. Fluid Mech. 164, 155.

60. Andersen D F, and Sturis J (1988) Chaotic structures in generic management models: pedagogical principles and examples, Sys. Dyn. Rev. 4, 218.

61. Andrade R F S, and Rauh A (1983) Nonlinear stability analysis of Lorenz model by Lyapunov's direct method, Z. Phys. B50, 151.

62. Andrey L (1986) The relationship between entropy production and K-entropy, Prog. Theor. Phys. 75, 1258.

63. Angelo P M, and Riela G (1981) A six-mode truncation of the Navier-Stokes equations on a two-dimensional torus: a numerical study, Nuovo Cimento, 64B, 207.

64. Anishchenko V S (1984a) Interaction of strange attractors and chaos – chaos intermittency, Sov. Tech. Phys. Lett. 10, 266; Russ. Orig. 10, 629.

65. Anishchenko V S (1984b) Effects of coupling on the dynamics of a system near the transition point to period-doubling behaviour, Sov. Phys. Tech. Phys. 29, 501; Russ. Orig. 54, 844.

66. Anishchenko V S (1986) Destruction of quasiperiodic oscillations leading to chaos in dissipative systems, Sov. Phys. Tech. Phys. 1, 137; Russ. Orig. 56, 225.

67. Anishchenko V S, Aranson I S, Postnov D E, and Rabinovich M I (1986) Spatial synchronization and bifurcation of the development of chaos in a chain of coupled generators, Sov. Phys. Dokl. 31, 169; Russ. Orig. 286, 1120.

68. Anishchenko V S, Astakov V V, and Letchford T E (1983) Experimental study of the structure of a strange attractor in a model generator with a nonlinear inertial term, Sov. Phys. Tech. Phys. 28, 91; Russ. Orig. 53, 152.

69. Anishchenko V S, Melekhin G V, Stepanov V A, and Chirkin M V (1986) Mechanism of onset and evolution of chaos in stratified positive column of a gas discharge, Radiophys. & Quantum Electr. 29, 692; Russ. Orig. 29, 903.

70. Anishchenko V S, and Astakhov V V (1983) Bifurcation phenomena in an autostochastic oscillator responding to a regular external signal, Sov. Phys. Tech. Phys. 28, 1326; Russ. Orig. 53, 2165.

71. Anishchenko V S, and Herzel H (1988) Noise-induced chaos in a system with homoclinic points, Z. Angew. Math. Mech. 68, 317.

72. Anishchenko V S, and Klimontovich Yu L (1984) Entropy evolution in an oscillator with an initial nonlinearity during the transition to chaos through a sequence of period-doubling bifurcations, Sov. Tech. Phys. Lett. 10, 368; Russ. Orig. 10, 876

73. Anishchenko V S, and Letchford T E (1986) Breakup of three-frequency oscillations and onset of chaos in a biharmonically excited oscillator, Sov. Phys. Tech. Phys. 31, 1347; Russ. Orig. 56, 2250.

74. Anishchenko V S, and Neiman A B (1987) Increase in the duration of correlations during a chaos-chaos intermittency, Sov. Tech. Phys. Lett. 1, 444; Russ. Orig. 13, 1063.

75. Anishchenko Y S, Letchford T E, and Safonova M A (1984) Stochasticity and the disruption of quasiperiodic motion due to doubling in a system of coupled generators, Radiophys. & Quantum Electr. 27, 381; Russ. Orig. 27, 565.

76. Ankiewicz A (1983) Regular and irregular motion: new mechanical results and firbre optics, J. Phys. A16, 3657.

77. Ankiewicz A, and Pask C (1984) Potential symmetry in dynamics and the onset of chaos, Phys. Lett. 101A, 239.

78. Antipin V A, Kazakov V P, and Parshin G S (1983) Phenomenon of 'chaos' in an oscillatory chemiluminescent reaction, Theor. & Exper. Chem. 19, 462; Russ. Orig. 19, 497.

79. Antorantz J C, and Mori H (1985) Intermittent incommensurate chaos, Physica 16D, 184.

80. Aoki K, Ikezawa O, and Yamamoto (1983) Firing wave instability and chaos of the current filament in a semiconductor, Phys. Lett. 98A, 217.

81. Aoki K, and Mugibayashi N (1986) Cellular automata and coupled chaos developed in a lattice chain of N equivalent switching elements, Phys. Lett. 114A, 425.

82. Aoki K, and Mugibayashi N (1988) Onset of turbulent patterns in a coupled map lattice. Case for soliton-like behavior, Phys. Lett. 128A, 349.

83. Aoki K, and Yamamoto K (1989) Nonlinear response and chaos in semiconductors induced by impact ionization, Appl. Phys. A48, 111.

84. Aranson I S, Gaponov-Grekhov A V, and Rabinovich M I (1985) The development of chaos in dynamic structure ensembles, Sov. Phys. JETP, 62, 52.

85. Arecchi F T (1987) Chaos in quantum optics: hyperchaos and long memory times, Phys. Scripta 36, 911.

86. Arecchi F T (1988a) Instabilities and chaos in optics, Phys. Scripta T23, 160.

87. Arecchi F T (1988b) Shilnikov chaos in lasers, in Zeifel, Gallavotti and Anile (B1988), 301.

88. Arecchi F T, Badii R, and Politi A (1984a) Low-frequency phenomena in dynamical systems with many attractors, Phys. Rev. A29, 1006.

89. Arecchi F T, Badii R, and Politi A (1984b) Scaling of first passage times for noise induced crises, Phys. Lett. 103A, 3.

90. Arecchi F T, Badii R, and Politi A (1985) Generalized multistability and noise-induced jumps in a nonlinear dynamical system, Phys. Rev. A32, 402.

91. Arecchi F T, Gadomski W, and Meucci R (1986) Generation of chaotic dynamics by feedback on a laser, Phys. Rev. A34, 1617.

92. Arecchi F T, Giliberto S, and Rubio M A (1984) Oscillations and chaos on the free surface of a heated fluid, Nuovo Cimento, 3D, 793.

93. Arecchi F T, Lapucci A, Meucci R, Roversi J A, and Coullet P H (1988) Experimental characterization of Shil'nikov chaos by statistics of return times, Europhys. Lett. 6, 677.

94. Arecchi F T, Lippi G L, Puccioni G P, and Tredicce J R (1984) Deterministic chaos in laser with injected signal, Opt. Commun. 51, 308.

95. Arecchi F T, Meucci R, Puccioni G, and Tredicce J (1982) Experimental evidence of subharmonic bifurcations, multistability, and turbulence in a Q-switched gas laser, Phys. Rev. Lett. 49, 1217.

96. Arecchi F T, and Califano A (1984) Low-frequency hopping phenomena in nonlinear systems with many attractors, Phys. Lett. 101A, 443.

97. Arecchi F T, and Califano A (1987) Noise-induced trapping at the boundary between two attractors: a source of $1/f$ spectra in nonlinear dynamics, Europhys. Lett. 3, 5.

98. Aref H (1983) Integrable, chaotic, and turbulent vortex motion in two-dimensional flows, Ann. Rev. Fluids Mech. 15, 345.

99. Argoul F, Arneodo A, Collet P, and Lesne A (1987) Transitions to chaos in the presence of an external periodic field: crossover effect in the measure of critical exponents, Europhys. Lett. 3, 643.

100. Argoul F, Arneodo A, Richetti P, Roux J C, and Swinney H L (1987) Chemical chaos: from hints to confirmation, Accounts for Chem. Research, 20, 436.

101. Argoul F, Arneodo A, Richetti P, and Roux J C (1987) From quasiperiodicity to chaos in the Belousov-Zhabotinskii reaction: I. Experiemnt, J. Chem. Phys. 86, 3325.

102. Argoul F, Arneodo A, and Richetti P (1987) Experimental evidence for homoclinic chaos in the Belousov-Zhabotinskii reaction, Phys. Lett. 120A, 269.

103. Argoul F, Arneodo A, and Richetti P (1988) A three-dimensional dissipative map modeling type-III intermittency, J. Physique, 49, 767.

104. Argoul F, and Arneodo A (1985) Scaling for periodic forcing at the onset of intermittency, J. Physique Lett. 46, L901.

105. Arneodo A (1984) Scaling for a periodic forcing of a period-doubling system, Phys. Rev. Lett. 53, 1240.

106. Arneodo A, Coullet P, Tresser C, Libchaber A, Maure J, and d'Humiéres D (1983) About the observation of the uncompleted cascade in Rayleigh-Bénard experiment, Physica 6D, 385.

107. Arneodo A, Coullet P, and Spiegel E A (1982) Chaos in a finite macroscopic system, Phys. Lett. 92A, 369.

108. Arneodo A, Coullet P, and Spiegel E A (1983) Cascade of period-doublings of tori, Phys. Lett. 94A, 1.

109. Arneodo A, Coullet P, and Spiegel E A (1985a) The dynamics of triple convection, Geophys. Astrophys. Fluid Dyn. 31, 1.

110. Arneodo A, Coullet P, and Spiegel E A (1985b) Asymptotic chaos, Physica 14D, 327.

111. Arneodo A, Coullet P, and Tresser C (1979) A renormalization group with periodic behavior, Phys. Lett. 70A, 74.

112. Arneodo A, Coullet P, and Tresser C (1980) Occurrence of strange attractors in three-dimensional Volterra equation, Phys. Lett. 79A, 59.

113. Arneodo A, Coullet P, and Tresser C (1981a) A possible new mechanism for the onset of turbulence, Phys. Lett. 81A, 197.

114. Arneodo A, Coullet P, and Tresser C (1981b) Possible new strange attractors with spiral structure, Commun. Math. Phys. 79, 573.

115. Arneodo A, Coullet P, and Tresser C (1982) Oscillators with chaotic bebehavior: an illustration of a theorem by Shilnikov, J. Stat. Phys. 27, 171.

116. Arneodo A, Grasseau G, and Kostelich E J (1987) Fractal dimension and $f(\alpha)$ spectrum of the Hénon attractor, Phys. Lett. 124A, 426.

117. Arneodo A, and Holschneider M (1987) Crossover effect in the $f(\alpha)$ spectrum for quasiperiodic trajectories at the onset of chaos, Phys. Rev. Lett. 58, 2007.

118. Arneodo A, and Holschneider M (1988) Fractal dimensions and homeomorphic conjugacies, J. Stat. Phys. 50, 995.

119. Arneodo A, and Thual O (1985) Direct numerical simulations of a triple convection problem versus normal form predictions, Phys. Lett. 109A, 367.

120. Arnold V I (1962) The classical theory of perturbations and the problem of stability of planetary systems, Sov. Math. Dokl. 3, 1008.

121. Arnold V I (1963a) Proof of a theorem of A. N. Kolmogorov on the invariance of periodic motions under small perturbations of the Hamiltonian, Russ. Math. Surveys, 18, 9; Russ. Orig. UMN 18:5, 13.

122. Arnold V I (1963b) Small denominators and problems of stability of motion in classical and celestial mechanics, Russ. Math. Surveys, 18, 85; Russ. Orig. UMN 18:6, 91.

123. Arnold V I (1964) Instability of dynamical systems with several degrees of freedom, Sov. Math. Dokl. 5, 581; Russ. Orig. 156, 9.

124. Arnold V I (1965) Small denominators. I: Mappings of the circumference onto itself, AMS Transl. Ser. 2, 46, 213.

125. Aronson D G, Chory M A, Hall G R, and McGehee R P (1982) Bifurcations from an invariant circle for two-parameter families of maps of the plane: A computer-assisted study, Commun. Math. Phys. 83, 303.

126. Aubry S (1983a) Devils' staircase and order without periodicity in classical condensed matter, J. Physique, 44, 147.

127. Aubry S (1983b) The twist map, the extended Frenkel-Kontorova model and the devil's staircase, Physica 7D, 240.

128. Auerbach D, Cvitanovic P, Eckmann J-P, Gunaratne G, and Procaccia I (1987) Exploring chaotic motion through periodic orbits, Phys. Rev. Lett. 58, 2387.

129. Auerbach D, O'Shaughnessy B, and Procaccia I (1988) Scaling structure of strange attractors, Phys. Rev. A37, 2234.

130. Aurell E (1987a) On the metric properties of the Feigenbaum attractor, J. Stat. Phys. 47, 439.

131. Aurell E (1987b) Feigenbaum attractor as a spin system, Phys. Rev. A35, 4016.

132. Aurell E (1988) An equivalence relation in some fractal sets, Phys. Lett. 130A, 449.

133. Aurich R, Sieber M, and Steiner F (1988) Quantum chaos of the Hadamard-Gutzwiller model, Phys. Rev. Lett. 61, 483.

134. Avraham B-M, Procaccia I, and Grassberger P (1984) Characterization of experimental (noisy) strange attractors, Phys. Rev. A29, 975.

135. Avraimovitch V, Bkov V, and Sil'nikov L (1977) On the origin and structure of the Lorenz attractor, Sov. Phys. Dokl. 22, 253; Russ. Orig. 234, 336.

136. Azzouz A, Duhr R, and Hasler M (1983) Transition to chaos in a simple nonlinear circuit driven by a sinusoidal voltage source, IEEE Trans. Cir. & Sys. CAS-30, 9113.

137. Babloyantz A, Salazar J M, and Nicolis C (1985) Evidence of chaotic dynamics of brain activity during the sleep cycle, Phys. Lett. 111A, 152.

138. Babloyantz A, and Destexhe A (1986) Low-dimensional chaos in an instance of epilepsy, Proc. Natl. Acad. Sci. USA, 83, 3513.

139. Badii R (1988) Generalized thermodynamic ensembles for fractal measures, Nuovo Cimento, 10D, 819.

140. Badii R (1989) Conservation laws and thermodynamic formalism for dissipative dynamical systems, Riv. Nuovo Cimento, 12, No.3, 1.

141. Badii R, Broggi G, Derighetti B, Ravani M, Ciliberto S, Politi A, and Rubio M A (1988) Dimension increase in filtered chaotic signals, Phys. Rev. Lett. 60, 979.

142. Badii R, and Broggi G (1988) Measurement of the dimensions spectrum $f(\alpha)$: fixed mass approach, Phys. Lett. 131A, 339.

143. Badii R, and Politi A (1984) Intrinsic oscillations in measuring the fractal dimension, Phys. Lett. 104A, 303.

144. Badii R, and Politi A (1985) Statistical description of chaotic attractors: the dimension function, J. Stat. Phys. 40, 725.

145. Badii R, and Politi A (1987a) Dimension function and phase transition-like behavior in strange attractors, Phys. Scripta 35, 2456.

146. Badii R, and Politi A (1987b) Renyi dimensions from local expansion rates, Phys. Rev. A35, 1288.

147. Baesens C, and Nicolis G (1983) Complex bifurcations in a periodically forced normal form, Z. Phys. B52, 345.

148. Bagley R J, Mayer-Kress J, and Farmer J D (1986) Mode locking, the Belousov-Zhabotinsky reaction, and one-dimensional mappings, Phys. Lett. 114A, 419.

149. Baive D, and Franceschini V (1981) Symmetry breaking on a model of five-mode truncated Navier-Stokes equations, J. Stat. Phys. 26, 471.

150. Bajkowski J, and Müller P C (1988) Lyapunov exponents and fractal dimensions in nonlinear mechanical systems, Z. Angew. Math. Mech. 68, 49.

151. Bak P (1981) Chaotic behavior and incommensurate phases in the anisotropic Ising model with competing interactions, Phys. Rev. Lett. 46, 791.

152. Bak P (1982) Commensurate phases ,incommensurate phases and the devil's staircase, Repts. Prog. Phys. 45, 587.

153. Bak P (1986) The devil's staircase, Phys. Today, December 1986, 38.

154. Bak P, Bohr T, Jensen M H, and Christiansen P V (1984) Josephson junctions and circle maps, Solid State Commun. 51, 231.

155. Bak P, and Bruinsma R (1982) One-dimensional Ising model and the complete devil's staircase, Phys. Rev. Lett. 49, 249.

156. Bak P, and Jensen M H (1982) Bifurcations and chaos in the $\phi^4$ theory on a lattice, J. Phys. A15, 1983.

157. Balakirev V A, and Buts V A (1984) Influence of a periodic external force on three-wave decay, Sov. Phys. Tech. Phys. 29, 507; Russ. Orig. 54, 854.

158. Balazs N L, and Voros A (1986) Chaos on the pseudosphere, Phys. Reports, 14, 109.

159. Bamon R, Malta I P, Pacifico M J, and Takens F (1984) Rotation intervals of endomorphisms of the circle, Ergod. Th. & Dyn. Sys. 4, 493.

160. Baranger M, and Devies K T R (1987) Periodic trajectories for a two-dimensional nonintegrable Hamiltonian, Ann. Phys. 177, 330.

161. Bardsley J N, Sundaram B, Pinnaduwage L A, and Bayfield J E (1986) Quantum dynamics for driven weakly bound electrons near the shreshold for classical chaos, Phys. Rev. Lett. 56, 1007.

162. Bardsley J N, and Sundaram B (1985) Microwave absorption by hydrogen atoms in high Rydberg states, Phys. Rev. A32, 689.

163. Barkley D (1988) Near-critical behavior for one-parameter families of circle maps, Phys. Lett. 129A, 219.

164. Barnett W, and Chen P (1988) Deterministic chaos and fractal attractors as tools for nonparametric dynamical econometric inference: with an application to the divisia monetary aggregates, Mathl. Comput. Modelling, 10, 275.

165. Barrow J D (1982a) General relativistic chaos and nonlinear mechanics, Gen. Relativ. and Gravit. 14, 523.

166. Barrow J D (1982b) Chaotic behavior in general relativity, Phys. Reports, 85, 1.

167. Bartuccelli M, Christiansen P L, Muto V, and Soerensen P (1987) Chaotic behavior of va pendulum with viriable length, Nuovo Cimento, 100B, 229.

168. Bartuccelli M, Christiansen P L, Pedersen N F, and Salerno M (1986) 'Horseshoe chaos' in the space-dependent double sine-Gordon system, Wave Motion, 8, 581.

169. Barugola A, Cathala J C, and Mira C (1986) Annular chaotic areas, Nonlin. Anal. Th. Meth. & Appl. 10, 1223.

170. Bassett M R, and Hudson J L (1989) Experimental evidence of period doubling of tori during an electrochemical reaction, Physica, 35D, 289.

171. Bassetti B, Butera P, Raciti M, and Sparpaglione M (1984) Complex poles, spatial intermittency, and energy transfer in a classical nonlinear string, Phys. Rev. A30, 1033.

172. Bayfield J E (1987b) Atomic physics and the quantum dynamics of classically chaotic nonintegrable systems, Comments At. & Mol. Phys. 20, 245.

173. Bayfield J E, and Pinnaduwage L A (1985) Diffusionlike aspects of multiphoton absorption in electrically polarized highly excited hydrogen atoms, Phys. Rev. Lett. 54, 313.

174. Beasley M R, and Huberman B A (1982) Chaos in Josephson junctions, Comments on Sol. Stat. Phys. 10, 155.

175. Becker W, Scully M O, Wodkiewicz K, and Zubairy M S (1984) Relativistic charged-particle interactions in a chaotic laser field, Phys. Rev. A30, 2245.

176. Beddington J R, Free C A, and Lauton J H (1975) Dynamic complexity in predator-pray models framed in difference equations, Nature, 255, 58.

177. Beiersdorfer P, and Wersinger J-M (1983) Topology of the invariant manifolds of period-doubling attractors for some forced nonlinear oscillators, Phys. Lett. 96A, 269.

178. Bélair J, and Glass L (1983) Self-similarity in periodically forced oscillators, Phys. Lett. 96A, 113.

179. Bélair J, and Glass L (1985) Universality and self-similarity in the bifurcations of circle maps, Physica 16D, 143.

180. Bellissard J, Besis D, and Moussa P (1982) Chaotic states of almost periodic Schrödinger operator, Phys. Rev. Lett. 49, 701.

181. Belmonte A L, Vinson M J, Glazier J A, Gunaratne G H, and Kenny A G (1988) Trajectory scaling functions at the onset of chaos: experimental results, Phys. Rev. Lett. 61, 539.

182. Belobrov P I, Beloshapkin V V, Zaslavsky G M, and Tretyakov A G (1987) Devil's staircase in double helics self-organization, Phys. Lett. 122A, 323.

183. Beloshapkin V V, and Zaslavsky G M (1983) On the spectral properties of dynamical systems in the transition region from order to chaos, Phys. Lett. 97A, 121.

184. Belyaev Yu N, Monakhov A A, Scherbakov S A, and Yavoskaya I (1979) Onset of turbulence in rotating fluid, Sov. Phys. JETP Lett. 29, 295; Russ. Orig. 29, 329.

185. Ben-Jacob E, Braiman Y, and Shainsky R, and Imry Y (1981) Microwave induced "devil's staircase" structure and "chaotic" behavior in current-fed Josephson junctions, Appl. Phys. Lett. 38, 822.

186. Ben-Jacob E, Goldhirsch I, Imry Y, and Fishman S (1982) Intermittent chaos in Josephson junctions, Phys. Rev. Lett. 49, 1599.

187. Ben-Mizrachi A, Procaccia I, and Grassberger P (1984) Characterization of experimental (noisy)' strange attractors, Phys. Rev. A29, 975.

188. Ben-Mizrachi A, Procaccia I, Rosenberg N, and Schmidt A (1985) Real and apparent divergencies in low-frequency spectra of nonlinear dynamical systems, Phys. Rev. A31, 1830.

189. Ben-Mizrachi A, and Procaccia I (1984) Universal power law for the dimension of strange attractors near the onset of chaos, Phys. Rev. Lett. 53, 1704.

190. Ben-Mizrachi A, and Procaccia I (1985) Wrinkling of mode-locked tori in the transition to chaos, Phys. Rev. A31, 3990.

191. Benatti F, and Narnhofer H (1988) Entropic dimension for completely positive maps, J. Stat. Phys. 53, 1273.

192. Bene J, and Szépfalusy P (1988) Properties of fully developed chaotic one-dimensional maps in the presence of external noise, Phys. Rev. A37, 871.

193. Benettin G (1984) Power-law behavior of Lyapunov exponents in some conservative dynamical systems, Physica 13D, 211.

194. Benettin G, Casartelli M, Galgani L, Giorgilli A, and Strelcyn J-M (1978-9) On the reliability of numerical studies of stochasticity: I. Existence of time averages; II. Identification of time averages, Nuovo Cimento, 44B, 183; 50B, 211.

195. Benettin G, Casati G, Galgani L, Giorgilli A, and Sironi L (1986) Apparent fractal dimensions in conservative dynamical systems, Phys. Lett. 118A, 325.

196. Benettin G, Cercignani C, Galgani L, and Giorgilli A (1980) Universal properties in conservative dynamical systems, Lett. Nuovo Cimento, 28, 1.

197. Benettin G, Froeschle C, and Scheidecker H P (1979) Kolmogorov entropy of a dynamical system with an increasing number of degrees of freedom, Phys. Rev. A19, 454.

198. Benettin G, Galgani L, and Giorgilli A (1980) Further results on universal properties in conservative dynamical systems, Lett. Nuovo Cimento, 29, 163.

199. Benettin G, Galgani L, and Strelcyn J-M (1976) Kolmogorov entropy and numerical experiments, Phys. Rev. 14A, 2338.

200. Benettin G, and Gallavotti G (1986) Stability of motions near resonances in quasi-integrable Hamiltonian systems, J. Stat. Phys. 44, 293.

201. Benettin G, and Galgani L (1982) Transition to stochasticity in a one-dimensional model of a radiant cavity, J. Stat. Phys. 27, 153.

202. Benettin G, and Strelcyn J-M (1978) Numerical experiments on the free motion of a point mass moving in a plane convex region: stochastic transition entropy, Phys. Rev. A17, 773.

203. Benettin G, and Tenenbaum A (1983) Ordered and stochastic behavior in a two-dimensional Lenard-Jones system, Phys. Rev. A28, 3020.

204. Benhabib J, and Day R H (1981) Rational choice and erratic behaviour, Rev. Economic Studies, 48, 459.

205. Benjamin T B, and Mullin T (1981) Anomalous modes in the Taylor experiment, Proc. R. Soc. Lond. A377, 221.

206. Bennett D, Bishop A R, and Trullinger S E (1982) Coherence and chaos in the driven damped sine-Gordon chain, Z. Phys. B47, 265.

207. Bensimon D, Jensen M H, and Kadanoff L P (1986) Renormalization-group analysis of the global structure of the period-doubling attractor, Phys. Rev. A33, 3622.

208. Bensimon D, and Kadanoff L P (1984) Extended chaos and disappearance of KAM trajectories, Physica 13D, 82.

209. Benza V G, and Koch S W (1987) Symmetry breaking and metastable chaos in a coherently driven superradiant system, Phys. Rev. A35, 174.

210. Benzi R, Paladin G, Parisi G, and Vulpiani A (1984) On the multifractal nature of fully developed turbulence and chaotic system, J. Phys. A18, 3521.

211. Benzi R, Paladin G, Parisi G, and Vulpiani A (1985) Characterization of intermittency in chaotic systems, J. Phys. A18, 2157.

212. Benzinger H E, Burns S A, and Palmore J I (1987) Chaotic complex dynamics and Newton's method, Phys. Lett. 119A, 441.

213. Berge P (1982) Study of the phase-space diagrams through experimental Poincaré sections in prechaotic and chaotic regimes, Phys. Scripta T1, 71.

214. Berge P, and Dubois M (1979) Time-dependent velocity in Rayleigh-Bénard convection: a transition to turbulence, Opt. Commun. 19, 129.

215. Berge P, and Dubois M (1983) Transient reemergent order in convective spatial chaos, Phys. Lett. 93A, 365.

216. Berker A N, and McKay S R (1984) Hierarchical models and chaotic spin glasses, J. Stat. Phys. 36, 787.

217. Berman G P, Kolovsky A R, and Izrailev F M (1988) Quantum chaos and peculiarities of diffusion in Wigner representation, Physica 152A, 273.

218. Berman G P, and Iomin A M (1985a) Structural order and chaos in a one-dimensional quantum atomic chain, Phys. Lett. 107A, 324.

219. Berman G P, and Iomin A M (1985b) Nonlinear dynamics and spatial structure of quantum one-dimensional chains, Sov. Phys. JETP, 62, 534; Russ. Orig. 89, 946.

220. Berman G P, and Kagansky A M (1983) On stochasticity for a system with a finite region of interaction of particles, Physica 9D, 225.

221. Berman G P, and Kolovsky A R (1983) Correlation function behavior in quantum systems which are classically chaotic, Physica 8D, 117.

222. Berman G P, and Kolovsky A R (1984) The limit of stochasticity for a one-dimensional chain of interacting oscillators, Sov. Phys. JETP, 60, 1116; Russ. Orig. 87, 1938.

223. Berman G P, and Kolovsky A R (1985) Dynamics of classically chaotic quantum systems in Wigner representation, Physica 17D, 183.

224. Berman G P, and Zaslavsky G M (1982) Quantum mappings and the problems of stochasticity in quantum systems, Physica 111A, 17.

225. Bernhardt C (1982) Rotation intervals of a class of endomorphisms of the circle, Proc. Lond. Math. Soc. III Ser. 45, part 2.

226. Bernhardt C (1987) The ordering on permutations induced by continuous maps of the real line, Ergod. Th. & Dyn. Sys. 7, 155.

227. Berry M V (1987) Quantum physics on the edge of chaos, New Scientist, 116, 44.

228. Berry M V, and Robnik M (1986) Statistics of energy levels without time – reversal symmetry: Aharonov – Bohm chaotic billiards, J. Phys. A19, 649.

229. Bershadskii A G (1984) Preturbulence (transitional laminar-turbulent flows), Magnetohydrodynamics, 20, 151; Russ. Orig. 20, 54.

230. Besis D, Fournier J D, Servizi G, Turchetti G, and Vaienti S (1987) Mellin transforms of correlation integrals and generalized dimensions of strange sets, Phys. Rev. A36, 920.

231. Besis D, Paladin G, Turchetti G, and Vaienti (1988) Generalized dimensions, entropies and Liapunov exponents from the pressure function for strange sets, J. Stat. Phys. 51, 109.

232. Besis D, Serrizi G, Turchetti G, and Vaienti S (1987) Mellin transforms and correlation dimensions, Phys. Lett. 119A, 345.

233. Beyer W A, Mauldin R D, and Stein P R (1986) Shift-maximal sequences in function iteration: existence, uniqueness, and multiplicity, J. Math. Anal. and Appl. 115, 305.

234. Bezruchko B P, Gulyaev Yu V, Kuznetsov S P, and Seleznev E P (1986) New type of critical behavior of coupled systems at the transition to chaos, Sov. Phys. Dokl. 31, 258; Russ. Orig. 287, 619.

235. Bhatia N B, and Egerland W O (1986) On the existence of Li-Yorke points in the theory of chaos, Nonlin. Anal. Th. Meth. & Appl. 10, 541.

236. Bhattacharjee J K (1986) Fractal dimension of Feigenbaum attrractors for a class of one-dimensional maps, Phys. Lett. 117A, 339.

237. Bhattacharjee J K, and Banerjee K (1984) Intermittency in the presence of control-parameter modulation, Phys. Rev. A29, 2301.

238. Bhattacharjee J K, and Banerjee K (1987) The one-dimensional map $1 - Cx^{2\mu}$ in the large $\mu$ limit, J. Phys. A20, L269.

239. Bialek J, Schmidt G, and Wang B H (1985) Bifurcation sequences and tree interactions in some Hamiltonian systems, Physica 14D, 265.

240. Bier M, and Bountis T C (1984) Remerging Feigenbaum trees in dynamical systems, Phys. Lett. 104A, 29.

241. Bindal V N, Saksena T K, and Singh G (1980) Subharmonic emission produced by high power ultrasonic waves in air, Acoust. Lett. 4, 89.

242. Binder P M, and Jensen R V (1986) Simulating chaotic behavior with finite-state machines, Phys. Rev. A34, 4460.

243. Bird N, and Vivaldi F (1988) Periodic orbits of the sawtooth maps, Physica 30D, 164.

244. Birnir B (1986) Chaotic perturbation of KdV. I. Rational solutions, Physica 19D, 238.

245. Bishop A R (1986) Space-time complexity in solid state models, Helv. Phys. Acta, 59, 811.

246. Bishop A R, Fesser K, and Lomdahl P S (1983) Coherent spatial structure versus time chaos in a perturbed Sine-Gordon system, Phys. Rev. Lett. 50, 1095.

247. Bishop A R, Forest M G, McLaughlin D W, and Overman II E A (1986) A quasi-periodic route to chaos in a near-integrable partial differential equation, Physica 2D, 293.

248. Bishop A R, McLaughlin D W, Forest M G, and Overman II E A (1988) Quasiperiodic route to chaos in a near-integrable PDE: homoclinic crossings, Phys. Lett. 127A, 335.

249. Biskamp D, and He K-F (1985) Three-drift-wave interaction at finite parallel wavelength: bifurcations and transition to turbulence, Phys. Fluids, 28, 2172.

250. Blackburn J A, Yang Z-J, Vik S, Smith H J T, Nerenberg M A H (1987) Experimental study of chaos in a driven pendulum, Physica 26D, 385.

251. Blackmore D (1986) The mathematical theory of chaos, Comput. & Math. with Appl. 12B, 1039.

252. Bleher S, Grebogi C, Ott E, and Brown R (1988) Fractal boundaries for exit in Hamiltonian dynamics, Phys. Rev. A38, 930.

253. Block A, Scheunhuber H J, and Unbschat H (1987) Analytic fractal dimension of Cantori, Phys. Rev. Lett. 58, 1046.

254. Block L (1976) Homoclinic points of mappings of the interval, Proc. AMS, 72, 576.

255. Block L, and Hart D (1987) Orbit types for maps of the interval, Ergod. Th. & Dyn. Sys. 7, 161.

256. Blumel R, Goldberg J, and Smilansky U (1988) Features of the quasienergy spectrum of the hydrogen atom in a microwave field, Z. Phys. 9D, 95.

257. Blumel R, and Smilansky U (1984) Suppression of classical stochasticity by quantum-mechanical effects in the dynamics of periodically perturbed surface-state electrons, Phys. Rev. A30, 1040.

258. Bogomolny E B (1988) Smoothed wave functions of chaotic quantum systems, Physica 31D, 169.

259. Bogoyavlenskii O I (1981) Geometrical methods of the qualitative theory of dynamical systems in problems of theoretical physics, in *Mathematical Physics Review*, ed. by S. P. Novikov, Vol. 2, p. 117.

260. Bohigas O, Giannoni M J, and Schmit C (1984) Characterization of chaotic quantum spectra and universality of level fluctuation laws, Phys. Rev. Lett. 52, 1.

261. Bohr T (1985) Destruction of invariant tori as an eigenvalue problem, Phys. Rev. Lett. 54, 1737.

262. Bohr T (1986) Precursors of chaos or how to draw the phase diagram of a swing, Phys. Scripta T13, 124.

263. Bohr T, Bak P, and Jensen M H (1984) Transition to chaos by interaction of resonances in dissipative systems. II. Josephson junctions, charge-density waves,and standard maps, Phys. Rev. A30, 1970.

264. Bohr T, Grinstein G, Yu He, and Jayaprakash C (1987) Coherence, chaos, and broken symmetry in classical, many-body dynamical systems, Phys. Rev. Lett. 58, 2155.

265. Bohr T, and Gunaratne G (1985) Scaling for supercritical circle-maps: numerical investigation of the onset of bistability and period doubling, Phys. Lett. 113A, 55.

266. Bohr T, and Rand D (1987) The entropy function for characteristic exponents, Physica 25D, 387.

267. Boldrighini C, and Franceschini V (1979) A five-dimensional truncation of the plane incompressible Navier-Stokes equations, Commun. Math. Phys. 64, 159.

268. Bolotin Yu L, Gonchar V Yu, and Inopin E V (1987) Chaos and catastrophies in quadrupole oscillations of nuclei, Sov. J. Nucl. Phys. 45, 220; Russ. Orig. 45, 350.

269. Bondeson A, Ott E, and Antonsen M Jr (1985) Quasiperiodically forced pendula and Schrödinger equations with quasiperiodic potentials: implications of their equivalence, Phys. Rev. Lett. 55, 2103.

270. Bonetti M, Meynart R, Boon J-P, and Olivari D (1985) Chaotic dynamics in a periodically excited air jet, Phys. Rev. Lett. 55, 492.

271. Booty M, Gibbon J D, and Fowler A C (1982) A study of the effect of mode truncation on an exact periodic solution of an infinite set of Lorenz equations, Phys. Lett. 87A, 261.

272. Born M (1955) Is classical mechanics in fact deterministic ? (reprinted in *Physics in my Generation*, p. 164, Springer-Verlag, 1969), Phys. Bllater. 11:9, 49.

273. Boswell R W (1985) Experimental measurements of bifurcations, chaos and three cycle behavior in a neutralized electron beam, Plasma Phys. & Control. Fus. 27, 405.

274. Bountis T C (1981) Period doubling, bifurcations, and universality in conservative systems, Physica 3D, 577.

275. Bountis T, Papageorgion V, and Bier M (1987) On the singularity analysis of intersecting separatrices in near-integrable dynamical systems, Physica 24D, 292.

276. Bountis T, and Helleman R H (1981) On the stability of periodic orbits of two-dimensional mappings, J. Math. Phys. 22, 1867.

277. Bowen R (1973) Symbolic dynamics for hyperbolic flows, Amer. J. Math. 95, 429.

278. Boyarsky A (1984) On the significance of absolutely continuous invariant measure, Physica 11D, 130.

279. Boyarsky A (1987) Singuler perturbations of piecewise monotonic maps of the interval, J. Stat. Phys. 48, 561.

280. Boyarsky A (1988) A matrix method for estimating the Liapunov exponent of one-dimensional systems, J. Stat. Phys. 50, 213.

281. Boyarsky A, Byers W, and Gauthier P (1987) Higher dimensional analogues of the tent map, Nonlin. Anal. Th. Meth. Appl. 11, 1317.

282. Boyland P L (1986) Bifurcations of circle maps: Arnol'd tongues, bistability and rotation intervals, Commun. Math. Phys. 106, 353.

283. Brachet M E, Coullet P, and Fauve S (1987) Propagative phase dynamics in temporarily intermittent systems, Europhys. Lett. 4, 1017.

284. Bramberg L, and Rechester A B (1988) Gibbs-type partitions in chaotic dynamics, Phys. Rev. A37, 1708.

285. Brambilla R, Casartelli M (1985) Local order and onset of chaos for a family of two-dimensional mappings, Nouvo Cimento, 88B, 102.

286. Brandl A, Geisel T, and Prettl W (1987) Oscillations and chaotic current fluctuations in n-GaAs, Europhys. Lett. 3, 401.

287. Brandstäter A, Swift J, Swinney H L, Wolf A, Farmer J D, Jen E, and Crutchfield P J (1983) Low-dimensional chaos in a hydrodynamic system, Phys. Rev. Lett. 51, 1442.

288. Brandstäter A, and Swinney H L (1987) Strange attractors in weakly turbulent Couette-Taylor flow, Phys. Rev. A35, 2207.

289. Braun T, Lisboa J A, and Francke R E (1987) Observation of deterministic choas in electrical discharges in gases, Phys. Rev. Lett. 59, 613.

290. Bray A J, and Moore M A (1987) Chaotic nature of spin-glass phase, Phys. Rev. Lett. 58, 57.

291. Bremer H (1987) Chaos in mechanical systems: demonstration models, Z. Angew. Math. & Mech. 67, T63.

292. Brenig L, and Banai N (1984) The logarithmic link between the solvable population growth mapping and the quadratic mapping, Phys. Lett. 191A, 479.

293. Brickmann J, Pfeiffer R, and Schmidt P C (1984) The transition between regular and chaotic dynamics and its influence on the vibrational energy transfer in molecules after local preparation, Ber. Bunsenges. Phys. Chem. 88, 382.

294. Briggs K (1987) Simple experiments in chaotic dynamics, Am. J. Phys. 55, 1083.

295. Brindley J, and Moroz I M (1980) Lorenz attractor behavior in a continuously stratified baroclinic fluid, Phys. Lett. 77A, 441.

296. Brock W A (1986) Distinguishing random and deterministic systems: abridged version, J. Economic Theory, 40, 168.

297. Brolin (1965) Invariant sets under iteration of rational functions, Arkiv Mat. 6, 103.

298. Broomhead D S, Jones R, and King G P (1987) Topological dimension and local coordinates from time series data, J. Phys. A20, L563.

299. Broomhead D S, McCreadie G, and Rowlands G (1981) On the analytic derivation of Poincaré maps — the forced Brusselator problem, Phys. Lett. 84A, 229.

300. Broomhead D S, and King G P (1986) Extracting qualitative dynamics from experimental data, Physica 20D, 217.

301. Broomhead D S, and Rowlands G (1982) A simple derivation of the Mel'nikov condition for the appearance of homoclinic points, Phys. Lett. 89A, 63.

302. Broomhead D S, and Rowlands G (1983) On the analytic treatment of nonintegrable difference equations, J. Phys. A16, 9.

303. Broomhead D S, and Rowlands G (1984) On the use of perturbation theory in the calculation of the fractal dimension of strange attractors, Physica 10D, 340.

304. Brorson S D, Dewey D, and Linsay P S (1983) Self-replicating attractor of a driven semiconductor oscillator, Phys. Rev. A28, 1201.

305. Brown R, Grebogi C, and Ott E (1986) Broadening of spectral peaks at the merging of chaotic bands in period-doubling systems, Phys. Rev. A34, 2248.

306. Brown R, Ott E, and Grebogi C (1987a) The goodness of ergodic adiabatic invariants, J. Stat. Phys. 49, 511.

307. Brown R, Ott E, and Grebogi C (1987b) Ergodic adiabatic invariants of chaotic systems, Phys. Rev. Lett. 59, 1173.

308. Bruhn B (1987) Chaos and order in weakly coupled systems of nonlinear oscillators, Phys. Scripta 35, 7.

309. Bruhn B (1988) Transient chaos and perturbed large range interactions, Phys. Scripta 37, 193.

310. Bruhn B, and Leven R W (1985) Existence of periodic and homoclinic trajectories for weakly coupled systems of nonlinear ordinary differential equations, Phys. Scripta 32, 486.

311. Brundson V, and Holmes P (1987) Power spectra of strange attractors near homoclinic orbits, Phys. Rev. Lett. 58, 1699.

312. Brunner W, Paul H (1983) Regular and chaotic behavior of multimode gas lasers, Opt. Quantum Electron. 15, 87.

313. Brunowsky P (1974) Generic properties of the rotation number of one dimensional diffeomorphisms of the circle, Czech. Math. J. 24, 74.

314. Bryant P, and Jeffries C (1984) Bifurcations of a forced magnetic oscillator near points of resonance, Phys. Rev. Lett. 53, 250.

315. Bryant P, and Wiesenfeld K (1986) Suppression of period-doubling and nonlinear parametric effects in periodically perturbed systems, Phys. Rev. A33, 2525.

316. Bucher M (1986) Universal scaling of windows from one-dimensional maps, Phys. Rev. A33, 3544.

317. Buchler J R, Goupil M-J, and Kovacs G (1988) Tangent bifurcation to chaos in stellar pulsations, Phys. Lett. 128A, 177.

318. Buchner J, and Zeleny L (1986) Deterministic chaos in the dynamics of charged particles near a magnetic field reversal, Phys. Lett. 118A, 395.

319. Budinsky N, and Bountis T (1983) Stability of nonlinear modes and chaotic properties of 1D Fermi-Ulam-Pasta lattices, Physica 8D, 445.

320. Bugalho M H, Rica da Silva A, and Sousa Ramos J (1986) The order of chaos on a Bianchi-IX cosmological model, Gen. Relativ. & Gravit. 18, 1263.

321. Bullet S (1986) Invariant circles of the piecewise linear standard map, Commun. Math. Phys. 107, 241.

322. Bullet S (1988) Dynamics of quadratic correspondences, Nonlinearity, 1, 27.

323. Bullet S, Osbaldestin A H, and Percival I (1986) An iterated implicit complex map, Physica 19D, 290.

324. Bunimovich L A (1985) Decay of correlations in dynamical systems with chaotic behavior, Sov. Phys. JETP, 62, 842; Russ. Orig. 89, 1452.

325. Bunow B, and Weiss G H (1979) How chaotic is chaos? Chaotic and other "noisy" dynamics in the frequency domain, Math. Biosciences, 47, 221.

326. Bunz H, Ohno H, and Haken H (1984) Subcritical period doubling in the Duffing equation — type-3 intermittency, attractor crisis, Z. Phys. 56B, 345.

327. Buskirk B V, and Jeffries C (1985) Observation of chaotic dynamics of complex nonlinear oscillators, Phys. Rev. A31, 3332.

328. Bussac M N, and Meunier C (1982) Statistical properties of type I intermittency, J. Physique, 43, 585.

329. Campanino M (1980) Two remarks on the computer study of differentiable dynamical systems, Commun. Math. Phys. 74, 15.

330. Campanino M, Epstein H, and Ruelle D (1982) On Feigenbaum's functional equation $g \circ g(\lambda x) + \lambda g(x) = 0$, Topology, 21, 125.

331. Campanino M, and Epstein H (1981) On the existence of Feigenbaum's fixed point, Commun. Math. Phys. 79, 261.

332. Caputo J G, and Atten P (1987) Metric entropy: an experimental means for characterizing and quantifying chaos, Phys. Rev. A35, 1311.

333. Caranicolas N, and Vizikis C (1987) Chaos in a quartic dynamical model (celestial mechanics), Celes. Mech. 40, 35.

334. Carmichael H J, Savage C M, and Walls P F (1983) From optical tristability to chaos, Phys. Rev. Lett. 50, 163.

335. Carnegie A, and Percival I C (1984) Regular and chaotic motion in some quartic potentials, J. Phys. A17, 801.

336. Caroli B, Caroli C, and Roulet B (1983) Effect of a small slow modulation on Pomeau-Manneville intermittencies, Phys. Lett. 94A, 117.

337. Carroll J L, Pecora L M, and Rachford F J (1987) Chaotic transients and multiple attractors in spin-wave experiments, Phys. Rev. Lett. 59, 2891.

338. Cartwright M L (1948) Forced oscillations in nearly sinusoidal systems, J. Inst. Elec. Eng. 95, 88.

339. Cartwright M L, and Littlewood J E (1945) On nonlinear differential equations of the second order, J. London Math. Soc. 20, 180.

340. Casartelli M, Diana E, Galgani L, and Scotti A (1976) Numerical computation on a stochastic parameter related to the Kolmogorov entropy, Phys. Rev. A13, 1921.

341. Casartelli M, and Sello S (1985) Low stochasticity and relaxation in Hénon-Heiles model, Phys. Lett. 112A, 249.

342. Casati G, Chirikov B V, Guarneri I, and Shepelyansky D L (1986) Dynamical stability of quantum 'chaotic' motion in a hydrogen atom, Phys. Rev. Lett. 56, 2437.

343. Casati G, Ford J, Vivaldi F, and Visscher W M (1984) One-dimensional classical many-body system having a normal thermal conductivity, Phys. Rev. Lett. 52, 1861.

344. Casati G, and Comparin G (1982) Decay of correlations in certain hyperbolic systems, Phys. Rev. A26, 1.

345. Casati G, and Guarneri I (1983) Chaos and special features of quantum systems under external noise, Phys. Rev. Lett. 50, 640.

346. Cascais J, Dilao R, and Costa A N (1983) Chaos and reverse bifurcation in a RCL circuit, Phys. Lett. 93A, 213.

347. Casdagli M (1987) Periodic orbits for dissipative twist maps, Ergod. Th. & Dyn. Sys. 7, 165.

348. Casdagli M (1988) Rotational chaos in dissipative systems, Physica 29D, 365.

349. Caurier E, and Grammaticos B (1986) Quantum chaos with nonergodic Hamiltonians, Europhys. Lett. 2, 417.

350. Celaschi S, and Zimmermann R L (1987) Evolution of a two-parameter chaotic dynamics from universal attractors, Phys. Lett. 120A, 447.

351. Celletti A, and Falcolini C (1988) A remark on the KAM theorem applied to a four-vortex system, J. Stat. Phys. 52, 471.

352. Cenys A, and Pyragas K (1988) Estimation of the number of degrees of freedom from chaotic series, Phys. Lett. 129A, 227.

353. Chang S-J (1984) Classical Yang-Mills solutions and iterated maps, Phys. Rev. D29, 259.

354. Chang S-J, Wortis M, and Wright J (1981) Iterative properties of a one dimensional quartic map, Phys. Rev. A24, 2669.

355. Chang S-J, and Fendley P R (1986) Scaling and universal behavior in the bifurcation attractor, Phys. Rev. A33, 4092.

356. Chang S-J, and McCown J (1984) Universal exponents and fractal dimensions of Feigenbaum attractors, Phys.Rev. A30, 1149.

357. Chang S-J, and McCown J (1985) Universality behaviors and fractal dimensions associated with M-furcations, Phys. Rev. A31, 3791.

358. Chang S-J, and Shi K-J (1985) Time evolution and eigenstates of a quantum iterative system, Phys. Rev. Lett. 55, 269.

359. Chang S-J, and Wright J (1981) Transitions and distribution functions for chaotic systems, Phys. Rev. A23, 1419.

360. Chang Y F, Tabor M, Weiss J, and Corliss G (1981) On the analytic structure of the Hénon-Heiles system, Phys. Lett. 85A, 211.

361. Chavoya-Aceves O, Angulo-Brown F, and Piña E (1985) Symbolic dynamics of the cubic map, Physica 14D, 374.

362. Chay T R, and Rinzel J (1985) Bursting, beating, and chaos in an excitable membrane model, Biophys. J. 47, 357.

363. Chen C, Gyorgyi G, and Schmidt G (1986) Universal transition between Hamiltonian and dissipative chaos, Phys. Rev. A34, 2568.

364. Chen C, Gyorgyi G, and Schmidt G (1987a) Universal scaling in dissipative systems, Phys. Rev. A35, 2660.

365. Chen C, Gyorgyi G, and Schmidt G (1987b) Rapid convergence to the universal dissipative sequence in dynamical systems, Phys. Rev. A36, 5502.

366. Chen C, Gyorgyi G, and Schmidt G (1987c) Scaling in the circle map above criticality, Phys. Lett. 122A, 89.

367. Chen L-X, Li C-F, and Hong J (1984) Periodic and chaotic behaviors in optical bistability, Chinese Phys. Lett. 1, 85.

368. Chen P (1988a) Multiperiodicity and irregularity in growth cycle: a continuous model of monetary attractors, Mathl. Comput. Modelling, 10, 647.

369. Chen P (1988b) Empirical and theoretical evidence of economic chaos, Sys. Dyn. Rev. 81.

370. Chen Q, Meiss J D, and Percival I C (1987) Orbit extension method for finding unstable orbits, Physica 29D, 143.

371. Chen S-G, and Wang G-R (1988) The critical properties and scaling laws for circle map, Commun. Theor. Phys. 10,11.

372. Chen X-L, Wang Y-Q, and Chen S-G (1984) Period-doubling bifurcation and chaotic behaviour in non-equilibrium superconducting film, Solid State Commun. 52, 551.

373. Chernikov A A, Natenzon M Ya, Petrovichev B A, Sagdeev R Z, and Zaslavsky G M (1987) Some pecularities of stochastic layer and stochastic web formation, Phys. Lett. 122A, 39.

374. Chernikov A A, Sagdeev R Z, Usikov D A, Zakharov M Yu, and Zaslavsky G M (1987) Minimal chaos and stochastic webs, Nature, 326, 559.

375. Chernoff D F, and Barrow J D (1983) Chaos in the mixmaster universe, Phys. Rev. Lett. 50, 134.

376. Chillingworth D R J, and Holmes P J (1980) Dynamical systems and models for reversals of the Earth's magnetic field, J. Internat. Assoc. Math. Geol, 12, 41.

377. Chirikov B V (1979) A universal instability of many oscillator systems, Phys. Reports 52, 265.

378. Chirikov B V (1985) Intrinsic stochasticity, in *Proceedings of the 1984 International Conference on PLasma Physics*, vol. 2, 761.

379. Chirikov B V (1986) Transient chaos in quantum and classical mechanics, Found. Phys. 16, 39.

380. Chirikov B V, Israilev F M, and Shepelyansky D L (1981) Dynamical stochasticity in classical and quantum mechanics, in *Mathematical Physics Reviews*, ed. by S. P. Novikov, vol. 2, p. 209.

381. Chirikov B V, and Shepelyansky D L (1981) Stochastic oscillations of classical Yang-Mills fields, Pis'ma JETP, 34, 171.

382. Chirikov B V, and Shepelyansky D L (1982) Diffusion during multiple passage through a nonlinear resonance, Sov. Phys. Tech. Phys. 27, 156; Russ. Orig. 52, 238.

383. Chirikov B V, and Shepelyansky D L (1984) Correlation properties of dynamical chaos in Hamiltonian systems, Physica 13D, 395.

384. Chirikov B V, and Shepelyansky D L (1986) Localization of dynamical chaos in quantum systems, Radiophys. & Quantum Electr. 29, 787; Russ. Orig. 29, 1041.

385. Choi M Y, and Huberman B A (1983) Dynamical behavior of nonlinear networks, Phys. Rev. A28, 1204.

386. Christiansen P L (1984) Aspects of modern nonlinear dynamics: soliton and chaos phenomena, Radio Sci. (USA), 19, 1124.

387. Christoffel K M, and Brumer P (1985a) Vertical motion and 'scarred' eigenfunctions in the stadium billiard, Phys. Rev. A31, 3466.

388. Christoffel K M, and Brumer P (1985b) Quantum and classical dynamics in the stadium billiard, Phys. Rev. A33, 1309.

389. Chu S I (1987) Quasiperiodic and chaotic motions in intense field multiphoton processes, AIP Conf. Proc. 160, 282.

390. Chua L O, and Lin T (1988) Chaos in digital filters, IEEE Trans. on Cir. & Sys. CAS-35, 638.

391. Chua L O, Komuro M, and Matsumoto T (1986) The double scroll family, IEEE Trans. on Cir. & Sys. CAS-33, 1073.

392. Chui S T, and Ma K B (1982) Nature of some chaotic states for Duffing's equation, Phys. Rev. A26, 2262.

393. Cicogna G (1987) A theoretical prediction of the threshold for chaos in a Josephson junction, Phys. Lett. 121A, 403.

394. Cicogna G (1988) The onset of chaos, a soluble model, and the Josephson-junction equation, Phys. Lett. 131A, 98.

395. Cicogna G, and Papoff F (1987) Asymptotic Duffing equation and the appearance of 'chaos', Europhys. Lett. 3, 963.

396. Ciliberto S, Gollub J P (1985) Chaotic mode competition in parametrically forced surface waves, J. Fluid Mech. 158, 381.

397. Ciliberto S, and Bigazzi P (1988) Spatio-temporal intermittency in Raileigh-Bénard convection, Phys. Rev. Lett. 60, 286.

398. Ciliberto S, and Rubio M A (1987) Chaos and order in the temperature field of Rayleigh-Bénard convection, Phys. Scripta 36, 920.

399. Ciliberto S, and Simonelli F (1986) Spatial structures of temporal chaos in Rayleigh-Bénard convection, Europhys. Lett. 2, 285.

400. Coffman K G, McCormick W D, and Swinney H L (1986) Multiplicity in a chemical reaction with one-dimensional dynamics, Phys. Rev. Lett. 56, 999.

401. Cohen A, and Fishmen S (1988) Classical diffusion and quantal localization of a kicked particle in a well, Int. J. Mod. Phys. B2, 103.

402. Cohen A, and Procaccia I (1985) Computing the Kolmogorov entropy from time signals of dissipative and conservative dynamical systems, Phys. Rev. A31, 1872.

403. Collet P, Coullet P, and Tresser C (1985) Scenarios under constraint, J. Physique Lett. 46, L-143.

404. Collet P, Crutchfield J P, and Echmann J-P (1983) Computing the topological entropy of maps, Commun. Math. Phys. 88, 257.

405. Collet P, Eckmann J-P, and Koch H (1981a) Period doubling bifurcations for families of maps on $R^n$, J. Stat. Phys. 25, 1.

406. Collet P, Eckmann J-P, and Koch H (1981b) On universality for area-preserving maps of the plane, Physica 3D, 457.

407. Collet P, Eckmann J-P, and Lanford O E (1980) Universal properties of maps on an interval, Commun. Math. Phys. 76, 211.

408. Collet P, Eckmann J-P, and Thomas L (1981) A note on the power spectrum of the iterates of Feigenbaum's function, Commun. Math. Phys. 81, 261.

409. Collet P, Lebowitz J L, and Parzio A (1987) The dimension spectrum of some dynamical systems, J. Stat. Phys. 47, 609.

410. Collet P, and Eckmann J-P (1980) On the abundance of aperiodic behavior for maps on the interval, Commun. Math. Phys. 73, 115.

411. Collet P, and Eckmann J-P (1983) Positive Liapunov exponents and absolute continuity for maps of the interval, Ergod. Th. & Dyn. Sys. 3, 13.

412. Conte R, and Dubois M, Lyapunov exponents of experimental systems, in Leon, B1988, 767.

413. Coon D D, Ma S N, and Perera A G U (1987) Farey-fraction frequency modulation in the neuronlike output of silicon p-i-n diodes at 4.2K, Phys. Rev. Lett, 58, 1139.

414. Corbet A B (1988) Suppression of chaos in 1D maps, Phys. Lett. 130A, 267.

415. Coste J (1980) Iterations of transformation on the unit interval: approach to aperiodic attractor, J. Stat. Phys. 23, 521.

416. Coste J, and Peyraud N (1981) Two dimensional convection, oscillation of rolls with reversal of their sign of rotation, Phys. Lett. 84A, 17.

417. Coste J, and Peyraud N (1982) A new type of period-doubling bifurcations in one-dimensional transformations with two extrema, Physica 5D, 415.

418. Coullet P (1984) Chaotic behaviors in the unfolding of singular vector fields, Phys. Reports, 103, 95.

419. Coullet P, Elphick C, and Repaux D (1987) Nature of spatial chaos, Phys. Rev. Lett. 58, 431.

420. Coullet P, Tresser C, and Arneodo A (1979) Transition to stochasticity for a class of forced oscillators, Phys. Lett. 72A, 268.

421. Coullet P, and Elphick C (1987) Topological defects dynamics and Melnikov's theory, Phys. Lett. 121A, 233.

422. Coullet P, and Repaux (1987) Strong resonances of periodic patterns, Europhys. Lett. 33, 573.

423. Coullet P, and Tresser C (1978) Iterations d'endomorphismes et groupe de renormalisation, C. R. Acad. Sci. Paris, 287, 577; J. Physique, 39, Coll. C5-25.

424. Coullet P, and Tresser C (1980) Critical transition to stochasticity for some dynamical systems, J. Physique Lett. 41, L255.

425. Coullet P, and Vanneste C (1983) Scenarios for the onset of chaos, Helv. Phys. Acta, 56, 813.

426. Courtemanche M, Glass L, Bélair J, Scagliotti D, and Gordon D (1989) A circle map in a human heart, Physica 40D, 299.

427. Cramer E, and Lauterborn W (1982) Acoustica cavitation noise spectra, Appl. Scientific Res. 38, 209.

428. Crawford J D, and Knobloch E (1988) Symmetry-breaking bifurcations in $O(2)$ maps, Phys. Lett. 28A, 27.

429. Crawford J D, and Omohundro S (1984) On the global structure of period doubling flows, Physica 13D, 161.

430. Cremers J, and Hubler A (1987) Construction of differential equations from experimental data, Z. Naturforsch. 42a, 797.

431. Crutchfield J P, Farmer J D, Packard N H, and Shaw R S (1980) Power spectral analysis of a dynamical system, Phys. Lett. 76A, 1.

432. Crisanti A, Paladin G, and Vulpiani A (1988) Generalized Lyapunov exponents in high-dimensional chaotic dynamics and products of large random matrices, J. Stat. Phys. 53, 583.

433. Cronemeyer D C, Chi C C, Davidson A, and Pedersen N F (1985) Chaos, noise,and tails on the I-V curve steps of rf-driven Josephson junctions, Phys. Rev. B31, 2667.Ω

434. Croquette V, and Poitou C (1981) Cascade of period doubling bifurcations and large stochasticity in the motions of a compass, J. Physique Lett. 42, 537.

435. Crutchfield J P, Farmer J D, Packard N H, and Shaw R S (1986) Chaos, Sci. Amer. 255, No.6, 38.

436. Crutchfield J P, Farmer J D, and Huberman B A (1982) Fluctuations and simple chaotic dynamics, Phys. Reports, 92, 45.

437. Crutchfield J P, Nauenberg M, and Rudnick J (1981) Scaling for external noise at the onset of chaos, Phys. Rev. Lett. 46, 933.

438. Crutchfield J P, and Huberman B A (1980) Fluctuations and the onset of chaos, Phys. Lett. 77A, 407.

439. Crutchfield J P, and Kaneko K (1988) Are attractors relevant to turbulence? Phys. Rev. Lett. 60, 2715.

440. Crutchfield J P, and Packard N H (1982) Symbolic dynamics of 1D maps: entropies, finite precision, and noise, Int. J. Theor. Phys. 21, 433.

441. Csordas A, and Szépfalusy P (1988) Generalized entropy decay rates of one-dimensional maps, Phys. Rev. A38, 2582.

442. Cumming A, and Linsay P S (1987) Deviations from universality in the transition from quasiperiodicity to chaos, Phys. Rev. Lett. 59, 1633.

443. Cumming A, and Linsay P S (1988) Quasiperiodicity and chaos in a system with three competing frequencies, Phys. Rev. Lett. 60, 2719.

444. Curry J H (1978) A generalized Lorenz system, Commun. Math. Phys. 60, 193.

445. Curry J H (1979a) On the Hénon transformation, Commun. Math. Phys. 68, 129.

446. Curry J H (1979b) Chaotic response to periodic modulation of a model of a convecting fluid, Phys. Rev. Lett. 4, 1013.

447. Curry J H (1980) On some systems motivated by the Lorenz equations: numerical results, Lect. Notes in Phys. 116, 316.

448. Curry J H (1981) On computing the entropy of the Hénon attractor, J. Stat. Phys. 26, 683.

449. Curry J H, and Johnson J R (1982) On the rate of approach to homoclinic tangency, Phys. Lett. 92, 5.

450. Cvitanovic P (1988) Invariant measurement of strange sets in terms of cycles, Phys. Rev. Lett. 61, 2729.

451. Cvitanovic P, Gunaratne G H, and Procaccia I (1988) Topological and metric properties of Hénon-type strange attractors, Phys. Rev. A38, 1503.

452. Cvitanovic P, Jensen M H, Kadanoff L P, and Procaccia I (1985) Renormalization, unstable manifolds, and the fractal structure of mode locking, Phys. Rev. Lett. 55, 343.

453. Cvitanovic P, Shraiman B, and Soderberg B (1985) Scaling laws for mode-locking in circle maps, Phys. Scripta 32, 263.

454. Cvitanovic P, and Myrheim J (1983) Universality for period n-tuplings in complex mappings, Phys. Lett. 94A, 329.

455. D'Humieres D, Beasley M R, Huberman B A, and Libchaber A (1982) Chaotic states and routes to chaos in the forced pendulum, Phys. Rev. A26, 3483.

456. Da Costa L N, Knobloch E, and Weiss N O (1981) Oscillations in double diffusive convection, J. Fluid Mech, 109, 25.

457. Da Silva Ritter, Ozorio de Almeida, and Douady R (1987) Analytical determination of unstable periodic orbits in area-preserving maps, Physica 29D, 181.

458. Daido H (1980) Analytic conditions for the appearance of homoclinic and heteroclinic points of a 2-dimensional mapping: the case of the Hénon mapping, Prog. Theor. Phys, 63, 1190; 1831.

459. Daido H (1981a) Theory of the period-doubling phenomenon of one-dimensional mappings based on the parameter dependence, Phys. Lett, 83A, 246.

460. Daido H (1981b) Universal relation of a band splitting sequence to a preceding period-doubling one, Phys. Lett, 86A, 259.

461. Daido H (1982a) Period-doubling bifurcations and associated universal properties including parameter dependence, Prog. Theor. Phys, 67, 1698.

462. Daido H (1982b) On the scaling behavior in a map of a circle onto itself, Prog. Theor. Phys. 68, 1935.

463. Daido H (1983a) Nonuniversal accumulation of bifurcations leading to homoclinic tangency, Prog. Theor. Phys, 69, 1304.

464. Daido H (1983b) Resonance and intermittent transition from torus to chaos in periodically forced systems near intermittency threshold, Prog. Theor. Phys, 70, 879.

465. Daido H (1984a) Onset of intermittency from torus, Prog. Theor. Phys. 71, 402.

466. Daido H (1984b) Coupling sensitivity of chaos, Prog. Theor. Phys. 72, 853.

467. Daido H (1985a) Cliff: sudden enhancement or enfeeblement of chaos in dissipative dynamical systems, Phys. Lett. 108A, 233.

468. Daido H (1985b) Coupling sensitivity of chaos and the Lyapunov dimension: the case of coupled two-dimensional map, Phys. Lett. 110A, 5.

469. Daido H (1987) Coupling sensitivity of chaos: theory and further numerical evidence, Phys. Lett. 121A, 60.

470. Daido H, and Haken H (1985) Cliffs and its inversion as rapid decay of fully developed chaos, Phys. Lett. 111A, 211.

471. Dana I, and Fishman S (1985) Diffusion in the standard map, Physica 17D, 63.

472. Dana I, and Reinhardt W P (1987) Adiabatic invariance in the standard map, Physica 28D, 115.

473. Dangoisse D, Bekkali A, Papoff F, and Glorieux P (1988) Shilnikov dynamics in a passive Q-switching laser, Europhys. Lett. 6, 335.

474. Davis P, and Ikeda K (1984) $T^3$ in a model of a nonlinear optical resonantor, Phys. Lett. 100A, 455.

475. Day R H (1983) The emergence of chaos from classical economic growth, Quart. J. Econ. 201.

476. De Aguiar M A M, Malta C P, Baranger M, and Davies K T R (1987) Bifurcations of periodic trajectories in nonintegrable Hamiltonian systems with two degrees of freedom: numerical and analytical results, Ann. Phys. 180, 167.

477. De Aguiar M A M, and Baranger M (1988) Invariant tori for two-dimensional nonintegrable Hamiltonians, Ann. Phys. 186, 355.

478. De Fillipo S, and Fusco Girard M (1988) Numerical evidence of a sharp order window in a Hamiltonian system, Physica 29D, 421.

479. De Leon N, and Berne B J (1981) Intramoleular rate process: isomerization dynamics and the transition to chaos, J. Chem. Phys, 75, 3495.

480. De Leon N, and Berne B J (1982a) Reaction dynamics in an ergodic system: the Siamese stadium billiard, Chem. Phys. Lett, 93, 162.

481. De Leon N, and Berne B J (1982b) Reaction dynamics in a non-ergodic system: the Siamese stadium billiard, Chem. Phys. Lett, 93, 169.

482. De Oliveira C R (1988) Attractors and charateristic exponents, J. Stat. Phys. 53, 603.

483. De Oliveira C R, and Malta C P (1987) Bifurcations in a class of time-delay equations, Phys. Rev. A36, 3997.

484. De Sousa Vieira M C (1988) Scaling factors associated with M-furcations of the $1 - \mu|x|^z$ map, J. Stat. Phys. 53, 1315.

485. De Sousa Vieira M C, Lazo E and Tsallis C (1987) New road to chaos, Phys. Rev. A35, 945.

486. Decroly O, and Goldbeter A (1982) Birthythmicity, chaos, and other patterns of temporal self-organization in a multiply regulated biochemical system, Proc. Natl. Acad. Sci. USA, 79, 6917.

487. Decroly O, and Goldbeter A (1985) Selection between multiple periodic regimes in a biochemical system: complex dynamic behaviour resolved by use of one-dimensional maps, J. Theor. Biol. 113, 649.

488. Decroly O, and Goldbeter A (1987) From simple to complex oscillatory behaviour: analysis of bursting in a multiply regulated biochemical system, J. Theor. Biol. 124, 219.

489. Deissler R J (1984a) Turbulent solutions of the equations of fluid motion, Rev. Mod. Phys. 56, 223.

490. Deissler R J (1984b) One-dimensional strings, random fluctuations, and complex chaotic structures, Phys. Lett. 100A, 451.

491. Deissler R J (1986) Is Navier-Stokes turbulence chaotic? Phys. Fluids, 29, 1453.

492. Deissler R J (1987) Spatially growing waves, intermittency, and convective chaos in an open-flow system, Physica 25D, 23.

493. Deissler R J, and Kaneko K (1987) Velocity-dependent Liyapunov exponents as a measure of chaos for open-flow systems, Phys. Lett. 119, 397.

494. Dekker H (1986a) Coherent tunnelling, squeezing, chaotic behavior and fractal dimension in a bistable potential, Phys Lett. 119A, 10.

495. Dekker H (1986b) The coherent tunneling propagator and chaotic bistability, J. Phys. A19, L1137.

496. Delande D, and Gay J C (1986) The quantum analog of chaos in the diamagnetic Kepler problem, Comments At. & Mol. Phys. 19, 35.

497. Delande D, and Gay J C (1987) Scars of symmetries in quantum chaos, Phys. Rev. Lett. 59, 1809.

498. Delbourgo R, Hart W, ank Kenny B G (1985) Dependence of universal constants upon multiplication periods in nonlinear maps, Phys. Rev. A31, 514.

499. Delbourgo R, Hughs P, and Kenny B G (1987) Islands of stability and complex universality relations, J. Math. Phys. 28, 60.

500. Delbourgo R, and Kenny B G (1985) Universal features of tangent bifurcation, Aust. J. Phys. 38, 1.

501. Delbourgo R, and Kenny B G (1986) Universality relations, Phys. Rev. A33, 3292.

502. Demaret J, De Rop Y, and Henneaux M (1988) Chaos in non-diagonal spatially homogeneous cosmological models in spacetime dimension $\leq$ 10, Phys. Lett. 211B, 37.

503. Deneckere R, and Pelikan S (1986) Competitive chaos, J. Economic Theory, 40, 14.

504. Deng Z, and Hioe F T (1985) Chaos-order-chaos transitions in a two-dimensional Hamiltonian system, Phys. Rev. Lett. 55, 1539.

505. Denker M, and Keller G (1986) Rigorous statistical procedures for data from dynamical systems, J. Stat. Phys. 44, 67.

506. Derighetti B, Ravani M, Stoop R, Meier P F, Brun E, and Badii R (1985) Period-doubling lasers as small-signal detectors, Phys. Rev. Lett. 55, 1746.

507. Derrida B, Gervois A, and Pomeau Y (1978) Iteration of endomorphisms on the real axis and representation of numbers, Ann. Inst. Henri Poincaré, 29A, 305.

508. Derrida B, Gervois A, and Pomeau Y (1979) Universal metric properties of bifurcations of endomorphisms, J. Phys. A12, 269.

509. Derrida B, and Pomeau Y (1980) Feigenbaum's ratios of two-dimensional area preserving maps, Phys. Lett, 80A, 217.

510. Destexhe A, Sepulchre J A, and Babloyantz A (1988) A comparative study of the experimental quantification of deterministic chaos, Phys. Lett. 132A, 101.

511. Deutsch J M (1985) The effect of noise on iterated maps, J. Phys. A18, 1457.

512. Devaney R L (1984a) A piecewise linear model for the zones of instability of an area-perserving map, Physica 10D, 387.

513. Devaney R L (1984b) Bursts into chaos, Phys. Lett. 104A, 385.

514. Devaney R L, and Nitecki Z (1979) Shift automorphisms in the Hénon mapping, Commun. Math. Phys. 67, 137.

515. Diamond P (1987) Nonlinear integral operators and chaos in Banach spaces, Bull. Austral. Math. Soc. 35, 275.

516. Dias De Deus J, Dilao R, and Taborta Durate J (1982) Topological entropy and approaches to chaos in dynamics of the interval, Phys. Lett, 90A, 1.

517. Dias De Deus J, Dilao R, and Taborta Durate J (1983) Topological entropy, characteristic exponents, and scaling behavior in dynamics of the interval, Phys. Lett, 93A, 1.

518. Dias de Deus J, Dilao R, and Noronha da Costa A (1984) Intermittency and sequence of periodic regimes in one-dimensional maps of the interval, Phys. Lett. 101A, 459.

519. Dias de Deus J, Dilao R, and Noronha da Costa A (1987) Scaling behavior of windows and intermittency in one-dimensional maps, Phys. Lett. 124A, 433.

520. Dias de Deus J, and Norouha da Costa A (1987) Symbolic approach to intermittency, Phys. Lett. 120A, 19.

521. Dias de Deus J, and Taborda Duarte J M (1984) Evolution patterns and iterative maps, Phys. Lett. 102A, 149.

522. Dias de Deus J, and Taborta Duarte J (1982) On the approach to the final aperiodic regime in maps of the interval, Commun. Math. Phys. 84, 251.

523. Diener M (1984) The canard unchained, or how fast/slow dynamical systems bifurcate, Math. Intelligencer, 6, No.3, 38.

524. Ding E-J (1986a) Exact treatment of a nonlinear driven oscillator, Phys. Scripta, T14, 89.

525. Ding E-J (1986b) Analytic treatment of periodic orbit: systematics for a nonlinear driven oscillator, Phys. Rev, A34, 3547.

526. Ding E-J (1987a) Analytic treatment of a driven oscillator with a limit cycle, Phys. Rev. A35, 2669.

527. Ding E-J (1987b) Structure of parameter space for a prototype nonlinear oscillator, Phys. Rev, A36, 1488.

528. Ding E-J (1987c) Scaling behavior in the supercritical sine circle map, Phys. Rev. Lett, 58, 1059.

529. Ding E-J (1988a) Structure of the parameter space 'for the van der Pol oscillator, Phys. Scripta 38, 9.

530. Ding E-J (1988b) Wave numbers for unimodal maps, Phys. Rev. A37, 1827.

531. Ding E-J, and Hemmer P C (1987) Exact treatment of mode locking for a piecewise linear map, J. Stat. Phys, 46, 99.

532. Ding E-J, and Hemmer P C (1988) Winding numbers for the supercritical sine circle map, Physica 32D, 153.

533. Ding M-Z, Grebogi C, and Ott E (1989a) Evolution of attractors in quasiperiodically forced systems: from quasiperiodic to strange nonchaotic to chaotic, Phys. Rev. A39, 2593.

534. Ding M-Z, Grebogi C, and Ott E (1989b) Dimensions of strange nonchaotic attractors, Phys. Lett. A137, 167.

535. Ding M-Z, Hao B-L, and Hao X (1985) Power spectrum analysis and the nomenclature of periods in the Lorenz model, Chinese Phys. Lett. 2, 1.

536. Ding M-Z, and Hao B-L (1988) Systematics of the periodic windows in the Lorenz model and its relation with the antisymmetric cubic map, Commun. Theor. Phys, 9, 375.

537. Dmitriev A S, Gulyaev Yu V, Kislov V Yu, and Panas A I (1988) Mode pulling and competition in a system with chaotic dynamics, Phys. Lett. 128A, 172.

538. Dmitriev A S, and Panas A I (1986) Strange attractors in self-excited ring systems with initial links, Sov. Phys. Tech. Phys. 31, 460; Russ. Orig. 56, 759.

539. Dmitriev A S, and Panas A I (1987) Mode competition and pulling in a system with strange attractors, Sov. Tech. Phys. Lett. 13, 295; Russ. Orig. 13, 713.

540. Doering C R, Gibbon J D, Holm D D, and Nicolaenko B (1987) Exact Lyapunov dimension of the universal attractor for the complex Ginzberg-Landau equation, Phys. Rev. Lett, 59, 2911.

541. Dolnik M, Schreiber I, and Marek M (1986) Dynamic regimes in a periodically forced reaction cell with oscillating chemical reaction, Physica 21D, 78.

542. Dombre T, Frisch U, Greene J M, Hénon M, Mehr A, and Soward A M (1986) Chaotic streamlines in the ABC flows, J. Fluid Mech. 167, 35.

543. Donnelly R J, Park I, Shaw R S, and Walden R W (1980) Early nonperiodic transitions in Couette flow, Phys. Rev. Lett, 44, 987.

544. Doolen G D, DuBois D F, Rose H A, and Hafizi B (1983) Coherence in chaos and caviton turbulence, Phys. Rev. Lett, 51, 335.

545. Dorfle M (1985) Spectrum and eigenfunctions of the Frobenius – Perron operator of the tent map, J. Stat. Phys. 40, 93.

546. Dorizzi B, Grammaticos B, Le Berre M, Pomeau Y, Ressayre E, and Tallet A (1987) Statistics and dimension of chaos in difference delay systems, Phys. Rev. A35, 328.

547. Dorizzi B, Grammaticos B, and Pomeau Y (1984) The periodically kicked rotator: recurrence and/or energy growth, J. Stat. Phys. 37, 93.

548. Doron E, and Fishmen S (1988) Anderson localization for a two-dimensional rotor, Phys. Rev. Lett. 60, 867.

549. Doron E, and Fishmen S (1988) Perturbative calculation of the diffusion coefficient for a multidimensional kicked rotor, Phys. Rev. A37, 2144.

550. Doveil F, and Escande D F (1981) Destabilization of an enumerable set of cycles in a Hamiltionian system, Phys. Lett, 84A, 399.

551. Doveil F, and Escande D F (1982) Fractal diagrams for non-integrable Hamiltonians, Phys. Lett, 90A, 226.

552. Dowell E H (1982) Flutter of a buckled plate as an example of chaotic motion of a deterministic autonomous system, J. Sound and Vib. 85, 333.

553. Dowell E H (1984) Observation and evolution of chaos for an autonomous system, Trans. ASME J. Appl. Mech. 51, 664.

554. Dowell E H (1988) Chaotic oscillations in mechanical systems, Comput. Struct. 30, 171.

555. Dowell E H, and Pezeshki C (1986) On the understanding of chaos in Duffing's equations including a comparison with experiments, Trans. ASME J. Appl. Mech. 53, 5.

556. Du B-S (1987) Topological entropy and chaos of iterated maps, Nonlin. Anal. Th. Meth. & Appl. 11, 105.

557. Duan J C (1987) Asymptotic decay rates from the growth properties of Liapunov function near singular attractors, J. Math. Anal. & Appl. 125, 6.

558. Dubois M, Berge P, and Croquette V (1981) Study of non-steady convection regimes using Poincaré sections, J. Physique Lett. 43, L295.

559. Dubois M, Rubio M A, and Berge P (1983) Experimental evidence of intermittencies associated with a subharmonic bifurcation, Phys. Rev. Lett, 51, 1446-9; Erratum: 51, 2345.

560. Dubois M, and Berge P (1986) Rotation number dependence at onset of chaos in free Rayleigh-Bénard convection, Phys. Scripta 33, 159.

561. Dumont R S, and Brumer P (1988) Characteristics of power spectra for regular and chaotic systems, J. Chem. Phys. 88, 1481.

562. Eckhardt B (1988) Quantum mechanics of classically non-integrable systems, Phys. Reports, 163, 205.

563. Eckmann J-P (1981) Roads to turbulence in dissipative dynamical systems, Rev. Mod. Phys. 53, 643.

564. Eckmann J-P, Oliffson Kamphorst S, Ruelle D, and Ciliberto S (1986) Liapunov exponents from time series, Phys. Rev. A34, 4971.

565. Eckmann J-P, Oliffson Kamphorst S, and Ruelle D (1987) Recurrence plots of dynamical systems, Europhys. Lett. 4, 973.

566. Eckmann J-P, Koch H, and Wittwer P (1982) Existence of a fixed point of the doubling transformation for area-preserving maps of the plane, Phys. Rev. A26, 720.

567. Eckmann J-P, Thomas L E, and Wittwer P (1981) Intermittency in the presence of noise, J. Phys. A14, 3153.

568. Eckmann J-P, and Procaccia I (1986) Fluctuations of dynamical scaling indices in nonlinear systems, Phys. Rev. A34, 659.

569. Eckmann J-P, and Ruelle D (1985) Ergodic theory of chaos and strange attractors, Rev. Mod. Phys. 57, 617.

570. Eckmann J-P, and Thomas L E (1982) Remarks on stochastic resonances, J. Phys. A15, L261.

571. Eckmann J-P, and Wayne C E (1988) Liapunov spectra for infinite chains of nonlinear oscillators, J. Stat. Phys. 50, 853.

572. Eckmann J-P, and Wittwer P (1987) A complete proof of the Feigenbaum conjectures, J. Stat. Phys. 46, 455.

573. Edelen D G (1985) Semi-inverse methods for obtaining partial differential equations with chaotic solutions, Int. J. Eng. Sci. 23, 331.

574. Eidson J, Flynn S, Holm C, Weeks D , and Fox R F (1986) Elementary explanation of boundary shading in chaotic-attractor plots for the Feigenbaum map and the circle map, Phys. Rev. A33, 2809.

575. Eidson J, and Fox R F (1986) Quantum chaos in a two-level system in a semiclassical radiation field, Phys. Rev. A34, 3288.

576. Elgin J N, and Forster D (1983) Mechanism for chaos in the Duffing's equation, Phys. Lett. 94A, 195.

577. El Naschie M.S, and Al Athel S (1989) On the connection between statical and dynamical chaos, Z. Naturfor. 44a, 449.

578. Elskens Y, and Henneaux M (1987) Chaos in Kaluza-Klein models, Class. & Quantum Grav. 4, L161.

579. Endo T, and Chua L O (1988) Chaos from phase-locked loop, IEEE Trans. on Cir. & Sys. CAS-35, 987.

580. Epstein H, and Lascoux J (1981) Analyticity properties of the Feigenbaum equation, Commun. Math. Phys. 81, 437.

581. Epstein I R (1983) Oscillations and chaos in chemical systems, Physica 7D, 47.

582. Erber T, Johnson P, and Everett P (1981) Cebysev mixing and harmonic oscillator models, Phys. Lett. 85A, 61.

583. Ermentrout G B (1984) Period doublings and possible chaos in neutral models, SIAM J. Appl. Math. 44, 80.

584. Escande D F (1982a) Large-scale stochasticity in Hamiltonian systems, Phys. Scripta T2, 126.

585. Escande D F (1982b) Renormalization approach to nonintegrable Hamiltonians, Phys. Lett. 91A, 327.

586. Escande D F (1982c) Renormalization for stochastic layers, Physica 6D, 119.

587. Escande D F (1985) Stochasticity in classical Hamiltonian systems: universal aspects, Phys. Reports, 121, 165.

588. Escande D F, Mohamed-Benkadda M S, and Doveil F (1984) Threshold of global stochasticity and universality in Hamiltonian systems, Phys. Lett. 101A, 309.

589. Escande D F, and Doveil F (1981a) Renormalization method for computing the threshold of the large-scale stochastic instability in two degrees of freedom Hamiltonian systems, J. Stat. Phys. 26, 257.

590. Escande D F, and Doveil F (1981b) Renormalization method for the onset of stochasticity in a Hamiltonian system, Phys. Lett. 83A, 307.

591. Escande D F, and Mehr A (1982) Link between KAM tori and nearby cycles, Phys. Lett. 91A, 327.

592. Essex C, Lookman T, and Nerenberg M A H (1987) The climate attractor over short time scale, Nature, 326, 64.

593. Everson R M (1986) Chaotic dynamics of a bouncing ball, Physica 19D, 355.

594. Everson R M (1987) Scaling of intermittency period with dimension of a partition boundary, Phys. Lett. 122A, 471.

595. Eykholt R, and Umberger D K (1986) Characterization of fat fractals in nonlinear dynamical systems, Phys. Rev. Lett. 57, 2333.

596. Eykholt R, and Umberger D K (1988) Relating the various scaling exponents used to characterize fat fractals in nonlinear dynamical systems, Physica 30D, 43.

597. Fairlie D B, and Siegwart D K (1988) Classical billiards in a rotating boundary, J. Phys. A21, 1157.

598. Falcioni M, Paladin G, and Vulpiani A (1988) Regular and chaotic motion of fluid particles in a two-dimensional fluid, J. Phys. A21, 3451.

599. Falconer K J, and Mash D T (1988) The dimension of affine-invariant fractals, J. Phys. A21, L121.

600. Falk H (1984) Evolution of the density for a chaotic map, Phys. Lett. 105A, 101.

601. Farantos S C, and Tennyson J (1985) Quantum and classical vibrational chaos in floppy molecules, J. Chem. Phys. 82, 800.

602. Farmer J D (1981) Spectral broadening of period doubling bifurcation sequences, Phys. Rev. Lett. 47, 179.

603. Farmer J D (1982a) Chaotic attractors of an infinite dimensional systems, Physica 4D, 366.

604. Farmer J D (1982b) Information dimension and the probabilistic structure of chaos, Z. Naturforsch. 37a, 1304.

605. Farmer J D (1985) Sensitive dependence on parameters in nonlinear dynamics, Phys. Rev. Lett. 55, 351.

606. Farmer J D, Hart J, and Weidman P (1982) A phase space analysis of baroclinic flow, Phys. Lett. 91A, 22.

607. Farmer J D, and Satija I I (1985) Renormalization of the quasiperiodic transition to chaos for arbitrary winding numbers, Phys. Rev. A31, 3520.

608. Farmer J D, and Sidorowich J J (1987) Predicting chaotic time series, Phys. Rev. Lett. 59, 845.

609. Fauve S, Laroche C, Libchaber A, and Perrin B (1984) Chaotic phases and magnetic order in a convective fluid, Phys. Rev. Lett. 52, 1774.

610. Fauve S, Laroche C, and Libchaber A (1981) Effect of a horizontal magnetic field on convective instabilities in mercury, J. Physique Lett. 42, 455.

611. Feigenbaum M J (1978) Quantitative universality for a class of nonlinear transformations, J. Stat. Phys. 19, 25.

612. Feigenbaum M J (1979) The universal metric properties of nonlinear transformations, J. Stat. Phys. 21, 69.

613. Feigenbaum M J (1980a) Universal behavior in nonlinear systems, Los Alamos Sci. Summer, 4.

614. Feigenbaum M J (1980b) The onset spectrum of turbulence, Phys. Lett. 74A, 375.

615. Feigenbaum M J (1980c) The transition to aperiodic behavior in turbulent systems, Commun. Math. Phys. 77, 65.

616. Feigenbaum M J (1987a) Scaling spectra and return times of dynamical systems, J. Stat. Phys. 46, 925.

617. Feigenbaum M J (1987b) Some characterizations of strange sets, J. Stat. Phys. 46, 919.

618. Feigenbaum M J (1988) Presentation functions, fixed points, and a theory of scaling function dynamics, J. Stat. Phys. 52, 527.

619. Feigenbaum M J, Jensen M H, and Procaccia I (1986) Time ordering and thermodynamics of strange sets: theory and experimental tests, Phys. Rev. Lett. 57, 1503.

620. Feigenbaum M J, Kadanoff L P, and Shenker S J (1982) Quasiperiodicity in dissipative systems: a renomalization group analysis, Physica 5D, 370.

621. Feigenbaum M J, and Hasslacher B (1982) Irrational decimations and path intergrals for external noise, Phys. Rev. Lett. 49, 605.

622. Feingold M, Kadanoff L P, Piro O (1988) Passive scalars, three-dimensional volume-preserving maps, and chaos, J. Stat. Phys. 50, 529.

623. Feingold M, Moiseyev N, and Peres A (1984) Ergodicity and mixing in quantum theory. II. Phys. Rev. A30, 509.

624. Feingold M, Moiseyev N, and Peres A (1985) Classical limit of quantum chaos, Chem. Phys. Lett. 117, 344.

625. Feingold M, and Fishman S (1987) Statistics of quasi-energies in chaotic and random systems, Physica 25D, 181.

626. Feingold M, and Peres A (1983) Regular and chaotic motion of coupled rotators, Physica 9D, 433.

627. Feingold M, and Peres A (1985) Regular and chaotic propagators in quantum theory, Phys. Rev. A31, 2472.

628. Feingold M, and Peres A (1986) Distribution of matrix elements of chaotic systems, Phys. Rev. A34, 591.

629. Feit S D (1978) Characterisic exponents snd strange attractors, Commun. Math. Phys. 61, 249.

630. Feit S D, and Fleck J A (1984) Wave packet dynamics and chaos in the Hénon-Heils system, J. Chem. Phys. 80, 2578.

631. Fenstermacher P R, Swinney H. L, and Gollub J P (1979) Dynamical insbilities and the transition to chaotic Taylor vortex flow, J. Fluid Mech. 94, 103.

632. Feroe J A (1982) Existence and stability of multiple impulse solutions of a nerve equation, SIAM J. Appl. Math. 42, 235.

633. Ferritti A, and Rahman N K (1987) Coupled logistic maps in physico-chemical processes: coexisting attractors and their implications, Chem Phys. Lett. 140, 71.

634. Fesser K, McLaughlin D W, Bishop A R, and Holian B L (1985) Chaos and nonlinear modes in a perturbed Toda chain, Phys. Rev. A31, 2728.

635. Fiel D (1987) Scaling for period doubling sequences with correlated noise, J. Phys. A20, 3209.

636. Finn J M, and Ott E (1988) Chaotic flows and magnetic dynamics, Phys. Rev. Lett. 60, 760.

637. Firth W J (1987) Optically bistable arrays and chaotic dynamics, Phys. Lett. 125A, 375.

638. Firth W J (1988) Optical memory and spatial chaos, Phys. Rev. Lett. 61, 329.

639. Firth W J, and Wrigth E M (1982) Oscillation and chaos in a Fabry-Perot bistable cavity with Gaussian input beam, Phys. Lett. 92, 211.

640. Fishman S, Grempel D R, and Prange R E (1982) Chaos, quantum recurrences, and Anderson localization, Phys. Rev. Lett. 49, 509.

641. Fiszden W, and Sen M (1988) Initial behaviour of solutions to the Lorenz equations, Int. J. Nonlin. Mech. 23, 53.

642. Flaherty J E, and Hoppensteadt F C (1978) Frequency entrainment of a forced van der Pol oscillator, Stud. Appl. Math. 58, 5.

643. Foias C, Jolly M S, Kevrekidis I G, Sell G R, and Titi E S (1988) On the computation of inertial manifold, Phys. Lett. 131A, 433.

644. Foias C, Manley O, and Temam R (1987) Attractors for the Bénard problem: existence and physical bounds on their fractal dimension, Nonlin. Anal. Th. Meth. & Appl. 11, 939.

645. Foias C, Sell G R, and Teman R (1988) Inertial manifold for nonlinear evolution equations, J. Diff. Eqs. 73, 309.

646. Ford J (1973) The transition form analytic dynamics to statistical mechanics, Adv. Chem. Phys. 24, 155.

647. Ford J (1986) What is chaos, that we should be mindful of it? in *The New Physics*, ed. by S. Capelin and P. C. M. Devies, Cambridge University Press.

648. Ford J, and Lunsford G H (1970) Stochastic behavior of resonant nearly linear oscillator systems as the nonlinear coupling approaches zero, Phys. Rev. A1, 59.

649. Ford J, and Lunsford G H (1972) On the stability of periodic orbits for nonlinear oscillator systems in regions exhibiting stochastic behavior, J. Math. Phys. 13, 700.

650. Fournier J D, Levine G, and Tabor M (1988) Singularity clustering in the Duffing oscillator, J. Phys. A21, 33.

651. Fowler A C, Gibbon J D, and McGuinness M J (1982) The complex Lorenz equations, Physica 4D, 139.

652. Fowler A C, and McGuinness M J (1982a) A description of the Lorenz attractor at high Prandtl number, Physica 5D, 149.

653. Fowler A C, and McGuinness M J (1982b) Hysteresis in the Lorenz equations, Phys. Lett. 92A, 103.

654. Fowler A G (1984) The use of the method of averaging in predicting chaotic motion, Phys. Lett. 100A, 1.

655. Fox R F, and Eidson J C (1987) Systematic corrections to the rotating-wave approximation and quantum chaos, Phys. Rev. A36, 4321.

656. Fradkin E, Hernandez O, Huberman B A, and Pandit R (1983) Periodic, incommensurate, and chaotic states in a continuum statistical mechanics model, Nucl. Phys. B215(FS7), 137.

657. Frahm H, and Mikeska H J (1985) On the dynamics of a quantum system which is classically chaotic, Z. Phys. B60, 117.

658. Frahm H, and Mikeska H J (1988) Quantum suppression of irregularity in the spectral properties of the kicked rotator, Phys. Rev. Lett. 60, 3.

659. Franaszek M (1984) Effect of random noise on the deterministic choas in a dissipative system, Phys. Lett. 105A, 383.

660. Franaszek M (1987) Chaotic, nonstrange attractors in the presence of external, random noise, Phys. Rev. A35, 3162.

661. Franaszek M, and Kowalik Z J (1986) Measurements of the dimension of the strange attractor for the Fermi-Ulam problem, Phys. Rev. A33, 3508.

662. Franaszek M, and Pieranski P (1985) Jumping particale model. Critical slowing down near the bifurcation point, Can. J. Phys. 63, 488.

663. Franceschini V (1980) A Feigenbaum sequence of bifurcations in the Lorenz model, J. Stat. Phys. 22, 397.

664. Franceschini V (1983) Bifurcations of tori and phase locking in a dissipative system of differential equations, Physica 6D, 285.

665. Franceschini V, Giberti C, and Nicolini M (1988) Common periodic behavior in larger and larger truncations of the Navier-Stokes equations, J. Stat. Phys. 50, 879.

666. Franceschini V, and Russo L (1981) Stable and unstable manifolds of the Hénon mapping, J. Stat. Phys. 25, 757.

667. Franceschini V, and Tebaldi C (1979) Sequences of infinite bifurcations and turbulence in a five-mode truncation of the Navier-Stokes equations, J. Stat. Phys. 21, 707.

668. Franceschini V, and Tebaldi C (1981) A seven-mode truncation of the plane incompressible Navier-Stokes equations, J. Stat. Phys. 25, 397.

669. Franceschini V, and Tebaldi C (1984) Breaking and disappearance of tori, Commun. Math. Phys. 94, 17.

670. Franciosi C (1986) Oscillations of torus and collison torus-chaos in a delayed circle map, Prog. Theor. Phys. 76, 302.

671. Fraser A M (1989a) Reconstructing attractors from scalar time series: a comparison of singular system and redundancy criteria, Physica 34D, 391-404.

672. Fraser A M (1989b) Information and entropy in strange attractors, IEEE Trans. Information Th. 35, 245-62.

673. Fraser A M (1989c) Hidden Markov models and dynamical systems, IEEE Trans. Acoustics, Speech and Signal Processing, 37, 962.

674. Fraser A M, and Swinney H L (1986) Independent coordinates for strange attractors from mutual information, Phys. Rev. A33, 1134.

675. Fraser S, Celarier E, and Karpal R (1983) Stochastic dynamics of the cubic map: a study of noise-induced transition phenomena, J. Stat. Phys. 33, 341.

676. Fraser S, and Kapral R (1981) A resonance model for chaos near period three, Phys. Rev. A23, 3303.

677. Fraser S, and Kapral R (1982a) Behavior of the Lyapunov number near period three, Phys. Rev. A25, 2827.

678. Fraser S, and Kapral R (1982b) Analysis of flow hysteresis by a one-dimensional map, Phys. Rev. A25, 3223.

679. Fraser S, and Kapral R (1984) Universal vector scaling in one-dimensional maps, Phys. Rev. A30, 1017.

680. Fraser S, and Kapral R (1985) Mass and dimension of Feigenbaum attractors, Phys. Rev. A31, 1687.

681. Fraujione J G, and Ottino J M (1987) Feasibility of numerical tracking of material lines and surfaces in chaotic flows, Phys. Fluids, 30, 3641.

682. Froehling H, Crutchfield J P, Farmer J D, Packard N H, and Shaw R (1981) On determining the dimension of chaotic flows, Physica 3D, 605.

683. Frøyland J (1983a) Some symmetric two-dimensional dissipative maps, Physica 8D, 423.

684. Frøyland J (1983b) Lyapunov exponents for multidimensional orbits, Phys. Lett. 97A, 2.

685. Frøyland J, and Alfsen K H (1984) Lyapunov exponent spectra for the Lorenz model, Phys. Rev. A29, 2928.

686. Fujisaka H (1982) Multiperiodic flows, chaos and Lyapunov exponents, Prog. Theor. Phys. 68, 1105.

687. Fujisaka H (1983) Statistical dynamics generated by fluctuations of local Lyapunov exponents, Prog. Theor. Phys. 70, 1264.

688. Fujisaka H (1984) Theory of diffusion and intermittency in chaotic systems, Prog. Theor. Phys. 71, 513.

689. Fujisaka H, Grassmann S, and Thomae S (1986) Chaos-induced diffusion, analogues to nonlinear Fokker-Planck equations, Z. Naturforsch. 40a, 867.

690. Fujisaka H, Inoue M, and Uchimura H (1984) Scaling behaviour of characteristic exponents near chaotic transition points, Prog. Theor. Phys. 72, 23.

691. Fujisaka H, Ishii H, Inoue M, and Yamada T (1986) Intermittency caused by chaotic modulation II. Lyapunov exponent, fractal structure and power spectrum, Prog. Theor. Phys. 76, 1198.

692. Fujisaka H, Kamifukumoto H, and Inoue M (1983) Intermittency associated with the breakdown of the chaos symmetry, Prog. Theor. Phys. 69, 333.

693. Fujisaka H, and Grossmann S (1982) Chaos-induced diffusion in nonlinear discrete dynamics, Z. Phys. B48, 261.

694. Fujisaka H, and Inoue M (1985) Theory of diffusion and intermittency in chaotic systems. II. Physical picture for non-perturbative non-diffusive motion, Prog. Theor. Phys. 74, 20.

695. Fujisaka H, and Inoue M (1987) Statistical-thermodynamics formalism of self-similarity, Prog. Theor. Phys. 77, 1334.

696. Fujisaka H, and Inoue M (1987) Theory of diffusion and intermittency in chaotic systems III. New approach to temporal correlations, Prog. Theor. Phys. 78, 268.

697. Fujisaka H, and Yamada T (1977) Theoretical study of a chemical turbulence, Prog. Theor. Phys. 57, 734.

698. Fujisaka H, and Yamada T (1978a) Trajectory instability and strange attractors in a discrete model exhibiting chaotic behavior, Phys. Lett. 66A, 450.

699. Fujisaka H, and Yamada T (1978b) Theoretical study of time correlation functions in a discrete chaotic process, Z. Naturforsch. 33a, 1455.

700. Fujisaka H, and Yamada T (1980) Limit cycles and chaos in realistic models of the Belousov – Zhabotinskii reaction system, Z. Phys. B37, 265.

701. Fujisaka H, and Yamada T (1983) Stability theory of synchronized motion in coupled oscillator systems, Prog. Theor. Phys. 69, 32.

702. Fujisaka H, and Yamada T (1985) A new intermittency in coupled dynamical systems, Prog. Theor. Phys. 74, 918.

703. Fujisaka H, and Yamada T (1986a) Stability theory of synchronized motion in coupled-oscillator systems IV. Instability of synchronized chaos and new intermittency, Prog. Theor. Phys. 75, 1087.

704. Fujisaka H, and Yamada T (1986b) Intermittency caused by chaotic modulation I. Analysis with a multiplicative noise model, Prog. Theor. Phys. 76, 582.

705. Fujisaka H, and Yamada T (1987) Intermittency caused by chaotic modulation III. Self-similarity and high order correlation functions, Prog. Theor. Phys. 77, 1045.

706. Fujita S, and Watanabe M (1986) Transition from periodic to non-periodic oscillation observed in a mathematical model of biconvection by motile micro-organisms, Physica 20D, 435.

707. Fukuda W, and Katsura S (1986) Exactly solvable models showing chaotic behavior II, Physica 136A, 588.

708. Fukushima K, and Yamada T (1986-7) Chaos for pinned sine-Gordon soliton. I and II. J. Phys. Soc. Japan, 55, 2581; 56, 467.

709. Fulinski A, Karczmarczuk J, and Rosciszewski K (1987) On the phase transition analogy in the Feigenbaum cascade, Acta Phys. Pol. A71, 861.

710. Furusawa T (1986) Quantum chaos of mixmaster universe, Prog. Theor. Phys. 75, 59.

711. Furusawa T, and Hosoya A (1985) Is the anisotropic Kaluza – Klein model of the Universe chaotic? Prog. Theor. Phys. 73, 467.

712. Furuya K, and de Almeida A M O (1987) Soliton energies in the standard map beyond the chaotic threshold, J. Phys. A20, 6211.

713. Galgani L (1982) Planck's formula for classical oscillator with stochacticity thresholds, Lett. Nuovo Cimento, 35, 93.

714. Gallavotti G (1982) A criterion of integrability for perturbed nonresonant harmonic oscillators. 'Wick ordering' of the perturbations in classical mechanics and invariance of the frequency spectrum, Commun. Math. Phys. 87, 365.

715. Gambaudo J-M, Glendinning P, and Tresser C (1984) The rotation interval as a computable measure of chaos, Phys. Lett. 105A, 97.

716. Gambaudo J-M, Glendinning P, and Tresser C (1985) Stable cycles with complicated structure, J. Physique Lett. 46, L-653.

717. Gambaudo J M, Glendinning P, and Tresser C (1988) The gluing bifurcation: I. Symbolic dynamics of closed curves, Nonlinearity, 1, 203.

718. Gambaudo J-M, Lanford III O, and Tresser C (1984) Dynamique symbolique des rotations, C. R. Acad. Sc. Paris, Sr. I, 299, 823.

719. Gambaudo J M, Los L E, and Tresser C (1987) A horseshoe for the doubling operator: topological dynamics for metric universality, Phys. Lett. 123A, 60.

720. Gambaudo J-M, Procaccia I, Thomae S,and Tresser C (1986) New universal scenarios for the onset of chaos in Lorenz-type flows, Phys. Rev. Lett. 57, 925.

721. Gambaudo J-M, and Tresser C (1983) Simple models for bifurcations creating horseshoes, J. Stat. Phys. 32, 455.

722. Gao J-Y, Narducci L M, Sadiky H, Squicciarini M, and Yuan J-M (1984) Higher order bifurcations in a bistable system with delay, Phys. Rev. A30, 901.

723. Gao J-Y, Narducci L M, Schulman L S, Squicciarini M,Yuan J-M (1983) Route to chaos in a hybrid bistable system with delay, Phys. Rev. A28, 2910.

724. Gao J-Y, Yuan J-M, and Narducci L M (1983) Instabilities and chaotic behavior in a hybrid bistable system with a short delay, Opt. Commun. 44, 201.

725. Gao J-Y, and Narducci L M (1986) The effect of modulation in a bistable system with delay, Opt. Commun. 58, 360.

726. Gapanov-Grekhov A V, Rabinovich M I, and Strobinets I M (1984) Appearance of multidimensional chaos in active arrays, Sov. Phys. Dokl. 29, 914; Russ. Orig. 279, 596.

727. Gardini L (1985) Hopf bifurcations and period-doubling transitions in Rössler model, Nuovo Cimento, 89B, 139.

728. Gardini L, Lupini R, Mammana C, and Messia M G (1987) Bifurcations and transitions to chaos in the three-dimensional Lotka-Volterra map, SIAM J. Appl. Math. 47, 455.

729. Gardini L, and Lupini R (1986) Chaotic attractors in a three-mode model of forced, dissipative, rotating fluid, Nuovo Cimento, 93B, 7.

730. Gaspard P (1983) Generation of a countable set of homoclinic flows through bifurcation, Phys. Lett. 97A, 1.

731. Gaspard P (1984) Generation of a countable set of homoclinic flows through bifurcation in multidimensional systems, Bull. Sci. Acad. Roy. Belgique, 70, 61.

732. Gaspard P, Kapral R, and Nicolis G (1984) Bifurcation phenomena near homoclinic systems: a two-parameter analysis, J. Stat. Phys. 35, 697.

733. Gaspard P, and Nicolis G (1983) What can we learn from homoclinic orbits in chaotic dynamics? J. Stat. Phys. 31, 499.

734. Gaspard P, and Wang X-J (1987) Homoclinic orbits and mixed-mode oscillations in far from equilibrium systems, J. Stat. Phys. 48, 151.

735. Gavrilov N K, and Shilnikov L P (1972-3) On three dimensional dynamical systems close to systems with a structurally unstable homoclinic curve, Math. USSR Sb. 88, 467; 90, 139.

736. Geisel T (1984) Deterministic diffusion: a chaotic phenomenon, Europhys. News, 15, No.5, 5.

737. Geisel T, Heldstab I, and Thomas H (1984) Linear and nonlinear response of discrete dynamical systems II. Chaotic attractors, Z. Phys. B55, 165.

738. Geisel T, Nierwetberg J, and Keller J (1981) Critical behavior of the Lyapunov number at the period-doubling onset of chaos, Phys. Lett. 86A, 75.

739. Geisel T, Nierwetberg J, and Zacherl A (1985) Accelerated diffusion in Josephson junctions and related chaotic systems, Phys. Rev. Lett. 54, 616.

740. Geisel T, Radons G, and Rubner J (1986) KAM barriers in the quantum dynamics of chaotic systems, Phys. Rev. Lett. 57, 2883.

741. Geisel T, Zacherl A, and Radons G (1988) Chaotic diffusion and 1/f noise of particles in two-dimensional solids, Z. Phys. B71, 117.

742. Geisel T, and Fairen V (1984) Statistical properties of chaos in Chebyshev maps, Phys. Lett. 105A, 263.

743. Geisel T, and Nierwetberg J (1981) A universal fine structure of the chaotic region in period-doubling systems, Phys. Rev. Lett. 47, 975.

744. Geisel T, and Nierwetberg J (1982) Onset of diffusion and universal scaling in chaotic systems, Phys. Rev. Lett. 48, 7.

745. Geisel T, and Nierwetberg J (1984a) Intermittent diffusion: a chaotic scenario in unbounded systems, Phys. Rev. A29, 2305.

746. Geisel T, and Nierwetberg J (1984b) Statistical properties of intermittent diffusion in chaotic systems, Z. Phys. 56B, 59.

747. Geisel T, and Thomae S (1984) Anomalous diffusion in intermittent chaotic systems, Phys. Rev. Lett. 52, 1936.

748. Geist K, and Lauterborn W (1988) The nonlinear dynamics of the damped and driven Toda chain. I. Energy bifurcation diagram, Physica 31D, 103.

749. Gell Y, and Nakach R (1986) Route to chaos in a nonlinear conservative oscillatory system, Phys. Rev. A34, 4276.

750. Gencher Z D, Ivanov Z G, and Todorov B N (1983) Effect of a periodic perturbation on radio frequency model of Josephson junction, IEEE Trans. Cir. & Sys. CAS-30, 633.

751. George D P (1986) Bifurcations in a piecewise linear system, Phys. Lett. 118A, 17.

752. Gertsberg V L, and Sivashinsky G I (1981) Large cells in nonlinear Rayleigh-Bénard convection, Prog. Theor. Phys. 66, 1219.

753. Ghendrih P (1986) Universality in the transition to chaos of dissipative systems, Physica 19D, 440.

754. Ghikas D P (1983) A method of approximation of invariant measures for maps on the unit interval, Lett. Math. Phys. 7, 91.

755. Ghikas D P (1984) Polynomial invariant for maps on the unit interval, Z. Phys. B55, 145.

756. Ghikas D P, and Nicolis G (1982) Stochasticity from deterministic dynamics: an explicit example of generation of Markovian strings, Z. Phys. B47, 279.

757. Giansanti A (1988) Onset of dynamical chaos in topologically massive gauge theories, Phys. Rev. D38, 1352.

758. Gibbon J D, and McGuinness M (1980) A derivation of the Lorenz equations for some unstable dispersive physical systems, Phys. Lett. 77A, 295.

759. Gibbon J D, and McGuinness M J (1982) The real and complex Lorenz equations in rotating fluids and lasers, Physica 5D, 8.

760. Gibbs H M, Hopf F A, Kaplan D L, and Shoemaker R L (1981) Observation of chaos in optical bistability, Phys. Rev. Lett. 46, 474.

761. Gibson C, and Jeffries C (1984) Observation of period doubling and chaos in spin-wave instabilities in YIG, Phys. Rev. 29, 811.

762. Giglio M, Musazzi S, and Perini U (1981) Transition to chaotic behavior via a reproducible sequence of period-doubling bifurcations, Phys. Rev. Lett. 47, 243.

763. Gilpin M E (1979) Spiral chaos in a predator-prey model, Am. Naturalist, 113, 306.

764. Gioggia R S, and Abraham N B (1983) Routes to chaotic output from a single-mode, dc-excited laser, Phys. Rev. Lett. 51, 650.

765. Gioggia R S, and Abraham N B (1984) Anomalous mode pulling, instabilities and chaos in a single-mode, standing-wave $3.39\mu m$ He-Ne laser, Phys. Rev. A29, 1304.

766. Giorgilli A, Casati G, Sironi L, and Galgani L (1986) An efficient procedure to compute fractal dimensions by box counting, Phys. Lett. 115A, 202.

767. Glass L, Guevara M R, Bélair J, and Shrier A (1984) Global bifurcations of a periodically forced biological oscillator, Phys. Rev. A29, 1348.

768. Glass L, and Perez R (1982) Fine structure of phase locking, Phys. Rev. Lett. 48, 1772.

769. Glazier J A, Gunaratne G H, and Libchaber A (1988) $f(\alpha)$ curves: experimental results, Phys. Rev. A37, 523.

770. Glazier J A, Jensen M H, Libchaber A, and Stavans J (1986) Structure of Arnold tongues and the $f(\alpha)$ spectrum for period doubling: experimental results, Phys. Rev. A34, 1621.

771. Glazier J A, and Libchaber A (1988) Quasi-periodicity and dynamical systems: an experimentalist's view, IEEE Trans. Cir. and Sys. CAS-35, 790.

772. Glendinning P (1984a) Local and global behaviour near homoclinic orbits, J. Stat. Phys. 35, 645.

773. Glendinning P (1984b) Bifurcations near homoclinic orbits with symmetry, Phys. Lett. 103A, 163.

774. Glendinning P (1985) Heteroclinic loops leading to hyperchaos, J. Physique Lett. 46, 347.

775. Glendinning P, and Sparrow C (1986) T-points: a codimension two heteroclinic bifurcation, J. Stat. Phys. 43, 749.

776. Goggin M E, and Milonni P W (1988) Driven Morse oscillator: classical chaos, quantum theory, and photodissociation, Phys. Rev. A37, 796.

777. Golberg A I, Sinai Ya G, and Khanin K M (1983) Universal properties of sequences of period-triplings, Usp. Mat. Nauk. 38, 159.

778. Goldberger A L (1987) Nonlinear dnamics, fractals, cardiac physiology, and sudden death, in *Temporal Disorder in Human Oscillatory Systems*, ed. L. Rensing, Springer-verlag.

779. Goldberger A L *et al.* (1984) Nonlinear dynamics in heart failure: implications of long-wavelength cardiopulmonary oscillations, Am. Heart J. 107, 612.

780. Goldberger A L, Bhargava V, West B J, and Mandell A J (1985a) Nonlinear dynamics of heartbeat II. subharmonic bifurcations of the cardiac interbeat interval in sinus mode disease, Physica 17D, 207.

781. Goldberger A L, Bhargava V, West B J, and Mandell A J (1985b) On a mechanism of cardiac electrical stability, Biophys. J. 98, 525.

782. Goldberger A L, and West B J (1987), Applications of nonlinear dynamics to clinic cardiology, Ann. N. Y. Acad. Sci. 504, 195.

783. Goldbeter A, and Decroly O (1983) Temporal self-organization in biochemical systems: periodic behavior versus chaos, Amer. J. Physiology, 245, R478.

784. Golden K, and Goldstein S (1988) Arbitrarily slow decay of correlations in quasiperiodic systems, J. Stat. Phys. 52, 1113.

785. Gollub J P (1983) What causes noise in a convecting fluid, Physica 118A, 28.

786. Gollub J P, Brunner T O, and Danly B G (1978) Periodicity and chaos in coupled nonlinear oscillators, Science, 200, 48.

787. Gollub J P, Romer E J, and Socolar J E (1980) Trajectory divergence for coupled relaxation oscillators: measurements and models, J. Stat. Phys. 23, 321.

788. Gollub J P, and Benson S V (1978) Chaotic response to periodic perturbation of a convecting fluid, Phys. Rev. Lett. 41, 948.

789. Gollub J P, and Benson S V (1980) Many routes to turbulent convection, J. Fluid Mech. 100, 449.

790. Gollub J P, and McCarriar A R (1982) Convection patterns in Fourier space, Phys. Rev. A26, 3470.

791. Gollub J P, and Swinney H L (1975) Onset of turbulence in a rotating fluid, Phys. Rev. Lett. 35, 927.

792. Gomez Liorente J M, and Pollak E (1987) Order out of chaos in the $H_3^+$ molecule, Chem. Phys. Lett. 138, 125.

793. Gong D-C, Qin G-R, Li R, Hu G, Mao J-Y, and Zhang L (1986) Experimental observation of the road from quasiperiodicity to chaos, Kexeu Tongbao (Sci. Bull. China), 31, 1601.

794. Gontis V, and Kaulakys B (1987) Stochastic dynamics of hydrogen atoms in the microwave field: modelling by maps and quantum description, J. Phys. B19, 5051.

795. Gonzalez D L, and Piro O (1983) Chaos in a nonlinear driven oscillator with exact solution, Phys. Rev. Lett. 50, 870.

796. Gonzalez D L, and Piro O (1984a) Disappearance of chaos and integrability in an externally modulated nonlinear oscillator, Phys. Rev, A30, 2788.

797. Gonzalez D L, and Piro O (1984b) One-dimensional Poincaré map for a nonlinear driven oscillator: analytical derivation and geometrical properties, Phys. Lett. 101A, 455.

798. Gonzalez D L, and Piro O (1985) Symmetric kicked self-oscillators: iterated maps, strange attractors, and symmetry of the phase-locking Farey hierarchy, Phys. Rev. Lett, 55, 17.

799. Gora P, Boyarsky A, and Proppe H (1988) Constructive approximations to densities invariant under nonexpanding transformations, J. Stat. Phys. 51, 179.

800. Gorman M, Swinney H L, and Rand D A (1981) Doubly periodic circular Couette flow: experiments compared with prediction from dynamics and symmetry, Phys. Rev. Lett. 46, 992.

801. Gorman M, Widmann P J, and Robbins K A (1984) Chaotic flow regimes in a convection loop, Phys. Rev. Lett. 52, 2241.

802. Gorman M, Widmann P J, and Robbins K A (1986) Nonlinear dynamics of a convection loop: a qualitative comparison of experiment with theory, Physica 19D, 255.

803. Gorshkov V G, Danileiko Yu K, Lebedeva T P, and Nesterov D A (1987) Transition from harmonic behavior to chaos in the interference of plane waves in a nonlinear medium, JETP Lett. 45, 243-6; Russ. Orig. 45, 196.

804. Gouesbet G, and Weill M E (1984) Complexities and entropies of periodic series with application to the transition to turbulence in the logistic map, Phys. Rev. A30, 1442.

805. Grabec I (1986) Chaos generated by the cutting process, Phys. Lett. 117A, 384.

806. Graffi S, Paul T, and Silverstone H J (1987) Classical resonance overlapping and quantum avoided crossings, Phys. Rev. Lett. 59, 255.

807. Graham R (1976) Onset of self-pulsing in lasers and the Lorenz model, Phys. Lett. 58A, 440.

808. Graham R (1983a) Exact solution of some discrete stochastic models with chaos, Phys. Rev. A28, 1679.

809. Graham R (1983b) Quantization of a two-dimensional map with a strange attractor, Phys. Lett. 99A, 131.

810. Graham R (1984a) Wigner distribution of the quantized Lorenz model, Phys. Rev. Lett. 53, 2020.

811. Graham R (1984b) Quantum chaos of the two-level atom, Acta Phys. Austriaca, 56, 45.

812. Graham R (1984c) Quantum noise and strange attractors, Phys. Reports, 1033, 143.

813. Graham R (1985) Global and local dissipation in a quantum map, Z. Phys. B59, 75.

814. Graham R (1987a) Period doubling in dissipative quantum systems, Europhys. Lett. 3, 259.

815. Graham R (1987b) Dissipative quantum maps, Phys. Scripta 335, 111.

816. Graham R, Isermann S, and Tél T (1988) Quantization of Hénon's map with dissipation II. Numerical results, Z. Phys. 71B, 237.

817. Graham R, and Scholz H J (1980) Analytic approximation of the Lorenz attractor by invariant manifolds, Phys. Rev. A22, 1198.

818. Graham R, and Tél T (1985) Quantization of Hénon's map with dissipation, Z. Phys. B60, 127.

819. Grappin R, and Léorat J (1987) Computation of the dimension of two-dimensional turbulence, Phys. Rev. Lett. 59, 1100.

820. Grasman J, Nijmeijer H, and Veling E J M (1984) Singular perturbation and a mapping on an interval for the forced Van der Pol relaxation oscillator, Physica 13D, 195.

821. Grasman J, Veling E J M, and Willems G M (1976) Relaxation oscillations governed by a van der Pol equation with periodic forcing term, SIAM J. Appl. Math. 31, 667.

822. Grasman J, and Boerdink J B T M (1989) Stochastic and chaotic relaxation oscillations, J. Stat. Phys. 54, 949.

823. Grassberger P (1981) On the Hausdorff dimension of fractal attractors, J. Stat. Phys. 26, 173.

824. Grassberger P (1983a) New mechanism for deterministic diffusion, Phys. Rev. A28, 3666.

825. Grassberger P (1983b) On the fractal dimension of the Hénon attractor, Phys. Lett. 97A, 224.

826. Grassberger P (1983c) Generalized dimensions of strange attractors, Phys. Lett. 97A, 227.

827. Grassberger P (1984) Chaos and diffusion in deterministic cellular automata, Physica 10D, 52.

828. Grassberger P (1985) Information flow and maximum entropy measures for one-dimensional maps, Physica 14D, 365.

829. Grassberger P (1986a) How to measure self-generated complexity, Physica, 140A, 319.

830. Grassberger P (1986b) Do climatic attractors exist? Nature, 323, 609.

831. Grassberger P (1988) On symbolic dynamics of one-humped maps of the interval, Z. Naturforsch. 43a, 671.

832. Grassberger P, Badii R, and Politi A (1988) Scaling laws for invariant measures on hyperbolic and nonhyperbolic attractors, J. Stat. Phys. 51, 135.

833. Grassberger P, and Kantz H (1985a) Universal scaling of long-time tails in Hamiltonian systems? Phys. Lett. 113A, 167.

834. Grassberger P, and Kantz H (1985b) Generating partitions for the dissipative Hénon map, Phys. Lett. 113A, 235.

835. Grassberger P, and Procaccia I (1983a) Measuring the strangeness of strange attractors, Physica 9D, 189.

836. Grassberger P, and Procaccia I (1983b) Characterization of strange attractors, Phys. Rev. Lett. 50, 346.

837. Grassberger P, and Procaccia I (1983c) Estimation of the Kolmogorov entropy from a chaotic signal, Phys. Rev. A28, 2591.

838. Grassberger P, and Procaccia I (1984) Dimensions and entropies of strange attractors from a fluctuating dynamics approach, Physica 13D, 34.

839. Grassberger P, and Scheunert M (1981) Some more universal scaling laws for critical mappings, J. Stat. Phys. 26, 697.

840. Grebogi C, Kostelich E, Ott E, and Yorke J A (1987) Multi-dimension intertwined basin boundaries and basin structure of the kicked double rotor, Phys. Lett. 118A, 448.

841. Grebogi C, McDonald S W, Ott E, and Yorke J A (1983) Final state sensitivity: an obstruction to predictability, Phys. Lett. 99A, 415.

842. Grebogi C, McDonald S W, Ott E, and Yorke J A (1985) Exterior dimension of fat fractals, Phys. Lett. 110A, 1.

843. Grebogi C, Ott E, Pelikan S, and Yorke J A (1984) Strange attractors that are not chaotic, Physica 13D, 261.

844. Grebogi C, Ott E, Romeiras F, and Yorke J A (1987) Critical exponents for crisis-induced intermittency, Phys. Rev. A36, 5365.

845. Grebogi C, Ott E, and Yorke J A (1982) Chaotic attractors in crisis, Phys. Rev. Lett. 48, 1507.

846. Grebogi C, Ott E, and Yorke J A (1983a) Are three-frequency quasiperiodic orbits to be expected in typical nonlinear dynamical systems? Phys. Rev. Lett. 51, 339.

847. Grebogi C, Ott E, and Yorke J A (1983b) Fractal basin boundaries, long-lived chaotic transients, and unstable-unstable pair bifurcation, Phys. Rev. Lett. 50, 935.

848. Grebogi C, Ott E, and Yorke J A (1985a) Super persistent chaotic transients, Ergod. Th. & Dyn. Sys. 5, 341.

849. Grebogi C, Ott E, and Yorke J A (1985b) Attractors on an N-torus: quasiperiodicity versus chaos, Physica 15D, 354.

850. Grebogi C, Ott E, and Yorke J A (1986a) Critical exponent of chaotic transients in nonlinear dynamical systems, Phys. Rev. Lett. 57, 1284.

851. Grebogi C, Ott E, and Yorke J A (1986b) Metamorphoses of basin boundaries in nonlinear dynamical systems, Phys. Rev. Lett. 56, 1011.

852. Grebogi C, Ott E, and Yorke J A (1987a) Basin boundary metamorphoses: changes in accessible boundary orbits, Physica 24D, 243.

853. Grebogi C, Ott E, and Yorke J A (1987b) Chaos, strange attractors, and fractal basin boundaries in nonlinear dynamics, Science, 238, 632.

854. Grebogi C, Ott E, and Yorke J A (1987c) Unstable periodic orbits and the dimension of chaotic attractors, Phys. Rev. A36, 3522.

855. Grebogi C, Ott E, and Yorke J A (1988) Unstable periodic orbits and the dimensions of multifractal chaotic attractors, Phys. Rev. A37, 1711.

856. Grebogi C, and Kaufman A N (1981) Decay of statistical dependence in chaotic orbits of deterministic mappings, Phys. Rev. A24, 2829.

857. Greene J M (1968) Two-dimensional measure-preserving mappings, J. Math. Phys. 9, 760.

858. Greene J M (1979) A method for determining a stochastic transition, J. Math. Phys. 20, 1183.

859. Greene J M, Mackay R S, Vivaldi F, and Feigenbaum M J (1981) Universal behaviour in families of area preserving maps, Physica 3D, 468.

860. Greene J M, and Kim J-S (1987) The calculation of Lyapunov spectra, Physica 24D, 213.

861. Greene J M, and Percival I (1981) Hamiltonian maps in the complex plane, Physica 3D, 530.

862. Greenside H S, Ahlers G, Hohenberg P C, and Walden R W (1982) A simple stochastic model for the onset of turbulence in Rayleigh-Bénard convection, Physica 5D, 322.

863. Greenside H S, Wolf A, Swift J, and Pignataro T (1982) Impracticality of a box-counting algorithm for calculating the dimensionality of strange attractors, Phys. Rev. A25, 3453.

864. Greenspan B D, and Holmes P J (1984) Repeated resonance and homoclinic bifurcations in a periodically forced family of oscillators, SIAM J. Math. Anal. 15, No.1.

865. Gregorio S, Scoppola E, and Tirozzi B (1983) A rigorous study of periodic orbits by means of a computer, J. Stat. Phys. 32, 25.

866. Grempel D R, Prange R E, and Fishman S (1984) Quantum dynamics of an nonintegrable system, Phys. Rev. A29, 1639.

867. Grossmann S (1983) Shape dependence of correlation times in chaos-induced diffusion, Phys. Lett. 97A, 263.

868. Grossmann S (1984) Linear response in chaotic states of discrete dynamics, Z. Phys. B57, 77.

869. Grossmann S, Schnedler E, and Thomae S (1985) Ergodicity and entraiment in a system of electrical relaxation oscillators, Ann. Physik, 42, 307.

870. Grossmann S, and Fujisaka H (1982) Diffusion in discrete nonlinear dynamical systems, Phys. Rev. A26, 1779.

871. Grossmann S, and Horner H (1985) Long time tail correlation in discrete chaotic dynamics, Z. Phys. B60, 79.

872. Grossmann S, and Mayer-Kress G (1989) Chaos in the international arms rate, Nature, 377, 701.

873. Grossmann S, and Thomae S (1977) Invariant distributions and stationary correlation functions of the one-dimensional discrete processes, Z. Naturforsch. 32a, 1353.

874. Gruendler J (1985) The existence of homoclinic orbits and the method of Melnikov for systems in $R^n$, SIAM J. Math. Anal. 16, 907.

875. Gu Y (1987) Most stable manifolds and destruction of tori in dissipative dynamical systems, Phys. Lett. 124A, 340.

876. Gu Y, Bandy D K, Yuan J-M, and Narducci L M (1985) Bifurcation routes in a laser with injected signal, Phys. Rev. A31, 354.

877. Gu Y, Tung M, Yuan J-M, Feng D H, and Narducci L M (1984) Crises and hysteresis in coupled logistic maps, Phys. Rev. Lett. 52, 701.

878. Gu Y, and Yuan J-M (1987) Classical dynamics and resonance structures in laser-induced dissociation of a Morse oscillator, Phys. Rev. 36A, 3788.

879. Gubakov V N, Ziglin S L, Konstantinyan K I, Koshelets V P, and Ovsyannikov (1984) Stochastic oscillations in Josephson tunnel junctions, Sov. Phys. – JETP 59, 198; Russ. Orig. 86, 343.

880. Guckenheimer J (1977) Bifurcations of maps of the interval, Inventiones Math. 39, 165.

881. Guckenheimer J (1979) Sensitive dependence on initial conditions for one-dimensional maps, Commun. Math. Phys. 70, 133.

882. Guckenheimer J (1980a) Dynamics of the van der Pol equation, IEEE Trans. Cir. & Sys. CAS-27, 983.

883. Guckenheimer J (1980b) Symbolic dynamics and relaxation oscillations, Physica 1D, 227.

884. Guckenheimer J (1982) Noise in chaotic systems, Nature, 298, 358.

885. Guckenheimer J (1984) Multiple bifurcation problems of codimension 2, SIAM J. Math. Anal. 15, 1.

886. Guckenheimer J, and Buzyna G (1983) Dimension measurements for geostrophic turbulence, Phys. Rev. Lett. 51, 1438.

887. Guckenheimer J, and Williams R (1979) Structural stability of the Lorenz attractor, Publ. Math. IHES, 50, 307.

888. Guevara M R, Glass L, Mackey M C, and Shirer L (1983) Chaos in neurobiology, IEEE Trans. Sys. Man & Cyb. SMC-13, 790.

889. Guevara M R, Glass L, and Shrier A (1981) Phase locking, period-doubling bifurcations, and irregular dynamics in periodically stimulated cardiac cells, Science, 214, 1350.

890. Guevara M R, Ward G, Shrier A, and Glass L (1984) Electrical alternans and period-doubling bifurcations, in *IEEE Computers in Cardiology*, IEEE Computer Society Press, 167.

891. Guevara M R, and Glass L (1982) Phase locking, period-doubling bifurcations and chaos, J. Math. Biol. 14, 1.

892. Gulyaev Yu V, Dmitriev A S, and Kislov V D (1985) Starnge attractors in ring-type self-oscillating systems, Sov. Phys. Dokl. 30, 360; Russ. Orig. 282, 53.

893. Gunaratne G H (1987) Trajectory scaling for period tripling in near conformal mappings, Phys. Rev. A36, 1834.

894. Gunaratne G H, Jensen M H, and Procaccia I (1988) Universal strange attractors on wrinkled tori, Nonlinearity, 1, 157.

895. Gunaratne G H, Linsay P S, and Vinson J (1989) Chaos beyond onset: a comparison of theory and experiment, Phys. rev. Lett. 63, 1.

896. Gunaratne G H, and Procaccia I (1987) The organization of chaos, Phys. Rev. Lett. 59, 1377.

897. Gurzadyan V G, and Kocharyan A A (1987) Relative chaos in stellar systems, Astrophys. & Space Sci. 135, 307.

898. Gutkin E (1983) Propagation of chaos and the Burgers equation, SIAM J. Appl. Math. 43, 971.

899. Gutzwiller M C, and Mandelbrot B B (1988) Invariant multifractal measures in chaotic Hamiltonian systems and related structures, Phys. Rev. Lett. 60, 673.

900. Gwinn E G, and Westervelt R M (1985) Intermittent chaos and low-frequency noise in the driven damped pendulum, Phys. Rev. Lett, 54, 1613.

901. Gwinn E G, and Westervelt R M (1986) Fractal basin boundaries and intermittency in the driven damped pendulum, Phys. Rev. A33, 4143.

902. Gwinn E G, and Westervelt R M (1987) Scaling structure of attractors at the transition from quasiperiodicity to chaos in electron transport in Ge, Phys. Rev. Lett. 59, 157.

903. Gyorgyi G, and Tishby N (1987a) Scaling in stochastic Hamiltonian systems: a renormalization approach, Phys. Rev. Lett. 58, 527.

904. Gyorgyi G, and Tishby N (1987b) Destabilization of islands in noisy Hamiltonian systems, Phys. Rev. A36, 4957.

905. Gyorgyi G, and Szépfalusy P (1984) Properties of fully developed chaos in one-dimensional maps, J. Stat. Phys. 34, 451.

906. Gyorgyi G, and Szépfalusy P (1985) Calculation of entropy in chaotic systems, Phys. Rev. A31, 3477.

907. Haake F, Kus M, and Scharf R (1987) Classical and quantum chaos for a kicked top, Z. Phys. 65B, 381.

908. Haken H (1975) Analogy between higher instabilities in fluids and lasers, Phys. Lett. 53A, 77.

909. Haken H (1983) At least one Lyapunov exponent vanishes if the trajectory of an attractor does not contain a fixed point, Phys. Lett. 94A, 71.

910. Haken H, and Mayer-Kress G (1981a) Chapman-Kolmogorov equation for discrete chaos, Phys. Lett. 84A, 159.

911. Haken H, and Mayer-Kress G (1981b) Chapman-Kolmogorov equation and path integrals for discrete chaos in presence of noise, Z. Phys. B43, 185.

912. Haken H, and Wunderlin A (1977) New interpretation and size of strange attractor of the Lorenz model of turbulence, Phys. Lett. 62A, 133.

913. Haken H, and Wunderlin A (1982) Some exact results on discrete noisy maps, Z. Phys. B46, 181.

914. Halas N J, Liu S-N, and Abraham N B (1983) Route to mode locking in a three-mode He-Ne 3.39 $\mu m$ laser including chaos in the secondary beat frequency, Phys. Rev. A28, 2915.

915. Hale J K, and Sternberg N (1988) Onset of chaos in differential delay equations, J. Comput. Phys. 77, 221.

916. Haller, Koppel H, and Cederbaum L S (1984) Uncovering the transition from regularity to irregularity in a quantum system, Phys. Rev. Lett. 52, 1665.

917. Halpern P (1987) Chaos in the long-term behavior of some Bianchi-type VIII models, Gen Relativ. & Grav. 19, 73.

918. Halschneider M (1988) On the wavelet transformation of fractal objects, J. Stat. Phys. 50, 963.

919. Halsey T C, Jensen M H, Kadanoff L P, Procaccia I, and Shraiman B I (1986) Fractal measures and their singularities: the generalization of strange sets, Phys. Rev. A33, 1141.

920. Hamilton I, and Brumer P (1982) Relaxation rates for two-dimensional deterministic mappings, Phys. Rev. A25, 3457.

921. Hammel S M, Yorke J A, and Grebogi C (1987) Do numerical orbits of chaotic dynamical processes represent true orbits? J. Complexity, 3, 136.

922. Hansen A (1988) A connection between the percolation transition and the onset of chaos in the Kaufman model, J. Phys. A21, 2481.

923. Hanson J D (1987) Fat-fractal scaling exponent of area-preserving maps, Phys. Rev. A35, 1470.

924. Hanson J D, Ott E, and Antonsen T M (1984) Influence of finite wavelength on the quantum kicked rotator in the semiclassical regime, Phys. Rev. A29, 819.

925. Hao B-L (1981) Universal slowing-down exponent near period-doubling bifurcation points, Phys. Lett. 86A, 267.

926. Hao B-L (1982) Two kinds of entrainment-beating transitions in a driven limit cycle oscillator, J. Theor. Biol. 98, 9.

927. Hao B-L (1985) Bifurcations and chaos in a periodically forced limit cycle oscillator, in *Advances in Science of China: Physics*, vol. 1, ed. Zhu Hong-yuan, Zhou Guang-zhao, and Fang Li-zhi, Science Press, 113.

928. Hao B-L (1986) Symbolic dynamics and systematics of periodic windows, Physica 140A, 85.

929. Hao B-L (1987) Bifurcation and chaos in the periodically forced Brusselator, in *Collected Papers Dedicated to Professor Kazuhisa Tomita on the Occasion of his Retirement from Kyoto University*, Kyoto University, 82.

930. Hao B-L, Wang G-R, and Zhang S-Y (1983) U-sequences in the periodically forced Brusselator, Commun. Theor. Phys. 2, 1075.

931. Hao B-L, and Zeng W-Z (1986) Information dimensions in unimodal mappings, in *Proceedings of the Sino-Japan Bilateral Workshop on Statistical Physics and Condensed Matter Theory*, ed. Xie Xi-de, World Scientific, 24.

932. Hao B-L, and Zeng W-Z (1987) Number of periodic windows in one-dimensional mappings, in *The XV International Colloquium on Group Theoretical Methods in Physics*, ed. R. Gilmore, World Scientific, 199.

933. Hao B-L, and Zhang S-Y (1982a) Subharmonic stroboscopy as a method to study period-doubling bifurcations, Phys. Lett. 87A, 267.

934. Hao B-L, and Zhang S-Y (1982b) Hierarchy of chaotic bands, J. Stat. Phys. 28, 769.

935. Hao B-L, and Zheng W-M (1989) Symbolic dynamics of unimodal maps revisited, Int. J. Mod. Phys. B3, 235.

936. Hardas B R, and Scheeline A (1984) Stability and chaos in a voltage-thresholded high-voltage spark source, Anal. Chem. 56, 169.

937. Harding R H, and Ross J (1987) The effect of different types of noise on some methods of distinguishing chaos from periodic oscillations, J. Chim. Phys. Phys.-Chim. Biol. 84, 1305.

938. Harikrishnan K P (1987) Universal behavior in a 'modulated' logistic map, Phys. Lett. 125A, 465.

939. Harms A A, Bilanovic Z, and Leung H K Y (1986) The existence of fission-driven nonlinear maps, Ann. Nucl. Energy, 13, 341.

940. Harnhofer H, Thirring W, and Wiklicky H (1988) Transitivity and ergodicity of quantum systems, J. Stat. Phys. 52, 1097.

941. Harrison R G, Firth W J, Emshary C A, and Al-Saidi I A (1983) Observation of period-doubling in an all-optical resonator containing $NH_3$ gas, Phys. Rev. Lett. 51, 562.

942. Harrison R G, Firth W J, and Al-Saidi I A (1984) Observation of bifurcation to chaos in an all-optical Fabry-Perot resonator, Phys. Rev. Lett. 53, 258.

943. Harrison R G, and Biswas D J (1985) Pulsatting instabilities and chaos in lasers, Prog. Quant. Electr. 10, 147.

944. Harrison R G, and Biswas D J (1986) Chaos in light, Nature, 321, 394.

945. Harrison R G, and Uppal J S (1988) Instabilities and chaos in lasers, Europhys. News, 19, No.6, 84.

946. Hart J E (1986) A model for the transition to baroclinic chaos, Physica 20D, 350.

947. Harth E (1983) Order and chaos in neural systems: an approach to the dynamics of higher brain functions, IEEE Trans. Sys. Man & Cyb. SMC-13, 782.

948. Hasegawa H, Harada A, and Okazaki Y (1984) Analytic simulation of the Poincaré surface of sections for the diamagnetic Kepler problem, J. Phys. A16, L883.

949. Hata H, Morita T, Tomita K, and Mori H (1987) Spectra of singularities for the Lozi and Hénon maps, Prog. Theor. Phys. 78, 721.

950. Hatori T, and Irie H (1987) Long-time correlation for the chaotic orbit in the two-wave Hamiltonian, Prog. Theor. Phys. 78, 249.

951. Haubs G, and Haken H (1985) Quantities describing local properties of chaotic attractors, Z. Phys. B59, 459.

952. Hauser P R, Curado E M F, and Tsallis C (1985) On the universality classes of the Hénon map, Phys. Lett. 108A, 308.

953. Hauser P R, Tsallis C, and Curado E M F (1984) Criticality of the routes to chaos of the $1 - a|x|^z$ map, Phys. Rev. A30, 2074.

954. Hayashi C, Ishizuka S, and Hirakawa K (1983) Transition to chaos via intermittency in the Onchidium pacemaker neuron, Phys. Lett. 98A, 474.

955. He D-R, Yeh W J, and Kao Y H (1984) Transition from quasiperiodicity to chaos in a Josephson-junction analog, Phys. Rev. B30, 172.

956. He D-R, Yeh W-J, and Kao Y H (1985) Studies of return maps, chaos, and phase-locked states in a current driven Josephson-junction simulator, Phys. Rev. B31, 1359.

957. He K-F (1987) Statistical model for a chaotic system of the kinds of interacting modes, Commun. Theor. Phys. 7, 15.

958. Held G A, Jeffries C, and Haller E E (1984) Observation of chaotic behavior in an electron-hole plasma in Ge, Phys. Rev. Lett. 52, 1037.

959. Held G A, and Jeffries C (1985) Spatial and temporal structure of chaotic instabilities in an electron-hole Plasma in Ge, Phys. Rev. Lett. 55, 887.

960. Heldstab J, Thomas H, Geisel T, and Radons G (1983) Linear and nonlinear response of discrete dynamical systems. I. Periodic attractor, Z. Phys. B50, 141.

961. Heller E J (1984) Bound-state eigenfunctions of classically chaotic Hamiltonian systems: scars of periodic orbits, Phys. Rev. Lett. 53, 1575.

962. Helton T W, and Tabor M (1985) On classical and quantal Kolmogorov entropies, J. Phys. A18, 2743.

963. Hemmer P C (1984) The exact invariant density for a cusp-shaped return map, J. Phys. A17, L247.

964. Hénon M (1976) A two-dimensional mapping with a strange attractor, Commun. Math. Phys. 50, 69.

965. Hénon M (1982) On the numerical computation of Poincaré maps, Physica 5D, 412.

966. Hénon M, and Heiles C (1964) The applicability of the third integral of the motion: some numerical experiments, Astron. J. 69, 73.

967. Hénon M, and Pomeau Y (1977) Two strange attractors with a simple structure, Lect. Notes in Math. 565, 29.

968. Hense A (1986) On the possible existence of a strange attractor for the Southern oscillation, Beitr. Phys. Atmosph. 60, 35.

969. Hentschel H G E, and Procaccia I (1983) The infinite number of generalized dimensions of fractals and strange attractors, Physica 8D, 435.

970. Henyey F S, and Pomphrey N (1982) The autocorrelation function of a pseudointegrable system, Physica 6D, 78.

971. Herring C, and Huberman B A (1980) Dislocation motion and solid-state turbulence, Appl. Phys. Lett. 36, 975.

972. Herzel H P, and Pompe B (1987) Effects of noise on a nonuniform chaotic map, Phys. Lett. 122A, 121.

973. Herzel H-P, and Ebeling W (1985) The decay of correlations in chaotic maps, Phys. Lett. 111A, 1.

974. Hioe F T, and Deng Z (1987) Stability-instability transitions in Hamiltonian systems of n dimensions, Phys. Rev. A35, 847.

975. Hirooka H, Kurokawa M, and Saito N (1985) How to extract regularity from disguised irregular levels of integrable Hamiltonian system, J. Phys. Soc. Japan, 54, 3209.

976. Hirooka H, Saito N, and Ford J (1984) Chaos around hyperbolic fixed points, J. Phys. Soc. Japan, 53, 895.

977. Hirsch J E, Huberman B A, and Scalapino D J (1982) A theory of intermittence, Phys. Rev. A25, 519.

978. Hirsch J E, Nauenberg M, and Scalapino D J (1982) Intermittency in the presence of noise: a renormalization group formulation, Phys. Lett. 87A, 391.

979. Hitzl D L (1981) Numerical determination of the capture escape boundary for the Hénon attractor, Physica 2D, 370.

980. Hitzl D L, and Zele F (1985a) A three-dimensional dissipative map with three routes to chaos, Phys. Lett. 110A, 181.

981. Hitzl D L, and Zele F (1985b) An exploration of the Hénon quadratic map, Physica 14D, 305.

982. Hnilo A A (1985) Chaotic (as the logistic map) laser cavity, Opt. Commun. 53, 194.

983. Hockett K, and Holmes P (1986) Josephson's junction, annulus maps, Birkhoff attractors, horseshoes and rotation sets, Ergod. Th. & Dyn. Sys. 6, 205.

984. Hoffmann K H (1982) The Birkhoff renormalization procedure and the reductive perturbation approach: two equivalent methods to discuss the Hopf bifurcations, Phys. Lett. 92A, 163.

985. Hoffnagle J, DeVoe R G, Reyna L, and Brewer R G (1988) Order-chaos transition of two trapped ions, Phys. Rev. Lett. 61, 255.

986. Hofstadter D R (1981) Strange attractors: mathematical patterns delicately poised between order and chaos, Sci. Amer. 245, 22.

987. Hogg T, and Huberman B A (1982) Recurrence phenomena in quantum dynamics, Phys. Rev. Lett. 48, 711.

988. Hogg T, and Huberman B A (1983) Quantum dynamics and nonintegrability, Phys. Rev. A28, 22.

989. Hogg T, and Huberman B A (1984a) Generic behavior of coupled oscillators, Phys. Rev. A29, 275.

990. Hogg T, and Huberman B A (1984b) Chaos and the classical limit of quantum systems, Phys. Scripta 30, 225.

991. Hogg T, and Huberman B A (1985) Attractors on finite sets: the dissipative dynamics of computing structures, Phys. Rev. A32, 2338.

992. Holden A V, and Winlow W (1983) Neuronal activity as the behavior of a differential system, IEEE Trans. Sys. Man & Cyb. SMC-13, 711.

993. Holmes P (1979a) A nonlinear oscillator with a strange attractor, Phil. Trans. Roy. Soc. A292, 419.

994. Holmes P (1979b) Domains of stability in a wind induced oscillation problem, Trans. ASME. J. Appl. Mech. 46, 672.

995. Holmes P (1980) Averaging and chaotic motions in forced oscillations, SIAM J. Appl. Math. 38, 65; 40, 167.

996. Holmes P (1982a) Proof of non-integrability for the Hénon-Heile's Hamiltonian near an exceptional integrable case, Physica 5D, 335.

997. Holmes P (1982b) The dynamics of repeated impacts with a sinusoidally vibrating table, J. Sound Vib. 84, 173.

998. Holmes P (1984) Bifurcation sequences in horseshoe maps: infinitely many routes to chaos, Phys. Lett. 104A, 299.

999. Holmes P (1986) Chaotic motion in a weakly nonlinear model for surface waves, J. Fluid Mech. 162, 3365.

1000. Holmes C, and Holmes P (1981) Second order averaging and bifurcations to subharmonics in Duffing's equation. J. Sound and Vib. 78, 161.

1001. Holmes P, and Marsden J E (1983) Horseshoes and Arnold diffusion for Hamiltonian systems on Lie groups, Indiana U. Math. J. 32, 273.

1002. Holmes P, and Rand D A (1978) Bifurcations of the forced van der Pol oscillator, Quart. Appl. Math. 35, 495.

1003. Holmes P, and Whitley D (1984) Bifurcations of one- and two-dimensional maps, Phil. Trans. Soc. Lond. A311, 43.

1004. Holt R G, and Schwartz I B (1984) Newton's method as a dynamical system: global convergence and predictibility, Phys. Lett. 105A, 327.

1005. Holtfort J, Mohring W, and Vogel H (1988) Computing the invariant density for bistable noisy maps, Z. Phys. B72, 115.

1006. Holzfuss J, and Lauterborn W (1989) Liapunov exponents from a time series of acoustic chaos, Phys. Rev. A39, 2146.

1007. Hongler M-O, and Streit L (1988) On the origin of chaos in gearbox models, Physica 29D, 402.

1008. Hoover W G, Tull C G, and Posch H A (1988) Negative Lyapunov exponents for dissipative systems, Phys. Lett. 131A, 211.

1009. Hopf F (1948) A mathematical example displaying features of turbulence, Commun. on Pure Appl. Math. 1, 303.

1010. Hopf F A, Kaplan D L, Gibbs H M, and Shoemaker R L (1982) Bifurcations to chaos in optical bistability, Phys. Rev. A25, 172.

1011. Hopf F A, Kaplan D L, Rose M H, Sanders L D, and Derstine M W (1986) Characterization of chaos in a hybrid optical bistability device, Phys. Rev. Lett. 57, 1394.

1012. Howard J E, and Hohs S M (1984) Stochasticity and reconnection in Hamiltonian systems, Phys. Rev. A29, 418.

1013. Hsu C S, and Kim M C (1984) Method of constructing generating partitions for entropy evaluation, Phys. Rev. A30, 3351.

1014. Hsu C S, and Kim M C (1985a) Construction of maps with generating partitions for entropy eveluation, Phys. Rev. A31, 3253.

1015. Hsu C S, and Kim M C (1985b) Statistics of strange attractors by generalized cell mapping, J. Stat. Phys. 38, 735.

1016. Hsu G-H, Ott E, and Grebogi C (1988) Strange saddles and the dimensions of their invariant manifolds, Phys. Lett. 127A, 199.

1017. Hu B (1981) Dissipative bifurcation ratio in the area-preserving Hénon map, J. Phys. A14, L423.

1018. Hu B (1982a) A two dimensional scaling theory of intermittency, Phys. Lett. 91A, 375.

1019. Hu B (1982b) Introduction to real space renormalization group methods in critical and chaotic phenomena, Phys. Reports, 91, 233.

1020. Hu B (1983) A simple derivation of the stochastic eigenvalue equation in the transition from quasiperiodicity to chaos, Phys. Lett. 98A, 79.

1021. Hu B, and Mao J-M (1982a) Third order renormalization-group calculation of the Feigenbaum universal bifurcation ratio in the transition to chaotic behavior, Phys. Rev. A25, 1196.

1022. Hu B, and Mao J-M (1982b) Period doubling: universality and critical-point order, Phys. Rev. A25, 3259.

1023. Hu B, and Mao J-M (1983) Universal metric properties of an approximate Poincaré map for Duffing's equation with negative stiffness, Phys. Rev. A27, 1700.

1024. Hu B, and Mao J-M (1984) Universal scaling of the power spectrum in area-preserving maps, Phys. Rev. A29, 1564.

1025. Hu B, and Mao J-M (1985) The eigenvalue-matching renormalization group, Phys. Lett. 108A, 305.

1026. Hu B, and Rudnick J (1982a) Exact solutions to the Feigenbaum renormalization group equations for intermittency, Phys. Rev. Lett. 48, 1645.

1027. Hu B, and Rudnick J (1982b) Exact solutions to the renormalization-group fixed-point equations for intermittency in two-dimensional maps, Phys. Rev. A26, 3035.

1028. Hu B, and Rudnick J (1986) Differential-equation approach to functional equations: exact solutions for intermittency, Phys. Rev. A34, 2453.

1029. Hu B, and Satija I I (1983) A spectrum of universality classes in period doubling and period tripling, Phys. Lett. 98A, 143.

1030. Hu G (1986) The invariat distribution of non-fully developed chaos, Chinese Phys. Lett. 3, 357.

1031. Hu G, and Hao B-L (1983) A scaling relation for the Hausdorff dimension of the limiting sets in one-dimensional mappings, Commun. Theor. Phys. 2, 1473.

1032. Huang Y-N (1985) Determination of the stable periodic orbits for the Hénon map by analytical method, Chinese Phys. Lett. 2, 98.

1033. Huang Y-N (1986) An algebraic analytical method for exploring periodic orbits of the Hénon map, Scientia Sinica (Series A), 29, 1302.

1034. Huberman B A (1983) Mostly chaos, Physica 118A, 323.

1035. Huberman B A, Crutchfield J P, and Packard N H (1980) Noise phenomena in Josephson junctions, Appl. Phys. Lett. 37, 750.

1036. Huberman B A, and Crutchfield J P (1979) Chaotic states of anharmonic systems in periodic fields, Phys. Rev. Lett. 43, 1743.

1037. Huberman B A, and Rudnick J (1980) Scaling behavior of chaotic flows, Phys. Rev. Lett. 45, 154.

1038. Huberman B A, and Zisook A B (1981) Power spectra of strange attractors, Phys. Rev. Lett. 46, 626.

1039. Hudson J L, Hart M, and Marinko D (1979) An experimental study of multiple peak periodic and nonperiodic oscillators in the Belousov-Zhabotinskii reaction, J. Chem. Phys. 71, 1601.

1040. Hudson J L, Rössler O E, and Killory H (1986) A four-variable chaotic chemical reaction, Chem. Eng. Commun. 46, 159.

1041. Hudson J L, and Mankin J C (1981) Chaos in the Belousov-Zhabotinskii reaction, J. Chem. Phys. 74, 6171.

1042. Hudson J L, and Rössler O E (1984) A piecewise-linear invertible noodle map, Physica 11D, 293.

1043. Huebener R P, Peinke J, and Parisi J (1989) Experimental progress in the nonlinear behavior of semiconductors, Appl. Phys. A48, 107.

1044. Hunt E R, and Rollins R W (1984) Exactly solvable model of a physical system exhibiting multidimensional chaotic behavior, Phys. Rev. A29, 1000.

1045. Hurley M (1986) Multiple attractors in Newton's method, Ergod. Th. & Dyn. Sys. 6, 561.

1046. Hurley M, and Martin C (1984) Newton's algorithm and chaotic dynamical systems, SIAM J. Math. Anal. 15, 238.

1047. Ibanez J L, Pomeau Y (1978) A simple case of non-periodic (strange) attractor, J. Non-Equil. Thermodyn. 3, 135.

1048. Ichikawa Y H, Kamimura T, and Hatori T (1987) Stochastic diffusion in the standard map, Physica 29D, 247.

1049. Ikeda K (1979) Multiple-valued stationary state and its instability of the transmitted light by a ring cavity system, Opt. Commun. 30, 257.

1050. Ikeda K, Daido H, and Akimoto O (1980) Optical turbulence: chaotic behavior of transmitted light from a ring cavity, Phys. Rev. Lett. 45, 709.

1051. Ikeda K, and Akimoto O (1982) Instability leading to periodic and chaotic self-pulsations in a bistable optical cavity, Phys. Rev. Lett. 48, 617.

1052. Ikeda K, and Kondo K (1982) Successive higher-harmonic bifurcations in systems with delayed feedback, Phys. Rev. Lett. 49, 1467.

1053. Ikeda K, and Matsumoto K (1987) High-dimensional chaotic behavior in systems with time-delayed feedback, Physica 29D, 223.

1054. Ikezi H, De Grassie J S, and Jensen T H (1983) Observation of multivalued attractors and crises in a driven nonlinear circuit, Phys. Rev. A28, 1207.

1055. Imada M (1983) Chaos caused by the soliton-solition interaction, J. Phys. Soc. Japan, 52, 1946.

1056. Imaeda K, Yamamoto Y, and Yamaguchi T (1987) Bifurcation of oscillation modes obtained by a shift type mapping simulating the Oregonator responding to an external periodic force, J. Phys. Soc. Japan, 56, 3832.

1057. Innanen K A (1985) The threshold for chaos for Hénon-Heiles and related potentials, Astron. J. 90, 2377.

1058. Inoue M, Kawaguchi T, and Fujisaka H (1986) Simultaneous onset of diffusion motion and intermittent chaos. A scaling property of characteristic exponents, Phys. Lett. 115A, 139.

1059. Inoue M, and Fujisaka H (1985) Chaos-induced diffusive motion in a modeled-dislocation system: a scaling property of characteristic exponents, Phys. Rev. B32, 277.

1060. Inoue M, and Fujisaka H (1987) Analytic properties of characteristic exponents for chaotic dynamical systems, Prog. Theor. Phys. 77, 1077.

1061. Inoue M, and Fujisaka H (1988) A fluctuation theory of local fractal dimensions, Prog. Theor. Phys. Lett. 79, 1251.

1062. Inoue M, and Kamifukumoto H (1984) Scenarios leading to chaos in a forced Lotka-Volterra model, Prog. Theor. Phs. 71, 930.

1063. Inoue M, and Koga H (1982) Chaos and diffusion in a sinusoidal potential with a periodic external field, Prog. Theor. Phys. 68, 2184.

1064. Inoue M, and Koga H (1983) Chaotic response of a self-interacting pseudo-spin model, Prog. Theor. Phys. 69, 1403.

1065. Iooss G, and Los J E (1988) Quasi-genericity of bifurcations to high dimensional tori for maps, Commun. Math. Phys. 119, 453.

1066. Ishii H, Fujisaka H, and Inoue M (1986) Breakdown of chaos symmetry and intermittency in the double-well potential system, Phys. Lett. 116A, 257.

1067. Isola S (1988) Resonances in chaotic dynamics, Commun. Math. Phys. 116, 343.

1068. Isomaki H M, von Boehm J, and Raty R (1985) Devil's attractor and chaos of a driven impact oscillator, Phys. Lett. 107A, 343.

1069. Isomaki H M, von Boehm J, and Raty R (1988) Fractal basin boundaries of an impacting particle, Phys. Lett. 126A, 484.

1070. Ito A (1979) Successive subharmonic bifurcations and chaos in a nonlinear Mathieu equation, Prog. Theor. Phys. 61, 815.

1071. Ito H M (1984) Ergodicity of randomly perturbed Lorenz model, J. Stat. Phys. 5, 151.

1072. Ito R (1981) Rotation sets are closed, Math. Proc. Camb. Phil. Soc. 89, 107.

1073. Izrailev F M (1987) Chaotic structure of eigenfunctions in systems with maximal quantum chaos, Phys. Lett. 125A, 250.

1074. Izrailev F M (1988) Transient chaos in a generalized Hénon map on the torus, Phys. Lett. 126A, 405.

1075. Izrailev F M, Rabinovich M J, and Ugodnikov A D (1981) Approximate description of three-dimensional dissipative systems with stochastic behaviour, Phys. Lett. 86A, 321.

1076. Jackson E A (1985) The Lorenz system. I. The global structure of the stable manifolds; II. The homoclinic convolution of the stable manifolds, Phys. Scripta 32, 469; 476.

1077. Jaffe C, and Reinhardt W P (1982) Uniform semiclassical quantization of regular and chaotic classical dynamics on the Hénon-Heiles surface, J. Chem. Phys. 77, 5191.

1078. Jakobson M K (1981) Absolutely continuous invariant measure for one parameter families of one dimensional maps, Commun. Math. Phys. 81, 39.

1079. Jannussis A, Theodoropoulou M, and Brodimas G (1984) Finite-difference approach in chemical reactions and their chaotic behaviour, Lett. Nuovo Cimento, 41, 145.

1080. Jefferies D J (1986) Bifurcation to chaos in clocked digital systems containing autonomous timing circuits, Phys. Lett. 115A, 89.

1081. Jeffries C, and Perez J (1982) Observation of a Pomeau-Manneville intermittent route to chaos in a nonlinear oscillator, Phys. Rev. A26, 2117.

1082. Jeffries C, and Perez J (1983) Direct observation of crises of the chaotic attractor in a nonlinear oscillator, Phys. Rev. A27, 601.

1083. Jeffries C, and Usher A (1983) Frequency division using diodes in resonant systems, Phys. Lett. 99A, 427.

1084. Jeffries C, and Wiesenfeld K (1985) Observation of noisy precursors of dynamical instabilities, Phys. Rev. A31, 1077.

1085. Jensen J H, Christiansen P L, Scott A C, and Skovgaard O (1984) Chaos in the Beeler-Reuter system for the action potential of ventricular myocardial fibres, Physica 13D, 269.

1086. Jensen M H, Bak P, and Bohr T (1983) Complete devil's staircase, fractal dimension, and universality of mode-locking structure in the circle map, Phys. Rev. Lett. 50, 1637.

1087. Jensen M H, Bak P, and Bohr T (1984) Transition to chaos by interaction of resonances in dissipative systems. I. Circle maps, Phys. Rev. A30, 1960.

1088. Jensen M H, Kadanoff L P, Libchaber A, Procaccia I, and Stavans J (1985) Global universality at the onset of chaos: results of a forced Rayleigh-Bénard experiment, Phys. Rev. Lett. 55, 2798.

1089. Jensen M H, Kadanoff L P, and Procaccia I (1987) Scaling structure and thermodynamics of strange sets, Phys. Rev. A36, 1409.

1090. Jensen M H, and Procaccia I (1985) Chaos via quasiperiodicity: universal scaling laws in the chaotic regime, Phys. Rev. A32, 1225.

1091. Jensen R V (1984) Stochastic ionization of surface-state electrons: classical theory, Phys. Rev. A30, 386.

1092. Jensen R V (1987a) Chaos in atomic physics, At. Phys. 10, 319.

1093. Jensen R V (1987b) Classical chaos, Am. Sci. 75, 168.

1094. Jensen R V, Susskind S M, and Sanders M M (1989) Microwave ionization of highly excited hydrogen atoms: a test of the correspondence principle, Phys. Rev. Lett. 62, 1476.

1095. Jensen R V, and Jessup E R (1986) Statistical properties of the circle map, J. Stat. Phys. 43, 369.

1096. Jensen R V, and Myers C R (1985) Images of the critical points of nonlinear maps, Phys. Rev. A32, 1222.

1097. Jensen R V, and Oberman C R (1982) Statistical properties of chaotic dynamical systems which exhibit strange attractors, Physica 4D, 183.

1098. Jensen R V, and Shanker R (1985) Statistical behaviour in deterministic quantum systems with few degrees of freedom, Phys. Rev. Lett. 54, 1879.

1099. Jetschke G, and Stiewe Ch (1985) Intermittency for tent maps is exactly calculable, Phys. Lett. 112A, 265.

1100. Jetschke G, and Stiewe Ch (1987) An ergodic theorem for intermittency of piecewise linear iterated maps, J. Phys. A20, 3185.

1101. Jiang L-Y, and Peng S-L (1987) Rigorous bounds on the power spectra of arbitrary prime $\eta$-order renormalization group equations, J. Phys. A20, 2325.

1102. Jinz S J, and Lucke M (1986) Effect of additive and multiplicative noise on the first bifurcation of the logistic model, Phys. Rev. A33, 2694.

1103. Jones C A, Weiss N O, and Cattaneo F (1985) Nonlinear dynamos: a complex generalization of the Lorenz equations, Physica 14D, 161.

1104. Jones S W, and Aref H (1988) Chaotic advection in pulsed source-sink system, Phys. Fluids, 31, 469.

1105. Jonker L, and Rand D (1980) The periodic orbits and entropy of certain maps of the unit interval, J. London Math. Soc. (2), 22, 175.

1106. José J V (1986) Study of a quantum Fermi-accelaration model, Phys. Rev. Lett. 56, 290.

1107. Jung C, and Scholz H J (1987) Cantor set structure in the singularities of classical potential scattering, J. Phys. A20, 3607.

1108. Jung C, and Scholz H J (1988) Chaotic scattering off the magnetic dipole, J. Phys. A21, 2300.

1109. Kadanoff L P (1981) Scaling for a critical Kolmogorov-Arnold-Moser trajectory, Phys. Rev. Lett. 47, 1641.

1110. Kadanoff L P (1983a) Supercritical behavior of an ordered trajectory, J. Stat. Phys. 31, 1.

1111. Kadanoff L P (1983b) Roads to chaos, Phys. Today, December, 46.

1112. Kadanoff L P (1985) Simulating hydrodynamics: a pedestrian model, J. Stat. Phys. 39, 267.

1113. Kadanoff L P (1986) Renormalization group analysis of the global properties of a strange attractor, J. Stat. Phys. 43, 395.

1114. Kadanoff L P (1987) Dimensional calculations for Julia sets, Phys. Scripta T19A, 19.

1115. Kadanoff L P, and Tang C (1984) Escape from strange repellers, Proc. Natl. Acad. Sci. USA, 81, 1276.

1116. Kahlert C, and Rössler O E (1984) Chaos as a limit in a boundary value problem, Z. Naturforsch. 39a, 1200.

1117. Kahlert C, and Rössler O E (1987) Analogues to a Julia boundary away from analyticity, Z. Naturforsch. 42a, 24.

1118. Kai T (1981) Universaility of power spectra of a dynamical system with an infinite sequence of period-doubling bifurcations, Phys. Lett. 86A, 263.

1119. Kai T (1982) Lyapunov number for a noisy $2^n$ Cycle, J. Stat. Phys. 29, 329.

1120. Kai T, and Tomita K (1979) Stroboscopic phase portrait of a forced nonlinear oscillator, Prog. Theor. Phys. 61, 54.

1121. Kajanto M J, and Salomaa M M (1985) Effects of external noise on the circle map and the transition to chaos in Josephson junctions, Solid State Commun. 53, 99.

1122. Kalafati Yu D, and Malakhov B A (1983) Dynamical chaos and spontaneous symmetry breaking in anharmonic systems excited by a periodic external force, JETP Lett. 7, 577; Russ. Orig. 37, 486.

1123. Kaneko K (1982) On the period-adding phenomena at the frequency locking in a one-dimensional mapping, Prog. Theor. Phys. 68, 669.

1124. Kaneko K (1983a) Similarity structure and scaling property of the period-adding phenomena, Prog. Theor. Phys. 69, 403.

1125. Kaneko K (1983b) Transition from torus to chaos accompanied by frequency lockings with symmetry breaking, Prog. Theor. Phys. 69, 1427.

1126. Kaneko K (1983c) Doubling of torus, Prog. Theor. Phys. 69, 1806.

1127. Kaneko K (1984a) Fates of three-torus. I. Double devil's staircase in lockings, Prog. Theor. Phys. 71, 282.

1128. Kaneko K (1984b) Fractalization of a torus, Prog. Theor. Phys. 71, 1112.

1129. Kaneko K (1984c) Oscillation and doubling of torus, Prog. Theor. Phys. 72, 202.

1130. Kaneko K (1984d) Period-doubling of kink-antikink patterns, quasiperiodicity in antiferro-like structures and spatial intermittency in coupled logistic lattice: towards a prelude of a "field theory of chaos", Prog. Theor. Phys. 72, 480.

1131. Kaneko K (1984e) Supercritical behaviour of disordered orbits of a circle map, Prog. Theor. Phys. 72, 1089.

1132. Kaneko K (1985a) Spatial period-doubling in open flow, Phys. Lett. 111A, 17.

1133. Kaneko K (1985b) Spatio-temporal intermittency in coupled map lattices, Prog. Theor. Phys. 74, 1033.

1134. Kaneko K (1987) Pattern competition intermittency and selective flicker noise in spatio-temporal chaos, Phys. Lett. 125A, 25.

1135. Kaneko K (1988a) Symplectic cellular automata, Phys. Lett. 129A, 9.

1136. Kaneko K (1988b) Chaotic diffusion of localized defect and pattern selection in spatio-temporal chaos, Europhys. Lett. 6, 193.

1137. Kaneko K, and Bagley R J (1985) Arnold diffusion, ergodicity and intermittency in a coupled standard mapping, Phys. Lett. 110A, 435.

1138. Kaneko K, and Konishi T (1987) Transition, ergodicity and Lyapunov spectra of Hamiltonian dynamical systems, J. Phys. Soc. Japan, 56, 2993.

1139. Kantz H, and Grassberger P (1985) Repellers, semi-attractors, and long-lived chaotic transients, Physica 17D, 75.

1140. Kantz H, and Grassberger P (1987) Chaos in four-dimensional Hamiltonian maps, Phys. Lett. 123A, 437.

1141. Kantz H, and Grassberger P (1988) Internal Arnold diffusion and chaos threshold in coupled symplectic maps, J. Phys. A21, L127.

1142. Kao Y H, Huang J C, and Gou Y S (1986) Direct observation of crises and the related low frequency noise in a Josephson analog, Phys. Rev. A34, 1628.

1143. Kao Y H, Huang J C, and Gou Y S (1987) Persistent properties of chaos in a Duffing oscillator, Phys. Rev. A35, 5228.

1144. Kao Y H, Huang J C, and Gou Y S (1988) Routes to chaos in the Duffing oscillator with a simple potential well, Phys. Lett. 131A, 91.

1145. Kapitaniak T (1986) Chaotic distribution of nonlinear systems perturbed by random noise, Phys. Lett. 116A, 251.

1146. Kapitaniak T (1987) Chaotic behaviour of anharmonic oscillations with time delay, J. Phys. Soc. Japan, 56, 1951.

1147. Kapitaniak T, Awrejcewicz J, and Steeb W-H (1987) Chaotic behavior of an anharmonic oscillator with almost periodic excitation, J. Phys. A20, L355.

1148. Kapitaniak T, and Wojewoda J (1988) Chaos in a limit cycle system with almost periodic excitation, J. Phys. A21, L843.

1149. Kaplan H (1983) New method for calculating stable and unstable periodic orbits, Phys. Lett. 97A, 365.

1150. Kaplan J L, and Yorke J A (1979a) Preturbulence: a regime observed in a fluid flow model of Lorenz, Commun. Math. Phys. 67, 93.

1151. Kaplan J L, and Yorke J A (1979b) Chaotic behavior of multidimensional difference equations, Lect. Notes in Math. 730, 228.

1152. Kapral R, Fraser S (1984) Bistable oscillating states in dissipative dynamical systems: scaling properties and one-dimensional maps, J. Phys. Chem. 88, 4845.

1153. Kapral R, Schell M, and Fraser S (1982) Chaos and fluctuations in nonlinear dissipative systems, J. Phys. Chem. 86, 2205.

1154. Kapral R, and Mandel P (1985) Bifurcation structure of the nonautonomous quadratic map, Phys. Rev. A32, 1076.

1155. Karney C F F (1983) Long-time correlations in the stochastic regime, Physica 8D, 360.

1156. Karney C F F, Rechester A B, and White R B (1982) Effect of noise on the standard mapping, Physica 4D, 425.

1157. Kaspar F, and Schuster H G (1986) Scaling at the onset of spatial disorder in coupled piecewise linear maps, Phys. Lett. 113A, 451.

1158. Kaspar F, and Schuster H G (1987) Easily calculable measure for the complexity of spatio-temporal patterns, Phys. Rev. A36, 842.

1159. Katok A B (1980) Lyapunov exponents, entropy and periodic points for diffeomorphisms, Publ. Math. IHES, 51, 137.

1160. Kats V A (1984) Experimental demonstration of the universal properties of a sequence of Feigenbaum period-doubling bifurcations in the onset of chaos in a distributed oscillator with retardation, Sov. Tech. Phys. Lett. 10, 288; Russ. Orig. 10, 684.

1161. Kats V A (1985) Appearance of chaos and its evolution in a distributed oscillator with delay (experiment), Radiophys. & Quantum Electr. 28, 107; Russ. Orig. 28, 161.

1162. Kats V A, and Kuznetsov S R (1987) Transition to multimode chaos in a simple model of an oscillator with a delay, Sov. Tech. Phys. Lett. 13, 302; Russ. Orig. 13, 727.

1163. Katsura S, and Fukuda W (1985) Exactly solvable models showing chaotic behavior, Physica 130A, 597.

1164. Katzen D, and Procaccia I (1987) Phase transitions in thermodynamic formalism of multifractals, Phys. Rev. Lett, 58, 1169.

1165. Kautz R L (1981) Chaotic states of RF-biased Josephson junctions, J. Appl. Phys. 52, 6241.

1166. Kautz R L, and Macfarlane J C (1986) Onset of chaos in the RF-biased Josephson junction, Phys. Rev. A33, 498.

1167. Kawai H, and Tye S-H H (1984) Approach to chaos: universal quantitative properties of one-dimensional maps, Phys. Rev. A30, 2005.

1168. Kay K G, and Ramachandran B (1988) Classical and quantal pseudo-ergodic regimes of the Hénon-Heiles system, J. Chem. Phys. 88, 5688.

1169. Kazarinoff N D, and Seydel R (1986) Bifurcations in Lorenz's symmetric fourth-order system, Phys. Rev. A4, 3387.

1170. Keener J P, and Glass L (1984) Global bifurcations of a periodically forced nonlinear oscillator, J. Math. Biol. 21, 175.

1171. Keolian R, Putterman S J, Turkevich L A, Rudnick I, and Rudnick J (1981) Subharmonic sequences in the Faraday experiment: departures from period-doubling, Phys. Rev. Lett. 47, 1133.

1172. Kerr W C, Williams M B, Bishop A R, Fesser K, Lomdahl P S, and Trullinger S E (1985) Symmetry and chaos in the motion of the damped driven pendulum, Z. Phys. B59, 103.

1173. Ketoja J A, and Kurkijarvi J (1986) Universality of the window structure and the density of aperiodic solutions in dissipative dynamical systems, Phys. Rev. A33, 2845.

1174. Khanin K M, and Sinai Ya G (1987) A new proof of M. Herman's theorem, Commun. Math. Phys. 112, 89.

1175. Khantha M (1987) Intermittency in Fibonacci chains, J. Phys. A20, L945.

1176. Killory H, Rössler O E, and Hudson J L (1987) Higher chaos in a four-variable chemical reaction model, Phys. Lett. 122A, 341.

1177. Kim M C, and Hsu C S (1986a) Computation of the largest Liapunov exponent by the generalized cell mapping: classification of persistent groups, J. Stat. Phys. 45, 49.

1178. Kim M C, and Hsu C S (1986b) Symmetry-breaking bifurcations for the standard mapping, Phys. Rev. A34, 4464.

1179. Kim S-H, Ostlund S, and Yu G (1988) Fourier analysis of multi-frequency dynamical systems, Physica 31D, 117.

1180. Kim S-H, and Ostlund S (1985) Renormalization of mappings of the two-torus, Phys. Rev. Lett. 55, 1165.

1181. Kim S-H, and Ostlund S (1986) Simultaneous rational approximants for physicists, Phys. Rev. A34, 3426.

1182. Kim S-Y, and B. Hu (1988) Singularity spectrum for period-n-tupling in area-preserving maps, Phys. Rev. A38, 1534.

1183. Kimura S, Schubert G, and Strans J M (1986) Route to chaos in porous-medium thermal convection, J. Fluid Mech. 166, 305.

1184. King R B (1983) Chemical applications of topology and group theory: 14. Topological aspects of chaotic chemical reactions, Theor. Chim. Acta (Berlin), 63, 323.

1185. King R, Barchas J D, and Huberman B A (1984) Chaotic behavior in dopamine neurodynamics, Proc. Natl. Acad. Sci. USA, 81, 1244.

1186. Kitano M, Yabuzaki T, and Ogawa T (1983) Chaos and period-doubling bifurcations in a simple acoustic system, Phys. Rev. Lett. 50, 713.

1187. Kitano M, Yabuzaki T, and Ogawa T (1984) Symmetry-recovering crises of chaos in polarization-related optical bistability, Phys. Rev. A29, 1288.

1188. Klar H (1986) Periodicity and chaos in strongly perturbed classical orbits for Coulomb interactions, Few-Body Systems (Austria), 1, 123.

1189. Klauder J R (1987) Semiclassical quantization of classically chaotic systems, Phys. Rev. Lett. 59, 748.

1190. Klinker T, Meyer-Ilse W, and Lauterborn W (1984) Period doubling and chaotic behavior in a driven Toda oscillator, Phys. Lett. 101A, 371.

1191. Knobloch E (1979) On the statistical dynamics of the Lorenz model, J. Stat. Phys. 20, 695.

1192. Knobloch E (1981) Chaos in a segmented disk dynamo, Phys. Lett. 82A, 439.

1193. Knobloch E, and Weiss N O (1981) Bifurcations in a model of double-diffusive convection, Phys. Lett. 85A, 127.

1194. Kocarev L (1987a) The basin boundaries of one-dimensional maps, Phys. Lett. 121A, 274.

1195. Kocarev L (1987b) Quasifractal metamorphoses of one dimensional maps, Phys. Lett. 125A, 3389.

1196. Kocarev L (1988) On a class of symmetrical chaotic attractors, Phys. Lett. 130A, 7.

1197. Koch B P, and Bruhn B (1988) Transient chaos in weakly coupled Josephson junctions, J. Physique, 49, 35.

1198. Koch B P, and Leven R W (1985) Subharmonic and homoclinic bifurcations in a parametrically forced pendulum, Physica 16D, 1.

1199. Koga H, Fujisaka H, and Inoue M (1983) Anomalous enhancement of the diffusion coefficient near the intermittency transition, Phys. Rev. A28, 2370.

1200. Koga S (1986) Phase description method to time averages in the Lorenz system, Prog. Theor. Phys. 76, 335.

1201. Koga S (1987) Phase description method to time averages in dissipative chaos. Construction of modified maps, Prog. Theor. Phys. 77, 1057.

1202. Kohyama T (1984) Non-stationarity of chaotic motions in an area preserving mapping, Prog. Theor. Phys. 71, 1104.

1203. Kohyama T, and Aizawa Y (1983) Orbital stability in a piece-wise linear map of the circle onto itself, Prog. Theor. Phys. 70, 1002.

1204. Kohyama T, and Aizawa Y (1984) Theory of the intermittent chaos: 1/f spectrum and the Pareto-Zipf law, Prog. Theor. Phys. 71, 917.

1205. Koiller J (1984) A mechanical system with a 'wild' horseshoe, J. Math. Phys. 25, 1599.

1206. Koiller J, Balthazar J M, and Yokoyama T (1987) Relaxation-chaos phenomena in celestial mechanics. I. On Wisdom's model for the 3/1 Kirkwood gap, Physica 26D, 85.

1207. Koiller J, de Mello Neto J R, and Damiao Soares I (1985) Homoclinic phenomena in the gravitational collapse, Phys. Lett. 110A, 260.

1208. Kolmogorov A N (1954) Preservation of conditionally periodic movements with small change in the Hamilton function, Akad. Nayk SSSR Dokl. 98, 527.

1209. Konno H, and Soneda H (1988) Chaotic behaviour of soliton-like pulses in a driven modified Kuramoto-Sivashinsky equation, J. Phys. Soc. Japan, 57, 1163.

1210. Konno K, Irie H, and Shimada I (1986) Generating function and its formal derivatives for dynamical systems, Prog. Theor. Phys. 76, 561.

1211. Konno K, and Tateno H (1984) Duffing's equation in complex times and chaos, Prog. Theor. Phys. 72, 1047.

1212. Kornev V K, and Likharev K K (1986) Chaos in a superconducting quantum interferometer, Sov. J. Commun. Tech. & Electr. 31, 113.

1213. Kostelich E J, and Yorke J A (1987) Lorenz cross sections of the chaotic attractor of the double rotor, Physica 24D, 263.

1214. Kostelich E J, and Yorke J A (1988) Noise reduction in dynamical systems, Phys. Rev. A38, 1649.

1215. Kottalam J, West B J, and Lindenberg K (1987) Analogy bewteen the Lorenz strange attractor and a bistable stochastic oscillator, J. Stat. Phys. 46, 119.

1216. Kowalik Z J, Franaszek M, and Pieranski P (1988) Self-reanimated chaos in the bouncing-ball system, Phys. Rev. A37, 4016.

1217. Kozak J J, Musho M K, and Hatlee M D (1982) Chaos, periodic chaos, and the random-walk problem, Phys. Rev. Lett. 49, 1801.

1218. Krug J (1987) Optical analog of a kicked quantum oscillator, Phys. Rev. Lett. 59, 2133.

1219. Kruscha K J G, and Pompe B (1988) Information flow in one-dimensional maps, Z. Naturforsch. 43a, 933.

1220. Kryukov B I, and Seredovich G I (1981) The 'strange' behavior of solutions of the Duffing equation, Sov. Phys. Dokl. 26, 501; Russ. Orig. 258, 311.

1221. Kubicek M, and Holodniok M (1987) Algorithms for determining period doubling bifurcation points in ordinary differential equations, J. Comput. Phys. 70, 203.

1222. Kumar K, Agarwal A K, Bhattacharjee J K, and Banerjee K (1987) Precursor transition in dynamical systems undergoing period doubling, Phys. Rev. A35, 2334.

1223. Kunick A, and Steeb W-H (1985) Coupled chaotic oscillators, J. Phys. Soc. Japan, 54, 1220.

1224. Kuramoto Y, and Koga S (1982) Anomalous period-doubling bifurcations leading to chemical turbulence, Phys. Lett. 92A, 1.

1225. Kus M (1983) Integrals of motion for the Lorenz system, J. Phys. A16, L689.

1226. Kus M (1985) Statistical properties of the spectrum of the two-level system, Phys. Rev. Lett. 54, 1343.

1227. Kus M, Scharf R, and Haake F (1987) Symmetry versus degree of level repulsion for kicked quantum systems, Z. Phys. 66B, 129.

1228. Kuzmin M V, Nemov I V, Struchebrukhov A A, Bagratashvili V N, and Letokhov V S (1986) Chaotic non-ergodic vibrational motion in a polyatomic molecule, Chem. Phys. Lett. 124, 522.

1229. Kuznetsov S P (1984) Effect of a periodic external perturbation in a system which exhibits an order-chaos transition through period doubling bifurcations, JETP Lett. 39, 133; Russ. Orig. 9, 113.

1230. Kuznetsov S P (1985) Universality and scaling in the behavior of coupled Feigenbaum systems, Radiophys. & Quantum Electron, 28, 681; Russ. Orig. 28, 991.

1231. Kuznetsov S P (1986) Renormalization group, universality, and scaling in the dynamics of one-dimensional autowave media, Radiophys. & Quantum Electr. 29, 679; Russ. Orig. 29, 889.

1232. Kuznetsov S P, and Pikovsky A S (1986) Universality and scaling of period doubling bifurcations in a dissipative distributed medium, Physica 19D, 384.

1233. Lafon A, Rossi A, and Vidal C (1983) The power of chaos measured through the spectral analysis of experimental data, J. Physigue, 44, 505.

1234. Lahiri A (1988) Spatially inhomogeneous structures in a one-dimensional array of Brusselators, J. Chem. Phys. 88, 7459.

1235. Landa P S, and Stratonovich R L (1982) Stationary probability distribution for one of the simplest strange attractors, Sov. Phys. Dokl. 27, 1032; Russ. Orig. 267, 832.

1236. Landa P S, and Stratonovich R L (1987) Theory of intermittency, Radiophys. Quantum Electron. 30, 53.

1237. Landau L D (1944) On the problem of turbulence, C. R. Acad. Sci. USSR: 44, 311; in *Collected Papers of Landau*, ed. by D. ter Haar, 387.

1238. Landsberg P T, Scholl E, and Shukla P (1988) A simple model for the origin of chaos in semiconductors, Physica 30D, 235.

1239. Lanford O E (1977a) Computer pictures of the Lorenz attractor, Appendix to Williams (1977).

1240. Lanford O E (1977b) An introduction to the Lorenz system, in *1976 Duke University Turbulence Conference*, Duke University Math. Series, III.

1241. Lanford O E (1982a) The strange attractor theory of turbulence, Ann. Rev. Fluid Mech. 14, 347.

1242. Lanford O E (1982b) A computer-assisted proof of the Feigenbaum conjectures, Bull. Amer. Math. Soc. 6, 427.

1243. Lanwerier H A (1986) The structure of a strange attractor, Physica 21D, 146.

1244. Lathrop D P, and Kostelich E J (1989) Characterization of an experimental strange attractor by periodic orbits, Phys. Rev. A40, 4028.

1245. Laufer J (1983) Deterministic and stochastic aspects of turbulence, Trans. ASME J. Appl. Mech. 50, No.4B, 1079.

1246. Laurien E, and Fasel H (1988) Numerical investigation of the onset of chaos in the flow between rotating cylinders, Z. Angew. Math. Mech. 68, 311.

1247. Lauterborn W, and Cramer E (1981) Subharmonic route to chaos observed in acoustics, Phys. Rev. Lett. 47, 1445.

1248. Lauterborn W, and Parlitz U (1988) Methods of chaos physics and their application to acoustics, J. Acoust. Soc. Am. 84, 1975.

1249. Le Berre M, Ressayre E, Gibbs H M, Kaplan D L, and Rose M (1987) Conjecture on the dimensions of chaotic attractors of delayed-feedback dynamical systems, Phys. Rev. A35, 4020.

1250. Le Berre M, Ressayre E, Tallet A, and Gibbs H M (1986) High-dimension chaotic attractors of a nonlinear ring cavity, Phys. Rev. Lett. 56, 274.

1251. Le Berre M, Ressayre E, and Tallet A (1989) Lyapunov analysis of the Ruelle-Takens route to chaos i an optical retarded differential system, Optics Commun. 72, 123.

1252. Le Treut H, and Ghil M (1983) Orbital forcing, climatic interactions, and glaciation cycles, J. Geophys. Res. 88, 5167.

1253. Ledrappier F (1981) Some relations between dimension and Lyapunov exponents, Commun. Math. Phys. 81, 229.

1254. Lee C-K, and Moon F C (1986) An optical technique for measuring fractal dimension of planar Poincaré maps, Phys. Lett. 114A, 222.

1255. Lee K-C (1983) The universality of period-doubling bifurcations in certain 2D reversible area-preserving mappings with guadratic nonlinearity, J. Phys. A16, L137.

1256. Lee K-C (1988) Long-time tails in a chaotic system, Phys. Rev. Lett. 60, 1991.

1257. Lee K-C, Kim S-Y, and Choi D-I (1984) Universality of $k \times 3^n$ and $k \times 4^n$ bifurcations in area-perserving maps, Phys. Lett. 103A, 225.

1258. Lehtihet H E, and Miller B N (1986) Numerical study of a billiard in a gravitational field, Physica 21D, 93.

1259. Leiber Th, and Risken H (1988) Stability of parametrically excited dissipative systems, Phys. Lett. 129A, 214.

1260. Leipnik R B, and Newton T A (1981) Double strange attractor in rigid body motion with linear feedback control, Phys. Lett. 86A, 63.

1261. Leopold J G, and Richards D (1988) A study of quantum dynamics in the classically chaotic regime, J. Phys. A21, 2179.

1262. Leven R W, Pompe B, Wilke C, and Koch B P (1985) Experiments on periodic and chaotic motions of a parametrically forced pendulum, Physica 16D, 371.

1263. Leven R W, and Koch B P (1981) Chaotic behaviour of a parametrically excited damped pendulum, Phys. Lett. 86A, 71.

1264. Levenson M T (1982) Even and odd harmonic frequencies and chaos in Josephson junctions: impact on parametric amplifiers? J. Appl. Phys. 53, 4294.

1265. Levi B G (1986) New global fractal formalism describes paths to turbulence, Phys. Today, 39, April, 17.

1266. Levi M (1981) Qualitative analysis of the periodically forced relaxation oscillations, Mem. AMS, 214, 1.

1267. Levi M, Hoppensteadt F, and Miranker W (1978) Dynamics of the Josephson junction, Quart. Appl. Math. 35, 167.

1268. Levinson N (1949) A second order differential equation with singular solutions, Ann. Math. 50, 127.

1269. Levy Y E (1982) Some remarks about computer studies of dynamical systems, Phys. Lett. 88A, 1.

1270. Lewenstein M, and Tél T (1985) On the dynamics of ensemble averages in chaotic maps, Phys. Lett. 109A, 411.

1271. Li J-B, and Liu Z-R (1985) Chaotic behaviour in planar quardratic Hamiltonian system with periodic perturbation, Kexue Tongbao (Sci. Bull. China), 30, 1285.

1272. Li J-N (1985) Period-doubling bifurcation for a delay-differential equation related to optical bistability, Chinese Phys. Lett. 2, 497.

1273. Li J-N, and Hao B-L (1989) Bifurcation spectrum in a delay-differential system related to optical bistability, Commun. Theor. Phys. 11, 265.

1274. Li T Y, Misiurewicz M, Pianigiani G, and Yorke J A (1982) Odd chaos, Phys. Lett. 87A, 271.

1275. Li T Y, and Yorke J A (1975) Period three implies chaos, Am. Math. Monthly, 82, 985.

1276. Li W-T (1986) Fractal dimension of Cantori, Phys. Rev. Lett. 57, 655.

1277. Li Y-X, Ding D-F, and Xu J-H (1984) Chaos and other temporal self-organization patterns in coupled enzyme-catalyzed systems, Commun. Theor. Phys. 3, 629.

1278. Liaw C Y (1987) Subharmonic response of offshore structures, ASCE J. Eng. Mech. March, 366.

1279. Liaw C Y (1988a) Bifurcations of subharmonics and chaotic motions of articulated towers, Eng. Struct. 10, 117.

1280. Liaw C Y (1988b) Chaotic and periodic responses of a coupled wave-force and structure system, Computers & Structures, 30, 985.

1281. Libchaber A (1987) From chaos to turbulence in Bénard convection, Proc. Roy. Soc. London, A413, 633.

1282. Libchaber A, Laroche C, and Fauve S (1982) Period doubling cascade in mercury: quantitative measurement, J. Physique Lett. 43, L211.

1283. Lie G C, and Yuan J -M (1986) Bistable and chaotic behavior in a damped driven Morse oscillator: a classical approach, J. Chem. Phys. 84, 5486.

1284. Lieberman M A, and Tsang K Y (1985) Transient chaos in dissipative perturbed near-integrable Hamiltonian systems, Phys. Rev. Lett. 55, 908.

1285. Lima R, and Ruffo S (1988) Scaling laws for all Liapunov exponents: models and measurements, J. Stat. Phys. 52, 259.

1286. Lin C A, and Lian B H (1986) Chaotic behaviour in a low-order unforced, inviscid barotropic model, Pure & Appl. Geophys. 124, 1087.

1287. Lin I, Wu M-S (1987) Spatio-temporal chaos in weakly ionized magneto-plasmas, Phys. Lett. 124A, 271.

1288. Lin W A, and Reichl L E (1986) External field induced chaos in an infinite square well potential, Physica 19D, 145.

1289. Linde A D (1988) Chaotic inflation with constrained fields, Phys. Lett. 202B, 194.

1290. Ling F-H (1985) A numerical method for determining bifurcation curves of mappings, Phys. Lett. 110A, 116.

1291. Ling F-H (1987) A numerical study of the applicability of Melnikov's method, Phys. Lett. 119A, 447.

1292. Ling F-H (1988a) Bifurcation curves of the Hénon map determined by a multiple shooting technique, Chinese Phys. Lett. 5, 121.

1293. Ling F-H (1988b) On fractal attracting basin boundaries and their consequences, Kexue Tongbao (Sci. Bull. China), 33, 783.

1294. Ling F-H, and Bao G-W (1987) A numerical implementation of Melnikov's method, Phys. Lett. 122A, 413.

1295. Linsay P S (1981) Period doubling and chaotic behavior in a driven anharmonic oscillator, Phys. Rev. Lett. 47, 1349.

1296. Linsay P S (1985) Approximate scaling of period doubling windows, Phys. Lett. 108A, 431.

1297. Lippi G L, Tredicce J R, Abraham N B, and Arecchi F T (1985) Deterministic mode alternation, giant pulses and chaos in a bidirectional $CO_2$ ring laser, Opt. Commun. 53, 129.

1298. Liu E, and Yuan J-M (1984) A chaotic attractor with hysteresis in laser-driven molecules, Phys. Rev. A29, 2257.

1299. Liu K L, Lo W S, and Young K (1984) Generalized renormalization group equation for period-doubling bifurcations, Phys. Lett. 105A, 103.

1300. Liu K L, Lo W S, and Young K (1987) Entrainment and chaos in a discrete map with commensurate external forcing, Nuovo Cimento, 97, 170.

1301. Liu K L, and Young K (1985) Stability of forced nonlinear oscillators via Poincaré map, J. Math. Phys. 27, 502.

1302. Liu S-D (1986) Chaos in internal-wave dynamics and onset of atmospheric turbulence, Scientia Sinica B29, 1201.

1303. Liverani C, and Turchetti G (1986) Improved KAM estimates for the Siegel radius, J. Stat. Phys. 45, 1071.

1304. Livi R, Pettini M, Ruffo S, Vulpiani A (1987) Chaotic behavior in nonlinear Hamiltonian systems and equilibrium statistical mechanics, J. Stat. Phys. 48, 539.

1305. Livi R, Politi A, Ruffo S, and Vulpiani A (1987) Liapunov exponents in high-dimensional symplectic dynamics, J. Stat. Phys. 46, 147.

1306. Livi R, Politi A, and Ruffo S (1986) Distribution of characteristic exponents in the thermodynamic limit, J. Phys. A19, 2033.

1307. Lohofer G, and Mayer D (1985) Correlation functions of a time-continuous dissipative system with a strange attractor, Phys. Lett. 113A, 105.

1308. Lombardi M, Labastie P, Bordas M C, and Broyer M (1988) Molecular Rydberg states: classical chaos and its correspondence in quantum mechanics, J. Chem. Phys. 89, 3479.

1309. Longcope D W, and Sudan R N (1987) Arnold diffusion in 11/2 dimensions, Phys. Rev. Lett. 59, 1500.

1310. Lorenz E N (1963) Deterministic nonperodic flow, J. Atmos. Sci. 20, 130.

1311. Lorenz E N (1979) On the prevalence of aperiodicity in simple systems, Lect. Notes in Math. 755, 53.

1312. Lorenz E N (1984a) The local structure of a chaotic attractor in four dimensions, Physica 13D, 90.

1313. Lorenz E N (1984b) Irregularity: a fundamental property of the atmosphere, Tellus, 36A, 98.

1314. Lorenz E N (1984c) A very narrow spectral band, J. Stat. Phys. 36, 1.

1315. Lorenz E N (1985) Lyapunov number and the local structure of attractors, Physica 17D, 279.

1316. Lozi R (1978) Un attracteur etrange (?) du type attracteur de Hénon, J. Physique, 39, Coll. C5, 9.

1317. Lozi R, and Ushiki S (1988) Organized confinors and anti-confinors in constrained 'Lorenz system', Ann. Telecommun. 43, 187.

1318. Lucke M (1976) Statistical dynamics of the Lorenz model, J. Stat. Phys. 15, 455.

1319. Lugiato L A, Narducci L M, Bandy D K, and Pennise C A (1983) Breathing, spiking and chaos in a laser with injected signal, Opt. Commun. 46, 64.

1320. Lunsford G H, and Ford J (1972) On the stability of periodic orbits for nonlinear oscillator systems in regions exhibiting stochastic behavior, J. Math. Phys. 13, 700.

1321. Lutzky M (1988) Reverse multifurcation and universal constants, Phys. Lett. 128A, 332.

1322. Lvov V S, Predtechenskii A A, and Chernykh A (1981) Bifurcation and chaos in a system of Taylor vortices: a natural and numerical experiment, Sov. Phys. JETP, 53, 562.

1323. Lvov V S, and Predtechensky A A (1981) On Landau and stochastic attractor pictures in the problem of transition to turbulence, Physica 2D, 38.

1324. Lyubimov D V, and Zaks M A (1983) Two mechanisms of the transition to chaos in finite-dimensional models of convection, Physica 9D, 52.

1325. MacDonald A H, and Plischke M (1983) Study of the driven damped pendulum: application to Josephson junctions and charge-density-wave systems, Phys. Rev. B27, 210.

1326. MacKay R S (1982) Islets of stability beyond period doubling, Phys. Lett. 87A, 7.

1327. MacKay R S (1987) Rotation interval from a time series, J. Phys. A20, 587.

1328. MacKay R S, Meiss J D, and Percival I C (1984a) Transport in Hamiltonian systems, Physica 13D, 55.

1329. MacKay R S, Meiss J D, and Percival I C (1984b) Stochasticity and transport in Hamiltonian systems, Phys. Rev. Lett. 52, 697.

1330. MacKay R S, Meiss J D, and Percival I C (1987) Resonances in area-preserving maps, Physica 27D, 1.

1331. MacKay R S, and Percival I C (1985) Converse KAM: theory and practice, Commun. Math. Phys. 98, 469.

1332. MacKay R S, and Tresser C (1984a) Transition to chaos for two-frequency systems, J. Physique Lett. 45, L741.

1333. MacKay R S, and Tresser C (1984b) Badly ordered orbits of circle maps, Math. Proc. Camb. Phil. Soc. 96, 447.

1334. MacKay R S, and Tresser C (1986) Transition to topological chaos for circle maps, Physica 19D, 206-37; Errata: 29D(1988) 427.

1335. MacKay R S, and Tresser C (1987) Some flesh on the skeleton: the bifurcation structure of bimodal maps, Physica 27D, 412.

1336. MacKay R S, and Van Zeijts J B (1988) Period doubling for bimodal maps: a horseshoe for a renormalization operator, Nonlinearity, 1, 253.

1337. Machacek M (1986) Invariant measure of dissipative dynamical systems, Czech. J. Phys. B336, 651.

1338. Mackey M C, and Glass L (1977) Oscillation and chaos in physiological control systems, Science, 197, 287.

1339. Maganza C, Causse R, and Laloe F (1986) Bifurcations, period doublings and chaos in clarinetlike systems, Europhys. Lett. 1, 295.

1340. Malagoli A, Paladin G, and Vulpiani A (1986) Transitions to stochasticity in Hamiltonian systems: some numerical results, Phys. Rev. A34, 1550.

1341. Malomed A B (1983) A simple dynamical system with stochastic behavior, Physica 8D, 343.

1342. Malraison B, Atten P, Berge P, and Dubois M (1983) Dimension of strange attractors: an experimental determination for the chaotic regime of two convective systems, J. Physique Lett. 44, L897.

1343. Malraison B, and Atten P (1982) Chaotic behavior of instability due to unipolar ion injection in a dielectric liquid, Phys. Rev. Lett. 49, 723.

1344. Mandel P, and Kapral R (1983) Subharmonic and chaotic bifurcation structure in optical bistability, Opt. Commun. 47, 151.

1345. Mandell A J (1987) Dynamical complexity and pathological order in the cardiac manitoring problem, Physica 27D, 235.

1346. Mankin J C, and Hudson J L (1984) Oscillatory and chaotic behaviour of a forced exothermic chemical reaction, Chem. Engin. Science, 39, 1907.

1347. Mankin J C, and Hudson J L (1986) The dynamics of coupled nonisothermal continuous stirred tank reactors, Chem. Eng. Sci. 41, 2651.

1348. Manneville P (1980a) Intermittency in dissipative dynamical systems, Phys. Lett. 79A, 33.

1349. Manneville P (1980b) Intermittency, self-similarity and $1/f$ spectrum in dissipative dynamical systems, J. Physique, 41, 1235.

1350. Manneville P (1982) On the statistics of turbulent transients in dissipative systems, Phys. Lett. 90A, 327.

1351. Manneville P, and Piquemal J M (1982) Transverse phase diffusion in Rayleigh-Bénard convection, J. Physique Lett. 43, 253.

1352. Manneville P, and Pomeau Y (1979) Intermittency and the Lorenz model, Phys. Lett. 75A, 1.

1353. Manneville P, and Pomeau Y (1980) Different ways to turbulence in dissipative systems, Physica 1D, 219.

1354. Manton N S, and Nauenberg M (1983) Universal scaling behavior for iterated maps in the complex plane, Commun. Math. Phys. 89, 555.

1355. Mao J-M (1988) Period doubling in six-dimensional symmetric volume preserving maps, J. Phys. A21, 3079.

1356. Mao J-M, Satija I I, and Hu B (1985) Evidence for a new period-doubling sequence in four-dimensional symplectic maps, Phys. Rev. A32, 1927.

1357. Mao J-M, and Greene J M (1987) Renormalization of period-doubling in symmetric four-dimensional volume-preserving maps, Phys. Rev. A35, 3911.

1358. Mao J-M, and Helleman R H G (1988) Nonsymmetric four-dimensional volume-preserving maps: universality classes of period doubling, Phys. Rev. A37, 3475.

1359. Mao J-M, and Hu B (1987) Corrections to scaling for period doubling, J. Stat. Phys. 46, 111.

1360. Mao J-M, and Hu B (1988) Multiple scaling and the fine structure of period doubling, Int. J. Mod. Phys. B2, 65.

1361. Markus M, Kuschmitz D, and Hess B (1984) Chaotic dynamics in yeast glycolysis under periodic substrate input flux, FEBS Lett. 172, 235.

1362. Markus M, Kuschmitz D, and Hess B (1985) Properties of strange attractors in yeast glycolysis, Biophys. Chem. 22, 9.

1363. Markus M, Müller S C, and Hess B (1985) Observation of entrainment, quasiperiodicity and chaos in glycolyzing yeast extracts under periodic glucose input, Ber. Bunsenges. Phys. Chem. 89, 651.

1364. Marotto F R (1978) Snap-back repellers imply chaos in $R^n$, J. Math. Anal. Appl. 63, 199.

1365. Marotto F R (1979) Chaotic behavior in the Hénon mapping, Commun. Math. Phys. 68, 187.

1366. Martiel J L, and Goldbeter A (1985) Autonomous chaotic behaviour of the slime mould Dictyostelium discoideum predicted by a model for cyclic AMP signalling, Nature, 313, 590.

1367. Martien P, Pope S C, Scott P L, and Shaw R S (1985) The chaotic behavior of the leaky faucet, Phys. Lett. 110A, 399.

1368. Martin P C (1976) Instabilities, oscillations, and chaos, J. Physique, 37, Coll. C1-57.

1369. Martin S, Leber H, and Martienssen W (1984) Oscillatory and chaotic states of the electrical conduction in barium sodium niobate crystals, Phys. Rev. Lett. 53, 303-6.

1370. Martin S, and Martienssen W (1986) Circle maps and mode locking in the driven electrical conductivity of barium sodium niobate crystals, Phys. Rev. Lett. 56, 1522.

1371. Martinez-Mekler G C, Mondragon R, and Perez R (1986) Basin-structure invariance of circle maps with bistable dynamics, Phys. Rev. A33, 2143.

1372. Marzec C J, and Spiegel E A (1980) Ordinary differential equations with strange attractors, SIAM J. Appl. Math. 38, 403.

1373. Maselko J and Epstein I R (1984) Chemical chaos in the chlorite-thiosulfate reaction, J. Chem. Phys. 80, 3175.

1374. Maselko J, and Swinney H L (1986) Complex periodic oscillations and Farey arithmetic in the Belousov-Zhabotinskii reaction, J. Chem. Phys. 85, 6430.

1375. Maselko J, and Swinney H L (1987) A Farey triangle in the Belousov-Zhabotinskii reaction, Phys. Lett. A119, 403-6.

1376. Mashiyama K T, Takayoshi K, and Mori H (1981) Anomalous fluctuations near nonequilibrium soft transitions. I. Homogeneous transitions, Prog. Theor. Phys. 65, 1820.

1377. Matinyan S G (1985) Dynamic chaos of non-Abelian gauge fields, Sov. J. Part. & Nucl. 16, 226; Russ. Orig. 16, 522.

1378. Matinyan S G, Prokhorenko E B, and Savvidy G K (1988) Non-integrability of time-dependent spherically symmetric Yang-Mills equations, Nucl. Phys. B298, 414.

1379. Matsumoto G, Aihara K, Hanyu Y, Takahashi N, and Yoshizawa S (1987) Chaos and phase locking in normal squid axons, Phys. Lett. 123A, 162.

1380. Matsumoto K, and Tsuda I (1983) Noise-induced order, J. Stat. Phys. 31, 87; 33, 757.

1381. Matsumoto K, and Tsuda I (1987) Extended information in one-dimensional maps, Physica 26D, 347.

1382. Matsumoto K, and Tsuda I (1988) Calculation of information flow rate from mutual information, J. Phys. A21, 1405.

1383. Matsumoto S, and Yasui Y (1988) Chaos on the super Riemann surface, Prog. Theor. Phys. 79, 1022.

1384. Matsumoto T (1984) A chaotic attractor from Chuo's circuit, IEEE Trans. on Cir. & Sys. CAS-31, 1055.

1385. Matsumoto T, Chua L O, and Komuro M (1985) The double scroll, IEEE Trans. Cir. and Sys. CAS-32, 798.

1386. Matsumoto T, Chua L O, and Komuro M (1986) The double scroll bifurcations, Int. J. Circuit Theory Appl. 14, 117.

1387. Matsumoto T, Chua L O, and Komuro M (1987) Birth and death of the double scroll, Physica 24D, 97.i

1388. Matsumoto T, Chua L O, and Takumasu K (1986) Double scroll via a two-transister circuit, IEEE Trans. on Cir. & Sys. CAS-33, 828.

1389. Matsushita T, and Terasaka T (1983) Mass dependence of the KAM stability and low-order resonances in the kinetically coupled two-degrees-of-freedom Morse system, Chem. Phys. Lett. 100, 138.

1390. Matsushita T, and Terasaka T (1984) A connection between classical chaos and the quantized energy spectrum: level spacing distribution in a kinetically coupled quantum Morse system with two degrees of freedom, Chem. Phys. Lett. 105, 511.

1391. Maurer J, and Libchaber A (1979) Rayleigh-Bénard experiment in liquid He: frequency locking and the onset of turbulence, J. Physique Lett. 40, L419.

1392. Maurer J, and Libchaber A (1980) Effect of Prandtl number on the onset of turbulence, J. Physique Lett. 41, L515.

1393. May R M (1974) Biological populations with nonoverlapping generations, stable points, stable cycles and chaos, Science, 186, 645.

1394. May R M (1976) Simple mathematical models with very complicated dynamics, Nature, 261, 459.

1395. May R M, and Oster G F (1976) Bifurcations and dynamic complexity in simple ecological models, Amer. Natur. 110, 573.

1396. May R M, and Oster G F (1980) Period-doubling and the onset of turbulence: an analytic estimate of the Feigenbaum ratio, Phys. Lett. 78A, 1.

1397. Mayer-Kress G, and Haken H (1981a) Intermittent behavior of the logistic system, Phys. Lett. 82A, 151.

1398. Mayer-Kress G, and Haken H (1981b) The influence of noise on the logistic model, J. Stat. Phys. 26, 149.

1399. Mayer-Kress G, and Haken H (1984) Attractors of convex maps with positive Schwarzian derivative in the presence of noise, Physica 10D, 329.

1400. Mayer-Kress G, and Haken H (1987) An explicit construction of a class of suspensions and autonomous differential equations for diffeomorphisms in the plane, Commun. Math. Phys. 111, 63.

1401. McAvity D M, Enns R H, and Rangnekar S S (1988) Bistable solitons and the route to chaos, Phys. Rev. A38, 4647.

1402. McCauley J L (1987) Chaotic dynamical systems as automata, Z. Naturforsch. 42a, 547.

1403. McCauley J L (1988) An introduction to nonlinear dynamics and chaos theory, Phys. Scripta T20, 5.

1404. McCauley J L, and Palmore J I (1986) Computable chaotic orbits, Phys. Lett. 115A, 433.

1405. McCreadie G A, and Rowlands G (1982) An analytical approximation to the Lyapunov number for 1D maps, Phys. Lett. 91A, 146.

1406. McDonald S W, Grebogi C, Ott E, and Yorke J A (1985a) Fractal basin boundaries, Physica 17D, 125.

1407. McDonald S W, Grebogi C, Ott E, and Yorke J A (1985b) Structure and crises of fractal basin boundaries, Phys. Lett. 107A, 51.

1408. McDonald S W, Grebogi C, and Kaufman A N (1985) Locally coupled evolution of wave and particle distribution in general magnetoplasma geometry, Phys. Lett. 111A, 19.

1409. McDonald S W, and Kaufman A N (1988) Wave chaos in the studium: statistical properties of short-wave solutions of the Helmholtz equation, J. Phys. A37, 3067.

1410. McGarr P R, and Percival I C (1984) The transition to chaos for a special solution of the area-preserving quadratic map, Physica 14D, 49.

1411. McGuinness M J (1983) The fractal dimension of the Lorenz attractor, Phys. Lett. 99A, 5.

1412. McGuire J B, and Thompson C J (1980) Distribution of iterates of first order difference equations, Bull. Austral. Math. Soc. 22, 133.

1413. McGuire J B, and Thompson C J (1981) On the universality and computation of Feigenbaum's $\delta$, Phys. Lett. 84A, 9.

1414. McGuire J B, and Thompson C J (1982) Asymptotic properties of iterates of nonlinear transformations, J. Stat. Phys. 27, 183.

1415. McKay S R, Berker A N, and Kirkpatrick S (1982) Spin-glass behavior in frustrated Ising models with chaotic renormalization-group trajectories, Phys. Rev. Lett. 48, 767.

1416. McLaughlin J B (1976) Successive bifurcations leading to stochastic behavior, J. Stat. Phys. 15, 307.

1417. McLaughlin J B (1979a) Stochastic behavior in slightly dissipative systems, Phys. Rev. A20, 2114.

1418. McLaughlin J B (1979b) The role of dissipation in a truncation of Hénon's map, Phys. Lett. 72A, 271.

1419. McLaughlin J B (1980) Connection between dissipative and resonant conservative nonlinear oscillators, J. Stat. Phys. 19, 587.

1420. McLaughlin J B (1981) Period-doubling bifurcations and chaotic motion of a parametrically forced pendulum, J. Stat. Phys. 24, 375.

1421. McLaughlin J B, and Orszag S A (1982) Transition from periodic to chaotic thermal convection, J. Fluid Mech. 122, 123.

1422. Mees A I, Rapp E P, and Jennings L S (1987) Singular value decomposition and embedding dimension, Phys. Rev. A36, 340.

1423. Meijaard J P, and de Pater A D (1989) Railway vehicle systems dynamics and chaotic vibrations, Int. J. Nonlin. Mech. 24, 1.

1424. Meiss J D (1986) Class renormalization: islands around islands, Phys. Rev. A34, 2375.

1425. Meiss J D, and Cary J R (1983) Correlations of periodic area-preserving maps, Physica 6D, 375.

1426. Meiss J D, and Ott E (1986) Markov tree model for transport in area-preserving maps, Physica 20D, 387.

1427. Meiss J D, Cary J R, Grebogi C, Crawford J D, Kaufman A N, and Abarbanel H D I (1983) Correlations of periodic, area-preserving maps, Physica 6D, 375.

1428. Meissner H, and Schmidt G (1985) A simple experiment for studying the transition from order to chaos, Am. J. Phys. 54, 800.

1429. Mello T M, and Tufillaro N B (1987) Strange attractors of a bouncing ball, Am. J. Phys. 55, 316.

1430. Melnikov V K (1963) On the stability of the center for time periodic perturbations, Trans. Moscow Math. Soc. 12, 1.

1431. Mendoza L (1985) The entropy of $C^2$ surface diffeomorphisms in terms of Hausdorff dimension and a Lyapunov exponent, Ergod. Th. & Dyn. Sys. 5, 273.

1432. Meredith D C, and Koonin S E (1988) Quantum chaos in a schematic shell model, Phys. Rev. A37, 3499.

1433. Meron E, and Procaccia I (1986) Theory of chaos in surface waves: the reduction from hydrodynamics to few-dimensional dynamics, Phys. Rev. Lett, 56, 1323.

1434. Meron E, and Procaccia I (1987) Gluing bifurcations in critical flows: the route to chaos in parametrically excited surfacec waves, Phys. Rev. A35, 4008.

1435. Mestel B, and Percival I C (1987) Newton method for highly unstable orbits, Physica 24D, 172.

1436. Metropolis N, Stein M L, and Stein P R (1973) On finite limit sets for transformations on the unit interval, J. Comb. Theor. 15, 25.

1437. Metzler W, Beau W, Frees W, and Ueberla A (1987) Symmetry and self-similarity with coupled logistic maps, Z. Naturforsch. 42a, 310.

1438. Meunier C (1984) Continuity of type-I intermittency from a measure-theoretical point of view, J. Stat. Phys. 6, 321.

1439. Meunier C, and Verga A D (1988) Noise and bifurcations, J. Stat. Phys. 50, 345.

1440. Meyer H-D (1986) Theory of the Liapunov exponents of Hamiltonian systems and a numerical study on the transition from regular to irregular classical motion, J. Chem. Phys. 84, 3147.

1441. Miao G-Q, Wang B-R, and Wei R-J (1987) Multi-mode and chaotic oscillations in a nonlinear system of acoustic hybrid with single peak frequency response, Chinese Phys. Lett. 4, 457.

1442. Miles J (1984a) Chaotic motion of a weakly nonlinear, modulated oscillator, Proc. Natl. Acad. Sci. USA, 81, 3919.

1443. Miles J (1984b) Resonant forced motion of two quadratically coupled oscillators, Physica 13D, 247.

1444. Miles J (1984c) Strange attractors in hydrodynamics, Advances in Appl. Mech. 24, 189.

1445. Miles J (1988a) Resonance and symmetry breaking for a nonlinear oscillator, Phys. Lett. 130A, 276.

1446. Miles J (1988b) Resonance and symmetry breaking of the pendulum, Physica 31D, 252.

1447. Milnor J (1985) On the concept of attractor, Commun. Math. Phys. 99, 177.

1448. Milnor J, and Thurston W (1988) On iterated maps of the interval, Lect. Notes in Math. 1342, 465.

1449. Milonni P W, Ackerhalt J R, and Galbraith H W (1983) Chaos and nonlinear optics: a chaotic Raman attractor, Phys. Rev. A28, 887.

1450. Milonni P W, Ackerhalt J R, and Goggin M E (1987) Quasiperiodically kicked quantum systems, Phys. Rev. A35, 1714.

1451. Minowa H (1988) Smale's horseshoe map in a Hamiltonian system around a separatrix, Z. Phys. 70B, 125.

1452. Miracky R F, Clarke J, and Koch R H (1983) Chaotic noise observed in a resistively shunted self resonant Josephson tunnel junction, Phys. Rev. Lett. 50, 856.

1453. Mishina T, Kohmoto T, and Hashi T (1985) Simple electronic circuit for the demonstration of chaotic phenomena, Am. J. Phys. 53, 332.

1454. Misiurewicz M (1979) Horseshoes for mappings of the interval, Bull. Acad. Pol. Ser. Sci. Math. 27, 167.

1455. Misiurewicz M (1980) Strange attractor for the Lozi mapping, Ann. N. Y. Acad. Sci. 357, 348.

1456. Misiurewicz M (1981a) The structure of mapping of an interval with zero entropy. Publ. Math. IHES, 53, 5.

1457. Misiurewicz M (1981b) Absolutely continuous measures for certain maps of an interval, Publ. Math. IHES, 53, 17.

1458. Misiurewicz M (1986) Rotation intervals for a class of maps of the real line into itself, Ergod. Th. & Dyn. Sys. 6, 117.

1459. Misiurewicz M, and Szewc B (1980) Existence of a homoclinic point for the Hénon map, Commun. Math. Phys. 75, 285.

1460. Misiurewicz M, and Szlenk W (1980) Entropy of piecewise monotone mappings, Studia Mathematica, 67, 45.

1461. Mistriotis A D, and Jackson E A (1987) Transition to stochasticity for a periodically perturbed area-perserving system, Phys. Scripta 35, 97.

1462. Mitschke F, Moller M, and Lange W (1988) Measuring filtered chaotic signals, Phys. Rev. A37, 4518.

1463. Miura T, and Kai T (1984) Chaotic behavior of a system of three disk dynamos, Phys. Lett. 101A, 450.

1464. Miura T, and Kai T (1986) A strange attractor of a system of three disk-dynamos and a geomagnetic attractor: their dimensions and $K_2$ entropies, J. Phys. Soc. Japan, 55, 2562.

1465. Moiseyev N, Brown R C, Wyatt R E, and Tsidoni E (1986) Analysis of chaotic eigenfunctions by the natural expansion method, Chem. Phys. Lett. 127, 37.

1466. Moiseyev N, and Perez A (1983) Motion of wave packets in regular and chaotic systems, J. Chem. Phys. 79, 5945.

1467. Moloney J V (1984) Coexistent attractors and nonperiodic cycles in a bistable ring cavity, Opt. Commun. 48, 435.

1468. Moloney J V (1986) Many-parameter roùtes to optical turbulence, Phys. Rev. A3, 4061.

1469. Moloney J V, Hopf F A, and Gibbs H M (1982) Effects of transverse beam variation on bifurcations in an intrinsic bistable ring cavity, Phys. Rev. 25A, 3442.

1470. Monin A S (1978) On the nature of turbulence, Sov. Phys. Usp. 21, 429; Russ. Orig. 125, 94.

1471. Moon F C (1980) Experiments on chaotic motion of forced nonlinear oscillator strange attractors, Trans. ASME J. Appl. Mech. 47, 638.

1472. Moon F C (1984) Fractal boundary for chaos in a two-state mechanical oscillator, Phys. Rev. Lett. 53, 962.

1473. Moon F C, Cusumano J, and Holmes P (1987) Evidence for homoclinic orbits as a precursor to chaos in a magnetic pendulum, Physica 24D, 383.

1474. Moon F C, and G-X Li (1985) The fractal dimension of the two-well potential strange attractor, Physica 17D, 99.

1475. Moon F C, and Holmes P (1979) A magnetoelastic strange attractor. J. Sound Vib. 65, 285; Errata: 69(1980), 339.

1476. Moon F C, and Holmes W T (1985) Double Poincaré sections of a quasiperiodically forced, chaotic attractors, Phys. Lett. 111A, 157.

1477. Moon F C, and Shaw S W (1983) Chaotic vibrations of a beam with nonlinear boundary conditions, Int. J. Non-Lin. Mech. 18, 465.

1478. Moon H T, Huerre P, and Redekopp L G (1982) Three frequency motion and chaos in the Ginzburg-Landau equation, Phys. Rev. Lett. 49, 458.

1479. Moon H T, Huerre P, and Redekopp L G (1983) Transtions to chaos in the Landau-Ginzburg equation, Physica 7D, 135.

1480. Moore D R, Toomre J, Knobloch E, and Weiss N O (1983) Period doubling and chaos in partial differential equations for thermosolutal convection, Nature, 303, 663.

1481. Mori H (1980) Fractal dimensions of chaotic flows of autonomous dissipative Systems, Prog. Theor. Phys. 63, 1044.

1482. Mori H, Okamoto H, So B C, and Kuroki S (1986) Global spectral structures of intermittent chaos, Prog. Theor. Phys. 76, 784.

1483. Mori H, Okamoto H, and Ogasawara M (1984) Self-similar cascades of band splittings of linear mod 1 maps, Prog. Theor. Phys. 71, 499.

1484. Mori H, So B C, and Kuroki S (1986) Spectral structure of intermittent chaos, Physica 21D, 355.

1485. Mori H, So B C, and Ose T (1981) Time-correlation functions of one-dimensional transformations, Prog. Theor. Phys. 66, 4.

1486. Mori H, and Fujisaka H (1980) Statistical dynamics of chaotic flows, Prog. Theor. Phys. 63, 1931.

1487. Mori N, Kuroki S, and Mori H (1988) Power spectra of intermittent chaos due to the collapse of period 3 windows, Prog. Theor. Phys. 79, 1260.

1488. Morimoto Y (1984) Bifurcation diagram of recurrence equation $x(t+1) = Ax(t)(1 - x(t)) - x(t-1) - x(t-2)$, J. Phys. Soc. Japan, 5, 2460.

1489. Morioka N, and Shimizu T (1978) Transition between turbulent and periodic states in the Lorenz model, Phys. Lett. 66A, 447.

1490. Morita T, Hata H, Mori H, Horita T, and Tomita K (1987) On partial dimensions and spectra of singularities of strange attractors, Prog. Theor. Phys. 78, 511.

1491. Morita T, Hata H, Mori H, Horita T, and Tomita K (1988) Spatial and temporal scaling properties of strange attractors and their presentations by unstable periodic orbits, Prog. Theor. Phys. 79, 296.

1492. Morris B, and Moss F (1986) Postponed bifurcations of a quadratic map with a swept parameter, Phys. Lett. 118A, 117.

1493. Mosekilde E, Aracil J, and Allen P M (1988) Instabilities and chaos in nonlinear dynamical systems, Sys. Dyn. Rev. 4, 14.

1494. Mosekilde E, and Larsen E R (1988) Deterministic chaos in the beer production-distribution model, Sys. Dyn. Rev. 4, 131.

1495. Moser J (1962) On invariant curves of area-preserving mappings of an annulus, Nachr. Akad. Wiss. Gottingen Math, 2, 1.

1496. Moser J (1978) Is the solar system stable? Math. Interlligencer, 1, 65.

1497. Müller G (1986) Nature of quantum chaos in spin systems, Phys. Rev. A34, 3345.

1498. Mullhaupt A P (1987) Boolean delay equations, DeBruijn oscillations and intermittency, Phys. Lett. 124A, 151.

1499. Murray N W, Lieberman M A, and Lichtenberg A J (1985) Corrections to quasilinear diffusion in area-preserving maps, Phys. Rev. A32, 2413.

1500. Nagai Y (1988) Pearson-walk visualization of the characteristic function of the invariant measure for one-dimensional chaos, Physica 150A, 40.

1501. Nagai Y, Hara R, Tsuchiya T, and Saito N (1988) Existence of low dimensional chaos in pamecium membrane potential suggested by the correlation integral method, J. Phys. Soc. Japan, 57, 3305.

1502. Nagashima H (1984) Chaos in a nonlinear wave equation with higher order dispersion, Phys. Lett. 105A, 439.

1503. Nagashima T, and Haken H (1983) Chaotic modulation of correlation functions, Phys. Lett. 96A, 385.

1504. Nagashima T, and Shimada I (1977) On the C-system-like property of the Lorenz system, Prog. Theor. Phys. 58, 1318.

1505. Nakamura K (1978) Numerical experiments on trajectory instabilities, Prog. Theor. Phys. 59, 64.

1506. Nakamura K (1979) Stochastic instabilities and turbulence in nonlinear dissipative systems, Proc. Inst. Nat. Sci. Nihon Un, 14, 9.

1507. Nakamura K, Okazaki Y, and Bishop A R (1986) Fat fractals in quantum chaos, Phys. Lett. 117A, 459.

1508. Nakamura K, and Lakshmanan M (1986) Complete integrability in a quantum description of chaotic systems, Phys. Rev. Lett. 57, 1661.

1509. Nakamura K, and Mikeska H J (1987) Quantum chaos of periodically pulsed systems. Underlying complete integrability. Phys. Rev. A35, 5294.

1510. Nakamura K, and Thomas H (1988) Quantum billiard in a magnetic field: chaos and diamagnetism, Phys. Rev. Lett. 61, 247.

1511. Nakamura M, Azumai K, Petrosky T Y, and Mishima N (1986) Irreversibility in chaotic regime of a conservative nonlinear system with a few degrees of freedom, Nuovo Cimento, 94B, 37.

1512. Nakatsuka H, Asaka S, Itoh H, Ikeda K, and Matzuoka M (1983) Observation of bifurcation to chaos in an all-optical bistable system, Phys. Rev. Lett. 50, 109.

1513. Namajunas A, Pozela J, and Tamasevicius A (1988) An electronic technique for measuring phase space dimension from chaotic time series, Phys. Lett. 131A, 85.

1514. Napiorkowski M (1985a) Average dynamics of noisy maps, Phys. Lett. 112A, 357.

1515. Napiorkowski M (1985b) Generalized final state sensitivity of the logistic map, Phys. Lett. 113A, 111.

1516. Napiorkowski M, and Thomae S (1987) Final state sensitivity in one-hump maps, Europhys. Lett. 4, 1247.

1517. Napiorkowski M, and Zaus U (1986) Average trajectories and fluctuation from noisy nonlinear map, J. Stat. Phys. 43, 349.

1518. Nardone P, Mandel P. and Kapral R (1986) Analysis of a delay-differential equation in optical bistability, Phys. Rev. A33, 2465.

1519. Nase J M, Dutton J A, and Wells R (1987) Calculated attractor dimensions for low-order spectral model (fluid dynamics), J. Atmos. Sci. 44, 1950.

1520. Nasuno S, Sano M, and Sawada Y (1988) Spatial instabilities and onset of chaos in Rayleigh-Bénard system with intermediate aspect ratio, J. Phys. Soc. Japan, 57, 3357.

1521. Nath A, and Ray D S (1986) Chaos in an exciton-biexciton two-oscillator model, Phys. Rev. A34, 4472.

1522. Nath A, and Ray D S (1987a) Period-doubling bifurcations in stimulated Raman scattering, Phys. Rev. A35, 1959.

1523. Nath A, and Ray D S (1987b) Horseshoe-shaped maps in chaotic dynamics of atom-field interaction, Phys. Rev. A36, 431.

1524. Nauenberg M (1982) On the fixed points for circle maps, Phys. Lett. 92A, 7.

1525. Nauenberg M (1987) Fractal boundary of domain of analyticity of the Feigenbaum function and relation to the Mandelbrot set, J. Stat. Phys. 47, 459.

1526. Nauenberg M, and Rudnick J (1981) Universality and the power spectrum at the onset of chaos, Phys. Rev. B24, 493.

1527. Nerenberg M A H, Baskey J H, and Blackburn J A (1987) Chaotic behavior in an array of coupled Josephson weak links, Phys. Rev. B36, 8333.

1528. Nerenberg M A H, Blackburn J A and Vik S (1984) Chaotic behavior in coupled superconducting weak links, Phys. Rev. B30, 5084.

1529. Neu J C (1979) Coupled chemical oscillators, SIAM J. Appl. Math. 37, 307.

1530. Neumann R, Koch S W, Schmidt H E, and Hang H (1984) Deterministic chaos and noise in optical bistability, Z. Phys. B55, 155.

1531. Newhouse S E, Ruelle D, and Takens F (1978) Occurrence of strange axiom A attractors near quasi-periodic flows on $T^m$ ($m = 3$ or more), Commun. Math. Phys. 64, 35.

1532. Newhouse S, Palis J, and Takens F (1983) Bifurcations and stability of families of diffeomorphisms, Publ. Math. IHES, 57, 5.

1533. Newton P K (1988) Chaos in Rayleigh-Bénard convection with external driving, Phys. Rev. A37, 932.

1534. Newton P K, and Sirovich L (1986) Instabilities of the Ginzburg-Landau equation: periodic solutions, Quart. Appl. Math. 44, 49.

1535. Ni W-S (1986) The period-adding phenomena in a two-dimensional mapping with three parameters, Chinese Phys. Lett. 3, 573.

1536. Ni W-S, Tong P-Q, and Hao B-L (1989) Homoclinic and heteroclinic intersections in the periodically forced Brusselator, Int. J. Mod. Phys. B3, 643.

1537. Nicolaenko B, Scheurer B, and Temam R (1985) Some global dynamical properties of the Kuramoto-Sivashinsky equations: nonlinear stability and attractors, Physica 16D, 155.

1538. Nicolis C, and Nicolis G (1984) Is there a climatic attractor? Nature, 311, 529.

1539. Nicolis C, and Nicolis G (1986) Effective noise of the Lorenz attractor, Phys. Rev. A34, 2384.

1540. Nicolis G, and Nicolis C (1988) Master equation approach to deterministic chaos, Phys. Rev. A8, 427.

1541. Nicolis J S (1986) Chaotic dynamics applied to information processing, Repts. Prog. Phys. 49, 1109.

1542. Nicolis J S (1987) Chaotic dynamics in biological information processing: a heuristic outline, Nuovo Cimento, 9D, 1359.

1543. Nicolis J S, Mayer-Kress G, and Haubs H (1983) Nonuniform chaotic dynamics with implications to information processing, Z. Naturforsch. 38a, 1157.

1544. Nicolis J S, and Tsuda I (1985) Chaotic dynamics of information processing: the 'magic number seven plus-minus two' resivited, Bull. Math. Biology, 47, 343.

1545. Normand C Y, Pomeau Y, and Velarde M G (1977) Convective instability: a physicist's approach, Rev. Mod. Phys. 49, 581.

1546. Novak F, Kosloff R, Tannor D J, Lorinzc A, Smith D D, and Rice S A (1985) Wave packet evolution in isolated pyrazine molecules: coherence triumphs over chaos, J. Chem. Phys. 82, 1073.

1547. Novak S, and Frehlich R G (1982) Transition to chaos in the Duffing oscillator, Phys. Rev. A26, 3660.

1548. Nozaki K, and Bekki N (1983a) Chaos in a perturbed nonlinear Schrödinger equation, Phys. Rev. Lett. 50, 1226.

1549. Nozaki K, and Bekki N (1983b) Pattern selection and spatio-temporal transition to chaos in the Ginzburg-Landau equation, Phys. Rev. Lett. 51, 2171.

1550. Nozaki K, and Bekki N (1985) Choatic solitons in a plasma driven by an RF field, J. Phys. Soc. Japan, 24, 2363.

1551. Nozaki K, and Bekki N (1986) Low-dimensional chaos in a driven damped nonlinear Schrödinger equation, Physica 21D, 381.

1552. Nusse H E (1986) Persistence of order and structure in chaos, Physica 20D, 374.

1553. Nusse H E (1987) Asymptotically periodic behavior in the dynamics of chaotic mappings, SIAM J. Appl. Math. 47, 498.

1554. Nusse H E, and Yorke J A (1988) Is every approximate trajectory of some process near an exact trajectory of a nearby process? Commun. Math. Phys. 363.

1555. Nusse H E, and Yorke J A (1988) Period doubling for $x_{n+1} = MF(x_n)$ where $F$ has negative Schwarzian derivative, Phys. Lett. 127A, 28.

1556. Oblow E M (1988) Supertracks, supertrack function and chaos in the quadratic map, Phys. Lett. 128A, 406.

1557. Octavio M, Da Costa A, and Aponte J (1986) Nonuniversality and metric properties of a forced nonlinear oscillator, Phys. Rev. A34, 1512.

1558. Octavio M, and Nasser C R (1984) Chaos in a dc-bias Josephson junction in the presence of microwave radiation, Phys. Rev. B30, 1586.

1559. Ogura H, Ueda Y, and Yoshida Y (1981) Periodic stationarity of a chaotic motion in the system governed by Duffing's equation, Prog. Theor. Phys. 66, 2280.

1560. Okada M, and Takizawa K (1981) Instability of an electro-optic bistable device with a delayed feedback, IEEE J. Quantum Electr. QE-17, 2135.

1561. Okamoto H, Mori H, and Kuroki S (1988) Global spectral structures of type III intermittent chaos, Prog. Theor. Phys. 79, 581.

1562. Okninski A (1988) Chaos in discrete maps, deterministic scattering, and nondifferentiable functions, J. Stat. Phys. 52, 577.

1563. Okubo A, Andreasen V, and Mitchell J (1984) Chaos-induced turbulent diffusion, Phys. Lett. 105A, 169.

1564. Olinger D J, and Screenivasan K R (1988) Nonlinear dynamics of the wake of an oscillating cylinder, Phys. Rev. Lett. 60, 797.

1565. Olsen L F, Truty G L, and Schaffer W M (1988) Oscillations and chaos in epidemics: a nonlinear dynamics study of six childhood diseases in Copenhagen, Denmark, Theor. Pop. Biol. 33, 344.

1566. Olsen L F, and Degn H (1979) Chaos in an enzyme reaction, Nature, 267, 177.

1567. Olsen L F, and Degn H (1985) Chaos in biological systems, Quarterly Rev. Biophys. 18, 165.

1568. Olsen O H, and Samuelsen M R (1987) Pattern competition and origin of intermittency-type chaos, Phys. Lett. 119A, 391.

1569. Olvera A, and Simó C (1987) An obstruction method for the destruction of invariant curves, Physica 26D, 181.

1570. Oono Y (1978a) A heuristic approach to the Kolmogorov entropy as a disorder parameter, Prog. Theor. Phys. 60, 1944.

1571. Oono Y (1978b) Period $\neq 2^n$ implies chaos, Prog. Theor. Phys. 59, 1029.

1572. Oono Y, Kohda T, and Yamazaki H (1980) Disorder parameter for chaos, J. Phys. Soc. Japan, 48, 738.

1573. Oono Y, and Osikawa M (1980) Chaos in nonlinear differential equations: I. Qualitative study of (formal) chaos, Prog. Theor. Phys. 64, 54.

1574. Oono Y, and Takahashi Y (1980) Chaos, external noise, and Fredholm theory, Prog. Theor. Phys. 63, 1804.

1575. Oppo G L, and Politi A (1984) Collision of Feigenbaum Cascades, Phys. Rev. A30, 435.

1576. Orban M, and Epstein I R (1982) Complex periodic and aperiodic oscillation in the chlorite-thiosulfate reaction, J. Phys. Chem. 86, 3907.

1577. Orszag S A, and McLaughlin J B (1980) Evidence that random behavior is generic for nonlinear differential equations, Physica 1D, 68.

1578. Osada H (1986) Propagation of chaos for the two-dimensional Navier-Stokes equation, Proc. Japan Acad. Ser. A, 62, 8.

1579. Osbaldestin A H, and Sarkis M Y (1987) Singularity spectrum of a critical KAM torus, J. Phys. A20, L953.

1580. Osborne A R, Kirwan A D, Provenzale A, and Bergamasco L (1986) A search for chaotic behavior in large and mesoscale motions in the Pacific Ocean, Physica 23D, 75.

1581. Oseledec V I (1968) A multiplicative ergodic theorem: Lyapunov characteristic numbers for dynamical systems, Trans. Moscow Math. Soc. 19, 197.

1582. Ostlund S, Rand D, Sethna J, and Siggia E (1983) Universal properties of the transition from quasiperiodicity to chaos in dissipative systems, Physica 8D, 303.

1583. Otsuka K, and Iwamura H (1983) Theory of optical multistability and chaos in a resonant-type semiconductor laser amplifier, Phys. Rev. A28, 3153.

1584. Ott E (1981) Strange attractors and chaotic motions of dynamical systems, Rev. Mod. Phys. 53, 655.

1585. Ott E, Antosen T M Jr, and Hanson J D (1984) Effect of noise on time-dependent quantum chaos, Phys. Rev. Lett. 53, 2187.

1586. Ott E, Withers W D, Yorke J A (1984) Is the dimension of chaotic attractors invariant under coordinate changes? J. Stat. Phys. 36, 687.

1587. Ott E, Yorke E D, and Yorke J A (1985) A scaling law: how an attractor's volume depends on noise level, Physica 16D, 62.

1588. Ott E, and Hanson J D (1981) The effect of noise on the structure of strange attractors, Phys. Lett. 85A, 20.

1589. Overman II E A, McLaughlin D W, and Bishop A R (1986) Coherence and chaos in the driven damped sine-Gordon equation: measurement of the soliton spectrum, Physica 19D, 1.

1590. Packard N H, Crutchfield J P, Farmer J D, and Shaw R S (1980) Geometry from a time series, Phys. Rev. Lett. 45, 712.

1591. Pade J, Rauh R, and Tsarouhas G (1987) Application of normal forms to the Lorenz model in the subcritical region, Physica 29D, 236.

1592. Pakarinen P, and Nieminen R M (1983) Period-multiplying bifurcations and multifurcations in conservative mappings, J. Phys. A16, 2105.

1593. Paladin G, Peliti L, and Vulpiani A (1986) Intermittency as multifractality in history space, J. Phys. A19, L991.

1594. Paladin G, and Vulpiani A (1984) Characterization of strange attractors as inhomogeneous fractals, Lett. Nuovo Cimento, 41, 82.

1595. Paladin G, and Vulpiani A (1985) The role of connectance on the chaoticity of Hamiltonian systems, Phys. Lett. 111A, 333.

1596. Paladin G, and Vulpiani A (1986a) Scaling law and asymptotic distribution of Lyapunov exponents in conservative dynamical systems with many degrees of freedom, J. Phys. A19, 1881.

1597. Paladin G, and Vulpiani A (1986b) Intermittency in chaotic systems and Renyi entropies, J. Phys. A19, L997.

1598. Paladin G, and Vulpiani A (1987) Anomalous scaling laws in multifractal objects, Phys. Reports, 156, 147.

1599. Palmore J J, and McCauley J L (1987) Shadowing by computable chaotic orbits, Phys. Lett. 122A, 399.

1600. Paramio M (1988) Invariant directions in the Hénon map, Phys. Lett. 132A, 98.

1601. Paramio M, and Sesma J (1987) Breakdown of KAM tori in the standard map, Phys. Lett. 124A, 345.

1602. Park K, and Crawford G L (1983) Deterministic transition in Taylor wavy-vortex flow, Phys. Rev. Lett. 50, 343.

1603. Park K. Crawford G L, and Donnelly R J (1981) Determination of transition in Couette flow in finite geometries, Phys. Rev. Lett. 47, 1448.

1604. Parker T S, and Chua L O (1987) Chaos: a tutorial for engineers, Proc. IEEE 982.

1605. Parlitz U, and Lauterborn W (1987) Period-doubling cascades and devil's staircase of the driven van der Pol oscillator, Phys. Rev. A36, 1428.

1606. Parry W (1976) Symbolic dynamics and transformation of the unit interval, Trans. Amer. Math. Soc. 122, 368.

1607. Pawelzik P, and Schuster H G (1987) Generalized dimensions and entropies from a measured time series, Phys. Rev. A35, 481.

1608. Pechukas P (1984) Remarks on "quantum chaos", J. Phys. Chem. 88, 4823.

1609. Pedersen N F (1988) Chaos in Josephson junctions and SQUIDs, Europhys. News, 19, No.4, 53.

1610. Pedersen N F, Soerenson O H, Dueholm B, and Mygind J (1980) Half-harmonic parametric oscillations in Josephson junctions, J. Low Temp. Phys. 38, 1.

1611. Pedersen N F, and Davidson A (1981) Chaos and noise rise in Josephson junctions, Appl. Phys. Lett. 39, 830.

1612. Pedlovsky J, and Frenzen C (1980) Chaotic and periodic behavior of finite amplitude baroclinic waves, J. Atmosph. Sci. 37, 1177.

1613. Pei L-Q, Guo F, Wu S-X, and Chua L O (1986) Experimental confirmation of the period-adding route to chaos in a nonlinear circuit, IEEE Trans. Cir. & Sys. CAS-33, 438.

1614. Peinke J, Parisi J, Rohricht B, Rössler O E, and Metzler W (1988) Smooth decomposition of generalized Fatou set explains smooth structure in generalized Mandelbrot set, Z. Naturforsch. 43a, 14.

1615. Peng S-L, Wang Y-X, and He Y-S (1986) Rigorous bound of the power spectrum for the renormalization group equation of a trifurcation sequence, Kexue Tongbao (Sci. Bull. China), 31, 1673.

1616. Peng S-L, and Cao K-F (1988) A new global regularity of fractal dimensions on critical points of transitions to chaos, Phys. Lett. 131A, 261.

1617. Peng S-L, and Qu C-C (1987) Existence of non-bijective solution for generalized Feigenbaum's functional equation, Kexue Tongbao (Sci. Bull. China), 32, 371.

1618. Percival I C (1982) Chaotic boundary of a Hamiltonian map, Physica 6D, 67.

1619. Percival I C (1987) Chaos in Hamiltonian systems, Proc. Roy. Soc. London, A413, 131.

1620. Percival I, and Vivaldi F (1987) Arithmetical properties of strongly chaotic motion, Physica 25D, 105.

1621. Percival I, and Vivaldi F (1987) A linear code for the sawtooth and cat maps, Physica 27D, 373.

1622. Peres A (1984a) Ergodicity and mixing in quantum theory. I. Phys. Rev. A30, 504.

1623. Peres A (1984b) Stability of quantum motion in chaotic and regular systems, Phys. Rev. A30, 1610.

1624. Perez J, and Jeffries C (1982a) Effects of additive noise on a nonlinear oscillator exhibiting period-doubling and chaotic behavior, Phys. Rev. B26, 3460.

1625. Perez Pascual R, and Lomnitz-Adler J (1988) Coupled relaxation oscillators and circle maps, Physica 30D, 61.

1626. Perez R, and Glass L (1982) Bistability, period doubling bifurcations, and chaos in a periodically forced oscillator, Phys. Lett. 90A, 441.

1627. Pesin Ya B (1976) Lyapunov characteristic exponent and ergodic properties of smooth dynamical systems with an invariant measure, Sov. Math. Dokl. 17, 196.

1628. Pesin Ya B (1977) Characteristic Lyapunov exponents and smooth ergodic theory, Russ. Math. Surv. 32:4, 55.

1629. Pesin Ya B, and Sinai Ya G (1981) Hyperbolicity and stochasticity of dynamical systems, in *Mathematical Physics Review*, ed. by S. P. Novikov, vol. 2, p. 53.

1630. Peters H (1982) Chaos in a time-delayed differential equation, Z. Angew. Math. Mech. 62, 297.

1631. Petrosky T Y (1984) Chaos and irreversibility in a conservative nonlinear dynamical system with a few degrees of freedom, Phys. Rev. A29, 2078.

1632. Petrosky T Y (1985) Chaos and nonunitary evolution in nonintegrable Hamiltonian systems, Phys. Rev. A32, 3716.

1633. Petrosky T Y (1986) Chaos and cometary clouds in the solar system, Phys. Lett. A117, 326.

1634. Petrosky T Y, and Schieve W C (1985) Limit of classical chaos in quantum systems, Phys. Rev. A31, 3907.

1635. Pfeiffer F (1988) Chaos in gearing, Z. Angew. Math. Mech. 68, 100.

1636. Pfenniger D (1985) Numerical study of complex instability. I. Mappings, Astron. & Astrophys. 150, No.1, Pt.1, 97.

1637. Phillipson P E (1988) The transition from simple to fractal basin boundary structure displayed by unimodal maps, Phys. Lett. 128A, 413.

1638. Pianigiani G (1981) Conditionally invariant measure and exponential decay, J. Math. Anal. and Appl. 82, 75.

1639. Pianigiani G, and Yorke J A (1979) Expanding maps on sets which are almost invariant: decay and chaos, Trans. Am. Math. Soc. 252, 351.

1640. Picard G, and Johnston T W (1982) Instability cascades; Lotka-Volterra population equations and Hamiltonian chaos, Phys. Rev. Lett. 48, 23.

1641. Pickover C A (1988a) Overrelaxation and chaos, Phys. Lett. 130A, 125.

1642. Pickover C A (1988b) Pattern formation and chaos in networks, Commun. ACM, 31, 136.

1643. Pieranski P (1983) Jumping particle model: period doubling cascade in an experimental system, J. Physique, 44, 573.

1644. Pieranski P (1988) Direct evidence for the suppression of period doubling in the bouncing-ball model, Phys. Rev. A37, 1782.

1645. Pieranski P, and Malecki J (1987) Noise-sensitive hysteresis loops around period doubling bifurcation points, Nuovo Cimento, 9D, 757.

1646. Pike E R, Sakar S, and Satchell J S (1988) A comparison of exact and semiclassical analysis of diagonal occupation probabilities for the Feingold-Perez chaotic Hamiltonian, J. Phys. A21, 1571.

1647. Pikovsky A S (1983) A new type of intermittent transition to chaos, J. Phys. A16, L109.

1648. Pikovsky A S (1984) On the interaction of strange attractors, Z. Phys. B55, 149.

1649. Pikovsky A S (1985) Chaotic autowaves, Sov. Tech. Phys. Lett. 11, 279; Russ. Orig. 11, 11.

1650. Pikovsky A S (1986) Evolution of the power spectrum in the period-doubling route to chaos, Radiophys. & Quantum Electr. 29, 1076; Russ. Orig. 29, 1438.

1651. Pikovsky A S, and Rabinovich M I (1981a) Stochastic oscillations in dissipative systems, Physica 2D, 8.

1652. Pikovsky A S, and Rabinovich M I (1981b) Stochastic behavior in dissipative systems, in *Mathematical Physics Review*, ed. S. P. Novikov, vol. 2, 165.

1653. Piña E, and Lara L J (1987) On the symmetry lines of the standard mapping, Physica 26D, 369.

1654. Pippard B (1982) Instability and chaos: physical models of everyday life, Interdisc. Sci. Rev. 7, 92.

1655. Pismen L M (1982) Bifurcation sequences in a third-order system with a folded slow maniford, Phys. Lett. 89A, 59.

1656. Pismen L M (1987) Homoclinic explosion in the vicinity of bifurcation at the triple-zero eigenvalue, Phys. Rev. A35, 2709.

1657. Poddar B, Moon F C, and Mukherjee S (1988) Chaotic motion of an elastic-plastic beam, Trans. ASME J. Appl. Mech. 55, 185.

1658. Pokrovskii L A (1986) Solutions of the system of Lorenz equations in the asymptotic limit of large Rayleigh numbers. II. Description of trajectories near a separatrix by the matching method, Theor. & Math. Phys. 67, 490; Russ. Orig. 67, 263.

1659. Politi A, Badii R, and Grassberger P (1988) On the geometric structure of non-hyperbolic attractors, J. Phys. A21, L763.

1660. Pollock M D (1987) On the possibility of chaotic inflation from a softly-broken superconformal invariance, Phys. Lett. 194B, 518.

1661. Pomeau Y, Pumir A, Pelce P (1984) Intrinsic stochasticity with many degrees of freedom, J. Stat. Phys. 37, 39.

1662. Pomeau Y, Roux J C, Rossi A, Bachelart, and Vidal C (1981) Intermittent behaviour in the Belousov-Zhabotionsky reaction, J. Physique Lett. 42, 271.

1663. Pomeau Y, and Manneville P (1980) Intermittent transition to turbulence in dissipative dynamical systems, Commun. Math. Phys. 74, 189.

1664. Pompe B, Kruscha J, and Leven B W (1986) State predictability and information flow in simple chaotic systems, Z. Naturforsch. 41a, 801.

1665. Pompe B, and Leven R W (1986) Transformation of chaotic systems, Phys. Scripta 34, 8.

1666. Pool R (1989a) Is it chaos, or is it just noise? Science, 243, 25.

1667. Pool R (1989b) Ecologists flirt with chaos, Science, 243, 310.

1668. Pool R (1989c) Is it healthy to be chaotic? Science, 243, 604.

1669. Pool R (1989d) Quantum chaos: enigma wrapped in a mystery, Science, 243, 893.

1670. Pool R (1989e) Is something strange about the weather? Science, 243, 1290.

1671. Poppe D, and Korsch J (1987) Stochastic and deterministic dissociation dynamics of a Morse oscillator driven by impulsive interactions, Physica 24D, 367.

1672. Posch H A, Hoover W G, and Vesely F J (1986) Canonical dynamics of the Nose oscillator: stability, order and chaos, Phys. Rev. A33, 4258.

1673. Pounder J R, and Rogers T D (1980) The geometry of chaos: dynamics of nonlinear second-order difference equation, Bull. Math. Biol. 42, 551.

1674. Pozela J, Namajunas A, Tamasevicius A, and Ulbikas J (1989) Quantitative characterization of chaotic instabilities in semiconductors, Appl. Phys. A48, 181.$\Omega$

1675. Procaccia I (1988) Universalities of dynamically complex systems: the organization of chaos, Nature, 333, 618.

1676. Procaccia I, Thomae S, and Tresser C (1987) First return maps as a unified renormalization scheme for dynamical systems, Phys. Rev. A35, 1884.

1677. Puccioni G P, Poggi A, Gadomski W, Tredicce J R, and Arecchi F T (1985) Measurement of the formation and evolution of a strange attractor in a laser, Phys. Rev. Lett. 55, 339.

1678. Qian M, and Yan Y (1986) Transversal heteroclinic cycle and its application to Hénon mapping, Kexue Tongbao (Sci. Bull. China), 31, 10.

1679. Qin G-R, Gong D-C, Yang C-Y, Hu G, Mao J-Y, and Zhang L (1985) Division of frequency and chaos, Chinese Phys. Lett. 2, 35.

1680. Qiu X-M, and Wang X-G (1986) A new strange attractor in MHD flow in sheared magnetic field, Chinese Phys. Lett. 3, 105.

1681. Qiu X-M, and Wang X-G (1987) Chaotic attractor of MHD flow in a sheared magnetic field, Chinese Phys. Lett. 4, 49.

1682. Quispel G R (1985) Scaling of the superstable fraction of the two-dimensional period-doubling interval, Phys. Lett. 112A, 353.

1683. Rabinovich M I, and Fabrikant A C (1979) Stchastic self-modulation of waves in nonequilibrium media, Sov. Phys. JETP, 50, 311; Russ. Orig. 77, 617.

1684. Rabinovitch A, and Thieberger R (1987) Time series analysis of chaotic signals, Physica 28D, 409.

1685. Rajasekar S, and Lakshmanan M (1988) Period-doubling bifurcations, chaos, phase-locking and devil's staircase in a Bonhoeffer-van der Pol oscillator, Physica 32D, 146.

1686. Ramachandran B, and Kay K G (1987) Local ergodicity as a probe for chaos in quantum systems: application to the Hénon-Heiles system, J. Chem. Phys. 86, 4628.

1687. Ramaswamy R, and Swaminathan S (1987) Fractal eigenfunctions in (classically) nonintegrable Hamiltonian systems, Europhys. Lett. 4, 27.

1688. Rand D (1978) The topological classification of Lorenz attractors, Math. Proc. Cambr. Phil. Soc. 83, 451.

1689. Rand D (1987) Fractal bifurcation sets, renormalization strange sets, and their universal invariants, Proc. Roy. Soc. London, A413, No.1844, 45.

1690. Rand D, Ostlund S, Sethna J, and Siggia E D (1982) A universal transition from quasi-periodicity to chaos in dissipative systems, Phys. Rev. Lett. 49, 132.

1691. Rapp P E, Zimmerman I D, Albano A M, deGuzman G C, Greenbaum N N (1985) Dynamics of spontaneous neural activity in the simian motor cortex: the dimension of chaotic neurons, Phys. Lett. 110A, 335.

1692. Rasmussen D R, and Bohr T (1987) Temporal chaos and spatial disorder, Phys. Lett. 125A, 107.

1693. Rasmussen S, Moskilde E, and Sterman J D (1985) Bifurcations and chaotic behavior in a simple model of the economic long wave, Sys. Dyn. Rev. 1, 92.

1694. Raty R, Von Boehm J, and Isomaki H M (1984) Absence of inversion-symmetric limit cycles of even periods and the chaotic motion of Duffing's oscillator, Phys. Lett. 103A, 289.

1695. Raty R, Von Boehm J, and Isomaki H M (1986) Chaotic motion of a periodically driven pendulum in an asymptotic potential well, Phys. Rev. A34, ,4310.

1696. Rechester A B, and White R B (1983) Invariant distribution on the attractors in the presence of noise, Phys. Rev. A27, 1203.

1697. Reichl J (1988) Statistical analysis of regular and irregular wave functions, Europhys. Lett. 6, 669.

1698. Reichl J, and Buttner H (1986) Stochastic and regular motion in a four-particle system, Phys. Rev. A33, 2184.

1699. Reichl J, and Buttner H (1987a) Energy spectra for nonlinear oscillators with broken symmetry, J. Phys. A20, 6321.

1700. Reichl J, and Buttner H (1987b) Quantum chaos in a nonlinear oscillator with three degrees of freedom, Europhys. Lett. 4, 1343.

1701. Reichl L E, and Lin W A (1986) Exact quantum model of field-induced resonance overlap, Phys. Rev. A33, 3598.

1702. Reichl L E, and Zheng W-M (1984) Chaos in the conservative Duffing system — renormalization group prediction, in *Fluctuations and Sensitivity in Nonequilibrium Systems*, ed. by W. Horsthemke and D. K. Kondepudi, Springer-Verlag, 228.

1703. Reichl L E, de Fainchtein R, and Petrosky T (1983) Field induced chaos in the Toda lattice, Phys. Rev. A28, 3051.

1704. Reiner R, Munz M, and Weidlich W (1988) Migratory dynamics of interacting sub-populations: regular and chaotic behavior, Sys. Dyn. Rev. 4, 179.

1705. Reis L H (1985) A model of the Universe. I. The theory of chaotons, Lett. Nuovo Cimento, 44, 637.

1706. Retzloff D G, Chan P C-H, Chicone C, Offin D, and Mohamed K (1987) Chaotic behavior in the dynamical system of a continuous stirred tank reactor, Physica 25D, 131.

1707. Richetti P, Argoul F, and Arneodo A (1986) Type-II intermittency in a periodically driven nonlinear oscillator, Phys. Rev. A34, 726.

1708. Richetti P, Roux J C, Argoul F, and Arneodo A (1987) From quasiperiodicity to chaos in the Belousov-Zhabotinskii reaction: II. Modelling and theory, J. Chem. Phys. 86, 3339.

1709. Richetti P, and Arneodo A (1985) The periodic-chaotic sequences in chemical reactions: a scenario close to homoclinic conditions? Phys. Lett. 109A, 359.

1710. Riedel U, Kuhn R, and van Hemmen J L (1988) Temporal sequences and chaos in neural nets, Phys. Rev. A38, 1105.

1711. Riela G (1982a) Universal spectral property in higher dimensional dynamical systems, Phys. Lett. 92A, 157.

1712. Riela G (1982b) A New six-mode truncation of the Navier-Stokes equations on a two-dimensional torus: a numerical study, Nuovo Cimento, 69B, 245.

1713. Riela G (1982c) Loss of stablity and disappearance of two-dimensional invariant tori in a dissipative dynamical system, Phys. Lett. 91A, 283.

1714. Riela G (1985) Transition tori-chaos through collisons with hyperbolic orbits, J. Stat. Phys. 41, 201.

1715. Ringland J, and Turner J S (1984) One-dimensinal behavior in a model of the Belousov-Zhabotinskii reaction, Phys. Lett. 105A, 93.

1716. Riste T, and Otnes K (1986) Chaos and neutron scattering, Physica 137B+C, 141.

1717. Robbins K A (1977) A new approach to subcritical instability and turbulent transitions in a simple dynamo, Math. Proc. Camb. Phil. Soc. 82, 309.

1718. Robbins K A (1979) Periodic solutions and bifurcation structure at high R in the Lorenz model, SIAM J. Appl. Math. 36, 457.

1719. Roberts J A G, and Tompson C J (1988) Dynamics of the classical Heisenberg spin chain, J. Phys. A21, 1769.

1720. Robinson A L (1982) Physicists try to find order in chaos, Science, 218, 554.

1721. Robinson A L (1983) How does fluid flow become turbulent? Science, 221, 140.

1722. Robnik M (1987) A note on the level spacings distribution of the Hamiltonians in the transition between integrability and chaos, J. Phys. A20, L495.

1723. Rogers T D, Pomder J R (1984) The evolution of crisis: bifurcation and demise of a snapback repeller, Physica 13D, 408.

1724. Rollins R W, and Hunt E R (1982) Exactly solvable model of a physical system exhibiting universal chaotic behavior, Phys. Rev. Lett. 49, 1295.

1725. Rollins R W, and Hunt E R (1984) Intermittent transient chaos at interior crises in the diode resonator, Phys. Rev. A29, 3327.

1726. Romanelli L, Figiola M A, Hirsch F A, and Radicella S M (1987) Chaotic behavior of solar radio flux, Sol. Phys. 110, 391.

1727. Romanelli L, Figliola M A, and Hirsch F A (1988) Deterministic chaos and natural phenomena, J. Stat. Phys. 53, 991.

1728. Romeiras F J, Bondeson A, Ott E, Antonsen T M, and Grebogi C (1987) Quasiperiodically forced dynamical systems with strange nonchaotic attractors, Physica 26D, 277.

1729. Romeiras F J, and Ott E (1987) Strange nonchaotic attractors of the damped pendulum with quasiperiodic forcing, Phys. Rev. A35, 4404.

1730. Romeiras F, Grebogi C, and Ott E (1988) Critical exponents for power-spectra scaling at mergings of chaotic bands, Phys. Rev. A38, 463.

1731. Rössler O E (1976a) An equation for continuous chaos, Phys. Lett. 57A, 397.

1732. Rössler O E (1976b) Chaotic behavior in simple reaction systems, Z. Naturforsch. 31a, 259.

1733. Rössler O E (1976c) Chemical turbulence: chaos in a simple reaction-diffusion system, Z. Naturforsch. 31a, 1168.

1734. Rössler O E (1976d) Different types of chaos in two simple differential equations, Z. Naturforsch. 31a, 1664.

1735. Rössler O E (1977a) Chaos in abstract kinetics: two prototypes, Bull. Math. Biol. 39, 275.

716

1736. Rössler O E (1977b) Horseshoe-map chaos in the Lorenz equation, Phys. Lett. 60A, 392.

1737. Rössler O E (1979a) Chaotic oscillations — an example of hyperchaos, Lect. Notes in Appl. Math. 17, 141.

1738. Rössler O E (1979b) An equation for hyperchaos, Phys. Lett. 71A, 155.

1739. Rössler O E (1983) The chaotic hierarchy, Z. Naturforsch. 38a, 788.

1740. Rössler O E, Hudson J L, and Yorke J A (1986) Cloud attractors and time-inversed Julia boundaries, Z. Naturforsch. 41a, 979.

1741. Rössler O E, Kahlert C, Parisi J, Peinke J, and Rohricht B (1986) Hyperchaos and Julia sets, Z. Naturforsch. 41a, 819.

1742. Rössler O E, and Hoffmann M (1987) Quasiperiodization in classical hyperchaos, J. Comput. Chem. 8, 510.

1743. Rössler O E, and Wegmann K (1978) Chaos in the Zhabotinskii reaction, Nature, 271, 89.

1744. Rotaru A Kh (1986) Optical turbulence in a system of coherent excitatons, photons, and biexcitons, Sov. Phys. Solid State, 28, 1383; Russ. Orig. 28, 2492.

1745. Rotenberg M (1988) Life on a regular polygon: maps induced by spatially extended prey-predator equations, Physica 30D, 192.

1746. Roux J C, Rossi A, Bachelart S, and Vidal C (1980) Representation of a strange attractor from an experimental study of chemical turbulence, Phys. Lett. 77A, 391.

1747. Roux J C, Rossi A, Bachelart S, and Vidal C (1981) Experimental observations of complex dynamical behavior during a chemical reaction, Physica 2D, 395.

1748. Roux J C, Simoyi R H, and Swinney H L (1983) Observation of a strange attractor, Physica 8D, 257.

1749. Ruelle D (1976) The Lorenz attractor and the problem of turbulence, Lect. Notes in Math. 565, 146.

1750. Ruelle D (1977a) Dynamical systems with turbulent behavior, Lect. Notes. in Phys. 80, 341.$\Omega$

1751. Ruelle D (1977b) Statistical mechanics and dynamical systems, Duke Univ. Math. Series, III, 1.

1752. Ruelle $\bar{\text{D}}$ (1979) Ergodic theory of differentiable dynamical systems, Publ. Math. IHES. 50, 275.$\Omega$

1753. Ruelle D (1980a) Recent results on differentiable dynamical systems, Lect. Notes in Phys. 116, 321.

1754. Ruelle D (1980b) Strange attractors, Math. Intelligencer, 2, 126.

1755. Ruelle D (1981a) Small random perturbations of dynamical systems and the definition of attractors, Commun. Math. Phys. 82, 137.

1756. Ruelle D (1981b) Differentiable dynamical systems and the problem of turbulence, Bull. Am. Math. Soc. 5, 29.

1757. Ruelle D (1982) Do turbulent crystals exit ? Physica 113A, 619.

1758. Ruelle D (1984) Characteristic exponents for a viscous fluid subjected to time dependent forces, Commun. Math. Phys. 93, 285.

1759. Ruelle D (1986) Resonances of chaotic dynamical systems, Phys. Rev. Lett. 56, 405.

1760. Ruelle D, and Takens F (1971) On the Nature of turbulence, Commun. Math. Phys. 20, 167; 23, 343.

1761. Russell D A, Hanson J D, and Ott E (1980) Dimension of strange attractors, Phys. Rev. Lett. 45, 1175.

1762. Russell D A, and Ott E (1981) Chaotic (strange) and periodic behavior in instability saturation by the oscillating two-stream instability, Phys. Fluids, 24, 1976.

1763. Saari D G, and Urenko J B (1985) Newton's method, circle maps, and chaotic motion, Am. Math. Monthly, 91, 3.

1764. Sagdeev R Z, and Zaslavsky G M (1987) Stochasticity in the Kepler problem and a model of possible dynamics of comets in the Oort cloud, Nuovo Cimento, 97B, 119.

1765. Sakaguchi H (1988) Phase transitions in coupled Bernoulli maps, Prog. Theor. Phys. 80, 7.

1766. Sakaguchi H, and Tomita K (1987) Bifurcations of the coupled logistic map, Prog. Theor. Phys. 78, 305.

1767. Salam F M A (1987) The Melnikov technique for highly dissipative systems, SIAM J. Appl. Math. 47, 232.

1768. Salam F M A, Marsden J E, and Varaiya P P (1983) Chaos and Arnold diffusion in dynamical systems, IEEE Trans. Cir. and Sys. CAS-32, 697.

1769. Salam F M A, and Bai S (1988) Complicated dynamics of a prototype continuous-time adaptive control system, IEEE Trans. Cir. & Sys. CAS-35, 842.

1770. Salam F M A, and Sastry S S (1985) Dynamics of the forced Josephson junction circuit: the regions of chaos, IEEE Trans. Cir. and Sys. CAS-32, 784.

1771. Saltzman B (1962) Finite amplitude convection as an initial value problem. I, J. Atmosph. Sci. 19, 329.

1772. Samuelides M, Feckinger R, Tonsiller L, and Bellisard J (1986) Instabilities of the quantum rotator and transition in the quasi-energy spectrum, Europhys. Lett. 1, 203.

1773. Sanders J A (1982) Melnikov's method and averaging, Celest. Mech. 28, 171.

1774. Sano M, Sato S, and Sawada Y (1986) Global spectral characterization of chaotic dynamics, Prog. Theor. Phys. 76, 945.

1775. Sano M, and Sawada Y (1982) Unavoidable pretransitional macroscopic fluid motion and its effect on a possible critical fluctuation in a finite Bénard system, Phys. Rev. 25A, 990.

1776. Sano M, and Sawada Y (1983) Transition from quasiperiodicity to chaos in a system of coupled nonlinear oscillators, Phys. Lett. 97A, 73.

1777. Sano M, and Sawada Y (1985) Measurement of the Lyapunov spectrum from a chaotic time series, Phys. Rev. Lett. 55, 1082.

1778. Saperstein A M (1984) Chaos — a model for the outbreak of war, Nature, 309, 303.

1779. Saravanan R, Narayan O, Banerjee K, and Bhattacharjee J K (1985) Chaos in a periodically forced Lorenz system, Phys. Rev. A31, 520.

1780. Sarkar S (1988) The response of a quantum field to classical chaos, J. Phys. A21, 971.

1781. Sarkar S K (1984) Information dimension of quasiperiodic trajectories with quadratically irrational winding number, Phys. Lett. 106A, 95.

1782. Sarkar S, Satchell J S, and Carmichael H J (1986) Quantum fluctuations and the Lorenz equations, J. Phys. A19, 2751.

1783. Sarkar S, and Satchell J S (1987a) The effect of measurement on the quantum kicked ratoator, Europhys. Lett. 4, 133.

1784. Sarkar S, and Satchell J S (1987b) Quantum measurement as dissipation in chaotic atomic systems, J. Phys. A20, L437.

1785. Sarkar S, and Satchell J S (1988) Measurements on quantum chaotic systems, Physica 29D, 343.

1786. Sarkovskii A N (1964) Coexistence of cycles of a continuous map of a line into itself, Ukranian Math. J. 16, 61 (in Russian).

1787. Satchell J S, and Sarkar S (1987) Stochastic Shilnikov maps, J. Phys. A20, 1333.

1788. Satija I I (1987) Universal strange attractor underlying Hamiltonian stochasticity, Phys. Rev. Lett. 58, 623.

1789. Satija I I, Bishop A R, and Fesser F (1985) Chaos in a damped and driven Toda system, Phys. Lett. 112A, 183.

1790. Sato S, Sano M, and Sawada Y (1983) Universal scaling property in bifurcation structure of Duffing's and of generalized Duffing's equations, Phys. Rev. A28, 1654.

1791. Sato S, Sano M, and Sawada Y (1987) Practical methods of measuring the generalized dimension and the largest Lyapunov exponent in high dimensional chaotic systems, Prog. Theor. Phys. Lett. 77, 1.

1792. Savarge C M, Carmichael H J, and Walls D F (1982) Optical multistability and self-oscillations in three level systems, Opt. Commun. 42, 211.

1793. Schaffer W M (1984) Stretching and folding in lynx fur returns: evidence for a strange attractor in Nature, The Am. Naturalist, 124, 798.

1794. Schaffer W M (1985) Order and chaos in ecological systems, Ecology, 66, 93.

1795. Schaffer W M (1986) Chaos in ecological systems: the coals that Newcastle forgot, Trends in Ecol. Sys. 1, 63.

1796. Schaffer W M, Ellner S, and Kot M (1986) Effects of noise on some dynamical models in ecology, J. Math. Biol. 24, 379.

1797. Schaffer W M, and Kot M (1985a) Nearly one-dimensional dynamics in an epidemic, J. Theor. Biol. 112, 403.

1798. Schaffer W M, and Kot M (1985b) Do strange attractors govern ecological systems? Bioscience, 35, 349.

1799. Schecter S (1987) Melnikov's method at a saddle-node and the dynamis of the forced Josephson junction, SIAM J. Math. Anal. 18, 1699.

1800. Scheeline A, and Kuhns D W (1985) Mode locking and instability in a voltage thresholded spark source, Anal. Chemistry, 57, 73.

1801. Schell M, Fraser S, and Kapral R (1982) Diffusive dynamics in systems with translational symmetry: a one-dimensional-map model, Phys. Rev. A26, 504.

1802. Schell M, Fraser S, and Kapral R (1983) Subharmonic bifurcation in the sine map: an infinite hierachy of cusp bistabilities, Phys. Rev. A28, 373.

1803. Schell M, and Albahadily F N (1989) Mixed-mode oscillations in an electrochemical system.II. A periodic-chaotic sequence, J. Chem. Phys. 90, 822.$\Omega$

1804. Scheurle J (1986) Chaotic solutions of systems with almost periodic forcing, Z. Angew. Math. & Phys. 37, 12.

1805. Schiehlen W (1987) On the loading of chaotically vibrating mechanical systems, Z. Angew. Math. & Mech. 67, T140.

1806. Schilling R (1984) Tunneling levels and specific heat of one-dimensional chaotic configurations, Phys. Rev. Lett. 53, 2258.

1807. Schlögl F (1987) The variance of information loss as a characteristic quantify of dynamical chaos, J. Stat. Phys. 46, 135.

1808. Schlögl F, and Scholl E (1988) Generalized specific heat as a characteristic measure in chaos, Z. Phys. 71B, 231.

1809. Schmidt G, and Bialek J (1982) Fractal diagrams for Hamiltonian stochasticity, Physica 5D, 397.

1810. Schmidt G, and Wang B-H (1985) Dissipative standard map, Phys. Rev. A32, 2994.

1811. Schmitz R A, Graziani K R, and Hudson J L (1977) Experimental evidence of chaotic states in the Belousov-Zhabotinskii reaction, J. Chem. Phys. 67, 3040.

1812. Schmutz M, and Rueff M (1984) Bifurcation schemes of the Lorenz model, Physica 11D, 167.

1813. Schneider F W (1985) Periodic perturbations of chemical oscillators: experiments, Ann. Rev. Phys. Chem. 36, 347.

1814. Schnellhuber H J, Urbschat H, and Block A (1983) Calculation of Cantori, Phys. Rev. A33, 2856.

1815. Scholl E (1989) Theoretical approaches to nonlinear and chaotic dynamics of generation recombination processes in semiconductors, Appl. Phys. A48, 95.$\Omega$

1816. Scholl E, Parisi J, Rohricht B, Reinke J, and Huebener R P (1987) Spatial corraltions of chaotic oscillations in the post-breakdown regime of p-Ge, Phys. Lett. 119A, 419.

1817. Scholz H J, Yamada Y, Brand H, and Graham R (1981) Intermittency and chaos in a laser system with modulated inversion, Phys. Lett. 82A, 321.

1818. Schreiber I, Dolnik M, Choc P, and Marek M (1988) Resonance behavior in two-parameter families of periodically forced oscillators, Phys. Lett. 128A, 66.

1819. Schreiber I, Holodniok M, Kubicek M, and Marek M (1986) Periodic and aperiodic regimes in coupled dissipative chemical oscillators, J. Stat. Phys. 43, 489.

1820. Schreiber I, and Marek M (1982a) Strange attractors in coupled reaction-diffusion cells, Physica 5D, 258.

1821. Schreiber I, and Marek M (1982b) Transition to chaos via two-torus in coupled reaction-diffusion cells, Phys. Lett. 91, 263.

1822. Schulmann N (1983) Chaos in piecewise-linear systems, Phys. Rev. A28, 477.

1823. Schulmeister Th, and Herzel H (1986) Chaos in forced Selkov system, Z. Angew. Math. & Mech. 66, 375.

1824. Schult R L, Cremer D B, Henyey F S, and Wright J A (1987) Symmetric and nonsymmetric coupled logistic maps, Phys. Rev. A35, 3115.

1825. Schuster H G (1987) Estimating the strength of chaos from the power spectrum, Z. Phys. 68B, 251.

1826. Schuster H G, Martin S, and Martienssen W (1986) New method for determining the largest Liapunov exponent of simple nonlinear systems, Phys. Rev. A33, 3547.

1827. Schutle M, Stiefelhagen W, and Demme E S (1987) Period in the chaotic phase of Q2R automata, J. Phys. A20, L1023.

1828. Schwartz I B (1988a) Infinite primary Saddle-node bifurcation in periodically forced systems, Phys. Lett. 126A, 411.

1829. Schwartz I B (1988b) Sequential horseshoe formation in the birth and death of chaotic attractors, Phys. Rev. Lett. 60, 1359.

1830. Seifert (1983) Intermittent chaos in Josephson junctions represented by stroboscopic maps, Phys. Lett. 98A, 43.

1831. Seisl M, Steindl A, and Troger H (1988) Chaos in a discrete mechanical model of the sine-Gordon equation, Z. Angew. Math. Mech. 68, 120.

1832. Seligman T H, Verbaarschot J J M, and Zirnbaum M R (1984) Quantum spectra and transition from regular to chaotic classical motion, Phys. Rev. Lett. 53, 215.

1833. Seligman T H, Verbaarschot J J M, and Zirnbaum M R (1985) Spectral fluctuation properties of Hamiltonian systems: the transition region between order and chaos, J. Phys. A18, 2751.

1834. Seligman T H, and Verbaarschot J J M (1985a) Quantum spectra of classically chaotic systems with time reversal invariance, Phys. Lett. 108A, 183.

1835. Seligman T H, and Verbaarschot J J M (1985b) Fluctuations of quantum spectra and their semiclassical limit in the transition between order and chaos, J. Phys. A18, 2227.

1836. Series C (1987) Some geometrical models of chaotic dynamics, Proc. Roy. Soc. London, A413, No.1844, 171.

1837. Sethna J P, and Siggia D E (1984) Universal transition in a dynamical system forced at two incommensurate frequencies, Physica 11D, 193.

1838. Seydel R (1985) Attractors of a Duffing equation — dependence on the exciting frequency, Physica 17D, 308.

1839. Shapiro M, Goelman G (1984) Onset of chaos in an isolated energy eigenstate, Phys. Rev. Lett. 53, 1714.

1840. Shapiro M, Ronkin J, and Brumer P (1988) Scaling laws and correlation lengths of quantum and classical ergodic states, Chem. Phys. Lett. 148, 177.

1841. Shapiro M, Taylor R D, and Brumer P (1984) Regularity of low-lying quantum eigenstates in a classically mixing system, Chem. Phys. Lett. 106, 325.

1842. Shapiro M, and Child M S. (1982) Quantum stochasticity and unimolecular decay, J. Chem. Phys. 76, 6176.

1843. Shaw R S (1981) Strange attractors, chaotic behavior, and information flow, Z. Naturforsch. 36a, 80.

1844. Shaw S W, and Holmes P (1983) Periodically forced linear oscillator with impacts: chaos and long-period motions, Phys. Rev. Lett. 51, 8.

1845. Shaw S W, and Rand R H (1989) The transition to chaos in a simple mechanical system, Int. J. Nonlin. Mech. 24, 41.

1846. Shaw S W, and Wiggins S (1988) Chaotic dynamics of a whirling pendulum, Physica 31D, 190.

1847. Shenker S J (1982) Scaling behavior in a map of a circle onto itself: Empirical results, Physica 5D, 405.

1848. Shenker S J, and Kadanoff L P (1981) Band to band hopping in one-dimensional maps, J. Phys. A14, L23.

1849. Shenker S J, and Kadanoff L P (1982) Critical behavior of a KAM surface: I. Empirical results, J. Stat. Phys. 27, 631.

1850. Shepelyansky D L (1981) Dynamic stochasticity in nonlinear quantum systems, Theor. Math. Phys. 49, 925; Russ. Orig. TMF 49, 36.

1851. Shepelyansky D L (1983) Some statistical properties of simple classically stochastic quantum systems, Physica 8D, 208.

1852. Shigematsu H, Mori H, Yoshida T, and Okamoto H (1983) Analytic study of the power spectra of the tent maps near band-splitting transitions, J. Stat. Phys. 30, 649.

1853. Shilnikov L P (1965) A case of the existence of denumerable set of periodic motions, Sov. Math. Dokl. 6, 163; Russ. Orig. 160, 588.

1854. Shilnikov L P (1967) The existence of a denumerable set of periodic motions in four-dimensional space in an extended neighbourhood of a saddle-focus, Sov. Math. Dokl. 8, 54.

1855. Shilnikov L P (1969) On a new type of bifurcation of multidimensional dynamical systems, Sov. Math. Dokl. 10, 1368.

1856. Shilnikov L P (1970) A contribution to the problem of the structure of an extended neighbourhood of a rough equilibrium state of saddle-focus type, Math. USSR Sb. 10, 91.

1857. Shilnikov L P (1976) Theory of the bifurcation of dynamical systems and dangerous boundaries, Sov. Phys. Dokl. 20, 674.

1858. Shimada T (1979) Gibbsian distribution on the Lorenz attractor, Prog. Theor. Phys. 62, 61.

1859. Shimada T, and Nagashima (1977) On the C-system like property of the Lorenz system, Prog. Theor. Phys. 58, 1318.

1860. Shimada T, and Nagashima T (1978) The iterative transition phenomenon between periodic and turbulent states in a dissipative dynamical system, Prog. Theor. Phys. 59, 1033.

1861. Shimada T, and Nagashima T (1979) A numerical approach to ergodic problem of dissipative systems, Prog. Theor. Phys. 61, 1605.

1862. Shimizu T (1979) Analytic form of the simplest limit cycle in the Lorenz model, Physica 97A, 383.

1863. Shimizu T (1981) Asymptotic form of a strange attractor, Phys. Lett. 84A, 85.

1864. Shimizu T (1987) Perturbation theory analysis of chaos. I. Transition from quasiperiodicity to chaos; II. Chaos in a time-dependent Hamiltonian system, Physica 142A, 75; 145A, 341.

1865. Shimizu T, and Ichimura A (1982) Asymptotic solution of a chaotic motion, Phys. Lett. 91A, 52.

1866. Shimizu T, and Morioka N (1978a) Chaos and limit cycles in the Lorenz model, Phys. Lett. 66A, 182.

1867. Shimizu T, and Morioka N (1978b) Transient behavior in periodic regions of the Lorenz model, Phys. Lett. 66A, 447.

1868. Shimizu T, and Morioka N (1978c) Transitions between turbulent and periodic states in the Lorenz model, Phys. Lett. 69A, 148.

1869. Shimizu T, and Morioka N (1981) Period-duobling bifurcations in a simple model, Phys. Lett. 83A, 243.

1870. Shlesinger M F, Klafter J, and West B J (1986) Levy walks with application to turbulence and chaos, Physica 140A, 212.

1871. Shobu K, Ose T, and Mori H (1984) Shapes of the power spectrum of intermittent turbulence near its onset point, Prog. Theor. Phys. 71, 458.

1872. Showalter K, Noyes R M. and Bar-Eli K (1978) A modified Oregonator model exhibiting complicated limit cycle behavior in a flow system, J. Chem. Phys. 69, 2514.

1873. Shraiman B I (1984) Transition from quasiperiodicity to chaos: a perturbative renormalization-group approach, Phys. Rev. A29, 3464.

1874. Shraiman B, Wayne C E, and Martin P C (1981) A scaling theory for noisy period-doubling transitions to chaos, Phys. Rev. Lett. 46, 935.

1875. Shtern V N (1983a) Attractor dimension for the generalized baker's transformation, Phys. Lett. 99A, 268.

722

1876. Shtern V N (1983b) On the dimensionality of attractors of turbulent motion, Sov. Phys. Dokl. 28, 384; Russ. Orig. 270, 582.

1877. Shtern V N, and Shumova L V (1984) Metamorphoses of preturbulence, Phys. Lett. 103A, 167.

1878. Shudo A, and Saito N (1987) Level spacing distribution and avoiding crossing in quantum chaos, J. Phys. Soc. Japan, 56, 2641.

1879. Shulga N F, Bolotin Yu L, Gonchar V Yu, and Truten V I (1987) Dynamical chaos in the motion of fast particles in crystals, Phys. Lett. 123A, 357.

1880. Siemens D P, and Bucher M (1986) New graphical method for the iteration of one-dimensional maps, Physica 20D, 336.

1881. Simm C W, Sawley M L, Skiff F, and Pochelon A (1987) On the analysis of experimental signals for evidence of deterministic chaos, Helvetica Phys. Acta, 60, 510

1882. Simó C (1979) On the Hénon-Pomeau attractor, J. Stat. Phys. 21, 465.

1883. Simoyi R H, Wolf A, and Swinney H L (1982) One-dimensional dynamics in a multi-component chemical reaction, Phys. Rev. Lett. 49, 245.

1884. Sinai Ya G, and Khanin K M (1988) Renormalization group method in the theory of dynamical systems, Int. J. Mod. Phys. B2, 147.

1885. Sinai Ya G, and Vul E B (1980) Discovery of closed orbits of dynamical systems with the use of computers, J. Stat. Phys. 23, 27.

1886. Sinai Ya G, and Vul E B (1981) Hyperbolicity conditions for the Lorenz model, Physica 2D, 3.

1887. Singer D (1978) Stable orbits and bifurcations of maps of the interval, SIAM J. Appl. Math. 35, 260.

1888. Singh S, and Agarwal G S (1983) Chaos in two-phonon coherent processes in a ring cavity, Opt. Commun. 47, 73.

1889. Singh V A, Bhattacharjee J K (1987) Fractal characteristics of classically chaotic quantum systems, Phys. Rev. A35, 3119.

1890. Sirovich L, and Rodriguez J D (1987) Coherent structures and chaos: a model problem, Phys. Lett. 120A, 211.

1891. Skarda C A, and Freeman W J (1987) How brains make chaos in order to make sense of the world, Behav. & Brain Sci. 10, 161.

1892. Skjolding H, Branner-Jorgensen B, Christiansen P L, and Jensen H E (1983) Bifurcations in discrete dynamical systems with cubic maps, SIAM J. Appl. Math. 43, 520.

1893. Smereka P, Birnir B, and Benerjee S (1987) Regular and chaotic bubble oscillations in periodically driven pressure fields, Phys. Fluids, 30, 3342.

1894. Smith C W, Tejwani M J, and Farris D A (1982) Bifurcation universality for first-sound subharmonic generation in superfluid helium-4, Phys. Rev. Lett. 48, 492.

1895. Smith C W, and Tejwani M J (1983) Bifurcation and the universal sequence for first-sound subharmonic generation in superfluid He$^4$, Physica 7D, 85.

1896. Snapp R R, Carmichael H J, and Schieve W C (1981) The path to turbulence: optical bistability and universality in the ring cavity, Opt. Commun. 40, 1.

1897. So B C, Yoshitake N, Okamoto H, and Mori H (1984) Correlations and spectra of an intermittent chaos near its onset point, J. Stat. Phys. 6, 367.

1898. So B C, and Mori H (1984) Power spectra of the intermittent chaos generated by the quadratic tangent bifurcation, Prog. Theor. Phys. 72, 1258.

1899. So B C, and Mori H (1986) Asymptotic shapes of power spectra of intermittent chaos near its onset point, Physica 21D, 126.

1900. Solari H G, and Gilmore R (1988) Relative rotation ratios for driven dynamical systems, Phys. Rev. A37, 3096.

1901. Sompolinsky H, Crisanti A, and Sommers H J (1988) Chaos in random neural networks, Phys. Rev. Lett. 61, 259.

1902. Sornette D, and Arneodo A (1984) Chaos, pseudo-random number generators and the random walk problem, J. Physique, 45, 1843.

1903. Spiegel E A (1987) Chaos: a mixed metaphor for turbulence, Proc. Roy. Soc. London, A413, 87.

1904. Sporns O, Roth S, and Seeling F F (1987) Chaotic dynamics of two coupled biochemical oscillators, Physica 26D, 215.

1905. Srivastava N, Kaufman C, Müller G, Weber R, and Thomas H (1988) Integrable and non-integrable classical spin clusters, Z. Phys. B70, 251.

1906. Steeb W-H, Erig W, and Kunick A (1983) Chaotic behavior and limit cycle behavior of anharmonic systems with periodic external perturbations, Phys. Lett. 93A, 267.

1907. Steeb W-H, Louw J A, Leach P G, and Mahomed F M (1986) Hamiltonian systems with three degrees of freedom, singular-point analysis, and chaotic behavior, Phys. Rev. A33, 211.

1908. Steeb W-H, Louw J A, and Kunick A (1987) Quantum chaos of an exciton-phonon system, Found. Phys. 17, 173.

1909. Steeb W-H, Louw J A, and Villet C M (1986a) Chaos in Yang-Mills equations, Phys. Rev. D33, 1174.

1910. Steeb W-H, Louw J A, and Villet C M (1986b) Maximal one-dimensional Lyapunov exponent and singular-point analysis for a quartic Hamiltonian, Phys. Rev. A34, 3489.

1911. Steeb W-H, Villet C M, and Kunick A (1985a) Quantum chaos and two exactly solvable second-quantized models, Phys. Rev. A32, 1232.

1912. Steeb W-H, Villet C M, and Kunick A (1985b) Chaotic behavior of a Hamiltonian with a quartic potential, J. Phys. A18, 3269.

1913. Steeb W-H, and Kunick A (1985) Instability of trajectories of a class of Hamiltonian systems, Lett. Nuovo Cimento, 42, 89.

1914. Steeb W-H, and Louw J A (1986) Chaotic behavior of an anharmonic oscillator with two external periodic forces, J. Phys. Soc. Japan, 55, 3279.

1915. Stefan P (1977) A theorem of Sarkovskii on the existence of periodic orbits of continuous endomorphisms of the real line, Commun. Math. Phys. 54, 237.

1916. Stefanski K (1982) Fluctuations and structure of attractors — simple tests on the Hénon mapping, Phys. Lett. 92A, 315.

1917. Stein P R, and Ulam S M (1964) Non-linear transformation studies on electronic computers, Rozprawy Matematyczne, 39, 1.

1918. Sterman J D (1988) Deterministic chaos in models of human behavior: methodological issues and experimental results, Sys. Dyn. Rev. 4, 148.

1919. Stoker J J (1980) Periodic forced vibrations of systems of relaxation oscillators, Comm. Pure Appl. Math. 33, 215.

1920. Stoop R, Peinke J, Parisi J, Rohricht B, and Huebener R P (1989) A p-Ge semiconductor experiment showing chaos and hyperchaos, Physica 35D, 425.

1921. Stratonovich R L (1982) Correlators of processes in very simple systems with strange attractors, Sov. Phys. Dokl. 27, 942; Russ. Orig. 267, 355.

724

1922. Stratonovich R L, and Nikolaevskii E S (1984) On the sequence of bifurcations of the motion of strange attractors as they split up in the case of simple systems, Sov. Phys. - Dokl. 29, 406; Russ. Orig. 276, 363.

1923. Sturis J, and Mosekilde E (1988) Bifurcation sequence in a simple model of migratory dynamics, Sys. Dyn. Rev. 4, 208.

1924. Stutzer M J (1980) Chaotic dynamics and bifurcation in a macro model, J. Econ. Dyn. Control, 2, 353.

1925. Sun G-Z, and Liu Z-R (1987) Chaotic state of soft spring Duffing system, Kexue Tongbao (Sci. Bull. China), 2, 1464.

1926. Sun Y-S (1983) On the measure-preserving mappings with odd dimension, Celest. Mech. 30, 7.

1927. Sun Y-S (1984a) An extension of quadratic area-preserving mappings to the measure-preserving mappings with three dimensions, Scientia Sinica, 27, 174.

1928. Sun Y-S (1984b) Invariant manifords in the measure-preserving mappings with three dimension, Celest. Mech. 33, 111.

1929. Sun Y-S (1985) Attractors in a dissipative dynamical system with three dimensions, Celestial Mech. 37, 171.

1930. Sun Y-S, and Froeschle C (1982) On the Kolmogorov entropy of two-dimensional area-preserving mappings, Scientia Sinica, 25, 750.

1931. Sun Y-S, and Yan Z-M (1988) A perturbed extension of hyperbolic twist mappings, Celest. Mech. 42, 369.

1932. Susman G J, and Wisdom J (1988) Numerical evidence that the motion of Pluto is chaotic, Science, 241, 433.

1933. Svensmark H, and Samuelsen M R (1987) Influence of perturbations on period-doubling bifurcations, Phys. Rev. A36, 2413.

1934. Swiatecki W J (1987) Order, chaos and nuclear dynamics, J. Physique Colloq. 48, C-2, 247.

1935. Swift J W, and Wiesenfeld K (1984) Suppression of period doubling in symmetric systems, Phys. Rev. Lett. 52, 705.

1936. Swinney H L (1985) Observations of complex dynamics and chaos, in *Fundamental Problems in Statistical Mechanics VI*, North-Holland, 253.

1937. Swinney H L (1988) Instabilities and chaos in rotating fluids, in Zweifel, Gallavotti, and Anile (B1988), 319.

1938. Swinney H L, and Gollub J P (1978) Transition to turbulence, Phys. Today, 31, 41.

1939. Szemplinska-Stupnicka W (1988) Bifurcations of harmonic solutions leading to chaotic motion in the softening type Duffing's oscillator, Int. J. Nonlin. Mech. 23, 257.

1940. Szemplinska-Stupnicka W, and Bajkowski J (1986) The 1/2 subharmonic resonance and its transition to chaotic motion in a nonlinear oscillator, Int. J. Nonlin. Mech. 21, 401.

1941. Szemplinska-Stupnicka W, and Bajkowski J (1987) Period doubling and chaotic motion in a nonlinear oscillator, Z. Angew. Math. Mech. 67, T153.

1942. Szépfalusy P, Tél T, Csordas A, and Kovaks Z (1987) Phase transitions associated with dynamical properties of chaotic systems, Phys. Rev. A36, 3525.

1943. Szépfalusy P, and Gyorgyi G (1986) Entropy decay as a measure of stochasticity in chaotic systems, Phys. Rev. A33, 2852.

1944. Szépfalusy P, and Tél T (1982) Fluctuations in the limit cycle state and the problem of phase chaos, Physica 112A, 146.

1945. Szépfalusy P, and Tél T (1985) Properties of maps related to flows around a saddle point, Physica 16D, 252.

1946. Szépfalusy P, and Tél T (1986) New approach to the problem of chaotic repellers, Phys. Rev. A34, 2520.

1947. Szépfalusy P, and Tél T (1987) Dynamical fractal properties of one-dimensional maps, Phys. Rev. A35, 477.

1948. Szydlowski M (1987) The problem of chaotic behavior in a homogeneous arbitrarily dimensional cosmology, Phys. Lett. 195B, 31.

1949. Tabeling P (1985) Sudden increase of the fractal dimension in a hydrodynamical system, Phys. Rev. A31, 3460.

1950. Tabor M (1981) The onset of chaotic motion in dynamical systems, Adv. in Chem. Phys. 46, 73.

1951. Tabor M, and Weiss J (1981) Analytical structure of the Lorenz system, Phys. Rev. A24, 2157.

1952. Takahashi K, and Saito N (1985) Chaos and Husimi distribution function in quantum mechanics, Phys. Rev. Lett. 55, 645.

1953. Takahashi T, Ichimura A, Hirooka H, and Saito N (1985) External disturbance to non-integrable Hamiltonian systems, J. Phys. Soc. Japan, 54, 500.

1954. Takahashi Y, and Oono Y (1984) Towards the statistical mechanics of chaos, Prog. Thoer. Phys. 71, 851.

1955. Takens F (1981) Detecting strange attractors in turbulence, Lect. Notes in Math. 898, 366.

1956. Takesue S (1986) Fractal dimension of strange repellers in one-dimensional interval mappings, Prog. Theor. Phys. 76, 11.

1957. Takeyama K (1978, 1980) Dynamics of the Lorenz model of convective instabilities. I and II, Prog. Theor. Phys. 60, 613; 63, 91.

1958. Takeyama K (1983) A parameter renormalization for orbit-splittings and band-mergings of iterated maps of an interval, Z. Phys. B52, 253.

1959. Takeyama K, and Satoh K K (1983) A finite-order splitting-merging bifurcation of bands in iterated maps of an interval with a cusp maximum, Phys. Lett. 98A, 313.

1960. Takimoto N, and Yamashida H (1987) The variational approach to the theory of subharmonic bifurcations, Physica 26D, 251.

1961. Tam W Y, Vastano J A, Swinney H L, and Horthemke W (1988) Regular and chaotic chemical spatiotemporal patterns — experiemnt and theory, Phys. Rev. Lett. 61, 2163.$\Omega$

1962. Tang D M, and Dowell E H (1988) On the threshold force for chaotic motion for a forced bucked beam, Trans. ASME J. Appl. Mech. 55, 190.

1963. Tang Y S, Mees A I, and Chua L O (1983) Synchronization and chaos, IEEE Trans. Cir. and Sys. CAS-30, 620.

1964. Tao H-J, Lu L, and Yang Q-S (1986) Observation of chaotic noise in Nb-In point contact, Chinese Phys. Lett. 3, 349.

1965. Tavakol R K, and Tworkowski A S (1984a) On the occurrence of quasiperodic motion on three tori, Phys. Lett. 100A, 65.

1966. Tavakol R K, and Tworkowski A S (1984b) An example of quasiperiodic motion on 4-torus, Phys. Lett. 100A, 273.

1967. Tavakol R K, and Tworkowski A S (1988) Fluid intermittency in low dimensional deterministic systems, Phys. Lett. 126A, 318.

1968. Tedeschini-Lalli L (1982) Truncated Navier-Stokes equations: continuos transition from a five-mode to a seven-mode model, J. Stat. Phys. 27, 365.

1969. Tedeschini-Lalli L, and Yorke J A (1986) How often do simple dynamical processes have infinitely many coexisting sinks? Commun. Math. Phys. 106, 635.

1970. Teitsworth S W, Westervelt R M and Haller E E (1983) Nonlinear oscillations and chaos in electrical breakdown in Ge, Phys. Rev. Lett. 51, 825.

1971. Teitsworth S W, and Westervelt R M (1984) Chaos and broadband noise in extrinsic photoconductors, Phys. Rev. Lett. 53, 2587.

1972. Tél T (1982a) On the construction of invariant curves of period-two points in two-dimensional maps, Phys. Lett. 94A, 334.

1973. Tél T (1982b) On the construction of stable and unstable manifolds of two-dimensional invertible maps, Z. Phys. B49, 157.

1974. Tél T (1983a) Fractal dimension of the strange attractor in a piecewise linear two-dimensional map, Phys. Lett. 97A, 219.

1975. Tél T (1983b) Invariant curves, attractors, and phase diagram of a piecewise linear map with chaos, J. Stat. Phys. 33, 195.

1976. Tél T (1986) Characteristic exponents of chaotic repellers as eigenvalues, Phys. Lett. 119A, 65.

1977. Tél T (1987a) Escape rate from strange sets as an eigenvalue, Phys. Rev. A36, 1502.

1978. Tél T (1987b) Dynamical spectrum and thermodynamic functions of strange sets from an eigenvalue problem, Phys. Rev. A36, 2507.

1979. Tél T (1988) On the stationary distribution of self-sustained oscillators around bifurcation points, J. Stat. Phys. 50, 897.

1980. Termonia Y (1984a) Random walk on a strange attractor, Phys. Rev. Lett. 53, 1356.

1981. Termonia Y (1984b) Kolmogorov entropy from a time series, Phys. Rev. A29, 1612.

1982. Termonia Y, and Alexandrowicz Z (1983) Fractal dimension of strange attrators from radius versus size of arbitary clusters, Phys. Rev. Lett. 51, 1265.

1983. Testa J (1985) Fractal dimension at chaos of a quasiperiodic driven tunnel diode, Phys. Lett. 111A, 243.

1984. Testa J, Perez J, and Jeffries C (1982) Evidence for universal chaotic behavior of a driven nonlinear oscillator, Phys. Rev. Lett. 48, 714.

1985. Testa J, and Held G A (1982) Study of a one-dimensional map with multiple basins, Phys. Rev. A28, 3085.

1986. Testa J, and Held G A (1983) Period doubling, bifurcations, chaos and periodic windows of the cubic map, Phys. Rev. A28, 3085.

1987. Theile E, and Stone J (1984) A measure of quantum chaos, J. Chem. Phys. 80, 5187.

1988. Theiler J (1986) Spurious dimension from correlation algorithms applied to limited time series data, Phys. Rev. A34, 2427.

1989. Theiler J (1987) Efficient algorithm for estimating the correlation dimension from a set of discrete points, Phys. Rev. A36, 4456.

1990. Thomae S, and Grossmann S (1981a) A scaling property in critical spectra of discrete systems, Phys. Lett. 83A, 181.

1991. Thomae S, and Grossmann S (1981b) Correlations and spectra of periodic chaos generated by the logistic parabola, J. Stat. Phys. 26, 485.

1992. Thompson C J, and McGuire J B (1988) Asymptotic and essentially singular solutions of the Feigenbaum equations, J. Stat. Phys. 51, 991.

1993. Thompson J M T (1983) Complex dynamics of compliant off-shore structures, Proc. Roy. Soc. London, A87, 407.

1994. Thompson J M T, Bishop S R, and Lenug L M (1987) Fractal basins and chaotic bifurcations prior to escape from a potential well, Phys. Lett. 121A, 116.

1995. Thompson J M T, Bokaian A R, and Ghaffari R (1983) Subharmonic resonances and chaotic motions of a bilinear oscillator, IMA J. Appl. Math. 31, 207.

1996. Thompson J M T, and Ghaffari R (1982) Chaos after period-doubling bifurcations in the resonance of an impact oscillator, Phys. Lett. 91A, 5.

1997. Thompson J M T, and Ghaffari R (1983) Chaotic dynamics of an impact oscillator, Phys. Rev. A27, 1741.

1998. Thompson J M T, and Stewart H B (1984) Folding and mixing in the Birkhoff-Shaw chaotic attractor, Phys. Lett. 103A, 229.

1999. Thompson J M T, and Virgin L N (1988) Spatial chaos and localization phenomenon in nonlinear elasticity, Phys. Lett. 126A, 491.

2000. Thompson M, and Bishop S (1988) From Newton to chaos, Phys. Bull. 39, 232.

2001. Thummel O (1987) Oscillations in the determination of Renyi dimension, Ann. Phys. 44, 601.

2002. Toda M, and Ikeda K (1985) Quantum version of global stochastization due to overlap of resonances, Phys. Lett. 110A, 235.

2003. Toda M, and Ikeda K (1987a) Quantal version of resonance overlap, J. Phys. A20, 83.

2004. Toda M, and Ikeda K (1987b) Quantal Lyapunov exponent, Phys. Lett. 124A, 165.

2005. Toh S (1987) Statistical model with localized structures describing the spatio-temporal chaos of Kuramoto-Shivashisky equation, J. Phys. Soc. Japan, 56, 949.

2006. Tomita K (1982a) Chaotic response of nonlinear oscillators, Phys. Reports, 86, 113.

2007. Tomita K (1982b) Structure and macroscopic chaos in biology, J. Theor. Biol. 99, 111.

2008. Tomita K (1987) Conjugate pair of representations in chaos and quantum mechanics, Found. Phys. 17, 699.

2009. Tomita K, Kai T, and Hikami F (1977) Entrainment of a limit cycle by a periodic external excitation, Prog. Theor. Phys. 57, 1159.

2010. Tomita K, and Daido H (1980) Possibility of chaotic behavior and multi-basins in forced glycolytic oscillations, Phys. Lett. 79A, 133.

2011. Tomita K, and Kai T (1978) Stroboscopic phase portrait and strange attractors, Phys. Lett. 66A, 91.

2012. Tomita K, and Kai T (1979a) Stroboscopic phase portrait of a forced nonlinear oscillator, Prog. Theor. Phys. 61, 54.

2013. Tomita K, and Kai T (1979b) Chaotic response of a nonlinear oscillator, J. Stat. Phys. 21, 65.

2014. Tomita K, and Tsuda I (1979) Chaos in Belousov-Zhabotinskii reaction in a flow system, Phys. Lett. 71A, 489.

2015. Tomita K, and Tsuda I (1980) Towards the interpretation of Hudson's experiment on the Belousov-Zhabotinskii reaction: chaos due to delocalization, Prog. Theor. Phys. 64, 1138.

2016. Toro M, and Aracil J (1988) Qualitative analysis of system dynamics ecological models, Sys. Dyn. Rev. 4, 16.

2017. Tredicce J R, Abraham N B, Puccioni G P and Arecchi F T (1985) On chaos in lasers with modulated parameters: a comparative analysis, Opt. Commun. 55, 131.

2018. Tredicce J R, Arecchi F T, Puccioni G P, and Gadomski W (1986) Dynamic behavior and onset of low-dimensional chaos in a modulated homogeneously broadened single-mode laser: experiments and theory, Phys. Rev. A34, 2073.

2019. Tresser C (1984) Homoclinic orbits for flows in $R^3$, J. Physique, 45, 837.

2020. Tresser C, Coullet P, and Arneodo A (1980a) Topological horseshoe and numerically observed chaotic behavior in the Hénon mapping, J. Phys. 13A, L123.

2021. Tresser C, Coullet P, and Arneodo A (1980b) Transition to turbulence for doubly periodic flows, Phys. Lett. 77A, 327.

2022. Tresser C, Coullet P, and Arneodo A (1980c) On the existence of hysteresis in a transition to chaos after a single bifurcation, J. Physique Lett. 41, L243.

2023. Tresser C, and Coullet P (1978) Iterations de endomorphismes et groupe de renormalisation, C. R. Acad. Sc. Paris, 287A, 577.

2024. Tresser C, and Coullet P (1980) Preturbulent states and renormalization group for simple models, Repts. Math. Phys. 17, 189.

2025. Tritton D (1986) Chaos in the swing of a pendulum, New Scientist, 111, No. 1518, 37.

2026. Troger H, Kacani V, and Stribersky A (1985) Definig 'strange attractors', Z. Angew. Math. und Mech. 65, T109.

2027. Trubetskov D I, and Dmitriev A Yu (1985) Dynamical chaos in a system of beam-electron oscillators and an electromagnetic wave (experiment), Sov. Tech. Phys. Lett. 11, 519-21; Russ. Orig. 11, 19.

2028. Tsang K Y (1986) Dimensionality of strange attractors determined analytically, Phys. Rev. Lett. 57, 1390.

2029. Tsang K Y (1987) Determining the spectrum of scaling indices for strange attractors in two-dimensional maps, Phys. Rev. A36, 4567.

2030. Tsang K Y, and Lieberman M A (1984a) Invariant distribution on strange attractors in highly dissipative systems, Phys. Lett. 103A, 175.

2031. Tsang K Y, and Lieberman M A (1984b) Analytical calculation of invariant distribution on strange attractors, Physica 11D, 147.

2032. Tsang K Y, and Lieberman M A (1986) Transient chaotic distribution in dissipative systems, Physica 21D, 401.

2033. Tsuchiya T (1984) An exactly solvable difference equation that gives pure chaos for a continuous range of a parameter, Z. Naturforsch. 39a, 80.

2034. Tsuchiya T, Szabo A, and Saito N (1983) Exact solutions of simple difference equation systems that show chaotic behavior, Z. Naturforsch. 38a, 1035.

2035. Tsuchiya T, and Ichimura A (1984) 'Pure chaotic' orbits of one-dimensional maps have third-order correlation, Phys. Lett. 101A, 447.

2036. Tufillaro N B (1985) Motions of a swinging Atwood's machine, J. Physique, 46, 1495.

2037. Tufillaro N B, and Albano A M (1986) Chaotic dynamics of a bouncing ball, Am. J. Phys. 54, 939.

2038. Tung M W, Yuan J-M (1987) Dissipative quantum mechanics: driven molecular vibrations, Phys. Rev. A36, 4463.

2039. Turner J S, Roux J C, McCormick W D, and Swinney H L (1981) Alternating preiodic and chaotic regimes in a chemical reaction — experiment and theory, Phys. Lett. 85A, 9.

2040. Tyson J J (1978) On the appearance of chaos in a model of the Belousov reaction, J. Math. Biology, 5, 351.

2041. Uchimura H, Fujisaka H, and Inoue M (1987) Breakdown of chaos symmetry and intermittency in band-splitting phenomena, Prog. Theor. Phys. 77, 1344.

2042. Ueda Y (1979) Randomly transitional phenomena in the system governed by Duffuing's equation, J. Stat. Phys. 20, 181.

2043. Ueda Y (1985) Random phenomena resulting from nonlinearity in the system described by Duffing's equation, Int. J. Nonlin. Mech. 20, 481.

2044. Ueda Y, Akamatsu N (1981) Chaotically transitional phenomena in the forced negative-resistance oscillator, IEEE Trans. Cir. & Sys. CAS-28, 217.

2045. Ueda Y, and Noguch A (1983) A model that realizes the soliton and the chaos simultaneously, J. Phys. Soc. Japan, 52, 713.

2046. Uezu T, and Aizawa Y (1982a) Some routes to chaos from limit cycle in the forced Lorenz system, Prog. Theor. Phys. 68, 1543.

2047. Uezu T, and Aizawa Y (1982b) Topological character of periodic solution in three-dimensional ODE system, Prog. Theor. Phys. 68, 1907.

2048. Ulam S M, and von Neumann J (1947) On combinations of stochastic and deterministic processes, Bull. Amer. Math. Soc. 53, 1120.

2049. Umberger D K, Farmer J D, and Satija I I (1986) A universal strange attractor underlying the quasiperiodic transition to chaos, Phys. Lett. 114A, 41.

2050. Urata K (1987) Low dimensional chaos in Boussinesq convection, Fluid Dyn. Res. 1, 257.

2051. Ushiki S (1982) Central difference scheme and chaos, Physica 4D, 407.

2052. Ushio T, and Hirai K (1983) Chaos in nonlinear sampled-data control systems, Int. J. Control, 38, 1023.

2053. Ushio T, and Hsu C S (1986) Cell simplex degeneracy, Liapunov functions and stability of simple cell mapping systems, Int. J. Nonl. Mech. 21, 183.

2054. Valkering T P (1987) The shape of the basin of attraction and transients via transformations to normal form, Physica 27D, 213.

2055. Vallee R, Delisle C, and Chrostowski J (1984) Noise versus chaos in acousto-optic bistability, Phys. Rev. A30, 336.

2056. Vallee R, and Delisle C (1986) Periodicity windows in a dynamical system with delayed feedback, Phys. Rev. A34, 309.

2057. Vallos G K (1986) El Nino: a chaotic dynamical system, Science, 232, 243.

2058. Van Buskirk R, and Jeffries C (1985) Observation of chaotic dynamics of coupled nonlinear oscillators, Phys. Rev. A31, 3332.

2059. Van Exter M, and Lagendijk A (1983) Observation of fine structure in the phase locking of a nonlinear oscillator, Phys. Lett. 99A, 1.

2060. Van Leeuwen K A H, Oppen G V, Renwick S, Bowlin J B, and Koch P M (1985) Microwave ionization of hydrogen atom: experiment versus classical dynamics, Phys. Rev. Lett. 55, 2231.

2061. Van der Pol B, and Van der Mark J (1928) The heart beat considered as a relaxation oscillation, and an electrical model of the heart, Phil. Mag. (7), 6, 763.

2062. Van der Water W, and Schram P (1988) Generalized dimensions from near-neighbour information, Phys. Rev. A37, 3118.

2063. Van der Weele J P, Capel H W, Post T, and Calkoen C J (1986) Crossover from dissipative to conservative behaviour in period doubling systems, Physica 137A, 1.

2064. Van der Weele J P, Capel H W, and Calkoen C J (1985) Crossover from dissipative to conservative systems, Phys. Lett. 111A, 5.

2065. Van der Weele J P, Capel H W, and Kluving R (1986) On the scaling factors $\alpha(z)$ and $\delta(z)$, Phys. Lett. 119A, 15.

2066. Varghese M, and Thorp J S (1988) Truncated-fractal basin boundaries in forced pendulum systems, Phys. Rev. Lett. 60, 665.

2067. Varosi F, Grebogi C, and Yorke J A (1987) Simplicial approximation of Poincaré maps of differential equations, Phys. Lett. 124A, 59.

2068. Vastano J A, and Swinney H L (1988) Information transport in spatio-temporal systems, Phys. Rev. Lett. 60, 1773.

2069. Veerman P (1986) Symbolic dynamics and rotation numbers, Physica 134A, 543.

2070. Veerman P (1987) Symbolic dynamics of order-preserving orbits, Physica 29D, 191.

2071. Velarde M G, and Antoranz J C (1981) Strange attractor (optical turbulence) in a model problem for the laser with saturable absorber and the two-component Bénard convection, Prog. Theor. Phys. 66, 717.

2072. Viet O, Westfreid J E, and Guyon E (1983) Kinetic art and mechanical chaos, Eur. J. Phys. 4, 72.

2073. Vilela Mendes R (1981) Critical-point dependence of universality in maps of the interval, Phys. Lett. 84A, 1.

2074. Vilela Mendes R (1984) A variational formulation for dissipative maps, Phys. Lett. 104A, 391.

2075. Vincent A P, and Yuen D A (1988) Thermal attractor in chaotic convection with high-Prandtl-number fluids, Phys. Rev. A38, 328.

2076. Vinson M J, and Gunaratne G H (1987) Trajectory scaling function for bifurcations in four-dimensional maps, Phys. Rev. A36, 2418.

2077. Visscher W M (1987) Class of classically chaotic time-periodic Hamiltoians which lead to separable evolution operators, Phys. Rev. A36, 5031.

2078. Vivaldi F, Casati G, and Guarneri I (1983) Origin of long-time tails in strongly chaotic systems, Phys. Rev. Lett. 51, 727.

2079. Vladimirskii K V, and Norvaishas A A (1984) On the dimensionality of a metastable strange attractor, Sov. Phys. Dokl. 29, 827; Russ. Orig. 278, 1106.

2080. Vlasova O F, and Zaslavsky G M (1983) Dissipation effect on chaos onset in a two-resonance overlap case, Phys. Lett. 99A, 405.

2081. Vlasova O F, and Zaslavsky G M (1984) Nonergodic regions in the standard dissipative mapping, Phys. Lett. 105A, 1.

2082. Vorobev P A, and Zaslavsky G M (1987) Quantum chaos and level distribution in the model of two coupled oscillators, Sov. Phys. JETP, 65, 877; Russ. Orig. 92, 1564.

2083. Vulovic V, and Prange R E (1986) Randomness of a true coin tossing, Phys. Rev. A33, 576.

2084. Walden R W, and Donnelly R J (1979) Reemergent order of chaotic circular Couette flow, Phys. Rev. Lett. 42, 301.

2085. Walden R W, Kolodner P, Passner A, and Surko C M (1984) Nonchaotic Rayleigh-Bénard convection with four and five incommensurable frequencies, Phys. Rev. Lett. 53, 242.

2086. Walker G H, Ford J (1969) Amplitude instability and ergodic behavior for conservative nonlinear oscillator systems, Phys. Rev. 188, 416.

2087. Waller I, and Kapral R (1984) Synchronization and chaos in coupled nonlinear oscillators, Phys. Lett. 105A, 163.

2088. Wang G-R, Chen S-G, and Hao B-L (1983) Intermittent chaos in the forced Brusselator, Chinese Phys. 4, 282.

2089. Wang G-R, Chen S-G, and Hao B-L (1984) On the nonconvergence problem in computing the capacity of strange attractors, Chinese Phys. Lett. 1, 11.

2090. Wang X-J, and Mou C Y (1985) A thermokinetic model of complex oscillations in gaseous hydrocarbon oxidation, J. Chem. Phys. 83, 4554.

2091. Wang Y-X, He Y-S, and Peng S-L (1984) Frequency sweep method for study of period-doubling bifurcation and chaotic behaviour, Chinese Phys. 4, 822.

2092. Warnock R L, and Ruth R D (1987) Invariant tori through direct solution of the Hamilton-Jacobi equation, Physica 26D, 1.

2093. Wedding B, Gasch A, and Jaeger D (1984) Chaos observed in a Fabry-Perot interferometer with quardratic nonlinear medium, Phys. Lett. 105A, 105.

2094. Wegmann K, and Rössler O E (1978) Differenct kinds of chaotic oscillations in the Belousov – Zhabotinskii reaction, Z. Naturforsch. 33a, 1179.

2095. Wei R-J, Tao Q-T, and Ni W-S (1986) Bifurcation and chaos of direct radiation loudspeaker, Chinese Phys. Lett. 3, 469.

2096. Weidenmüller H A (1986) Statistics and the shell model for chaotic motion in atomic nuclei, Comments Nucl. & Part. Phys. 16, 199.

2097. Weiss C O (1984) Chaotic (turbulent) behaviour of laser oscillation, Laser & Optoelektron. 16, 41.

2098. Weiss C O, Abraham N B, and Hubner U (1988) Homoclinic and heteroclinic chaos in a single-mode laser, Phys. Rev. Lett. 61, 1587.

2099. Weiss C O, Godone A, and Olafsson A (1983) Routes to chaotic emission in a CW He-Ne laser, Phys. Rev. A28, 892.

2100. Weiss C O, and Brock J (1986) Evidence for Lorenz-type chaos in a laser, Phys. Rev. Lett. 57, 2804.

2101. Weiss C O, and King H (1982) Oscillation period-doubling chaos in a laser, Opt. Commun. 44, 59.

2102. Weiss C O, and Kische W (1984) On observability of Lorenz instability in lasers, Opt. Commun. 51, 47.

2103. Weiss N O (1985) Chaotic behavior in stellar dynamos, J. Stat. Phys. 39, 477.

2104. Weissman Y, and Jortner J (1982) Quantum manifestations of classical stochasticity: I. Energetics of some nonlinear systems; II. Dynamics of wave packets of bound states, J. Chem. Phys. 77, 1486; 1496.

2105. Wersinger J-M, Finn J M, and Ott E (1980) Bifurcations and strange behavior in instability saturation by nonlinear mode coupling, Phys. Rev. Lett. 44, 453.

2106. West B J, Goldberger A L, Rovner G, and Bhargava V (1985) Nonlinear dynamics of the heartbeat I. The AV junctions: passive conduit or active oscillator? Physica 17D, 198.

2107. West B J, and Mandell A J (1985) On a mechanism of cardiac electrical stability: the fractal hypothesis, Biophys. J. 48, 525.

2108. Whitehead R R (1984) A chaotic mapping that displays its own homoclinic structure, Physica 13D, 401.

2109. Whitehead R R, and MacDonald N (1984) A transition between two kinds of chaos in the Burgers mapping, Phys. Lett. 105A, 433.

2110. Widmann P J, Gorman M, and Robbins K A (1989) Nonlinear dynamics of a convection loop II. Chaos in laminar and turbulent flows, Physica 36D, 157.

2111. Widom M, Bensimon D, Kadanoff L P, and Shenker S J (1983) Strange objects in the complex plane, J. Stat. Phys. 32, 443.

2112. Widom M, and Kadanoff L P (1982) Renormalization group analysis of area-preserving maps, Physica 5D, 287.

2113. Wiesenfeld K (1985a) Virtual Hopf phenomenon: a new precursor of period-doubling bifurcations, Phys. Rev. A32, 1744.

2114. Wiesenfeld K (1985b) Noisy precursors of nonlinear instabilities, J. Stat. Phys. 38, 1071.

2115. Wiesenfeld K (1986) Parametric amplification in semiconductor lasers: a dynamical perspective, Phys. Rev. A33, 4026.

2116. Wiesenfeld K, and McNamara B (1985) Period-doubling systems as small-signal amplifiers, Phys. Rev. Lett. 55, 13.

2117. Wiesenfeld K, and McNamara B (1986) Small-signal amplification in bifurcating dynamical systems, Phys. Rev. A33, 629.

2118. Wiesenfeld K, and Pederson N F (1987) Amplitude calculation near a period-doubling bifurcation: an example, Phys. Rev. A36, 1440.

2119. Wiesenfeld K, and Tufillaro N B (1987) Suppression of period doubling in the dynamics of a bouncing ball, Physica 26D, 321.

2120. Wiessefeld K, Knobloch E, Miracky R F, and Clarke J (1984) Calculation of period doubling in a Josephson circuit, Phys. Rev. A29, 2102.

2121. Wiggins S (1987) Chaos in the quasiperiodically forced Duffing oscillation, Phys. Lett. 124A, 138.

2122. Wiggins S (1988) Adiabatic chaos, Phys. Lett. 128A, 339.

2123. Wiggins S, and Holmes P (1987) Homoclinic orbits in slowly varying oscillators, SIAM J. Math. Anal. 18, 612.

2124. Wilbrink J (1987) Erratic behavior of invariant circles in standard-like mappings, Physica 26D, 358.

2125. Wilbrink J (1988) New fixed points in the renormalization operator for invariant circles, Phys. Lett. 131A, 251.

2126. Wilkinson M (1987) A semiclassical sum rule for matrix elements of classically chaotic systems, J. Phys. A20, 2415.

2127. Wilkinson M (1988) Random matrix theory in semiclassical quantum mechanics of chaotic systems, J. Phys. A21, 1173.

2128. Williams R F (1979) The structure of Lorenz attractors, Publ. Math. IHES, 50, 321.

2129. Wintgen D (1987a) Calculation of the Liapunov exponent for periodic orbits of the magnetised hydrogen atom, J. Phys. A20, L511.

2130. Wintgen D (1987b) Connection between long-range correlations in quantum spectra and classical periodic orbits, Phys. Rev. Lett. 58, 1589.

2131. Wintgen D, and Friedrich H (1987) Classical and quantum-mechanical transition between regularity and irregularity in a Hamiltonian system, Phys. Rev. A35, 1464.

2132. Wintgen D, and Marxer H (1988) Level statistics of a quantized Cantori system, Phys. Rev. Lett. 60, 971.

2133. Wisdom J (1983) Chaotic behavior and the origin of the 3/1 Kirkwood gap, Icarus, 56, 51.

2134. Wisdom J (1985) Meteorites may follow a chaotic route to Earth, Nature, 315, 731.

2135. Wisdom J (1987) Urey Prize lecture: Chaotic dynamics in the solar system, Icarus, 72, 241.

2136. Withers W D (1987) Calculating derivatives with respect to parameters in iterated function systems, Physica 28D, 206.

2137. Wolf A (1983) Simplicity and universality in the transition to chaos, Nature, 305, 182.

2138. Wolf A, Swift J B, Swinney H L, and Vastano J A (1985) Determining Lyapunov exponents from a time series, Physica 16D, 285.

2139. Wolf A, and Swift J (1981) Universal power spectra for the reverse bifurcation sequence, Phys. Lett. 83A, 184.

2140. Wolf M (1988) Example of order and disorder: $x_{n+1} = (Ax_n + B) \mathrm{mod}\ C$, Int. J. Theor. Phys. 27, 73.

2141. Wolfe R, and Morris H C (1987) Chaotic solutions of systems of first order partial differential equations, SIAM J. Math. Anal. 18, 1040.

2142. Wolfram S (1985) Origin of randomness in physical systems, Phys. Rev. Lett. 55, 449.

2143. Wong S (1984) Newton's method and symbolic dynamics, Proc. Amer. Math. Soc. 91, 245.

2144. Woo C H (1984) Temporary order amid chaos, Phys. Rev. A29, 2866.

2145. Woo C H (1989) Chaos, ineffectiveness, and the contrast between classical and quantal physics, Found. Phys. 19, 17.

2146. Wright J (1984) Method for calculating a Lyapunov exponent, Phys. Rev. A29, 2924.

2147. Wu S-X, Pei L-Q, and Guo F (1985) U-sequence and period-tripling phenomena in a forced nonlinear oscillator, Chinese Phys. Lett. 2, 213.

2148. Xie B-M (1988) Molecular spectra in chaotic regime expressed by the Wigner function, J. Chem. Phys. 89, 1351.

2149. Xu J-H, and Li W (1986) The dynamics of large scale neuron-glia network and its relation to the brain function, Commun. Theor. Phys. 5, 339.

2150. Yabuzaki T, Okamoto T, Kitano M, and Ogawa T (1984) Optical bistability with symmetry breaking, Phys. Rev. A29, 1694.

2151. Yahata H (1983) Period-doubling cascade in the Rayleigh-Bénard convection, Prog. Theor. Phys. 69, 1802.

2152. Yamada M, and Ohkitani K (1987) Lyapunov spectrum of a chaotic model of three-dimensional turbulence, J. Phys. Soc. Japan, 56, 4210.

2153. Yamada M, and Ohkitani K (1988) Lyapunov spectrum of a model of two-dimensional turbulence, Phys. Rev. Lett. 60, 983.

2154. Yamada T, and Fujisaka H (1977) A discrete model exhibiting successive bifurcations leading to the onset of chaos, Z. Phys. B28, 239.

2155. Yamada T, and Fujisaka H (1986) Intermittency caused by chaotic modulation. I. Analysis with a multiplicative noise model, Prog. Theor. Phys. 76, 582.

2156. Yamada T, and Fujisaka H (1987) Effect of inhomogeneity on intermittent chaos in a coupled system, Phys. Lett. 124A, 421.

2157. Yamaguchi Y (1983) Chaotic behavior of magnetization in superfluid He$^3$ driven by external periodic field, Prog. Theor. Phys. 69, 1377.

2158. Yamaguchi Y (1985) Structure of the stochastic layer of a perturbed double-well potential system, Phys. Lett. 109A, 191.

2159. Yamaguchi Y (1986) Breakup of an invariant curve in a dissipative standard mapping, Phys. Lett. 116A, 307.

2160. Yamaguchi Y, Kometani K, and Shimizu H (1978) Self-synchronization of nonlinear oscillations in the presence of fluctuations, J. Stat. Phys. 27, 719.

2161. Yamaguchi Y, Tanikawa K, and Mishima N (1988) Fractal basin boundary in dynamical systems and the Weierstrass-Takagi functions, Phys. Lett. 128A, 470.

2162. Yamaguchi Y, and Minowa H (1985) Power spectrum of chaotic motion near a crisis of a one-dimensional map, Phys. Rev. A32, 3758.

2163. Yamaguchi Y, and Mishima N (1984) Structure of strange attractor and homoclinic bifurcation of two-dimensional cubic map, Phys. Lett. 104A, 179.

2164. Yamaguchi Y, and Sakai K (1983) New type of 'crisis' showing hysteresis, Phys. Rev. A27, 2755.

2165. Yamaguchi Y, and Sakai K (1986) $1/f$ noise spectrum of the chaotic motion in a whisker mapping, Phys. Lett. 117A, 387.

2166. Yamaguchi Y, and Sakai K (1988) Structure change of basins by crisis in a two-dimensional map, Phys. Lett. 131A, 499.

2167. Yamaguti M, and Ushiki S (1981) Chaos in numerical analysis of ODE's, Physica, 3D, 618.

2168. Yamazaki H, Oono Y, and Hirakawa K (1978-9) Experimental study on chemical turbulence: I and II, J. Phys. Soc. Japan, 44, 335; 46, 721.

2169. Yang W-M, and Hao B-L (1987) How the Arnold tongues become sausages in a piecewise linear circle map, Commun. Theor. Phys. 8, 1.

2170. Yazaki Y, Takashima S, and Mizutani F (1987) Complex quasiperiodic and chaotic states observed in thermally induced oscillations of gas columns, Phys. Rev. Lett. 58, 1108.

2171. Yeh W-J, He D-R, and Kao Y H (1984) Fractal dimension and self-similarity of the devil's staircase in a Josephson-junction simulator, Phys. Rev. Lett. 52, 480.

2172. Yeh W-J, and Kao Y H (1982) Universal scaling and chaotic behavior of a Josephson-junction analog, Phys. Rev. Lett. 49, 1888.

2173. Yeh W-J, and Kao Y H (1983) Intermittency in Josephson junctions, Appl. Phys. Lett. 42, 299.

2174. Yokoi C S O, de Oliveira M J, and Salinas S R (1985) Strange attractors in the Ising model with competing interactions on the Cayley tree, Phys. Rev. Lett. 54, 163.

2175. Yokoyama J, Sato K, and Kodama H (1987) Baryogenesis in chaotic inflationary cosmology, Phys. Lett. 196B, 129.

2176. Yoon T-H, Song J-W, Shin S-Y, and Ra J-W (1984) One-dimensional map and its modification for periodic-chaotic sequence in a driven nonlinear oscillator, Phys. Rev. A30, 3347.

2177. Yorke J A, Grebogi C, Ott E, and Tedeschini-Lalli L (1985) Scaling behavior of windows in dissipative dynamical systems, Phys. Rev. Lett. 54, 1095.

2178. Yorke J A, Yorke E D, and Mallet-Paret J (1987) Lorenz-like chaos in a partial differential equation for a heated flow loop, Physica 24D, 279.

2179. Yorke J A, and Alligood K T (1985) Period doubling cascades of attractors: a prerequisite for horseshoes, Commun. Math. Phys. 101, 305.

2180. Yorke J A, and Yorke E D (1979) Metastable chaos: the transition to sustained chaotic behavior in the Lorenz model, J. Stat. Phys. 21, 263.

2181. Yoshida H (1986) Existence of exponentially unstable periodic solutions and the non-integrability of homogeneous Hamiltonian systems, Physica 21D, 167.

2182. Yoshida T, Mori H, and Shigematsu H (1983) Analytic study of chaos of the tent map: band structures, power spectra and critical behavior, J. Stat. Phys. 31, 279.

2183. Yoshida T, and So B-C (1988) Spectra of singularities and entropies for local expansion rate, Prog. Theor. Phys. 79, 1.

2184. Yoshida T, and Tomita K (1987) Characteristic structures of power spectra in periodic chaos. II. The Hénon map, the driven damped pendulum and the Rössler system, Prog. Theor. Phys. 77, 239.

2185. Yoshimura K, and Watanabe S (1982) Chaotic behavior of nonlinear evolution equation with fifth power dispersion, J. Phys. Soc. Japan, 51, 3028.

2186. Yosida T (1984) Analytic study of periodic chaos I. Method and spectral characteristics, Prog. Theor. Phys. 72, 895.

2187. Yosida T (1985) Analytic study of periodic chaos II. Recursion relations for the power spectra, Prog. Theor. Phys. 73, 3499.

2188. Young K (1985) Further regularities in period-doubling bifurcations, Phys. Lett. 111A, 161.

2189. Young L-S (1981) Capacity of attractors, Ergod. Th. & Dyn. Sys. 1, 381.

2190. Young L-S (1982) Dimension, entropy and Lyapunov exponents, Ergod. Th. & Dyn. Sys. 2, 109.

2191. Young L-S (1984) Dimension, entropy and Lyapunov exponents in differentiable dynamical systems, Physica 124A, 639.

2192. Yu J, and Hu G (1986) The invariant distribution of non-fully developed chaos for the tent map, Chinese Phys. Lett. 3, 357.

2193. Yu J, and Hu G (1988) Exact solutions of the invariant density for piecewise linear approximation to cubic maps, J. Phys. A21, 2717.

2194. Yuan J-M, Liu E, and Tung M (1983) Bistability and hysteresis in laser-driven polyatomic molecules, J. Chem. Phys. 79, 5034.

2195. Yuan J-M, Tung M, Feng D H, and Narducci L M (1983) Instability and irregular behavior of coupled logistic equations, Phys. Rev. A28, 1662.

2196. Zak M (1985) Deterministic representation of chaos in classical dynamics, Phys. Lett. 107A, 125.

2197. Zak M (1986) Criteria of chaos in nonlinear mechanics, Int. J. Nonlin. Mech. 21, 175.

2198. Zak M (1987) Deterministic representation of chaos with application to turbulence, Math. Modelling, 8, 430.

2199. Zaks M A, and Lyubimov D V (1984) Anomalously fast convergence of a doubling-type bifurcation chain in systems with two saddle equilibria, Sov. Phys. JETP, 60, 974; Russ. Orig. 87, 1696.

2200. Zaleski S, and Lallemand P (1985) Scaling laws and intermittency in phase turbulence, J. Physique Lett. 46, L793.

2201. Zardecki A (1982) Noisy Ikeda attractor, Phys. Lett. 90A, 6.

2202. Zaslavsky G M (1978) The simplest case of a strange attractor, Phys. Lett. 69A, 145.

2203. Zaslavsky G M (1981) Stochasticity in quantum systems, Phys. Reports, 80, 157.

2204. Zaslavsky G M, Sagdeev R Z, and Chernikov A A (1988) Stochastic nature of streamlines in steady-state flows, Sov. Phys. JETP, 67, 270; Russ. Orig. 94, 102.

2205. Zaslavsky G M, and Chernikov A A (1985) Group response, universal mapping, and stochastic particle dynamics, Sov. Phys. JETP, 62, 945; Russ. Orig. 89, 1633.

2206. Zaslavsky G M, and Chirikov B V (1972) Stochastic instabilities in nonlinear oscillations, Sov. Phys. Usp. 14, 549; Russ. Orig. 105 (1973), 3.

2207. Zaslavsky G M, and Rachko Kh Ya (1978) Singularties of the transition to a turbulent motion, Sov. Phys. JETP, 49, 1039; Russ. Orig. 76, 2052.

2208. Zeeman E C (1988) Stability of dynamical systems, Nonlinearity, 1, 115.

2209. Zeng W-Z (1985) A recursion formula for the number of stable orbits in the cubic map, Chinese Phys. Lett. 2, 429.

2210. Zeng W-Z (1987) On the number of stable cycles in the cubic map, Commun. Theor. Phys. 8, 273.

2211. Zeng W-Z, Ding M-Z, and Li J-N (1985) Symbolic description of periodic windows in the antisymmetric cubic map, Chinese Phys. Lett. 2, 293.

2212. Zeng W-Z, Ding M-Z, and Li J-N (1988) Symbolic dynamics for one-dimensional mappings with multiple critical points, Commun. Theor. Phys. 9, 141.

2213. Zeng W-Z, Hao B-L, Wang G-R, and Chen S-G (1984) Scaling property of period-n-tupling sequences in one-dimensional mappings, Commun. Theor. Phys. 3, 283.Ω

2214. Zeng W-Z, and Glass L (1989) Symbolic dynamics and skeletons of circle maps, Physica 40D, 218.

2215. Zeng W-Z, and Hao B-L (1986) Dimensions of the limiting sets of period-n-tupling sequences, Chinese Phys. Lett. 3, 285.

2216. Zeng W-Z, and Hao B-L (1987) The derivation of a sum rule detrmining the q-th order information dimension, Commun. Theor. Phys. 8, 295.

2217. Zhang H-J, Dai J-H, Wang P-Y, Jin C-D, and Hao B-L (1987) Analytical study of a bimodal map related to optical bistability, Commun. Theor. Phys. 8, 281.

2218. Zhang H-J, Wang P-Y, Dai J-H, Jin C-D, and Hao B-L (1985) Analytical study of a bimodal mapping related to a hybrid optical bistable device using liquid crystal, Chinese Phys. Lett. 2, 5.

2219. Zhang T-L, and Sun Y-S (1988) Measure-preserving mapping in the Couette-Taylor system, Scientia Sinica, 31, 87.

2220. Zhao Y, Wang H-D, and Huo Y-P (1988) Anomalous behavior of bifurcation in a system with delayed feedback, Phys. Lett. 131A, 353.

2221. Zhao Y, and Huo Y-P (1987) Dynamical behavior of a system with delayed feedback, Chin. Phys. (AIP), 7, 625.

2222. Zhao Y, and Huo Y-P (1988) New behavior of bifurcations and chaos in a nonlinear optical ring cavity, Chin. Phys. (AIP), 8, 326.

2223. Zheng W-M (1984a) Analysis of the Poincaré surface of section for the double-resonance Hamiltonian, Phys. Rev. A29, 2204.

2224. Zheng W-M (1984b) Renormalization calculation of the bifurcation limit point for the whisker mapping, Phys. Lett. 101A, 246.

2225. Zheng W-M (1986a) Derivation of the spectrum for the standard mapping: a simple renormalization procedure, Phys. Rev. A33, 2850.

2226. Zheng W-M (1986b) Renormalization procedure for incommensurate self-dual lattices, Phys. Rev. A34, 1625.

2227. Zheng W-M (1986c) Simple renormalization procedure for quasiperiodic maps, Phys. Rev. A34, 2336.

2228. Zheng W-M (1987a) "Momentum-space" suppression technique for Josephson junctions, Phys. Lett. 121A, 451.

2229. Zheng W-M (1987b) Global scaling properties of the spectrum for the Fibonacci chains, Phys. Rev. A35, 1467.

2230. Zheng W-M (1989a) Generalized composition law for symbolic itineraries, J. Phys. A22, 3307.

2231. Zheng W-M (1989b) The W-sequence for circle maps, J. Phys. A22, 3647.

2232. Zheng W-M (1989c) Symbolic dynamics for the gap map, Phys. Rev. A39, 6608.

2233. Zheng W-M, and Hao B-L (1989) Symbolic dynamics analysis of symmetry breaking and restoration, Int. J. Mod. Phys. B3, 1183.

2234. Zheng W-M, and Reichl L E (1987) Level mixing in the stretched hydrogen atom, Phys. Rev. A35, 474.

2235. Zhong G-Q, and Ayrom F (1985) Experimental confirmation of chaos from Chua's circuit, Circuit Theory and Appl. 13, 93.

2236. Zimmermann T, Koppel H, Haller E, Meyer H-D, and Cederbaum L S (1987) Energy level statistics of coupled oscillators, Phys. Scripta 5, 125.

2237. Zimmermann Th, Meyer H-D, Koppel H, and Cederbaum L S (1986) Manifestation of classical chaos in the statistics of quantum energy levels, Phys. Rev. A33, 4334.

2238. Zippelius A, and Lucke M (1981) The effect of external noise in the Lorenz model of the Bénard Problem, J. Stat. Phys. 24, 345.

2239. Zisook A B (1981) Universal effects of dissipation in two-dimensional mappings, Phys. Rev. A24, 1640.

2240. Zisook A B (1982) Intermittency in area-preserving mappings, Phys. Rev. A25, 2289.

2241. Zisook A B, and Shenker S J (1982) Renormalization group for intermittency in area-preserving mappings, Phys. Rev. A25, 2824.

2242. Zong Y-T, and Wei R-J (1984) Bifurcation in air aciustic systems, Chinese Phys. Lett. 4, 457.Ω

2243. Zou C-Z (1986) Amalgamation of periodic and chaotic solutions in a double parallel-connected Lorenz system, Chinese Phys. Lett. 3, 161.

2244. Zyczkowski K (1987) Quantum chaotic system in the generalized Husimi representation, Phys. Rev. A35, 3546.